Encyclopedia of Chemical Physics and Physical Chemistry

Volume I: Fundamentals

Online version at www.ecppc.iop.org

Encyclopedia of Chemical Physics and Physical Chemistry

Volume I: Fundamentals

Edited by

John H Moore

University of Maryland

and

Nicholas D Spencer

ETH-Zürich

Institute of Physics Publishing
Bristol and Philadelphia

British Library Cataloguing-in-Publication Data

A catalogue record for this book is available from the British Library.

ISBN 0 7503 0798 6 (Vol. I)
 0 7503 0799 4 (Vol. II)
 0 7503 0800 1 (Vol. III)
 0 7503 0313 1 (3 Vol. set)

Library of Congress Cataloging-in-Publication Data are available

Online version of encyclopedia at www.ecppc.iop.org

Jacket illustration. Parametric two-photon down conversion of photons by a nonlinear crystal. The photo shows the conversion of 351 nm laser radiation by a KDP crystal into pairs of correlated photons. When the input photons are split evenly, two photons at 702 nm (red) are produced at an angle of $4°$ to the incident photon direction. Other pairs at shorter and longer wavelengths than 702 nm are also produced, giving the other colours. Because momentum is conserved the different wavelengths are spatially separated. (From the laboratory of Alan Migdall, Optical Technology Division, NIST.)

Publisher: Nicki Dennis
Editorial Assistant: Bret Tobias
Production Editor: Simon Laurenson
Production Control: Sarah Plenty
Cover Design: Frédérique Swist

Published by Institute of Physics Publishing, wholly owned by The Institute of Physics, London

Institute of Physics Publishing, Dirac House, Temple Back, Bristol BS1 6BE, UK

US Office: Institute of Physics Publishing, The Public Ledger Building, Suite 1035, 150 South Independence Mall West, Philadelphia, PA 19106, USA

Typeset in TEX using the IOP Bookmaker Macros
Printed in the UK by MPG Books Ltd, Bodmin, Cornwall

Contents

Contents

Editors

John H Moore

Department of Chemistry and Biochemistry,
University of Maryland,
College Park, MD 20742,
USA
jm89@umail.umd.edu

Nicholas D Spencer

Laboratory for Surface Science and Technology,
Department of Materials,
ETH-Zürich,
CH-8092, Zürich,
Switzerland
nspencer@surface.mat.ethz.ch

Scientific Advisory Board

List of contributors

A C Albrecht (B1.3)

Department of Chemistry,
G-62 Olin Lab,
Cornell University,
Ithaca, NY 14853-1301,
USA
ACA7@CORNELL.EDU

M P Allen (B3.3)

H H Wills Physics Laboratory,
University of Bristol,
Royal Fort,
Tyndall Avenue,
Bristol, BS8 1TL,
UK
m.p.allen@bristol.ac.uk

S M Anderson (B3.4)

Department of Chemistry,
University of California,
Los Angeles, CA 90095-1569,
USA
sybil@chem.ucla.edu

D L Andrews (C3.4)

School of Chemical Sciences,
University of East Anglia,
Norwich, NR4 7TJ,
UK
D.L.Andrews@uea.ac.uk

A D Bain (B2.4)

Department of Chemistry,
McMaster University,
1280 Main Street West,
Hamilton,
Ontario, L8S 4M1,
Canada
bain@mcmaster.ca

J Baker (B3.5)

Chemistry Department,
University of Arkansas,
Fayetteville, AR 72701,
USA
baker@uafchem1.uark.edu

W F Beck (B2.1)

Department of Chemistry,
3 Chemistry Building,
Michigan State University,
East Lansing, MI 48824-1322,
USA
beck@cem.msu.edu

K Becker (C2.13)

Department of Physics,
Stevens Institute of Technology,
Castle Point Station,
Hoboken, NJ 07030,
USA
kbecker@stevens-tech.edu

D N Beratan (C3.2)

Department of Chemistry,
University of Pittsburgh,
Pittsburgh, PA 15260,
USA
beratan@pitt.edu

G A Blake (B1.4)

Division of Geological and Planetary Sciences,
California Institute of Technology,
Mail Stop 150-21,
Pasadena, CA 91125,
USA
gab@gps.caltech.edu

P R Bunker (A1.4)

Steacie Institute for Molecular Science,
Theory and Computational Program,
National Research Council of Canada,
Ottawa,
Ontario, K1A 0R6,
Canada
Philip.Bunker@nrc.ca

E A Carter (B3.2)

Department of Chemistry and Biochemistry,
University of California,
Los Angeles, CA 90095-1569,
USA
eac@chem.ucla.edu

J R Chelikowsky (A1.3)

Chemical Engineering and Materials Science,
University of Minnesota,
Minneapolis, MN 55455,
USA
jrc@msi.umn.edu

B Chu (B1.9)

Department of Chemistry,
SUNY at Stony Brook,
Stony Brook, NY 11794-3400,
USA
BCHU@ccmail.sunysb.edu

V L Colvin (C2.17)

Chemistry Department,
Rice University,
PO Box 1892,
Houston, TX 77251,
USA
colvin@ruf.rice.edu

M A Coplan (B1.10)

Insitute for Physical Science and Technology,
University of Maryland,
College Park, MD 20742,
USA
mc45@umail.umd.edu

L Coulier (B1.25)

Schuit Institute of Catalysis,
Eindhoven University of Technology,
5600 MB, Eindhoven,
Netherlands
L.Coulier@tue.nl

J I Dadap (B1.5)

Department of Physics,
Columbia University,
538 West 120th St,
New York, NY 10027,
USA
dadap@phys.columbia.edu

P J Dagdigian (B2.3)

Department of Chemistry,
Johns Hopkins University,
Baltimore, MD 21218,
USA
pjdagdigian@jhu.edu

G R Darling (A3.9)

Surface Science Research Centre,
University of Liverpool,
Liverpool, L69 3BX,
UK
darling@ssci.liv.ac.uk

A A Demidov (C3.4)

Physics Department,
Northeastern University,
111 Dana Research Center,
Boston, MA 02115,
USA
ademidov@lynx.dac.neu.edu

R C Desai (A2.2, A3.3)

Department of Physics,
Rm 1001,
University of Toronto,
60 George St,
Toronto,
Ontario, M5S 1A7,
Canada
desai@krishna.physics.utoronto.ca

D D Dlott (C3.5)

Department of Chemistry,
Noyes Laboratory,
University of Illinois,
Urbana, IL 61801-3364,
USA
D-Dlott@uiuc.edu

J R Dorfman (A3.1)

Institute for Physical Science and Technology,
University of Maryland,
College Park, MD 20742,
USA
jrd@ipst.umd.edu

R Dupree (B1.12)

Department of Physics,
University of Warwick,
Coventry, CV4 7AL,
UK
nmrd@spec.warwick.ac.uk

S K Estreicher (C2.16)

Texas Technical University,
Lubbock, TX 79409,
USA

K G Ewsuk (C2.11)

Advanced Materials Laboratory,
Ceramic Materials Department, 1843,
Sandia National Laboratories, MS 1349,
1001 University Blvd, SE,
Albuquerque, NM 87106,
USA
kgewsuk@sandia.gov

M R Flannery (B2.2)

School of Physics,
Georgia Institute of Technology,
Atlanta, GA 30331-0430,
USA
ray.flannery@physics.gatech.edu

G W Flynn (C3.3)

Department of Chemistry,
Columbia University,
3000 Broadway,
Mail code 3109,
New York, NY 10027,
USA
flynn@chem.columbia.edu

M D E Forbes (B1.16)

Department of Chemistry, CB #3290,
University of North Carolina at Chapel Hill,
Chapel Hill, NC 27599,
USA
mdef@unc.edu

R F Fox (A3.2)

School of Physics,
Georgia Institute of Technology,
837 State St,
Atlanta, GA 30332-0430,
USA
ron.fox@physics.gatech.edu

S J Fraser (C3.6)

Lash Miller Chemical Laboratories,
University of Toronto,
80 St George St,
Toronto,
Ontario, M5S 1A1,
Canada
sfraser@chem.utoronto.ca

B C Gates (C2.7)

Department of Chemical Engineering,
University of California,
Davis, CA 95616,
USA
bcgates@ucdavis.edu

A J Gellman (C2.9)

Department of Chemical Engineering,
Carnegie Mellon University,
Pittsburgh, PA 15213,
USA
aj4b@andrew.cmu.edu

H P Gillis (C2.18)

Department of Materials Science and Engineering,
University of California,
Los Angeles, CA 90095-1595,
USA
hpgillis@ucla.edu

U Goerke (B1.14)

Stephanstr. 1a,
D-04103, Leipzig,
Germany
goerke@cns.mpg.de

R A Goldbeck (C3.1)

Department of Chemistry and Biochemistry,
University of California,
Santa Cruz, CA 95064,
USA
goldbeck@hydrogen.UCSC.EDU

D W Goodman (A3.10)

Texas A&M University,
College Station, TX 77843,
USA
goodman@chemvx.chem.tamu.edu

M J Graham (C2.8)

National Research Council of Canada,
Institute for Microstructural Studies,
Montreal Road, Bldg M-50,
Ottawa,
Ontario, K1A 0R6,
Canada
mike.graham@nrc.ca

D M Guldi (C1.2)

Radiation Laboratory,
University of Notre Dame,
Notre Dame, IN 46556,
USA
guldi@hertz.rad.nd.edu

G Hähner (C2.4)

Department of Physics and Astronomy,
(Condensed Matter and Material Physics),
University College London,
Gower Street,
London, WC1E 6BT,
UK
g.haehner@ucl.ac.uk

I W Hamley (C2.2)

School of Chemistry,
University of Leeds,
Leeds, LS2 9JT,
UK
I.W.Hamley@chem.leeds.ac.uk

A Hamnett (A2.4)

Department of Chemistry,
University of Newcastle,
Newcastle upon Tyne, NE1 7RU,
UK
Andrew.Hamnett@ncl.ac.uk

E J Harbron (B1.16)

Department of Chemistry, CB #3290,
University of North Carolina at Chapel Hill,
Chapel Hill, NC 27599,
USA

W L Hase (A3.12)

Department of Chemistry,
Wayne State University,
Detroit, MI 48202,
USA
hase@sun.chem.wayne.edu

T F Heinz (B1.5)

Department of Physics,
Columbia University,
538 West 120th St,
New York, NY 10027,
USA
tony.heinz@columbia.edu

M Heuberger (B1.20)

Laboratory for Surface Science and Technology,
Department of Materials,
ETH-Zürich,
CH-8092, Zürich,
Switzerland
heuberger@surface.mat.ethz.ch

A W E Hodgson (B1.28)

Institute of Materials Chemistry and Corrosion, and
Department of Materials,
ETH-Zürich,
CH-8093, Zürich,
Switzerland
hodgson@lbwk.baug.ethz.ch

S Holloway (A3.9)

Surface Science Research Centre,
University of Liverpool,
Liverpool, L69 3BX,
UK
stephen@ssci.liv.ac.uk

O W Howarth (B1.11)

Department of Chemistry,
University of Warwick,
Coventry, CV4 7AL,
UK
msrhs@warwick.ac.uk

B S Hsiao (B1.9)

Department of Chemistry,
SUNY at Stony Brook,
Stony Brook, NY 11794-3400,
USA
BHSIAO@ccmail.sunysb.edu

J M Hutson (C1.3)

Department of Chemistry,
University of Durham,
Durham, DH1 3LE,
UK
J.M.Hutson@durham.ac.uk

L K Iwaki (C3.5)

Department of Chemistry,
Noyes Laboratory,
University of Illinois,
Urbana, IL 61801-3364,
USA
lawrence.iwaki@nist.gov

S P Jarvis (B1.19)

Nanotechnology Research Institute,
National Institute for Advanced Industrial Science and
Technology,
1-1-4 Higashi,
Tsukuba, Ibaraki 305-0046,
Japan
spjarvis@jrcat.or.jp

P Jensen (A1.4)

FB 9 – Theoretische Chemie,
Bergische Universität,
Gesamthochschule Wuppertal,
D-42097, Wuppertal,
Germany
jensen@wrcs3.urz.uni-wuppertal.de

R Kapral (C3.6)

Lash Miller Chemical Laboratories,
University of Toronto,
80 St George St,
Toronto,
Ontario, M5S 1A1,
Canada
rkapral@gatto.chem.utoronto.ca

H Keiss (B1.18)

im unteren Tollacker 11,
CH-8162, Steinmaur,
Switzerland

M E Kellman (A1.2)

Department of Chemistry,
University of Oregon,
Klamath Hall,
1370 Franklin Blvd,
Eugene, OR 97403,
USA
kellman@oregon.uoregon.edu

R Kimmich (B1.14)

Sektion Kernresonanz,
University of Ulm,
Albert-Einstein-Allee 11,
D-89069, Ulm,
Germany
rainer.kimmich@physik.uni-ulm.de

J C Kirkwood (B1.3)

Department of Chemistry,
Cornell University.
Ithaca, NY 14853-1301,
USA
jason.kirkwood@kla-tencor.com

D S Kliger (C3.1)

Department of Chemistry and Biochemistry,
389 Thimann Labs,
University of California,
Santa Cruz, CA 95064,
USA
kliger@chemistry.ucsc.edu

D K Klimov (C2.5)

Institute for Physical Science and Technology,
University of Maryland,
College Park, MD 20742,
USA
klimov@glue.umd.edu

A Kogelbauer (C2.12)

Laboratory for Technical Chemistry,
ETH-Zürich,
CH-8092, Zürich,
Switzerland
kogelbauer@tech.chem.ethz.ch

J Kowalewski (B1.13)

Arrhenius Laboratory of Physical Chemistry,
Stockholm University,
S-10691, Stockholm,
Sweden
jk@physc.su.se

D Luckhaus (A3.4, B2.5)

Laboratorium für Physikalische Chemie,
ETH-Zürich,
CH-8092, Zürich,
Switzerland
luckhaus@ir.phys.chem.ethz.ch

D Lützenkirchen-Hecht (C2.10)

Institut für Angewandte Physik,
Heinrich-Heine-Universität Düsseldorf,
Universitätsstr. 1,
D-40225, Düsseldorf
Germany
hechtd@uni-duesseldorf.de

R Marquardt (A3.13)

Laboratoire de Chimie Théorique,
Université de Marne-la-Vallée,
5, Bd Descartes,
Champs-sur-Marne,
F-77454, Marne-la-Vallée,
France
roberto.marquardt@univ-mlv.fr

K N Marsh (B1.27)

Department of Chemical and Process Engineering,
University of Canterbury,
Private Bag 4800,
Christchurch,
New Zealand
k.marsh@cape.canterbury.ac.nz

J W Mayer (B1.24)

Center for Solid State Sciences,
Arizona State University,
Tempe, AZ 85287-1704,
USA
IFJWM@asuvm.inre.asu.edu

P M Mayer (B1.7)

Chemistry Department,
University of Ottawa,
10 Marie Curie,
Ottawa, K1N 6N5,
Canada
pmayer@science.uottawa.ca

P J McDonald (B1.14)

Physics Department,
University of Surrey,
Guildford, GU2 5XH,
UK
p.mcdonald@surrey.ac.uk

T M Miller (A3.5)

Air Force Research Laboratory,
Space Vehicles Directorate (VSBP),
29 Randolph Rd,
Hanscom AFB, MA 01731-3010,
USA
tmiller@plh.af.mil

D M Mittleman (C2.17)

Electrical and Computer Engineering Department,
Rice University,
Houston, TX 77005,
USA
daniel@rice.edu

J H Moore (B1.6)

Department of Chemistry and Biochemistry,
University of Maryland,
College Park, MD 20742,
USA
jm89@umail.umd.edu

M Müller (B1.17)

Laboratory for Electron Microscopy,
ETH-Zürich,
CH-8092, Zürich,
Switzerland
mueller@em.biol.ethz.ch

M Müller (B3.6)

Staudinger Weg 7,
Universität Mainz,
D-55099, Mainz,
Germany
mueller@plato.physik.uni-mainz.de

A Myers Kelley (C1.5)

Department of Chemistry,
Kansas State University,
Manhattan, KS 66506-3701,
USA
amkelley@ksu.edu

D Neuhauser (B3.4)

Department of Chemistry and Biochemistry,
University of California,
Los Angeles, CA 90095-1569,
USA
dxn@chem.ucla.edu

D M Neumark (A3.7)

Department of Chemistry,
University of California,
Berkeley, CA 94720,
USA
dan@radon.cchem.berkeley.edu

M F Nicol (B1.29)

Department of Physics,
University of Nevada,
PO Box 454002,
Las Vegas, NV 89154-4002,
USA
nicol@physics.unlv.edu

J W Niemantsverdriet (B1.25)

Schuit Institute of Catalysis,
Eindhoven University of Technology,
PO Box 513,
5600 MB, Eindhoven,
Netherlands,
tgtahn@chem.tue.nl

E Prince (B1.8)

NIST Center for Neutron Research,
Reactor Building (235), Room B107,
Gaithersburg, MD 20899,
USA
PRINCE@enh.nist.gov

R Prins (C2.12)

Laboratory for Technical Chemistry,
ETH-Zürich,
CH-8092, Zürich,
Switzerland
prins@tech.chem.ethz.ch

V M Prozesky (B1.24)

SAGE International Ltd,
1 Love Lane,
London, EC2V 7JN,
UK
vprozesky@sage-intl.co.uk

P Pulay (B3.5)

Chemistry Department,
University of Arkansas,
Fayetteville, AR 72701,
USA

M Quack (A3.4, A3.13, B2.5)

Laboratorium für Physikalische Chemie,
ETH-Zürich,
CH-8092, Zürich,
Switzerland
martin@quack.ch

J W Rabalais (B1.23)

Department of Chemistry,
University of Houston,
4800 Calhoun,
Houston, TX 77204-5641,
USA
rabalais@jetson.uh.edu

J J Ramsden (C2.14)

Institut für Anorganische Chemie,
Universität Basel,
Spitalstrasse 51,
CH-4056, Basel,
Switzerland
J.Ramsden@unibas.ch

J C Rasaiah (A2.3)

Department of Chemistry,
University of Maine,
Orono, ME 04469,
USA
rasaiah@maine.maine.edu

C Rettner (A3.9)

Research Division, Almaden Research Center,
IBM Almaden,
650 Harry Road,
San Jose, CA 95120,
USA
rettner@almadan.ibm.com

P Robyr (C2.1)

Department of Materials,
Institute of Polymers,
ETH-Zürich,
CH-8092, Zürich,
Switzerland
probyr@ifp.mat.ethz.ch

R G Sadygov (B3.4)

Department of Chemistry,
University of California,
Los Angeles, CA 90095-1569,
USA

T P St Clair (A3.10)

Department of Chemistry,
Goodman Laboratory,
Texas A&M University,
PO Box 30012,
College Station, TX 77842-3012,
USA
clair@mail.chem.tamu.edu

G C Schatz (A3.11)

Department of Chemistry,
Northwestern University,
Evanston, IL 60208-3113,
USA
schatz@chem.nwu.edu

M Schmidt (C2.13)

Institut fur Niedertemperatur Plasmaphysik e.V.,
Ernst-Moritz-Arndt Universität Greifswald,
Robert Blum Strasse 8–10,
D-17489, Greifswald,
Germany
schmidtm@INP-GREIFSWALD.DE

P Schmuki (C2.8)

Department of Materials Science,
Institute for Surface Science and Corrosion,
Friedrich-Alexander Universität,
Martensstrasse 7,
D-91058, Erlangen,
Germany
schmuki@ww.uni-erlangen.de

C Schmuttenmaer (B1.2)

Department of Chemistry,
Yale University,
PO Box 208107,
New Haven, CT 06520-8107,
USA
charles.schmuttenmaer@yale.edu

R R Schröder (B1.17)

Max-Planck-Institut für Medizinische Forschung,
Jahnstrasse 29,
D-69120, Heidelberg,
Germany
rasmus@mpimf-heidelberg.mpg.de

J Schroeder (A3.6)

Abteilung Spektroskopie und Photochemische Kinetik,
(Abt. 010),
Max-Planck-Institut für Biophysikalische Chemie,
Postfach 2481,
D-37018, Göttingen,
Germany
jschroe2@gwdg.de

R L Scott (A2.1, A2.5)

Department of Chemistry and Biochemistry,
University of California,
Los Angeles, CA 90095-1569,
USA
scott@chem.ucla.edu

S K Scott (A3.14)

School of Chemistry,
University of Leeds,
Leeds, LS2 9JT,
UK
S.K.Scott@chemistry.leeds.ac.uk

J Simons (B3.1)

Department of Chemistry,
Henry Eyring Building,
University of Utah,
Salt Lake City, UT 84112,
USA
simons@chemistry.chem.utah.edu

M E Smith (B1.12)

Department of Physics,
University of Warwick,
Coventry, CV4 7AL,
UK
M.E.Smith.1@warwick.ac.uk

N D Spencer (B1.19)

Laboratory for Surface Science and Technology,
Department of Materials,
ETH-Zürich,
CH-8092, Zürich,
Switzerland
nspencer@surface.mat.ethz.ch

J F Stanton (A1.1)

Department of Chemistry,
University of Texas,
Austin, TX 78712,
USA
stanton@jfs1.cm.utexas.edu

F Starrost (B3.2)

Department of Chemistry and Biochemistry,
University of California,
Los Angeles, CA 90095-1569,
USA
fstar@chem.ucla.edu

H-H Strehblow (C2.10)

Institut für Physikalische Chemie und Elektrochemie,
Heinrich-Heine-Universität Düsseldorf,
Universitätsstr. 1, Geb. 2642,
D-40225, Düsseldorf,
Germany
henning@uni-duesseldorf.de

S J Strickler (B1.1)

Department of Chemistry,
University of Colorado,
Boulder, CO 80309,
USA
Stewart.Strickler@Colorado.edu

D J Tannor (A1.6)

Chemical Physics Department,
Weizmann Institute of Science,
Rehovot, 76100,
Israel
tannor@quantum.weizmann.ac.il

H Temkin (C2.16)

Electrical Engineering Department,
Texas Technical University,
Lubbock, TX 79409,
USA
htemkin@coe2.coe.TTU.edu

J Texter (C2.3)

Strider Research Corp.,
265 Clover St,
Rochester, NY 14610-2246,
USA
texter@striderresearch.com

A J Thakkar (A1.5)

Department of Chemistry,
University of New Brunswick,
Fredericton,
New Brunswick, E3B 5A3,
Canada
ajit@unb.ca

C C Theron (B1.24)

National Accelarator Centre,
PO Box 72,
Faure, 7131,
South Africa
ctheron@nac.ac.za

D Thirumalai (C2.5)

Institute for Physical Sciences and Technology,
University of Maryland,
College Park, MD 20742,
USA
thirum@glue.umd.edu

W T Tysoe (B1.26)

Department of Chemistry,
University of Wisconsin,
Milwaukee, WI 53201,
USA
wtt@csd.uwm.edu

D J Ulness (B1.3)

Department of Chemistry,
Concordia College,
Moorhead, MN 56562,
USA
ulnessd@cord.edu

J S van Duijneveldt (C2.6)

School of Chemistry,
University of Bristol,
Bristol, BS8 1TS,
UK
J.S.van-Duijneveldt@bristol.ac.uk

M A Van Hove (B1.21)

Bldg 66,
Lawrence Berkeley Laboratory,
Berkeley, CA 94720,
USA
vanhove@lbl.gov

A A Viggiano (A3.5)

Air Force Research Laboratory,
Space Vehicles Directorate (VSBP),
29 Randolph Road,
Hanscom AFB, MA 01731-3010,
USA
viggiano@plh.af.mil

G A Voth (A3.8)

Department of Chemistry,
University of Utah,
315 South, 1400 East DOC,
Salt Lake City, UT 84112-0850,
USA
voth@atlas.chem.utah.edu

G C Walker (C3.2)

Department of Chemistry,
University of Pittsburgh,
Pittsburgh, PA 15260,
USA
gilbertw@vms.cis.pitt.edu

L-S Wang (C1.1)

Department of Physics,
Washington State University,
2710 University Drive,
Richland, WA 99352,
USA
LS.WANG@PNL.GOV

S Weber (B1.15)

Fachbereich Physik,
Institut fur Experimentalphysik,
Freie Universität Berlin,
Arnimallee 14,
D-14195, Berlin,
Germany
stefan.weber@physik.fu-berlin.de

J Weiner (C1.4)

IRSAMC/LCAR,
Université Paul Sabatier,
118 route de Narbonne,
Toulouse, 310062,
France
jweiner@irsamc.ups-tlse.fr

W L Wilson (C2.15)

Photonic Materials Research Department,
Bell Laboratories, Lucent Technologies,
600–700 Mountain Ave,
Murray Hill, NJ 07974-0636,
USA
WLW@lucent.com

G Wu (B1.26)

Department of Chemistry,
University of Wisconsin,
Milwaukee, WI 53201,
USA
gefei@uwm.edu

J A Yarmoff (A1.7)

Department of Physics,
University of California,
Riverside, CA 92507,
USA
jory.yarmoff@ucr.edu

F Zaera (B1.22)

Department of Chemistry,
University of California,
Riverside, CA 95251,
USA
francisco.zaera@ucr.edu

Acknowledgments

The support of the following agencies is gratefully acknowledged.

A1.4: PJ is grateful to the Steacie Institute for Molecular Sciences for hospitality and financial support; PRB acknowledges an Alexander von Humboldt Award during the tenure of which, at the University of Wuppertal, further parts of the paper were written. A2.3: The author gratefully acknowledges the support of the National Science Foundation under grant No CHE-9610288. A3.11: This work was supported by NSF grant CHE-9527677. A3.12: The author wishes to thank the National Science Foundation for support of his research on unimolecular reaction dynamics. B1.3: We are most grateful for support during the preparation of this chapter from the National Science Foundation through grant CHE-9616635. B1.5: The authors acknowledge support from the National Science Foundation under the EMSI programme at Columbia University. B1.9: The authors acknowledge the financial support of two grants from NSF (DMR9732653 for BH and DMR9612386 for BC). B1.13: I wish to express my gratitude to the Swedish Natural Science Research Council for their generous and long lasting support of my own research in the field of nuclear spin relaxation. B1.21: This work was supported by the Director, Office of Energy Research, Office of Basic Energy Sciences, Materials Science Division, of the US Department of Energy under contract No DE-AC03-76SF00098. B1.23: This material is based on work supported by the National Science Foundation under grant No CHE-9700665, the R A Welch Foundation under grant No E-656, and the Texas Advanced Research Programme under grant No 3652-683. B3.2: EAC is grateful for support from the Air Force Office of Scientific Research, the National Science Foundation, and the Army Research Office; FS thanks the Deutsche Forschungsgemeinschaft for partial funding. B3.5: The authors gratefully acknowledge support from the National Science Foundation (under grant No CHE-9707202) and the US Air Force office for scientific research (under grant No F49620-98-1-0082) during the preparation of this article. C1.2: This work was supported by the Office of Basic Energy Sciences of the US Department of Energy (contribution No NDRL-4119 from the Notre Dame Radiation Laboratory). C2.11: This work was supported by the United States Department of Energy under contract DE-AC04-94AL85000; Sandia is a multiprogramme laboratory operated by Sandia Corporation, a Lockheed Martin company, for the United States Department of Energy. C2.17: VLC was supported during the writing of this chapter by NSF (CHE-9702520) and the R A Welch Foundation (C-1342). C3.2: We thank NSF, NIH, ONR and AFOSR/DARPA for supporting our studies of ET reactions. C3.3: Work performed at Columbia University was supported by the Department of Energy (DE-FG02-88-ER13937), with equipment support provided by the National Science Foundation (CHE-97-27205) and the Joint Services Electronics Program (US Army, US Navy and US Air Force; DAAG55-97-1-0166).

Foreword

The principles that govern the behaviour of molecular systems, and the methods of measuring the structure, reactivity and dynamics of molecules are what physical chemists and chemical physicists seek in their research. All the scientific fields in which molecules play an essential role, which include chemistry and physics, the biomedical sciences, atmospheric, interstellar and the earth sciences, continue to benefit from these fundamental discoveries. But how does one go about defining the underlying structure of a subject whose research frontiers change so radically with time? The Editors of the *Encyclopedia of Chemical Physics and Physical Chemistry* have found the answer by enlisting the scientists who create the frontiers, putting together an amazing collection of truly distinguished contributors, representing both the ultra-modern and the traditional in our subject. The result is a single source of the experimental and theoretical tools that form the basis of physical chemistry and chemical physics.

Physical chemistry research is now beginning to embrace more complex systems, structure determination at shorter timescales and larger length scales, tackling essential questions from genomics to materials science. As physical chemistry turns to face these new challenges of the 21st century, its graduate students will continue to seek deep understanding of the phenomena they uncover in their research. This encyclopedia will help to guide them into these new subjects.

The Scientific Advisory Board expresses its indebtedness to all the experts who have contributed chapters to this encyclopedia. Theirs is a wonderful contribution to our field from which we will all benefit. Last, but not least, the Editors are to be congratulated for their vision in planning the unique structure of the encyclopedia.

Robin M Hochstrasser
Donner Professor of Science
University of Pennsylvania
Philadelphia, PA 19104
USA

Introduction

Chemical physics and physical chemistry are not distinctly separate fields of study. Nevertheless, the areas covered by these two disciplines together are distinguished from others by the incredible range of problems addressed by their practitioners. An effective physical chemist or chemical physicist is a 'jack-of-all-trades', able to apply the principles and techniques of the field to everything from high-tech materials to biology. Furthermore, it is nearly impossible to specialize in one aspect of physical chemistry or chemical physics throughout one's professional career. Even within the confines of a single area, real progress often requires drawing on the ideas and techniques developed in quite disparate fields. As a consequence, scientists in the areas that bridge chemistry and physics are always needing to learn and apply unfamiliar aspects of their discipline.

The modern library is filled with texts, and even whole series of volumes, on every imaginable aspect of physics and chemistry. We perceive a need, however, for a more compact source of information that covers the basic aspects of the application of physics to chemistry along with the modern manifestations of these applications—an encyclopedia to which the chemist and physicist can turn for an introduction to an unfamiliar area, an explanation of important experimental and computational techniques, and a description of modern endeavours in the field. For the practicing physicist or chemist, the encyclopedia would be the place to start when confronted with a new problem or when the techniques of an unfamiliar area might be exploited. The encyclopedia quickly provides the basics, defines the scope of each subdiscipline, and indicates where to go for a more complete and detailed explanation. For a graduate student in chemistry or physics the encyclopedia gives a synopsis of the fundamentals and an overview of the range of activities in which physical principles are applied to chemical problems.

The *Encyclopedia of Chemical Physics and Physical Chemistry* is intended for the reader with a solid undergraduate education in physics or chemistry including mechanics, electricity and magnetism, differential equations, simple matrix algebra, chemical nomenclature and the basics of chemical spectroscopy. In addition, the reader has knowledge of quantum mechanics, thermodynamics, and statistical mechanics at the level of an undergraduate physical chemistry text. The encyclopedia is divided into three major sections:

FUNDAMENTALS—the mechanics of atoms and molecules and their interactions, the macroscopic and statistical description of systems at equilibrium, and the basic ways of treating reacting systems. The contributions in this section assume a somewhat less sophisticated audience than the two subsequent sections.

METHODS—the instrumentation and fundamental theory employed in the major spectroscopic techniques, the experimental means for characterizing materials, the instrumentation and basic theory employed in the study of chemical kinetics, and the computational techniques used to predict the static and dynamic properties of materials.

APPLICATIONS—specific topics of current interest and intensive research.

John H Moore
Nicholas D Spencer

PART A1

MICROSCOPICS

A1.1
The quantum mechanics of atoms and molecules

John F Stanton

A1.1.1 Introduction

At the turn of the 19th century, it was generally believed that the great distance between earth and the stars would forever limit what could be learned about the universe. Apart from their approximate size and distance from earth, there seemed to be no hope of determining intensive properties of stars, such as temperature and composition. While this pessimistic attitude may seem quaint from a modern perspective, it should be remembered that all knowledge gained in these areas has been obtained by exploiting a scientific technique that did not exist 200 years ago—spectroscopy.

In 1859, Kirchoff made a breakthrough discovery about the nearest star—our sun. It had been known for some time that a number of narrow dark lines are found when sunlight is bent through a prism. These absences had been studied systematically by Fraunhofer, who also noted that dark lines can be found in the spectrum of other stars; furthermore, many of these absences are found at the same wavelengths as those in the solar spectrum. By burning substances in the laboratory, Kirchoff was able to show that some of the features are due to the presence of sodium atoms in the solar atmosphere. For the first time, it had been demonstrated that an element found on our planet is not unique, but exists elsewhere in the universe. Perhaps most important, the field of modern spectroscopy was born.

Armed with the empirical knowledge that each element in the periodic table has a characteristic spectrum, and that heating materials to a sufficiently high temperature disrupts all interatomic interactions, Bunsen and Kirchoff invented the spectroscope, an instrument that atomizes substances in a flame and then records their emission spectrum. Using this instrument, the elemental composition of several compounds and minerals were deduced by measuring the wavelength of radiation that they emit. In addition, this new science led to the discovery of elements, notably caesium and rubidium.

Despite the enormous benefits of the fledgling field of spectroscopy for chemistry, the underlying physical processes were completely unknown a century ago. It was believed that the characteristic frequencies of elements were caused by (nebulously defined) vibrations of the atoms, but even a remotely satisfactory quantitative theory proved to be elusive. In 1885, the Swiss mathematician Balmer noted that wavelengths in the visible region of the hydrogen atom emission spectrum could be fitted by the empirical equation

$$\lambda = b \left(\frac{n^2}{n^2 - m^2} \right) \tag{A1.1.1}$$

where $m = 2$ and n is an integer. Subsequent study showed that frequencies in other regions of the hydrogen spectrum could be fitted to this equation by assigning different integer values to m, albeit with a different value of the constant b. Ritz noted that a simple modification of Balmer's formula

$$\frac{1}{\lambda} = R_H \left(\frac{1}{m^2} - \frac{1}{n^2} \right) \tag{A1.1.2}$$

succeeds in fitting all the line spectra corresponding to different values of m with only the single constant R_H. Although this formula provides an important clue regarding the underlying processes involved in spectroscopy, more than two decades passed before a theory of atomic structure succeeded in deriving this equation from first principles.

The origins of line spectra as well as other unexplained phenomena such as radioactivity and the intensity profile in the emission spectrum of hot objects eventually led to a realization that the physics of the day was incomplete. New ideas were clearly needed before a detailed understanding of the submicroscopic world of atoms and molecules could be gained. At the turn of the 20th century, Planck succeeded in deriving an equation that gave a correct description of the radiation emitted by an idealized isolated solid (blackbody radiation). In the derivation, Planck assumed that the energy of electromagnetic radiation emitted by the vibrating atoms of the solid cannot have just any energy, but must be an integral multiple of $h\nu$, where ν is the frequency of the radiation and h is now known as Planck's constant. The resulting formula matched the experimental blackbody spectrum perfectly.

Another phenomenon that could not be explained by classical physics involved what is now known as the photoelectric effect. When light impinges on a metal, ionization leading to ejection of electrons happens only at wavelengths ($\lambda = c/\nu$, where c is the speed of light) below a certain threshold. At shorter wavelengths (higher frequency), the kinetic energy of the photoelectrons depends linearly on the frequency of the applied radiation field and is independent of its intensity. These findings were inconsistent with conventional electromagnetic theory. A brilliant analysis of this phenomenon by Einstein convincingly demonstrated that electromagnetic energy is indeed absorbed in bundles, or quanta (now called photons), each with energy $h\nu$ where h is precisely the same quantity that appears in Planck's formula for the blackbody emission spectrum.

While the revolutionary ideas of Planck and Einstein forged the beginnings of the quantum theory, the physics governing the structure and properties of atoms and molecules remained unknown. Independent experiments by Thomson, Weichert and Kaufmann had established that atoms are not the indivisible entities postulated by Democritus 2000 years ago and assumed in Dalton's atomic theory. Rather, it had become clear that all atoms contain identical negative charges called electrons. At first, this was viewed as a rather esoteric feature of matter, the electron being an entity that 'would never be of any use to anyone'. With time, however, the importance of the electron and its role in the structure of atoms came to be understood. Perhaps the most significant advance was Rutherford's interpretation of the scattering of alpha particles from a thin gold foil in terms of atoms containing a very small, dense, positively charged core surrounded by a cloud of electrons. This picture of atoms is fundamentally correct, and is now learned each year by millions of elementary school students.

Like the photoelectric effect, the atomic model developed by Rutherford in 1911 is not consistent with the classical theory of electromagnetism. In the hydrogen atom, the force due to Coulomb attraction between the nucleus and the electron results in acceleration of the electron (Newton's first law). Classical electromagnetic theory mandates that all accelerated bodies bearing charge must emit radiation. Since emission of radiation necessarily results in a loss of energy, the electron should eventually be captured by the nucleus. But this catastrophe does not occur. Two years after Rutherford's gold-foil experiment, the first quantitatively successful theory of an atom was developed by Bohr. This model was based on a combination of purely classical ideas, use of Planck's constant h and the bold assumption that radiative loss of energy does not occur provided the electron adheres to certain special orbits, or 'stationary states'. Specifically, electrons that move in a circular path about the nucleus with a classical angular momentum mvr equal to an integral multiple of Planck's constant divided by 2π (a quantity of sufficient general use that it is designated by the simple symbol $\hbar$) are immune from energy loss in the Bohr model. By simply writing the classical energy of the orbiting electron in terms of its mass m, velocity v, distance r from the nucleus and charge e,

$$E = \frac{1}{2}mv^2 - \frac{e^2}{r} \tag{A1.1.3}$$

invoking the (again classical) virial theorem that relates the average kinetic ($\langle T \rangle$) and potential ($\langle V \rangle$) energy of a system governed by a potential that depends on pairwise interactions of the form r^k via

$$\langle T \rangle = \frac{k}{2} \langle V \rangle \qquad (A1.1.4)$$

and using Bohr's criterion for stable orbits

$$r = \frac{n\hbar}{mv} \qquad (A1.1.5)$$

it is relatively easy to demonstrate that energies associated with orbits having angular momentum $n\hbar$ in the hydrogen atom are given by

$$E_n = -\frac{me^4}{2n^2\hbar^2} \qquad (A1.1.6)$$

with corresponding radii

$$r_n = \frac{n^2\hbar^2}{me^2}. \qquad (A1.1.7)$$

Bohr further postulated that *quantum jumps* between the different allowed energy levels are always accompanied by absorption or emission of a photon, as required by energy conservation, *viz.*

$$\Delta E \equiv E_n - E_m = \frac{me^4}{2\hbar^2} \left(\frac{1}{m^2} - \frac{1}{n^2} \right) = h\nu_{\text{photon}} \qquad (A1.1.8)$$

or perhaps more illustratively

$$\frac{1}{\lambda_{\text{photon}}} = \frac{me^4}{4\pi\hbar^3 c} \left(\frac{1}{m^2} - \frac{1}{n^2} \right) \qquad (A1.1.9)$$

precisely the form of the equation deduced by Ritz. The constant term of equation (A1.1.2) calculated from Bohr's equation did not exactly reproduce the experimental value at first. However, this situation was quickly remedied when it was realized that a proper treatment of the two-particle problem involved use of the reduced mass of the system $\mu \equiv mm_{\text{proton}}/(m + m_{\text{proton}})$, a minor modification that gives striking agreement with experiment.

Despite its success in reproducing the hydrogen atom spectrum, the Bohr model of the atom rapidly encountered difficulties. Advances in the resolution obtained in spectroscopic experiments had shown that the spectral features of the hydrogen atom are actually composed of several closely spaced lines; these are not accounted for by quantum jumps between Bohr's allowed orbits. However, by modifying the Bohr model to allow for elliptical orbits and to include the special theory of relativity, Sommerfeld was able to account for some of the *fine structure* of spectral lines. More serious problems arose when the planetary model was applied to systems that contained more than one electron. Efforts to calculate the spectrum of helium were completely unsuccessful, as was a calculation of the spectrum of the hydrogen molecule ion (H_2^+) that used a generalization of the Bohr model to treat a problem involving two nuclei. This latter work formed the basis of the PhD thesis of Pauli, who was to become one of the principal players in the development of a more mature and comprehensive theory of atoms and molecules.

In retrospect, the Bohr model of the hydrogen atom contains several flaws. Perhaps most prominent among these is that the angular momentum of the hydrogen ground state ($n = 1$) given by the model is $\hbar$; it is now known that the correct value is zero. Efforts to remedy the Bohr model for its insufficiencies, pursued doggedly by Sommerfeld and others, were ultimately unsuccessful. This 'old' quantum theory was replaced in the 1920s by a considerably more abstract framework that forms the basis for our current understanding of the detailed physics governing chemical processes. The modern quantum theory, unlike Bohr's, does not involve classical ideas coupled with an *ad hoc* incorporation of Planck's quantum hypothesis. It is instead founded upon a limited number of fundamental principles that cannot be proven, but must be regarded as laws of nature. While the modern theory of quantum mechanics is exceedingly complex and fraught with certain philosophical paradoxes (which will not be discussed), it has withstood the test of time; no contradiction between predictions of the theory and actual atomic or molecular phenomena has ever been observed.

The purpose of this chapter is to provide an introduction to the basic framework of quantum mechanics, with an emphasis on aspects that are most relevant for the study of atoms and molecules. After summarizing the basic principles of the subject that represent required knowledge for all students of physical chemistry, the independent-particle approximation so important in molecular quantum mechanics is introduced. A significant effort is made to describe this approach in detail and to communicate how it is used as a foundation for qualitative understanding and as a basis for more accurate treatments. Following this, the basic techniques used in accurate calculations that go beyond the independent-particle picture (variational method and perturbation theory) are described, with some attention given to how they are actually used in practical calculations.

It is clearly impossible to present a comprehensive discussion of quantum mechanics in a chapter of this length. Instead, one is forced to present cursory overviews of many topics or to limit the scope and provide a more rigorous treatment of a select group of subjects. The latter alternative has been followed here. Consequently, many areas of quantum mechanics are largely ignored. For the most part, however, the areas lightly touched upon or completely absent from this chapter are specifically dealt with elsewhere in the encyclopedia. Notable among these are the interaction between matter and radiation, spin and magnetism, techniques of quantum chemistry including the Born–Oppenheimer approximation, the Hartree–Fock method and electron correlation, scattering theory and the treatment of internal nuclear motion (rotation and vibration) in molecules.

A1.1.2   Concepts of quantum mechanics

A1.1.2.1   Beginnings and fundamental postulates

The modern quantum theory derives from work done independently by Heisenberg and Schrödinger in the mid-1920s. Superficially, the mathematical formalisms developed by these individuals appear very different; the quantum mechanics of Heisenberg is based on the properties of matrices, while that of Schrödinger is founded upon a differential equation that bears similarities to those used in the classical theory of waves. Schrödinger's formulation was strongly influenced by the work of de Broglie, who made the revolutionary hypothesis that entities previously thought to be strictly particle-like (electrons) can exhibit wavelike behaviour (such as diffraction) with particle 'wavelength' and momentum (p) related by the equation $\lambda = h/p$. This truly startling premise was subsequently verified independently by Davisson and Germer as well as by Thomson, who showed that electrons exhibit diffraction patterns when passed through crystals and very small circular apertures, respectively. Both the treatment of Heisenberg, which did not make use of wave theory concepts, and that of Schrödinger were successfully applied to the calculation of the hydrogen atom spectrum. It was ultimately proven by both Pauli and Schrödinger that the 'matrix mechanics' of Heisenberg and the 'wave mechanics' of Schrödinger are mathematically equivalent. Connections between the two methods were further clarified by the transformation theory of Dirac and Jordan. The importance of this new quantum theory was recognized immediately and Heisenberg, Schrödinger and Dirac shared the 1932 Nobel Prize in physics for their work.

While not unique, the Schrödinger picture of quantum mechanics is the most familiar to chemists principally because it has proven to be the simplest to use in practical calculations. Hence, the remainder of this section will focus on the Schrödinger formulation and its associated wavefunctions, operators and eigenvalues. Moreover, effects associated with the special theory of relativity (which include spin) will be ignored in this subsection. Treatments of alternative formulations of quantum mechanics and discussions of relativistic effects can be found in the reading list that accompanies this chapter.

Like the geometry of Euclid and the mechanics of Newton, quantum mechanics is an axiomatic subject. By making several assertions, or postulates, about the mathematical properties of and physical interpretation associated with solutions to the Schrödinger equation, the subject of quantum mechanics can be applied to understand behaviour in atomic and molecular systems. The first of these postulates is:

1. Corresponding to any collection of n particles, there exists a time-dependent function $\Psi(q_1, q_2, \ldots, q_n; t)$ that comprises all information that can be known about the system. This function must be continuous and single valued, and have continuous first derivatives at all points where the classical force has a finite magnitude.

In classical mechanics, the state of the system may be completely specified by the set of Cartesian particle coordinates r_i and velocities dr_i/dt at any given time. These evolve according to Newton's equations of motion. In principle, one can write down equations involving the state variables and forces acting on the particles which can be solved to give the location and velocity of each particle at any later (or earlier) time t', provided one knows the precise state of the classical system at time t. In quantum mechanics, the state of the system at time t is instead described by a well behaved mathematical function of the particle coordinates q_i rather than a simple list of positions and velocities. The relationship between this *wavefunction* (sometimes called *state function*) and the location of particles in the system forms the basis for a second postulate:

2. The product of $\Psi(q_1, q_2, \ldots, q_n; t)$ and its complex conjugate has the following physical interpretation. The probability of finding the n particles of the system in the regions bounded by the coordinates $q_1', q_2', \ldots, q_n'$ and $q_1'', q_2'', \ldots, q_n''$ at time t is proportional to the integral

$$\int_{q_1'}^{q_1''} \int_{q_2'}^{q_2''} \cdots \int_{q_n'}^{q_n''} \Psi^*(q_1, q_2, \ldots, q_n; t)\Psi(q_1, q_2, \ldots q_n; t)\,dq_1\,dq_2 \cdots dq_n. \qquad (A1.1.10)$$

The proportionality between the integral and the probability can be replaced by an equivalence if the wavefunction is scaled appropriately. Specifically, since the probability that the n particles will be found somewhere must be unity, the wavefunction can be scaled so that the equality

$$\int \Psi^*(q_1, q_2, \ldots, q_n; t)\Psi(q_1, q_2, \ldots, q_n; t)\,d\tau = 1 \qquad (A1.1.11)$$

is satisfied. The symbol $d\tau$ introduced here and used throughout the remainder of this section indicates that the integral is to be taken over the full range of all particle coordinates. Any wavefunction that satisfies equation (A1.1.11) is said to be *normalized*. The product $\Psi^*\Psi$ corresponding to a normalized wavefunction is sometimes called a probability, but this is an imprecise use of the word. It is instead a *probability density*, which must be integrated to find the chance that a given measurement will find the particles in a certain region of space. This distinction can be understood by considering the classical counterpart of $\Psi^*\Psi$ for a single particle moving on the x-axis. In classical mechanics, the probability at time t for finding the particle at the coordinate (x') obtained by propagating Newton's equations of motion from some set of initial conditions is exactly equal to one; it is zero for any other value of x. What is the corresponding probability density function, $P(x; t)$? Clearly, $P(x; t)$ vanishes at all points other than x' since its integral over any interval that does not include x' must equal zero. At x', the value of $P(x; t)$ must be chosen so that the normalization condition

$$\int_{-\infty}^{\infty} P(x; t)\,dx = 1 \qquad (A1.1.12)$$

is satisfied. Functions such as this play a useful role in quantum mechanics. They are known as *Dirac delta functions*, and are designated by $\delta(r - r_0)$. These functions have the properties

$$\int \delta(r - r_0)\,d\tau = 1 \qquad (A1.1.13)$$

$$\int f(r)\delta(r - r_0)\, d\tau = f(r_0) \tag{A1.1.14}$$

$$\delta(r - r_0) = 0 \qquad \text{for } r \neq r_0. \tag{A1.1.15}$$

Although a seemingly odd mathematical entity, it is not hard to appreciate that a simple one-dimensional realization of the classical $P(x; t)$ can be constructed from the familiar Gaussian distribution centred about x' by letting the standard deviation (σ) go to zero,

$$P(x; t) = \lim_{\sigma \to 0} \frac{1}{\sqrt{2\pi\sigma^2}} \exp\left[\frac{-(x - x')^2}{2\sigma^2}\right]. \tag{A1.1.16}$$

Hence, although the probability for finding the particle at x' is equal to one, the corresponding probability density function is infinitely large. In quantum mechanics, the probability density is generally nonzero for all values of the coordinates, and its magnitude can be used to determine which regions are most likely to contain particles. However, because the number of possible coordinates is infinite, the probability associated with any precisely specified choice is zero. The discussion above shows a clear distinction between classical and quantum mechanics; given a set of initial conditions, the locations of the particles are determined exactly at all future times in the former, while one generally can speak only about the probability associated with a given range of coordinates in quantum mechanics.

To extract information from the wavefunction about properties other than the probability density, additional postulates are needed. All of these rely upon the mathematical concepts of operators, eigenvalues and eigenfunctions. An extensive discussion of these important elements of the formalism of quantum mechanics is precluded by space limitations. For further details, the reader is referred to the reading list supplied at the end of this chapter.

In quantum mechanics, the classical notions of position, momentum, energy etc are replaced by mathematical operators that act upon the wavefunction to provide information about the system. The third postulate relates to certain properties of these operators:

3. Associated with each system property A is a linear, Hermitian operator $\hat{A}$.

Although not a unique prescription, the quantum-mechanical operators $\hat{A}$ can be obtained from their classical counterparts A by making the substitutions $x \to x$ (coordinates); $t \to t$ (time); $p_q \to -i\hbar\partial/\partial q$ (component of momentum). Hence, the quantum-mechanical operators of greatest relevance to the dynamics of an n-particle system such as an atom or molecule are:

Dynamical variable A	Classical quantity	Quantum-mechanical operator $\hat{A}$
Time	t	t
Position of particle i	r_i	r_i
Momentum of particle i	$m_i v_i$	$-i\hbar\nabla_i$
Angular momentum of particle i	$m_i v_i \times r_i$	$-i\hbar\nabla_i \times r_i$
Kinetic energy of particle i	$\frac{p_i \cdot p_i}{2m_i}$	$-\frac{\hbar^2}{2m_i}\nabla_i^2$
Potential energy	$V(q, t)$	$V(q, t)$

where the gradient

$$\nabla_i \equiv \frac{\partial}{\partial x_i}\boldsymbol{i} + \frac{\partial}{\partial y_i}\boldsymbol{j} + \frac{\partial}{\partial z_i}\boldsymbol{k} \tag{A1.1.17}$$

and Laplacian

$$\nabla_i^2 \equiv \nabla \cdot \nabla = \frac{\partial^2}{\partial x_i^2} + \frac{\partial^2}{\partial y_i^2} + \frac{\partial^2}{\partial z_i^2} \tag{A1.1.18}$$

operators have been introduced. Note that a potential energy which depends upon only particle coordinates and time has exactly the same form in classical and quantum mechanics. A particularly useful operator in quantum mechanics is that which corresponds to the total energy. This *Hamiltonian* operator is obtained by simply adding the potential and kinetic energy operators

$$\hat{H} \equiv \hat{T} + \hat{V} = - \sum_{\substack{\text{particles} \\ i}} \frac{\hbar^2}{2m_i} \left[\frac{\partial^2}{\partial x_i^2} + \frac{\partial^2}{\partial y_i^2} + \frac{\partial^2}{\partial z_i^2} \right] + V(q, t). \qquad \text{(A1.1.19)}$$

The relationship between the abstract quantum-mechanical operators $\hat{A}$ and the corresponding physical quantities A is the subject of the fourth postulate, which states:

4. If the system property A is measured, the only values that can possibly be observed are those that correspond to eigenvalues of the quantum-mechanical operator $\hat{A}$.

An illustrative example is provided by investigating the possible momenta for a single particle travelling in the x-direction, p_x. First, one writes the equation that defines the eigenvalue condition

$$\hat{p}_x f(x) = -i\hbar \frac{\mathrm{d}f(x)}{\mathrm{d}x} = \lambda f(x) \qquad \text{(A1.1.20)}$$

where λ is an eigenvalue of the momentum operator and $f(x)$ is the associated eigenfunction. It is easily verified that this differential equation has an infinite number of solutions of the form

$$f_k(x) = A \exp(ikx) \qquad \text{(A1.1.21)}$$

with corresponding eigenvalues

$$\lambda_k = \hbar k \qquad \text{(A1.1.22)}$$

in which k can assume any value. Hence, nature places no restrictions on allowed values of the linear momentum. Does this mean that a quantum-mechanical particle in a particular state $\Psi(x; t)$ is allowed to have any value of p_x? The answer to this question is 'yes', but the interpretation of its consequences rather subtle. Eventually a fifth postulate will be required to establish the connection between the quantum-mechanical wavefunction Ψ and the possible outcomes associated with measuring properties of the system. It turns out that the set of possible momenta for our particle depends entirely on its wavefunction, as might be expected from the first postulate given above. The infinite set of solutions to equation (A1.1.20) means only that no values of the momentum are excluded, in the sense that they can be associated with a particle described by an *appropriately chosen* wavefunction. However, the choice of a *specific* function might (or might not) impose restrictions on which values of p_x are allowed.

The rather complicated issues raised in the preceding paragraph are central to the subject of quantum mechanics, and their resolution forms the basis of one of the most important postulates associated with the Schrödinger formulation of the subject. In the example above, discussion focuses entirely on the eigenvalues of the momentum operator. What significance, if any, can be attached to the eigenfunctions of quantum-mechanical operators? In the interest of simplicity, the remainder of this subsection will focus entirely on the quantum mechanics associated with operators that have a finite number of eigenvalues. These are said to have a *discrete spectrum*, in contrast to those such as the linear momentum, which have a *continuous spectrum*. Discrete spectra of eigenvalues arise whenever boundaries limit the region of space in which a system can be. Examples are particles in hard-walled boxes, or soft-walled shells and particles attached to springs. The results developed below can all be generalized to the continuous case, but at the expense of increased mathematical complexity. Readers interested in these details should consult chapter 1 of Landau and Lifschitz (see additional reading).

It can be shown that the eigenfunctions of Hermitian operators necessarily exhibit a number of useful mathematical properties. First, if all eigenvalues are distinct, the set of eigenfunctions $\{f_1, f_2 \cdots f_n\}$ are *orthogonal* in the sense that the integral of the product formed from the complex conjugate of eigenfunction j (f_j^*) and eigenfunction k (f_k) vanishes unless $j = k$,

$$\int f_j^* f_k \, d\tau = 0 \text{ if } j \neq k. \tag{A1.1.23}$$

If there are identical eigenvalues (a common occurrence in atomic and molecular quantum mechanics), it is permissible to form linear combinations of the eigenfunctions corresponding to these *degenerate* eigenvalues, as these must also be eigenfunctions of the operator. By making a judicious choice of the expansion coefficients, the degenerate eigenfunctions can also be made orthogonal to one another. Another useful property is that the set of eigenfunctions is said to be *complete*. This means that any function of the coordinates that appear in the operator can be written as a linear combination of its eigenfunctions, provided that the function obeys the same *boundary conditions* as the eigenfunctions and shares any fundamental symmetry property that is common to all of them. If, for example, all of the eigenfunctions vanish at some point in space, then only functions that vanish at the same point can be written as linear combinations of the eigenfunctions. Similarly, if the eigenfunctions of a particular operator in one dimension are all odd functions of the coordinate, then all linear combinations of them must also be odd. It is clearly impossible in the latter case to expand functions such as $\cos(x)$, $\exp(-x^2)$ etc in terms of odd functions. This qualification is omitted in some elementary treatments of quantum mechanics, but it is one that turns out to be important for systems containing several identical particles. Nevertheless, if these criteria are met by a suitable function g, then it is always possible to find coefficients c_k such that

$$g = \sum_k c_k f_k \tag{A1.1.24}$$

where the coefficient c_j is given by

$$c_j = \frac{\int f_j^* g \, d\tau}{\int f_j^* f_j \, d\tau}. \tag{A1.1.25}$$

If the eigenfunctions are normalized, this expression reduces to

$$c_j = \int f_j^* g \, d\tau. \tag{A1.1.26}$$

When normalized, the eigenfunctions corresponding to a Hermitian operator are said to represent an *orthonormal* set.

The mathematical properties discussed above are central to the next postulate:

5. In any experiment, the probability of observing a particular non-degenerate value for the system property A can be determined by the following procedure. First, expand the wavefunction in terms of the complete set of normalized eigenfunctions of the quantum-mechanical operator, $\hat{A}$,

$$\Psi = \sum_j c_j \phi_j. \tag{A1.1.27}$$

The probability of measuring $A = \lambda_k$, where λ_k is the eigenvalue associated with the normalized eigenfunction ϕ_k, is precisely equal to $|c_k|^2$ $(\equiv c_k^* c_k)$. For degenerate eigenvalues, the probability of observation is given by $\sum |c_k|^2$, where the sum is taken over all of the eigenfunctions ϕ_k that correspond to the degenerate eigenvalue λ_k.

At this point, it is appropriate to mention an elementary concept from the theory of probability. If there are n possible numerical outcomes (ξ_n) associated with a particular process, the average value $\langle \xi \rangle$ can be calculated by summing up all of the outcomes, each weighted by its corresponding probability

$$\langle \xi \rangle = \sum_i P_i \xi_i. \qquad (A1.1.28)$$

As an example, the possible outcomes and associated probabilities for rolling a pair of six-sided dice are

Sum	Probability
2	1/36
3	1/18
4	1/12
5	1/9
6	5/36
7	1/6
8	5/36
9	1/9
10	1/12
11	1/18
12	1/36

The average value is therefore given by the sum $\frac{1}{36}(2) + \frac{1}{18}(3) + \frac{1}{12}(4) + \frac{1}{9}(5) + \frac{5}{36}(6) + \frac{1}{6}(7) + \frac{5}{36}(8) + \frac{1}{9}(9) + \frac{1}{12}(10) + \frac{1}{18}(11) + \frac{1}{36}(12) = 7$.

What does this have to do with quantum mechanics? To establish a connection, it is necessary to first expand the wavefunction in terms of the eigenfunctions of a quantum-mechanical operator $\hat{A}$,

$$\Psi = \sum_k c_k \phi_k. \qquad (A1.1.29)$$

We will assume that both the wavefunction and the orthogonal eigenfunctions are normalized, which implies that

$$\sum_j c_j^* c_j = \sum_j |c_j|^2 = 1. \qquad (A1.1.30)$$

Now, the operator $\hat{A}$ is applied to both sides of equation (A1.1.29), which because of its *linearity*, gives

$$\hat{A}\Psi = \hat{A}\sum_k c_k \phi_k = \sum_k c_k \hat{A}\phi_k = \sum_k c_k \lambda_k \phi_k \qquad (A1.1.31)$$

where λ_k represents the eigenvalue associated with the eigenfunction ϕ_k. Next, both sides of the preceding equation are multiplied from the left by the complex conjugate of the wavefunction and integrated over all space

$$\int \Psi^* \hat{A}\Psi \, d\tau = \int \sum_j \sum_k c_j^* c_k \lambda_k \phi_j^* \phi_k \, d\tau \qquad (A1.1.32)$$

$$= \sum_j \sum_k c_j^* c_k \lambda_k \int \phi_j^* \phi_k \, d\tau \qquad (A1.1.33)$$

$$= \sum_k c_k^* c_k \lambda_k = \sum_k |c_k|^2 \lambda_k. \qquad (A1.1.34)$$

The last identity follows from the orthogonality property of eigenfunctions and the assumption of normalization. The right-hand side in the final result is simply equal to the sum over all eigenvalues of the operator (possible results of the measurement) multiplied by the respective probabilities. Hence, an important corollary to the fifth postulate is established:

$$\langle A \rangle = \int \Psi^* \hat{A} \Psi \, d\tau. \tag{A1.1.35}$$

This provides a recipe for calculating the average value of the system property associated with the quantum-mechanical operator $\hat{A}$, for a specific but arbitrary choice of the wavefunction Ψ, notably those choices which are not eigenfunctions of $\hat{A}$.

The fifth postulate and its corollary are extremely important concepts. Unlike classical mechanics, where everything can in principle be known with precision, one can generally talk only about the probabilities associated with each member of a set of possible outcomes in quantum mechanics. By making a measurement of the quantity A, all that can be said with certainty is that *one* of the eigenvalues of $\hat{A}$ will be observed, and its probability can be calculated precisely. However, if it happens that the wavefunction corresponds to one of the eigenfunctions of the operator $\hat{A}$, then and only then is the outcome of the experiment certain: the measured value of A will be the corresponding eigenvalue.

Up until now, little has been said about time. In classical mechanics, complete knowledge about the system at any time t suffices to predict with absolute certainty the properties of the system at any other time t'. The situation is quite different in quantum mechanics, however, as it is not possible to know everything about the system at *any* time t. Nevertheless, the temporal behavior of a quantum-mechanical system evolves in a well defined way that depends on the Hamiltonian operator and the wavefunction Ψ according to the last postulate

6. The time evolution of the wavefunction is described by the differential equation

$$i\hbar \frac{d}{dt} \Psi(q_1, q_2, \ldots, q_n; t) = \hat{H} \Psi(q_1, q_2, \ldots, q_n; t). \tag{A1.1.36}$$

The differential equation above is known as the time-dependent Schrödinger equation. There is an interesting and intimate connection between this equation and the classical expression for a travelling wave

$$A(x, t) = A \exp\left(2\pi i \left\{ \frac{x}{\lambda} - \nu t \right\}\right). \tag{A1.1.37}$$

To convert (A1.1.37) into a quantum-mechanical form that describes the 'matter wave' associated with a free particle travelling through space, one might be tempted to simply make the substitutions $\nu = E/h$ (Planck's hypothesis) and $\lambda = h/p$ (de Broglie's hypothesis). It is relatively easy to verify that the resulting expression satisfies the time-dependent Schrödinger equation. However, it should be emphasized that this is not a derivation, as there is no compelling reason to believe that this *ad hoc* procedure should yield one of the fundamental equations of physics. Indeed, the time-dependent Schrödinger equation cannot be derived in a rigorous way and therefore must be regarded as a postulate.

The time-dependent Schrödinger equation allows the precise determination of the wavefunction at any time t from knowledge of the wavefunction at some initial time, provided that the forces acting within the system are known (these are required to construct the Hamiltonian). While this suggests that quantum mechanics has a deterministic component, it must be emphasized that it is not the observable system properties that evolve in a precisely specified way, but rather the probabilities associated with values that might be found for them in a measurement.

A1.1.2.2   Stationary states, superposition and uncertainty

From the very beginning of the 20th century, the concept of energy conservation has made it abundantly clear that electromagnetic energy emitted from and absorbed by material substances must be accompanied by compensating energy changes within the material. Hence, the discrete nature of atomic line spectra suggested that only certain energies are allowed by nature for each kind of atom. The wavelengths of radiation emitted or absorbed must therefore be related to the *difference* between energy levels via Planck's hypothesis, $\Delta E = h\nu = hc/\lambda$.

The Schrödinger picture of quantum mechanics summarized in the previous subsection allows an important deduction to be made that bears directly on the subject of energy levels and spectroscopy. Specifically, the energies of spectroscopic transitions must correspond precisely to differences between distinct eigenvalues of the Hamiltonian operator, as these correspond to the allowed energy levels of the system. Hence, the set of eigenvalues of the Hamiltonian operator are of central importance in chemistry. These can be determined by solving the so-called *time-independent Schrödinger equation*,

$$H\psi_k(q_1, q_2, \ldots, q_n) = E_k\psi_k(q_1, q_2, \ldots, q_n) \tag{A1.1.38}$$

for the eigenvalues E_k and eigenfunctions ψ_k. It should be clear that the set of eigenfunctions and eigenvalues does not evolve with time provided the Hamiltonian operator itself is time independent. Moreover, since the eigenfunctions of the Hamiltonian (like those of any other operator) form a complete set, it is always possible to expand the exact wavefunction of the system at any time in terms of them:

$$\Psi(q_1, q_2, \ldots, q_n; t) = \sum_j c_j(t)\psi_j(q_1, q_2, \ldots, q_n). \tag{A1.1.39}$$

It is important to point out that this expansion is valid even if time-dependent terms are added to the Hamiltonian (as, for example, when an electric field is turned on). If there is more than one nonzero value of c_j at any time t, then the system is said to be in a *superposition* of the energy eigenstates ψ_k associated with non-vanishing expansion coefficients, c_k. If it were possible to measure energies directly, then the fifth postulate of the previous section tells us that the probability of finding energy E_k in a given measurement would be $c_k^* c_k$.

When a molecule is isolated from external fields, the Hamiltonian contains only kinetic energy operators for all of the electrons and nuclei as well as terms that account for repulsion and attraction between all distinct pairs of like and unlike charges, respectively. In such a case, the Hamiltonian is constant in time. When this condition is satisfied, the representation of the time-dependent wavefunction as a superposition of Hamiltonian eigenfunctions can be used to determine the time dependence of the expansion coefficients. If equation (A1.1.39) is substituted into the time-dependent Schrödinger equation

$$i\hbar\frac{d}{dt}\sum_k c_k(t)\psi_k = H\sum_k c_k(t)\psi_k \tag{A1.1.40}$$

the simplification

$$i\hbar\sum_k \psi_k\frac{d}{dt}c_k(t) = \sum_k E_k c_k(t)\psi_k \tag{A1.1.41}$$

can be made to the right-hand side since the restriction of a time-independent Hamiltonian means that ψ_k is *always* an eigenfunction of H. By simply equating the coefficients of the ψ_k, it is easy to show that the choice

$$c_k(t) = c_k(0)\exp\left(\frac{iE_k t}{\hbar}\right) \tag{A1.1.42}$$

for the time-dependent expansion coefficients satisfies equation (A1.1.41). Like any differential equation, there are an infinite number of solutions from which a choice must be made to satisfy some set of initial conditions. The state of the quantum-mechanical system at time $t = 0$ is used to fix the arbitrary multipliers $c_k(0)$, which can always be chosen as real numbers. Hence, the wavefunction Ψ becomes

$$\Psi = \sum_k c_k(0) \exp\left(\frac{iE_k t}{\hbar}\right) \psi_k. \qquad (A1.1.43)$$

Suppose that the system property A is of interest, and that it corresponds to the quantum-mechanical operator $\hat{A}$. The average value of A obtained in a series of measurements can be calculated by exploiting the corollary to the fifth postulate

$$\langle A \rangle = \int \Psi^* \hat{A} \Psi \, d\tau = \sum_j \sum_k c_j(0) c_k(0) \int \exp\left(\frac{-iE_j t}{\hbar}\right) \psi_j^* \hat{A} \exp\left(\frac{iE_k t}{\hbar}\right) \psi_k \, d\tau. \qquad (A1.1.44)$$

Now consider the case where $\hat{A}$ is itself a time-independent operator, such as that for the position, momentum or angular momentum of a particle or even the energy of the benzene molecule. In these cases, the time-dependent expansion coefficients are unaffected by application of the operator, and one obtains

$$\langle A \rangle = \sum_j \sum_k c_j(0) c_k(0) \exp\left[\frac{i(E_k - E_j)t}{\hbar}\right] \int \psi_j^* \hat{A} \psi_k \, d\tau$$

$$= \sum_j |c_j(0)|^2 \int \psi_j \hat{A} \psi_j + \sum_j \sum_{k \neq j} c_j(0) c_k(0) \cos\left[\frac{(E_j - E_k)t}{\hbar}\right] \int \psi_j^* \hat{A} \psi_k \, d\tau. \qquad (A1.1.45)$$

As one might expect, the first term that contributes to the expectation value of A is simply its value at $t = 0$, while the second term exhibits an oscillatory time dependence. If the superposition initially includes large contributions from states of widely varying energy, then the oscillations in $\langle A \rangle$ will be rapid. If the states that are strongly mixed have similar energies, then the timescale for oscillation in the properties will be slower. However, there is one special class of system properties A that exhibit no time dependence whatsoever. If (and only if) every one of the states ψ_k is an eigenfunction of $\hat{A}$, then the property of orthogonality can be used to show that every contribution to the second term vanishes. An obvious example is the Hamiltonian operator itself; it turns out that the expectation value for the energy of a system subjected to forces that do not vary with time is a constant. Are there other operators that share the same set of eigenfunctions ψ_k with $\hat{H}$, and if so, how can they be recognized? It can be shown that any two operators which satisfy the property

$$\hat{A}\hat{B}f = \hat{B}\hat{A}f \Rightarrow [\hat{A}, \hat{B}]f = 0 \qquad (A1.1.46)$$

for all functions f share a common set of eigenfunctions, and A and B are said to *commute*. (The symbol $[\hat{A}, \hat{B}]$ meaning $\hat{A}\hat{B} - \hat{B}\hat{A}$, is called the *commutator* of the operators $\hat{A}$ and $\hat{B}$.) Hence, there is no time dependence for the expectation value of any system property that corresponds to a quantum-mechanical operator that commutes with the Hamiltonian. Accordingly, these quantities are known as *constants of the motion*: their average values will not vary, provided the environment of the system does not change (as it would, for example, if an electromagnetic field were suddenly turned on). In nonrelativistic quantum mechanics, two examples of constants of the motion are the square of the total angular momentum, as well as its projection along an arbitrarily chosen axis. Other operators, such as that for the dipole moment, do not commute with the Hamiltonian and the expectation value associated with the corresponding properties can indeed oscillate with time. It is important to note that the frequency of these oscillations is given by differences between the allowed energies of the system divided by Planck's constant. These are the so-called *Bohr frequencies*, and it is perhaps not surprising that these are exactly the frequencies of electromagnetic radiation that cause transitions between the corresponding energy levels.

Close inspection of equation (A1.1.45) reveals that, under very special circumstances, the expectation value does not change with time for *any* system properties that correspond to fixed (static) operator representations. Specifically, if the spatial part of the time-dependent wavefunction is the exact eigenfunction ψ_j of the Hamiltonian, then $c_j(0) = 1$ (the zero of time can be chosen arbitrarily) and all other $c_k(0) = 0$. The second term clearly vanishes in these cases, which are known as *stationary states*. As the name implies, all observable properties of these states do not vary with time. In a stationary state, the energy of the system has a precise value (the corresponding eigenvalue of $\hat{H}$) as do observables that are associated with operators that commute with $\hat{H}$. For all other properties (such as the position and momentum), one can speak only about average values or probabilities associated with a given measurement, but these quantities themselves do not depend on time. When an external perturbation such as an electric field is applied or a collision with another atom or molecule occurs, however, the system and its properties generally will evolve with time. The energies that can be absorbed or emitted in these processes correspond precisely to differences in the stationary state energies, so it should be clear that solving the time-independent Schrödinger equation for the stationary state wavefunctions and eigenvalues provides a wealth of spectroscopic information. The importance of stationary state solutions is so great that it is common to refer to equation (A1.1.38) as 'the Schrödinger equation', while the qualified name 'time-dependent Schrödinger equation' is generally used for equation (A1.1.36). Indeed, the subsequent subsections are devoted entirely to discussions that centre on the former and its exact and approximate solutions, and the qualifier 'time independent' will be omitted.

Starting with the quantum-mechanical postulate regarding a one-to-one correspondence between system properties and Hermitian operators, and the mathematical result that only operators which commute have a common set of eigenfunctions, a rather remarkable property of nature can be demonstrated. Suppose that one desires to determine the values of the two quantities A and B, and that the corresponding quantum-mechanical operators do not commute. In addition, the properties are to be measured simultaneously so that both reflect the same quantum-mechanical state of the system. If the wavefunction is neither an eigenfunction of $\hat{A}$ nor $\hat{B}$, then there is necessarily some uncertainty associated with the measurement. To see this, simply expand the wavefunction ψ in terms of the eigenfunctions of the relevant operators

$$\psi = \sum_k a_k f_k^A \tag{A1.1.47}$$

$$\psi = \sum_k b_k f_k^B \tag{A1.1.48}$$

where the eigenfunctions f_k^A and f_k^B of operators $\hat{A}$ and $\hat{B}$, respectively, are associated with corresponding eigenvalues λ_k^A and λ_k^B. Given that ψ is not an eigenfunction of either operator, at least two of the coefficients a_k and two of the b_k must be nonzero. Since the probability of observing a particular eigenvalue is proportional to the square of the expansion coefficient corresponding to the associated eigenfunction, there will be no less than four possible outcomes for the set of values A and B. Clearly, they both cannot be determined precisely. Indeed, under these conditions, neither of them can be!

In a more favourable case, the wavefunction ψ might indeed correspond to an eigenfunction of one of the operators. If $\psi = f_m^A$, then a measurement of A necessarily yields λ_m^A, and this is an unambiguous result. What can be said about the measurement of B in this case? It has already been said that the eigenfunctions of two commuting operators are identical, but here the pertinent issue concerns eigenfunctions of two operators that do not commute. Suppose f_m^A is an eigenfunction of $\hat{A}$. Then, it must be true that

$$\hat{A} f_m^A = \lambda_m^A f_m^A$$
$$\hat{B}\hat{A} f_m^A = \lambda_m^A \hat{B} f_m^A. \tag{A1.1.49}$$

If f_m^A is also an eigenfunction of $\hat{B}$, then it follows that $\hat{A}\hat{B} f_m^A = \hat{B}\hat{A} f_m^A = \lambda_m^A \lambda_m^B f_m^A$, which contradicts the assumption that $\hat{A}$ and $\hat{B}$ do not commute. Hence, no nontrivial eigenfunction of $\hat{A}$ can also be an eigenfunction of $\hat{B}$. Therefore, if measurement of A yields a precise result, then some uncertainty must be associated with B. That is, the expansion of ψ in terms of eigenfunctions of $\hat{B}$ (equation (A1.1.48)) must have at least two non-vanishing coefficients; the corresponding eigenvalues therefore represent distinct possible outcomes of the experiment, each having probability $b_k^* b_k$. A physical interpretation of $\hat{A} f_m^A$ is the process of measuring the value of A for a system in a state with a unique value for this property λ_m^A. However $\hat{B} f_m^A$ represents a measurement that changes the state of the system, so that if after we measure B and then measure A, we would no longer find λ_m^A as its value: $\hat{B} \hat{A} f_m^A = \lambda_m^A \hat{B} f_m^A \neq \hat{A} \hat{B} f_m^A$.

The *Heisenberg uncertainty principle* offers a rigorous treatment of the qualitative picture sketched above. If several measurements of A and B are made for a system in a particular quantum state, then quantitative uncertainties are provided by standard deviations in the corresponding measurements. Denoting these as σ_A and σ_B, respectively, it can be shown that

$$\sigma_A \sigma_B \geq \tfrac{1}{2} |\langle [\hat{A}, \hat{B}] \rangle|. \tag{A1.1.50}$$

One feature of this inequality warrants special attention. In the previous paragraph it was shown that the precise measurement of A made possible when ψ is an eigenfunction of $\hat{A}$ necessarily results in some uncertainty in a simultaneous measurement of B when the operators $\hat{A}$ and $\hat{B}$ do not commute. However, the mathematical statement of the uncertainty principle tells us that measurement of B is in fact completely uncertain: one can say nothing at all about B apart from the fact that any and all values of B are equally probable! A specific example is provided by associating A and B with the position and momentum of a particle moving along the x-axis. It is rather easy to demonstrate that $[p_x, x] = -i\hbar$, so that $\sigma_{p_x} \sigma_x \geq \hbar/2$. If the system happens to be described by a Dirac delta function at the point x_0 (which is an eigenfunction of the position operator corresponding to eigenvalue x_0), then the probabilities associated with possible momenta can be determined by expanding $\delta(x - x_0)$ in terms of the momentum eigenfunctions $A \exp(ikx)$. Carrying out such a calculation shows that all of the infinite number of possible momenta (the momentum operator has a continuous spectrum) appear in the wavefunction expansion, all with precisely the same weight. Hence, no particular momentum or (more properly in this case) range bounded by $p_x + dp_x$ is more likely to be observed than any other.

A1.1.2.3 *Some qualitative features of stationary states*

A great number of qualitative features associated with the stationary states that correspond to solutions of the time-independent Schrödinger can be worked out from rather general mathematical considerations and use of the postulates of quantum mechanics. Mastering these concepts and the qualifications that may apply to them is essential if one is to obtain an intuitive feeling for the subject. In general, the systems of interest to chemists are atoms and molecules, both in isolation as well as how they interact with each other or with an externally applied field. In all of these cases, the forces acting upon the particles in the system give rise to a potential energy function that varies with the positions of the particles, strength of the applied fields etc. In general, the potential is a smoothly varying function of the coordinates, either growing without bound for large values of the coordinates or tending asymptotically towards a finite value. In these cases, there is necessarily a minimum value at what is known as the global equilibrium position (there may be several global minima that are equivalent by symmetry). In many cases, there are also other minima (meaning that the matrix of second derivatives with respect to the coordinates has only non-negative eigenvalues) that have higher energies, which are called local minima. If the potential becomes infinitely large for infinite values of the coordinates (as it does, for example, when the force on a particle varies linearly with its displacement from equilibrium) then all solutions to the Schrödinger equation are known as *bound states*; that with the smallest eigenvalue is called the *ground state* while the others are called *excited states*. In other cases, such as potential functions that represent realistic models for diatomic molecules by approaching a constant finite value at large separation (zero force on the particles, with a finite dissociation energy), there are two classes of solutions. Those associated with eigenvalues that are below the asymptotic value of the potential energy

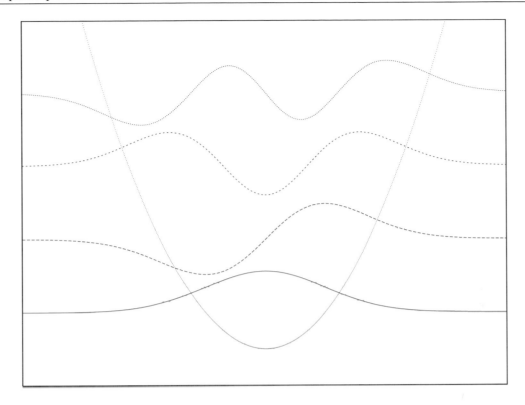

Figure A1.1.1. Wavefunctions for the four lowest states of the harmonic oscillator, ordered from the $n = 0$ ground state (at the bottom) to the $n = 3$ state (at the top). The vertical displacement of the plots is chosen so that the location of the classical turning points are those that coincide with the superimposed potential function (dotted line). Note that the number of nodes in each state corresponds to the associated quantum number.

are the bound states, of which there is usually a finite number; those having higher energies are called the *scattering* (or *continuum*) states and form a continuous spectrum. The latter are dealt with in section A3.11 of the encyclopedia and will be mentioned here only when necessary for mathematical reasons.

Bound state solutions to the Schrödinger equation decay to zero for infinite values of the coordinates, and are therefore integrable since they are continuous functions in accordance with the first postulate. The solutions may assume zero values elsewhere in space and these regions—which may be a point, a plane or a three- or higher-dimensional hypersurface—are known as *nodes*. From the mathematical theory of differential eigenvalue equations, it can be demonstrated that the lowest eigenvalue is always associated with an eigenfunction that has the same sign at all points in space. From this result, which can be derived from the calculus of variations, it follows that the wavefunction corresponding to the smallest eigenvalue of the Hamiltonian must have no nodes. It turns out, however, that relativistic considerations require that this statement be qualified. For systems that contain more than two identical particles of a specific type, not all solutions to the Schrödinger equation are allowed by nature. Because of this restriction, which is described in subsection A1.1.3.3, it turns out that the ground states of lithium, all larger atoms and all molecules other than H_2^+, H_2 and isoelectronic species have nodes. Nevertheless, our conceptual understanding of electronic structure as well as the basis for almost all highly accurate calculations is ultimately rooted in a single-particle approximation. The quantum mechanics of one-particle systems is therefore important in chemistry.

Shapes of the ground- and first three excited-state wavefunctions are shown in figure A1.1.1 for a particle in one dimension subject to the potential $V = \frac{1}{2}kx^2$, which corresponds to the case where the force acting on the particle is proportional in magnitude and opposite in direction to its displacement from equilibrium $(f \equiv -\nabla V = -kx)$. The corresponding Schrödinger equation

$$-\frac{\hbar^2}{2m}\frac{\mathrm{d}^2}{\mathrm{d}x^2}\psi + \frac{1}{2}kx^2 = E\psi \qquad (A1.1.51)$$

can be solved analytically, and this problem (probably familiar to most readers) is that of the quantum harmonic oscillator. As expected, the ground-state wavefunction has no nodes. The first excited state has a single node, the second two nodes and so on, with the number of nodes growing with increasing magnitude of the eigenvalue. From the form of the kinetic energy operator, one can infer that regions where the slope of the wavefunction is changing rapidly (large second derivatives) are associated with large kinetic energy. It is quite reasonable to accept that wavefunctions with regions of large curvature (where the function itself has appreciable magnitude) describe states with high energy, an expectation that can be made rigorous by applying a quantum-mechanical version of the virial theorem.

Classically, a particle with fixed energy E described by a quadratic potential will move back and forth between the points where $V = E$, known as the *classical turning points*. Movement beyond the classical turning points is forbidden, because energy conservation implies that the particle will have a negative kinetic energy in these regions, and imaginary velocities are clearly inconsistent with the Newtonian picture of the universe. Inside the turning points, the particle will have its maximum kinetic energy as it passes through the minimum, slowing in its climb until it comes to rest and subsequently changes direction at the turning points (imagine a marble rolling in a parabola). Therefore, if a camera were to take snapshots of the particle at random intervals, most of the pictures would show the particle near the turning points (the equilibrium position is actually the least likely location for the particle). A more detailed analysis of the problem shows that the probability of seeing the classical particle in the neighbourhood of a given position x is proportional to $\frac{1}{\sqrt{E-V(x)}}$. Note that the situation found for the ground state described by quantum mechanics bears very little resemblance to the classical situation. The particle is most likely to be found at the equilibrium position and, within the classically allowed region, least likely to be seen at the turning points. However, the situation is even stranger than this: the probability of finding the particle outside the turning points is non-zero! This phenomenon, known as *tunnelling*, is not unique to the harmonic oscillator. Indeed, it occurs for bound states described by every potential that tends asymptotically to a finite value since the wavefunction and its derivatives must approach zero in a smooth fashion for large values of the coordinates where (by the definition of a bound state) V must exceed E. However, at large energies (see the 29th excited state probability density in figure A1.1.2), the situation is more consistent with expectations based on classical theory: the probability density has its largest value near the turning points, the general appearance is as implied by the classical formula (if one ignores the oscillations) and its magnitude in the classically forbidden region is reduced dramatically with respect to that found for the low-lying states. This merging of the quantum-mechanical picture with expectations based on classical theory always occurs for highly excited states and is the basis of the *correspondence principle*.

The energy level spectrum of the harmonic oscillator is completely regular. The ground state energy is given by $\frac{1}{2}h\nu$, where ν is the classical frequency of oscillation given by

$$\nu = \frac{1}{2\pi}\sqrt{\frac{k}{m}} \qquad (A1.1.52)$$

although it must be emphasized that our inspection of the wavefunction shows that the motion of the particle cannot be literally thought of in this way. The energy of the first excited state is $h\nu$ above that of the ground state and precisely the same difference separates each excited state from those immediately above and below. A different example is provided by a particle trapped in the *Morse potential*

$$V(x) = D_e[\exp(-ax) - 1]^2, \qquad (A1.1.53)$$

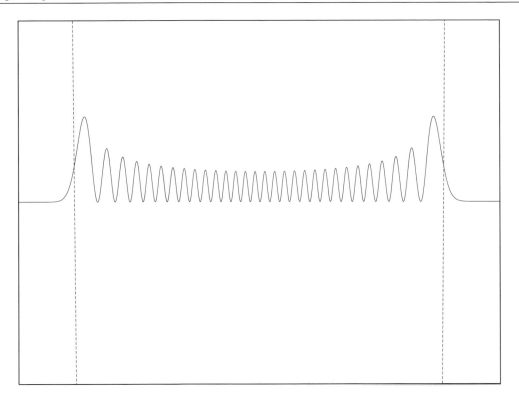

Figure A1.1.2. Probability density ($\Psi^*\Psi$) for the $n = 29$ state of the harmonic oscillator. The vertical state is chosen as in figure A1.1.1, so that the locations of the turning points coincide with the superimposed potential function.

originally suggested as a realistic model for the vibrational motion of diatomic molecules. Although the wavefunctions associated with the Morse levels exhibit largely the same qualitative features as the harmonic oscillator functions and are not shown here, the energy level structures associated with the two systems are qualitatively different. Since $V(x)$ tends to a finite value (D_e) for large x, there are only a limited number of bound state solutions, and the spacing between them decreases with increasing eigenvalue. This is another general feature; energy level spacings for states associated with potentials that tend towards asymptotic values at infinity tend to decrease with increasing quantum number.

The one-dimensional cases discussed above illustrate many of the qualitative features of quantum mechanics, and their relative simplicity makes them quite easy to study. Motion in more than one dimension and (especially) that of more than one particle is considerably more complicated, but many of the general features of these systems can be understood from simple considerations. While one relatively common feature of multidimensional problems in quantum mechanics is degeneracy, it turns out that the ground state must be non-degenerate. To prove this, simply assume the opposite to be true, i.e.

$$H\psi_1 = E_0\psi_1 \tag{A1.1.54}$$
$$H\psi_2 = E_0\psi_2 \tag{A1.1.55}$$

where E_0 is the ground state energy, and

$$\int \psi_1^*\psi_2 \, d\tau = 0. \tag{A1.1.56}$$

In order to satisfy equation (A1.1.56), the two functions must have identical signs at some points in space and different signs elsewhere. It follows that at least one of them must have at least one node. However, this is incompatible with the nodeless property of ground-state eigenfunctions.

Having established that the ground state of a single-particle system is non-degenerate and nodeless, it is straightforward to prove that the wavefunctions associated with every excited state must contain at least one node (though they need not be degenerate!), just as seen in the example problems. It follows from the orthogonality of eigenfunctions corresponding to a Hermitian operator that

$$\int \psi_g^* \psi_x \, d\tau = 0 \tag{A1.1.57}$$

for all excited states ψ_x. In order for this equality to be satisfied, it is necessary that the integrand either vanishes at all points in space (which contradicts the assumption that both ψ_g and ψ_x are nodeless) or is positive in some regions of space and negative in others. Given that the ground state has no nodes, the latter condition can be satisfied only if the excited-state wavefunction changes sign at one or more points in space. Since the first postulate states that all wavefunctions are continuous, it is therefore necessary that ψ_x has at least one node.

In classical mechanics, it is certainly possible for a system subject to dissipative forces such as friction to come to rest. For example, a marble rolling in a parabola lined with sandpaper will eventually lose its kinetic energy and come to rest at the bottom. Rather remarkably, making a measurement of E that coincides with V_{min} (as would be found classically for our stationary marble) is incompatible with quantum mechanics. Turning back to our example, the ground-state energy is indeed larger than the minimum value of the potential energy for the harmonic oscillator. That this property of *zero-point energy* is guaranteed in quantum mechanics can be demonstrated by straightforward application of the basic principles of the subject. Unlike nodal features of the wavefunction, the arguments developed here also hold for many-particle systems. Suppose the total energy of a stationary state is E. Since the energy is the sum of kinetic and potential energies, it must be true that expectation values of the kinetic and potential energies are related according to

$$E = \langle T \rangle + \langle V \rangle. \tag{A1.1.58}$$

If the total energy associated with the state is equal to the potential energy at the equilibrium position, it follows that

$$V_{min} - \langle V \rangle = \langle T \rangle. \tag{A1.1.59}$$

Two cases must be considered. In the first, it will be assumed that the wavefunction is nonzero at one or more points for which $V > V_{min}$ (for the physically relevant case of a smoothly varying and continuous potential, this includes all possibilities other than that in which the wavefunction is a Dirac delta function at the equilibrium position). This means that $\langle V \rangle$ must also be greater than V_{min} thereby forcing the average kinetic energy to be negative. This is not possible. The kinetic energy operator for a quantum-mechanical particle moving in the x-direction has the (unnormalized) eigenfunctions

$$f = \exp(ikx) \tag{A1.1.60}$$

where

$$k = \left(\frac{2m\alpha}{\hbar^2} \right)^{\frac{1}{2}} \tag{A1.1.61}$$

and α are the corresponding eigenvalues. It can be seen that negative values of α give rise to real arguments of the exponential and correspondingly divergent eigenfunctions. Zero and non-negative values are associated with constant and oscillatory solutions in which the argument of the exponential vanishes or is imaginary, respectively. Since divergence of the actual wavefunction is incompatible with its probabilistic interpretation, no contribution from negative α eigenfunctions can appear when the wavefunction is expanded in terms of kinetic energy eigenfunctions. It follows from the fifth postulate that the kinetic energy of each particle in the system (and therefore the total kinetic energy) is restricted to non-negative values. Therefore, the expectation value of the kinetic energy cannot be negative.

The other possibility is that the wavefunction is non-vanishing only when $V = V_{min}$. For the case of a smoothly varying, continuous potential, this corresponds to a state described by a Dirac delta function at the equilibrium position, which is the quantum-mechanical equivalent of a particle at rest. In any event, the fact that the wavefunction vanishes at all points for which $V \neq V_{min}$ means that the expectation value of the kinetic energy operator must also vanish if there is to be no zeropoint energy. Considering the discussion above, this can occur only when the wavefunction is the same as the zero-kinetic-energy eigenfunction ($\psi = $ constant). This contradicts the assumption used in this case, where the wavefunction is a delta function. Following the general arguments used in both cases above, it is easily shown that E can only be larger than V_{min}, which means that any measurement of E for a particle in a stationary or non-stationary state must give a result that satisfies the inequality $E > V_{min}$.

A1.1.3 Quantum mechanics of many-particle systems

A1.1.3.1 The hydrogen atom

It is admittedly inconsistent to begin a section on many-particle quantum mechanics by discussing a problem that can be treated as a single particle. However, the hydrogen atom and atomic ions in which only one electron remains (He^+, Li^{2+} etc) are the only atoms for which exact analytic solutions to the Schrödinger equation can be obtained. In no cases are exact solutions possible for molecules, even after the Born–Oppenheimer approximation (see section B3.1.1.1) is made to allow for separate treatment of electrons and nuclei. Despite the limited interest of hydrogen atoms and hydrogen-like ions to chemistry, the quantum mechanics of these systems is both highly instructive and provides a basis for treatments of more complex atoms and molecules. Comprehensive discussions of one-electron atoms can be found in many textbooks; the emphasis here is on qualitative aspects of the solutions.

The Schrödinger equation for a one-electron atom with nuclear charge Z is

$$\frac{-\hbar^2}{2\mu}\nabla^2 - \frac{Ze^2}{r}\psi = E\psi \tag{A1.1.62}$$

where μ is the reduced mass of the electron–nucleus system and the Laplacian is most conveniently expressed in spherical polar coordinates. While not trivial, this differential equation can be solved analytically. Some of the solutions are normalizable, and others are not. The former are those that describe the bound states of one-electron atoms, and can be written in the form

$$\psi_{nlm} = N R_{nl}(r) Y_{l,m_l}(\theta, \phi) \tag{A1.1.63}$$

where N is a normalization constant, and $R_{nl}(r)$ and $Y_{l,m_l}(\theta, \phi)$ are specific functions that depend on the *quantum numbers n, l* and m_l. The first of these is called the principal quantum number, while l is known as the angular momentum, or azimuthal, quantum number, and m_l the magnetic quantum number. The quantum numbers that allow for normalizable wavefunctions are limited to integers that run over the ranges

$$n = 1, 2, 3, \ldots \tag{A1.1.64}$$

$$l = -n + 1, -n + 2, \ldots, 0, 1, 2, \ldots n - 1 \tag{A1.1.65}$$

$$m_l = -l, -l + 1, \ldots, l - 1, l. \tag{A1.1.66}$$

The fact that there is no restriction on n apart from being a positive integer means that there are an infinite number of bound-state solutions to the hydrogen atom, a peculiarity that is due to the form of the Coulomb potential. Unlike most bound state problems, the range of the potential is infinite (it goes to zero at large r, but diverges to negative infinity at $r = 0$). The eigenvalues of the Hamiltonian depend only on the principal quantum number and are (in attojoules (10^{-18} J))

$$E_n = -2.18\frac{Z^2}{n^2} \qquad (A1.1.67)$$

where it should be noted that the zero of energy corresponds to infinite separation of the particles. For each value of n, the Schrödinger equation predicts that all states are degenerate, regardless of the choice of l and m_l. Hence, any linear combination of wavefunctions corresponding to some specific value of n is also an eigenfunction of the Hamiltonian with eigenvalue E_n. States of hydrogen are usually characterized as ns, np, nd etc where n is the principal quantum number and s is associated with $l = 0$, p with $l = 1$ and so on. The functions $R_{nl}(r)$ describe the radial part of the wavefunctions and can all be written in the form

$$R_{nl}(r) = \exp(-\rho/2)\rho^l L_{nl}(\rho) \qquad (A1.1.68)$$

where ρ is proportional to the electron–nucleus separation r and the atomic number Z. L_{nl} is a polynomial of order $n - l - 1$ that has zeros (where the wavefunction, and therefore the probability of finding the electron, vanishes—a *radial node*) only for positive values of ρ. The functions $Y_{l,m_l}(\theta, \phi)$ are the *spherical harmonics*. The first few members of this series are familiar to everyone who has studied physical chemistry: Y_{00} is a constant, leading to a spherically symmetric wavefunction, while $Y_{1,0}$, and specific linear combinations of $Y_{1,1}$ and $Y_{1,-1}$, vanish (have an *angular node*) in the xy, xz and yz planes, respectively. In general, these functions exhibit l nodes, meaning that the number of overall nodes corresponding to a particular ψ_{nlm_l} is equal to $n - 1$. For example, the 4d state has two angular nodes ($l = 2$) and one radial node ($L_{nl}(\rho)$ has one zero for positive ρ). In passing, it should be noted that many of the ubiquitous qualitative features of quantum mechanics are illustrated by the wavefunctions and energy levels of the hydrogen atom. First, the system has a zero-point energy, meaning that the ground-state energy is larger than the lowest value of the potential ($-\infty$) and the spacing between the energy levels decreases with increasing energy. Second, the ground state of the system is nodeless (the electron may be found at any point in space), while the number of nodes exhibited by the excited states increases with energy. Finally, there is a finite probability that the electron is found in a classically forbidden region in all bound states. For the hydrogen atom ground state, this corresponds to all electron–proton separations larger than 105.8 pm, where the electron is found 23.8% of the time. As usual, this tunnelling phenomenon is less pronounced in excited states: the corresponding values for the 3s state are 1904 pm and 16.0%.

The Hamiltonian commutes with the angular momentum operator $\hat{L}_z$ as well as that for the square of the angular momentum $\hat{L}^2$. The wavefunctions above are also eigenfunctions of these operators, with eigenvalues $m_l\hbar$ ($\hat{L}_z$) and $l(l + 1)\hbar^2$ ($\hat{L}^2$). It should be emphasized that the total angular momentum is $L = \sqrt{l(l + 1)}\hbar$, and not a simple integral multiple of $\hbar$ as assumed in the Bohr model. In particular, the ground state of hydrogen has zero angular momentum, while the Bohr atom ground state has $L = \hbar$. The meaning associated with the m_l quantum number is more difficult to grasp. The choice of z instead of x or y seems to be (and is) arbitrary and it is illogical that a specific value of the angular momentum projection along one coordinate must be observed in any experiment, while those associated with x and y are not similarly restricted. However, the states with a given l are degenerate, and the wavefunction at any particular time will in general be some linear combination of the m_l eigenfunctions. The only way to isolate a specific ψ_{nlm_l} (and therefore ensure the result of measuring L_z) is to apply a magnetic field that lifts the degeneracy and breaks the symmetry of the problem. The z axis then corresponds to the magnetic field direction, and it is the projection of the angular momentum vector on this axis that must be equal to $m_l\hbar$.

The quantum-mechanical treatment of hydrogen outlined above does not provide a completely satisfactory description of the atomic spectrum, even in the absence of a magnetic field. Relativistic effects cause both a scalar shifting in all energy levels as well as splittings caused by the magnetic fields associated with both motion and intrinsic properties of the charges within the atom. The features of this *fine structure* in the energy spectrum were successfully (and miraculously, given that it preceded modern quantum mechanics by a decade and was based on a two-dimensional picture of the hydrogen atom) predicted by a formula developed by Sommerfeld in 1915. These interactions, while small for hydrogen, become very large indeed for larger atoms where very strong electron–nucleus attractive potentials cause electrons to move at velocities close to the speed of light. In these cases, quantitative calculations are extremely difficult and even the separability of orbital and intrinsic angular momenta breaks down.

A1.1.3.2 *The independent-particle approximation*

Applications of quantum mechanics to chemistry invariably deal with systems (atoms and molecules) that contain more than one particle. Apart from the hydrogen atom, the stationary-state energies cannot be calculated exactly, and compromises must be made in order to estimate them. Perhaps the most useful and widely used approximation in chemistry is the *independent-particle approximation*, which can take several forms. Common to all of these is the assumption that the Hamiltonian operator for a system consisting of n particles is approximated by the sum

$$\hat{H}_0 = \hat{h}_1 + \hat{h}_2 + \cdots + \hat{h}_n \tag{A1.1.69}$$

where the single-particle Hamiltonians $\hat{h}_i$ consist of the kinetic energy operator plus a potential ($\hat{v}_i$) that does not explicitly depend on the coordinates of the other $n - 1$ particles in the system. Of course, the simplest realization of this model is to completely neglect forces due to the other particles, but this is often too severe an approximation to be useful. In any event, the quantum mechanics of a system described by a Hamiltonian of the form given by equation (A1.1.69) is worthy of discussion simply because the independent-particle approximation is the foundation for molecular orbital theory, which is the central paradigm of descriptive chemistry.

Let the orthonormal functions $\chi_i(1), \chi_j(2), \ldots, \chi_p(n)$ be selected eigenfunctions of the corresponding single-particle Hamiltonians $\hat{h}_1, \hat{h}_2, \ldots, \hat{h}_n$, with eigenvalues $\lambda_i, \lambda_j, \ldots, \lambda_p$. It is easily verified that the product of these *single-particle wavefunctions* (which are often called *orbitals* when the particles are electrons in atoms and molecules)

$$\phi = \chi_i(1)\chi_j(2) \cdots \chi_p(n) \tag{A1.1.70}$$

satisfies the approximate Schrödinger equation for the system

$$\hat{H}_0\phi = E_0\phi \tag{A1.1.71}$$

with the corresponding energy

$$E_0 = \lambda_i + \lambda_j + \cdots + \lambda_p. \tag{A1.1.72}$$

Hence, if the Hamiltonian can be written as a sum of terms that individually depend only on the coordinates of one of the particles in the system, then the wavefunction of the system can be written as a product of functions, each of which is an eigenfunction of one of the single-particle Hamiltonians, h_i. The corresponding eigenvalue is then given by the sum of eigenvalues associated with each single-particle wavefunction χ appearing in the product.

The approximation embodied by equations (A1.1.69)–(A1.1.72) presents a conceptually appealing picture of many-particle systems. The behaviour and energetics of each particle can be determined from a simple function of three coordinates and the eigenvalue of a differential equation considerably simpler than the one that explicitly accounts for all interactions. It is precisely this simplification that is invoked in qualitative interpretations of chemical phenomena such as the inert nature of noble gases and the strongly reducing property of the alkali metals. The price paid is that the model is only approximate, meaning that properties predicted from it (for example, absolute ionization potentials rather than just trends within the periodic table) are not as accurate as one might like. However, as will be demonstrated in the latter parts of this section, a carefully chosen independent-particle description of a many-particle system provides a starting point for performing more accurate calculations. It should be mentioned that even qualitative features might be predicted incorrectly by independent-particle models in extreme cases. One should always be aware of this possibility and the oft-misunderstood fact that there really is no such thing as an orbital. Fortunately, however, it turns out that qualitative errors are uncommon for electronic properties of atoms and molecules when the best independent-particle models are used.

One important feature of many-particle systems has been neglected in the preceding discussion. Identical particles in quantum mechanics must be indistinguishable, which implies that the exact wavefunctions ψ which describe them must satisfy certain symmetry properties. In particular, interchanging the coordinates of any two particles in the mathematical form of the wavefunction cannot lead to a different prediction of the system properties. Since any rearrangement of particle coordinates can be achieved by successive pairwise permutations, it is sufficient to consider the case of a single permutation in analysing the symmetry properties that wavefunctions must obey. In the following, it will be assumed that the wavefunction is real. This is not restrictive, as stationary state wavefunctions for isolated atoms and molecules can always be written in this way. If the operator P_{ij} is that which permutes the coordinates of particles i and j, then indistinguishability requires that

$$\left(\int P_{ij}\psi \right)^* \hat{A} P_{ij}\psi \, d\tau = \int \psi^* \hat{A} \psi \, d\tau \tag{A1.1.73}$$

for any operator $\hat{A}$ (including the identity) and choice of i and j. Clearly, a wavefunction that is symmetric with respect to the interchange of coordinates for any two particles

$$P_{ij}\psi = \psi \tag{A1.1.74}$$

satisfies the indistinguishability criterion. However, equation (A1.1.73) is also satisfied if the permutation of particle coordinates results in an overall sign change of the wavefunction, i.e.

$$P_{ij}\psi = -\psi. \tag{A1.1.75}$$

Without further considerations, the only acceptable real quantum-mechanical wavefunctions for an n-particle system would appear to be those for which

$$P_{ij}\psi = \pm\psi \tag{A1.1.76}$$

where i and j are any pair of identical particles. For example, if the system comprises two protons, a neutron and two electrons, the relevant permutations are that which interchanges the proton coordinates and that which interchanges the electron coordinates. The other possible pairs involve distinct particles and the action of the corresponding P_{ij} operators on the wavefunction will in general result in something quite different. Since indistinguishability is a necessary property of exact wavefunctions, it is reasonable to impose the same constraint on the approximate wavefunctions ϕ formed from products of single-particle solutions. However, if two or more of the χ_i in the product are different, it is necessary to form linear combinations if the condition $P_{ij}\psi = \pm\psi$ is to be met. An additional consequence of indistinguishability is that the h_i operators corresponding to identical particles must also be identical and therefore have precisely the same eigenfunctions. It should be noted that there is nothing mysterious about this perfectly reasonable restriction placed on the mathematical form of wavefunctions.

For the sake of simplicity, consider a system of two electrons for which the corresponding single-particle states are $\chi_i, \chi_j, \chi_k, \ldots, \chi_n$, with eigenvalues $\lambda_i, \lambda_j, \lambda_k, \ldots, \lambda_n$. Clearly, the two-electron wavefunction $\phi = \chi_i(1)\chi_i(2)$ satisfies the indistinguishability criterion and describes a stationary state with energy $E_0 = 2\lambda_i$. However, the state $\chi_i(1)\chi_j(2)$ is not satisfactory. While it is a solution to the Schrödinger equation, it is neither symmetric nor antisymmetric with respect to particle interchange. However, two such states can be formed by taking the linear combinations

$$\phi_S = \sqrt{\tfrac{1}{2}}[\chi_i(1)\chi_j(2) + \chi_i(2)\chi_j(1)] \tag{A1.1.77}$$

$$\phi_A = \sqrt{\tfrac{1}{2}}[\chi_i(1)\chi_j(2) - \chi_i(2)\chi_j(1)] \tag{A1.1.78}$$

which are symmetric and antisymmetric with respect to particle interchange, respectively. Because the functions χ are orthonormal, the energies calculated from ϕ_S and ϕ_A are the same as that corresponding to the unsymmetrized product state $\chi_i(1)\chi_j(2)$, as demonstrated explicitly for ϕ_S:

$$\int \phi_S \hat{H} \phi_S \, d\tau = \frac{1}{2}\Bigg[\int \chi_i(1)\chi_j(2)\hat{H}\chi_i(1)\chi_j(2)\, d\tau_1\, d\tau_2 + \int \chi_i(1)\chi_j(2)\hat{H}\chi_i(2)\chi_j(1)\, d\tau_1\, d\tau_2$$

$$+ \int \chi_i(2)\chi_j(1)\hat{H}\chi_i(1)\chi_j(2)\, d\tau_1\, d\tau_2 + \int \chi_i(2)\chi_j(1)\hat{H}\chi_i(2)\chi_j(1)\, d\tau_1\, d\tau_2 \Bigg]$$

$$= \frac{1}{2}(\lambda_i + \lambda_j)\Bigg[\int \chi_i(1)\chi_i(1)\, d\tau_1 \int \chi_j(2)\chi_j(2)\, d\tau_2 + \int \chi_i(1)\chi_j(1)\, d\tau_1 \int \chi_j(2)\chi_i(2)\, d\tau_2$$

$$+ \int \chi_j(1)\chi_i(1)\, d\tau_1 \int \chi_i(2)\chi_j(2)\, d\tau_2 + \int \chi_j(1)\chi_j(1)\, d\tau_1 \int \chi_i(2)\chi_i(2)\, d\tau_2 \Bigg]$$

$$= \frac{1}{2}(\lambda_i + \lambda_j)[1 + 0 + 0 + 1] = \lambda_i + \lambda_j. \tag{A1.1.79}$$

It should be mentioned that the single-particle Hamiltonians in general have an infinite number of solutions, so that an uncountable number of wavefunctions ψ can be generated from them. Very often, interest is focused on the ground state of many-particle systems. Within the independent-particle approximation, this state can be represented by simply assigning each particle to the lowest-lying energy level. If a calculation is performed on the lithium atom in which interelectronic repulsion is ignored completely, the single-particle Schrödinger equations are precisely the same as those for the hydrogen atom, apart from the difference in nuclear charge. The following lithium atom wavefunction could then be constructed from single-particle orbitals

$$\phi = N\chi_{1s}(1)\chi_{1s}(2)\chi_{1s}(3) \tag{A1.1.80}$$

a form that is obviously symmetric with respect to interchange of particle coordinates. If this wavefunction is used to calculate the expectation value of the energy using the exact Hamiltonian (which includes the explicit electron–electron repulsion terms),

$$\epsilon = \int \psi^* H \psi \, d\tau \tag{A1.1.81}$$

one obtains an energy lower than the actual result, which (see A1.1.4.1) suggests that there are serious problems with this form of the wavefunction. Moreover, a relatively simple analysis shows that ionization potentials of atoms would increase monotonically—approximately linearly for small atoms and quadratically for large atoms—if the independent-particle picture discussed thus far has any validity. Using a relatively simple model that assumes that the lowest lying orbital is a simple exponential, ionization potentials of 13.6, 23.1, 33.7 and 45.5 electron volts (eV) are predicted for hydrogen, helium, lithium and beryllium, respectively. The value for hydrogen (a one-electron system) is exact and that for helium is in relatively good agreement with the experimental value of 24.8 eV. However, the other values are well above the actual ionization energies of Li and Be (5.4 and 9.3 eV, respectively), both of which are smaller than those of H and He! All freshman chemistry students learn that ionization potentials do not increase monotonically with atomic number, and that there are in fact many pronounced and more subtle decreases that appear when this property is plotted as a function of atomic number.

There is evidently a grave problem here. The wavefunction proposed above for the lithium atom contains all of the particle coordinates, adheres to the boundary conditions (it decays to zero when the particles are removed to infinity) and obeys the restrictions $P_{12}\phi = P_{13}\phi = P_{23}\phi = \pm\phi$ that govern the behaviour of the exact wavefunctions. Therefore, if no other restrictions are placed on the wavefunctions of multiparticle systems, the product wavefunction for lithium must lie in the space spanned by the exact wavefunctions. However, it clearly does not, because it is proven in subsection A1.1.4.1 that any function expressible as a linear combination of Hamiltonian eigenfunctions cannot have an energy lower than that of the exact ground state. This means that there is at least one additional symmetry obeyed by all of the exact wavefunctions that is not satisfied by the product form given for lithium in equation (A1.1.80).

This missing symmetry provided a great puzzle to theorists in the early part days of quantum mechanics. Taken together, ionization potentials of the first four elements in the periodic table indicate that wavefunctions which assign two electrons to the same single-particle functions such as

$$\phi = \chi_a(1)\chi_a(2) \tag{A1.1.82}$$

(helium) and

$$\phi = S\chi_a(1)\chi_a(2)\chi_b(3)\chi_b(4) \tag{A1.1.83}$$

(beryllium, the operator $\hat{S}$ produces the labelled $\chi_a\chi_a\chi_b\chi_b$ product that is symmetric with respect to interchange of particle indices) are somehow acceptable but that those involving three or more electrons in one state are not! The resolution of this *zweideutigkeit* (two-valuedness) puzzle was made possible only by the discovery of electron spin, which is discussed below.

A1.1.3.3 Spin and the Pauli principle

In the early 1920s, spectroscopic experiments on the hydrogen atom revealed a striking inconsistency with the Bohr model, as adapted by Sommerfeld to account for relativistic effects. Studies of the fine structure associated with the $n = 4 \rightarrow n = 3$ transition revealed five distinct peaks, while six were expected from arguments based on the theory of interaction between matter and radiation. The problem was ultimately reconciled by Uhlenbeck and Goudsmit, who reinterpreted one of the quantum numbers appearing in Sommerfeld's fine structure formula based on a startling assertion that the electron has an intrinsic angular momentum independent of that associated with its motion. This idea was also supported by previous experiments of Stern and Gerlach, and is now known as *electron spin*. Spin is a mysterious phenomenon with a rather unfortunate name. Electrons are fundamental particles, and it is no more appropriate to think of them as charges that resemble extremely small billiard balls than as waves. Although they exhibit behaviour characteristic of both, they are in fact neither. Elementary textbooks often depict spin in terms of spherical electrons whirling about their axis (a compelling idea in many ways, since it reinforces the Bohr model by introducing a spinning planet), but this is a purely classical perspective on electron spin that should not be taken literally.

Electrons and most other fundamental particles have two distinct spin wavefunctions that are degenerate in the absence of an external magnetic field. Associated with these are two abstract states which are eigenfunctions of the intrinsic spin angular momentum operator $\hat{S}_z$

$$S_z\sigma = m_s\hbar\sigma. \tag{A1.1.84}$$

The allowed quantum numbers m_s are $\frac{1}{2}$ and $-\frac{1}{2}$, and the corresponding eigenfunctions are usually written as α and β, respectively. The associated eigenvalues $\frac{\hbar}{2}$ and $-\frac{\hbar}{2}$ give the projection of the intrinsic angular momentum vector along the direction of a magnetic field that can be applied to resolve the degeneracy. The overall spin angular momentum of the electron is given in terms of the quantum number s by $\sqrt{s(s+1)}\hbar$. For an electron, $s = \frac{1}{2}$. For a collection of particles, the overall spin and its projection are given in terms of the spin quantum numbers S and M_S (which are equal to the corresponding lower-case quantities for single particles) by $\sqrt{S(S+1)}\hbar$ and $M_S\hbar$, respectively. S must be positive and can assume either integral or half-integral values, and the M_S quantum numbers lie in the interval

$$M_S = -S, -S+1, -S+2, \ldots, 0, \ldots, S-1, S \qquad \text{(A1.1.85)}$$

where a correspondence to the properties of orbital angular momentum should be noted. The *multiplicity* of a state is given by $2S+1$ (the number of possible M_S values) and it is customary to associate the terms *singlet* with $S = 0$, doublet with $S = \frac{1}{2}$, triplet with $S = 1$ and so on.

In the non-relativistic quantum mechanics discussed in this chapter, spin does not appear naturally. Although Dirac showed in 1928 that a fourth quantum number associated with intrinsic angular momentum appears in a relativistic treatment of the free electron, it is customary to treat spin heuristically. In general, the wavefunction of an electron is written as the product of the usual spatial part (which corresponds to a solution of the non-relativistic Schrödinger equation and involves only the Cartesian coordinates of the particle) and a *spin part* σ, where σ is either α or β. A common shorthand notation is often used, whereby

$$\psi \equiv \psi_{\text{spatial}}\alpha \qquad \text{(A1.1.86)}$$
$$\bar{\psi} \equiv \psi_{\text{spatial}}\beta. \qquad \text{(A1.1.87)}$$

In the context of electronic structure theory, the composite functions above are often referred to as *spin orbitals*.

When spin is taken into account, one finds that the ground state of the hydrogen atom is actually doubly degenerate. The spatial part of the wavefunction is the Schrödinger equation solution discussed in section A1.1.3.1, but the possibility of either spin α or β means that there are two distinct overall wavefunctions. The same may be said for any of the excited states of hydrogen (all of which are, however, already degenerate in the nonrelativistic theory), as the level of degeneracy is doubled by the introduction of spin. Spin may be thought of as a fourth coordinate associated with each particle. Unlike Cartesian coordinates, for which there is a continuous distribution of possible values, there are only two possible values of the spin coordinate available to each particle. This has important consequences for our discussion of indistinguishability and symmetry properties of the wavefunction, as the concept of coordinate permutation must be amended to include the spin variable of the particles. As an example, the independent-particle ground state of the helium atom based on hydrogenic wavefunctions

$$\chi_{1s}(1)\chi_{1s}(2) \qquad \text{(A1.1.88)}$$

must be replaced by the four possibilities

$$\chi_{1s}(1)\chi_{1s}(2) \qquad \text{(A1.1.89)}$$
$$\bar{\chi}_{1s}(1)\chi_{1s}(2) \qquad \text{(A1.1.90)}$$
$$\chi_{1s}(1)\bar{\chi}_{1s}(2) \qquad \text{(A1.1.91)}$$
$$\bar{\chi}_{1s}(1)\bar{\chi}_{1s}(2). \qquad \text{(A1.1.92)}$$

While the first and fourth of these are symmetric with respect to particle interchange and thereby satisfy the indistinguishability criterion, the other two are not and appropriate linear combinations must be formed.

Doing so, one finds the following four wavefunctions

$$\phi_{S1} = \chi_{1s}(1)\chi_{1s}(2) \tag{A1.1.93}$$

$$\phi_{S2} = \sqrt{\tfrac{1}{2}}[\chi_{1s}(1)\bar{\chi}_{1s}(2) + \chi_{1s}(2)\bar{\chi}_{1s}(1)] \tag{A1.1.94}$$

$$\phi_{S3} = \bar{\chi}_{1s}(1)\bar{\chi}_{1s}(2) \tag{A1.1.95}$$

$$\phi_A = \sqrt{\tfrac{1}{2}}[\chi_{1s}(1)\bar{\chi}_{1s}(2) - \chi_{1s}(2)\bar{\chi}_{1s}(1)] \tag{A1.1.96}$$

where the first three are symmetric with respect to particle interchange and the last is antisymmetric. This suggests that under the influence of a magnetic field, the ground state of helium might be resolved into components that differ in terms of overall spin, but this is not observed. For the lithium example, there are eight possible ways of assigning the spin coordinates, only two of which

$$\phi = \chi_{1s}(1)\chi_{1s}(2)\chi_{1s}(3) \tag{A1.1.97}$$

$$\phi = \bar{\chi}_{1s}(1)\bar{\chi}_{1s}(2)\bar{\chi}_{1s}(3) \tag{A1.1.98}$$

satisfy the criterion $P_{ij}\phi = \pm\phi$. The other six must be mixed in appropriate linear combinations. However, there is an important difference between lithium and helium. In the former case, all assignments of the spin variable to the state given by equation (A1.1.88) produce a product function in which the same state (in terms of both spatial and spin coordinates) appears at least twice. A little reflection shows that it is not possible to generate a linear combination of such functions that is antisymmetric with respect to all possible interchanges; only symmetric combinations such as

$$\phi = \sqrt{\tfrac{1}{3}}[\chi_{1s}(1)\chi_{1s}(2)\bar{\chi}_{1s}(3) + \chi_{1s}(1)\bar{\chi}_{1s}(2)\chi_{1s}(3) + \bar{\chi}_{1s}(1)\chi_{1s}(2)\chi_{1s}(3)] \tag{A1.1.99}$$

can be constructed.

The fact that antisymmetric combinations appear for helium (where the independent-particle ground state made up of hydrogen 1s functions is qualitatively consistent with experiment) and not for lithium (where it is not) raises the interesting possibility that the exact wavefunction satisfies a condition more restrictive than $P_{ij}\psi = \pm\psi$, namely $P_{ij}\psi = -\psi$. For reasons that are not at all obvious, or even intuitive, nature does indeed enforce this restriction, which is one statement of the *Pauli exclusion principle*. When this idea is first met with, one usually learns an equivalent but less general statement that applies only within the independent-particle approximation: *no two electrons can have the same quantum numbers*. What does this mean? Within the independent-particle picture of an atom, each single-particle wavefunction, or orbital, is described by the quantum numbers n, l, m_l and (when spin is considered) m_s. Since it is not possible to generate antisymmetric combinations of products if the same spin orbital appears twice in each term, it follows that states which assign the same set of four quantum numbers twice cannot possibly satisfy the requirement $P_{ij}\psi = -\psi$, so this statement of the exclusion principle is consistent with the more general symmetry requirement. An even more general statement of the exclusion principle, which can be regarded as an additional postulate of quantum mechanics, is

> The wavefunction of a system must be antisymmetric with respect to interchange of the coordinates of identical particles γ and δ if they are *fermions*, and symmetric with respect to interchange of γ and δ if they are *bosons*.

Electrons, protons and neutrons and all other particles that have $s = \tfrac{1}{2}$ are known as fermions. Other particles are restricted to $s = 0$ or 1 and are known as bosons. There are thus profound differences in the quantum-mechanical properties of fermions and bosons, which have important implications in fields ranging from statistical mechanics to spectroscopic selection rules. It can be shown that the spin quantum number S

associated with an even number of fermions must be integral, while that for an odd number of them must be half-integral. The resulting composite particles behave collectively like bosons and fermions, respectively, so the wavefunction symmetry properties associated with bosons can be relevant in chemical physics. One prominent example is the treatment of nuclei, which are typically considered as composite particles rather than interacting protons and neutrons. Nuclei with even atomic number therefore behave like individual bosons and those with odd atomic number as fermions, a distinction that plays an important role in rotational spectroscopy of polyatomic molecules.

A1.1.3.4 *Independent-particle models in electronic structure*

At this point, it is appropriate to make some comments on the construction of approximate wavefunctions for the many-electron problems associated with atoms and molecules. The Hamiltonian operator for a molecule is given by the general form

$$\hat{H} = -\frac{\hbar^2}{2}\left[\sum_{\substack{\text{nuclei}\\\alpha}} \frac{\nabla_\alpha^2}{M_\alpha} + \sum_{\substack{\text{electrons}\\i}} \frac{\nabla_i^2}{m_e}\right] + \sum_{\substack{\text{nuclei}\\\alpha<\beta}} \frac{Z_\alpha Z_\beta e^2}{r_{\alpha\beta}} + \sum_{\substack{\text{electrons}\\i<j}} \frac{e^2}{r_{ij}} - \sum_{\substack{\text{electrons}\\i}} \sum_{\substack{\text{nuclei}\\\alpha}} \frac{eZ_\alpha}{r_{i\alpha}}. \tag{A1.1.100}$$

It should be noted that nuclei and electrons are treated equivalently in $\hat{H}$, which is clearly inconsistent with the way that we tend to think about them. Our understanding of chemical processes is strongly rooted in the concept of a potential energy surface which determines the forces that act upon the nuclei. The potential energy surface governs all behaviour associated with nuclear motion, such as vibrational frequencies, mean and equilibrium internuclear separations and preferences for specific conformations in molecules as complex as proteins and nucleic acids. In addition, the potential energy surface provides the transition state and activation energy concepts that are at the heart of the theory of chemical reactions. Electronic motion, however, is never discussed in these terms. All of the important and useful ideas discussed above derive from the Born–Oppenheimer approximation, which is discussed in some detail in section B3.1. Within this model, the *electronic states* are solutions to the equation

$$\left[-\frac{\hbar^2}{2m}\sum_{\substack{\text{electrons}\\i}} \nabla_i^2 - \sum_{\substack{\text{electrons}\\i}} \sum_{\substack{\text{nuclei}\\\alpha}} \frac{eZ_\alpha}{r_{i\alpha}} + \sum_{\substack{\text{electrons}\\i<j}} \frac{e^2}{r_{ij}}\right]\psi = \lambda\psi \tag{A1.1.101}$$

where the nuclei are assumed to be stationary. The *electronic energies* are given by the eigenvalues (usually augmented by the wavefunction-independent internuclear repulsion energy) of $\hat{H}$. The functions obtained by plotting the electronic energy as a function of nuclear position are the potential energy surfaces described above. The latter are different for every electronic state; their shape gives the usual information about molecular structure, barrier heights, isomerism and so on. The Born–Oppenheimer separation is also made in the study of electronic structure in atoms. However, this is a rather subtle point and is not terribly important in applications since the only assumption made is that the nucleus has infinite mass.

Although a separation of electronic and nuclear motion provides an important simplification and appealing qualitative model for chemistry, the electronic Schrödinger equation is still formidable. Efforts to solve it approximately and apply these solutions to the study of spectroscopy, structure and chemical reactions form the subject of what is usually called *electronic structure theory* or *quantum chemistry*. The starting point for most calculations and the foundation of molecular orbital theory is the independent-particle approximation.

For many-electron systems such as atoms and molecules, it is obviously important that approximate wavefunctions obey the same boundary conditions and symmetry properties as the exact solutions. Therefore, they should be antisymmetric with respect to interchange of each pair of electrons. Such states can always be constructed as linear combinations of products such as

$$\chi_i(1)\chi_j(2)\chi_k(3)\ldots\chi_q(n). \tag{A1.1.102}$$

The χ are assumed to be spin orbitals (which include both the spatial and spin parts) and each term in the product differs in the way that the electrons are assigned to them. Of course, it does not matter how the electrons are distributed amongst the χ in equation (A1.1.102), as the necessary subsequent antisymmetrization makes all choices equivalent apart from an overall sign (which has no physical significance). Hence, the product form is usually written without assigning electrons to the individual orbitals, and the set of unlabelled χ included in the product represents an *electron configuration*. It should be noted that all of the single-particle orbitals χ in the product are distinct. A very convenient method for constructing antisymmetrized combinations corresponding to products of particular single-particle states is to form the *Slater determinant*

$$\phi = \begin{vmatrix} \chi_i(1) & \chi_i(2) & \chi_i(3) & \cdots & \chi_i(n) \\ \chi_j(1) & \chi_j(2) & \chi_j(3) & \cdots & \chi_j(n) \\ \chi_k(1) & \chi_k(2) & \chi_k(3) & \cdots & \chi_k(n) \\ \vdots & \vdots & \vdots & \ddots & \vdots \\ \chi_q(1) & \chi_q(2) & \chi_q(3) & \cdots & \chi_q(n) \end{vmatrix} \Big/ \sqrt{n!} \tag{A1.1.103}$$

where the nominal electron configuration can be determined by simply scanning along the main diagonal of the matrix. A fundamental result of linear algebra is that the determinant of a matrix changes sign when any two rows or columns are interchanged. Inspection of equation (A1.1.103) shows that interchanging any two columns of the Slater determinant corresponds to interchanging the labels of two electrons, so the Pauli exclusion principle is automatically incorporated into this convenient representation. Whether all orbitals in a given row are identical and all particle labels the same in each column (as above) or *vice versa* is not important, as determinants are invariant with respect to transposition. In particular, it should be noted that the Slater determinant necessarily vanishes when two of the spin orbitals are identical, reflecting the alternative statement of the Pauli principle—no two electrons can have the same quantum number. One qualification which should be stated here is that Slater determinants are not necessarily eigenfunctions of the S^2 operator, and it is often advantageous to form linear combinations of those corresponding to electron configurations that differ only in the assignment of the spin variable to the spatial orbitals. The resulting functions ϕ are sometimes known as *spin-adapted configurations*.

Within an independent-particle picture, there are a very large number of single-particle wavefunctions χ available to each particle in the system. If the single-particle Schrödinger equations can be solved exactly, then there are often an infinite number of solutions. Approximate solutions are, however, necessitated in most applications, and some subtleties must be considered in this case. The description of electrons in atoms and molecules is often based on the Hartree–Fock approximation, which is discussed in section B3.1 of this encyclopedia. In the Hartree–Fock method, only briefly outlined here, the orbitals are chosen in such a way that the total energy of a state described by the Slater determinant that comprises them is minimized. There are cogent reasons for using an energy minimization strategy that are based on the variational principle discussed later in this section. The Hartree–Fock method derives from an earlier treatment of Hartree, in which indistinguishability and the Pauli principle were ignored and the wavefunction expressed as in equation (A1.1.102). However, that approach is not satisfactory because it can lead to pathological solutions such as that discussed earlier for lithium. In Hartree–Fock theory, the orbital optimization is achieved at the expense of introducing a very complicated single-particle potential term v_i. This potential depends on all of the other orbitals in which electrons reside, requires the evaluation of difficult integrals and necessitates a self-consistent (iterative) solution. The resulting one-electron Hamiltonian is known as the Fock operator, and it has (in principle) an infinite number of eigenfunctions; a subset of these are exactly the same as the χ that correspond to the occupied orbitals upon which it is parametrized. The resulting equations cannot be solved analytically; for atoms, exact solutions for the occupied orbitals can be determined by numerical methods, but the infinite number of unoccupied functions are unknown apart from the fact that they must be orthogonal to the occupied ones. In molecular calculations, it is customary to assume that the orbitals χ can be written as linear combinations of a fixed set of N *basis functions*, where N is typically of the order of tens to a few hundred. Iterative solution of a set of matrix equations provides approximations for the orbitals describing the n electrons of the molecule and $N - n$ unoccupied orbitals.

The choice of basis functions is straightforward in atomic calculations. It can be demonstrated that all solutions to an independent-particle Hamiltonian have the symmetry properties of the hydrogenic wavefunctions. Each is, or can be written as, an eigenfunction of the $\hat{L}_z$ and $\hat{L}^2$ operators and involves a radial part multiplied by a spherical harmonic. Atomic calculations that use basis sets (not all of them do) typically choose functions that are similar to those that solve the Schrödinger equation for hydrogen. If the complete set of hydrogenic functions is used, the solution to the basis set equations are the exact Hartree–Fock solutions. However, practical considerations require the use of finite basis sets; the corresponding solutions are therefore only approximate. Although the distinction is rarely made, it is preferable to refer to these as *self-consistent field* (SCF) solutions and energies in order to distinguish them from the exact Hartree–Fock results. As the quality of a basis is improved, the energy approaches that of the Hartree–Fock solution from above.

In molecules, things are a great deal more complicated. In principle, one can always choose a subset of all the hydrogenic wavefunctions centred at some point in space. Since the resulting basis functions include all possible electron coordinates and Slater determinants constructed from them vanish at infinity and satisfy the Pauli principle, the corresponding approximate solutions must lie in the space spanned by the exact solutions and be qualitatively acceptable. In particular, use of enough basis functions will result in convergence to the exact Hartree–Fock solution. Because of the difficulties involved with evaluating integrals involving exponential hydrogenic functions centred at more than one point in space, such *single-centre expansions* were used in the early days of quantum chemistry. The main drawback is that convergence to the exact Hartree–Fock result is extraordinarily slow. The states of the hydrogen molecule are reasonably well approximated by linear combinations of hydrogenic functions centred on each of the two nuclei. Hence, a more practical strategy is to construct a basis by choosing a set of hydrogenic functions for each atom in the molecule (the same functions are usually used for identical atoms, whether or not they are equivalent by symmetry). Linear combinations of a relatively small number of these functions are capable of describing the electronic distribution in molecules much better than is possible with a corresponding number of functions in a single-centre expansion. This approach is often called the *linear combination of atomic orbitals* (LCAO) approximation, and is used in virtually all molecular SCF calculations performed today. The problems associated with evaluation of multicentre integrals alluded to above was solved more than a quarter-century ago by the introduction of Gaussian—rather than exponential—basis functions, which permit all of the integrals appearing in the Fock operator to be calculated analytically. Although Gaussian functions are not hydrogenic functions (and are inferior basis functions), the latter can certainly be approximated well by linear combinations of the former. The ease of integral evaluation using Gaussian functions makes them the standard choice for practical calculations. The importance of selecting an appropriate basis set is of great practical importance in quantum chemistry and many other aspects of atomic and molecular quantum mechanics. An illustrative example of basis set selection and its effect on calculated energies is given in subsection A1.1.4.2. While the problem studied there involves only the motion of a single particle in one dimension, an analogy with the LCAO and single-centre expansion methods should be apparent, with the desirable features of the former clearly illustrated.

Even Hartree–Fock calculations are difficult and expensive to apply to large molecules. As a result, further simplifications are often made. Parts of the Fock operator are ignored or replaced by parameters chosen by some sort of statistical procedure to account, in an average way, for the known properties of selected compounds. While calculating properties that have already been measured experimentally is of limited interest to anyone other than theorists trying to establish the accuracy of a method, the hope of these approximate

Hartree–Fock procedures (which include the well known Hückel approximation and are collectively known as *semiempirical methods*) is that the parametrization works just as well for both unmeasured properties of known molecules (such as transition state structures) and the structure and properties of transient or unknown species. No further discussion of these approaches is given here (more details are given in sections B3.1 and B3.2); it should only be emphasized that all of these methods are based on the independent-particle approximation.

Regardless of how many single-particle wavefunctions χ are available, this number is overwhelmed by the number of n-particle wavefunctions ϕ (Slater determinants) that can be constructed from them. For example, if a two-electron system is treated within the Hartree–Fock approximation using 100 basis functions, both of the electrons can be assigned to any of the χ obtained in the calculation, resulting in 10 000 two-electron wavefunctions. For water, which has ten electrons, the number of electronic wavefunctions with equal numbers of α and β spin electrons that can be constructed from 100 single-particle wavefunctions is roughly 10^{15}! The significance of these other solutions may be hard to grasp. If one is interested solely in the electronic ground state and its associated potential energy surface (the focus of investigation in most quantum chemistry studies), these solutions play no role whatsoever within the HF–SCF approximation. Moreover, one might think (correctly) that solutions obtained by putting an electron in one of the unoccupied orbitals offers a poor treatment of excited states since only the occupied orbitals are optimized. However, there is one very important feature of the extra solutions. If the HF solution has been obtained and all (an infinite number) of virtual orbitals available, then the basic principles of quantum mechanics imply that the exact wavefunction can be written as the sum of Slater determinants

$$\psi_{\text{exact}} = \sum_k c_k \phi_k \qquad (A1.1.104)$$

where the ϕ_k correspond to all possible electron configurations. The individual Slater determinants are thus seen to play a role in the representation of the exact wavefunction that is analogous to that played by the hydrogenic (or LCAO) functions in the expansion of the Hartree–Fock orbitals. The Slater determinants are sometimes said to form an *n-electron basis*, while the hydrogenic (LCAO) functions are the *one-electron basis*.

A similar expansion can be made in practical finite-basis calculations, except that limitations of the basis set preclude the possibility that the exact wavefunction lies in the space spanned by the available ϕ. However, it should be clear that the formation of linear combinations of the finite number of ϕ_k offers a way to better approximate the exact solution. In fact, it is possible to obtain by this means a wavefunction that exactly satisfies the electronic Schrödinger equation when the assumption is made that the solution must lie in the space spanned by the n-electron basis functions ϕ. However, even this is usually impossible, and only a select number of the ϕ_k are used. The general principle of writing n-electron wavefunctions as linear combinations of Slater determinants is known as *configuration interaction*, and the resultant improvement in the wavefunction is said to account for *electron correlation*. The origin of this term is easily understood. Returning to helium, an inspection of the Hartree–Fock wavefunction

$$\psi = \sqrt{\tfrac{1}{2}}[\chi_{1s}(1)\bar{\chi}_{1s}(2) - \chi_{1s}(2)\bar{\chi}_{1s}(1)] \qquad (A1.1.105)$$

exhibits some rather unphysical behaviour: the probability of finding one electron at a particular point in space is entirely independent of where the other electron is! In particular, the probability does not vanish when the two particles are coincident, the associated singularity in the interelectronic repulsion potential notwithstanding. Of course, electrons do not behave in this way, and do indeed tend to avoid each other. Hence, their motion is correlated, and this qualitative feature is absent from the Hartree–Fock approximation when the electrons have different spins. When they are of like spin, then the implicit incorporation of the Pauli principle into the form of the Slater determinant allows for some measure of correlation (although these like-spin effects are characteristically overestimated) since the wavefunction vanishes when the coordinates of the two electrons coincide. Treatments of electron correlation and the related concept of *correlation energy* (the difference between the Hartree–Fock and exact non-relativistic results) take a number of different forms that differ in the strategies used to determine the expansion coefficients c_k and the energy (which is not always given by the expectation value of the Hamiltonian over a function of the form equation (A1.1.104)). The basic theories underlying the most popular choices are the variational principle and perturbation theory, which are discussed in a general way in the remainder of this section. Specific application of these tools in electronic structure theory is dealt with in section B3.1.

Before leaving this discussion, it should also be mentioned that a concept very similar to the independent-particle approximation is used in the quantum-mechanical treatment of molecular vibrations. In that case, it is always possible to solve the Schrödinger equation for nuclear motion exactly if the potential energy function is assumed to be quadratic. The corresponding functions χ_i, χ_j etc. then define what are known as the *normal coordinates* of vibration, and the Hamiltonian can be written in terms of these in precisely the form given by equation (A1.1.69), with the caveat that each term refers not to the coordinates of a single particle, but rather to independent coordinates that involve the collective motion of many particles. An additional distinction is that treatment of the vibrational problem does not involve the complications of antisymmetry associated with identical fermions and the Pauli exclusion principle. Products of the normal coordinate functions nevertheless describe all vibrational states of the molecule (both ground and excited) in very much the same way that the product states of single-electron functions describe the electronic states, although it must be emphasized that one model is based on independent motion and the other on collective motion, which are qualitatively very different. Neither model faithfully represents reality, but each serves as an extremely useful conceptual model and a basis for more accurate calculations.

A1.1.4 Approximating eigenvalues of the Hamiltonian

Since its eigenvalues correspond to the allowed energy states of a quantum-mechanical system, the time-independent Schrödinger equation plays an important role in the theoretical foundation of atomic and molecular spectroscopy. For cases of chemical interest, the equation is always easy to write down but impossible to solve exactly. Approximation techniques are needed for the application of quantum mechanics to atoms and molecules. The purpose of this subsection is to outline two distinct procedures—the variational principle and perturbation theory—that form the theoretical basis for most methods used to approximate solutions to the Schrödinger equation. Although some tangible connections are made with ideas of quantum chemistry and the independent-particle approximation, the presentation in the next two sections (and example problem) is intended to be entirely general so that the scope of applicability of these approaches is not underestimated by the reader.

A1.1.4.1 *The variational principle*

Although it may be impossible to solve the Schrödinger equation for a specific choice of the Hamiltonian, it is always possible to guess! While randomly chosen functions are unlikely to be good approximations to the exact quantum-mechanical wavefunction, an educated guess can usually be made. For example, if one is interested in the ground state of a single particle subjected to some potential energy function, the qualitative features discussed in subsection A1.1.2.3 can be used as a guide in constructing a guess. Specifically, an appropriate choice would be one that decays to zero at positive and negative infinity, has its largest values in regions where the potential is deepest, and has no nodes. For more complicated problems—especially those involving several identical particles—it is not so easy to intuit the form of the wavefunction. Nevertheless, guesses can be based on solutions to a (perhaps grossly) simplified Schrödinger equation, such as the Slater determinants associated with independent-particle models.

In general, approaches based on guessing the form of the wavefunction fall into two categories. In the first, the ground-state wavefunction is approximated by a function that contains one or more nonlinear parameters. For example, if $\exp(ax)$ is a solution to a simplified Schrödinger equation, then the function $\exp(bx)$ provides a plausible guess for the actual problem. The parameter b can then be varied to obtain the most accurate description of the exact ground state. However, there is an apparent contradiction here. If the exact ground-state wavefunction and energy are not known (and indeed impossible to obtain analytically), then how is one to determine the best choice for the parameter b?

The answer to the question that closes the preceding paragraph is the essence of the *variational principle* in quantum mechanics. If a guessed or *trial wavefunction* ϕ is chosen, the energy ϵ obtained by taking the expectation value of the Hamiltonian (it must be emphasized that the actual Hamiltonian is used to evaluate the expectation value, rather than the approximate Hamiltonian that may have been used to generate the form of the trial function) over ϕ must be higher than the exact ground-state energy. It seems rather remarkable that the mathematics seems to know precisely where the exact eigenvalue lies, even though the problem cannot be solved exactly. However, it is not difficult to prove that this assertion is true. The property of mathematical completeness tells us that our trial function can be written as a linear combination of the exact wavefunctions (so long as our guess obeys the boundary conditions and fundamental symmetries of the problem), even when the latter cannot be obtained. Therefore one can always write

$$\phi = \sum_k c_k \psi_k \tag{A1.1.106}$$

where ψ_k is the exact Hamiltonian eigenfunction corresponding to eigenvalue λ_k, and ordered so that $\lambda_0 \leq \lambda_1 \leq \lambda_2 \cdots$. Assuming normalization of both the exact wavefunctions and the trial function, the expectation value of the Hamiltonian is

$$\epsilon = \int \phi^* H \phi \, d\tau$$

$$= \int \left(\sum_j c_j^* \psi_j^* \right) H \left(\sum_k c_k \psi_k \right) d\tau = \sum_k \sum_j c_j^* c_k \int \psi_j^* H \psi_k. \tag{A1.1.107}$$

Since the ψ_k represent exact eigenfunctions of the Hamiltonian, equation (A1.1.107) simplifies to

$$\epsilon = \sum_j \sum_k c_j^* c_k \lambda_k \int \psi_j^* \psi_k = \sum_k c_k^* c_k \lambda_k = \sum_k |c_k|^2 \lambda_k. \tag{A1.1.108}$$

The assumption of normalization imposed on the trial function means that

$$\sum_k |c_k|^2 = 1 \tag{A1.1.109}$$

hence

$$|c_0|^2 = 1 - |c_1|^2 - |c_2|^2 - |c_3|^2 - \cdots. \tag{A1.1.110}$$

Inserting equation (A1.1.110) into equation (A1.1.108) yields

$$\epsilon = \lambda_0 + |c_1|^2 (\lambda_1 - \lambda_0) + |c_2|^2 (\lambda_2 - \lambda_0) + \cdots. \tag{A1.1.111}$$

The first term on the right-hand side of the equation for ϵ is the exact ground-state energy. All of the remaining contributions involve norms of the expansion coefficients and the differences $\lambda_k - \lambda_0$, both of which must be either positive or zero. Therefore, ϵ is equal to the ground-state energy plus a number that cannot be negative. In the case where the trial function is precisely equal to the ground-state wavefunction, then $\epsilon = \lambda_0$; otherwise

$\epsilon > \lambda_0$. Hence, the expectation value of the Hamiltonian with respect to any arbitrarily chosen trial function provides an upper bound to the exact ground-state energy. The dilemma raised earlier—how to define the best value of the variational parameter b—has a rather straightforward answer, namely the choice that minimizes the value of ϵ, known as the *variational energy*.

A concrete example of the variational principle is provided by the Hartree–Fock approximation. This method asserts that the electrons can be treated independently, and that the n-electron wavefunction of the atom or molecule can be written as a Slater determinant made up of orbitals. These orbitals are defined to be those which minimize the expectation value of the energy. Since the general mathematical form of these orbitals is not known (especially in molecules), then the resulting problem is highly nonlinear and formidably difficult to solve. However, as mentioned in subsection A1.1.3.2, a common approach is to assume that the orbitals can be written as linear combinations of one-electron basis functions. If the basis functions are fixed, then the optimization problem reduces to that of finding the best set of coefficients for each orbital. This tremendous simplification provided a revolutionary advance for the application of the Hartree–Fock method to molecules, and was originally proposed by Roothaan in 1951.

A similar form of the trial function occurs when it is assumed that the exact (as opposed to Hartree–Fock) wavefunction can be written as a linear combination of Slater determinants (see equation (A1.1.104)). In the conceptually simpler latter case, the objective is to minimize an expression of the form

$$\epsilon = \int \phi^* \hat{H} \phi \, d\tau \qquad (A1.1.112)$$

where ϕ is parametrized as

$$\phi = \sum_{k=0}^{N} c_k \chi_k \qquad (A1.1.113)$$

and both the (fixed functions) χ_k and ϕ are assumed to be normalized.

The representation of trial functions as linear combinations of fixed basis functions is perhaps the most common approach used in variational calculations; optimization of the coefficients c_k is often said to be an application of the *linear variational principle*. Although some very accurate work on small atoms (notably helium and lithium) has been based on complicated trial functions with several nonlinear parameters, attempts to extend these calculations to larger atoms and molecules quickly runs into formidable difficulties (not the least of which is how to choose the form of the trial function). Basis set expansions like that given by equation (A1.1.113) are much simpler to design, and the procedures required to obtain the coefficients that minimize ϵ are all easily carried out by computers.

For the example discussed above, where $\exp(ax)$ is the solution to a simpler problem, a trial function using five basis functions

$$\phi = c_1 e^{(a-2)x} + c_2 e^{(a-1)x} + c_3 e^{ax} + c_4 e^{(a+1)x} + c_5 e^{(a+2)x} \qquad (A1.1.114)$$

could be used instead of $\exp(bx)$ if the exact function is not expected to deviate too much from $\exp(ax)$. What is gained from replacing a trial function containing a single parameter by one that contains five? To see, consider the problem of how coefficients can be chosen to minimize the variational energy ϵ,

$$\epsilon = \frac{\int \phi^* \hat{H} \phi \, d\tau}{\int \phi^* \phi \, d\tau}. \qquad (A1.1.115)$$

The denominator is included in equation (A1.1.115) because it is impossible to ensure that the trial function is normalized for arbitrarily chosen coefficients c_k. In order to minimize the value of ϵ for the trial function

$$\phi = \sum_{k=0}^{N} c_k \chi_k \qquad (A1.1.116)$$

it is necessary (but not sufficient) that its first partial derivatives with respect to all expansion coefficients vanish, *viz*

$$\frac{\partial \epsilon}{\partial c_1} = \frac{\partial \epsilon}{\partial c_2} = \frac{\partial \epsilon}{\partial c_3} = \cdots = 0. \tag{A1.1.117}$$

It is worthwhile, albeit tedious, to work out the condition that must satisfied in order for equation (A1.1.117) to hold true. Expanding the trial function according to equation (A1.1.113), assuming that the basis functions and expansion coefficients are real and making use of the technique of implicit differentiation, one finds

$$\frac{\partial \epsilon}{\partial c_k} \left[\sum_{i=0}^{N} \sum_{j=0}^{N} c_i c_j S_{ij} \right] + 2\epsilon \left[\sum_{j=0}^{N} c_j S_{jk} \right] = 2 \sum_{j=0}^{N} c_j H_{jk}$$

$$\frac{\partial \epsilon}{\partial c_k} \left[\sum_{i=0}^{N} \sum_{j=0}^{N} c_i c_j S_{ij} \right] = 2 \sum_{j=0}^{N} c_j (H_{jk} - \epsilon S_{jk}) \tag{A1.1.118}$$

where shorthand notations for the *overlap matrix elements*

$$S_{jk} \equiv \int \chi_j \chi_k \, d\tau \tag{A1.1.119}$$

and *Hamiltonian matrix elements*

$$H_{jk} \equiv \int \chi_j H \chi_k \, d\tau \tag{A1.1.120}$$

have been introduced. Since the term multiplying the derivative of the expansion coefficient is simply the norm of the wavefunction, the variational condition equation (A1.1.117) is satisfied if the term on the right-hand side of equation (A1.1.118) vanishes for all values of k. Specifically, the set of homogeneous linear equations corresponding to the matrix expression

$$(c_1 \quad c_2 \quad \cdots \quad c_N) \begin{pmatrix} H_{00} - \epsilon S_{00} & H_{01} - \epsilon S_{01} & \cdots & H_{0N} - \epsilon S_{0N} \\ H_{10} - \epsilon S_{10} & H_{11} - \epsilon S_{11} & \cdots & H_{1N} - \epsilon S_{1N} \\ \vdots & \vdots & \ddots & \vdots \\ H_{N0} - \epsilon S_{N0} & H_{N1} - \epsilon S_{N1} & \cdots & H_{NN} - \epsilon S_{NN} \end{pmatrix} = \begin{pmatrix} 0 \\ 0 \\ \vdots \\ 0 \end{pmatrix} \tag{A1.1.121}$$

must be satisfied. It is a fundamental principle of linear algebra that systems of equations of this general type are satisfied only for certain choices of ϵ, namely those for which the *determinant*

$$\begin{vmatrix} H_{00} - \epsilon S_{00} & H_{01} - \epsilon S_{01} & \cdots & H_{0N} - \epsilon S_{1N} \\ H_{10} - \epsilon S_{10} & H_{11} - \epsilon S_{11} & \cdots & H_{1N} - \epsilon S_{2N} \\ \vdots & \vdots & \ddots & \vdots \\ H_{N0} - \epsilon S_{N0} & H_{N1} - \epsilon S_{N1} & \cdots & H_{NN} - \epsilon S_{NN} \end{vmatrix} \tag{A1.1.122}$$

is identically equal to zero. There are precisely N values of ϵ that satisfy this condition, some of which might be degenerate, and their determination constitutes what is known as the generalized eigenvalue problem. While this is reasonably well suited to computation, a further simplification is usually made. When suited to the problem under consideration, the basis functions are usually chosen to be members of an orthonormal set. In other cases (for example, in the LCAO treatment of molecules) where this is not possible, the original basis functions χ_k' corresponding to the overlap matrix $\mathbf{S}'$ can be subjected to the orthonormalizing transformation

$$\chi_k = \sum_l \chi_l' X_{lk} \tag{A1.1.123}$$

where $\mathbf{X}$ is the reciprocal square root of the overlap matrix in the primed basis,

$$\mathbf{X} \equiv \mathbf{S}'^{-1/2}. \tag{A1.1.124}$$

The simplest way to obtain $\mathbf{X}$ is to diagonalize $\mathbf{S}'$, take the reciprocal square roots of the eigenvalues and then transform the matrix back to its original representation, i.e.

$$\mathbf{X} = \mathbf{C}' \mathbf{s}^{-1/2} \mathbf{C}'^{\dagger} \tag{A1.1.125}$$

where $\mathbf{s}$ is the diagonal matrix of reciprocal square roots of the eigenvalues, and $\mathbf{C}'$ is the matrix of eigenvectors for the original $\mathbf{S}'$ matrix. Doing this, one finds that the transformed basis functions are orthonormal. In terms of implementation, elements of the Hamiltonian are usually first evaluated in the primed basis, and the resulting *matrix representation* of $\mathbf{H}$ is then transformed to the orthogonal basis ($\mathbf{H} = \mathbf{X}^{\dagger}\mathbf{H}'\mathbf{X}$).

In an orthonormal basis, $S_{kj} = 1$ if $k = j$, and vanishes otherwise. The problem of finding the variational energy of the ground state then reduces to that of determining the smallest value of ϵ that satisfies

$$\begin{vmatrix} H_{00} - \epsilon & H_{01} & \cdots & H_{0N} \\ H_{10} & H_{11} - \epsilon & \cdots & H_{1N} \\ \vdots & \vdots & \ddots & \vdots \\ H_{N0} & H_{N1} & \cdots & H_{NN} - \epsilon \end{vmatrix} = 0 \tag{A1.1.126}$$

a task that modern digital computers can perform very efficiently. Given an orthonormal basis, the variational problem can be solved by *diagonalizing* the matrix representation of the Hamiltonian, $\mathbf{H}$. Associated with each eigenvalue ϵ is an eigenvector $(c_0, c_1, c_2, \ldots, c_N)$ that tells how the basis functions are combined in the corresponding approximate wavefunction ϕ as parametrized by equation (A1.1.116). That the lowest eigenvalue ϵ of $\mathbf{H}$ provides an upper bound to the exact ground-state energy has already been proven; it is also true (but will not be proved here) that the first excited state of the actual system must lie below the next largest eigenvalue λ_1, and indeed all remaining eigenvalues provide upper bounds to the corresponding excited states. That is,

$$\epsilon_0 \geq \lambda_0, \epsilon_1 \geq \lambda_1, \epsilon_2 \geq \lambda_2, \ldots, \epsilon_N \geq \lambda_N. \tag{A1.1.127}$$

The equivalence between variational energies and the exact eigenvalues of the Hamiltonian is achieved only in the case where the corresponding exact wavefunctions can be written as linear combinations of the basis functions.

Suppose that the Schrödinger equation for the problem of interest cannot be solved, but another simpler problem that involves precisely the same set of coordinates lends itself to an analytic solution. In practice, this can often be achieved by ignoring certain interaction terms in the Hamiltonian, as discussed earlier. Since the eigenfunctions of the simplified Hamiltonian form a complete set, they provide a conceptually useful basis since all of the eigenfunctions of the intractable Hamiltonian can be written as linear combinations of them (for example, Slater determinants for electrons or products of normal mode wavefunctions for vibrational states). In this case, diagonalization of $\mathbf{H}$ in this basis of functions provides an exact solution to the Schrödinger equation. It is worth pausing for a moment to analyse what is meant by this rather remarkable statement. One simply needs to ignore interaction terms in the Hamiltonian that preclude an analytic determination of the stationary states and energies of the system. The corresponding Schrödinger equation can then be solved to provide a set of orthonormal basis functions, and the integrals that represent the matrix elements of $\mathbf{H}$

$$H_{ij} = \int \chi_i^* H \chi_j \, d\tau \tag{A1.1.128}$$

computed. Diagonalization of the resulting matrix provides the sought-after solution to the quantum-mechanical problem. Although this process replaces an intractable differential equation by a problem in linear algebra, the latter offers its own insurmountable hurdle: the *dimension of the matrix* (equal to the number of rows or columns) is equal to the number of functions included in the complete set of solutions to the simplified Schrödinger equation. Regrettably, this number is usually infinite. At present, special algorithms can be used with modern computers to obtain eigenvalues of matrices with dimensions of about 100 million relatively routinely, but this still falls far short of infinity. Therefore, while it seems attractive (and much simpler) to do away with the differential equation in favour of a matrix diagonalization, it is not a magic bullet that makes exact quantum-mechanical calculations a possibility.

In order to apply the linear variational principle, it is necessary to work with a matrix sufficiently small that it can be diagonalized by a computer; such calculations are said to employ a *finite basis*. Use of a finite basis means that the eigenvalues of **H** are not exact unless the basis chosen for the problem has the miraculous (and extraordinarily unlikely) property of being sufficiently flexible to allow one or more of the exact solutions to be written as linear combinations of them. For example, if the intractable system Hamiltonian contains only small interaction terms that are ignored in the simplified Hamiltonian used to obtain the basis functions, then χ_0 is probably a reasonably good approximation to the exact ground-state wavefunction. At the very least, one can be relatively certain that it is closer to ψ_0 than are those that correspond to the thousandth, millionth and billionth excited states of the simplified system. Hence, if the objective of a variational calculation is to determine the ground-state energy of the system, it is important to include χ_0 and other solutions to the simplified problem with relatively low lying energies, while $\chi_{1\,000\,000}$ and other high lying solutions can be excluded more safely.

A1.1.4.2   Example problem: the double-well oscillator

To illustrate the use of the variational principle, results are presented here for calculations of the five lowest energy states (the ground state and the first four excited states) of a particle subject to the potential

$$V(q) = 0.05q^4 - q^2 \tag{A1.1.129}$$

which is shown in figure A1.1.3. The potential goes asymptotically to infinity (like that for the harmonic oscillator), but exhibits two symmetrical minima at $q = \pm\sqrt{10}$ and a maximum at the origin. This function is known as a double well, and provides a qualitative description of the potential governing a variety of quantum-mechanical processes, such as motion involved in the inversion mode of ammonia (where the minima play the role of the two equivalent pyramidal structures and the maximum that of planar NH_3). For simplicity, the potential is written in terms of the dimensionless coordinate q defined by

$$q \equiv \alpha x \equiv \left(\frac{mk}{\hbar^2}\right)^{\frac{1}{4}} x \tag{A1.1.130}$$

where x is a Cartesian displacement and k is a constant with units of (mass)(time)$^{-2}$. The corresponding Schrödinger equation can be written as

$$\left[-\frac{1}{2}\frac{\mathrm{d}}{\mathrm{d}q^2} + 0.05q^4 - q^2\right]\psi = E\psi \tag{A1.1.131}$$

where the energy is given as a multiple of $\hbar^2\alpha^2/m$. This value corresponds to $h\nu$ where ν is the frequency corresponding to a quantum harmonic oscillator with force constant k.

It is not possible to solve this equation analytically, and two different calculations based on the linear variational principle are used here to obtain the approximate energy levels for this system. In the first, eigenfunctions corresponding to the potential $V = 2q^2$ (this corresponds to the shape of the double-well potential in the vicinity of its minima) are used as a basis. It should be noted at the outset that these functions

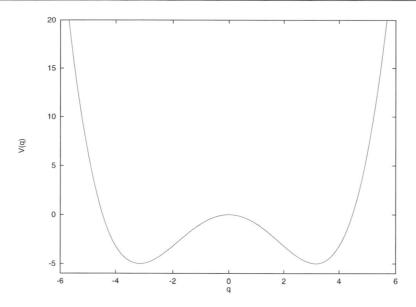

Figure A1.1.3. Potential function used in the variational calculations. Note that the energies of all states lie above the lowest point of the potential ($V = -5$), which occurs at $q = \pm\sqrt{10}$.

form a complete set, and it is therefore possible to write exact solutions to the double-well oscillator problem in terms of them. However, since we expect the ground-state wavefunction to have maximum amplitude in the regions around $q = \pm\sqrt{10}$, it is unlikely that the first few harmonic oscillator functions (which have maxima closer to the origin) are going to provide a good representation of the exact ground state. The first four eigenvalues of the potential are given in the table below, where N indicates the size of the variational basis which includes the N lowest energy harmonic oscillator functions centred at the origin.

N	λ_1	λ_2	λ_3	λ_4
2	0.259 37	0.796 87	—	—
4	−0.467 37	−0.358 63	1.989 99	3.248 62
6	−1.414 39	−1.051 71	0.689 35	1.434 57
8	−2.225 97	−1.850 67	0.097 07	0.396 14
10	−2.891 74	−2.580 94	−0.358 57	−0.339 30
20	−4.021 22	−4.012 89	−2.162 21	−2.125 38
30	−4.026 63	−4.026 60	−2.204 11	−2.200 79
40	−4.026 63	−4.026 60	−2.204 11	−2.200 79
50	−4.026 63	−4.026 60	−2.204 11	−2.200 79

Note that the energies decrease with increasing size of the basis set, as expected from the variational principle. With 30 or more functions, the energies of the four states are well converged (to about one part in 100 000). In figure A1.1.4, the wavefunctions of the ground and first excited states of the system calculated with 40 basis functions are shown. As expected, the probability density is localized near the symmetrically disposed minima on the potential. The ground state has no nodes and the first excited state has a single node. The ground-state wavefunction calculated with only eight basis functions (shown in figure A1.1.5) is clearly imperfect. The rapid oscillations in the wavefunction are not real, but rather an artifact of the incomplete basis used in the calculation. A larger number of functions is required to reduce the amplitude of the oscillations.

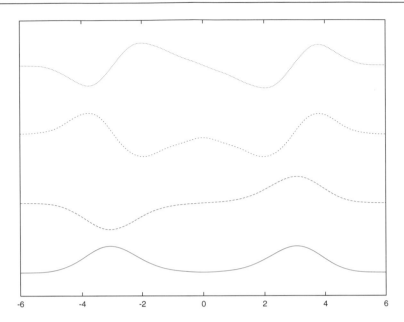

Figure A1.1.4. Wavefunctions for the four lowest states of the double-well oscillator. The ground-state wavefunction is at the bottom and the others are ordered from bottom to top in terms of increasing energy.

The form of the approximate wavefunctions suggests another choice of basis for this problem, namely one comprising some harmonic oscillator functions centred about one minimum and additional harmonic oscillator functions centred about the other minimum. The only minor difficulty in this calculation is that the basis set is not orthogonal (which should be clear simply by inspecting the overlap of the ground-state harmonic oscillator wavefunctions centred at the two points) and an orthonormalization based on equations (A1.1.123)–(A1.1.125) is necessary. Placing an equal number of $V = 2q^2$ harmonic oscillator functions at the position of each minimum (these correspond to solutions of the harmonic oscillator problems with $V = 2(q - \sqrt{10})^2$ and $V = 2(q + \sqrt{10})^2$, respectively) yields the eigenvalues given below for the four lowest states (in each case, there are $N/2$ functions centred at each point).

N	λ_1	λ_2	λ_3	λ_4
2	−3.990 62	−3.990 62	—	—
4	−4.017 87	−4.017 87	−1.925 88	−1.925 88
6	−4.018 51	−4.018 51	−2.125 22	−2.125 21
8	−4.025 23	−4.025 23	−2.142 47	−2.142 45
10	−4.026 32	−4.026 32	−2.176 90	−2.176 80
20	−4.026 63	−4.026 60	−2.202 90	−2.200 64
30	−4.026 63	−4.026 60	−2.204 11	−2.200 79
40	−4.026 63	−4.026 60	−2.204 11	−2.200 79
50	−4.026 63	−4.026 60	−2.204 11	−2.200 79

These results may be compared to those obtained with the basis centred at $q = 0$. The rate of convergence is faster in the present case, which attests to the importance of a carefully chosen basis. It should be pointed out that there is a clear correspondence between the two approaches used here and the single-centre and

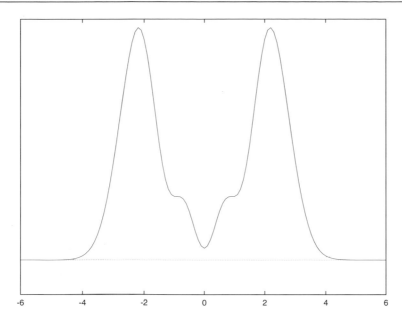

Figure A1.1.5. Ground state wavefunction of the double-well oscillator, as obtained in a variational calculation using eight basis functions centred at the origin. Note the spurious oscillatory behaviour near the origin and the location of the peak maxima, both of which are well inside the potential minima.

LCAO expansions used in molecular orbital theory; the reader should appreciate the advantages of choosing an appropriately designed multicentre basis set in achieving rapid convergence in some calculations. Finally, in figure A1.1.6, the ground-state wavefunctions calculated with a mixed basis of eight functions (four centred about each of the two minima) are displayed. Note that oscillations seen in the single-centre basis calculation using the same number of functions are completely missing in the non-orthogonal basis calculation.

A1.1.4.3 Perturbation theory

Calculations that employ the linear variational principle can be viewed as those that obtain the exact solution to an approximate problem. The problem is approximate because the basis necessarily chosen for practical calculations is not sufficiently flexible to describe the exact states of the quantum-mechanical system. Nevertheless, within this finite basis, the problem is indeed solved exactly: the variational principle provides a recipe to obtain the best possible solution in the *space spanned by the basis functions*. In this section, a somewhat different approach is taken for obtaining approximate solutions to the Schrödinger equation. Instead of obtaining exact eigenvalues of **H** in a finite basis, a strategy is developed for determining approximate eigenvalues of the exact matrix representation of $\hat{H}$. It can also be used (and almost always is in practical calculations) to obtain approximate eigenvalues to approximate (incomplete basis) Hamiltonian matrices that are nevertheless much larger in dimension than those that can be diagonalized exactly. The standard textbook presentation of this technique, which is known as perturbation theory, generally uses the Schrödinger differential equation as the starting point. However, some of the generality and usefulness of the technique can be lost in the treatment. Students may not come away with an appreciation for the role of linear algebra in perturbation theory, nor do they usually grasp the (approximate problem, exact answer)/(right—or at least less approximate—problem/approximate answer) distinction between matrix diagonalization in the linear variational principle and the use of perturbation theory.

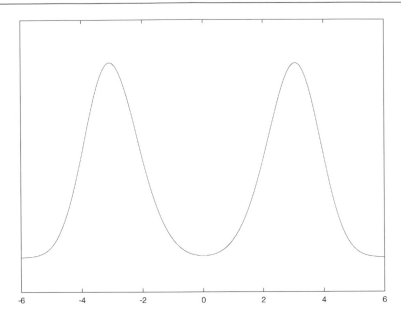

Figure A1.1.6. Ground-state wavefunction of the double-well oscillator, as obtained in a variational calculation using four basis functions centred at $q = \sqrt{10}$ and four centred at $q = -\sqrt{10}$. Note the absence of a node at the origin.

In perturbation theory, the Hamiltonian is divided into two parts. One of these corresponds to a Schrödinger equation that can be solved exactly

$$\hat{H}_0 \chi_k^{(0)} = \lambda_k^{(0)} \chi_k^{(0)} \qquad (A1.1.132)$$

while the remainder of the Hamiltonian is designated here as $\hat{V}$. The orthonormal eigenfunctions $\chi_k^{(0)}$ of the *unperturbed*, or *zeroth-order Hamiltonian* $\hat{H}_0$ form a convenient basis for a matrix representation of the Hamiltonian $\hat{H}$. Diagonalization of **H** gives the exact quantum-mechanical energy levels if the complete set of $\chi_k^{(0)}$ is used, and approximate solutions if the basis is truncated. Instead of focusing on the exact eigenvalues of **H**, however, the objective of perturbation theory is to approximate them. The starting point is the matrix representation of H_0 and V, which will be designated as **h**

$$\mathbf{h} = \begin{pmatrix} h_{00} & 0 & 0 & \cdots & 0 \\ 0 & h_{11} & 0 & \cdots & 0 \\ 0 & 0 & h_{22} & \cdots & 0 \\ \vdots & \vdots & \vdots & \ddots & \vdots \\ 0 & 0 & 0 & \cdots & h_{NN} \end{pmatrix} \qquad (A1.1.133)$$

and **v**

$$\mathbf{v} = \begin{pmatrix} v_{00} & v_{01} & v_{02} & \cdots & v_{0N} \\ v_{10} & v_{11} & v_{12} & \cdots & v_{1N} \\ v_{20} & v_{21} & v_{22} & \cdots & v_{2N} \\ \vdots & \vdots & \vdots & \ddots & \vdots \\ v_{N0} & v_{N1} & v_{N2} & \cdots & v_{NN} \end{pmatrix} \qquad (A1.1.134)$$

respectively, where the matrix elements h_{ii} and v_{ij} are given by the integrals

$$h_{ii} \equiv \int \chi_i^{(0)*} H_0 \chi_i^{(0)} \, d\tau = \lambda_i^{(0)} \qquad (A1.1.135)$$

and

$$v_{ij} \equiv \int \chi_i^{(0)*} V \chi_j^{(0)} \, d\tau. \tag{A1.1.136}$$

Note that $\mathbf{h}$ is simply the diagonal matrix of zeroth-order eigenvalues $\lambda_k^{(0)}$. In the following, it will be assumed that the zeroth-order eigenfunction $\chi_0^{(0)}$ is a reasonably good approximation to the exact ground-state wavefunction (meaning that $\lambda_0^{(0)} \sim \lambda_0$), and $\mathbf{h}$ and $\mathbf{v}$ will be written in the compact representations

$$\mathbf{h} = \begin{pmatrix} \lambda_0^{(0)} & \mathbf{0} \\ \mathbf{0} & \Lambda_q^{(0)} \end{pmatrix} \tag{A1.1.137}$$

$$\mathbf{v} = \begin{pmatrix} v_{00} & \mathbf{v}_{0q} \\ \mathbf{v}_{q0} & \mathbf{v}_{qq} \end{pmatrix}. \tag{A1.1.138}$$

It is important to realize that while the uppermost diagonal elements of these matrices are numbers, the other diagonal element is a matrix of dimension N. Specifically, these are the matrix representations of H_0 and V in the basis q which consists of all $\chi_k^{(0)}$ in the original set, apart from $\chi_0^{(0)}$, i.e.

$$q = \{\chi_1^{(0)}, \chi_2^{(0)} \cdots \chi_N^{(0)}\}. \tag{A1.1.139}$$

The off-diagonal elements in this representation of $\mathbf{h}$ and $\mathbf{v}$ are the zero vector of length N (for $\mathbf{h}$) and matrix elements which *couple* the zeroth-order ground-state eigenfunction $\chi_0^{(0)}$ to members of the set q (for $\mathbf{v}$):

$$\mathbf{v}_{q0} \ni v_{k0} \equiv \int \chi_k^{(0)*} V \chi_0^{(0)} \qquad (k \neq 0). \tag{A1.1.140}$$

The exact ground-state eigenvalue λ_0 and corresponding eigenvector

$$c \equiv \begin{pmatrix} c_0 \\ c_q \end{pmatrix} \tag{A1.1.141}$$

clearly satisfy the coupled equations

$$H_{00}c_0 + \mathbf{H}_{0q}c_q = c_0\lambda_0 \tag{A1.1.142}$$

$$\mathbf{H}_{q0}c_0 + \mathbf{H}_{qq}c_q = c_q\lambda_0. \tag{A1.1.143}$$

The latter of these can be solved for c_q

$$c_q = [\lambda_0 \mathbf{1} - \mathbf{H}_{qq}]^{-1} \mathbf{h}_{q0} c_0 \tag{A1.1.144}$$

(the N by N identity matrix is represented here and in the following by $\mathbf{1}$) and inserted into equation (A1.1.142) to yield the implicit equation

$$\lambda_0 = \{H_{00} + \mathbf{H}_{0q}[\lambda_0 \mathbf{1} - \mathbf{H}_{qq}]^{-1} \mathbf{H}_{q0}\}. \tag{A1.1.145}$$

Thus, one can solve for the eigenvalue iteratively, by guessing λ_0, evaluating the right-hand side of equation (A1.1.145), using the resulting value as the next guess and continuing in this manner until convergence is achieved. However, this is not a satisfactory method for solving the Schrödinger equation, because the problem of diagonalizing a matrix of dimension $N + 1$ is replaced by an iterative procedure in which a matrix of dimension N must be inverted for each successive improvement in the guessed eigenvalue. This is an even more computationally intensive problem than the straightforward diagonalization approach associated with the linear variational principle.

Nevertheless, equation (A1.1.145) forms the basis for the approximate diagonalization procedure provided by perturbation theory. To proceed, the exact ground-state eigenvalue and corresponding eigenvector are written as the sums

$$c = c^{(0)} + c^{(1)} + c^{(2)} + \cdots \tag{A1.1.146}$$

and

$$\lambda_0 = \lambda_0^{(0)} + \lambda_0^{(1)} + \lambda_0^{(2)} + \cdots \tag{A1.1.147}$$

where $c^{(k)}$ and $\lambda_0^{(k)}$ are said to be *kth-order contributions* in the *perturbation expansion*. What is meant here by order? Ultimately, the various contributions to c and λ_0 will be written as matrix products involving the unperturbed Hamiltonian matrix $\mathbf{h}$ and the matrix representation of the perturbation $\mathbf{v}$. The order of a particular contribution is defined by the number of times $\mathbf{v}$ appears in the corresponding matrix product. Roughly speaking, if $\lambda_0^{(0)}$ is of order unity, and the matrix elements of $\mathbf{v}$ are an order of magnitude or two smaller, then the third-order energy contribution should be in the range 10^{-3}–10^{-6}. Therefore, one expects the low order contributions to be most important and the expansions given by equations (A1.1.146) and (A1.1.147) to converge rapidly, provided the zeroth-order description of the quantum-mechanical system is reasonably accurate.

To derive equations for the order-by-order contributions to the eigenvalue λ, the implicit equation for the eigenvalue is first rewritten as

$$
\begin{aligned}
\lambda_0^{(0)} + \Delta\lambda &= \{\lambda_0^{(0)} + v_{00} + \mathbf{v}_{0q}[\lambda_0^{(0)}\mathbf{1} + \Delta\lambda\mathbf{1} - \Lambda_q^{(0)} - \mathbf{v}_{qq}]^{-1}\mathbf{v}_{q0}\} \\
&= \{\lambda_0^{(0)} + v_{00} + \mathbf{v}_{0q}[\lambda_0^{(0)}\mathbf{1} - \Lambda_q^{(0)} + \Delta\lambda\mathbf{1} - \mathbf{v}_{qq}]^{-1}\mathbf{v}_{q0}\} \\
&= \{\lambda_0^{(0)} + v_{00} + \mathbf{v}_{0q}[(\lambda_0^{(0)}\mathbf{1} - \Lambda_q^{(0)})\{\mathbf{1} - (\lambda_0^{(0)}\mathbf{1} - \Lambda_q^{(0)})^{-1}(\mathbf{v}_{qq} - \Delta\lambda\mathbf{1})\}]^{-1}\mathbf{v}_{q0}\} \\
&= \{\lambda_0^{(0)} + v_{00} + \mathbf{v}_{0q}[\mathbf{1} - (\lambda_0^{(0)}\mathbf{1} - \Lambda_q^{(0)})^{-1}(\mathbf{v}_{qq} - \Delta\lambda\mathbf{1})]^{-1}(\lambda_0^{(0)}\mathbf{1} - \Lambda_q^{(0)})^{-1}\mathbf{v}_{q0}\}
\end{aligned}
\tag{A1.1.148}
$$

where $\Delta\lambda$ is a shorthand notation for the error in the zeroth-order eigenvalue λ

$$\Delta\lambda \equiv \lambda_0 - \lambda_0^{(0)} = \lambda_0^{(1)} + \lambda_0^{(2)} + \lambda_0^{(3)} + \cdots. \tag{A1.1.149}$$

There are two matrix inverses that appear on the right-hand side of these equations. One of these is trivial; the matrix $\lambda_0^{(0)}\mathbf{1} - \Lambda_q^{(0)}$ is diagonal. The other inverse

$$[\mathbf{1} - (\lambda_0^{(0)}\mathbf{1} - \Lambda_q^{(0)})^{-1}(\mathbf{v}_{qq} - \Delta\lambda\mathbf{1})]^{-1} \tag{A1.1.150}$$

is more involved because the matrix $\mathbf{v}_{qq}$ is not diagonal, and direct inversion is therefore problematic. However, if the zeroth-order ground-state energy is well separated from low lying excited states, the diagonal matrix hereafter designated as $\mathbf{R}_q$

$$\mathbf{R}_q \equiv (\lambda_0^{(0)}\mathbf{1} - \Lambda_q^{(0)})^{-1} \tag{A1.1.151}$$

that acts in equation (A1.1.150) to scale

$$(\mathbf{v}_{qq} - \Delta\lambda\mathbf{1}) \tag{A1.1.152}$$

will consist of only small elements. Thus, the matrix to be inverted can be considered as

$$\mathbf{1} - \mathbf{X} \tag{A1.1.153}$$

where $\mathbf{X}$ is, in the sense of matrices, small with respect to $\mathbf{1}$. It can be shown that the inverse of the matrix $\mathbf{1} - \mathbf{X}$ can be written as a series expansion

$$(\mathbf{1} - \mathbf{X})^{-1} = \mathbf{1} + \mathbf{X} + \mathbf{XX} + \mathbf{XXX} + \mathbf{XXXX} + \cdots \tag{A1.1.154}$$

that converges if all eigenvalues of $\mathbf{X}$ lie within the unit circle in the complex plane (complex numbers $a + bi$ such that $a^2 + b^2 < 1$). Applications of perturbation theory in quantum mechanics are predicated on the assumption that the series converges for the inverse given by equation (A1.1.150), but efforts are rarely made to verify that this is indeed the case. Use of the series representation of the inverse in equation (A1.1.148) gives the unwieldy formal equality

$$\lambda_0^{(0)} + \Delta\lambda = \lambda_0^{(0)} + v_{00} + \mathbf{v}_{0q}\mathbf{R}_q\mathbf{v}_{q0} + \mathbf{v}_{0q}\mathbf{R}_q(\mathbf{v}_{qq} - \Delta\lambda\mathbf{1})\mathbf{R}_q\mathbf{v}_{q0}$$
$$+ \mathbf{v}_{0q}\mathbf{R}_q(\mathbf{v}_{qq} - \Delta\lambda\mathbf{1})\mathbf{R}_q(\mathbf{v}_{qq} - \Delta\lambda\mathbf{1})\mathbf{R}_q\mathbf{v}_{q0} + \cdots \qquad \text{(A1.1.155)}$$

from which the error in the zeroth-order energy $\Delta\lambda$ is easily seen to be

$$\lambda_0^{(1)} + \lambda_0^{(2)} + \lambda_0^{(3)} + \cdots = \mathbf{v}_{00} + \mathbf{v}_{0q}\mathbf{R}_q\mathbf{v}_{q0} + \mathbf{v}_{0q}\mathbf{R}_q(\mathbf{v}_{qq} - \Delta\lambda\mathbf{1})\mathbf{R}_q\mathbf{v}_{q0}$$
$$+ \mathbf{v}_{0q}\mathbf{R}_q(\mathbf{v}_{qq} - \Delta\lambda\mathbf{1})\mathbf{R}_q(\mathbf{v}_{qq} - \Delta\lambda\mathbf{1})\mathbf{R}_q\mathbf{v}_{q0} + \cdots . \qquad \text{(A1.1.156)}$$

Each term on the right-hand side of the equation involves matrix products that contain $\mathbf{v}$ a specific number of times, either explicitly or implicitly (for the terms that involve $\Delta\lambda$). Recognizing that $\mathbf{R}_q$ is a zeroth-order quantity, it is straightforward to make the associations

$$\lambda_0^{(1)} = v_{00} \qquad \text{(A1.1.157)}$$

$$\lambda_0^{(2)} = \mathbf{v}_{0q}\mathbf{R}_q\mathbf{v}_{q0} \qquad \text{(A1.1.158)}$$

$$\lambda_0^{(3)} = \mathbf{v}_{0q}\mathbf{R}_q\mathbf{v}_{qq}\mathbf{R}_q\mathbf{v}_{q0} - \lambda_0^{(1)}\mathbf{v}_{0q}\mathbf{R}_q\mathbf{R}_q\mathbf{v}_{q0} \qquad \text{(A1.1.159)}$$

$$\lambda_0^{(4)} = \mathbf{v}_{0q}\mathbf{R}_q\mathbf{v}_{qq}\mathbf{R}_q\mathbf{v}_{qq}\mathbf{R}_q\mathbf{v}_{q0} - \lambda_0^{(2)}\mathbf{v}_{0q}\mathbf{R}_q\mathbf{R}_q\mathbf{v}_{q0}$$
$$- 2\lambda_0^{(1)}\mathbf{v}_{0q}\mathbf{R}_q\mathbf{R}_q\mathbf{v}_{qq}\mathbf{R}_q\mathbf{v}_{q0} + \lambda_0^{(1)}\lambda_0^{(1)}\mathbf{v}_{0q}\mathbf{R}_q\mathbf{R}_q\mathbf{R}_q\mathbf{v}_{q0} \cdots \qquad \text{(A1.1.160)}$$

which provide recipes for calculating corrections to the energy to fourth order. Similar analysis of equation (A1.1.146) provides successively ordered corrections to the zeroth-order eigenvector ($c_0^{(0)} = 1$, $c_q^{(0)} = 0$), specifically

$$c_q^{(1)} = \mathbf{R}_q\mathbf{v}_{q0} \qquad \text{(A1.1.161)}$$

$$c_q^{(2)} = \mathbf{R}_q\mathbf{v}_{qq}\mathbf{R}_q\mathbf{v}_{q0} - \lambda_0^{(1)}\mathbf{R}_q\mathbf{R}_q\mathbf{v}_{q0} \qquad \text{(A1.1.162)}$$

$$\vdots$$

At this point, it is appropriate to make some general comments about perturbation theory that relate to its use in qualitative aspects of chemical physics. Very often, our understanding of complex systems is based on some specific zeroth-order approximation that is then modified to allow for the effect of a perturbation. For example, chemical bonding is usually presented as a weak interaction between atoms in which the atomic orbitals interact to form bonds. Hence, the free atoms represent the zeroth-order picture, and the perturbation is the decrease in internuclear distance that accompanies bond formation. Many rationalizations for bonding trends traditionally taught in descriptive chemistry are ultimately rooted in perturbation theory. As a specific illustration, the decreasing bond strength of carbon–halogen bonds in the sequence C–F > C–Cl > C–Br > C–I (a similar trend is found in the sequence CO, CS, CSe, CTe) can be attributed to a 'mismatch' of the np halogen orbitals with the 2p orbitals of carbon as for larger values of n. From the point of perturbation theory, it is easily understood that the interaction between the bonding electrons is maximized when the corresponding energy levels are close (small denominators, large values of $\mathbf{R}_q$) while large energy mismatches (such as that between the valence orbitals of iodine and carbon) allow for less interaction and correspondingly weaker bonds.

For qualitative insight based on perturbation theory, the two lowest order energy corrections and the first-order wavefunction corrections are undoubtedly the most useful. The first-order energy corresponds to averaging the effects of the perturbation over the approximate wavefunction χ_0, and can usually be evaluated without difficulty. The sum of $\lambda_0^{(0)}$ and $\lambda_0^{(1)}$ is precisely equal to the expectation value of the Hamiltonian over the zeroth-order description χ_0, and is therefore the proper energy to associate with a simplified model. (It should be pointed out that it is this energy and not the zeroth-order energy obtained by summing up orbital eigenvalues that is used as the basis for orbital optimization in Hartree–Fock theory. It is often stated that the first-order correction to the Hartree–Fock energy vanishes, but this is misleading; the first-order energy is defined instead to be part of the Hartree–Fock energy.) The second-order correction allows for *interaction* between the zeroth-order wavefunction and all others, weighted by the reciprocal of the corresponding energy differences and the magnitude of the matrix elements $\mathbf{v}_{0q}$. The same interactions between $\chi_0^{(0)}$ and the $\chi_q^{(0)}$ determine the extent to which the latter are *mixed in* to the first-order perturbed wavefunction described by $c_q^{(1)}$. This is essentially the idea invoked in the theory of orbital hybridization. In the presence of four identical ligands approaching a carbon atom tetrahedrally, its valence s and p orbitals are mixed (through the corresponding $\mathbf{v}_{0q}$ elements, which vanish at infinite separation) and their first-order correction in the presence of the perturbation (the ligands) can be written as four equivalent linear combinations between the s and three p zeroth-order orbitals.

Some similarities and differences between perturbation theory and the linear variational principle need to be emphasized. First, neither approach can be used in practice to obtain exact solutions to the Schrödinger equation for intractable Hamiltonians. In either case, an infinite basis is required; neither the sums given by perturbation theory nor the matrix diagonalization of a variational calculation can be carried out. Hence, the strengths and weaknesses of the two approaches should be analysed from the point of view that the basis is necessarily truncated. Within this constraint, diagonalization of $\mathbf{H}$ represents the best solution that is possible in the space spanned by the basis set. In variational calculations, rather severe truncation of $\mathbf{H}$ is usually required, with the effect that its eigenvalues might be poor approximations to the exact values. The problem, of course, is that the basis is not sufficiently flexible to accurately represent the true quantum-mechanical wavefunction. In perturbation theory, one can include significantly more functions in the calculation. It turns out that the results of a low order perturbation calculation are often superior to a practical variational treatment of the same problem. Unlike variational methods, perturbation theory does not provide an upper bound to the energy (apart from a first-order treatment) and is not even guaranteed to converge. However, in chemistry, it is virtually always energy differences—and not absolute energies—that are of interest, and differences of energies obtained variationally are not themselves upper (or lower) bounds to the exact values. For example, suppose a spectroscopic transition energy between the states ψ_i and ψ_j is calculated from the difference $\lambda_i - \lambda_j$ obtained by diagonalizing $\mathbf{H}$ in a truncated basis. There is no way of knowing whether this value is above or below the exact answer, a situation no different than that associated with taking the difference between two approximate eigenvalues obtained from two separate calculations based on perturbation theory.

In the quantum mechanics of atoms and molecules, both perturbation theory and the variational principle are widely used. For some problems, one of the two classes of approach is clearly best suited to the task, and is thus an established choice. However, in many others, the situation is less clear cut, and calculations can be done with either of the methods or a combination of both.

Further Reading

Berry R S, Rice S A and Ross J R 1980 *Physical Chemistry* 6th edn (New York, NY: Wiley)

The introductory treatment of quantum mechanics presented in this textbook is excellent. Particularly appealing is the effort devoted to developing a qualitative understanding of quantum-mechanical principles.

Karplus M and Porter R N 1970 *Atoms and Molecules: an Introduction for Students of Physical Chemistry* (Reading, MA: Addison-Wesley)

An excellent treatment of molecular quantum mechanics, on a level comparable to that of Szabo and Ostlund. The scope of this book is quite different, however, as it focuses mainly on the basic principles of quantum mechanics and the theoretical treatment of spectroscopy.

Levine I N 1991 *Quantum Chemistry* 4th edn (Englewood Cliffs, NJ: Wiley)

A relatively complete survey of quantum chemistry, written on a level just below that of the Szabo and Ostlund text. Levine has done an excellent job in including up-to-date material in successive editions of this text, which makes for interesting as well as informative reading.

Szabo A and Ostlund N S 1996 *Modern Quantum Chemistry* (New York: Dover)

Although it is now somewhat dated, this book provides one of the best treatments of the Hartree–Fock approximation and the basic ideas involved in evaluating the correlation energy. An especially valuable feature of this book is that much attention is given to how these methods are actually implemented.

Pauling L and Wilson E B 1935 *Introduction to Quantum Mechanics* (New York: Dover)

This venerable book was written in 1935, shortly after the birth of modern quantum mechanics. Nevertheless, it remains one of the best sources for students seeking to gain an understanding of quantum-mechanical principles that are relevant in chemistry and chemical physics. Equally outstanding jobs are done in dealing with both quantitative and qualitative aspects of the subject. More accessible to most chemists than Landau and Lifschitz.

Landau L D and Lifschitz E M 1977 *Quantum Mechanics (Nonrelativistic Theory)* (Oxford: Pergamon)

A marvellous and rigorous treatment of non-relativistic quantum mechanics. Although best suited for readers with a fair degree of mathematical sophistication and a desire to understand the subject in great depth, the book contains all of the important ideas of the subject and many of the subtle details that are often missing from less advanced treatments. Unusual for a book of its type, highly detailed solutions are given for many illustrative example problems.

Simons J and Nichols J 1997 *Quantum Mechanics in Chemistry* (New York: Oxford)

A new text that provides a relatively broad view of quantum mechanics in chemistry ranging from electron correlation to time-dependent processes and scattering.

Parr R G and Yang W 1994 *Density-Functional Theory of Atoms and Molecules* (New York: Oxford)

A comprehensive treatment of density functional theory, an idea that is currently very popular in quantum chemistry.

Albright T A, Burdett J K and Whangbo M-H 1985 *Orbital Interactions in Chemistry* (New York: Wiley)

A superb treatment of applied molecular orbital theory and its application to organic, inorganic and solid state chemistry. Perhaps the best source for appreciating the power of the independent-particle approximation and its remarkable ability to account for qualitative behaviour in chemical systems.

Salem L 1966 *Molecular Orbital Theory of Conjugated Systems* (Reading, MA: Benjamin)

A highly readable account of early efforts to apply the independent-particle approximation to problems of organic chemistry. Although more accurate computational methods have since been developed for treating all of the problems discussed in the text, its discussion of approximate Hartree–Fock (semiempirical) methods and their accuracy is still useful. Moreover, the view supplied about what was understood and what was not understood in physical organic chemistry three decades ago is fascinating.

Pais A 1988 *Inward Bound: of Matter and Forces in the Physical World* (Oxford: Oxford University Press)

A good account of the historical development of quantum mechanics. While much of the book deals with quantum field theory and particle physics, the first third of the book focuses on the period 1850–1930 and the origins of quantum theory. An admirable job is done in placing events in a proper historical context.

A1.2
Internal molecular motions

Michael E Kellman

A1.2.1   Introduction

Ideas on internal molecular motions go back to the very beginnings of chemistry as a natural science, to the days of Robert Boyle and Isaac Newton [1]. Much of Boyle's interest in chemistry, apart from the 'bewitchment' he found in performing chemical experiments [2], arose from his desire to revive and transform the corpuscular philosophy favoured by some of the ancient Greeks, such as Epicurus [3]. This had lain dormant for centuries, overshadowed by the apparently better-founded Aristotelian cosmology [4], including the theory of the four elements. With the revolution in celestial mechanics that was taking place in modern Europe in the 17th century, Boyle was concerned to persuade natural philosophers that chemistry, then barely emerging from alchemy, was potentially of great value for investigating the corpuscular view, which was re-emerging as a result of the efforts of thinkers such as Francis Bacon and Descartes. This belief of Boyle's was based partly on the notion that the qualitative properties of real substances and their chemical changes could be explained by the joining together of elementary corpuscles, and the 'local motions' within these aggregates—what we now call the internal motions of molecules. Boyle influenced his younger colleague in the Royal Society, Isaac Newton. Despite immense efforts in chemical experimentation, Newton wrote only one paper in chemistry, in which he conjectured the existence of short-range forces in what we now recognize as molecules. Thus, in a true sense, with Boyle and Newton was born the science of chemical physics [1].

This was a child whose development was long delayed, however. Not until the time of John Dalton in the early 19th century, after the long interlude in which the phlogiston theory triumphed and then was overthrown in the chemistry of Lavoisier, did the nascent corpuscular view of Boyle and Newton really begin to grow into a useful atomic and molecular theory [1, 5]. It became apparent that it was necessary to think of the compound states of the elements of Lavoisier in terms of definite molecular formulae, to account for the facts that were becoming known about the physical properties of gases and the reactions of the elements, their joining into compounds and their separation again into elements.

However, it was still a long time even after Dalton before anything definite could be known about the internal motions in molecules. The reason was that the microscopic nature of atoms and molecules was a bar to any knowledge of their internal constituents. Furthermore, nothing at all was known about the physical laws that applied at the microscopic level. The first hints came in the late 19th century, with the classical Maxwell–Lorentz theory of the dynamics of charged particles interacting through the electromagnetic field. The electron was discovered by Thomson, and a little later the nuclear structure of the atom by Rutherford. This set the stage in the 20th century for a physical understanding in terms of quantum theory of the constituents of molecules, and the motions of which they partake.

This section will concentrate on the motions of atoms within molecules—'internal molecular motions'—as comprehended by the revolutionary quantum ideas of the 20th century. Necessarily, limitations of space prevent many topics from being treated in the detail they deserve. Some of these are treated in more detail in other articles in this Encyclopedia, or in references in the Bibliography. The emphasis is on treating certain key topics in sufficient depth to build a foundation for further exploration by the reader, and for branching off into related topics that cannot be treated in depth at all. There will not be much focus on molecules undergoing chemical reactions, except for unimolecular rearrangements, which are a rather extreme example of internal molecular motion. However, it must be emphasized that the distinctions between the internal motions of molecules, the motions of atoms in a molecule which is undergoing dissociation and the motion of atoms in two or more molecules undergoing reaction are somewhat artificial. Even the motions which are most properly called 'internal' play a central role in theories of reaction dynamics. In fact, their character in chemical reactions is one of the most important unsolved mysteries in molecular motion. Although we will not have anything directly to say about general theories of reaction [6], the internal motion of molecules undergoing isomerization and the importance of the internal motions in efforts to control reactions with sophisticated laser sources will be two of the topics considered.

A key theme of contemporary chemical physics and physical chemistry is 'ultrafast' molecular processes [7–9], including both reaction dynamics and internal molecular motions that do not involve reaction. The probing of ultrafast processes generally is thought of in terms of very short laser pulses, through the window of the time domain. However, most of the emphasis of this section is on probing molecules through the complementary window of the frequency domain, which usually is thought of as the realm of the time-independent processes, which is to say, the 'ultraslow'. One of the key themes of this section is that encrypted within the totality of the information which can be gathered on a molecule in the frequency domain is a vast store of information on ultrafast internal motions. The decoding of this information by new theoretical techniques for analysis of experimental spectra is a leading theme of recent work.

A1.2.2   Quantum theory of atomic and molecular structure and motion

The understanding of molecular motions is necessarily based on quantum mechanics, the theory of microscopic physical behaviour worked out in the first quarter of the 20th century. This is because molecules are microscopic systems in which it is impossible—or at least very dangerous!—to ignore the dual wave–particle nature of matter first recognized in quantum theory by Einstein (in the case of classical waves) and de Broglie (in the case of classical particles).

The understanding of the quantum mechanics of atoms was pioneered by Bohr, in his theory of the hydrogen atom. This combined the classical ideas on planetary motion—applicable to the atom because of the formal similarity of the gravitational potential to the Coulomb potential between an electron and nucleus—with the quantum ideas that had recently been introduced by Planck and Einstein. This led eventually to the formal theory of quantum mechanics, first discovered by Heisenberg, and most conveniently expressed by Schrödinger in the wave equation that bears his name.

However, the hydrogen atom is relatively a very simple quantum mechanical system, because it contains only two constituents, the electron and the nucleus. This situation is the quantum mechanical analogue of a single planet orbiting a sun. It might be thought that an atom with more than one electron is much like a solar system with more than one planet, in which the motion of each of the planets is more or less independent and regular. However, this is not the case, because the relative strength of the interaction between the electrons is much stronger than the attraction of the planets in our solar system. The problem of the internal dynamics of atoms—the internal motion when there is more than one electron—is still very far from a complete understanding. The electrons are not really independent, nor would their motion, if it were described by classical rather than quantum mechanics, be regular, unlike the annual orbits of the planets. Instead, in general, it would be *chaotic*. The corresponding complexity of the quantum mechanical atom with more than one electron, or even one electron in a field, is to this day a challenge [10–12]. (In fact, even in the solar system, despite the relative strengths of planetary attraction, there are constituents, the asteroids, with very irregular, chaotic behaviour. The issue of chaotic motion in molecules is an issue that will appear later with great salience.)

As we shall see, in molecules as well as atoms, the interplay between the quantum description of the internal motions and the corresponding classical analogue is a constant theme. However, when referring to the internal motions of molecules, we will be speaking, loosely, of the motion of the atoms in the molecule, rather than of the fundamental constituents, the electrons and nuclei. This is an extremely fundamental point to which we now turn.

A1.2.3   The molecular potential energy surface

One of the most salient facts about the structure of molecules is that the electrons are far lighter than the nuclei, by three orders of magnitude and more. This is extremely fortunate for our ability to attain a rational understanding of the internal motion of the electrons and nuclei. In fact, without this it might well be that not much progress would have been made at all! Soon after the discovery of quantum mechanics it was realized that the vast difference in the mass scales of the electrons and nuclei means that it is possible, in the main, to separate the problem into two parts, an electronic and a nuclear part. This is known as the Born–Oppenheimer separability or approximation [13]. The underlying physical idea is that the electrons move much faster than the nuclei, so they adjust rapidly to the relatively much slower nuclear motion. Therefore, the electrons are described by a quantum mechanical 'cloud' obtained by solving the Schrödinger wave equation. The nuclei then move slowly within this cloud, which in turn adjusts rapidly as the nuclei move.

The result is that, to a very good approximation, as treated elsewhere in this Encyclopedia, the nuclei move in a mechanical potential created by the much more rapid motion of the electrons. The electron cloud itself is described by the quantum mechanical theory of electronic structure. Since the electronic and nuclear motion are approximately separable, the electron cloud can be described mathematically by the quantum mechanical theory of electronic structure, in a framework where the nuclei are fixed. The resulting Born–Oppenheimer potential energy surface (PES) created by the electrons is the mechanical potential in which the nuclei move. When we speak of the internal motion of molecules, we therefore mean essentially the motion of the nuclei, which contain most of the mass, on the molecular potential energy surface, with the electron cloud rapidly adjusting to the relatively slow nuclear motion.

We will now treat the internal motion on the PES in cases of progressively increasing molecular complexity. We start with the simplest case of all, the diatomic molecule, where the notions of the Born–Oppenheimer PES and internal motion are particularly simple.

The potential energy surface for a diatomic molecule can be represented as in figure A1.2.1. The x-axis gives the internuclear separation R and the y-axis the potential function $V(R)$. At a given value of R, the potential $V(R)$ is determined by solving the quantum mechanical electronic structure problem in a framework with the nuclei fixed at the given value of R. (To reiterate the discussion above, it is only possible to regard the nuclei as fixed in this calculation because of the Born–Oppenheimer separability, and it is important to keep in mind that this is only an approximation. There can be subtle but important *non-adiabatic effects* [14, 15], due to the non-exactness of the separability of the nuclei and electrons. These are treated elsewhere in this Encyclopedia.) The potential function $V(R)$ is determined by repeatedly solving the quantum mechanical electronic problem at different values of R. Physically, the variation of $V(R)$ is due to the fact that the electronic cloud adjusts to different values of the internuclear separation R in a subtle interplay of mutual particle attractions and repulsions: electron–electron repulsions, nuclear–nuclear repulsions and electron–nuclear attractions.

The potential function in figure A1.2.1 has several crucial characteristics. It has a minimum at a certain value R_0 of the internuclear separation. This is the equilibrium internuclear distance. Near R_0, the function $V(R)$ rises as R increases or decreases. This means that there is an attractive mechanical force tending to restore the nuclei to R_0. At large values of R, $V(R)$ flattens out and asymptotically approaches a value which

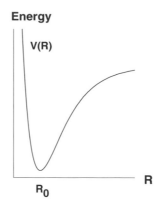

Figure A1.2.1. Potential $V(R)$ of a diatomic molecule as a function of the internuclear separation R. The equilibrium distance R_0 is at the potential minimum.

in figure A1.2.1 is arbitrarily chosen to be zero. This means that the molecule dissociates into separated atoms at large R. The difference between the equilibrium potential $V(R_0)$ and the asymptotic energy is the dissociation, or binding, energy. At values of R less than R_0, the potential $V(R)$ again rises, but now without limit. This represents the repulsion between nuclei as the molecule is compressed.

Classically, the nuclei vibrate in the potential $V(R)$, much like two steel balls connected by a spring which is stretched or compressed and then allowed to vibrate freely. This vibration along the nuclear coordinate R is our first example of internal molecular motion. Most of the rest of this section is concerned with different aspects of molecular vibrations in increasingly complicated situations.

Near the bottom of the potential well, $V(R)$ can be approximated by a parabola, so the function $V(R)$ is approximated as

$$V(R) = kR^2. \tag{A1.2.1}$$

This is the form of the potential for a harmonic oscillator, so near the bottom of the well, the nuclei undergo nearly harmonic vibrations. For a harmonic oscillator with potential as in (A1.2.1), the classical frequency of oscillation is independent of energy and is given by [16–18]

$$\omega_0 = 2\pi \nu_0 = \sqrt{k/\mu} \tag{A1.2.2}$$

where μ is the reduced mass. Quantum mechanically, the oscillator has a series of discrete energy levels, characterized by the number of quanta n in the oscillator. This is the quantum mechanical analogue for the oscillator of the quantized energy levels of the electron in a hydrogen atom. The energy levels of the harmonic oscillator are given by

$$E_n = \omega_0(n + \tfrac{1}{2}) \tag{A1.2.3}$$

where $\hbar$, i.e. Planck's constant h divided by 2π, has been omitted as a factor on the right-hand side, as is appropriate when the customary wavenumber (cm^{-1}) units are used [18].

A1.2.4 Anharmonicity

If the potential were exactly harmonic for all values of R, the vibrational motion would be extremely simple, consisting of vibrations with frequency ω_0 for any given amount of vibrational energy. The fact that this is a drastic oversimplification for a real molecule can be seen from the fact that such a molecule would never dissociate, lacking the flatness in the potential at large R that we saw in figure A1.2.1. As the internuclear separation departs from the bottom of the well at R_0, the harmonic approximation (A1.2.1) progressively becomes less accurate as a description of the potential. This is known as *anharmonicity* or *nonlinearity*. Anharmonicity introduces complications into the description of the vibrational motion. The frequency is no longer given by the simple harmonic formula (A1.2.2). Instead, it varies with the amount of energy in the oscillator. This variation of frequency with the number of quanta is the essence of the nonlinearity.

The variation of the frequency can be approximated by a series in the number of quanta, so the energy levels are given by

$$E_n = \omega_0(n + \tfrac{1}{2}) + \gamma_1(n + \tfrac{1}{2})^2 + \gamma_2(n + \tfrac{1}{2})^3 + \cdots. \qquad (A1.2.4)$$

Often, it is a fair approximation to truncate the series at the quadratic term with γ_1. The energy levels are then approximated as

$$E_n = \omega_0(n + \tfrac{1}{2}) + \gamma_1(n + \tfrac{1}{2})^2. \qquad (A1.2.5)$$

The first term is known as the harmonic contribution and the second term as the quadratic anharmonic correction.

Even with these complications due to anharmonicity, the vibrating diatomic molecule is a relatively simple mechanical system. In polyatomics, the problem is fundamentally more complicated with the presence of more than two atoms. The anharmonicity leads to many extremely interesting effects in the internal molecular motion, including the possibility of chaotic dynamics.

It must be pointed out that another type of internal motion is the overall rotation of the molecule. The vibration and rotation of the molecule are shown schematically in figure A1.2.2.

A1.2.5   Polyatomic molecules

In polyatomic molecules there are many more degrees of freedom, or independent ways in which the atoms of the molecule can move. With n atoms, there are a total of $3n$ degrees of freedom. Three of these are the motion of the centre of mass, leaving $(3n - 3)$ internal degrees of freedom [18]. Of these, except in linear polyatomics, three are rotational degrees of freedom, leaving $(3n - 6)$ vibrational degrees of freedom. (In linear molecules, there are only two rotational degrees of freedom, corresponding to the two individual orthogonal axes of rotation about the molecular axis, leaving $(3n - 5)$ vibrational degrees of freedom. For example, the diatomic has only one vibrational degree of freedom, the vibration along the coordinate R which we encountered above.)

Because of limitations of space, this section concentrates very little on rotational motion and its interaction with the vibrations of a molecule. However, this is an extremely important aspect of molecular dynamics of long-standing interest, and with development of new methods it is the focus of intense investigation [18–23]. One very interesting aspect of rotation–vibration dynamics involving *geometric phases* is addressed in section A1.2.20.

The $(3n - 6)$ degrees of vibrational motion again take place on a PES. This implies that the PES itself must be a function in a $(3n - 6)$ dimensional space, i.e. it is a function of $(3n - 6)$ *internal coordinates* $r_1 \cdots r_N$, where $N = (3n - 6)$, which depend on the positions of all the nuclei. The definition of the coordinates $r_1 \cdots r_N$ has a great deal of flexibility. To be concrete, for H_2O one choice is the set of internal coordinates illustrated in figure A1.2.3. These are a bending coordinate, i.e. the angular bending displacement from the equilibrium geometry, and two bond displacement coordinates, i.e. the stretching displacement of each O–H bond from its equilibrium length.

An equilibrium configuration for the molecule is any configuration $(r_{10} \cdots r_{n0})$ where the PES has a minimum, analogous to the minimum in the diatomic potential at R_0 in figure A1.2.1. In general, there can be a number of local equilibrium configurations in addition to the lowest equilibrium configuration, which is called the global equilibrium or minimum. We will refer to an equilibrium configuration in speaking of any of the local equilibria, and *the* equilibrium configuration when referring to the global minimum. In the

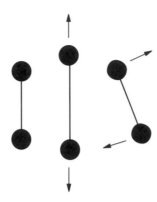

Figure A1.2.2. Internal nuclear motions of a diatomic molecule. Top: the molecule in its equilibrium configuration. Middle: vibration of the molecule. Bottom: rotation of the molecule.

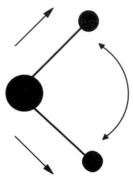

Figure A1.2.3. The internal coordinates of the H_2O molecule. There are two bond stretching coordinates and a bend coordinate.

very close vicinity of the equilibrium configuration, the molecule will execute harmonic vibrations. Since there are $(3n - 6)$ vibrational degrees of freedom, there must be $(3n - 6)$ harmonic *modes*, or independent vibrational motions. This means that on the multi-dimensional PES, there must be $(3n - 6)$ independent coordinates, along any of which the potential is harmonic, near the equilibrium configuration. We will denote these independent degrees of freedom as the *normal modes coordinates* $R_1 \cdots R_N$. Each of the R_i in general is some combination of the internal coordinates $r_1 \cdots r_N$ in terms of which the nuclear positions and PES were defined earlier. These are illustrated for the case of water in figure A1.2.4. One of the normal modes is a bend, very much like the internal bending coordinate in figure A1.2.3. The other two modes are a symmetric and antisymmetric stretch. Near the equilibrium configuration, given knowledge of the molecular potential, it is possible by the procedure of *normal mode analysis* [24] to calculate the frequencies of each of the normal modes and their exact expression in terms of the original internal coordinates $r_1 \cdots r_N$.

It is often very useful to describe classical vibrations in terms of a *trajectory* in the space of coordinates $r_1 \cdots r_N$. If the motion follows one of the normal modes, the trajectory is one in which the motion repeats itself along a closed curve. An example is shown in figure A1.2.5 for the symmetric and antisymmetric stretch modes. The x and y coordinates r_1, r_2 are the displacements of the two O–H bonds. (For each mode i there is a family of curves, one for each value of the energy, with the amplitude of vibration along the normal modes in figure A1.2.5 increasing with energy; the figure shows the trajectory of each mode for one value of the energy.)

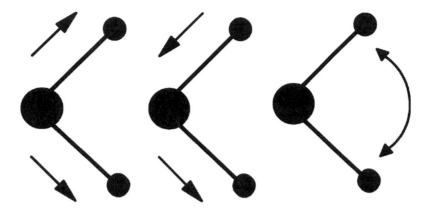

Figure A1.2.4. The normal vibrational coordinates of H_2O. Left: symmetric stretch. Middle: antisymmetric stretch. Right: bend.

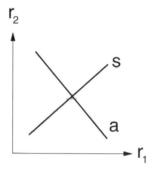

Figure A1.2.5. Harmonic stretch normal modes of a symmetric triatomic. The symmetric stretch s and antisymmetric stretch a are plotted as a function of the bond displacements r_1, r_2.

In general, each normal mode in a molecule has its own frequency, which is determined in the normal mode analysis [24]. However, this is subject to the constraints imposed by molecular symmetry [18, 25, 26]. For example, in the methane molecule CH_4, four of the normal modes can essentially be designated as normal stretch modes, i.e. consisting primarily of collective motions built from the four C–H bond displacements. The molecule has tetrahedral symmetry, and this constrains the stretch normal mode frequencies. One mode is the totally symmetric stretch, with its own characteristic frequency. The other three stretch normal modes are all constrained by symmetry to have the same frequency, and are referred to as being *triply-degenerate*.

The $(3n - 6)$ normal modes with coordinates $R_1 \cdots R_N$ are often designated $\nu_1 \cdots \nu_N$. (Not to be confused with the common usage of ν to denote a frequency, as in equation (A1.2.2), the last such usage in this section.) Quantum mechanically, each normal mode ν_i is characterized by the number of vibrational quanta n_i in the mode. Then the vibrational state of the molecule is designated or *assigned* by the number of quanta n_i in each of the modes, i.e. $(n_1 \cdots n_N)$. In the harmonic approximation in which each mode i is characterized by a frequency ω_i, the vibrational energy of a state assigned as $(n_1 \cdots n_N)$ is given by

$$E(n_1 \cdots n_N) = (n_1 + \tfrac{1}{2})\omega_1 + (n_2 + \tfrac{1}{2})\omega_2 + \cdots + (n_N + \tfrac{1}{2})\omega_N \qquad (A1.2.6)$$

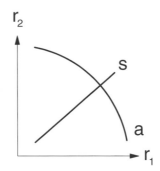

Figure A1.2.6. Anharmonic stretch normal modes of a symmetric triatomic. The plot is similar to figure A1.2.5, except the normal modes are now anharmonic and can be curvilinear in the bond displacement coordinates r_1, r_2. The antisymmetric stretch is curved, but the symmetric stretch is linear because of symmetry.

A1.2.6   Anharmonic normal modes

In the polyatomic molecule, just as in the diatomic, the PES must again be highly anharmonic away from the vicinity of the potential minimum, as seen from the fact that the polyatomic can dissociate; in fact in a multiplicity of ways, because in general there can be several dissociation products. In addition, the molecule can have complicated internal rearrangements in which it isomerizes. This means that motion takes place from one minimum in the PES, over a saddle, or 'pass', and into another minimum. We will have something to say about these internal rearrangements later. However, the fact of anharmonicity raises important questions about the normal modes even in the near vicinity of an equilibrium configuration. We saw above that anharmonicity in a diatomic means that the frequency of the vibrational motion varies with the amount of vibrational energy. An analogous variation of frequency of the normal modes occurs in polyatomics.

However, there is a much more profound prior issue concerning anharmonic normal modes. The existence of the normal vibrational modes, involving the collective motion of all the atoms in the molecule as illustrated for H_2O in figure A1.2.4, was predicated on the basis of the existence of a harmonic potential. But if the potential is not exactly harmonic, as is the case everywhere except right at the equilibrium configuration, are there still collective normal modes? And if so, since they cannot be harmonic, what is their nature and their relation to the harmonic modes?

The beginning of an answer comes from a theorem of Moser and Weinstein in mathematical nonlinear dynamics [27, 28]. This theorem states that in the vicinity of a potential minimum, a system with $(3n - 6)$ vibrational degrees of freedom has $(3n - 6)$ *anharmonic normal modes*. What is the difference between the harmonic normal modes and the *anharmonic* normal modes proven to exist by Moser and Weinstein? Figure A1.2.6 shows anharmonic stretch normal modes. The symmetric stretch looks the same as its harmonic counterpart in figure A1.2.5; this is necessarily so because of the symmetry of the problem. The antisymmetric stretch, however, is distinctly different, having a curvilinear appearance in the zero-order bond modes. The significance of the Moser–Weinstein theorem is that it guarantees that in the vicinity of a minimum in the PES, there must be a set of $(3n - 6)$ of these anharmonic modes.

It is sometimes very useful to look at a trajectory such as the symmetric or antisymmetric stretch of figures A1.2.5 and A1.2.6, not in the physical spatial coordinates $(r_1 \cdots r_N)$, but in the *phase space* of Hamiltonian mechanics [16, 29], which in addition to the coordinates $(r_1 \cdots r_N)$ also has as additional coordinates the set of conjugate momenta $(p_1 \cdots p_N)$. In phase space, a one-dimensional trajectory such as the antisymmetric stretch again appears as a one-dimensional curve, but now the curve closes on itself. Such a

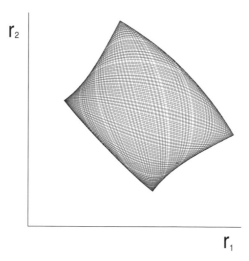

Figure A1.2.7. Trajectory of two coupled stretches, obtained by integrating Hamilton's equations for motion on a PES for the two modes. The system has stable anharmonic symmetric and antisymmetric stretch modes, like those illustrated in figure A1.2.6. In this trajectory, semiclassically there is one quantum of energy in each mode, so the trajectory corresponds to a combination state with quantum numbers $[n_s, n_a] = [1, 1]$. The 'woven' pattern shows that the trajectory is regular rather than chaotic, corresponding to motion in phase space on an invariant torus.

trajectory is referred to in nonlinear dynamics as a *periodic orbit* [29]. One says that the anharmonic normal modes of Moser and Weinstein are *stable* periodic orbits.

What does it mean to say the modes are stable? Suppose that one fixes the initial conditions—the initial values of the coordinates and momenta, for a given fixed value of the energy—so the trajectory does not lie entirely on one of the anharmonic modes. At any given time the position and momentum is some combination of each of the normal motions. An example of the kind of trajectory that can result is shown in figure A1.2.7. The trajectory lies in a box with extensions in each of the anharmonic normal modes, filling the box in a very regular, 'woven' pattern. In phase space, a regular trajectory in a box is no longer a one-dimensional closed curve, or periodic orbit. Instead, in phase space a box-filling trajectory lies on a surface which has the qualitative form, or topology, of a torus—the surface of a doughnut. The confinement of the trajectory to such a box indicates that the normal modes are stable. (Unstable modes do exist and will be of importance later.) Another quality of the trajectory in the box is its 'woven' pattern. Such a trajectory is called *regular*. We will consider other, *chaotic* types of trajectories later; the chaos and instability of modes are closely related. The issues of periodic orbits, stable modes and regular and chaotic motion have been studied in great depth in the theory of Hamiltonian or energy-preserving dynamical systems [29, 30]. We will return repeatedly to concepts of classical dynamical systems.

However, the reader may be wondering, what is the connection of all of these classical notions—stable normal modes, regular motion on an invariant torus—to the quantum spectrum of a molecule observed in a spectroscopic experiment? Recall that in the harmonic normal modes approximation, the quantum levels are defined by the set of quantum numbers $(n_1 \cdots n_N)$ giving the number of quanta in each of the normal modes.

Does it make sense to associate a definite quantum number n_i to each mode i in an anharmonic system? In general, this is an extremely difficult question! But remember that so far, we are speaking of the situation in some small vicinity of a minimum on the PES, where the Moser–Weinstein theorem *guarantees* the existence of the anharmonic normal modes. This essentially guarantees that quantum levels with low enough n_i values correspond to trajectories that lie on invariant tori. Since the levels are quantized, these must be special tori,

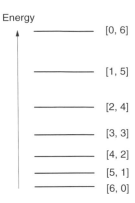

Figure A1.2.8. Typical energy level pattern of a sequence of levels with quantum numbers $[n_s, n_a]$ for the number of quanta in the symmetric and antisymmetric stretch. The bend quantum number is neglected and may be taken as fixed for the sequence. The total number of quanta $(n_s + n_a = 6)$ is the polyad number, which is the same for all levels. [6, 0] and [0, 6] are the overtones of the symmetric and antisymmetric stretch; the other levels are combination levels. The levels have a monotonic sequence of energy spacings from bottom to top.

each characterized by quantized values of the classical actions $I_i = (n_i + \frac{1}{2})\hbar$, which are constants of the motion on the invariant torus. As we shall see, the possibility of assigning a set of N quantum numbers n_i to a level, one for each mode, is a very special situation that holds only near the potential minimum, where the motion is described by the N anharmonic normal modes. However, let us continue for now with the region of the spectrum where this special situation applies.

If there are n_i quanta in mode i and zero quanta in all the other modes, the state is called an *overtone* of the normal mode i. What does such a state correspond to in terms of a classical trajectory? Consider the overtone of the antisymmetric stretch, again neglecting the bend. If all the energy in the overtone were in mode i, the trajectory would look like the anharmonic mode itself in figure A1.2.6. However, because of the unavoidable quantum mechanical zero-point energy associated with the action $\hbar/2$ in each mode, an overtone state actually has a certain amount of energy in all of the normal modes. Therefore, classically, the overtone of the antisymmetric stretch corresponds to a box-like trajectory, with most of the extension along the antisymmetric stretch, but with some extension along the symmetric stretch, and corresponding to the irreducible zero-point energy.

The other kind of quantum level we considered above is one with quanta in more than one mode, i.e. $(n_1 \cdots n_N)$ with more than one of the n_i not equal to zero. Such a state is called a *combination* level. This corresponds, classically, to a box-like trajectory with extension in each mode corresponding to the number of quanta; an example was seen in figure A1.2.7.

What does one actually observe in the experimental spectrum, when the levels are characterized by the set of quantum numbers $(n_1 \cdots n_N)$ for the normal modes? The most obvious spectral observation is simply the set of energies of the levels; another important observable quantity is the intensities. The latter depend very sensitively on the type of probe of the molecule used to obtain the spectrum; for example, the intensities in absorption spectroscopy are in general far different from those in Raman spectroscopy. From now on we will focus on the energy levels of the spectrum, although the intensities most certainly carry much additional information about the molecule, and are extremely interesting from the point of view of theoretical dynamics.

If the molecule really had harmonic normal modes, the energy formula (A1.2.6) would apply and the spectrum would be extremely simple. It is common to speak of a *progression* in a mode i; a progression consists of the series of levels containing the fundamental, with $n_i = 1$, along with the overtone levels $n_i > 1$.

Each progression of a harmonic system would consist of equally spaced levels, with the level spacing given by the frequency ω_i. It is also common to speak of sequences, in which the sum of the number of quanta in two modes is fixed. In a harmonic spectrum, the progressions and sequences would be immediately evident to the eye in a plot of the energy levels.

In a system with anharmonic normal modes, the spectral pattern is not so simple. Instead of the simple energy level formula (A1.2.6), in addition to the harmonic terms there are anharmonic terms, similar to the terms $\gamma_1(n+\frac{1}{2})^2$, $\gamma_2(n+\frac{1}{2})^3$, ... in (A1.2.4). For each mode i, there is a set of such terms $\gamma_{ii}(n_i+\frac{1}{2})^2$, $\gamma_{iii}(n_i+\frac{1}{2})^3$, etc, where now by common convention the i's in the subscript refer to mode i and the order of the subscript and superscript match, for example γ_{ii} with the quadratic power $(n_i+\frac{1}{2})^2$. However, there are also *cross terms* $\gamma_{ij}(n_i+\frac{1}{2})(n_j+\frac{1}{2})$, $\gamma_{iij}(n_i+\frac{1}{2})^2(n_j+\frac{1}{2})$, etc. As an example, the anharmonic energy level formula for just a symmetric and antisymmetric stretch is given to the second order in the quantum numbers by

$$E(n_s, n_a) = \omega_s(n_s + \tfrac{1}{2}) + \omega_a(n_a + \tfrac{1}{2}) + \omega_b(n_b + \tfrac{1}{2}) + \gamma_{ss}(n_s + \tfrac{1}{2})^2 + \gamma_{aa}(n_a + \tfrac{1}{2})^2 + \gamma_{bb}(n_b + \tfrac{1}{2})^2$$
$$+ \gamma_{sa}(n_s + \tfrac{1}{2})(n_a + \tfrac{1}{2}) + \gamma_{sb}(n_s + \tfrac{1}{2})(n_b + \tfrac{1}{2}) + \gamma_{ab}(n_a + \tfrac{1}{2})(n_b + \tfrac{1}{2}). \tag{A1.2.7}$$

An energy expression for a polyatomic in powers of the quantum numbers like (A1.2.7) is an example of an anharmonic expansion [18]. In the anharmonic spectrum, within a progression or sequence there will not be equal spacings between levels; rather, the spacings will depend on the quantum numbers of the adjacent levels. Nonetheless, the spectrum will appear very regular to the eye. Spectra that follow closely a formula such as (A1.2.7), perhaps including higher powers in the quantum numbers, are very common in the spectroscopy of polyatomic molecules at relatively low energy near the minimum of the PES. This regularity is not too surprising, when one recalls that it is associated with the existence of the good quantum numbers $(n_1 \cdots n_N)$, which themselves correspond classically to regular motion of the kind shown in figure A1.2.7.

A1.2.7 Spectra that are not so regular

If this was all there is to molecular spectra they would be essentially well understood by now and their power to give information on molecules nearly exhausted. However, this cannot be the case: consider that molecules dissociate—a very irregular type of motion!—while a molecule whose spectrum strictly followed a formula such as (A1.2.7) would have quantum levels all corresponding semiclassically to motion on invariant tori that are described by the N anharmonic normal modes. Motion as simple as this is expected only near a potential minimum, where the Weinstein–Moser theorem applies. How is the greater complexity of real molecules manifested in a spectrum? The spectrum is a reflection of the physical PES, since the vibrational spectrum is determined quantum mechanically by the PES. Since the PES contains the possibility of much less regular motion than that reflected in a Dunham formula such as (A1.2.7), how can a Dunham formula be modified so as to represent a real spectrum, including portions corresponding to less regular motion? We will consider first what these modifications must look like, then pose the following question: suppose we have a generalized spectral Hamiltonian and use this to represent experimental observations, how can we use this representation to *decode* the dynamical information on the internal molecular motions that is contained in the spectrum?

A1.2.8 Resonance couplings

The fact that terms in addition to those present in the energy level formula (A1.2.7) might arise in molecular spectra is already strongly suggested by one of the features already discussed; the cross-anharmonic terms such as $\gamma_{ij}(n_i+\frac{1}{2})(n_j+\frac{1}{2})$. These terms show that the anharmonicity arises not only from the normal modes themselves—the 'self-anharmonicity' terms like $\gamma_{ii}(n_i+\frac{1}{2})^2$—but also from *couplings between the normal modes*. The cross-anharmonic terms depend only on the vibrational quantum numbers—the Hamiltonian so far is diagonal in the normal mode quantum numbers. However, there are also terms in the generalized

Hamiltonian that are not diagonal in the quantum numbers. It is these that are responsible for profoundly greater complexity of the internal motion of a polyatomic, as compared to a diatomic.

Consider how these non-diagonal terms would arise in the analysis of an experimental spectrum. Given a set of spectral data, one would try to fit the data to a Hamiltonian of the form of (A1.2.7). The Hamiltonian then is to be regarded as a 'phenomenological' or 'effective' spectroscopic Hamiltonian, to be used to describe the results of experimental observations. The fitting consists of adjusting the parameters of the Hamiltonian, for example ω's, the γ's, etc, until the best match possible is obtained between the spectroscopic Hamiltonian and the data. If a good fit is not obtained with a given number of terms in the Dunham expansion, one could simply add terms of higher order in the quantum numbers. However, it is found in fitting the spectrum of the stretch modes of a molecule like H_2O that this does not work at all well. Instead, a large *resonance coupling* term which *exchanges quanta* between the modes is found to be necessary to obtain a good fit to the data, as was first discovered long ago by Darling and Dennison [31]. Specifically, the Darling–Dennison coupling takes two quanta out of the symmetric stretch, and places two into the antisymmetric stretch. There is also a coupling which does the reverse, taking two quanta from the antisymmetric stretch and placing them into the symmetric stretch. It is convenient to represent this coupling in terms of the raising and lowering operators [32] a_i^+, a_i. These, respectively, have the action of placing a quantum into or removing a quantum from an oscillator which originally has n quanta:

$$a^+|n\rangle = |n+1\rangle \qquad a|n\rangle = |n-1\rangle \qquad (A1.2.8)$$

The raising and lowering operators originated in the algebraic theory of the quantum mechanical oscillator, essentially by the path followed by Heisenberg in formulating quantum mechanics [33]. In terms of raising and lowering operators, the Darling–Dennison coupling operator is

$$\kappa_{DD}(a_s^+ a_s^+ a_a a_a + a_s a_s a_a^+ a_a^+) \qquad (A1.2.9)$$

where κ_{DD} is a parameter which defines the strength of the coupling; κ_{DD} is optimized to obtain the best possible fit between the data and the spectroscopic Hamiltonian.

Physically, why does a term like the Darling–Dennison coupling arise? We have said that the spectroscopic Hamiltonian is an abstract representation of the more concrete, physical Hamiltonian formed by letting the nuclei in the molecule move with specified initial conditions of displacement and momentum on the PES, with a given total kinetic plus potential energy. This is the sense in which the spectroscopic Hamiltonian is an 'effective' Hamiltonian, in the nomenclature used above. The concrete Hamiltonian that it mimics is expressed in terms of particle momenta and displacements, in the representation given by the normal coordinates. Then, in general, it may contain terms proportional to all the powers of the products of the normal coordinates $R_i^{\eta_i} R_j^{\eta_j}$. (It will also contain terms containing the momenta that arise from the kinetic energy; however, these latter kinetic energy terms are more restricted in form than the terms from the potential.) In the spectroscopic Hamiltonian, these will partly translate into expressions with terms proportional to the powers of the quantum numbers, as in (A1.2.7). However, there will also be resonance couplings, such as the Darling–Dennison coupling (A1.2.9). These arise directly from the fact that the oscillator raising and lowering operators (A1.2.8) have a close connection to the position and momentum operators of the oscillator [32], so the resonance couplings are implicit in the terms of the physical Hamiltonian such as $R_i^{\eta_i} R_j^{\eta_j}$.

Since all powers of the coordinates appear in the physical PES, and these give rise to resonance couplings, one might expect a large, in fact infinite, number of resonance couplings in the spectroscopic Hamiltonian. However, in practice, a small number of resonance couplings—and often none, especially at low energy—is sufficient to give a good fit to an experimental spectrum, so effectively the Hamiltonian has a rather simple form. To understand why a small number of resonance couplings is usually sufficient we will focus again on H_2O.

In fitting the H_2O stretch spectrum, it is found that the Darling–Dennison coupling is necessary to obtain a good fit, but *only* the Darling–Dennison and no other. (It turns out that a second coupling, between the symmetric stretch and bend, is necessary to obtain a good fit when significant numbers of bending quanta are involved; we will return to this point later.) If all resonance terms in principle are involved in the Hamiltonian, why it is that, empirically, only the Darling–Dennison coupling is important? To understand this, a very important notion, the *polyad quantum number*, is necessary.

A1.2.9   Polyad number

The characteristic of the Darling–Dennison coupling is that it exchanges two quanta between the symmetric and antisymmetric stretches. This means that the individual quantum numbers n_s, n_a are no longer good quantum numbers of the Hamiltonian containing V_{DD}. However, the *total* number of stretch quanta

$$n_{str} = (n_s + n_a) \tag{A1.2.10}$$

is left unchanged by V_{DD}. Thus, while it might appear that V_{DD} has destroyed two quantum numbers, corresponding to two constants of motion, it has in fact preserved n_{str} as a good quantum number, often referred to as a *polyad* quantum number. So, the Darling–Dennison term V_{DD} couples together a set of zero-order states with common values of the polyad number n_{str}. For example, the set with $n_{str} = 4$ contains zero-order states $[n_s, n_a] = [4, 0], [3, 1], [2, 2], [1, 3], [0, 4]$. These five, zero-order states are referred to as the zero-order polyad with $n_{str} = 4$.

If only zero-order states from the same polyad are coupled together, this constitutes a fantastic simplification in the Hamiltonian. Enormous computational economies result in fitting spectra, because the spectroscopic Hamiltonian is block diagonal in the polyad number. That is, only zero-order states within blocks with the same polyad number are coupled; the resulting small matrix diagonalization problem is vastly simpler than diagonalizing a matrix with all the zero-order states coupled to each other.

However, why should such a simplification be a realistic approximation? For example, why should not a coupling of the form

$$(a_s^+ a_s^+ a_s^+ a_a + a_s a_s a_s a_a^+) \tag{A1.2.11}$$

which would break the polyad number n_{str}, be just as important as V_{DD}? There is no reason *a priori* why it might not have just as large a contribution as V_{DD} when the coordinate representation of the PES is expressed in terms of the raising and lowering operators a_i^+, a_i. To see why it nonetheless is found empirically to be unimportant in the fit, and therefore is essentially negligible, consider again the molecule H_2O. A coupling like (A1.2.11), which removes three quanta from one mode but puts only one quantum in the other mode, is going to couple zero-order states with vastly different zero-order energy. For example, $[n_s, n_a] = [3, 0]$ will be coupled to $[0, 1]$, but these zero-order states are nowhere near each other in energy. By general quantum mechanical arguments of perturbation theory [32], the coupling of states which differ greatly in energy will have a correspondingly small effect on the wavefunctions and energies. In a molecule like H_2O, such a coupling can essentially be ignored in the fitting Hamiltonian.

This is why the coupling V_{DD} is often called a Darling–Dennison *resonance* coupling: it is significant precisely when it couples zero-order states that differ by a small number of quanta which are approximately degenerate with each other, which classically is to say that they are in resonance. The Darling–Dennison coupling, because it involves taking two quanta from one mode and placing two in another, is also called a 2:2 coupling. Other orders of coupling $n:m$ also arise in different situations (such as the stretch–bend coupling in H_2O), and these will be considered later.

However, if *only* the Darling–Dennison coupling is important for the coupled stretches, what is its importance telling us about the internal molecular motion? It turns out that the right kind of analysis of the spectroscopic fitting Hamiltonian reveals a vast amount about the dynamics of the molecule: it allows us to decipher the story encoded in the spectrum of what the molecule is 'really doing' in its internal motion. We will approach this 'spectral cryptology' from two complementary directions: the spectral pattern of the Darling–Dennison spectroscopic Hamiltonian; and, less directly, the analysis of a classical Hamiltonian corresponding to the spectroscopic quantum Hamiltonian. We will see that the Darling–Dennison coupling produces a pattern in the spectrum that is very distinctly different from the pattern of a 'pure normal modes Hamiltonian', without coupling, such as (A1.2.7). Then, when we look at the classical Hamiltonian corresponding to the Darling–Dennison quantum fitting Hamiltonian, we will subject it to the mathematical tool of bifurcation analysis [34]. From this, we will infer a dramatic birth in bifurcations of new 'natural motions' of the molecule, i.e. *local modes*. This will be directly connected with the distinctive quantum spectral pattern of the polyads. Some aspects of the pattern can be accounted for by the classical bifurcation analysis; while others give evidence of intrinsically non-classical effects in the quantum dynamics.

It should be emphasized here that while the discussion of contemporary techniques for decoding spectra for information on the internal molecular motions will largely concentrate on spectroscopic Hamiltonians and bifurcation analysis, there are distinct, but related, contemporary developments that show great promise for the future. For example approaches using advanced 'algebraic' techniques [35, 36] for alternative ways to build the spectroscopic Hamiltonian, and 'hierarchical analysis' using techniques related to general classification methods [37].

A1.2.10   Spectral pattern of the Darling–Dennison Hamiltonian

Consider the polyad $n_{str} = 6$ of the Hamiltonian (A1.2.7). This polyad contains the set of levels conventionally assigned as [6, 0,], [5, 1], . . . , [0, 6]. If a Hamiltonian such as (A1.2.7) described the spectrum, the polyad would have a pattern of levels with monotonically varying spacing, like that shown in figure A1.2.8. However, suppose the fit of the experimental spectrum requires the addition of a strong Darling–Dennison term V_{DD}, as empirically is found to be the case for the stretch spectrum of a molecule like H_2O. In general, because of symmetry, only certain levels may be spectroscopically allowed; for example, in absorption spectra, only levels with odd number of quanta n_a in the antisymmetric stretch. However, diagonalization of the polyad Hamiltonian gives all the levels of the polyad. When these are plotted for the Darling–Dennison Hamiltonian, including the spectroscopically unobserved levels with even n_a, a striking pattern, shown in figure A1.2.9, is immediately evident. At the top of the polyad the level spacing pattern is like that of the anharmonic normal modes, as in figure A1.2.8, but at the bottom of the polyad the levels come in near-degenerate doublets. What is this pattern telling us about the change in the internal molecular motion resulting from inclusion of the Darling–Dennison coupling?

This has been the subject of a great deal of work by many people over more than 20 years. Breakthroughs in the theoretical understanding of the basic physics began to accumulate in the early 1980s [38–41]. One approach that has a particularly close relation between experiment and theory uses bifurcation analysis of a classical analogue of the spectroscopic fitting Hamiltonian. The mathematical details are presented elsewhere [42–45]; the qualitative physical meaning is easily described.

A classical Hamiltonian is obtained from the spectroscopic fitting Hamiltonian by a method that has come to be known as the 'Heisenberg correspondence' [46], because it is closely related to the techniques used by Heisenberg in fabricating the form of quantum mechanics known as matrix mechanics.

Once the classical Hamiltonian has been obtained, it is subjected to bifurcation analysis. In a bifurcation, typically, a stable motion of the molecule—say, one of the Weinstein–Moser normal modes—suddenly becomes unstable; and new stable, anharmonic modes suddenly branch out from the normal mode. An illuminating example is presented in figure A1.2.10, which illustrates the results of the bifurcation analysis of the classical version of the Darling–Dennison Hamiltonian. One of the normal modes—it can be either the symmetric or antisymmetric stretch depending on the specific parameters found empirically in the fitting Hamiltonian—remains stable. Suppose it is the antisymmetric stretch that remains stable. At the bifurcation, the symmetric stretch suddenly becomes unstable. This happens at some critical value of the mathematical

Figure A1.2.9. Energy level pattern of polyad 6 of a spectroscopic Hamiltonian for coupled stretches with strong Darling–Dennison coupling. Within the polyad the transition from normal to local modes is evident. At the bottom of the polyad are two nearly degenerate pairs of levels. Semiclassically, the bottom pair derive from local mode overtone states. The levels are symmetrized mixtures of the individual local mode overtones. Semiclassically, they are exactly degenerate; quantum mechanically, a small splitting is present, due to tunnelling. The next highest pair are symmetrized local mode combination states. The tunnelling splitting is larger than in the bottom pair; above this pair, the levels have normal mode character, as evidenced by the energy level pattern.

Normal-Local Bifurcation

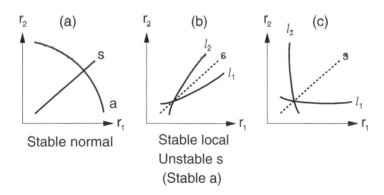

Figure A1.2.10. Birth of local modes in a bifurcation. In (a), before the bifurcation there are stable anharmonic symmetric and antisymmetric stretch modes, as in figure A1.2.6. At a critical value of the energy and polyad number, one of the modes, in this example the symmetric stretch, becomes unstable and new stable local modes are born in a bifurcation; the system is shown shortly after the bifurcation in (b), where the new modes have moved away from the unstable symmetric stretch. In (c), the new modes clearly have taken the character of the anharmonic local modes.

'control parameters' [34], which we may take to be some critical combination of the energy and polyad number. From the unstable symmetric stretch, there immediately emerge two new stable periodic orbits, or anharmonic modes. As the control parameter is increased, the new stable modes creep out from the symmetric stretch—which remains in 'fossilized' form as an unstable periodic orbit. Eventually, the new modes point more or less along the direction of the zero-order bond displacements, but as curvilinear trajectories. We can say that in this bifurcation, anharmonic local modes have been born.

It is the 'skeleton' of stable and unstable modes in figure A1.2.10(c) that explains the spectral pattern seen in figure A1.2.9. Some of the levels in the polyad, those in the upper part, have wavefunctions that are quantized in patterns that shadow the normal modes—the still-stable antisymmetric stretch and the now-unstable symmetric stretch. Other states, the lower ones in the polyad, are quantized along the local modes. These latter states, described by *local mode quantum numbers*, account for the pattern of near-degenerate doublets. First, why is the degeneracy there at all? The two classical local modes have exactly the same energy and frequency, by symmetry. In the simplest semiclassical [29] picture, there are two *exactly* degenerate local mode overtones, each pointed along one or the other of the local modes. There are also combination states possible with quanta in each of the local modes and, again, semiclassically these must come in exactly degenerate pairs.

The classical bifurcation analysis has succeeded in decoding the spectrum to reveal the existence of local and normal modes, and the local modes have accounted for the changeover from a normal mode spectral pattern to the pattern of degenerate doublets. But why the *splitting* of the near-degenerate doublets? Here, non-classical effects unique to quantum mechanics come into play. A trajectory in the box for one of the local modes is confined in phase space to an invariant torus, and classically will never leave its box. However, quantum mechanically, there is some probability for classically forbidden processes to take place in which the trajectory jumps from one box to the other! This may strike the reader as akin to the quantum mechanical phenomenon of tunnelling. In fact, this is more than an analogy. The effect has been called 'dynamical tunnelling' [41, 47], and it can be formulated rigorously as a mathematical tunnelling problem [40, 48]. The effect of the dynamical tunnelling on the energy levels comes through in another unique manifestation of quantum mechanics. The quantum eigenfunctions—the wavefunctions for the energy levels of the true quantum spectrum—are symmetrized combinations of the two semiclassical wavefunctions corresponding to the two classical boxes [38]. These wavefunctions come in pairs of $+$ and $-$ symmetry; the two levels of a near-degenerate pair are split into a $+$state and a $-$state. The amount of the splitting is directly related to the rate of the non-classical tunnelling process [49].

A1.2.11 Fermi resonances

In the example of H_2O, we saw that the Darling–Dennison coupling between the stretches led to a profound change in the internal dynamics; the birth of local modes in a bifurcation from one of the original low-energy normal modes. The question arises of the possibility of other types of couplings, if not between two identical stretch modes, then between other kinds of modes. We have seen that, effectively, only a very small subset of possible resonance couplings between the stretches is actually important; in the case of the H_2O stretches, only the 2:2 Darling–Dennison coupling. This great simplification came about because of the necessity to satisfy a condition of frequency resonance between the zero-order modes for the 2:2 Darling–Dennison coupling to be important. In H_2O, there is also an approximate 2:1 resonance condition satisfied between the stretch and bend frequencies. Not surprisingly, in fitting the H_2O spectrum, in particular when several bending quanta are present, it is necessary to consider a 2:1 coupling term between the symmetric stretch (s) and bend (b), of the form

$$\kappa_{sbb}(a_s^+ a_b a_b + a_s a_b^+ a_b^+). \tag{A1.2.12}$$

(The analogous coupling between the antisymmetric stretch and bend is forbidden in the H_2O Hamiltonian because of symmetry.) The 2:1 resonance is known as a 'Fermi resonance' after its introduction [50] in molecular spectroscopy. The 2:1 resonance is often very prominent in spectra, especially between stretch and bend modes, which often have approximate 2:1 frequency ratios. The 2:1 coupling leaves unchanged as a

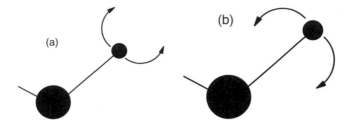

Figure A1.2.11. Resonant collective modes of the 2:1 Fermi resonance system of a coupled stretch and bend with an approximate 2:1 frequency ratio. Shown is one end of a symmetric triatomic such as H_2O. The normal stretch and bend modes are superseded by the horseshoe-shaped modes shown in (a) and (b). These two modes have different frequency, as further illustrated in figure A1.2.12.

polyad number the sum:

$$n_{sb} = (n_s + n_b/2). \tag{A1.2.13}$$

Other resonances, of order $n{:}m$, are possible in various systems. Another type of resonance is a 'multimode' resonance. For example, in C_2H_2 the coupling

$$\kappa_{2345}(a_3^+ a_2 a_4 a_5 + a_3 a_2^+ a_4^+ a_5^+) \tag{A1.2.14}$$

that transfers one quantum from the antisymmetric stretch ν_3 to the C–C stretch ν_2 and each of the bends ν_4 and ν_5 is important [51–53]. Situations where couplings such as the $n{:}m$ resonance and the 2345 multimode resonance need to be invoked are often referred to as 'Fermi resonances', though some authors restrict this term to the 2:1 resonance and use the term 'anharmonic resonance' to describe the more general $n{:}m$ or multimode cases. Here, we will use the terms 'Fermi' and 'anharmonic' resonances interchangeably.

It turns out that the language of 'normal and local modes' that emerged from the bifurcation analysis of the Darling–Dennison Hamiltonian is not sufficient to describe the general Fermi resonance case, because the bifurcations are qualitatively different from the normal-to-local bifurcation in figure A1.2.10. For example, in 2:1 Fermi systems, one type of bifurcation is that in which 'resonant collective modes' are born [54]. The resonant collective modes are illustrated in figure A1.2.11; their difference from the local modes of the Darling–Dennison system is evident. Other types of bifurcations are also possible in Fermi resonance systems; a detailed treatment of the 2:1 resonance can be found in [44].

A1.2.12 More subtle energy level patterns

The Darling–Dennison Hamiltonian displayed a striking energy level pattern associated with the bifurcation to local modes: approximately degenerate local mode doublets, split by dynamical tunnelling. In general Fermi resonance systems, the spectral hallmarks of bifurcations are not nearly as obvious. However, subtle, but clearly observable spectral markers of bifurcations do exist. For example, associated with the formation of resonant collective modes in the 2:1 Fermi system there is a pattern of a minimum in the spacing of adjacent energy levels within a polyad [55], as illustrated in figure A1.2.12. This pattern has been invoked [56, 57] in the analysis of 'isomerization spectra' of the molecule HCP, which will be discussed later. Other types of bifurcations have their own distinct, characteristic spectral patterns; for example, in 2:1 Fermi systems a second type of bifurcation has a pattern of alternating level spacings, of a 'fan' or a 'zigzag', which was predicted in [55] and subsequently observed in isomerization spectra [57].

Figure A1.2.12. Energy level pattern of a polyad with resonant collective modes. The top and bottom energy levels correspond to overtone motion along the two modes shown in figure A1.2.11, which have a different frequency. The spacing between adjacent levels decreases until it reaches a minimum between the third and fourth levels from the top. This minimum is the hallmark of a separatrix [29, 45] in phase space.

A1.2.13 Multiple resonances in polyatomics

Implicit in the discussion of the Darling–Dennison and Fermi resonances has been the assumption that we can isolate each individual resonance, and consider its bifurcations and associated spectral patterns separately from other resonances in the system. However, strictly speaking, this cannot be the case. Consider again H_2O. The Darling–Dennison resonance couples the symmetric and antisymmetric stretches; the Fermi resonance couples the symmetric stretch and bend. Indirectly, all three modes are coupled, and the two resonances are linked. It is no longer true that the stretch polyad number $(n_s + n_a)$ is conserved, because it is broken by the 2:1 Fermi coupling; nor is the Fermi polyad number $(n_s + n_b/2)$ preserved, because it is broken by the Darling–Dennison coupling. However, there is still a generalized 'total' polyad number

$$n_{\text{total}} = (n_s + n_a + n_b/2) \tag{A1.2.15}$$

that is conserved by both couplings, as may readily be verified. All told, the Hamiltonian with both couplings has two constants of motion, the energy and the polyad number (A1.2.15). A system with fewer constants than the number of degrees of freedom, in this case two constants and three degrees of freedom, is 'nonintegrable', in the language of classical mechanics [29]. This means that, in general, trajectories do not lie on higher-dimensional invariant tori; instead, they may be chaotic, and in fact this is often observed to be the case [58, 59] in trajectories of the semiclassical Hamiltonian for H_2O.

Nonetheless, it is still possible to perform the bifurcation analysis on the multiresonance Hamiltonian. In fact, the existence of the polyad number makes this almost as easy, despite the presence of chaos, as in the case of an isolated single Fermi or Darling–Dennison resonance. It is found [60] that most often (though not always), the same qualitative bifurcation behaviour is seen as in the single resonance case, explaining why the simplified individual resonance analysis very often is justified. The bifurcation analysis has now been performed for triatomics with two resonances [60] and for C_2H_2 with a number of resonances [61].

A1.2.14 Potential and experiment: closing the circle

We have alluded to the connection between the molecular PES and the spectroscopic Hamiltonian. These are two very different representations of the molecular Hamiltonian, yet both are supposed to describe the same molecular dynamics. Furthermore, the PES often is obtained via *ab initio* quantum mechanical calculations; while the spectroscopic Hamiltonian is most often obtained by an empirical fit to an experimental spectrum. Is there a direct link between these two seemingly very different ways of apprehending the molecular Hamiltonian and dynamics? And if so, how consistent are these two distinct ways of viewing the molecule?

There has been a great deal of work [62, 63] investigating how one can use perturbation theory to obtain an effective Hamiltonian like the spectroscopic Hamiltonian, starting from a given PES. It is found that one can readily obtain an effective Hamiltonian in terms of normal mode quantum numbers and coupling. Furthermore, the actual Hamiltonians obtained very closely match those obtained via the empirical fitting of spectra! This consistency lends great confidence that both approaches are complementary, mutually consistent ways of apprehending real information on molecules and their internal dynamics.

Is it possible to approach this problem the other way, from experiment to the molecular PES? This is difficult to answer in general, because 'inversion' of spectra is not a very well-posed question mathematically. Nonetheless, using spectra to gain information on potentials has been pursued with great vigor. Even for diatomics, surprising new, mathematically powerful methods are being developed [64]. For polyatomics, it has been shown [65] how the effective spectroscopic Hamiltonian is a very useful way-station on the road from experiment back to the PES. This closes the circle, because it shows that one can go from an assumed PES to the effective Hamiltonian derived via perturbation theory; or take the opposite path from the experimentally obtained effective spectroscopic Hamiltonian to the PES.

A1.2.15 Polyad quantum numbers in larger systems

We have seen that resonance couplings destroy quantum numbers as constants of the spectroscopic Hamiltonian. With both the Darling–Dennison stretch coupling and the Fermi stretch–bend coupling in H_2O, the individual quantum numbers n_s, n_a and n_b were destroyed, leaving the total polyad number $(n_s + n_a + n_b/2)$ as the only remaining quantum number. We can ask: (1) Is there also a good polyad number in larger molecules? (2) If so, how robust is this quantum number? For example, how high in energy does it persist as the molecule approaches dissociation or a barrier to isomerization? (3) Is the total polyad number the only good vibrational quantum number left over after the resonances have been taken into account, or can there be others?

It may be best to start with question (3). Given the set of resonance coupling operators found to be necessary to obtain a good fit of an experimental spectrum, it can be shown that the resonance couplings may be represented as vectors, which are not necessarily orthogonal. This leads to a simple but very powerful 'resonance vector analysis' [62, 66, 67]. The original vector space of the normal mode coordinates has N dimensions. The subspace spanned by the resonance vectors is the space of the vibrational quantum numbers that was destroyed; the complement of this space gives the quantities that remain as good quantum numbers. In general, there can be more than one such quantum number; we will encounter an example of this in C_2H_2, and see that it has important implications for the internal molecular dynamics. The set of good quantum numbers may contain one or more of the original individual normal mode quantum numbers; but in general, the good constants are combinations of the original quantum numbers. Examples of this are the polyad numbers that we have already encountered.

The resonance vector analysis has been used to explore all of the questions raised above on the fate of the polyad numbers in larger molecules, the most thoroughly investigated case so far probably being C_2H_2. This molecule has been very extensively probed by absorption as well as stimulated emission pumping and dispersed fluorescence techniques [52, 53, 68–71], the experimental spectra have been analysed in great detail and the fits to data have been carefully refined with each new experiment. A large number of resonance coupling operators has been found to be important, a good many more than the number of vibrational modes, which are seven in number: a symmetric C–H stretch ν_1, antisymmetric C–H stretch ν_3, C–C stretch ν_2 and two bends ν_4 and ν_5, each doubly degenerate. Despite the plethora of couplings, the resonance vector analysis shows that the total polyad number

$$N_{\text{total}} = (5n_1 + 3n_2 + 5n_3 + n_4 + n_5) \tag{A1.2.16}$$

is a good quantum number up to at least about $15\,000$ cm^{-1}. This is at or near the barrier to the formation of the isomer vinylidene! (The coefficients 5, 3, 5, 1 and 1 in (A1.2.16) are close to the frequency ratios of the zero-order normal modes, which is to say, the polyad number satisfies a resonance condition, as in the earlier examples for H_2O.) The polyad number N_{total} has been used with great effect to identify remarkable order [66–70] in the spectrum: groups of levels can clearly be identified that belong to distinct polyads. Furthermore, there are additional 'polyad' constants—that is, quantum numbers that are combinations of the original quantum numbers—in addition to the total polyad number (A1.2.16). These additional constants have great significance for the molecular dynamics. They imply the existence of energy transfer pathways [67]. For example, in dispersed fluorescence spectra in which pure bending motion is excited, it has been found that with as many as 22 quanta of bend, all of the vibrational excitation remains in the bends on the time scale associated with dispersed fluorescence spectroscopy, with no energy transfer to the stretches [72].

A1.2.16 Isomerization spectra

We have spoken of the simplicity of the bifurcation analysis when the spectroscopic Hamiltonian possesses a good polyad number, and also of the persistence of the polyad number in C_2H_2 as the molecule approaches the barrier to isomerization to the species vinylidene. This suggests that it might be possible to use detailed spectra to probe the dynamics of a system undergoing an internal rearrangement. Several groups [56, 57] have been investigating the rearrangement of HCP to the configuration CPH, through analysis of the 'isomerization spectrum'. Many of the tools described in this section, including decoding the dynamics through analysis of bifurcations and associated spectral patterns, have come into play. The various approaches all implicate an 'isomerization mode' in the rearrangement process, quite distinct from any of the low-energy normal modes of the system. An explanation has been provided [57] in terms of the abrupt birth of the isomerization mode. This occurs at a bifurcation, in which the HCP molecule suddenly acquires a stable motion that takes it along the isomerization pathway, thereby altering the geometry and with it the rotational constant.

It should be emphasized that isomerization is by no means the only process involving chemical reactions in which spectroscopy plays a key role as an experimental probe. A very exciting topic of recent interest is the observation and computation [73, 74] of the spectral properties of the transition state [6]—catching a molecule 'in the act' as it passes the point of no return from reactants to products. Furthermore, it has been discovered from spectroscopic observation [75] that molecules can have motions that are stable for long times even *above* the barrier to reaction.

A1.2.17 Breakdown of the polyad numbers

The polyad concept is evidently a very simple but powerful tool in the analysis and description of the internal dynamics of molecules. This is especially fortunate in larger molecules, where the intrinsic spectral complexity grows explosively with the number of atoms and degrees of freedom. Does the polyad number ever break down? Strictly speaking, it must: the polyad number is only an approximate property of a molecule's dynamics and spectrum. The actual molecular Hamiltonian contains resonance couplings of all forms, and these must destroy the polyad numbers at some level. This will show up by looking at high enough resolution at a spectrum which at lower resolution has a good polyad number. Levels will be observed of small intensity, which would be rigorously zero if the polyad numbers were exact. The fine detail in the spectrum corresponds to long-time dynamics, according to the time–energy uncertainty relation [49].

One reason the polyad-breaking couplings are of interest is because they govern the long-time intramolecular energy flow, which is important for theories on reaction dynamics. These are considered elsewhere in this Encyclopedia and in monographs [6] and will not be considered further here. The long-time energy flow may also be important for efforts of coherent control and for problems of energy flow from a molecule to a bath, such as a surrounding liquid. Both of these will be considered later.

Several questions arise on the internal dynamics associated with the breakdown of the polyad number. We can only speculate in what follows, awaiting the illumination of future research.

When the polyad number breaks down, as evidenced by the inclusion of polyad-breaking terms in the spectroscopic Hamiltonian, what is the residue left in the spectrum of the polyads as approximately conserved entities? There is already some indication [76] that the polyad organization of the spectrum will still be evident even with the inclusion of weak polyad-breaking terms. The identification of these polyad-breaking resonances will be a challenge, because each such resonance probably only couples a given polyad to a very small subset of 'dark' states of the molecule that lie outside those levels visible in the polyad spectrum. There will be a large number of such resonances, each of them coupling a polyad level to a small subset of dark levels.

Another question is the nature of the changes in the classical dynamics that occur with the breakdown of the polyad number. In all likelihood there are further bifurcations. Apart from the identification of the individual polyad-breaking resonances, the bifurcation analysis itself presents new challenges. This is partly because with the breakdown of the polyad number, the great computational simplicity afforded by the block-diagonalization of the Hamiltonian is lost. Another problem is that the bifurcation analysis is exactly solvable only when a polyad number is present [45], so approximate methods will be needed.

When the polyad number breaks down, the bifurcation analysis takes on a new kind of interest. The approximate polyad number can be thought of as a type of 'bottleneck' to energy flow, which is restricted to the phase space of the individual polyad; the polyad breakdown leads to energy flow in the full phase space. We can think of the goal as the search for the 'energy transfer modes' of long-time energy flow processes in the molecule, another step beyond the current use of bifurcation analysis to find the natural anharmonic modes that emerge within the polyad approximation.

The existence of the polyad number as a bottleneck to energy flow on short time scales is potentially important for efforts to control molecular reactivity using advanced laser techniques, discussed below in section A1.2.20. Efforts at control seek to intervene in the molecular dynamics to prevent the effects of widespread vibrational energy flow, the presence of which is one of the key assumptions of Rice–Ramsperger–Kassel–Marcus (RRKM) and other theories of reaction dynamics [6].

In connection with the energy transfer modes, an important question, to which we now turn, is the significance of classical chaos in the long-time energy flow process, in particular the relative importance of chaotic classical dynamics, versus classically forbidden processes involving 'dynamical tunnelling'.

A1.2.18   Classical versus non-classical effects

To understand the internal molecular motions, we have placed great store in classical mechanics to obtain a picture of the dynamics of the molecule and to predict associated patterns that can be observed in quantum spectra. Of course, the classical picture is at best an imprecise image, because the molecular dynamics are intrinsically quantum mechanical. Nonetheless, the classical metaphor must surely possess a large kernel of truth. The classical structure brought out by the bifurcation analysis has accounted for real patterns seen in wavefunctions and also for patterns observed in spectra, such as the existence of local mode doublets, and the more subtle level-spacing patterns seen in connection with Fermi resonance spectra.

However, we have also seen that some of the properties of quantum spectra are intrinsically non-classical, apart from the discreteness of quantum states and energy levels implied by the very existence of quanta. An example is the splitting of the local mode doublets, which was ascribed to dynamical tunnelling, i.e. processes which classically are forbidden. We can ask if non-classical effects are ubiquitous in spectra and, if so, are there manifestations accessible to observation other than those we have encountered so far? If there are such manifestations, it seems likely that they will constitute subtle peculiarities in spectral patterns, whose discernment and interpretation will be an important challenge.

The question of non-classical manifestations is particularly important in view of the chaos that we have seen is present in the classical dynamics of a multimode system, such as a polyatomic molecule, with more than one resonance coupling. Chaotic classical dynamics is expected to introduce its own peculiarities into quantum spectra [29, 77]. In H_2O, we noted that chaotic regions of phase space are readily seen in the classical dynamics corresponding to the spectroscopic Hamiltonian. How important are the effects of chaos in the observed spectrum, and in the wavefunctions of the molecule? In H_2O, there were some states whose wavefunctions appeared very disordered, in the region of the phase space where the two resonances should both be manifesting their effects strongly. This is precisely where chaos should be most pronounced, and indeed this was observed to be the case [58]. However, close examination of the states in question by Keshavamurthy and Ezra [78] showed that the disorder in the quantum wavefunction was due not primarily to chaos, but to dynamical tunnelling, the non-classical effect invoked earlier to explain the splitting of local mode doublets.

This demonstrated importance of the non-classical processes in systems with intact polyad numbers prompts us to consider again the breakdown of the polyad number. Will it be associated mainly with chaotic classical diffusion, or non-classical effects? It has been suggested [47] that high-resolution structure in spectra, which we have said is one of the manifestations of the polyad breakdown, may be predominantly due to non-classical, dynamical tunnelling processes, rather than chaotic diffusion. Independent, indirect support comes from the observation that energy flow from vibrationally excited diatomic molecules in a liquid bath is predominantly due to non-classical effects, to the extent of several orders of magnitude [79]. Whether dynamical tunnelling is a far more important energy transfer mechanism within molecules than is classical chaos is an important question for the future exploration of the interface of quantum and classical dynamics.

It should be emphasized that the existence of 'energy transfer modes' hypothesized earlier with the polyad breakdown is completely consistent with the energy transfer being due to non-classical, dynamical tunnelling processes. This is evident from the observation above that the disorder in the H_2O spectrum is attributable to *non-classical* effects which nonetheless are accompaniments of *classical* bifurcations.

The general question of the spectral manifestations of classical chaos and of non-classical processes, and their interplay in complex quantum systems, is a profound subject worthy of great current and future interest. Molecular spectra can provide an immensely important laboratory for the exploration of these questions. Molecules provide all the necessary elements: a mixture of regular and chaotic classical motion, with ample complexity for the salient phenomena to make their presence known and yet sufficient simplicity and control in the number of degrees of freedom to yield intelligible answers. In particular, the fantastic simplification afforded by the polyad constants, together with their gradual breakdown, may well make the spectroscopic study of internal molecular motions an ideal arena for a fundamental investigation of the quantum–classical correspondence.

A1.2.19   Molecules in condensed phase

So far we have considered internal motions mostly of isolated molecules, not interacting with an environment. This condition will be approximately met in a dilute gas. However, many of the issues raised may be of relevance in processes where the molecule is not isolated at all. An example already briefly noted is the transfer of vibrational energy from a molecule to a surrounding bath, for example a liquid. It has been found

[79] that when a diatomic molecule such as O_2 is vibrationally excited in a bath of liquid oxygen, the transfer of vibrational energy is extremely slow. This is due to the extreme mismatch between the energy of an O_2 vibrational quantum, and the far lower energy of the bath's phonon modes—vibrations involving large numbers of the bath molecules oscillating together. Classically, the energy transfer is practically non-existent; semiclassical approximations, however, show that quantum effects increase the rate by orders of magnitude.

The investigation of energy transfer in polyatomic molecules immersed in a bath is just beginning. One issue has to do with energy flow *from* the molecule to the bath. Another issue is the effect of the bath on energy flow processes *within* the molecule. Recent experimental work [80] using ultrafast laser probes of ClO_2 immersed in solvents points to the importance of bifurcations within the solute triatomic for the understanding of energy flow both within and from the molecule.

For a polyatomic, there are many questions on the role of the polyad number in energy flow from the molecule to the bath. Does polyad number conservation in the isolated molecule inhibit energy flow to the bath? Is polyad number breaking a facilitator or even a *prerequisite* for energy flow? Finally, does the energy flow to the bath increase the polyad number breaking in the molecule? One can only speculate until these questions become accessible to future research.

A1.2.20   Laser control of molecules

So far, we have talked about the internal motions of molecules which are exhibiting their 'natural' behaviour, either isolated in the gas phase or surrounded by a bath in a condensed phase. These natural motions are inferred from carefully designed spectroscopic experiments that are sufficiently mild that they simply probe what the molecule does when left to 'follow its own lights'. However, there is also a great deal of effort toward using high-intensity, carefully sculpted laser pulses which are anything but mild, in order to control the dynamics of molecules. In this quest, what role will be played by knowledge of their natural motions?

Surprisingly, a possible answer may be 'not much of a role at all'. One promising approach [81] using *coherent* light sources seeks to have the apparatus 'learn' how to control the molecule without knowing much at all about its internal properties in advance. Instead, a 'target' outcome is selected, and a large number of automated experiments performed, in which the control apparatus learns how to achieve the desired goal by rationally programmed trial and error in tailoring coherent light sources. It might not be necessary to learn much at all about the molecule's dynamics before, during or after, to make the control process work, even though the control apparatus might seem to all appearances to be following a cunning path to achieve its ends.

It can very well be objected that such a hit-or-miss approach, no matter how cleverly designed, is not likely to get very far in controlling polyatomic molecules with more than a very small number of atoms—in fact one will do much better by harnessing knowledge of the natural internal motions of molecules in tandem with the process of external control. The counter-argument can be made that in the trial and error approach, one will hit on the 'natural' way of controlling the molecule, even if one starts out with a method which at first tries nothing but brute force, even if one remains resolutely ignorant of why the molecule is responding to the evolving control procedure. Of course, if a good way is found to control the molecule, a retrospective explanation of how and why it worked almost certainly must invoke the natural motions of the molecule, about which much will perhaps have been learned along the way in implementing the process of control.

The view of this author is that knowledge of the internal molecular motions, perhaps as outlined in this chapter, is likely to be important in achieving successful control, in approaches that make use of coherent light sources and quantum mechanical coherence. However, at this point, opinions on these issues may not be much more than speculation.

There are also approaches [82–84] to control that have had marked success and which do not rely on quantum mechanical coherence. These approaches typically rely explicitly on a knowledge of the internal molecular dynamics, both in the design of the experiment and in the achievement of control. So far, these approaches have exploited only implicitly the very simplest types of bifurcation phenomena, such as the transition from local to normal stretch modes. If further success is achieved along these lines in larger molecules, it seems likely that deliberate knowledge and exploitation of more complicated bifurcation phenomena will be a matter of necessity.

As discussed in section A1.2.17, the existence of the approximate polyad numbers, corresponding to short-time bottlenecks to energy flow, could be very important in efforts for laser control, apart from the separate question of bifurcation phenomena.

Another aspect of laser control of molecular dynamics is the use of control techniques to *probe* the internal motions of molecules. A full account of this topic is far beyond the scope of this section, but one very interesting case in point has important relations to other branches of physics and mathematics. This is the phenomenon of 'geometric phases', which are closely related to gauge theories. The latter were originally introduced into quantum physics from the classical theory of electromagnetism by Weyl and others (see [85]). Quantum field theories with generalizations of the electromagnetic gauge invariance were developed in the 1950s and have since come to play a paramount role in the theory of elementary particles [86, 87]. Geometric phases were shown to have directly observable effects in quantum phenomena such as the Aharanov–Bohm effect [88]. It was later recognized that these phases are a general phenomenon in quantum systems [89]. One of the first concrete examples was pointed out [90] in molecular systems involving the coupling of rotation and vibration. A very systematic exposition of geometric phases and gauge ideas in molecular systems was presented in [91]. The possibility of the direct optical observation of the effects of the geometric phases in the time domain through coherent laser excitations has recently been explored [92].

A1.2.21   Larger molecules

This section has focused mainly on the internal dynamics of small molecules, where a coherent picture of the detailed internal motion has been emerging from intense efforts of many theoretical and experimental workers. A natural question is whether these kinds of issues will be important in the dynamics of larger molecules, and whether their investigation at the same level of detail will be profitable or tractable.

There will probably be some similarities, but also some fundamental differences. We have mainly considered small molecules with relatively rigid structures, in which the vibrational motions, although much different from the low-energy, near-harmonic normal modes, are nonetheless of relatively small amplitude and close to an equilibrium structure. (An important exception is the isomerization spectroscopy considered earlier, to which we shall return shortly.)

Molecules larger than those considered so far are formed by linking together several smaller components. A new kind of dynamics typical of these systems is already seen in a molecule such as C_2H_6, in which there is hindered rotation of the two methyl groups. Systems with hindered internal rotation have been studied in great depth [93], but there are still many unanswered questions. It seems likely that semiclassical techniques, using bifurcation analysis, could be brought to bear on these systems with great benefit.

The dynamics begin to take on a qualitatively different nature as the number of components, capable of mutual hindered rotation, starts to become only a little larger than in C_2H_6. The reason is that large-amplitude, very flexible twisting motions, such as those that start to be seen in a small polymer chain, become very important. These large scale 'wiggly motions' define a new class of dynamics and associated frequency scale as a characteristic internal motion of the system.

A hint that bifurcation techniques should be a powerful aid to the understanding of these problems comes from the example already considered in HCP isomerization [57]. Here the bifurcation techniques have given dramatic insights into the motions that stray very far from the equilibrium structure, in fact approaching the top of a barrier to the rearrangement to a different molecular isomer. It seems likely that similar approaches will be invaluable for molecules with internal rotors, including flexible polymer systems, but with an increase in complexity corresponding to the larger size of the systems. Probably, techniques to separate out the characteristic large-amplitude flexible motions from faster high-frequency vibrations, such as those of the individual bonds, will be necessary to unlock, along with the tools of the bifurcation analysis, the knowledge of the detailed anharmonic motions encrypted in the spectrum. This separation of time scales would be similar in some ways to the Born–Oppenheimer separability of nuclear and electronic motion.

Another class of problems in larger systems, also related to isomerization, is the question of large-amplitude motions in clusters of atoms and molecules. The phenomena of internal rearrangements, including processes akin to 'melting' and the seeking of minima on potential surfaces of very high dimensionality (due to the number of particles), have been extensively investigated [94]. The question of the usefulness of bifurcation techniques and the dynamical nature of large-amplitude natural motions in these systems has yet to be explored. These problems of large-amplitude motions and the seeking of potential minima in large clusters are conceptually related to the problem of protein folding, to which we now turn.

A1.2.22   Protein folding

An example of a kind of extreme challenge in the complexity of internal molecular dynamics comes with very complicated biological macromolecules. One of the major classes of these is proteins, very long biopolymers consisting of large numbers of amino acid residues [95]. They are very important in biological systems because they are the output of the translation of the genetic code: the DNA codes for the sequences of amino acid residues for each individual protein produced by the organism. A good sequence, i.e. one which forms a biologically useful protein, is one which folds to a more-or-less unique 'native' three-dimensional *tertiary* structure. (The sequence itself is the *primary* structure; subunits within the tertiary structure, consisting of chains of residues, fold to well defined *secondary* structures, which themselves are folded into the tertiary structure.) An outstanding problem, still very far from a complete understanding, is the connection between the sequence and the specific native structure, and even the prior question whether a given sequence has a reliable native structure at all. For sequences which do fold up into a unique structure, it is not yet possible to reliably predict what the structure will be, or what it is about the sequence that makes it a good folder. A solution of the sequence–structure problem would be very important, because it would make it possible to design sequences in the laboratory to fold to a definite, predictable structure, which then could be tailored for biological activity. A related question is the kinetic mechanism by which a good protein folds to its native structure.

Both the structural and kinetic aspects of the protein-folding problem are complicated by the fact that folding takes place within a bath of water molecules. In fact, hydrophobic interactions are almost certainly crucial for both the relation of the sequence and the native structure, and the process by which a good sequence folds to its native structure.

It is presently unknown whether the kind of detailed dynamical analysis of the natural motions of molecules outlined in this section will be useful for a problem as complicated as that of protein folding. The likely applicability of such methods to systems with several internal rotors strung together, and the incipient interest in bifurcation phenomena of small molecules immersed in a bath [80], suggests that dynamical analysis might also be useful for the much larger structures in proteins. In a protein, most of the molecular motion may be essentially irrelevant, i.e. the high-frequency, small-amplitude vibrations of the backbone of the amino acid sequence, and, also, probably much of the localized large-amplitude 'wiggly' motion. It is likely that there is a far smaller number of relevant large-amplitude, low-frequency motions that are crucial to the folding process. It will be of great interest to discover if techniques of dynamical systems such as bifurcation analysis can be used to reveal the 'folding modes' of proteins. For this to work, account must be taken of the complication of the bath of water molecules in which the folding process takes place. This introduces effects such as friction, for which there is little or no experience at present in applying bifurcation techniques in molecular systems. Proteins themselves interact with other proteins and with nucleic acids in biological processes of every conceivable kind considered at the molecular level.

A1.2.23 Outlook

Knowledge of internal molecular motions became a serious quest with Boyle and Newton, at the very dawn of modern natural science. However, real progress only became possible with the advent of quantum theory in the 20th century. The study of internal molecular motion for most of the century was concerned primarily with molecules near their equilibrium configuration on the PES. This gave an enormous amount of immensely valuable information, especially on the structural properties of molecules.

In recent years, especially the past two decades, the focus has changed dramatically to the study of highly-excited states. This came about because of a conjunction of powerful influences, often in mutually productive interaction with molecular science. Perhaps the first was the advent of lasers as revolutionary light sources for the probing of molecules. Coherent light of unprecedented intensities and spectral purity became available for studies in the traditional frequency domain of spectroscopy. This allowed previously inaccessible states of molecules to be reached, with new levels of resolution and detail. Later, the development of ultrafast laser pulses opened up the window of the ultrafast time domain as a spectroscopic complement to the new richness in the frequency domain. At the same time, revolutionary information technology made it possible to apply highly-sophisticated analytical methods, including new pattern recognition techniques, to process the wealth of new experimental information. The computational revolution also made possible the accurate investigation of highly-excited regions of molecular potential surfaces by means of quantum chemistry calculations. Finally, new mathematical developments in the study of nonlinear classical dynamics came to be appreciated by molecular scientists, with applications such as the bifurcation approaches stressed in this section.

With these radical advances in experimental technology, computational ability to handle complex systems, and new theoretical ideas, the kind of information being sought about molecules has undergone an equally profound change. Formerly, spectroscopic investigation, even of vibrations and rotations, had focused primarily on structural information. Now there is a marked drive toward dynamical information, including problems of energy flow, and internal molecular rearrangement. As emphasized in this section, a tremendous impetus to this was the recognition that other kinds of motion, such as local modes, could be just as important as the low-energy normal modes, in the understanding of the internal dynamics of highly-excited states. Ultrafast pulsed lasers have played a major role in these dynamical investigations. There is also a growing awareness of the immense potential for frequency domain spectroscopy to yield information on ultrafast processes in the time domain. This involves sophisticated measurements and data analysis of the very complex spectra of excited states; and equally sophisticated theoretical analysis to unlock the dynamical information encoded in the spectra. One of the primary tools is the bifurcation analysis of phenomenological Hamiltonians used directly to model experimental spectra. This gives information on the birth of new anharmonic motions in bifurcations of the low-energy normal modes. This kind of analysis is yielding information of startling detail about the internal molecular dynamics of high-energy molecules, including molecules undergoing isomerization. The ramifications are beginning to be explored for molecules in condensed phase. Here, ultrafast time-domain laser spectroscopy is usually necessary; but the requisite knowledge of internal molecular dynamics at the level of bifurcation analysis must be obtained from frequency-domain, gas phase experiments. Thus, a fruitful interplay is starting between gas and condensed phase experiments, and probes using sophisticated time- and frequency-domain techniques. Extension to much larger systems such as proteins is an exciting, largely unexplored future prospect. The interplay of research on internal molecular dynamics at the levels of small molecules, intermediate-size molecules, such as small polymer chains, and the hyper-complex scale of biological macromolecules is a frontier area of chemistry which surely will yield fascinating insights and discoveries for a long time to come.

References

[1]	Brock W H 1992 *The Norton History of Chemistry* (New York: Norton)
[2]	Hall M B 1965 *Robert Boyle on Natural Philosophy* (Bloomington, IN: Indiana University Press)
[3]	Schrödinger E 1996 *Nature and the Greeks* (Cambridge: Cambridge University Press)
	Schrödinger E 1996 *Science and Humanism* (Cambridge: Cambridge University Press)
[4]	Kuhn T S 1957 *The Copernican Revolution: Planetary Astronomy in the Development of Western Thought* (Cambridge, MA: Harvard University Press)
[5]	Ihde A J 1984 *The Development of Modern Chemistry* (New York: Dover)
[6]	Steinfeld J I, Francisco J S and Hase W L 1999 *Chemical Kinetics and Dynamics* (Upper Saddle River, NJ: Prentice-Hall)
[7]	Ball P 1994 *Designing the Molecular World: Chemistry at the Frontier* (Princeton, NJ: Princeton University Press)
[8]	Rosker M J, Dantus M and Zewail A H 1988 *Science* **241** 1200
[9]	Mokhtari A, Cong P, Herek J L and Zewail A H 1990 *Nature* **348** 225
[10]	Kellman M E 1994 *Phys. Rev. Lett.* **75** 2543
[11]	Blumel R and Reinhardt W P 1997 *Chaos in Atomic Physics* (Cambridge: Cambridge University Press)
[12]	Jaffe C, Farrelly D and Uzer T 2000 *Phys. Rev.* A **60** 3833
[13]	Berry R S, Rice S A and Ross J 1980 *Physical Chemistry* (New York: Wiley)
[14]	Herzberg G 1966 *Molecular Spectra and Molecular Structure III: Electronic Spectra and Electronic Structure of Polyatomic Molecules* (New York: Van Nostrand-Reinhold)
[15]	Nikitin E E 1999 *Ann. Rev. Phys. Chem.* **50** 1
[16]	Goldstein H 1980 *Classical Mechanics* (Reading, MA: Addison-Wesley)
[17]	Herzberg G 1950 *Molecular Spectra and Molecular Structure I: Spectra of Diatomic Molecules* (New York: Van Nostrand-Reinhold)
[18]	Herzberg G 1945 *Molecular Spectra and Molecular Structure II: Infrared and Raman Spectra of Polyatomic Molecules* (New York: Van Nostrand-Reinhold)
[19]	Papousek D and Aliev M R 1982 *Molecular Vibrational–Rotational Spectra* (Amsterdam: Elsevier)
[20]	Dai H L, Field R W and Kinsey J L 1985 *J. Chem. Phys.* **82** 2161
[21]	Frederick J H and McClelland G M 1987 *J. Chem. Phys.* **84** 4347
[22]	Littlejohn R G, Mitchell K A, Reinsch M, Aquilanti V and Cavalli S 1998 *Phys. Rev.* A **58** 3718
[23]	Sarkar P, Poulin N and Carrington T Jr 1999 *J. Chem. Phys.* **110** 10 269
[24]	Wilson E B Jr, Decius J C and Cross P C 1955 *Molecular Vibrations: The Theory of Infrared and Raman Vibrational Spectra* (New York: McGraw-Hill)
[25]	Bunker P R 1979 *Molecular Symmetry and Spectroscopy* (New York: Academic)
[26]	Harter W G 1993 *Principles of Symmetry, Dynamics, and Spectroscopy* (New York: Wiley)
[27]	Weinstein A 1973 Normal modes for nonlinear Hamiltonian systems *Inv. Math.* **20** 47
[28]	Moser J 1976 Periodic orbits near an equilibrium and a theorem by Alan Weinstein *Comm. Pure Appl. Math.* **29** 727
[29]	Tabor M 1989 *Chaos and Integrability in Nonlinear Dynamics: An Introduction* (New York: Wiley)
[30]	Lichtenberg A J and Lieberman M A 1983 *Regular and Stochastic Motion* (Berlin: Springer)
[31]	Darling B T and Dennison D M 1940 *Phys. Rev.* **57** 128
[32]	Messiah A 1961 *Quantum Mechanics* transl. G M Temmer (Amsterdam: North-Holland)
[33]	Heisenberg W 1925 *Z. Phyz.* **33** 879 (Engl. Transl. van der Waerden B L (ed) 1967 *Sources of Quantum Mechanics* (New York: Dover)
[34]	Golubitsky M and Schaeffer D G 1985 *Singularities and Groups in Bifurcation Theory* vol 1 (New York: Springer)
[35]	Iachello F and Levine R D 1995 *Algebraic Theory of Molecules* (Oxford: Oxford University Press)
[36]	Temsamani M A, Champion J-M and Oss S 1999 *J. Chem. Phys.* **110** 2893
[37]	Davis M J 1995 Trees from spectra: generation, analysis, and energy transfer information *Molecular Dynamics and Spectroscopy by Stimulated Emission Pumping* ed H-L Dai and R W Field (Singapore: World Scientific)
[38]	Lawton R T and Child M S 1980 *Mol. Phys.* **40** 773
[39]	Jaffe C and Brumer P 1980 *J. Chem. Phys.* **73** 5646
[40]	Sibert E L III, Hynes J T and Reinhardt W P 1982 *J. Chem. Phys.* **77** 3583
[41]	Davis M J and Heller E J 1981 *J. Chem. Phys.* **75** 246
[42]	Xiao L and Kellman M E 1989 *J. Chem. Phys.* **90** 6086
[43]	Li Z, Xiao L and Kellman M E 1990 *J. Chem. Phys.* **92** 2251
[44]	Xiao L and Kellman M E 1990 *J. Chem. Phys.* **93** 5805
[45]	Kellman M E 1995 Dynamical analysis of highly excited vibrational spectra: progress and prospects *Molecular Dynamics and Spectroscopy by Stimulated Emission Pumping* ed H-L Dai and R W Field (Singapore: World Scientific)
[46]	Clark A P, Dickinson A S and Richards D 1977 *Adv. Chem. Phys.* **36** 63
[47]	Heller E J 1995 *J. Phys. Chem.* **99** 2625
[48]	Kellman M E 1985 *J. Chem. Phys.* **83** 3843
[49]	Merzbacher E 1998 *Quantum Mechanics* 3rd edn (New York: Wiley)
[50]	Fermi E 1931 *Z. Phyzik* **71** 250

[51] Pliva J 1972 *J. Mol. Spec.* **44** 165
[52] Smith B C and Winn J S 1988 *J. Chem. Phys.* **89** 4638
[53] Smith B C and Winn J S 1991 *J. Chem. Phys.* **94** 4120
[54] Kellman M E and Xiao L 1990 *J. Chem. Phys.* **93** 5821
[55] Svitak J, Li Z, Rose J and Kellman M E 1995 *J. Chem. Phys.* **102** 4340
[56] Ishikawa H, Nagao C, Mikami N and Field R W 1998 *J. Chem. Phys.* **109** 492
[57] Joyeux M, Sugny D, Tyng V, Kellman M E, Ishikawa H and Field R W 2000 *J. Chem. Phys.* **112** 4162
[58] Lu Z-M and Kellman M E 1995 *Chem. Phys. Lett.* **247** 195–203
[59] Keshavamurthy S and Ezra G S 1997 *J. Chem. Phys.* **107** 156
[60] Lu Z-M and Kellman M E 1997 *J. Chem. Phys.* **107** 1–15
[61] Jacobson M P, Jung C, Taylor H S and Field R W 1999 *J. Chem. Phys.* **111** 66
[62] Fried L E and Ezra G S 1987 *J. Chem. Phys.* **86** 6270
[63] Sibert E L 1988 *J. Chem. Phys.* **88** 4378
[64] Herrick D R and O'Connor S 1998 *J. Chem. Phys.* **109** 2071
[65] Sibert E L and McCoy A B 1996 *J. Chem. Phys.* **105** 469
[66] Kellman M E 1990 *J. Chem. Phys.* **93** 6330
[67] Kellman M E and Chen G 1991 *J. Chem. Phys.* **95** 8671
[68] Solina S A B, O'Brien J P, Field R W and Polik W F 1996 *J. Phys. Chem.* **100** 7797
[69] Abbouti Temsamani M and Herman M 1995 *J. Chem. Phys.* **102** 6371
[70] El Idrissi M I, Lievin J, Campargue A and Herman M 1999 *J. Chem. Phys.* **110** 2074
[71] Jonas D M, Solina S A B, Rajaram B, Silbey R J, Field R W, Yamanouchi K and Tsuchiya S 1993 *J. Chem. Phys.* **99** 7350
[72] Jacobson M P, O'Brien J P, Silbey R J and Field R W 1998 *J. Chem. Phys.* **109** 121
[73] Waller I M, Kitsopoulos T M and Neumark D M 1990 *J. Phys. Chem.* **94** 2240
[74] Sadeghi R and Skodje R T 1996 *J. Chem. Phys.* **105** 7504
[75] Choi Y S and Moore C B 1991 *J. Chem. Phys.* **94** 5414
[76] Wu G 1998 *Chem. Phys. Lett.* **292** 369
[77] Stockmann H-J 1999 *Quantum Chaos: An Introduction* (Cambridge: Cambridge University Press)
[78] Keshavamurthy S and Ezra G S 1996 *Chem. Phys. Lett.* **259** 81
[79] Everitt K F, Egorov S A and Skinner J L 1998 *Chem. Phys.* **235** 115
[80] Hayes S C, Philpott M P, Mayer S G and Reid P J 1999 *J. Phys. Chem.* A **103** 5534
[81] Rabitz H 1997 *Adv. Chem. Phys.* **101** 315
[82] Hoffmann R 2000 *Am. Sci.* **88** 14
[83] VanderWal R L, Scott J L and Crim F F 1990 *J. Chem. Phys.* **92** 803
[84] Kandel S A and Zare R N 1998 *J. Chem. Phys.* **109** 9719
[85] O'Raifeartaigh L 1997 *The Dawning of Gauge Theory* (Princeton, NJ: Princeton University Press)
[86] Coughlan G D and Dodd J E 1991 *The Ideas of Particle Physics: An Introduction for Scientists* 2nd edn (Cambridge: Cambridge University Press)
[87] Aitchison I J R and Hey A J G 1996 *Gauge Theories in Particle Physics: A Practical Introduction* (Bristol: Institute of Physics Publishing)
[88] Aharonov Y and Bohm D 1959 *Phys. Rev.* **115** 485
[89] Berry M V 1984 *Proc. R. Soc.* A **392** 45
[90] Mead C A 1992 *Rev. Mod. Phys.* **64** 1
[91] Littlejohn R G and Reinsch M 1997 *Rev. Mod. Phys.* **69** 213
[92] Cina J A 2000 *J. Raman Spec.* **31** 95
[93] Ortigoso J, Kleiner I and Hougen J T 1999 *J. Chem. Phys.* **110** 11 688
[94] Berry R S 1999 Phases and phase changes of small systems *Theory of Atomic and Molecular Clusters* ed J Jellinek (Berlin: Springer)
[95] Creighton T E 1993 *Proteins* (New York: Freeman)

Further Reading

Herzberg G 1950 *Molecular Spectra and Molecular Structure I: Spectra of Diatomic Molecules* (New York: Van Nostrand-Reinhold)

Herzberg G 1945 *Molecular Spectra and Molecular Structure II: Infrared and Raman Spectra of Polyatomic Molecules* (New York: Van Nostrand-Reinhold)

Herzberg G 1966 *Molecular Spectra and Molecular Structure III: Electronic Spectra and Electronic Structure of Polyatomic Molecules* (New York: Van Nostrand-Reinhold)

The above three sources are a classic and comprehensive treatment of rotation, vibration, and electronic spectra of diatomic and polyatomic molecules.

Kellman M E 1995 Dynamical analysis of highly excited vibrational spectra: progress and prospects *Molecular Dynamics and Spectroscopy by Stimulated Emission Pumping* ed H-L Dai and R W Field (Singapore: World Scientific)

This is a didactic introduction to some of the techniques of bifurcation theory discussed in this article.

Steinfeld J I, Francisco J S and Hase W L 1999 *Chemical Kinetics and Dynamics* (Upper Saddle River, NJ: Prentice-Hall)
Papousek D and Aliev M R 1982 *Molecular Vibrational–Rotational Spectra* (Amsterdam: Elsevier)

This is a readable and fairly comprehensive treatment of rotation–vibration spectra and their interactions.

Tabor M 1989 *Chaos and Integrability in Nonlinear Dynamics: An Introduction* (New York: Wiley)
Lichtenberg A J and Lieberman M A 1983 Regular and Stochastic Motion (Berlin: Springer)

The above is a comprehensive, readable introduction to modern nonlinear classical dynamics, with quantum applications.

Iachello F and Levine R D 1995 *Algebraic Theory of Molecules* (Oxford: Oxford University Press)

This is a comprehensive survey of 'algebraic' methods for internal molecular motions.

Kellman M E 1995 Algebraic methods in spectroscopy *Ann. Rev. Phys. Chem.* **46** 395

This survey compares 'algebraic' methods with more standard approaches, and the bifurcation approach in this article.

Bunker P R 1979 *Molecular Symmetry and Spectroscopy* (New York: Academic)
Harter W G 1993 *Principles of Symmetry, Dynamics, and Spectroscopy* (New York: Wiley)

The above two references are comprehensive and individualistic surveys of symmetry, molecular structure and dynamics.

A1.3
Quantum mechanics of condensed phases

James R Chelikowsky

A1.3.1 Introduction

Traditionally one categorizes matter by phases such as gases, liquids and solids. Chemistry is usually concerned with matter in the gas and liquid phases, whereas physics is concerned with the solid phase. However, this distinction is not well defined: often chemists are concerned with the solid state and reactions between solid-state phases, and physicists often study atoms and molecular systems in the gas phase. The term *condensed phases* usually encompasses both the liquid state and the solid state, but not the gas state. In this section, the emphasis will be placed on the solid state with a brief discussion of liquids.

The solid phase of matter offers a very different environment to examine the chemical bond than does a gas or liquid [1–5]. The obvious difference involves describing the atomic positions. In a solid state, one can often describe atomic positions by a *static* configuration, whereas for liquid and gas phases this is not possible. The properties of the liquids and gases can be characterized only by considering some time-averaged ensemble. This difference between phases offers advantages in describing the solid phase, especially for crystalline matter. Crystals are characterized by a periodic symmetry that results in a system occupying all space [6]. Periodic, or translational, symmetry of crystalline phases greatly simplifies discussions of the solid state since knowledge of the atomic structure within a fundamental 'subunit' of the crystal, called the *unit cell*, is sufficient to describe the entire system encompassing all space. For example, if one is interested in the spatial distribution of electrons in a crystal, it is sufficient to know what this distribution is within a unit cell.

A related advantage of studying crystalline matter is that one can have symmetry-related operations that greatly expedite the discussion of a chemical bond. For example, in an elemental crystal of diamond, all the chemical bonds are equivalent. There are no terminating bonds and the characterization of one bond is sufficient to understand the entire system. If one were to know the binding energy or polarizability associated with one bond, then properties of the diamond crystal associated with all the bonds could be extracted. In contrast, molecular systems often contain different bonds and always have atoms at the boundary between the molecule and the vacuum.

Since solids do not exist as truly infinite systems, there are issues related to their termination (i.e. surfaces). However, in most cases, the existence of a surface does not strongly affect the properties of the crystal as a whole. The number of atoms in the interior of a cluster scale as the cube of the size of the specimen while the number of surface atoms scale as the square of the size of the specimen. For a sample of macroscopic size, the number of interior atoms vastly exceeds the number of atoms at the surface. On the other hand, there are interesting properties of the surface of condensed matter systems that have no analogue in atomic or molecular systems. For example, electronic states can exist that 'trap' electrons at the interface between a solid and the vacuum [1].

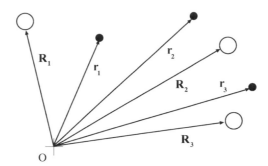

Figure A1.3.1. Atomic and electronic coordinates. The electrons are illustrated by filled circles; the nuclei by open circles.

Issues associated with order occupy a large area of study for crystalline matter [1, 7, 8]. For nearly perfect crystals, one can have systems with defects such as *point defects* and *extended defects* such as dislocations and grain boundaries. These defects occur in the growth process or can be mechanically induced. In contrast to molecular systems that can be characterized by 'perfect' molecular systems, solids always have defects. Individuals atoms that are missing from the ideal crystal structure, or extra atoms unneeded to characterize the ideal crystal are called point defects. The missing atoms correspond to vacancies; additional atoms are called interstitials. Extended defects are entire planes of atoms or interfaces that do not correspond to those of the ideal crystal. For example, edge dislocations occur when an extra half-plane of atoms is inserted in a perfect crystal and grain boundaries occur when a solid possesses regions of crystalline matter that have different structural orientations. In general, if a solid has no long-range order then one considers the phase to be an *amorphous* solid. The idea of atomic order and 'order parameters' is not usually considered for molecular systems, although for certain systems such as long molecular chains of atoms one might invoke a similar concept.

Another issue that distinguishes solids from atomic or molecular systems is the role of controlled defects or impurities. Often a pure, elemental crystal is not of great interest for technological applications; however, crystals with controlled additions of impurities are of great interest. The alteration of electronic properties with defects can be dramatic, involving changes in electrical conductivity by orders of magnitude. As an example, the addition of one boron atom for every 10^5 silicon atoms increases the conductivity of pure silicon by factor of 10^3 at room temperature [1]. Much of the electronic materials revolution is based on capitalizing on the dramatic changes in electronic properties via the controlled addition of electronically active dopants.

Of course, condensed phases also exhibit interesting physical properties such as electronic, magnetic, and mechanical phenomena that are not observed in the gas or liquid phase. Conductivity issues are generally not studied in isolated molecular species, but are actively examined in solids. Recent work in solids has focused on dramatic conductivity changes in superconducting solids. Superconducting solids have resistivities that are identically zero below some transition temperature [1, 9, 10]. These systems cannot be characterized by interactions over a few atomic species. Rather, the phenomenon involves a collective mode characterized by a phase representative of the entire solid.

A1.3.2 Many-body wavefunctions in condensed phases

One of the most significant achievements of the twentieth century is the description of the quantum mechanical laws that govern the properties of matter. It is relatively easy to write down the Hamiltonian for interacting fermions. Obtaining a solution to the problem that is sufficient to make predictions is another matter.

Let us consider N nucleons of charge Z_n at positions $\{\mathbf{R}_n\}$ for $n = 1, \ldots, N$ and M electrons at positions $\{\mathbf{r}_i\}$ for $i = 1, \ldots, M$. This is shown schematically in figure A1.3.1. The Hamiltonian for this system in its simplest form can be written as

$$\hat{\mathcal{H}}(\mathbf{R}_1, \mathbf{R}_2, \mathbf{R}_3, \ldots; \mathbf{r}_1, \mathbf{r}_2, \mathbf{r}_3 \ldots) = \sum_{n=1}^{N} \frac{-\hbar^2 \nabla_n^2}{2\mathcal{M}_n} + \frac{1}{2} \sum_{n,m=1,n \neq m}^{N} \frac{Z_n Z_m e^2}{|\mathbf{R}_n - \mathbf{R}_m|} + \sum_{i=1}^{M} \frac{-\hbar^2 \nabla_i^2}{2m}$$

$$- \sum_{n=1}^{N} \sum_{i=1}^{M} \frac{Z_n e^2}{|\mathbf{R}_n - \mathbf{r}_j|} + \frac{1}{2} \sum_{i,j=1,i \neq j}^{M} \frac{e^2}{|\mathbf{r}_i - \mathbf{r}_j|}. \qquad (A1.3.1)$$

$\mathcal{M}_n$ is the mass of the nucleon, $\hbar$ is Planck's constant divided by 2π, m is the mass of the electron. This expression omits some terms such as those involving relativistic interactions, but captures the essential features for most condensed matter phases.

Using the Hamiltonian in equation (A1.3.1), the quantum mechanical equation known as the Schrödinger equation for the electronic structure of the system can be written as

$$\hat{\mathcal{H}}(\mathbf{R}_1, \mathbf{R}_2, \mathbf{R}_3, \ldots; \mathbf{r}_1, \mathbf{r}_2, \mathbf{r}_3 \ldots) \Psi(\mathbf{R}_1, \mathbf{R}_2, \mathbf{R}_3, \ldots; \mathbf{r}_1, \mathbf{r}_2, \mathbf{r}_3 \ldots) = E \Psi(\mathbf{R}_1, \mathbf{R}_2, \mathbf{R}_3, \ldots; \mathbf{r}_1, \mathbf{r}_2, \mathbf{r}_3 \ldots)$$

$$(A1.3.2)$$

where E is the total electronic energy of the system, and Ψ is the many-body wavefunction. In the early part of the twentieth century, it was recognized that this equation provided the means of solving for the electronic and nuclear degrees of freedom. Using the variational principle, which states that an approximate wavefunction will always have a less favourable energy than the true ground-state energy, one had an equation and a method to test the solution. One can estimate the energy from

$$E = \frac{\int \Psi^* \hat{\mathcal{H}} \Psi \, d^3 R_1 \, d^3 R_2 \, d^3 R_3 \ldots d^3 r_1 \, d^3 r_2 \, d^3 r_3 \ldots}{\int \Psi^* \Psi \, d^3 R_1 \, d^3 R_2 \, d^3 R_3 \ldots d^3 r_1 \, d^3 r_2 \, d^3 r_3 \ldots}. \qquad (A1.3.3)$$

Solving equation (A1.3.2) for anything more complex than a few particles becomes problematic even with the most modern computers. Obtaining an approximate solution for condensed matter systems is difficult, but considerable progress has been made since the advent of digital computers. Several highly successful approximations have been made to solve for the ground-state energy. The nature of the approximations used is to remove as many degrees of freedom from the system as possible.

One common approximation is to separate the nuclear and electronic degrees of freedom. Since the nuclei are considerably more massive than the electrons, it can be assumed that the electrons will respond 'instantaneously' to the nuclear coordinates. This approximation is called the Born–Oppenheimer or adiabatic approximation. It allows one to treat the nuclear coordinates as classical parameters. For most condensed matter systems, this assumption is highly accurate [11, 12].

A1.3.2.1   The Hartree approximation

Another common approximation is to construct a specific form for the many-body wavefunction. If one can obtain an accurate estimate for the wavefunction, then, via the variational principle, a more accurate estimate for the energy will emerge. The most difficult part of this exercise is to use physical intuition to define a trial wavefunction.

One can utilize some very simple cases to illustrate this approach. Suppose one considers a solution for *non-interacting electrons*: i.e. in equation (A1.3.1) the last term in the Hamiltonian is ignored. In this limit, it is possible to write the many-body wavefunction as a sum of independent Hamiltonians. Using the adiabatic approximation, the *electronic* part of the Hamiltonian becomes

$$\hat{\mathcal{H}}_{\text{el}}(\mathbf{r}_1, \mathbf{r}_2, \mathbf{r}_3 \ldots) = \sum_{i=1}^{M} \frac{-\hbar^2 \nabla_i^2}{2m} - \sum_{n=1}^{N} \sum_{i=1}^{M} \frac{Z_n e^2}{|\mathbf{R}_n - \mathbf{r}_i|}. \tag{A1.3.4}$$

Let us define a nuclear potential, V_{N}, which the ith electron sees as

$$V_{\text{N}}(\mathbf{r}_i) = - \sum_{n=1}^{N} \frac{Z_n e^2}{|\mathbf{R}_n - \mathbf{r}_i|}. \tag{A1.3.5}$$

One can now rewrite a simplified Schrödinger equation as

$$\hat{\mathcal{H}}_{\text{el}}(\mathbf{r}_1, \mathbf{r}_2, \mathbf{r}_3 \ldots)\psi(\mathbf{r}_1, \mathbf{r}_2, \mathbf{r}_3 \ldots) = \sum_{i=1}^{M} \hat{H}^i \psi(\mathbf{r}_1, \mathbf{r}_2, \mathbf{r}_3 \ldots) \tag{A1.3.6}$$

where the Hamiltonian is now defined for the ith electron as

$$\hat{H}^i = \frac{-\hbar^2 \nabla_i^2}{2m} + V_{\text{N}}(\mathbf{r}_i). \tag{A1.3.7}$$

For this simple Hamiltonian, let us write the many-body wavefunction as

$$\psi(\mathbf{r}_1, \mathbf{r}_2, \mathbf{r}_3 \ldots) = \phi_1(\mathbf{r}_1)\phi_2(\mathbf{r}_2)\phi_3(\mathbf{r}_3) \ldots . \tag{A1.3.8}$$

The $\phi_i(\mathbf{r})$ orbitals can be determined from a 'one-electron' Hamiltonian

$$\hat{H}^i \phi_i(\mathbf{r}) = \left(\frac{-\hbar^2 \nabla^2}{2m} + V_{\text{N}}(\mathbf{r}) \right) \phi(\mathbf{r}) = E_i \phi_i(\mathbf{r}). \tag{A1.3.9}$$

The index i for the orbital $\phi_i(\mathbf{r})$ can be taken to include the spin of the electron plus any other relevant quantum numbers. The index i runs over the number of electrons, each electron being assigned a unique set of quantum numbers. This type of Schrödinger equation can be easily solved for fairly complex condensed matter systems. The many-body wavefunction in equation (A1.3.8) is known as the Hartree wavefunction. If one uses this form of the wavefunction as an approximation to solve the Hamiltonian *including* the electron–electron interactions, this is known as the Hartree approximation. By ignoring the electron–electron terms, the Hartree approximation simply reflects the electrons independently moving in the nuclear potential. The total energy of the system in this case is simply the sum of the eigenvalues, E_i.

To obtain a realistic Hamiltonian, the electron–electron interactions must be reinstated in equation (A1.3.6):

$$\hat{\mathcal{H}}_{\text{el}}(\mathbf{r}_1, \mathbf{r}_2, \mathbf{r}_3 \ldots)\psi(\mathbf{r}_1, \mathbf{r}_2, \mathbf{r}_3 \ldots) = \sum_{i=1}^{M} \left(\hat{H}^i + \frac{1}{2} \sum_{j=1, j \neq i}^{M} \frac{e^2}{|\mathbf{r}_i - \mathbf{r}_j|} \right) \psi(\mathbf{r}_1, \mathbf{r}_2, \mathbf{r}_3 \ldots). \tag{A1.3.10}$$

In this case, the individual orbitals, $\phi_i(\mathbf{r})$, can be determined by minimizing the total energy as per equation (A1.3.3), with the constraint that the wavefunction be normalized. This minimization procedure results in the following Hartree equation:

$$\hat{H}^i \phi_i(\mathbf{r}) = \left(\frac{-\hbar^2 \nabla^2}{2m} + V_{\text{N}}(\mathbf{r}) + \sum_{j=1, j \neq i}^{M} \int \frac{e^2 |\phi_j(\mathbf{r}')|^2}{|\mathbf{r} - \mathbf{r}'|} \, \mathrm{d}^3 r' \right) \phi_i(\mathbf{r}) = E_i \phi_i(\mathbf{r}). \tag{A1.3.11}$$

Using the orbitals, $\phi(\mathbf{r})$, from a solution of equation (A1.3.11), the Hartree many-body wavefunction can be constructed and the total energy determined from equation (A1.3.3).

The Hartree approximation is useful as an illustrative tool, but it is not a very accurate approximation. A significant deficiency of the Hartree wavefunction is that it does not reflect the anti-symmetric nature of the electrons as required by the Pauli principle [7]. Moreover, the Hartree equation is difficult to solve. The Hamiltonian is orbitally dependent because the summation in equation (A1.3.11) does not include the ith orbital. This means that if there are M electrons, then M Hamiltonians must be considered and equation (A1.3.11) solved for each orbital.

A1.3.2.2   The Hartree–Fock approximation

It is possible to write down a many-body wavefunction that will reflect the antisymmetric nature of the wavefunction. In this discussion, the spin coordinate of each electron needs to be explicitly treated. The coordinates of an electron may be specified by $\mathbf{r}_i s_i$ where s_i represents the spin coordinate. Starting with one-electron orbitals, $\phi_i(\mathbf{r}s)$, the following form can be invoked:

$$\Psi(\mathbf{r}_1 s_1, \mathbf{r}_1 s_2, \mathbf{r}_1 s_3, \dots) = \begin{vmatrix} \phi_1(\mathbf{r}_1 s_1) & \phi_1(\mathbf{r}_2 s_2) & \dots & \dots & \phi_1(\mathbf{r}_M s_M) \\ \phi_2(\mathbf{r}_1 s_1) & \phi_2(\mathbf{r}_2 s_2) & \dots & \dots & \dots \\ \dots & \dots & \dots & \dots & \dots \\ \phi_M(\mathbf{r}_1 s_1) & \dots & \dots & \dots & \phi_M(\mathbf{r}_M s_M) \end{vmatrix}. \tag{A1.3.12}$$

This form of the wavefunction is called a Slater determinant. It reflects the proper symmetry of the wavefunction and the Pauli principle. If two electrons occupy the same orbit, two rows of the determinant will be identical and the many-body wavefunction will have zero amplitude. Likewise, the determinant will vanish if two electrons occupy the same point in generalized space (i.e. $\mathbf{r}_i s_i = \mathbf{r}_j s_j$) as two columns of the determinant will be identical. If two particles are exchanged, this corresponds to a sign change in the determinant. The Slater determinant is a convenient representation. It is probably the simplest form that incorporates the required symmetry properties for fermions, or particles with non-integer spins.

If one uses a Slater determinant to evaluate the total electronic energy and maintains the orbital normalization, then the orbitals can be obtained from the following Hartree–Fock equations:

$$H^i \phi_i(\mathbf{r}) = \left(\frac{-\hbar^2 \nabla^2}{2m} + V_N(\mathbf{r}) + \sum_{j=1}^{M} \int \frac{e^2 |\phi_j(\mathbf{r}')|^2}{|\mathbf{r} - \mathbf{r}'|} \, \mathrm{d}^3 r' \right) \phi_i(\mathbf{r}) - \sum_{j=1}^{M} \int \frac{e^2}{|\mathbf{r} - \mathbf{r}'|} \phi_j^*(\mathbf{r}') \phi_i(\mathbf{r}') \, \mathrm{d}^3 r' \, \delta_{s_i, s_j} \, \phi_j(\mathbf{r})$$

$$= E_i \phi_i(\mathbf{r}). \tag{A1.3.13}$$

It is customary to simplify this expression by defining an electronic charge density, ρ:

$$\rho(\mathbf{r}) = \sum_{j=1}^{M} |\phi_j(\mathbf{r})|^2 \tag{A1.3.14}$$

and an orbitally dependent *exchange-charge density*, ρ_i^{HF} for the ith orbital:

$$\rho_i^{HF}(\mathbf{r}, \mathbf{r}') = \sum_{j=1}^{M} \frac{\phi_j^*(\mathbf{r}') \phi_i(\mathbf{r}') \phi_i^*(\mathbf{r}) \phi_j(\mathbf{r})}{\phi_i^*(\mathbf{r}) \phi_i(\mathbf{r})} \delta_{s_i, s_j}. \tag{A1.3.15}$$

This 'density' involves a spin-dependent factor which couples only states (i, j) with the same spin coordinates (s_i, s_j). It is not a true density in that it is dependent on $\mathbf{r}, \mathbf{r}'$; it has meaning only as defined below.

With these charge densities defined, it is possible to define corresponding potentials. The Coulomb or Hartree potential, V_H, is defined by

$$V_H(\mathbf{r}) = \int \rho(\mathbf{r}) \frac{e^2}{|\mathbf{r} - \mathbf{r}'|} \, d^3 r' \tag{A1.3.16}$$

and an *exchange* potential can be defined by

$$V_x^i(\mathbf{r}) = - \int \rho_i^{HF}(\mathbf{r}, \mathbf{r}') \frac{e^2}{|\mathbf{r} - \mathbf{r}'|} \, d^3 r'. \tag{A1.3.17}$$

This combination results in the following Hartree–Fock equation:

$$\left(\frac{-\hbar^2 \nabla^2}{2m} + V_N(\mathbf{r}) + V_H(\mathbf{r}) + V_x^i(\mathbf{r}) \right) \phi_i(\mathbf{r}) = E_i \phi_i(\mathbf{r}). \tag{A1.3.18}$$

Once the Hartree–Fock orbitals have been obtained, the total Hartree–Fock electronic energy of the system, E_{HF}, can be obtained from

$$E_{HF} = \sum_i^M E_i - \frac{1}{2} \int \rho(\mathbf{r}) V_H(\mathbf{r}) \, d^3 r - \frac{1}{2} \sum_i^M \int \phi_i^*(\mathbf{r}) \phi_i(\mathbf{r}) V_x^i(\mathbf{r}) \, d^3 r. \tag{A1.3.19}$$

E_{HF} is not a sum of the Hartree–Fock orbital energies, E_i. The factor of $\frac{1}{2}$ in the electron–electron terms arises because the electron–electron interactions have been double-counted in the Coulomb and exchange potentials. The Hartree–Fock Schrödinger equation is only slightly more complex than the Hartree equation. Again, the equations are difficult to solve because the exchange potential is orbitally dependent.

There is one notable difference between the Hartree–Fock summation and the Hartree summation. The Hartree–Fock sums include the $i = j$ terms in equation (A1.3.13). This difference arises because the exchange term corresponding to $i = j$ cancels an equivalent term in the Coulomb summation. The $i = j$ term in both the Coulomb and exchange term is interpreted as a 'self-screening' of the electron. Without a cancellation between Coulomb and exchange terms a 'self-energy' contribution to the total energy would occur. Approximate forms of the exchange potential often do not have this property. The total energy then contains a self-energy contribution which one needs to remove to obtain a correct Hartree–Fock energy.

The Hartree–Fock wavefunctions are approximations to the true ground-state many-body wavefunctions. Terms not included in the Hartree–Fock energy are referred to as *correlation* contributions. One definition for the correlation energy, E_{corr} is to write it as the difference between the correct total energy of the system and the Hartree–Fock energies: $E_{corr} = E_{exact} - E_{HF}$. Correlation energies are sometimes included by considering Slater determinants composed of orbitals which represent excited-state contributions. This method of including unoccupied orbitals in the many-body wavefunction is referred to as *configuration interaction* or 'CI'.

Applying Hartree–Fock wavefunctions to condensed matter systems is not routine. The resulting Hartree–Fock equations are usually too complex to be solved for extended systems. It has been argued that many-body wavefunction approaches to the condensed matter or large molecular systems do not represent a reasonable approach to the electronic structure problem of extended systems.

A1.3.3 Density functional approaches to quantum descriptions of condensed phases

Alternative descriptions of quantum states based on a knowledge of the electronic charge density equation (A1.3.14) have existed since the 1920s. For example, the Thomas–Fermi description of atoms based on a knowledge of $\rho(\mathbf{r})$ was reasonably successful [13–15]. The starting point for most discussions of condensed matter begins by considering a limiting case that may be appropriate for condensed matter systems, but not for small molecules. One often considers a free electron gas of uniform charge density. The justification for this approach comes from the observation that simple metals like aluminium and sodium have properties which appear to resemble those of a free electron gas. This model cannot be applied to systems with localized electrons such as highly covalent materials like carbon or highly ionic materials like sodium chloride. It is also not appropriate for very open structures. In these systems large variations of the electron distribution can occur.

A1.3.3.1   Free electron gas

Perhaps the simplest description of a condensed matter system is to imagine non-interacting electrons contained within a box of volume, Ω. The Schrödinger equation for this system is similar to equation (A1.3.9) with the potential set to zero:

$$\frac{-\hbar^2 \nabla^2}{2m} \phi(\mathbf{r}) = E\phi(\mathbf{r}). \tag{A1.3.20}$$

Ignoring spin for the moment, the solution of equation (A1.3.20) is

$$\phi(\mathbf{r}) = \frac{1}{\sqrt{\Omega}} \exp(i\mathbf{k} \cdot \mathbf{r}). \tag{A1.3.21}$$

The energy is given by $E(k) = \hbar^2 k^2/2m$ and the charge density by $\rho = 1/\Omega$. $\mathbf{k}$ is called a *wavevector*.

A key issue in describing condensed matter systems is to account properly for the number of states. Unlike a molecular system, the eigenvalues of condensed matter systems are closely spaced and essentially 'infinite' in number. For example, if one has 10^{23} electrons, then one can expect to have 10^{23} occupied states. In condensed matter systems, the number of states per energy unit is a more natural measure to describe the energy distribution of states.

It is easy to do this with *periodic boundary conditions*. Suppose one considers a one-dimensional specimen of length L. In this case the wavefunctions obey the rule $\phi(x + L) = \phi(x)$ as $x + L$ corresponds in all physical properties to x. For a free electron wavefunction, this requirement can be expressed as $\exp(ik(x + L)) = \exp(ikx)$ or as $\exp(ikL) = 1$ or $k = 2\pi n/L$ where n is an integer.

Periodic boundary conditions force k to be a discrete variable with allowed values occurring at intervals of $2\pi/L$. For very large systems, one can describe the system as continuous in the limit of $L \to \infty$. Electron states can be defined by a *density of states* defined as follows:

$$\begin{aligned} D(E) &= \lim_{\Delta E \to 0} \frac{N(E + \Delta E) - N(E)}{\Delta E} \\ &= \frac{dN}{dE} \end{aligned} \tag{A1.3.22}$$

where $N(E)$ is the number of states whose energy resides below E. For the one-dimensional case, $N(k) = 2k/(2\pi/L)$ (the factor of two coming from spin) and $dN/dE = (dN/dk) \cdot (dk/dE)$. Using $E(k) = \hbar^2 k^2/2m$, we have $k = \sqrt{2mE}/\hbar$ and $dk/dE = \frac{1}{2}\sqrt{2m/E}/\hbar$. This results in the one-dimensional density of states as

$$D(E) = \frac{L}{\pi\hbar}\sqrt{2m/E}. \tag{A1.3.23}$$

The density of states for a one-dimensional system diverges as $E \to 0$. This divergence of $D(E)$ is not a serious issue as the integral of the density of states remains finite. In three dimensions, it is straightforward to show that

$$D(E) = \frac{\Omega}{2\pi^2}\left(\frac{2m}{\hbar^2}\right)^{3/2}\sqrt{E}. \tag{A1.3.24}$$

The singularity is removed, although a discontinuity in the derivative exists as $E \to 0$.

One can determine the total number of electrons in the system by integrating the density of states up to the highest occupied energy level. The energy of the highest occupied state is called the *Fermi level* or *Fermi energy*, E_F:

$$N = \frac{\Omega}{2\pi^2}\left(\frac{2m}{\hbar^2}\right)^{3/2}\int_0^{E_F}\sqrt{E}\,\mathrm{d}E \tag{A1.3.25}$$

and

$$E_F = \frac{\hbar^2}{2m}\left(\frac{3\pi^2 N}{\Omega}\right)^{2/3}. \tag{A1.3.26}$$

By defining a *Fermi* wavevector as $k_F = (3\pi^2 n_{el})^{1/3}$ where n_{el} is the electron density, $n_{el} = N/\Omega$, of the system, one can write

$$E_F = \frac{\hbar^2 k_F^2}{2m}. \tag{A1.3.27}$$

It should be noted that typical values for E_F for simple metals like sodium or potassium are of the order of several electronvolts. If one defines a temperature, T_F, where $T_F = E_F/k_B$ and k_B is the Boltzmann constant, typical values for T_F might be 10^3–10^4 K. Thus, at ambient temperatures one can often neglect the role of temperature in determining the Fermi energy.

A1.3.3.2   Hartree–Fock exchange in a free electron gas

For a free electron gas, it is possible to evaluate the Hartree–Fock exchange energy directly [3, 16]. The Slater determinant is constructed using free electron orbitals. Each orbital is labelled by a $\mathbf{k}$ and a spin index. The Coulomb potential for an infinite free electron gas diverges, but this divergence can be removed by imposing a compensating uniform positive charge. The resulting Hartree–Fock eigenvalues can be written as

$$E_k = \frac{\hbar^2 k^2}{2m} - \frac{1}{\Omega}\sum_{k'<k_F}\frac{4\pi e^2}{|\mathbf{k}-\mathbf{k'}|^2} \tag{A1.3.28}$$

where the summation is over occupied $\mathbf{k}$-states. It is possible to evaluate the summation by transposing the summation to an integration. This transposition is often done for solid-state systems as the state density is so high that the system can be treated as a continuum:

$$\frac{1}{\Omega}\sum_{k'<k_F}\frac{4\pi e^2}{|\mathbf{k}-\mathbf{k'}|^2} = \frac{1}{(2\pi)^3}\int_{k'<k_F}\frac{4\pi e^2}{|\mathbf{k}-\mathbf{k'}|^2}\,\mathrm{d}^3 k. \tag{A1.3.29}$$

This integral can be solved analytically. The resulting eigenvalues are given by

$$E_k = \frac{\hbar^2 k^2}{2m} - \frac{e^2 k_F}{\pi}\left(1 + \frac{1-(k/k_F)^2}{2(k/k_F)}\ln\left|\frac{k+k_F}{k-k_F}\right|\right). \tag{A1.3.30}$$

Using the above expression and equation (A1.3.19), the total electron energy, E_{HF}^{FEG}, for a free electron gas within the Hartree–Fock approximation is given by

$$E_{HF}^{FEG} = 2\sum_{k<k_F}\frac{\hbar^2 k^2}{2m} - \frac{e^2 k_F}{\pi}\sum_{k<k_F}\left(1 + \frac{1-(k/k_F)^2}{2(k/k_F)}\ln\left|\frac{k+k_F}{k-k_F}\right|\right). \tag{A1.3.31}$$

The factor of 2 in the first term comes from spin. In the exchange term, there is no extra factor of 2 because one can subtract off a 'double-counting term' (see equation (A1.3.19)). The summations can be executed as per equation (A1.3.29) to yield

$$E_{\text{HF}}^{\text{FEG}}/N = \frac{3}{5}E_{\text{F}} - \frac{3e^2}{4\pi}k_{\text{F}}.$$ (A1.3.32)

The first term corresponds to the average energy per electron in a free electron gas. The second term corresponds to the exchange energy per electron. The exchange energy is attractive and scales with the cube root of the average density. This form provides a clue as to what form the exchange energy might take in an interacting electron gas or non-uniform electron gas.

Slater was one of the first to propose that one replace V_x^i in equation (A1.3.18) by a term that depends only on the cube root of the charge density [17–19]. In analogy to equation (A1.3.32), he suggested that V_x^i be replaced by

$$V_{\text{x}}^{\text{Slater}}[\rho(\mathbf{r})] = -\frac{3e^2}{2\pi}(3\pi\rho(\mathbf{r}))^{1/3}.$$ (A1.3.33)

This expression is not orbitally dependent. As such, a solution of the Hartree–Fock equation (equation (A1.3.18)) is much easier to implement. Although Slater exchange was not rigorously justified for non-uniform electron gases, it was quite successful in replicating the essential features of atomic and molecular systems as determined by Hartree–Fock calculations.

A1.3.3.3 The local density approximation

In a number of classic papers Hohenberg, Kohn and Sham established a theoretical framework for justifying the replacement of the many-body wavefunction by one-electron orbitals [15, 20, 21]. In particular, they proposed that the charge density plays a central role in describing the electronic structure of matter. A key aspect of their work was the *local density approximation* (LDA). Within this approximation, one can express the exchange energy as

$$E_{\text{x}}[\rho(\mathbf{r})] = \int \rho(\mathbf{r})\mathcal{E}_{\text{x}}[\rho(\mathbf{r})]\,\mathrm{d}^3r$$ (A1.3.34)

where $\mathcal{E}_{\text{x}}[\rho]$ is the exchange energy per particle of uniform gas at a density of ρ. Within this framework, the exchange potential in equation (A1.3.18) is replaced by a potential determined from the functional derivative of $E_x[\rho]$:

$$V_{\text{x}}[\rho] = \frac{\delta E_x[\rho]}{\delta \rho}.$$ (A1.3.35)

One serious issue is the determination of the exchange energy per particle, $\mathcal{E}_{\text{x}}$, or the corresponding exchange potential, V_{x}. The exact expression for either of these quantities is unknown, save for special cases. If one assumes the exchange energy is given by equation (A1.3.32), i.e. the Hartree–Fock expression for the exchange energy of the free electron gas, then one can write

$$E_{\text{x}}[\rho] = -\frac{3e^2}{4\pi}(3\pi^2)^{1/3}\int [\rho(\mathbf{r})]^{4/3}\,\mathrm{d}^3r$$ (A1.3.36)

and taking the functional derivative, one obtains

$$V_{\text{x}}[\rho] = -\frac{e^2}{\pi}(3\pi^2\rho(\mathbf{r}))^{1/3}.$$ (A1.3.37)

Comparing this to the form chosen by Slater, we note that this form, known as Kohn–Sham exchange, differs by a factor of $\frac{2}{3}$: i.e. $V_{\text{x}} = 2V_{\text{x}}^{\text{Slater}}/3$. For a number of years, some controversy existed as to whether the

Kohn–Sham or Slater exchange was more accurate for realistic systems [15]. Slater suggested that a parameter be introduced that would allow one to vary the exchange between the Slater and Kohn–Sham values [19]. The parameter, α, was often placed in front of the Slater exchange: $V_{x\alpha} = \alpha V_x^{\text{Slater}}$. α was often chosen to replicate some known feature of an exact Hartree–Fock calculation such as the total energy of an atom or ion. Acceptable values of α were viewed to range from $\alpha = \frac{2}{3}$ to $\alpha = 1$. Slater's so-called 'X_α' method was very successful in describing molecular systems [19]. Notable drawbacks of the X_α method centre on its ad hoc nature through the α parameter and the omission of an explicit treatment of correlation energies.

In contemporary theories, α is taken to be $\frac{2}{3}$, and correlation energies are explicitly included in the energy functionals [15]. Sophisticated numerical studies have been performed on uniform electron gases resulting in local density expressions of the form $V_{xc}[\rho(\mathbf{r})] = V_x[\rho(\mathbf{r})] + V_c[\rho(\mathbf{r})]$ where V_c represents contributions to the total energy beyond the Hartree–Fock limit [22]. It is also possible to describe the role of spin explicitly by considering the charge density for up and down spins: $\rho = \rho_\uparrow + \rho_\downarrow$. This approximation is called the *local spin density approximation* [15].

The Kohn–Sham equation [21] for the electronic structure of matter is given by

$$\left(\frac{-\hbar^2\nabla^2}{2m} + V_N(\mathbf{r}) + V_H(\mathbf{r}) + V_{xc}[\rho(\mathbf{r})]\right)\phi_i(\mathbf{r}) = E_i\phi_i(\mathbf{r}). \tag{A1.3.38}$$

This equation is usually solved 'self-consistently'. An approximate charge is assumed to estimate the exchange-correlation potential and to determine the Hartree potential from equation (A1.3.16). These approximate potentials are inserted in the Kohn–Sham equation and the total charge density is obtained from equation (A1.3.14). The 'output' charge density is used to construct new exchange-correlation and Hartree potentials. The process is repeated until the input and output charge densities or potentials are identical to within some prescribed tolerance.

Once a solution of the Kohn–Sham equation is obtained, the total energy can be computed from

$$E_{\text{KS}} = \sum_i^M E_i - \frac{1}{2}\int \rho(\mathbf{r})V_H(\mathbf{r})\,d^3r + \int \rho(\mathbf{r})(\mathcal{E}_{xc}[\rho(\mathbf{r})] - V_{xc}[\rho(\mathbf{r})])\,d^3r. \tag{A1.3.39}$$

The electronic energy, as determined from E_{KS}, must be added to the ion–ion interactions to obtain the structural energies. This is a straightforward calculation for confined systems. For extended systems such as crystals, the calculations can be done using Madelung summation techniques [2].

Owing to its ease of implementation and overall accuracy, the local density approximation is the current method of choice for describing the electronic structure of condensed matter. It is relatively easy to implement and surprisingly accurate. Moreover, recent developments have included so-called gradient corrections to the local density approximation. In this approach, the exchange-correlation energy depends on the local density the gradient of the density. This approach is called the generalized gradient approximation or GGA [23].

When first proposed, density functional theory was not widely accepted in the chemistry community. The theory is not 'rigorous' in the sense that it is not clear how to improve the estimates for the ground-state energies. For wavefunction-based methods, one can include more Slater determinants as in a configuration interaction approach. As the wavefunctions improve via the variational theorem, the energy is lowered. In density functional theory, there is no analogous procedure. The Kohn–Sham equations are also variational, but need not approach the true ground-state energy. This is not a problem provided that one is interested in *relative* energies and any inherent density functional errors cancel in the difference.

In some sense, density functional theory is an *a posteriori* theory. Given the transference of the exchange-correlation energies from an electron gas, it is not surprising that errors would arise in its implementation to highly non-uniform electron gas systems as found in realistic systems. However, the degree of error cancellations is rarely known *a priori*. The reliability of density functional theory has only been established by numerous calculations for a wide variety of condensed matter systems. For example, the cohesive energies, compressibility, structural parameters and vibrational spectra of elemental solids have been calculated within the density functional theory [24]. The accuracy of the method is best for systems in which the cancellation of errors is expected to be complete. Since cohesive energies involve the difference in energies between atoms in solids and atoms in free space, error cancellations are expected to be significant. This is reflected in the fact that historically cohesive energies have presented greater challenges for density functional theory: the errors between theory and experiment are typically $\sim$5–10%, depending on the nature of the density functional. In contrast, vibrational frequencies which involve small structural changes within a given crystalline environment are easily reproduced to within 1–2%.

A1.3.4 Electronic states in periodic potentials: Bloch's theorem

Crystalline matter serves as the testing ground for electronic structure methods applied to extended systems. Owing to the translational periodicity of the system, a knowledge of the charge density in part of the crystal is sufficient to understand the charge density throughout the crystal. This greatly simplifies quantum descriptions of condensed matter.

A1.3.4.1   The structure of crystalline matter

A key aspect in defining a crystal is the existence of a building block which, when translated by a precise prescription an infinite number of times, replicates the structure of interest. This building block is call a *unit cell*. The numbers of atoms required to define a unit cell can vary greatly from one solid to another. For simple metals such as sodium only one atom may be needed in defining the unit cell. Complex organic crystals can require thousands of atoms to define the building block.

The unit cell can be defined in terms of three *lattice vectors*: $(\mathbf{a}, \mathbf{b}, \mathbf{c})$. In a periodic system, the point $\mathbf{x}$ is equivalent to any point $\mathbf{x}'$, provided the two points are related as follows:

$$\mathbf{x} = \mathbf{x}' + n_1 \mathbf{a} + n_2 \mathbf{b} + n_3 \mathbf{c} \tag{A1.3.40}$$

where n_1, n_2, n_3 are arbitrary integers. This requirement can be used to define the translation vectors. Equation (A1.3.40) can also be written as

$$\mathbf{x} = \mathbf{x}' + \mathbf{R}_{n_1,n_2,n_3} \tag{A1.3.41}$$

where $\mathbf{R}_{n_1,n_2,n_3} = n_1 \mathbf{a} + n_2 \mathbf{b} + n_3 \mathbf{c}$ is called a *translation* vector. The set of points located by $\mathbf{R}_{n_1,n_2,n_3}$ formed by all possible combinations of (n_1, n_2, n_3) is called a *lattice*.

Knowing the lattice is usually not sufficient to reconstruct the crystal structure. A knowledge of the vectors $(\mathbf{a}, \mathbf{b}, \mathbf{c})$ does not specify the positions of the atoms within the unit cell. The positions of the atoms within the unit cell is given by a set of vectors: $\tau_i, i = 1, 2, 3 \ldots n$ where n is the number of atoms in the unit cell. The set of vectors, τ_i, is called the *basis*. For simple elemental structures, the unit cell may contain only one atom. The lattice sites in this case can be chosen to correspond to the atomic sites, and no basis exists.

The position of the ith atom in a crystal, $\mathbf{r}_i$, is given by

$$\mathbf{r}_i = \tau_j + \sum_{n_1,n_2,n_3} \mathbf{R}_{n_1,n_2,n_3} \tag{A1.3.42}$$

where the index j refers to the jth atom in the cell and the indices (n_1, n_2, n_3) refer to the cell. The construction of the unit cell, i.e. the lattice vectors $\mathbf{R}_{n_1,n_2,n_3}$ and the basis vector τ, is not unique. The choice of unit cell is usually dictated by convenience. The smallest possible unit cell which properly describes a crystal is called the *primitive unit cell*.

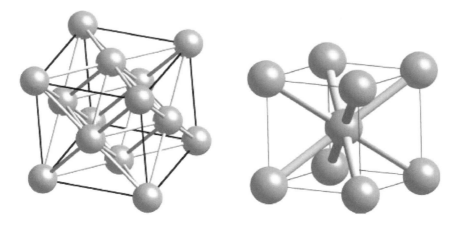

Figure A1.3.2. Structure of a FCC crystal.   **Figure A1.3.3.** Structure of a BCC crystal.

(a) Face-centred cubic (FCC) structure

The FCC structure is illustrated in figure A1.3.2. Metallic elements such as calcium, nickel, and copper form in the FCC structure, as well as some of the inert gases. The conventional unit cell of the FCC structure is cubic with the length of the edge given by the *lattice parameter, a*. There are four atoms in the conventional cell. In the primitive unit cell, there is only one atom. This atom coincides with the lattice points. The lattice vectors for the primitive cell are given by

$$\mathbf{a} = a(\hat{\mathbf{y}} + \hat{\mathbf{z}})/2 \qquad \mathbf{b} = a(\hat{\mathbf{x}} + \hat{\mathbf{z}})/2 \qquad \mathbf{c} = a(\hat{\mathbf{x}} + \hat{\mathbf{y}})/2. \tag{A1.3.43}$$

This structure is called 'close packed' because the number of atoms per unit volume is quite large compared with other simple crystal structures.

(b) Body-centred cubic (BCC) structure

The BCC structure is illustrated in figure A1.3.3. Elements such as sodium, tungsten and iron form in the BCC structure. The conventional unit cell of the BCC structure is cubic, like FCC, with the length of the edge given by the *lattice parameter, a*. There are two atoms in the conventional cell. In the primitive unit cell, there is only one atom and the lattice vectors are given by

$$\mathbf{a} = a(-\hat{\mathbf{x}} + \hat{\mathbf{y}} + \hat{\mathbf{z}})/2 \qquad \mathbf{b} = a(\hat{\mathbf{x}} - \hat{\mathbf{y}} + \hat{\mathbf{z}})/2 \qquad \mathbf{c} = a(\hat{\mathbf{x}} + \hat{\mathbf{y}} - \hat{\mathbf{z}})/2. \tag{A1.3.44}$$

(c) Diamond structure

The diamond structure is illustrated in figure A1.3.4. Elements such as carbon, silicon and germanium form in the diamond structure. The conventional unit cell of the diamond structure is cubic with the length of the edge given by the *lattice parameter, a*. There are eight atoms in the conventional cell. The diamond structure can be constructed by considering two interpenetrating FCC crystals displaced one-fourth of the body diagonal. For the primitive unit cell, the lattice vectors are the same as for the FCC crystal; however, each lattice point has a basis associated with it. The basis can be chosen as

$$\tau_1 = -a(1, 1, 1)/8 \qquad \tau_2 = a(1, 1, 1)/8. \tag{A1.3.45}$$

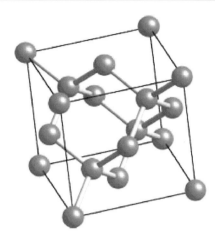 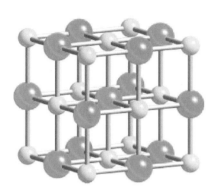

Figure A1.3.4. Structure of a diamond crystal.   **Figure A1.3.5.** Structure of a rocksalt crystal.

(d) Rocksalt structure

The rocksalt structure is illustrated in figure A1.3.5. This structure represents one of the simplest compound structures. Numerous ionic crystals form in the rocksalt structure, such as sodium chloride (NaCl). The conventional unit cell of the rocksalt structure is cubic. There are eight atoms in the conventional cell. For the primitive unit cell, the lattice vectors are the same as FCC. The basis consists of two atoms: one at the origin and one displaced by one-half the body diagonal of the conventional cell.

A1.3.4.2   Bloch's theorem

The periodic nature of crystalline matter can be utilized to construct wavefunctions which reflect the translational symmetry. Wavefunctions so constructed are called *Bloch functions* [1]. These functions greatly simplify the electronic structure problem and are applicable to any periodic system.

For example, consider a simple crystal with one atom per lattice point: the total ionic potential can be written as

$$V_{\text{ion}}^{\text{xtal}}(\mathbf{r}) = \sum_{\mathbf{R},\tau} V_{\text{ion}}^{\text{a}}(\mathbf{r} - \mathbf{R} - \tau). \tag{A1.3.46}$$

This ionic potential is periodic. A translation of $\mathbf{r}$ to $\mathbf{r} + \mathbf{R}$ can be accommodated by simply reordering the summation. Since the valence charge density is also periodic, the total potential is periodic as the Hartree and exchange-correlation potentials are functions of the charge density. In this situation, it can be shown that the wavefunctions for crystalline matter can be written as

$$\phi_{\mathbf{k}}(\mathbf{r}) = \exp(i\mathbf{k} \cdot \mathbf{r}) u_{\mathbf{k}}(\mathbf{r}) \tag{A1.3.47}$$

where $\mathbf{k}$ is a wavevector and $u_{\mathbf{k}}(\mathbf{r})$ is a periodic function, $u_{\mathbf{k}}(\mathbf{r} + \mathbf{R}) = u_{\mathbf{k}}(\mathbf{r})$. This is known as *Bloch's theorem*. In the limit of a free electron, $\mathbf{k}$ can be identified with the momentum of the electron and $u_{\mathbf{k}} = 1$.

The wavevector is a *good* quantum number: e.g., the orbitals of the Kohn–Sham equations [21] can be rigorously labelled by $\mathbf{k}$ and spin. In three dimensions, four quantum numbers are required to characterize an eigenstate. In spherically symmetric atoms, the numbers correspond to n, l, m, s, the principal, angular momentum, azimuthal and spin quantum numbers, respectively. Bloch's theorem states that the equivalent quantum numbers in a crystal are k_x, k_y, k_z and spin. The spin index is usually dropped for non-magnetic materials.

By taking the ϕ_k orbitals to be of the Bloch form, the Kohn–Sham equations can be written as

$$\left(\frac{(\mathbf{p} + \hbar\mathbf{k})^2}{2m} + V_N(\mathbf{r}) + V_H(\mathbf{r}) + V_{xc}[\rho(\mathbf{r})]\right)u_k(\mathbf{r}) = E(\mathbf{k})u_k(\mathbf{r}). \tag{A1.3.48}$$

Knowing the energy distributions of electrons, $E(\mathbf{k})$, and the spatial distribution of electrons, $\rho(\mathbf{r})$, is important in obtaining the structural and electronic properties of condensed matter systems.

A1.3.5   Energy bands for crystalline solids

A1.3.5.1   Kronig–Penney model

One of the first models to describe electronic states in a periodic potential was the *Kronig–Penney* model [1]. This model is commonly used to illustrate the fundamental features of Bloch's theorem and solutions of the Schrödinger equation for a periodic system.

This model considers the solution of wavefunctions for a one-dimensional Schrödinger equation:

$$\left[\frac{-\hbar^2\nabla^2}{2m} + V(x)\right]\psi(x) = E\psi(x). \tag{A1.3.49}$$

This Schrödinger equation has a particularly simple solution for a finite energy well: $V(x) = -V_0$ for $0 < x < a$ (region I) and $V(x) = 0$ elsewhere (region II) as indicated in figure A1.3.6. This is a standard problem in elementary quantum mechanics. For a bound state ($E < 0$) the wavefunctions have solutions in region I: $\psi_I(x) = B \exp(iKx) + C \exp(-iKx)$ and in region II: $\psi_{II}(x) = A \exp(-Q|x|)$. The wavefunctions are required to be continuous: $\psi_I(0) = \psi_{II}(0)$ and $\psi_I(a) = \psi_{II}(a)$ and have continuous first derivatives: $\psi_I'(0) = \psi_{II}'(0)$ and $\psi_I'(a) = \psi_{II}'(a)$. With these conditions imposed at $x = 0$

$$B/C = -(1 + iK/Q)^2/(1 + K^2/Q^2) \tag{A1.3.50}$$

and at $x = a$

$$B/C = -(1 - iK/Q)^2 \exp(-2iKa)/(1 + K^2/Q^2). \tag{A1.3.51}$$

A nontrivial solution will exist only if

$$(1 + iK/Q)^2 = (1 - iK/Q)^2 \exp(-2iKa) \tag{A1.3.52}$$

or

$$Q^2 - 2QK\cot(Ka) - K^2 = 0. \tag{A1.3.53}$$

This results in two solutions:

$$Q = -K\cot(Ka/2) \qquad \text{and} \qquad Q = K\tan(Ka/2). \tag{A1.3.54}$$

If ψ_I and ψ_{II} are inserted into the one-dimensional Schrödinger equation, one finds $E = \hbar^2K^2/2m - V_0$ or $K = \sqrt{2m(E + V_0)/\hbar^2}$ and $E = -\hbar^2Q^2/2m$. In the limit $V_0 \to \infty$, or $K \to \infty$, equation (A1.3.53) can result in a finite value for Q only if $\tan(Ka/2) \to 0$, or $\cot(Ka/2) \to 0$ (i.e. $Ka = n\pi$ where n is an integer). The energy levels in this limit correspond to the standard 'particle in a box' eigenvalues:

$$E_n = \frac{\hbar^2(2n\pi/a)^2}{2m}.$$

In the Kronig–Penney model, a periodic array of such potentials is considered as illustrated in figure A1.3.6. The width of the wells is a and the wells are separated by a distance b. Space can be divided to

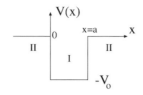

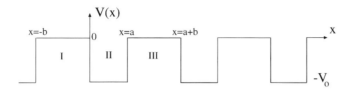

Figure A1.3.6. An isolated square well (top). A periodic array of square wells (bottom). This model is used in the Kronig–Penney description of energy bands in solids.

distinct regions: region I ($-b < x < 0$), region II ($0 < x < a$) and region III ($a < x < a + b$). In region I, the wavefunction can be taken as

$$\psi_I(x) = C \exp(Qx) + D \exp(-Qx). \tag{A1.3.55}$$

In region II, the wavefunction is

$$\psi_{II}(x) = A \exp(iKx) + B \exp(-iKx). \tag{A1.3.56}$$

Unlike an isolated well, there is no restriction on the sign on the exponentials, i.e. both $\exp(+Qx)$ and $\exp(-Qx)$ are allowed. For an isolated well, the sign was restricted so that the exponential vanished as $|x| \to \infty$. Either sign is allowed for the periodic array as the extent of the wavefunction within each region is finite.

Because our system is periodic, one need only consider the wavefunctions in I and II and apply the periodic boundary conditions for other regions of space. Bloch's theorem can be used in this case: $\psi(x+a) = \exp(ika)\psi(x)$ or $\psi(x + (a + b)) = \exp(ik(a + b))\psi(x)$. This relates ψ_{III} and ψ_I:

$$\psi_{III}(x) = \exp(ik(a + b))\psi_I(x) \tag{A1.3.57}$$

or

$$\psi_{III}(x) = \exp(ik(a + b))(C \exp(Q(x - a - b)) + D \exp(-Q(x - a - b))). \tag{A1.3.58}$$

k now serves to label the states in the same sense n serves to label states for a square well.

As in the case of the isolated well, one can impose continuity of the wavefunctions and the first derivatives of the wavefunctions at $x = 0$ and $x = a$. At $x = 0$,

$$A + B = C + D \qquad iK(A - B) = Q(C - D) \tag{A1.3.59}$$

and at $x = a$

$$A \exp(iKa) + B \exp(-iKa) = \exp(ik(a + b))(C \exp(-Qb) + D \exp(Qb)) \tag{A1.3.60}$$

$$iKa(A \exp(iKa) - B \exp(-iKa)) = Q \exp(ik(a + b))(C \exp(-Qb) + D \exp(Qb)). \tag{A1.3.61}$$

This results in four equations and four unknowns. Since the equations are homogeneous, a nontrivial solution exists only if the determinant formed by the coefficients of A, B, C and D vanishes. The solution to this equation is

$$\frac{(Q^2 - K^2)}{2QK} \sinh(Qb) \sin(Ka) + \cosh(Qb) \cos(Ka) = \cos(k(a + b)). \tag{A1.3.62}$$

Equation (A1.3.62) provides a relationship between the wavevector, k, and the energy, E, which is implicit in Q and K.

Before this result is explored in more detail, consider the limit where $b \rightarrow \infty$. In this limit, the wells become isolated and k has no meaning. As $b \rightarrow \infty$, $\sinh(Qb) \rightarrow \exp(Qb)/2$ and $\cosh(Qb) \rightarrow \exp(Qb)/2$. One can rewrite equation (A1.3.62) as

$$(\exp(Qb)/2)\left(\frac{(Q^2 - K^2)}{2QK} \sin(Ka) + \cos(Ka)\right) = \cos(k(a + b)). \tag{A1.3.63}$$

As $\exp(Qb)/2 \rightarrow \infty$, this equation can be valid if

$$\frac{(Q^2 - K^2)}{2QK} \sin(Ka) + \cos(Ka) \rightarrow 0 \tag{A1.3.64}$$

otherwise the rhs of equation (A1.3.63) would diverge. In this limit, equation (A1.3.64) reduces to the isolated well solution (equation (A1.3.53)):

$$Q^2 - 2QK \cot(Ka) - K^2 = 0. \tag{A1.3.65}$$

Since k does not appear in equation (A1.3.65) in this limit, it is undefined.

One can illustrate how the energy states evolve from discrete levels in an isolated well to states appropriate for periodic systems by varying the separation between wells. In figure A1.3.7, solutions for E versus k are shown for isolated wells and for strongly interacting wells. It is important to note that k is not defined except within a factor of $2\pi m/(a + b)$ where m is an integer as $\cos((k + 2\pi m/(a + b))(a + b)) = \cos(k(a + b))$. The E versus k plot need be displayed only for k between 0 and $\pi/(a + b)$ as larger values of k can be mapped into this interval by subtracting off values of $2\pi/(a + b)$.

In the case where the wells are far apart, the resulting energy levels are close to the isolated well. However, an interesting phenomenon occurs as the atoms are brought closer together. The energy levels cease being constant as a function of the wavevector, k. There are regions of allowed solutions and regions where no energy state occurs. The region of allowed energy values is called an *energy band*. The range of energies within the band is called the *band width*. As the width of the band increases, it is said that the band has greater *dispersion*.

The Kronig–Penney solution illustrates that, for *periodic* systems, gaps can exist between bands of energy states. As for the case of a free electron gas, each band can hold $2N$ electrons where N is the number of wells present. In one dimension, this implies that if a well contains an *odd* number, one will have *partially occupied* bands. If one has an *even* number of electrons per well, one will have *fully occupied energy* bands. This distinction between odd and even numbers of electrons per cell is of fundamental importance. The Kronig–Penney model implies that crystals with an odd number of electrons per unit cell are *always* metallic whereas an even number of electrons per unit cell implies an insulating state. This simple rule is valid for more realistic potentials and need be only slightly modified in three dimensions. In three dimensions, an even number of electrons per unit cells is a necessary condition for an insulating state, but not a sufficient condition.

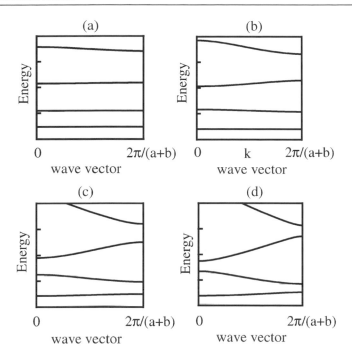

Figure A1.3.7. Evolution of energy bands in the Kronig–Penney model as the separation between wells, b (figure A1.3.6) is decreased from (a) to (d). In (a) the wells are separated by a large distance (large value of b) and the energy bands resemble discrete levels of an isolated well. In (d) the wells are quite close together (small value of b) and the energy bands are free-electron-like.

One of the major successes of *energy band* theory is that it can be used to predict whether a crystal exists as a metal or insulator. If a band is filled, the Pauli principle prevents electrons from changing their momentum in response to the electric field as all possible momentum states are occupied. In a metal this constraint is not present as an electron can change its momentum state by moving from a filled to an occupied state within a given band. The distinct types of energy bands for insulators, metals, semiconductors and semimetals are schematically illustrated in figure A1.3.8. In an insulator, energy bands are either completely empty or completely filled. The band gap between the highest occupied band and lowest empty band is large, e.g. above 5 eV. In a semiconductor, the bands are also completely filled or empty, but the gap is smaller, e.g. below 3 eV. In metals bands are not completely occupied and no gap between filled and empty states occurs. Semimetals are a special case. No gap exists, but one band is almost completely occupied; it overlaps with a band that is almost completely empty.

A1.3.5.2 Reciprocal space

Expressing $E(\mathbf{k})$ is complicated by the fact that $\mathbf{k}$ is not unique. In the Kronig–Penney model, if one replaced k by $k + 2\pi/(a + b)$, the energy remained unchanged. In three dimensions $\mathbf{k}$ is known only to within a *reciprocal lattice vector*, $\mathbf{G}$. One can define a set of reciprocal vectors, given by

$$\mathbf{G} = m_1\mathbf{A} + m_2\mathbf{B} + m_3\mathbf{C} \tag{A1.3.66}$$

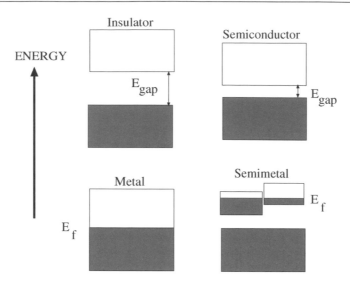

Figure A1.3.8. Schematic energy bands illustrating an insulator (large band gap), a semiconductor (small band gap), a metal (no gap) and a semimetal. In a semimetal, one band is almost filled and another band is almost empty.

where the set $(\mathbf{A}, \mathbf{B}, \mathbf{C})$ define a lattice in reciprocal space. These vectors can be defined by

$$\mathbf{A} = \frac{2\pi}{\Omega}\mathbf{b} \times \mathbf{c} \qquad \mathbf{B} = \frac{2\pi}{\Omega}\mathbf{c} \times \mathbf{b} \qquad \mathbf{C} = \frac{2\pi}{\Omega}\mathbf{a} \times \mathbf{b} \qquad (A1.3.67)$$

where Ω is defined as the unit cell volume. Note that $\Omega = |\mathbf{a} \cdot \mathbf{b} \times \mathbf{c}|$ from elementary vector analysis. It is easy to show that

$$\begin{aligned}
\mathbf{A} \cdot \mathbf{a} &= 2\pi & \mathbf{A} \cdot \mathbf{b} &= 0 & \mathbf{A} \cdot \mathbf{c} &= 0 \\
\mathbf{B} \cdot \mathbf{a} &= 0 & \mathbf{B} \cdot \mathbf{b} &= 2\pi & \mathbf{B} \cdot \mathbf{c} &= 0 \\
\mathbf{C} \cdot \mathbf{a} &= 0 & \mathbf{C} \cdot \mathbf{b} &= 0 & \mathbf{C} \cdot \mathbf{c} &= 2\pi.
\end{aligned} \qquad (A1.3.68)$$

It is apparent that

$$\mathbf{G} \cdot \mathbf{R} = 2\pi(n_1 m_1 + n_2 m_2 + n_3 m_3). \qquad (A1.3.69)$$

Reciprocal lattice vectors are useful in defining periodic functions. For example, the valence charge density, $\rho(\mathbf{r})$, can be expressed as

$$\rho(\mathbf{r}) = \sum_{\mathbf{G}} \rho(\mathbf{G}) \exp(i\mathbf{G} \cdot \mathbf{r}). \qquad (A1.3.70)$$

It is clear that $\rho(\mathbf{r} + \mathbf{R}) = \rho(\mathbf{r})$ from equation (A1.3.69). The Fourier coefficients, $\rho(\mathbf{G})$, can be determined from

$$\rho(\mathbf{G}) = \frac{1}{\Omega} \int \rho(\mathbf{r}) \exp(-i\mathbf{G} \cdot \mathbf{r}) \, \mathrm{d}^3 r. \qquad (A1.3.71)$$

Because $E(\mathbf{k}) = E(\mathbf{k} + \mathbf{G})$, a knowledge of $E(\mathbf{k})$ within a given volume called the *Brillouin zone* is sufficient to determine $E(\mathbf{k})$ for all $\mathbf{k}$. In one dimension, $G = 2\pi n/d$ where d is the lattice spacing between atoms. In this case, $E(k)$ is known once k is determined for $-\pi/d < k < \pi/d$. (For example, in the Kronig–Penney model (figure A1.3.6), $d = a + b$ and k was defined only to within a vector $2\pi/(a+b)$.) In three dimensions, this subspace can result in complex polyhedrons for the Brillouin zone.

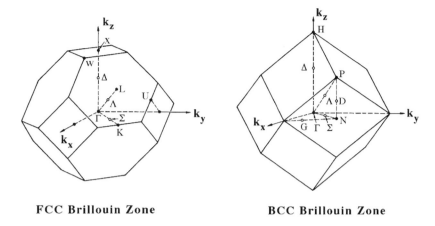

FCC Brillouin Zone   BCC Brillouin Zone

Figure A1.3.9. Brillouin zones for the FCC and BCC crystal structures.

In figure A1.3.9 the Brillouin zone for a FCC and a BCC crystal are illustrated. It is a common practice to label high-symmetry point and directions by letters or symbols. For example, the $\mathbf{k} = 0$ point is called the Γ point. For cubic crystals, there exist 48 symmetry operations and this symmetry is maintained in the energy bands: e.g., $E(k_x, k_y, k_z)$ is invariant under sign permutations of (x, y, z). As such, one need only have knowledge of $E(\mathbf{k})$ in $\frac{1}{48}$ of the zone to determine the energy band throughout the zone. The part of the zone which cannot be reduced by symmetry is called the *irreducible* Brillouin zone.

A1.3.5.3 *Realistic energy bands*

Since the electronic structure of a solid can be determined from a knowledge of the spatial and energetic distribution of electrons (i.e. from the charge density, $\rho(\mathbf{r})$, and the electronic density of states, $D(E)$), it is highly desirable to have the ability to determine the quantum states of crystal. The first successful electronic structure calculations for energy bands of crystalline matter were not performed from 'first principles'. Although elements of density functional theory were understood by the mid-1960s, it was not clear how reliable these methods were. Often, two seemingly identical calculations would yield very different results for simple issues such as whether a solid was a metal or an insulator. Consequently, some of the first reliable energy bands were constructed using *empirical pseudopotentials* [25]. These potentials were extracted from experimental data and not determined from first principles.

A1.3.5.4 *Empirical pseudopotentials*

The first reliable energy band theories were based on a powerful approximation, call the *pseudopotential approximation*. Within this approximation, the *all-electron potential* corresponding to interaction of a valence electron with the inner, core electrons and the nucleus is replaced by a pseudopotential. The pseudopotential reproduces only the properties of the outer electrons. There are rigorous theorems such as the *Phillips–Kleinman* cancellation theorem that can be used to justify the pseudopotential model [2, 3, 26]. The Phillips–Kleinman cancellation theorem states that the *orthogonality* requirement of the valence states to the core states can be described by an effective repulsive potential. This repulsive potential cancels the strong Coulombic potential within the core region. The cancellation theorem explains, in part, why valence electrons feel a less attractive potential than would be expected on the basis of the Coulombic part of the potential. For example, in alkali metals an 'empty' core pseudopotential approximation is often made. In this model pseudopotential, the valence electrons experience no Coulomb potential within the core region.

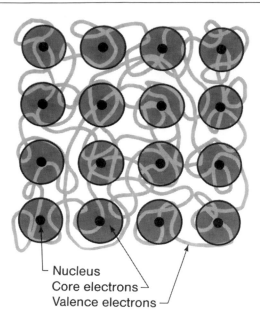

Figure A1.3.10. Pseudopotential model. The outer electrons (valence electrons) move in a fixed arrangement of chemically inert ion cores. The ion cores are composed of the nucleus and core electrons.

Since the pseudopotential does not bind the core states, it is a very weak potential. Simple basis functions can be used to describe the pseudo-wavefunctions. For example, a simple grid or plane wave basis will yield a converged solution [25]. The simplicity of the basis is important as it results in an unbiased, flexible description of the charge density. Also, since the nodal structure of the pseudo-wavefunctions has been removed, the charge density varies slowly in the core region. A schematic model of the pseudopotential model is illustrated in figure A1.3.10. The pseudopotential model describes a solid as a sea of valence electrons moving in a periodic background of cores (composed of nuclei and inert core electrons). In this model many of the complexities of all-electron calculations, calculations that include the core and valence electrons on an equal footing, are avoided. A group IV solid such as C with 6 electrons per atom is treated in a similar fashion to Sn with 50 electrons per atom since both have 4 valence electrons per atom. In addition, the focus of the calculation is only on the accuracy of the valence electron wavefunction in the spatial region away from the chemically inert core.

One can quantify the pseudopotential by writing the total crystalline potential for an elemental solid as

$$V_{\mathrm{p}}(\mathbf{r}) = \sum_{\mathbf{G}} S(\mathbf{G}) V_{\mathrm{p}}^{\mathrm{a}}(G) \exp(\mathrm{i}\mathbf{G} \cdot \mathbf{r}). \tag{A1.3.72}$$

$S(\mathbf{G})$ is the *structure factor* given by

$$S(\mathbf{G}) = \frac{1}{N_{\mathrm{a}}} \sum_{\tau} \exp(\mathrm{i}\mathbf{G} \cdot \boldsymbol{\tau}) \tag{A1.3.73}$$

where N_{a} is the number of atoms in the unit call and τ is a basis vector. $V_{\mathrm{p}}^{\mathrm{a}}(G)$ is the *form factor* given by

$$V_{\mathrm{p}}^{\mathrm{a}}(G) = \frac{1}{\Omega_{\mathrm{a}}} \int V_{\mathrm{p}}^{\mathrm{a}}(r) \exp(\mathrm{i}\mathbf{G} \cdot \mathbf{r}) \, \mathrm{d}^3 r \tag{A1.3.74}$$

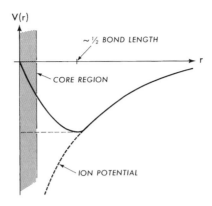

Figure A1.3.11. Schematic pseudopotential in real space.

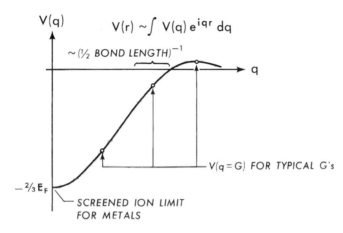

Figure A1.3.12. Schematic pseudopotential in reciprocal space.

where Ω_a is the volume per atom and $V_p^a(r)$ is a pseudopotential associated with an atom. Often this potential is assumed to be spherically symmetry. In this case, the form factor depends only on the magnitude of **G**: $V_p^a(\mathbf{G}) = V_p^a(|\mathbf{G}|)$. A schematic pseudopotential is illustrated in figure A1.3.11. Outside the core region the pseudopotential is commensurate with the all-electron potential. When this potential is transformed into Fourier space, it is often sufficient to keep just a few unique form factors to characterize the potential. These form factors are then treated as adjustable parameters which can be fitted to experimental data. This is illustrated in figure A1.3.12.

The empirical pseudopotential method can be illustrated by considering a specific semiconductor such as silicon. The crystal structure of Si is diamond. The structure is shown in figure A1.3.4. The lattice vectors and basis for a primitive cell have been defined in the section on crystal structures (A1.3.4.1). In Cartesian coordinates, one can write **G** for the diamond structure as

$$\mathbf{G} = \frac{2\pi}{a}(n, l, m) \tag{A1.3.75}$$

where the indices (n, l, m) must be either all odd or all even: e.g., $\mathbf{G} = \frac{2\pi}{a}(1, 0, 0)$ is not allowed, but $\mathbf{G} = \frac{2\pi}{a}(2, 0, 0)$ is permitted. It is convenient to organize **G**-vectors by their magnitude squared in units of $(2\pi/a)^2$. In this scheme: $G^2 = 0, 3, 4, 8, 11, 12, \ldots$. The structure factor for the diamond structure is

$S(\mathbf{G}) = \cos(\mathbf{G} \cdot \boldsymbol{\tau})$. For some values of G, this structure factor vanishes: e.g., if $\mathbf{G} = (2\pi/a)(2,0,0)$, then $\mathbf{G} \cdot \boldsymbol{\tau} = \pi/2$ and $S(\mathbf{G}) = 0$. If the structure factor vanishes, the corresponding form factor is irrelevant as it is multiplied by a zero structure factor. In the case of diamond structure, this eliminates the $G^2 = 4, 12$ form factors. Also, the $G^2 = 0$ factor is not important for spectroscopy as it corresponds to the average potential and serves to shift the energy bands by a constant. The rapid convergence of the pseudopotential in Fourier space coupled with the vanishing of the structure factor for certain $\mathbf{G}$ means that only three form factors are required to fix the energy bands for diamond semiconductors like Si and Ge: $V_p^a(G^2 = 3)$, $V_p^a(G^2 = 8)$ and $V_p^a(G^2 = 11)$. These form factors can be fixed by comparisons to reflectivity measurements or photoemission [25].

A1.3.5.5   Density functional pseudopotentials

Another realistic approach is to construct pseudopotentials using density functional theory. The implementation of the Kohn–Sham equations to condensed matter phases without the pseudopotential approximation is not easy owing to the dramatic span in length scales of the wavefunction and the energy range of the eigenvalues. The pseudopotential eliminates this problem by removing the core electrons from the problem and results in a much simpler problem [27].

In the pseudopotential construction, the atomic wavefunctions for the valence electrons are taken to be nodeless. The pseudo-wavefunction is taken to be identical to the appropriate all-electron wavefunction in the regions of interest for solid-state effects. For the core region, the wavefunction is extrapolated back to the origin in a manner consistent with the normalization condition. This type of construction was first introduced by Fermi to account for the shift in the wavefunctions of high-lying states of alkali atoms subject to perturbations from foreign atoms. In this remarkable paper, Fermi introduced the conceptual basis for both the pseudopotential and the scattering length [28].

With the density functional theory, the first step in the construction of a pseudopotential is to consider the solution for an isolated atom [27]. If the atomic wavefunctions are known, the pseudo-wavefunction can be constructed by removing the nodal structure of the wavefunction. For example, if one considers a valence wavefunction for the isolated atom, $\psi_v(r)$, then a pseudo-wavefunction, $\phi_p(r)$, might have the properties

$$\phi_p(r) = r^l \exp(-\alpha r^4 - \beta r^3 - \gamma r^2 - \delta) \qquad r < r_c$$
$$= \psi_v(r) \qquad r > r_c. \tag{A1.3.76}$$

The pseudo-wavefunction within this frame work is guaranteed to be nodeless. The parameters $(\alpha, \beta, \gamma, \delta)$ are fixed so that (1) ϕ_v and ϕ_p have the same eigenvalue, $\mathcal{E}_v$, and the same norm:

$$\int_0^{r_c} |\psi_v(r)|^2 r^2 \, dr = \int_0^{r_c} |\phi_p(r)|^2 r^2 \, dr. \tag{A1.3.77}$$

This ensures that $\phi_p(r) = \psi_v(r)$ for $r > r_c$ after the wavefunctions have been normalized. (2) The pseudo-wavefunction should be continuous and have continuous first and second derivatives at r_c. An example of a pseudo-wavefunction is given in figure A1.3.13.

Once the eigenvalue and pseudo-wavefunction are known for the atom, the Kohn–Sham equation can be inverted to yield the ionic pseudopotential:

$$V_{ion}^p(r) = -\mathcal{E}_v - V_H(r) - V_{xc}(r) + \frac{\hbar^2 \nabla^2 \phi_p}{2m\phi_p}. \tag{A1.3.78}$$

Since V_H and V_{xc} depend only on the valence charge densities, they can be determined once the valence pseudo-wavefunctions are known. Because the pseudo-wavefunctions are nodeless, the resulting pseudopotential is well defined despite the last term in equation (A1.3.78). Once the pseudopotential has been constructed

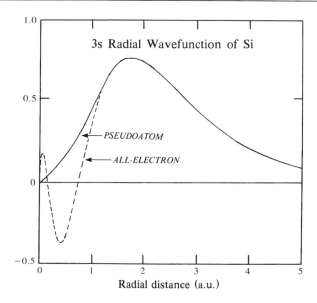

Figure A1.3.13. All-electron and pseudopotential wavefunction for the 3s state in silicon. The all-electron 3s state has nodes which arise because of an orthogonality requirement to the 1s and 2s core states.

from the atom, it can be transferred to the condensed matter system of interest. For example, the ionic pseudopotential defined by equation (A1.3.78) from an *atomistic* calculation can be transferred to *condensed matter phases* without any significant loss of accuracy.

There are complicating issues in defining pseudopotentials, e.g. the pseudopotential in equation (A1.3.78) is state dependent, orbitally dependent and the energy and spatial separations between valence and core electrons are sometimes not transparent. These are not insurmountable issues. The state dependence is usually weak and can be ignored. The orbital dependence requires different potentials for different angular momentum components. This can be incorporated via *non-local* operators. The distinction between valence and core states can be addressed by incorporating the core level in question as part of the valence shell. For example, in Zn one can treat the $3d^{10}$ shell as a valence shell. In this case, the valency of Zn is 12, not 2. There are also very reliable approximate methods for treating the outer core states without explicitly incorporating them in the valence shell.

A1.3.5.6   Other approaches

There are a variety of other approaches to understanding the electronic structure of crystals. Most of them rely on a density functional approach, with or without the pseudopotential, and use different bases. For example, instead of a plane wave basis, one might write a basis composed of atomic-like orbitals:

$$\psi_{\mathbf{k}}(\mathbf{r}) = \sum_{i,\mathbf{R}} \alpha_i(\mathbf{k}) \exp(i\mathbf{k} \cdot \mathbf{R}) \phi_i(\mathbf{r} - \mathbf{R}) \qquad (A1.3.79)$$

where the $\exp(i\mathbf{k} \cdot \mathbf{R})$ is explicitly written to illustrate the Bloch form of this wavefunction: i.e. $\psi_{\mathbf{k}}(\mathbf{r} + \mathbf{R}) = \exp(i\mathbf{k} \cdot \mathbf{R})\psi_{\mathbf{k}}(\mathbf{r})$. The orbitals ϕ_i can be taken from atomic structure solutions where i is a general index such as $lmns$, or ϕ_i can be taken to be a some localized function such as an exponential, called a Slater-type orbital, or a Gaussian orbital. Provided the basis functions are appropriately chosen, this approach works quite well for a wide variety of solids. This approach is called the *tight binding method* [2, 7].

An approach closely related to the pseudopotential is the *orthogonalized plane wave* method [29]. In this method, the basis is taken to be as follows:

$$\phi_{\mathbf{k}}^{\text{OPW}}(\mathbf{r}) = \exp(i\mathbf{k} \cdot \mathbf{r}) - \sum_i \beta_i \chi_{i,\mathbf{k}}(\mathbf{r}) \tag{A1.3.80}$$

and

$$\chi_{i,\mathbf{k}}(\mathbf{r}) = \sum_{\mathbf{R}} \exp(i\mathbf{k} \cdot \mathbf{R}) a_i(\mathbf{r} - \mathbf{R}) \tag{A1.3.81}$$

where $\chi_{i,\mathbf{k}}$ is a tight binding wavefunction composed of atomic core functions, a_i. As an example, one would take (a_{1s}, a_{2s}, a_{2p}) atomic orbitals for the core states of silicon. The form for $\phi_{\mathbf{k}}(\mathbf{r})$ is motivated by several factors. In the interstitial regions of a crystal, the potential should be weak and slowly varying. The wavefunction should look like a plane wave in this region. Near the nucleus, the wavefunction should look atomic-like. The basis reflects these different regimes by combining plane waves with atomic orbitals. Another important attribute of the wavefunction is an *orthogonality* condition. This condition arises from the form of the Schrödinger equation; higher-energy eigenvalues must have wavefunctions which are orthogonal to more tightly bound states of the same symmetry: e.g., the 2s wavefunction of an atom must be orthogonal to the 1s state. It is possible to choose β_i so that

$$\int \phi_{\mathbf{k}}^*(\mathbf{r}) \chi_{i,\mathbf{k}}(\mathbf{r}) \, d^3r = 0. \tag{A1.3.82}$$

The orthogonality condition assures one that the lowest energy state will not converge to core-like states, but valence states. The wavefunction for the solid can be written as

$$\psi_{\mathbf{k}}(\mathbf{r}) = \sum_{\mathbf{G}} \alpha(\mathbf{k}, \mathbf{G}) \phi_{\mathbf{k}}^{\text{OPW}}(\mathbf{r}). \tag{A1.3.83}$$

As with any basis method the $\alpha(\mathbf{k}, \mathbf{G})$ coefficients are determined by solving a secular equation.

Other methods for determining the energy band structure include cellular methods, Green function approaches and augmented plane waves [2, 3]. The choice of which method to use is often dictated by the particular system of interest. Details in applying these methods to condensed matter phases can be found elsewhere (see section B3.2).

A1.3.6 Examples for the electronic structure and energy bands of crystals

Many phenomena in solid-state physics can be understood by resort to energy band calculations. Conductivity trends, photoemission spectra, and optical properties can all be understood by examining the quantum states or energy bands of solids. In addition, electronic structure methods can be used to extract a wide variety of properties such as structural energies, mechanical properties and thermodynamic properties.

A1.3.6.1 *Semiconductors*

A prototypical semiconducting crystal is silicon. Historically, silicon has been the testing ground for quantum theories of condensed matter. This is not surprising given the importance of silicon for technological applications. The energy bands for Si are shown in figure A1.3.14. Each band can hold two electrons per unit cell. There are four electrons per silicon atom and two atoms in the unit cell. This would lead to four filled bands. It is customary to show the filled bands and the lowest few empty bands. In the case of silicon the bands are separated by a gap of approximately 1 eV. Semiconductors have *band gaps* that are less than a few electron-volts. Displaying the energy bands is not a routine matter as $E(\mathbf{k})$ is often a complex function. The bands

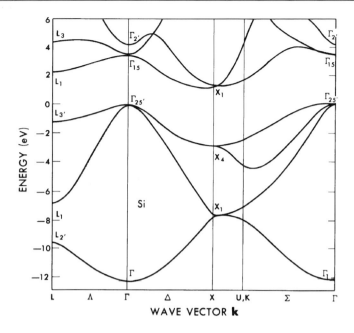

Figure A1.3.14. Band structure for silicon as calculated from empirical pseudopotentials [25].

are typically displayed only along high-symmetry directions in the Brillouin zone (see figure A1.3.9). For example, one might plot the energy bands along the (100) direction (the Δ direction).

The occupied bands are called *valence* bands; the empty bands are called *conduction* bands. The top of the valence band is usually taken as energy zero. The lowest conduction band has a minimum along the Δ direction; the highest occupied valence band has a maximum at Γ. Semiconductors which have the highest occupied k_v-state and lowest empty state k_c at different points are called *indirect* gap semiconductors. If $k_v = k_c$, the semiconductor is call *direct* gap semiconductor. Germanium is also an indirect gap semiconductor whereas GaAs has a direct gap. It is not easy to predict whether a given semiconductor will have a direct gap or not.

Electronic and optical excitations usually occur between the upper valence bands and lowest conduction band. In optical excitations, electrons are transferred from the valence band to the conduction band. This process leaves an empty state in the valence band. These empty states are called *holes*. Conservation of wavevectors must be obeyed in these transitions: $k_{photon} + k_v = k_c$ where k_{photon} is the wavevector of the photon, k_v is the wavevector of the electron in the initial valence band state and k_c is the wavevector of the electron in the final conduction band state. For optical excitations, $k_{photon} \approx 0$. This implies that the excitation must be *direct*: $k_v \approx k_c$. Because of this conservation rule, direct optical excitations are stronger than indirect excitations.

Semiconductors are poor conductors of electricity at low temperatures. Since the valence band is completely occupied, an applied electric field cannot change the total momentum of the valence electrons. This is a reflection of the Pauli principle. This would not be true for an electron that is excited into the conduction band. However, for a band gap of 1 eV or more, few electrons can be thermally excited into the conduction band at ambient temperatures. Conversely, the electronic properties of semiconductors at ambient temperatures can be profoundly altered by the addition of impurities. In silicon, each atom has four covalent bonds, one to each neighbouring atom. All the valence electrons are consumed in saturating these bonds. If a silicon atom is removed and replaced by an atom with a different number of valence electrons, there will be a mismatch between the number of electrons and the number of covalent bonds. For example, if one replaces a silicon atom by a phosphorous atom, then there will an extra electron that cannot be accommodated as phosphorous

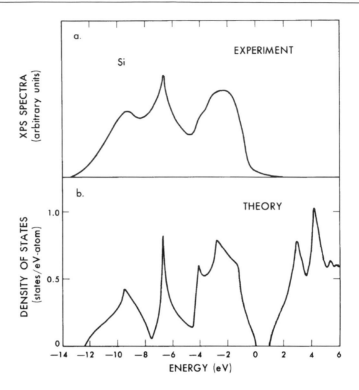

Figure A1.3.15. Density of states for silicon (bottom panel) as calculated from empirical pseudopotential [25]. The top panel represents the photoemission spectra as measured by x-ray photoemission spectroscopy [30]. The density of states is a measure of the photoemission spectra.

possesses five instead of four valence electrons. This extra electron is only loosely bound to the phosphorous atom and can be easily excited into the conduction band. Impurities with an 'extra' electron are called *donors*. Under the influence of an electric field, this donor electron can contribute to the electrical conductivity of silicon. If one were to replace a silicon atom by a boron atom, the opposite situation would occur. Boron has only three valence electrons and does not possess a sufficient number of electrons to saturate the bonds. In this case, an electron in the valence band can readily move into the unsaturated bond. Under the influence of an electric field, this unsaturated bond can propagate and contribute to the electrical conductivity as if it were a positively charged particle. The unsaturated bond corresponds to a *hole* excitation. Impurity atoms that have less than the number of valence electrons to saturate all the covalent bonds are called *acceptors*.

Several factors determine how efficient impurity atoms will be in altering the electronic properties of a semiconductor. For example, the size of the band gap, the shape of the energy bands near the gap and the ability of the valence electrons to screen the impurity atom are all important. The process of adding controlled impurity atoms to semiconductors is called *doping*. The ability to produce well defined doping levels in semiconductors is one reason for the revolutionary developments in the construction of solid-state electronic devices.

Another useful quantity is defining the electronic structure of a solid is the electronic *density of states*. In general the density of states can be defined as

$$D(E) = \frac{\Omega}{(2\pi)^3} \sum_n \int_{BZ} \delta(E - E_n(\mathbf{k})) \, d^3k. \qquad (A1.3.84)$$

Unlike the density of states defined in equation (A1.3.24), which was specific for the free electron gas, equation (A1.3.84) is a general expression. The sum in equation (A1.3.84) is over all energy bands and the integral is over all **k**-points in the Brillouin zone. The density of states is an extensive function that scales with the size of the sample. It is usually normalized to the number of electronic states per atom. In the case of silicon, the number of states contained by integrating $D(E)$ up to the highest occupied states is four states per atom. Since each state can hold two electrons with different spin coordinates, eight electrons can be accommodated within the valence bands. This corresponds to the number of electrons within the unit cell with the resulting valence bands being fully occupied.

The density of states for crystalline silicon is shown in figure A1.3.15. The density of states is a more general representation of the energetic distribution of electrons than the energy band structure. The distribution of states can be given without regard to the **k** wavevector. It is possible to compare the density of states from the energy band structure directly to experimental probes such as those obtained in photoemission. *Photoemission measurements* can be used to measure the distribution of binding electrons within a solid. In these measurements, a photon with a well defined energy impinges on the sample. If the photon carries sufficient energy, an electron can be excited from the valence state to a free electron state. By knowing the energy of the absorbed photon and the emitted electron, it is possible to determine the energy of the electron in the valence state. The number of electrons emitted is proportional to the number of electrons in the initial valence states; the density of states gives a measure of the number of photoemitted electrons for a given binding energy. In realistic calculations of the photoemission spectra, the probability of making a transition from the valence band to the vacuum must be included, but often the transition probabilities are similar over the entire valence band. This is illustrated in figure A1.3.15. Empty states cannot be measured using photoemission so these contributions are not observed.

By examining the spatial character of the wavefunctions, it is possible to attribute atomic character-istics to the density of states spectrum. For example, the lowest states, 8 to 12 eV below the top of the valence band, are s-like and arise from the atomic 3s states. From 4 to 6 eV below the top of the valence band are states that are also s-like, but change character very rapidly toward the valence band maximum. The states residing within 4 eV of the top of the valence band are p and arise from the 3p states.

A major achievement of the quantum theory of matter has been to explain the interaction of light and matter. For example, the first application of quantum theory, the Bohr model of the atom, accurately predicted the electronic excitations in the hydrogen atom. In atomic systems, the absorption and emission of light is characterized by sharp lines. Predicting the exact frequencies for atomic absorption and emission lines provides a great challenge and testing ground for any theory. This is in apparent contrast to the spectra of solids. The continuum of states in solids, i.e. energy bands, allows many possible transitions. A photon with energy well above the band gap can excite a number of different states corresponding to different bands and **k**-points. The resulting spectra correspond to broad excitation spectra without the sharp structures present in atomic transitions. This is illustrated in figure A1.3.16. The spectrum consists of three broad peaks with the central peak at about 4.5 eV.

The interpretation of solid-state spectra as featureless and lacking the information content of atomic spectra is misleading. If one modulates the reflectivity spectra of solids, the spectra are quite rich in struc-ture. This is especially the case at low temperatures where vibrational motions of the atoms are reduced. In figure A1.3.17 the spectra of silicon is differentiated. The process of measuring a differentiated spectra is called *modulation spectroscopy*. In modulated reflectivity spectra, broad undulating features are sup-pressed and sharp features are enhanced. It is possible to modulate the reflectivity spectrum in a variety of ways. For example, one can mechanically vibrate the crystal, apply an alternating electric field or modu-late the temperature of the sample. One of the most popular methods is to measure the reflectivity directly and then numerically differentiate the reflectivity data. This procedure has the advantage of being easily interpreted [25].

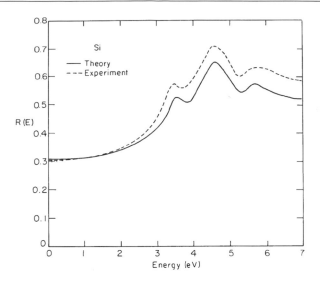

Figure A1.3.16. Reflectivity of silicon. The theoretical curve is from an empirical pseudopotential method calculation [25]. The experimental curve is from [31].

The structure in the reflectivity can be understood in terms of band structure features: i.e. from the quantum states of the crystal. The normal incident reflectivity from matter is given by

$$R = \left(\frac{I}{I_0}\right)^2 = \left|\frac{N-1}{N+1}\right|^2 \qquad (A1.3.85)$$

where I_0 is the incident intensity of the light and I is the reflected intensity. N is the complex index of refraction. The complex index of refraction, N, can be related to the dielectric function of matter by

$$N^2 = \epsilon_1 + i\epsilon_2 \qquad (A1.3.86)$$

where ϵ_1 is the real part of the dielectric function and ϵ_2 is the imaginary part of the dielectric function.

It is possible to make a connection between the quantum states of a solid and the resulting optical properties of a solid. In contrast to metals, most studies have concentrated on insulators and semiconductors where the optical structure readily lends itself to a straightforward interpretation. Within certain approximations, the imaginary part of the dielectric function for semiconducting or insulating crystals is given by

$$\epsilon_2(\omega) = \frac{4\pi e^2 \hbar}{3m^2\omega^2} \sum_{vc} \frac{2}{(2\pi)^3} \int_{BZ} \delta(\omega_{vc}(\mathbf{k}) - \omega)|M_{vc}(\mathbf{k})|^2 \, d^3k. \qquad (A1.3.87)$$

The matrix elements are given by

$$M_{vc}(\mathbf{k}) = \int u_{\mathbf{k},v}^*(\mathbf{r}) \nabla u_{\mathbf{k},c}(\mathbf{r}) \, d^3r \qquad (A1.3.88)$$

where $u_{\mathbf{k},c}$ is the periodic part of the Bloch wavefunction. The summation in equation (A1.3.87) is over all occupied to empty state transitions from valence (v) to conduction bands (c). The energy difference between occupied and empty states is given by $E_c(\mathbf{k}) - E_v(\mathbf{k})$ which can be defined as a frequency; $\omega_{vc}(\mathbf{k}) = (E_c(\mathbf{k}) - E_v(\mathbf{k}))/\hbar$. The delta function term, $\delta(\omega_{vc}(\mathbf{k}) - \omega)$, ensures conservation of energy. The matrix

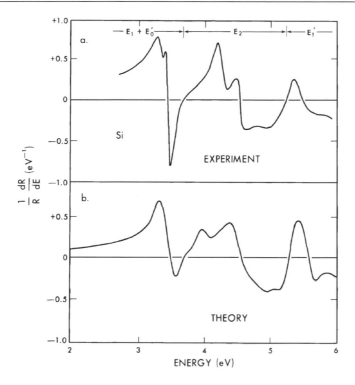

Figure A1.3.17. Modulated reflectivity spectrum of silicon. The theoretical curve is obtained from an empirical pseudopotential calculation [25]. The experimental curve is from a wavelength modulation experiment from [32].

elements, M_{vc}, control the oscillator strength. As an example, suppose that the v → c transition couples states which have similar parity. The matrix elements will be small because the momentum operator is odd. Although angular momentum is not a good quantum number in condensed matter phases, atomic selection rules remain approximately true.

This expression for ϵ_2 neglects the spatial variation of the perturbing electric field. The wavelength of light for optical excitations is between 4000–7000 Å and greatly exceeds a typical bond length of 1–2 Å. Thus, the assumption of a uniform field is usually a good approximation. Other effects ignored include many-body contributions such as correlation and electron–hole interactions.

Once the imaginary part of the dielectric function is known, the real part can be obtained from the Kramers–Kronig relation:

$$\epsilon_1(\omega) = 1 + \frac{2}{\pi} P \int_0^\infty \frac{\omega' \epsilon_2(\omega')}{\omega'^2 - \omega^2} \, d\omega'. \tag{A1.3.89}$$

The principal part of the integral is taken and the integration must be done over all frequencies. In practice, the integration is often terminated outside of the frequency range of interest. Once the full dielectric function is known, the reflectivity of the solid can be computed.

It is possible to understand the fine structure in the reflectivity spectrum by examining the contributions to the imaginary part of the dielectric function. If one considers transitions from two bands (v → c), equation (A1.3.87) can be written as

$$\epsilon_2(\omega)_{vc} = \frac{4\pi e^2 \hbar}{3m^2 \omega^2} \frac{2}{(2\pi)^3} |M_{vc}| \int_{BZ} \delta(\omega_{vc}(\mathbf{k}) - \omega) \, d^3k. \tag{A1.3.90}$$

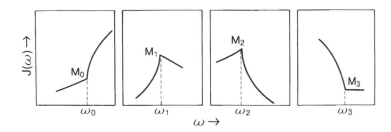

Figure A1.3.18. Typical critical point structure in the joint density of states.

Under the assumption that the matrix elements can be treated as constants, they can be factored out of the integral. This is a good approximation for most crystals. By comparison with equation (A1.3.84), it is possible to define a function similar to the density of states. In this case, since both valence and conduction band states are included, the function is called the *joint density of states*:

$$J_{vc}(\omega) = \frac{2}{(2\pi)^3} \int_{BZ} \delta(\omega_{vc}(\mathbf{k}) - \omega)\, d^3k. \tag{A1.3.91}$$

With this definition, one can write

$$\epsilon_2(\omega)_{vc} = \frac{4\pi e^2 \hbar}{3m^2\omega^2} |M_{vc}| J_{vc}(\omega). \tag{A1.3.92}$$

Within this approximation, the structure in $\epsilon_2(\omega)_{vc}$ can be related to structure in the joint density of states. The joint density of states can be written as a surface integral [1]:

$$J_{vc}(\omega) = \frac{2}{(2\pi)^3} \int_{\omega_{vc}=\omega} \frac{ds}{|\nabla_{\mathbf{k}}\omega_{vc}(\mathbf{k})|}. \tag{A1.3.93}$$

ds is a surface element defined by $\omega_{vc}(\mathbf{k}) = \omega$. The sharp structure in the joint density of states arises from zeros in the dominator. This occurs at *critical points* where

$$\nabla_{\mathbf{k}}\omega_{vc}(\mathbf{k}) = 0 \tag{A1.3.94}$$

or

$$\nabla_{\mathbf{k}}E_v(\mathbf{k}) = \nabla_{\mathbf{k}}E_c(\mathbf{k}) \tag{A1.3.95}$$

when the slopes of the valence band and conduction band are equal. The group velocity of an electron or hole is defined as $\mathbf{v}_g = \nabla_{\mathbf{k}}E(\mathbf{k})$. Thus, the critical points occur when the hole and electrons have the same group velocity.

The band energy difference or $\omega_{vc}(\mathbf{k})$ can be expanded around a critical point $\mathbf{k}_{cp}$ as

$$\omega_{vc}(\mathbf{k}) = \omega_{vc}(\mathbf{k}_{cp}) + \sum_{n=1}^{3} \alpha_n(\mathbf{k} - \mathbf{k}_{cp})_n^2 + \cdots. \tag{A1.3.96}$$

The expansion is done around the principal axes so only three terms occur in the summation. The nature of the critical point is determined by the signs of the α_n. If $\alpha_n > 0$ for all n, then the critical point corresponds to a local minimum. If $\alpha_n < 0$ for all n, then the critical point corresponds to a local maximum. Otherwise, the critical points correspond to saddle points.

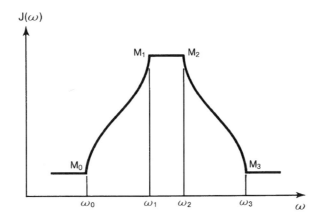

Figure A1.3.19. Simplest possible critical point structure in the joint density of states for a given energy band.

The types of critical points can be labelled by the number of α_n less than zero. Specifically, the critical points are labelled by M_i where i is the number of α_n which are negative: i.e. a local minimum critical point would be labelled by M_0, a local maximum by M_3 and the saddle points by (M_1, M_2). Each critical point has a characteristic line shape. For example, the M_0 critical point has a joint density of state which behaves as $J_{vc} = \text{constant} \times \sqrt{\omega - \omega_0}$ for $\omega > \omega_0$ and zero otherwise, where ω_0 corresponds to the M_0 critical point energy. At $\omega = \omega_0$, J_{vc} has a discontinuity in the first derivative. In figure A1.3.18, the characteristic structure of the joint density of states is presented for each type of critical point.

For a given pair of valence and conduction bands, there must be at least one M_0 and one M_3 critical points and at least three M_1 and three M_2 critical points. However, it is possible for the saddle critical points to be degenerate. In the simplest possible configuration of critical points, the joint density of states appears as in figure A1.3.19.

It is possible to identify particular spectral features in the modulated reflectivity spectra to band structure features. For example, in a direct band gap the joint density of states must resemble that of a M_0 critical point. One of the first applications of the *empirical pseudopotential method* was to calculate reflectivity spectra for a given energy band. Differences between the calculated and measured reflectivity spectra could be assigned to errors in the energy band structure. Such errors usually involve the incorrect placement or energy of a critical point feature. By making small adjustments in the pseudopotential, it is almost always possible to extract an energy band structure consistent with the measure reflectivity.

The critical point analysis performed for the joint density of states can also be applied to the density of states. By examining the photoemission spectrum compared with the calculated density of states, it is also possible to assess the quality of the energy band structure. Photoemission spectra are superior to reflectivity spectra in the sense of giving the band structure energies relative to a fixed energy reference, such as the vacuum level. Reflectivity measurements only give relative energy differences between energy bands.

In figures A1.3.20 and A1.3.21, the real and imaginary parts of the dielectric function are illustrated for silicon. There are some noticeable differences in the line shapes between theory and experiment. These differences can be attributed to issues outside of elementary band theory such as the interactions of electrons and holes. This issue will be discussed further in the following section on insulators. Qualitatively, the real part of the dielectric function appears as a simple harmonic oscillator with a resonance at about 4.5 eV. This energy corresponds approximately to the cohesive energy per atom of silicon.

It is possible to determine the spatial distributions, or charge densities, of electrons from a knowledge of the wavefunctions. The arrangement of the charge density is very useful in characterizing the bond in the solid. For example, if the charge is highly localized between neighbouring atoms, then the bond corresponds

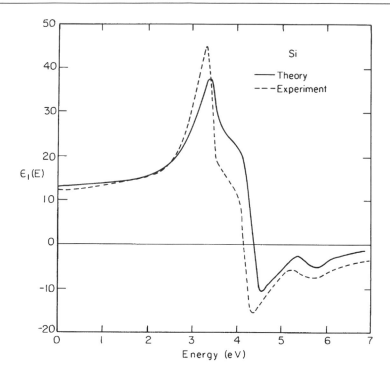

Figure A1.3.20. Real part of the dielectric function for silicon. The experimental work is from [31]. The theoretical work is from an empirical pseudopotential calculation [25].

to a covalent bond. The classical picture of the covalent bond is the sharing of electrons between two atoms. This picture is supported by quantum calculations. In figure A1.3.22, the electronic distribution charge is illustrated for crystalline carbon and silicon in the diamond structure. In carbon the midpoint between neighbouring atoms is a saddle point: this is typical of the covalent bond in organics, but not in silicon where the midpoint corresponds to a maximum of the charge of the density. X-ray measurements also support the existence of the covalent bonding charge as determined from quantum calculations [33].

Although empirical pseudopotentials present a reliable picture of the electronic structure of semiconductors, these potentials are not applicable for understanding structural properties. However, density-functional-derived pseudopotentials can be used for examining the structural properties of matter. Once a self-consistent field solution of the Kohn–Sham equations has been achieved, the total electronic energy of the system can be determined from equation (A1.3.39). One of the first applications of this method was to forms of crystalline silicon. Various structural forms of silicon were considered: diamond, hexagonal diamond, β-Sn, simple cubic, FCC, BCC and so on. For a given volume, the lattice parameters and any internal parameters can be optimized to achieve a ground-state energy. In figure A1.3.23, the total structural energy of the system is plotted for eight different forms of silicon. The lowest energy form of silicon is correctly predicted to be the diamond structure. By examining the change in the structural energy with respect to volume, it is possible to determine the equation of state for each form. It is possible to determine which phase is lowest in energy for a specified volume and to determine transition pressures between different phases. As an example, one can predict from this phase diagram the transition pressure to transform silicon in the diamond structure to the white tin (β-Sn) structure. This pressure is predicted to be approximately 90 MPa; the measured pressure is about 120 MPa [34]. The role of temperature has been neglected in the calculation of the structural energies. For most applications, this is not a serious issue as the role of temperature is often less than the inherent errors within density functional theory.

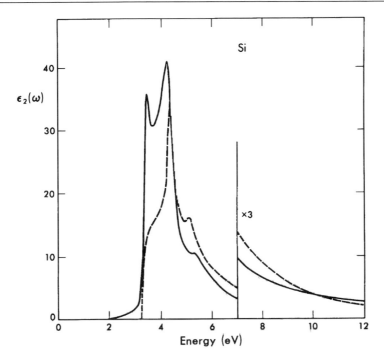

Figure A1.3.21. Imaginary part of the dielectric function for silicon. The experimental work is from [31]. The theoretical work is from an empirical pseudopotential calculation [25].

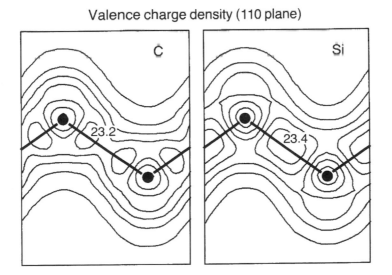

Figure A1.3.22. Spatial distributions or charge densities for carbon and silicon crystals in the diamond structure. The density is only for the valence electrons; the core electrons are omitted. This charge density is from an *ab initio* pseudopotential calculation [27].

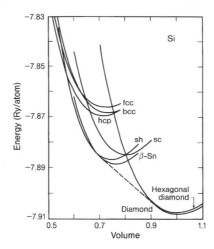

Figure A1.3.23. Phase diagram of silicon in various polymorphs from an *ab initio* pseudopotential calculation [34]. The volume is normalized to the experimental volume. The binding energy is the total electronic energy of the valence electrons. The slope of the dashed curve gives the pressure to transform silicon in the diamond structure to the β-Sn structure. Other polymorphs listed include face-centred cubic (fcc), body-centred cubic (bcc), simple hexagonal (sh), simple cubic (sc) and hexagonal close-packed (hcp) structures.

One notable consequence of the phase diagram in figure A1.3.23 was the prediction that high-pressure forms of silicon might be superconducting [35, 36]. This prediction was based on the observation that some high-pressure forms of silicon are metallic, but retain strong covalent-like bonds. It was later verified by high-pressure measurements that the predicted phase was a superconductor [36]. This success of the structural phase diagram of silicon helped verify the utility of the pseudopotential density functional method and has resulted in its widespread applicability to condensed phases.

A1.3.6.2 Insulators

Insulating solids have band gaps which are notably larger than semiconductors. It is not unusual for an alkali halide to have a band gap of ~ 10 eV or more. Electronic states in insulators are often highly localized around the atomic sites in insulating materials In most cases, this arises from a large transfer of electrons from one site to another. Exceptions are insulating materials like sulfur and carbon where the covalent bonds are so strong as to strongly localize charge between neighbouring atoms.

As an example of the energy band structures for an insulator, the energy bands for lithium fluoride are presented in figure A1.3.24. LiF is a highly ionic material which forms in the rocksalt structure (figure A1.3.25). The bonding in this crystal can be understood by transferring an electron from the highly electropositive Li to the electronegative F atoms: i.e. one can view crystalline LiF as consisting of Li^+F^- constituents. The highly localized nature of the electronic charge density results in very narrow, almost atomic-like, energy bands.

One challenge of modern electronic structure calculations has been to reproduce excited-state properties. Density functional theory is a ground-state theory. The eigenvalues for empty states do not have physical meaning in terms of giving excitation energies. If one were to estimate the band gap from density functional theory by taking the eigenvalue differences between the highest occupied and lowest empty states, the energy difference would badly underestimate the band gap. Contemporary approaches [37, 38] have resolved this issue by correctly including spatial variations in the electron–electron interactions and including self-energy terms (see section A1.3.2.2).

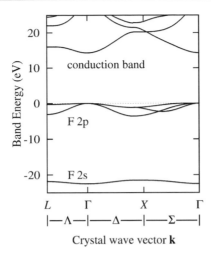

Figure A1.3.24. Band structure of LiF from *ab initio* pseudopotentials [39].

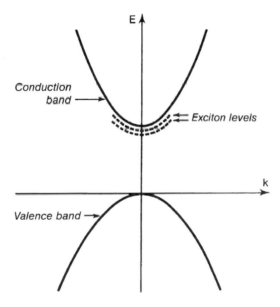

Figure A1.3.25. Schematic illustration of exciton binding energies in an insulator or semiconductor.

Because of the highly localized nature of electronic and hole states in insulators, it is difficult to describe the optical excitations. The excited electron is strongly affected by the presence of the hole state. One failure of the energy band picture concerns the interaction between the electron and hole. The excited electron and the hole can form a hydrogen atomic-like interaction resulting in the formation of an *exciton*, or a bound electron–hole pair. The exciton binding energy reduces the energy for an excitation below that of the conduction band and results in strong, discrete optical lines. This is illustrated in figure A1.3.25.

A simple model for the exciton is to assume a screened interaction between the electron and hole using a static dielectric function. In addition, it is common to treat the many-body interactions in a crystal by replacing the true mass of the electron and hole by a dynamical or *effective mass*. Unlike a hydrogen atom, where the

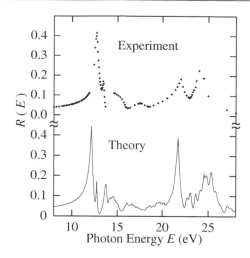

Figure A1.3.26. Reflectivity of LiF from *ab initio* pseudopotentials. (Courtesy of E L Shirley, see [39] and references therein.)

proton mass exceeds that of the electron by three orders of magnitude, the masses of the interacting electron and hole are almost equivalent. Using the reduced mass for this system, we have $1/\mu = 1/m_e + 1/m_h$. Within this model, the binding energy of the exciton can be found from

$$\left[\frac{-\hbar^2\nabla^2}{2\mu} - \frac{e^2}{\epsilon r}\right]\psi(\mathbf{r}) = E_b\psi(\mathbf{r}) \tag{A1.3.97}$$

where ϵ is the static dielectric function for the insulator of interest. The binding energy from this hydrogenic Schrödinger equation is given by

$$E_b = -\frac{\mu e^4}{2\hbar^2\epsilon^2 n^2} \tag{A1.3.98}$$

where $n = 1, 2, 3, \ldots$. Typical values for a semiconductor are μ and ϵ are $\mu = 0.1$ m and $\epsilon = 10$. This results in a binding energy of about 0.01 eV for the ground state, $n = 1$. For an insulator, the binding energy is much larger. For a material like silicon dioxide, one might have $\mu = 0.5$ m and $\epsilon = 3$ or a binding energy of roughly 1 eV. This estimate suggests that reflectivity spectra in insulators might be strongly altered by exciton interactions.

Even in semiconductors, where it might appear that the exciton binding energies would be of interest only for low temperature regimes, excitonic effects can strongly alter the line shape of excitations away from the band gap.

The size of the electron–hole pair can be estimated from the Bohr radius for this system:

$$r_0 = -\frac{\epsilon\hbar^2}{\mu e^2}. \tag{A1.3.99}$$

The size of the exciton is approximately 50 Å in a material like silicon, whereas for an insulator the size would be much smaller: for example, using our numbers above for silicon dioxide, one would obtain a radius of only ~3 Å or less. For excitons of this size, it becomes problematic to incorporate a static dielectric constant based on macroscopic crystalline values.

The reflectivity of LiF is illustrated in figure A1.3.26. The first large peak corresponds to an excitonic transition.

A1.3.6.3 Metals

Metals are fundamentally different from insulators as they possess no gap in the excitation spectra. Under the influence of an external field, electrons can respond by readily changing from one $\mathbf{k}$ state to another. The ease by which the ground-state configuration is changed accounts for the high conductivity of metals.

Arguments based on a free electron model can be made to explain the conductivity of a metal. It can be shown that the $\mathbf{k}$ will evolve following a Newtonian law [1]:

$$\hbar \frac{d\mathbf{k}}{dt} = -e\mathcal{E}. \tag{A1.3.100}$$

This can be integrated to yield

$$(\mathbf{k} - \mathbf{k}_0) = -e\mathcal{E}(t - t_0)/\hbar. \tag{A1.3.101}$$

After some typical time, τ, the electron will scatter off a lattice imperfection. This imperfection might be a lattice vibration or an impurity atom. If one assumes that no memory of the event resides after the scattering event, then on average one has $\Delta \mathbf{k} = -e\mathcal{E}\tau/\hbar$. In this picture, the conductivity of the metal, σ, can be extracted from Ohm's law: $\sigma = J/\mathcal{E}$ where J is the current density. The current density is given by

$$\mathbf{J} = -ne\Delta\mathbf{v} = -ne(-e\mathcal{E}\tau/m) = ne^2\tau\mathcal{E}/m \tag{A1.3.102}$$

or

$$\sigma = ne^2\tau/m. \tag{A1.3.103}$$

This expression for the conductivity is consistent with experimental trends.

Another important accomplishment of the free electron model concerns the heat capacity of a metal. At low temperatures, the heat capacity of a metal goes linearly with the temperature and vanishes at absolute zero. This behaviour is in contrast with classical statistical mechanics. According to classical theories, the equipartition theory predicts that a free particle should have a heat capacity of $\frac{3}{2}k_B$ where k_B is the Boltzmann constant. An ideal gas has a heat capacity consistent with this value. The electrical conductivity of a metal suggests that the conduction electrons behave like 'free particles' and might also have a heat capacity of $\frac{3}{2}k_B$, which would be strongly at variance with the observed behaviour and in violation of the third law of thermodynamics.

The resolution of this issue is based on the application of the Pauli exclusion principle and Fermi–Dirac statistics. From the free electron model, the total electronic energy, U, can be written as

$$U(T) = \int_0^\infty \epsilon f(\epsilon, T) D(\epsilon) \, d\epsilon \tag{A1.3.104}$$

where $f(\epsilon, T)$ is the Fermi–Dirac distribution function and $D(\epsilon)$ is the density of states. The Fermi–Dirac function gives the probability that a given orbital will be occupied:

$$f(\epsilon, T) = \frac{1}{\exp\left((\epsilon - E_F)/kT\right) + 1}. \tag{A1.3.105}$$

The value of E_F at zero temperature can be estimated from the electron density (equation (A1.3.26)). Typical values of the Fermi energy range from about 1.6 eV for Cs to 14.1 eV for Be. In terms of temperature ($T_F = E_F/k$), the range is approximately 2000–16 000 K. As a consequence, the Fermi energy is a very weak function of temperature under ambient conditions. The electronic contribution to the heat capacity, C, can be determined from

$$C = \frac{dU}{dT} = \int_0^\infty \epsilon D(\epsilon) \frac{df(\epsilon, T)}{dT} \, d\epsilon. \tag{A1.3.106}$$

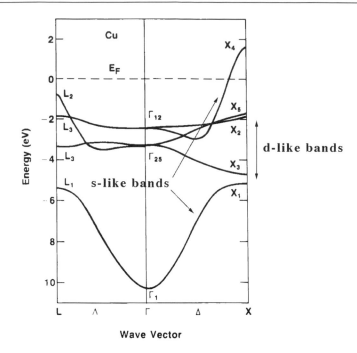

Figure A1.3.27. Energy bands of copper from *ab initio* pseudopotential calculations [40].

The integral can be approximated by noting that the derivative of the Fermi function is highly localized around E_F. To a very good approximation, the heat capacity is

$$C = \frac{\pi^2}{3} D(\epsilon_F) k^2 T. \qquad (A1.3.107)$$

The linear dependence of C with temperature agrees well with experiment, but the pre-factor can differ by a factor of two or more from the free electron value. The origin of the difference is thought to arise from several factors: the electrons are not truly free, they interact with each other and with the crystal lattice, and the dynamical behaviour the electrons interacting with the lattice results in an *effective mass* which differs from the free electron mass. For example, as the electron moves through the lattice, the lattice can distort and exert a dragging force.

Simple metals like alkalis, or ones with only s and p valence electrons, can often be described by a free electron gas model, whereas transition metals and rare earth metals which have d and f valence electrons cannot. Transition metal and rare earth metals do not have energy band structures which resemble free electron models. The formed bonds from d and f states often have some strong covalent character. This character strongly modulates the free-electron-like bands.

An example of metal with significant d-bonding is copper. The atomic configuration of copper is $1s^2 2s^2 2p^6 3s^2 3p^6 3d^{10} 4s^1$. If the 3d states were truly core states, then one might expect copper to resemble potassium as its atomic configuration is $1s^2 2s^2 2p^6 3s^2 3p^6 4s^1$. The strong differences between copper and potassium in terms of their chemical properties suggest that the 3d states interact strongly with the valence electrons. This is reflected in the energy band structure of copper (figure A1.3.27).

Copper has a FCC structure with one atom in the primitive unit cell. From simple orbital counting, one might expect the ten d electrons to occupy five d-like bands and the one s electron to $\frac{1}{2}$ occupy one s-like band. This is apparent in the figure, although the interpretation is not straightforward. The lowest band (L_1

to Γ_1 to X_4) is s-like, but it mixes strongly with the d-like bands (at Γ_{25} and Γ_{12}), these bands are triply and doubly degenerate at Γ. Were it not for the d-mixing, the s-like band would be continuous from Γ_1 to X_4. The d-mixing 'splits' the s bands. The Fermi level cuts the s-like band along the Δ direction, reflecting the partial occupation of the s levels.

A1.3.7 Non-crystalline matter

A1.3.7.1 Amorphous solids

Crystalline matter can be characterized by *long-range order*. For a perfect crystal, a prescription can be used to generate the positions of atoms arbitrarily far away from a specified origin. However, 'real crystals' always contain imperfections. They contain defects which can be characterized as *point defects* localized to an atomic site or *extended defects* spread over a number of sites. Vacancies on the lattice site or atoms of impurities are examples of point defects. Grain boundaries or dislocations are examples of extended defects. One might imagine starting from an ideal crystal and gradually introducing defects such as vacancies. At some point the number of defects will be so large as to ruin the long-range order of the crystal. Solid materials that lack long-range order are called *amorphous* solids or *glasses*. The precise definition of an amorphous material is somewhat problematic. Usually, any material which does not display a sharp x-ray pattern is considered to be 'amorphous'. Some text books [1] define amorphous solids as 'not crystalline on any significant scale'.

Glassy materials are usually characterized by an additional criterion. It is often possible to cool a liquid below the thermodynamic melting point (i.e. to supercool the liquid). In glasses, as one cools the liquid state significantly below the melting point, it is observed that at a temperature well below the melting point of the solid the viscosity of the supercooled liquid increases dramatically. This temperature is called the *glass transition* temperature, and labelled as T_g. This increase of viscosity delineates the supercooled liquid state from the glass state. Unlike thermodynamic transitions between the liquid and solid state, the liquid $\rightarrow$ glass transition is not well defined. Most amorphous materials such as tetrahedrally coordinated semiconductors like silicon and germanium do not exhibit a glass transformation.

Defining order in an amorphous solid is problematic at best. There are several 'qualitative concepts' that can be used to describe disorder [7]. In figure A1.3.28, a perfect crystal is illustrated. A simple form of disorder involves crystals containing more than one type of atom. Suppose one considers an alloy consisting of two different atoms (A and B). In an ordered crystal one might consider each A surrounded by B and *vice versa*. In a random alloy, one might consider the lattice sites to remain unaltered but randomly place A and B atoms. This type of disorder is called *compositional disorder*. Other forms of disorder may involve minor distortions of the lattice that destroy the long-range order of the solid, but retain the chemical ordering and short-range order of the solid. For example, in short-range ordered solids, the coordination number of each atom might be preserved. In a highly disordered solid, no short-range order is retained: the chemical ordering is random with a number of over- and under-coordinated species.

In general, it is difficult to quantify structural properties of disordered matter via experimental probes as with x-ray or neutron scattering. Such probes measure statistically averaged properties like the *pair-correlation function*, also called the *radial distribution function*. The pair-correlation function measures the average distribution of atoms from a particular site.

Several models have been proposed to describe amorphous solids, in particular glasses. The structure of glasses often focus on two archetypes [1]: the continuous random network and microcrystallite models. In the continuous random network model, short-range order is preserved. For example, in forms of crystalline silica each silicon atom is surrounded by four oxygen atoms. The SiO_4 tetrahedra are linked together in a regular way which establishes the crystal structure. In a continuous random network each SiO_4 tetrahedral unit is preserved, but the relative arrangement between tetrahedral units is random. In another model, the so-called microcrystallite model, small 'crystallites' of the perfect structure exist, but these crystallites are randomly

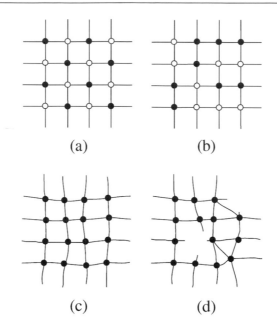

Figure A1.3.28. Examples of disorder: (a) perfect crystal, (b) compositional disorder, (c) positional disorder which retains the short-range order and (d) no long-range or short-range order.

arranged. The difference between the random network model and the crystallite model cannot be experimentally determined unless the crystallites are sufficiently large to be detected; this is usually not the situation.

Amorphous materials exhibit special quantum properties with respect to their electronic states. The loss of periodicity renders Bloch's theorem invalid; **k** is no longer a good quantum number. In crystals, structural features in the reflectivity can be associated with critical points in the joint density of states. Since amorphous materials cannot be described by **k**-states, selection rules associated with **k** are no longer appropriate. Reflectivity spectra and associated spectra are often featureless, or they may correspond to highly smoothed versions of the crystalline spectra.

One might suppose that optical gaps would not exist in amorphous solids, as the structural disorder would result in allowed energy states throughout the solid. However, this is not the case, as disordered insulating solids such as silica are quite transparent. This situation reflects the importance of local order in determining gaps in the excitation spectra. It is still possible to have gaps in the joint density of states without resort to a description of energy *versus* wavevector. For example, in silica the large energy gap arises from the existence of SiO_4 units. Disordering these units can cause states near the top of the occupied states and near the bottom of the empty states to tail into the gap region, but not remove the gap itself.

Disorder plays an important role in determining the extent of electronic states. In crystalline matter one can view states as existing throughout the crystal. For disordered matter, this is not the case: electronic states become localized near band edges. The effect of localization has profound effects on transport properties. Electrons and holes can still carry current in amorphous semiconductors, but the carriers can be strongly scattered by the disordered structure. For the localized states near the band edges, electrons can be propagated only by a thermally activated hopping process.

A1.3.7.2   Liquids

Unlike the solid state, the liquid state cannot be characterized by a static description. In a liquid, bonds break and reform continuously as a function of time. The quantum states in the liquid are similar to those

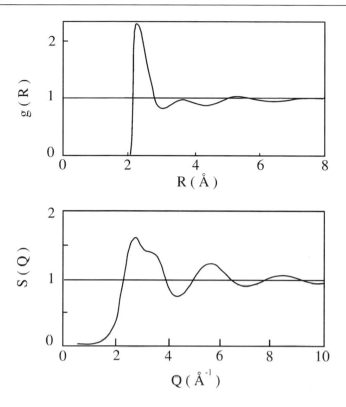

Figure A1.3.29. Pair correlation and structure factor for liquid silicon from experiment [41].

in amorphous solids in the sense that the system is also disordered. The liquid state can be quantified only by considering some ensemble averaging and using statistical measures. For example, consider an elemental liquid. Just as for amorphous solids, one can ask what is the distribution of atoms at a given distance from a reference atom on average, i.e. the *radial distribution function* or the *pair correlation function* can also be defined for a liquid. In scattering experiments on liquids, a *structure factor* is measured. The radial distribution function, $g(r)$, is related to the structure factor, $S(q)$, by

$$S(q) = 1 + \rho_0 \int [g(r) - 1] \exp(i\mathbf{q} \cdot \mathbf{r}) \, d^3 r \qquad (A1.3.108)$$

where ρ_0 is the average concentration density of the liquid. By taking the Fourier transform of the structure, it is possible to determine the radial distribution function of the liquid.

Typical results for a semiconducting liquid are illustrated in figure A1.3.29, where the experimental pair correlation and structure factors for silicon are presented. The radial distribution function shows a sharp first peak followed by oscillations. The structure in the radial distribution function reflects some local ordering. The nature and degree of this order depends on the chemical nature of the liquid state. For example, semiconductor liquids are especially interesting in this sense as they are believed to retain covalent bonding characteristics even in the melt.

One simple measure of the liquid structure is the average coordination number of an atom. For example, the average coordination of a silicon atom is four in the *solid phase* at ambient pressure and increases to six in high pressure forms of silicon. In the *liquid* state, the average coordination of silicon is six. The average coordination of the liquid can be determined from the radial distribution function. One common prescription

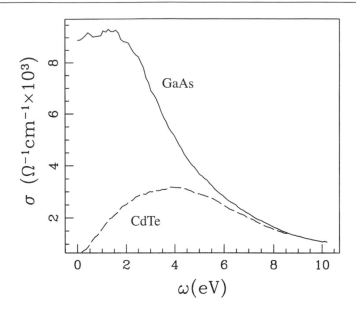

Figure A1.3.30. Theoretical frequency-dependent conductivity for GaAs and CdTe liquids from *ab initio* molecular dynamics simulations [42].

is to integrate the area under the first peak of the radial distribution function. The integration is terminated at the first local minimum after the first peak. For a crystalline case, this procedure gives the exact number of nearest neighbours. In general, coordination numbers greater than four correspond to metallic states of silicon. As such, the radial distribution function suggests that silicon is a metal in the liquid state. This is consistent with experimental values of the conductivity. Most tetrahedrally coordinated semiconductors, e.g. Ge, GaAs, InP and so on, become metallic upon melting.

It is possible to use the quantum states to predict the electronic properties of the melt. A typical procedure is to implement molecular dynamics simulations for the liquid, which permit the wavefunctions to be determined at each time step of the simulation. As an example, one can use the eigenpairs for a given atomic configuration to calculate the optical conductivity. The real part of the conductivity can be expressed as

$$\sigma_r(\omega) = \frac{2\pi e^2}{3m^2\omega\Omega} \sum_{n,m} \sum_{\alpha=x,y,z} |\langle \psi_m | p_\alpha | \psi_n \rangle|^2 \times \delta(E_n - E_m - \hbar\omega) \tag{A1.3.109}$$

where E_i and ψ_i are eigenvalues and eigenfunctions, and Ω is the volume of the supercell. The dipole transition elements, $\langle \psi_m | p_\alpha | \psi_n \rangle$, reflect the spatial resolution of the initial and final wavefunctions. If the initial and final states were to have an even parity, then the electromagnetic field would not couple to these states.

The conductivity can be calculated for each time step in a simulation and averaged over a long simulation time. This procedure can be used to distinguish the metallic and semiconducting behaviour of the liquid state. As an example, the calculated frequency dependence of the electrical conductivity of gallium arsenide and cadmium telluride are illustrated in figure A1.3.30. In the melt, gallium arsenide is a metal. As the temperature of the liquid is increased, its DC conductivity decreases. For cadmium telluride, the situation is reversed. As the temperature of the liquid is increased, the DC conductivity increases. This is similar to the behaviour of a semiconducting solid. As the temperature of the solid is increased, more carriers are thermally excited into the conduction bands and the conductivity increases. The relative conductivity of GaAs *versus* CdTe as determined via theoretical calculations agrees well with experiment.

References

[1]   Kittel C 1996 *Introduction to Solid State Physics* 7th edn (New York: Wiley)
[2]   Ziman J M 1986 *Principles of the Theory of Solids* 2nd edn (Cambridge: Cambridge University Press)
[3]   Kittel C 1987 *Quantum Theory of Solids* 2nd revn (New York: Wiley)
[4]   Callaway J 1974 *Quantum Theory of the Solid State* (Boston: Academic)
[5]   Yu P and Cardona M 1996 *Fundamentals of Semiconductors* (New York: Springer)
[6]   Wells A F 1984 *Structural Inorganic Chemistry* 5th edn (Oxford: Clarendon)
[7]   Madelung O 1996 *Introduction to Solid State Theory* (New York: Springer)
[8]   Tauc J (ed) 1974 *Amorphous and Liquid Semiconductors* (New York: Plenum)
[9]   Phillips J C 1989 *Physics of High-T_c Superconductors* (Boston: Academic)
[10] Anderson P W 1997 *The Theory of Superconductivity in the High-T_c Cuprates (Princeton Series in Physics)* (Princeton: Princeton University Press)
[11] Ziman J M 1960 *Electrons and Phonons* (Oxford: Oxford University Press)
[12] Haug A 1972 *Theoretical Solid State Physics* (New York: Pergamon)
[13] Thomas L H 1926 *Proc. Camb. Phil. Soc.* **23** 542
[14] Fermi E 1928 *Z. Phys.* **48** 73
[15] Lundqvist S and March N H (eds) 1983 *Theory of the Inhomogeneous Electron Gas* (New York: Plenum)
[16] Aschroft N W and Mermin N D 1976 *Solid State Physics* (New York: Holt, Rinehart and Winston)
[17] Slater J C 1951 *Phys. Rev.* **81** 385
[18] Slater J C 1964–74 *Quantum Theory of Molecules and Solids* vols 1–4 (New York: McGraw-Hill)
[19] Slater J C 1968 *Quantum Theory of Matter* (New York: McGraw-Hill)
[20] Hohenberg P and Kohn W 1964 *Phys. Rev.* B **136** 864
[21] Kohn W and Sham L 1965 *Phys. Rev.* A **140** 1133
[22] Ceperley D M and Alder B J 1980 *Phys. Rev. Lett.* **45** 566
[23] Perdew J P, Burke K and Wang Y 1996 *Phys. Rev.* B **54** 16 533 and references therein
[24] Chelikowsky J R and Louie S G (eds) 1996 *Quantum Theory of Real Materials* (Boston: Kluwer)
[25] Cohen M L and Chelikowsky J R 1989 *Electronic Structure and Optical Properties of Semiconductors* 2nd edn (Springer)
[26] Phillips J C and Kleinman L 1959 *Phys. Rev.* **116** 287
[27] Chelikowsky J R and Cohen M L 1992 *Ab initio* pseudopotentials for semiconductors *Handbook on Semiconductors* vol 1, ed P Landsberg (Amsterdam: Elsevier) p 59
[28] Fermi E 1934 *Nuovo Cimento* **11** 157
[29] Herring C 1940 *Phys. Rev.* **57** 1169
[30] Ley L, Kowalczyk S P, Pollack R A and Shirley D A 1972 *Phys. Rev. Lett.* **29** 1088
[31] Philipp H R and Ehrenreich H 1963 *Phys. Rev. Lett.* **127** 1550
[32] Zucca R R L and Shen Y R 1970 *Phys. Rev.* B **1** 2668
[33] Yang L W and Coppens P 1974 *Solid State Commun.* **15** 1555
[34] Yin M T and Cohen M L 1980 *Phys. Rev. Lett.* **45** 1004
[35] Chang K J, Dacorogna M M, Cohen M L, Mignot J M, Chouteau G and Martinez G 1985 *Phys. Rev. Lett.* **54** 2375
[36] Dacorogna M M, Chang K J and Cohen M L 1985 *Phys. Rev.* B **32** 1853
[37] Hybertsen M and Louie S G 1985 *Phys. Rev. Lett.* **55** 1418
[38] Hybertsen M and Louie S G 1986 *Phys. Rev.* B **34** 5390
[39] Benedict L X and Shirley E L 1999 *Phys. Rev.* B **59** 5441
[40] Chelikowsky J R and Chou M Y 1988 *Phys. Rev.* B **38** 7966
[41] Waseda Y 1980 *The Structure of Non-Crystalline Materials* (New York: McGraw-Hill)
[42] Godlevsky V, Derby J and Chelikowsky J R 1998 *Phys. Rev. Lett.* **81** 4959

Further Reading

Anderson P W 1963 *Concepts in Solids* (New York: Benjamin)

Cox P A 1987 *The Electronic Structure and Chemistry of Solids* (Oxford: Oxford University Press)

Harrison W A 1999 *Elementary Electronic Structure* (River Edge: World Scientific)

Harrison W A 1989 *Electronic Structure and the Properties of Solids: The Physics of the Chemical Bond* (New York: Dover)

Hummel R 1985 *Electronic Properties of Materials* (New York: Springer)

Jones W and March N 1973 *Theoretical Solid State Physics* (New York: Wiley)

Lerner R G and Trigg G L (eds) 1983 *Concise Encyclopedia of Solid State Physics* (Reading, MA: Addison-Wesley)

Myers H P 1997 *Introductory Solid State Physics* (London: Taylor and Francis)

Patterson J D 1971 *Introduction to the Theory of Solid State Physics* (Reading, MA: Addison-Wesley)
Peierls R 1955 *Quantum Theory of Solids* (Oxford: Clarendon)
Phillips J C 1973 *Bands and Bonds in Semiconductors* (New York: Academic)
Pines D 1963 *Elementary Excitations in Solids* (New York: Benjamin)
Seitz F 1948 *Modern Theory of Solids* (New York: McGraw-Hill)

A1.4
The symmetry of molecules

Per Jensen and P R Bunker

A1.4.1   Introduction

Unlike most words in a glossary of terms associated with the theoretical description of molecules, the word 'symmetry' has a meaning in every-day life. Many objects look exactly like their mirror image, and we say that they are symmetrical or, more precisely, that they have *reflection* symmetry. In addition to having reflection symmetry, a pencil (for example) is such that if we rotate it through any angle about its long axis it will look the same. We say it has *rotational* symmetry. The concepts of rotation and reflection symmetry are familiar to us all.

The ball-and-stick models used in elementary chemistry education to visualize molecular structure are frequently symmetrical in the sense discussed above. Reflections in certain planes, rotations by certain angles about certain axes, or more complicated symmetry operations involving both reflection and rotation, will leave them looking the same. One might initially think that this is 'the symmetry of molecules' discussed in the present chapter, but it is not. Ball-and-stick models represent molecules fixed at their equilibrium configuration, that is, at the minimum (or at one of the minima) of the potential energy function for the electronic state under consideration. A real molecule is not static and generally it does not possess the rotation–reflection symmetry of its equilibrium configuration. Anyway, the use we make of molecular symmetry in understanding molecules, their spectra and their dynamics, has its basis in considerations other than the appearance of the molecule at equilibrium.

The true basis for understanding molecular symmetry involves studying the operations that leave the energy of a molecule unchanged, rather than studying the rotations or reflections that leave a molecule in its equilibrium configuration looking the same. Symmetry is a general concept. Not only does it apply to molecules, but it also applies, for example, to atoms, to atomic nuclei and to the particles that make up atomic nuclei. Also, the concept of symmetry applies to *nonrigid* molecules such as ammonia NH_3, ethane C_2H_6, the hydrogen dimer $(H_2)_2$, the water trimer $(H_2O)_3$ and so on, that easily contort through structures that differ in the nature of their rotational and reflection symmetry. For a hydrogen molecule that is translating, rotating and vibrating in space, with the electrons orbiting, it is clear that the total energy of the molecule is unchanged if we interchange the coordinates and momenta of the two protons; the total kinetic energy is unchanged (since the two protons have the same mass), and the total electrostatic potential energy is unchanged (since the two protons have the same charge). However, the interchange of an electron and a proton will almost certainly not leave the molecular energy unchanged. Thus the permutation of identical particles is a symmetry operation and we will introduce others. In quantum mechanics the possible molecular energies are the eigenvalues of the molecular Hamiltonian and if the Hamiltonian is invariant to a particular operation (or, equivalently, if the Hamiltonian commutes with a particular operation) then that operation is a symmetry operation.

We collect symmetry operations into various 'symmetry groups', and this chapter is about the definition and use of such symmetry operations and symmetry groups. Symmetry groups are used to label molecular states and this labelling makes the states, and their possible interactions, much easier to understand. One important symmetry group that we describe is called *the molecular symmetry group* and the symmetry operations it contains are permutations of identical nuclei with and without the inversion of the molecule at its centre of mass. One fascinating outcome is that indeed for *rigid* molecules (i.e., molecules that do not undergo large amplitude contortions to become *nonrigid* as discussed above) we can obtain a group of rotation and reflection operations that describes the rotation and reflection symmetry of the equilibrium molecular structure from the molecular symmetry group. However, by following the energy-invariance route we can understand the generality of the concept of symmetry and can readily deduce the symmetry groups that are appropriate for nonrigid molecules as well.

This introductory section continues with a subsection that presents the general motivation for using symmetry and ends with a short subsection that lists the various types of molecular symmetry.

A1.4.1.1 *Motivation: rotational symmetry as an example*

Rotational symmetry is used here as an example to explain the motivation for using symmetry in molecular physics; it will be discussed in more detail in section A1.4.3.2.

We consider an isolated molecule in field-free space with Hamiltonian $\widehat{H}$. We let $\widehat{F}$ be the total angular momentum operator of the molecule, that is

$$\widehat{F} = \widehat{N} + \widehat{S} + \widehat{I} \tag{A1.4.1}$$

where $\widehat{N}$ is the operator for the rovibronic angular momentum that results from the rotational motion of the nuclei and the orbital motion of the electrons, $\widehat{S}$ is the total electron spin angular momentum operator and $\widehat{I}$ is the total nuclear spin angular momentum operator. We introduce a Cartesian axis system (X, Y, Z). The orientation of the (X, Y, Z) axis system is fixed in space (i.e., it is independent of the orientation in space of the molecule), but the origin is tied to the molecular centre of mass. It is well known that the molecular Hamiltonian $\widehat{H}$ commutes with the operators

$$\widehat{F}^2 = \widehat{F}_X^2 + \widehat{F}_Y^2 + \widehat{F}_Z^2 \tag{A1.4.2}$$

and $\widehat{F}_Z$, where this is the component of $\widehat{F}$ along the Z axis, i.e.,

$$[\widehat{F}^2, \widehat{H}] = \widehat{F}^2\widehat{H} - \widehat{H}\widehat{F}^2 = 0 \tag{A1.4.3}$$

and

$$[\widehat{F}_Z, \widehat{H}] = 0. \tag{A1.4.4}$$

It is also well known that $\widehat{F}^2$ and $\widehat{F}_Z$ have simultaneous eigenfunctions $|F, m_F\rangle$ and that

$$\widehat{F}^2|F, m_F\rangle = F(F+1)\hbar^2|F, m_F\rangle \tag{A1.4.5}$$

and

$$\widehat{F}_Z|F, m_F\rangle = m_F\hbar|F, m_F\rangle \tag{A1.4.6}$$

where, for a given molecule, F assumes non-negative values that are either integral ($=0, 1, 2, 3, \ldots$) or half-integral ($=1/2, 3/2, 5/2, \ldots$) and, for a given F value, m_F has the $2F+1$ values $-F, -F+1, \ldots, F-1, F$.

We can solve the molecular Schrödinger equation

$$\widehat{H}\Psi_j = E_j\Psi_j \tag{A1.4.7}$$

by representing the unknown wavefunction Ψ_j (where j is an index labelling the solutions) as a linear combination of known basis functions Ψ_n^0,

$$\Psi_j = \sum_n C_{jn} \Psi_n^0 \tag{A1.4.8}$$

where the C_{jn} are expansion coefficients and n is an index labelling the basis functions. As described, for example, in section 6.6 of Bunker and Jensen [1], the eigenvalues E_j and expansion coefficients C_{jn} can be determined from the 'Hamiltonian matrix' by solving the secular equation

$$|H_{mn} - \delta_{mn} E| = 0 \tag{A1.4.9}$$

where the Kronecker delta δ_{mn} has the value 1 for $m = n$ and the value 0 for $m \neq n$, and the Hamiltonian matrix elements H_{mn} are given by

$$H_{mn} = \int \Psi_m^0 {}^* \widehat{H} \Psi_n^0 \, d\tau \tag{A1.4.10}$$

with integration carried out over the configuration space of the molecule. This process is said to involve 'diagonalizing the Hamiltonian matrix'.

We now show what happens if we set up the Hamiltonian matrix using basis functions Ψ_n^0 that are eigenfunctions of $\widehat{F}^2$ and $\widehat{F}_Z$ with eigenvalues given by equations (A1.4.5) and (A1.4.6). We denote this particular choice of basis functions as Ψ_{n,F,m_F}^0. From equation (A1.4.3), equation (A1.4.5) and the fact that $\widehat{F}^2$ is a Hermitian operator, we derive

$$\langle \Psi_{m,F',m_F'}^0 | [\widehat{F}^2, \widehat{H}] | \Psi_{n,F'',m_F''}^0 \rangle = \langle \widehat{F}^2 \Psi_{m,F',m_F'}^0 | \widehat{H} | \Psi_{n,F'',m_F''}^0 \rangle - \langle \Psi_{m,F',m_F'}^0 | \widehat{H} | \widehat{F}^2 \Psi_{n,F'',m_F''}^0 \rangle$$

$$= (F'(F'+1) - F''(F''+1)) \hbar^2 \langle \Psi_{m,F',m_F'}^0 | \widehat{H} | \Psi_{n,F'',m_F''}^0 \rangle = 0 \tag{A1.4.11}$$

from which it follows that the matrix element $\langle \Psi_{m,F',m_F'}^0 | \widehat{H} | \Psi_{n,F'',m_F''}^0 \rangle$ must vanish if $F' \neq F''$. From equation (A1.4.4) it follows in a similar manner that the matrix element must also vanish if $m_F' \neq m_F''$. That is, in the basis Ψ_{n,F,m_F}^0 the Hamiltonian matrix is block diagonal in F and m_F, and we can rewrite equation (A1.4.8) as

$$\Psi_j^{(F,m_F)} = \sum_n C_{jn}^{(F,m_F)} \Psi_{n,F,m_F}^0 \tag{A1.4.12}$$

the eigenfunctions of $\widehat{H}$ are also eigenfunctions of $\widehat{F}^2$ and $\widehat{F}_Z$. We can further show that since m_F quantizes the molecular angular momentum along the arbitrarily chosen, space-fixed Z axis, the energy (i.e., the eigenvalue of $\widehat{H}$) associated with the function $\Psi_j^{(F,m_F)}$ is independent of m_F. That is, the $2F + 1$ states with common values of j and F and $m_F = -F, -F + 1, \ldots, F$, are degenerate.

In order to solve equation (A1.4.7) we do not have to choose the basis functions to be eigenfunctions of $\widehat{F}^2$ and $\widehat{F}_Z$, but there are obvious advantages in doing so:

• The Hamiltonian matrix factorizes into blocks for basis functions having common values of F and m_F. This reduces the numerical work involved in diagonalizing the matrix.

• The solutions can be labelled by their values of F and m_F. We say that F and m_F are *good quantum numbers*. With this labelling, it is easier to keep track of the solutions and we can use the good quantum numbers to express selection rules for molecular interactions and transitions. In field-free space only states having the same values of F and m_F can interact, and an electric dipole transition between states with $F = F'$ and F'' will take place if and only if

$$|F' - F''| \leq 1 \text{ and } F' + F'' \geq 1. \tag{A1.4.13}$$

At this point the reader may feel that we have done little in the way of explaining molecular symmetry. All we have done is to state basic results, normally treated in introductory courses on quantum mechanics, connected with the fact that it is possible to find a complete set of simultaneous eigenfunctions for two or more commuting operators. However, as we shall see in section A1.4.3.2, the fact that the molecular Hamiltonian $\widehat{H}$ commutes with $\widehat{F}^2$ and $\widehat{F}_Z$ is intimately connected to the fact that $\widehat{H}$ commutes with (or, equivalently, is invariant to) any rotation of the molecule about a space-fixed axis passing through the centre of mass of the molecule. As stated above, an operation that leaves the Hamiltonian invariant is a symmetry operation of the Hamiltonian. The infinite set of all possible rotations of the molecule about all possible axes that pass through the molecular centre of mass can be collected together to form a *group* (see below). Following the notation of Bunker and Jensen [1] we call this group K(spatial). Since all elements of K(spatial) are symmetry operations of $\widehat{H}$, we say that K(spatial) is a *symmetry group* of $\widehat{H}$. Any group has a set of *irreducible representations* and they define the way coordinates, wavefunctions and operators have to transform under the operations in the group; it so happens that the irreducible representations of K(spatial), $D^{(F)}$, are labelled by the angular momentum quantum number F. The $2F + 1$ functions $|F, m_F\rangle$ (or Ψ^0_{n,F,m_F} or $\Psi^{(F,m_F)}_j$) with a common value of F (and n or j) and $m_F = -F, -F+1, \ldots, F$ transform according to the irreducible representation $D^{(F)}$ of K(spatial). As a result, we can reformulate our procedure for solving the Schrödinger equation of a molecule as follows:

- For the Hamiltonian $\widehat{H}$ we identify a symmetry group, and this is a group of symmetry operations of $\widehat{H}$; a symmetry operation being defined as an operation that leaves $\widehat{H}$ invariant (i.e., that commutes with $\widehat{H}$). In our example, the symmetry group is K(spatial).

- Having done this we solve the Schrödinger equation for the molecule by diagonalizing the Hamiltonian matrix in a complete set of known basis functions. We choose the basis functions so that they transform according to the irreducible representations of the symmetry group.

- The Hamiltonian matrix will be block diagonal in this basis set. There will be one block for each irreducible representation of the symmetry group.

- As a result the eigenstates of $\widehat{H}$ can be labelled by the irreducible representations of the symmetry group and these irreducible representations can be used as 'good quantum numbers' for understanding interactions and transitions.

We have described here one particular type of molecular symmetry, *rotational symmetry*. On one hand, this example is complicated because the appropriate symmetry group, K(spatial), has infinitely many elements. On the other hand, it is simple because each irreducible representation of K(spatial) corresponds to a particular value of the quantum number F which is associated with a physically observable quantity, the angular momentum. Below we describe other types of molecular symmetry, some of which give rise to finite symmetry groups.

A1.4.1.2   A list of the various types of molecular symmetry

The possible types of symmetry for the Hamiltonian of an isolated molecule in field-free space (all of them are discussed in more detail later on in the article) can be listed as follows:

(i)   *Translational symmetry*. A translational symmetry operation displaces all nuclei and electrons in the molecule uniformly in space (i.e., all particles are moved in the same direction and by the same distance). This symmetry is a consequence of the uniformity of space.

(ii) *Rotational symmetry*. A rotational symmetry operation rotates all nuclei and electrons by the same angle about a space-fixed axis that passes through the molecular centre of mass. This symmetry is a consequence of the isotropy of space.

(iii) *Inversion symmetry*. The Hamiltonian that we customarily use to describe a molecule involves only the electromagnetic forces between the particles (nuclei and electrons) and these forces are invariant to the 'inversion operation' E^* which inverts all particle positions through the centre of mass of the molecule. Thus such a Hamiltonian commutes with E^*; the use of this operation leads (as we see in section A1.4.2.5) to the concept of *parity*, and parity can be + or −. This symmetry results from the fact that the electromagnetic force is invariant to inversion. It is not a property of space.

(iv) *Identical particle permutation symmetry*. The corresponding symmetry operations permute identical particles in a molecule. These particles can be electrons, or they can be identical nuclei. This symmetry results from the indistinguishability of identical particles.

(v) *Time reversal symmetry*. The time reversal symmetry operation T or $\widehat{\theta}$ reverses the direction of motion in a molecule by reversing the sign of all linear and angular momenta. This symmetry results from the properties of the Schrödinger equation of a system of particles moving under the influence of electromagnetic forces. It is not a property of space–time.

We hope that by now the reader has it firmly in mind that the way molecular symmetry is defined and used is based on energy invariance and not on considerations of the geometry of molecular equilibrium structures. Symmetry defined in this way leads to the idea of *conservation*. For example, the total angular momentum of an isolated molecule in field-free space is a conserved quantity (like the total energy) since there are no terms in the Hamiltonian that can mix states having different values of F. This point is discussed further in sections A1.4.3.1 and A1.4.3.2.

A1.4.2   Group theory

The use of symmetry involves the mathematical apparatus of *group theory*, and in this section we summarize the basics. We first define the concept of a *group* by considering the permutations of the protons in the phosphine molecule PH_3 (figure A1.4.1) as an example. This leads to the definition of the nuclear permutation group for PH_3. We briefly discuss point groups and then introduce *representations* of a group; in particular we define *irreducible representations*. We then go on to show how wavefunctions are transformed by symmetry operations, and how this enables molecular states to be labelled according to the irreducible representations of the applicable symmetry group. The final subsection explains *the vanishing integral rule* which is of major use in applying molecular symmetry in order to determine which transitions and interactions can and cannot occur.

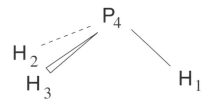

Figure A1.4.1. A PH_3 molecule at equilibrium. The protons are labelled 1, 2 and 3, respectively, and the phosphorus nucleus is labelled 4.

A1.4.2.1   Nuclear permutation groups

The three protons in PH_3 are identical and indistinguishable. Therefore the molecular Hamiltonian will commute with any operation that permutes them, where such a permutation interchanges the space and spin coordinates of the protons. Although this is a rather obvious symmetry, and a proof is hardly necessary, it can be proved by formal algebra as done in chapter 6 of [1].

How many distinct ways of permuting the three protons are there? For example, we can interchange protons 1 and 2. The corresponding symmetry operation is denoted (12) (pronounced 'one–two') and it is to be understood quite literally: protons 1 and 2 interchange their positions in space. There are obviously two further distinct operations of this type: (23) and (31)[1]. A permutation operation that interchanges just two nuclei is called a *transposition*. A more complicated symmetry operation is (123). Here, nucleus 1 is replaced by nucleus 2, nucleus 2 by nucleus 3 and nucleus 3 by nucleus 1. Thus, after (123) nucleus 2 ends up at the position in space initially occupied by nucleus 1, nucleus 3 ends up at the position in space initially occupied by nucleus 2 and nucleus 1 ends up at the position in space initially occupied by nucleus 3. Such an operation, which involves more than two nuclei, is called a *cycle*. A moment's thought will show that in the present case, there exists one other distinct cycle, namely (132). We could write further cycles like (231), (321) etc, but we discover that each of them has the same effect as (123) or (132). There are thus five distinct ways of permuting three protons: (123), (132), (12), (23) and (31).

We can apply permutations successively. For example, we can first apply (12), and then (123); the net effect of doing this is to interchange protons 1 and 3. Thus we have

$$(123)(12) = (31).\tag{A1.4.14}$$

When we apply permutations (or other symmetry operations) successively (this is commonly referred to as *multiplying* the operations so that (31) is the *product* of (123) and (12)), we write the operation to be applied first to the right in the manner done for general quantum mechanical operators. Permutations do not necessarily commute. For example,

$$(12)(123) = (23).\tag{A1.4.15}$$

If we apply the operation (12) twice, or the operation (123) three times, we obviously get back to the starting point. We write this as

$$(12)(12) = (123)(123)(123) = E\tag{A1.4.16}$$

where the *identity operation E* leaves the molecule unchanged by definition. Having defined E, we define the *reciprocal* (or *inverse*) R^{-1} of a symmetry operation R (which, in our present example, could be (123), (132), (12), (23) or (31)) by the equation

$$RR^{-1} = R^{-1}R = E.\tag{A1.4.17}$$

It is easy to verify that for example

$$(12)^{-1} = (12) \text{ and } (123)^{-1} = (132).\tag{A1.4.18}$$

The six operations

$$S_3 = \{E, (123), (132), (12), (23), (31)\}\tag{A1.4.19}$$

are said to form a *group* because they satisfy the following *group axioms*:

(i)   We can multiply (i.e., successively apply) the operations together in pairs and the result is a member of the group.

(ii) One of the operations in the group is the identity operation E.

(iii) The reciprocal of each operation is a member of the group.

---

[1] Clearly, the operation (21) has the same effect as (12), (13) has the same effect as (31) etc.

Table A1.4.1. The multiplication table of the S_3 group.

	E	(123)	(132)	(12)	(23)	(31)
E	E	(123)	(132)	(12)	(23)	(31)
(123)	(123)	(132)	E	(31)	(12)	(23)
(132)	(132)	E	(123)	(23)	(31)	(12)
(12)	(12)	(23)	(31)	E	(123)	(132)
(23)	(23)	(31)	(12)	(132)	E	(123)
(31)	(31)	(12)	(23)	(123)	(132)	E

Each entry is the product of first applying the permutation at the top of the column and then applying the permutation at the left end of the row.

(iv) Multiplication of the operations is associative; that is, in a multiple product the answer is independent of how the operations are associated in pairs, e.g.,

$$(12)(123)(23) = (12)\underbrace{[(123)(23)]}_{(12)} = \underbrace{[(12)(123)]}_{(23)}(23) = E. \tag{A1.4.20}$$

The fact that the group axioms (i), (ii), (iii) and (iv) are satisfied by the set in equation (A1.4.19) can be verified by inspecting the *multiplication table* of the group S_3 given in table A1.4.1; this table lists all products $R_1 R_2$ where R_1 and R_2 are members of S_3. The group S_3 is the permutation group (or symmetric group) of degree 3, and it consists of all permutations of three objects. There are six elements in S_3 and the group is said to have *order* six. In general, the permutation group S_n (all permutations of n objects) has order $n!$.

There is another way of developing the algebra of permutation multiplication, and we briefly explain it. In this approach for PH_3 three positions in space are introduced and labelled $\underline{1}$, $\underline{2}$ and $\underline{3}$; the three protons are labelled H_1, H_2 and H_3. The permutation $(12)^S$ (where S denotes space-fixed position labels) is defined in this approach as permuting the nuclei that are in positions $\underline{1}$ and $\underline{2}$, and the permutation $(123)^S$ as replacing the proton in position $\underline{1}$ by the proton in position $\underline{2}$ etc. With this definition the effect of first doing $(12)^S$ and then doing $(123)^S$ can be drawn as

$$\begin{array}{ccccc} 123 & \xrightarrow{\;(12)^S\;} & 213 & \xrightarrow{\;(123)^S\;} & 132 \\ 123 & & 123 & & 123 \\ \end{array}$$
$$\xrightarrow{\;(23)^S\;}$$

and we see that

$$(123)^S(12)^S = (23)^S. \tag{A1.4.21}$$

This is not the same as equation (A1.4.14). In fact, in this convention, which we can call the S-convention, the multiplication table is the transpose of that given in table A1.4.1. The convention we use and which leads to the multiplication table given in table A1.4.1, will be called the N-convention (where N denotes nuclear-fixed labels).

A1.4.2.2   Point groups

Having defined the concept of a group in section A1.4.2.1, we discuss in the present section a particular type of group that most readers will have heard about: the *point group*. We do this with some reluctance since point group operations do not commute with the complete molecular Hamiltonian and thus they are not true symmetry operations of the kind discussed in section A1.4.1.2. Also the actual effect that the operations have on molecular coordinates is not straightforward to explain. From a pedagogical and logical point of view it would be better to bring them into the discussion of molecular symmetry only after groups consisting of the true symmetry operations enumerated in section A1.4.1.2 have been thoroughly explained. However, because of their historical importance we have decided to introduce them early on. As explained in section A1.4.4 the operations of a molecular point group involve the rotation and/or reflection of vibrational displacement coordinates and electronic coordinates, within the molecular-fixed coordinate system which itself remains fixed in space. Thus the rotational variables (called Euler angles) that define the orientation of a molecule in space are not transformed and in particular the molecule is not subjected to an overall rotation by the operations that are called 'rotations' in the molecular point group. It turns out that the molecular point group is a symmetry group of use in the understanding of the vibrational and electronic states of molecules. However, because of centrifugal and Coriolis forces the vibrational and electronic motion is not completely separable from the rotational motion and, as we explain in section A1.4.5, the molecular point group is only a *near symmetry group* of the complete molecular Hamiltonian appropriate for the hypothetical, non-rotating molecule.

In general, a point group symmetry operation is defined as a rotation or reflection of a macroscopic object such that, after the operation has been carried out, the object looks the same as it did originally. The macroscopic objects we consider here are models of molecules in their equilibrium configuration; we could also consider idealized objects such as cubes, pyramids, spheres, cones, tetrahedra etc. in order to define the various possible point groups.

As an example, we again consider the PH_3 molecule. In its pyramidal equilibrium configuration PH_3 has all three P–H distances equal and all three bond angles $\angle(HPH)$ equal. This object has the point group symmetry C_{3v} where the operations of the group are

$$C_{3v} = \{E, C_3, C_3^2, \sigma_1, \sigma_2, \sigma_3\}. \tag{A1.4.22}$$

The operations in the group can be understood by referring to figure A1.4.2. In this figure the right-handed Cartesian (p, q, r) axis system has origin at the molecular centre of mass, the P nucleus is above the pq plane (the plane of the page), and the three H nuclei are below the pq plane[2]. The operations C_3 and C_3^2 in equation (A1.4.22) are right-handed rotations of $120°$ and $240°$, respectively, about the r axis. In general, we use the notation C_n for a rotation of $2\pi/n$ radians about an axis[3]. Somewhat unfortunately, it is customary to use the symbol C_n to denote not only the rotation operation, but also the rotation axis. That is, we say that the r axis in figure A1.4.2 is a C_3 axis. The operation σ_1 is a reflection in the pr plane (which, with the same unfortunate lack of distinction used in the case of the C_3 operation and the C_3 axis, we call the σ_1 plane), and σ_2 and σ_3 are reflections in the σ_2 and σ_3 planes; these planes are obtained by rotating by $120°$ and $240°$, respectively, about the r axis from the pr plane. As shown in figure A1.4.2, each of the H nuclei in the PH_3 molecule lies in a σ_k plane ($k = 1, 2, 3$) and the P nucleus lies on the C_3 axis. It is clear that the operations of C_{3v} as defined here leave the PH_3 molecule in its static equilibrium configuration looking unchanged. It is important to realize that when we apply the point group operations we do not move the (p, q, r) axes

---

[2] The axis labels (p, q, r) are chosen in order not to confuse this axis system with other systems, such as the molecule fixed axes (x, y, z) discussed below, used to describe molecular motion.

[3] For an observer viewing the pq plane from a point that has a positive r coordinate (figure A1.4.2), the positive right-handed direction of the C_3 and C_3^2 rotations is anticlockwise.

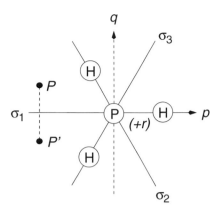

Figure A1.4.2. The PH$_3$ molecule at equilibrium. The symbol $(+r)$ indicates that the r axis points up, out of the plane of the page.

Table A1.4.2. The multiplication table of the C_{3v} point group using the space-fixed axis convention (see text).

	E	C_3	C_3^2	σ_1	σ_2	σ_3
E	E	C_3	C_3^2	σ_1	σ_2	σ_3
C_3	C_3	C_3^2	E	σ_3	σ_1	σ_2
C_3^2	C_3^2	E	C_3	σ_2	σ_3	σ_1
σ_1	σ_1	σ_2	σ_3	E	C_3	C_3^2
σ_2	σ_2	σ_3	σ_1	C_3^2	E	C_3
σ_3	σ_3	σ_1	σ_2	C_3	C_3^2	E

Each entry is the product of first applying the operation at the top of the column and then applying the operation at the left end of the row.

(we call this the 'space-fixed' axis convention) and we will now show how this aspect of the way point group operations are defined affects the construction of the group multiplication table.

Formally, we can say that the operations in C_{3v} act on points in space. For example, we show in figure A1.4.2 how a point P in the pq plane is transformed into another point P' by the operation σ_1; we can say that $P' = \sigma_1 P$. The reader can now show by geometrical considerations that if we first reflect a point P in the σ_1 plane to obtain $P' = \sigma_1 P$, and we then reflect P' in the σ_2 plane to obtain $P'' = \sigma_2 P' = \sigma_2 \sigma_1 P$, then P'' can be obtained directly from P by a 240° anticlockwise rotation about the r axis. Thus $P'' = C_3^2 P$ or generally

$$C_3^2 = \sigma_2 \sigma_1. \tag{A1.4.23}$$

We can also show that

$$C_3 = \sigma_3 \sigma_1. \tag{A1.4.24}$$

The complete multiplication table of the C_{3v} point group, worked out using arguments similar to those leading to equation (A1.4.23) and (A1.4.24), is given in table A1.4.2. It is left as an exercise for the reader to use this table to show that the elements of C_{3v} satisfy the group axioms given in section A1.4.2.1.

If we were to define the operations of the point group as also rotating and reflecting the (p, q, r) axis system (in which case the axes would be 'tied' to the positions of the nuclei), we would obtain a different multiplication table. We could call this the 'nuclear-fixed axis convention.' To implement this the protons in the σ_1, σ_2 and σ_3 planes in figure A1.4.2 would be numbered H_1, H_2 and H_3 respectively. With this convention the C_3 operation would move the σ_1 plane to the position in space originally occupied by the σ_2 plane. If we follow such a C_3 operation by the σ_1 reflection (in the plane containing H_1) we find that, in the nuclear-fixed axis convention:

$$\sigma_1 C_3 = \sigma_3. \tag{A1.4.25}$$

Similarly, with the nuclear-fixed axis convention, we determine that

$$C_3 = \sigma_1 \sigma_3 \tag{A1.4.26}$$

and this result also follows by multiplying equation (A1.4.25) on the left by σ_1. The multiplication table obtained using the nuclear-fixed axis convention is the transpose of the multiplication table obtained using the space-fixed axis convention (compare equations (A1.4.24) and (A1.4.26)). In dealing with point groups we will use the space-fixed axis convention. For defining the effect of permutation operations the S-convention (see equation (A1.4.21)) is related to the N-convention (see equation (A1.4.14)) in the same way that the space-fixed and nuclear-fixed axis conventions for point groups are related.

The operations in a point group are associated with so-called *symmetry elements*. Symmetry elements can be rotation axes (such as the C_3 axis that gives rise to the C_3 and C_3^2 operations in C_{3v}) or reflection planes (such as the planes σ_1, σ_2, σ_3; each of which gives rise to a reflection operation in C_{3v}). A third type of symmetry element not present in C_{3v} is the *rotation–reflection axis* or *improper axis*. For example, an allene molecule H_2CCCH_2 in its equilibrium configuration will be unchanged in appearance by a rotation of $90°$ about the CCC axis combined with a reflection in a plane perpendicular to this axis and containing the 'middle' C nucleus. This operation (a *rotation–reflection* or an *improper rotation*) is called S_4; it is an element of the point group of allene, D_{2d}. Allene is said to have as a symmetry element the rotation-reflection axis or improper axis S_4. It should be noted that neither the rotation of $90°$ about the CCC axis nor the reflection in the plane perpendicular to it are themselves in D_{2d}. For an arbitrary point group, all symmetry elements will intersect at the centre of mass of the object; this point is left unchanged by the group operations and hence the name point group. In order to determine the appropriate point group for a given static arrangement of nuclei, one first identifies the symmetry elements present. Cotton [2] gives in his section 3.14 a systematic procedure to select the appropriate point group from the symmetry elements found. The labels customarily used for point groups (such as C_{3v} and D_{2d}) are named *Schönflies symbols* after their inventor. The most important point groups (defined by their symmetry elements) are

C_n one n-fold rotation axis,
C_{nv} one n-fold rotation axis and n reflection planes containing this axis,
C_{nh} one n-fold rotation axis and one reflection plane perpendicular to this axis,
D_n one n-fold rotation axis and n twofold rotation axes perpendicular to it,
D_{nd} those of D_n plus n reflection planes containing the n-fold rotation axis and bisecting
          the angles between the n twofold rotation axes,
D_{nh} those of D_n plus a reflection plane perpendicular to the n-fold rotation axis,
S_n one alternating axis of symmetry (about which rotation by $2\pi/n$ radians followed
          by reflection in a plane perpendicular to the axis is a symmetry operation).

The point groups T_d, O_h and I_h consist of all rotation, reflection and rotation–reflection symmetry operations of a regular tetrahedron, cube and icosahedron, respectively.

Point groups are discussed briefly in sections 4.3 and 4.4 of [1] and very extensively in chapter 3 of Cotton [2]. We refer the reader to these literature sources for more details.

A1.4.2.3   Irreducible representations and character tables

If we have two groups A and B, of the same order h:

$$A = \{A_1, A_2, A_3, \ldots, A_h\} \tag{A1.4.27}$$

$$B = \{B_1, B_2, B_3, \ldots, B_h\} \tag{A1.4.28}$$

where $A_1 = B_1 = E$, the identity operation and if there is a one-to-one correspondence between the elements of A and B, $A_k \leftrightarrow B_k$, $k = 1, 2, 3, \ldots, h$, so that if

$$A_i A_j = A_m \tag{A1.4.29}$$

it can be inferred that

$$B_i B_j = B_m \tag{A1.4.30}$$

for all $i \le h$ and $j \le h$, then the two groups A and B are said to be *isomorphic*.

As an example we consider the group S_3 introduced in equation (A1.4.19) and the point group C_{3v} given in equation (A1.4.22). Inspection shows that the multiplication table of C_{3v} in table A1.4.2 can be obtained from the multiplication table of the group S_3 (table A1.4.1) by the following mapping:

$$
\begin{array}{llllllll}
S_3 : & E & (123) & (132) & (12) & (23) & (31) \\
C_{3v} : & E & C_3 & C_3^2 & \sigma_3 & \sigma_1 & \sigma_2.
\end{array} \tag{A1.4.31}
$$

Thus, C_{3v} and S_3 are isomorphic.

Homomorphism is analogous to isomorphism. Where an isomorphism is a one-to-one correspondence between elements of groups of the same order, homomorphism is a many-to-one correspondence between elements of groups having different orders. The larger group is said to be homomorphic onto the smaller group. For example, the point group C_{3v} is homomorphic onto $S_2 = \{E, (12)\}$ with the following correspondences:

$$
\begin{array}{lllllll}
C_{3v} : & E & C_3 & C_3^2 & \sigma_1 & \sigma_2 & \sigma_3 \\
S_2 : & & E & & & (12) & .
\end{array} \tag{A1.4.32}
$$

The multiplication table of S_2 has the entries $EE = E$, $E(12) = (12)E = (12)$ and $(12)(12) = E$. If, in the multiplication table of C_{3v} (table A1.4.2), the elements E, C_3 and C_3^2 are each replaced by E (of S_2) and σ_1, σ_2 and σ_3 each by (12), we obtain the multiplication table of S_2 nine times over.

We are particularly concerned with isomorphisms and homomorphisms, in which one of the groups involved is a *matrix group*. In this circumstance the matrix group is said to be a *representation* of the other group. The elements of a matrix group are square matrices, all of the same dimension. The 'successive application' of two matrix group elements (in the sense of group axiom (i) in section A1.4.2.1) is matrix multiplication. Thus, the identity operation E of a matrix group is the unit matrix of the appropriate dimension, and the inverse element of a matrix is its inverse matrix. Matrices and matrix groups are discussed in more detail in section 5.1 of [1].

For the group A in equation (A1.4.27) to be isomorphic to, or homomorphic onto, a matrix group containing matrices of dimension ℓ, say, each element A_k of A is mapped onto an $\ell \times \ell$ matrix $\mathbf{M}_k$, $k = 1, 2, 3, 4, \ldots, h$, and equations (A1.4.29) and (A1.4.30) can be rewritten in the form

$$\text{if } A_i A_j = A_m \text{ then } \mathbf{M}_i \mathbf{M}_j = \mathbf{M}_m \tag{A1.4.33}$$

for all $i \le h$ and $j \le h$. The latter part of this equation says that the $\ell \times \ell$ matrix $\mathbf{M}_m$ is the product of the two $\ell \times \ell$ matrices $\mathbf{M}_i$ and $\mathbf{M}_j$.

If we have found one representation of ℓ-dimensional matrices $\mathbf{M}_1$, $\mathbf{M}_2$, $\mathbf{M}_3$, ... of the group A, then, at least for $\ell > 1$, we can define infinitely many other *equivalent representations* consisting of the matrices

$$\mathbf{M}'_k = \mathbf{V}^{-1}\mathbf{M}_k\mathbf{V} \qquad k = 1, 2, 3, \ldots, h \tag{A1.4.34}$$

where $\mathbf{V}$ is an $\ell \times \ell$ matrix. The determinant of $\mathbf{V}$ must be nonvanishing, so that $\mathbf{V}^{-1}$ exists, but otherwise $\mathbf{V}$ is arbitrary. We say that $\mathbf{M}'_k$ is obtained from $\mathbf{M}_k$ by a *similarity transformation*. It is straightforward to show that the matrices $\mathbf{M}'_k$, $k = 1, 2, 3, \ldots, h$ form a representation of A since they satisfy an equation analogous to equation (A1.4.33).

It is well known that the *trace* of a square matrix (i.e., the sum of its diagonal elements) is unchanged by a similarity transformation. If we define the traces

$$\chi'_k = \sum_{p=1}^{\ell}(\mathbf{M}'_k)_{pp} \text{ and } \chi_k = \sum_{p=1}^{\ell}(\mathbf{M}_k)_{pp} \tag{A1.4.35}$$

we have

$$\chi'_k = \chi_k. \tag{A1.4.36}$$

The traces of the representation matrices are called the *characters* of the representation, and equation (A1.4.36) shows that all equivalent representations have the same characters. Thus, the characters serve to distinguish inequivalent representations.

If we select an element of A, A_j say, and determine the set of elements S given by forming all products

$$S = R^{-1}A_jR \tag{A1.4.37}$$

where R runs over all elements of A, then the set of distinct elements obtained, which will include A_j (since for $R = R^{-1} = E$ we have $S = A_j$), is said to form a *class* of A. For any group the identity operation E is always in a class of its own since for all R we have $S = R^{-1}ER = R^{-1}R = E$. The reader can use the multiplication table (table A1.4.1) to determine the classes of the group S_3 (equation (A1.4.19)); there are three classes $[E]$, $[(123), (132)]$ and $[(12), (23), (31)]$. Since the groups S_3 and C_{3v} (equation (A1.4.22)) are isomorphic, the class structure of C_{3v} can be immediately inferred from the class structure of S_3 together with equation (A1.4.31). C_{3v} has the classes $[E]$, $[C_3, C_3^2]$ and $[\sigma_1, \sigma_2, \sigma_3]$.

If two elements of A, A_i and A_j say, are in the same class, then there exists a third element of A, R, such that

$$A_i = R^{-1}A_jR. \tag{A1.4.38}$$

Then by equation (A1.4.33)

$$\mathbf{M}_i = \mathbf{M}_R^{-1}\mathbf{M}_j\mathbf{M}_R \tag{A1.4.39}$$

where $\mathbf{M}_i$, $\mathbf{M}_j$ and $\mathbf{M}_R$ are the representation matrices associated with A_i, A_j and R, respectively. That is, $\mathbf{M}_i$ is obtained from $\mathbf{M}_j$ in a similarity transformation, and these two matrices thus have the same trace or character. Consequently, all the elements in a given class of a group are represented by matrices with the same character.

If we start with an ℓ-dimensional representation of A consisting of the matrices $\mathbf{M}_1$, $\mathbf{M}_2$, $\mathbf{M}_3$, ..., it may be that we can find a matrix $\mathbf{V}$ such that when it is used with equation (A1.4.34) it produces an equivalent representation $\mathbf{M}'_1$, $\mathbf{M}'_2$, $\mathbf{M}'_3$, ... each of whose matrices is in the same *block diagonal form*. For example, the nonvanishing elements of each of the matrices $\mathbf{M}'_k$ could form an upper-left-corner $\ell_1 \times \ell_1$ block and a lower-right-corner $\ell_2 \times \ell_2$ block, where $\ell_1 + \ell_2 = \ell$. In this situation, a few moments' consideration of the rules of matrix multiplication shows that all the upper-left-corner $\ell_1 \times \ell_1$ blocks, taken on their own, form an ℓ_1-dimensional representation of A and all the lower-right-corner $\ell_2 \times \ell_2$ blocks, taken on their own, form an

Table A1.4.3. The character table of the S_3 group.

S_3	E	(123)	(12)
	1	2	3
A_1	1	1	1
A_2	1	1	-1
E	2	-1	0

One representative element in each class is given, and the number written below each element is the number of elements in the class.

ℓ_2-dimensional representation. In these circumstances the original representation Γ consisting of $\mathbf{M}_1, \mathbf{M}_2, \mathbf{M}_3$, ... is *reducible* and we have *reduced* it to the sum of the two representations, Γ_1 and Γ_2 say, of dimensions ℓ_1 and ℓ_2, respectively. We write this reduction as

$$\Gamma = \Gamma_1 \oplus \Gamma_2. \tag{A1.4.40}$$

Clearly, a one-dimensional representation (also called a *non-degenerate* representation) is of necessity *irreducible* in that it cannot be reduced to representations of lower dimensions. *Degenerate representations* (i.e., groups of matrices with dimension higher than 1) can also be irreducible, which means that there is no matrix that by a similarity transformation will bring all the matrices of the representation into the same block diagonal form. It can be shown that the number of irreducible representations of a given group is equal to the number of classes in the group. We have seen that the group S_3 has three classes $[E]$, $[(123), (132)]$ and $[(23), (31), (12)]$ and therefore it has three irreducible representations. For a general group with n irreducible representations with dimensions $\ell_1, \ell_2, \ell_3, \ldots, \ell_n$, it can also be shown that

$$\sum_{i=1}^{n} \ell_i^2 = h \tag{A1.4.41}$$

where h is the order of the group. For S_3 this equation yields

$$\ell_1^2 + \ell_2^2 + \ell_3^2 = 6 \tag{A1.4.42}$$

and, since the ℓ_i have to be positive integers, we obtain $\ell_1 = \ell_2 = 1$ and $\ell_3 = 2$. When developing general formulae we label the irreducible representations of a group as $\Gamma_1, \Gamma_2, \ldots, \Gamma_n$ and denote the characters associated with Γ_i as $\chi^{\Gamma_i}[R]$, where R is an element of the group under study. However, the irreducible representations of symmetry groups are denoted by various other special symbols such as A_2, Σ^- and $D^{(4)}$. The characters of the irreducible representations of a symmetry group are collected together into a *character table* and the character table of the group S_3 is given in table A1.4.3. The construction of character tables for finite groups is treated in section 4.4 of [2] and section 3-4 of [3].

For any Γ_i we have $\chi^{\Gamma_i}[E] = \ell_i$, the dimension of Γ_i. This is because the identity operation E is always represented by an $\ell_i \times \ell_i$ unit matrix whose trace obviously is ℓ_i. For any group there will be one irreducible representation (called the *totally symmetric representation* $\Gamma^{(s)}$) which has all $\chi^{\Gamma_i}[R] = 1$. Such a representation exists because any group is homomorphic onto the one-member matrix group $\{1\}$ (where the '1' is interpreted as a 1×1 matrix). The irreducible characters $\chi^{\Gamma_i}[R]$ satisfy several equations (see, for example, section 4.3 of [2] and section 3-3 of [3]), for example

$$\sum_R \chi^{\Gamma_i}[R]^* \chi^{\Gamma_j}[R] = h\delta_{ij} \tag{A1.4.43}$$

where the sum runs over all elements R of the group.

In applications of group theory we often obtain a reducible representation, and we then need to reduce it to its irreducible components. The way that a given representation of a group is reduced to its irreducible components depends only on the characters of the matrices in the representation and on the characters of the matrices in the irreducible representations of the group. Suppose that the reducible representation is Γ and that the group involved has irreducible representations that we label $\Gamma_1, \Gamma_2, \Gamma_3, \ldots$. What we mean by 'reducing' Γ is finding the integral coefficients a_i in the expression

$$\Gamma = a_1\Gamma_1 \oplus a_2\Gamma_2 \oplus a_3\Gamma_3 \oplus \cdots \tag{A1.4.44}$$

where

$$\chi^\Gamma[R] = \sum_j a_j \chi^{\Gamma_j}[R] \tag{A1.4.45}$$

with the sum running over all the irreducible representations of the group. Multiplying equation (A1.4.45) on the right by $\chi^{\Gamma_i}[R]^*$ and summing over R it follows from the character orthogonality relation (equation (A1.4.43)) that the required a_i are given by

$$a_i = \frac{1}{h} \sum_R \chi^\Gamma[R]\chi^{\Gamma_i}[R]^* \tag{A1.4.46}$$

where h is the order of the group and R runs over all the elements of the group.

A1.4.2.4 The effects of symmetry operations

For the PH_3 molecule, which we continue using as an example, we consider that proton i ($= 1, 2$ or 3) initially has the coordinates (X_i, Y_i, Z_i) in the (X, Y, Z) axis system, and the phosphorus nucleus has the coordinates (X_4, Y_4, Z_4). After applying the permutation operation (12) to the PH_3 molecule, nucleus 1 is where nucleus 2 was before. Consequently, nucleus 1 now has the coordinates (X_2, Y_2, Z_2). Nucleus 2 is where nucleus 1 was before and has the coordinates (X_1, Y_1, Z_1). Thus we can write

$$\begin{aligned}
(12)\,&[X_1, Y_1, Z_1, X_2, Y_2, Z_2, X_3, Y_3, Z_3, X_4, Y_4, Z_4] \\
&= [X_1', Y_1', Z_1', X_2', Y_2', Z_2', X_3', Y_3', Z_3', X_4', Y_4', Z_4'] \\
&= [X_2, Y_2, Z_2, X_1, Y_1, Z_1, X_3, Y_3, Z_3, X_4, Y_4, Z_4]
\end{aligned} \tag{A1.4.47}$$

where X_i', Y_i' and Z_i' are the X, Y and Z coordinates of nucleus i after applying the permutation (12). By convention we always give first the (X, Y, Z) coordinates of nucleus 1, then those of nucleus 2, then those of nucleus 3 etc.

Similarly, after applying the operation (123) to the PH_3 molecule, nucleus 2 is where nucleus 1 was before and has the coordinates (X_1, Y_1, Z_1). Nucleus 3 is where nucleus 2 was before and has the coordinates (X_2, Y_2, Z_2) and, finally, nucleus 1 is where nucleus 3 was before and has the coordinates (X_3, Y_3, Z_3). So

$$\begin{aligned}
(123)\,&[X_1, Y_1, Z_1, X_2, Y_2, Z_2, X_3, Y_3, Z_3, X_4, Y_4, Z_4] \\
&= [X_1', Y_1', Z_1', X_2', Y_2', Z_2', X_3', Y_3', Z_3', X_4', Y_4', Z_4'] \\
&= [X_3, Y_3, Z_3, X_1, Y_1, Z_1, X_2, Y_2, Z_2, X_4, Y_4, Z_4]
\end{aligned} \tag{A1.4.48}$$

where here X_i', Y_i' and Z_i' are the X, Y and Z coordinates of nucleus i after applying the permutation (123).

The procedure exemplified by equations (A1.4.47) and (A1.4.48) can be trivially generalized to define the effect of any symmetry operation, R say, on the coordinates (X_i, Y_i, Z_i) of any nucleus or electron i in any molecule by writing

$$R[X_i, Y_i, Z_i] = [RX_i, RY_i, RZ_i] = [X_i', Y_i', Z_i']. \tag{A1.4.49}$$

We can also write

$$R^{-1}[X_i', Y_i', Z_i'] = [R^{-1}X_i', R^{-1}Y_i', R^{-1}Z_i'] = [X_i, Y_i, Z_i]. \qquad (A1.4.50)$$

We use the nuclear permutation operations (123) and (12) to show what happens when we apply two operations in succession. We write the successive effect of these two permutations as (remember that we are using the N-convention; see equation (A1.4.14))

$$
\begin{aligned}
(123)(12) \, & [X_1, Y_1, Z_1, X_2, Y_2, Z_2, X_3, Y_3, Z_3, X_4, Y_4, Z_4] \\
= (123) \, & [X_1', Y_1', Z_1', X_2', Y_2', Z_2', X_3', Y_3', Z_3', X_4', Y_4', Z_4'] \\
= & [X_3', Y_3', Z_3', X_1', Y_1', Z_1', X_2', Y_2', Z_2', X_4', Y_4', Z_4'] \\
= & [X_3, Y_3, Z_3, X_2, Y_2, Z_2, X_1, Y_1, Z_1, X_4, Y_4, Z_4] \\
= (31) \, & [X_1, Y_1, Z_1, X_2, Y_2, Z_2, X_3, Y_3, Z_3, X_4, Y_4, Z_4]
\end{aligned}
\qquad (A1.4.51)
$$

where X_i', Y_i', Z_i' are the coordinates of the nuclei after applying the operation (12). The result in equation (A1.4.51) is in accord with equation (A1.4.14).

Molecular wavefunctions are functions of the coordinates of the nuclei and electrons in a molecule, and we are now going to consider how such functions can be transformed by the general symmetry operation R as defined in equation (A1.4.49). To do this we introduce three functions of the coordinates, $f(X_i, Y_i, Z_i)$, $f_N^R(X_i, Y_i, Z_i)$ and $f_S^R(X_i, Y_i, Z_i)$. The functions f_N^R and f_S^R are such that their values at any point in configuration space are each related to the value of the function f at another point in configuration space, where the coordinates of this 'other' point are defined by the effect of R as follows:

$$f_N^R(X_i, Y_i, Z_i) = f(X_i', Y_i', Z_i') \qquad (A1.4.52)$$

and

$$f_S^R(X_i', Y_i', Z_i') = f(X_i, Y_i, Z_i) \qquad (A1.4.53)$$

or equivalently,

$$f_N^R(X_i, Y_i, Z_i) = f(RX_i, RY_i, RZ_i) \qquad (A1.4.54)$$

and

$$f_S^R(X_i, Y_i, Z_i) = f(R^{-1}X_i, R^{-1}Y_i, R^{-1}Z_i), \qquad (A1.4.55)$$

This means that f_N^R is such that its value at any point (X_i, Y_i, Z_i) is the same as the value of f at the point (RX_i, RY_i, RZ_i), and that f_S^R is such that its value at any point (X_i, Y_i, Z_i) is the same as the value of f at the point $(R^{-1}X_i, R^{-1}Y_i, R^{-1}Z_i)$. Alternatively, for the latter we can say that f_S^R is such that its value at (RX_i, RY_i, RZ_i) is the same as the value of f at the point (X_i, Y_i, Z_i).

We define the effect of a symmetry operation on a wavefunction in two different ways depending on whether the symmetry operation concerned uses a moving or fixed 'reference frame' (see [4]). Either we define its effect using the equation

$$Rf(X_i, Y_i, Z_i) = f_N^R(X_i, Y_i, Z_i) = f(RX_i, RY_i, RZ_i), \qquad (A1.4.56)$$

or we define its effect using

$$Rf(X_i, Y_i, Z_i) = f_S^R(X_i, Y_i, Z_i) = f(R^{-1}X_i, R^{-1}Y_i, R^{-1}Z_i). \qquad (A1.4.57)$$

Nuclear permutations in the N-convention (which convention we always use for nuclear permutations) and rotation operations relative to a nuclear-fixed or molecule-fixed reference frame, are defined to transform wavefunctions according to equation (A1.4.56). These symmetry operations involve a moving reference frame. Nuclear permutations in the S-convention, point group operations in the space-fixed axis convention (which is the convention that is always used for point group operations; see section A1.4.2.2) and rotation operations relative to a space-fixed frame are defined to transform wavefunctions according to equation (A1.4.57). These operations involve a fixed reference frame.

Another distinction we make concerning symmetry operations involves the *active* and *passive* pictures. Below we consider translational and rotational symmetry operations. We describe these operations in a space-fixed axis system (X, Y, Z) with axes parallel to the (X, Y, Z) axes, but with the origin fixed in space. In the active picture, which we adopt here, a translational symmetry operation displaces all nuclei and electrons in the molecule along a vector A, say, and leaves the (X, Y, Z) axis system unaffected. In the passive picture, the molecule is left unaffected but the (X, Y, Z) axis system is displaced by $-A$. Similarly, in the active picture a rotational symmetry operation physically rotates the molecule, leaving the axis system unaffected, whereas in the passive picture the axis system is rotated and the molecule is unaffected. If we think about symmetry operations in the passive picture, it is immediately obvious that they must leave the Hamiltonian invariant (i.e., commute with it). The energy of an isolated molecule in field-free space is obviously unaffected if we translate or rotate the (X, Y, Z) axis system.

A1.4.2.5   The labelling of molecular energy levels

The irreducible representations of a symmetry group of a molecule are used to label its energy levels. The way we label the energy levels follows from an examination of the effect of a symmetry operation on the molecular Schrödinger equation.

$$\widehat{H}\Psi_n(X_i, Y_i, Z_i) = E_n\Psi_n(X_i, Y_i, Z_i) \tag{A1.4.58}$$

where $\Psi_n(X_i, Y_i, Z_i)$ is a molecular eigenfunction having eigenvalue E_n.

By definition, a symmetry operation R commutes with the molecular Hamiltonian $\widehat{H}$ and so we can write the operator equation:

$$\widehat{H}R = R\widehat{H}. \tag{A1.4.59}$$

If we act with each side of this equation on an eigenfunction $\Psi_n(X_i, Y_i, Z_i)$ from equation (A1.4.58) we derive

$$\widehat{H}R\Psi_n(X_i, Y_i, Z_i) = R\widehat{H}\Psi_n(X_i, Y_i, Z_i) = RE_n\Psi_n(X_i, Y_i, Z_i) = E_nR\Psi_n(X_i, Y_i, Z_i). \tag{A1.4.60}$$

The second equality follows from equation (A1.4.58)[4], and the third equality from the fact that E_n is a number and numbers are not affected by symmetry operations. We can rewrite the result of equation (A1.4.60) as

$$\widehat{H}[R\Psi_n(X_i, Y_i, Z_i)] = E_n[R\Psi_n(X_i, Y_i, Z_i)]. \tag{A1.4.61}$$

Thus

$$R\Psi_n(X_i, Y_i, Z_i) = \Psi_n^R(X_i, Y_i, Z_i) \tag{A1.4.62}$$

is an eigenfunction having the same eigenvalue as $\Psi_n(X_i, Y_i, Z_i)$. If E_n is a nondegenerate eigenvalue then Ψ_n^R cannot be linearly independent of Ψ_n, which means that we can only have

$$R\Psi_n(X_i, Y_i, Z_i) = c\Psi_n(X_i, Y_i, Z_i) \tag{A1.4.63}$$

where c is a constant. An arbitrary symmetry operation R is such that $R^m = E$ the identity, where m is an integer. From equation (A1.4.63) we deduce that

$$R^m\Psi_n(X_i, Y_i, Z_i) = c^m\Psi_n(X_i, Y_i, Z_i). \tag{A1.4.64}$$

---

[4] Equivalently, it follows if we apply R to both sides of equation (A1.4.58) and then use equation (A1.4.59) on the left hand side.

Table A1.4.4. The character table of a symmetry group for the H_2 molecule.

	E	(12)	E^*	$(12)^*$
$+s$	1	1	1	1
$-s$	1	1	-1	-1
$+a$	1	-1	1	-1
$-a$	1	-1	-1	1

Since $R^m = E$ we must have $c^m = 1$ in equation (A1.4.64), which gives

$$c = \sqrt[m]{1}. \tag{A1.4.65}$$

Thus, for example, for the PH_3 molecule any nondegenerate eigenfunction can only be multiplied by $+1$, $\omega = \exp(2\pi \, i/3)$, or $\omega^2 = \exp(4\pi \, i/3)$ by the symmetry operation (123) since $(123)^3 = E$ (so that $m = 3$ in equation (A1.4.65)). In addition, such a function can only be multiplied by $+1$ or -1 by the symmetry operations (12), (23) or (31) since each of these operations is self-reciprocal (so that $m = 2$ in equation (A1.4.65)).

We will apply this result to the H_2 molecule as a way of introducing the fact that nondegenerate molecular energy levels can be labelled according to the one-dimensional irreducible representations of a symmetry group of the molecular Hamiltonian. The Hamiltonian for the H_2 molecule commutes with E^* and with the operation (12) that permutes the protons. Thus, the eigenfunction of any nondegenerate molecular energy level is either invariant, or changed in sign, by the inversion operation E^* since $(E^*)^2 = E$ (i.e., $m = 2$ for $R = E^*$ in equation (A1.4.65)); invariant states are said to have positive parity (+) and states that are changed in sign by E^* to have negative parity ($-$). Similarly, any nondegenerate energy level will be invariant or changed in sign by the proton permutation operation (12); states that are invariant are said to be symmetric (s) with respect to (12) and states that are changed in sign are said to be antisymmetric (a). This enables us to label nondegenerate energy levels of the H_2 molecule as being $(+s)$, $(-s)$, $(+a)$ or $(-a)$ according to the effect of the operations E^* and (12). For the H_2 molecule we can form a symmetry group using these elements: $\{E, (12), E^*, (12)^*\}$, where

$$(12)^* = (12)E^* = E^*(12) \tag{A1.4.66}$$

and the character table of the group is given in table A1.4.4. The effect of the operation $(12)^*$ on a wavefunction is simply the product of the effects of (12) and E^*. The labelling of the states as $(+s)$, $(-s)$, $(+a)$ or $(-a)$ is thus according to the irreducible representations of the symmetry group and the nondegenerate energy levels of the H_2 molecule are of four different symmetry types in this group.

The energy level of an an l-fold degenerate eigenstate can be labelled according to an l-fold degenerate irreducible representation of the symmetry group, as we now show. Suppose the l orthonormal[5] eigenfunctions $\Psi_{n1}, \Psi_{n2}, \ldots, \Psi_{nl}$ all have the same eigenvalue E_n of the molecular Hamiltonian. If we apply a symmetry operation R to one of these functions the resulting function will also be an eigenfunction of the Hamiltonian with eigenvalue E_n (see equation (A1.4.61) and the sentence after it) and the most general function of this type is a linear combination of the l functions Ψ_{ni} given above. Thus, using matrix notation, we can write the effect of R as[6]

$$R\Psi_{ni} = \sum_{j=1}^{l} D[R]_{ij} \Psi_{nj} \tag{A1.4.67}$$

---

[5] Two functions Ψ_{ni} and Ψ_{nj} are orthogonal if the product $\Psi_{ni}^* \Psi_{nj}$, integrated over all configuration space, vanishes. A function Ψ is normalized if the product $\Psi^* \Psi$ integrated over all configuration space is unity. An orthonormal set contains functions that are normalized and orthogonal to each other.

[6] Note the order of the subscripts on $D[R]$ which follows from the fact that we use the N-convention of equation (A1.4.56) to define the effect of a permutation on a function.

where $i = 1, 2, \ldots, l$. For example, choosing $i = 1$, we have the effect of R on Ψ_{n1} as:

$$R\Psi_{n1} = D[R]_{11}\Psi_{n1} + D[R]_{12}\Psi_{n2} + \cdots + D[R]_{1l}\Psi_{nl}. \tag{A1.4.68}$$

The $D[R]_{ij}$ are numbers and $D[R]$ is a matrix of these numbers; the matrix $D[R]$ is *generated* by the effect of R on the l functions Ψ_{ni}. We can visualize equation (A1.4.67) as the effect of R acting on a column matrix Ψ_n being equal to the product of a square matrix $D[R]$ and a column matrix Ψ_n, i.e.,

$$R\left[\Psi_n\right] = \left[D[R]\right]\left[\Psi_n\right]. \tag{A1.4.69}$$

Each operation in a symmetry group of the Hamiltonian will generate such an $l \times l$ matrix, and it can be shown (see, for example, appendix 6-1 of [1]) that if three operations of the group P_1, P_2 and P_{12} are related by

$$P_1 P_2 = P_{12} \tag{A1.4.70}$$

then the matrices generated by application of them to the Ψ_{ni} (as described by equation (A1.4.67)) will satisfy

$$D[P_1]D[P_2] = D[P_{12}]. \tag{A1.4.71}$$

Thus, the matrices will have a multiplication table with the same structure as the multiplication table of the symmetry group and hence will form an l-dimensional representation of the group.

A given l-fold degenerate state can generate a reducible or an irreducible l-dimensional representation of the symmetry group considered. If the representation is irreducible then the degeneracy is said to be *necessary*, i.e., imposed by the symmetry of the Hamiltonian. However, if the representation is reducible then the degeneracy between the different states is said to be accidental and it is not imposed by the symmetry of the Hamiltonian. The occurrence of accidental degeneracy can indicate that some other symmetry operation has been forgotten, or paradoxically it can indicate that too many symmetry operations (called *unfeasible* symmetry operations in section A1.4.4) have been introduced.

These considerations mean that, for example, using the symmetry group S_3 for the PH$_3$ molecule (see table A1.4.3), the energy levels are determined to be of symmetry type A_1, A_2 or E. In molecular physics the labelling of molecular energy levels according to the irreducible representations of a symmetry group is mainly what we use symmetry for. Once we have labelled the energy levels of a molecule, we can use the labels to determine which of the levels can *interact* with each other as the result of adding a term $\widehat{H}'$ to the molecular Hamiltonian. This term could be the result of applying an external perturbation such as an electric or magnetic field, it could be the result of including a previously unconsidered term from the Hamiltonian, or this term could result from the effect of shining electromagnetic radiation through a gas of the molecules. In this latter case the symmetry labels enable us to determine the selection rules for allowed transitions in the spectrum of the molecule. All this becomes possible by making use of the *vanishing integral rule*.

A1.4.2.6   The vanishing integral rule

To explain the vanishing integral rule we first have to explain how we determine the symmetry of a product. Given an s-fold degenerate state of energy E_n and symmetry Γ_n, with eigenfunctions $\Phi_{n1}, \Phi_{n2}, \ldots, \Phi_{ns}$, and an r-fold degenerate state of energy E_m and symmetry Γ_m, with eigenfunctions $\Phi_{m1}, \Phi_{m2}, \ldots, \Phi_{mr}$, we wish to determine the symmetry Γ_{mn} of the set of functions $\Psi_{ij} = \Phi_{ni}\Phi_{mj}$, where $i = 1, 2, \ldots, s$ and $j = 1, 2, \ldots, r$. There will be $s \times r$ functions of the type Ψ_{ij}. The matrices D^{Γ_n} and D^{Γ_m} in the representations Γ_n and Γ_m, respectively, are obtained from (see equation (A1.4.67))

$$R\Phi_{ni} = \sum_{k=1}^{s} D^{\Gamma_n}[R]_{ik}\Phi_{nk} \tag{A1.4.72}$$

and

$$R\Phi_{mj} = \sum_{l=1}^{r} D^{\Gamma_m}[R]_{jl}\Phi_{ml} \tag{A1.4.73}$$

where R is an operation of the symmetry group. To obtain the matrices in the representation Γ_{nm} we write

$$R[\Phi_{ni}\Phi_{mj}] = \sum_{k=1}^{s}\sum_{l=1}^{r} D^{\Gamma_n}[R]_{ik}D^{\Gamma_m}[R]_{jl}\Phi_{nk}\Phi_{ml} \tag{A1.4.74}$$

and we can write this as

$$R\Psi_{ij} = \sum_{k=1}^{s}\sum_{l=1}^{r} D^{\Gamma_{nm}}[R]_{ij,kl}\Psi_{kl}. \tag{A1.4.75}$$

From this we see that the $s \times r$ dimensional representation Γ_{nm} generated by the $s \times r$ functions Ψ_{ij} has matrices with elements given by

$$D^{\Gamma_{nm}}[R]_{ij,kl} = D^{\Gamma_n}[R]_{ik}D^{\Gamma_m}[R]_{jl} \tag{A1.4.76}$$

where each element of $D^{\Gamma_{nm}}$ is indexed by a row label ij and a column label kl, each of which runs over $s \times r$ values. The ij, ij diagonal element is given by

$$D^{\Gamma_{nm}}[R]_{ij,ij} = D^{\Gamma_n}[R]_{ii}D^{\Gamma_m}[R]_{jj} \tag{A1.4.77}$$

and the character of the matrix is given by

$$\chi^{\Gamma_{nm}}[R] = \sum_{k=1}^{s}\sum_{l=1}^{r} D^{\Gamma_{nm}}[R]_{ij,ij} = \sum_{k=1}^{s}\sum_{l=1}^{r} D^{\Gamma_n}[R]_{ii}D^{\Gamma_m}[R]_{jj} = \chi^{\Gamma_n}[R]\chi^{\Gamma_m}[R]. \tag{A1.4.78}$$

We can therefore calculate the character, under a symmetry operation R, in the representation generated by the product of two sets of functions, by multiplying together the characters under R in the representations generated by each of the sets of functions. We write Γ_{nm} symbolically as

$$\Gamma_{nm} = \Gamma_n \otimes \Gamma_m \tag{A1.4.79}$$

where the characters satisfy equation (A1.4.78) in which usual algebraic multiplication is used. Knowing the character in Γ_{nm} from equation (A1.4.78) we can then reduce the representation to its irreducible components using equation (A1.4.47). Suppose Γ_{nm} can be reduced to irreducible representations Γ_1, Γ_2 and Γ_3 according to

$$\Gamma_n \otimes \Gamma_m = 3\Gamma_1 \oplus \Gamma_2 \oplus 2\Gamma_3. \tag{A1.4.80}$$

In this circumstance we say that Γ_{nm} *contains* Γ_1, Γ_2 and Γ_3; since $\Gamma_n \otimes \Gamma_m$ contains Γ_1, for example, we write

$$\Gamma_n \otimes \Gamma_m \supset \Gamma_1. \tag{A1.4.81}$$

Suppose that we can write the total Hamiltonian as $\widehat{H} = \widehat{H}^0 + \widehat{H}'$, where $\widehat{H}'$ is a perturbation. Let us further suppose that the Hamiltonian $\widehat{H}^0$ ($\widehat{H}'$ having been neglected) has normalized eigenfunctions Ψ_m^0 and Ψ_n^0, with eigenvalues E_m^0 and E_n^0, respectively, and that $\widehat{H}^0$ commutes with the group of symmetry operations $G = \{R_1, R_2, \ldots, R_h\}$. $\widehat{H}^0$ will transform as the totally symmetric representation $\Gamma^{(s)}$ of G, and we let Ψ_m^0, Ψ_n^0 and $\widehat{H}'$ generate the representations Γ_m, Γ_n and Γ' of G, respectively. The complete set of eigenfunctions of $\widehat{H}^0$ forms a basis set for determining the eigenfunctions and eigenvalues of the Hamiltonian

$\widehat{H} = \widehat{H}^0 + \widehat{H}'$ and the Hamiltonian matrix H in this basis set is a matrix with elements H_{mn} given by the integrals

$$H_{mn} = \int \Psi_m^0{}^* (\widehat{H}^0 + \widehat{H}') \Psi_n^0 \, d\tau = \delta_{mn} E_n^0 + H'_{mn} \qquad (A1.4.82)$$

where

$$H'_{mn} = \int \Psi_m^0{}^* \widehat{H}' \Psi_n^0 \, d\tau. \qquad (A1.4.83)$$

The eigenvalues E of $\widehat{H}$ can be determined from the Hamiltonian matrix by solving the secular equation

$$|H_{mn} - \delta_{mn} E| = 0. \qquad (A1.4.84)$$

In solving the secular equation it is important to know which of the off-diagonal matrix elements H'_{mn} vanish since this will enable us to simplify the equation.

We can use the symmetry labels Γ_m and Γ_n on the levels E_m^0 and E_n^0, together with the symmetry Γ' of $\widehat{H}'$, to determine which H'_{mn} elements must vanish. The function $\Psi_m^0{}^* \widehat{H}' \Psi_n^0$ generates the product representation $\Gamma_m{}^* \otimes \Gamma' \otimes \Gamma_n = \Gamma'_{mn}$ ($\Psi_m^0{}^*$ has symmetry $\Gamma_m{}^*$). We can now state *the vanishing integral rule*[7]: the matrix element

$$\int \Psi_m^0{}^* \widehat{H}' \Psi_n^0 \, d\tau = 0 \qquad (A1.4.85)$$

if

$$\Gamma_m{}^* \otimes \Gamma' \otimes \Gamma_n \not\supset \Gamma^{(s)} \qquad (A1.4.86)$$

where $\Gamma^{(s)}$ is the totally symmetric representation. If $\widehat{H}'$ is totally symmetric in G then H'_{mn} will vanish if

$$\Gamma_m{}^* \otimes \Gamma_n \not\supset \Gamma^{(s)} \qquad (A1.4.87)$$

i.e., if

$$\Gamma_m \neq \Gamma_n. \qquad (A1.4.88)$$

It would be an accident if H'_{mn} vanished even though $\Gamma'_{mn} \supset \Gamma^{(s)}$, but if this were the case it might well indicate that there is some extra unconsidered symmetry present.

The value of the vanishing integral rule is that it allows the matrix H to be block diagonalized. This occurs if we order the eigenfunctions Ψ_n^0 according to their symmetry when we set up H. Let us initially consider the case when $\Gamma' = \Gamma^{(s)}$. In this case all off-diagonal matrix elements between Ψ_n^0 basis functions of different symmetry will vanish, and the Hamiltonian matrix will block diagonalize with there being one block for each symmetry type of Ψ_n^0 function. Each eigenfunction of $\widehat{H}$ will only be a linear combination of Ψ_n^0 functions having the same symmetry in G (G being the symmetry group of $\widehat{H}^0$). Thus the symmetry of each eigenfunction Ψ_j of $\widehat{H}$ in the group G will be the same as the symmetry of the Ψ_n^0 basis functions that make it up (G is a symmetry group of $\widehat{H}$ when $\Gamma' = \Gamma^{(s)}$) and each block of a block diagonal matrix can be diagonalized separately, which is a great simplification. The symmetry of the Ψ_j functions can be obtained from the symmetry of the Ψ_n^0 functions without worrying about the details of $\widehat{H}'$ and this is frequently very useful. When $\Gamma' \neq \Gamma^{(s)}$ all off-diagonal matrix elements between Ψ^0 functions of symmetry Γ_m and Γ_n will vanish if equation (A1.4.87) is satisfied, and there will also be a block diagonalization of H (it will be necessary to rearrange the rows or columns of H, i.e., to rearrange the order of the Ψ_n^0 functions, to obtain H in block diagonal form). However, now nonvanishing matrix elements occur in H that connect Ψ_n^0 functions of different symmetry in G and as a result the eigenfunctions of $\widehat{H}$ may not contain only functions of one symmetry type of G; when $\Gamma' \neq \Gamma^{(s)}$ the group G is not a symmetry group of $\widehat{H}$ and its eigenfunctions

---

[7] Proved, for example, in section 6.5 of [1].

Ψ_j cannot be classified in G. However, the classification of the basis functions Ψ_n^0 in G will still allow a simplification of the Hamiltonian matrix.

The vanishing integral rule is not only useful in determining the nonvanishing elements of the Hamiltonian matrix H. Another important application is the derivation of *selection rules* for transitions between molecular states. For example, the intensity of an electric dipole transition from a state with wavefunction $\Psi_{j'}^{(F',m_F')}$ to a state with wavefunction $\Psi_{j''}^{(F'',m_F'')}$ (see equation (A1.4.12)) is proportional to the quantity

$$|T|^2 = \left| \int \Psi_{j'}^{(F',m_F')*} \mu_A \Psi_{j''}^{(F'',m_F'')} \, d\tau \right|^2 \tag{A1.4.89}$$

where μ_A, $A = X, Y, Z$, is the component of the molecular dipole moment operator along the A axis. If $\Psi_{j'}^{(F',m_F')}$ and $\Psi_{j''}^{(F'',m_F'')}$ belong to the irreducible representations $\Gamma_{j'}^{(F',m_F')}$ and $\Gamma_{j''}^{(F'',m_F'')}$, respectively and μ_A has the symmetry $\Gamma(\mu_A)$, then $|T|^2$, and thus the intensity of the transition, vanishes unless

$$\Gamma_{j'}^{(F',m_F')*} \otimes \Gamma(\mu_A) \otimes \Gamma_{j''}^{(F'',m_F'')} \supset \Gamma^{(s)}. \tag{A1.4.90}$$

In the rotational symmetry group K(spatial) discussed in section A1.4.1.1, we have $\Gamma_{j'}^{(F',m_F')} = D^{(F')}$, $\Gamma_{j''}^{(F'',m_F'')} = D^{(F'')}$ and $\Gamma(\mu_A) = D^{(1)}$. In this case the application of the vanishing integral rule leads to the selection rule given in equation (A1.4.13) (see section 7.3.2, in particular equation (7-47), of [1]).

A1.4.3   Symmetry operations and symmetry groups

The various types of symmetry enumerated in section A1.4.1.2 are discussed in detail here and the symmetry groups containing such symmetry operations are presented.

A1.4.3.1   *Translational symmetry*

In the active picture adopted here the (X,Y,Z) axis system remains fixed in space and a translational symmetry operation changes the (X,Y,Z) coordinates of all nuclei and electrons in the molecule by constant amounts, $(\Delta X, \Delta Y, \Delta Z)$ say,

$$(X_i, Y_i, Z_i) \rightarrow (X_i + \Delta X, Y_i + \Delta Y, Z_i + \Delta Z). \tag{A1.4.91}$$

We obtain a coordinate set more suitable for describing translational symmetry by introducing the centre of mass coordinates

$$(X_0, Y_0, Z_0) = \left(\frac{1}{M} \sum_{i=1}^{l} m_i X_i, \frac{1}{M} \sum_{i=1}^{l} m_i Y_i, \frac{1}{M} \sum_{i=1}^{l} m_i Z_i \right) \tag{A1.4.92}$$

together with

$$(X_i, Y_i, Z_i) = (X_i - X_0, Y_i - Y_0, Z_i - Z_0) \tag{A1.4.93}$$

for each particle i, where there are l particles in the molecule (N nuclei and $l - N$ electrons), m_i is the mass of particle i and $M = \sum_{i=1}^{l} m_i$ is the total mass of the molecule. In this manner we have introduced a new axis system (X, Y, Z) with axes parallel to the (X,Y,Z) axes but with origin at the molecular centre of mass. The molecule is described by the $3l$ coordinates

$$X_0, Y_0, Z_0, X_2, Y_2, Z_2, X_3, Y_3, Z_3, \ldots, X_l, Y_l, Z_l.$$

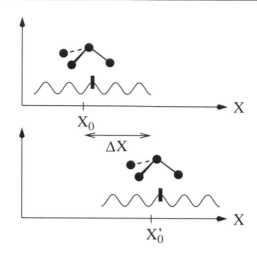

Figure A1.4.3. A PH$_3$ molecule and its wavefunction, symbolized by a sine wave, before (top) and after (bottom) a translational symmetry operation.

The coordinates (X_1, Y_1, Z_1) are redundant since they can be determined from the condition that the (X, Y, Z) axis system has origin at the molecular centre of mass. Obviously, the translational symmetry operation discussed above has the effect of changing the centre of mass coordinates

$$(X_0, Y_0, Z_0) \rightarrow (X_0 + \Delta X, Y_0 + \Delta Y, Z_0 + \Delta Z) \tag{A1.4.94}$$

whereas the coordinates $X_2, Y_2, Z_2, X_3, Y_3, Z_3, \ldots, X_l, Y_l, Z_l$ are unchanged by this operation.

We now define the effect of a translational symmetry operation on a function. Figure A1.4.3 shows how a PH$_3$ molecule is displaced a distance ΔX along the X axis by the translational symmetry operation that changes X_0 to $X'_0 = X_0 + \Delta X$. Together with the molecule, we have drawn a sine wave symbolizing the molecular wavefunction, Ψ_j say. We have marked one wavecrest to better keep track of the way the function is displaced by the symmetry operation. For the physical situation to be unchanged by the symmetry operation, the marked wavecrest and thus the entire wavefunction, is displaced by ΔX along the X axis as shown in figure A1.4.3. Thus, an operator $R_T^{(\Delta X, \Delta Y, \Delta Z)}$, which describes the effect of the translational symmetry operation on a wavefunction, is defined according to the S-convention (see equation (A1.4.57))

$$R_T^{(\Delta X, \Delta Y, \Delta Z)} \Psi_j (X_0, Y_0, Z_0, X_2, Y_2, Z_2, X_3, Y_3, Z_3, \ldots, X_l, Y_l, Z_l)$$
$$= \Psi_j (X_0 - \Delta X, Y_0 - \Delta Y, Z_0 - \Delta Z, X_2, Y_2, Z_2, X_3, Y_3, Z_3, \ldots, X_l, Y_l, Z_l). \tag{A1.4.95}$$

This definition causes the wavefunction to 'move with the molecule' as shown for the X direction in figure A1.4.3. The set of all translation symmetry operations $R_T^{(\Delta X, \Delta Y, \Delta Z)}$ constitutes a group which we call the translational group G_T. Because of the uniformity of space, G_T is a symmetry group of the molecular Hamiltonian $\widehat{H}$ in that all its elements commute with $\widehat{H}$:

$$[R_T^{(\Delta X, \Delta Y, \Delta Z)}, \widehat{H}] = 0. \tag{A1.4.96}$$

We could stop here in the discussion of the translational group. However, for the purpose of understanding the relation between translational symmetry and the conservation of linear momentum, we now show how the operator $R_T^{(\Delta X, \Delta Y, \Delta Z)}$ can be expressed in terms of the quantum mechanical operators representing the translational linear momentum of the molecule; these operators are defined as

$$(\widehat{P}_X, \widehat{P}_Y, \widehat{P}_Z) = \left(-i\hbar \frac{\partial}{\partial X_0}, -i\hbar \frac{\partial}{\partial Y_0}, -i\hbar \frac{\partial}{\partial Z_0} \right). \tag{A1.4.97}$$

The translational linear momentum is conserved for an isolated molecule in field free space and, as we see below, this is closely related to the fact that the molecular Hamiltonian commutes with $R_T^{(\Delta X, \Delta Y, \Delta Z)}$ for all values of $(\Delta X, \Delta Y, \Delta Z)$. The conservation of linear momentum and translational symmetry are directly related.

In order to determine the relationship between $R_T^{(\Delta X, \Delta Y, \Delta Z)}$ and the $(\widehat{P}_X, \widehat{P}_Y, \widehat{P}_Z)$ operators, we consider a translation $R_T^{(\delta X, 0, 0)}$ where δX is infinitesimally small. In this case we can approximate the right hand side of equation (A1.4.95) by a first-order Taylor expansion:

$$R_T^{(\delta X, 0, 0)} \Psi_j (X_0, Y_0, Z_0, X_2, Y_2, Z_2, \ldots, X_l, Y_l, Z_l)$$
$$= \Psi_j (X_0 - \delta X, Y_0, Z_0, X_2, Y_2, Z_2, \ldots, X_l, Y_l, Z_l)$$
$$= \Psi_j (X_0, Y_0, Z_0, X_2, Y_2, Z_2, \ldots, X_l, Y_l, Z_l) - \frac{\partial \Psi_j}{\partial X_0} \delta X. \qquad (A1.4.98)$$

From the definition of the translational linear momentum operator $\widehat{P}_X$ (in equation (A1.4.97)) we see that

$$\frac{\partial \Psi_j}{\partial X_0} = \frac{i}{\hbar} \widehat{P}_X \Psi_j \qquad (A1.4.99)$$

and by introducing this identity in equation (A1.4.98) we obtain

$$R_T^{(\delta X, 0, 0)} \Psi_j = \Psi_j - \frac{i}{\hbar} \delta X \widehat{P}_X \Psi_j \qquad (A1.4.100)$$

where we have omitted the coordinate arguments for brevity. Since the function Ψ_j in equation (A1.4.100) is arbitrary, it follows that we can write the symmetry operation as

$$R_T^{(\delta X, 0, 0)} = 1 - \frac{i}{\hbar} \delta X \widehat{P}_X. \qquad (A1.4.101)$$

The operation $R_T^{(\Delta X, 0, 0)}$, for which ΔX is an arbitrary finite length, obviously has the same effect on a wavefunction as the operation $R_T^{(\delta X, 0, 0)}$ applied to the wavefunction $\Delta X / \delta X$ times. We simply divide the translation by ΔX into $\Delta X / \delta X$ steps, each step of length δX. This remains true in the limit of $\delta X \to 0$. Thus

$$R_T^{(\Delta X, 0, 0)} = \lim_{\delta X \to 0} (R_T^{(\delta X, 0, 0)})^{\frac{\Delta X}{\delta X}} = \lim_{\delta X \to 0} \left(1 - \frac{i}{\hbar} \delta X \widehat{P}_X\right)^{\frac{\Delta X}{\delta X}} = \exp\left(-\frac{i}{\hbar} \Delta X \widehat{P}_X\right) \qquad (A1.4.102)$$

where we have used the general identity

$$\lim_{x \to 0} (1 + ax)^{y/x} = \exp(ay). \qquad (A1.4.103)$$

We can derive expressions analogous to equation (A1.4.102) for $R_T^{(0, \Delta Y, 0)}$ and $R_T^{(0, 0, \Delta Z)}$ and we can resolve a general translation $R_T^{(\Delta X, \Delta Y, \Delta Z)}$ as

$$R_T^{(\Delta X, \Delta Y, \Delta Z)} = R_T^{(\Delta X, 0, 0)} R_T^{(0, \Delta Y, 0)} R_T^{(0, 0, \Delta Z)}. \qquad (A1.4.104)$$

Consequently,

$$R_T^{(\Delta X, \Delta Y, \Delta Z)} = \exp\left[-\frac{i}{\hbar} (\Delta X \widehat{P}_X + \Delta Y \widehat{P}_Y + \Delta Z \widehat{P}_Z)\right]. \qquad (A1.4.105)$$

We deal with the exponentials in equation (A1.4.102) and (A1.4.105) whose arguments are operators by using their Taylor expansion

$$\exp(i\widehat{O}) = 1 + i\widehat{O} + \frac{1}{2!}(i\widehat{O})^2 + \cdots \qquad (A1.4.106)$$

where $\widehat{O}$ is a Hermitian operator.

It follows from equation (A1.4.105) that equation (A1.4.96) is satisfied for arbitrary $\Delta X, \Delta Y, \Delta Z$ if and only if

$$[\widehat{H}, \widehat{P}_X] = [\widehat{H}, \widehat{P}_Y] = [\widehat{H}, \widehat{P}_Z] = 0. \qquad (A1.4.107)$$

From the fact that $\widehat{H}$ commutes with the operators $(\widehat{P}_X, \widehat{P}_Y, \widehat{P}_Z)$ it is possible to show that the linear momentum of a molecule in free space must be conserved. First we note that the time-dependent wavefunction $\Psi(t)$ of a molecule fulfills the time-dependent Schrödinger equation

$$i\hbar\frac{\partial \Psi(t)}{\partial t} = \widehat{H}\Psi(t). \qquad (A1.4.108)$$

For $A = X, Y,$ or Z, we use this identity to derive an expression for

$$\frac{\partial}{\partial t}\langle \Psi(t)|\widehat{P}_A|\Psi(t)\rangle = \left\langle\frac{\partial\Psi(t)}{\partial t}\Big|\widehat{P}_A\Big|\Psi(t)\right\rangle + \left\langle\Psi(t)\Big|\frac{\partial}{\partial t}(\widehat{P}_A\Psi(t))\right\rangle = \left\langle\frac{\partial\Psi(t)}{\partial t}\Big|\widehat{P}_A\Big|\Psi(t)\right\rangle + \left\langle\Psi(t)\Big|\widehat{P}_A\Big|\frac{\partial\Psi(t)}{\partial t}\right\rangle \qquad (A1.4.109)$$

where, in the last equality, we have used the fact that $\widehat{P}_A$ does not depend explicitly on t. We obtain $\partial\Psi(t)/\partial t$ from equation (A1.4.108) and insert the resulting expression in equation (A1.4.109); this yields

$$\frac{\partial}{\partial t}\langle\Psi(t)|\widehat{P}_A|\Psi(t)\rangle = \frac{i}{\hbar}(\langle\widehat{H}\Psi(t)|\widehat{P}_A|\Psi(t)\rangle - \langle\Psi(t)|\widehat{P}_A|\widehat{H}\Psi(t)\rangle) = \frac{i}{\hbar}\langle\Psi(t)|[\widehat{H}, \widehat{P}_A]|\Psi(t)\rangle = 0 \qquad (A1.4.110)$$

where we have used equation (A1.4.107) in conjunction with the fact that $\widehat{H}$ is Hermitian. Equation (A1.4.110) shows that the expectation value of each linear momentum operator is conserved in time and thus the conservation of linear momentum directly follows from the translational invariance of the molecular Hamiltonian (equation (A1.4.96)).

Because of equation (A1.4.107) and because of the fact that $\widehat{P}_X, \widehat{P}_Y$ and $\widehat{P}_Z$ commute with each other, we know that there exists a complete set of simultaneous eigenfunctions of $\widehat{P}_X, \widehat{P}_Y, \widehat{P}_Z$ and $\widehat{H}$. An eigenfunction of $\widehat{P}_X, \widehat{P}_Y$ and $\widehat{P}_Z$ has the form

$$\Psi_T(X_0, Y_0, Z_0) = \exp[i(k_X X_0 + k_Y Y_0 + k_Z Z_0)] \qquad (A1.4.111)$$

where

$$\widehat{P}_A\Psi_T(X_0, Y_0, Z_0) = \hbar k_A \Psi_T(X_0, Y_0, Z_0) \qquad (A1.4.112)$$

with $A = X, Y$ or Z, so that (equation (A1.4.105))

$$R_T^{(\Delta X, \Delta Y, \Delta Z)}\Psi_T(X_0, Y_0, Z_0) = \exp[-i(\Delta X k_X + \Delta Y k_Y + \Delta Z k_Z)]\Psi_T(X_0, Y_0, Z_0). \qquad (A1.4.113)$$

That is, the effect of a translational operation is determined solely by the vector with components (k_X, k_Y, k_Z) which defines the linear momentum.

For a molecular wavefunction $\Psi_j(X_0, Y_0, Z_0, X_2, Y_2, Z_2, \ldots, X_l, Y_l, Z_l)$ to be a simultaneous eigenfunction of $\widehat{P}_X, \widehat{P}_Y, \widehat{P}_Z$ and $\widehat{H}$ it must have the form

$$\Psi_j(X_0, Y_0, Z_0, X_2, Y_2, Z_2, \ldots, X_l, Y_l, Z_l) = \Psi_T(X_0, Y_0, Z_0)\Psi_{int}(X_2, Y_2, Z_2, \ldots, X_l, Y_l, Z_l) \qquad (A1.4.114)$$

where Ψ_{int} describes the *internal* motion of the molecule (see also section 7.3.1 of [1]).

We can describe the conservation of linear momentum by noting the analogy between the time-dependent Schrödinger equation, equation (A1.4.108), and equation (A1.4.99). For an isolated molecule, $\widehat{H}$ does not depend explicitly on t and we can repeat the arguments expressed in equations (A1.4.98)–(A1.4.102) with X replaced by t and $\widehat{P}_X$ replaced by $-\widehat{H}$ to show that

$$\Psi(t) = \exp\left(\frac{i}{\hbar}t\widehat{H}\right)\Psi(t=0). \qquad (A1.4.115)$$

If the wavefunction at $t = 0$, $\Psi(t = 0)$, is an eigenfunction of $\widehat{P}_X$, $\widehat{P}_Y$, $\widehat{P}_Z$ and $\widehat{H}$ so that it can be expressed as given in equation (A1.4.114), it follows from equation (A1.4.115) that at any other time t,

$$\Psi(t) = \exp\left(\frac{i}{\hbar}tE\right)\Psi(t=0) \qquad (A1.4.116)$$

where E is the energy (i.e., the eigenvalue of $\widehat{H}$ associated with the eigenfunction $\Psi(t = 0)$). It is straight-forward to show that this function is an eigenfunction of $\widehat{P}_X$, $\widehat{P}_Y$, $\widehat{P}_Z$ and $\widehat{H}$ with the same eigenvalues as $\Psi(t = 0)$. This is another way of proving that linear momentum and energy are conserved in time.

A1.4.3.2 Rotational symmetry

In order to discuss rotational symmetry, we must first introduce the rotational and vibrational coordinates customarily used in molecular theory. We define a set of (x, y, z) axes with an orientation relative to the (X, Y, Z) axes discussed above that is defined by the positions of the nuclei. These axes are called 'molecule fixed' axes; their orientation is determined by the coordinates of the nuclei only and the coordinates of the electrons are not involved. The (x, y, z) and (X, Y, Z) axis systems are always chosen to be right handed. For any placement of the N nuclei in space (i.e., any set of values for the $3N-3$ independent coordinates X_i, Y_i and Z_i of the nuclei) there is an unambiguous way of specifying the orientation of the (x, y, z) axes with respect to the (X, Y, Z) axes. Three equations are required to define the three Euler angles (θ, ϕ, χ) (see figure A1.4.4) that specify this orientation and the equations used are the Eckart equations [5]. The Eckart equations minimize the angular momentum in the (x, y, z) axis system and so they optimize the separation of the rotational and vibrational degrees of freedom in the rotation–vibration Schrödinger equation. It is described in detail in chapter 10 of [1] how, by introducing the Eckart equations, we can define the (x, y, z) axis system and thus the Euler angles (θ, ϕ, χ). Suffice it to say that we describe the internal motion of a nonlinear molecule[8] by $3l - 3$ coordinates, where the first three are the Euler angles (θ, ϕ, χ) describing rotation, the next $3N - 6$ are *normal coordinates* $Q_1, Q_2, Q_3, \ldots, Q_{3N-6}$ describing the vibration of the nuclei and the remaining $3(l - N)$ are electronic coordinates $x_{N+1}, y_{N+1}\ z_{N+1}, x_{N+2}, y_{N+2}\ z_{N+2}, \ldots, x_l, y_l\ z_l$, simply chosen as the Cartesian coordinates of the electrons in the (x, y, z) axis system.

We consider rotations of the molecule about space-fixed axes in the active picture. Such a rotation causes the (x, y, z) axis system to rotate so that the Euler angles change

$$(\theta, \phi, \chi) \rightarrow (\theta + \Delta\theta, \phi + \Delta\phi, \chi + \Delta\chi). \qquad (A1.4.117)$$

The normal coordinates Q_r, $r = 1, 2, \ldots, 3N - 6$, and the electronic coordinates $x_i, y_i\ z_i, i = N + 1$, $N + 2, \ldots, l$, all describe motion relative to the (x, y, z) axis system and are invariant to rotations.

Initially, we neglect terms depending on the electron spin $\widehat{S}$ and the nuclear spin $\widehat{I}$ in the molecular Hamiltonian $\widehat{H}$. In this approximation, we can take the total angular momentum to be $\widehat{N}$ (see equation (A1.4.1))

---

[8] That is, a molecule for which the minimum of the Born–Oppenheimer potential energy function corresponds to a nonlinear geometry. The theory of linear molecules is explained in chapter 17 of [1].

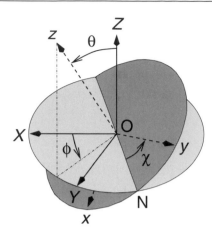

Figure A1.4.4. The definition of the Euler angles (θ, ϕ, χ) that relate the orientation of the molecule fixed (x, y, z) axes to the (X, Y, Z) axes. The origin of both axis systems is at the nuclear centre of mass O, and the node line ON is directed so that a right handed screw is driven along ON in its positive direction by twisting it from Z to z through θ where $0 \leq \theta \leq \pi$. ϕ and χ have the ranges 0 to 2π. χ is measured from the node line.

which results from the rotational motion of the nuclei and the orbital motion of the electrons. The components of $\widehat{N}$ in the (X, Y, Z) axis system are given by:

$$\widehat{N}_X = -i\hbar \left(-\sin\phi \frac{\partial}{\partial\theta} + \operatorname{cosec}\theta \cos\phi \frac{\partial}{\partial\chi} - \cot\theta \cos\phi \frac{\partial}{\partial\phi} \right) \qquad (A1.4.118)$$

$$\widehat{N}_Y = -i\hbar \left(\cos\phi \frac{\partial}{\partial\theta} + \operatorname{cosec}\theta \sin\phi \frac{\partial}{\partial\chi} - \cot\theta \sin\phi \frac{\partial}{\partial\phi} \right) \qquad (A1.4.119)$$

and

$$\widehat{N}_Z = -i\hbar \frac{\partial}{\partial\phi}. \qquad (A1.4.120)$$

By analogy with our treatment of translation symmetry, we aim to derive an operator $R_R^{(\Delta\theta, \Delta\phi, \Delta\chi)}$ which, when applied to a wavefunction, describes the effect of a general symmetry operation that causes the change in the Euler angles given in equation (A1.4.117). Because of the analogy between equation (A1.4.120) and the definition of $\widehat{P}_X$ in equation (A1.4.97), we can repeat the arguments expressed in equations (A1.4.98)–(A1.4.102) with X replaced by ϕ to show that

$$R_R^{(0, \Delta\phi, 0)} = \exp\left(-\frac{i}{\hbar} \Delta\phi \widehat{N}_Z \right). \qquad (A1.4.121)$$

A more involved derivation (see, for example, section 3.2 of Zare [6]) shows that for a general rotation

$$R_R^{(\Delta\theta, \Delta\phi, \Delta\chi)} = \exp\left(-\frac{i}{\hbar} \Delta\phi \widehat{N}_Z \right) \exp\left(-\frac{i}{\hbar} \Delta\theta \widehat{N}_Y \right) \exp\left(-\frac{i}{\hbar} \Delta\chi \widehat{N}_Z \right). \qquad (A1.4.122)$$

The operators $\widehat{N}_Y$ and $\widehat{N}_Z$ in equation (A1.4.122) do not commute and we have (see equation (10-90) of [1])

$$[\widehat{N}_Y, \widehat{N}_Z] = i\hbar \widehat{N}_X. \qquad (A1.4.123)$$

The commutators $[\widehat{N}_X, \widehat{N}_Y]$ and $[\widehat{N}_Z, \widehat{N}_X]$ are obtained by replacing XYZ by ZXY and YZX, respectively, in equation (A1.4.123). It is, therefore, important in using equation (A1.4.122) that the exponential factors be applied in the correct order.

The set of all rotation operations $R_R^{(\Delta\theta, \Delta\phi, \Delta\chi)}$ forms a group which we call the rotational group K(spatial). Since space is isotropic, K(spatial) is a symmetry group of the molecular Hamiltonian $\widehat{H}$ in that all its elements commute with $\widehat{H}$:

$$[R_R^{(\Delta\theta, \Delta\phi, \Delta\chi)} \widehat{H}] = 0. \qquad (A1.4.124)$$

It follows from equation (A1.4.122) that equation (A1.4.124) is satisfied for arbitrary $(\Delta\theta, \Delta\phi, \Delta\chi)$ if and only if $\widehat{H}$ commutes with $\widehat{N}_Y$ and $\widehat{N}_Z$. But then $\widehat{H}$ also commutes with $\widehat{N}_X$ because of equation (A1.4.123). That is

$$[\widehat{H}, \widehat{N}_X] = [\widehat{H}, \widehat{N}_Y] = [\widehat{H}, \widehat{N}_Z] = 0 \qquad (A1.4.125)$$

this equation is analogous to equation (A1.4.107). We discussed above (in connection with equations (A1.4.108)–(A1.4.110)) how the invariance of the molecular Hamiltonian to translation is related to the conservation of linear momentum. We now see that, in a completely analogous manner, the invariance of the molecular Hamiltonian to rotation is related to the conservation of angular momentum.

The (X, Y, Z) components of $\widehat{N}$ do not commute and so we cannot find simultaneous eigenfunctions of all the four operators occurring in equation (A1.4.125). It is straightforwardly shown from the commutation relations in equation (A1.4.123) that the operator

$$\widehat{N}^2 = \widehat{N}_X^2 + \widehat{N}_Y^2 + \widehat{N}_Z^2 \qquad (A1.4.126)$$

commutes with $\widehat{N}_X$, $\widehat{N}_Y$, and $\widehat{N}_Z$. Because of equation (A1.4.125), this operator also commutes with $\widehat{H}$. As a consequence, we can find simultaneous eigenfunctions of $\widehat{H}$, $\widehat{N}^2$ and one component of $\widehat{N}$, customarily chosen as $\widehat{N}_Z$. We can use this result to simplify the diagonalization of the matrix representation of the molecular Hamiltonian. We choose the basis functions as $\Psi^0_{n,N,m}$. They are eigenfunctions of $\widehat{N}^2$ (with eigenvalues $N(N+1)\hbar^2$, $N = 0, 1, 2, 3, 4, \ldots$) and $\widehat{N}_Z$ (with eigenvalues $m\hbar$, $m = -N, -N+1, \ldots, N-1, N$). The functions $\Psi^0_{n,N,m}$, $m = -N, -N+1, \ldots, N-1, N$, transform according to the irreducible representation $D^{(N)}$ of K(spatial) (see section A1.4.1.1). With these basis functions, the matrix representation of the molecular Hamiltonian will be block diagonal in N and m in the manner described for the quantum numbers F and m_F in section A1.4.1.1.

If we allow for the terms in the molecular Hamiltonian depending on the electron spin $\widehat{S}$ (see chapter 7 of [1]), the resulting Hamiltonian no longer commutes with the components of $\widehat{N}$ as given in equation (A1.4.125), but with the components of

$$\widehat{J} = \widehat{N} + \widehat{S}. \qquad (A1.4.127)$$

In this case, we choose the basis functions Ψ^0_{n,J,m_J}, that is, the eigenfunctions of $\widehat{J}^2$ (with eigenvalues $J(J+1)\hbar^2$, $J = |N - S|, |N - S| + 1, \ldots, N + S - 1, N + S$) and $\widehat{N}_Z$ (with eigenvalues $m_J\hbar$, $m_J = -J, -J+1, \ldots, J-1, J$). These functions are linear combinations of products $\Psi^0_{n,N,m} \Psi^0_{e,S,m_S}$, where the function $\Psi^0_{n,N,m}$ is an eigenfunction of $\widehat{N}^2$ and $\widehat{N}_Z$ as described above, and Ψ^0_{e,S,m_S} is an eigenfunction of $\widehat{S}^2$ (with eigenvalues $S(S+1)\hbar^2$, $S = 0, 1/2, 1, 3/2, 2, 5/2, 3, \ldots$) and $\widehat{S}_Z$ (with eigenvalues $m_S\hbar$, $m_S = -S, -S+1, \ldots, S-1, S$). In this basis, the matrix representation of the molecular Hamiltonian is block diagonal in J and m_J. The functions Ψ^0_{e,S,m_S}, $m_S = -S, -S+1, \ldots, S-1, S$, transform according to the irreducible representation $D^{(S)}$ of K(spatial) and the functions Ψ^0_{n,J,m_J}, $m_J = -J, -J+1, \ldots, J-1, J$, have $D^{(J)}$ symmetry in K(spatial). Singlet states have $S = 0$ and for them $\widehat{J} = \widehat{N}$, $J = N$ and $m_J = m$.

Finally, we consider the complete molecular Hamiltonian which contains not only terms depending on the electron spin, but also terms depending on the nuclear spin $\widehat{I}$ (see chapter 7 of [1]). This Hamiltonian commutes

Table A1.4.5. The character table of the inversion group $\mathcal{E}$

	E	E^*
$+$	1	1
$-$	1	-1

with the components of $\widehat{F}$ given in equation (A1.4.1). The diagonalization of the matrix representation of the complete molecular Hamiltonian proceeds as described in section A1.4.1.1.

The theory of rotational symmetry is an extensive subject and we have only scratched the surface here. A relatively new book, which is concerned with molecules, is by Zare [6] (see [7] for the solutions to all the problems in [6] and a list of the errors). This book describes, for example, the method for obtaining the functions Ψ^0_{n,J,m_J} from $\Psi^0_{n,N,m}$ and Ψ^0_{e,S,m_S}, and for obtaining the functions Ψ^0_{n,F,m_F} (section A1.4.1.1) from the Ψ^0_{n,J,m_J} combined with eigenfunctions of $\widehat{I}^2$ and $\widehat{I}_Z$.

A1.4.3.3   Inversion symmetry

We have already discussed inversion symmetry and how it leads to the parity label in sections A1.4.1.2 and A1.4.2.5. For any molecule in field-free space, if we neglect terms arising from the weak interaction force (see the next paragraph), the molecular Hamiltonian commutes with the inversion operation E^* and thus for such a Hamiltonian the *inversion group* $\mathcal{E} = \{E, E^*\}$ is a symmetry group. The character table of the inversion group is given in table A1.4.5 and the irreducible representations are labelled $+$ and $-$ to give the parity.

Often molecular energy levels occur in closely spaced doublets having opposite parity. This is of particular interest when there are symmetrically equivalent minima, separated by a barrier, in the potential energy function of the electronic state under investigation. This happens in the PH_3 molecule and such pairs of levels are called 'inversion doublets'; the splitting between such parity doublet levels depends on the extent of the quantum mechanical tunnelling through the barrier that separates the two minima. This is discussed further in section A1.4.4.

The Hamiltonian considered above, which commutes with E^*, involves the electromagnetic forces between the nuclei and electrons. However, there is another force between particles, the weak interaction force, that is not invariant to inversion. The weak charged current interaction force is responsible for the beta decay of nuclei, and the related weak neutral current interaction force has an effect in atomic and molecular systems. If we include this force between the nuclei and electrons in the molecular Hamiltonian (as we should because of electroweak unification) then the Hamiltonian will not commute with E^*, and states of opposite parity will be mixed. However, the effect of the weak neutral current interaction force is incredibly small (and it is a very short range force), although its effect has been detected in extremely precise experiments on atoms (see, for example, Wood *et al* [8], who detect that a small part ($\sim 10^{-11}$) of a P state of caesium is mixed into an S state by this force). Its effect has not been detected in a molecule and, thus, for practical purposes we can neglect it and consider E^* to be a symmetry operation. Note that inversion symmetry is not a universal symmetry like translational or rotational symmetry and it does not derive from a general property of space. In the theoretical physics community, when dealing with particle symmetry, the inversion operation is called the 'parity operator' P.

An optically active molecule is a particular type of molecule in which there are two equivalent minima separated by an insuperable barrier in the potential energy surface and for which the molecular structures at these two minima are not identical (as they are in PH_3) but are mirror images of one another. The two forms of the molecule are called the dextrorotatory (D) and laevorotatory (L) forms and they can be separated. The D and L wavefunctions are not eigenfunctions of E^* and E^* interconverts them. In the general case eigenstates of the Hamiltonian are eigenstates of E^* and they have a definite parity. In the laboratory, when one makes an optically active molecule one obtains a racemic 50/50 mixture of the D and L forms, but in living organisms use is made of only one isomer; natural proteins, for example, are composed exclusively of L-amino acids, whereas nucleic acids contain only D-sugars. This fact is unexplained but it has been pointed out (see [9] and references therein) that in the molecular Hamiltonian the weak neutral current interaction term $\widehat{H}_{WI}$ would give rise to a small energy difference between the energy levels of the D and L forms, and this small energy difference could have acted to select one isomer over the long time of prebiotic evolution. The experimental determination of the energy difference between the D and L forms of any optically active molecule has yet to be achieved. However, see Daussy C, Marrel T, Amy-Klein A, Nguyen C T, Bordé C J and Chardonnet C 1999 *Phys. Rev. Lett.* **83** 1554 for a recent determination of an upper bound of 13 Hz on the energy difference between CHFClBr enantiomers.

A very recent paper concerning the search for a parity-violating energy difference between enantiomers of a chiral molecule is by Lahamer A S, Mahurin S M, Compton R N, House D, Laerdahl J K, Lein M and Schwerdtfeger P 2000 *Phys. Rev. Lett.* **85** 4470. The importance of the parity-violating energy difference in leading to prebiotic asymmetric synthesis is discussed in Frank P, Bonner W A and Zare R N 2000 On one hand but not the other: the challenge of the origin and survival of homochirality in prebiotic chemistry *Chemistry for the 21st Century* ed E Keinan and I Schechter (Weinheim: Wiley-VCH) pp 175–208.

A1.4.3.4 Identical particle permutation symmetry

If there are n electrons in a molecule there are $n!$ ways of permuting them and we can form the permutation group (or symmetric group) $S_n^{(e)}$ of degree n and order $n!$ that contains all the electron permutations. The molecular Hamiltonian is invariant to the elements of this group. Similarly, there can be sets of identical nuclei in a molecule and the Hamiltonian is invariant to the relevant identical-nucleus permutation groups. For example, the ethanol molecule CH_3CH_2OH consists of 26 electrons, a set of six identical hydrogen nuclei, a set of two identical carbon nuclei and a lone oxygen nucleus. The molecular Hamiltonian of ethanol is therefore invariant to the 26! ($\sim 4 \times 10^{26}$) elements of the electron permutation group $S_{26}^{(e)}$, the 6! = 720 possible permutations of the hydrogen nuclei in the group $S_6^{(H)}$ and the two possible permutations of the C nuclei (E and their exchange) in the group $S_2^{(C)}$. The group of all possible permutations of identical nuclei in a molecule is called the *complete nuclear permutation* (CNP) group of the molecule G^{CNP}. For ethanol G^{CNP} consists of all 6! elements of $S_6^{(H)}$ *and* of all these elements taken in combination with the exchange of the two C nuclei; $2 \times 6!$ elements in all. This CNP group is called the *direct product* of the groups $S_6^{(H)}$ and $S_2^{(C)}$ and is written

$$G^{CNP} = S_6^{(H)} \otimes S_2^{(C)}. \tag{A1.4.128}$$

The CNP group of a molecule containing l identical nuclei of one type, m of another, n of another and so on is the direct product group

$$G^{CNP} = S_l \otimes S_m \otimes S_n \ldots \tag{A1.4.129}$$

and the order of the group is $l! \times m! \times n! \ldots$. It would seem that we have a very rich set of irreducible representation labels with which we can label the molecular energy levels of a molecule using the electron permutation group and the CNP group. But this is not the case for internal states described by Ψ_{int} (see equation (A1.4.114)) because there is fundamentally no observable difference between states that differ merely in the permutation of identical particles. The environment of a molecule (e.g. an external electric or magnetic field, or the effect of a neighbouring molecule) affects whether the Hamiltonian of that molecule is invariant to a rotation operation or the inversion operation; states having different symmetry labels from the rotation or inversion groups can be mixed and transitions can occur between such differently labelled states. However, the Hamiltonian of a molecule *regardless of the environment of the molecule* is invariant to any identical particle permutation. Two Ψ_{int} states that differ only in the permutation of identical particles are observationally indistinguishable and there is only one state. Since there is only one state it can only transform as one set of irreducible representations of the various identical particle permutation groups that apply for the particular molecule under investigation. It is an experimental fact that particles with half integral spin (called *fermions*), such as electrons and protons, transform as that one-dimensional irreducible representation of their permutation group that has character +1 for all even permutations[9] and character −1 for all odd permutations. Nuclei that have integral spin (called *bosons*), such as ^{12}C nuclei and deuterons, transform as the totally symmetric representation of their permutation group (having character +1 for all permutations). Thus fermion wavefunctions are changed in sign by an odd permutation but boson wavefunctions are invariant. This simple experimental observation has defied simple theoretical proof but there is a complicated proof [10] that we cannot recommend any reader of the present article to look at.

The fact that allowed fermion states have to be antisymmetric, i.e., changed in sign by any odd permutation of the fermions, leads to an interesting result concerning the allowed states. Let us write a state wavefunction for a system of n noninteracting fermions as

$$|X\rangle = |a_1\rangle|b_2\rangle|c_3\rangle \ldots |q_n\rangle \tag{A1.4.130}$$

where this indicates that particle 1 is in state a, particle 2 in state b and so on. Clearly this does not correspond to an allowed (i.e., antisymmetric) state since making an odd permutation of the indices, such as (12), does not give −1 times $|X\rangle$. But we can get an antisymmetric function by making all permutations of the indices in $|X\rangle$ and adding the results with the coefficient −1 for those functions obtained by making an odd permutation, i.e.,

$$|F\rangle = \sum_P \pm P|a_1\rangle|b_2\rangle|c_3\rangle \ldots |q_n\rangle \tag{A1.4.131}$$

where the sum over all permutations involves a + or − sign as the permutation P is even or odd respectively. We can write equation (A1.4.131) as the determinant

$$|F\rangle = \begin{vmatrix} |a_1\rangle & |b_1\rangle & |c_1\rangle & \ldots & |q_1\rangle \\ |a_2\rangle & |b_2\rangle & |c_2\rangle & \ldots & |q_2\rangle \\ \vdots & \vdots & \vdots & \vdots & \vdots \\ |a_n\rangle & |b_n\rangle & |c_n\rangle & \ldots & |q_n\rangle \end{vmatrix}. \tag{A1.4.132}$$

The state $|F\rangle$ is such that the particle states $a, b, c, \ldots, q$ are occupied and each particle is equally likely to be in any one of the particle states. However, if two of the particle states $a, b, c, \ldots, q$ are the same then $|F\rangle$ vanishes; it does not correspond to an allowed state of the assembly. This is a characteristic of antisymmetric states and it is called 'the Pauli exclusion principle': no two identical fermions can be in the same particle state. The general function for an assembly of bosons is

$$|B\rangle = \sum_P P|a_1\rangle|b_2\rangle|c_3\rangle \ldots |q_n\rangle \tag{A1.4.133}$$

where the sum over all permutations involves just '+' signs. In such a state it is possible for two or more of the particles to be in the same particle state.

It would appear that identical particle permutation groups are not of help in providing distinguishing symmetry labels on molecular energy levels as are the other groups we have considered. However, they do provide very useful restrictions on the way we can build up the complete molecular wavefunction from basis functions. Molecular wavefunctions are usually built up from basis functions that are products of electronic

[9] An even (odd) permutation is one that when expressed as the product of pair exchanges involves an even (odd) number of such exchanges. Thus (123) = (12)(23) and (12345) = (12)(23)(34)(45) are even permutations, whereas (12), (1234) = (12)(23)(34) and (123456) = (12)(23)(34)(45)(56) are odd permutations.

Table A1.4.6. The character table of the group $S_2^{(e)}$.

	E	(ab)
$\Gamma_1^{(e)}$	1	1
$\Gamma_2^{(e)}$	1	-1

Table A1.4.7. The character table of the group G^{CNP} for H_2.

	E	(12)
$\Gamma_1^{(\mathrm{CNP})}$	1	1
$\Gamma_2^{(\mathrm{CNP})}$	1	-1

and nuclear parts. Each of these parts is further built up from products of separate 'uncoupled' coordinate (or orbital) and spin basis functions. When we combine these separate functions, the final overall product states must conform to the permutation symmetry rules that we stated above. This leads to restrictions in the way that we can combine the uncoupled basis functions.

We explain this by considering the H_2 molecule. For the H_2 molecule we label the electrons a and b, and the hydrogen nuclei 1 and 2. The electron permutation group is $S_2^{(e)} = \{E, (ab)\}$, and the CNP group $G^{\mathrm{CNP}} = \{E, (12)\}$. The character tables of these groups are given in tables A1.4.6 and A1.4.7. If there were no restriction on permutation symmetry we might think that the energy levels of the H_2 molecule could be of any one of the following four symmetry types using these two groups: $(\Gamma_1^{(e)}, \Gamma_1^{(\mathrm{CNP})})$, $(\Gamma_1^{(e)}, \Gamma_2^{(\mathrm{CNP})})$, $(\Gamma_2^{(e)}, \Gamma_1^{(\mathrm{CNP})})$ and $(\Gamma_2^{(e)}, \Gamma_2^{(\mathrm{CNP})})$. However, both electrons and protons are fermions (having a spin of 1/2) and so, from the above rules, the wavefunctions of the H_2 molecule must be multiplied by -1 by both (ab) and (12). Thus the energy levels of the H_2 molecule can only be of symmetry $(\Gamma_2^{(e)}, \Gamma_2^{(\mathrm{CNP})})$.

These limitations lead to electron spin multiplicity restrictions and to differing nuclear spin statistical weights for the rotational levels. Writing the electronic wavefunction as the product of an orbital function Ψ_e and a spin function Ψ_{es}, there are restrictions on how these functions can be combined. The restrictions are imposed by the fact that the complete function $\Psi_e \Psi_{es}$ has to be of symmetry $\Gamma_2^{(e)}$ in the group $S_2^{(e)}$. The orbital function Ψ_e can be of symmetry $\Gamma_1^{(e)}$ or $\Gamma_2^{(e)}$ and, for example, Ψ_e for the ground electronic state of H_2 has symmetry $\Gamma_1^{(e)}$. For a two electron system there are four possible electron spin functions[10]: $\alpha\alpha$, $\alpha\beta$, $\beta\alpha$ and $\beta\beta$, where α is a 'spin-up' function having $m_S = +1/2$ and β is a 'spin-down' function having $m_S = -1/2$. The functions $\Psi_{es}^{(1)} = \alpha\alpha$ and $\Psi_{es}^{(2)} = \beta\beta$ are invariant to the operation (ab) and therefore have symmetry $\Gamma_1^{(e)}$. The functions $\alpha\beta$ and $\beta\alpha$ are interchanged by (ab) and do not transform irreducibly, but it is easy to see that their sum and difference, $\Psi_{es}^{(3)} = (\alpha\beta + \beta\alpha)/\sqrt{2}$ and $\Psi_{es}^{(4)} = (\alpha\beta - \beta\alpha)/\sqrt{2}$, transform as $\Gamma_1^{(e)}$ and $\Gamma_2^{(e)}$ respectively. The three functions $\Psi_{es}^{(1)}$, $\Psi_{es}^{(2)}$ and $\Psi_{es}^{(3)}$, each of symmetry $\Gamma_1^{(e)}$, form a triplet electron spin state (with $m_S = 1, -1$ and 0, for $S = 1$) and the function $\Psi_{es}^{(1)}$, having symmetry $\Gamma_2^{(e)}$ is a singlet state (with $S = 0$). The ground electronic state cannot be a triplet state since if it were then the symmetry of both Ψ_e and Ψ_{es} would be $\Gamma_1^{(e)}$ and the product would therefore be of symmetry $\Gamma_1^{(e)}$ which is not allowed. Hence the ground electronic state of H_2 has to be a singlet electronic state.

The way we combine the nuclear spin basis functions Ψ_{ns} with the rotation–vibration–electronic basis functions Ψ_{rve} in H_2 follows the same type of argument using the nuclear permutation group G^{CNP}. Rovibronic states of symmetry $\Gamma_1^{(\mathrm{CNP})}$ can only be combined with Ψ_{ns} of species $\Gamma_2^{(\mathrm{CNP})}$ (of which there is one with $I = 0$),

---

[10] We give the spin of electron a first and of electron b second.

and rovibronic states of symmetry $\Gamma_2^{(\text{CNP})}$ can only be combined with Ψ_{ns} of species $\Gamma_1^{(\text{CNP})}$ (of which there are three with $I = 1$). Thus rovibronic states of symmetry $\Gamma_1^{(\text{CNP})}$ have a *nuclear spin statistical weight* of 1, and rovibronic states of symmetry $\Gamma_2^{(\text{CNP})}$ have a nuclear spin statistical weight of 3. An interesting result of these considerations follows for the $^{16}\text{O}_2$ molecule by using the G^{CNP} group. Labelling the O nuclei 1 and 2 this group is as in table A1.4.7. The spin of ^{16}O nuclei is 0 and so the nuclear spin wavefunction is of species $\Gamma_1^{(\text{CNP})}$. There is no nuclear spin wavefunction of species $\Gamma_2^{(\text{CNP})}$ in $^{16}\text{O}_2$. Since ^{16}O nuclei are bosons the complete wavefunction must be of symmetry $\Gamma_1^{(\text{CNP})}$ and thus rovibronic states of species $\Gamma_2^{(\text{CNP})}$ (which can only be combined with a nuclear spin wavefunction of species $\Gamma_2^{(\text{CNP})}$) have no nuclear spin partner with which to combine. Thus these states cannot occur and are 'missing.' This means that half the rotational levels of every vibronic state are missing in this molecule. Missing levels arise in other molecules and can also involve nuclei with nonzero spin; they arise for the ammonia molecule NH_3.

The Pauli exclusion principle follows from the indistinguishability of electrons and the rules of fermion permutation. It prevents the occurrence of states that have two or more electrons in the same particle state. As a result of the indistinguishability of nuclei, and the rules of fermion and boson permutation, there are missing levels. Both of these results can be tested experimentally. A negative result from trying to put an electron into the 1S state of Cu (this state already having two electrons of opposite spin in it) was reported by Ramberg and Snow [11] and by analysing their results they determined an upper limit for the violation of the Pauli exclusion principle of 1.7×10^{-26}; this means that at this level the electrons are indistinguishable. Attempts to observe spectral lines that would arise from transitions between 'missing' levels have been made in order to see whether the levels are truly missing. Such missing levels would arise if the nuclei involved are not completely identical. Such a situation is conceivable. Three negative attempts at a sensitivity level of only about 10^{-6} have been reported [12, 13, 14].

A1.4.3.5   Time reversal symmetry

The time reversal symmetry operation $\widehat{\theta}$ (or T) is the operation of reversing the direction of motion; it reverses all momenta, including spin angular momenta, but not the coordinates (see [15] for a good general account of this symmetry operation). As with the inversion operation E^* the weak interaction force is not invariant to time reversal and we discuss this further in the next subsection. However, for all practical purposes in molecular physics we can take this to be a symmetry operation. This symmetry operation has the property, unlike the other symmetry operations discussed here, of being *antiunitary*. Also, time reversal invariance does not lead to any conservation law and molecular states are not eigenstates of $\widehat{\theta}$. However, this symmetry operator constrains the form of the Hamiltonian, an example being that no term in the Hamiltonian can contain the product of an odd number of momenta. Also, it is sometimes a useful tool in determining whether certain matrix elements vanish (see, for example, [16]) and it can be responsible for extra degeneracies. In particular, if a symmetry group has a pair of irreducible representations, Γ and Γ^* say, whose characters are the complex conjugates of each other, then energy levels of symmetry Γ and Γ^* will always coincide in pairs and be degenerate because of time reversal symmetry. Such a pair of irreducible representations of a symmetry group are called 'separably degenerate'. The irreducible representations E_+ and E_- of the point group C_3 (see table A1.4.8) are separably degenerate. Such a character table can be condensed by adding the characters of the separably degenerate irreducible representations and this is done for the C_3 group in table A1.4.9. In the condensed character table the separably degenerate representations are marked 'sep'.

Apart from the degeneracy of separably degenerate states, time reversal symmetry leads to *Kramers' degeneracy* or *Kramers' theorem*: all energy levels of a system containing an odd number of particles with half-integral spin (i.e., fermions) must be at least doubly degenerate. One generally only considers systems having an odd number of electrons, but if nuclei with half integral spin cause the degeneracy then one must resolve the nuclear hyperfine structure for the degeneracy to be revealed.

Table A1.4.8. The character table of the point group C_3.

	E	C_3	C_3^2
A_1	1	1	1
E_+	1	ω	ω^2
E_-	1	ω^2	ω

$\omega = \exp(2\pi i/3)$.

Table A1.4.9. The condensed character table of the C_3 group.

	E	C_3, C_3^2	
A_1	1	1	
E	2	-1	sep

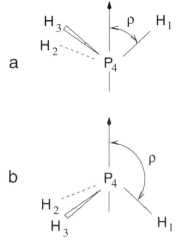

Figure A1.4.5. PH_3 inversion.

A1.4.3.6   Concluding remarks about symmetries

In the above we have discussed several different symmetry groups: the translation group G_T, the rotation group K(spatial), the inversion group $\mathcal{E}$, the electron permutation group S_n^e and the complete nuclear permutation group G^{CNP}. We have also discussed the time reversal symmetry operation $\hat{\theta}$. The translational states Φ_{CM} can be classified according to their linear momentum using G_T, but we rarely worry about the translational state of a molecule. The internal states Φ_{int} can be labelled with their angular momentum (F, m_F) using K(spatial), and their parity ($\pm$) using $\mathcal{E}$. The symmetry in the group S_n^e leads to restrictions on the electron spin multiplicities (the Pauli exclusion principle) and the symmetry in G^{CNP} leads to nuclear spin statistical weights.

One might think that we should form a 'full' symmetry group of the molecular Hamiltonian, G_{FULL} say, describing all symmetry types simultaneously and symmetry classify our basis functions and eigenfunctions in this group. If we neglect time reversal symmetry (which requires special consideration because the operator

$\widehat{\theta}$ is antiunitary), we have

$$G_{\text{FULL}} = G_{\text{T}} \otimes K(\text{spatial}) \otimes \mathcal{E} \otimes S_n^e \otimes G^{\text{CNP}} \tag{A1.4.134}$$

that is, the full symmetry group for an isolated molecule in field-free space is the direct product of the groups describing the individual symmetry types. However, it can be shown that it is completely equivalent and easier, to treat each type of symmetry and each symmetry group, separately. In order to transform irreducibly in G_{FULL}, a wavefunction must transform irreducibly in each of the groups G_{T}, $K(\text{spatial})$, $\mathcal{E}$, S_n^e and G^{CNP}. This is discussed in section 7.3 of [1]. Watson [17] has shown that for a molecule in an external electric field the full symmetry group cannot be factorized in the simple manner of equation (A1.4.134). In this case, instead of the three separate groups $K(\text{spatial})$, $\mathcal{E}$ and G^{CNP}, it is necessary to consider a more complicated group containing selected elements of their direct product group. In the following section we show how the direct product of the groups $\mathcal{E}$ and G^{CNP}, called the complete nuclear permutation–inversion (CNPI) group G^{CNPI}, is used in molecular physics; it leads to the definition of the molecular symmetry (MS) group. In the final section we show how the molecular point group emerges from the molecular symmetry group as a near symmetry group of the molecular Hamiltonian.

As a postscript to this section we consider the operation of charge conjugation symmetry. This operation is not used in molecular physics but it is an important symmetry in nature, and it does lead to an important implication about the probable breakdown of time reversal symmetry. Classical electrodynamic forces are invariant if we change the signs of the charges. In elementary particle physics the 'charge conjugation operation' C is introduced as a generalization of this changing-the-sign-of-the-charge operation: it is the operation of changing every particle (including uncharged particles like the neutron) into its antiparticle. Weak interactions are not invariant to the operation C just as they are not invariant to the inversion operation P. One might hope to preserve the exact 'mirror symmetry' of nature if invariance to the product CP were a fact. Unfortunately, CP symmetry is not universal [18], although its violation is a small effect that has never been observed outside the neutral K meson (kaon) system and the extent of its violation cannot be calculated (unlike the situation with parity violation, which by comparison is a big effect). CP violation permits unequal treatment of particles and antiparticles and it may be responsible for the domination of matter over antimatter in the universe [19]. Very recent considerations concerning CP violation are summarized in [20]; in particular, this reference points out that the study of CP violation in neutral B mesons will probe the physics behind the 'standard model', which does not predict sufficient CP violation to account, by itself, for the predominance of matter over antimatter in the universe. In the light of the fact that C was introduced as a generalization of the changing-the-sign-of-the-charge operation, it is appropriate that CP violation provides an unambiguous 'convention-free' definition of positive charge: *it is the charge carried by the lepton preferentially produced in the decay of the long-lived neutral K meson* [21]. Although CP violation is a fact there is one invariance in nature involving C that is believed to be universal (based on quantum field theory) and that is invariance under the triple operation TCP, which also involves the time reversal operation T. TCP symmetry implies that every particle has the same mass and lifetime as its antiparticle. However, now, if TCP symmetry is true the observation of CP violation in experiments on neutral K mesons must mean that there is a compensating violation of time reversal symmetry at the same time. A direct experimental measure of the violation of time reversal symmetry has not been made, mainly because the degree of violation is very small.

A1.4.4 The molecular symmetry group

The complete nuclear permutation inversion (CNPI) group of the PH_3 molecule is the direct product of the complete nuclear permutation (CNP) group S_3 (see equation (A1.4.19)) and the inversion group $\mathcal{E} = \{E, E^*\}$. This is a group of 12 elements that we call G_{12}:

$$G_{12} = \{E, (123), (132), (12), (23), (31), E^*, (123)^*, (132)^*, (12)^*, (23)^*, (31)^*\}. \tag{A1.4.135}$$

Table A1.4.10. The character table of the CNPI group G_{12}.

	E	(123) (132)	(12) (23) (31)	E^*	(123)* (132)*	(12)* (23)* (31)*
A_1^+	1	1	1	1	1	1
A_1^-	1	1	1	-1	-1	-1
A_2^+	1	1	-1	1	1	-1
A_2^-	1	1	-1	-1	-1	1
E^+	2	-1	0	2	-1	0
E^-	2	-1	0	-2	1	0

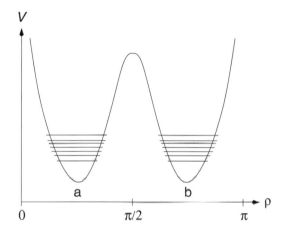

Figure A1.4.6. A cross-section of the potential energy surface of PH$_3$. The coordinate ρ is defined in figure A1.4.5.

The rotation–vibration–electronic energy levels of the PH$_3$ molecule (neglecting nuclear spin) can be labelled with the irreducible representation labels of the group G_{12}. The character table of this group is given in table A1.4.10.

Before we consider the results of this symmetry labelling, we should consider the effect of the inversion motion in PH$_3$. In figure A1.4.5 we depict the two *versions* (see [22] for a discussion of this term) of the numbered equilibrium structure of the molecule and call them a and b. The inversion coordinate ρ is also indicated in this figure. In figure A1.4.6 we schematically indicate the cross-section in the potential energy surface of the PH$_3$ molecule that contains the two minima and the barrier between them. In this figure we also indicate several vibrational energy levels of the molecule. The barrier to inversion is so high ($\approx$11 300 cm^{-1}; see [23]) that there is no observable inversion tunnelling splitting. Thus, the energy levels can be calculated by just considering the motion in one of the two minima and we do not need to consider both minima. The 'single minimum' calculation is represented in figure A1.4.7; each minimum has a duplicate set of energy levels.

If we were to calculate the vibrational energy levels using the double minimum potential energy surface, we would find that well below the barrier, every energy level would be doubly degenerate to within measurement accuracy for PH$_3$. If we symmetry classified the levels using the group G_{12} we would find that there were three types of energy level: $A_1^+ + A_1^-$, $A_2^+ + A_2^-$ or $E^+ + E^-$. This double degeneracy would be resolved by inversion tunnelling and it is an accidental degeneracy not forced by the symmetry group G_{12}.

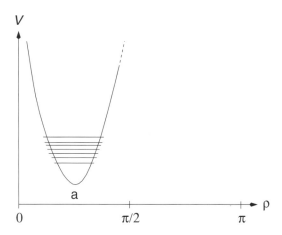

Figure A1.4.7. A cross-section of the potential energy surface of PH$_3$ obtained by ignoring the version b (see figure A1.4.6). The coordinate ρ is defined in figure A1.4.5.

Table A1.4.11. The character table of the molecular symmetry group $C_{3v}(M)$.

	E	(123) (132)	$(12)^*$ $(23)^*$ $(31)^*$
A_1	1	1	1
A_2	1	1	-1
E	2	-1	0

If the inversion tunnelling is not resolved we have actually done too much work here. There are only three distinct types of level and yet we have used a symmetry group with six irreducible representations. However, Longuet-Higgins [24] showed how to obtain the appropriate subgroup of G_{12} that avoids the unnecessary double labels. This is achieved by just using the elements of G_{12} that are appropriate for a single minimum; we delete elements such as E^* and (12) that interconvert the a and b forms. Longuet-Higgins termed the deleted elements 'unfeasible.' The group obtained is 'the molecular symmetry (MS) group'. In the case of PH$_3$, we obtain the particular MS group

$$C_{3v}(M) = \{E, (123), (132), (12)^*, (23)^*, (31)^*\}; \qquad (A1.4.136)$$

its character table (with the class structure indicated) is given in table A1.4.11. Using this group, we achieve a sufficient symmetry labelling of the levels as being either A_1, A_2 or E. All possible interactions can be understood using this group (apart from the effect of inversion tunnelling).

For PH$_3$ the labour saved by using the MS group rather than the CNPI group is not very great, but for larger molecules, such as the water trimer for example, a great saving is achieved if all unfeasible elements of the CNPI group are eliminated from consideration. An unfeasible element of the CNPI group is one that takes the molecule between versions that are separated by an insuperable energy barrier in the potential energy function. For the water trimer the CNPI group has $6! \times 3! \times 2 = 8640$ elements. The MS group that is used to interpret the spectrum has 48 elements [25].

Ammonia (NH$_3$) is pyramidal like PH$_3$ and in its electronic ground state there are two versions of the numbered equilibrium structure exactly as shown for PH$_3$ in figure A1.4.5. The potential barrier between the two versions, however, is around 2000 cm^{-1} for ammonia [26] and thus much lower than in PH_3. This barrier is so low that the molecule will tunnel through it on the time scale of a typical spectroscopic experiment, and the tunnelling motion gives rise to energy level splittings that can be resolved experimentally (see, for example, figure 15-3 of [1]). Thus, for NH_3, all elements of the group G_{12} are feasible, and the molecular symmetry group of NH_3 in its electronic ground state is G_{12}. This group is isomorphic to the point group D_{3h} and in the literature it is customarily called $D_{3h}(M)$.

A1.4.5   The molecular point group

The MS group is introduced by deleting unfeasible elements from the CNPI group. It can be applied to symmetry label the rotational, vibrational, electronic and spin wavefunctions of a molecule, regardless of whether the molecule is rigid or nonrigid. It is a *true* symmetry group and no terms in the Hamiltonian can violate the symmetry labels obtained (with the exception of the as yet undetected effect of the weak neutral current interaction). The MS group can be used to determine nuclear spin statistical weights, to determine which states can and cannot interact as a result of considering previously neglected higher order terms in the Hamiltonian, or the effect of externally applied magnetic or electric fields and it can be used to determine the selection rules for allowed electric and magnetic dipole transitions. What then of the molecular point group?

For a molecule that has no observable tunnelling between minima on the potential energy surface (i.e., for a rigid molecule) and for which the equilibrium structure is nonlinear[11], it turns out that the MS group is isomorphic to the point group of the equilibrium structure. For example, PH_3 has the molecular symmetry group $C_{3v}(M)$ given in equation (A1.4.136) and its equilibrium structure has the point group C_{3v} given in equation (A1.4.22). It is easy to show from equation (A1.4.31) (using the fact that $E^*E^* = E$) that these two groups are isomorphic with the following mapping:

$$C_{3v}(M): \quad E \quad (123) \quad (132) \quad (12)^* \quad (23)^* \quad (31)^*$$
$$C_{3v}: \quad\quad E \quad C_3 \quad C_3^2 \quad \sigma_3 \quad \sigma_1 \quad \sigma_2. \tag{A1.4.137}$$

Obviously, we have chosen the name $C_{3v}(M)$ for the molecular symmetry group of PH_3 because this group is isomorphic to C_{3v}.

Quite remarkably, if we neglect the effect of the MS group elements on the rotational variables (the Euler angles θ, ϕ and χ) then each element of the MS group rotates and/or reflects the vibrational displacements and electronic coordinates in the manner described by its partner in the point group. In fact, for the purpose of classifying vibrational and electronic wavefunctions this *defines* what the elements of the molecular point group actually do to the molecular coordinates for a rigid nonlinear molecule. By starting with the fundamental definition of symmetry in terms of energy invariance, by considering the operations of inversion and identical nuclei permutation and, finally, by deleting unfeasible elements of the CNPI group, we recover the simple description of molecular symmetry in terms of rotations and reflections, but the rotations and reflections are of the vibrational displacements and the electronic coordinates—not of the entire molecule at its equilibrium configuration. Such operations are not symmetry operations of the full Hamiltonian (unlike the elements of the MS group) since the transformation of the rotational variables is neglected. This means that such effects as Coriolis coupling for example, which involve a coupling of rotation and vibration, will mix vibrational states of different point group symmetry. The molecular point group is a *near* symmetry group of the full Hamiltonian. However, the molecular point group is a symmetry group of the vibration–electronic Hamiltonian of a rigid molecule and in practice it is always used for labelling the vibration–electronic states of such molecules. Its use enables one, for example, to classify the normal vibration coordinates and to study the transformation properties of the electronic wavefunction without having to bother about molecular rotation. This is a useful

---

[11] Rigid linear molecules are a special case in which an *extended* MS group, rather than the MS group, is isomorphic to the point group of the equilibrium structure; see chapter 17 of [1].

simplification, but the reader must be aware that the rotation and/or reflection operations of the molecular point group do not rotate and/or reflect the molecule in space; they rotate and/or reflect the vibrational displacements and electronic coordinates[12]. To study the effect of molecular rotation (as one needs to do if one is interested in understanding high resolution rotationally resolved molecular spectra), or to study nonrigid molecules such as the water trimer, the point group is of no use and one must employ the appropriate MS group.

Acknowledgments

We thank Antonio Fernandez-Ramos for critically reading the manuscript. Part of this paper was written while PJ worked as a guest at the Steacie Institute for Molecular Sciences.

References

[1]   Bunker P R and Jensen P 1998 *Molecular Symmetry and Spectroscopy* 2nd edn (Ottawa: NRC)
[2]   Cotton F A 1990 *Chemical Applications of Group Theory* 3rd edn (New York: Wiley)
[3]   Tinkham M 1964 *Group Theory and Quantum Mechanics* (New York: McGraw-Hill)
[4]   Bunker P R and Howard B J 1983 *Symmetries and Properties of Non-Rigid Molecules: a Comprehensive Survey (Studies in Physical and Theoretical Chemistry 23)* ed J Maruani and J Serre (Amsterdam: Elsevier) p 29
[5]   Eckart C 1935 *Phys. Rev.* **47** 552
[6]   Zare R N 1988 *Angular Momentum* (New York: Wiley)
[7]   Kleiman V, Gordon R J, Park H and Zare R N 1998 *Companion to Angular Momentum* (New York: Wiley)
[8]   Wood C S, Bennett S C, Cho D, Masterson B P, Roberts J L, Tanner C E and Wieman C E 1997 Measurement of parity nonconservation and an anapole moment in cesium *Science* **275** 1759–63
[9]   Hegstrom R A, Rein D W and Sandars P G H 1980 *J. Chem. Phys.* **73** 2329
[10] Pauli W 1940 *Phys. Rev.* **58** 716
[11] Ramberg E and Snow G A 1990 *Phys. Lett.* B **238** 438
[12] de Angelis M, Gagliardi G, Gianfrani L and Tino G M 1996 Test of the symmetrization postulate for spin-0 particles *Phys. Rev. Lett.* **76** 2840
[13] Hilborn R C and Yuca C L 1996 Spectroscopic test of the symmetrization postulate for spin-0 nuclei *Phys. Rev. Lett.* **76** 2844
[14] Naus H, de Lange A and Ubachs W 1997 *Phys. Rev.* A **56** 4755
[15] Overseth O E 1969 *Sci. Am.* October
[16] Watson J K G 1974 *J. Mol. Spectrosc.* **50** 281
[17] Watson J K G 1975 *Can. J. Phys.* **53** 2210
[18] Christenson J H, Cronin J W, Fitch V L and Turlay R 1964 *Phys. Rev. Lett.* **13** 138
[19] Wilczek F 1980 *Sci. Am.* December
[20] Schwarzschild B 1999 *Phys. Today* January 22
[21] Griffiths D 1987 *Introduction to Elementary Particles* (New York: Harper and Row)
[22] Bone R G A, Rowlands T W, Handy N C and Stone A J 1991 *Mol. Phys.* **72** 33
[23] Špirko V, Civiš S, Ebert M and Danielis V 1986 *J. Mol. Spectrosc.* **119** 426
[24] Longuet-Higgins H C 1963 *Mol. Phys.* **6** 445
[25] van der Avoird A, Olthof E H T and Wormer P E S 1996 *J. Chem. Phys.* **105** 8034
[26] Špirko V and Kraemer W P 1989 *J. Mol. Spectrosc.* **133** 331

Further Reading

Bunker P R and Jensen P 1998 *Molecular Symmetry and Spectroscopy* 2nd edn (Ottawa: NRC)
Cotton F A 1990 *Chemical Applications of Group Theory* 3rd edn (New York: Wiley)
Griffiths D 1987 *Introduction to Elementary Particles* (New York: Harper and Row)
Kleiman V, Gordon R J, Park H and Zare R N 1998 *Companion to Angular Momentum* (New York: Wiley)
Tinkham M 1964 *Group Theory and Quantum Mechanics* (New York: McGraw-Hill)
Zare R N 1988 *Angular Momentum* (New York: Wiley)

---

[12] A detailed discussion of the relation between MS group operations and point group operations is given in section 4.5 of [1].

A1.5
Intermolecular interactions

Ajit J Thakkar

A1.5.1 Introduction

The existence of intermolecular interactions is apparent from elementary experimental observations. There must be attractive forces because otherwise condensed phases would not form, gases would not liquefy, and liquids would not solidify. There must be short-range repulsive interactions because otherwise solids and liquids could be compressed to much smaller volumes with ease. The kernel of these notions was formulated in the late eighteenth century, and Clausius made a clear statement along the lines of this paragraph as early as 1857 [1].

Since the interaction energy V between a pair of molecules must have an attractive region at large intermolecular separations r and a steeply repulsive region at short distances, it is evident that $V(r)$ must have the schematic form illustrated in figure A1.5.1. It is conventional to denote the distance at which the interaction energy is a minimum by either r_m or r_e and to refer to this distance as the equilibrium distance. Similarly it is common to denote the shorter distance at which the interaction energy is zero by σ and refer to it as the slow collision diameter. The net potential energy of attraction at the minimum is $V(r_m) = -\varepsilon$, and ε is called the well depth.

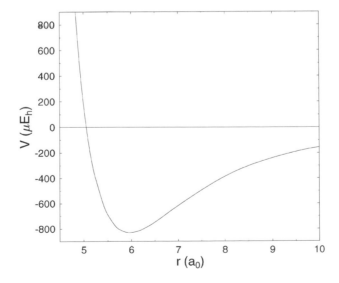

Figure A1.5.1. Potential energy curve for NeF$^-$ based on *ab initio* calculations of Archibong *et al* [88].

In 1873, van der Waals [2] first used these ideas to account for the deviation of real gases from the ideal gas law $P\bar{V} = RT$ in which P, $\bar{V}$ and T are the pressure, molar volume and temperature of the gas and R is the gas constant. He argued that the incompressible molecules occupied a volume b leaving only the volume $\bar{V} - b$ free for the molecules to move in. He further argued that the attractive forces between the molecules reduced the pressure they exerted on the container by $a/\bar{V}^2$; thus the pressure appropriate for the gas law is $P + a/\bar{V}^2$ rather than P. These ideas led him to the van der Waals equation of state:

$$(P + a/\bar{V}^2)(\bar{V} - b) = RT. \tag{A1.5.1}$$

The importance of the van der Waals equation is that, unlike the ideal gas equation, it predicts a gas–liquid transition and a critical point for a pure substance. Even though this simple equation has been superseded, its remarkable success led to the custom of referring to the attractive and repulsive forces between molecules as van der Waals forces.

The feature that distinguishes intermolecular interaction potentials from intramolecular ones is their relative strength. Most typical single bonds have a dissociation energy in the 150–500 kJ mol^{-1} range but the strength of the interactions between small molecules, as characterized by the well depth, is in the 1–25 kJ mol^{-1} range.

A1.5.1.1   Many-body expansion

The total energy of an assembly of molecules can be written as

$$E = \sum_i E_i + \sum_{i>j} V_{ij} + \sum_{i>j>k} V_{ijk} + \cdots \tag{A1.5.2}$$

in which E_i is the energy of isolated molecule i, V_{ij} is the energy of interaction between molecules i and j in the absence of any others, V_{ijk} is the *non-additive* energy of interaction among the three molecules i, j and k in the absence of any others, and so on. The interaction energy is then

$$V = E - \sum_i E_i = \sum_{i>j} V_{ij} + \sum_{i>j>k} V_{ijk} + \cdots. \tag{A1.5.3}$$

For example, if there are three molecules A, B and C, then equation (A1.5.3) can be written as

$$V = V_{AB} + V_{BC} + V_{CA} + V_{ABC}. \tag{A1.5.4}$$

V_{AB} is the interaction energy of molecules A and B in the *absence* of molecule C. The interaction between molecules A and B will be different in the presence of molecule C, and so on. The non-additive, three-body term V_{ABC} is the total correction for these errors in the three pair interactions. When there are four molecules, a three-body correction is included for each distinct triplet of molecules and the remaining error is corrected by the non-additive four-body term.

In many cases, it is reasonable to expect that the sum of two-body interactions will be much greater than the sum of the three-body terms which in turn will be greater than the sum of the four-body terms and so on. Retaining only the two-body terms in equation (A1.5.3) is called the *pairwise additivity* approximation. This approximation is quite good so the bulk of our attention can be focused on describing the two-body interactions. However, it is now known that the many-body terms cannot be neglected altogether, and they are considered briefly in sections A1.5.2.6 and A1.5.3.5.

A1.5.1.2 Types of intermolecular interactions

It is useful to classify various contributions to intermolecular forces on the basis of the physical phenomena that give rise to them. The first level of classification is into long-range forces that vary as inverse powers of the distance r^{-n}, and short-range forces that decrease exponentially with distance as in $\exp(-\alpha r)$.

There are three important varieties of long-range forces: *electrostatic*, *induction* and *dispersion*. Electrostatic forces are due to classical Coulombic interactions between the static charge distributions of the two molecules. They are strictly pairwise additive, highly anisotropic, and can be either repulsive or attractive. The distortions of a molecule's charge distribution induced by the electric field of all the other molecules leads to induction forces that are always attractive and highly non-additive. Dispersion forces are always present, always attractive, nearly pairwise additive, and arise from the instantaneous fluctuations of the electron distributions of the interacting molecules. If the molecules are in closed-shell ground states, then there are no other important long-range interactions. However, if one or more of the molecules are in degenerate states, then non-additive, resonance interactions of either sign can arise. Long-range forces are discussed in greater detail in section A1.5.2.

The most important short-range forces are exchange and repulsion; they are very often taken together and referred to simply as exchange–repulsion. They are both non-additive and of opposing sign, but the repulsion dominates at short distances. The overlap between the electron densities of molecules when they are close to one another leads to modifications of the long-range terms and thence to short-range penetration, charge transfer and damping effects. All these effects are discussed in greater detail in section A1.5.3.

A1.5.1.3 Potential energy surfaces

Only the interactions between a pair of atoms can be described as a simple function $V(r)$ of the distance between them. For nonlinear molecules, several coordinates are required to describe the relative orientation of the interacting species. Thus it is necessary to think of the interaction energy as a 'potential energy surface' (PES) that depends on many variables. There are usually several points of minimum energy on this surface; many of these will be 'local minima' and at least one will be the 'global minimum'. The interaction energy at a local minimum is lower than at any point in its neighbourhood but there can be lower energy minima further away. If there is more than one global minimum, then these are located at symmetry equivalent points on the surface, corresponding to the same minimum energy.

For the interaction between a nonlinear molecule and an atom, one can place the coordinate system at the centre of mass of the molecule so that the PES is a function of the three spherical polar coordinates r, θ, ϕ needed to specify the location of the atom. If the molecule is linear, V does not depend on ϕ and the PES is a function of only two variables. In the general case of two nonlinear molecules, the interaction energy depends on the distance between the centres of mass, and five of the six Euler angles needed to specify the relative orientation of the molecular axes with respect to the global or 'space-fixed' coordinate axes.

A1.5.2 Long-range forces

A1.5.2.1 Long-range perturbation theory

Perturbation theory is a natural tool for the description of intermolecular forces because they are relatively weak. If the interacting molecules (A and B) are far enough apart, then the theory becomes relatively simple because the overlap between the wavefunctions of the two molecules can be neglected. This is called the polarization approximation. Such a theory was first formulated by London [3,4], and then reformulated by several others [5–7].

Each electron in the system is assigned to either molecule A or B, and Hamiltonian operators $\mathcal{H}^A$ and $\mathcal{H}^B$ for each molecule defined in terms of its assigned electrons. The unperturbed Hamiltonian for the system is then $\mathcal{H}^0 = \mathcal{H}^A + \mathcal{H}^B$, and the perturbation $\lambda\mathcal{H}'$ consists of the Coulomb interactions between the nuclei and electrons of A and those of B. The unperturbed states, eigenfunctions of $\mathcal{H}^0$, are simple product functions $\Psi_m^A\Psi_n^B$. For closed-shell molecules, non-degenerate, Rayleigh–Schrödinger, perturbation theory gives the energy of the ground state of the interacting system. The first-order interaction energy is the electrostatic energy, and the second-order energy is partitioned into induction and dispersion energies. The induction energy consists of all terms that involve excited states of only one molecule at a time, whereas the dispersion energy includes all the remaining terms that involve excited states of both molecules simultaneously.

Long-range forces are most conveniently expressed as a power series in $1/r$, the reciprocal of the intermolecular distance. This series is called the multipole expansion. It is so common to use the multipole expansion that the electrostatic, induction and dispersion energies are referred to as 'non-expanded' if the expansion is not used. In early work it was noted that the multipole expansion did not converge in a conventional way and doubt was cast upon its use in the description of long-range electrostatic, induction and dispersion interactions. However, it is now established [8–13] that the series is asymptotic in Poincaré's sense. The interaction energy can be written as

$$V(r) = \sum_{n=0}^{N} V_n/r^n + \mathrm{O}(1/r^{N+1}) \tag{A1.5.5}$$

with the assurance that the remainder left upon truncation after some chosen term in r^{-N} tends to zero in the limit as $r \to \infty$. In other words, the multipole expansion can be made as accurate as one desires for large enough intermolecular separations, even though it cannot be demonstrated to converge at any given value of r and, in some cases, diverges for all r!

Some electric properties of molecules are described in section A1.5.2.2 because the coefficients of the powers of $1/r$ turn out to be related to them. The electrostatic, induction and dispersion energies are considered in turn in sections A1.5.2.3–A1.5.2.5, respectively.

A1.5.2.2   Multipole moments and polarizabilities

The long-range interactions between a pair of molecules are determined by electric multipole moments and polarizabilities of the individual molecules. *Multipole moments* are measures that describe the non-sphericity of the charge distribution of a molecule. The zeroth-order moment is the total charge of the molecule: $Q = \sum_i q_i$ where q_i is the charge of particle i and the sum is over all electrons and nuclei in the molecule. The first-order moment is the dipole moment vector with Cartesian components given by

$$\mu_\alpha = \int \rho(\mathbf{r})r_\alpha \, \mathrm{d}^3r \qquad \alpha \in \{x, y, z\} \tag{A1.5.6}$$

in which $\rho(\mathbf{r})$ is the total (electronic plus nuclear) charge density of the molecule. The direction of the dipole moment is from negative to positive. Dipole moments have been measured for a vast variety of molecules [14–16].

Next in order is the quadrupole moment tensor Θ with components:

$$\Theta_{\alpha\beta} = \tfrac{1}{2} \int \rho(\mathbf{r})(3r_\alpha r_\beta - r^2\delta_{\alpha\beta}) \, \mathrm{d}^3r \qquad \alpha, \beta \in \{x, y, z\} \tag{A1.5.7}$$

where the 'Kronecker delta' $\delta_{\alpha\beta} = 1$ for $\alpha = \beta$ and $\delta_{\alpha\beta} = 0$ for $\alpha \neq \beta$. The quadrupole moment is a symmetric ($\Theta_{\alpha\beta} = \Theta_{\beta\alpha}$) second-rank tensor. Moreover, it is traceless:

$$\Theta_{xx} + \Theta_{yy} + \Theta_{zz} = 0. \tag{A1.5.8}$$

Therefore, it has at most five independent components, and fewer if the molecule has some symmetry. Symmetric top molecules have only one independent component of Θ, and, in such cases, the axial component is often referred to as the quadrupole moment. A quadrupolar distribution can be created from four charges of the same magnitude, two positive and two negative, by arranging them in the form of two dipole moments parallel to each other but pointing in opposite directions. Centro-symmetric molecules, like CO_2, have a zero dipole moment but a non-zero quadrupole moment.

The multipole moment of rank n is sometimes called the 2^n-pole moment. The first non-zero multipole moment of a molecule is origin independent but the higher-order ones depend on the choice of origin. Quadrupole moments are difficult to measure and experimental data are scarce [17–19]. The octopole and hexadecapole moments have been measured only for a few highly symmetric molecules whose lower multipole moments vanish. *Ab initio* calculations are probably the most reliable way to obtain quadrupole and higher multipole moments [20–22].

The charge redistribution that occurs when a molecule is exposed to an electric field is characterized by a set of constants called *polarizabilities*. In a uniform electric field F, a component of the dipole moment is

$$\mu_\alpha = \mu_\alpha^0 + \alpha_{\alpha\beta} F_\beta + \frac{1}{2}\beta_{\alpha\beta\gamma} F_\beta F_\gamma + \frac{1}{3!}\Gamma_{\alpha\beta\gamma\delta} F_\beta F_\gamma F_\delta + \cdots \tag{A1.5.9}$$

in which $\alpha_{\alpha\beta}$, $\beta_{\alpha\beta\gamma}$ and $\Gamma_{\alpha\beta\gamma\delta}$, respectively, are components of the dipole polarizability, *hyperpolarizability* and second hyperpolarizability tensors, and a summation is implied over repeated subscripts.

The dipole polarizability tensor characterizes the lowest-order dipole moment *induced* by a uniform field. The α tensor is symmetric and has no more than six independent components, less if the molecule has some symmetry. The scalar or mean dipole polarizability

$$\bar{\alpha} = \frac{1}{3}\text{Tr}\,\alpha = \frac{1}{3}\sum_i \alpha_{ii} \tag{A1.5.10}$$

is invariant to the choice of coordinate system and is often referred to simply as 'the polarizability'. It is related to many important bulk properties of an ensemble of molecules including the dielectric constant, the refractive index, the extinction coefficient, and the electric susceptibility. The polarizability is a measure of the softness of the molecule's electron density, and correlates directly with molecular size, and inversely with the ionization potential and HOMO–LUMO gap. Another scalar polarizability invariant commonly encountered is the polarizability anisotropy:

$$(\Delta\alpha)^2 = \tfrac{1}{2}[3\text{Tr}\,\alpha^2 - (\text{Tr}\,\alpha)^2]. \tag{A1.5.11}$$

In linear, spherical and symmetric tops the components of α along and perpendicular to the principal axis of symmetry are often denoted by $\alpha_\parallel$ and $\alpha_\perp$, respectively. In such cases, the anisotropy is simply $\Delta\alpha = \alpha_\parallel - \alpha_\perp$. If the applied field is oscillating at a frequency ω, then the dipole polarizability is frequency dependent as well $\alpha(\omega)$. The zero frequency limit of the 'dynamic' polarizability $\alpha(\omega)$ is the static polarizability described above.

There are higher multipole polarizabilities that describe higher-order multipole moments induced by non-uniform fields. For example, the quadrupole polarizability is a fourth-rank tensor **C** that characterizes the lowest-order quadrupole moment induced by an applied field gradient. There are also mixed polarizabilities such as the third-rank dipole–quadrupole polarizability tensor **A** that describes the lowest-order response of the dipole moment to a field gradient and of the quadrupole moment to a dipolar field. All polarizabilities of order higher than dipole depend on the choice of origin. Experimental values are basically restricted to the dipole polarizability and hyperpolarizability [23–25]. *Ab initio* calculations are an important source of both dipole and higher polarizabilities [20]; some recent examples include [26, 27].

A1.5.2.3 Electrostatic interactions

The electrostatic potential generated by a molecule A at a distant point B can be expanded in inverse powers of the distance r between B and the centre of mass (CM) of A. This series is called the *multipole expansion* because the coefficients can be expressed in terms of the multipole moments of the molecule. With this expansion in hand, it is straightforward to write the electrostatic interaction between molecule A and another molecule with its CM at B as a multipole expansion. The formal expression [7, 28] for this electrostatic interaction, in terms of '**T** tensors', is intimidating to all but the experts. However, explicit expressions for individual terms in this expansion are easily understood.

Consider the case of two neutral, linear, dipolar molecules, such as HCN and KCl, in a coordinate system with its origin at the CM of molecule A and the z-axis aligned with the intermolecular vector r pointing from the CM of A to the CM of B. The relative orientation of the two molecules is uniquely specified by their spherical polar angles θ_A, θ_B and the difference $\phi = \phi_A - \phi_B$ between their azimuthal angles. The leading term in the multipole expansion of the electrostatic interaction energy is the dipole–dipole term

$$V_{dd}(r, \theta_A, \theta_B, \phi) = -\frac{\mu_A \mu_B}{4\pi\varepsilon_0 r^3}(2\cos\theta_A\cos\theta_B - \sin\theta_A\sin\theta_B\cos\phi) \qquad (A1.5.12)$$

in which ε_0 is the vacuum permittivity, and μ_A and μ_B are the magnitudes of the dipole moments of A and B. This expression is also applicable to the dipole–dipole interaction between any pair of neutral molecules provided that the angles are taken to specify the relative orientation of the dipole moment vectors of the molecules.

The leading term in the electrostatic interaction between a pair of linear, quadrupolar molecules, such as HCCH and CO_2 is

$$V_{qq} = \frac{3\Theta_A\Theta_B}{16\pi\varepsilon_0 r^5}[1 - 5\cos^2\theta_A - 5\cos^2\theta_B - 15\cos^2\theta_A\cos^2\theta_B + 2(4\cos\theta_A\cos\theta_B - \sin\theta_A\sin\theta_B\cos\phi)^2]$$

$$(A1.5.13)$$

in which Θ_A and Θ_B are the axial quadrupole moments of A and B. This expression is also applicable to the quadrupole–quadrupole interaction between any pair of spherical or symmetric top molecules provided that the angles are taken to specify the relative orientation of the axial component of the quadrupole moment tensors of the molecules.

The leading term in the electrostatic interaction between the dipole moment of molecule A and the axial quadrupole moment of a linear, spherical or symmetric top B is

$$V_{dq} = \frac{3\mu_A\Theta_B}{8\pi\varepsilon_0 r^4}[\cos\theta_A(3\cos^2\theta_B - 1) - \sin\theta_A\sin 2\theta_B\cos\phi]. \qquad (A1.5.14)$$

Note the r dependence of these three terms: the dipole–dipole interaction varies as r^{-3}, the dipole–quadrupole as r^{-4} and the quadrupole–quadrupole as r^{-5}. In general, the interaction between a 2^ℓ-pole moment and a 2^L-pole moment varies as $r^{-(\ell+L+1)}$. Thus, the dipole–octopole interaction also varies as r^{-5}. At large enough r, only the term involving the lowest-rank, non-vanishing, multipole moment is important. Higher terms begin to play a role as r decreases.

The angular variation of the electrostatic interaction is much greater than that of the induction and dispersion. Hence, electrostatic forces often determine the geometry of a van der Waals complex even when they do not constitute the dominant contribution to the overall interaction.

At a fixed distance r, the angular factor in equation (A1.5.12) leads to the greatest attraction when the dipoles are lined up in a linear head-to-tail arrangement, $\theta_A = \theta_B = 0$, whereas the linear tail-to-tail geometry, $\theta_A = \pi, \theta_B = 0$, is the most repulsive. A head-to-tail, parallel arrangement, $\theta_A = \theta_B = \pi/2, \phi = \pi$, is attractive but less so than the linear head-to-tail geometry. Nevertheless, if the molecules are linear, the head-to-tail, parallel geometry may be more stable because it allows the molecules to get closer and thus increases the r^{-3} factor. For example, the HCN dimer takes the linear head-to-tail geometry in the gas phase [29], but the crystal structure shows a parallel, head-to-tail packing [30].

For interactions between two quadrupolar molecules which have Θ_A and Θ_B of the same sign, at a fixed separation r, the angular factor in equation (A1.5.13) leads to a planar, T-shaped structure, $\theta_A = 0, \theta_B = \pi/2, \phi = 0$, being preferred. This geometry is often seen for nearly spherical quadrupolar molecules. There are other planar ($\phi = 0$) configurations with $\theta_A = \pi/2 - \theta_B$ that are also attractive. A planar, 'slipped parallel' structure, $\theta_A = \theta_B \approx \pi/4, \phi = 0$ is often preferred by planar molecules, and long and narrow molecules because it allows them to approach closer thereby increasing the radial factor. For example, benzene, naphthalene and many other planar quadrupolar molecules have crystal structures consisting of stacks of tilted parallel molecules.

For interactions between two quadrupolar molecules which have Θ_A and Θ_B of the opposite sign, at a fixed separation r, the angular factor in equation (A1.5.13) leads to a linear structure, $\theta_A = \theta_B = 0$, being the most attractive. Linear molecules may also prefer a C_{2v} rectangular or non-planar 'cross' arrangement with $\theta_A = \theta_B = \pi/2$, which allows them to approach closer and increase the radial factor.

Although such structural arguments based purely on electrostatic arguments are greatly appealing, they are also grossly over-simplified because all other interactions, such as exchange–repulsion and dispersion, are neglected, and there are serious shortcomings of the multipole expansion at smaller intermolecular separations.

A1.5.2.4 Induction interactions

If the long-range interaction between a pair of molecules is treated by quantum mechanical perturbation theory, then the electrostatic interactions considered in section A1.5.2.3 arise in first order, whereas induction and dispersion effects appear in second order. The multipole expansion of the induction energy in its full generality [7, 28] is quite complex. Here we consider only explicit expressions for individual terms in the multipole expansion that can be understood readily.

Consider the interaction of a neutral, dipolar molecule A with a neutral, S-state atom B. There are no electrostatic interactions because all the multipole moments of the atom are zero. However, the electric field of A distorts the charge distribution of B and induces multipole moments in B. The leading induction term is the interaction between the permanent dipole moment of A and the dipole moment induced in B. The latter can be expressed in terms of the polarizability of B, see equation (A1.5.9), and the dipole–induced-dipole interaction is given by

$$V_{\text{did}} = -\frac{\mu_A^2 \alpha_B}{2(4\pi \varepsilon_0)^2 r^6}(3\cos^2\theta_A + 1) \tag{A1.5.15}$$

in which θ_A is the angle between the dipole moment vector of A and the intermolecular vector, and α_B is the mean dipole polarizability of B. Since B is a spherical atom, its polarizability tensor is diagonal with the three diagonal components equal to one another and to the mean.

If molecule A is a linear, spherical or symmetric top that has a zero dipole moment like benzene, then the leading induction term is the quadrupole–induced-dipole interaction

$$V_{\text{qid}} = -\frac{9\Theta_A^2 \alpha_B}{8(4\pi \varepsilon_0)^2 r^8}(4\cos^4\theta_A + \sin^4\theta_A) \tag{A1.5.16}$$

in which θ_A is the angle between the axial component of the quadrupole moment tensor of A and the intermolecular vector.

If the molecule is an ion bearing a charge Q_A, then the leading induction term is the isotropic, charge–induced-dipole interaction

$$V_{\text{cid}} = -\frac{Q_A^2 \alpha_B}{2(4\pi \varepsilon_0)^2 r^4}. \tag{A1.5.17}$$

For example, this is the dominant long-range interaction between a neon atom and a fluoride anion F^-.

Note the r dependence of these terms: the charge–induced-dipole interaction varies as r^{-4}, the dipole–induced-dipole as r^{-6} and the quadrupole–induced-dipole as r^{-8}. In general, the interaction between a permanent 2^ℓ-pole moment and an induced 2^L-pole moment varies as $r^{-2(\ell+L+1)}$. At large enough r, only the leading term is important, with higher terms increasing in importance as r decreases. The induction forces are clearly non-additive because a third molecule will induce another set of multipole moments in the first two, and these will then interact. Induction forces are almost never dominant since dispersion is usually more important.

A1.5.2.5 Dispersion interactions

The most important second-order forces are dispersion forces. London [3, 31, 32] showed that they are caused by a correlation of the electron distribution in one molecule with that in the other, and pointed out that the electrons contributing most strongly to these forces are the same as those responsible for the dispersion of light. Since then, these forces have been called London or dispersion forces. Dispersion interactions are always present, even between S-state atoms such as neon and krypton, although there are no electrostatic or induction interaction terms since all the multipole moments of both species are zero.

Dispersion forces cannot be explained classically but a semiclassical description is possible. Consider the electronic charge cloud of an atom to be the time average of the motion of its electrons around the nucleus. The average cloud is spherically symmetric with respect to the nucleus, but at any instant of time there may be a polarization of charge giving rise to an instantaneous dipole moment. This instantaneous dipole induces a corresponding instantaneous dipole in the other atom and there is an interaction between the instantaneous dipoles. The dipole of either atom averages to zero over time, but the interaction energy does not because the instantaneous and induced dipoles are correlated and they stay in phase. The average interaction energy falls off as r^{-6} just as the dipole–induced-dipole energy of equation (A1.5.15). Higher-order instantaneous multipole moments are also involved, giving rise to higher-order dispersion terms. This picture is visually appealing but it should not be taken too literally. The actual effect is not time dependent in the sense of classical fluctuations taking place.

The multipole expansion of the dispersion interaction can be written as

$$V(r) = -C_6/r^6 - C_8/r^8 - C_{10}/r^{10} - \cdots \tag{A1.5.18}$$

where the dispersion coefficients C_6, C_8 and C_{10} are positive, and depend on the electronic properties of the interacting species. The first term is the interaction between the induced-dipole moments on the atoms, the second is the induced-dipole–induced-quadrupole term and the third consists of the induced-dipole–induced-octopole term as well as the interaction between induced quadrupoles. In general, the interaction between an induced 2^ℓ-pole moment and an induced 2^L-pole moment varies as $r^{-2(\ell+L+1)}$. The dispersion coefficients are constants for atoms but, for non-spherical molecules, they depend upon the five angles describing the relative orientation of the molecules. For example, the dispersion coefficients for the interactions between an S-state atom and a Σ_g^+-state diatomic molecule can be expressed as

$$C_{2n}(\theta) = \sum_{L=0}^{n-2} C_{2n}^{2L} P_{2L}(\cos\theta) \tag{A1.5.19}$$

where the C_{2n}^{2L} are dispersion constants, the $P_{2L}(\cos\theta)$ are Legendre polynomials, and θ is the angle between the symmetry axis of the diatomic and the intermolecular vector. Note that C_{2n}^0 is the spherical average of $C_{2n}(\theta)$ and is the appropriate quantity to use in equation (A1.5.18) if the orientation dependence is being neglected. Purely anisotropic dispersion terms varying as $r^{-7}, r^{-9}, \ldots$ arise if at least one of the interacting species lacks inversion symmetry.

Perturbation theory yields a sum-over-states formula for each of the dispersion coefficients. For example, the isotropic C_6^{AB} coefficient for the interaction between molecules A and B is given by

$$C_6^{AB} = \frac{3e^4\hbar^4}{2m_e^2(4\pi\varepsilon_0)^2} \sum_{m,n\neq0} \frac{f_{Am}f_{Bn}}{\Delta E_{Am}\Delta E_{Bn}(\Delta E_{Am}+\Delta E_{Bn})} \qquad (A1.5.20)$$

in which $\hbar$ is the Planck–Dirac constant, $\Delta E_{Am} = E_{Am} - E_{A0}$ is the excitation energy from the ground state $m = 0$ to state m for molecule A and f_{Am} is the corresponding dipole oscillator strength averaged over degenerate final states. Similarly, the sum-over-states formula for the mean, frequency-dependent, polarizability can be written as

$$\bar{\alpha}(\omega) = \frac{e^2}{m_e} \sum_{m\neq0} \frac{f_{Am}}{\omega_{Am}^2 - \omega^2} \qquad (A1.5.21)$$

where $\omega_{Am} = \Delta E_{Am}/\hbar$ is the mth excitation frequency. An important advance consisted in the realization [33–35] that use of the Feynman identity

$$[ab(a+b)]^{-1} = (2/\pi)\int_0^\infty \frac{du}{(a^2+u^2)(b^2+u^2)} \qquad \text{for } a>0, b>0 \qquad (A1.5.22)$$

together with equations (A1.5.20) and (A1.5.21) leads to

$$C_6^{AB} = \frac{3\hbar}{\pi(4\pi\varepsilon_0)^2} \int_0^\infty \bar{\alpha}_A(i\omega)\bar{\alpha}_B(i\omega)\,d\omega \qquad (A1.5.23)$$

where $\bar{\alpha}_A(i\omega)$ is the analytic continuation of the dynamic dipole polarizability to the imaginary axis. The significance of equation (A1.5.23) is that it expresses an interaction coefficient in terms of properties of the individual, interacting molecules. The anisotropic components of C_6 can be written as similar integrals involving $\Delta\alpha(i\omega)$, and the higher dispersion coefficients as integrals involving components of the higher-order, dynamic polarizability tensors at imaginary frequency.

Many methods for the evaluation of C_6 from equation (A1.5.20) use moments of the dipole oscillator strength distribution (DOSD) defined, for molecule A, by

$$S_A(k) = (a_0/e^2)^k \sum_{m\neq0} f_{Am}\Delta E_{Am}^k \qquad \text{for } k = 2, 1, 0, -1, -2, \ldots. \qquad (A1.5.24)$$

These moments are related to many physical properties. The Thomas–Kuhn–Reiche sum rule says that $S(0)$ equals the number of electrons in the molecule. Other sum rules [36] relate $S(2)$, $S(1)$ and $S(-1)$ to ground state expectation values. The mean static dipole polarizability is $\bar{\alpha}(0) = e^2S(-2)/m_e$. The Cauchy expansion of the refractive index n at low frequencies ω is given by

$$n^2 - 1 = K_0[S(-2) + \omega^2 K_1 S(-4) + \omega^4 K_2 S(-6) + \cdots] \qquad (A1.5.25)$$

where the K_n are known constants. One approach is to use experimental photoabsorption, refractive index and Verdet constant data, together with known sum rules to construct a constrained DOSD from which dipole properties including C_6 can be calculated. This approach was pioneered by Margenau [5, 37], extended by Dalgarno and coworkers [38, 39], and refined and exploited by Meath and coworkers [40–42] who also generalized it to anisotropic properties [43, 44]. Many methods for bounding C_6 in terms of a few DOSD moments have been explored, and the best of these have been identified by an extensive comparative study [45]. *Ab initio* calculations are the only route to the higher-order dispersion coefficients, and Wormer and his colleagues [46–50] have led the field in this area. The dimensionless ratio $C_{10}C_6/C_8^2$ is predicted to be a constant for all interactions by simple models [51], and this ratio still serves as a useful check on *ab initio* computations [48]. Dispersion coefficients of even higher order can be estimated from simple models as well [52, 53].

The dispersion coefficient for interactions C_6^{AB} between molecules A and B can be estimated to an average accuracy of 0.5% [45] from those of the A–A and B–B interactions using the Moelwyn-Hughes [54] combining rule:

$$C_6^{AB} = \frac{2C_6^{AA}C_6^{BB}\alpha_A\alpha_B}{C_6^{AA}\alpha_B^2 + C_6^{BB}\alpha_A^2} \tag{A1.5.26}$$

where α_A and α_B are the static dipole polarizabilities of A and B, respectively. This rule has a sound theoretical basis [55, 56].

A1.5.2.6   *Many-body long-range forces*

The induction energy is inherently non-additive. In fact, the non-additivity is displayed elegantly in a distributed polarizability approach [28]. Non-additive induction energies have been found to stabilize what appear to be highly improbable crystal structures of the alkaline earth halides [57].

In the third order of long-range perturbation theory for a system of three atoms A, B and C, the leading non-additive dispersion term is the Axilrod–Teller–Mutō triple–dipole interaction [58, 59]

$$V_{ddd} = C_9\frac{(1 + 3\cos\theta_A\cos\theta_B\cos\theta_C)}{(r_{AB}r_{BC}r_{CA})^3} \tag{A1.5.27}$$

where r_{AB}, r_{BC} and r_{CA} are the sides of the triangle formed by the atoms, and θ_A, θ_B and θ_C are its internal angles, and the C_9 coefficient can be written [60] in terms of the dynamic polarizabilities of the monomers as

$$C_9^{ABC} = \frac{3\hbar}{\pi(4\pi\varepsilon_0)^2}\int_0^{\infty}\bar{\alpha}_A(i\omega)\bar{\alpha}_B(i\omega)\bar{\alpha}_C(i\omega)\,d\omega. \tag{A1.5.28}$$

Hence, the same techniques used to calculate C_6 are also used for C_9. Note that equation (A1.5.28) has a geometrical factor whose sign depends upon the geometry, and that, unlike the case of the two-body dispersion interaction, the triple–dipole dispersion energy has no minus sign in front of the positive coefficient C_9. For example, for an equilateral triangle configuration the triple–dipole dispersion is repulsive and varies as $+(11/8)C_9r^{-9}$. There are strongly anisotropic, non-additive dispersion interactions arising from higher-order polarizabilities as well [61], and the relevant coefficients for rare gas atoms have been calculated *ab initio* [48].

A1.5.3   Short- and intermediate-range forces

A1.5.3.1   *Exchange perturbation theories*

The perturbation theory described in section A1.5.2.1 fails completely at short range. One reason for the failure is that the multipole expansion breaks down, but this is not a fundamental limitation because it is feasible to construct a 'non-expanded', long-range, perturbation theory which does not use the multipole expansion [6]. A more profound reason for the failure is that the polarization approximation of zero overlap is no longer valid at short range.

When the overlap between the wavefunctions of the interacting molecules cannot be neglected, the zeroth-order wavefunction must be anti-symmetrized with respect to all the electrons. The requirement of anti-symmetrization brings with it some difficult problems. If electrons have been assigned to individual molecules in order to partition the Hamiltonian into an unperturbed part $\mathcal{H}^0$ and a perturbation $\lambda\mathcal{H}'$, as described in section A1.5.2.1, then these parts do not commute with the antisymmetrization operator $\mathcal{A}^{AB}$ for the full system

$$[\mathcal{A}^{AB}, \mathcal{H}^0] \neq 0, \qquad [\mathcal{A}^{AB}, \lambda\mathcal{H}'] \neq 0. \qquad (A1.5.29)$$

On the other hand, the system Hamiltonian $\mathcal{H}^{AB} = \mathcal{H}^0 + \lambda\mathcal{H}'$ is symmetric with respect to all the electrons and commutes with $\mathcal{A}^{AB}$

$$[\mathcal{A}^{AB}, \mathcal{H}^0 + \lambda\mathcal{H}'] = 0. \qquad (A1.5.30)$$

Combining these commutation relations, we find

$$[\mathcal{A}^{AB}, \mathcal{H}^0] = -[\mathcal{A}^{AB}, \lambda\mathcal{H}'] \neq 0 \qquad (A1.5.31)$$

which indicates that a zeroth-order quantity is equal to a non-zero, first-order quantity. Unfortunately, this means that there will be no unique definition of the order of a term in our perturbation expansion. Moreover, antisymmetrized products of the wavefunctions of A and B will be non-orthogonal, and therefore they will not be eigenfunctions of any Hermitian, zeroth-order Hamiltonian.

Given these difficulties, it is natural to ask whether we really need to antisymmetrize the zeroth-order wavefunction. If we start with the product function, can we reasonably expect that the system wavefunction obtained by perturbation theory will converge to a properly antisymmetric one? Unfortunately, in that case, the series barely converges [62, 63]. Moreover, there are an infinite number of non-physical states with bosonic character that lie below the physical ground state [64] for most systems of interest—all those containing at least one atom with atomic number greater than two [13]. Claverie [64] has argued that if perturbation theory converges at all, it will converge to one of these unphysical states.

Clearly, standard Rayleigh–Schrödinger perturbation theory is not applicable and other perturbation methods have to be devised. Excellent surveys of the large and confusing variety of methods, usually called 'exchange perturbation theories', that have been developed are available [28, 65]. Here it is sufficient to note that the methods can be classified as either 'symmetric' or 'symmetry-adapted'. Symmetric methods start with antisymmetrized product functions in zeroth order and deal with the non-orthogonality problem in various ways. Symmetry-adapted methods start with non-antisymmetrized product functions and deal with the antisymmetry problem in some other way, such as antisymmetrization at each order of perturbation theory.

A further difficulty arises because the exact wavefunctions of the isolated molecules are not known, except for one-electron systems. A common starting point is the Hartree–Fock wavefunctions of the individual molecules. It is then necessary to include the effects of intramolecular electron correlation by considering them as additional perturbations. Jeziorski and coworkers [66] have developed and computationally implemented a triple perturbation theory of the symmetry-adapted type. They have applied their method, dubbed SAPT, to many interactions with more success than might have been expected given the fundamental doubts [67] raised about the method. SAPT is currently both useful and practical. A recent application [68] to the CO_2 dimer is illustrative of what can be achieved with SAPT, and a rich source of references to previous SAPT work.

A1.5.3.2   First-order interactions

In all methods, the first-order interaction energy is just the difference between the expectation value of the system Hamiltonian for the antisymmetrized product function and the zeroth-order energy

$$E^{(1)} = \frac{\langle \mathcal{A}^{AB}\Psi_0^A\Psi_0^B | \mathcal{H}^{AB} | \Psi_0^A\Psi_0^B \rangle}{\langle \mathcal{A}^{AB}\Psi_0^A\Psi_0^B | \Psi_0^A\Psi_0^B \rangle} - (E_0^A + E_0^B) \qquad (A1.5.32)$$

in which E_0^A and E_0^B are the ground-state energies of isolated molecules A and B. An electrostatic part is usually separated out from the first-order energy, also called the Heitler–London energy, and the remainder is called the exchange–repulsion part:

$$E^{(1)} = E_c^{(1)} + E_{xr}^{(1)}. \qquad (A1.5.33)$$

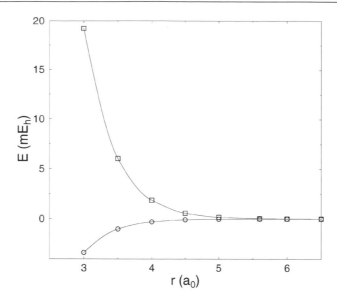

Figure A1.5.2. First-order Coulomb ($\bigcirc$) and exchange-repulsion ($\square$) energies for He–He. Based on data from Komasa and Thakkar [70].

The 'non-expanded' form of the electrostatic or 'Coulomb' energy is

$$E_c^{(1)} = \int\int \frac{\rho_A(r_1)\rho_B(r_2)}{|r_1 - r_2|}\, d^3 r_1\, d^3 r_2 \tag{A1.5.34}$$

where ρ_A and ρ_B are the total (nuclear plus electronic) charge densities of A and B, respectively. A multipole expansion of equation (A1.5.34) leads to the long-range electrostatic energy discussed in section A1.5.2.3.

The difference between the converged multipole expansion of the electrostatic energy and $E_c^{(1)}$ is sometimes called the first-order penetration energy. The exchange–repulsion is often simply called the exchange energy. For Hartree–Fock monomer wavefunctions, $E_{xr}^{(1)}$ can be divided cleanly [69] into attractive exchange and dominant repulsion parts. The exchange part arises because the electrons of one molecule can extend over the entire system, whereas the repulsion arises because the Pauli principle does not allow electrons of the same spin to be in the same place.

Figure A1.5.2 shows $E_c^{(1)}$ and $E_{xr}^{(1)}$ for the He–He interaction computed from accurate monomer wavefunctions [70]. Figure A1.5.3 shows that, as in interactions between other species, the first-order energy $E^{(1)}$ for He–He decays exponentially with interatomic distance. It can be fitted [70] within 0.6% by a function of the form

$$E^{(1)} = (A/r)\, e^{-br - cr^2} \tag{A1.5.35}$$

where A, b, c are fitted parameters.

The exchange–repulsion energy is approximately proportional to the overlap of the charge densities of the interacting molecules [71–73]

$$E_{xr}^{(1)} \approx k\left[\int \rho_A(r)\rho_B(r)\, d^3 r\right]^n \tag{A1.5.36}$$

where $n \approx 1$.

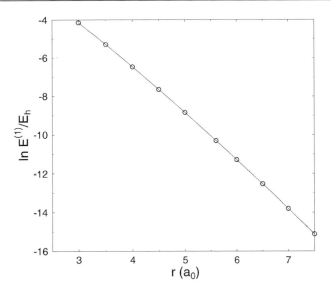

Figure A1.5.3. First-order interaction energy for He–He. Based on data from Komasa and Thakkar [70].

A1.5.3.3 Second-order interactions

The details of the second-order energy depend on the form of exchange perturbation theory used. Most known results are numerical. However, there are some common features that can be described qualitatively. The short-range induction and dispersion energies appear in a non-expanded form and the differences between these and their multipole expansion counterparts are called penetration terms.

The non-expanded dispersion energy can be written as

$$V_{\text{disp}}(r) = -f_6(r)C_6/r^6 - f_8(r)C_8/r^8 - f_{10}(r)C_{10}/r^{10} - \cdots \qquad (A1.5.37)$$

where the $f_6(r)$, $f_8(r)$, ... are 'damping' functions. The damping functions tend to unity as $r \to \infty$ so that the long-range form of equation (A1.5.18) is recovered. As $r \to 0$, the damping functions tend to zero as r^n so that they suppress the spurious r^{-n} singularity of the undamped dispersion, equation (A1.5.18). Meath and coworkers [74–79] have performed *ab initio* calculations of these damping functions for interactions between small species. The general form is shown in figure A1.5.4. Observe that the distance at which the damping functions begin to decrease significantly below unity increases with n. The orientation dependence of the damping functions is not known. Similar damping functions also arise for the induction energy [74, 76, 79].

A 'charge transfer' contribution is often identified in perturbative descriptions of intermolecular forces. This, however, is not a new effect but a part of the short-range induction energy. It is possible to separate the charge transfer part from the rest of the induction energy [80]. It turns out to be relatively small and often negligible. Stone [28] has explained clearly how charge transfer has often been a source of confusion and error.

A1.5.3.4 Supermolecule calculations

The conceptually simplest way to calculate potential energy surfaces for weakly interacting species is to treat the interacting system AB as a 'supermolecule', use the Schrödinger equation to compute its energy as a function of the relative coordinates of the interacting molecules, and subtract off similarly computed energies

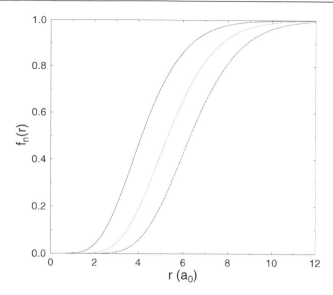

Figure A1.5.4. Dispersion damping functions, f_6: ——, f_8: $\cdots\cdots$ and f_{10}: - - - - for H–H based on data from [74].

of the isolated molecules. This scheme permits one to use any available method for solving the Schrödinger equation.

Unfortunately, the supermolecule approach [81, 82] is full of technical difficulties, which stem chiefly from the very small magnitude of the interaction energy relative to the energy of the supermolecule. Even today, a novice would be ill-advised to attempt such a computation using one of the 'black-box' computer programs available for performing *ab initio* calculations.

That said, the remarkable advances in computer hardware have made *ab initio* calculations feasible for small systems, provided that various technical details are carefully treated. A few examples of recent computations include potential energy surfaces for He–He [83], Ne–Ne and Ar–Ar [84], Ar–H_2, Ar–HF and Ar–NH_3 [85], N_2–He [86, 87], He–F^- and Ne–F^- [88]. Density-functional theory [89] is currently unsuitable for the calculation of van der Waals interactions [90], but the situation could change.

A1.5.3.5   *Many-body short-range forces*

A few *ab initio* calculations are the main source of our current, very meagre knowledge of non-additive contributions to the short-range energy [91]. It is unclear whether the short-range non-additivity is more or less important than the long-range, dispersion non-additivity in the rare-gas solids [28, 92].

A1.5.4   **Experimental information**

Despite the recent successes of *ab initio* calculations, many of the most accurate potential energy surfaces for van der Waals interactions have been obtained by fitting to a combination of experimental and theoretical data. The future is likely to see many more potential energy surfaces obtained by starting with an *ab initio* surface, fitting it to a functional form and then allowing it to vary by small amounts so as to obtain a good fit to many experimental properties simultaneously; see, for example, a recent study on 'morphing' an *ab initio* potential energy surface for Ne–HF [93].

This section discusses how spectroscopy, molecular beam scattering, pressure virial coefficients, measurements on transport phenomena and even condensed phase data can help determine a potential energy surface.

A1.5.4.1 Spectroscopy

Spectroscopy is the most important experimental source of information on intermolecular interactions. A wide range of spectroscopic techniques is being brought to bear on the problem of weakly bound or 'van der Waals' complexes [94, 95]. Molecular beam microwave spectroscopy, pioneered by Klemperer and refined by Flygare, has been used to determine the microwave spectra of a large number of weakly bound complexes and obtain structural information averaged over the vibrational ground state. With the development of tunable far-infrared lasers and sophisticated detectors, far-infrared 'vibration–rotation–tunnelling' spectroscopy has enabled Saykally and others to measure data that probes portions of the potential energy surface further from the minimum. Other techniques including vacuum ultraviolet spectroscopy and conventional gas-phase absorption spectroscopy with very long path lengths have also been used.

Spectroscopic data for a complex formed from two atoms can be inverted by the Rydberg–Klein–Rees procedure to determine the interatomic potential in the region probed by the data. The classical turning points r_L and r_R corresponding to a specific energy level $E(v, J)$ with vibrational and rotational quantum numbers v and J can be determined from a knowledge of all the vibrational and rotational energy level spacings between the bottom of the well and the given energy level. The standard equations are [96]

$$f(v, J) = r_R - r_L = \frac{2\hbar}{\sqrt{2\mu}} \int_{-1/2}^{v} \frac{dv'}{[E(v, J) - E(v', J)]^{1/2}} \qquad (A1.5.38)$$

and

$$g(v, J) = 1/r_{\hat{L}} - 1/r_{\hat{R}} = \frac{2\sqrt{2\mu}}{\hbar} \int_{-1/2}^{v} \frac{B(v', J)\,dv'}{[E(v, J) - E(v', J)]^{1/2}} \qquad (A1.5.39)$$

where $B(v, J) = (2J + 1)^{-1}(\partial E/\partial J)_v$ is a generalized rotational constant. If the rotational structure has not been resolved, then the vibrational spacings alone can be used to determine the well-width function $f(v, 0)$. Similar methods have been developed which enable a spherically averaged potential function to be obtained by inversion of rotational levels, measured precisely enough to yield information on centrifugal distortion, for a single vibrational state. However, most van der Waals complexes are too floppy for a radial potential energy function to be a useful representation of the full PES.

Determination of a PES from spectroscopic data generally requires fitting a parameterized surface to the observed energy levels together with theoretical and other experimental data. This is a difficult process because it is not easy to devise realistic functional representations of a PES with parameters that are not strongly correlated, and because calculation of the vibrational and rotational energy levels from a PES is not straightforward and is an area of current research. The former issue will be discussed further in section A1.5.5.3. The approaches available for the latter currently include numerical integration of a truncated set of 'close-coupled' equations, methods based on the discrete variable representation and diffusion Monte Carlo techniques [28]. Some early and fine examples of potential energy surfaces determined in this manner include the H_2–rare gas surfaces of LeRoy and coworkers [97–99], and the hydrogen halide–rare gas potential energy surfaces of Hutson [100–102]. More recent work is reviewed by van der Avoird et al [103].

A1.5.4.2 Molecular beam scattering

One direct way to study molecular interactions is to cross two molecular beams, one for each of the interacting species, and to study how the molecules scatter after elastic collisions at the crossing point of the two beams.

A collision of two atoms depends upon the relative kinetic energy E of collision and the impact parameter b, which is the distance by which the centres of mass would miss each other in the absence of interatomic interaction. Collimated beams with well defined initial velocities can be used, and the scattering measured as a function of deflection angle χ. However, it is not possible to restrict the collisions to a single impact parameter, and results are therefore reported in the form of differential cross sections $\sigma(\chi, E)$ which are measures of the observed scattering intensity. The integral cross section

$$Q(E) = \int \sigma(\chi, E)\, d\Omega \qquad (A1.5.40)$$

is simply the integral of the differential cross section over all solid angles.

The situation is much the same as with spectroscopic measurements. In the case of interactions between monatomic species, if all the oscillations in the measured differential cross sections are fully resolved, then an inversion procedure can be applied to obtain the interatomic potential [104, 105]. No formal inversion procedures exist for the determination of a PES from measured cross sections for polyatomic molecules, and it is necessary to fit a parametrized surface to the observed cross sections.

A1.5.4.3   Gas imperfections

The virial equation of state, first advocated by Kamerlingh Onnes in 1901, expresses the compressibility factor of a gas as a power series in the number density:

$$P\bar{V}/RT = 1 + B(T)/\bar{V} + C(T)/\bar{V}^2 + \cdots \qquad (A1.5.41)$$

in which $B(T)$, $C(T)$, ... are called the second, third, ... virial coefficients. The importance of this equation in the study of intermolecular forces stems from the statistical mechanical proof that the second virial coefficient depends only on the pair potential, even if the total interaction contains significant many-body contributions. For spherically symmetric interactions the relationship between $B(T)$ and $V(r)$ was well established by 1908, and first Keesom in 1912, and then Jones (later known as Lennard-Jones) in the 1920s exploited it as a tool for the determination of intermolecular potentials from experiment [106, 107]. The relationship is simply [108]:

$$B(T) = -2\pi N_A \int_0^\infty [\exp(-V(r)/kT) - 1]r^2\, dr. \qquad (A1.5.42)$$

In the repulsive region ($r < \sigma$) there is a one-to-one correspondence between the interaction energy and the intermolecular distance. Hence it is possible, in principle at least, to obtain $V(r)$ for $r < \sigma$ by inverting $B(T)$. However, in the region of the potential well ($r > \sigma$), both the inner and outer turning points of the classical motion correspond to the same V and hence it is impossible to obtain $V(r)$ uniquely by inverting $B(T)$. In fact [109, 110], inversion of $B(T)$ can only yield the width of the well as a function of its depth. For light species, equation (A1.5.42) is the first term in a semi-classical expansion, and the following terms are called the quantum corrections [106, 107, 111]. For nonlinear molecules, the classical relationship is analogous to equation (A1.5.42) except that the integral is six dimensional since five angles are required to specify the relative orientation of the molecules. In such cases, inversion of $B(T)$ is a hopeless task. Nevertheless, virial coefficient data provide an important test of a proposed potential function.

The third virial coefficient $C(T)$ depends upon three-body interactions, both additive and non-additive. The relationship is well understood [106, 107, 111]. If the pair potential is known precisely, then $C(T)$ ought to serve as a good probe of the non-additive, three-body interaction energy. The importance of the non-additive contribution has been confirmed by $C(T)$ measurements. Unfortunately, large experimental uncertainties in $C(T)$ have precluded unequivocal tests of details of the non-additive, three-body interaction.

A1.5.4.4   Transport properties

The viscosity, thermal conductivity and diffusion coefficient of a monatomic gas at low pressure depend only on the pair potential but through a more involved sequence of integrations than the second virial coefficient. The transport properties can be expressed in terms of 'collision integrals' defined [111] by

$$\bar{\Omega}^{(\ell,s)}(T) = [(s+1)!(kT)^{s+2}]^{-1} \int_0^\infty Q^{(\ell)}(E)\,\mathrm{e}^{-E/kT}E^{s+1}\,\mathrm{d}E \tag{A1.5.43}$$

where k is the Boltzmann constant and E is the relative kinetic energy of the collision. The collision integral is a thermal average of the transport cross section

$$Q^{(\ell)}(E) = 2\pi\left[1 - \frac{1+(-1)^\ell}{2(1+\ell)}\right]^{-1}\int_0^\infty (1-\cos^\ell\chi)b\,\mathrm{d}b \tag{A1.5.44}$$

in which b is the impact parameter of the collision, and χ is the deflection angle given by

$$\chi(E,b) = \pi - 2b\int_{r_0}^\infty \frac{\mathrm{d}r}{r^2(1-b^2/r^2-V(r)/E)^{1/2}} \tag{A1.5.45}$$

where r_0, the distance of closest approach in the collision, is the outermost classical turning point of the effective potential. The latter is the sum of the true potential and the centrifugal potential so that $V_{\mathrm{eff}}(L,r) = V(r) + L^2/(2\mu r^2) = V(r) + Eb^2/r^2$ in which L is the angular momentum and μ the reduced mass. Hence r_0 is the outermost solution of $E = V_{\mathrm{eff}}(L, r_0)$.

The Chapman–Enskog solution of the Boltzmann equation [112] leads to the following expressions for the transport coefficients. The viscosity of a pure, monatomic gas can be written as

$$\eta(T) = \frac{5(m\pi kT)^{1/2}}{16\bar{\Omega}^{(2,2)}(T)}f_\eta \tag{A1.5.46}$$

and the thermal conductivity as

$$\lambda(T) = \frac{75}{64}\left(\frac{\pi k^3 T}{m}\right)^{1/2}\frac{1}{\bar{\Omega}^{(2,2)}(T)}f_\lambda \tag{A1.5.47}$$

where m is the molecular mass. f_η and f_λ are higher-order correction factors that differ from unity by only 1 or 2% over a wide temperature range, and can be expressed in terms of collision integrals with different values of ℓ and s. Expressions (A1.5.46) and (A1.5.47) imply that

$$\frac{\lambda(T)}{\eta(T)} = \frac{15kf_\lambda}{4mf_\eta} \tag{A1.5.48}$$

and this is borne out experimentally [111] with the ratio of correction factors being a gentle function of temperature: $f_\lambda/f_\eta \approx 1 + 0.0042(1 - \mathrm{e}^{0.33(1-T^*)})$ for $1 < T^* < 90$ with $T^* = kT/\varepsilon$. The self-diffusion coefficient can be written in a similar fashion:

$$D(T) = \frac{3}{8n}\left(\frac{\pi kT}{m}\right)^{1/2}\frac{1}{\bar{\Omega}^{(1,1)}(T)}f_D \tag{A1.5.49}$$

where n is the number density. The higher-order correction factor f_D differs from unity by only a few per cent and can also be expressed in terms of other collision integrals.

Despite the complexity of these expressions, it is possible to invert transport coefficients to obtain information about the intermolecular potential by an iterative procedure [111] that converges rapidly, provided that the initial guess for $V(r)$ has the right well depth.

The theory connecting transport coefficients with the intermolecular potential is much more complicated for polyatomic molecules because the internal states of the molecules must be accounted for. Both quantum mechanical and semi-classical theories have been developed. McCourt and his coworkers [113, 114] have brought these theories to computational fruition and transport properties now constitute a valuable test of proposed potential energy surfaces that can be performed routinely. Electric and magnetic field effects on transport properties [113, 114] depend primarily on the non-spherical part of the interaction, and serve as stringent checks on the anisotropy of potential energy surfaces.

A1.5.5   Model interaction potentials

There are many large molecules whose interactions we have little hope of determining in detail. In these cases we turn to models based on simple mathematical representations of the interaction potential with empirically determined parameters. Even for smaller molecules where a detailed interaction potential has been obtained by an *ab initio* calculation or by a numerical inversion of experimental data, it is useful to fit the calculated points to a functional form which then serves as a computationally inexpensive interpolation and extrapolation tool for use in further work such as molecular simulation studies or predictive scattering computations. There are a very large number of such models in use, and only a small sample is considered here. The most frequently used simple spherical models are described in section A1.5.5.1 and some of the more common elaborate models are discussed in sections A1.5.5.2–A1.5.5.4.

A1.5.5.1   Simple spherical models

The hard sphere model considers each molecule to be an impenetrable sphere of diameter σ so that

$$V(r) = \begin{cases} \infty & r \leq \sigma \\ 0 & r > \sigma. \end{cases} \qquad (A1.5.50)$$

This simple model is adequate for some properties of rare gas fluids. When it is combined with an accurate description of the electrostatic interactions, it can rationalize the structures of a large variety of van der Waals complexes [115–117].

The venerable bireciprocal potential consists of a repulsive term A/r^n and an attractive term $-B/r^m$ with $n > m$. This potential function was introduced by Mie [118] but is usually named after Lennard-Jones who used it extensively. Almost invariably, $m = 6$ is chosen so that the attractive term represents the leading dispersion term. Many different choices of n have been used, but the most common is $n = 12$ because of its computational convenience. The 'Lennard-Jones (12,6)' potential can be written in terms of the well depth (ε) and either the minimum position (r_{m}) or the zero potential location (σ) as

$$V(r) = 4\varepsilon[(\sigma/r)^{12} - (\sigma/r)^{6}] = \varepsilon[(r_{\mathrm{m}}/r)^{12} - 2(r_{\mathrm{m}}/r)^{6}] \qquad (A1.5.51)$$

in which the relationship $\sigma = 2^{-1/6}r_{\mathrm{m}}$ is a consequence of having only two parameters. Fitted values of the coefficient $4\varepsilon\sigma^6$ of the r^{-6} term are often twice as large as the true C_6 value because the attractive term has to compensate for the absence of the higher-order dispersion terms. It is remarkable that this simple model continues to be used almost a century after its introduction.

Morse [119] introduced a potential energy model for the vibrations of bound molecules

$$V(r) = \varepsilon[e^{-2(c/\sigma)(r-r_{\mathrm{m}})} - 2\,e^{-(c/\sigma)(r-r_{\mathrm{m}})}] \qquad (A1.5.52)$$

where c is a dimensionless parameter related to the curvature of the potential at its minimum. This function has a more realistic repulsion than the Lennard-Jones potential, but has incorrect long-range behaviour. It has the merit that its vibrational and rotational energy levels are known analytically [119, 120].

The 'exp-6' potential replaces the inverse power repulsion in the Lennard-Jones $(12, 6)$ function by a more realistic exponential form:

$$V(r) = \begin{cases} \varepsilon(1 - 6/a)^{-1}[(6/a)\,e^{a(1-r/r_{\mathrm{m}})} - (r_{\mathrm{m}}/r)^6] & r > r_{\max} \\ \infty & r \le r_{\max}. \end{cases} \tag{A1.5.53}$$

The potential has a spurious maximum at $r_{\max}$ where the r^{-6} term again starts to dominate. The dimensionless parameter a is a measure of the steepness of the repulsion and is often assigned a value of 14 or 15. The ideas of an exponential repulsion and of its combination with an r^{-6} attraction were introduced by Slater and Kirkwood [121], and the cut-off at $r_{\max}$ by Buckingham [122]. An exponential repulsion, $A\,e^{-br}$, is commonly referred to as a Born–Mayer form, perhaps because their work [123] is better known than that of Slater and Kirkwood.

The parameters in simple potential models for interactions between unlike molecules A and B are often deduced from the corresponding parameters for the A–A and B–B interactions using 'combination rules'. For example, the σ and ε parameters are often estimated from the 'Lorentz–Berthelot' rules:

$$\sigma_{\mathrm{AB}} = (\sigma_{\mathrm{A}} + \sigma_{\mathrm{B}})/2 \tag{A1.5.54}$$

$$\varepsilon_{\mathrm{AB}} = (\varepsilon_{\mathrm{A}}\varepsilon_{\mathrm{B}})^{1/2}. \tag{A1.5.55}$$

The former is useful but the latter tends to overestimate the well depth. A harmonic mean rule

$$\varepsilon_{\mathrm{AB}} = 2\varepsilon_{\mathrm{A}}\varepsilon_{\mathrm{B}}/(\varepsilon_{\mathrm{A}} + \varepsilon_{\mathrm{B}}) \tag{A1.5.56}$$

proposed by Fender and Halsey [124] is generally better than the geometric mean of equation (A1.5.55). Combination rules for the steepness parameter in the exp-6 model include the arithmetic mean

$$a_{\mathrm{AB}} = (a_{\mathrm{A}} + a_{\mathrm{B}})/2 \tag{A1.5.57}$$

and the somewhat more accurate harmonic mean

$$a_{\mathrm{AB}} = 2a_{\mathrm{A}}a_{\mathrm{B}}/(a_{\mathrm{A}} + a_{\mathrm{B}}). \tag{A1.5.58}$$

Many other rules, some of which are rather more elaborate, have been proposed [111], but these rules have insubstantial theoretical underpinnings and they continue to be used only because there is often no better way to proceed.

A1.5.5.2 *Elaborate spherical models*

The potential functions for the interactions between pairs of rare-gas atoms are known to a high degree of accuracy [125]. However, many of them use *ad hoc* functional forms parametrized to give the best possible fit to a wide range of experimental data. They will not be considered because it is more instructive to consider representations that are more firmly rooted in theory and could be used for a wide range of interactions with confidence.

Slater and Kirkwood's idea [121] of an exponential repulsion plus dispersion needs only one concept, damping functions, see section A1.5.3.3, to lead to a working template for contemporary work. Buckingham and Corner [126] suggested such a potential with an empirical damping function more than 50 years ago:

$$V(r) = A\,e^{-br} - (C_6/r^6 + C_8/r^8)f(r) \tag{A1.5.59}$$

where the damping function is

$$f(r) = \begin{cases} \exp[4(1 - r/r_\mathrm{m})^3] & r < r_\mathrm{m} \\ 1 & r \geq r_\mathrm{m}. \end{cases} \tag{A1.5.60}$$

Modern versions of this approach use a more elaborate exponential function for the repulsion, more dispersion terms, induction terms if necessary, and individual damping functions for each of the dispersion, and sometimes induction, terms as in equation (A1.5.37).

Functional forms used for the repulsion include the simple exponential multiplied by a linear combination of powers (possibly non-integer) of r, a generalized exponential function $\exp(-b(r))$, where $b(r)$ is typically a polynomial in r, and a combination of these two ideas.

Parametrized representations of individual damping dispersion functions were first obtained [127] by fitting *ab initio* damping functions [74] for H–H interactions. The one-parameter damping functions of Douketis *et al* are [127]:

$$f_n(r) = [1 - \exp(-2.1s/n - 0.109s^2/\sqrt{n})]^n \tag{A1.5.61}$$

where $s = \rho r$, and ρ is a scale parameter (defined to be $\rho = 1/a_0$ for H–H) that enables the damping functions to be used for any interaction. Meath and coworkers [78, 128] prefer the more elaborate form

$$f_n(r) = [1 - \exp(-a_n s - b_n s^2 - d_n s^3)]^n \tag{A1.5.62}$$

in which the a_n, b_n, d_n ($n = 6, 8, \ldots, 20$) are parameters obtained by fitting to *ab initio* damping functions for H–H. A one-parameter damping function of the incomplete gamma form, based on asymptotic arguments and the H–H interaction, is advocated by Tang and Toennies [129]:

$$f_n(r) = 1 - \exp(-br) \sum_{k=0}^{n} (br)^k / k! \tag{A1.5.63}$$

where b is a scale parameter which is often set equal to the corresponding steepness parameter in the Born–Mayer repulsion.

Functional forms based on the above ideas are used in the HFD [127] and Tang–Toennies models [129], where the repulsion term is obtained by fitting to Hartree–Fock calculations, and in the XC model [92] where the repulsion is modelled by an *ab initio* Coulomb term $E_c^{(1)}$ and a semi-empirical exchange–repulsion term $E_{xr}^{(1)}$. Current versions of all these models employ an individually damped dispersion series for the attractive term.

An example of a potential energy function based on all these ideas is provided by the 10-parameter function used [88] as a representation of *ab initio* potential energy curves for He–F$^-$ and Ne–F$^-$

$$V(r) = A \exp[-b(r)] - \sum_{n=2}^{5} f_{2n}(r) C_{2n} / r^{2n} \tag{A1.5.64}$$

where $b(r) = (b_0 + b_1 z + b_2 z^2) r$ with $z = (r - r_s)/(r + r_s)$, the damping functions $f_n(r)$ are those of equation (A1.5.61), the r^{-4} term is a pure induction term, and the higher r^{-2n} terms contain both dispersion and induction. Note that this representation implicitly assumes that the dispersion damping functions are applicable to induction without change.

A1.5.5.3 Model non-spherical intermolecular potentials

The complete intermolecular potential energy surface depends upon the intermolecular distance and up to five angles, as discussed in section A1.5.1.3.

The interaction energy can be written as an expansion employing Wigner rotation matrices and spherical harmonics of the angles [28, 130]. As a simple example, the interaction between an atom and a diatomic molecule can be expanded in Legendre polynomials as

$$V(r, \theta) = \sum_{L=0}^{N} V_L(r) P_L(\cos \theta). \tag{A1.5.65}$$

This Legendre expansion converges rapidly only for weakly anisotropic potentials. Nonetheless, truncated expansions of this sort are used more often than justified because of their computational advantages.

A more natural way to account for the anisotropy is to treat the parameters in an interatomic potential, such as equation (A1.5.64), as functions of the relative orientation of the interacting molecules. Corner [131] was perhaps the first to use such an approach. Pack [132] pointed out that Legendre expansions of the well depth ε and equilibrium location r_m of the interaction potential converge more rapidly than Legendre expansions of the potential itself.

As an illustration, consider the function used to fit an *ab initio* surface for N_2–He [86, 87]. It includes a repulsive term of the form

$$V_{\text{rep}}(r, \theta) = \exp[A(\theta) - b(\theta)R + \gamma(\theta) \ln r] \tag{A1.5.66}$$

in which

$$A(\theta) = A_0 + A_2 P_2(\cos \theta) + A_4 P_4(\cos \theta) \tag{A1.5.67}$$

and similar three-term Legendre expansions are used for $b(\theta)$ and $\gamma(\theta)$. The same surface includes an anisotropic attractive term consisting of damped dispersion and induction terms:

$$V_{\text{att}}(r, \theta) = -\sum_{n=3}^{5} f_{2n}(r, \theta) C_{2n}(\theta)/r^{2n} \tag{A1.5.68}$$

in which the combined dispersion and induction coefficients $C_{2n}(\theta)$ are given by Legendre series as in equation (A1.5.19), and the damping functions are given by a version of equation (A1.5.61) modified so that the scale factor has a weak angle dependence

$$\rho(\theta) = \rho_0 + \rho_2 P_2(\cos \theta). \tag{A1.5.69}$$

To improve the description of the short-range anisotropy, the surface also includes a repulsive 'site–site' term

$$V_{\text{ssr}} = \sqrt{r_A}\, e^{Z - \zeta r_A} + \sqrt{r_B}\, e^{Z - \zeta r_B} \tag{A1.5.70}$$

where r_A and r_B are distances between the nitrogen atoms and the helium atom.

A1.5.5.4   Site–site intermolecular potentials

The approach described in section A1.5.5.3 is best suited for accurate representations of the PES for interactions between small molecules. Interactions between large molecules are usually handled with an atom–atom or site–site approach. For example, an atom–atom, exp-6 potential for the interaction between molecules A and B can be written as

$$V_{\text{ss}} = \sum_{a \in A} \sum_{b \in B} [A_{ab} \exp(-b_{ab} r_{ab}) - C_6^{ab}/r_{ab}^6] \tag{A1.5.71}$$

where the sums are over the atoms of each molecule, and there are three parameters A_{ab}, b_{ab} and C_6^{ab} for each distinct type of atom pair. A set of parameters was developed by Filippini and Gavezzotti [133, 134] for describing crystal structures and another set for hydrogen bonding.

A more accurate approach is to begin with a model of the charge distribution for each of the molecules. Various prescriptions for obtaining point charge models, such as fitting to the electrostatic potential of the molecule [135, 136], are currently in use. Unfortunately, these point charge models are insufficiently accurate if only atom-centred charges are used [137]. Hence, additional charges are sometimes placed at off-atom sites. This increases the accuracy of the point charge model at the expense of arbitrariness in the choice of off-atom sites and an added computational burden. A less popular but sounder procedure is to use a distributed multipole model [28, 138, 139] instead of a point charge model.

Once the models for the charge distributions are in hand, the electrostatic interaction is computed as the interaction between the sets of point charges or distributed multipoles, and added to an atom–atom, exp-6 form that represents the repulsion and dispersion interactions. Different exp-6 parameters, often from [140–142], are used in this case. The induction interaction is frequently omitted because it is small, or it is modelled by a single site polarizability on each molecule interacting with the point charges or distributed multipoles on the other.

A further refinement [143, 144] is to treat the atoms as being non-spherical by rewriting the repulsive part of the atom–atom exp-6 model, equation (A1.5.71), as

$$V_{\text{rep}} = V_{\text{ref}} \sum_{a \in A} \sum_{b \in B} \exp[-b_{ab}(\Omega_{ab})(r_{ab} - \rho_{ab}(\Omega_{ab}))] \qquad (A1.5.72)$$

where Ω_{ab} is used as a generic designation for all the angles required to specify the relative orientation of the molecules, and V_{ref} is an energy unit. The $\rho_{ab}(\Omega_{ab})$ functions describe the shape of the contour on which the repulsion energy between atoms a and b equals V_{ref}. The spherical harmonic expansions used to represent the angular variation of the steepness $b_{ab}(\Omega_{ab})$ and shape $\rho_{ab}(\Omega_{ab})$ functions are quite rapidly convergent.

References

[1]   Clausius R 1857 Über die Art von Bewegegung, die wir Wärme nennen *Ann. Phys. Chem.* **100** 353
[2]   van der Waals J D 1873 Over de Continuïteit van den Gas- en Vloeistoftoestand *PhD Thesis* Leiden
[3]   London F 1930 Zur theorie und systematik der molekularkräfte *Z. Phys.* **63** 245
[4]   London F 1937 The general theory of molecular forces *Trans. Faraday Soc.* **33** 8
[5]   Margenau H 1939 van der Waals forces *Rev. Mod. Phys.* **11** 1
[6]   Longuet-Higgins H C 1956 The electronic states of composite systems *Proc. R. Soc.* A **235** 537
[7]   Buckingham A D 1967 Permanent and induced molecular moments and long-range intermolecular forces *Adv. Chem. Phys.* **12** 107
[8]   Brooks F C 1952 Convergence of intermolecular force series *Phys. Rev.* **86** 92
[9]   Roe G M 1952 Convergence of intermolecular force series *Phys. Rev.* **88** 659
[10] Dalgarno A and Lewis J T 1956 The representation of long-range forces by series expansions. I. The divergence of the series *Proc. Phys. Soc.* A **69** 57
[11] Dalgarno A and Lewis J T 1956 The representation of long-range forces by series expansions. II. The complete perturbation calculation of long-range forces *Proc. Phys. Soc.* A **69** 59
[12] Ahlrichs R 1976 Convergence properties of the intermolecular force series (1/r expansion) *Theor. Chim. Acta* **41** 7
[13] Morgan J D III and Simon B 1980 Behavior of molecular potential energy curves for large nuclear separations *Int. J. Quantum Chem.* **17** 1143
[14] McClellan A L 1963 *Tables of Experimental Dipole Moments* vol 1 (New York: Freeman)
[15] McClellan A L 1974 *Tables of Experimental Dipole Moments* vol 2 (El Cerrito, CA: Rahara Enterprises)
[16] McClellan A L 1989 *Tables of Experimental Dipole Moments* vol 3 (El Cerrito, CA: Rahara Enterprises)
[17] Sutter D H and Flygare W H 1976 The molecular Zeeman effect *Topics Curr. Chem.* **63** 89
[18] Gray C G and Gubbins K E 1984 *Theory of Molecular Fluids. 1. Fundamentals* (Oxford: Clarendon)
[19] Spackman M A 1992 Molecular electric moments from X-ray diffraction data *Chem. Rev.* **92** 1769
[20] Dykstra C E 1988 Ab initio *Calculation of the Structures and Properties of Molecules* (Amsterdam: Elsevier)
[21] Bündgen P, Grein F and Thakkar A J 1995 Dipole and quadrupole moments of small molecules. An *ab initio* study using perturbatively corrected, multi-reference, configuration interaction wavefunctions *J. Mol. Struct. (Theochem)* **334** 7
[22] Doerksen R J and Thakkar A J 1999 Quadrupole and octopole moments of heteroaromatic rings *J. Phys. Chem.* A **103** 10 009
[23] Miller T M and Bederson B 1988 Electric dipole polarizability measurements *Adv. At. Mol. Phys.* **25** 37

[24] Shelton D P and Rice J E 1994 Measurements and calculations of the hyperpolarizabilities of atoms and small molecules in the gas phase *Chem. Rev.* **94** 3
[25] Bonin K D and Kresin V V 1997 *Electric-dipole Polarizabilities of Atoms, Molecules and Clusters* (Singapore: World Scientific)
[26] Doerksen R J and Thakkar A J 1999 Structures, vibrational frequencies and polarizabilities of diazaborinines, triazadiborinines, azaboroles and oxazaboroles *J. Phys. Chem.* A **103** 2141
[27] Maroulis G 1999 On the accurate theoretical determination of the static hyperpolarizability of trans-butadiene *J. Chem. Phys.* **111** 583
[28] Stone A J 1996 *The Theory of Intermolecular Forces* (New York: Oxford)
[29] Legon A C, Millen D J and Mjöberg P J 1977 The hydrogen cyanide dimer: identification and structure from microwave spectroscopy *Chem. Phys. Lett.* **47** 589
[30] Dulmage W J and Lipscomb W N 1951 The crystal structures of hydrogen cyanide, HCN *Acta Crystallogr.* **4** 330
[31] Eisenschitz R and London F 1930 Über das verhältnis der van der Waalschen kräften zu density homöopolaren bindungskräften *Z. Phys.* **60** 491
[32] London F 1930 Über einige eigenschaften und anwendungen der molekularkräfte *Z. Phys. Chem.* B **11** 222
[33] Casimir H B G and Polder D 1948 The influence of retardation on the London–van der Waals forces *Phys. Rev.* **73** 360
[34] Mavroyannis C and Stephen M J 1962 Dispersion forces *Mol. Phys.* **5** 629
[35] McLachlan A D 1963 Retarded dispersion forces between molecules *Proc. R. Soc.* A **271** 387
[36] Bethe H A and Salpeter E E 1957 *Quantum Mechanics of One- and Two-electron Atoms* (Berlin: Springer)
[37] Margenau H 1931 Note on the calculation of van der Waals forces *Phys. Rev.* **37** 1425
[38] Dalgarno A and Lynn N 1957 Properties of the helium atom *Proc. Phys. Soc. London* **70** 802
[39] Dalgarno A and Kingston A E 1961 van der Waals forces for hydrogen and the inert gases *Proc. Phys. Soc. London* **78** 607
[40] Zeiss G D, Meath W J, MacDonald J C F and Dawson D J 1977 Dipole oscillator strength distributions, sums, and some related properties for Li, N, O, H_2, N_2, O_2, NH_3, H_2O, NO and N_2O *Can. J. Phys.* **55** 2080
[41] Kumar A, Fairley G R G and Meath W J 1985 Dipole properties, dispersion energy coefficients and integrated oscillator strengths for SF_6 *J. Chem. Phys.* **83** 70
[42] Kumar A and Meath W J 1992 Dipole oscillator strength properties and dispersion energies for acetylene and benzene *Mol. Phys.* **75** 311
[43] Meath W J and Kumar A 1990 Reliable isotropic and anisotropic dipole dispersion energies, evaluated using constrained dipole oscillator strength techniques, with application to interactions involving H_2, N_2 and the rare gases *Int. J. Quantum Chem. Symp.* **24** 501
[44] Kumar A, Meath W J, Bündgen P and Thakkar A J 1996 Reliable anisotropic dipole properties and dispersion energy coefficients for O_2, evaluated using constrained dipole oscillator strength techniques *J. Chem. Phys.* **105** 4927
[45] Thakkar A J 1984 Bounding and estimation of van der Waals coefficients *J. Chem. Phys.* **81** 1919
[46] Rijks W and Wormer P E S 1989 Correlated van der Waals coefficients. II. Dimers consisting of CO, HF, H_2O and NH_3 *J. Chem. Phys.* **90** 6507
[47] Rijks W and Wormer P E S 1990 *Erratum*: correlated van der Waals coefficients. II. Dimers consisting of CO, HF, H_2O and NH_3 *J. Chem. Phys.* **92** 5754
[48] Thakkar A J, Hettema H and Wormer P E S 1992 *Ab initio* dispersion coefficients for interactions involving rare-gas atoms *J. Chem. Phys.* **97** 3252
[49] Wormer P E S and Hettema H 1992 Many-body perturbation theory of frequency-dependent polarizabilities and van der Waals coefficients: application to $H_2O \cdots H_2O$ and $Ar \cdots NH_3$ *J. Chem. Phys.* **97** 5592
[50] Hettema H, Wormer P E S and Thakkar A J 1993 Intramolecular bond length dependence of the anisotropic dispersion coefficients for interactions of rare gas atoms with N_2, CO, Cl_2, HCl and HBr *Mol. Phys.* **80** 533
[51] Thakkar A J and Smith V H Jr 1974 On a representation of the long range interatomic interaction potential *J. Phys. B: At. Mol. Phys.* **7** L321
[52] Tang K T and Toennies J P 1978 A simple model of the van der Waals potential at intermediate distances. II. Anisotropic potential of $He \cdots H_2$ and $Ne \cdots H_2$ *J. Chem. Phys.* **68** 5501
[53] Thakkar A J 1988 Higher dispersion coefficients: accurate values for hydrogen atoms and simple estimates for other systems *J. Chem. Phys.* **89** 2092
[54] Moelwyn-Hughes E A 1957 *Physical Chemistry* (New York: Pergamon) p 332
[55] Tang K T 1969 Dynamic polarizabilities and van der Waals coefficients *Phys. Rev.* **177** 108
[56] Kutzelnigg W and Maeder F 1978 Natural states of interacting systems and their use for the calculation of intermolecular forces. III. One-term approximations of oscillator strength sums and dynamic polarizabilities *Chem. Phys.* **35** 397
[57] Wilson M and Madden P A 1994 Anion polarization and the stability of layered structures in MX_2 systems *J. Phys.: Condens. Matter* **6** 159
[58] Axilrod P M and Teller E 1943 Interaction of the van der Waals type between three atoms *J. Chem. Phys.* **11** 299
[59] Mutō Y 1943 Force between non-polar molecules *J. Phys. Math. Soc. Japan* **17** 629
[60] McLachlan A D 1963 Three-body dispersion forces *Mol. Phys.* **6** 423
[61] Bell R J 1970 Multipolar expansion for the non-additive third-order interaction energy of three atoms *J. Phys. B: At. Mol. Phys.* **3** 751

[62] Kutzelnigg W 1992 Does the polarization approximation converge for large-r to a primitive or a symmetry-adapted wavefunction? *Chem. Phys. Lett.* **195** 77

[63] Cwiok T, Jeziorski B, Kołos W, Moszynski R, Rychlewski J and Szalewicz K 1992 Convergence properties and large-order behavior of the polarization expansion for the interaction energy of hydrogen atoms *Chem. Phys. Lett.* **195** 67

[64] Claverie P 1971 Theory of intermolecular forces. I. On the inadequacy of the usual Rayleigh–Schrödinger perturbation method for the treatment of intermolecular forces *Int. J. Quantum Chem.* **5** 273

[65] Claverie P 1978 Elaboration of approximate formulas for the interactions between large molecules: applications in organic chemistry *Intermolecular Interactions: From Diatomics to Biopolymers* ed B Pullman (New York: Wiley) p 69

[66] Jeziorski B, Moszynski R and Szalewicz K 1994 Perturbation theory approach to intermolecular potential energy surfaces of van der Waals complexes *Chem. Rev.* **94** 1887

[67] Adams W H 1994 The polarization approximation and the Amos–Musher intermolecular perturbation theories compared to infinite order at finite separation *Chem. Phys. Lett.* **229** 472

[68] Bukowski R, Sadlej J, Jeziorski B, Jankowski P, Szalewicz K, Kucharski S A, Williams H L and Rice B M 1999 Intermolecular potential of carbon dioxide dimer from symmetry-adapted perturbation theory *J. Chem. Phys.* **110** 3785

[69] Hayes I C and Stone A J 1984 An intermolecular perturbation theory for the region of moderate overlap *Mol. Phys.* **53** 83

[70] Komasa J and Thakkar A J 1995 Accurate Heitler–London interaction energy for He_2 *J. Mol. Struct. (Theochem)* **343** 43

[71] Kita S, Noda K and Inouye H 1976 Repulsion potentials for Cl^-–R and Br^-–R (R = He, Ne and Ar) derived from beam experiments *J. Chem. Phys.* **64** 3446

[72] Kim Y S, Kim S K and Lee W D 1981 Dependence of the closed-shell repulsive interaction on the overlap of the electron densities *Chem. Phys. Lett.* **80** 574

[73] Wheatley R J and Price S L 1990 An overlap model for estimating the anisotropy of repulsion *Mol. Phys.* **69** 507

[74] Kreek H and Meath W J 1969 Charge-overlap effects. Dispersion and induction forces *J. Chem. Phys.* **50** 2289

[75] Knowles P J and Meath W J 1986 Non-expanded dispersion energies and damping functions for Ar_2 and Li_2 *Chem. Phys. Lett.* **124** 164

[76] Knowles P J and Meath W J 1986 Non-expanded dispersion and induction energies, and damping functions, for molecular interactions with application to $HF \cdots He$ *Mol. Phys.* **59** 965

[77] Knowles P J and Meath W J 1987 A separable method for the calculation of dispersion and induction energy damping functions with applications to the dimers arising from He, Ne and HF *Mol. Phys.* **60** 1143

[78] Wheatley R J and Meath W J 1993 Dispersion energy damping functions, and their relative scale with interatomic separation, for (H,He,Li)–(H,He,Li) interactions *Mol. Phys.* **80** 25

[79] Wheatley R J and Meath W J 1994 Induction and dispersion damping functions, and their relative scale with interspecies distance, for (H^+,He^+,Li^+)–(H,He,Li) interactions *Chem. Phys.* **179** 341

[80] Stone A J 1993 Computation of charge-transfer energies by perturbation theory *Chem. Phys. Lett.* **211** 101

[81] van Lenthe J H, van Duijneveldt-van de Rijdt J G C M and van Duijneveldt F B 1987 Weakly bonded systems. *Adv. Chem. Phys.* **69** 521

[82] van Duijneveldt F B, van Duijneveldt-van de Rijdt J G C M and van Lenthe J H 1994 State of the art in counterpoise theory *Chem. Rev.* **94** 1873

[83] Anderson J B, Traynor C A and Boghosian B M 1993 An exact quantum Monte-Carlo calculation of the helium–helium intermolecular potential *J. Chem. Phys.* **99** 345

[84] Woon D E 1994 Benchmark calculations with correlated molecular wavefunctions. 5. The determination of accurate *ab initio* intermolecular potentials for He_2, Ne_2, and Ar_2 *J. Chem. Phys.* **100** 2838

[85] Tao F M and Klemperer W 1994 Accurate *ab initio* potential energy surfaces of Ar–HF, Ar–H_2O, and Ar–NH_3 *J. Chem. Phys.* **101** 1129–45

[86] Hu C H and Thakkar A J 1996 Potential energy surface for interactions between N_2 and He: *ab initio* calculations, analytic fits, and second virial coefficients *J. Chem. Phys.* **104** 2541

[87] Reid J P, Thakkar A J, Barnes P W, Archibong E F, Quiney H M and Simpson C J S M 1997 Vibrational deactivation of N_2 ($v = 1$) by inelastic collisions with ^3He and ^4He: an experimental and theoretical study *J. Chem. Phys.* **107** 2329

[88] Archibong E F, Hu C H and Thakkar A J 1998 Interaction potentials for He–F^- and Ne–F^- *J. Chem. Phys.* **109** 3072

[89] Parr R G and Yang W 1989 *Density-functional Theory of Atoms and Molecules* (Oxford: Clarendon)

[90] Pérez-Jordá J M and Becke A D 1995 A density functional study of van der Waals forces: rare gas diatomics *Chem. Phys. Lett.* **233** 134

[91] Elrod M J and Saykally R J 1994 Many-body effects in intermolecular forces *Chem. Rev.* **94** 1975

[92] Meath W J and Koulis M 1991 On the construction and use of reliable two- and many-body interatomic and intermolecular potentials *J. Mol. Struct. (Theochem)* **226** 1

[93] Meuwly M and Hutson J M 1999 Morphing *ab initio* potentials: a systematic study of Ne-HF *J. Chem. Phys.* **110** 8338

[94] 1994 van der Waals molecules *Chem. Rev.* **94**

[95] 1994 Structure and dynamics of van der Waals complexes *Faraday Disc.* **97**

[96] Child M S 1991 *Semiclassical Mechanics with Molecular Applications* (Oxford: Clarendon)

[97] LeRoy R J and van Kranendonk J 1974 Anisotropic intermolecular potentials from an analysis of spectra of H_2- and D_2-inert gas complexes *J. Chem. Phys.* **61** 4750

[98] LeRoy R J and Carley J S 1980 Spectroscopy and potential energy surfaces of van der Waals molecules *Adv. Chem. Phys.* **42** 353

[99] LeRoy R J and Hutson J M 1987 Improved potential energy surfaces for the interaction of H_2 with Ar, Kr and Xe *J. Chem. Phys.* **86** 837

[100] Hutson J M and Howard B J 1980 Spectroscopic properties and potential surfaces for atom–diatom van der Waals molecules *Mol. Phys.* **41** 1123

[101] Hutson J M 1989 The intermolecular potential of Ne–HCl: determination from high-resolution spectroscopy *J. Chem. Phys.* **91** 4448

[102] Hutson J M 1990 Intermolecular forces from the spectroscopy of van der Waals molecules *Ann. Rev. Phys. Chem.* **41** 123

[103] van der Avoird A, Wormer P E S and Moszynski R 1994 From intermolecular potentials to the spectra of van der Waals molecules and vice versa *Chem. Rev.* **94** 1931

[104] Buck U 1974 Inversion of molecular scattering data *Rev. Mod. Phys.* **46** 369

[105] Buck U 1975 Elastic scattering *Adv. Chem. Phys.* **30** 313

[106] Hirschfelder J O, Curtiss C F and Bird R B 1954 *Molecular Theory of Gases and Liquids* (New York: Wiley)

[107] Mason E A and Spurling T H 1969 *The Virial Equation of State* (Oxford: Pergamon)

[108] McQuarrie D A 1973 *Statistical Thermodynamics* (Mill Valley, CA: University Science Books)

[109] Keller J B and Zumino B 1959 Determination of intermolecular potentials from thermodynamic data and the law of corresponding states *J. Chem. Phys.* **30** 1351

[110] Frisch H L and Helfand E 1960 Conditions imposed by gross properties on the intermolecular potential *J. Chem. Phys.* **32** 269

[111] Maitland G C, Rigby M, Smith E B and Wakeham W A 1981 *Intermolecular Forces: Their Origin and Determination* (Oxford: Clarendon)

[112] Chapman S and Cowling T G 1970 *The Mathematical Theory of Non-uniform Gases* 3rd edn (London: Cambridge University Press)

[113] McCourt F R, Beenakker J, Köhler W E and Kuščer I 1990 *Nonequilibrium Phenomena in Polyatomic Gases. 1. Dilute Gases* (Oxford: Clarendon)

[114] McCourt F R, Beenakker J, Köhler W E and Kuščer I 1991 *Nonequilibrium Phenomena in Polyatomic Gases. 2. Cross-sections, Scattering and Rarefied Gases* (Oxford: Clarendon)

[115] Buckingham A D and Fowler P W 1983 Do electrostatic interactions predict structures of van der Waals molecules? *J. Chem. Phys.* **79** 6426

[116] Buckingham A D and Fowler P W 1985 A model for the geometries of van der Waals complexes *Can. J. Chem.* **63** 2018

[117] Buckingham A D, Fowler P W and Stone A J 1986 Electrostatic predictions of shapes and properties of van der Waals molecules *Int. Rev. Phys. Chem.* **5** 107

[118] Mie G 1903 Zur kinetischen theorie der einatomigen körper *Ann. Phys., Lpz* **11** 657

[119] Morse P M 1929 Diatomic molecules according to the wave mechanics: II. Vibrational levels *Phys. Rev.* **34** 57

[120] Pekeris C L 1934 The rotation–vibration coupling in diatomic molecules *Phys. Rev.* **45** 98

[121] Slater J C and Kirkwood J G 1931 The van der Waals forces in gases *Phys. Rev.* **37** 682

[122] Buckingham R A 1938 The classical equation of state of gaseous helium, neon and argon *Proc. R. Soc.* A **168** 264

[123] Born M and Mayer J E 1932 Zur gittertheorie der ionenkristalle *Z. Phys.* **75** 1

[124] Fender B E F and Halsey G D Jr 1962 Second virial coefficients of argon, krypton and argon–krypton mixtures at low temperatures *J. Chem. Phys.* **36** 1881

[125] Aziz R A 1984 Interatomic potentials for rare-gases: pure and mixed interactions *Inert Gases: Potentials, Dynamics and Energy Transfer in Doped Crystals* ed M L Klein (Berlin: Springer) ch 2, pp 5–86

[126] Buckingham R A and Corner J 1947 Tables of second virial and low-pressure Joule–Thompson coefficients for intermolecular potentials with exponential repulsion *Proc. R. Soc.* A **189** 118

[127] Douketis C, Scoles G, Marchetti S, Zen M and Thakkar A J 1982 Intermolecular forces via hybrid Hartree–Fock SCF plus damped dispersion (HFD) energy calculations. An improved spherical model *J. Chem. Phys.* **76** 3057

[128] Koide A, Meath W J and Allnatt A R 1981 Second-order charge overlap effects and damping functions for isotropic atomic and molecular interactions *Chem. Phys.* **58** 105

[129] Tang K T and Toennies J P 1984 An improved simple model for the van der Waals potential based on universal damping functions for the dispersion coefficients *J. Chem. Phys.* **80** 3726

[130] van der Avoird A, Wormer P E S, Mulder F and Berns R M 1980 *Ab initio* studies of the interactions in van der Waals molecules *Topics Curr. Chem.* **93** 1

[131] Corner J 1948 The second virial coefficient of a gas of non-spherical molecules *Proc. R. Soc.* A **192** 275

[132] Pack R T 1978 Anisotropic potentials and the damping of rainbow and diffraction oscillations in differential cross-sections *Chem. Phys. Lett.* **55** 197

[133] Filippini G and Gavezzotti A 1993 Empirical intermolecular potentials for organic crystals: the 6-exp approximation revisited *Acta Crystallogr.* B **49** 868

[134] Gavezzotti A and Filippini G 1994 Geometry of the intermolecular XH$\cdots$Y (X, Y = N, O) hydrogen bond and the calibration of empirical hydrogen-bond potentials *J. Phys. Chem.* **98** 4831

[135] Momany F A 1978 Determination of partial atomic charges from *ab initio* molecular electrostatic potentials. Application to formamide, methanol and formic acid *J. Phys. Chem.* **82** 592

[136] Singh U C and Kollman P A 1984 An approach to computing electrostatic charges for molecules *J. Comput. Chem.* **5** 129

[137] Wiberg K B and Rablen P R 1993 Comparison of atomic charges by different procedures *J. Comput. Chem.* **14** 1504

[138] Stone A J 1981 Distributed multipole analysis; or how to describe a molecular charge distribution *Chem. Phys. Lett.* **83** 233

[139] Stone A J and Alderton M 1985 Distributed multipole analysis—methods and applications *Mol. Phys.* **56** 1047

[140] Williams D E 1965 Non-bonded potential parameters derived from crystalline aromatic hydrocarbons *J. Chem. Phys.* **45** 3770

[141] Williams D E 1967 Non-bonded potential parameters derived from crystalline hydrocarbons *J. Chem. Phys.* **47** 4680

[142] Mirsky K 1978 The determination of the intermolecular interaction energy by empirical methods *Computing in Crystallography* ed R Schenk *et al* (Delft, The Netherlands: Delft University) p 169

[143] Stone A J 1979 Intermolecular forces *The Molecular Physics of Liquid Crystals* ed G R Luckhurst and G W Gray (New York: Academic) pp 31–50

[144] Price S L and Stone A J 1980 Evaluation of anisotropic model intermolecular pair potentials using an *ab initio* SCF-CI surface *Mol. Phys.* **40** 805

Further Reading

Stone A J 1996 *The Theory of Intermolecular Forces* (New York: Oxford)

A fine text suitable for both graduate students and researchers. Emphasizes theory of long-range forces.

Maitland G C, Rigby M, Smith E B and Wakeham W A 1981 *Intermolecular Forces: Their Origin and Determination* (Oxford: Clarendon)

A thorough reference work with emphasis on the determination of intermolecular potentials from experimental data.

Scheiner S 1997 *Hydrogen Bonding: A Theoretical Perspective* (New York: Oxford)

A survey of research on hydrogen bonding with emphasis on theoretical calculations.

1994 van der Waals molecules *Chem. Rev.* **94** 1721

A special issue devoted to review articles on various aspects of van der Waals molecules.

Müller-Dethlefs K and Hobza P 2000 Noncovalent interactions: a challenge for experiment and theory *Chem. Rev.* **100** 143

A survey of challenges that have yet to be met.

Pyykkö P 1997 Strong closed-shell interactions in inorganic chemistry *Chem. Rev.* **97** 597

A review of fertile ground for further research.

Margenau H and Kestner N R 1971 *Theory of Intermolecular Forces* 2nd edn (New York: Pergamon)

An older treatment that contains a wealth of references to the earlier literature, and an interesting history of the subject beginning with the work of Clairault in the mid-eighteenth century.

A1.6
Interaction of light with matter: a coherent perspective

David J Tannor

A1.6.1 The basic matter–field interaction

There has been phenomenal expansion in the range of experiments connected with light–molecule interactions. If one thinks of light as an electromagnetic (EM) wave, like any wave it has an amplitude, a frequency and a phase. The advent of the laser in 1960 completely revolutionized the control over all three of these factors. The amplitude of the EM wave is related to its intensity; current laser capabilities allow intensities up to about 10^{20} W cm^{-2}, fifteen orders of magnitude larger than prelaser technology allowed. Laser beams can be made extremely monochromatic. Finally, it is increasingly possible to control the absolute phase of the laser light. There have also been remarkable advances in the ability to construct ultrashort pulses. Currently it is possible to construct pulses of the order of 10^{-15} s (several femtoseconds), a time scale short compared with typical vibrational periods of molecules. These short pulses consist of a *coherent* superposition of many frequencies of the light; the word coherent implies a precise phase relationship between the different frequency components. When these coherent ultrashort pulse interact with a molecule they excite coherently many frequency components in the molecule. Such coherent excitation, whether it is with short pulses or with monochromatic light, introduces new concepts in thinking about the light–matter interaction. These new concepts can be used passively, to learn about molecular properties via new coherent spectroscopies, or actively, to control chemical reactions using light, or to use light to cool atoms and molecules to temperatures orders of magnitude lower than 1 K.

A theme which will run through this section is the complementarity of light and the molecule with which it interacts. The simplest example is energy: when a photon of energy $E = \hbar\omega$ is absorbed by a molecule it disappears, transferring the identical quantity of energy $E = \hbar(\omega_f - \omega_i)$ to the molecule. But this is only one of a complete set of such complementary relations: the amplitude of the EM field determines the amplitude of the excitation; the phase of the EM phase determines the phase of the excitation; and the time of the interaction with the photon determines the time of excitation of the molecule. Moreover, both the magnitude and direction of the momentum of the photon are imparted to the molecules, an observation which plays a crucial role in translational cooling. Finally, because of the conservation or increase in entropy in the universe, any entropy change in the system has to be compensated for by an entropy change in the light; specifically, coherent light has zero or low entropy while incoherent light has high entropy. Entropy exchange between the system and the light plays a fundamental role in laser cooling, where entropy from the system is carried off by the light via incoherent, spontaneous emission, as well in lasing itself where entropy from incoherent light must be transferred to the system.

This section begins with a brief description of the basic light–molecule interaction. As already indicated, coherent light pulses excite coherent superpositions of molecular eigenstates, known as 'wavepackets', and we will give a description of their motion, their coherence properties, and their interplay with the light. Then we will turn to linear and nonlinear spectroscopy, and, finally, to a brief account of coherent control of molecular motion.

A1.6.1.1 Electromagnetic fields

The material in this section can be found in many textbooks and monographs. Our treatment follows that in [1–3].

(a) Maxwell's equations and electromagnetic potentials

The central equations of electromagnetic theory are elegantly written in the form of four coupled equations for the electric and magnetic fields. These are known as Maxwell's equations. In free space, these equations take the form:

$$\boldsymbol{\nabla} \times \boldsymbol{E} = -\frac{1}{c}\frac{\mathrm{d}\boldsymbol{B}}{\mathrm{d}t} \tag{A1.6.1}$$

$$\boldsymbol{\nabla} \times \boldsymbol{B} = \frac{1}{c}\frac{\mathrm{d}\boldsymbol{E}}{\mathrm{d}t} + 4\frac{\pi}{c}\boldsymbol{J} \tag{A1.6.2}$$

$$\boldsymbol{\nabla} \cdot \boldsymbol{E} = 4\pi\rho \tag{A1.6.3}$$

$$\boldsymbol{\nabla} \cdot \boldsymbol{B} = 0 \tag{A1.6.4}$$

where $\boldsymbol{E}$ is the electric field vector, $\boldsymbol{B}$ is the magnetic field vector, $\boldsymbol{J}$ is the current density, ρ is the charge density and c is the speed of light. It is convenient to define two potentials, a scalar potential ϕ and a vector potential $\boldsymbol{A}$, such that the electric and magnetic fields are defined in terms of derivatives of these potentials. The four Maxwell equations are then replaced by two equations which define the fields in terms of the potentials,

$$\boldsymbol{E} = -\boldsymbol{\nabla}\phi - \frac{1}{c}\frac{\mathrm{d}\boldsymbol{A}}{\mathrm{d}t} \tag{A1.6.5}$$

$$\boldsymbol{B} = \boldsymbol{\nabla} \times \boldsymbol{A} \tag{A1.6.6}$$

together with two equations for the vector and scalar fields themselves. Note that there is a certain amount of flexibility in the choice of $\boldsymbol{A}$ and ϕ, such that the same values for $\boldsymbol{E}$ and $\boldsymbol{B}$ are obtained (called gauge invariance). We will adopt below the Coulomb gauge, in which $\boldsymbol{\nabla} \cdot \boldsymbol{A} = 0$.

In free space ($\rho = 0$, $\boldsymbol{J} = 0$, $\phi = $ constant), the equations for the potentials decouple and take the following simple form:

$$\boldsymbol{\nabla}^2\phi = 0 \tag{A1.6.7}$$

$$\boldsymbol{\nabla}^2\boldsymbol{A} = \frac{1}{c^2}\frac{\mathrm{d}^2\boldsymbol{A}}{\mathrm{d}t^2}. \tag{A1.6.8}$$

Equation (A1.6.8), along with the definitions (A1.6.5) and (A1.6.6) constitute the central equation for the propagation of electromagnetic waves in free space. The form of section A1.6.4 admits harmonic solutions of the form

$$\boldsymbol{A} \equiv \boldsymbol{A}_0 \cos(\boldsymbol{k} \cdot \boldsymbol{r} - \omega t) \tag{A1.6.9}$$

from which it follows that

$$\boldsymbol{E} = -\frac{\omega}{c}\boldsymbol{A}_0 \sin(\boldsymbol{k} \cdot \boldsymbol{r} - \omega t) \equiv E_0 e \sin(kr - \omega t) \tag{A1.6.10}$$

$$\boldsymbol{B} = -\boldsymbol{A}_0(\boldsymbol{k} \times e) \sin(\boldsymbol{k} \cdot \boldsymbol{r} - \omega t) \tag{A1.6.11}$$

(e is a unit vector in the direction of $\boldsymbol{E}$ and $\omega = |\boldsymbol{k}|c$).

(b) Energy and photon number density

In what follows it will be convenient to convert between field strength and numbers of photons in the field. According to classical electromagnetism, the energy E in the field is given by

$$E = \int d^3r \frac{E^2 + B^2}{8\pi}.$$
(A1.6.12)

If we assume a single angular frequency of the field, ω, and a constant magnitude of the vector potential, A_0, in the volume V, we obtain, using equations (A1.6.10) and (A1.6.11), and noting that the average value of $\sin^2(x) = 1/2$,

$$E = V \frac{E_0^2}{8\pi}.$$
(A1.6.13)

But by the Einstein relation we know that the energy of a single photon on frequency ω is given by $\hbar\omega$, and hence the total energy in the field is

$$E = N\hbar\omega$$
(A1.6.14)

where N is the number of photons. Combining equations (A1.6.13) and (A1.6.14) we find that

$$E_0 = \left(\frac{8\pi N\hbar\omega}{V}\right)^{1/2},$$
(A1.6.15)

Equation (A1.6.15) provides the desired relationship between field strength and the number of photons.

A1.6.1.2 Interaction between field and matter

(a) Classical theory

To this point, we have considered only the radiation field. We now turn to the interaction between the matter and the field. According to classical electromagnetic theory, the force on a particle with charge e due to the electric and magnetic fields is

$$F = e\left(E + \frac{v \times B}{c}\right).$$
(A1.6.16)

This interaction can also be expressed in terms of a Hamiltonian:

$$H(p, A) = \frac{1}{2m}\left(p - \frac{e}{c}A\right)^2$$
(A1.6.17)

where $A = A(x)$ and where p and x are the conjugate variables that obey the canonical Hamilton equations. (Verifying that equation (A1.6.17) reduces to equation (A1.6.16) is non-trivial (cf [3])). Throughout the remainder of this section the radiation field will be treated using classical electromagnetic theory, while the matter will be treated quantum mechanically, that is, a 'semiclassical' treatment. The Hamiltonian form for the interaction, equation (A1.6.17), provides a convenient starting point for this semiclassical treatment.

(b) Quantum Hamiltonian for a particle in an electromagnetic field

To convert the Hamiltonian for the material from a classical to a quantum form, we simply replace p with $-i\hbar\nabla$. This gives:

$$H = \frac{1}{2m}\left[-i\hbar\nabla - \frac{e}{c}A\right]^2 + V_s \qquad (A1.6.18)$$

$$= \underbrace{-\frac{\hbar^2\nabla^2}{2m} + V_s} + \underbrace{\frac{i\hbar e}{2mc}(\nabla\cdot A + A\cdot\nabla) + \frac{e^2}{2mc}A\cdot A} \qquad (A1.6.19)$$

$$= \qquad H_0 \qquad + \qquad V \qquad (A1.6.20)$$

where H_0 is the Hamiltonian of the bare system and V is the part of the Hamiltonian that comes from the radiation field and the radiation–matter interaction. Note that an additional term, V_s, has been included in the system Hamiltonian, to allow for internal potential energy of the system. V_s contains all the interesting features that make different atoms and molecules distinct from one another, and will play a significant role in later sections.

We now make the following observations.

(i)   For many charged particles

$$V = \sum_i \frac{i\hbar e}{2mc}(\nabla\cdot A + A\cdot\nabla) + \frac{e^2}{2mc}A\cdot A.$$

(ii) In the Coulomb gauge $\nabla\cdot A = 0$. This implies that $\nabla\cdot(A\psi) = A\cdot\nabla\psi$ for any ψ, and hence the terms linear in A can be combined:

$$\nabla\cdot A + A\cdot\nabla = 2A\cdot\nabla.$$

(iii) The quadratic term in A,

$$\frac{e^2}{2m}c^2 A\cdot A$$

can be neglected except for very strong fields, on the order of 10^{15} W cm^{-2} [4].

(iv) For isolated molecules, it is generally the case that the wavelength of light is much larger than the molecular dimensions. In this case it is a good approximation to make the replacement $e^{ik\cdot r} \approx 1$, which allows the replacement [3]

$$V = -\frac{e}{mc}A\cdot\hat{p} = -E\cdot e\hat{r}.$$

For many electrons and nuclei, V takes the following form:

$$V = -E\cdot\sum_i Z_i e\hat{r}_i = -E\cdot\bar{\hat{\mu}} \qquad (A1.6.21)$$

where we have defined the dipole operator, $\bar{\hat{\mu}} \equiv \sum_i Z_i e\hat{r}_i$. The dipole moment is seen to be a product of charge and distance, and has the physical interpretation of the degree of charge separation in the atom or molecule. Note that for not-too-intense fields, equation (A1.6.21) is the dominant term in the radiation–matter interaction; this is the dipole approximation.

A1.6.1.3   Absorption, stimulated emission and spontaneous emission of light

Consider a quantum system with two levels, a and b, with energy levels E_a and E_b. Furthermore, let the perturbation between these levels be of the form equation (A1.6.21), with monochromatic light, that is, $\mathbf{E} = \mathbf{E}_0 \cos(\omega t)$ resonant to the transition frequency between the levels, so $\omega = (E_b - E_a)/\hbar \equiv E_{ba}/\hbar$. The perturbation matrix element between a and b is then given by

$$V_{ba} = \mathbf{E}_0 \cos(\omega t) \cdot \langle b|\bar{\hat{\mu}}|a \rangle = \frac{\mathbf{E}}{2}(\mathrm{e}^{i\omega t} + \mathrm{e}^{-i\omega t}) \cdot \bar{\mu}_{ba} \tag{A1.6.22}$$

where

$$\bar{\mu}_{ab} \equiv \langle b|\bar{\hat{\mu}}|a \rangle = \bar{\mu}_{ba}$$

is the dipole matrix element. There are three fundamental possible kinds of transitions connected by the dipole interaction: absorption ($a \rightarrow b$), corresponding to the second term in equation (A1.6.22); stimulated emission ($b \rightarrow a$) governed by the first term in equation (A1.6.22); and spontaneous emission (also $b \rightarrow a$), for which there is no term in the classical radiation field. For a microscopic description of the latter, a quantum mechanical treatment of the radiation field is required. Nevertheless, there is a simple prescription for taking spontaneous emission into account, which was derived by Einstein during the period of the old quantum theory on the basis of considerations of thermal equilibrium between the matter and the radiation. Although for most of the remainder of this section the assumption of thermal equilibrium will not be satisfied, it is convenient to invoke it here to quantify spontaneous emission.

Fermi's Golden Rule expresses the rate of transitions between b and a as

$$W = \frac{2\pi}{\hbar}|V_{ba}|^2 \rho(E_{ba}) \tag{A1.6.23}$$

where $\rho(E_{ba})$ is the density of final states for both the system and the light. As described above, we will consider the special case of both the matter and light at thermal equilibrium. The system final state is by assumption non-degenerate, but there is a frequency dependent degeneracy factor for thermal light, $\rho(E)\,\mathrm{d}E$, where

$$\rho(E) = \frac{V}{(2\pi c)^3}\frac{\omega^2}{\hbar}\,\mathrm{d}\Omega \tag{A1.6.24}$$

and V is the volume of the 'box' and Ω is an element of solid angle.

The thermal light induces transitions from $a \rightarrow b$ and from $b \rightarrow a$ in proportion to the number of photons present. The number of transitions per second induced by absorption is

$$W_{\mathrm{abs}}(a \rightarrow b) = \frac{2\pi}{\hbar}|V_{ab}|^2 \rho(E_b - \hbar\omega) \tag{A1.6.25}$$

$$= \frac{2\pi}{\hbar}\frac{E_0^2}{4}|e \cdot \bar{\mu}_{ab}|^2 \frac{V}{(2\pi c)^3}\frac{\omega^2}{\hbar}\,\mathrm{d}\Omega. \tag{A1.6.26}$$

Integrating over all solid angles and using equations (A1.6.15) and (A1.6.10) we find

$$W_{\mathrm{abs}}(a \rightarrow b) = \frac{4}{3\hbar}N\frac{\omega^3}{c^3}|\mu_{ab}|^2. \tag{A1.6.27}$$

For thermal light, the number of transitions per second induced by stimulated emission integrated over solid angles, W_{stim}, is equal to W_{abs}. The total emission, which is the sum of the stimulated and spontaneous emission, may be obtained by letting $N \rightarrow N + 1$ in the expression for stimulated emission, giving

$$W_{\mathrm{em}}(b \rightarrow a) = \frac{4}{3\hbar}(N + 1)\frac{\omega^3}{c^3}|\mu_{ab}|^2. \tag{A1.6.28}$$

Einstein's original treatment [5] used a somewhat different notation, which is still in common use:

$$W_{\text{stim}}(b \to a) \equiv B(b \to a)\bar{\rho} = W_{\text{abs}}(a \to b)$$

where

$$\bar{\rho} = \frac{2hN\hbar\omega}{V}\rho(E) = \frac{2N\omega^3\hbar}{\pi c^3}$$

is the energy in the field per unit volume between frequencies ν and $\nu + d\nu$ (the 'radiation density') (the factor of 2 comes from the two polarizations of the light and the factor h from the scaling between energy and frequency). Comparing with equation (A1.6.27) leads to the identification

$$B(b \to a) = B(a \to b) = \frac{2\pi}{3\hbar^2}|\mu_{ab}|^2.$$

Moreover, in Einstein's treatment

$$W_{\text{spont}}(b \to a) \equiv A(b \to a) = \frac{4}{3\hbar}\left(\frac{\omega}{c}\right)^3|\mu_{ab}|^2$$

leading to the following ratio of the Einstein A and B coefficients:

$$\frac{A(b \to a)}{B(b \to a)} = \frac{2\hbar}{\pi}\left(\frac{\omega}{c}\right)^3. \tag{A1.6.29}$$

The argument is sometimes given that equation (A1.6.29) implies that the ratio of spontaneous to stimulated emission goes as the cube of the emitted photon frequency. This argument must be used with some care: recall that for light at thermal equilibrium, W_{stim} goes as $B\bar{\rho}$, and hence the rate of stimulated emission has a factor of $(\omega/c)^3$ coming from $\bar{\rho}$. The ratio of the spontaneous to the stimulated emission rates is therefore frequency independent! However, for non-thermal light sources (e.g. lasers), only a small number of energetically accessible states of the field are occupied, and the $\bar{\rho}$ factor is on the order of unity. The rate of spontaneous emission still goes as ω^3, but the rate of stimulated emission goes as ω, and hence the ratio of spontaneous to stimulated emission goes as ω^2. Thus, for typical light sources, spontaneous emission dominates at frequencies in the UV region and above, while stimulated emission dominates at frequencies in the far-IR region and below, with both processes participating at intermediate frequencies.

A1.6.1.4   Interaction between matter and field

In the previous sections we have described the interaction of the electromagnetic field with matter, that is, the way the material is affected by the presence of the field. But there is a second, reciprocal perspective: the excitation of the material by the electromagnetic field generates a dipole (polarization) where none existed previously. Over a sample of finite size this dipole is macroscopic, and serves as a new source term in Maxwell's equations. For weak fields, the source term, P, is linear in the field strength. Thus,

$$P = \chi E \tag{A1.6.30}$$

where the proportionality constant χ, called the (linear) susceptibility, is generally frequency dependent and complex. As we shall see below, the imaginary part of the linear susceptibility determines the absorption spectrum while the real part determines the dispersion, or refractive index of the material. There is a universal relationship between the real part and the imaginary part of the linear susceptibility, known as the Kramers–Kronig relation, which establishes a relationship between the absorption spectrum and the frequency-dependent refractive index. With the addition of the source term P, Maxwell's equations still have wavelike solutions, but the relation between frequency and wavevector in equation (A1.6.10) must be generalized as follows:

$$\left(\frac{kc}{\omega}\right)^2 = 1 \rightarrow \left(\frac{kc}{\omega}\right)^2 = 1 + \chi. \tag{A1.6.31}$$

The quantity $1 + \chi$ is known as the dielectric constant, ϵ; it is constant only in the sense of being independent of E, but is generally dependent on the frequency of E. Since χ is generally complex so is the wavevector k. It is customary to write

$$\frac{kc}{\omega} = \eta + i\kappa \tag{A1.6.32}$$

where η and κ are the refractive index and extinction coefficient, respectively. The travelling wave solutions to Maxwell's equations, propagating in the z-direction now take the form

$$\exp(i(kz - \omega t)) = \exp\left[i\omega\left(\frac{\eta z}{c} - t\right) - \left(\frac{\omega \kappa z}{c}\right)\right]. \tag{A1.6.33}$$

In this form it is clear that κ leads to an attenuation of the electric field amplitude with distance (i.e. absorption).

For stronger fields the relationship between the macroscopic polarization and the incident field is non-linear. The general relation between P and E is written as

$$P = \chi^{(1)} E + \chi^{(2)} : E^2 + \chi^{(3)} : E^3 + \cdots \equiv P^{(1)} + P^{(2)} + P^{(3)} + \cdots. \tag{A1.6.34}$$

The microscopic origin of χ and hence of P is the non-uniformity of the charge distribution in the medium. To lowest order this is given by the dipole moment, which in turn can be related to the dipole moments of the component molecules in the sample. Thus, on a microscopic quantum mechanical level we have the relation

$$P = \langle \psi | \bar{\bar{\mu}} | \psi \rangle. \tag{A1.6.35}$$

Assuming that the material has no permanent dipole moment, P originates from changes in the wavefunction ψ that are induced by the field; this will be our starting point in section A1.6.4.

A1.6.2 Coherence properties of light and matter

In the previous section we discussed light and matter at equilibrium in a two-level quantum system. For the remainder of this section we will be interested in light and matter which are not at equilibrium. In particular, laser light is completely different from the thermal radiation described at the end of the previous section. In the first place, only one, or a small number of states of the field are occupied, in contrast with the Planck distribution of occupation numbers in thermal radiation. Second, the field state can have a precise *phase*; in thermal radiation this phase is assumed to be random. If multiple field states are occupied in a laser they can have a precise phase relationship, something which is achieved in lasers by a technique called 'mode-locking'. Multiple frequencies with a precise phase relation give rise to laser pulses in time. Nanosecond experiments have been very useful in probing, for example, radiationless transitions, intramolecular dynamics and radiative lifetimes of single vibronic levels in molecules. Picosecond experiments have been useful in probing, for example, collisional relaxation times and rotational reorientation times in solutions. Femtosecond experiments have been useful in observing the real time breaking and formation of chemical bonds; such experiments will be described in the next section. Any time that the phase is precisely correlated in time over the duration of an experiment, or there is a superposition of frequencies with well-defined relative phases, the process is called coherent. Single frequency coherent processes will be the major subject of section A1.6.2, while multifrequency coherent processes will be the focus for the remainder of the section.

A1.6.2.1   Wavepackets: solutions of the time-dependent Schrödinger equation

The central equation of (non-relativistic) quantum mechanics, governing an isolated atom or molecule, is the time-dependent Schrödinger equation (TDSE):

$$i\hbar \frac{\partial \psi(x,t)}{\partial t} = H\psi(x,t). \tag{A1.6.36}$$

In this equation H is the Hamiltonian (developed in the previous section) which consists of the bare system Hamiltonian and a term coming from the interaction between the system and the light. That is,

$$H = -\frac{\hbar^2}{2m}\nabla^2 + V_s(x) - E(t)\mu. \tag{A1.6.37}$$

Since we are now interested in the possibility of coherent light, we have taken the interaction between the radiation and matter to be some general time-dependent interaction, $V = -E(t)\mu$, which could in principle contain many frequency components. At the same time, for simplicity, we neglect the vector character of the electric field in what follows. The vector character will be reintroduced in section A1.6.4, in the context of nonlinear spectroscopy.

   Real molecules in general have many quantum levels, and the TDSE can exhibit complicated behaviour even in the absence of a field. To simplify matters, it is worthwhile discussing some properties of the solutions of the TDSE in the absence of a field and then reintroducing the field. First let us consider

$$H = -\frac{\hbar^2}{2m}\nabla^2 + V_s(x). \tag{A1.6.38}$$

Since in this case the Hamiltonian is time independent, the general solution can be written as

$$\Psi(x,t) = \sum_{n=1}^{\infty} a_n \psi_n(x) e^{-(i/\hbar)E_n t}. \tag{A1.6.39}$$

(This expression assumes a system with a discrete level structure; for systems with both a discrete and a continuous portion to their spectrum the expression consists of a sum over the discrete states and an integral over the continuous states.) Here, $\psi_n(x)$ is a solution of the time-independent Schrödinger equation,

$$H\psi_n(x) = E_n \psi(x),$$

with eigenvalue E_n. The coefficients, a_n, satisfy the normalization condition $\sum_n |a_n|^2 = 1$, and are time independent in this case. Equation (A1.6.39) describes a moving *wavepacket*, that is, a state whose average values in coordinate and momentum change with time. To see this, note that according to quantum mechanics $|\Psi(x,t)|^2 \, dx$ is the probability to find the particle between x and $x + dx$ at time t. Using equation (A1.6.39) we see that

$$|\Psi(x,t)|^2 = \sum_{m,n=1}^{\infty} a_m^* a_n \psi_m^*(x)\psi_n(x) e^{-(i/\hbar)(E_n - E_m)t}$$

in other words, the probability density has a non-vanishing time dependence so long as there are components of two or more different energy eigenstates.

   One of the remarkable features of time evolution of wavepackets is the close connection they exhibit with the motion of a classical particle. Specifically, Ehrenfest's theorem indicates that for potentials up to quadratic, the average value of position and momentum of the quantum wavepacket as a function of time is exactly the same as that of a classical particle on the same potential that begins with the corresponding

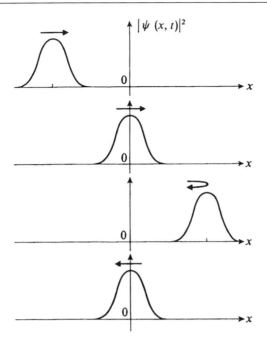

Figure A1.6.1. Gaussian wavepacket in a harmonic oscillator. Note that the average position and momentum change according to the classical equations of motion (adapted from [6]).

initial conditions in position and momentum. This classical-like behaviour is illustrated in figure A1.6.1 for a displaced Gaussian wavepacket in a harmonic potential. For the case shown, the initial width is the same as the ground-state width, a 'coherent state', and hence the Gaussian moves without spreading. By way of contrast, if the initial Gaussian has a different width parameter, the centre of the Gaussian still satisfies the classical equations of motion; however, the width will spread and contract periodically in time, twice per period.

A1.6.2.2   Coherence in a two-level system: the Rabi solution

We now add the field back into the Hamiltonian, and examine the simplest case of a two-level system coupled to coherent, monochromatic radiation. This material is included in many textbooks (e.g. [6–11]). The system is described by a Hamiltonian H_0 having only two eigenstates, ψ_a and ψ_b, with energies $E_a = \hbar\omega_a$ and $E_b = \hbar\omega_b$. Define $\omega_0 = \omega_b - \omega_a$. The most general wavefunction for this system may be written as

$$\Psi(t) = a(t)\mathrm{e}^{-\mathrm{i}\omega_a t}\psi_a + b(t)\,\mathrm{e}^{-\mathrm{i}\omega_b t}\psi_b. \tag{A1.6.40}$$

The coefficients $a(t)$ and $b(t)$ are subject to the constraint that $|a(t)|^2 + |b(t)|^2 = 1$. If we couple this system to a light field, represented as $V = -\mu_{ab}E\cos(\omega t)$, then we may write the TDSE in matrix form as

$$\mathrm{i}\hbar\frac{\mathrm{d}}{\mathrm{d}t}\begin{pmatrix} a(t)\mathrm{e}^{-\mathrm{i}\omega_a t} \\ b(t)\mathrm{e}^{-\mathrm{i}\omega_b t} \end{pmatrix} = \begin{pmatrix} E_a & -\mu_{ab}E\cos(\omega t) \\ -\mu_{ab}E\cos(\omega t) & E_b \end{pmatrix}\begin{pmatrix} a(t)\mathrm{e}^{-\mathrm{i}\omega_a t} \\ b(t)\mathrm{e}^{-\mathrm{i}\omega_b t} \end{pmatrix}. \tag{A1.6.41}$$

To continue we define a detuning parameter, $\Delta \equiv \omega - \omega_0$. If $\Delta \ll \omega_0$ then $\exp(-\mathrm{i}(\omega - \omega_0)t)$ is slowly varying while $\exp(-\mathrm{i}(\omega+\omega_0)t)$ is rapidly varying and cannot transfer much population from state A to state B.

We therefore ignore the latter term; this is known as the 'rotating wave approximation'. If we choose as initial conditions $|a(0)|^2 = 1$ and $|b(0)|^2 = 0$ then the solution of equation (A1.6.41) is

$$a(t) = e^{+\frac{i}{2}\Delta t}\left(\cos\left(\frac{1}{2}\Omega t\right) - i\frac{\Delta}{\Omega}\sin\left(\frac{1}{2}\Omega t\right)\right) \tag{A1.6.42}$$

$$b(t) = e^{-\frac{i}{2}\Delta t}\left(\frac{\mu E}{2\hbar\Omega}\right)\left(2i\sin\left(\frac{1}{2}\Omega t\right)\right). \tag{A1.6.43}$$

where the Rabi frequency, Ω, is defined as

$$\Omega = \sqrt{\Delta^2 + \left(\frac{\mu E}{\hbar}\right)^2}. \tag{A1.6.44}$$

The populations as functions of time are then

$$|a(t)|^2 = \left(\frac{\Delta}{\Omega}\right)^2 + \left(\frac{\mu E}{\hbar\Omega}\right)^2\cos^2\left(\frac{1}{2}\Omega t\right) \tag{A1.6.45}$$

$$|b(t)|^2 = \left(\frac{\mu E}{\hbar\Omega}\right)^2\sin^2\left(\frac{1}{2}\Omega t\right). \tag{A1.6.46}$$

The population in the upper state as a function of time is shown in figure A1.6.2. There are several important things to note. At early times, resonant and non-resonant excitation produce the same population in the upper state because, for short times, the population in the upper state is independent of the Rabi frequency:

$$|b(t)|^2 = \left(\frac{\mu E}{\hbar\Omega}\right)^2\sin^2\left(\frac{1}{2}\Omega t\right) \xrightarrow{t\ \text{small}} \left(\frac{\mu E}{2\hbar}\right)^2 t^2. \tag{A1.6.47}$$

One should also notice that resonant excitation completely cycles the population between the lower and upper state with a period of $2\pi/\Omega$. Non-resonant excitation also cycles population between the states but never completely depopulates the lower state. Finally, one should notice that non-resonant excitation cycles population between the two states at a faster rate than resonant excitation.

A1.6.2.3   Geometrical representation of the evolution of a two-level system

A more intuitive, and more general, approach to the study of two-level systems is provided by the Feynman–Vernon–Hellwarth geometrical picture. To understand this approach we need to first introduce the density matrix.

In the Rabi solution of the previous section we considered a wavefunction $\Psi(t)$ of the form

$$\Psi(t) = a(t)e^{-i\omega_a t}\psi_a + b(t)e^{-i\omega_b t}\psi_b. \tag{A1.6.48}$$

We saw that the time-dependent populations in each of the two levels is given by $P_a = |a(t)|^2$ and $P_b = |b(t)|^2$. So long as the field is on, these populations continue to change; however, once the external field is turned off, these populations remain constant (discounting relaxation processes, which will be introduced below). Yet the *amplitudes* in the states ψ_a and ψ_b do continue to change with time, due to the accumulation of time-dependent phase factors during the field-free evolution. We can obtain a convenient separation of the time-dependent and the time-independent quantities by defining a density matrix, ρ. For the case of the wavefunction $|\Psi\rangle$, ρ is given as the 'outer product' of $|\Psi\rangle$ with itself,

$$\rho \equiv |\Psi\rangle\langle\Psi|. \tag{A1.6.49}$$

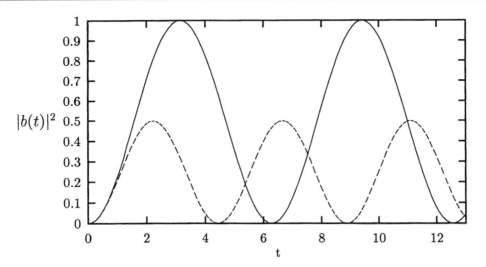

Figure A1.6.2. The population in the upper state as a function of time for resonant excitation (full curve) and for non-resonant excitation (dashed curve).

This outer product gives four terms, which may be arranged in matrix form as

$$\rho = \begin{pmatrix} |a|^2 & a^*b\,\mathrm{e}^{-\mathrm{i}(\omega_b-\omega_a)t} \\ ab^*\,\mathrm{e}^{\mathrm{i}(\omega_b-\omega_a)t/\hbar} & |b|^2 \end{pmatrix}. \tag{A1.6.50}$$

Note that the diagonal elements of the matrix, $|a|^2$ and $|b|^2$, correspond to the *populations* in the energy levels, a and b, and contain no time dependence, while the off-diagonal elements, called the *coherences*, contain all the time dependence.

   A differential equation for the time evolution of the density operator may be derived by taking the time derivative of equation (A1.6.49) and using the TDSE to replace the time derivative of the wavefunction with the Hamiltonian operating on the wavefunction. The result is called the Liouville equation, that is,

$$\mathrm{i}\hbar\frac{\partial\rho}{\partial t} = [H,\rho]. \tag{A1.6.51}$$

The strategy for representing this differential equation geometrically is to expand both H and ρ in terms of the three Pauli spin matrices, σ_1, σ_2 and σ_3 and then view the coefficients of these matrices as time-dependent vectors in three-dimensional space. We begin by writing the the two-level system Hamiltonian in the following general form,

$$H = \begin{pmatrix} E_b & V_{ba} \\ V_{ab} & E_a \end{pmatrix} \tag{A1.6.52}$$

where we take the radiation–matter interaction to be of the dipole form, but allow for arbitrary time-dependent electric fields:

$$V_{ba} = -\mu_{ba}E(t). \tag{A1.6.53}$$

Moreover, we will write the density matrix for the system as

$$\rho = \begin{pmatrix} bb^* & a^*b \\ ab^* & aa^* \end{pmatrix} \tag{A1.6.54}$$

where a and b now contain the bare system evolution phase factors. We proceed to express both the Hamiltonian and the density matrix in terms of the standard Pauli spin matrices:

$$H = \overbrace{(V_{ab} + V_{ba})}^{E_1} \sigma_1 + \overbrace{\mathrm{i}(V_{ba} - V_{ab})}^{E_2} \sigma_2 + \overbrace{(E_b - E_a)}^{E_3} \sigma_3$$

$$\rho = \underbrace{(ab^* + a^*b)}_{r_1} \sigma_1 + \underbrace{\mathrm{i}(a^*b - ab^*)}_{r_2} \sigma_2 + \underbrace{(bb^* - aa^*)}_{r_3} \sigma_3.$$

We now define the three-dimensional vectors, $\bar{r}$ and $\bar{\Omega}$, consisting of the coefficients of the Pauli matrices in the expansion of ρ and H, respectively:

$$\bar{r} = (r_1, r_2, r_3) \tag{A1.6.55}$$

$$\bar{\Omega} = \frac{1}{\hbar}(E_1, E_2, E_3). \tag{A1.6.56}$$

Using these vectors, we can rewrite the Liouville equation for the two-level system as

$$\frac{\mathrm{d}}{\mathrm{d}t}\bar{r} = \bar{\Omega} \times \bar{r}. \tag{A1.6.57}$$

Note that r_3 is the population difference between the upper and lower states: having all the population in the lower state corresponds to $r_3 = -1$ while having a completely inverted population (i.e. no population in the lower state) corresponds to $r_3 = +1$.

This representation is slightly inconvenient since E_1 and E_2 in equation (A1.6.56) are explicitly time-dependent. For a monochromatic light field of frequency ω, we can transform to a frame of reference rotating at the frequency of the light field so that the vector $\bar{\Omega}$ is a constant. To completely remove the time dependence we make the rotating wave approximation (RWA) as before: $E\cos(\omega t) = \frac{1}{2}(E\,\mathrm{e}^{-\mathrm{i}\omega t} + E\,\mathrm{e}^{\mathrm{i}\omega t}) \rightarrow \frac{1}{2}E\,\mathrm{e}^{-\mathrm{i}\omega t}$. In the rotating frame, the Liouville equation for the system is

$$\frac{\mathrm{d}}{\mathrm{d}t}\bar{r}' = \bar{\Omega}' \times \bar{r}' \tag{A1.6.58}$$

where $\bar{\Omega}'$ is now time independent. The geometrical interpretation of this equation is that the pseudospin vector, r', precesses around the field vector, Ω', in exactly the same way that the angular momentum vector precesses around a body fixed axis of a rigid object in classical mechanics. This representation of the two-level system is called the Feynman–Vernon–Hellwarth, or FVH representation; it gives a unified, pictorial view with which one can understand the effect of a wide variety of optical pulse effects in two-level systems. For example, the geometrical picture of Rabi cycling within the FVH picture is shown in figure A1.6.3. Assuming that at $t = 0$ all the population is in the ground-state then the initial position of the $\bar{r}'$ vector is $(0, 0, -1)$, and so $\bar{r}'$ points along the negative z-axis. For a resonant field, $\omega_0 - \omega = 0$ and so the $\bar{\Omega}'$ vector points along the x-axis. Equation (A1.6.58) then says that the population vector simply precesses about the x-axis. It then periodically points along the positive z-axis, which corresponds to having all the population in the upper state. If the field is non-resonant, then $\bar{\Omega}'$ no longer points along the x-axis but along some other direction in the xz-plane. The population vector still precesses about the field vector, but now at some angle to the z-axis. Thus, the projection onto the z-axis of $\bar{r}'$ never equals one and so there is never a complete population inversion.

The FVH representation allows us to visualize the results of more complicated laser pulse sequences. A laser pulse which takes $\bar{r}'$ from $(0, 0, -1)$ to $(0, 0, 1)$ is called a π-pulse since the $\bar{r}'$ vector precesses π radians about the field vector. Similarly, a pulse which takes $\bar{r}'$ from $(0, 0, -1)$ to $(+1, 0, 0)$ is called a $\pi/2$-pulse. The state represented by the vector $(+1, 0, 0)$ is a coherent superposition of the upper and lower states of the system.

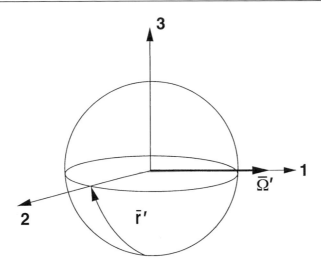

Figure A1.6.3. FVH diagram, exploiting the isomorphism between the two-level system and a pseudospin vector precessing on the unit sphere. The pseudospin vector, $\bar{r}'$, precesses around the field vector, $\bar{\Omega}'$, according to the equation $d\bar{r}'/dt = \bar{\Omega} \times \bar{r}$. The z-component of the $\bar{r}$ vector is the population difference between the two levels, while the x- and y-components refer to the polarization, that is, the real and imaginary parts of the coherence between the amplitude in the two levels. In the frame of reference rotating at the carrier frequency, the z-component of the $\bar{\Omega}$ vector is the detuning of the field from resonance, while the x- and y-components indicate the field amplitude. In the rotating frame, the y-component of $\bar{\Omega}$ may be set equal to zero (since the overall phase of the field is irrelevant, assuming no coherence of the levels at $t = 0$), unless there is non-uniform change in phase in the field during the process.

One interesting experiment is to apply a $\pi/2$-pulse followed by a $\pi/2$ phase shift of the field. This phase shift will bring $\bar{\Omega}'$ parallel to $\bar{r}'$. Since now $\bar{\Omega}' \times \bar{r}' = 0$, the population is fixed in time in a coherent superposition between the ground and excited states. This is called photon locking.

A second interesting experiment is to begin with a pulse which is far below resonance and slowly and continuously sweep the frequency until the pulse is far above resonance. At $t = -\infty$ the field vector is pointing nearly along the $-z$-axis, and is therefore almost parallel to the state vector. As the field vector slowly moves from $z = -1$ to $z = +1$ the state vector adiabatically follows it, precessing about the instantaneous direction of the field vector (figure A1.6.4). When, at $t \to +\infty$, the field vector is directed nearly along the $+z$-axis, the state vector is directed there as well, signifying complete population inversion. The remarkable feature of 'adiabatic following', as this effect is known, is its robustness—there is almost no sensitivity to either the field strength or the exact schedule of changing the frequency, provided the conditions for adiabaticity are met.

A1.6.2.4 Relaxation of the density operator to equilibrium

In real physical systems, the populations $|a(t)|^2$ and $|b(t)|^2$ are not truly constant in time, even in the absence of a field, because of relaxation processes. These relaxation processes lead, at sufficiently long times, to thermal equilibrium, characterized by the populations $P_a = e^{-\beta E_a}/Q$, $P_b = e^{-\beta E_b}/Q$, where Q is the canonical partition function which serves as a normalization factor and $\beta = 1/kT$, where k is the Boltzmann's constant and T is the temperature. The thermal equilibrium state for a two-level system, written as a density matrix,

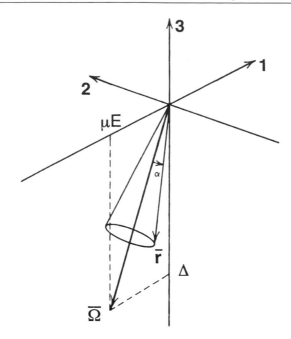

Figure A1.6.4. FVH diagram, showing the concept of adiabatic following. The Bloch vector, $\bar{r}'$, precesses in a narrow cone about the rotating frame torque vector, $\bar{\Omega}'$. As the detuning, Δ, changes from negative to positive, the field vector, $\bar{\Omega}'$, becomes inverted. If the change in $\bar{\Omega}'$ is adiabatic the Bloch vector follows the field vector in this inversion process, corresponding to complete population transfer to the excited state.

takes the following form:

$$\rho = \begin{pmatrix} e^{-\beta E_a}/Q & 0 \\ 0 & e^{-\beta E_b}/Q \end{pmatrix}. \qquad (A1.6.59)$$

The populations, $e^{-\beta E_n}/Q$, appear on the diagonal as expected, but note that there are no off-diagonal elements—no coherences; this is reasonable since we expect the equilibrium state to be time-independent, and we have associated the coherences with time.

It follows that there are two kinds of processes required for an arbitrary initial state to relax to an equilibrium state: the diagonal elements must redistribute to a Boltzmann distribution and the off-diagonal elements must decay to zero. The first of these processes is called population decay; in two-level systems this time scale is called T_1. The second of these processes is called dephasing, or coherence decay; in two-level systems there is a single time scale for this process called T_2. There is a well-known relationship in two level systems, valid for weak system–bath coupling, that

$$\frac{1}{T_2} = \frac{1}{2T_1} + \frac{1}{T_2^*} \qquad (A1.6.60)$$

where T_2^* is the time scale for so-called pure dephasing. Equation (A1.6.60) has the following significance: even without pure dephasing there is still a minimal dephasing rate that accompanies population relaxation.

In the presence of some form of relaxation the equations of motion must be supplemented by a term involving a relaxation *superoperator*—superoperator because it maps one operator into another operator. The literature on the correct form of such a superoperator is large, contradictory and incomplete. In brief, the extant theories can be divided into two kinds, those without memory relaxation (Markovian) $\Gamma\rho$ and those with memory relaxation (non-Markovian) $\int_{-\infty}^{t} \Gamma(t - t')\rho(t')\, dt'$. The Markovian theories can be further subdivided into those that preserve positivity of the density matrix (all $p_n > 0$ in equation (A1.6.66) for all admissible ρ) and those that do not. For example, the following widely used Markovian equation of motion is guaranteed to preserve positivity of the density operator for any choice of $\{V_i\}$:

$$\frac{\partial \rho}{\partial t} \equiv \left[\frac{H}{i\hbar}, \rho\right] + \Gamma\rho = \left[\frac{H}{i\hbar}, \rho\right] + \sum_i V_i \rho V_i^\dagger - \frac{1}{2}\sum_i [V_i^\dagger V_i \rho + \rho V_i^\dagger V_i]. \tag{A1.6.61}$$

As an example, consider the two-level system, with relaxation that arises from spontaneous emission. In this case there is just a single V_i:

$$V = \gamma^{1/2}\begin{pmatrix} 0 & 1 \\ 0 & 0 \end{pmatrix} \qquad V^\dagger = \gamma^{1/2}\begin{pmatrix} 0 & 0 \\ 1 & 0 \end{pmatrix}. \tag{A1.6.62}$$

It is easy to verify that the dissipative contribution is given by

$$\Gamma\rho = \gamma\begin{pmatrix} \rho_{22} & -\rho_{12}/2 \\ -\rho_{21}/2 & -\rho_{22} \end{pmatrix}. \tag{A1.6.63}$$

We now make two connections with topics discussed earlier. First, at the beginning of this section we defined $1/T_1$ as the rate constant for population decay and $1/T_2$ as the rate constant for coherence decay. Equation (A1.6.63) shows that for spontaneous emission $1/T_1 = \gamma$, while $1/T_2 = \gamma/2$; comparing with equation (A1.6.60) we see that for spontaneous emission, $1/T_2^* = 0$. Second, note that γ is the rate constant for population transfer due to spontaneous emission; it is identical to the Einstein A coefficient which we defined in equation (A1.6.3).

For the two-level system, the evolution equation for ρ may also be expressed, as before, in terms of the three-vector $\bar{r}$:

$$\frac{d}{dt}\bar{r} = \bar{\Omega} \times \bar{r} - \bar{\Gamma} \cdot \bar{r} \tag{A1.6.64}$$

where

$$\bar{\Gamma} = \gamma\left(\frac{1}{2}, \frac{1}{2}, 1\right) \equiv \left(\frac{1}{T_2}, \frac{1}{T_2}, \frac{1}{T_1}\right). \tag{A1.6.65}$$

Equation (A1.6.64) describes the relaxation to equilibrium of a two-level system in terms of a vector equation. It is the analogue of the Bloch equation, originally developed for magnetic resonance, in the optical regime and hence is called the optical Bloch equation.

In the above discussion of relaxation to equilibrium, the density matrix was implicitly cast in the energy representation. However, the density operator can be cast in a variety of representations other than the energy representation. Two of the most commonly used are the coordinate representation and the Wigner phase space representation. In addition, there is the diagonal representation of the density operator; in this representation, the most general form of ρ takes the form

$$\rho = \sum_i p_i |\Psi_i\rangle\langle\Psi_i| \tag{A1.6.66}$$

where the p_i are real numbers, $0 \leq p_i \leq 1$ and $\sum p_i = 1$. This equation expresses ρ as an *incoherent* superposition of fundamental density operators, $|\Psi_i\rangle\langle\Psi_i|$, where $|\Psi_i\rangle$ is a wavefunction but not necessarily an eigenstate. In equation (A1.6.66), the p_i are the *probabilities* (not amplitudes) of finding the system in state $|\Psi_i\rangle$. Note that in addition to the usual probabilistic interpretation for finding the particle described by a particular wavefunction at a specified location, there is now a probability distribution for being in different eigenstates! If one of the $p_i = 1$ and all the others are zero, the density operator takes the form equation (A1.6.49) and corresponds to a single wavefunction; we say the system is in a *pure state*. If more than one of the $p_i > 0$ we say the system is in a *mixed state*.

A measure of the purity or coherence of a system is given by $\sum_i p_i^2$: $\sum_i p_i^2 = 1$ for a pure state and $\sum p_n^2 \leq 1$ for a mixed state; the greater the degree of mixture the lower will be the purity. A general expression for the purity, which reduces to the above definition but is representation free, is given by $\mathrm{Tr}(\rho^2)$: $\mathrm{Tr}(\rho^2) < 1$ for a mixed state and $\mathrm{Tr}(\rho^2) = 1$ for a pure state. Note that in the absence of dissipation, the purity of the system, as measured by $\mathrm{Tr}(\rho^2)$, is conserved in time. To see this, take the equation of motion for ρ to be purely Hamiltonian, that is,

$$\dot{\rho} = -\frac{i}{\hbar}[H, \rho].\tag{A1.6.67}$$

Then:

$$\frac{d}{dt}\mathrm{Tr}(\rho^2) = 2\mathrm{Tr}\rho\dot{\rho} = \frac{2}{i\hbar}\mathrm{Tr}(\rho[H, \rho]) = \frac{2}{i\hbar}\mathrm{Tr}(\rho(H\rho - \rho H)) = 0\tag{A1.6.68}$$

where in the last step we have used the cyclic invariance of the trace. This invariance of the purity to Hamiltonian manipulations is essentially equivalent to the invariance of phase space density, or entropy, to Hamiltonian manipulations. Including the dissipative part to the equations of motion gives

$$\dot{\rho} = -\frac{i}{\hbar}[H, \rho] + \Gamma\rho \qquad \text{and} \qquad \frac{d}{dt}\mathrm{Tr}(\rho^2) = 2\mathrm{Tr}(\rho\Gamma\rho).\tag{A1.6.69}$$

In concluding this section, we note the complementarity of the light and matter, this time in terms of coherence properties (i.e. phase relations). The FVH geometrical picture shows explicitly how the *phase* of the field is inseparably intertwined with the phase change in the matter; in the next section, in the context of short pulses, we shall see how the *time* of interaction with the pulse is similarly intertwined with the time of the response of the molecule, although in general an integration over all such times must be performed. But both these forms of complementarity are on the level of the Hamiltonian portion of the evolution only. The complementarity of the *dissipation* will appear at the end of this section, in the context of laser cooling.

A1.6.3   The field transfers its coherence to the matter

Much of the previous section dealt with two-level systems. Real molecules, however, are not two-level systems: for many purposes there are only two electronic states that participate, but each of these electronic states has many states corresponding to different quantum levels for vibration and rotation. A coherent femtosecond pulse has a bandwidth which may span many vibrational levels; when the pulse impinges on the molecule it excites a coherent superposition of all these vibrational states—a vibrational wavepacket. In this section we deal with excitation by one or two femtosecond optical pulses, as well as continuous wave excitation; in section A1.6.4 we will use the concepts developed here to understand nonlinear molecular electronic spectroscopy.

The pioneering use of wavepackets for describing absorption, photodissociation and resonance Raman spectra is due to Heller [12–16]. The application to pulsed excitation, coherent control and nonlinear spectroscopy was initiated by Tannor and Rice ([17] and references therein).

A1.6.3.1   First-order amplitude: wavepacket interferometry

Consider a system governed by Hamiltonian $H \equiv H_0 + H_1$, where H_0 is the bare molecular Hamiltonian and H_1 is the perturbation, taken to be the $-\mu E(t)$ as we have seen earlier. Adopting the Born–Oppenheimer (BO) approximation and specializing to two BO states, H_0 can be written as

$$H_0 = \begin{pmatrix} H_a & 0 \\ 0 & H_b \end{pmatrix}\tag{A1.6.70}$$

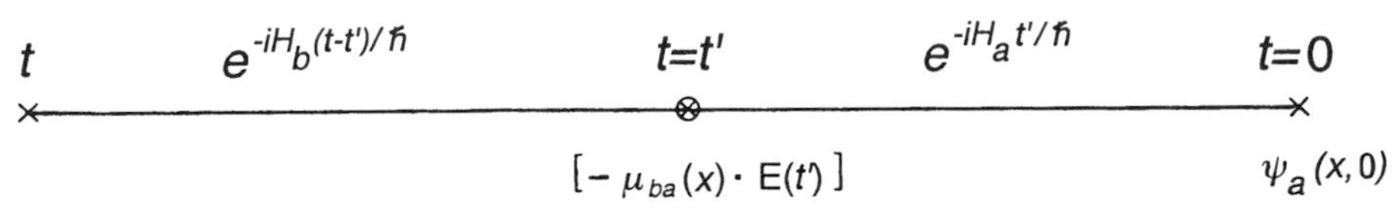

$$t \qquad e^{-iH_b(t-t')/\hbar} \qquad t=t' \qquad e^{-iH_a t'/\hbar} \qquad t=0$$

$$\underset{\otimes}{\times} \xleftarrow{\hspace{6cm}} \times$$

$$[-\mu_{ba}(x)\cdot E(t')] \qquad\qquad \psi_a(x,0)$$

Figure A1.6.5. Feynman diagram for the first-order process described in the text.

and H_1 as

$$H_1 = \begin{pmatrix} 0 & -\mu_{ab}E^*(t) \\ -\mu_{ba}E(t) & 0 \end{pmatrix}. \tag{A1.6.71}$$

The TDSE in matrix form reads:

$$i\hbar \frac{\partial}{\partial t}\begin{pmatrix} \psi_a(t) \\ \psi_b(t) \end{pmatrix} = \begin{pmatrix} H_a & -\mu_{ab}E^*(t) \\ -\mu_{ba}E(t) & H_b \end{pmatrix}\begin{pmatrix} \psi_a(t) \\ \psi_b(t) \end{pmatrix}. \tag{A1.6.72}$$

Note the structural similarity between equations (A1.6.72) and equation (A1.6.41), with E_a and E_b being replaced by H_a and H_b, the BO Hamiltonians governing the quantum mechanical evolution in electronic states a and b, respectively. These Hamiltonians consist of a nuclear kinetic energy part and a potential energy part which derives from nuclear–electron attraction and nuclear–nuclear repulsion, which differs in the two electronic states.

If H_1 is small compared with H_0 we may treat H_1 by perturbation theory. The first-order perturbation theory formula takes the form [18–21]:

$$\psi^{(1)}(x,t) = \frac{1}{i\hbar}\int_0^t e^{-(i/\hbar)H_b(t-t')}\{-\mu_{ba}E(t')\}e^{-(i/\hbar)H_a t'}\psi_a(x,0)\,dt' \tag{A1.6.73}$$

where we have assumed that all the amplitude starts on the ground electronic state. This formula has a very appealing physical interpretation. At $t=0$ the wavefunction is in, say, $v=0$ of the ground electronic state. The wavefunction evolves from $t=0$ until time t' under the ground electronic state, Hamiltonian, H_a. If we assume that the initial state is a vibrational eigenstate of H_a, $(H_a\psi_v = E_v\psi_v)$, there is no spatial evolution, just the accumulation of an overall phase factor; that is the action of $e^{-(i/\hbar)H_a t'}\psi_a(x,0)$ can be replaced by $e^{-(i/\hbar)E_v t'}\psi_v(x,0)\,dt'$. For concreteness, in what follows we will take $v=0$, which is the most common case of interest. At $t=t'$ the electric field, of amplitude $E(t')$, interacts with the transition dipole moment, promoting amplitude to the excited electronic state. This amplitude evolves under the influence of H_b from time t' until time t. The integral dt' indicates that one must take into account all instants in time t' at which the interaction with the field could have taken place. In general, if the field has some envelope of finite duration in time, the promotion to the excited state can take place at any instant under this envelope, and there will be interference from portions of the amplitude that are excited at one instant and portions that are excited at another. The various steps in the process may be visualized schematically with the use of Feynman diagrams. The Feynman diagram for the process just described is shown in figure A1.6.5.

We will now proceed to work through some applications of this formula to different pulse sequences. Perhaps the simplest is to consider the case of a δ-function excitation by light. That is,

$$E(t') = \delta(t'-t_1). \tag{A1.6.74}$$

In this case, the first-order amplitude reduces to

$$\psi^{(1)}(x,t) = \frac{1}{i\hbar}e^{-(i/\hbar)H_b(t-t_1)}\{-\mu_{ba}\}e^{-(i/\hbar)E_0 t_1}\psi_0(x,0). \tag{A1.6.75}$$

Within the Condon approximation (μ_{ba} independent of x), the first-order amplitude is simply a constant times the initial vibrational state, propagated on the excited-state potential energy surface! This process can be visualized by drawing the ground-state vibrational wavefunction displaced vertically to the excited-state potential. The region of the excited-state potential which is accessed by the vertical transition is called the Franck–Condon region, and the vertical displacement is the Franck–Condon energy. Although the initial vibrational state was an eigenstate of H_a, in general it is not an eigenstate of H_b, and starts to evolve as a coherent wavepacket. For example, if the excited-state potential energy surface is repulsive, the wavepacket will evolve away from the Franck–Condon region toward the asymptotic region of the potential, corresponding to separated atomic or molecular fragments (see figure A1.6.6). If the excited-state potential is bound, the wavepacket will leave the Franck–Condon region, but after half a period reach a classical turning point and return to the Franck–Condon region for a complete or partial revival.

An alternative perspective is as follows. A δ-function pulse in time has an infinitely broad frequency range. Thus, the pulse promotes transitions to all the excited-state vibrational eigenstates having good overlap (Franck–Condon factors) with the initial vibrational state. The pulse, by virtue of its coherence, in fact prepares a coherent superposition of all these excited-state vibrational eigenstates. From the earlier sections, we know that each of these eigenstates evolves with a different time-dependent phase factor, leading to coherent spatial translation of the wavepacket.

The δ-function excitation is not only the simplest case to consider; it is the fundamental building block, in the sense that the more complicated pulse sequences can be interpreted as superpositions of δ-functions, giving rise to *superpositions of wavepackets* which can in principle interfere.

The simplest case of this interference is the case of two δ-function pulses [22–24]:

$$E(t') = \delta(t' - t_1)e^{-i\omega_L t_1} + \delta(t' - t_2)e^{-i\omega_L t_2}e^{i\phi}. \tag{A1.6.76}$$

We will explore the effect of three parameters: $t_2 - t_1$, ω_L and ϕ, that is, the time delay between the pulses, the tuning or detuning of the carrier frequency from resonance with an excited-state vibrational transition and the relative phase of the two pulses. We follow closely the development of [22]. Using equation (A1.6.73),

$$\psi^{(1)}(x, t) = \frac{1}{i\hbar}e^{-(i/\hbar)H_b(t-t_1)}\{-\mu_{ba}e^{-i\omega_L t_1}\}e^{-(i/\hbar)E_0 t_1}\psi(x, 0)$$

$$+ \frac{1}{i\hbar}e^{-(i/\hbar)H_b(t-t_2)}\{-\mu_{ba}e^{-i\omega_L t_2}e^{i\phi}\}e^{-(i/\hbar)E_0 t_2}\psi(x, 0) \tag{A1.6.77}$$

$$= \frac{1}{i\hbar}(e^{(i/\hbar)H_b(t_2-t_1)}e^{-i\omega_L(t_2-t_1)}e^{-i\omega_0(t_2-t_1)}e^{i\phi} + 1)e^{-(i/\hbar)H_b(t-t_1)}e^{-(i/\hbar)E_0 t_1}\{-\mu_{ba}e^{-i\omega_L t_1}\}\psi(x, 0). \tag{A1.6.78}$$

To simplify the notation, we define $\tilde{H}_b = H_b - E_{00} - E_{0b}$, $\tilde{\omega}_L = \omega_L + \omega_0 - E_{00}/\hbar - E_{0b}/\hbar$, where E_{00} is the vertical displacement between the minimum of the ground electronic state and the minimum of the excited electronic state, E_{0b} is the zero point energy on the excited-state surface and $\omega_0 = E_0/\hbar$. Specializing to the harmonic oscillator, $e^{-(i/\hbar)\tilde{H}_b \tau}\psi(x, 0) = \psi(x, 0)$, where $\tau = 2\pi/\omega$ is the excited-state vibrational period, that is, any wavefunction in the harmonic oscillator returns exactly to its original spatial distribution after one period. To highlight the effect of detuning we write $\omega_L = n\omega + \Delta$, where Δ is the detuning from an excited-state vibrational eigenstate, and we examine time delays equal to the vibrational period $t_2 - t_1 = \tau$. We obtain:

$$\psi^{(1)}(x, t) = \frac{1}{i\hbar}(e^{(i/\hbar)\tilde{H}_b \tau}e^{-i(n\omega_0+\Delta)\tau}e^{i\phi} + 1)e^{-(i/\hbar)\tilde{H}_b(t-t_1)}\{-\mu_{ba}e^{-i\tilde{\omega}_L t_1}\}\psi(x, 0) \tag{A1.6.79}$$

$$= \frac{1}{i\hbar}(e^{-(i\Delta\tau-\phi)} + 1)e^{-(i/\hbar)\tilde{H}_b(t-t_1)}\{-\mu_{ba}e^{-i\tilde{\omega}_L t_1}\}\psi(x, 0). \tag{A1.6.80}$$

To illustrate the dependence on detuning, Δ, time delay, τ, and phase difference, ϕ, we consider some special cases. (i) If $\Delta = \phi = 0$ then the term in parentheses gives $1 + 1 = 2$. In this case, the two pulses create two wavepackets which add constructively, giving two units of amplitude or four units of excited-state population. (ii) If $\Delta = 0$ and $\phi = \pm\pi$ then the term in parentheses gives $-1 + 1 = 0$. In this case, the two pulses create two wavepackets which add destructively, giving no excited-state population! Viewed from the point of view of the light, this is stimulated emission. Emission against absorption is therefore controlled by the relative phase of the second pulse relative to the first. (iii) If $\Delta = 0$ and $\phi = \pm(\pi/2)$ then the term in parentheses gives $\pm i + 1$. In this case, the excited-state population, $\langle \psi^{(1)} | \psi^{(1)} \rangle$, is governed by the factor $(-i + 1)(i + 1) = 2$. The amplitude created by the two pulses overlap, but have no net interference contribution. This result is related to the phenomenon of 'photon locking', which was be discussed in section A1.6.2. (iv) If $\Delta = \omega/2$ and $\phi = 0$ then the term in parentheses gives $-1 + 1 = 0$. This is the ultrashort excitation counterpart of tuning the excitation frequency between vibrational resonances in a single frequency excitation: no net excited-state population is produced. As in the case above, of the two pulses π out of phase, the two wavepackets destructively interfere. In this case, the destructive interference comes from the offset of the carrier frequency from resonance, leading to a phase factor of $(\omega/2)\tau = \pi$. For time delays that are significantly different from τ the first wavepacket is not in the Franck–Condon region when the second packet is promoted to the excited state, and the packets do not interfere; two units of population are prepared on the excited state, as in the case of a $\pm(\pi/2)$ phase shift. These different cases are summarized in figure A1.6.7

Figure A1.6.8 shows the experimental results of Scherer *et al* of excitation of I_2 using pairs of phase locked pulses. By the use of heterodyne detection, those authors were able to measure just the interference contribution to the total excited-state fluorescence (i.e. the difference in excited-state population from the two units of population which would be prepared if there were no interference). The basic qualitative dependence on time delay and phase is the same as that predicted by the harmonic model: significant interference is observed only at multiples of the excited-state vibrational frequency, and the relative phase of the two pulses determines whether that interference is constructive or destructive.

There is a good analogy between the effects of pulse pairs and pulse shapes, and Fresnel and Fraunhofer diffraction in optics. Fresnel diffraction refers to the interference pattern obtained by light passing through two slits; interference from the wavefronts passing through the two slits is the spatial analogue of the interference from the two pulses in time discussed above. Fraunhofer diffraction refers to interference arising from the finite width of a single slit. The different subportions of a single slit can be thought of as independent slits that happen to adjoin; wavefronts passing through each of these subslits will interfere. This is the analogue of a single pulse with finite duration: there is interference from excitation coming from different subportions of the pulse, which may be insignificant if the pulse is short but can be important for longer pulse durations.

There is an alternative, and equally instructive, way of viewing the effect of different pulse sequences, by Fourier transforming the pulse train to the frequency domain. In the time domain, the wavefunction produced is the convolution of the pulse sequence with the excited-state dynamics; in frequency it is simply the product of the frequency envelope with the Franck–Condon spectrum (the latter is simply the spectrum of overlap factors between the initial vibrational state and each of the excited vibrational states). The Fourier transform of δ-function excitation is simply a constant excitation in frequency, which excites the entire Franck–Condon spectrum. The Fourier transform of a sequence of two δ-functions in time with spacing τ is a spectrum having peaks with a spacing of $2\pi/\tau$. If the carrier frequency of the pulses is resonant and the relative phase between the pulses is zero, the frequency spectrum of the pulses will lie on top of the Franck–Condon spectrum and the product will be non-zero; if, on the other hand, the carrier frequency is between resonances, or the relative phase is π, the frequency spectrum of the pulses will lie in between the features of the Franck–Condon spectrum, signifying zero net absorption. Similarly, a single pulse of finite duration may have a frequency envelope which is smaller than that of the entire Franck–Condon spectrum. The absorption process will depend on the overlap of the frequency spectrum with the Franck–Condon spectrum, and hence on both pulse shape and carrier frequency.

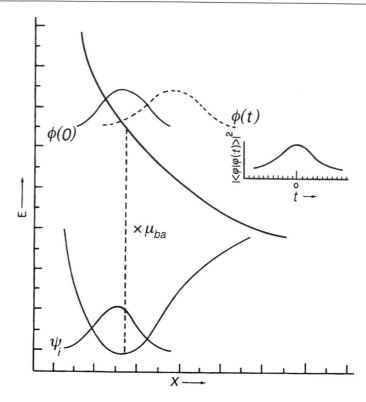

Figure A1.6.6. The wavepacket picture corresponding to the first-order process described in the text. The wavepacket propagates on the ground-state surface until time t_1, but since it is an eigenstate of this surface it only develops a phase factor. At time t_1 a photon impinges and promotes the initial vibrational state to an excited electronic state, for which it is not an eigenstate. The state is now a *wavepacket* and begins to move according to the TDSE. Often the ensuing motion is very classical-like, the initial motion being along the gradient of the excited-state potential, with recurrences at multiples of the excited-state vibrational period (adapted from [32]).

A1.6.3.2   Second-order amplitude: clocking chemical reactions

We now turn to the second-order amplitude. This quantity is given by [18–21]

$$\psi^{(2)}(x,t) = \left(\frac{1}{i\hbar}\right)^2 \int_0^t \int_0^{t'} e^{-(i/\hbar)H_c(t-t')}\{-\mu_{cb}E(t')\}e^{-(i/\hbar)H_b(t'-t'')}\{-\mu_{ba}E(t'')\}e^{-(i/\hbar)H_a t''}\psi(x,0)\,dt'\,dt''.$$

(A1.6.81)

This expression may be interpreted in a very similar spirit to that given above for one-photon processes. Now there is a second interaction with the electric field and the subsequent evolution is taken to be on a third surface, with Hamiltonian H_c. In general, there is also a second-order interaction with the electric field through μ_{ab} which returns a portion of the excited-state amplitude to surface a, with subsequent evolution on surface a. The Feynman diagram for this second-order interaction is shown in figure A1.6.9.

Second-order effects include experiments designed to 'clock' chemical reactions, pioneered by Zewail and co-workers [25]. The experiments are shown schematically in figure A1.6.10. An initial 100–150 fs pulse moves population from the bound ground state to the dissociative first excited state in ICN. A second

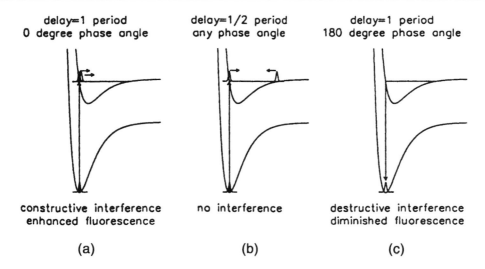

Figure A1.6.7. Schematic diagram illustrating the different possibilities of interference between a pair of wavepackets, as described in the text. The diagram illustrates the role of phase ((a) and (c)), as well as the role of time delay (b). These cases provide the interpretation for the experimental results shown in figure A1.6.8. Reprinted from [22].

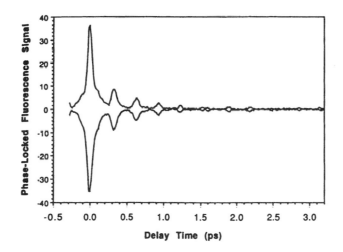

Figure A1.6.8. Wavepacket interferometry. The *interference contribution* to the excited-state fluorescence of I_2 as a function of the time delay between a pair of ultrashort pulses. The interference contribution is isolated by heterodyne detection. Note that the structure in the interferogram occurs only at multiples of 300 fs, the excited-state vibrational period of I_2: it is only at these times that the wavepacket promoted by the first pulse is back in the Franck–Condon region. For a phase shift of 0 between the pulses the returning wavepacket and the newly promoted wavepacket are in phase, leading to constructive interference (upper trace), while for a phase shift of π the two wavepackets are out of phase, and interfere destructively (lower trace). Reprinted from Scherer N F *et al* 1991 *J. Chem. Phys.* **95** 1487.

pulse, time delayed from the first then moves population from the first excited state to the second excited state, which is also dissociative. By noting the frequency of light absorbed from the second pulse, Zewail can estimate the distance between the two excited-state surfaces and thus infer the motion of the initially prepared wavepacket on the first excited state (figure A1.6.10).

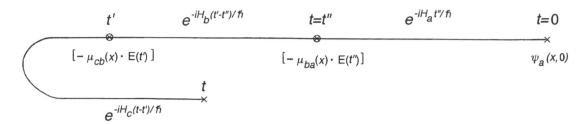

Figure A1.6.9. Feynman diagram for the second-order process described in the text.

A dramatic set of experiments by Zewail involves the use of femtosecond pulse pairs to probe the wavepacket dynamics at the crossing between covalent and ionic states of NaI [25]. A first pulse promotes wavepacket amplitude from the ionic to the covalent potential curve. The packet begins to move out, but most of the amplitude is reflected back from the crossing between the covalent and ionic curves, that is, the adiabatic potential changes character to ionic at large distances, and this curve is bound, leading to wavepacket reflection back to the FC region. The result is a long progression of wavepacket revivals, with a slow overall decay coming from amplitude which dissociates on the diabatic curve every period.

Femtosecond pump–probe experiments have burgeoned in the last ten years, and this field is now commonly referred to as 'laser femtochemistry' [26–29].

A1.6.3.3 *Spectroscopy as the rate of absorption of monochromatic radiation*

In this section we will discuss more conventional spectroscopies: absorption, emission and resonance Raman scattering. These spectroscopies are generally measured under single frequency conditions, and therefore our formulation will be tailored accordingly: we will insert monochromatic perturbations of the form $e^{i\omega t}$ into the perturbation theory formulae used earlier in the section. We will then *define* the spectrum as the time rate of change of the population in the final level. The same formulae apply with only minor modifications to electronic absorption, emission, photoionization and photodetachment/transition state spectroscopy. If the CW perturbation is inserted into the second-order perturbation theory one obtains the formulae for resonance Raman scattering, two-photon absorption and dispersed fluorescence spectroscopy. The spectroscopies of this section are to be contrasted with coherent nonlinear spectroscopies, such as coherent anti-Stokes Raman spectroscopy (CARS) or photon echoes, in which the signal is directional, which will be described in section A1.6.4.

(a) Electronic absorption and emission spectroscopy

Consider the radiation–matter Hamiltonian, equation (A1.6.73), with the interaction term of the form:

$$H_1(t) = -\mu E(t) = \begin{cases} \frac{-\mu E_0}{2}e^{-i\omega_I t} & \text{absorption} \\ \frac{-\mu E_0}{2}e^{+i\omega_S t} & \text{emission} \end{cases} \tag{A1.6.82}$$

where the incident (scattered) light has frequency $\omega_I (\omega_S)$ and μ is the (possibly coordinate-dependent) transition dipole moment for going from the lower state to the upper state. This form for the matter–radiation interaction Hamiltonian represents a light field that is 'on' all the time from $-\infty$ to ∞. This interaction will continuously move population from the lower state to the upper state. The propagating packets on the upper states will interfere with one another: constructively, if the incident light is resonant with a transition from an eigenstate of the lower surface to an eigenstate of the upper surface, destructively if not. Since for a one-photon process we have two potential energy surfaces we have, in effect, two different H_0's: one for

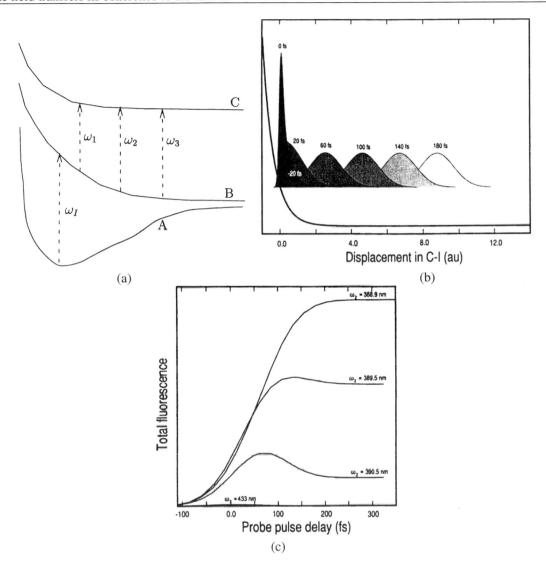

(a)

(b)

(c)

Figure A1.6.10. (a) Schematic representation of the three potential energy surfaces of ICN in the Zewail experiments.
(b) Theoretical quantum mechanical simulations for the reaction ICN $\rightarrow$ ICN* $\rightarrow$ [I$---$CN]†* $\rightarrow$ I+CN. Wavepacket
moves and spreads in time, with its centre evolving about 5 Å in 200 fs. Wavepacket dynamics refers to motion on the
intermediate potential energy surface B. Reprinted from Williams S O and Imre D G 1988 *J. Phys. Chem.* **92** 6648.
(c) Calculated FTS signal (total fluorescence from state C) as a function of the time delay between the first excitation
pulse (A $\rightarrow$ B) and the second excitation pulse (B $\rightarrow$ C). Reprinted from Williams S O and Imre D G, as above.

before excitation (call it H_a) and one for after (call it H_b). With this in mind, we can use the results of
section A1.6.2 to write down the first-order correction to the unperturbed wavefunction. If $|\psi_i(-\infty)\rangle$ is an
eigenstate of the ground-state Hamiltonian, H_a, then

$$|\psi^{(1)}(t)\rangle = \frac{-1}{2i\hbar} \int_{-\infty}^{t} e^{-(i/\hbar)H_b(t-t')} \mu E_0 e^{-i\omega_1 t'} e^{-(i/\hbar)E_i t'} |\psi_i(-\infty)\rangle \, dt'. \qquad (A1.6.83)$$

Defining $\tilde{\omega} = E_i/\hbar + \omega_I$, replacing $\psi(-\infty)$ by $\psi(0)$, since the difference is only a phase factor, which exactly cancels in the bra and ket, and assuming that the electric field vector is time independent, we find

$$\frac{\mathrm{d}}{\mathrm{d}t}\langle\psi^{(1)}(t)|\psi^{(1)}(t)\rangle = \frac{1}{4\hbar^2}\int_{-\infty}^{\infty}\langle\psi_i(0)|\mu E_0 \mathrm{e}^{-(\mathrm{i}/\hbar)H_b t}E_0\mu|\psi_i(0)\rangle\mathrm{e}^{\mathrm{i}\tilde{\omega}t}\,\mathrm{d}t. \tag{A1.6.84}$$

The absorption spectrum, $\sigma(\omega)$, is the ratio of transition probability per unit time/incident photon flux. The incident photon flux is the number of photons per unit area per unit time passing a particular location, and is given by

$$\frac{Nc}{V} = \frac{E_0^2 c}{8\pi\hbar\omega}$$

where we have used equation (A1.6.15). Finally, we obtain [12, 13]:

$$\sigma(\omega) = \frac{2\pi\hbar\omega}{E_0^2 c}\frac{\mathrm{d}}{\mathrm{d}t}\langle\psi^{(1)}(t)|\psi^{(1)}(t)\rangle \tag{A1.6.85}$$

$$= \frac{2\pi\omega}{\hbar c}\int_{-\infty}^{\infty}\langle\psi_i(0)|\mu\mathrm{e}^{-\mathrm{i}Ht/\hbar}\mu|\psi_i(0)\rangle\mathrm{e}^{\mathrm{i}\tilde{\omega}t}\,\mathrm{d}t. \tag{A1.6.86}$$

Rotational averaging yields

$$\sigma(\omega) = \frac{2\pi\omega}{3\hbar c}\int_{-\infty}^{\infty}\langle\phi_i(0)|\phi_i(t)\rangle\mathrm{e}^{\mathrm{i}\tilde{\omega}t}\,\mathrm{d}t \tag{A1.6.87}$$

where in the last equation we have defined $|\phi_i(0)\rangle \equiv \mu|\psi_i(0)\rangle$ and $|\phi_i(t)\rangle = \mathrm{e}^{-H_b t/\hbar}|\phi_i(0)\rangle$.

Since the absorption spectrum is a ratio it is amenable to other interpretations. One such interpretation is that the absorption spectrum is the ratio of energy absorbed to energy incident. From this perspective, the quantity $\hbar\omega(\mathrm{d}/\mathrm{d}t)\langle\psi^{(1)}(t)|\psi^{(1)}(t)\rangle$ is interpreted as the rate of energy absorption (per unit volume), since $\mathrm{d}E/\mathrm{d}t = \hbar\omega(\mathrm{d}N/\mathrm{d}t)$ while the quantity $E_0^2 c/\hbar\omega$ is interpreted as the incident energy flux, which depends only on the field intensity and is independent of frequency.

Equation (A1.6.87) expresses the absorption spectrum as the Fourier transform of a wavepacket correlation function. This is a result of central importance. The Fourier transform relationship between the wavepacket autocorrelation function and the absorption spectrum provides a powerful tool for interpreting absorption spectra in terms of the underlying nuclear wavepacket dynamics that follows the optically induced transition. The relevant correlation function is that of the moving wavepacket on the excited-state potential energy surface (or more generally, on the potential energy surface accessed by interaction with the light) with the stationary wavepacket on the ground-state surface (more generally, the initial wavepacket on the potential surface of origin), and thus the spectrum is a probe of excited-state dynamics, particularly in the Franck–Condon region (i.e. the region accessed by the packet undergoing a vertical transition at $t = 0$). Since often only short or intermediate dynamics enter in the spectrum (e.g. because of photodissociation or radiationless transitions to other electronic states) computation of the time correlation function can be much simpler than evaluation of the spectrum in terms of Franck–Condon overlaps, which formally can involve millions of eigenstates for an intermediate sized molecule.

We now proceed to some examples of this Fourier transform view of optical spectroscopy. Consider, for example, the UV absorption spectrum of CO_2, shown in figure A1.6.11, The spectrum is seen to have a long progression of vibrational features, each with fairly uniform shape and width. What is the physical interpretation of this vibrational progression and what is the origin of the width of the features? The goal is to come up with a dynamical model that leads to a wavepacket autocorrelation function whose Fourier transform agrees with the spectrum in figure A1.6.11. Figure A1.6.12 gives a plausible dynamical model leading to such an autocorrelation function. In (a), equipotential contours of the excited-state potential energy surface of CO_2 are shown, as a function of the two bond lengths, R_1 and R_2, or, equivalently, as a function of the symmetric and antisymmetric stretch coordinates, v and u (the latter are linear combinations of the former). Along the axis $u = 0$ the potential has a minimum; along the axis $v = 0$ (the local 'reaction path') the potential has a maximum. Thus, the potential in the region $u = 0$, $v = 0$ has a 'saddle-point'. There are two symmetrically related exit channels, for large values of R_1 and R_2, respectively, corresponding to the formation of OC + O versus O + CO. Figure A1.6.12(a) also shows the initial wavepacket, which is approximately a two-dimensional Gaussian. Its centre is displaced from the minimum in the symmetric stretch coordinate. Figures A1.6.12(b)–(f) show the subsequent dynamics of the wavepacket. It moves downhill along the v coordinate, while at the same time spreading. After one vibrational period in the v coordinate the centre of the wavepacket comes back to its starting point in v, but has spread in u (figure A1.6.12(e)). The resulting wavepacket autocorrelation function is shown in figure A1.6.12(right) (a). At $t = 0$ the autocorrelation function is 1. On a time scale τ_b the correlation function has decayed to nearly 0, reflecting the fact that the wavepacket has moved away from its initial Franck–Condon location (figure A1.6.12(b)). At time τ_e the wavepacket has come back to the Franck–Condon region in the v coordinate, and the autocorrelation function has a recurrence. However, the magnitude of the recurrence is much smaller than the initial value, since there is irreversible spreading of the wavepacket in the u coordinate. Note there are further, smaller recurrences at multiples of τ_e.

The spectrum obtained by Fourier transform of figure A1.6.12 (right) (a) is shown in figure A1.6.12 (right) (b). Qualitatively, it has all the features of the spectrum in figure A1.6.11: a broad envelope with resolved vibrational structure underneath, but with an ultimate, unresolvable linewidth. Note that the shortest time decay, δ, determines the overall envelope in frequency, $1/\delta$; the recurrence time, T, determines the vibrational frequency spacing, $2\pi/T$; the overall decay time determines the width of the vibrational features. Moreover, note that decays in time correspond to widths in frequency, while recurrences in time correspond to spacings in frequency.

Perhaps the more conventional approach to electronic absorption spectroscopy is cast in the energy, rather than in the time domain. It is straightforward to show that equation (A1.6.87) can be rewritten as

$$\sigma(\omega) = \frac{4\pi^2\omega}{3c\hbar} \sum_n |\langle\psi_n|\mu|\psi_i\rangle|^2 \delta(\bar{\omega} - \omega_n). \tag{A1.6.88}$$

Note that if we identify the sum over δ-functions with the density of states, then equation (A1.6.88) is just Fermi's Golden Rule, which we employed in section A1.6.1. This is consistent with the interpretation of the absorption spectrum as the transition rate from state i to state n.

The coefficients of the δ-function in the sum are called Franck–Condon factors, and reflect the overlap of the initial state with the excited-state ψ_n at energy $E_n = \hbar\omega_n$ (see figure A1.6.13). Formally, equation (A1.6.88) gives a 'stick' spectrum of the type shown in figure A1.6.13(b); generally, however, the experimental absorption spectrum is diffuse, as in figure A1.6.11. This highlights one of the advantages of the time domain approach: that the broadening of the stick spectrum need not be introduced artificially, but arises naturally from the decay of the wavepacket correlation function, as we have seen in figure A1.6.11.

The above formulae for the absorption spectrum can be applied, with minor modifications, to other one-photon spectroscopies, for example, emission spectroscopy, photoionization spectroscopy and photodetachment spectroscopy (photoionization of a negative ion). For stimulated emission spectroscopy, the factor of ω_I is simply replaced by ω_S, the stimulated light frequency; however, for spontaneous emission spectroscopy, the prefactor ω_I is replaced by the prefactor ω_S^3. The extra factor of ω_S^2 is due to the density of states of vacuum field states which induce the spontaneous emission, which increase quadratically with frequency. Note that in emission spectroscopy the roles of the ground- and excited-state potential energy surfaces are reversed: the initial wavepacket starts from the vibrational ground state of the *excited* electronic state and its spectrum has information on the vibrational eigenstates and potential energy surface of the *ground* electronic state.

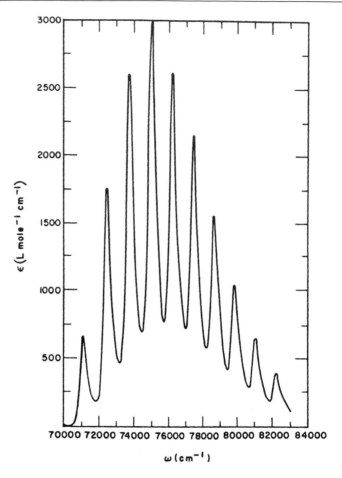

Figure A1.6.11. Idealized UV absorption spectrum of CO_2. Note the regular progression of intermediate resolution vibrational progression. In the frequency regime this structure is interpreted as a Franck–Condon progression in the symmetric stretch, with broadening of each of the lines due to predissociation. Reprinted from [31].

(b) Resonance Raman spectroscopy

We will now look at two-photon processes. We will concentrate on Raman scattering although two-photon absorption can be handled using the same approach. In Raman scattering, absorption of an incident photon of frequency ω_I carries the initial wavefunction, ψ_i, from the lower potential to the upper. The emission of a photon of frequency ω_S returns the system to the lower potential, to state ψ_f. If $\omega_S = \omega_I$ then the scattering is elastic and the process is called Rayleigh scattering. Raman scattering occurs when $\omega_S \neq \omega_I$ and in that case $\psi_f \neq \psi_i$. The measured quantity is the Raman intensity, $I(\omega_I; \omega_S)$. The amplitudes of the incident and emitted fields are taken as E_I and E_S; for simplicity, we begin with the case of stimulated Raman scattering, and then discuss the modifications for spontaneous Raman scattering at the end.

We start from the expression for the second-order wavefunction:

$$|\psi^{(2)}(t)\rangle = -\frac{1}{\hbar^2} \int_{-\infty}^{t} dt' \int_{-\infty}^{t'} dt'' \, e^{-(i/\hbar)H_a(t-t')} E_S \mu e^{+i\omega_S t'} \} e^{-(i/\hbar)H_b(t'-t'')} e^{-\frac{\gamma}{2}(t'-t'')} E_I \mu e^{-i\bar{\omega}_I t''} \}$$
$$\times \, |\psi_i(-\infty)\rangle + NRT$$

(A1.6.89)

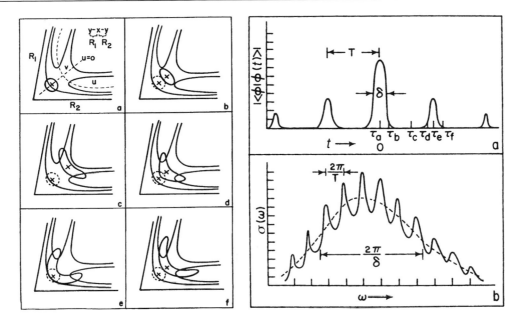

Figure A1.6.12. Left: A qualitative diagram showing evolution of $\phi(t)$ on the upper potential surface. Note the oscillation along the v (symmetric stretch) coordinate, and the spreading along the u (antisymmetric stretch) coordinate. Reprinted from [32]. Right: (a) The absolute value of the correlation function, $|\langle\phi|\phi(t)\rangle|$ versus t for the dynamical situation shown in figure A1.6.12. (b) The Fourier transform of $\langle\phi|\phi(t)\rangle$, giving the absorption spectrum. Note that the central lobe in the correlation function, with decay constant δ, gives rise to the overall width of the absorption spectrum, on the order of $2\pi/\delta$. Furthermore, the recurrences in the correlation on the time scale T give rise to the oscillations in the spectrum on the time scale $2\pi/T$. Reprinted from [32].

where H_a (H_b) is the Hamiltonian for the lower (upper) potential energy surface and, as before, $\tilde{\omega}_I = \omega_I + \omega_i$. In words, equation (A1.6.89) is saying that the second-order wavefunction is obtained by propagating the initial wavefunction on the ground-state surface until time t'', at which time it is excited up to the excited state, upon which it evolves until it is returned to the ground state at time t', where it propagates until time t. *NRT* stands for non-resonant term: it is obtained by $E_I \leftrightarrow E_S$ and $\omega_I \leftrightarrow -\omega_S$, and its physical interpretation is the physically counterintuitive possibility that the emitted photon precedes the incident photon. γ is the spontaneous emission rate.

If we define $\tilde{\omega}_S = \omega_S - \tilde{\omega}_I$, then we can follow the same approach as in the one-photon case. We now take the time derivative of the norm of $|\psi^{(2)}(t)\rangle$, with the result:

$$\omega_I\omega_S\frac{d}{dt}\langle\psi^{(2)}(t)|\psi^{(2)}(t)\rangle/|E|^2 = \omega_I\omega_S\frac{1}{\hbar^4}\sum_f|\alpha_{fi}(\omega_I)|^2\delta(\omega_f + \omega_S - (\omega_I + \omega_i)). \tag{A1.6.90}$$

where

$$\alpha_{fi}(\omega_I) = \int_0^\infty \langle\psi_f|\mu e^{-(i/\hbar)H_b t}\mu|\psi_i\rangle e^{-\frac{\gamma}{2}t}e^{i\tilde{\omega}_I t}dt + NRT. \tag{A1.6.91}$$

Again, NRT is obtained from the first term by the replacement $\omega_I \to -\omega_S$. If we define $|\phi_f\rangle = \mu|\psi_f\rangle$ and $|\phi_i(t)\rangle = e^{-(i/\hbar)H_b t}\mu|\psi_i\rangle$, then we see that the frequency-dependent polarizability, $\alpha_{fi}(\omega_I)$, can be written in the following compact form [14]:

$$\alpha_{fi}(\omega_I) = \int_0^\infty \langle\phi_f|\phi_i(t)\rangle e^{-\frac{\gamma}{2}t}e^{i\tilde{\omega}_I t}\,dt + NRT. \tag{A1.6.92}$$

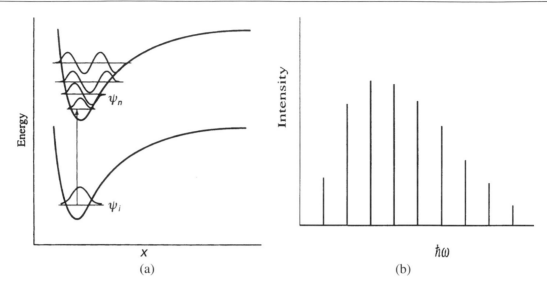

Figure A1.6.13. (a) Potential energy curves for two electronic states. The vibrational wavefunctions of the excited electronic state and for the lowest level of the ground electronic state are shown superimposed. (b) Stick spectrum representing the Franck–Condon factors (the square of overlap integral) between the vibrational wavefunction of the ground electronic state and the vibrational wavefunctions of the excited electronic state (adapted from [3]).

The only modification of equation (A1.6.90) for spontaneous Raman scattering is the multiplication by the density of states of the cavity, equation (A1.6.24), leading to a prefactor of the form $\omega_I \omega_S^3$.

Equation (A1.6.92) has a simple physical interpretation. At $t = 0$ the initial state, ψ_i is multiplied by μ (which may be thought of as approximately constant in many cases, the Condon approximation). This product, denoted ϕ_i, constitutes an initial wavepacket which begins to propagate on the excited-state potential energy surface (figure A1.6.14). Initially, the wavepacket will have overlap only with ψ_i, and will be orthogonal to all other ψ_f on the ground-state surface. As the wavepacket begins to move on the excited state, however, it will develop overlap with ground vibrational states of ever-increasing quantum number. Eventually, the wavepacket will reach a turning point and begin moving back towards the Franck–Condon region of the excited-state surface, now overlapping ground vibrational states in decreasing order of their quantum number. These time-dependent overlaps determine the Raman intensities via equations (A1.6.92) and (A1.6.90). If the excited state is dissociative, then the wavepacket never returns to the Franck–Condon region and the Raman spectrum has a monotonically decaying envelope. If the wavepacket bifurcates on the excited state due to a bistable potential, then it will only have non-zero overlaps with ground vibrational states which are of even parity; the Raman spectrum will then have 'missing' lines. In multidimensional systems, there are ground vibrational states corresponding to each mode of vibration. The Raman intensities then contain information about the extent to which different coordinates participate in the wavepacket motion, to what extent, and even in what sequence [15, 33]. Clearly, resonance Raman intensities can be a sensitive probe of wavepacket dynamics on the excited-state potential.

One of the most interesting features of the Raman spectrum is its dependence on the incident light frequency, ω_I. When ω_I is on resonance with the excited electronic state, the scattering process closely resembles a process of absorption followed by emission. However, as ω_I is detuned from resonance there are no longer any nearby eigenstates, and thus no absorption: the transition from the initial state i to the final state f is a 'scattering' process. In the older literature the non-existent intermediate state was called a 'virtual' state.

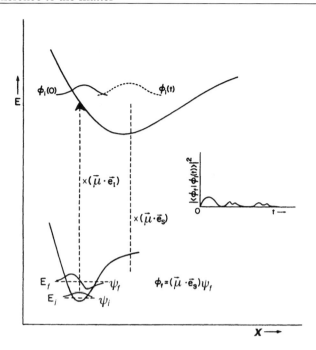

Figure A1.6.14. Schematic diagram showing the promotion of the initial wavepacket to the excited electronic state, followed by free evolution. Cross-correlation functions with the excited vibrational states of the ground-state surface (shown in the inset) determine the resonance Raman amplitude to those final states (adapted from [14]).

There can be no completely rigorous separation between absorption–emission and Raman scattering. This is clear from the time-domain expression, equation (A1.6.92), in which the physical meaning of the variable t is the time *interval* between incident and emitted photons. If the second photon is emitted long after the first photon was incident the process is called absorption/emission. If the second photon is emitted almost immediately after the first photon is incident the process is called scattering. The limits on the integral in (A1.6.92) imply that the Raman amplitude has contributions from all values of this interval ranging from 0 (scattering) to ∞ (absorption/emission). However, the regions that contribute most depend on the incident light frequency. In particular, as the incident frequency is detuned from resonance there can be no absorption and the transition becomes dominated by scattering. This implies that as the detuning is increased, the relative contribution to the integral from small values of t is greater.

Mathematically, the above observation suggests a time–energy uncertainty principle [15]. If the incident frequency is detuned by an amount $\Delta\omega$ from resonance with the excited electronic state, the wavepacket can 'live' on the excited state only for a time $\tau \approx 1/\Delta\omega$ (see figure A1.6.15). This follows from inspection of the integral in equation (A1.6.92): if the incident light frequency is mismatched from the intrinsic frequencies of the evolution operator, there will be a rapidly oscillating phase to the integrand. Normally, such a rapidly oscillating phase would kill the integral completely, but there is a special effect that comes into play here, since the lower bound of the integral is 0 and not $-\infty$. The absence of contributions from negative t leads to an incomplete cancellation of the portions of the integral around $t = 0$. The size of the region around $t = 0$ is inversely proportional to the mismatch in frequencies, $\Delta\omega$. Since the physical significance of t is time delay between incident and scattered photons, and this time delay is the effective wavepacket lifetime in the excited state, we are led to conclude that the effective lifetime decreases as the incident frequency is detuned from resonance.

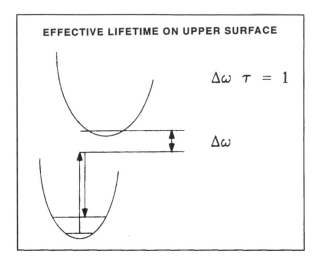

Figure A1.6.15. Schematic diagram, showing the time–energy uncertainty principle operative in resonance Raman scattering. If the incident light is detuned from resonance by an amount $\Delta\omega$, the effective lifetime on the excited-state is $\tau \approx 1/\Delta\omega$ (adapted from [15]).

Because of the two frequencies, ω_I and ω_S, that enter into the Raman spectrum, Raman spectroscopy may be thought of as a 'two-dimensional' form of spectroscopy. Normally, one fixes ω_I and looks at the intensity as a function of ω_S; however, one may vary ω_I and probe the intensity as a function of $\omega_I - \omega_S$. This is called a Raman excitation profile.

The more conventional, energy domain formula for resonance Raman scattering is the expression by Kramers–Heisenberg–Dirac (KHD). The differential cross section for Raman scattering into a solid angle $d\Omega$ can be written in the form

$$\frac{d\sigma_{fi}(\omega_I)}{d\Omega} = \frac{\omega_I \omega_S^3}{c^4} \langle |(\alpha_{fi})_{\rho\lambda}(\omega_I)|^2 \rangle \tag{A1.6.93}$$

where

$$(\alpha_{fi})_{\rho\lambda}(\omega_i) = \sum_j \frac{\langle \psi_f | e_S \cdot \mu_\rho | \psi_j \rangle \langle \psi_j | e_I \cdot \mu_\lambda | \psi_i \rangle}{E_i + \hbar\omega_I - E_j - i\gamma/2} + \frac{\langle \psi_f | e_S \cdot \mu_\rho | \psi_j \rangle \langle \psi_j | e_I \cdot \mu_\lambda | \psi_i \rangle}{E_i - \hbar\omega_S - E_j - i\gamma/2} \tag{A1.6.94}$$

and the angular brackets indicate orientational averaging. The labels $e_{I,S}$ refer to the direction of polarization of the incident and scattered light, respectively, while the subscripts ρ and λ refer to x, y and z components of the vector $\bar{\mu}$. Integrated over all directions and polarizations one obtains [33, 34]:

$$\sigma_{fi}(\omega_I) = \frac{8\pi \omega_I \omega_S^3}{9c^4} \sum_{\rho\lambda} |\alpha_{\rho\lambda}|^2. \tag{A1.6.95}$$

Equation (A1.6.94) is called the KHD expression for the polarizability, α. Inspection of the denominators indicates that the first term is the resonant term and the second term is the non-resonant term. Note the product of Franck–Condon factors in the numerator: one corresponding to the amplitude for excitation and the other to the amplitude for emission. The KHD formula is sometimes called the 'sum-over-states' formula, since formally it requires a sum over all intermediate states j, each intermediate state participating according to how far it is from resonance and the size of the matrix elements that connect it to the states ψ_i and ψ_f. The KHD formula is fully equivalent to the time domain formula, equation (A1.6.92), and can be derived from the latter in a straightforward way. However, the time domain formula can be much more convenient, particularly as one detunes from resonance, since one can exploit the fact that the effective dynamic becomes shorter and shorter as the detuning is increased.

A1.6.4 Coherent nonlinear spectroscopy

As described at the end of section A1.6.1, in nonlinear spectroscopy a polarization is created in the material which depends in a nonlinear way on the strength of the electric field. As we shall now see, the microscopic description of this nonlinear polarization involves multiple interactions of the material with the electric field. The multiple interactions in principle contain information on both the ground electronic state and excited electronic state dynamics, and for a molecule in the presence of solvent, information on the molecule–solvent interactions. Excellent general introductions to nonlinear spectroscopy may be found in [35–37]. Raman spectroscopy, described at the end of the previous section, is also a nonlinear spectroscopy, in the sense that it involves more than one interaction of light with the material, but it is a pathological example since the second interaction is through spontaneous emission and therefore not proportional to a driving field and not directional; at the end of this section we will connect the present formulation with Raman spectroscopy [38].

What information is contained in nonlinear spectroscopy? For gas-phase experiments, that is, experiments in which the state of the system undergoes little or no dissipation, the goal of nonlinear spectroscopy is generally as in linear spectroscopy, that is, revealing the quantum energy level structure of the molecule, both in the ground and the excited electronic state(s). For example, two-photon spectroscopy allows transitions that are forbidden due to symmetry with one photon; thus the two-photon spectrum allows the spectroscopic study of many systems that are otherwise dark. Moreover, nonlinear spectroscopy allows one to access highly excited vibrational levels that cannot be accessed by ordinary spectroscopies, as in the example of time-dependent CARS spectroscopy below. Moreover, nonlinear spectroscopy has emerged as a powerful probe of molecules in anisotropic environments, for example, molecules at interfaces, where there is a $P^{(2)}$ signal which is absent for molecules in an isotropic environment.

A feature of nonlinear spectroscopy which is perhaps unique is the ability to probe not only energy levels and their populations, but to probe directly coherences, be they electronic or vibrational, via specially designed pulse sequences. For an isolated molecule this is generally uninteresting, since in the absence of relaxation the coherences are completely determined by the populations; however, for a molecule in solution the decay of the coherence is an indicator of molecule–solvent interactions. One normally distinguishes two sources of decay of the coherence: inhomogeneous decay, which represents static differences in the environment of different molecules; and homogeneous decay, which represents the dynamics interaction with the surroundings and is the same for all molecules. Both these sources of decay contribute to the *linewidth* of spectral lines; in many cases the inhomogenous decay is faster than the homogeneous decay, masking the latter. In echo spectroscopies, which are related to a particular subset of diagrams in $P^{(3)}$, one can at least partially discriminate between homogeneous and inhomogeneous decay.

From the experimental point of view, nonlinear spectroscopy has the attractive feature of giving a directional signal (in a direction other than that of any of the incident beams), and hence a background free signal (figure A1.6.16). A significant amount of attention is given in the literature on nonlinear spectroscopy to the directionality of the signals that are emitted in different directions, and their dynamical interpretation. As we shall see, many dynamical pathways can contribute to the signal in each direction, and the dynamical interpretation of the signal depends on sorting out these contributions or designing an experiment which selects for just a single dynamical pathway.

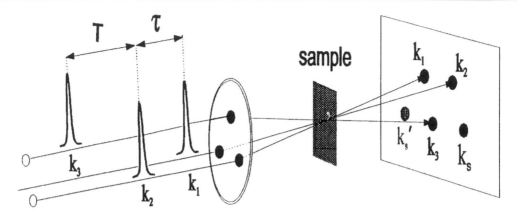

Figure A1.6.16. Diagram showing the directionality of the signal in coherent spectroscopy. Associated with the carrier frequency of each interaction with the light is a wavevector, k. The output signal in coherent spectroscopies is determined from the direction of each of the input signals via momentum conservation (after [48a]).

A1.6.4.1 General development

As discussed in section A1.6.1, on a microscopic quantum mechanical level, within the dipole approximation, the polarization, $P(t)$, is given by

$$P(t) \equiv \langle \psi | \mu | \psi \rangle. \tag{A1.6.96}$$

Assuming that the system has no permanent dipole moment, the existence of $P(t)$ depends on a non-stationary ψ induced by an external electric field. For weak fields, we may expand the polarization in orders of the perturbation,

$$P(t) \equiv \langle \psi | \mu | \psi \rangle = P^{(0)}(t) + P^{(1)}(t) + P^{(2)}(t) + P^{(3)}(t) + \cdots. \tag{A1.6.97}$$

We can then identify each term in the expansion with one or more terms in the perturbative expansion of

$$P^{(0)}(t) \equiv \langle \psi^{(0)}(t) | \mu | \psi^{(0)}(t) \rangle \tag{A1.6.98}$$

$$P^{(1)}(t) \equiv \langle \psi^{(0)}(t) | \mu | \psi^{(1)}(t) \rangle + \text{cc} \tag{A1.6.99}$$

$$P^{(2)}(t) \equiv \langle \psi^{(0)}(t) | \mu | \psi^{(2)}(t) \rangle + \text{cc} + \langle \psi^{(1)}(t) | \mu | \psi^{(1)}(t) \rangle \tag{A1.6.100}$$

and

$$P^{(3)}(t) \equiv \langle \psi^{(0)}(t) | \mu | \psi^{(3)}(t) \rangle + \text{cc} + \langle \psi^{(1)}(t) | \mu | \psi^{(2)}(t) \rangle + \text{cc} \tag{A1.6.101}$$

etc. Note that for an isotropic medium, terms of the form $P^{(2n)}(t)$ ($P^{(0)}(t)$, $P^{(2)}(t)$, etc) do not survive orientational averaging. For example, the first term, $\langle \psi^{(0)} | \mu | \psi^{(0)} \rangle$, is the permanent dipole moment, which gives zero when averaged over an isotropic medium. At an interface, however (e.g. between air and water), these even orders of $P(t)$ do not vanish, and in fact are sensitive probes of interface structure and dynamics.

The central dynamical object that enters into the polarization are the coherences of the form $\langle \psi^{(0)}(t) | \mu | \psi^{(1)}(t) \rangle$ and $\langle \psi^{(1)}(t) | \mu | \psi^{(2)}(t) \rangle$, etc. These quantities are overlaps between wavepackets moving on different potential energy surfaces [40–42, 52]: the instantaneous overlap of the wavepackets creates a non-vanishing transition dipole moment which interacts with the light. This view is appropriate both in the regime of weak fields, where perturbation theory is valid, and for strong fields, where perturbation theory is no longer valid. Note that in the previous sections we saw that the absorption and Raman spectra were related to $\frac{\mathrm{d}}{\mathrm{d}t} \langle \psi^{(1)}(t) | \psi^{(1)}(t) \rangle$ and $\frac{\mathrm{d}}{\mathrm{d}t} \langle \psi^{(2)}(t) | \psi^{(2)}(t) \rangle$. The coherences that appear in equations (A1.6.99) and (A1.6.101) are precisely equivalent to these derivatives: the rate of change of a population is proportional to

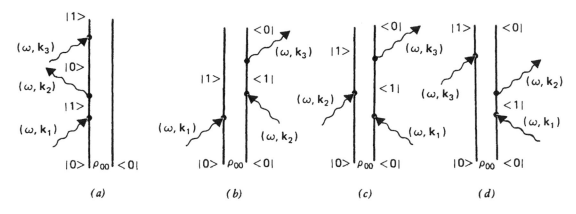

Figure A1.6.17. Double-sided Feynman diagrams, showing the interaction time with the ket (left) and the bra (right). Time moves forward from down to up (adapted from [36]).

the instantaneous coherence, a relationship which can be observed already in the vector precession model of the two-level system (section A1.6.2.3).

The coherences can be written compactly using the language of density matrices. The total polarization is given by

$$P = \text{Tr}(\rho\mu) = P^{(0)}(t) + P^{(1)}(t) + P^{(2)}(t) + P^{(3)}(t) + \cdots. \tag{A1.6.102}$$

where the different terms in the perturbative expansion of P are accordingly as follows:

$$P^{(1)} = \text{Tr}(\rho^{(1)}\mu) \qquad P^{(2)} = \text{Tr}(\rho^{(2)}\mu) \qquad P^{(3)} = \text{Tr}(\rho^{(3)}\mu)\,\text{etc.} \tag{A1.6.103}$$

In the absence of dissipation and pure state initial conditions, equations (A1.6.102) and (A1.6.103) are equivalent to equations (A1.6.97)–(A1.6.101). But equations (A1.6.102) and (A1.6.103) are more general, allowing for the possibility of dissipation, and hence for describing nonlinear spectroscopy in the presence of an environment. There is an important caveat however. In the presence of an environment, it is customary to define a reduced density matrix which describes the system, in which the environment degrees of freedom have been traced out. The tracing out of the environment should be performed only at the end, after all the interactions of the system environment with the field, otherwise important parts of the nonlinear spectrum (e.g. phonon sidebands) will be missing. The tracing of the environment at the end can be done analytically if the system is a two-level system and the environment is harmonic, the so-called spin-boson or Brownian oscillator model. However, in general the dynamics in the full system-environment degrees of freedom must be calculated, which essentially entails a return to a wavefunction description, equations (A1.6.97)–(A1.6.101), but in a larger space.

The total of three interactions of the material with the field can be distributed in several different ways between the ket and the bra (or more generally, between left and right interactions of the field with the density operator). For example, the first term in equation (A1.6.101) corresponds to all three interactions being with the ket, while the second term corresponds to two interactions with the ket and one with the bra. The second term can be further subdivided into three possibilities: that the single interaction with the bra is before, between or after the two interactions with the ket (or correspondingly, left/right interactions of the field with the density operator) [37]. These different contributions to $P^{(3)}$ (or, equivalently, to $\rho^{(3)}$) are represented conveniently using double-sided Feynman diagrams, a generalization of the single-sided Feynman diagrams introduced in section A1.6.3, as shown in figures A1.6.17(a)–(d).

The subdivision of the second term into three possibilities has an interesting physical interpretation. The ordering of the interactions determines whether diagonal vs off-diagonal elements of the density matrix are

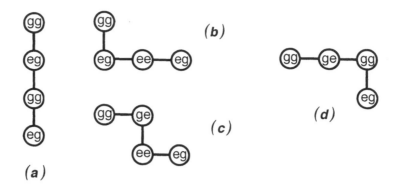

Figure A1.6.18. Liouville space lattice representation in one-to-one correspondence with the diagrams in figure A1.6.17. Interactions of the density matrix with the field from the left (right) is signified by a vertical (horizontal) step. The advantage to the Liouville lattice representation is that populations are clearly identified as diagonal lattice points, while coherences are off-diagonal points. This allows immediate identification of the processes subject to population decay processes (adapted from [37]).

produced: populations versus coherences. In the presence of relaxation processes (dephasing and population relaxation) the order of the interactions and the duration between them determines the duration for which population versus coherence relaxation mechanisms are in effect. This can be shown schematically using a Liouville space diagram, figure A1.6.18 [37]. The different pathways in Liouville space are drawn on a lattice, where ket interactions are horizontal steps and bra interactions are vertical. The diagonal vertices represent populations and the off-diagonal vertices are coherences. The three different time orderings for contributions to $|\psi^{(2)}\rangle\langle\psi^{(1)}|$ correspond to the three Liouville pathways shown in figure A1.6.18. From such a diagram one sees at a glance which pathways pass through intermediate populations (i.e. diagonal vertices) and hence are governed by population decay processes, and which pathways do not.

As a first application of the lattice representation of Liouville pathways, it is interesting to re-examine the process of electromagnetic spontaneous light emission, discussed in the previous section. Note that formally, diagrams A1.6.18(*b*)–(*d*) all contribute to the Kramers–Heisenberg–Dirac formula for resonance Raman scattering. However, diagrams (*b*) and (*c*) produce an excited electronic state population (both the bra and ket are excited in the first two interactions) and hence are subject to excited-state vibrational population relaxation processes, while diagram (*d*) does not. Typically, in the condensed phase, the fluorescence spectrum consists of sharp lines against a broad background. Qualitatively speaking, the sharp lines are associated with diagram (*d*), and are called the resonance Raman spectrum, while the broad background is associated with diagrams (*a*) and (*b*), and is called the resonance fluorescence spectrum [38]. Indeed, the emission frequency of the sharp lines changes with the excitation frequency, indicating no excited electronic state population relaxation, while the broad background is independent of excitation frequency, indicating vibrationally relaxed fluorescence.

There is an aspect of nonlinear spectroscopy which we have so far neglected, namely the spatial dependence of the signal. In general, three incident beams, described by k-vectors k_1, k_2 and k_3 will produce an outgoing beam at each of the directions:

$$k_{\text{out}} = \pm k_1 \pm k_2 \pm k_3. \tag{A1.6.104}$$

Figure A1.6.19 shows eight out of the 48 Feynman diagrams that contribute to an outgoing k-vector at $-k_1 + k_2 + k_3$. The spatial dependence is represented by the wavevector k on each of the arrows in figure A1.6.19. Absorption (emission) by the ket corresponds to a plus (minus) sign of k; absorption (emission) by the

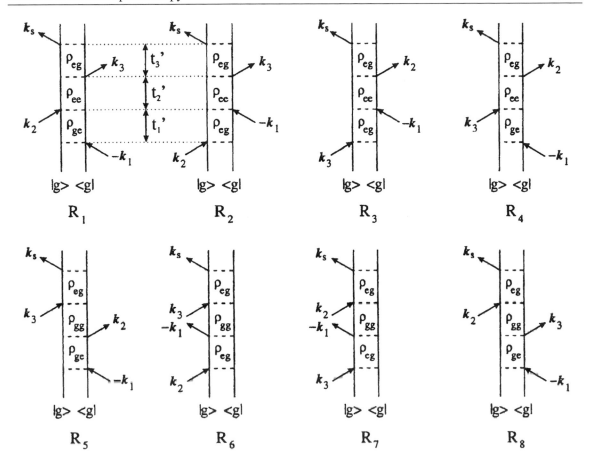

Figure A1.6.19. Eight double-sided Feynman diagrams corresponding to electronic resonance and emission at $-k_1 + k_2 + k_3$. Absorption is shown by an incoming arrow, while emission is indicated by an outgoing arrow. Note that if an arrow is moved from (to) the ket to (from) the bra while changing from (to) absorption to (from) emission, the slope of the arrow and therefore its k-vector will be unchanged. The eight diagrams are for arbitrary time ordering of the interactions; with a fixed time ordering of the interactions, as in the case of non-overlapping pulses, only two of the diagrams survive (adapted from [48]).

bra corresponds to a minus (plus) sign of k. The eight diagrams shown dominate under conditions of electronic resonance; the other 40 diagrams correspond to non-resonant contributions, involving emission before absorption. The reason there are eight resonant diagrams now, instead of the four in figure A1.6.17, is a result of the fact that the introduction of the k-dependence makes the order of the interactions distinguishable. At the same time, the k-dependence of the detection eliminates many additional processes that might otherwise contribute; for example, detection at $-k_1 + k_2 + k_3$ eliminates processes in which k_1 and k_2 are interchanged, as well as processes representing two or more interactions with a single beam. Under conditions in which the interactions have a well-defined temporal sequence, just two diagrams dominate, while two of the diagrams in figure A1.6.17 are eliminated since they emit to $k_1 - k_2 + k_3$. Below we will see that in resonant CARS, where in addition to the electronic resonance there is a vibrational resonance after the second interaction, there is only a single resonant diagram. All else being equal, the existence of multiple diagrams complicates the interpretation of the signal, and experiments that can isolate the contribution of individual diagrams have a better chance for complete interpretation and should be applauded.

A1.6.4.2   Linear response

We now proceed to the spectrum, or frequency-dependent response [41, 42]. The power, or rate of energy absorption, is given by

$$P = \frac{\mathrm{d}E}{\mathrm{d}t} = \frac{\mathrm{d}\langle\psi|H|\psi\rangle}{\mathrm{d}t} = -2\mathrm{Re}\{\langle\psi_a|\mu|\psi_b\rangle\dot{E}^*(t)\}. \tag{A1.6.105}$$

(In the second step we have used equation (A1.6.72) and noted that the terms involving $\partial\psi/\partial t$ cancel.) To lowest order, this gives

$$P = -2\mathrm{Re}\{P_{01}^{(1)}\dot{E}^*\}. \tag{A1.6.106}$$

The total energy absorbed, ΔE, is the integral of the power over time. Keeping just the lowest order terms we find

$$\Delta E = \int_{-\infty}^{\infty} P\,\mathrm{d}t = -2\mathrm{Re}\int_{-\infty}^{\infty} P_{01}^{(1)}(t)\dot{E}^*(t)\,\mathrm{d}t = 2\mathrm{Im}\int_{-\infty}^{\infty} \omega\tilde{P}_{01}^{(1)}(\omega)\tilde{E}^*(\omega)\,\mathrm{d}\omega \tag{A1.6.107}$$

where

$$\tilde{P}_{01}^{(1)}(\omega) \equiv \int_{-\infty}^{\infty} P_{01}^{(1)}(t)\mathrm{e}^{\mathrm{i}\omega t}\,\mathrm{d}t \tag{A1.6.108}$$

and

$$\tilde{E}(\omega) \equiv \int_{-\infty}^{\infty} E(t)\mathrm{e}^{\mathrm{i}\omega t}\,\mathrm{d}t. \tag{A1.6.109}$$

The last relation in equation (A1.6.107) follows from the Fourier convolution theorem and the property of the Fourier transform of a derivative; we have also assumed that $E(\omega) = E(-\omega)$. The absorption spectrum is defined as the total energy absorbed at frequency ω, normalized by the energy of the incident field at that frequency. Identifying the integrand on the right-hand side of equation (A1.6.107) with the total energy absorbed at frequency ω, we have

$$\sigma(\omega) = \frac{|\tilde{E}'(\omega)|^2}{|\tilde{E}(\omega)|^2} = \frac{4\pi\omega}{3c\hbar}\frac{\mathrm{Im}(\tilde{P}_{01}^{(1)}(\omega)\tilde{E}^*(\omega))}{|\tilde{E}(\omega)|^2}. \tag{A1.6.110}$$

Note the presence of the ω prefactor in the absorption spectrum, as in equation (A1.6.87); again its origin is essentially the faster rate of the change of the phase of higher frequency light, which in turn is related to a higher rate of energy absorption. The equivalence between the other factors in equation (A1.6.110) and equation (A1.6.87) under linear response will now be established.

    In the perturbative regime one may decompose these coherences into the contribution from the field and a part which is intrinsic to the matter, the *response* function. For example, note that the expression $P_{01}^{(1)}(t) = \langle\psi^{(0)}(t)|\mu|\psi^{(1)}(t)\rangle$ is not simply an intrinsic function of the molecule: it depends on the functional form of the field, since $\psi^{(1)}(t)$ does. However, since the dependence on the field is linear it is possible to write $P_{01}^{(1)}$ as a *convolution* of the field with a response function which depends on the material. Using the definition of $\psi^{(1)}$,

$$\psi^{(1)} = \frac{1}{\mathrm{i}\hbar}\int_{-\infty}^{t} \mathrm{e}^{-\mathrm{i}H_b(t-t')}\{-\mu E(t')\}\mathrm{e}^{-\mathrm{i}E_0 t'}\psi^{(0)}\,\mathrm{d}t' \tag{A1.6.111}$$

we find that

$$P_{01}^{(1)}(t) = \frac{1}{\mathrm{i}\hbar}\int_{-\infty}^{t} \langle\psi^{(0)}(t)|\mu\mathrm{e}^{-\mathrm{i}H_b(t-t')}\{-\mu E(t')\}\mathrm{e}^{-\mathrm{i}E_0 t'}|\psi^{(0)}\rangle\,\mathrm{d}t' \tag{A1.6.112}$$

$$= \frac{\mathrm{i}}{\hbar}\int_{0}^{\infty} \langle\psi^{(0)}(t)|\mu\mathrm{e}^{-\mathrm{i}H_b\tau}\mu|\psi^{(0)}\rangle E(t-\tau)\,\mathrm{d}\tau \tag{A1.6.113}$$

$$= \frac{\mathrm{i}}{\hbar}\{E(t)\otimes S_{00}(t)\} \tag{A1.6.114}$$

where $S_{00}(t)$ is the half or *causal* form of the autocorrelation function:

$$S_{00}(t) = \begin{cases} C_{00}(t) & t > 0 \\ 0 & t < 0 \end{cases} \qquad \begin{array}{l} \text{(A1.6.115)} \\ \text{(A1.6.116)} \end{array}$$

and $\otimes$ signifies convolution. We have defined the wavepacket autocorrelation function

$$C_{00}(t) \equiv \langle \psi^{(0)} | \mu e^{-iH_b t/\hbar} \mu | \psi^{(0)} \rangle. \qquad \text{(A1.6.117)}$$

where $C_{00}(t)$ is just the wavepacket autocorrelation function we encountered in section A1.6.3.3. There we saw that the Fourier transform of $C_{00}(t)$ is proportional to the linear absorption spectrum. The same result appears here but with a different interpretation. There, the correlation function governed the rate of excited-state population change. Here, the expectation value of the dipole moment operator with the correlation function is viewed as the *response* function of the molecule.

By the Fourier convolution theorem

$$P_{01}^{(1)}(\omega) \equiv \int_{-\infty}^{\infty} P_{01}^{(1)}(t) e^{i\omega t} \, dt = \left\{ \frac{1}{i\hbar} \tilde{E}(\omega) \tilde{S}_{00}(\omega) \right\}. \qquad \text{(A1.6.118)}$$

Using the definition of the susceptibility, χ (equation (A1.6.30)) we see that

$$\chi^{(1)}(\omega) = \left\{ \frac{1}{i\hbar} \tilde{S}_{00}(\omega) \right\}. \qquad \text{(A1.6.119)}$$

Substituting $P_{01}^{(1)}(\omega)$ into equation (A1.6.110) we find that the linear absorption spectrum is given by

$$\sigma(\omega) = \frac{4\pi\omega}{3c\hbar} \text{Re}\{\tilde{S}_{00}(\omega)\} \qquad \text{(A1.6.120)}$$

$$= \frac{2\pi\omega}{3c\hbar} \int_{-\infty}^{\infty} C_{00}(t) e^{i\omega t} \, dt \qquad \text{(A1.6.121)}$$

in agreement with equation (A1.6.87). We also find that

$$\sigma(\omega) = \frac{4\pi\omega}{3c\hbar} \text{Im}\{\chi^{(1)}(\omega)\} \qquad \text{(A1.6.122)}$$

establishing the result in section A1.6.1.4 that the absorption spectrum is related to the imaginary part of the susceptibility χ at frequency ω.

A1.6.4.3   *Nonlinear response: isolated systems*

As discussed above, the nonlinear material response, $P^{(3)}(t)$ is the most commonly encountered nonlinear term since $P^{(2)}$ vanishes in an isotropic medium. Because of the special importance of $P^{(3)}$ we will discuss it in some detail. We will now focus on a few examples of $P^{(3)}$ spectroscopy where just one or two of the 48 double-sided Feynman diagrams are important, and will stress the dynamical interpretation of the signal. A pictorial interpretation of all the different resonant diagrams in terms of wavepacket dynamics is given in [41].

Coherent anti-Stokes Raman spectroscopy (CARS)

Our first example of a $P^{(3)}$ signal is coherent anti-Stokes Raman spectroscopy, or CARS. Formally, the emission signal into direction $k = k_1 - k_2 + k_3$ has 48 Feynman diagrams that contribute. However, if the frequency ω_1 is resonant with the electronic transition from the ground to the excited electronic state, and the mismatch between frequencies ω_1 and ω_2 is resonant with a ground-state vibrational transition or transitions, only one diagram is resonant, namely, the one corresponding to R_6 in figure A1.6.19 (with the interchange of labels k_1 and k_2).

To arrive at a dynamical interpretation of this diagram it is instructive to write the formula for the dominant term in $P^{(3)}$ explicitly:

$$P^{(3)}(t) = \langle \psi^{(0)}(t)|\mu|\psi^{(3)}(t)\rangle \tag{A1.6.123}$$

$$= \frac{(-)^3}{(i\hbar)^3} \int_{-\infty}^{t_4} dt_3 \int_{-\infty}^{t_3} dt_2 \int_{-\infty}^{t_2} dt_1 \langle \psi^{(0)}(t')|\{\mu\}e^{-iH_b(t-t_3)/\hbar}\{\mu E_3(t_3)\}e^{-iH_a(t_3-t_2)/\hbar}$$
$$\times \{\mu E_2(t_2)\}e^{-iH_b(t_2-t_1)/\hbar}\{\mu E_1(t_1)\}e^{-iH_a t_1}|\psi^{(0)}\rangle \tag{A1.6.124}$$

where in the second line we have substituted explicitly for the third-order wavefunction, $\psi^{(3)}(t)$. This formula, although slightly longer than the formulae for the first- and second-order amplitude discussed in the previous section, has the same type of simple dynamical interpretation. The initial wavepacket, $\psi^{(0)}$ interacts with the field at time t_1 and propagates on surface b for time $t_2 - t_1$; at time t_2 it interacts a second time with the field and propagates on the ground surface a for time $t_3 - t_2$; at time t_3 it interacts a third time with the field and propagates on surface b until variable time t. The third-order wavepacket on surface b is projected onto the initial wavepacket on the ground state; this overlap is a measure of the coherence which determines both the magnitude and phase of the CARS signal. Formally, the expression involves an integral over three time variables, reflecting the coherent contribution of all possible instants at which the interaction with the light took place, for each of the three interactions. However, if the interaction is with pulses that are short compared with a vibrational period, as we saw in equation (A1.6.76), one can approximate the pulses by δ-functions in time, eliminate the three integrals and the simple dynamical interpretation above becomes precise.

Qualitatively, the delay between interaction 1 and 2 is a probe of excited-state dynamics, while the delay between interaction 2 and 3 reflects ground-state dynamics. If pulses 1 and 2 are coincident, the combination of the first two pulses prepares a vibrationally excited wavepacket on the ground-state potential energy surface; the time delay between pulses 2 and 3 then determines the time interval for which the wavepacket evolves on the ground-state potential, and is thus a probe of ground-state dynamics [43, 45, 52]. If a second delay, the delay between pulses 1 and 2, is introduced this allows large wavepacket excursions on the excited state before coming back to the ground state. The delay between pulses 1 and 2 can be used in a very precise way to tune the level of ground-state vibrational excitation, and can prepare ground vibrational wavepackets with extremely high energy content [44]. The sequence of pulses involving one against two time delays is shown in figures A1.6.20(a) and (b). The control over the vibrational energy content in the ground electronic state via the delay between pulses 1 and 2 is illustrated in figure A1.6.20(right).

(b) Stimulated Raman and dynamic absorption spectroscopy

In CARS spectroscopy, $\omega_1 = \omega_3$, and ω_2 is generally different and of lower frequency. If $\omega_1 = \omega_2 = \omega_3$ the process is called degenerate four-wave mixing (DFWM). Now, instead of a single diagram dominating, two diagrams participate if the pulses are non-overlapping, four dominate if two pulses overlap and all eight resonant diagrams contribute if all three pulses overlap (e.g., in continuous wave excitation) [43, 46]. The additional diagrams correspond to terms of the form $\langle \psi^{(1)}(t)|\mu|\psi^{(2)}(t)\rangle$ discussed above; this is the overlap of a second-order wavepacket on the ground-state surface with a first-order wavepacket on the excited-state

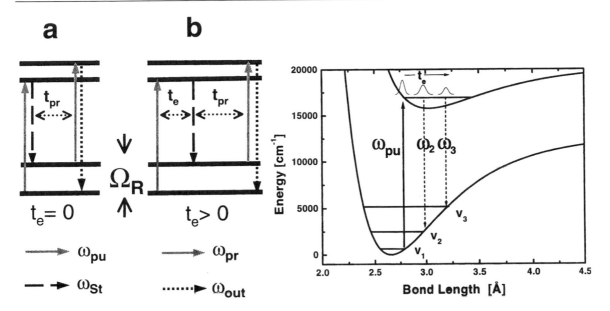

Figure A1.6.20. (Left) Level scheme and nomenclature used in (a) single time-delay CARS. (b) Two-time delay CARS ((TD)^{2}CARS). The wavepacket is excited by ω_{pu}, then transferred back to the ground state by ω_{st} with Raman shift Ω_R. Its evolution is then monitored by ω_{pr} (after [44]). (Right) Relevant potential energy surfaces for the iodine molecule. The creation of the wavepacket in the excited state is done by ω_{pu}. The transfer to the final state is shown by the dashed arrows according to the state one wants to populate (after [44]).

surface. These new diagrams come in for two reasons. First, even if the pulses are non-overlapping, the degeneracy of the first two interactions allows the second interaction to produce an absorption, not just emission. If the pulses are overlapping there is the additional flexibility of interchanging the order of pulses 1 and 2 (at the same time exchanging their role in producing absorption versus emission). The contribution of these additional diagrams to the $P^{(3)}$ signal is not simply additive, but there are interference terms among all the contributions, considerably complicating the interpretation of the signal. Diagrams R_1–R_4 are commonly referred to as stimulated Raman scattering: the first two interactions produce an excited-state population while the last interaction produces stimulated emission back to the ground electronic state.

A process which is related diagrammatically to stimulated Raman scattering is transient absorption spectroscopy. In an ordinary absorption spectrum, the initial state is typically the ground vibrational eigenstate of the ground electronic state. Dynamic absorption spectroscopy refers to the excitation of a vibrational wavepacket to an electronic state b via a first pulse, and then the measurement of the spectrum of that moving wavepacket on a third electronic state c as function of time delay between the pump and the probe. The time delay controls the instantaneous wavepacket on state b whose spectrum is being measured with the second pulse; in an ideal situation, one may obtain 'snapshots' of the wavepacket on electronic b as a function of time, by observing its shadow onto surface c. This form of spectroscopy is very similar in spirit to the pump–probe experiments of Zewail *et al* [25], described in section A1.6.3.2, but there are two differences. First, the signal in a dynamic absorption spectrum is a coherent signal in the direction of the probe pulse (pulse 3), as opposed to measuring fluorescence from state c, which is non-directional. Second, field intensity in the direction of the probe pulse can be frequency resolved to give simultaneous time and frequency resolution of the transient absorption. Although in principle the fluorescence from state c can also be frequency resolved, this fluorescence takes place over a time which is orders of magnitude longer than the vibrational dynamics

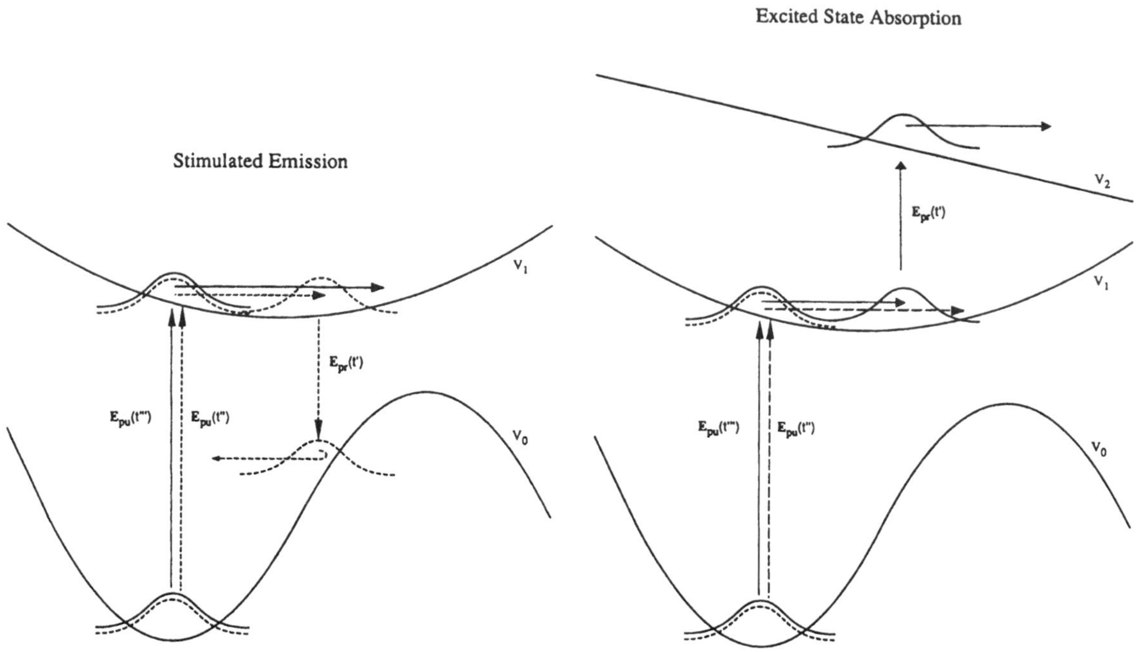

Excited State Absorption

Stimulated Emission

Figure A1.6.21. Bra and ket wavepacket dynamics which determine the coherence overlap, $\langle \phi^{(1)} | \phi^{(2)} \rangle$. Vertical arrows mark the transitions between electronic states and horizontal arrows indicate free propagation on the potential surface. Full curves are used for the ket wavepacket, while dashed curves indicate the bra wavepacket. (a) Stimulated emission. (b) Excited state (transient) absorption (from [41]).

of interest and the signal contains a complicated combination of all excited- and ground-state frequency differences.

The dynamic absorption signal, $P^{(3)}$, can be written in a form which looks analogous to the linear absorption signal $P^{(1)}$ (see equation (A1.6.113)),

$$P^{(3)}(t) = \frac{i}{\hbar} \int_{-\infty}^{\infty} \langle \psi^{(1)}(t') | \mu e^{-iH_c(t-t')/\hbar} \mu | \psi^{(1)}(t') \rangle E(t') \, dt'. \qquad (A1.6.125)$$

However, because of the t' dependence in $\psi^{(1)}(t')$ one cannot write that $P^{(3)} = E(t) \otimes S_{11}(t)$. For the latter to hold, it is necessary to go to the limit of a probe pulse which is short compared with the dynamics on surface 1. In this case, $\psi^{(1)}(t')$ is essentially frozen and we can write $\psi^{(1)} \approx \psi_\tau^{(1)}$, where we have indicated explicitly the parametric dependence on the pump–probe delay time, τ. In this case, equation (A1.6.125) is isomorphic with equation (A1.6.113), indicating that under conditions of impulsive excitation, dynamic absorption spectroscopy is just first-order spectroscopy on the frozen state, $\phi_\tau^{(1)}$, on surface c. Note the residual dependence of the frozen state on τ, the pump–probe delay, and thus variation of the variables (ω, τ) generates a two-dimensional dynamic absorption spectrum. Note that the pair of variables (ω, τ) are not limited by some form of time–energy uncertainty principle. This is because, although the absorption is finished when the probe pulse is finished, the spectral analysis of which frequency components were absorbed depends on the full time evolution of the system, beyond its interaction with the probe pulse. Thus, the dynamic absorption signal can give high resolution both in time (i.e. time delay between pump and probe pulses) and frequency, simultaneously.

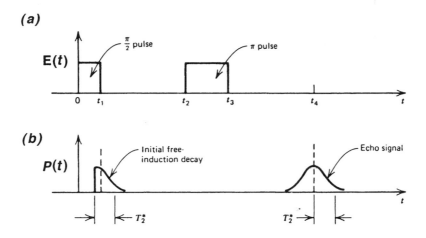

Figure A1.6.22. (*a*) Sequence of pulses in the canonical echo experiment. (*b*) Polarization versus time for the pulse sequence in (*a*), showing an echo at a time delay equal to the delay between the excitation pulses.

A1.6.4.4 Nonlinear response: systems coupled to an environment

(a) Echo spectroscopy

In discussing spectroscopy in condensed phase environments, one normally distinguishes two sources of decay of the coherence: Inhomogeneous decay, which represents static differences in the environment of different molecules, and homogeneous decay, which represents the dynamics interaction with the surroundings and is the same for all molecules. Both these sources of decay contribute to the *linewidth* of spectral lines; in many cases the inhomogenous decay is faster than the homogeneous decay, masking the latter. In echo spectroscopies, which are related to a particular subset of diagrams in $P^{(3)}$, one can at least partially discriminate between homogeneous and inhomogeneous decay.

Historically, photon echoes grew up as optical analogues of spin echoes in NMR. Thus, the earliest photon echo experiments were based on a sequence of two excitation pulses, a $\pi/2$ pulse followed by a π pulse, analogous to the pulse sequence used in NMR. Conceptually, the $\pi/2$ pulse prepares an optical coherence, which will proceed to dephase due to both homogeneous and inhomogeneous mechanisms. After a delay time τ, the π pulse reverses the role of the excited and ground electronic states, which causes the inhomogeneous contribution to the dephasing to reverse itself but does not affect the homogeneous decay. The reversal of phases generated by the π-pulse has been described in many colourful ways over the years (see figure A1.6.23).

Fundamentally, the above description of photon echoes is based on a two-level description of the system. As we have seen throughout this article, much of molecular electronic spectroscopy is described using two electronic states, albeit with a vibrational manifold in each of these electronic states. This suggests that photon echoes can be generalized to include these vibrational manifolds, provided that the echo signal is now defined in terms of a wavepacket overlap (or density matrix coherence) involving the coherent superposition of all the participating vibrational levels. This is shown schematically in figure A1.6.24. The $\pi/2$ pulse transfers 50% of the wavepacket amplitude to the excited electronic state. This creates a non-stationary vibrational wavepacket in the excited electronic state (and generally, the remaining amplitude in the ground electronic state is non-stationary as well). After a time delay τ a π pulse comes in, exchanging the wavepackets on the ground and excited electronic states. The wavepackets continue to evolve on their new respective surfaces. At some later time, when the wavepackets overlap, an echo will be observed. This sequence is shown in figure A1.6.24. Note that this description refers only to the isolated molecule; if there are dephasing

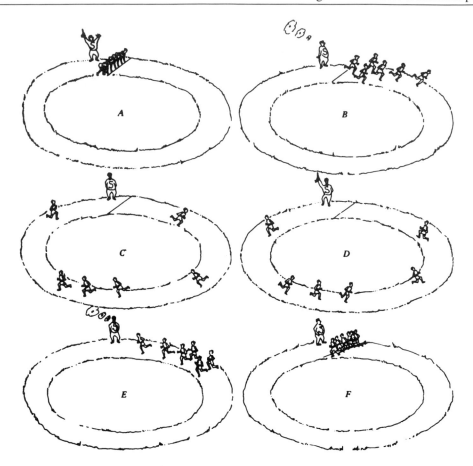

Figure A1.6.23. Schematic representation of dephasing and reversal on a race track, leading to coherent rephasing and an 'echo' of the starting configuration. From *Phys. Today*, (Nov. 1953), front cover.

mechanisms due to the environment as well, the echo requires the rephasing in both the intramolecular and the environmental degrees of freedom.

Although the early photon echo experiments were cast in terms of $\pi/2$ and π pulses, these precise inversions of the population are by no means necessary [36]. In fact echoes can be observed using sequences of weak pulses, and can be described within the perturbative $P^{(3)}$ formalism which we have used throughout section A1.6.4. Specifically, the diagrams R_1, R_4, R_5 and R_8 in figure A1.6.19 correspond to echo diagrams, while the diagrams R_2, R_3, R_6 and R_7 do not. In the widely used Brownian oscillator model for the relaxation of the system [37, 48], the central dynamical object is the electronic frequency correlation,

$$M(t) = \frac{\langle \Delta\omega(0)\Delta\omega(t)\rangle}{\langle \Delta\omega^2\rangle} \tag{A1.6.126}$$

where $\Delta\omega(t) = \langle\omega_{eg}\rangle - \omega(t)$. Here $\langle\omega_{eg}\rangle$ is the average transition frequency, $\omega(t)$ is the transition frequency at time t, and the brackets denote an ensemble average. It can be shown that as long as $M(t)$ is a monotonically decaying function, the diagrams R_1, R_4, R_5 and R_8 can cause rephasing of $P^{(3)}$ while the diagrams R_2, R_3, R_6 and R_7 cannot (see figure A1.6.25).

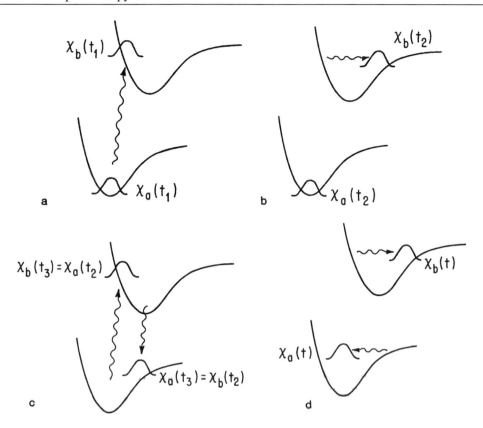

Figure A1.6.24. Schematic representation of a photon echo in an isolated, multilevel molecule. (a) The initial pulse prepares a superposition of ground- and excited-state amplitude. (b) The subsequent motion on the ground and excited electronic states. The ground-state amplitude is shown as stationary (which in general it will not be for strong pulses), while the excited-state amplitude is non-stationary. (c) The second pulse exchanges ground- and excited-state amplitude. (d) Subsequent evolution of the wavepackets on the ground and excited electronic states. When they overlap, an echo occurs (after [40]).

It is instructive to contrast echo spectroscopy with single time-delayed CARS spectroscopy, discussed above. Schematically, TD-CARS spectroscopy involves the interaction between pulses 1 and 2 being close in time, creating a ground-state coherence, and then varying the delay before interaction 3 to study ground-state dynamics. In contrast, echo spectroscopy involves an isolated interaction 1 creating an electronic coherence between the ground and the excited electronic state, followed by a pair of interactions 2 and 3, one of which operates on the bra and the other on the ket. The pair of interactions 2,3 essentially reverses the role of the ground and the excited electronic states. If there is any inhomogeneous broadening, or more generally any bath motions that are slow compared with the time intervals between the pulses, these modes will show up as echo signal after the third pulse is turned off [47].

We close with three comments. First, there is preliminary work on retrieving not only the amplitude but also the phase of photon echoes [49]. This appears to be a promising avenue to acquire complete 2-dimensional time and frequency information on the dynamics, analogous to methods that have been used in NMR. Second, we note that there is a growing literature on non-perturbative, numerical simulation of nonlinear spectroscopies. In these methods, the consistency of the order of interaction with the field and the appropriate

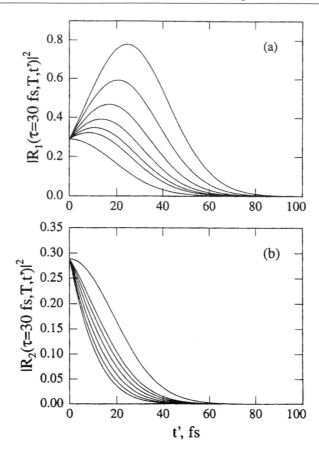

Figure A1.6.25. Modulus squared of the rephasing, $|R_1|^2$, (a), and non-rephasing, $|R_2|^2$, (b), response functions versus final time t for a near-critically overdamped Brownian oscillator model $M(t)$. The time delay between the second and third pulse, T, is varied as follows: (a) from top to bottom, $T = 0, 20, 40, 60, 80, 100, \infty$ fs; (b) from bottom to top, $T = 0, 20, 40, 60, 80, 100, \infty$ fs. Note that $|R_1|^2$ and $|R_2|^2$ are identical at $T = \infty$. After [48].

relaxation process is achieved automatically, and thus these methods may become a useful alternative to the perturbative formalism [50, 51]. Third, there is a growing field of single molecule spectroscopy. If the optical response from individual molecules in a condensed phase environment is detected, then one has a more direct approach than echo spectroscopy for removing the effect of environmental inhomogeneity. Moreover, the spectral change of individual molecules can be followed in time, giving data that are masked in even the best echo spectrum.

A1.6.5   Coherent control of molecular dynamics

Not only has there been great progress in making femtosecond pulses in recent years, but also progress has been made in the shaping of these pulses, that is, giving each component frequency any desired amplitude and phase. Given the great experimental progress in shaping and sequencing femtosecond pulses, the inexorable question is: How is it possible to take advantage of this wide possible range of coherent excitations to bring about selective and energetically efficient photochemical reactions? Many intuitive approaches to laser selective chemistry have been tried since 1980. Most of these approaches have focused on depositing energy in a sustained manner, using monochromatic radiation, into a particular state or mode of the molecule. Virtually all such schemes have failed, due to rapid intramolecular energy redistribution.

The design of pulse sequences to selectively control chemical bond breaking is naturally formulated as a problem in the calculus of variations [17, 52]. This is the mathematical apparatus for finding the best shape, subject to certain constraints. For example, the shape which encloses the maximum area for a given perimeter; the minimum distance between two points on a sphere subject to the constraint that the connecting path be on the sphere; the shape of a cable of fixed length and fixed endpoints which minimizes the potential energy; the trajectory of least time; the path of least action; all these are searches for the best shape, and are problems in the classical calculus of variations. In our case, we are searching for the best shape of laser pulse intensity against time. If we admit complex pulses this involves an optimization over the real and imaginary parts of the pulse shape. We may be interested in the optimal pulse subject to some constraints, for example for a fixed total energy in the pulse.

It turns out that there is another branch of mathematics, closely related to the calculus of variations, although historically the two fields grew up somewhat separately, known as optimal control theory (OCT). Although the boundary between these two fields is somewhat blurred, in practice one may view optimal control theory as the application of the calculus of variations to problems with differential equation constraints. OCT is used in chemical, electrical, and aeronautical engineering; where the differential equation constraints may be chemical kinetic equations, electrical circuit equations, the Navier–Stokes equations for air flow, or Newton's equations. In our case, the differential equation constraint is the TDSE in the presence of the control, which is the electric field interacting with the dipole (permanent or transition dipole moment) of the molecule [53–56]. From the point of view of control theory, this application presents many new features relative to conventional applications; perhaps most interesting mathematically is the admission of a complex state variable and a complex control; conceptually, the application of control techniques to steer the *microscopic* equations of motion is both a novel and potentially very important new direction.

A very exciting approach adopted more recently involves letting the laser learn to design its own optimal pulse shape in the laboratory [59–63]. This is achieved by having a feedback loop, such that the increase or decrease in yield from making a change in the pulse is fed back to the pulse shaper, guiding the design of the next trial pulse. A particular implementation of this approach is the 'genetic algorithm', in which large set of initial pulses are generated; those giving the highest yield are used as 'parents' to produce a new 'generation' of pulses, by allowing segments of the parent pulses to combine in random new combinations.

The various approaches to laser control of chemical reactions have been discussed in detail in several recent reviews [64, 65].

A1.6.5.1 *Intuitive control concepts*

Consider the ground electronic state potential energy surface in figure A1.6.26. This potential energy surface, corresponding to collinear ABC, has a region of stable ABC and two exit channels, one corresponding to A + BC and one to AB + C. This system is the simplest paradigm for control of chemical product formation: a two degree of freedom system is the minimum that can display two distinct chemical products. The objective is, starting out in a well-defined initial state ($v = 0$ for the ABC molecule) to design an electric field as a function of time which will steer the wavepacket out of channel 1, with no amplitude going out of channel 2, and *vice versa* [19, 52].

We introduce a single excited electronic state surface at this point. The motivation is severalfold. (i) Transition dipole moments are generally much stronger than permanent dipole moments. (ii) The difference in functional form of the excited and ground potential energy surface will be our dynamical kernel; with a single surface one must make use of the (generally weak) coordinate dependence of the dipole. Moreover, the use of excited electronic states facilitates large changes in force on the molecule, effectively instantaneously,

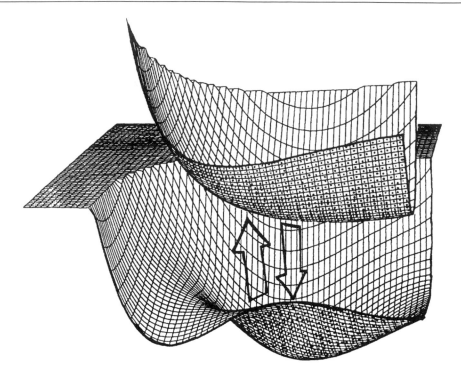

Figure A1.6.26. Stereoscopic view of ground- and excited-state potential energy surfaces for a model collinear ABC system with the masses of HHD. The ground-state surface has a minimum, corresponding to the stable ABC molecule. This minimum is separated by saddle points from two distinct exit channels, one leading to AB+C the other to A+BC. The object is to use optical excitation and stimulated emission between the two surfaces to 'steer' the wavepacket selectively out of one of the exit channels (reprinted from [54]).

without necessarily using strong fields. (iii) The technology for amplitude and phase control of optical pulses is significantly ahead of the corresponding technology in the infrared.

The object now will be to steer the wavefunction out of a specific exit channel on the ground electronic state, using the excited electronic state as an intermediate. Insofar as the control is achieved by transferring amplitude between two electronic states, all the concepts regarding the central quantity μ_{eg} introduced above will now come into play.

(a) Pump–dump scheme

Consider the following intuitive scheme, in which the timing between a pair of pulses is used to control the identity of products [52]. The scheme is based on the close correspondence between the centre of a wavepacket in time and that of a classical trajectory (Ehrenfest's theorem). The first pulse produces an excited electronic state wavepacket. The time delay between the pulses controls the time that the wavepacket evolves on the excited electronic state. The second pulse stimulates emission. By the Franck–Condon principle, the second step prepares a wavepacket on the ground electronic state with the same position and momentum, instantaneously, as the excited-state wavepacket. By controlling the position and momentum of the wavepacket produced on the ground state through the second step, one can gain some measure of control over product formation on the ground state. This 'pump–dump' scheme is illustrated classically in figure A1.6.27. The trajectory originates at the ground-state surface minimum (the equilibrium geometry).

At $t = 0$ it is promoted to the excited-state potential surface (a two-dimensional harmonic oscillator in this model) where it originates at the Condon point, that is, vertically above the ground-state minimum. Since this position is displaced from equilibrium on the excited state, the trajectory begins to evolve, executing a two-dimensional Lissajous motion. After some time delay, the trajectory is brought down vertically to the ground state (keeping both the instantaneous position and momentum it had on the excited state) and allowed to continue to evolve on the ground-state. Figure A1.6.27 shows that for one choice of time delay it will exit into channel 1, for a second choice of time delay it will exit into channel 2. Note how the position and momentum of the trajectory on the ground state, immediately after it comes down from the excited state, are both consistent with the values it had when it left the excited state, and at the same time are ideally suited for exiting out their respective channels.

A full quantum mechanical calculation based on these classical ideas is shown in figures A1.6.28 and A1.6.29 [19]. The dynamics of the two-electronic-state model was solved, starting in the lowest vibrational eigenstate of the ground electronic state, in the presence of a pair of femtosecond pulses that couple the states. Because the pulses were taken to be much shorter than a vibrational period, the effect of the pulses is to prepare a wavepacket on the excited/ground state which is almost an exact replica of the instantaneous wavefunction on the other surface. Thus, the first pulse prepares an initial wavepacket which is almost a perfect Gaussian, and which begins to evolve on the excited-state surface. The second pulse transfers the instantaneous wavepacket at the arrival time of the pulse back to the ground state, where it continues to evolve on the ground-state surface, given its position and momentum at the time of arrival from the excited state. For one choice of time delay the exit out of channel 1 is almost completely selective (figure A1.6.28), while for a second choice of time delay the exit out of channel 2 is almost completely selective (A1.6.29). Note the close correspondence with the classical model: the wavepacket on the excited state is executing a Lissajous motion almost identical with that of the classical trajectory (the wavepacket is a nearly Gaussian wavepacket on a two-dimensional harmonic oscillator). On the groundstate, the wavepacket becomes spatially extended but its exit channel, as well as the partitioning of energy into translation and vibration (i.e. parallel and perpendicular to the exit direction) are seen to be in close agreement with the corresponding classical trajectory.

This scheme is significant for three reasons: (i) it shows that control is possible, (ii) it gives a starting point for the design of optimal pulse shapes, and (iii) it gives a framework for interpreting the action of two pulse and more complicated pulse sequences. Nevertheless, the approach is limited: in general with the best choice of time delay and central frequency of the pulses one may achieve only partial selectivity. Perhaps most importantly, this scheme does not exploit the phase of the light. Intuition breaks down for more complicated processes and classical pictures cannot adequately describe the role of the phase of the light and the wavefunction. Hence, attempts were made to develop a systematic procedure for improving an initial pulse sequence.

Before turning to these more systematic procedures for designing shaped pulses, we point out an interesting alternative perspective on pump–dump control. A central tenet of Feynman's approach to quantum mechanics was to think of quantum interference as arising from multiple dynamical paths that lead to the same final state. The simple example of this interference involves an initial state, two intermediate states and a single final state, although if the objective is to control some branching ratio at the final energy then at least two final states are necessary. By controlling the phase with which each of the two intermediate states contributes to the final state, one may control constructive versus destructive interference in the final states. This is the basis of the Brumer–Shapiro approach to coherent control [57, 58]. It is interesting to note that pump–dump control can be viewed entirely from this perspective. Now, however, instead of two intermediate states there are many, corresponding to the vibrational levels of the excited electronic state (see figure A1.6.31). The control of the phase which determines how each of these intermediate levels contributes to the final state is achieved via the time delay between the excitation and the stimulated emission pulse. This 'interfering pathways' interpretation of pump–dump control is shown in figure A1.6.30.

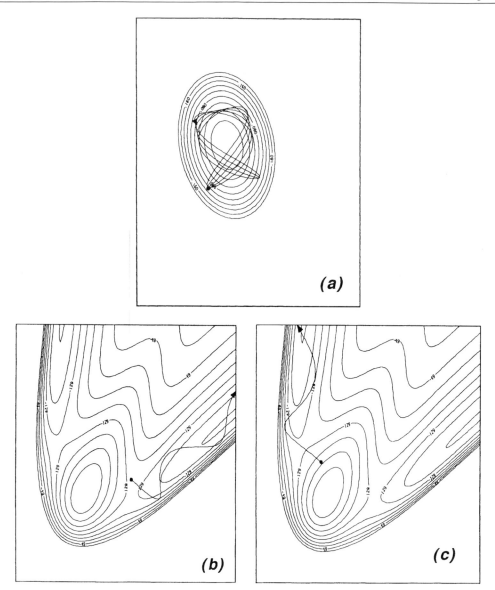

Figure A1.6.27. Equipotential contour plots of (a) the excited- and (b), (c) ground-state potential energy surfaces. (Here a harmonic excited state is used because that is the way the first calculations were performed.) (a) The classical trajectory that originates from rest on the ground-state surface makes a vertical transition to the excited state, and subsequently undergoes Lissajous motion, which is shown superimposed. (b) Assuming a vertical transition down at time t_1 (position and momentum conserved) the trajectory continues to evolve on the ground-state surface and exits from channel 1. (c) If the transition down is at time t_2 the classical trajectory exits from channel 2 (reprinted from [52]).

A1.6.5.2 Variational formulation of control of product formation

The next step, therefore, is to address the question: how is it possible to take advantage of the many additional available parameters: pulse shaping, multiple pulse sequences, etc—in general an $E(t)$ with arbitrary

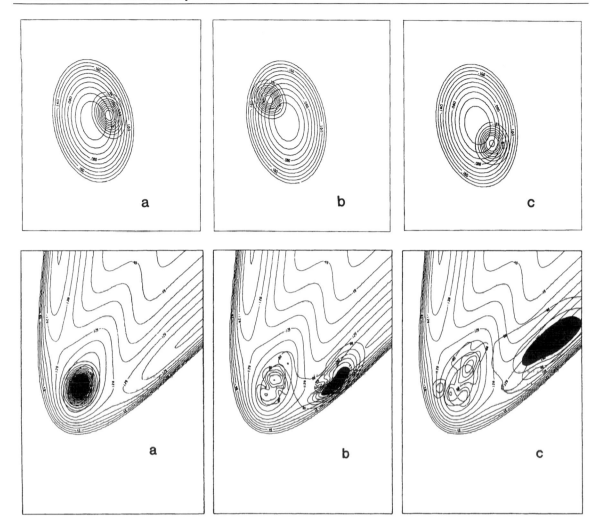

Figure A1.6.28. Magnitude of the excited-state wavefunction for a pulse sequence of two Gaussians with time delay of 610 a.u. = 15 fs. (a) $t = 200$ a.u., (b) $t = 400$ a.u., (c) $t = 600$ a.u. Note the close correspondence with the results obtained for the classical trajectory (figure A1.6.27(a) and (b)). Magnitude of the ground-state wavefunction for the same pulse sequence, at (a) $t = 0$, (b) $t = 800$ a.u., (c) $t = 1000$ a.u. Note the close correspondence with the classical trajectory of figure A1.6.27(c)). Although some of the amplitude remains in the bound region, that which does exit does so exclusively from channel 1 (reprinted from [52]).

complexity—to maximize and perhaps obtain perfect selectivity? Posing the problem mathematically, one seeks to maximize

$$J \equiv \lim_{T \to \infty} \langle \psi(T) | P_\alpha | \psi(T) \rangle \tag{A1.6.127}$$

where P_α is a projection operator for chemical channel α (here, α takes on two values, referring to arrangement channels A + BC and AB + C; in general, in a triatomic molecule ABC, α takes on three values, 1,2,3, referring to arrangement channels A + BC, AB + C and AC + B). The time T is understood to be longer than the duration of the pulse sequence, $E(t)$; the yield, J, is defined as $T \to \infty$, that is, after the wavepacket amplitude has time to reach its asymptotic arrangement. The key observation is that the quantity J is a *functional* of $E(t)$,

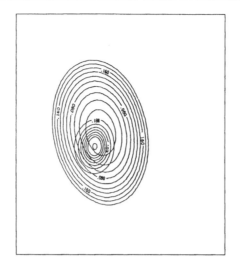

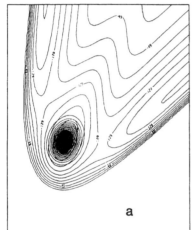
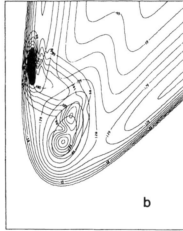
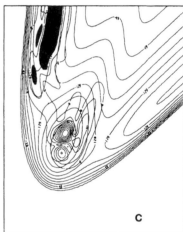

a   b   c

Figure A1.6.29. Magnitude of the ground- and excited-state wavefunctions for a sequence of two Gaussian pulses with time delay of 810 a.u. (upper diagram) excited-state wavefunction at 800 a.u., before the second pulse. (a) Ground-state wavefunction at 0 a.u. (b) Ground-state wavefunction at 1000 a.u. (c) Ground-state wavefunction at 1200 a.u. That amplitude which does exit does so exclusively from channel 2. Note the close correspondence with the classical trajectory of figure A1.6.27(c) (reprinted from [52]).

that is, J is a function of a function, because $\psi(T)$ depends on the whole history of $E(t)$. To make this dependence on $E(T)$ explicit we may write

$$J[E(t)] \equiv \lim_{T \to \infty} \langle \psi[E(t)](T) | P_\alpha | \psi[E(t)](T) \rangle \qquad (A1.6.128)$$

where square brackets are used to indicate functional dependence. The problem of maximizing a function of a function has a rich history in mathematical physics, and falls into the class of problems belonging to the calculus of variations.

In the OCT formulation, the TDSE written as a 2×2 matrix in a BO basis set, equation (A1.6.72), is introduced into the objective functional with a Lagrange multiplier, $\chi(x, t)$ [54]. The modified objective

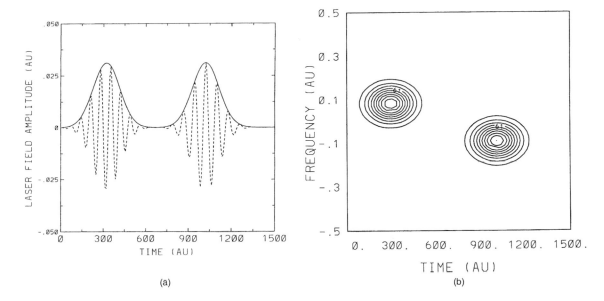

Figure A1.6.30. (a) Two pulse sequence used in the Tannor–Rice pump–dump scheme. (b) The Husimi time–frequency distribution corresponding to the two pump sequence in (a), constructed by taking the overlap of the pulse sequence with a two-parameter family of Gaussians, characterized by different centres in time and carrier frequency, and plotting the overlap as a function of these two parameters. Note that the Husimi distribution allows one to visualize both the time delay and the frequency offset of pump and dump simultaneously (after [52a]).

functional may now be written as

$$\bar{J} \equiv \lim_{T\to\infty} \langle\psi(T)|P_\alpha|\psi(T)\rangle + 2\mathrm{Re}\int_0^T dt \langle\chi(t)\left|\frac{\partial}{\partial t} - \frac{H}{i\hbar}\right|\psi(t)\rangle - \lambda\int_0^T dt |E(t)|^2 \qquad (A1.6.129)$$

where a constraint (or penalty) on the time integral of the energy in the electric field has also been added. It is clear that as long as ψ satisfies the TDSE the new term in $\bar{J}$ will vanish for any $\chi(x,t)$. The function of the new term is to make the variations of $\bar{J}$ with respect to E and with respect to ψ independent, to first-order in δE (i.e. to 'deconstrain' ψ and E).

The requirement that $\delta\bar{J}/\delta\psi = 0$ leads to the following equations:

$$i\hbar\frac{\partial\chi}{\partial t} = H\chi \qquad (A1.6.130)$$

$$\chi(x,T) = P_\alpha\psi(x,T) \qquad (A1.6.131)$$

that is, the Lagrange multiplier must obey the TDSE, subject to the boundary condition at the *final* time T that χ be equal to the projection operator operating on the Schrödinger wavefunction. These conditions 'conspire', so that a change in E, which would ordinarily change $\bar{J}$ through the dependence of $\psi(T)$ on E, does not do so to first-order in the field. For a physically meaningful solution it is required that

$$i\hbar\frac{\partial\psi}{\partial t} = H\psi \qquad (A1.6.132)$$

$$\psi(x,0) = \psi_0(x). \qquad (A1.6.133)$$

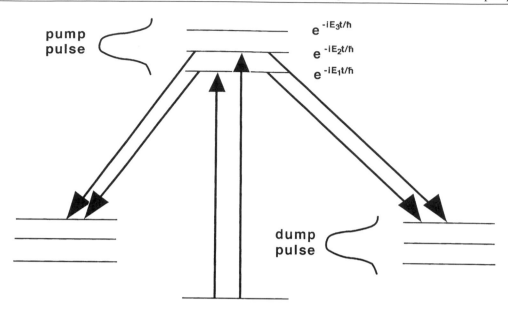

Figure A1.6.31. Multiple pathway interference interpretation of pump–dump control. Since each of the pair of pulses contains many frequency components, there are an infinite number of combination frequencies which lead to the same final energy state, which generally interfere. The time delay between the pump and dump pulses controls the relative phase among these pathways, and hence determines whether the interference is constructive or destructive. The frequency domain interpretation highlights two important features of coherent control. First, if final products are to be controlled there must be degeneracy in the dissociative continuum. Second, a single interaction with the light, no matter how it is shaped, cannot produce control of final products: at least two interactions with the field are needed to obtain interfering pathways.

Finally, the optimal $E(t)$ is given by the condition that $\delta \bar{J}/\delta E = 0$ which leads to the equation

$$E(t) = \frac{-i}{\hbar \lambda} [\langle \chi_a | \mu | \psi_b \rangle - \langle \psi_a | \mu | \chi_b \rangle]. \tag{A1.6.134}$$

The interested reader is referred to [54] for the details of the derivation.

Equations (A1.6.129)–(A1.6.133) form the basis for a double-ended boundary value problem. ψ is known at $t = 0$, while χ is known at $t = T$. Taking a guess for $E(t)$ one can propagate ψ forward in time to obtain $\psi(t)$; at time T the projection operator P_α may be applied to obtain $\chi(T)$, which may be propagated backwards in time to obtain $\chi(t)$. Note, however, that the above description is not self-consistent: the guess of $E(t)$ used to propagate $\psi(t)$ forward in time and to propagate $\chi(t)$ backwards in time is not, in general, equal to the value of $E(t)$ given by equation (A1.6.133). Thus, in general, one has to solve these equations iteratively until self-consistency is achieved. Optimal control theory has become a widely used tool for designing laser pulses with specific objectives. The interested reader can consult the review in [65] for further examples.

A1.6.5.3   Optimal control and laser cooling of molecules

The use of lasers to cool atomic translational motion has been one of the most exciting developments in atomic physics in the last 15 years. For excellent reviews, see [66, 67]. Here we give a non-orthodox presentation, based on [68].

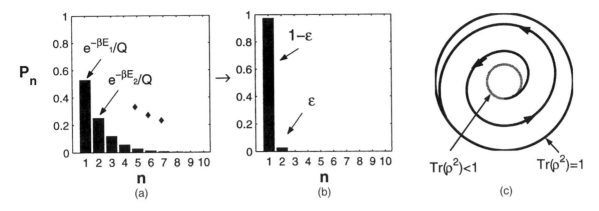

Figure A1.6.32. (a) Initial and (b) final population distributions corresponding to cooling. (c) Geometrical interpretation of cooling. The density matrix is represented as a point on generalized Bloch sphere of radius $R = \text{Tr}(\rho^2)$. For an initially thermal state the radius $R < 1$, while for a pure state $R = 1$. The object of cooling is to manipulate the density matrix onto spheres of increasingly larger radius.

(a) Calibration of cooling: the zeroth law

Consider, figure A1.6.32, in which a system is initially populated with an incoherent distribution of populations with Boltzmann probabilities, P_n, $\sum_n P_n = 1$. The simple-minded definition of cooling is to manipulate all the population into the lowest energy quantum state, i.e. to make $P_0 = 1$ and all the other $P_n = 0$. Cooling can then be measured by the quantity $\sum_n P_n^2$: for the initial, incoherent distribution $\sum_n P_n^2 < 1$ while for the final distribution $\sum_n P_n^2 = 1$. However, adoption of this definition of cooling implies that if all the population is put into *any* single quantum state, not necessarily the lowest energy state, the degree of cooling is identical. Although this seems surprising at first, it is in fact quite an appealing definition of cooling. It highlights the fact that the essence of cooling is the creation of a pure state starting from a mixed state; once the state is pure then coherent manipulations, which are relatively straightforward, can transfer this population to the ground state. As described in section A1.6.2.4, the conventional measure of the degree of purity of a system in quantum mechanics is $\text{Tr}(\rho^2)$, where ρ is the system's density matrix, and thus we have here defined cooling as the process of bringing $\text{Tr}(\rho^2)$ from its initial value less than 1 to unity. The definition of cooling in terms of $\text{Tr}(\rho^2)$ leads to an additional surprise, namely, that the single quantum state need not even be an eigenstate: it can, in principle, be a superposition consisting of a coherent superposition of many eigenstates. So long as the state is pure (i.e. can be described by a single Schrödinger wavefunction) it can manipulated into the lowest energy state by a unitary transformation, and in a very real sense is already cold! Figure A1.6.32 gives a geometrical interpretation of cooling. The density matrix is represented as a point on a generalized Bloch sphere of radius $R = \text{Tr}(\rho^2)$. For an initially thermal state the radius $R < 1$, while for a pure state $R = 1$. Thus, the object of cooling, that is, increasing the purity of the density matrix, corresponds to manipulating the density matrix onto spheres of increasingly larger radius.

We have seen in section A1.6.2.4 that external fields alone cannot change the value of $\text{Tr}(\rho^2)$! Changes in the purity can arise only from the spontaneous emission, which is inherently uncontrollable. Where then is the control?

A first glimmer of the resolution to the paradox of how control fields can control purity content is obtained by noting that the second derivative, $\text{Tr}(\ddot{\rho^2})$, does depend on the external field. Loosely speaking, the independence of the first derivative and the dependence of the second derivative on the control field indicates that the control of cooling is achieved only in two-stages: preparation of the initial state by the control field, followed by spontaneous emission into that recipient state. This two-stage interpretation will now be quantified.

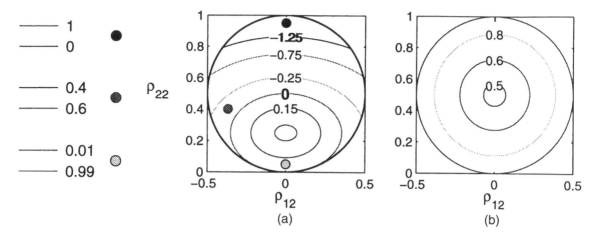

Figure A1.6.33. (a) Contour map of $\frac{d}{dt}\mathrm{Tr}(\rho^2)$ as a function of the parameters ρ_{22} (z) and $|\rho_{12}|$ (x). The dark region corresponds to $\frac{d}{dt}\mathrm{Tr}(\rho^2) < 0$, i.e. cooling while the light region corresponds to $\frac{d}{dt}\mathrm{Tr}(\rho^2) > 0$, i.e. heating. For fixed z, the maximum occurs along the line $x = 0$. (b) Isopurity, or isocoherence contours (contours of fixed $\mathrm{Tr}(\rho^2)$) as a function of ρ_{22} (z) and $|\rho_{12}|$ (x) for the two-level system. The contour takes its maximum value of 1, corresponding to a pure state, along the outermost circle, while the function takes its minimum value of 1/2, representing the most impure state, at the centre.

To find the boundary between heating and cooling we set $\mathrm{Tr}(\rho^2) = 0$. Figure A1.6.33 shows isocontours of $\mathrm{Tr}(\rho^2)$ as a function of the parameters ρ_{22} (z) and $|\rho_{12}|$ (x). The dark region corresponds to $\frac{d}{dt}\mathrm{Tr}(\rho^2) < 0$; that is, cooling, while the light region corresponds to $\frac{d}{dt}\mathrm{Tr}(\rho^2) > 0$, (i.e. heating). Note that the cooling region fills part, but not all of the lower hemisphere. For fixed z, the maximum occurs along the line $x = 0$, with the global maximum at $z = 1/4$, $x = 0$.

To gain a qualitative understanding for the heating and cooling regions we consider three representative points (top to bottom, figure A1.6.33(a)). (i) Spontaneous emission will lead from 1:99 to 0:100 and hence purity increase. (ii) Spontaneous emission will lead from 100:0 to 99:1 and hence purity decrease. (iii) Spontaneous emission will lead from 40:60 to 30:70 which suggests a purity increase; however, if there is purity stored in the coherences ρ_{12}, spontaneous emission will force these to decay at a rate $\Gamma_2 = 1/2\Gamma_1$; this leads to a decrease in purity which is greater than the increase in purity brought about by the population transfer.

The manipulations allowed by the external field are those that move the system along a contour of constant value of $\mathrm{Tr}(\rho^2)$, an isocoherence contour; it is clear from figure A1.6.33 that the location on this contour has a profound affect on $\mathrm{Tr}(\rho^2)$. This gives a second perspective on how the external field cannot directly change $\mathrm{Tr}(\rho^2)$, but can still affect the rate of change of $\mathrm{Tr}(\rho^2)$. If we imagine that at every instant in time the external field moves the system along the instantaneous isocoherence contour until it intersects the curve of maximum $\mathrm{Tr}(\rho^2)$, that would provide an optimal cooling strategy. This last observation is the crux of our cooling theory and puts into sharp perspective the role played by the external field: while the external field cannot itself change the purity of the system it can perform purity-preserving transformations which subsequently affect the rate of change of purity.

To summarize, we have the following chain of dependence: $(\rho_{22}, |\rho_{12}|) \rightarrow \mathrm{Tr}(\rho^2) \rightarrow (\tilde{\rho}_{22}, |\tilde{\rho}_{12}|) \rightarrow \mathrm{Tr}(\dot{\rho}^2)$. This chain of dependence gives $\mathrm{Tr}(\dot{\rho}^2)$ as a function of $\mathrm{Tr}(\rho^2)$, which is a *differential equation for the optimal trajectory* $\mathrm{Tr}(\rho^2)(t)$. By studying the rate of approach of the optimal trajectory to absolute zero (i.e. to a pure state) we will have found an inviolable limitation on cooling rate with the status of a third law of thermodynamics.

Note that the differential equation obtained from this approach will never agree perfectly with the results of a simulation. The above formulation is essentially an adiabatic formulation of the process: the spontaneous emission is considered to be slow compared with the time scale for the purity-preserving transformations generated by the external field, which is what allows us to assume in the theory that the external field manipulation along the isocoherence contour is instantaneous. If the external field is sufficiently intense, the population transfer may become nearly instantaneous relative to the spontaneous emission, and the adiabatic approximation will be excellent.

(b) Cooling and lasing as complementary processes

It is interesting to consider the regions of heating, that is, regions where $\text{Tr}(\dot{\rho^2}) < 0$. We conjecture that these regions correspond to regions where lasing can occur. The conjecture is based on the following considerations:

(i)   Note that for the two-level system with no coherence ($\rho_{12} = 0$), the region where $\text{Tr}(\dot{\rho^2}) < 0$ corresponds to $\rho_{22} > \frac{1}{2}$. This corresponds to the conventional population inversion criterion for lasing: that population in the excited state be larger than in the ground state.

(ii) The fact that in this region the system coherence is decreasing, leaves open the possibility that coherence elsewhere can increase. In particular, excitation with incoherent light can lead to emission of coherent light. This is precisely the reverse situation as with laser cooling, where coherent light is transformed to incoherent light (spontaneous emission), increasing the level of coherence of the system.

(iii) The regions with $\text{Tr}(\dot{\rho^2}) < 0$ and $d < \frac{1}{2}$ necessarily imply $\gamma > 0$, that is, coherences between the ground and excited state. This may correspond to lasing without population inversion, an effect which has attracted a great deal of attention in recent years, and is made possible by coherences between the ground and excited states. Indeed, in the three-level Λ system the boundary between heating and cooling is in exact agreement with the boundary between lasing and non-lasing.

Fundamentally, the conditions for lasing are determined unambiguously once the populations and coherences of the system density matrix are known. Yet, we have been unable to find in the literature any simple criterion for lasing in multilevel systems in terms of the system density matrix alone. Our conjecture is that entropy, as expressed by the purity content $\text{Tr}(\rho^2)$, is the unifying condition; the fact that such a simple criterion could have escaped previous observation may be understood, given the absence of thermodynamic considerations in conventional descriptions of lasing.

References

[1]   Jackson J D 1975 *Classical Electrodynamics* (New York: Wiley)
[2]   Loudon R 1983 *The Quantum Theory of Light* (Oxford: Oxford University Press)
[3]   Schatz G C and Ratner M A 1993 *Quantum Mechanics in Chemistry* (Englewood Cliffs, NJ: Prentice-Hall) ch 5
[4]   We follow [3] here. However, see van Kranendonk J and Sipe J E 1976 *Can. J. Phys.* **54** 471
[5]   Einstein A 1917 On the quantum theory of radiation *Phys. Z.* **18** 121
      Reprinted ter Haar D 1967 *The Old Quantum Theory* (New York: Pergamon)
[6]   Cohen-Tannoudji C, Diu B and Laloë F 1977 *Quantum Mechanics* vol 1 (New York: Wiley) ch 4
[7]   Allen L and Eberly J H 1987 *Optical Resonance and Two-Level Atoms* (New York: Dover)
[8]   Steinfeld J I 1986 *Molecules and Radiation* (Cambridge, MA: MIT)
[9]   Siegman A E 1986 *Lasers* (Mill Valley, CA: University Science Books)
[10] Sargent III M, Scully M O and Lamb W E Jr 1974 *Laser Physics* (Reading, MA: Addison-Wesley)
[11] Cohen-Tannoudji C, Dupont-Roc J and Grynberg G 1992 *Atom–Photon Interaction* (New York: Wiley)
[12] Heller E J 1978 Quantum corrections to classical photodissociation models *J. Chem. Phys.* **68** 2066
[13] Kulander K C and Heller E J 1978 Time-dependent formulation of polyatomic photofragmentation: application to H_3^+ *J. Chem. Phys.* **69** 2439
[14] Lee S-Y and Heller E J 1979 Time-dependent theory of Raman scattering *J. Chem. Phys.* **71** 4777
[15] Heller E J, Sundberg R L and Tannor D J 1982 Simple aspects of Raman scattering *J. Phys. Chem.* **86** 1822–33

[16]   Heller E J 1981 The semiclassical way to molecular spectroscopy *Acc. Chem. Res.* **14** 368

[17]   Tannor D J and Rice S A 1988 Coherent pulse sequence control of product formation in chemical reactions *Adv. Chem. Phys.* **70** 441–524

[18]   Lee S-Y and Heller E J 1979 *op. cit.* [14] equations (2.9) and (2.10)

[19]   Tannor D J, Kosloff R and Rice S A 1986 Coherent pulse sequence induced control of selectivity of reactions: exact quantum mechanical calculations *J. Chem. Phys.* **85** 5805–20, equations (1)–(6)

[20]   Tannor D J and Rice S A 1988 Coherent pulse sequence control of product formation in chemical reactions *Adv. Chem. Phys.* **70** 441–524, equations (5.1)–(5.6)

[21]   Tannor D J 2001 *Introduction to Quantum Mechanics: A Time Dependent Perspective* (Mill Valley, CA: University Science Books)

[22]   Scherer N F, Carlson R J, Matro A, Du M, Ruggiero A J, Romero-Rochin V, Cina J A, Fleming G R and Rice S A 1991 Fluorescence-detected wave packet interferometry: time resolved molecular spectroscopy with sequences of femtosecond phase-locked pulses *J. Chem. Phys.* **95** 1487

[23]   Scherer N F, Matro A, Ziegler L D, Du M, Cina J A and Fleming G R 1992 Fluorescence-detected wave packet interferometry. 2. Role of rotations and determination of the susceptibility *J. Chem. Phys.* **96** 4180

[24]   Engel V and Metiu H 1994 2-Photon wave-packet interferometry *J. Chem. Phys.* **100** 5448

[25]   Zewail A H 1988 Laser femtochemistry *Science* **242** 1645

[26]   Zewail A H (ed) 1994 *Femtochemistry* vols 1 and 2 (Singapore: World Scientific)

[27]   Manz J and Wöste L (eds) 1995 *Femtosecond Chemistry* (Heidelberg: VCH)

[28]   Baumert T, Engel V, Meier Ch and Gerber G 1992 High laser field effects in multiphoton ionization of Na_2 – experiment and quantum calculations *Chem. Phys. Lett.* **200** 488

[29]   Wang Q, Schoenlein R W, Peteanu L A, Mathies R A and Shank C V 1994 Vibrationally coherent photochemistry in the femtosecond primary event of vision *Science* **266** 422

[30]   Pugliano N, Gnanakaran S and Hochstrasser R M 1996 The dynamics of photodissociation reactions in solution *J. Photochem. and Photobiol. A—Chemistry* **102** 21–8

[31]   Pack R T 1976 Simple theory of diffuse vibrational structure in continuous UV spectra of polyatomic molecules. I. Collinear photodissociation of symmetric triatomics *J. Chem. Phys.* **65** 4765

[32]   Heller E J 1978 Photofragmentation of symmetric triatomic molecules: Time dependent picture *J. Chem. Phys.* **68** 3891

[33]   Myers A B and Mathies R A 1987 Resonance Raman intensities: A probe of excited-state structure and dynamics *Biological Applications of Raman Spectroscopy* vol 2, ed T G Spiro (New York: Wiley-Interscience) pp 1–58

[34]   Albrecht A C 1961 On the theory of Raman intensities *J. Chem. Phys.* **34** 1476

[35]   Bloembergen N 1965 *Nonlinear Optics* (Reading, MA: Benjamin-Cummings)

[36]   Shen Y R 1984 *The Principles of Nonlinear Optics* (New York: Wiley)

[37]   Mukamel S 1995 *Principles of Non-linear Optical Spectroscopy* (New York: Oxford University Press)

[38]   Lee D and Albrecht A C 1985 *Advances in Infrared and Raman Spectroscopy* **12** 179

[39]   Tannor D J, Rice S A and Weber P M 1985 Picosecond CARS as a probe of ground electronic state intramolecular vibrational redistribution *J. Chem. Phys.* **83** 6158

[40]   Tannor D J and Rice S A 1987 Photon echoes in multilevel systems *Understanding Molecular Properties* ed J Avery *et al* (Dordrecht: Reidel) p 205

[41]   Pollard W T, Lee S-Y and Mathies R A 1990 Wavepacket theory of dynamic absorption spectra in femtosecond pump–probe experiments *J. Chem. Phys.* **92** 4012

[42]   Lee S-Y 1995 Wave-packet model of dynamic dispersed and integrated pump–probe signals in femtosecond transition state spectroscopy *Femtosecond Chemistry* ed J Manz and L Wöste (Heidelberg: VCH)

[43]   Meyer S and Engel V 2000 Femtosecond time-resolved CARS and DFWM spectroscopy on gas-phase I_2: a wave-packet description *J. Raman Spectrosc.* **31** 33

[44]   Knopp G, Pinkas I and Prior Y 2000 Two-dimensional time-delayed coherent anti-Stokes Raman spectroscopy and wavepacket dynamics of high ground-state vibrations *J. Raman Spectrosc.* **31** 51

[45]   Pausch R, Heid M, Chen T, Schwoerer H and Kiefer W 2000 Quantum control by stimulated Raman scattering *J. Raman Spectrosc.* **31** 7

[46]   Pastirk I, Brown E J, Grimberg B I, Lozovoy V V and Dantus M 1999 Sequences for controlling laser excitation with femtosecond three-pulse four-wave mixing *Faraday Discuss.* **113** 401

[47]   Shen Y R 1984 *The Principles of Nonlinear Optics* (New York: Wiley) ch 21 for a clear discussion of the connection between the perturbative and nonperturbative treatment of photon echoes

[48]   Joo T, Jia Y, Yu J-Y, Lang M J and Fleming G R 1996 Third-order nonlinear time domain probes of solvation dynamics *J. Chem. Phys.* **104** 6089

[48a] Passino S A, Nagasawa Y, Joo T and Fleming G R 1997 Three pulse photon echoes *J. Phys. Chem. A* **101** 725

[49]   Gallagher Faeder S M and Jonas D 1999 Two-dimensional electronic correlation and relaxation spectra: theory and model calculations *J. Phys. Chem. A* **102** 10 489–505

[50]   Seidner L, Stock G and Domcke W 1995 Nonperturbative approach to femtosecond spectroscopy – general theory and application to multidimensional nonadiabatic photoisomerization processes *J. Chem. Phys.* **103** 4002

[51] Ashkenazi G, Banin U, Bartana A, Ruhman S and Kosloff R 1997 Quantum description of the impulsive photodissociation dynamics of I_3^- in solution *Adv. Chem. Phys.* **100** 229

[52] Tannor D J and Rice S A 1985 Control of selectivity of chemical reaction via control of wave packet evolution *J. Chem. Phys.* **83** 5013–18

[52a] Tannor D J 1994 Design of femtosecond optical pulses to control photochemical products *Molecules in Laser Fields* ed A Bandrauk (New York: Dekker) p 403

[53] Peirce A P, Dahleh M A and Rabitz H 1988 Optimal control of quantum mechanical systems – Existence, numerical approximations and applications *Phys. Rev.* A **37** 4950

[54] Kosloff R, Rice S A, Gaspard P, Tersigni S and Tannor D J 1989 Wavepacket dancing: achieving chemical selectivity by shaping light pulses *Chem. Phys.* **139** 201–20

[55] Warren W S, Rabitz H and Dahleh M 1993 Coherent control of quantum dynamics: the dream is alive *Science* **259** 1581

[56] Yan Y J, Gillilan R E, Whitnell R M, Wilson K R and Mukamel S 1993 Optimal control of molecular dynamics – Liouville space theory *J. Chem. Phys.* **97** 2320

[57] Shapiro M and Brumer P 1986 Laser control of product quantum state populations in unimolecular reactions *J. Chem. Phys.* **84** 4103

[58] Shapiro M and Brumer P 1989 Coherent chemistry—Controlling chemical reactions with lasers *Acc. Chem. Res.* **22** 407

[59] Judson R S and Rabitz H 1992 Teaching lasers to control molecules *Phys. Rev. Lett.* **68** 1500

[60] Bardeen C J, Yakovlev V V, Wilson K R, Carpenter S D, Weber P M and Warren W S 1997 Feedback quantum control of molecular electronic population transfer *Chem. Phys. Lett.* **280** 151

[61] Yelin D, Meshulach D and Silberberg Y 1997 Adaptive femtosecond pulse compression *Opt. Lett.* **22** 1793–5

[62] Assion A, Baumert T, Bergt M, Brixner T, Kiefer B, Seyfried V, Strehle M and Gerber G 1998 Control of chemical reactions by feedback-optimized phase-shaped femtosecond laser pulses *Science* **282** 919

[63] Weinacht T C, White J L and Bucksbaum P H 1999 Toward strong field mode-selective chemistry *J. Phys. Chem.* A **103** 10 166–8

[64] Gordon R J and Rice S A 1997 Active control of the dynamics of atoms and molecules *Annu. Rev. Phys. Chem.* **48** 601

[65] Rice S A and Zhao M 2000 *Optical Control of Molecular Dynamics* (New York: Wiley)

[66] Cohen-Tannoudji C N and Phillips W D 1990 New mechanisms for laser cooling *Phys. Today* **43** 33–40

[67] Cohen-Tannoudji C 1991 Atomic motion in laser light *Fundamental Systems in Quantum Optics* ed J Dalibard *et al* (Oxford: Elsevier)

[68] Tannor D J and Bartana A 1999 On the interplay of control fields and spontaneous emission in laser cooling *J. Phys. Chem.* A **103** 10 359–63

Further Reading

Loudon R 1983 *The Quantum Theory of Light* (Oxford: Oxford University Press)

An excellent and readable discussion of all aspects of the interaction of light with matter, from blackbody radiation to lasers and nonlinear optics.

Herzberg G 1939, 1945, 1966 *Molecular Spectra and Molecular Structure* (New York: van Nostrand) 3 vols

This is the classic work on molecular rotational, vibrational and electronic spectroscopy. It provides a comprehensive coverage of all aspects of infrared and optical spectroscopy of molecules from the traditional viewpoint and, both for perspective and scope, is an invaluable supplement to this section.

Steinfeld J I 1986 *Molecules and Radiation* (Cambridge, MA: MIT)

A good introduction to the use of coherent optical techniques and their use to probe molecular spectra.

Shen Y R 1984 *The Principles of Non-linear Optics* (New York: Wiley)

A clear, comprehensive discussion of the many facets of nonlinear optics. The emphasis is on optical effects, such as harmonic generation. The treatment of nonlinear spectroscopy, although occupying only a fraction of the book, is clear and physically well-motivated.

Mukamel S 1995 *Principles of Non-linear Optical Spectroscopy* (New York: Oxford University Press)

A valuable handbook describing the many uses of nonlinear optics for spectroscopy. The focus of the book is a unified treatment of $P^{(3)}$, and methods for modelling the $P^{(3)}$ signal.

Heller E J 1981 The semiclassical way to molecular spectroscopy *Acc. Chem. Res.* **14** 368

A beautiful, easy-to-read introduction to wavepackets and their use in interpreting molecular absorption and resonance Raman spectra.

Rice S A and Zhao M 2000 *Optical Control of Molecular Dynamics* (New York: Wiley)

A valuable resource, reviewing both theoretical and experimental progress on coherent control to date.

Tannor D J 2001 *Introduction to Quantum Mechanics: A Time Dependent Perspective* (Mill Valley, CA: University Science Books)

A comprehensive discussion of wavepackets, classical-quantum correspondence, optical spectroscopy, coherent control and reactive scattering from a unified, time dependent perspective.

A1.7
Surfaces and interfaces

J A Yarmoff

A1.7.1   Introduction

Some of the most interesting and important chemical and physical interactions occur when dissimilar materials meet, i.e. at an interface. The understanding of the physics and chemistry at interfaces is one of the most challenging and important endeavors in modern science.

Perhaps the most intensely studied interface is that between a solid and vacuum, i.e. a surface. There are a number of reasons for this. For one, it is more experimentally accessible than other interfaces. In addition, it is conceptually simple, as compared to interfaces between two solids or between a solid and a liquid, so that the vacuum–solid interface is more accessible to fundamental theoretical investigation. Finally, it is the interface most easily accessible for modification, for example by photons or charged particle beams that must be propagated in vacuum.

Studies of surfaces and surface properties can be traced to the early 1800s [1]. Processes that involved surfaces and surface chemistry, such as heterogeneous catalysis and Daguerre photography, were first discovered at that time. Since then, there has been a continual interest in catalysis, corrosion and other chemical reactions that involve surfaces. The modern era of surface science began in the late 1950s, when instrumentation that could be used to investigate surface processes on the molecular level started to become available.

Since the modern era began, the study of solid surfaces has been one of the fastest growing areas in solid-state research. The geometric, electronic and chemical structure at the surface of a solid is generally quite different from that of the bulk material. It is now possible to measure the properties of a surface on the atomic scale and, in fact, to image individual atoms on a surface. The theoretical understanding of the chemistry and physics at surfaces is also improving dramatically. Much of the theoretical work has been motivated by the experimental results, as well as by the vast improvements in computer technology that are required to carry out complex numerical calculations.

Surface studies address important issues in basic physics and chemistry, but are also relevant to a variety of applications. One of the most important uses of a surface, for example, is in heterogeneous catalysis. Catalysis occurs via adsorption, diffusion and reaction on a solid surface, so that delineation of surface chemical mechanisms is critical to the understanding of catalysis. Microelectronic devices are manufactured by processing of single-crystal semiconductor surfaces. Most dry processes that occur during device manufacture involve surface etching or deposition. Thus, understanding how molecules adsorb and react on surfaces and how electron and ion beams modify surfaces is crucial to the development of manufacturing techniques for semiconductor and, more recently, micro-electromechanical (MEMS), devices. Surfaces are also the active component in tribology, i.e. solid lubrication. In order to design lubricants that will stick to one surface, yet have minimal contact with another, one must understand the fundamental surface interactions involved. In addition, the movement of pollutants through the environment is controlled by the interactions of chemicals with the surfaces encountered in the soil. Thus, a fundamental understanding of the surface chemistry of metal oxide materials is needed in order to properly evaluate and solve environmental problems.

Surfaces are found to exhibit properties that are different from those of the bulk material. In the bulk, each atom is bonded to other atoms in all three dimensions. In fact, it is this infinite periodicity in three dimensions that gives rise to the power of condensed matter physics. At a surface, however, the three-dimensional periodicity is broken. This causes the surface atoms to respond to this change in their local environment by adjusting their geometric and electronic structures. The physics and chemistry of clean surfaces is discussed in section A1.7.2.

The importance of surface science is most often exhibited in studies of adsorption on surfaces, especially in regards to technological applications. Adsorption is the first step in any surface chemical reaction or film-growth process. The mechanisms of adsorption and the properties of adsorbate-covered surfaces are discussed in section A1.7.3.

Most fundamental surface science investigations employ single-crystal samples cut along a low-index plane. The single-crystal surface is prepared to be nearly atomically flat. The surface may also be modified in vacuum. For example, it may be exposed to a gas that adsorbs (sticks) to the surface, or a film can be grown onto a sample by evaporation of material. In addition to single-crystal surfaces, many researchers have investigated vicinal, i.e. stepped, surfaces as well as the surfaces of polycrystalline and disordered materials. In section A1.7.4, methods for the preparation of surfaces are discussed.

Surfaces are investigated with surface-sensitive techniques in order to elucidate fundamental information. The approach most often used is to employ a variety of techniques to investigate a particular materials system. As each technique provides only a limited amount of information, results from many techniques must be correlated in order to obtain a comprehensive understanding of surface properties. In section A1.7.5, methods for the experimental analysis of surfaces in vacuum are outlined. Note that the interactions of various kinds of particles with surfaces are a critical component of these techniques. In addition, one of the more interesting aspects of surface science is to use the tools available, such as electron, ion or laser beams, or even the tip of a scanning probe instrument, to modify a surface at the atomic scale. The physics of the interactions of particles with surfaces and the kinds of modifications that can be made to surfaces are an integral part of this section.

The liquid–solid interface, which is the interface that is involved in many chemical and environmental applications, is described in section A1.7.6. This interface is more complex than the solid–vacuum interface, and can only be probed by a limited number of experimental techniques. Thus, obtaining a fundamental understanding of its properties represents a challenging frontier for surface science.

A1.7.2   Clean surfaces

The study of clean surfaces encompassed a lot of interest in the early days of surface science. From this, we now have a reasonable idea of the geometric and electronic structure of many clean surfaces, and the tools are readily available for obtaining this information from other systems, as needed.

When discussing geometric structure, the macroscopic morphology must be distinguished from the microscopic atomic structure. The morphology is the macroscopic shape of the material, which is a collective property of groups of atoms determined largely by surface and interfacial tension. The following discussion, however, will concentrate on the structure at the atomic level. Note that the atomic structure often plays a role in determining the ultimate morphology of the surface. What is most important about the atomic structure, however, is that it affects the manner in which chemistry occurs on a surface at the molecular level.

A1.7.2.1   Surface crystallography

To first approximation, a single-crystal surface is atomically flat and uniform, and is composed of a regular array of atoms positioned at well defined lattice sites. Materials generally have of the order of 10^{15} atoms positioned at the outermost atomic layer of each square centimetre of exposed surface. A bulk crystalline material has virtually infinite periodicity in three dimensions, but infinite periodicity remains in only two dimensions when a solid is cut to expose a surface. In the third dimension, i.e. normal to the surface, the periodicity abruptly ends. Thus, the surface crystal structure is described in terms of a two-dimensional unit cell parallel to the surface.

In describing a particular surface, the first important parameter is the Miller index that corresponds to the orientation of the sample. Miller indices are used to describe directions with respect to the three-dimensional bulk unit cell [2]. The Miller index indicating a particular surface orientation is the one that points in the direction of the surface normal. For example, a Ni crystal cut perpendicular to the [100] direction would be labelled Ni(100).

The second important parameter to consider is the size of the surface unit cell. A surface unit cell cannot be smaller than the projection of the bulk cell onto the surface. However, the surface unit cell is often bigger than it would be if the bulk unit cell were simply truncated at the surface. The symmetry of the surface unit cell is easily determined by visual inspection of a low-energy electron diffraction (LEED) pattern (LEED is discussed in sections A1.7.5.1 and B1.21).

There is a well defined nomenclature employed to describe the symmetry of any particular surface [1]. The standard notation for describing surface symmetry is in the form

$$M(hkl)\text{-}(p \times q)\text{–A}$$

where M is the chemical symbol of the substrate material, h, k, and l are the Miller indices that indicate the surface orientation, p and q relate the size of the surface unit cell to that of the substrate unit cell and A is the chemical symbol for an adsorbate (if applicable). For example, atomically clean Ni cut perpendicular to the [100] direction would be notated as Ni(100)-(1 × 1), since this surface has a bulk-terminated structure. If the unit cell were bigger than that of the substrate in one direction or the other, then p and/or q would be larger than one. For example, if a Si single crystal is cleaved perpendicular to the [111] direction, a Si(111)-(2 × 1) surface is produced. Note that p and q are often, but are not necessarily, integers. If an adsorbate is involved in forming the reconstruction, then it is explicitly part of the nomenclature. For example, when silver is adsorbed on Si(111) under the proper conditions, the Si(111)-($\sqrt{3} \times \sqrt{3}$)–Ag structure is formed.

In addition, the surface unit cell may be rotated with respect to the bulk cell. Such a rotated unit cell is notated as
$$M(hkl)\text{-}(p \times q)\text{R}r°\text{–A}$$

where r is the angle in degrees between the surface and bulk unit cells. For example, when iodine is adsorbed onto the (111) face of silver, the Ag(111)-($\sqrt{3} \times \sqrt{3}$)R30°–I structure can be produced.

Finally, there is an abbreviation 'c', which stands for 'centred', that is used to indicate certain common symmetries. In a centred structure, although the primitive unit cell is rotated from the substrate unit cell, the structure can also be considered as a non-rotated unit cell with an additional atom placed in the centre. For example, a common adsorbate structure involves placing an atom at every other surface site of a square lattice. This has the effect of rotating the primitive unit cell by 45°, so that such a structure would ordinarily be notated as ($\sqrt{2} \times \sqrt{2}$)R45°. However, the unit cell can also be thought of as a (2 × 2) in registry with the substrate with an additional atom placed in the centre of the cell. Thus, in order to simplify the nomenclature, this structure is equivalently called a c(2 × 2). Note that the abbreviation 'p', which stands for 'primitive', is sometimes used for a unit cell that is in registry with the substrate in order to distinguish it from a centred symmetry. Thus, p(2 × 2) is just an unambiguous way of representing a (2 × 2) unit cell.

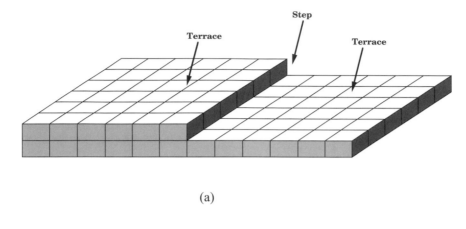

(a)

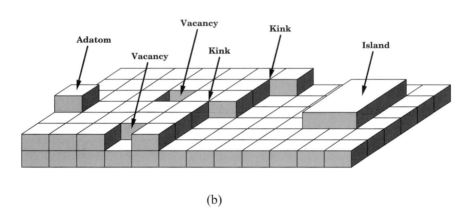

(b)

Figure A1.7.1. Schematic diagram illustrating terraces, steps, and defects. (a) Perfect flat terraces separated by a straight, monoatomic step. (b) A surface containing various defects.

A1.7.2.2   Terraces and steps

For many studies of single-crystal surfaces, it is sufficient to consider the surface as consisting of a single domain of a uniform, well ordered atomic structure based on a particular low-Miller-index orientation. However, real materials are not so flawless. It is therefore useful to consider how real surfaces differ from the ideal case, so that the behaviour that is intrinsic to a single domain of the well ordered orientation can be distinguished from that caused by defects.

Real, clean, single-crystal surfaces are composed of terraces, steps and defects, as illustrated in figure A1.7.1. This arrangement is called the TLK, or terrace-ledge-kink, model. A terrace is a large region in which the surface has a well-defined orientation and is atomically flat. Note that a singular surface is defined as one that is composed solely of one such terrace. It is impossible to orient an actual single-crystal surface to precise atomic flatness, however, and steps provide the means to reconcile the macroscopic surface plane with the microscopic orientation. A step separates singular terraces, or domains, from each other. Most steps are single atomic height steps, although for certain surfaces a double-height step is required in order that each terrace is equivalent. Figure A1.7.1(a) illustrates two perfect terraces separated by a perfect monoatomic step. The overall number and arrangement of the steps on any actual surface is determined by the misorientation, which

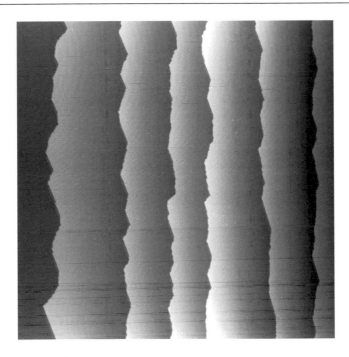

Figure A1.7.2. Large-scale (5000 Å × 5000 Å) scanning tunnelling microscope image of a stepped Si(111)-(7 × 7) surface showing flat terraces separated by step edges (courtesy of Alison Baski).

is the angle between the nominal crystal axis direction and the actual surface normal. If the misorientation is not along a low-index direction, then there will be kinks in the steps to adjust for this, as illustrated in figure A1.7.1(b).

A surface that differs from a singular orientation by a finite amount is called vicinal. Vicinal surfaces are composed of well oriented singular domains separated by steps. Figure A1.7.2 shows a large-scale scanning tunnel microscope (STM) image of a stepped Si(111) surface (STM instruments are described in sections A1.7.5.3 and B1.20). In this image, flat terraces separated by well defined steps are easily visible. It can be seen that the steps are all pointing along the same general direction.

Although all real surfaces have steps, they are not usually labelled as vicinal unless they are purposely misoriented in order to create a regular array of steps. Vicinal surfaces have unique properties, which make them useful for many types of experiments. For example, steps are often more chemically reactive than terraces, so that vicinal surfaces provide a means for investigating reactions at step edges. Also, it is possible to grow 'nanowires' by deposition of a metal onto a surface of another metal in such a way that the deposited metal diffuses to and attaches at the step edges [3].

Many surfaces have additional defects other than steps, however, some of which are illustrated in figure A1.7.1(b). For example, steps are usually not flat, i.e. they do not lie along a single low-index direction, but instead have kinks. Terraces are also not always perfectly flat, and often contain defects such as adatoms or vacancies. An *adatom* is an isolated atom adsorbed on top of a terrace, while a *vacancy* is an atom or group of atoms missing from an otherwise perfect terrace. In addition, a group of atoms called an *island* may form on a terrace, as illustrated.

Much surface work is concerned with the local atomic structure associated with a single domain. Some surfaces are essentially bulk-terminated, i.e. the atomic positions are basically unchanged from those of the bulk as if the atomic bonds in the crystal were simply cut. More common, however, are deviations from the

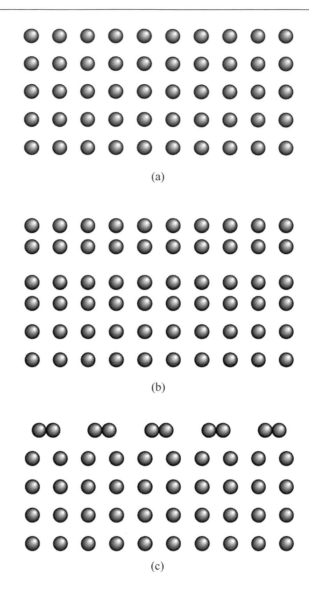

Figure A1.7.3. Schematic illustration showing side views of (a) a bulk-terminated surface, (b) a relaxed surface with oscillatory behaviour, and (c) a reconstructed surface.

bulk atomic structure. These structural adjustments can be classified as either relaxations or reconstructions. To illustrate the various classifications of surface structures, figure A1.7.3(a) shows a side-view of a bulk-terminated surface, figure A1.7.3(b) shows an oscillatory relaxation and figure A1.7.3(c) shows a reconstructed surface.

A1.7.2.3 Relaxation

Most metal surfaces have the same atomic structure as in the bulk, except that the interlayer spacings of the outermost few atomic layers differ from the bulk values. In other words, entire atomic layers are shifted as

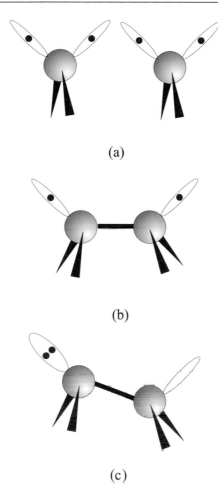

Figure A1.7.4. Schematic illustration of two Si atoms as they would be oriented on the (100) surface. (a) Bulk-terminated structure showing two dangling bonds (lone electrons) per atom. (b) Symmetric dimer, in which two electrons are shared and each atom has one remaining dangling bond. (c) Asymmetric dimer in which two electrons pair up on one atom and the other has an empty orbital.

a whole in a direction perpendicular to the surface. This is called *relaxation*, and it can be either inward or outward. Relaxation is usually reported as a percentage of the value of the bulk interlayer spacing. Relaxation does not affect the two-dimensional surface unit cell symmetry, so surfaces that are purely relaxed have (1×1) symmetry.

The reason that relaxation occurs can be understood in terms of the free electron character of a metal. Because the electrons are free, they are relatively unperturbed by the periodic ion cores. Thus, the electron density is homogeneous parallel to the surface. At the surface of a metal the solid abruptly stops, so that there is a net dipole perpendicular to the surface. This dipole field acts to attract electrons to the surface and is, in fact, responsible for the surface work function. The dipole field also interacts with the ion cores of the outermost atomic layer, however, causing them to move perpendicular to the surface. Note that some metals are also reconstructed since the assumption of perfectly free electrons unperturbed by the ion cores is not completely valid.

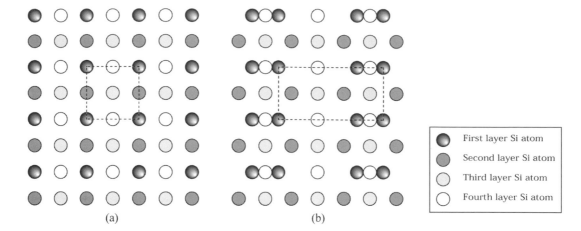

Figure A1.7.5. Schematic illustration showing the top view of the Si(100) surface. (a) Bulk-terminated structure. (b) Dimerized Si(100)-(2 × 1) structure. The dashed boxes show the two-dimensional surface unit cells.

In many materials, the relaxations between the layers oscillate. For example, if the first-to-second layer spacing is reduced by a few percent, the second-to-third layer spacing would be increased, but by a smaller amount, as illustrated in figure A1.7.3(b). These oscillatory relaxations have been measured with LEED [4, 5] and ion scattering [6, 7] to extend to at least the fifth atomic layer into the material. The oscillatory nature of the relaxations results from oscillations in the electron density perpendicular to the surface, which are called Friedel oscillations [8]. The Friedel oscillations arise from Fermi–Dirac statistics and impart oscillatory forces to the ion cores.

A1.7.2.4   Reconstruction

The three-dimensional symmetry that is present in the bulk of a crystalline solid is abruptly lost at the surface. In order to minimize the surface energy, the thermodynamically stable surface atomic structures of many materials differ considerably from the structure of the bulk. These materials are still crystalline at the surface, in that one can define a two-dimensional surface unit cell parallel to the surface, but the atomic positions in the unit cell differ from those of the bulk structure. Such a change in the local structure at the surface is called a *reconstruction*.

For covalently bonded semiconductors, the largest driving force behind reconstructions is the need to pair up electrons. For example, as shown in figure A1.7.4(a), if a Si(100) surface were to be bulk-terminated, each surface atom would have two lone electrons pointing away from the surface (assuming that each atom remains in a tetrahedral configuration). Lone electrons protruding into the vacuum are referred to as *dangling bonds*. Instead of maintaining two dangling bonds at each surface atom, however, dimers can form in which electrons are shared by two neighbouring atoms. Figure A1.7.4(b) shows two symmetrically dimerized Si atoms, in which two dangling bonds have been eliminated, although the atoms still have one dangling bond each. Figure A1.7.4(c) shows the asymmetric arrangement that further lowers the energy by pairing up two lone electrons onto one atom. In this arrangement, the electrons at any instant are associated with one Si atom, while the other has an empty orbital. This distorts the crystal structure, as the upper atom is essentially sp^3 hybridized, i.e. tetrahedral, while the other is sp^2, i.e. flat.

Figure A1.7.5(a) shows a larger scale schematic of the Si(100) surface if it were to be bulk-terminated, while figure A1.7.5(b) shows the arrangement after the dimers have been formed. The dashed boxes outline the two-dimensional surface unit cells. The reconstructed Si(100) surface has a unit cell that is two times

Si(1 1 1)-7×7

Top view

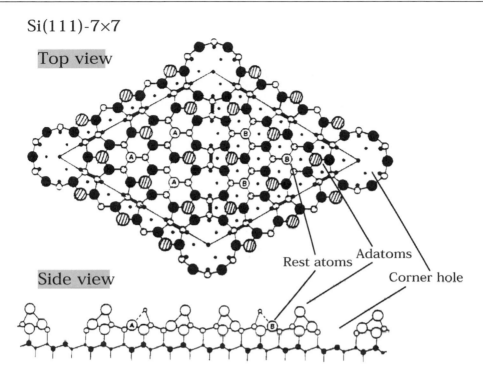

Side view

Rest atoms   Adatoms

Corner hole

Figure A1.7.6. Schematic diagrams of the DAS model of the Si(111)-(7 × 7) surface structure. There are 12 'adatoms' per unit cell in the outermost layer, which each have one dangling bond perpendicular to the surface. The second layer, called the rest layer, also has six 'rest' atoms per unit cell, each with a perpendicular dangling bond. The 'corner holes' at the edges of the unit cells also contain one atom with a dangling bond.

larger than the bulk unit cell in one direction and the same in the other. Thus, it has a (2 × 1) symmetry and the surface is labelled as Si(100)-(2 × 1). Note that in actuality, however, any real Si(100) surface is composed of a mixture of (2 × 1) and (1 × 2) domains. This is because the dimer direction rotates by 90° at each step edge.

The surface unit cell of a reconstructed surface is usually, but not necessarily, larger than the corresponding bulk-terminated two-dimensional unit cell would be. The LEED pattern is therefore usually the first indication that a reconstruction exists. However, certain surfaces, such as GaAs(110), have a reconstruction with a surface unit cell that is still (1 × 1). At the GaAs(110) surface, Ga atoms are moved inward perpendicular to the surface, while As atoms are moved outward.

The most celebrated surface reconstruction is probably that of Si(111)-(7 × 7). The fact that this surface has such a large unit cell had been known for some time from LEED, but the detailed atomic structure took many person-years of work to elucidate. Photoelectron spectroscopy [9], STM [10] and many other techniques were applied to the determination of this structure. It was transmission electron diffraction (TED), however, that provided the final information enabling the structure to be determined [11]. The structure now accepted is the so-called DAS, or dimer adatom stacking-fault, model, as shown in figure A1.7.6. In this structure, there are a total of 19 dangling bonds per unit cell, which can be compared to the 49 dangling bonds that the bulk-terminated surface would have. Figure A1.7.7 shows an atomic resolution STM image of the Si(111)-(7 × 7) surface. The bright spots in the image represent individual Si adatoms.

Although most metal surfaces exhibit only relaxation, some do have reconstructions. For example, the fcc metals, Pt(110), Au(110) and Ir(110), each have a (1 × 2) surface unit cell. The accepted structure of

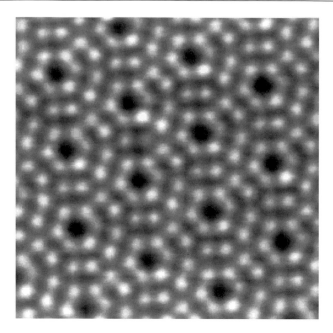

Figure A1.7.7. Atomic-resolution, empty-state STM image (100 Å × 100 Å) of the reconstructed Si(111)-7 × 7 surface. The bright spots correspond to a top layer of adatoms, with 12 adatoms per unit cell (courtesy of Alison Baski).

these surfaces is a missing row model, in which every other surface row is missing. Also, as discussed below, when an adsorbate attaches to a metal surface, a reconstruction of the underlying substrate may be induced.

Reliable tables that list many known surface structures can be found in [1]. Also, the National Institute of Standards and Technology (NIST) maintains databases of surface structures and other surface-related information, which can be found at *http://www.nist.gov/srd/surface.htm*.

A1.7.2.5   Self-diffusion

The atoms on the outermost surface of a solid are not necessarily static, particularly as the surface temperature is raised. There has been much theoretical [12, 13] and experimental work (described below) undertaken to investigate surface self-diffusion. These studies have shown that surfaces actually have dynamic, changing structures. For example, atoms can diffuse along a terrace to or from step edges. When atoms diffuse across a surface, they may move by hopping from one surface site to the next, or by exchanging places with second layer atoms.

The field ion microscope (FIM) has been used to monitor surface self-diffusion in real time. In the FIM, a sharp, crystalline tip is placed in a large electric field in a chamber filled with He gas [14]. At the tip, He ions are formed, and then accelerated away from the tip. The angular distribution of the He ions provides a picture of the atoms at the tip with atomic resolution. In these images, it has been possible to monitor the diffusion of a single adatom on a surface in real time [15]. The limitations of FIM, however, include its applicability only to metals, and the fact that the surfaces are limited to those that exist on a sharp tip, i.e. diffusion along a large terrace cannot be observed.

More recently, studies employing STM have been able to address surface self-diffusion across a terrace [16–19]. It is possible to image the same area on a surface as a function of time, and 'watch' the movement of individual atoms. These studies are limited only by the speed of the instrument. Note that the performance of STM instruments is constantly improving, and has now surpassed the 1 ps time resolution mark [20]. Not only has self-diffusion of surface atoms been studied, but the diffusion of vacancy defects on surfaces has also been observed with STM [18].

It has also been shown that sufficient surface self-diffusion can occur so that entire step edges move in a concerted manner. Although it does not achieve atomic resolution, the low-energy electron microscopy (LEEM) technique allows for the observation of the movement of step edges in real time [21]. LEEM has also been useful for studies of epitaxial growth and surface modifications due to chemical reactions.

A1.7.2.6 *Surface electronic structure*

At a surface, not only can the atomic structure differ from the bulk, but electronic energy levels are present that do not exist in the bulk band structure. These are referred to as 'surface states'. If the states are occupied, they can easily be measured with photoelectron spectroscopy (described in sections A1.7.5.1 and B1.25.2). If the states are unoccupied, a technique such as inverse photoemission or x-ray absorption is required [22, 23]. Also, note that STM has been used to measure surface states by monitoring the tunnelling current as a function of the bias voltage [24] (see section B1.20). This is sometimes called scanning tunnelling spectroscopy (STS).

Surface states can be divided into those that are intrinsic to a well ordered crystal surface with two-dimensional periodicity, and those that are extrinsic [25]. Intrinsic states include those that are associated with relaxation and reconstruction. Note, however, that even in a bulk-terminated surface, the outermost atoms are in a different electronic environment than the substrate atoms, which can also lead to intrinsic surface states. Extrinsic surface states are associated with imperfections in the perfect order of the surface region. Extrinsic states can also be formed by an adsorbate, as discussed below.

Note that in core-level photoelectron spectroscopy, it is often found that the surface atoms have a different binding energy than the bulk atoms. These are called surface core-level shifts (SCLS), and should not be confused with intrinsic surface states. An SCLS is observed because the atom is in a chemically different environment than the bulk atoms, but the core-level state that is being monitored is one that is present in all of the atoms in the material. A surface state, on the other hand, exists only at the particular surface.

A1.7.3 Adsorption

When a surface is exposed to a gas, the molecules can adsorb, or stick, to the surface. Adsorption is an extremely important process, as it is the first step in any surface chemical reaction. Some of the aspects of adsorption that surface science is concerned with include the mechanisms and kinetics of adsorption, the atomic bonding sites of adsorbates and the chemical reactions that occur with adsorbed molecules.

The coverage of adsorbates on a given substrate is usually reported in monolayers (ML). Most often, 1 ML is defined as the number of atoms in the outermost atomic layer of the unreconstructed, i.e. bulk-terminated, substrate. Sometimes, however, 1 ML is defined as the maximum number of adsorbate atoms that can stick to a particular surface, which is termed the saturation coverage. The saturation coverage can be much smaller than the number of surface atoms, particularly with large adsorbates. Thus, in reading the literature, care must be taken to understand how a particular author defines 1 ML.

Molecular adsorbates usually cover a substrate with a single layer, after which the surface becomes passive with respect to further adsorption. The actual saturation coverage varies from system to system, and is often determined by the strength of the repulsive interactions between neighbouring adsorbates. Some molecules will remain intact upon adsorption, while others will adsorb dissociatively. This is often a function of the surface temperature and composition. There are also often multiple adsorption states, in which the stronger, more tightly bound states fill first, and the more weakly bound states fill last. The factors that control adsorbate behaviour depend on the complex interactions between adsorbates and the substrate, and between the adsorbates themselves.

The probability for sticking is known as the sticking coefficient, S. Usually, S decreases with coverage. Thus, the sticking coefficient at zero coverage, the so-called initial sticking coefficient, S_0, reflects the interaction of a molecule with the bare surface.

In order to calibrate the sticking coefficient, one needs to determine the exposure, i.e. how many molecules have initially impacted a surface. The Langmuir (L) is a unit of exposure that is defined as 10^{-6} Torr s. An exposure of 1 L is approximately the number of incident molecules such that each outermost surface atom is impacted once. Thus, a 1 L exposure would produce 1 ML of adsorbates if the sticking coefficient were unity. Note that a quantitative calculation of the exposure per surface atom depends on the molecular weight of the gas molecules and on the actual density of surface atoms, but the approximations inherent in the definition of the Langmuir are often inconsequential.

A1.7.3.1   Physisorption

Adsorbates can physisorb onto a surface into a shallow potential well, typically 0.25 eV or less [25]. In physisorption, or physical adsorption, the electronic structure of the system is barely perturbed by the interaction, and the physisorbed species are held onto a surface by weak van der Waals forces. This attractive force is due to charge fluctuations in the surface and adsorbed molecules, such as mutually induced dipole moments. Because of the weak nature of this interaction, the equilibrium distance at which physisorbed molecules reside above a surface is relatively large, of the order of 3 Å or so. Physisorbed species can be induced to remain adsorbed for a long period of time if the sample temperature is held sufficiently low. Thus, most studies of physisorption are carried out with the sample cooled by liquid nitrogen or helium.

Note that the van der Waals forces that hold a physisorbed molecule to a surface exist for all atoms and molecules interacting with a surface. The physisorption energy is usually insignificant if the particle is attached to the surface by a much stronger chemisorption bond, as discussed below. Often, however, just before a molecule forms a strong chemical bond to a surface, it exists in a physisorbed precursor state for a short period of time, as discussed below in section A1.7.3.3.

A1.7.3.2   Chemisorption

Chemisorption occurs when the attractive potential well is large so that upon adsorption a strong chemical bond to a surface is formed. Chemisorption involves changes to both the molecule and surface electronic states. For example, when oxygen adsorbs onto a metal surface, a partially ionic bond is created as charge transfers from the substrate to the oxygen atom. Other chemisorbed species interact in a more covalent manner by sharing electrons, but this still involves perturbations to the electronic system.

Chemisorption is always an exothermic process. By convention, the heat of adsorption, ΔH_{ads}, has a positive sign, which is opposite to the normal thermodynamic convention [1]. Although the heat of adsorption has been directly measured with the use of a very sensitive microcalorimeter [26], it is more commonly measured via adsorption isotherms [1]. An isotherm is generated by measuring the coverage of adsorbates obtained by reaction at a fixed temperature as a function of the flux of incoming gas molecules. The flux is adjusted by regulating the pressure used during exposure. An analysis of the data then allows H_{ads} and other parameters to be determined. Heats of adsorption can also be determined from temperature programmed desorption (TPD) if the adsorption is reversible (TPD is discussed in sections A1.7.5.4 and B1.25).

When a molecule adsorbs to a surface, it can remain intact or it may dissociate. Dissociative chemisorption is common for many types of molecules, particularly if all of the electrons in the molecule are tied up so that there are no electrons available for bonding to the surface without dissociation. Often, a molecule will dissociate upon adsorption, and then recombine and desorb intact when the sample is heated. In this case, dissociative chemisorption can be detected with TPD by employing isotopically labelled molecules.

If mixing occurs during the adsorption/desorption sequence, it indicates that the initial adsorption was dissociative.

Atom abstraction occurs when a dissociation reaction occurs on a surface in which one of the dissociation
products sticks to the surface, while another is emitted. If the chemisorption reaction is particularly exothermic,
the excess energy generated by chemical bond formation can be channelled into the kinetic energy of the
desorbed dissociation fragment. An example of atom abstraction involves the reaction of molecular halogens
with Si surfaces [27, 28]. In this case, one halogen atom chemisorbs while the other atom is ejected from the
surface.

A1.7.3.3   Adsorption kinetics

When an atom or molecule approaches a surface, it feels an attractive force. The interaction potential between
the atom or molecule and the surface, which depends on the distance between the molecule and the surface
and on the lateral position above the surface, determines the strength of this force. The incoming molecule
feels this potential, and upon adsorption becomes trapped near the minimum in the well. Often the molecule
has to overcome an activation barrier, E_{act}, before adsorption can occur.

It is the relationship between the bound potential energy surface of an adsorbate and the vibrational states
of the molecule that determine whether an adsorbate remains on the surface, or whether it desorbs after a
period of time. The lifetime of the adsorbed state, τ, depends on the size of the well relative to the vibrational
energy inherent in the system, and can be written as

$$\tau = \tau_0 \exp(\Delta H_{ads}/kT). \qquad (A1.7.1)$$

Such lifetimes vary from less than a picosecond to times greater than the age of the universe [29]. Thus,
adsorbed states with short lifetimes can occur during a surface chemical reaction, or long-lived adsorbed
states exist in which atoms or molecules remain attached to a surface indefinitely.

In this manner, it can also be seen that molecules will desorb as the surface temperature is raised. This
is the phenomenon employed for TPD spectroscopy (see sections A1.7.5.4 and B1.25). Note that some
adsorbates may adsorb and desorb reversibly, i.e. the heats of adsorption and desorption are equal. Other
adsorbates, however, will adsorb and desorb via different pathways.

Note that chemisorption often begins with physisorption into a weakly bound precursor state. While in
this state, the molecule can diffuse along the surface to find a likely site for chemisorption. This is particularly
important in the case of dissociative chemisorption, as the precursor state can involve physisorption of the
intact molecule. If a precursor state is involved in adsorption, a negative temperature dependence to the
adsorption probability will be found. A higher surface temperature reduces the lifetime of the physisorbed
precursor state, since a weakly bound species will not remain on the surface in the presence of thermal
excitation. Thus, the sticking probability will be reduced at higher surface temperatures.

The kinetics of the adsorption process are important in determining the value and behaviour of S for any
given system. There are several factors that come into play in determining S [25].

(a)   The activation barrier must be overcome in order for a molecule to adsorb. Thus, only the fraction of the
incident particles whose energy exceeds E_{act} will actually stick.

(b)   The electronic orbitals of the incoming molecule must have the correct orientation with respect to the
orbitals of the surface. Thus, only a fraction of the incoming molecules will immediately stick to the
surface. Some of the incoming molecules may, however, diffuse across the surface while in a precursor
state until they achieve the proper orientation. Thus, the details of how the potential energy varies across
the surface are critical in determining the adsorption kinetics.

(c)   Upon adsorption, a molecule must effectively lose the remaining part of its kinetic energy, and possibly
the excess energy liberated by an exothermic reaction, in a time period smaller than one vibrational

period. Thus, excitations of the surface that can carry away this excess energy, such as plasmons or phonons, play a role in the adsorption kinetics.

(d) Adsorption sites must be available for reaction. Thus, the kinetics may depend critically on the coverage of adsorbates already present on the surface, as these adsorbates may block or modify the remaining adsorption sites.

A1.7.3.4 Adsorption models

The most basic model for chemisorption is that developed by Langmuir. In the Langmuir model, it is assumed that there is a finite number of adsorption sites available on a surface, and each has an equal probability for reaction. Once a particular site is occupied, however, the adsorption probability at that site goes to zero. Furthermore, it is assumed that the adsorbates do not diffuse, so that once a site is occupied it remains unreactive until the adsorbate desorbs from the surface. Thus, the sticking probability S goes to zero when the coverage, θ, reaches the saturation coverage, θ_0. These assumptions lead to the following relationship between the sticking coefficient and the surface coverage,

$$S = S_0(1 - \theta/\theta_0). \tag{A1.7.2}$$

The straight line in figure A1.7.8 shows the relationships between S and θ expected for various models, with the straight line indicating Langmuir adsorption.

Adsorbate atoms have a finite lifetime, τ, for remaining on a surface. Thus, there will always be a flux of molecules leaving the surface even as additional molecules are being adsorbed. If the desorption rate is equal to the rate of adsorption, then an isotherm can be collected by measuring the equilibrium coverage at a fixed temperature as a function of pressure, p. From the assumptions of the Langmuir model, one can derive the following expression relating the equilibrium coverage to pressure [29].

$$\theta = \frac{\chi p}{1 + \chi p} \tag{A1.7.3}$$

where χ is a constant that depends on the adsorbate lifetime and surface temperature, T, as

$$\chi \propto \tau T^{1/2}. \tag{A1.7.4}$$

If Langmuir adsorption occurs, then a plot of θ versus p for a particular isotherm will display the form of equation (A1.7.3). Measurements of isotherms are routinely employed in this manner in order to determine adsorption kinetics.

Langmuir adsorption adequately describes the behaviour of many systems in which strong chemisorption takes place, but it has limitations. For one, the sticking at surface sites actually does depend on the occupancy of neighbouring sites. Thus, sticking probability usually changes with coverage. A common observation, for example, is that the sticking probability is reduced exponentially with coverage, i.e.

$$S \propto \exp(-\alpha\theta/kT) \tag{A1.7.5}$$

which is called the Elovich equation [25]. This is compared to the Langmuir model in figure A1.7.8.

If adsorption occurs via a physisorbed precursor, then the sticking probability at low coverages will be enhanced due to the ability of the precursor to diffuse and find a lattice site [30]. The details depend on parameters such as strength of the lateral interactions between the adsorbates and the relative rates of desorption and reaction of the precursor. In figure A1.7.8, an example of a plot of S versus θ for precursor mediated adsorption is presented.

Another limitation of the Langmuir model is that it does not account for multilayer adsorption. The Braunauer, Emmett and Teller (BET) model is a refinement of Langmuir adsorption in which multiple layers

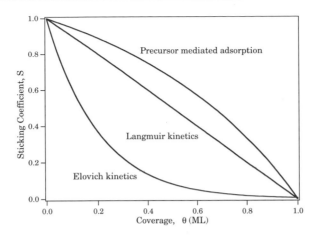

Figure A1.7.8. Sticking probability as a function of surface coverage for three different adsorption models.

of adsorbates are allowed [29, 31]. In the BET model, the particles in each layer act as the adsorption sites for the subsequent layers. There are many refinements to this approach, in which parameters such as sticking coefficient, activation energy, etc, are considered to be different for each layer.

A1.7.3.5 Adsorption sites

When atoms, molecules, or molecular fragments adsorb onto a single-crystal surface, they often arrange themselves into an ordered pattern. Generally, the size of the adsorbate-induced two-dimensional surface unit cell is larger than that of the clean surface. The same nomenclature is used to describe the surface unit cell of an adsorbate system as is used to describe a reconstructed surface, i.e. the symmetry is given with respect to the bulk terminated (unreconstructed) two-dimensional surface unit cell.

When chemisorption takes place, there is a strong interaction between the adsorbate and the substrate. The details of this interaction determine the local bonding site, particularly at the lowest coverages. At higher coverages, adsorbate–adsorbate interactions begin to also play a role. Most non-metallic atoms will adsorb above the surface at specific lattice sites. Some systems have multiple bonding sites. In this case, one site will usually dominate at low coverage, but a second, less stable site will be filled at higher coverages. Some adsorbates will interact with only one surface atom, i.e. be singly coordinated, while others prefer multiple coordinated adsorption sites. Other systems may form alloys or intermix during adsorption.

Local adsorption sites can be roughly classified either as on-top, bridge or hollow, as illustrated for a four-fold symmetric surface in figure A1.7.9. In the on-top configuration, a singly coordinated adsorbate is attached directly on top of a substrate atom. A bridge site is the two-fold site between two neighbouring surface atoms. A hollow site is positioned between three or four surface atoms, for surfaces with three- or four-fold symmetry, respectively.

There are interactions between the adsorbates themselves, which greatly affect the structure of the adsorbates [32]. If surface diffusion is sufficiently facile during or following the adsorption step, attractive interactions can induce the adsorbates to form islands in which the local adsorbate concentration is quite high. Other adsorbates may repel each other at low coverages forming structures in which the distance between adsorbates is maximized. Certain co-adsorption systems form complex ordered overlayer structures. The driving force in forming ordered overlayers are these adsorbate–adsorbate interactions. These interactions dominate the long-range structure of the surface in the same way that long-range interactions cause the formation of three-dimensional solid crystals.

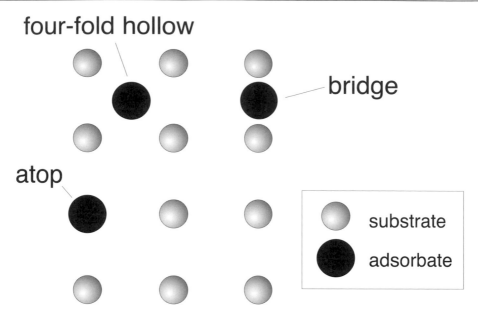

Figure A1.7.9. Schematic diagram illustrating three types of adsorption sites.

Adsorbed atoms and molecules can also diffuse across terraces from one adsorption site to another [33]. On a perfect terrace, adatom diffusion could be considered as a 'random walk' between adsorption sites, with a diffusivity that depends on the barrier height between neighbouring sites and the surface temperature [29]. The diffusion of adsorbates has been studied with FIM [14], STM [34, 35] and laser-induced thermal desorption [36].

A1.7.3.6 *Adsorption-induced reconstruction*

When an adsorbate attaches to a surface, the substrate itself may respond to the perturbation by either losing its relaxation or reconstruction, or by forming a new reconstruction. This is not surprising, considering the strength of a chemisorption bond. Chemisorption bonds can provide electrons to satisfy the requirements for charge neutrality or electron pairing that may otherwise be missing at a surface.

For a reconstructed surface, the effect of an adsorbate can be to provide a more bulk-like environment for the outermost layer of substrate atoms, thereby lifting the reconstruction. An example of this is As adsorbed onto Si(111)-(7 × 7) [37]. Arsenic atoms have one less valence electron than Si. Thus, if an As atom were to replace each outermost Si atom in the bulk-terminated structure, a smooth surface with no unpaired electrons would be produced, with a second layer consisting of Si atoms in their bulk positions. Arsenic adsorption has, in fact, been found to remove the reconstruction and form a Si(111)-(1 × 1)–As structure. This surface has a particularly high stability due to the absence of dangling bonds.

An example of the formation of a new reconstruction is given by certain fcc (110) metal surfaces. The clean surfaces have (1 × 1) symmetry, but become (2 × 1) upon adsorption of oxygen [16, 38]. The (2 × 1) symmetry is not just due to oxygen being adsorbed into a (2 × 1) surface unit cell, but also because the substrate atoms rearrange themselves into a new configuration. The reconstruction that occurs is sometimes called the 'missing-row' structure because every other row of surface atoms along the 2× direction is missing. A more correct terminology, however, is the 'added-row' structure, as STM studies have shown that it is formed by metal atoms diffusing away from a step edge and onto a terrace to create a new first layer, rather than by atoms

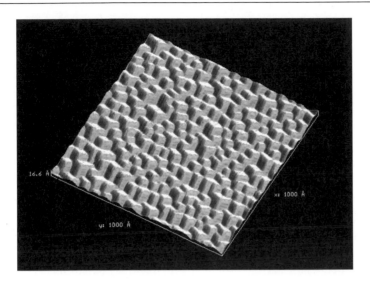

Figure A1.7.10. STM image (1000 Å × 1000 Å) of the (111) surface of a tungsten single crystal, after it had been coated with a very thin film of palladium and heated to about 800 K (courtesy of Ted Madey).

being removed [16]. In this case, the (2 × 1) symmetry results not just from the long-range structure of the adsorbed layer, but also from a rearrangement of the substrate atoms.

A more dramatic type of restructuring occurs with the adsorption of alkali metals onto certain fcc metal surfaces [39]. In this case, multilayer composite surfaces are formed in which the alkali and metal atoms are intermixed in an ordered structure. These structures involve the substitution of alkali atoms into substrate sites, and the details of the structures are found to be coverage-dependent. The structures are influenced by the repulsion between the dipoles formed by neighbouring alkali adsorbates and by the interactions of the alkalis with the substrate itself [40].

There is also an interesting phenomenon that has been observed following the deposition of the order of 1 ML of a metal onto another metallic substrate. For certain systems, this small coverage is sufficient to alter the surface energy so that a large-scale faceting of the surface occurs [41]. The morphology of such a faceted surface can be seen in the STM image of figure A1.7.10, which was collected from an annealed W(111) surface onto which a small amount of Pd had been deposited.

A1.7.3.7 *Work function changes induced by adsorbates*

The surface work function is formally defined as the minimum energy needed in order to remove an electron from a solid. It is often described as being the difference in energy between the Fermi level and the vacuum level of a solid. The work function is a sensitive measure of the surface electronic structure, and can be measured in a number of ways, as described in section B1.26.4. Many processes, such as catalytic surface reactions or resonant charge transfer between ions and surfaces, are critically dependent on the work function.

When an electropositive or electronegative adsorbate attaches itself to a surface, there is usually a change in the surface dipole, which, in turn, affects the surface work function. Thus, very small coverages of adsorbates can be used to modify the surface work function in order to ascertain the role that the work function plays in a given process. Conversely, work function measurements can be used to accurately determine the coverage of these adsorbates.

For example, alkali ions adsorbed onto surfaces donate some or all of their valence electron to the solid, thereby producing dipoles pointing away from the surface [40, 42]. This has the effect of substantially lowering the work function for coverages as small as 0.01 ML. When the alkali coverage is increased to the point at which the alkali adsorbates can interact with each other, they tend to depolarize. Thus, the work function initially decreases as alkali atoms are adsorbed until a minimum in the work function is attained. At higher alkali coverages, the work function may increase slightly due to the adsorbate–adsorbate interactions. Note that it is very common to use alkali adsorption as a means of modifying the surface work function.

A1.7.3.8 *Surface chemical reactions*

Surface chemical reactions can be classified into three major categories [29]:

(a) corrosion reactions,

(b) crystal growth reactions,

(c) catalytic reactions.

All three types of reactions begin with adsorption of species onto a surface from the gas phase.

In corrosion, adsorbates react directly with the substrate atoms to form new chemical species. The products may desorb from the surface (volatilization reaction) or may remain adsorbed in forming a corrosion layer. Corrosion reactions have many industrial applications, such as dry etching of semiconductor surfaces. An example of a volatilization reaction is the etching of Si by fluorine [43]. In this case, fluorine reacts with the Si surface to form SiF_4 gas. Note that the crystallinity of the remaining surface is also severely disrupted by this reaction. An example of corrosion layer formation is the oxidation of Fe metal to form rust. In this case, none of the products are volatile, but the crystallinity of the surface is disrupted as the bulk oxide forms. Corrosion and etching reactions are discussed in more detail in sections A3.10, C2.9 and C2.20.

The growth of solid films onto solid substrates allows for the production of artificial structures that can be used for many purposes. For example, film growth is used to create pn junctions and metal–semiconductor contacts during semiconductor manufacture, and to produce catalytic surfaces with properties that are not found in any single material. Lubrication can be applied to solid surfaces by the appropriate growth of a solid lubricating film. Film growth is also used to fabricate quantum-wells and other types of layered structures that have unique electronic properties. These reactions may involve dissociative or non-dissociative adsorption as the first step. The three basic types of film growth reactions are physical vapour deposition (PVD), chemical vapour deposition (CVD) and molecular beam epitaxy (MBE). In PVD, an atomic gas is condensed onto a surface forming a solid. In CVD, a molecular gas dissociates upon adsorption. Some of the dissociation fragments solidify to form the material, while other dissociation fragments are evolved back into the gas phase. In MBE, carefully controlled atomic and/or molecular beams are condensed onto a surface in the proper stoichiometry in order to grow a desired material [44]. MBE is particularly important in the growth of III–V semiconductor materials.

In crystal growth reactions, material is deposited onto a surface in order to extend the surface crystal structure, or to grow a new material, without disruption of the underlying substrate. Growth mechanisms can be roughly divided into three categories. If the film grows one atomic layer at a time such that a smooth, uniform film is created, it is called *Frank von der Merwe* growth. Such layer-by-layer growth will occur if the surface energy of the overlayer is lower than that of the substrate. If the film grows in a von der Merwe growth mode such that it forms a single crystal in registry with the substrate, it is referred to as epitaxial. The smaller the lattice mismatch between the overlayer and the substrate, the more likely it is that epitaxial growth can be achieved. If the first ML is deposited uniformly, but subsequent layers agglomerate into islands, it is called *Stranski–Krastanov* growth. In this case, the surface energy of the first layer is lower than that of the substrate, but the surface energy of the bulk overlayer material is higher. If the adsorbate agglomerates into islands immediately, without even wetting the surface, it is referred to as *Vollmer–Weber* growth. In this case, the surface energy of the substrate is lower than that of the overlayer. Growth reactions are discussed in more detail in sections A3.10 and C2.20.

The desire to understand catalytic chemistry was one of the motivating forces underlying the development of surface science. In a catalytic reaction, the reactants first adsorb onto the surface and then react with each other to form volatile product(s). The substrate itself is not affected by the reaction, but the reaction would not occur without its presence. Types of catalytic reactions include exchange, recombination, unimolecular decomposition, and bimolecular reactions. A reaction would be considered to be of the *Langmuir–Hinshelwood* type if both reactants first adsorbed onto the surface, and then reacted to form the products. If one reactant first adsorbs, and the other then reacts with it directly from the gas phase, the reaction is of the *Eley–Ridel* type. Catalytic reactions are discussed in more detail in sections A3.10 and C2.8.

A tremendous amount of work has been done to delineate the detailed reaction mechanisms for many catalytic reactions on well characterized surfaces [1, 45]. Many of these studies involved impinging molecules onto surfaces at relatively low pressures, and then interrogating the surfaces in vacuum with surface science techniques. For example, a useful technique for catalytic studies is TPD, as the reactants can be adsorbed onto the sample in one step, and the products formed in a second step when the sample is heated. Note that catalytic surface studies have also been performed by reacting samples in a high-pressure cell, and then returning them to vacuum for measurement.

Recently, *in situ* studies of catalytic surface chemical reactions at high pressures have been undertaken [46, 47]. These studies employed sum frequency generation (SFG) and STM in order to probe the surfaces as the reactions are occurring under conditions similar to those employed for industrial catalysis (SFG is a laser-based technique that is described in sections A1.7.5.5 and B1.22). These studies have shown that the highly stable adsorbate sites that are probed under vacuum conditions are not necessarily the same sites that are active in high-pressure catalysis. Instead, less stable sites that are only occupied at high pressures are often responsible for catalysis. Because the active adsorption sites are not populated at low pressures, they are not seen in vacuum surface science experiments. Despite this, however, the low-pressure experiments are necessary in order to calibrate the spectroscopy so that the high-pressure results can be properly interpreted.

A1.7.4 Preparation of clean surfaces

The exact methods employed to prepare any particular surface for study vary from material to material, and are usually determined empirically. In some respects, sample preparation is more of an art than a science. Thus, it is always best to consult the literature to look for preparation methods before starting with a new material.

Most samples require some initial *ex situ* preparation before insertion into a vacuum chamber [45]. A bulk single crystal must first be oriented [48], which is usually done with back-reflection Laue x-ray diffraction, and then cut to expose the desired crystal plane. Samples are routinely prepared to be within $\pm 1°$ of the desired orientation, but an accuracy of $\pm 1/4°$ or better can be routinely obtained. Cutting is often done using an electric discharge machine (spark cutter) for metals or a diamond saw or slurry drill for semiconductors. The surface must then be polished. Most polishing is done mechanically, with alumina or diamond paste, by polishing with finer and finer grits until the finest available grit is employed, which is usually of the order of 0.5 μm. Often, as a final step, the surface is electrochemically or chemi-mechanically polished. In addition, some samples are chemically reacted in solution in order to remove a large portion of the oxide layer that is present due to reaction with the atmosphere. Note that this layer is referred to as the *native oxide*.

In order to maintain the cleanliness of a surface at the atomic level, investigations must be carried out in ultra-high vacuum (UHV). UHV is usually considered to be a pressure of the order of 1×10^{-10} Torr or below. Surface science techniques are often sensitive to adsorbate levels as small as 1% of ML or less, so that great care must be taken to keep the surface contamination to a minimum. Even at moderate pressures, many contaminants will easily adsorb onto a surface. For example, at 1×10^{-6} Torr, which is a typical pressure realized by many diffusion-pumped systems, a 1 L exposure to the background gases will occur in 1 s. Thus, any molecule that is present in the background and has a high sticking probability, such as water or oxygen, will cover the surface within seconds. It is for this reason that extremely low pressures are necessary in order to keep surfaces contaminant-free at the atomic level.

Once a sample is properly oriented and polished, it is placed into a UHV chamber for the final preparation steps. Samples are processed *in situ* by a variety of methods in order to produce an atomically clean and flat surface. Ion bombardment and annealing (IBA) is the most common method used. Other methods include cleaving and film growth.

In IBA, the samples are first irradiated for a period of time with noble gas ions, such as Ar^+ or Ne^+, that have kinetic energies in the range of 0.5–2.0 keV. This removes the outermost layers of adsorbed contaminants and oxides by the process of sputtering. In sputtering, ions directly collide with the atoms at the surface of the sample, physically knocking out material. Usually the sample is at room temperature during sputtering and the ion beam is incident normal to the surface. Certain materials, however, are better prepared by sputtering at elevated temperature or with different incidence directions.

Because keV ions penetrate several layers deep into a solid, a side effect of sputtering is that it destroys the crystallinity of the surface region. In the preparation of a single-crystal surface, the damage is removed by annealing (heating) the surface in UHV in order to re-crystallize it. Care must be taken to not overheat the sample for (at least) two reasons. First, surfaces will melt and/or sublime well below the melting point of the bulk material. Second, contaminants sometimes diffuse to the surface from the bulk at high temperatures. If the annealing temperature is not high enough, however, the material will not be sufficiently well ordered. Thus, care must be taken to determine the optimal annealing temperature for any given material.

After a sample has been sputtered to remove the contaminants and then annealed at the proper temperature to re-crystallize the surface region, a clean, atomically smooth and homogeneous surface can be produced. Note, however, that it usually takes many cycles of IBA to produce a good surface. This is because a side effect of annealing is that the chamber pressure is raised as adsorbed gases are emitted from the sample holder, which causes additional contaminants to be deposited on the surface. Also, contaminants may have diffused to the surface from the bulk during annealing. Another round of sputtering is then needed to remove these additional contaminants. After a sufficient number of cycles, the contaminants in either the sample holder or the bulk solid are depleted to the point that annealing does not significantly contaminate the surface.

For some materials, the most notable being silicon, heating alone suffices to clean the surface. Commercial Si wafers are produced with a thin layer of silicon dioxide covering the surface. This native oxide is inert to reaction with the atmosphere, and therefore keeps the underlying Si material clean. The native oxide layer is desorbed, i.e. removed into the gas phase, by heating the wafer in UHV to a temperature above approximately 1100 °C. This procedure directly forms a clean, well ordered Si surface.

At times, *in situ* chemical treatments are used to remove particular contaminants. This is done by introducing a low pressure ($\sim 10^6$ Torr) of gas to the vacuum chamber, which causes it to adsorb (stick) to the sample surface, followed by heating the sample to remove the adsorbates. The purpose is to induce a chemical reaction between the contaminants and the adsorbed gas to form a volatile product. For example, carbon can be removed by exposing a surface to hydrogen gas and then heating it. This procedure produces methane gas, which desorbs from the surface into the vacuum. Similarly, hydrogen adsorption can be used to remove oxygen by forming gaseous water molecules.

Certain materials, most notably semiconductors, can be mechanically cleaved along a low-index crystal plane *in situ* in a UHV chamber to produce an ordered surface without contamination. This is done using a sharp blade to slice the sample along its preferred cleavage direction. For example, Si cleaves along the (111) plane, while III–V semiconductors cleave along the (110) plane. Note that the atomic structure of a cleaved surface is not necessarily the same as that of the same crystal face following treatment by IBA.

In addition, ultra-pure films are often grown *in situ* by evaporation of material from a filament or crucible, by molecular beam epitaxy (MBE), or with the use of chemical methods. Since the films are grown in UHV, the surfaces as grown will be atomically clean. Film growth has the advantage of producing a much cleaner and/or more highly ordered surface than could be obtained with IBA. In addition, certain structures can be formed with MBE that cannot be produced by any other preparation method. Film growth is discussed more explicitly above in section A1.7.3.8 and in section A3.10.

A1.7.5   Techniques for the investigation of surfaces

Because surface science employs a multitude of techniques, it is necessary that any worker in the field be acquainted with at least the basic principles underlying the most popular ones. These will be briefly described here. For a more detailed discussion of the physics underlying the major surface analysis techniques, see the appropriate chapter in this encyclopedia, or [49].

With the exception of the scanning probe microscopies, most surface analysis techniques involve scattering of one type or another, as illustrated in figure A1.7.11. A particle is incident onto a surface, and its interaction with the surface either causes a change to the particles' energy and/or trajectory, or the interaction induces the emission of a secondary particle(s). The particles that interact with the surface can be electrons, ions, photons or even heat. An analysis of the mass, energy and/or trajectory of the emitted particles, or the dependence of the emitted particle yield on a property of the incident particles, is used to infer information about the surface. Although these probes are indirect, they do provide reliable information about the surface composition and structure.

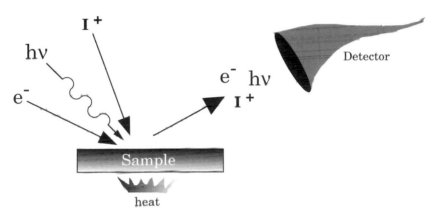

Figure A1.7.11. Schematic diagram of a generic surface science experiment. Particles, such as photons, electrons, or ions, are incident onto a solid surface, while the particles emitted from the surface are collected and measured by the detector.

Energetic particles interacting can also modify the structure and/or stimulate chemical processes on a surface. Absorbed particles excite electronic and/or vibrational (phonon) states in the near-surface region. Some surface scientists investigate the fundamental details of particle–surface interactions, while others are concerned about monitoring the changes to the surface induced by such interactions. Because of the importance of these interactions, the physics involved in both surface analysis and surface modification are discussed in this section.

The instrumentation employed for these studies is almost always housed inside a stainless-steel UHV chamber. One UHV chamber usually contains equipment for performing many individual techniques, each mounted on a different port, so that they can all be applied to the same sample. The sample is mounted onto a manipulator that allows for movement of the sample from one port to another, as well as for *in situ* heating and often cooling with liquid nitrogen (or helium). The chamber contains facilities for sample preparation, such as sputtering and annealing, as well as the possibility for gaseous exposures and/or film growth. Many instruments also contain facilities for the transfer of the sample from one chamber to another while maintaining UHV. This allows for the incorporation of even more techniques, as well as the easy introduction of new samples into the chamber via a load-lock mechanism. Sample transfer into a reaction chamber also allows for the exposure of samples at high pressures or with corrosive gases or liquids that could not otherwise be introduced into a UHV chamber.

Below are brief descriptions of some of the particle–surface interactions important in surface science. The descriptions are intended to provide a basic understanding of how surfaces are probed, as most of the information that we have about surfaces was obtained through the use of techniques that are based on such interactions. The section is divided into some general categories, and the important physics of the interactions used for analysis are emphasized. All of these techniques are described in greater detail in subsequent sections of the encyclopaedia. Also, note that there are many more techniques than just those discussed here. These particular techniques were chosen not to be comprehensive, but instead to illustrate the kind of information that can be obtained from surfaces and interfaces.

A1.7.5.1   Electron spectroscopy

Electrons are extremely useful as surface probes because the distances that they travel within a solid before scattering are rather short. This implies that any electrons that are created deep within a sample do not escape into vacuum. Any technique that relies on measurements of low-energy electrons emitted from a solid therefore provides information from just the outermost few atomic layers. Because of this inherent surface sensitivity, the various electron spectroscopies are probably the most useful and popular techniques in surface science.

Electrons interact with solid surfaces by elastic and inelastic scattering, and these interactions are employed in electron spectroscopy. For example, electrons that elastically scatter will diffract from a single-crystal lattice. The diffraction pattern can be used as a means of structural determination, as in LEED. Electrons scatter inelastically by inducing electronic and vibrational excitations in the surface region. These losses form the basis of electron energy loss spectroscopy (EELS). An incident electron can also knock out an inner-shell, or core, electron from an atom in the solid that will, in turn, initiate an Auger process. Electrons can also be used to induce stimulated desorption, as described in section A1.7.5.6.

Figure A1.7.12 shows the scattered electron kinetic energy distribution produced when a monoenergetic electron beam is incident on an Al surface. Some of the electrons are elastically backscattered with essentially no energy loss, as evidenced by the elastic peak. Others lose energy inelastically, however, by inducing particular excitations in the solid, but are then emitted from the surface by elastic backscattering. The plasmon loss features seen in figure A1.7.12 represent scattered electrons that have lost energy inelastically by excitation of surface plasmons. A plasmon is a collective excitation of substrate electrons, and a single plasmon excitation typically has an energy in the range of 5–25 eV. A small feature due to the emission of Auger electrons is also seen in the figure. Finally, the largest feature in the spectrum is the inelastic tail. The result of all of the electronic excitations is the production of a cascade of secondary electrons that are ejected from the surface. The intensity of the secondary electron 'tail' increases as the kinetic energy is reduced, until the cutoff energy is reached. The exact position of the cutoff is determined by the surface work function, and, in fact, is often used to measure the work function changes as the surface composition is modified.

The inelastic mean free path (IMFP) is often used to quantify the surface sensitivity of electron spectroscopy. The IMFP is the average distance that an electron travels through a solid before it is annihilated by inelastic scattering. The minimum in the IMFP for electrons travelling in a solid occurs just above the plasmon energy, as these electrons have the highest probability for excitation. Thus, for most materials, the electrons with the smallest mean free path are those with approximately 25–50 eV of kinetic energy [50]. When performing electron spectroscopy for quantitative analysis, it is necessary to define the mean escape depth (MED), rather then just use the IMFP [51]. The MED is the average depth below the surface from which electrons have originated, and includes losses by all possible elastic and inelastic mechanisms. Typical values

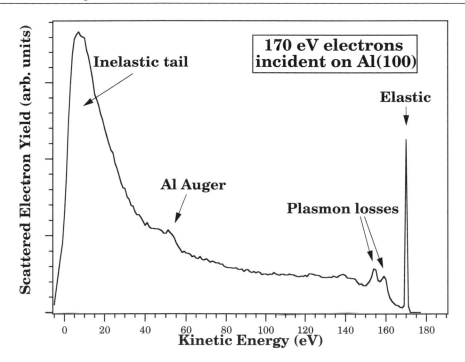

Figure A1.7.12. Secondary electron kinetic energy distribution, obtained by measuring the scattered electrons produced by bombardment of Al(100) with a 170 eV electron beam. The spectrum shows the elastic peak, loss features due to the excitation of plasmons, a signal due to the emission of Al LMM Auger electrons and the inelastic tail. The exact position of the cutoff at 0 eV depends on the surface work function.

of the MED for 10–1000 eV electrons are in the range of 4–10 Å, which is of the order of the interlayer spacings of a solid [52, 53]. Electron attenuation is modelled by assuming that the yield of electrons originating from a particular depth within the sample decreases exponentially with increasing depth, i.e.,

$$\text{Number of electrons} = \exp\left(-d/\lambda\right). \qquad (A1.7.6)$$

Where λ is the MED for the particular material and d is the distance below the surface from which the electron originated. This consideration allows measurements of depth distributions by changing either the electron kinetic energy or the emission angle in order to vary λ.

A popular electron-based technique is Auger electron spectroscopy (AES), which is described in section B1.25.2.2. In AES, a 3–5 keV electron beam is used to knock out inner-shell, or core, electrons from atoms in the near-surface region of the material. Core holes are unstable, and are soon filled by either fluorescence or Auger decay. In the Auger process, one valence, or lower-lying core, electron fills the hole while another is emitted from the sample, in order to satisfy conservation of energy. The emitted Auger electrons have kinetic energies that are characteristic of a particular element. The Perkin–Elmer Auger handbook contains sample spectra of each element, along with information on the relative sensitivity of each Auger line [54]. AES is most useful as a quantitative measure of the surface atomic composition, and is a standard technique employed to determine sample cleanliness. The ratio of the AES signal from an adsorbate to that of the substrate is also commonly used to quantify the coverage of an adsorbate.

LEED is used primarily to ascertain the crystallinity and symmetry of a single-crystal surface, but can also be used to obtain detailed structural information [55, 56]. LEED is described in detail in section B1.21.

In LEED, a 20–200 eV electron beam is incident upon a single-crystal surface along the sample normal. The angular distribution of the elastically scattered electrons is then measured, usually by viewing a phosphorescent screen. At certain angles, there are spots that result from the diffraction of electrons. The symmetry of the pattern of spots is representative of the two-dimensional unit cell of the surface. Note, however, that the spacings between LEED spots provide distances in inverse space, i.e. more densely packed LEED spots correspond to larger surface unit cells. The sharpness of the spots is an indication of the average size of the ordered domains on the surface. In order to extract detailed atomic positions from LEED, the intensity of the spots as a function of the electron energy, or intensity–voltage ($I–V$) curves, are collected and then compared to theoretical predictions for various surface structures [55, 56]. LEED $I–V$ analysis is capable of providing structural details to an accuracy of 0.01 Å. LEED is probably the most accurate structural technique available, but it will only work for structures that are not overly complex.

The excitation of surface quanta can be monitored directly with EELS, as discussed in sections B1.7 and B1.25.5. In EELS, a monoenergetic electron beam is incident onto a surface and the kinetic energy distribution of the scattered electrons is collected. The kinetic energy distribution will display peaks corresponding to electrons that have lost energy by exciting transitions in the near-surface region, such as the plasmon loss peaks shown in figure A1.7.12. EELS can be used to probe electronic transitions, in which case incident electron energies in the range of 10–100 eV are used. More commonly, however, EELS is used to probe low-energy excitations, such as molecular vibrations or phonon modes [57]. In this case, very low incident electron energies (< 10 eV) are employed and a very high-energy resolution is required. When EELS is performed in this manner, the technique is known as high-resolution electron energy loss spectroscopy (HREELS).

Photoelectron spectroscopy provides a direct measure of the filled density of states of a solid. The kinetic energy distribution of the electrons that are emitted via the photoelectric effect when a sample is exposed to a monochromatic ultraviolet (UV) or x-ray beam yields a photoelectron spectrum. Photoelectron spectroscopy not only provides the atomic composition, but also information concerning the chemical environment of the atoms in the near-surface region. Thus, it is probably the most popular and useful surface analysis technique. There are a number of forms of photoelectron spectroscopy in common use.

X-ray photoelectron spectroscopy (XPS), also called electron spectroscopy for chemical analysis (ESCA), is described in section B1.25.2.1. The most commonly employed x-rays are the Mg Kα (1253.6 eV) and the Al Kα (1486.6 eV) lines, which are produced from a standard x-ray tube. Peaks are seen in XPS spectra that correspond to the bound core-level electrons in the material. The intensity of each peak is proportional to the abundance of the emitting atoms in the near-surface region, while the precise binding energy of each peak depends on the chemical oxidation state and local environment of the emitting atoms. The Perkin–Elmer XPS handbook contains sample spectra of each element and binding energies for certain compounds [58].

XPS is also often performed employing synchrotron radiation as the excitation source [59]. This technique is sometimes called soft x-ray photoelectron spectroscopy (SXPS) to distinguish it from laboratory XPS. The use of synchrotron radiation has two major advantages: (1) a much higher spectral resolution can be achieved and (2) the photon energy of the excitation can be adjusted which, in turn, allows for a particular electron kinetic energy to be selected.

One of the more recent advances in XPS is the development of photoelectron microscopy [60]. By either focusing the incident x-ray beam, or by using electrostatic lenses to image a small spot on the sample, spatially-resolved XPS has become feasible. The limits to the spatial resolution are currently of the order of 1 μm, but are expected to improve. This technique has many technological applications. For example, the chemical makeup of micromechanical and microelectronic devices can be monitored on the scale of the device dimensions.

Ultraviolet photoelectron spectroscopy (UPS) is a variety of photoelectron spectroscopy that is aimed at measuring the valence band, as described in section B1.25.2.3. Valence band spectroscopy is best performed with photon energies in the range of 20–50 eV. A He discharge lamp, which can produce 21.2 or

40.8 eV photons, is commonly used as the excitation source in the laboratory, or UPS can be performed with synchrotron radiation. Note that UPS is sometimes just referred to as photoelectron spectroscopy (PES), or simply valence band photoemission.

A particularly useful variety of UPS is angle-resolved photoelectron spectroscopy (ARPES), also called angle-resolved ultraviolet photoelectron spectroscopy (ARUPS) [61, 62]. In this technique, measurements are made of the valence band photoelectrons emitted into a small angle as the electron emission angle or photon energy is varied. This allows for the simultaneous determination of the kinetic energy and momentum of the photoelectrons with respect to the two-dimensional surface Brillouin zone. From this information, the electronic band structure of a single-crystal material can be experimentally determined.

The diffraction of photoelectrons (or Auger electrons) is also used as a structural tool [63, 64]. When electrons of a well defined energy are created at a particular atomic site, such as in XPS or AES, then the emitted electrons interact with other atoms in the crystal structure prior to leaving the surface. The largest effect is 'forward scattering', in which the intensity of an electron wave emitted from one atom is enhanced when it passes through another atom. Thus, the angular distribution of the emitted electron intensity provides a 'map' of the surface crystal structure. More generally, however, there is a complex multiple scattering behaviour, which produces variations of the emitted electron intensity with respect to both angle and energy such that the intensity modulations do not necessarily relate to the atomic bond directions. In order to determine a surface structure from such diffraction data, the measured angular and/or energy distributions of the Auger or photoelectrons is compared to a theoretical prediction for a given structure. Similar to LEED analysis, the structure employed for the calculation is varied until the best fit to the data is found.

A1.7.5.2   Ion spectroscopy

Ions scattered from solid surfaces are useful probes for elemental identification of surface species and for measurements of the three-dimensional atomic structure of a single-crystal surface. Ions used for surface studies can be roughly divided into low (0.5–10 keV), medium (10–100 keV) and high (100 keV–1 MeV) energy regimes. In each regime, ions have distinct interactions with solid material and each regime is used for different types of measurements. The use of particle scattering for surface structure determination is described in detail in section B1.23.

The fundamental interactions between ions and surfaces can be separated into elastic and inelastic processes. When an ion undergoes a direct collision with a single atom in a solid, it loses energy elastically by transferring momentum to the target atom. As an ion travels through a material, it also loses energy inelastically by initiating various electronic and vibrational excitations. The elastic and inelastic energy losses can usually be treated independently from each other.

Elastic losses result from binary collisions between the ions and unbound target atoms positioned at the lattice sites. For keV and higher energy ions, the cross sections for collisions are small enough that the ions essentially 'see' each atom in the solid individually, i.e. the trajectory can be considered as a sequence of events in which the ion interacts with one target atom at a time. This is the so-called binary collision approximation (BCA). The energy of a scattered particle is determined by conservation of energy and momentum during the single collision (the binding energy of the target atom to the surface can be neglected since it is considerably smaller than the energy of the ions). The smaller the mass of the target atom relative to the projectile, the more the energy that is lost during an elastic collision and the lower the scattered energy. Peaks are seen in scattered ion energy spectra, called single scattering peaks (SSP), or quasi-single (QS) scattering peaks, that result from these binary collisions. In this manner, ion scattering produces a mass spectrum of the surface region, as the position of each SSP indicates the mass of the target atom.

Ions in the low-energy range have reasonably short penetration depths, and therefore provide a surface-sensitive means for probing a material. Low-energy ion scattering (LEIS), often called ion scattering spectroscopy (ISS), is generally used as a measure of the surface composition. The surface sensitivity when using noble gas ions for standard ISS results from the high probability for neutralization for any ions that have penetrated past the first atomic layer. The intensity of an SSP is related to the surface concentration of the particular element, but care must be taken in performing quantitative analysis to properly account for ion neutralization. Energy losses due to inelastic excitations further modify the ion energies and charge states of scattered particles. In the low-energy regime, these effects are often neglected, as they only slightly alter the shapes of the SSP and shift it to a lower energy. In the high-energy regime, however, inelastic excitations are dominant in determining the shape of the scattered ion energy spectrum, as in Rutherford backscattering spectroscopy (RBS) [65, 66], which is discussed in section B1.24.

Measurements of the angular distributions of scattered ions are often used as a structural tool, as they depend strongly on the relative positions of the atoms in the near-surface region. Ion scattering is used for structure determination by consideration of the shadow cones and blocking cones. These 'cones' are the regions behind each atom from which incoming ions are excluded because of scattering. A shadow cone is formed when an ion is incident onto the surface, while a blocking cone is formed when an ion that has scattered from a deeply-lying atom interacts with a surface atom along the outgoing trajectory. The ion flux is increased at the edges of the cones. Thus, rotating the ion beam or detector relative to the sample alters the flux of ions that scatter from any particular atom. The angular distributions are usually analysed by comparing the measured distributions to those obtained by computer simulation for a given geometry. Shadow/blocking cone analysis is used in both low- and medium-energy ion scattering to provide the atomic structure, and is accurate to about 0.1 Å [67, 68].

In the high-energy ion regime, ion channelling is used for surface structure determination [65, 66]. In this technique, the incident ion beam is aligned along a low-index direction in the crystal. Thus, most of the ions will penetrate into 'channels' created by the crystal structure. Those few ions that do backscatter from a surface atom are collected. The number of these scattering events is dependent on the detailed atomic structure. For performing a structure determination, the data is usually collected as 'rocking curves' in which the backscattered ion yield is collected as the crystal is precisely rotated about the channelling direction. The measured rocking curves are then compared to the results of computer simulations performed for particular model surface structures. As in LEED I–V analysis, the structure employed for the simulation that most closely matches the experimental data is deemed to be correct.

Ions are also used to initiate secondary ion mass spectrometry (SIMS) [69], as described in section B1.25.3. In SIMS, the ions sputtered from the surface are measured with a mass spectrometer. SIMS provides an accurate measure of the surface composition with extremely good sensitivity. SIMS can be collected in the 'static' mode in which the surface is only minimally disrupted, or in the 'dynamic' mode in which material is removed so that the composition can be determined as a function of depth below the surface. SIMS has also been used along with a shadow and blocking cone analysis as a probe of surface structure [70].

A1.7.5.3   Scanning probe methods

Scanning probe microscopies have become the most conspicuous surface analysis techniques since their invention in the mid-1980s and the awarding of the 1986 Nobel Prize in Physics [71, 72]. The basic idea behind these techniques is to move an extremely fine tip close to a surface and to monitor a signal as a function of the tip's position above the surface. The tip is moved with the use of piezoelectric materials, which can control the position of a tip to a sub-Ångstrøm accuracy, while a signal is measured that is indicative of the surface topography. These techniques are described in detail in section B1.20.

The most popular of the scanning probe techniques are STM and atomic force microscopy (AFM). STM and AFM provide images of the outermost layer of a surface with atomic resolution. STM measures the spatial distribution of the surface electronic density by monitoring the tunnelling of electrons either from the sample to the tip or from the tip to the sample. This provides a map of the density of filled or empty electronic states, respectively. The variations in surface electron density are generally correlated with the atomic positions. AFM measures the spatial distribution of the forces between an ultrafine tip and the sample. This distribution of these forces is also highly correlated with the atomic structure. STM is able to image many semiconductor and metal surfaces with atomic resolution. AFM is necessary for insulating materials, however, as electron conduction is required for STM in order to achieve tunnelling. Note that there are many modes of operation for these instruments, and many variations in use. In addition, there are other types of scanning probe microscopies under development.

Scanning probe microscopies have afforded incredible insight into surface processes. They have provided visual images of surfaces on the atomic scale, from which the atomic structure can be observed in real time. All of the other surface techniques discussed above involve averaging over a macroscopic region of the surface. From STM images, it is seen that many surfaces are actually not composed of an ideal single domain, but rather contain a mixture of domains. STM has been able to provide direct information on the structure of atoms in each domain, and at steps and defects on surfaces. Furthermore, STM has been used to monitor the movement of single atoms on a surface. Refinements to the instruments now allow images to be collected over temperatures ranging from 4 to 1200 K, so that dynamical processes can be directly investigated. An STM has also been adapted for performing single-atom vibrational spectroscopy [73].

One of the more interesting new areas of surface science involves manipulation of adsorbates with the tip of an STM. This allows for the formation of artificial structures on a surface at the atomic level. In fact, STM tips are being investigated for possible use in lithography as part of the production of very small features on microcomputer chips [74].

Some of the most interesting work in this area has involved physisorbed molecules at temperatures as low as 4 K [75]. Note that it takes a specialized instrument to be able to operate at these low temperatures. An STM tip is brought into contact with the physisorbed species by lightly pushing down on it. Then, the STM tip is translated parallel to the surface while pressure is maintained on the adsorbate. In this manner, the adsorbates can be moved to any location on the surface. Manipulation of this type has led to the writing of 'IBM' with single atoms [76], as well as to the formation of structures such as the 'quantum corral' [77]. The quantum corral is so named, as it is an oval-shaped enclosure made from adsorbate atoms that provides a barrier for the free electrons of the metal substrate. Inside the corral, standing wave patterns are set up that can be imaged with the STM.

There are many other experiments in which surface atoms have been purposely moved, removed or chemically modified with a scanning probe tip. For example, atoms on a surface have been induced to move via interaction with the large electric field associated with an STM tip [78]. A scanning force microscope has been used to create three-dimensional nanostructures by 'pushing' adsorbed particles with the tip [79]. In addition, the electrons that are tunnelling from an STM tip to the sample can be used as sources of electrons for stimulated desorption [80]. The tunnelling electrons have also been used to promote dissociation of adsorbed O_2 molecules on metal or semiconductor surfaces [81, 82].

A1.7.5.4   Thermal desorption

Temperature programmed desorption (TPD), also called thermal desorption spectroscopy (TDS), provides information about the surface chemistry such as surface coverage and the activation energy for desorption [49]. TPD is discussed in detail in section B1.25. In TPD, a clean surface is first exposed to a gaseous molecule that adsorbs. The surface is then quickly heated (on the order of 10 K s^{-1}), while the desorbed molecules are measured with a mass spectrometer. An analysis of TPD spectra basically provides three types of information: (1) The identities of the desorbed product(s) are obtained directly from the mass spectrometer. (2) The area of a TPD peak provides a good measure of the surface coverage. In cases where there are multiple species desorbed, the ratios of the TPD peaks provide the stoichiometry. (3) The shapes of the peaks, and how they change with surface coverage, provide detailed information on the kinetics of desorption. For example, the shapes of TPD curves differ for zeroth-, first- or second-order processes.

A1.7.5.5   Laser–surface interactions

Lasers have been used to both modify and probe surfaces. When operated at low fluxes, lasers can excite electronic and vibrational states, which can lead to photochemical modification of surfaces. At higher fluxes, the laser can heat the surface to extremely high temperatures in a region localized at the very surface. A high-power laser beam produces a very non-equilibrium situation in the near-surface region, during which the effective electron temperature can be extremely high. Thus, lasers can also be used to initiate thermal desorption. Laser-induced thermal desorption (LITD) has some advantages over TPD as an analytical technique [36]. When a laser is used to heat the surface, the heat is localized in the surface region and the temperature rise is extremely fast. It is also possible to produce excitations that involve multiple photons because of the high flux available with lasers. Furthermore, there are nonlinear effects that occur with laser irradiation of surfaces that allow for surface sensitive probes that do not require UHV, such as second harmonic generation (SHG) and sum frequency generation (SFG) [83, 84]. Optical techniques in surface science are discussed in section B1.22.

Surface photochemistry can drive a surface chemical reaction in the presence of laser irradiation that would not otherwise occur. The types of excitations that initiate surface photochemistry can be roughly divided into those that occur due to direct excitations of the adsorbates and those that are mediated by the substrate. In a direct excitation, the adsorbed molecules are excited by the laser light, and will directly convert into products, much as they would in the gas phase. In substrate-mediated processes, however, the laser light acts to excite electrons from the substrate, which are often referred to as 'hot electrons'. These hot electrons then interact with the adsorbates to initiate a chemical reaction.

Femtosecond lasers represent the state-of-the-art in laser technology. These lasers can have pulse widths of the order of 100 fm s. This is the same time scale as many processes that occur on surfaces, such as desorption or diffusion. Thus, femtosecond lasers can be used to directly measure surface dynamics through techniques such as two-photon photoemission [85]. Femtochemistry occurs when the laser imparts energy over an extremely short time period so as to directly induce a surface chemical reaction [86].

A1.7.5.6   Stimulated desorption

An electron or photon incident on a surface can induce an electronic excitation. When the electronic excitation decays, an ion or neutral particle can be emitted from the surface as a result of the excitation. Such processes are known as desorption induced by electronic transitions (DIET) [87]. The specific techniques are known as electron-stimulated desorption (ESD) and photon-stimulated desorption (PSD), depending on the method of excitation.

A DIET process involves three steps: (1) an initial electronic excitation, (2) an electronic rearrangement to form a repulsive state and (3) emission of a particle from the surface. The first step can be a direct excitation to an antibonding state, but more frequently it is simply the removal of a bound electron. In the second step, the surface electronic structure rearranges itself to form a repulsive state. This rearrangement could be, for example, the decay of a valence band electron to fill a hole created in step (1). The repulsive state must have a sufficiently long lifetime that the products can desorb from the surface before the state decays. Finally, during the emission step, the particle can interact with the surface in ways that perturb its trajectory.

There are two main theoretical descriptions applied to stimulated desorption. The Menzel–Gomer–Redhead (MGR) model is used to describe low-energy valence excitations, while the Knotek–Feibelman mechanism is used to describe a type of desorption that occurs with ionically-bound species. In the MGR model, it is assumed that the initial excitation occurs by absorption of a photon or electron to directly create an excited, repulsive state. This excited state can be neutral or ionic. It simply needs to have a sufficient lifetime so that desorption can occur before the system relaxes to the ground state. Thus, the MGR mechanism can be applied to positive or negative ion emission, or to the emission of a neutral atom. The Knotek–Feibelman

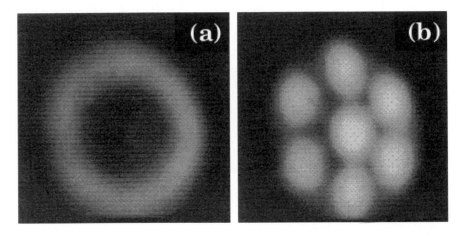

Figure A1.7.13. ESDIAD patterns showing the angular distributions of F^+ emitted from PF_3 adsorbed on Ru(0001) under electron bombardment. (a) 0.25 ML coverage, (b) the same surface following electron beam damage.

mechanism applies when there is an ionic bond at the surface. In this case, the incident electron kicks out an inner-shell electron, and an Auger process then fills the resulting core hole. In the Auger process, one electron drops down to fill the hole, while another electron is emitted from the surface in order to satisfy conservation of energy. Thus, the system has lost at least two electrons, which is sufficient to turn the negatively charged anion into a positive ion. Finally, Coulomb repulsion between this positive ion and the cation leads to the emission of a positive ion from the surface. Although this mechanism was originally proposed for maximally valent bonding, it has since been observed to occur in a variety of systems providing that there is at least a moderate amount of charge transfer involved in the bonding. Note that this mechanism is often referred to as Auger-stimulated desorption (ASD).

Electron stimulated desorption angular distributions (ESDIAD) [88] provide a quick measure of the bond angles for a lightly bound adsorbate. ESDIAD patterns are recorded by impinging an electron beam onto a surface and then measuring the angular distributions of the desorbed ions with an imaging analyser. The measured ion emission angles are related to the original surface bond angles. The initial excitation responsible for ESD is normally directly along the bond axis. As an ion is exiting from a surface, however, there are two effects that act to alter the ion's trajectory. First, the ion is attracted to its image charge, which tends to spread out the ESDIAD pattern. Second, however, is that there is inhomogeneous neutralization of the emitted ions, in that the ions emitted at more grazing angles are preferentially neutralized. This acts to compress the observed pattern. Thus, a balance between these competing effects produces the measured angular distribution, and it is therefore difficult, although not impossible, to quantitatively determine the bond angle.

The ESDIAD pattern does, however, provide very useful information on the nature and symmetry of an adsorbate. As an example, figure A1.7.13(a) shows the ESDIAD pattern of desorbed F^+ collected from a 0.25 ML coverage of PF_3 on Ru(0001) [89]. The F^+ pattern displays a ring of emission, which indicates that the molecule adsorbs intact and is bonded through the P end. It freely rotates about the P–Ru bond so that the F^+ emission occurs at all azimuthal angles, regardless of the substrate structure. In figure A1.7.13(b), the ESDIAD pattern is shown following sufficient e^--beam damage to remove much of the fluorine and produce adsorbed PF_2 and PF. Now, the F^+ emission shows six lobes along particular azimuths and one lobe along the surface normal. The off-normal lobes arise from PF_2, and indicate that PF_2 adsorbs in registry with the substrate, with the F atoms pointing away from the surface at an off-normal angle. The centre lobe arises from PF and indicates that the PF moiety is bonded through the P end, with F pointing normal to the surface.

Some recent advances in stimulated desorption were made with the use of femtosecond lasers. For example, it was shown by using a femtosecond laser to initiate the desorption of CO from Cu while probing the surface with SHG, that the entire process is completed in less than 325 fs [90]. The mechanism for this kind of laser-induced desorption has been termed desorption induced by multiple electronic transitions (DIMET) [91]. Note that the mechanism must involve a multiphoton process, as a single photon at the laser frequency has insufficient energy to directly induce desorption. DIMET is a modification of the MGR mechanism in which each photon excites the adsorbate to a higher vibrational level, until a sufficient amount of vibrational energy has been amassed so that the particle can escape the surface.

A1.7.6   Liquid–solid interface

One of the less explored frontiers in atomic-scale surface science is the study of the liquid–solid interface. This interface is critically important in many applications, as well as in biological systems. For example, the movement of pollutants through the environment involves a series of chemical reactions of aqueous groundwater solutions with mineral surfaces. Although the liquid–solid interface has been studied for many years, it is only recently that the tools have been developed for interrogating this interface at the atomic level. This interface is particularly complex, as the interactions of ions dissolved in solution with a surface are affected not only by the surface structure, but also by the solution chemistry and by the effects of the electrical double layer [31]. It has been found, for example, that some surface reconstructions present in UHV persist under solution, while others do not.

The electrical double layer basically acts as a capacitor by storing charge at the surface that is balanced by ions in solution [92]. The capacitance of the double layer is a function of the electrochemical potential of the solution, and has a maximum at the potential of zero-charge (pzc). The pzc in solution is essentially equivalent to the work function of that surface in vacuum. In solution, however, the electrode potential can be used to vary the surface charge in much the same way that alkali adsorbates are used to vary the work function of a surface in vacuum. The difference is that in solution the surface charge can be varied, while the surface composition is unchanged. The surface energy, which effects the atomic structure and reactivity, is directly related to the surface charge. It has been shown, for example, that by adjusting the electrode potential the reconstructions of certain surfaces in solution can be altered in a reversible manner. Electrochemistry can also be used to deposit and remove adsorbates from solution in a manner that is controlled by the electrode potential.

Studies of the liquid–solid interface can be divided into those that are performed *ex situ* and those performed *in situ*. In an *ex situ* experiment, a surface is first reacted in solution, and then removed from the solution and transferred into a UHV spectrometer for measurement. There has recently been, however, much work aimed at interrogating the liquid–solid interface *in situ*, i.e. while chemistry is occurring rather than after the fact.

In performing *ex situ* surface analysis, the transfer from solution to the spectrometer sometimes occurs either through the air or within a glove bag filled with an inert atmosphere. Many *ex situ* studies of chemical reactions at the liquid–solid interface, however, have been carried out using special wet cells that are directly attached to a UHV chamber [93, 94]. With this apparatus, the samples can be reacted and then immediately transferred to UHV without encountering air. Note that some designs enable complete immersion of the sample into solution, while others only allow the sample surface to interact with a meniscus. Although these investigations do not probe the liquid–solid interface directly, they can provide much information on the surface chemistry that has taken place.

One of the main uses of these wet cells is to investigate surface electrochemistry [94, 95]. In these experiments, a single-crystal surface is prepared by UHV techniques and then transferred into an electrochemical cell. An electrochemical reaction is then run and characterized using cyclic voltammetry, with the sample itself being one of the electrodes. In order to be sure that the electrochemical measurements all involved the same crystal face, for some experiments a single-crystal cube was actually oriented and polished on all six

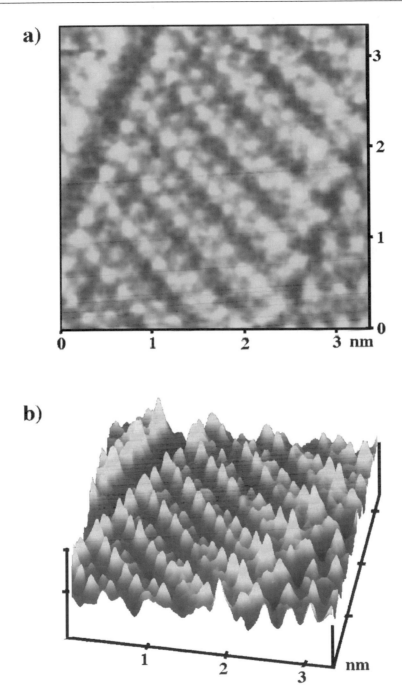

Figure A1.7.14. 3.4 nm × 3.4 nm STM images of 1-docosanol physisorbed onto a graphite surface in solution. This image reveals the hydrogen-bonding alcohol molecules assembled in lamellar fashion at the liquid–solid interface. Each 'bright' circular region is attributed to the location of an individual hydrogen atom protruding upward out of the plane of the all-*trans* hydrocarbon backbone, which is lying flat on the surface. (a) Top view, and (b) a perspective image (courtesy of Leanna Giancarlo and George Flynn).

sides! Following surface modification by electrochemistry, the sample is returned to UHV for measurement with standard techniques, such as AES and LEED. It has been found that the chemisorbed layers that are deposited by electrochemical reactions are stable and remain adsorbed after removal from solution. These studies have enabled the determination of the role that surface structure plays in electrochemistry.

The force between two adjacent surfaces can be measured directly with the surface force apparatus (SFA), as described in section B1.20 [96]. The SFA can be employed in solution to provide an *in situ* determination of the forces. Although this instrument does not directly involve an atomically resolved measurement, it has provided considerable insight into the microscopic origins of surface friction and the effects of electrolytes and lubricants [97].

Scanning probe microscopies are atomically resolved techniques that have been successfully applied to measurements of the liquid–solid interface *in situ* [98–102]. The STM has provided atomically resolved images of surface reconstructions and adsorption geometry under controlled conditions in solution, and the dependence of these structures on solution composition and electrode potential. Note that in order to perform STM under solution, a special tip coated with a dielectric must be used in order to reduce the Faradaic current that would otherwise transmit through the solution. As an example, figure A1.7.14 shows an STM image collected in solution from docosanol physisorbed on a graphite surface. The graphite lattice and the individual atoms in the adsorbed molecules can be imaged with atomic resolution. In addition, scanning probe microscopies have been used to image the surfaces of biological molecules and even living cells in solution [103].

Since water is transparent to visible light, optical techniques can be used to interrogate the liquid–solid interface *in situ* [104]. For example, SFG has been used to perform IR spectroscopy directly at the liquid–solid interface [105, 106]. The surface sensitivity of SFG arises from the breaking of centrosymmetry at the interface, rather than from electron attenuation as in more traditional surface techniques, so that the information obtained is relevant to atomic-scale processes at the solid–liquid interface. This allows for the identification of the adsorbed species while a reaction is occurring. Note that these techniques can be extended to the liquid–liquid interface, as well [107]. In addition, x-ray scattering employing synchrotron radiation is being developed for use at the liquid–solid interface. For example, an *in situ* electrochemical cell for x-ray scattering has been designed [108].

References

[1] Somorjai G A 1994 *Introduction to Surface Chemistry and Catalysis* (New York: Wiley)

[2] Kittel C 1996 *Introduction to Solid State Physics* 7th edn (New York: Wiley)

[3] Himpsel F J, Jung T and Ortega J E 1997 Nanowires on stepped metal surfaces *Surf. Rev. Lett.* **4** 371

[4] Noonan J R and Davis H L 1984 Truncation-induced multilayer relaxation of the Al(110) surface *Phys. Rev.* B **29** 4349

[5] Adams D L, Jensen V, Sun X F and Vollesen J H 1988 Multilayer relaxation of the Al(210) surface *Phys. Rev.* B **38** 7913

[6] Holub-Krappe E, Horn K, Frenken J W M, Krans R L and van der Veen J F 1987 Multilayer relaxation at the Ag(110) surface *Surf. Sci.* **188** 335

[7] Busch B W and Gustafsson T 1998 Oscillatory relaxation of Al(110) reinvestigated by using medium-energy ion scattering *Surf. Sci.* **415** L1074

[8] Cho J-H, Ismail, Zhang Z and Plummer E W 1999 Oscillatory lattice relaxation at metal surfaces *Phys. Rev.* B **59** 1677

[9] Himpsel F J, McFeely F R, Morar J F, Taleb-Ibrahimi A and Yarmoff J A 1990 Core level spectroscopy at silicon surfaces and interfaces *Proc. Enrico Fermi School on 'Photoemission and Adsorption Spectroscopy and Interfaces with Synchrotron Radiation'* vol course CVIII, eds M Campagna and R Rosei (Amsterdam: Elsevier) p 203

[10] Hamers R J, Tromp R M and Demuth J M 1986 Surface electronic structure of Si(111)-7 × 7 resolved in real space *Phys. Rev. Lett.* **56** 1972

[11] Takayanagi K, Tanishiro Y, Takahashi M and Takahashi S 1985 Structural analysis of Si(111)-7 × 7 by UHV-transmission electron diffraction and microscopy *J. Vac. Sci. Technol.* A **3** 1502

[12] Liu C L, Cohen J M, Adams J B and Voter A F 1991 EAM study of surface self-diffusion of single adatoms of fcc metals Ni, Cu, Al, Ag, Au, Pd, and Pt *Surf. Sci.* **253** 334

[13] Bonig L, Liu S and Metiu H 1996 An effective medium theory study of Au islands on the Au(100) surface: reconstruction, adatom diffusion, and island formation *Surf. Sci.* **365** 87

[14] Tsong T T 1988 Experimental studies of the behaviour of single adsorbed atoms on solid surfaces *Rep. Prog. Phys.* **51** 759

[15] Chen C-L and Tsong T T 1991 Self-diffusion on the reconstructed and nonreconstructed Ir(110) surfaces *Phys. Rev. Lett.* **66** 1610

[16] Jensen F, Besenbacher F, Laesgaard E and Stensgaard I 1990 Surface reconstruction of Cu (110) induced by oxygen chemisorption *Phys. Rev.* B **41** 10 233

[17] Besenbacher F, Jensen F, Laegsgaard E, Mortensen K and Stensgaard I 1991 Visualization of the dynamics in surface reconstructions *J. Vac. Sci. Technol.* B **9** 874

[18] Kitamura N, Lagally M G and Webb M B 1993 Real-time observations of vacancy diffusion on Si(100)-(2 × 1) by scanning tunneling microscopy *Phys. Rev. Lett.* **71** 2082

[19] Linderoth T R, Horsch S, Laesgaard E, Stensgaard I and Besenbacher F 1997 Surface diffusion of Pt on Pt(110): Arrhenius behavior of long jumps *Phys. Rev. Lett.* **78** 4978

[20] Botkin D, Glass J, Chemla D S, Ogletree D F, Salmeron M and Weiss S 1996 Advances in ultrafast scanning tunneling microscopy *Appl. Phys. Lett.* **69** 1321

[21] Bauer E 1994 Low energy electron microscopy *Rep. Prog. Phys.* **57** 895

[22] Himpsel F J 1990 Inverse photoemission from semiconductors *Surf. Sci. Rep.* **12** 1

[23] Himpsel F J 1991 Unoccupied electronic states at surfaces *Surface Physics and Related Topics. Festschrift for Xide Xie* ed F-J Yang, G-J Ni, X Wang, K-M Zhang and D Lu (Singapore: World Scientific) p 179

[24] Avouris P and Wolkow R 1989 Atom-resolved chemistry studied by scanning tunneling microscopy and spectroscopy *Phys. Rev.* B **39** 5091

[25] Lüth H 1995 *Surfaces and Interfaces of Solid Materials* 3rd edn (Berlin: Springer)

[26] Borroni-Bird C E, Al-Sarraf N, Andersson S and King D A 1991 Single crystal adsorption microcalorimetry *Chem. Phys. Lett.* **183** 516

[27] Li Y L *et al* 1995 Experimental verification of a new mechanism for dissociative chemisorption: atom abstraction *Phys. Rev. Lett.* **74** 2603

[28] Jensen J A, Yan C and Kummel A C 1996 Direct chemisorption site selectivity for molecular halogens on the Si(111)-(7 × 7) surface *Phys. Rev. Lett.* **76** 1388

[29] Hudson J B 1992 *Surface Science: An Introduction* (Boston: Butterworth-Heinemann)

[30] Kang H C and Weinberg W H 1994 Kinetic modeling of surface rate processes *Surf. Sci.* **299–300** 755

[31] Adamson A W and Gast A P 1997 *Physical Chemistry of Surfaces* 6th edn (New York: Wiley-Interscience)

[32] Over H 1998 Crystallographic study of interaction between adspecies on metal surfaces *Prog. Surf. Sci.* **58** 249

[33] Gomer R 1990 Diffusion of adsorbates on metal surfaces *Rep. Prog. Phys.* **53** 917

[34] Lagally M G 1993 Atom motion on surfaces *Physics Today* **46** 24

[35] Dunphy J C, Sautet P, Ogletree D F, Dabbousi O and Salmeron M B 1993 Scanning-tunneling-microscopy study of the surface diffusion of sulfur on Re(0001) *Phys. Rev.* B **47** 2320

[36] George S M, DeSantolo A M and Hall R B 1985 Surface diffusion of hydrogen on Ni(100) studied using laser-induced thermal desorption *Surf. Sci.* **159** L425

[37] Olmstead M A, Bringans R D, Uhrberg R I G and Bachrach R Z 1986 Arsenic overlayer on Si(111): removal of surface reconstruction *Phys. Rev.* B **34** 6041

[38] Yarmoff J A, Cyr D M, Huang J H, Kim S and Williams R S 1986 Impact-collision ion-scattering spectroscopy of Cu(110) and Cu(110)-(2 × 1)–O using 5-keV ^{6}Li$^+$ *Phys. Rev.* B **33** 3856

[39] Tochihara H and Mizuno S 1998 Composite surface structures formed by restructuring-type adsorption of alkali-metals on FCC metals *Prog. Surf. Sci.* **58** 1

[40] Diehl R D and McGrath R 1996 Structural studies of alkali metal adsorption and coadsorption on metal surfaces *Surf. Sci. Rep.* **23** 43

[41] Madey T E, Guan J, Nien C-H, Dong C-Z, Tao H-S and Campbell R A 1996 Faceting induced by ultrathin metal films on W(111) and Mo(111): structure, reactivity, and electronic properties *Surf. Rev. Lett.* **3** 1315

[42] Bonzel H P, Bradshaw A M and Ertl G 1989 *Physics and Chemistry of Alkali Metal Adsorption* (Amsterdam: Elsevier)

[43] Winters H F and Coburn J W 1992 Surface science aspects of etching reactions *Surf. Sci. Rep.* **14** 161

[44] Herman M A and Sitter H 1996 *Molecular Beam Epitaxy: Fundamentals and Current Status* (Berlin: Springer)

[45] Somorjai G A 1981 *Chemistry in Two Dimensions: Surfaces* (Ithaca: Cornell University Press)

[46] Somorjai G A 1996 Surface science at high pressures *Z. Phys. Chem.* **197** 1

[47] Somorjai G A 1998 Molecular concepts of heterogeneous catalysis *J. Mol. Struct. (Theochem)* **424** 101

[48] Wood E A 1963 *Crystal Orientation Manual* (New York: Columbia University Press)

[49] Woodruff D P and Delchar T A 1994 *Modern Techniques of Surface Science* 2nd edn (Cambridge: Cambridge University Press)

[50] Seah M P and Dench W A 1979 Quantitative electron spectroscopy of surfaces: a standard data base for electron inelastic mean free paths in solids *Surf. Interface Anal.* **1** 2

[51] Powell C J, Jablonski A, Tilinin I S, Tanuma S and Penn D R 1999 Surface sensitivity of Auger-electron spectroscopy and x-ray photoelectron spectroscopy *J. Electron Spec. Relat. Phenom.* **98–9** 1

[52] Duke C B 1994 Interaction of electrons and positrons with solids: from bulk to surface in thirty years *Surf. Sci.* **299–300** 24

[53] Powell C J 1994 Inelastic interactions of electrons with surfaces: applications to Auger-electron spectroscopy and x-ray photoelectron spectroscopy *Surf. Sci.* **299–300** 34

[54] Davis L E, MacDonald N C, Palmberg P W, Riach G E and Weber R E 1976 *Handbook of Auger Electron Spectroscopy* 2nd edn (Eden Prairie, MN: Perkin-Elmer Corporation)

[55] Pendry J B 1974 *Low Energy Electron Diffraction: The Theory and its Application to Determination of Surface Structure* (London: Academic)
[56] van Hove M A, Weinberg W H and Chan C-M 1986 *Low-Energy Electron Diffraction: Experiment, Theory, and Surface Structure Determination* (Berlin: Springer)
[57] Ibach H and Mills D L 1982 *Electron Energy Loss Spectroscopy and Surface Vibrations* (New York: Academic)
[58] Wagner C D, Riggs W M, Davis L E, Moulder J F and Muilenberg G E (eds) 1979 *Handbook of X-ray Photoelectron Spectroscopy* (Eden Prairie, MN: Perkin-Elmer Corporation)
[59] Margaritondo G 1988 *Introduction to Synchrotron Radiation* (New York: Oxford University Press)
[60] Tonner B P, Dunham D, Droubay T, Kikuma J, Denlinger J, Rotenberg E and Warwick A 1995 The development of electron spectromicroscopy *J. Electron Spectrosc.* **75** 309
[61] Smith N V and Himpsel F J 1983 Photoelectron spectroscopy *Handbook on Synchrotron Radiation* ed E E Koch (Amsterdam: North-Holland)
[62] Plummer E W and Eberhardt W 1982 Angle-resolved photoemission as a tool for the study of surfaces *Adv. Chem. Phys.* **49** 533
[63] Egelhoff W F Jr 1990 X-ray photoelectron and Auger electron forward scattering: a new tool for surface crystallography *CRC Crit. Rev. Solid State Mater. Sci.* **16** 213
[64] Fadley C S 1993 Diffraction and holography with photoelectrons and Auger electrons: some new directions *Surf. Sci. Rep.* **19** 231
[65] Chu W-K, Mayer J W and Nicolet M-A 1978 *Backscattering Spectrometry* (New York: Academic)
[66] Feldman L C, Mayer J W and Picraux S T 1982 *Materials Analysis by Ion Channeling: Submicron Crystallography* (New York: Academic)
[67] Niehus H, Heiland W and Taglauer E 1993 Low-energy ion scattering at surfaces *Surf. Sci. Rep.* **17** 213
[68] Fauster T 1988 Surface geometry determination by large-angle ion scattering *Vacuum* **38** 129
[69] Benninghoven A, Rüdenauer F G and Werner H W 1987 *Secondary Ion Mass Spectrometry: Basic Concepts, Instrumental Aspects, Applications, and Trends* (New York: Wiley)
[70] Chang C-C and Winograd N 1989 Shadow-cone-enhanced secondary-ion mass-spectrometry studies of Ag(110) *Phys. Rev. B* **39** 3467
[71] Binnig G and Rohrer H 1987 Scanning tunneling microscopy—from birth to adolescence *Rev. Mod. Phys.* **59** 615
[72] Wiesendanger R 1994 *Scanning Probe Microscopy and Spectroscopy: Methods and Applications* (New York: Cambridge University Press)
[73] Stipe B C, Rezaei M A and Ho W 1998 Single-molecule vibrational spectroscopy and microscopy *Science* **280** 1732
[74] Marrian C R K, Perkins F K, Brandow S L, Koloski T S, Dobisz E A and Calvert J M 1994 Low voltage electron beam lithography in self-assembled ultrathin films with the scanning tunneling microscope *Appl. Phys. Lett.* **64** 390
[75] Stroscio J A and Eigler D M 1991 Atomic and molecular manipulation with the scanning tunneling microscope *Science* **254** 319
[76] Eigler D M and Schweizer E K 1990 Positioning single atoms with a scanning tunneling microscope *Nature* **344** 524
[77] Crommie M F, Lutz C P and Eigler D M 1993 Confinement of electrons to quantum corrals on a metal surface *Science* **262** 218
[78] Boland J J 1993 Manipulating chlorine atom bonding on the Si(100)-(2 × 1) surface with the STM *Science* **262** 1703
[79] Resch R, Baur C, Bugacov A, Koel B E, Madhukar A, Requicha A A G and Will P 1998 Building and manipulating three-dimensional and linked two-dimensional structures of nanoparticles using scanning force microscopy *Langmuir* **14** 6613
[80] Shen T-C, Wang C, Abeln G C, Tucker J R, Lyding J W, Avouris P and Walkup R E 1995 Atomic-scale desorption through electronic and vibrational excitation mechanisms *Science* **268** 1590
[81] Martel R, Avouris Ph and Lyo I-W 1996 Molecularly adsorbed oxygen species on Si(111)-(7 × 7): STM-induced dissociative attachment studies *Science* **272** 385
[82] Stipe B C, Rezaei M A, Ho W, Gao S, Persson M and Lundqvist B I 1997 Single-molecule dissociation by tunneling electrons *Phys. Rev. Lett.* **78** 4410
[83] Shen Y R 1994 Nonlinear optical studies of surfaces *Appl. Phys.* A **59** 541
[84] Shen Y R 1994 Surfaces probed by nonlinear optics *Surf. Sci.* **299–300** 551
[85] Petek H and Ogawa S 1997 Femtosecond time-resolved two-photon photoemission studies of electron dynamics in metals *Prog. Surf. Sci.* **56** 239
[86] Her T-H, Finlay R J, Wu C and Mazur E 1998 Surface femtochemistry of CO/O2/Pt(111): the importance of nonthermalized substrate electrons *J. Chem. Phys.* **108** 8595
[87] Ramsier R D and Yates J T Jr 1991 Electron-stimulated desorption: principles and applications *Surf. Sci. Rep.* **12** 243
[88] Madey T E 1986 Electron- and photon-stimulated desorption: probes of structure and bonding at surfaces *Science* **234** 316
[89] Madey T E *et al* 1993 Structure and kinetics of electron beam damage in a chemisorbed monolayer: PF3 on Ru(0001) *Desorption Induced by Electronic Transitions DIET V* vol 31, ed A R Burns, E B Stechel and D R Jennison (Berlin: Springer)
[90] Prybyla J A, Tom H W K and Aumiller G D 1992 Femtosecond time-resolved surface reaction: desorption of Co from Cu(111) in <325 fsec *Phys. Rev. Lett.* **68** 503
[91] Misewich J A, Heinz T F and Newns D M 1992 Desorption induced by multiple electronic transitions *Phys. Rev. Lett.* **68** 3737
[92] Kolb D M 1996 Reconstruction phenomena at metal–electrolyte interfaces *Prog. Surf. Sci.* **51** 109
[93] Chusuei C C, Murrell T S, Corneille J S, Nooney M G, Vesecky S M, Hossner L R and Goodman D W 1999 Liquid reaction apparatus for surface analysis *Rev. Sci. Instrum.* **70** 2462

[94]   Soriaga M P 1992 Ultra-high vacuum techniques in the study of single-crystal electrode surfaces *Prog. Surf. Sci.* **39** 325

[95]   Hubbard A T 1990 Surface electrochemistry *Langmuir* **6** 97

[96]   Craig V S J 1997 An historical review of surface force measurement techniques *Colloids Surf. A: Physicochem. Eng. Aspects* **129–30** 75

[97]   Kumacheva E 1998 Interfacial friction measurements in surface force apparatus *Prog. Surf. Sci.* **58** 75

[98]   Itaya K 1998 *In situ* scanning tunneling microscopy in electrolyte solutions *Prog. Surf. Sci.* **58** 121

[99]   Cyr D M, Venkataraman B and Flynn G W 1996 STM investigations of organic molecules physisorbed at the liquid–solid interface *Chem. Mater.* **8** 1600

[100] Drake B, Sonnenfeld R, Schneir J and Hansma P K 1987 Scanning tunneling microscopy of process at liquid–solid interfaces *Surf. Sci.* **181** 92

[101] Giancarlo L C and Flynn G W 1988 Scanning tunneling and atomic force microscopy probes of self-assembled, physisorbed monolayers *Ann. Rev. Phys. Chem.* **49** 297

[102] Schneir J, Harary H H, Dagata J A, Hansma P K and Sonnenfeld R 1989 Scanning tunneling microscopy and fabrication of nanometer scale structure at the liquid–gold interface *Scanning Microsc.* **3** 719

[103] Vansteenkiste S O, Davies M C, Roberts C J, Tendler S J B and Williams P M 1998 Scanning probe microscopy of biomedical interfaces *Prog. Surf. Sci.* **57** 95

[104] Iwasita T and Nart F C 1997 *In situ* infrared spectroscopy at electrochemical interfaces *Prog. Surf. Sci.* **55** 271

[105] Raduge C, Pflumio V and Shen Y R 1997 Surface vibrational spectroscopy of sulfuric acid–water mixtures at the liquid–vapor interface *Chem. Phys. Lett.* **274** 140

[106] Shen Y R 1998 Sum frequency generation for vibrational spectroscopy: applications to water interfaces and films of water and ice *Solid State Commun.* **108** 399

[107] Gragson D E and Richmond G I 1998 Investigations of the structure and hydrogen bonding of water molecules at liquid surfaces by vibrational sum frequency spectroscopy *J. Phys. Chem.* **102** 3847

[108] Koop T, Schindler W, Kazimirov A, Scherb G, Zegenhagen J, Schulz T, Feidenhans'l R and Kirschner J 1998 Electrochemical cell for *in situ* x-ray diffraction under ultrapure conditions *Rev. Sci. Instrum.* **69** 1840

PART A2

THERMODYNAMICS AND STATISTICAL MECHANICS

A2.1
Classical thermodynamics

Robert L Scott

A2.1.1 Introduction

Thermodynamics is a powerful tool in physics, chemistry and engineering and, by extension, to substantially all other sciences. However, its power is narrow, since it says nothing whatsoever about time-dependent phenomena. It can demonstrate that certain processes are impossible, but it cannot predict whether thermo-dynamically allowed processes will actually take place.

It is important to recognize that thermodynamic laws are generalizations of experimental observations on systems of macroscopic size; for such bulk systems the equations are exact (at least within the limits of the best experimental precision). The validity and applicability of the relations are independent of the correctness of any model of molecular behaviour adduced to explain them. Moreover, the usefulness of thermodynamic relations depends crucially on *measurability*; unless an experimenter can keep the constraints on a system and its surroundings under control, the measurements may be worthless.

The approach that will be outlined here is due to Carathéodory [1] and Born [2] and should present fresh insights to those familiar only with the usual development in many chemistry, physics or engineering textbooks. However, while the formulations differ somewhat, the equations that finally result are, of course, identical.

A2.1.2 The zeroth law

A2.1.2.1 The state of a system

First, a few definitions: a system is any region of space, any amount of material for which the boundaries are clearly specified. At least for thermodynamic purposes it must be of macroscopic size and have a topological integrity. It may not be only part of the matter in a given region, e.g. all the sucrose in an aqueous solution. A system could consist of two non-contiguous parts, but such a specification would rarely be useful.

To define the thermodynamic state of a system one must specify the values of a minimum number of variables, enough to reproduce the system with all its macroscopic properties. If special forces (surface effects, external fields—electric, magnetic, gravitational, etc) are absent, or if the bulk properties are insensitive to these forces, e.g. the weak terrestrial magnetic field, it ordinarily suffices—for a one-component system—to specify three variables, e.g. the temperature T, the pressure p and the number of moles n, or an equivalent set. For example, if the volume of a surface layer is negligible in comparison with the total volume, surface effects usually contribute negligibly to bulk thermodynamic properties.

In order to specify the size of the system, at least one of these variables ought to be *extensive* (one that is proportional to the size of the system, like n or the total volume V). In the special case of several phases in equilibrium several extensive properties, e.g. n and V for two phases, may be required to determine the relative amounts of the two phases. The rest of the variables can be *intensive* (independent of the size of the system) like T, p, the molar volume $\bar{V} = V/n$, or the density ρ. For multicomponent systems, additional variables, e.g. several ns, are needed to specify composition.

For example, the definition of a system as 10.0 g H_2O at 10.0 °C at an applied pressure $p = 1.00$ atm is sufficient to specify that the water is liquid and that its other properties (energy, density, refractive index, even non-thermodynamic properties like the coefficients of viscosity and thermal conductivity) are uniquely fixed.

Although classical thermodynamics says nothing about time effects, one must recognize that nearly all thermodynamic systems are *metastable* in the sense that over long periods of time—much longer than the time to perform experiments—they may change their properties, e.g. perhaps by a very slow chemical reaction. Moreover, the time scale is merely relative; if a thermodynamic measurement can be carried out fast enough that it is finished before some other reaction can perturb the system, but slow enough for the system to come to internal equilibrium, it will be valid.

A2.1.2.2   *Walls and equations of state*

Of special importance is the nature of the boundary of a system, i.e. the wall or walls enclosing it and separating it from its *surroundings*. The concept of 'surroundings' can be somewhat ambiguous, and its thermodynamic usefulness needs to be clarified. It is not the rest of the universe, but only the external neighbourhood with which the system may interact. Moreover, unless this neighbourhood is substantially at internal equilibrium, its thermodynamic properties cannot be exactly specified. Examples of 'surroundings' are a thermostatic bath or the external atmosphere.

If neither matter nor energy can cross the boundary, the system is described as *isolated*; if only energy (but not matter) can cross the boundary, the system is *closed*; if both matter and energy can cross the boundary, the system is *open*.

(Sometimes, when defining a system, one must be careful to clarify whether the walls are part of the system or part of the surroundings. Usually the contribution of the wall to the thermodynamic properties is trivial by comparison with the bulk of the system and hence can be ignored.)

Consider two distinct closed thermodynamic systems each consisting of n moles of a specific substance in a volume V and at a pressure p. These two distinct systems are separated by an idealized wall that may be either *adiabatic* (heat-impermeable) or *diathermic* (heat-conducting). However, because the concept of heat has not yet been introduced, the definitions of adiabatic and diathermic need to be considered carefully. Both kinds of walls are impermeable to matter; a *permeable* wall will be introduced later.

If a system at equilibrium is enclosed by an adiabatic wall, the only way the system can be disturbed is by moving part of the wall; i.e. the only coupling between the system and its surroundings is by work, normally mechanical. (The adiabatic wall is an idealized concept; no real wall can prevent any conduction of heat over a long time. However, heat transfer must be negligible over the time period of an experiment.)

The diathermic wall is defined by the fact that two systems separated by such a wall cannot be at equilibrium at arbitrary values of their variables of state, p^α, V^α, p^β and V^β. (The superscripts are not exponents; they symbolize different systems, subsystems or phases; numerical subscripts are reserved for components in a mixture.) Instead there must be a relation between the four variables, which can be called an *equation of state*:

$$F(p^\alpha, V^\alpha, p^\beta, V^\beta) = 0. \qquad (A2.1.1)$$

Equation (A2.1.1) is essentially an expression of the concept of *thermal equilibrium*. Note, however, that, in this formulation, this concept precedes the notion of temperature.

To make the differences between the two kinds of walls clearer, consider the situation where both are ideal gases, each satisfying the ideal-gas law $pV = nRT$. If the two were separated by a diathermic wall,

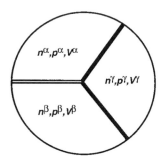

Figure A2.1.1. Illustration of the zeroth law. Three systems with two diathermic walls (solid) and one adiabatic wall (open).

one would observe experimentally that $p^\alpha V^\alpha / p^\beta V^\beta = C$ where the constant C would be n^α/n^β. If the wall were adiabatic, the two pV products could be varied independently.

A2.1.2.3 Temperature and the zeroth law

The concept of temperature derives from a fact of common experience, sometimes called the 'zeroth law of thermodynamics', namely, *if two systems are each in thermal equilibrium with a third, they are in thermal equilibrium with each other.* To clarify this point, consider the three systems shown schematically in figure A2.1.1, in which there are diathermic walls between systems α and γ and between systems β and γ, but an adiabatic wall between systems α and β.

Equation (A2.1.1) governs the diathermic walls, so one may write

$$F_A(p^\alpha, V^\alpha, p^\gamma, V^\gamma) = 0 \qquad (A2.1.2a)$$
$$F_B(p^\beta, V^\beta, p^\gamma, V^\gamma) = 0. \qquad (A2.1.2b)$$

It is a universal experimental observation, i.e. a 'law of nature', that the equations of state of systems 1 and 2 are then coupled as if the wall separating them were diathermic rather than adiabatic. In other words, there is a relation

$$F_C(p^\alpha, V^\alpha, p^\beta, V^\beta) = 0. \qquad (A2.1.2c)$$

It may seem that equation (A2.1.2c) is just a mathematical consequence of equations (2.1.2a, b), but it is not; it conveys new physical information. If one rewrites equations (2.1.2a, b) in the form

$$p^\gamma = \phi_\alpha(p^\alpha, V^\alpha, V^\gamma) = \phi_\beta(p^\beta, V^\beta, V^\gamma)$$

it is evident that this does not reduce to equation (A2.1.2c) unless one can separate V^γ out of the equation. This is not possible unless $\phi_\alpha = f_\alpha(p^\alpha, V^\alpha)g(V^\gamma) + h(V^\alpha)$ and $\phi_\beta = f_\beta(p^\beta, V^\beta)g(V^\gamma) + h(V^\beta)$. If equation (A2.1.2c) is a statement of a general experimental result, then $f_\alpha(p^\alpha, V^\alpha) = f_\beta(p^\beta, V^\beta)$ and the symmetry of equations (2.1.2a, b, c) extends the equality to $f_\gamma(p^\gamma, V^\gamma)$:

$$f_\alpha(p^\alpha, V^\alpha) = f_\beta(p^\beta, V^\beta) = f_\gamma(p^\gamma, V^\gamma) = \theta. \qquad (A2.1.3)$$

The three systems share a common property θ, the numerical value of the three functions f_α, f_β and f_γ, which can be called the *empirical temperature*. The equations (A2.1.3) are *equations of state* for the various systems, but the choice of θ is entirely arbitrary, since any function of f (e.g. f^2, $\log f$, $\cos^2 f - 3f^3$, etc) will satisfy equation (A2.1.3) and could serve as 'temperature'.

Redlich [3] has criticized the 'so-called zeroth law' on the grounds that the argument applies equally well for the introduction of any generalized force, mechanical (pressure), electrical (voltage), or otherwise. The difference seems to be that the physical nature of these other forces has already been clearly defined or postulated (at least in the conventional development of physics) while in classical thermodynamics, especially in the Born–Carathéodory approach, the existence of temperature has to be inferred from experiment.

For convenience, one of the systems will be taken as an ideal gas whose equation of state follows *Boyle's law*,

$$pV = nf(\theta) = nC\theta_{ig} \qquad (A2.1.4)$$

and which defines an ideal-gas temperature θ_{ig} proportional to pV/n. Later this will be identified with the thermodynamic temperature T. It is now possible to use the pair of variables V and θ instead of p and V to define the state of the system (for fixed n). [The pair p and θ would also do unless there is more than one phase present, in which case some variable or variables (in addition to n) must be extensive.]

A2.1.3 The first law

A2.1.3.1 *Work*

There are several different forms of work, all ultimately reducible to the basic definition of the infinitesimal work $Dw = f\, dl$ where f is the force acting to produce movement along the distance dl. Strictly speaking, both f and dl are vectors, so Dw is positive when the extension dl of the system is in the same direction as the applied force; if they are in opposite directions Dw is negative. Moreover, this definition assumes (as do all the equations that follow in this section) that there is a substantially equal and opposite force resisting the movement. Otherwise the actual work done on the system or by the system on the surroundings will be less or even zero. As will be shown later, the maximum work is obtained when the process is essentially 'reversible'.

The work depends on the detailed path, so Dw is an 'inexact differential' as symbolized by the capitalization. (There is no established convention about this symbolism; some books—and all mathematicians—use the same symbol for all differentials; some use δ for an inexact differential; others use a bar through the d; still others—as in this article—use D.) The difference between an *exact* and an *inexact* differential is crucial in thermodynamics. In general, the integral of a differential depends on the path taken from the initial to the final state. However, for some special but important cases, the integral is independent of the path; then and only then can one write

$$\int_i^f dF = F_f - F_i = \Delta F.$$

One then speaks of F as a 'state function' because it is a function only of those variables that define the state of the system, and not of the path by which the state was reached. An especially important feature of such functions is that if one writes DF as a function of several variables, say x, y, z,

$$DF = X(x, y, z)\, dx + Y(x, y, z)\, dy + Z(x, y, z)\, dz$$

then, for exact differentials *only*, $X = (\partial F/\partial x)_{y,z}$, $Y = (\partial F/\partial y)_{x,z}$ and $Z = (\partial F/\partial z)_{x,y}$. Since these exact differentials are path-independent, the order of differentiation is immaterial and one can then write

$$(\partial^2 F/\partial x\partial y)_z = (\partial X/\partial y)_{x,z} = (\partial Y/\partial x)_{y,z} \qquad \text{etc.}$$

One way of verifying the exactness of a differential is to check the validity of expressions like that above.

(a) Gravitational work

What is probably the simplest form of work to understand occurs when a force is used to raise the system in a gravitational field:

$$Dw_{grav} = mg\,dh$$

where m is the mass of the system, g is the acceleration of gravity, and dh is the infinitesimal increase in height. Gravitational work is rarely significant in most thermodynamic applications except when a falling weight outside the system drives a paddle wheel inside the system, as in one of the famous experiments in which Joule (1849) compared the work done with the increase in temperature of the system, and determined the 'mechanical equivalent of heat'. Note that, in this example, positive work is done on the system as the potential energy of the falling weight decreases. Note also that, in free fall, the potential energy of the weight decreases, but no work is done.

(b) One-dimensional work

When a spring is stretched or compressed, work is done. If the spring is the system, then the work done on it is simply

$$Dw_1 = f\,dl.$$

Note that a displacement from the initial equilibrium, either by compression or by stretching, produces positive work on the system. A situation analogous to the stretching of a spring is the stretching of a chain polymer.

(c) Two-dimensional (surface) work

When a surface is compressed by a force $f = \pi L$, the 'surface pressure' $\pi = f/L$ is the force per unit width L producing a decrease in length dl. (Note that L and l are not the same; indeed they are orthogonal.) The work is then

$$Dw_2 = -\pi\,dA$$

where $dA = L\,dl$ is the change in the surface area. This kind of work and the related thermodynamic functions for surfaces are important in dealing with monolayers in a Langmuir trough, and with membranes and other materials that are quasi-two-dimensional.

(d) Three-dimensional (pressure-volume) work

When a piston of area A, driven by a force $f = pA$, moves a distance $dl = -dV/A$, it produces a compression of the system by a volume dV. The work is then

$$Dw_3 = -p\,dV. \tag{A2.1.5}$$

It is this type of work that is ubiquitous in chemical thermodynamics, principally because of changes of the volume of the system under the external pressure of the atmosphere. The negative sign of the work done *on* the system is, of course, because the application of excess pressure produces a decrease in volume. (The negative sign in the two-dimensional case is analogous.)

(e) Other mechanical work

One can also do work by stirring, e.g. by driving a paddle wheel as in the Joule experiment above. If the paddle is taken as part of the system, the energy input (as work) is determined by appropriate measurements on the electric motor, falling weights or whatever drives the paddle.

(f) Electrical work

When a battery (or a generator or other power supply) outside the system drives current, i.e. a flow of electric charge, through a wire that passes through the system, work is done on the system:

$$\mathrm{D}w_{\mathrm{elec}} = \mathcal{E}\,\mathrm{d}\mathcal{Q}$$

where $\mathrm{d}\mathcal{Q}$ is the infinitesimal charge that crosses the boundary of the system and $\mathcal{E}$ is the electric potential (voltage) across the system, i.e. between the point where the wire enters and the point where it leaves. Converting to current $\mathcal{I} = \mathrm{d}\mathcal{Q}/\mathrm{d}t$ where $\mathrm{d}t$ is an infinitesimal time interval and to resistance $\mathcal{R} = \mathcal{E}/\mathcal{I}$ one can rewrite this equation in the form

$$\mathrm{D}w_{\mathrm{elec}} = \mathcal{E}\mathcal{I}\,\mathrm{d}t = (\mathcal{E}^2/\mathcal{R})\,\mathrm{d}t.$$

Such a resistance device is usually called an 'electrical heater' but, since there is no means of measurement at the boundary between the resistance and the material in contact with it, it is easier to regard the resistance as being inside the system, i.e. a part of it. Energy enters the system in the form of work where the wire breaches the wall, i.e. enters the container.

(g) Electrochemical work

A special example of electrical work occurs when work is done on an electrochemical cell or by such a cell on the surroundings ($-w$ in the convention of this article). Thermodynamics applies to such a cell when it is at equilibrium with its surroundings, i.e. when the electrical potential (*electromotive force* emf) of the cell is balanced by an external potential.

(h) Electromagnetic work

This poses a special problem because the source of the electromagnetic field may lie outside the defined boundaries of the system. A detailed discussion of this is outside the scope of this section, but the basic features can be briefly summarized.

When a specimen is moved in or out of an electric field or when the field is increased or decreased, the total work done on the whole system (charged condenser + field + specimen) in an infinitesimal change is

$$\mathrm{D}w_{\mathrm{el}} = \int \mathrm{d}V(\boldsymbol{E} \cdot \mathrm{d}\boldsymbol{D}),$$

where $\boldsymbol{E}$ is the electric field vector, $\boldsymbol{D} = \varepsilon\boldsymbol{E}$ is the electric displacement vector, and ε is the electric susceptibility tensor. The integration is over the whole volume encompassed by the total system, which must in principle extend as far as measurable fields exist.

Similarly, when a specimen is moved in or out of a magnetic field or when the magnetic field is increased or decreased, the total work done on the whole system (coil + field + specimen) in an infinitesimal change is

$$\mathrm{D}w_{\mathrm{mag}} = \int \mathrm{d}V(\boldsymbol{H} \cdot \mathrm{d}\boldsymbol{B})$$

where $\boldsymbol{H}$ is the magnetic field vector, $\boldsymbol{B} = \mu\boldsymbol{H}$ is the magnetic induction vector and μ is the magnetic permeability tensor. (Some modern discussions of magnetism regard $\boldsymbol{B}$ as the fundamental magnetic field vector, but usually fail to give a new name to $\boldsymbol{H}$.) As before the integration is over the whole volume.

For the special but familiar case of an isotropic specimen in a uniform external field E_0 or B_0, it can be shown [4] that

$$Dw_{el} = \int dV(\varepsilon_0 E_0 \cdot dE_0 - P \cdot dE_0) \tag{A2.1.6}$$

$$Dw_{mag} = \int dV(B_0 \cdot dB_0/\mu_0 + B_0 \cdot dM) \tag{A2.1.7}$$

where P is the polarization vector and M the magnetization vector; ε_0 and μ_0 are the susceptibility and permeability of the vacuum in the absence of the specimen. The vector notation could now be dropped since the external field and the induced field are parallel and the scalar product of two vectors oriented identically is simply the product of their scalar magnitudes; this will not be done in this article to avoid confusion with other thermodynamic quantities. (Note that equation (A2.1.7) is not the analogue of equation (A2.1.6).)

The work done increases the energy of the total system and one must now decide how to divide this energy between the field and the specimen. This separation is not measurably significant, so the division can be made arbitrarily; several self-consistent systems exist. The first term on the right-hand side of equation (A2.1.6) is obviously the work of creating the electric field, e.g. charging the plates of a condenser in the absence of the specimen, so it appears logical to consider the second term as the work done on the specimen.

By analogy, one is tempted to make the same division in equation (A2.1.7), regarding the first term as the work of creating the magnetic field in the absence of the specimen and the second, $\int dV(B_0 \cdot dM)$, as the work done on the specimen. This is the way most books on thermodynamics present the problem and it is an acceptable convention, except that it is inconsistent with the measured spectroscopic energy levels and with one's intuitive idea of work. For example, equation (A2.1.7) says that the work done in moving a permanent magnet (constant magnetization M) into or out of an electromagnet of constant B_0 is exactly zero! This is actually correct if one considers the extra electrochemical work done on the battery driving the current through the electromagnet while the permanent magnet is moving; this exactly balances the mechanical work. A careful analysis [5, 6] shows that, if one writes equation (A2.1.7) in the following form:

$$Dw_{mag} = \int dV[\underbrace{B_0 \cdot dB_0/\mu_0}_{A} - \underbrace{M\, dB_0}_{B} + \underbrace{d(B_0 \cdot M)}_{C}]$$

then term A is the work of creating the field in the absence of the specimen; term B is the work done on the specimen by 'ponderable forces', e.g. by a spring or by a physical push or pull; this is directly reflected in a change of the kinetic energy of the electrons; and term C is the work done by the electromotive force in the coil in creating the interaction field between B_0 and M. We elect to consider term B as the only work done on the specimen and write for the electromagnetic work

$$Dw_{electromag} = \int dV(-P \cdot dE_0 - M \cdot dB_0).$$

If in addition the specimen is assumed to be spherical as well as isotropic, so that P and M are uniform throughout the volume V, one can then write for the electromagnetic work

$$Dw_{electromag} = V(-P \cdot dE_0 - M \cdot dB_0). \tag{A2.1.8}$$

Equation (A2.1.8) turns out to be consistent with the changes of the energy levels measured spectroscopically, so the energy produced by work defined this way is frequently called the 'spectroscopic energy'. Note that the electric and magnetic parts of the equations are now symmetrical.

A2.1.3.2   Adiabatic work

One may now consider how changes can be made in a system across an adiabatic wall. The first law of thermodynamics can now be stated as another generalization of experimental observation, but in an unfamiliar form: *the work required to transform an adiabatic (thermally insulated) system from a completely specified initial state to a completely specified final state is independent of the source of the work (mechanical, electrical, etc.) and independent of the nature of the adiabatic path.* This is exactly what Joule observed; the same amount of work, mechanical or electrical, was always required to bring an adiabatically enclosed volume of water from one temperature θ_1 to another θ_2.

This can be illustrated by showing the net work involved in various adiabatic paths by which one mole of helium gas (4.00 g) is brought from an initial state in which $p = 1.000$ atm, $V = 24.62$ l [$T = 300.0$ K], to a final state in which $p = 1.200$ atm, $V = 30.779$ l [$T = 450.0$ K]. Ideal-gas behaviour is assumed (actual experimental measurements on a slightly non-ideal real gas would be slightly different). Information shown in brackets could be measured or calculated, but is not essential to the experimental verification of the first law.

Path I   (a)   Do electrical work on the system at constant $V = 24.62$ l until the
                   pressure has risen to 1.500 atm. [$\Delta T = 150.0$ K, $w = (3/2)R\Delta T$] $w_{elec} = 1871$ J
             (b)   Expand the gas into a vacuum (i.e. against zero external pressure)
                   until the total volume V is 30.77 l and $p = 1.200$ atm. [$\Delta T = 0$] $w_{exp} = 0$ J
                                                                                                        $w_{tot} = 1871$ J

Path II (a)   Compress the gas reversibly and adiabatically from 1.000 atm to
                   1.200 atm. [At the end of the compression $T = 322.7$ K,
                   $V = 22.07$ l, $w = (3/2)R\Delta T$] $w_{comp} = 283$ J
             (b)   Do electrical work on the system, holding the pressure constant
                   at 1.200 atm, until the volume V has increased to 30.77 l; under
                   these circumstances the system also does expansion work against
                   the external pressure.
                   [Electrical work $= (5/2)R\Delta T$] $w_{elec} = 2646$ J
                   [Expansion work $= -p\Delta V = -10.45$ l atm] $w_{exp} = -1058$ J
                                                                                                        $w_{tot} = 1871$ J

Path III (a)   Do electrical work on the system, holding the pressure constant at
                   1.000 atm, until the volume V has increased to 34.33 l; under these
                   circumstances, the system also does expansion work against the
                   external pressure.
                   [Final $T = 418.4$ K]
                   [Electrical work $= (5/2)RT$] $w_{elec} = 2460$ J
                   [Expansion work $= -p\Delta V = -9.71$ l atm] $w_{exp} = -984$ J
             (b)   Compress the gas reversibly and adiabatically from 1.000 atm to
                   1.200 atm. [At the end of the compression $T = 450.0$ K, $V = $ $w_{comp} = 395$ J
                   30.77 l, $\Delta T = 31.65$ K, $w = (3/2)RT$]
                                                                                                        $w_{tot} = 1871$ J

For all of these adiabatic processes, the total (net) work is exactly the same.

(As we shall see, because of the limitations that the second law of thermodynamics imposes, it may be impossible to find any adiabatic paths from a particular state A to another state B because $S_A - S_B < 0$. In this situation, however, there will be several adiabatic paths from state B to state A.)

If the adiabatic work is independent of the path, it is the integral of an exact differential and suffices to define a change in a function of the state of the system, the *energy U*. (Some thermodynamicists call this the 'internal energy', so as to exclude any kinetic energy of the motion of the system as a whole.)

$$dU = dw_{\text{adiabatic}}$$

or

$$\Delta U = U_f(V_f, \theta_f) - U_i(V_i, \theta_i) = \int dw_{\text{adiabatic}} = w_{\text{adiabatic}}. \tag{A2.1.9}$$

Here the subscripts i and f refer to the initial and final states of the system and the work w is defined as the work performed on the system (the opposite sign convention—with w as work done by the system on the surroundings—is also in common use). Note that a cyclic process (one in which the system is returned to its initial state) is *not* introduced; as will be seen later, a cyclic adiabatic process is possible only if every step is reversible. Equation (A2.1.9), i.e. the introduction of U as a state function, is an expression of the law of conservation of energy.

A2.1.3.3   *Non-adiabatic processes. Heat*

Not all processes are adiabatic, so when a system is coupled to its environment by diathermic walls, the heat q absorbed by the system is defined as the difference between the actual work performed and that which would have been required had the change occurred adiabatically.

$$Dq = dw_{\text{adiabatic}} - Dw = dU - Dw$$

or

$$q = w_{\text{adiabatic}} - w = \Delta U - w. \tag{A2.1.10}$$

Note that, since Dw is inexact, so also must be Dq.

This definition may appear eccentric because many people have an intuitive feeling for 'heat' as a certain kind of energy flow. However, thoughtful reconsideration supports a suspicion that the intuitive feeling is for the heat absorbed in a particular kind of process, e.g. constant pressure, for which, as we shall see, the heat q_p is equal to the change in a state function, the enthalpy change ΔH. For another example, the 'heats' measured in modern calorimeters are usually determined either by a *measurement* of electrical or mechanical work or by comparing one process with another so calibrated (as in an ice calorimeter). Indeed one can argue that one never measures q directly, that all 'measurements' require equation (A2.1.10); one always infers q from other measurements.

A2.1.4   The second law

In this and nearly all subsequent sections, the work Dw will be restricted to pressure–volume work, $-p\,dV$, and the fact that the 'heat' Dq may in some cases be electrical work will be ignored.

A2.1.4.1   *Reversible processes*

A particular path from a given initial state to a given final state is the reversible process, one in which after each infinitesimal step the system is in equilibrium with its surroundings, and one in which an infinitesimal change in the conditions (constraints) would reverse the direction of the change.

A simple example (figure A2.1.2) consists of a gas confined by a movable piston supporting a pile of sand whose weight produces a downward force per unit area equal to the pressure of the gas. Removal of a grain of sand decreases the downward pressure by an amount δp and the piston rises with an increase of volume δV sufficient to decrease the gas pressure by the *same* δp; the system is now again at equilibrium. Restoration of the grain of sand will drive the piston and the gas back to their initial states. Conversely, the successive removal of additional grains of sand will produce additional small decreases in pressure and small increases in volume; the sum of a very large number of such small steps can produce substantial changes in the thermodynamic properties of the system. Strictly speaking, such experimental processes are never quite

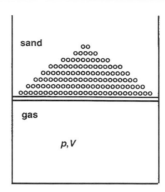

Figure A2.1.2. Reversible expansion of a gas with the removal one-by-one of grains of sand atop a piston.

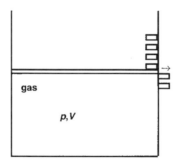

Figure A2.1.3. Irreversible expansion of a gas as stops are removed.

reversible because one can never make the small changes in pressure and volume infinitesimally small (in such a case there would be no tendency for change and the process would take place only at an infinitely slow rate). The true reversible process is an idealized concept; however, one can usually devise processes sufficiently close to reversibility that no measurable differences will be observed.

The mere fact that a substantial change can be broken down into a very large number of small steps, with equilibrium (with respect to any applied constraints) at the end of each step, does not guarantee that the process is reversible. One can modify the gas expansion discussed above by restraining the piston, not by a pile of sand, but by the series of stops (pins that one can withdraw one-by-one) shown in figure A2.1.3. Each successive state is indeed an equilibrium one, but the pressures on opposite sides of the piston are not equal, and pushing the pins back in one-by-one will not drive the piston back down to its initial position. The two processes are, in fact, quite different even in the infinitesimal limit of their small steps; in the first case work is done by the gas to raise the sand pile, while in the second case there is no such work. Both the processes may be called 'quasi-static' but only the first is anywhere near reversible. (Some thermodynamics texts restrict the term 'quasi-static' to a more restrictive meaning equivalent to 'reversible', but this then leaves no term for the slow irreversible process.)

If a system is coupled with its environment through an adiabatic wall free to move without constraints (such as the stops of the second example above), mechanical equilibrium, as discussed above, requires equality of the pressure p on opposite sides of the wall. With a diathermic wall, thermal equilibrium requires that the temperature θ of the system equal that of its surroundings. Moreover, it will be shown later that, if the wall is permeable and permits exchange of matter, material equilibrium (no tendency for mass flow) requires equality of a chemical potential μ.

Obviously the first law is not all there is to the structure of thermodynamics, since some adiabatic changes occur spontaneously while the reverse process never occurs. An aspect of the second law is that a state function, the *entropy S*, is found that increases in a spontaneous adiabatic process and remains unchanged in a reversible adiabatic process; it cannot decrease in any adiabatic process.

The next few sections deal with the way these experimental results can be developed into a mathematical system. A reader prepared to accept the second law on faith, and who is interested primarily in applications, may skip sections A2.1.4.2–A2.1.4.6 and perhaps even A2.1.4.7, and go to the final statement in section A2.1.4.8.

A2.1.4.2 *Adiabatic reversible processes and integrability*

In the example of the previous section, the release of the stop always leads to the motion of the piston in one direction, to a final state in which the pressures are equal, never in the other direction. This obvious experimental observation turns out to be related to a mathematical problem, the integrability of differentials in thermodynamics. The differential Dq, even Dq_{rev}, is inexact, but in mathematics many such expressions can be converted into exact differentials with the aid of an integrating factor.

In the example of pressure–volume work in the previous section, the adiabatic reversible process consisted simply of the sufficiently slow motion of an adiabatic wall as a result of an infinitesimal pressure difference. The work done on the system during an infinitesimal reversible change in volume is then $-p\,dV$ and one can write equation (A2.1.11) in the form

$$Dq_{rev} = dU + p\,dV = 0. \tag{A2.1.11}$$

If U is expressed as a function of two variables of state, e.g. V and θ, one can write $dU = (\partial U/\partial V)_\theta\,dV + (\partial U/\partial\theta)_V\,d\theta$ and transform equation (A2.1.11) into the following:

$$Dq_{rev} = [(\partial U/\partial V)_\theta + p]\,dV + (\partial U/\partial\theta)_V\,d\theta = Y\,dV + Z\,d\theta = 0. \tag{A2.1.12}$$

The coefficients Y and Z are, of course, functions of V and θ and therefore state functions. However, since in general $(\partial p/\partial\theta)_V$ is not zero, $\partial Y/\partial\theta$ is not equal to $\partial Z/\partial V$, so Dq_{rev} is not the differential of a state function but rather an inexact differential.

For a system composed of two subsystems α and β separated from each other by a diathermic wall and from the surroundings by adiabatic walls, the equation corresponding to equation (A2.1.12) is

$$\begin{aligned}
Dq_{rev} &= Dq^\alpha + Dq^\beta \\
&= [(\partial U^\alpha/\partial V^\alpha)_\theta + p^\alpha]\,dV^\alpha + [(\partial U^\beta/\partial V^\beta)_\theta + p^\beta]\,dV^\beta + [(\partial U^\alpha/\partial\theta)_{V^\alpha} + (\partial U^\beta/\partial\theta)_{V^\beta}]\,d\theta \\
&= X\,dV^\alpha + Y\,dV^\beta + Z\,d\theta = 0.
\end{aligned} \tag{A2.1.13}$$

One must now examine the integrability of the differentials in equations (A2.1.12) and (A2.1.13), which are examples of what mathematicians call *Pfaff differential equations*. If the equation is integrable, one can find an integrating denominator λ, a function of the variables of state, such that $Dq_{rev}/\lambda = d\phi$ where $d\phi$ is the exact differential of a function ϕ that defines a surface (line in the case of equation (A2.1.12)) in which the reversible adiabatic path must lie.

All equations of two variables, such as equation (A2.1.12), are necessarily integrable because they can be written in the form $dy/dx = f(x, y)$, which determines a unique value of the slope of the line through any point (x, y). Figure A2.1.4 shows a set of non-intersecting lines in V–θ space representing solutions of equation (A2.1.12).

For equations such as (A2.1.13) involving more than two variables the problem is no longer trivial. Most such equations are not integrable.

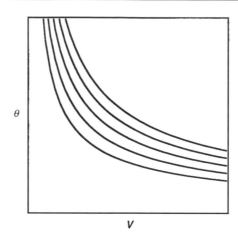

Figure A2.1.4. Adiabatic reversible (isentropic) paths that do not intersect. (The curves have been calculated for the isentropic expansion of a monatomic ideal gas.)

(Born [2] cites as an example of a simple expression for which no integrating factor exists

$$\mathrm{D}F = \mathrm{d}y + x\,\mathrm{d}z \overset{?}{=} \lambda(x, y, z)\,\mathrm{d}\phi.$$

If an integrating factor exists $\partial\phi/\partial x = 0$, $\partial\phi/\mathrm{d}y = 1/\lambda$ and $\partial\phi/\partial z = x/\lambda$. From the first of these relations one concludes that ϕ depends only on y and z. Using this result in the second relation one concludes that λ depends only on y and z. Given that ϕ and λ are both functions only of y and z, the third relation is a contradiction, so no factor λ can exist.)

There are now various adiabatic reversible paths because one can choose to vary $\mathrm{d}V^{\alpha}$ or $\mathrm{d}V^{\beta}$ in any combination of steps. The paths can cross and interconnect. The question of integrability is tied to the question of whether all regions of V^{α}, V^{β}, θ space are accessible by a series of connected adiabatic reversible paths or whether all such paths lie in a series of non-crossing surfaces. To distinguish, one must use a theorem of Carathéodory (the proof can be found in [1] and [2] and in books on differential equations):

If a Pfaff differential expression $\mathrm{D}F = X\,\mathrm{d}x + Y\,\mathrm{d}y + Z\,\mathrm{d}z$ *has the property that every arbitrary neighbourhood of a point* $P(x, y, z)$ *contains points that are inaccessible along a path corresponding to a solution of the equation* $\mathrm{D}F = 0$, *then an integrating denominator exists.* Physically this means that there are two mutually exclusive possibilities: either (a) a hierarchy of non-intersecting surfaces $\phi(x, y, z) = C$, each with a different value of the constant C, represents the solutions $\mathrm{D}F = 0$, in which case a point on one surface is inaccessible by a path that is confined to another, or (b) any two points can be connected by a path, each infinitesimal segment of which satisfies the condition $\mathrm{D}F = 0$. One must perform some experiments to determine which situation prevails in the physical world.

It suffices to carry out *one* such experiment, such as the expansion or compression of a gas, to establish that there are states inaccessible by adiabatic reversible paths, indeed even by any adiabatic irreversible path. For example, if one takes one mole of N_2 gas in a volume of 24 litres at a pressure of 1.00 atm (i.e. at $25\,°\mathrm{C}$), there is no combination of adiabatic reversible paths that can bring the system to a final state with the same volume and a different temperature. A higher temperature (on the ideal-gas scale θ_{ig}) can be reached by an adiabatic irreversible path, e.g. by doing electrical work on the system, but a state with the same volume and a lower temperature θ_{ig} is inaccessible by *any* adiabatic path.

A2.1.4.3   Entropy and temperature

One concludes, therefore, that equation (A2.1.13) is integrable and there exists an integrating factor λ. For the general case $\mathrm{D}q_{\mathrm{rev}} = \lambda \, \mathrm{d}\phi$ it can be shown [1, 2] that

$$\ln \lambda = \int g(\theta) \, \mathrm{d}\theta + \ln I(\phi)$$

where $I(\phi)$ is a constant of integration. It then follows that one may define two new quantities by the relations:

$$\ln(T/C) = \int g(\theta) \, \mathrm{d}\theta \qquad S = (1/C) \int I(\phi) \, \mathrm{d}\phi.$$

and one can now write

$$\mathrm{D}dq_{\mathrm{rev}} = \lambda \, \mathrm{d}\phi = T \, \mathrm{d}S. \tag{A2.1.14}$$

There are an infinite number of other integrating factors λ with corresponding functions ϕ; the new quantities T and S are chosen for convenience. S is, of course, the *entropy* and T, a function of θ only, is the 'absolute temperature', which will turn out to be the ideal-gas temperature, θ_{ig}. The constant C is just a scale factor determining the size of the degree.

The surfaces in which the paths satisfying the condition $\mathrm{D}q_{\mathrm{rev}} = 0$ must lie are, thus, surfaces of constant entropy; they do not intersect and can be arranged in an order of increasing or decreasing numerical value of the constant S. One half of the second law of thermodynamics, namely that for reversible changes, is now established.

Since $\mathrm{D}w_{\mathrm{rev}} = -p \, \mathrm{d}V$, one can utilize the relation $\mathrm{d}U = \mathrm{D}q_{\mathrm{rev}} + \mathrm{D}w_{\mathrm{rev}}$ and write

$$\mathrm{d}U = T \, \mathrm{d}S - p \, \mathrm{d}V. \tag{A2.1.15}$$

Equation (A2.1.15) involves only state functions, so it applies to any infinitesimal change in state whether the actual process is reversible or not (although, as equation (A2.1.14) suggests, $\mathrm{d}S$ is not experimentally accessible unless some reversible path exists).

A2.1.4.4   Thermodynamic temperature and the ideal-gas thermometer

So far, the thermodynamic temperature T has appeared only as an integrating denominator, a function of the empirical temperature θ. One now can show that T is, except for an arbitrary proportionality factor, the *same* as the empirical ideal-gas temperature θ_{ig} introduced earlier. Equation (A2.1.15) can be rewritten in the form

$$T \, \mathrm{d}S = \mathrm{d}U + p \, \mathrm{d}V = (\partial U/\partial \theta)_V \, \mathrm{d}\theta + [(\partial U/\partial V)_\theta + p] \, \mathrm{d}V. \tag{A2.1.16}$$

One assumes the existence of a fluid that obeys Boyle's law (equation (A2.1.4)) and that, on adiabatic expansion into a vacuum, shows no change in temperature, i.e. for which $pV = f(\theta)$ and $(\partial U/\partial V)_\theta = 0$. (All real gases satisfy this condition in the limit of zero pressure.) Equation (A2.1.16) then simplifies to

$$T \, \mathrm{d}S = (\mathrm{d}U/\mathrm{d}\theta) \, \mathrm{d}\theta + [f(\theta)/V] \, \mathrm{d}V = f(\theta)\{[(\mathrm{d}U/\mathrm{d}\theta)/f(\theta)] \, \mathrm{d}\theta + \mathrm{d}V/V\}.$$

The factor in wavy brackets is obviously an exact differential because the coefficient of $\mathrm{d}\theta$ is a function only of θ and the coefficient of $\mathrm{d}V$ is a function only of V. (The cross-derivatives vanish.) Manifestly then

$$\left. \begin{aligned} T &= Cf(\theta) = C(pV) \\ \mathrm{d}S &= \frac{\mathrm{d}U/\mathrm{d}\theta}{Cf(\theta)} \, \mathrm{d}\theta + \frac{1}{CV} \, \mathrm{d}V = \frac{\mathrm{d}U}{T} + \frac{\mathrm{d}V}{CV} \end{aligned} \right\} \quad \text{ideal gas only.}$$

If the arbitrary constant C is set equal to $(nR)^{-1}$ where n is the number of moles in the system and R is the gas constant per mole, then the thermodynamic temperature $T = \theta_{ig}$ where θ_{ig} is the temperature measured by the ideal-gas thermometer depending on the equation of state

$$pV = nR\theta_{ig} = nRT. \tag{A2.1.17}$$

Now that the identity has been proved θ_{ig} need not be used again.

A2.1.4.5   Irreversible changes and the second law

It is still necessary to consider the role of entropy in irreversible changes. To do this we return to the system considered earlier in section A2.1.4.2, the one composed of two subsystems in thermal contact, each coupled with the outside through movable adiabatic walls. Earlier this system was described as a function of three independent variables, V^α, V^β and θ (or T). Now, instead of the temperature, the entropy $S = S^\alpha + S^\beta$ will be used as the third variable. A final state $V^{\alpha'}$, $V^{\beta'}$, S' can always be reached from an initial state $V^{\alpha 0}$, $V^{\beta 0}$, S^0 by a two-step process.

(1)   The volumes are changed adiabatically and reversibly from $V^{\alpha 0}$ and $V^{\beta 0}$ to $V^{\alpha'}$ and $V^{\beta'}$, during which change the entropy remains constant at S^0.

(2)   At constant volumes $V^{\alpha'}$ and $V^{\beta'}$, the state is changed by the adiabatic performance of work (stirring, rubbing, electrical 'heating') until the entropy is changed from S^0 to S'.

If the entropy change in step (2) could be at times greater than zero and at other times less than zero, every neighbouring state $V^{\alpha'}$, $V^{\beta'}$, S' would be accessible, for there is no restriction on the adjustment of volumes in step (1). This contradicts the experimental fact that allowed the integration of equation (A2.1.13) and established the entropy S as a state function. It must, therefore, be true that either $S' > S^0$ always or that $S' < S^0$ always. One experiment demonstrates that the former is the correct alternative; if one takes the absolute temperature as a positive number, one finds that the entropy cannot decrease in an adiabatic process. This completes the specification of temperature, entropy and part of the second law of thermodynamics. One statement of the second law of thermodynamics is therefore:

$$\text{for any adiabatic process } (Dq = 0) \, dS \geq 0. \tag{A2.1.18}$$

(This is frequently stated for an isolated system, but the same statement about an adiabatic system is broader.)

A2.1.4.6   Irreversible changes and the measurement of entropy

Thermodynamic measurements are possible only when both the initial state and the final state are essentially at equilibrium, i.e. internally and with respect to the surroundings. Consequently, for a spontaneous thermodynamic change to take place, some constraint—internal or external—must be changed or released. For example, the expansion of a gas requires the release of a pin holding a piston in place or the opening of a stopcock, while a chemical reaction can be initiated by mixing the reactants or by adding a catalyst. One often finds statements that 'at equilibrium in an isolated system (constant U, V, n), the entropy is maximized'. What does this mean?

Consider two ideal-gas subsystems α and β coupled by a movable diathermic wall (piston) as shown in figure A2.1.5. The wall is held in place at a fixed position l by a stop (pin) that can be removed; then the wall is free to move to a new position l'. The total system $(\alpha + \beta)$ is adiabatically enclosed, indeed isolated $(q = w = 0)$, so the total energy, volume and number of moles are fixed.

When the pin is released, the wall will either (a) move to the right, or (b) move to the left, or (c) remain at the original position l. It is evident that these three cases correspond to initial situations in which $p^\alpha > p^\beta$,

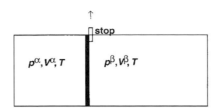

Figure A2.1.5. Irreversible changes. Two gases at different pressures separated by a diathermic wall, a piston that can be released by removing a stop (pin).

$p^\alpha < p^\beta$ and $p^\alpha = p^\beta$, respectively; if there are no other stops, the piston will come to rest in a final state where $p^{\alpha'} = p^{\beta'}$. For the two spontaneous adiabatic changes (a) and (b), the second law requires that $\Delta S > 0$, but one does not yet know the magnitude. (Nothing happens in case (c), so $\Delta S = 0$.)

The change of case (a) can be carried out in a series of small steps by having a large number of stops separated by successive distances Δl. For any intermediate step, $p^{\alpha'} > p^{\alpha''} > p^{\beta''} > p^{\beta'}$, but since the steps, no matter how small, are never reversible, one still has no information about ΔS.

The only way to determine the entropy change is to drive the system back from the final state to the initial state along a reversible path. One reimposes a constraint, not with simple stops, but with a gear system that permits one to do mechanical work driving the piston back to its original position l_0 along a reversible path; this work can be measured in various conventional ways. During this reverse change the system is no longer isolated; the total V and the total n remain unchanged, but the work done on the system adds energy. To keep the total energy constant, an equal amount of energy must leave the system in the form of heat:

$$dU = \mathrm{D}q_{rev} + \mathrm{D}w_{rev} = 0$$

or

$$-\Delta S_{forward} = \Delta S_{reverse} = \int \frac{\mathrm{D}q_{rev}}{T} = -\int \frac{\mathrm{D}w_{rev}}{T}.$$

For an ideal gas and a diathermic piston, the condition of constant energy means constant temperature. The reverse change can then be carried out simply by relaxing the adiabatic constraint on the external walls and immersing the system in a thermostatic bath. More generally the initial state and the final state may be at different temperatures so that one may have to have a series of temperature baths to ensure that the entire series of steps is reversible.

Note that although the change in state has been reversed, the system has not returned along the same detailed path. The forward spontaneous process was adiabatic, unlike the driven process and, since it was not reversible, surely involved some transient temperature and pressure gradients in the system. Even for a series of small steps ('quasi-static' changes), the infinitesimal forward and reverse paths must be different in detail. Moreover, because q and w are different, there are changes in the surroundings; although the system has returned to its initial state, the surroundings have not.

One can, in fact, drive the piston in both directions from the equilibrium value $l = l_e$ ($p^\alpha = p^\beta$) and construct a curve of entropy S (with an arbitrary zero) as a function of the piston position l (figure A2.1.6). If there is a series of stops, releasing the piston will cause l to change in the direction of increasing entropy until the piston is constrained by another stop or until l reaches l_e. It follows that at $l = l_e$, $dS/dl = 0$ and $d^2S/dl^2 < 0$; i.e. S is maximized when l is free to seek its own value. Were this not so, one could find spontaneous processes to jump from the final state to one of still higher entropy.

Thus, the spontaneous process involves the release of a constraint while the driven reverse process involves the imposition of a constraint. The details of the reverse process are irrelevant; any series of reversible steps by which one can go from the final state back to the initial state will do to measure ΔS.

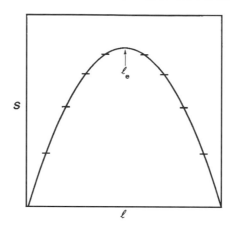

Figure A2.1.6. Entropy as a function of piston position l (the piston held by stops). The horizontal lines mark possible positions of stops, whose release produces an increase in entropy, the amount of which can be measured by driving the piston back reversibly.

A2.1.4.7   Irreversible processes: work, heat and entropy creation

One has seen that thermodynamic measurements can yield information about the change ΔS in an irreversible process (and thereby the changes in other state functions as well). What does thermodynamics tell one about work and heat in irreversible processes? Not much, in spite of the assertion in many thermodynamics books that

$$Dw = -p_{ext}\,dV = p_{ext}\,dV_{ext} \tag{A2.1.19}$$

and

$$Dq = -T_{ext}\,dS_{ext} = -dU_{ext} - Dw \tag{A2.1.20}$$

where p_{ext} and T_{ext} are the external pressure and temperature, i.e. those of the surroundings in which the changes $dV_{ext} = -dV$ and dS_{ext} occur.

Consider the situation illustrated in figure A2.1.5, with the modification that the piston is now an adiabatic wall, so the two temperatures need not be equal. Energy is transmitted from subsystem α to subsystem β only in the form of work; obviously $dV^{\alpha} = -dV^{\beta}$ so, in applying equation (A2.1.20), is $dU^{\alpha \to \beta}$ equal to $-p^{\alpha}\,dV^{\beta} = p^{\alpha}\,dV^{\alpha}$ or equal to $p^{\beta}\,dV^{\alpha}$, or is it something else entirely? One can measure the changes in temperature, $T^{\alpha'} - T^{\alpha}$ and $T^{\beta'} - T^{\beta}$ and thus determine $\Delta U^{\alpha \to \beta}$ after the fact, but could it have been predicted in advance, at least for ideal gases? If the piston were a diathermic wall so the final temperatures are equal, the energy transfer $\Delta U^{\alpha \to \beta}$ would be calculable, but even in this case it is unclear how this transfer should be divided between heat and work.

In general, the answers to these questions are ambiguous. When the pin in figure A2.1.5 is released, the potential energy inherent in the pressure difference imparts a kinetic energy to the piston. Unless there is a mechanism for the dissipation of this kinetic energy, the piston will oscillate like a spring; frictional forces, of course, dissipate this energy, but the extent to which the dissipation takes place in subsystem α or subsystem β depends on details of the experimental design not uniquely fixed by specifying the initial thermodynamic state. (For example, one can introduce braking mechanisms that dissipate the energy largely in subsystem α or, conversely, largely in subsystem β.) Only in one special case is there a clear prediction: if one subsystem (β) is empty no work can be done by α on β; for expansion into a vacuum necessarily $w = 0$. A more detailed discussion of the work involved in irreversible expansion has been given by Kivelson and Oppenheim [7].

The paradox involved here can be made more understandable by introducing the concept of entropy creation. Unlike the energy, the volume or the number of moles, the entropy is not conserved. The entropy of a system (in the example, subsystems α or β) may change in two ways: first, by the transport of entropy across the boundary (in this case, from α to β or *vice versa*) when energy is transferred in the form of heat, and second, by the creation of entropy within the subsystem during an irreversible process. Thus one can write for the change in the entropy of subsystem α in which some process is occurring

$$dS^\alpha = d_t S^\alpha + d_i S^\alpha$$

where $d_t S^\alpha = -d_t S^\beta$ is the change in entropy due to heat transfer to subsystem α and $d_i S^\alpha$ is the irreversible entropy creation inside subsystem α. (In the adiabatic example the dissipation of the kinetic energy of the piston by friction creates entropy, but no entropy is transferred because the piston is an adiabatic wall.)

The total change dS^α can be determined, as has been seen, by driving the subsystem α back to its initial state, but the separation into $d_i S^\alpha$ and $d_t S^\alpha$ is sometimes ambiguous. Any statistical mechanical interpretation of the second law requires that, at least for any volume element of macroscopic size, $d_i S \geq 0$. However, the total entropy change dS^α can be either positive or negative since the second law places no limitation on either the sign or the magnitude of $d_t S^\alpha$. (In the example above, the piston's adiabatic wall requires that $d_t S^\alpha = d_t S^\beta = 0$.)

In an irreversible process the temperature and pressure of the system (and other properties such as the chemical potentials μ_v to be defined later) are not necessarily definable at some intermediate time between the equilibrium initial state and the equilibrium final state; they may vary greatly from one point to another. One can usually define T and p for each small volume element. (These volume elements must not be too small; e.g. for gases, it is impossible to define T, p, S, etc for volume elements smaller than the cube of the mean free path.) Then, for each such sub-subsystem, $d_i S$ (but not the total dS) must not be negative. It follows that $d_i S^\alpha$, the sum of all the $d_i S$s for the small volume elements, is zero or positive. A detailed analysis of such irreversible processes is beyond the scope of classical thermodynamics, but is the basis for the important field of 'irreversible thermodynamics'.

The assumption (frequently unstated) underlying equations (A2.1.19) and (A2.1.20) for the measurement of irreversible work and heat is this: in the surroundings, which will be called subsystem β, internal equilibrium (uniform T^β, p^β and μ_i^β throughout the subsystem; i.e. no temperature, pressure or concentration gradients) is maintained throughout the period of time in which the irreversible changes are taking place in subsystem α. If this condition is satisfied $d_i S^\beta = 0$ and all the entropy creation takes place entirely in α. In any thermodynamic measurement that purports to yield values of q or w for an irreversible change, one must ensure that this condition is very nearly met. (Obviously, in the expansion depicted in figure A2.1.5 neither subsystem α nor subsystem β satisfied this requirement.)

Essentially this requirement means that, during the irreversible process, immediately inside the boundary, i.e. on the system side, the pressure and/or the temperature are only infinitesimally different from that outside, although substantial pressure or temperature gradients may be found outside the vicinity of the boundary. Thus an infinitesimal change in p_{ext} or T_{ext} would instantly reverse the direction of the energy flow, i.e. the sign of w or q. That part of the total process occurring *at the boundary* is then 'reversible'.

Subsystem β may now be called the 'surroundings' or as Callen (see further reading at the end of this article) does, in an excellent discussion of this problem, a 'source'. To formulate this mathematically one notes that, if $d_i S^\beta = 0$, one can then write

$$d_t S^\beta = Dq^\beta / T^\beta$$

and thus

$$d_t S^\alpha = -d_t S^\beta = -Dq^\beta / T^\beta = Dq^\alpha / T^\beta$$

because Dq^α, the energy received by α in the form of heat, must be the negative of that lost by β. Note, however, that the temperature specified is still T^β, since only in the β subsystem has no entropy creation been assumed ($d_i S^\beta = 0$). Then

$$d_i S^\alpha = dS^\alpha - d_t S^\alpha = dS^\alpha - Dq^\alpha / T^\beta \geq 0.$$

If one adds Dq^α / T^β to both sides of the inequality one has

$$dS^\alpha \geq Dq^\alpha / T^\beta.$$

A2.1.4.8   Final statement

If one now considers α as the 'system' and β as the 'surroundings' the second law can be reformulated in the form:

There exists a state function S, called the entropy of a system, related to the heat Dq absorbed from the surroundings during an infinitesimal change by the relations

$$\text{spontaneous process:}\quad dS > Dq/T_{\text{surr}} \quad \text{or}\quad \Delta S > \int (Dq/T_{\text{surr}})$$

$$\text{reversible process:}\quad dS = Dq/T_{\text{surr}} \quad \text{or}\quad \Delta S = \int (Dq/T_{\text{surr}})$$

$$\text{impossible process}\quad dS < Dq/T_{\text{surr}} \quad \text{or}\quad \Delta S < \int (Dq/T_{\text{surr}})$$

where T_{surr} is a positive quantity depending only on the (empirical) temperature of the surroundings. It is understood that for the surroundings $d_i S_{\text{surr}} = 0$. For the integral to have any meaning T_{surr} must be constant, or one must change the surroundings in each step. The above equations can be written in the more compact form

$$dS \geq Dq/T_{\text{surr}} \tag{A2.1.21}$$

where, in this and subsequent similar expressions, the symbol $\geq$ ('greater than or equal to') implies the equality for a reversible process and the inequality for a spontaneous (irreversible) process.

Equation (A2.1.21) includes, as a special case, the statement $dS \geq 0$ for adiabatic processes (for which $Dq = 0$) and, *a fortiori*, the same statement about processes that may occur in an isolated system ($Dq = Dw = 0$). If the universe is an isolated system (an assumption that, however plausible, is not yet subject to experimental verification), the first and second laws lead to the famous statement of Clausius: 'The energy of the universe is constant; the entropy of the universe tends always toward a maximum.'

It must be emphasized that equation (A2.1.21) permits the entropy of a particular system to decrease; this can occur if more entropy is transferred to the surroundings than is created within the system. The entropy of the system cannot decrease, however, without an equal or greater increase in entropy somewhere else.

There are many equivalent statements of the second law, some of which involve statements about heat engines and 'perpetual motion machines of the second kind' that appear superficially quite different from equation (A2.1.21). They will not be dealt with here, but two variant forms of equation (A2.1.21) may be noted: in view of the definition $dS = Dq_{\text{rev}}/T_{\text{surr}}$ one can also write for an infinitesimal change

$$Dq_{\text{rev}} \geq Dq$$

and, because $dU = Dq_{\text{rev}} + Dw_{\text{rev}} = Dq + Dw$,

$$Dw_{\text{rev}} \leq Dw.$$

Since w is defined as work done on the system, the minimum amount of work necessary to produce a given change in the system is that in a reversible process. Conversely, the amount of work done by the system on the surroundings is maximal when the process is reversible.

One may note, in concluding this discussion of the second law, that in a sense the zeroth law (thermal equilibrium) presupposes the second. Were there no irreversible processes, no tendency to move toward equilibrium rather than away from it, the concepts of thermal equilibrium and of temperature would be meaningless.

A2.1.5   Open systems

A2.1.5.1   Permeable walls and the chemical potential

We now turn to a new kind of boundary for a system, a wall permeable to matter. Molecules that pass through a wall carry energy with them, so equation (A2.1.15) must be generalized to include the change of the energy with a change in the number of moles dn:

$$dU = T\,dS - p\,dV + \mu\,dn. \tag{A2.1.22}$$

Here μ is the 'chemical potential' just as the pressure p is a mechanical potential and the temperature T is a thermal potential. A difference in chemical potential $\Delta\mu$ is a driving 'force' that results in the transfer of molecules through a permeable wall, just as a pressure difference Δp results in a change in position of a movable wall and a temperature difference ΔT produces a transfer of energy in the form of heat across a diathermic wall. Similarly equilibrium between two systems separated by a permeable wall must require equality of the chemical potential on the two sides. For a multicomponent system, the obvious extension of equation (A2.1.22) can be written

$$dU = T\,dS - p\,dV + \sum_i \mu_i\,dn_i \tag{A2.1.23}$$

where μ_i and n_i are the chemical potential and number of moles of the ith species. Equation (A2.1.23) can also be generalized to include various forms of work (such as gravitational, electrochemical, electromagnetic, surface formation, etc., as well as the familiar pressure–volume work), in which a generalized force X_j produces a displacement dx_j along the coordinate x_j, by writing

$$dU = T\,dS + \sum_j X_j\,dx_j + \sum_i \mu_i\,dn_i.$$

As a particular example, one may take the electromagnetic work terms of equation (A2.1.8) and write

$$dU = T\,dS - V(\mathbf{P}\cdot d\mathbf{E}_0 + \mathbf{M}\cdot d\mathbf{B}_0) + \sum_i \mu_i\,dn_i. \tag{A2.1.24}$$

The chemical potential now includes any such effects, and one refers to the *gravochemical potential*, the *electrochemical potential*, etc. For example, if the system consists of a gas extending over a substantial difference in height, it is the gravochemical potential (which includes a term mgh) that is the same at all levels, not the pressure. The electrochemical potential will be considered later.

A2.1.5.2   Internal equilibrium

Two subsystems α and β, in each of which the potentials T, p, and all the μ_is are uniform, are permitted to interact and come to equilibrium. At equilibrium all infinitesimal processes are reversible, so for the overall system $(\alpha + \beta)$, which may be regarded as isolated, the quantities conserved include not only energy, volume and numbers of moles, but also entropy, i.e. there is no entropy creation in a system at equilibrium. One now considers an infinitesimal reversible process in which small amounts of entropy $dS^{\alpha\to\beta}$, volume $dV^{\alpha\to\beta}$ and numbers of moles $dn_i^{\alpha \to \beta}$ are transferred from subsystem α to subsystem β. For this reversible change, one may use equation ((A2.1.23)) and write for dU^α and dU^β

$$dU^\alpha = -T^\alpha \, dS^{\alpha \to \beta} + p^\alpha \, dV^{\alpha \to \beta} - \sum_i \mu_i^\alpha \, dn_i^{\alpha \to \beta}$$

$$dU^\beta = T^\beta \, dS^{\alpha \to \beta} - p^\beta \, dV^{\alpha \to \beta} + \sum_i \mu_i^\beta \, dn_i^{\alpha \to \beta}.$$

Combining, one obtains for dU

$$dU = dU^\alpha + dU^\beta = 0 = (T^\beta - T^\alpha) \, dS^{\alpha \to \beta} - (p^\beta - p^\alpha) \, dV^{\alpha \to \beta} + \sum_i (\mu_i^\beta - \mu_i^\alpha) \, dn_i^{\alpha \to \beta}.$$

Thermal equilibrium means free transfer (exchange) of energy in the form of heat, mechanical (hydrostatic) equilibrium means free transfer of energy in the form of pressure–volume work, and material equilibrium means free transfer of energy by the motion of molecules across the boundary. Thus it follows that at equilibrium our choices of $dS^{\alpha \to \beta}$, $dV^{\alpha \to \beta}$, and $dn_i^{\alpha \to \beta}$ are independent and arbitrary. Yet the total energy must be kept unchanged, so the conclusion that the coefficients of $dS^{\alpha \to \beta}$, $dV^{\alpha \to \beta}$ and $dn_i^{\alpha \to \beta}$ must vanish is inescapable.

$$T^\alpha = T^\beta \qquad p^\alpha = p^\beta \qquad \mu_i^\alpha = \mu_i^\beta.$$

If there are more than two subsystems in equilibrium in the large isolated system, the transfers of S, V and n_i between any pair can be chosen arbitrarily; so it follows that at equilibrium all the subsystems must have the same temperature, pressure and chemical potentials. The subsystems can be chosen as *very small* volume elements, so it is evident that the criterion of internal equilibrium within a system (asserted earlier, but without proof) is uniformity of temperature, pressure and chemical potentials throughout. It has now been demonstrated conclusively that T, p and μ_i are *potentials*; they are intensive properties that measure 'levels'; they behave like the (equal) heights of the water surface in two interconnected reservoirs at equilibrium.

A2.1.5.3   Integration of dU

Equation (A2.1.23) can be integrated by the following trick: One keeps T, p, and all the chemical potentials μ_i constant and increases the number of moles n_i of each species by an amount $n_i \, d\xi$ where $d\xi$ is the same fractional increment for each. Obviously one is increasing the size of the system by a factor $(1 + d\xi)$, increasing all the extensive properties (U, S, V, n_i) by this factor and leaving the relative compositions (as measured by the mole fractions) and all other intensive properties unchanged. Therefore, $dS = S \, d\xi$, $dV = V \, d\xi$, $dn_i = n_i \, d\xi$, etc, and

$$dU = U \, d\xi = TS \, d\xi - pV \, d\xi + \sum_i \mu_i n_i \, d\xi.$$

Dividing by $d\xi$ one obtains

$$U = TS - pV + \sum_i \mu_i n_i. \tag{A2.1.25}$$

Mathematically equation (A2.1.25) is the direct result of the statement that U is homogeneous and of first degree in the extensive properties S, V and n_i. It follows, from a theorem of Euler, that

$$U = (\partial U / \partial S)_{V,n_i} S + (\partial U / \partial V)_{S,n_i} V + \sum_i (\partial U / \partial n_i)_{V,S,n_j} n_i. \tag{A2.1.26}$$

(The expression $(\partial U / \partial n_i)_{S,V,n_j}$ signifies, by common convention, the partial derivative of U with respect to the number of moles n_i of a particular species, holding S, V and the number of moles n_j of all other species $(j \neq i)$ constant.)

Equation (A2.1.26) is equivalent to equation (A2.1.25) and serves to identify T, p, and μ_i as appropriate partial derivatives of the energy U, a result that also follows directly from equation (A2.1.23) and the fact that dU is an exact differential.

$$T = (\partial U/\partial S)_{V,n_i} \qquad p = -(\partial U/\partial V)_{S,n_i} \qquad \mu = (\partial U/\partial n_i)_{V,S,n_j}.$$

If equation (A2.1.25) is differentiated, one obtains

$$dU = T\,dS + S\,dT - p\,dV - V\,dp + \sum_l \mu_i\,dn_i + \sum_i n_i\,d\mu_i$$

which, on combination with equation (A2.1.23), yields a very important relation between the differentials of the potentials:

$$S\,dT - V\,dp + \sum_i n_i\,d\mu_i = 0. \qquad (A2.1.27)$$

The special case of equation (A2.1.27) when T and p are constant ($dT = 0$, $dp = 0$) is called the Gibbs–Duhem equation, so equation (A2.1.27) is sometimes called the 'generalized Gibbs–Duhem equation'.

A2.1.5.4   *Additional functions and differing constraints*

The preceding sections provide a substantially complete *summary* of the fundamental concepts of classical thermodynamics. The basic equations, however, can be expressed in terms of other variables that are frequently more convenient in dealing with experimental situations under which different constraints are applied. It is often not convenient to use S and V as independent variables, so it is useful to define other quantities that are also functions of the thermodynamic state of the system. These include the enthalpy (or sometimes unfortunately called 'heat content') $H = U + pV$, the Helmholtz free energy (or 'work content') $A = U - TS$ and the Gibbs free energy (or 'Lewis free energy', frequently just called the 'free energy') $G = A + pV$. The usefulness of these will become apparent as some special situations are considered. In what follows it shall be assumed that there is no entropy creation in the surroundings, whose temperature and pressure can be controlled, so that equations (A2.1.19) and (A2.1.20) can be used to determine dw and dq. Moreover, for simplicity, the equations will be restricted to include only pressure–volume work; i.e. to equation (A2.1.5); the extension to other forms of work should be obvious.

(a) Constant-volume (isochoric) processes

If there is no volume change ($dV = 0$), then obviously there is no pressure–volume work done ($dw = 0$) irrespective of the pressure, and it follows from equation (A2.1.10) that the change in energy is due entirely to the heat absorbed, which can be designated as q_V:

$$(\Delta V = 0) \qquad dU = dq \qquad \Delta U = q_V. \qquad (A2.1.28)$$

Note that in this special case, the heat absorbed directly measures a state function. One still has to consider how this constant-volume 'heat' is measured, perhaps by an 'electric heater', but then is this not really work? Conventionally, however, if work is restricted to pressure–volume work, any remaining contribution to the energy transfers can be called 'heat'.

(b) Constant-pressure (isobaric) processes

For such a process the pressure p_{ext} of the surroundings remains constant and is equal to that of the system in its initial and final states. (If there are transient pressure changes within the system, they do not cause changes in the surroundings.) One may then write

$$dw = -p_{ext}\, dV$$
$$dq = dU + p_{ext}\, dV.$$

However, since $dp_{ext} = 0$ and the initial and final pressures inside equal p_{ext}, i.e. $\Delta p = 0$ for the change in state,

$$(\Delta p = 0) \qquad dq = d(U + p_{ext}\, dV) = d(U + pV) = dH$$
$$\Delta(U + pV) = \Delta H = q_p. \tag{A2.1.29}$$

Thus for isobaric processes a new function, the *enthalpy* H, has been introduced and its change ΔH is more directly related to the heat that must have been absorbed than is the energy change ΔU. The same reservations about the meaning of heat absorbed apply in this process as in the constant-volume process.

(c) Constant-temperature constant-volume (isothermal–isochoric) processes

In analogy to the constant-pressure process, constant temperature is defined as meaning that the temperature T of the surroundings remains constant and equal to that of the system in its initial and final (equilibrium) states. First to be considered are constant-temperature constant-volume processes (again $dw = 0$). For a reversible process

$$(\Delta T = 0, \Delta V = 0) \qquad dU = dq_{rev} = T\, dS.$$

For an irreversible process, invoking the notion of entropy transfer and entropy creation, one can write

$$dU = dq = T\, d_t S < T(d_t S + d_i S) = T\, dS = d(TS) = dq_{rev} \tag{A2.1.30}$$

which includes the inequality of equation (A2.1.21). Expressed this way the inequality $dU < T\, dS$ looks like a contradiction of equation (A2.1.15) until one realizes that the right-hand side of equation (A2.1.30) refers to the measurement of the entropy by a totally different process, a reverse (driven) process in which some work must be done on the system. If equation (A2.1.30) is integrated to obtain the isothermal change in state one obtains

$$(\Delta T = 0, \Delta V = 0) \qquad \Delta U = q < q_{rev} = T\Delta S = \Delta(TS)$$

or, rearranging the inequality,

$$(\Delta T = 0, \Delta V = 0) \qquad \Delta U - \Delta(TS) = \Delta(U - TS) = \Delta A < 0. \tag{A2.1.31}$$

Thus, for spontaneous processes at constant temperature and volume a new quantity, the *Helmholtz free energy* A, decreases. At equilibrium under such restrictions $dA = 0$.

(d) Constant-temperature constant-pressure (isothermal–isobaric) processes

The constant-temperature constant-pressure situation yields an analogous result. One can write for the reversible process

$$(\Delta T = 0, \Delta p = 0) \qquad dU = dq_{rev} + dw_{rev} = T\, dS - p\, dV$$

and for the irreversible process

$$dU = dq + dw = T\, d_t S - p\, dV < T\, dS - p\, dV$$

which integrated becomes

$$(\Delta T = 0, \Delta p = 0) \qquad \Delta U < T\Delta S - p\Delta V = \Delta(TS - pV)$$
$$\Delta(U + pV - TS) = \Delta G < 0. \qquad \text{(A2.1.32)}$$

For spontaneous processes at constant temperature and pressure it is the *Gibbs free energy* G that decreases, while at equilibrium under such conditions $dG = 0$.

More generally, without considering the various possible kinds of work, one can write for an isothermal change in a closed system ($dn_i = 0$)

$$\Delta U = q + w = T\Delta S + \Delta A.$$

Now, as has been shown, $q = T\Delta S$ for an isothermal *reversible* process only; for an isothermal *irreversible* process $\Delta S = \Delta_t S + \Delta_i S$, and $q = T\Delta_t S$. Since $\Delta_i S$ is positive for irreversible changes and zero only for reversible processes, one concludes

$$q = T\Delta S \qquad w = \Delta A \qquad \text{(isothermal reversible changes)}$$
$$q < T\Delta S \qquad w > \Delta A \qquad \text{(isothermal irreversible changes)}.$$

Another statement of the second law would be: 'The maximum work from (i.e. $-w$) a given isothermal change in thermodynamic state is obtained when the change in state is carried out reversibly; for irreversible isothermal changes, the work obtained is less'. Thus, in the expression $U = TS + A$, one may regard the TS term as that part of the energy of a system that is unavailable for conversion into work in an isothermal process, while A measures the 'free' energy that is available for isothermal conversion into work to be done on the surroundings. In isothermal changes some of A may be transferred quantitatively from one subsystem to another, or it may spontaneously decrease (be destroyed), but it cannot be created. Thus one may transfer the available part of the energy of an isothermal system (its 'free' energy) to a more convenient container, but one cannot increase its amount. In an irreversible process some of this 'free' energy is lost in the creation of entropy; some capacity for doing work is now irretrievably lost.

The usefulness of the Gibbs free energy G is, of course, that most changes of chemical interest are carried out under constant atmospheric pressure where work done on (or by) the atmosphere is not under the experimenter's control. In an isothermal–isobaric process (constant T and p), the *maximum* available 'useful' work, i.e. work other than pressure–volume work, is $-\Delta G$; indeed Guggenheim (1950) suggested the term 'useful energy' for G to distinguish it from the Helmholtz 'free energy' A. (Another suggested term for G is 'free enthalpy' from $G = H - TS$.) An international recommendation is that A and G simply be called the 'Helmholtz function' and the 'Gibbs function', respectively.

A2.1.5.5   Useful interrelations

By differentiating the defining equations for H, A and G and combining the results with equations (A2.1.25) and (A2.1.27) for dU and U (which are repeated here) one obtains general expressions for the differentials dH, dA, dG and others. One differentiates the defined quantities on the left-hand side of equations (A2.1.34)–(A2.1.39) and then substitutes the right-hand side of equation (A2.1.33) to obtain the appropriate differential. These are examples of *Legendre transformations*:

$$U = TS - pV + \sum_i \mu_i n_i \qquad dU = T\,dS - p\,dV + \sum_i \mu_i\,dn_i. \qquad \text{(A2.1.33)}$$

$$H = U + pV = TS + \sum_i \mu_i n_i \qquad dH = T\,dS + V\,dp + \sum_i \mu_i\,dn_i. \qquad \text{(A2.1.34)}$$

$$A = U - TS = -pV + \sum_i \mu_i n_i \qquad \mathrm{d}A = -S\,\mathrm{d}T - p\,\mathrm{d}V + \sum_i \mu_i\,\mathrm{d}n_i. \qquad (A2.1.35)$$

$$G = U + pV - TS = \sum_i \mu_i n_i \qquad \mathrm{d}G = -S\,\mathrm{d}T + V\,\mathrm{d}p + \sum_i \mu_i\,\mathrm{d}n_i. \qquad (A2.1.36)$$

$$-pV = U - TS - \sum_i \mu_i n_i \qquad -\mathrm{d}(pV) = -S\,\mathrm{d}T - p\,\mathrm{d}V - \sum_i n_i\,\mathrm{d}\mu_i. \qquad (A2.1.37)$$

$$TS = U + pV - \sum_i \mu_i n_i \qquad -\mathrm{d}(TS) = T\,\mathrm{d}S + V\,\mathrm{d}p - \sum_i n_i\,\mathrm{d}\mu_i. \qquad (A2.1.38)$$

$$0 = U - TS + pV - \sum_i \mu_i n_i \qquad -\mathrm{d}(0) = -S\,\mathrm{d}T + V\,\mathrm{d}p - \sum_i n_i\,\mathrm{d}\mu_i. \qquad (A2.1.39)$$

Equation (A2.1.39) is the 'generalized Gibbs–Duhem equation' previously presented (equation (A2.1.27)). Note that the Gibbs free energy is just the sum over the chemical potentials.

If there are other kinds of work, similar expressions apply. For example, with electromagnetic work (equation (A2.1.8)) instead of pressure–volume work, one can write for the Helmholtz free energy

$$\mathrm{d}A = -S\,\mathrm{d}T - V(\boldsymbol{P} \cdot \mathrm{d}\boldsymbol{E}_0 + \boldsymbol{M} \cdot \mathrm{d}\boldsymbol{B}_0) + \sum_i \mu_i\,\mathrm{d}n_i. \qquad (A2.1.40)$$

It should be noted that the differential expressions on the right-hand side of equations (A2.1.33)–(A2.1.40) express for each function the appropriate independent variables for that function, i.e. the variables—read constraints—that are kept constant during a spontaneous process.

All of these quantities are state functions, i.e. the differentials are exact, so each of the coefficients is a partial derivative. For example, from equation (A2.1.35) $p = -(\partial A/\partial V)_{T,n_i}$, while from equation (A2.1.36) $S = -(\partial G/\partial T)_{p,n_i}$. Moreover, because the order of partial differentiation is immaterial, one obtains as cross-differentiation identities from equations (A2.1.33)–(A2.1.40) a whole series of useful equations usually known as 'Maxwell relations'. A few of these are: from equation (A2.1.33):

$$(\partial^2 U/\partial S\,\partial V)_{n_i} = \partial(\partial U/\partial V)/\partial S = \partial(\partial U/\partial S)/\partial V$$
$$= -(\partial p/\partial S)_{V,n_i} = (\partial T/\partial V)_{S,n_i}$$

from equation (A2.1.35):

$$(\partial^2 A/\partial T\,\partial V)_{n_i} = -(\partial S/\partial V)_{T,n_i} = -(\partial p/\partial T)_{V,n_i} \qquad (A2.1.41)$$

from equation (A2.1.36):

$$(\partial^2 G/\partial T\,\partial p)_{n_i} = -(\partial S/\partial p)_{T,n_i} = (\partial V/\partial T)_{p,n_i} \qquad (A2.1.42)$$

and from equation (A2.1.40)

$$(\partial^2 A/\partial T\,\partial \boldsymbol{E}_0)_n = -(\partial S/\partial \boldsymbol{E}_0)_{T,n} = -V(\partial \boldsymbol{P}_0/\partial T)_{\boldsymbol{E}_0,n}$$
$$(\partial^2 A/\partial T\,\partial \boldsymbol{B}_0)_n = -(\partial S/\partial \boldsymbol{B}_0)_{T,n} = -V(\partial \boldsymbol{M}_0/\partial T)_{\boldsymbol{B}_0,n}. \qquad (A2.1.43)$$

(Strictly speaking, differentiation with respect to a vector quantity is not allowed. However for the isotropic spherical samples for which equation (A2.1.8) is appropriate, the two vectors have the same direction and could have been written as scalars; the vector notation was kept to avoid confusion with other thermodynamic quantities such as energy, pressure, etc. It should also be noted that the Maxwell equations above are correct for either of the choices for electromagnetic work discussed earlier; under the other convention A is replaced by a generalized G.)

A2.1.5.6 Features of equilibrium

Earlier in this section it was shown that, when a constraint, e.g. fixed l, was released in a system for which U, V and n were held constant, the entropy would seek a maximum value consistent with the remaining restrictions (e.g. $dS/dl = 0$ and $d^2S/dl^2 < 0$). One refers to this, a result of equation (A2.1.33), as a 'feature of equilibrium'. We can obtain similar features of equilibrium under other conditions from equations (A2.1.34)–(A2.1.39). Since at equilibrium all processes are reversible, all these equations are valid at equilibrium. Each equation is a linear relation between differentials; so, if all but one are fixed equal to zero, at equilibrium the remaining differential quantity must also be zero. That is to say, the function of which it is the differential must have an equilibrium value that is either maximized or minimized and it is fairly easy, in any particular instance, to decide between these two possibilities. To summarize the more important of these equilibrium features:

for fixed U, V, n_i	S is a maximum
for fixed H, p, n_i	S is a maximum
for fixed S, V, n_i	U is a minimum
for fixed S, p, n_i	H is a minimum
for fixed T, V, n_i	A is a minimum
for fixed T, p, n_i	G is a minimum.

Of these the last condition, minimum Gibbs free energy at constant temperature, pressure and composition, is probably the one of greatest practical importance in chemical systems. (This list does not exhaust the mathematical possibilities; thus one can also derive other apparently unimportant conditions such as that at constant U, S and n_i, V is a minimum.) However, an experimentalist will wonder how one can hold the entropy constant and release a constraint so that some other state function seeks a minimum.

A2.1.5.7 The chemical potential and partial molar quantities

From equations (A2.1.33)–(A2.1.36) it follows that the chemical potential may be defined by any of the following relations:

$$\mu_i = (\partial U/\partial n_i)_{S,V,n_j} = (\partial H/\partial n_i)_{S,p,n_j}$$
$$= (\partial A/\partial n_i)_{T,V,n_j} = (\partial G/\partial n_i)_{S,V,n_j} = \bar{G}_i. \tag{A2.1.44}$$

In experimental work it is usually most convenient to regard temperature and pressure as the independent variables, and for this reason the term *partial molar quantity* (denoted by a bar above the quantity) is always restricted to the derivative with respect to n_i holding T, p, and all the other n_j constant. (Thus $\bar{V}_i = (\partial V/\partial n_i)_{T,p,n_j}$.) From the right-hand side of equation (A2.1.44) it is apparent that the chemical potential is the same as the partial molar Gibbs free energy $\bar{G}_i$ and, therefore, some books on thermodynamics, e.g. Lewis and Randall (1923), do not give it a special symbol. Note that the partial molar Helmholtz free energy is *not* the chemical potential; it is

$$\bar{A}_i = (\partial A/\partial n_i)_{T,p,n_j} = [\partial(G - pV)/\partial n_i]_{T,p,n_j} = \mu_i - p\bar{V}_i.$$

On the other hand, in the theoretical calculations of statistical mechanics, it is frequently more convenient to use volume as an independent variable, so it is important to preserve the general importance of the chemical potential as something more than a quantity $\bar{G}_i$ whose usefulness is restricted to conditions of constant temperature and pressure.

From cross-differentiation identities one can derive some additional Maxwell relations for partial molar quantities:

$$(\partial^2 G/\partial T \partial n_i)_{p,n_j} = (\partial \mu_i/\partial T)_{p,n_i} = -(\partial S/\partial n_i)_{p,n_j} = -\bar{S}_i$$
$$(\partial^2 G/\partial p \partial n_i)_{T,n_j} = (\partial \mu_i/\partial p)_{T,n_i} = (\partial V/\partial n_i)_{T,p,n_j} = \bar{V}_i.$$

In passing one should note that the method of expressing the chemical potential is arbitrary. The amount of matter of species i in this article, as in most thermodynamics books, is expressed by the number of moles n_i; it can, however, be expressed equally well by the number of molecules N_i (convenient in statistical mechanics) or by the mass m_i (Gibbs' original treatment).

A2.1.5.8 Some additional important quantities

As one raises the temperature of the system along a particular path, one may define a *heat capacity* $C_{\text{path}} = Dq_{\text{path}}/dT$. (The term 'heat capacity' is almost as unfortunate a name as the obsolescent 'heat content' for H; alas, no alternative exists.) However several such paths define state functions, e.g. equations (A2.1.28) and (A2.1.29). Thus we can define the heat capacity at constant volume C_V and the heat capacity at constant pressure C_p as

$$C_V = (\partial U/\partial T)_{V,n_i} = T(\partial S/\partial T)_{V,n_i} \tag{A2.1.45}$$
$$C_p = (\partial H/\partial T)_{p,n_i} = T(\partial S/\partial T)_{p,n_i}. \tag{A2.1.46}$$

The right-hand equalities in these two equations arise directly from equations (A2.1.33) and (A2.1.34).

Two other important quantities are the *isobaric expansivity* ('coefficient of thermal expansion') α_p and the *isothermal compressibility* κ_T, defined as

$$\alpha_p = (1/V)(\partial V/\partial T)_{p,n_i}$$
$$\kappa_T = -(1/V)(\partial V/\partial p)_{T,n_i}.$$

The adjectives 'isobaric' and 'isothermal' and the corresponding subscripts are frequently omitted, but it is important to distinguish between the isothermal compressibility and the *adiabatic compressibility* $\kappa_S = -(1/V)(\partial V/\partial p)_{S,n_i}$.

A relation between C_p and C_V can be obtained by writing

$$(\partial S/\partial T)_{p,n_i} = (\partial S/\partial T)_{V,n_i} + (\partial S/\partial V)_{T,n_i}(\partial V/\partial T)_{p,n_i}$$
$$= (\partial S/\partial T)_{V,n_i} + (\partial p/\partial T)_{V,n_i}(\partial V/\partial T)_{p,n_i}$$
$$(\partial p/\partial T)_{V,n_i} = -(\partial p/\partial V)_{T,n_i}(\partial V/\partial T)_{p,n_i} = \alpha_p/\kappa_T \qquad \text{(from the cyclic rule)}.$$

Combining these, we have

$$(\partial H/\partial T)_{p,n_i} - (\partial U/\partial T)_{V,n_i} = T(\partial p/\partial T)_{V,n_i}(\partial V/\partial T)_{p,n_i}$$

or

$$C_p = C_V = TV\alpha_p^2/\kappa_T. \tag{A2.1.47}$$

For the special case of the ideal gas (equation (A2.1.17)), $\alpha_p = 1/T$ and $\kappa_T = 1/p$,

$$C_p - C_V = TVp/T^2 = nR \qquad \text{(ideal gas only)}.$$

A similar derivation leads to the difference between the isothermal and adiabatic compressibilities:

$$\kappa_T - \kappa_S = TV\alpha_p^2/C_p. \tag{A2.1.48}$$

A2.1.5.9   Thermodynamic equations of state

Two exact equations of state can be derived from equations (A2.1.33) and (A2.1.34)

$$(\partial U/\partial V)_{T,n_i} = -p + T(\partial S/V)_{T,n_i} = -p + T(\partial p/\partial T)_{V,n_i} \quad \text{or} \quad p = T(\partial p/\partial T)_{V,n_i} - (\partial U/\partial V)_{T,n_i}.$$
$$\text{(A2.1.49)}$$

$$(\partial H/\partial p)_{T,n_i} = V + T(\partial S/\partial p)_{T,n_i} = V - T(\partial V/\partial T)_{p,n_i} \quad \text{or} \quad V = T(\partial V/\partial T)_{p,n_i} + (\partial H/\partial p)_{T,n_i}.$$
$$\text{(A2.1.50)}$$

It is interesting to note that, when the van der Waals equation for a fluid,

$$p = nRT/(V - nb) - n^2a/V^2,$$

is compared with equation (A2.1.49), the right-hand sides separate in the same way:

$$T(\partial p/\partial T)_{V,n} = nRT/(V - nb) \quad \text{and} \quad (\partial U/\partial V)_{T,n} = n^2a/V^2.$$

A2.1.6   Applications

A2.1.6.1   Phase equilibria

When two or more phases, e.g. gas, liquid or solid, are in equilibrium, the principles of internal equilibrium developed in section A2.1.5.2 apply. If transfers between two phases α and β can take place, the appropriate potentials must be equal, even though densities and other properties can be quite different.

$$T^\alpha = T^\beta \qquad p^\alpha = p^\beta \qquad \mu_i^\alpha = \mu_i^\beta.$$

As shown in preceding sections, one can have equilibrium of some kinds while inhibiting others. Thus, it is possible to have thermal equilibrium ($T^\alpha = T^\beta$) through a fixed impermeable diathermic wall; in such a case p^α need not equal p^β, nor need μ_i^α equal μ_i^β. It is possible to achieve mechanical equilibrium ($p^\alpha = p^\beta$) through a movable impermeable adiabatic wall; in such a case the transfer of heat or matter is prevented, so T and μ_i can be different on opposite sides. It is possible to have both thermal and mechanical equilibrium ($p^\alpha = p^\beta$, $T^\alpha = T^\beta$) through a movable diathermic wall. For a one-component system $\mu = f(T, p)$, so $\mu^\alpha = \mu^\beta$ even if the wall is impermeable. However, for a system of two or more components one can have $p^\alpha = p^\beta$ and $T^\alpha = T^\beta$, but the chemical potential is now also a function of composition, so μ_i^α need not equal μ_i^β. It does not seem experimentally possible to permit material equilibrium ($\mu_i^\alpha = \mu_i^\beta$) without simultaneously achieving thermal equilibrium ($T^\alpha = T^\beta$).

Finally, in membrane equilibria, where the wall is permeable to some species, e.g. the solvent, but not others, thermal equilibrium ($T^\alpha = T^\beta$) and solvent equilibrium ($\mu_i^\alpha = \mu_i^\beta$) are found, but $\mu_j^\alpha \neq \mu_j^\beta$ and $p^\alpha \neq p^\beta$; the difference $p^\beta - p^\alpha$ is the osmotic pressure.

For a one-component system, $\Delta G^{\alpha \to \beta} = \mu^\beta - \mu^\alpha = 0$, so one may write

$$\Delta G^{\alpha \to \beta} = \Delta H^{\alpha \to \beta} - T\Delta S^{\alpha \to \beta} = 0 \quad \text{or} \quad \Delta S^{\alpha \to \beta} = \Delta H^{\alpha \to \beta}/T. \qquad \text{(A2.1.51)}$$

The Clapeyron equation

Moreover, using the generalized Gibbs–Duhem equations (A2.1.27) for each of the two one-component phases,

$$S^\alpha \, dT - V^\alpha \, dp + n^\alpha \, d\mu = 0$$

or

$$d\mu = \bar{V}^\alpha \, dp - \bar{S}^\alpha \, dT = \bar{V}^\beta \, dp - \bar{S}^\beta \, dT$$

one obtains the Clapeyron equation for the change of pressure with temperature as the two phases continue to coexist:

$$\mathrm{d}p/\mathrm{d}T = \Delta \bar{S}^{\alpha \to \beta}/\Delta \bar{V}^{\alpha \to \beta} = \Delta \bar{H}^{\alpha \to \beta}/T\Delta \bar{V}^{\alpha \to \beta}. \tag{A2.1.52}$$

The analogue of the Clapeyron equation for multicomponent systems can be derived by a complex procedure of systematically eliminating the various chemical potentials, but an alternative derivation uses the Maxwell relation (A2.1.41)

$$(\partial^2 A/\partial T\,\partial V)_{n_i} = -(\partial S/\partial V)_{T,n_i} = -(\partial p/\partial T)_{V,n_i}. \tag{A2.1.41}$$

Applied to a two-phase system, this says that the change in pressure with temperature is equal to the change in entropy at constant temperature as the total volume of the system ($\alpha + \beta$) is increased, which can only take place if some α is converted to β:

$$\mathrm{d}p/\mathrm{d}T = \Delta S^{\alpha \to \beta}/\Delta V^{\alpha \to \beta} = \Delta H^{\alpha \to \beta}/T\Delta V^{\alpha \to \beta}.$$

In this case, whatever n_i moles of each species are required to accomplish the ΔV are the same n_is that determine ΔS or ΔH. Note that this general equation includes the special one-component case of equation (A2.1.52).

When, for a one-component system, one of the two phases in equilibrium is a sufficiently dilute gas, i.e. is at a pressure well below 1 atm, one can obtain a very useful approximate equation from equation (A2.1.52). The molar volume of the gas is at least two orders of magnitude larger than that of the liquid or solid, and is very nearly an ideal gas. Then one can write

$$\Delta \bar{V}^{l \to g} \approx \bar{V}^g \approx RT/p$$

which can be substituted into equation (A2.1.52) to obtain

$$\mathrm{d}p/\mathrm{d}T \approx \Delta \bar{H}^{l \to g}(p/RT^2)$$

or

$$\mathrm{d}\ln p/\mathrm{d}T \approx \Delta \bar{H}^{l \to g}/RT^2 = \Delta \bar{H}_{\mathrm{vap}}/RT^2 \tag{A2.1.53}$$

or

$$\mathrm{d}\ln p/\mathrm{d}(1/T) \approx -\Delta \bar{H}_{\mathrm{vap}}/R,$$

where $\Delta \bar{H}_{\mathrm{vap}} = \Delta \bar{H}^{l \to g}$ is the molar enthalpy of vaporization at the temperature T. The corresponding equation for the vapour pressure of the solid is identical except for the replacement of the enthalpy of vaporization by the enthalpy of sublimation.

(Equation (A2.1.53) is frequently called the *Clausius–Clapeyron equation*, although this name is sometimes applied to equation (A2.1.52). Apparently Clapeyron first proposed equation (A2.1.52) in 1834, but it was derived properly from thermodynamics decades later by Clausius, who also obtained the approximate equation (A2.1.53).)

It is interesting and surprising to note that, although the molar enthalpy $\Delta \bar{H}_{\mathrm{vap}}$ and the molar volume of vaporization $\Delta \bar{V}_{\mathrm{vap}}$ both decrease to zero at the critical temperature of the fluid (where the fluid is very non-ideal), a plot of $\ln p$ against $1/T$ for most fluids is very nearly a straight line all the way from the melting point to the critical point. For example, for krypton, the slope $\mathrm{d}\ln p/\mathrm{d}(1/T)$ varies by less than 1% over the entire range of temperatures; even for water the maximum variation of the slope is only about 15%.

The phase rule

Finally one can utilize the generalized Gibbs–Duhem equations (A2.1.27) for each phase

$$S^\alpha \, dT - V^\alpha \, dp + \sum_i n_i^\alpha \, d\mu_i = 0$$

$$S^\beta \, dT - V^\beta \, dp + \sum_i n_i^\beta \, d\mu_i = 0$$

etc to obtain the 'Gibbs phase rule'. The number of variables (potentials) equals the number of components $\mathcal{C}$ plus two (temperature and pressure), and these are connected by an equation for each of the $\mathcal{P}$ phases. It follows that the number of potentials that can be varied independently (the 'degrees of freedom' $\mathcal{F}$) is the number of variables minus the number of equations:

$$\mathcal{F} = \mathcal{C} + 2 - \mathcal{P}.$$

From this equation one concludes that the maximum number of phases that can coexist in a one-component system ($\mathcal{C} = 1$) is three, at a unique temperature and pressure ($\mathcal{F} = 0$). When two phases coexist ($\mathcal{F} = 1$), selecting a temperature fixes the pressure. Conclusions for other situations should be obvious.

A2.1.6.2 Real and ideal gases

Real gases follow the ideal-gas equation (A2.1.17) only in the limit of zero pressure, so it is important to be able to handle the thermodynamics of real gases at non-zero pressures. There are many semi-empirical equations with parameters that purport to represent the physical interactions between gas molecules, the simplest of which is the van der Waals equation (A2.1.50). However, a completely general form for expressing gas non-ideality is the series expansion first suggested by Kamerlingh Onnes (1901) and known as the *virial equation of state*:

$$pV/nRT = 1 + B(n/V) + C(n/V)^2 + D(n/V)^3 + \cdots.$$

The equation is more conventionally written expressing the variable n/V as the inverse of the molar volume, $1/\bar{V}$, although n/V is just the molar concentration c, and one could equally well write the equation as

$$p/RT = c + Bc^2 + Cc^3 + Dc^4 + \cdots. \tag{A2.1.54}$$

The coefficients B, C, D, etc for each particular gas are termed its second, third, fourth, etc. *virial coefficients*, and are functions of the temperature only. It can be shown, by statistical mechanics, that B is a function of the interaction of an isolated pair of molecules, C is a function of the simultaneous interaction of three molecules, D, of four molecules, etc., a feature suggested by the form of equation (A2.1.54).

While volume is a convenient variable for the calculations of theoreticians, the pressure is normally the variable of choice for experimentalists, so there is a corresponding equation in which the equation of state is expanded in powers of p:

$$pV/n = RT + B'p + C'p^2 + D'p^3 + \cdots. \tag{A2.1.55}$$

The pressure coefficients can be related to the volume coefficients by reverting the series and one finds that

$$B' = B \qquad C' = (C - B^2)/RT \qquad D' = (D - 3BC + 2B^3)/(RT)^2 \qquad \text{etc.}$$

According to equation (A2.1.39) $(\partial\mu/\partial p)_T = V/n$, so equation (A2.1.55) can be integrated to obtain the chemical potential:

$$\mu(T, p) - \mu^0(T, p^0) = RT \ln(p/p^0) + B'p + C'p^2/2 + D'p^3/3 + \cdots. \tag{A2.1.56}$$

Note that a constant of integration μ^0 has come into the equation; this is the chemical potential of the hypothetical *ideal gas* at a reference pressure p^0, usually taken to be one atmosphere. In principle this involves a process of taking the real gas down to zero pressure and bringing it back to the reference pressure as an ideal gas. Thus, since $d\mu = (V/n)\,dp$, one may write

$$\mu(T, p^0) - \mu^0(T, p^0) = \int_0^{p^0} [(V/n) - (RT/p)]\,dp = B'p^0 + C'(p^0)^2 + \cdots .$$

If $p^0 = 1$ atm, it is sufficient to retain only the first term on the right. However, one does not need to know the virial coefficients; one may simply use volumetric data to evaluate the integral.

The molar entropy and the molar enthalpy, also with constants of integration, can be obtained, either by differentiating equation (A2.1.56) or by integrating equations (A2.1.42) or (A2.1.50):

$$\bar{S}(T, p) - \bar{S}^0(T, p^0) = -R\ln(p/p^0) - (dB'/dT)p - (dC'/dT)p^2/2 - (dD'/dT)p^3/3 - \cdots$$
$$\bar{H}(T, p) - \bar{H}^0(T, p^0) = [B' - T(dB'/dT)]p + [C' - T(dC'/dT)]p^2/2 + [D' - T(dD'/dT)]p^3/3 - \cdots$$

$$(A2.1.57)$$

where, as in the case of the chemical potential, the reference molar entropy $\bar{S}^0$ and reference molar enthalpy $\bar{H}^0$ are for the hypothetical ideal gas at a pressure p^0.

It is sometimes convenient to retain the generality of the limiting ideal-gas equations by introducing the *activity a*, an 'effective' pressure (or, as we shall see later in the case of solutions, an effective mole fraction, concentration, or molality). For gases, after Lewis (1901), this is usually called the *fugacity* and symbolized by f rather than by a. One can then write

$$\mu(T, p) - \mu^0(T, p^0) = RT\ln(f/p^0).$$

One can also define an *activity coefficient* or *fugacity coefficient* $\gamma = f/p$; obviously

$$RT\ln\gamma = B'p + C'p^2/2 + D'p^3/3 + \cdots .$$

Temperature dependence of the second virial coefficient

Figure A2.1.7 shows schematically the variation of $B = B'$ with temperature. It starts strongly negative (theoretically at minus infinity for zero temperature, but of course unmeasurable) and decreases in magnitude until it changes sign at the *Boyle temperature* ($B = 0$, where the gas is more nearly ideal to higher pressures). The slope dB/dT remains positive, but decreases in magnitude until very high temperatures. Theory requires the virial coefficient finally to reach a maximum and then slowly decrease, but this has been experimentally observed only for helium.

It is widely believed that gases are virtually ideal at a pressure of one atmosphere. This is more nearly true at relatively high temperatures, but at the normal boiling point (roughly 20% of the Boyle temperature), typical gases have values of pV/nRT that are 5 to 15% lower than the ideal value of unity.

The Joule–Thomson effect

One of the classic experiments on gases was the measurement by Joule and Thomson (1853) of the change in temperature when a gas flows through a porous plug (throttling valve) from a pressure p_1 to a pressure p_2 (figure (A2.1.8)). The total system is jacketed in such a way that the process is adiabatic ($q = 0$), and the pressures are constant (other than an infinitesimal δp) in the two parts. The work done on the gas in the right-hand region to bring it through is $p_2\,dV_2$, while that in the left-hand region is $-p_1\,dV_1$ (because dV_1 is

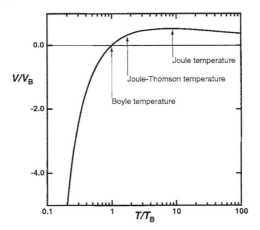

Figure A2.1.7. The second virial coefficient B as a function of temperature T/T_B. (Calculated for a gas satisfying the Lennard-Jones potential [8].)

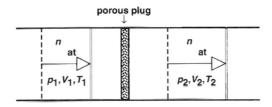

Figure A2.1.8. The Joule–Thomson experiment. The arrows indicate the motion of the two pistons from their initial positions (dashed lines) to their final positions (double lines).

negative). The two volumes are of course unequal, but no assumption about the ideality of the gas is necessary (or even desirable). The total energy change can then be written as the loss of energy from the left-hand region plus the gain in energy by the right-hand region

$$dU = -dU_1 + dU_2 = p_1 \, dV_1 - p_2 \, dV_2$$

or

$$dU_2 + p_2 \, dV_2 = dH_2 = dH_1 = dU_1 + p_1 \, dV_1.$$

This establishes the process as isenthalpic ($\Delta H = 0$), so that one can write the Joule–Thomson coefficient μ_{JT} (not to be confused with the chemical potential) as

$$\mu_{JT} = (\partial T/\partial p)_H = -(\partial H/\partial p)_T/(\partial H/\partial T)_p \qquad (A2.1.58)$$

where the right-hand side of the equation comes from the cyclic rule of calculus.

The denominator of equation (A2.1.58) is just the heat capacity C_p, while the numerator can be obtained by differentiating the enthalpy expression, equation (A2.1.57), with respect to pressure:

$$\mu_{JT} = -\{[B' - T(dB'/dT)] + [C' - T(dC'/dT)]p + \cdots\}/C_p.$$

It is important to observe that the Joule–Thomson coefficient of a real gas is non-zero even in the limit of zero pressure. (Note that C_p is also a power series in the pressure, which can be obtained by differentiating the enthalpy expression, equation (A2.1.57), with respect to temperature.)

The Joule–Thomson inversion temperature T_{JT}, at which $\mu_{JT} = 0$, is shown in figure A2.1.7 and is approximately 1.9 times the Boyle temperature T_B. Below this temperature (which is above room temperature for all gases except hydrogen and helium), μ_{JT} is positive.

The Joule effect

The change in temperature on the adiabatic free expansion of a gas ($\Delta U = 0$) was measured by Joule in 1845 and, with the precision available then, found to be zero. One may write for the infinitesimal change

$$(\partial T/\partial V)_U = -(\partial U/\partial V)_T/(\partial U/\partial T)_V = -RT^2[(dB/dT)(n/V)^2 + \cdots]/C_V.$$

Unlike $(\partial H/\partial p)_T$, $(\partial U/\partial V)_T$ does indeed vanish for real gases as the pressure goes to zero, but this is because the derivative is with respect to V, not because of the difference between U and H. At appreciable pressures $(\partial T/\partial V)_U$ is almost invariably negative, because the Joule temperature, at which dB/dT becomes negative, is extremely high (see figure A2.1.7).

A2.1.6.3 Gaseous mixtures

According to Dalton's *law of partial pressures*, observed experimentally at sufficiently low pressures, the pressure of a gas mixture in a given volume V is the sum of the pressures that each gas would exert alone in the same volume at the same temperature. Expressed in terms of moles n_i

$$p(n_1, n_2, \ldots, n_j, V, T) = \sum_{i=1}^{j} p_i(n_i, V, T),$$

or, given the validity of the ideal-gas law (equation (A2.1.18)) at these pressures,

$$p(n_1, n_2, \ldots, n_j, V, T) = \sum_i (n_i RT/V) = \left(\sum_i n_i\right)(RT/V).$$

The *partial pressure* p_i of a component in an ideal-gas mixture is thus

$$p_i = \left(n_i \Big/ \sum_i n_i\right) p = x_i p \tag{A2.1.59}$$

where $x_i = n_i/n$ is the mole fraction of species i in the mixture. (The partial pressure is always defined by equation (A2.1.59) even at higher pressures where Dalton's law is no longer valid.)

Given this experimental result, it is plausible to assume (and is easily shown by statistical mechanics) that the chemical potential of a substance with partial pressure p_i in an ideal-gas mixture is equal to that in the one-component ideal gas at pressure $p' = p_i$

$$\mu_i(p, T, x_i) = \mu_i'(p' = x_i p, T, x_i' = 1). \tag{A2.1.60}$$

What thermodynamic experiments can be cited to support such an assumption? There are several:

(1) There are a few semipermeable membranes that separate a gas mixture from a pure component gas. One is palladium, which is permeable to hydrogen, but not (in any reasonable length of time) to other gases. Another is rubber, through which carbon dioxide or ammonia diffuses rapidly, while gases like nitrogen or argon diffuse much more slowly. In such cases, at equilibrium (when the chemical potential of the diffusing gas must be the same on both sides of the membrane) the pressure of the one-component gas

on one side of the membrane is found to be equal to its partial pressure in the gas mixture on the other side.

(2) In the phase equilibrium between a pure solid (or a liquid) and its vapour, the addition of other gases, as long as they are insoluble in the solid or liquid, has negligible effect on the partial pressure of the vapour.

(3) In electrochemical cells (to be discussed later), if a particular gas participates in a chemical reaction at an electrode, the observed electromotive force is a function of the partial pressure of the reactive gas and not of the partial pressures of any other gases present.

For precise measurements, there is a slight correction for the effect of the slightly different pressure on the chemical potentials of the solid or of the components of the solution. More important, corrections must be made for the non-ideality of the pure gas and of the gaseous mixture. With these corrections, equation (A2.1.60) can be verified within experimental error.

Given equation (A2.1.60) one can now write for an ideal-gas mixture

$$
\begin{aligned}
\mu_i(p, T, x_i) = \mu_i' &= \mu_i^0(p^0, T) + RT \ln(p_i/p^0) \\
&= \mu_i^0(p^0, T) + RT \ln(x_i p/p^0) \\
&= \mu_i^0(p^0, T) + RT \ln(p/p^0) + RT \ln x_i.
\end{aligned}
\tag{A2.1.61}
$$

Note that this has resulted in the separation of pressure and composition contributions to chemical potentials in the ideal-gas mixture. Moreover, the thermodynamic functions for ideal-gas mixing at constant pressure can now be obtained:

$$
\left.
\begin{aligned}
\Delta G_m(T, p) &= \sum_i n_i[\mu_i(T, p, x_i) - \mu_i(T, p, 1)] = RT \sum_i n_i \ln x_i \\
\Delta S_m(T, p) &= -(\partial \Delta G_m/\partial T)_{p,n_i} = -R \sum_i n_i \ln x_i \\
\Delta H_m(T, p) &= \Delta G_m(T, p) + T \Delta S_m(T, p) = 0.
\end{aligned}
\right\} \text{ ideal gas only}
$$

Gas mixtures are subject to the same degree of non-ideality as the one-component ('pure') gases that were discussed in the previous section. In particular, the second virial coefficient for a gas mixture can be written as a quadratic average

$$
B(T, x_1, \ldots, x_k) = \sum_{i=1}^{k} \sum_{j=1}^{k} x_i x_j B_{ij}.
\tag{A2.1.62}
$$

where B_{ij}, a function of temperature only, depends on the interaction of an i, j pair. Thus, for a binary mixture of gases, one has B_{11} and B_{22} from measurements on the pure gases, but one needs to determine B_{12} as well. The corresponding third virial coefficient is a cubic average over the C_{ijk}s, but this is rarely needed. Appropriate differentiation of equation (A2.1.62) will lead to the non-ideal corrections to the equations for the chemical potentials and the mixing functions.

A2.1.6.4   Dilute solutions and Henry's law

Experiments on sufficiently dilute solutions of non-electrolytes yield *Henry's law*, that the vapour pressure of a volatile solute, i.e. its partial pressure in a gas mixture in equilibrium with the solution, is directly proportional to its concentration, expressed in *any* units (molar concentrations, molality, mole fraction, weight fraction, etc.) because in sufficiently dilute solution these are all proportional to each other.

$$
p_i = k_c c_i = k_m m_i = k_x x_i
$$

where c_i is the molar concentration of species i (conventionally, but not necessarily, expressed in units of moles per litre of *solution*), m_i is its *molality* (conventionally expressed as moles per kilogram of *solvent*), and x_i is its mole fraction. The Henry's law constants k_c, k_m and k_x differ, of course, with the choice of units.

It follows that, because phase equilibrium requires that the chemical potential μ_i be the same in the solution as in the gas phase, one may write for the chemical potential in the solution:

$$\mu_i(T, c_i) - \mu_i(T, c^0) = RT \ln(c_i/c^0). \tag{A2.1.63}$$

Here the composition is expressed as concentration c_i and the reference state is for unit concentration c^0 (conventionally 1 mol l^{-1}) but it could have been expressed using any other composition variable and the corresponding reference state.

It seems appropriate to assume the applicability of equation (A2.1.63) to sufficiently dilute solutions of non-volatile solutes and, indeed, to electrolyte species. This assumption can be validated by other experimental methods (e.g. by electrochemical measurements) and by statistical mechanical theory.

Just as increasing the pressure of a gas or a gas mixture introduces non-ideal corrections, so does increasing the concentration. As before, one can introduce an activity a_i and an activity coefficient γ_i and write $a_i = c_i \gamma_i$ and

$$\mu_i(T, c_i) - \mu_i(T, c^0) = RT \ln(a_i/c^0) = RT \ln(c_i/c^0) + RT \ln \gamma_i.$$

In analogy to the gas, the reference state is for the ideally dilute solution at c^0, although at c^0 the real solution may be far from ideal. (Technically, since this has now been extended to non-volatile solutes, it is defined at the reference pressure p^0 rather than at the vapour pressure; however, because $(\partial \mu_i / \partial p)_T = \bar{V}_i$, and molar volumes are small in condensed systems, this is rarely of any consequence.)

Using the Gibbs–Duhem equation ((A2.1.27) with $dT = 0$, $dp = 0$), one can show that the solvent must obey *Raoult's law* over the same concentration range where Henry's law is valid for the solute (or solutes):

$$p_0 = p_0^0 x_0.$$

where x_0 is the mole fraction of solvent, p_0 is its vapour pressure, and p_0^0 is the vapour pressure of pure solvent, i.e. at $x_0 = 1$. A more careful expression of Raoult's law might be

$$\lim_{x_0 \to 1} (\partial p_0 / \partial x_0) = p_0^0.$$

It should be noted that, whatever the form of Henry's law (i.e. in whatever composition units), Raoult's law must necessarily be expressed in mole fraction. This says nothing about the appropriateness of mole fractions in condensed systems, e.g. in equilibrium expressions; it arises simply from the fact that it is a statement about the *gas phase*.

The reference state for the solvent is normally the pure solvent, so one may write

$$\mu_0(T, p^0, x_0) - \mu_0^0(T, p^0, 1) = RT \ln a_0 = RT \ln x_0 + RT \ln \gamma_0.$$

Finally, a brief summary of the known behaviour of activity coefficients:
Binary non-electrolyte mixtures:

$$\begin{aligned} \text{solvent:} \quad & \ln \gamma_0 = k c_1^2 + O(c_1^3) \\ \text{solute:} \quad & \ln \gamma_1 = k' c_1 + O(c_1^2). \end{aligned}$$

(Theory shows that these equations must be simple power series in the concentration (or an alternative composition variable) and experimental data can always be fitted this way.)
Single electrolyte solution:

$$\begin{aligned} \text{solvent:} \quad & \ln \gamma_0 = k'' m_1^{3/2} + O(m_1^2) \\ \text{solute:} \quad & \ln \gamma_1 = k''' m_1^{1/2} + O(m_1). \end{aligned}$$

(The situation for electrolyte solutions is more complex; theory confirms the limiting expressions (originally from Debye–Hückel theory), but, because of the long-range interactions, the resulting equations are non-analytic rather than simple power series.) It is evident that electrolyte solutions are 'ideally dilute' only at extremely low concentrations. Further details about these activity coefficients will be found in other articles.

A2.1.6.5   Chemical equilibrium

If a thermodynamic system includes species that may undergo chemical reactions, one must allow for the fact that, even in a closed system, the number of moles of a particular species can change. If a chemical reaction (e.g. $N_2 + 3H_2 \rightarrow 2NH_3$) is represented by the symbolic equation

$$\nu_A A + \nu_B B + \cdots \rightarrow \nu_Y Y + \nu_Z Z + \cdots \qquad (A2.1.64)$$

it is obvious that any changes in the numbers of moles of the species must be proportional to the coefficients ν. Thus if n_A^0, n_B^0, etc, are the numbers of moles of the species at some initial point, we may write for the number of moles at some subsequent point

$$n_A = n_A^0 - \nu_A \xi \qquad dn_A = -\nu_A \, d\xi$$
$$n_B = n_B^0 - \nu_B \xi \qquad dn_B = -\nu_B \, d\xi$$
$$\cdots \qquad\qquad \cdots$$
$$n_Y = n_Y^0 + \nu_Y \xi \qquad dn_Y = \nu_Y \, d\xi$$
$$n_Z = n_Z^0 + \nu_Z \xi \qquad dn_Z = \nu_Z \, d\xi$$
$$\cdots \qquad\qquad \cdots$$

where the parameter ξ is called the 'degree of advancement' of the reaction. (If the variable ξ goes from 0 to 1, one unit of the reaction represented by equation (A2.1.64) takes place, but $\xi = 0$ does not necessarily mean that only reactants are present, nor does $\xi = 1$ mean that only products remain.) More generally one can write

$$n_i = n_i^0 + \nu_i \xi \qquad dn_i = \nu_i \, d\xi \qquad (A2.1.65)$$

where positive values of ν_i designate products and negative values of ν_i designate reactants. Equations (A2.1.33) to (A2.1.36) can be rewritten in new forms appropriate for these closed, but chemically reacting systems. Substitution from equation (A2.1.65) yields

$$dU = T \, dS - p \, dV + \sum_i \mu_i \, dn_i = T \, dS - p \, dV + \left(\sum_i \nu_i \mu_i \right) d\xi$$

$$dH = T \, dS + V \, dp + \sum_i \mu_i \, dn_i = T \, dS + V \, dp + \left(\sum_i \nu_i \mu_i \right) d\xi$$

$$dA = -S \, dT - p \, dV + \sum_i \mu_i \, dn_i = -S \, dT - p \, dV + \left(\sum_i \nu_i \mu_i \right) d\xi$$

$$dG = -S \, dT + V \, dp + \sum_i \mu_i \, dn_i = -S \, dT + V \, dp + \left(\sum_i \nu_i \mu_i \right) d\xi.$$

We have seen that equilibrium in an isolated system ($dU = 0$, $dV = 0$) requires that the entropy S be a maximum, i.e. that $(\partial S / \partial \xi)_{U,V} = 0$. Examination of the first equation above shows that this can only be true if $\sum_i \nu_i \mu_i$ vanishes. Exactly the same conclusion applies for equilibrium under the other constraints. Thus, for constant temperature and pressure, minimization of the Gibbs free energy requires that $(\partial G / \partial \xi)_{T,p} = \sum_i \nu_i \mu_i = 0$.

The affinity

This new quantity $\sum_i \nu_i \mu_i$, the negative of which De Donder (1920) has called the 'affinity' and given the symbol of a script $\mathcal{A}$, is obviously the important thermodynamic function for chemical equilibrium:

$$-\mathcal{A} = \sum_i \nu_i \mu_i = -T(\partial S/\partial \xi)_{U,V} = (\partial U/\partial \xi)_{S,V} = (\partial H/\partial \xi)_{S,p} = (\partial A/\partial \xi)_{T,V} = (\partial G/\partial \xi)_{T,p}.$$

Figure A2.1.9 illustrates how the entropy S and the affinity $\mathcal{A}$ vary with ξ in a constant U, V system. It is apparent that when the slope $(\partial S/\partial \xi)_{U,V}$ is positive (positive affinity), ξ will spontaneously increase; when it is negative, ξ will spontaneously decrease; when it is zero, ξ has no tendency to change and the system is at equilibrium. Moreover, one should note the feature that $\mathcal{A} = 0$ is the criterion for equilibrium for all these sets of constraints, whether U and V are fixed, or T and p are fixed.

Instead of using the chemical potential μ_i one can use the *absolute activity* $\lambda_i = \exp(\mu_i/RT)$. Since at equilibrium $\mathcal{A} = 0$,

$$-\mathcal{A} = \sum_i \nu_i \mu_i = RT \sum_i \nu_i \ln \lambda_i = 0 \qquad \text{or} \qquad \prod_i \lambda_i^{\nu_i} = 1.$$

It is convenient to define a *relative activity* a_i in terms of the standard states of the reactants and products at the same temperature and pressure, where $\lambda_i = \lambda_i^0$, $\mu_i = \mu_i^0$

$$a_i = \lambda_i/\lambda_i^0 = \lambda_i/\exp(\mu_i^0/RT).$$

Thus, at equilibrium

$$-\mathcal{A} = \sum_i \nu_i \mu_i = \sum_i \nu_i \ln(\lambda_i^0 a_i) = \sum_i \nu_i \mu_i^0 + RT \sum_i \nu_i \ln a_i = 0.$$

If we define an equilibrium constant K as

$$K = \prod_i a_i^{\nu_i} = \prod_i (\lambda_i^0)^{-\nu_i} \tag{A2.1.66}$$

it can now be related directly to $\mathcal{A}^0$ or ΔG^0:

$$RT \ln K = RT \sum_i \nu_i \ln a_i = -RT \sum_i \nu_i \ln \lambda_i^0 = -\sum_i \nu_i \mu_i^0$$

$$= -\Delta G^0 = \mathcal{A}^0. \tag{A2.1.67}$$

To proceed further, to evaluate the standard free energy ΔG^0, we need information (experimental or theoretical) about the particular reaction. One source of information is the equilibrium constant for a chemical reaction involving gases. Previous sections have shown how the chemical potential for a species in a gaseous mixture or in a dilute solution (and the corresponding activities) can be defined and measured. Thus, if one can determine (by some kind of analysis) the partial pressures of the reacting gases in an equilibrium mixture or the concentrations of reacting species in a dilute solution equilibrium (and, where possible, adjust them to activities so one allows for non-ideality), one can obtain the thermodynamic equilibrium constant K and the standard free energy of reaction ΔG^0 from equations (A2.1.66) and (A2.1.67).

A cautionary word about units: equilibrium constants are usually expressed in units, because pressures and concentrations have units. Yet the argument of a logarithm must be dimensionless, so the activities in equation (A2.1.66), defined in terms of the absolute activities (which are dimensionless) are dimensionless.

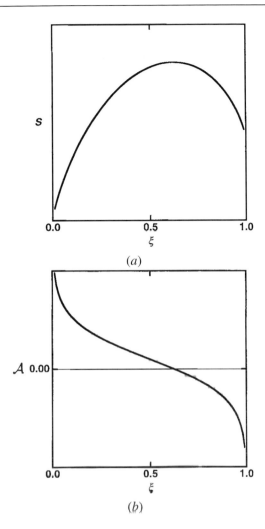

Figure A2.1.9. Chemically reacting systems. (a) The entropy S as a function of the degree of advancement ξ of the reaction at constant U and V. (b) The affinity $\mathcal{A}$ as a function of ξ for the same reacting system. Equilibrium is reached at $\xi = 0.623$ where S is a maximum and $\mathcal{A} = 0$.

The value of the standard free energy ΔG^0 depends on the choice of reference state, as does the equilibrium constant. Thus it would be safer to write the equilibrium constant K for a gaseous reaction as

$$K = \prod_i (p_i/p^0)^{v_i} = (p^0)^{\Sigma v_i} \prod_i p^{v_i} = (p^0)^{\Sigma v_i} K_p.$$

Here K is dimensionless, but K_p is not. Conversely, the factor $(p^0)^{\Sigma v_i}$ has units (unless $\Sigma v_i = 0$) but the value unity if $p^0 = 1$ atm. Similar considerations apply to equilibrium constants expressed in concentrations or molalities.

A2.1.6.6   Reversible galvanic cells

A second source of standard free energies comes from the measurement of the electromotive force of a galvanic cell. Electrochemistry is the subject of other articles (A2.4 and B1.28), so only the basics of a reversible chemical cell will be presented here. For example, consider the cell conventionally written as

$$\text{Pt, } H_2(1 \text{ atm})|HCl(m)|AgCl(s), Ag(s)$$

for which the electrode reactions are oxidation at the left electrode, the anode,

$$\tfrac{1}{2}H_2(1 \text{ atm}) \leftrightarrow H^+(m) + e^-(\psi_{\text{left}})$$

and reduction at the right electrode, the cathode,

$$AgCl(s) + e^-(\psi_{\text{right}}) \leftrightarrow Cl^-(m) + Ag(s)$$

which can be rewritten as two concurrent reactions:

$$\tfrac{1}{2}H_2(1 \text{ atm}) + AgCl(s) \leftrightarrow H^+(m) + Cl^-(m) + Ag(s) \tag{I}$$

$$e^-(\psi_{\text{right}}) \leftrightarrow e^-(\psi_{\text{left}}). \tag{II}$$

The chemical reaction (I) cannot come to equilibrium directly; it can come to equilibrium only if the two electrodes are connected so that electrons can flow. One can use this feature to determine the affinity (or the ΔG) of reaction (I) by determining the affinity of reaction (II) which balances it.

In these equations the electrostatic potential ψ might be thought to be the potential at the actual electrodes, the platinum on the left and the silver on the right. However, electrons are not the hypothetical test particles of physics, and the electrostatic potential difference at a junction between two metals is unmeasurable. What *is* measurable is the difference in the *electrochemical* potential μ of the electron, which at equilibrium must be the same in any two wires that are in electrical contact. One assumes that the electrochemical potential can be written as the combination of two terms, a 'chemical' potential minus the electrical potential ($-\psi$ because of the negative charge on the electron). When two copper wires are connected to the two electrodes, the 'chemical' part of the electrochemical potential is assumed to be the same in both wires; then the potentiometer measures, under conditions of zero current flow, the electrostatic potential difference $\Delta\psi = \psi_{\text{right}} - \psi_{\text{left}}$ between the two copper wires, which is called the *electromotive force* (emf) $\mathcal{E}$ of the cell.

For reaction (I) the two solids and the hydrogen gas are in their customary standard states ($a = 1$), so

$$\Delta G_{\text{I}} = \Delta G_{\text{I}}^0 + RT \ln \left(\frac{a_{\text{Ag}} a_{\text{H}^+} a_{\text{Cl}^-}}{a_{\text{H}_2}^{1/2} a_{\text{AgCl}}} \right) = \Delta G_{\text{I}}^0 + RT \ln(a_{\text{H}^+} a_{\text{Cl}^-})$$

while for the electrons

$$\Delta G_{\text{II}} = \mathcal{F}(\psi_{\text{right}} - \psi_{\text{left}}) = \mathcal{F}\mathcal{E}$$

where $\mathcal{F}$ is the Faraday constant (the amount of charge in one mole of electrons).

When no current flows, there is a constrained equilibrium in which the chemical reaction cannot proceed in either direction, and $\mathcal{E}$ can be measured. With this constraint, for the *overall reaction* $\Delta G = \Delta G_{\text{I}} + \Delta G_{\text{II}} = 0$, so

$$\Delta G_{\text{I}}^0 + RT \ln(a_{\text{H}^+} a_{\text{Cl}^-}) = -\mathcal{F}\mathcal{E}.$$

Were the HCl in its standard state, ΔG_I^0 would equal $-\mathcal{F}\mathcal{E}^0$, where $\mathcal{E}^0$ is the standard emf for the reaction. In general, for any reversible chemical cell without transference, i.e. one with a single electrolyte solution, not one with any kind of junction between two solutions,

$$\Delta G_I^0 + RT \ln \left(\sum_i a_i^{\nu_i} \right) = -n\mathcal{F}\mathcal{E} \tag{A2.1.68}$$

$$\Delta G_I^0 = -n\mathcal{F}\mathcal{E}^0 \tag{A2.1.69}$$

where n is the number of electrons associated with the cell reaction as written. By combining equations (A2.1.68) and (A2.1.69) one obtains the *Nernst equation*

$$\mathcal{E} = \mathcal{E}^0 - (RT/n\mathcal{F}) \ln \left(\sum_i a_i^{\nu_i} \right).$$

Thus, if the activities of the various species can be determined or if one can extrapolate to infinite dilution, the measurement of the emf yields the standard free energy of the reaction.

A2.1.6.7   Standard states and standard free energies of formation

With several experimental methods for determining the ΔG^0s of chemical reactions, one can start to organize the information in a systematic way. (To complete this satisfactorily, or at least efficiently and precisely, one needs the third law to add third-law entropies to calorimetrically determined ΔH^0s. Discussion of this is deferred to the next section, but it will be assumed for the purpose of this section that all necessary information is available.)

Standard states

Conventions about standard states (the reference states introduced earlier) are necessary because otherwise the meaning of the standard free energy of a reaction would be ambiguous. We summarize the principal ones:

(1) All standard states, both for pure substances and for components in mixtures and solutions, are defined for a pressure of exactly 1 atmosphere. However the temperature must be specified. (There is some movement towards metricating this to a pressure of 1 bar $= 100$ kPa $= 0.986\,924$ atm. This would make a significant difference only for gases; at $T = 298$ K, this would decrease a μ^0 by 32.6 J mol^{-1}.)

(2) As noted earlier, the standard state of a gas is the hypothetical ideal gas at 1 atmosphere and the specified temperature T.

(3) The standard state of a substance in a condensed phase is the real liquid or solid at 1 atm and T.

(4) The standard state of an electrolyte is the hypothetical ideally dilute solution (Henry's law) at a molarity of 1 mol kg^{-1}. (Actually, as will be seen, electrolyte data are conventionally reported as for the formation of individual ions.) Standard states for non-electrolytes in dilute solution are rarely invoked.

(5) For a free energy of formation, the preferred standard state of the element should be the thermodynamically stable (lowest chemical potential) form of it; e.g. at room temperature, graphite for carbon, the orthorhombic crystal for sulfur.

Compounds that are products in reactions are sometimes reported in standard states for phases that are not the most stable at the temperature in question. The stable standard state of H_2O at 298 K (and 1 atm) is, of course, the liquid, but ΔG^0s are sometimes reported for reactions leading to gaseous H_2O at 298 K. Moreover the standard functions for the formation of some metastable states, e.g. C(diamond) or S(monoclinic) at 298 K, are sometimes reported in tables.

The useful thermodynamic functions (e.g. G, H, S, C_p, etc) are all state functions, so their values in any particular state are independent of the path by which the state is reached. Consequently, one can combine (by addition or subtraction) the ΔG^0s for several chemical reactions at the same temperature, to obtain the ΔG^0 of another reaction that is not directly measurable. (Indeed one experimentalist has commented that the principal usefulness of thermodynamics arises 'because some quantities are easier to measure than others'.)

In particular, one can combine reactions to yield the ΔG_f^0, ΔH_f^0 and ΔS_f^0 for formation of compounds from their elements, quantities rarely measurable directly. (Many ΔH_f^0s for formation of substances are easily calculated from the calorimetric measurement of the enthalpies of combustion of the compound and of its constituent elements.) For example, consider the dimerization of NO_2 at 298 K. In appropriate tables one finds

$$\tfrac{1}{2}N_2(g) + O_2(g) \rightarrow NO_2(g) \qquad \Delta G_f^0 = 51.840 \text{ kJ mol}^{-1}$$
$$N_2(g) + 2O_2(g) \rightarrow N_2O_4(g) \qquad \Delta G_f^0 = 98.286 \text{ kJ mol}^{-1}$$
$$2NO_2(g) \rightarrow N_2O_4(g) \qquad \Delta G_f^0 = (98.286 - 2 \times 51.840) \text{ kJ mol}^{-1}$$
$$= -5.394 \text{ kJ mol}^{-1}.$$

With this information one can now use equation (A2.1.67) to calculate the equilibrium constant at 298 K. One finds $K = 75.7$ or, using the dimensional constant, $K_p = 75.7 \text{ atm}^{-1}$. (In fact, the free energies of formation were surely calculated using the experimental data on the partial pressures of the gases in the equilibrium. One might also note that this is one of the very few equilibria involving only gaseous species at room temperature that have constants K anywhere near unity.)

Thermodynamic tables usually report at least three quantities: almost invariably the standard enthalpy of formation at 298 K, $\Delta H_f^0(298 \text{ K})$; usually the standard entropy at 298 K, $S^0(298 \text{ K})$ (not $\Delta S_f^0(298 \text{ K})$, but the entropy based on the third-law convention (see subsequent section) that $S^0(0 \text{ K}) = 0$); and some form of the standard free energy of formation, usually either $\Delta G_f^0(298 \text{ K})$ or $\log_{10} K_f$. Many tables will include these quantities at a series of temperatures, as well as the standard heat capacity C_p^0, and enthalpies and entropies of various transitions (phase changes).

The standard free energy of formation of ions

A special convention exists concerning the free energies of ions in aqueous solution. Most thermodynamic information about strong (fully dissociated) electrolytes in aqueous solutions comes, as has been seen, from measurements of the emf of reversible cells. Since the ions in very dilute solution (or in the hypothetical ideally dilute solution at $m = 1 \text{ mol kg}^{-1}$) are essentially independent, one would like to assign free energy values to individual ions and add together the values for the anion and the cation to get that for the electrolyte. Unfortunately the emf of a half cell is unmeasurable, although there have been some attempts to estimate it theoretically. Consequently, the convention that the standard half-cell emf of the hydrogen electrode is exactly zero has been adopted.

$$\tfrac{1}{2}H_2(\text{ideal gas, } T) \rightarrow H^+(\text{aq, ideal, } m = 1 \text{ kg mol}^{-1}, T) + e^- \qquad \mathcal{E}^0 = 0, \qquad \Delta G_f^0 = 0.$$

Thus, when tables report the standard emf or standard free energy of the chloride ion,

$$\tfrac{1}{2}Cl_2(\text{ideal gas, } T) + e^- \rightarrow Cl^-(\text{aq, ideal, } m = 1 \text{ kg mol}^{-1}, T)$$

it is really that of the reaction

$$\tfrac{1}{2}H_2(\text{ideal gas, } T) + \tfrac{1}{2}Cl_2(\text{ideal gas, } T)$$
$$\rightarrow H^+(\text{aq, ideal, } m = 1 \text{ kg mol}^{-1}, T) + Cl^-(\text{aq, ideal, } m = 1 \text{ kg mol}^{-1}, T).$$

Similarly, the standard free energy or standard emf of the sodium ion, reported for

$$Na(s, T) \rightarrow Na^+(aq, \text{ideal}, m = 1 \text{ kg mol}^{-1}, T) + e^-$$

is really that for

$$H^+(aq, \text{ideal}, m = 1 \text{ kg mol}^{-1}, T) + Na(s, T)$$
$$\rightarrow \tfrac{1}{2}H_2(\text{ideal gas}, T) + Na^+(aq, \text{ideal}, m = 1 \text{ kg mol}^{-1}, T).$$

Temperature dependence of the equilibrium constant

Since equation (A2.1.67) relates the equilibrium constant K to the standard free energy ΔG^0 of the reaction, one can rewrite the equation as

$$\ln K = -\Delta G^0/RT$$

and differentiate with respect to temperature

$$d \ln K/dT = -(d\Delta G^0/dT)/RT + \Delta G^0/RT^2 = (T\Delta S^0 + \Delta G^0)/RT^2 = \Delta H^0/RT^2. \qquad (A2.1.70)$$

This important relation between the temperature derivative of the equilibrium constant K and the standard enthalpy of the reaction ΔH^0 is sometimes known as the *van't Hoff equation*. (Note that the derivatives are not expressed as partial derivatives at constant pressure, because the quantities involved are all defined for the standard pressure p^0. Note also that in this derivation one has not assumed—as is sometimes alleged—that ΔH^0 and ΔS^0 are independent of temperature.)

The validity of equation (A2.1.70) has sometimes been questioned when enthalpies of reaction determined from calorimetric experiments fail to agree with those determined from the temperature dependence of the equilibrium constant. The thermodynamic equation is rigorously correct, so doubters should instead examine the experimental uncertainties and whether the two methods actually relate to exactly the same reaction.

A2.1.7   The third law

A2.1.7.1   *History; the Nernst heat theorem*

The enthalpy, entropy and free energy changes for an isothermal reaction near 0 K cannot be measured directly because of the impossibility of carrying out the reaction reversibly in a reasonable time. One can, however, by a suitable combination of measured values, calculate them indirectly. In particular, if the value of ΔH^0, ΔS^0 or ΔG^0 is known at a specified temperature T, say 298 K, its value at another temperature T' can be computed using this value and the changes involved in bringing the products and the reactants separately from T' to T. If these measurements can be extrapolated to 0 K, the isothermal changes for the reaction at 0 K can be calculated.

If, in going from 0 K to T, a substance undergoes phase changes (fusion, vaporization, etc) at T_A and T_B with molar enthalpies of transition $\Delta \bar{H}_A$ and $\Delta \bar{H}_B$, one can write

$$\bar{H}^0(T) - \bar{H}^0(0) = \int_0^{T_A} \bar{C}_p^0 \, dT + \Delta \bar{H}_A + \int_{T_A}^{T_B} \bar{C}_p^0 \, dT + \Delta \bar{H}_B + \int_{T_B}^{T} \bar{C}_p^0 \, dT$$

$$\bar{S}^0(T) - \bar{S}^0(0) = \int_0^{T_A} (\bar{C}_p^0/T) \, dT + \Delta \bar{H}_A/T_A + \int_{T_A}^{T_B} (\bar{C}_p^0/T) \, dT + \Delta \bar{H}_B/T_B + \int_{T_B}^{T} (\bar{C}_p^0/T) \, dT.$$

$$(A2.1.71)$$

It is manifestly impossible to measure heat capacities down to exactly 0 K, so some kind of extrapolation is necessary. Unless C_p were to approach zero as T approaches zero, the limiting value of C_p/T would not be finite and the first integral in equation (A2.1.71) would be infinite. Experiments suggested that C_p might approach zero and Nernst (1906) noted that computed values of the entropy change ΔS^0 for various reactions appeared to approach zero as the temperature approached 0 K. This empirical discovery, known as the *Nernst heat theorem*, can be expressed mathematically in various forms as

$$\lim_{T \to 0} (d\Delta G^0/dT) = \lim_{T \to 0} \Delta S^0 = \lim_{T \to 0} (d\Delta H^0/dT) = 0. \tag{A2.1.72}$$

However, the possibility that C_p might not go to zero could not be excluded before the development of the quantum theory of the heat capacity of solids. When Debye (1912) showed that, at sufficiently low temperatures, C_p is proportional to T^3, this uncertainty was removed, and a reliable method of extrapolation for most crystalline substances could be developed. (For metals there is an additional term, proportional to T, a contribution from the electrons.) If the temperature T' is low enough that $C_p = \alpha T^3$, one may write

$$\bar{S}^0(T') - \bar{S}^0(0) = \int_0^{T'} (\bar{C}_p^0/T)\, dT = \int_0^{T'} \alpha T^2\, dT = \alpha T'^3/3 = \bar{C}_p^0(T')/3. \tag{A2.1.73}$$

With this addition, better entropy determinations, e.g. measurements plus extrapolations to 0 K, became available.

The evidence in support of equation (A2.1.72) is of several kinds:

(1) Many substances exist in two or more solid allotropic forms. At 0 K, the thermodynamically stable form is of course the one of lowest energy, but in many cases it is possible to make thermodynamic measurements on another (metastable) form down to very low temperatures. Using the measured entropy of transition at equilibrium, the measured heat capacities of both forms and equation (A2.1.73) to extrapolate to 0 K, one can obtain the entropy of transition at 0 K. Within experimental error ΔS^0 is zero for the transitions between β- and γ-phosphine, between orthorhombic and monoclinic sulfur and between different forms of cyclohexanol.

(2) As seen in previous sections, the standard entropy ΔS^0 of a chemical reaction can be determined from the equilibrium constant K and its temperature derivative, or equivalently from the temperature derivative of the standard emf of a reversible electrochemical cell. As in the previous case, calorimetric measurements on the separate reactants and products, plus the usual extrapolation, will yield $\Delta S^0(0)$.

The limiting ΔS^0 so calculated is usually zero within experimental error, but there are some disturbing exceptions. Not only must solutions and some inorganic and organic glasses be excluded, but also crystalline CO, NO, N_2O and H_2O. It may be easy to see, given the most rudimentary statistical ideas of entropy, that solutions and glasses have some positional disorder frozen in, and one is driven to conclude that the same situation must occur with these few simple crystals as well. For these substances in the gaseous state at temperature T there is a disagreement between the 'calorimetric' entropy calculated using equation (A2.1.71) and the 'spectroscopic' entropy calculated by statistical mechanics using the rotational and vibrational constants of the gas molecule; this difference is sometimes called 'residual entropy'. However, it can be argued that, because such a substance or mixture is frozen into a particular disordered state, its entropy is in fact zero. In any case, it is not in internal equilibrium (unless some special hypothetical constraints are applied), and it cannot be reached along a reversible path.

It is beyond the scope of this article to discuss the detailed explanation of these exceptions; suffice it to say that there are reasonable explanations in terms of the structure of each crystal.

A2.1.7.2   First statement ($\Delta S^0 \to 0$)

Because it is necessary to exclude some substances, including some crystals, from the Nernst heat theorem, Lewis and Gibson (1920) introduced the concept of a 'perfect crystal' and proposed the following modification as a definitive statement of the 'third law of thermodynamics' (exact wording due to Lewis and Randall (1923)):

> *If the entropy of each element in some crystalline state be taken as zero at the absolute zero of temperature, every substance has a finite positive entropy, but at the absolute zero of temperature the entropy may become zero, and does so become in the case of perfect crystalline substances.*

Because of the Nernst heat theorem and the third law, standard thermodynamic tables usually do not report entropies of formation of compounds; instead they report the molar entropy $\bar{S}^0(T)$ for each element and compound. The entropies reported for those substances that show 'residual entropy' (the 'imperfect' crystalline substances) are 'spectroscopic' entropies, not 'calorimetric' entropies.

For those who are familiar with the statistical mechanical interpretation of entropy, which asserts that at 0 K substances are normally restricted to a single quantum state, and hence have zero entropy, it should be pointed out that the conventional thermodynamic zero of entropy is not quite that, since most elements and compounds are mixtures of isotopic species that in principle should separate at 0 K, but of course do not. The thermodynamic entropies reported in tables ignore the entropy of isotopic mixing, and in some cases ignore other complications as well, e.g. ortho- and para-hydrogen.

A2.1.7.3   Second statement (unattainability of 0 K)

In the Lewis and Gibson statement of the third law, the notion of 'a perfect crystalline substance', while understandable, strays far from the macroscopic logic of classical thermodynamics and some scientists have been reluctant to place this statement in the same category as the first and second laws of thermodynamics. Fowler and Guggenheim (1939), noting that the first and second laws both state universal limitations on processes that are experimentally possible, have pointed out that the principle of the unattainability of absolute zero, first enunciated by Nernst (1912) expresses a similar universal limitation:

> *It is impossible by any procedure, no matter how idealized, to reduce the temperature of any system to the absolute zero of temperature in a finite number of operations.*

No one doubts the correctness of either of these statements of the third law and they are universally accepted as equivalent. However, there seems to have been no completely satisfactory proof of their equivalence; some additional, but very plausible, assumption appears necessary in making the connection.

Consider how the change of a system from a thermodynamic state α to a thermodynamic state β could decrease the temperature. (The change in state $\alpha \to \beta$ could be a chemical reaction, a phase transition, or just a change of volume, pressure, magnetic field, etc). Initially assume that α and β are always in complete internal equilibrium, i.e. neither has been cooled so rapidly that any disorder is frozen in. Then the Nernst heat theorem requires that $S^\alpha(0) = S^\beta(0)$ and the plot of entropy versus temperature must look something like the sketch in figure A2.1.10(a).

The most effective cooling process will be an adiabatic reversible one ($\Delta S = 0$). In any non-adiabatic process, heat will be absorbed from the surroundings (which are at a higher temperature), thus defeating the cooling process. Moreover, according to the second law, for an irreversible adiabatic process $\Delta S > 0$; it is obvious from figure A2.1.10(a) that the reversible process gets to the lower temperature. It is equally obvious that the process must end with β at a non-zero temperature.

But what if the thermodynamic state β is not in 'complete internal equilibrium', but has some 'residual entropy' frozen in? One might then imagine a diagram like figure A2.1.10(b) with a path for β leading to a positive entropy at 0 K. But β is not the true internal equilibrium situation at low temperature; it was obtained by freezing in what was equilibrium at a much higher temperature. In a process that generates β at a much

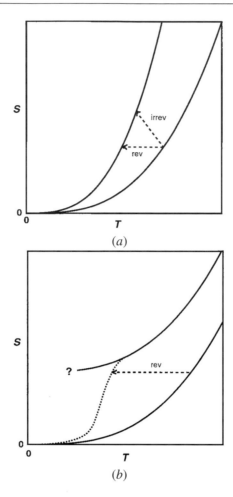

Figure A2.1.10. The impossibility of reaching absolute zero. (*a*) Both states α and β in complete internal equilibrium. Reversible and irreversible paths (dashed) are shown. (*b*) State β not in internal equilibrium and with 'residual entropy'. The true equilibrium situation for β is shown dotted.

lower temperature, one will not get this same frozen disorder; one will end on something more nearly like the true internal equilibrium curve (shown dotted). This inconceivability of the low-temperature process yielding the higher temperature's frozen disorder is the added assumption needed to prove the equivalence of the two statements. (Most ordinary processes become increasingly unlikely at low temperatures; only processes with essentially zero activation energy can occur and these are hardly the kinds of processes that could generate 'frozen' situations.)

The principle of the unattainability of absolute zero in no way limits one's ingenuity in trying to obtain lower and lower thermodynamic temperatures. The third law, in its statistical interpretation, essentially asserts that the ground quantum level of a system is ultimately non-degenerate, that some energy difference $\Delta\varepsilon$ must exist between states, so that at equilibrium at 0 K the system is certainly in that non-degenerate ground state with zero entropy. However, the $\Delta\varepsilon$ may be very small and temperatures of the order of $\Delta\varepsilon/k$ (where k is the Boltzmann constant, the gas constant per molecule) may be obtainable.

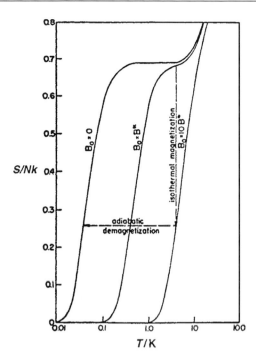

Figure A2.1.11. Magnetic cooling: isothermal magnetization at 4 K followed by adiabatic demagnetization to 0.04 K. (Constructed for a hypothetical magnetic substance with two magnetic states with an energy separation $\Delta\varepsilon = k(0.1$ K$)$ at $B_0 = 0$ and B^* (the field at which the separation is $10\Delta\varepsilon$ or 7400 gauss); the crystalline Stark effect has been ignored. The entropy above $S/Nk = 0.69 = \ln 2$ is due to the vibration of a Debye crystal.)

Magnetic cooling

A standard method of attaining very low temperatures is *adiabatic demagnetization*, a procedure suggested independently by Debye and by Giauque in 1926. A paramagnetic solid is cooled to a low temperature (one can reach about 1 K by the vaporization of liquid helium) and the solid is then magnetized isothermally in a high magnetic field B_0. (Any heat developed is carried away by contact with dilute helium gas.) As shown in figure A2.1.11, the entropy obviously decreases (compare equation (A2.1.43)).

Now the system is thermally insulated and the magnetic field is decreased to zero; in this adiabatic, essentially reversible (isentropic) process, the temperature necessarily decreases since

$$(\partial T/\partial B_0)_S = -(\partial S/\partial B_0)_T/(\partial S/\partial T)_{B_0}.$$

$((\partial S/\partial B_0)_T$ is negative and $(\partial S/\partial T)_{B_0}$, a heat capacity divided by temperature, is surely positive, so $(\partial T/\partial B_0)_S$ is positive.)

A2.1.7.4   Third statement (simple limits and statistical thermodynamics)

As we have seen, the third law of thermodynamics is closely tied to a statistical view of entropy. It is hard to discuss its implications from the exclusively macroscopic view of classical thermodynamics, but the problems become almost trivial when the molecular view of statistical thermodynamics is introduced. Guggenheim (1949) has noted that the usefulness of a molecular view is not unique to the situation of substances at low temperatures, that there are other limiting situations where molecular ideas are helpful in interpreting general experimental results:

(1) Substances at high dilution, e.g. a gas at low pressure or a solute in dilute solution, show simple behaviour. The ideal-gas law and Henry's law for dilute solutions antedate the development of the formalism of classical thermodynamics. Earlier sections in this article have shown how these experimental laws lead to simple thermodynamic equations, but these results are added to thermodynamics; they are not part of the formalism. Simple molecular theories, even if they are not always recognized as statistical mechanics, e.g. the 'kinetic theory of gases', make the experimental results seem trivially obvious.

(2) The entropy of mixing of very similar substances, i.e. the ideal solution law, can be derived from the simplest of statistical considerations. It too is a limiting law, of which the most nearly perfect example is the entropy of mixing of two isotopic species.

With this in mind Guggenheim suggested still another statement of the 'third law of thermodynamics':

> By the standard methods of statistical thermodynamics it is possible to derive for certain entropy changes general formulas that cannot be derived from the zeroth, first, and second laws of classical thermodynamics. In particular one can obtain formulae for entropy changes in highly disperse systems, for those in very cold systems, and for those associated with the mixing of very similar substances.

A2.1.8   Thermodynamics and statistical mechanics

Any detailed discussion of statistical mechanics would be inappropriate for this section, especially since other sections (A2.2 and A2.3) treat this in detail. However, a few aspects that relate to classical thermodynamics deserve brief mention.

A2.1.8.1   Ensembles and the constraints of classical thermodynamics

It is customary in statistical mechanics to obtain the average properties of members of an *ensemble*, an essentially infinite set of systems subject to the same constraints. Of course each of the systems contains the same substance or group of substances, but in addition the constraints placed on a particular ensemble are parallel to those encountered in classical thermodynamics.

The *microcanonical ensemble* is a set of systems each having the same number of molecules N, the same volume V and the same energy U. In such an ensemble of isolated systems, any allowed quantum state is equally probable. In classical thermodynamics at equilibrium at constant n (or equivalently, N), V, and U, it is the entropy S that is a maximum. For the microcanonical ensemble, the entropy is directly related to the number of allowed quantum states $\Omega(N, V, U)$:

$$S(N, V, U) = k \ln \Omega(N, V, U).$$

The *canonical ensemble* is a set of systems each having the same number of molecules N, the same volume V and the same temperature T. This corresponds to putting the systems in a thermostatic bath or, since the number of systems is essentially infinite, simply separating them by diathermic walls and letting them equilibrate. In such an ensemble, the probability of finding the system in a particular quantum state l is proportional to $\mathrm{e}^{-U_l/kT}$ where $U_l(N, V)$ is the energy of the lth quantum state and k, as before, is the Boltzmann constant. In classical thermodynamics, the appropriate function for fixed N, V and T is the Helmholtz free energy A, which is at a minimum at equilibrium and in statistical mechanics it is A that is directly related to the *canonical partition function* Q for the canonical ensemble.

$$Q(N, V, T) = \sum_l \mathrm{e}^{-U_l(N,V)/kT}$$

$$A(N, V, T) = -kT \ln Q(N, V, T).$$

The *grand canonical ensemble* is a set of systems each with the same volume V, the same temperature T and the same chemical potential μ (or if there is more than one substance present, the same set of μ_is). This corresponds to a set of systems separated by diathermic and permeable walls and allowed to equilibrate. In classical thermodynamics, the appropriate function for fixed μ, V, and T is the product pV (see equation (A2.1.37)) and statistical mechanics relates pV directly to the *grand canonical partition function* Ξ.

$$\Xi(\mu, V, T) = \sum_N Q_N(N, V, T) e^{-N\mu/kT} = \sum_N \lambda^N Q_N(N, V, T)$$

$$pV = kT \ln \Xi(\mu, V, T).$$

where λ is the absolute activity of section 2.1.6.5.

Since other sets of constraints can be used, there are other ensembles and other partition functions, but these three are the most important.

A2.1.8.2   *Fluctuations; the 'exactness' of thermodynamics*

In defining the thermodynamic state of a system in terms of fixed macroscopic constraints, classical thermodynamics appears to assume the identical character of two states subject to the identical set of constraints. However, any consideration of the fact that such systems are composed of many molecules in constant motion suggests that this must be untrue. Surely, fixing the number of gas molecules N in volume V at temperature T does not guarantee that the molecules striking the wall create exactly the same pressure at all times and in all such systems. If the pressure p is just an average, what can one say about the magnitude of fluctuations about this average?

According to statistical mechanics, for the canonical ensemble one may calculate $\langle U \rangle$, the average energy of all the members of the ensemble, while for the grand canonical ensemble one can calculate two averages, $\langle N \rangle$ and $\langle U \rangle$. Of crucial importance, however, is the probability of observing significant variations (fluctuations) from these averages in any particular member of the ensemble. Fortunately, statistical mechanics yields an answer to these questions.

Probability theory shows that the standard deviation σ_x of a quantity x can be written as

$$\sigma_x^2 = \langle x^2 \rangle - \langle x \rangle^2$$

and statistical mechanics can relate these averages to thermodynamic quantities. In particular, for the canonical ensemble

$$\sigma_U^2 = kT^2 C_V$$

while for the grand canonical ensemble

$$\sigma_N^2 = kT\kappa_T(N^2/V).$$

All the quantities in these equations are intensive (independent of the size of the system) except C_V, N, and V, which are extensive (proportional to the size of the system, i.e. to N). It follows that σ_N^2 is of the order of N, so σ_N/N is of the order of $N^{-1/2}$, as is σ_U/U. Since a macroscopic system described by thermodynamics probably has at least about 10^{20} molecules, the uncertainty, i.e. the typical fluctuation, of a measured thermodynamic quantity must be of the order of 10^{-10} times that quantity, orders of magnitude below the precision of any current experimental measurement. Consequently we may describe thermodynamic laws and equations as 'exact'.

(An exception to this conclusion is found in the immediate vicinity of critical points, where fluctuations become much more significant, although—with present experimental precision—still not of the order of N.)

References

[1] Carathéodory C 1909 Untersuchungen über die Grundlagen der Thermodynamik *Math. Ann.* **67** 355–86
[2] Born M 1921 Kritische Betrachtungen zur traditionellen Darstellung der Thermodynamik *Physik. Z.* **22** 218–24, 249–54, 282–6
[3] Redlich O 1970 The so-called zeroth law of thermodynamics *J. Chem. Educ.* **47** 740
[4] Guggenheim E A 1957 *Thermodynamics* 3rd edn (Amsterdam: North-Holland) p 441
[5] Fokker A D 1939 Remark on the fundamental relations of thermomagnetics *Physica* **6** 791–6
[6] Broer L J F 1946 On the statistical mechanics in a magnetic field *Physica* **12** 49–60
[7] Kivelson D and Oppenheim I 1966 Work in irreversible expansions *J. Chem. Educ.* **43** 233–5
[8] Hirschfelder J O Curtiss C F and Bird R B 1954 *Molecular Theory of Gases and Liquids* (New York: Wiley) p 1114

Further Reading

Callen H B 1960 *Thermodynamics, an Introduction to the Physical Theories of Equilibrium Thermostatics and Irreversible Thermodynamics* (New York: Wiley)

General principles, representative applications, fluctuations and irreversible thermodynamics. Chapter 4 discusses quasi-static processes, reversible work and heat sources, and thermodynamic engines.

Guggenheim E A 1967 *Thermodynamics, An Advanced Treatment for Chemists and Physicists* 5th edn (Amsterdam: North-Holland, New York: Interscience)

Basic thermodynamics, statistical thermodynamics, third-law entropies, phase transitions, mixtures and solutions, electrochemical systems, surfaces, gravitation, electrostatic and magnetic fields. (In some ways the 3rd and 4th editions (1957 and 1960) are preferable, being less idiosyncratic.)

Lewis G N and Randall M 1961 *Thermodynamics* 2nd edn, ed K S Pitzer and L Brewer (New York: McGraw-Hill)

Classical thermodynamics with many applications and much experimental information. Tables and graphs. Results of statistical mechanical calculations. The first edition (1923) was the first real presentation of thermodynamics for chemists.

McGlashan M L 1979 *Chemical Thermodynamics* (London: Academic)

An idiosyncratic book with comparatively little emphasis on irreversible processes, but with much emphasis upon experimental procedures and experimental results.

Pippard A E 1957 (reprinted with corrections 1964) *The Elements of Classical Thermodynamics* (Cambridge: Cambridge University Press)

Fundamentals of thermodynamics. Applications to phase transitions. Primarily directed at physicists rather than chemists.

Reid C E 1990 *Chemical Thermodynamics* (New York: McGraw-Hill)

Basic laws, using the Carathéodory approach. Applications to gases, mixtures and solutions, chemical and phase equilibria, electrochemical systems, surfaces.

Reiss H 1965 *Methods of Thermodynamics* (New York: Blaisdell) (reissued unabridged with a few corrections by Dover) (Mineola, NY: 1996)

A careful analysis of the fundamentals of classical thermodynamics, using the Born–Carathéodory approach. Emphasis on constraints, chemical potentials. Discussion of difficulties with the third law. Few applications.

A2.2
Statistical mechanics of weakly interacting systems

Rashmi C Desai

A2.2.1   Introduction

Thermodynamics is a phenomenological theory based upon a small number of fundamental laws, which are deduced from the generalization and idealization of experimental observations on macroscopic systems. The goal of statistical mechanics is to deduce the macroscopic laws of thermodynamics and other macroscopic theories (e.g. hydrodynamics and electromagnetism) starting from mechanics at a microscopic level and combining it with the rules of probability and statistics. As a branch of theoretical physics, statistical mechanics has extensive applications in physics, chemistry, biology, astronomy, materials science and engineering. Applications have been made to systems which are in thermodynamic equilibrium, to systems in steady state and also to non-equilibrium systems. Even though the scope of statistical mechanics is quite broad, this section is mostly limited to basics relevant to equilibrium systems.

At its foundation level, statistical mechanics involves some profound and difficult questions which are not fully understood, even for systems in equilibrium. At the level of its applications, however, the rules of calculation that have been developed over more than a century have been very successful.

The approach outlined here will describe a viewpoint which leads to the standard calculational rules used in various applications to systems in thermodynamic (thermal, mechanical and chemical) equilibrium. Some applications to ideal and weakly interacting systems will be made, to illustrate how one needs to think in applying statistical considerations to physical problems.

Equilibrium is a macroscopic phenomenon which implies a description on a length and time scale much larger than those appropriate to the molecular motion. The concept of 'absolute equilibrium' is an idealization and refers to a state of an absolutely isolated system and an infinitely long observation time. In non-equilibrium systems with slowly varying macroscopic properties, it is often useful to consider 'local equilibrium' where the 'macroscopic' time and length scales are determined in the context of an observation, or an experiment, and the system. A typical value of an experimentally measured property corresponds to the average value over the observation time of the corresponding physical observable; the physical properties of a system in equilibrium are time invariant and should be independent of the observation time. The observation time of an equilibrium state is typically quite long compared to the time characteristic of molecular motions.

Conservation laws at a microscopic level of molecular interactions play an important role. In particular, energy as a conserved variable plays a central role in statistical mechanics. Another important concept for equilibrium systems is the law of detailed balance. Molecular motion can be viewed as a sequence of collisions, each of which is akin to a reaction. Most often it is the momentum, energy and angular momentum of each of the constituents that is changed during a collision; if the molecular structure is altered, one has a chemical reaction. The law of detailed balance implies that, in equilibrium, the number of each reaction in the forward direction is the same as that in the reverse direction; i.e. each microscopic reaction is in equilibrium. This is a consequence of the time reversal symmetry of mechanics.

A2.2.2 Mechanics, microstates and the degeneracy function

Macroscopic systems contain a large number, N, of microscopic constituents. Typically N is of the order of 10^{20}–10^{25}. Thus many aspects of statistical mechanics involve techniques appropriate to systems with large N. In this respect, even the non-interacting systems are instructive and lead to non-trivial calculations. The degeneracy function that is considered in this subsection is an essential ingredient of the formal and general methods of statistical mechanics. The degeneracy function is often referred to as the density of states.

We first consider three examples as a prelude to the general discussion of basic statistical mechanics. These are: (i) N non-interacting spin-$\frac{1}{2}$ particles in a magnetic field, (ii) N non-interacting point particles in a box, and (iii) N non-interacting harmonic oscillators. For each example the results of quantum mechanics are used to enumerate the microstates of the N-particle system and then obtain the degeneracy function (density of states) of the system's energy levels. Even though these three examples are for ideal non-interacting systems, there are many realistic systems which turn out to be well approximated by them.

A microstate (or a microscopic state) is one of the quantum states determined from $\mathcal{H}\phi_l = E_l\phi_l$, $(l = 1, 2, \ldots)$, where $\mathcal{H}$ is the Hamiltonian of the system, E_l is the energy of the quantum state l and ϕ_l is the wavefunction representing the quantum state l. The large-N behaviour of the degeneracy function is of great relevance. The calculation of the degeneracy function in these three examples is a useful precursor to the conceptual use of the density of states of an arbitrary interacting system in the general framework of statistical mechanics.

(i) N non-interacting spin-$\frac{1}{2}$ particles in a magnetic field. Each particle can be considered as an elementary magnet (its magnetic moment has a magnitude equal to μ) which can point along two possible directions in space ($+z$ or 'up' and $-z$ or 'down'). A microstate of such a model system is given by giving the orientation ($+$ or $-$) of each magnet. It is obvious that for N such independent magnets, there are 2^N different microstates for this system. Note that this number grows exponentially with N. The total magnetic moment $\mathcal{M}$ of the model system is the vector sum of the magnetic moments of its N constituents. The component of $\mathcal{M}$ along the z direction varies between $-N\mu$ and $+N\mu$, and can take any of the $(N+1)$ possible values $N\mu$, $(N-2)\mu$, $(N-4)\mu$, $\ldots$, $-N\mu$. This number of possible values of $\mathcal{M}$ is much less than the total number of microstates 2^N, for large N. The number of microstates for a given $\mathcal{M}$ is the degeneracy function. In each of these microstates, there will be $\frac{1}{2}N + m$ spins up and $\frac{1}{2}N - m$ spins down, such that the difference between the two is $2m$, which is called the spin excess and equals $\mathcal{M}/\mu$. If x is the probability for a particle to have its spin up and $y = (1 - x)$ is the probability for its spin down, the degeneracy function $g(N, m)$ can be obtained by inspection from the binomial expansion

$$(x + y)^N = \sum_{m=-\frac{1}{2}N}^{\frac{1}{2}N} \frac{N!}{(\frac{1}{2}N + m)!(\frac{1}{2}N - m)!} x^{\frac{1}{2}N+m} y^{\frac{1}{2}N-m}.$$

That is

$$g(N, m) = \frac{N!}{(\frac{1}{2}N + m)!(\frac{1}{2}N - m)!}. \tag{A2.2.1}$$

By setting $x = y = \frac{1}{2}$, one can see that $\sum_m g(N, m) = 2^N$. For typical macroscopic systems, the number N of constituent molecules is very large: $N \sim 10^{23}$. For large N, $g(N, m)$ is a very sharply peaked function of m. In order to see this one needs to use the Stirling approximation for $N!$ which is valid when $N \gg 1$:

$$N! \approx (2\pi N)^{\frac{1}{2}} N^N \exp[-N + 1/(12N) + \cdots]. \tag{A2.2.2}$$

For sufficiently large N, the terms $1/(12N) + \cdots$ can be neglected in comparison with N and one obtains

$$\log N! \approx \frac{1}{2}\log(2\pi) + (N + \frac{1}{2})\log N - N \tag{A2.2.3}$$

$$\approx N \log N - N \tag{A2.2.4}$$

since both $\frac{1}{2}\log(2\pi)$ and $\frac{1}{2}$ are negligible compared to N. Using (A2.2.3), for $\log g(N, m)$ one obtains

$$\log g \approx \frac{1}{2} \log\left(\frac{2}{\pi N}\right) + N\log 2 - \frac{2m^2}{N}$$

which reduces to a Gaussian distribution for $g(N, m)$:

$$g(N, m) \approx g(N, 0)\exp(-2m^2/N) \tag{A2.2.5}$$

with

$$g(N, 0) = \frac{N!}{(\frac{1}{2}N)!(\frac{1}{2}N)!} \approx \left(\frac{2}{\pi N}\right)^{\frac{1}{2}} 2^N. \tag{A2.2.6}$$

When $m^2 = N/2$, the value of g is decreased by a factor of e from its maximum at $m = 0$. Thus the fractional width of the distribution is $\Delta(m/N) \sim (1/N)^{\frac{1}{2}}$. For $N \sim 10^{22}$ the fractional width is of the order of 10^{-11}. It is the sharply peaked behaviour of the degeneracy functions that leads to the prediction that the thermodynamic properties of macroscopic systems are well defined.

For this model system the magnetic potential energy in the presence of a uniform magnetic field $\vec{H}$ is given by $-\vec{\mathcal{M}} \cdot \vec{H}$ and, for $\vec{H}$ pointing in $+z$ direction, it is $-\mathcal{M}H$ or $-2m\mu H$. A fixed magnetic potential energy thus implies a fixed value of m. For a given H, the magnetic potential energy of the system is bounded from above and below. This is not the case for the next two examples. Besides the example of an ideal paramagnet that this model explicitly represents, there are many other systems which can be modelled as effective two-state systems. These include the lattice gas model of an ideal gas, binary alloy and a simple two-state model of a linear polymer. The time dependence of the mean square displacement of a Brownian particle can also be analysed using such a model.

(ii) N non-interacting point particles in a box. The microstate (orbital) of a free particle of mass M confined in a cube of volume L^3 is specified by three integers (quantum numbers): $(n_x, n_y, n_z) \equiv n$; $n_z, n_y, n_x = 1, 2, 3, \ldots$. Its wavefunction is

$$\phi_n = (2/L)^{3/2} \sin(xp_x) \sin(yp_y) \sin(zp_z)$$

with $p = (\hbar\pi/L)n$, and the energy $\epsilon = p^2/(2M) = (\hbar\pi/L)^2 n^2/(2M)$. The energy grows quadratically with n, and without bounds. One can enumerate the orbitals by considering the positive octant of a sphere in the space defined by n_x, n_y and n_z for free particle orbitals. With every unit volume $\Delta n_x \Delta n_y \Delta n_z = 1$, there is one orbital per spin orientation of the particle. For particles of spin I, there are $\gamma = (2I + 1)$ independent spin orientations. The energy of an orbital on the surface of a sphere of radius n_0 in the n space is $\epsilon_0 = (\hbar\pi/L)^2 n_0^2/(2M)$. The degeneracy function, or equivalently, the number of orbitals in the allowed (positive) octant of a spherical shell of thickness Δn is $\gamma \frac{1}{8} 4\pi n_0^2 \Delta n = \frac{1}{2}\gamma\pi n_0^2 \Delta n$. This is an approximate result valid asymptotically for large n_0. Often one needs the number of orbitals with energy between ϵ and $\epsilon + d\epsilon$. If it is denoted by $\mathcal{D}(\epsilon)\, d\epsilon$, it is easy to show by using

$$\mathcal{D}(\epsilon)\, d\epsilon = \frac{1}{2}\gamma\pi n^2 \frac{dn}{d\epsilon}\, d\epsilon \tag{A2.2.7}$$

that

$$\mathcal{D}(\epsilon) = \frac{\gamma V}{4\pi^2}\left(\frac{2M}{\hbar^2}\right)^{\frac{3}{2}} \epsilon^{\frac{1}{2}}. \tag{A2.2.8}$$

This is the density of microstates for one free particle in volume $V = L^3$.

For N non-interacting particles in a box, the result depends on the particle statistics: Fermi, Bose of Boltzmann. The state of a quantum system can be specified by the wavefunction for that state,

$\Psi_\nu(q_1, q_2, \ldots q_N)$. Ψ_ν is the νth eigensolution to the Schrödinger equation for an N-particle system. If the particles are non-interacting, then the wavefunction can be expressed in terms of the single-particle wavefunctions (ϕ_n given above is an example). Let these be denoted by $\phi_1(q), \phi_2(q), \ldots, \phi_j(q), \ldots$. For a specific state ν, $\Psi_\nu(q_1, q_2, \ldots, q_N)$ will be the appropriately symmetrized product containing n_1 particles with the single-particle wavefunction $\phi_1(q)$, n_2 particles with $\phi_2(q)$, etc. For Fermi particles (with half integral spin) the product is antisymmetric and for Bose particles (with integer spin) it is symmetric. The antisymmetry of the Fermi particle wavefunction implies that fermions obey the Pauli exclusion principle. The numbers $n_1, n_2, \ldots, n_j, \ldots$ are the occupation numbers of the respective single-particle states, and this set of occupation numbers $\{n_j\}$ completely specify the state ν of the system. If there are N_ν particles in this state then $N_\nu = \sum_j n_j$, and if the jth single-particle state has energy ϵ_j, then the energy of the system in the state ν is $E_\nu = \sum_j \epsilon_j n_j$.

In an ideal molecular gas, each molecule typically has translational, rotational and vibrational degrees of freedom. The example of 'one free particle in a box' is appropriate for the translational motion. The next example of oscillators can be used for the vibrational motion of molecules.

(iii) N non-interacting harmonic oscillators. Energy levels of a harmonic oscillator are non-degenerate, characterized by a quantum number l and are given by the expression $\epsilon_l = (l + \frac{1}{2})\hbar\omega, l = 0, 1, 2, 3, \ldots, \infty$. For a system of N independent oscillators, if the ith oscillator is in the state n_i, the set $\{n_i\}$ gives the microstate of the system, and its total energy E is given by $E = \frac{1}{2}N\hbar\omega + \sum_{i=1}^{N} n_i\hbar\omega$. Consider the case when the energy of the system, above its zero point energy $\frac{1}{2}N\hbar\omega$, is a fixed amount $E_o \equiv (E - \frac{1}{2}N\hbar\omega)$. Define n such that $E_o = n\hbar\omega$. Then, one needs to find the number of ways in which a set of $\{n_i\}$ can be chosen such that $\sum_{i=1}^{N} n_i = n$. This number is the degeneracy function $g(N, n)$ for this system. For a single-oscillator case, $n_1 = n$ and $g(1, n) = 1$, for all n. Consider a sum $\sum_{n=0}^{\infty} g(1, n)t^n$; it is called a generating function. Since $g(1, n) = 1$, it sums to $(1-t)^{-1}$, if $|t| < 1$. For N independent oscillators, one would therefore use $(1-t)^{-N}$, rewriting it as

$$\left(\frac{1}{1-t}\right)^N = \left(\sum_{s=0}^{\infty} t^s\right)^N \equiv \sum_{m=0}^{\infty} g(N, m)t^m.$$

Now since in general,

$$g(N, n) = \lim_{t\to\infty} \frac{1}{n!}\left(\frac{d}{dt}\right)^n \sum_{m=0}^{\infty} g(N, m)t^m$$

its use for the specific example of N oscillators gives

$$g(N, n) = \lim_{t\to\infty} \frac{1}{n!}\left(\frac{d}{dt}\right)^n (1-t)^{-N}$$

with the final result that the degeneracy function for the N-oscillator system is

$$g(N, n) = \frac{(N+n-1)!}{n!(N-1)!}. \tag{A2.2.9}$$

The model of non-interacting harmonic oscillators has a broad range of applicability. Besides vibrational motion of molecules, it is appropriate for phonons in harmonic crystals and photons in a cavity (black-body radiation).

A2.2.2.1   Classical mechanics

The set of microstates of a finite system in quantum statistical mechanics is a finite, discrete denumerable set of quantum states each characterized by an appropriate collection of quantum numbers. In classical statistical mechanics, the set of microstates form a continuous (and therefore infinite) set of points in Γ space (also called phase space). Following Gibbs, the Γ space is defined as a $2f$-dimensional space for a system with f degrees of freedom: $(p_1, p_2, \ldots, p_f; q_1, q_2, \ldots, q_f)$, abbreviated as (p, q). Here $(p_i, q_i), i = 1, \ldots, f$ are the canonical momenta and canonical coordinates of the f degrees of freedom of the system. Given a precise initial state (p°, q°), a system with the Hamiltonian $\mathcal{H}(p, q)$ evolves deterministically according to the canonical equations of motion:

$$\frac{\partial \mathcal{H}(p, q)}{\partial p_i} = \dot{q}_i \qquad \frac{\partial \mathcal{H}(p, q)}{\partial q_i} = -\dot{p}_i. \qquad (A2.2.10)$$

Now, if D/Dt represents time differentiation along the deterministic trajectory of the system in the Γ space, it follows that

$$\dot{\mathcal{H}} = \frac{D\mathcal{H}}{Dt} = \frac{\partial \mathcal{H}}{\partial t} + \sum_k \left(\dot{q}_k \frac{\partial \mathcal{H}}{\partial q_k} + \dot{p}_k \frac{\partial \mathcal{H}}{\partial p_k} \right) = \frac{\partial \mathcal{H}}{\partial t} \qquad (A2.2.11)$$

where the last equality is obtained using the equations of motion. Thus, when $\mathcal{H}$ does not depend on time explicitly, i.e. when $\partial \mathcal{H}/\partial t = 0$, the above equation implies that $\mathcal{H}(p, q) = E = $ constant. The locus of points in Γ space satisfying this condition defines a $(2f - 1)$-dimensional energy hypersurface S, and the trajectory of such a system in Γ space would lie on this hypersurface. Furthermore, since a given trajectory is uniquely determined by the equations of motion and the initial conditions, two trajectories in Γ space can never intersect.

A2.2.2.2   Liouville's theorem

The volume of a Γ-space-volume-element does not change in the course of time if each of its points traces out a trajectory in Γ space determined by the equations of motion. Equivalently, the Jacobian

$$J(t, t_0) \equiv \frac{\partial(p, q)}{\partial(p^\circ, q^\circ)} = 1. \qquad (A2.2.12)$$

Liouville's theorem is a restatement of mechanics. The proof of the theorem consists of two steps.

(1) Expand $J(t, t_0)$ around t_0 to obtain:

$$J(t, t_0) = 1 + (t - t_0) \left[\frac{\partial J}{\partial t}(t, t_0) \right]_{t=t_0} + \cdots = 1 + (t - t_0) \sum_i \left(\frac{\partial \dot{p}_i}{\partial p_i^\circ} + \frac{\partial \dot{q}_i}{\partial q_i^\circ} \right)_{t=t_0} + \cdots.$$

Hence,

$$\left[\frac{\partial J}{\partial t}(t, t_0) \right]_{t=t_0} = \sum_i \left(\frac{\partial \dot{p}_i}{\partial p_i^\circ} + \frac{\partial \dot{q}_i}{\partial q_i^\circ} \right)_{t=t_0}$$

$$= \sum_i \left(-\frac{\partial^2 \mathcal{H}}{\partial p_i^\circ \partial q_i} + \frac{\partial^2 \mathcal{H}}{\partial q_i^\circ \partial p_i} \right)_{t=t_0}$$

$$= \sum_i \left(-\frac{\partial^2 \mathcal{H}}{\partial p_i^\circ \partial q_i^\circ} + \frac{\partial^2 \mathcal{H}}{\partial q_i^\circ \partial p_i^\circ} \right)$$

$$= 0.$$

(2) From the multiplication rule of Jacobians, one has for *any* t_1 between t_0 and t,

$$\frac{\partial J(t, t_0)}{\partial t} = \frac{\partial J(t, t_1)}{\partial t} J(t_1, t_0).$$

Now let t_1 approach t. Then, by the result of the first step, the first factor on the right-hand side vanishes. Hence,

$$\frac{\partial J(t, t_o)}{\partial t} = 0 \qquad \text{and} \qquad J(t, t_o) = \text{constant}.$$

Finally, since $J(t_o, t_o) = 1$, one obtains the result $J(t, t_o) = 1$, which concludes the proof of Liouville's theorem.

Geometrically, Liouville's theorem means that if one follows the motion of a small phase volume in Γ space, it may change its shape but its volume is invariant. In other words the motion of this volume in Γ space is like that of an incompressible fluid. Liouville's theorem, being a restatement of mechanics, is an important ingredient in the formulation of the theory of statistical ensembles, which is considered next.

A2.2.3   Statistical ensembles

In equilibrium statistical mechanics, one is concerned with the thermodynamic and other macroscopic properties of matter. The aim is to derive these properties from the laws of molecular dynamics and thus create a link between microscopic molecular motion and thermodynamic behaviour. A typical macroscopic system is composed of a large number N of molecules occupying a volume V which is large compared to that occupied by a molecule:

$$N \approx 10^{23} \text{ molecules} \qquad V \approx 10^{23} \text{ molecular volumes}.$$

Due to such large numbers, it is useful to consider the limiting case of the *thermodynamic limit*, which is defined as

$$N \to \infty \qquad V \to \infty \qquad \frac{V}{N} = v \qquad\qquad (\text{A2.2.13})$$

where the specific volume v is a given finite number. For a three-dimensional system with N *point* particles, the total number of degrees of freedom $f = 3N$.

A statistical ensemble can be viewed as a description of how an experiment is repeated. In order to describe a macroscopic system in equilibrium, its thermodynamic state needs to be specified first. From this, one can infer the macroscopic constraints on the system, i.e. which macroscopic (thermodynamic) quantities are held fixed. One can also deduce, from this, what are the corresponding microscopic variables which will be constants of motion. A macroscopic system held in a specific thermodynamic equilibrium state is typically consistent with a very large number (classically infinite) of microstates. Each of the repeated experimental measurements on such a system, under ideal conditions and with identical macroscopic constraints, would correspond to the system in a different accessible microstate which satisfies the macroscopic constraints. It is natural to represent such a collection of microstates by an ensemble (a mental construct) of systems, which are identical in composition and macroscopic conditions (constraints), but each corresponding to a different microstate. For a properly constructed ensemble, each of its member systems satisfies the macroscopic constraints appropriate to the experimental conditions. Collectively the ensemble then consists of all the microstates that satisfy the macroscopic constraints (all accessible states). The simplest assumption that one can make in order to represent the repeated set of experimental measurements by the ensemble of accessible microstates is to give each an equal weight. The fundamental assumption in the ensemble theory is then the *Postulate of 'Equal a priori probabilities'*. It states that 'when a macroscopic system is in thermodynamic equilibrium, its microstate is equally likely to be any of the accessible states, each of which satisfy the macroscopic constraints on the system'.

Such an ensemble of systems can be geometrically represented by a distribution of representative points in the Γ space (classically a continuous distribution). It is described by an ensemble density function $\rho(p, q, t)$ such that $\rho(p, q, t) \, d^{2f}\Omega$ is the number of representative points which at time t are within the infinitesimal phase volume element $d^f p \, d^f q$ (denoted by $d^{2f}\Omega$) around the point (p, q) in the Γ space.

Let us consider the consequence of mechanics for the ensemble density. As in subsection A2.2.2.1, let D/Dt represent differentiation along the trajectory in Γ space. By definition,

$$\frac{D}{Dt}(\rho \, d^{2f}\Omega) = 0.$$

According to Liouville's theorem,

$$\frac{D}{Dt}(d^{2f}\Omega) = 0.$$

Therefore,

$$\frac{D\rho}{Dt} = 0$$

or, equivalently,

$$\frac{\partial \rho}{\partial t} + \sum_k \left(\dot{q}_k \frac{\partial \rho}{\partial q_k} + \dot{p}_k \frac{\partial \rho}{\partial p_k} \right) = 0$$

which can be rewritten in terms of Poisson brackets using the equations of motion, (A2.2.10):

$$\frac{\partial \rho}{\partial t} + \sum_k \left(\frac{\partial \mathcal{H}}{\partial p_k} \frac{\partial \rho}{\partial q_k} - \frac{\partial \mathcal{H}}{\partial q_k} \frac{\partial \rho}{\partial p_k} \right) = 0.$$

This is same as

$$\frac{\partial \rho}{\partial t} + [\mathcal{H}, \rho]_{\text{P.B.}} = 0. \qquad (A2.2.14)$$

For the quantum mechanical case, ρ and $\mathcal{H}$ are operators (or matrices in appropriate representation) and the Poisson bracket is replaced by the commutator $[\mathcal{H}, \rho]_-$. If the distribution is stationary, as for the systems in equilibrium, then $\partial \rho / \partial t = 0$, which implies

$$[\mathcal{H}, \rho]_{\text{P.B.}} = 0 \quad \text{classically} \qquad \text{and} \quad [\mathcal{H}, \rho]_- = 0 \quad \text{quantum mechanically.} \qquad (A2.2.15)$$

A stationary ensemble density distribution is constrained to be a functional of the constants of motion (globally conserved quantities). In particular, a simple choice is $\rho(p, q) = \rho^*(\mathcal{H}(p, q))$, where $\rho^*(\mathcal{H})$ is some functional (function of a function) of $\mathcal{H}$. Any such functional has a vanishing Poisson bracket (or a commutator) with $\mathcal{H}$ and is thus a stationary distribution. Its dependence on (p, q) through $\mathcal{H}(p, q) = E$ is expected to be reasonably smooth. Quantum mechanically, $\rho^*(\mathcal{H})$ is the density operator which has some functional dependence on the Hamiltonian $\mathcal{H}$ depending on the ensemble. It is also normalized: $\text{Tr}\rho = 1$. The density matrix is the matrix representation of the density operator in some chosen representation of a complete orthonormal set of states. If the complete orthonormal set of eigenstates of the Hamiltonian is known:

$$\mathcal{H}|\nu\rangle = E_\nu|\nu\rangle \qquad \langle \nu|\nu'\rangle = \delta_{\nu\nu'}$$

then the density operator is

$$\rho = \sum_\nu |\nu\rangle\langle\nu|\rho^*(E_\nu).$$

Often the eigenstates of the Hamiltonian are not known. Then one uses an appropriate set of states $|\nu\rangle$ which are complete and orthonormal. In any such representation the density matrix, given as $\langle\nu|\rho^*(\mathcal{H})|\nu\rangle$, is not diagonal.

A2.2.3.1 Microcanonical ensemble

An explicit example of an equilibrium ensemble is the microcanonical ensemble, which describes closed systems with adiabatic walls. Such systems have constraints of fixed N, V and $E < \mathcal{H} < E + dE$. dE is very small compared to E, and corresponds to the assumed very weak interaction of the 'isolated' system with the surroundings. dE has to be chosen such that it is larger than $(\delta E)_{qu} \sim h/t_{ob}$ where h is the Planck's constant and t_{ob} is the duration of the observation time. In such a case, even though dE may be small, there will be a great number of microstates for a macroscopic size system. For a microcanonical ensemble, the 'equal *a priori* probability' postulate gives its density distribution as:

classically,

$$\rho(p, q) = \begin{cases} \text{constant} & \text{if } E < \mathcal{H}(p, q) < E + dE \\ 0 & \text{otherwise} \end{cases} \tag{A2.2.16}$$

quantum mechanically, if the system microstate is denoted by l, then

$$\rho_l = \begin{cases} \text{constant} & \text{if } E < E_l < E + dE \\ 0 & \text{otherwise.} \end{cases} \tag{A2.2.17}$$

One considers systems for which the energy shell is a closed (or at least finite) hypersurface $\mathcal{S}$. Then the energy shell has a finite volume:

$$\int_{E < \mathcal{H} < E + dE} d^{2f}\Omega.$$

For each degree of freedom, classical states within a small volume $\Delta p_i \Delta q_i \sim h$ merge into a single quantum state which cannot be further distinguished on account of the uncertainty principle. For a system with f degrees of freedom, this volume is h^f. Furthermore, due to the indistinguishability of identical particles in quantum mechanics, there are $N!$ distinguishable classical states for each quantum mechanical state, which are obtained by simple permutations of the N particles in the system. Then the number of microstates $\Gamma(E)$ in Γ space occupied by the microcanonical ensemble is given by

$$\Gamma(E) \equiv \mathcal{D}(E)\, d(E) = [h^f N!]^{-1} \int_{E < \mathcal{H} < E + dE} d^{2f}\Omega \tag{A2.2.18}$$

where f is the total number of degrees of freedom for the N-particle system, and $\mathcal{D}(E)$ is the density of states of the system at energy E. If the system of N particles is made up of N_A particles of type A, N_B particles of type B, ..., then $N!$ is replaced by $N_A! N_B!$, ... Even though dE is conceptually essential, it does not affect the thermodynamic properties of macroscopic systems. In order that the ensemble density ρ is normalized, the 'constant' above in (A2.2.16) has to be $[\Gamma(E)]^{-1}$. Quantum mechanically $\Gamma(E)$ is simply the total number of microstates within the energy interval $E < E_l < E + dE$, and fixes the 'constant' in (A2.2.17). $\Gamma(E)$ is the microcanonical partition function; in addition to its indicated dependence on E, it also depends on N and V.

Consider a measurable property $\mathcal{B}(p, q)$ of the system, such as its energy or momentum. When a system is in equilibrium, according to Boltzmann, what is observed macroscopically are the time averages of the form

$$\bar{\mathcal{B}}^t = \lim_{T \to \infty} \frac{1}{2T} \int_{-T}^{T} \mathcal{B}(p_t, q_t)\, dt. \tag{A2.2.19}$$

It was assumed that, apart from a vanishingly small number of exceptions, the initial conditions do not have an effect on these averages. However, since the limiting value of the time averages cannot be computed, an *ergodic hypothesis* was introduced: *time averages are identical with statistical averages over a microcanonical ensemble, for reasonable functions $\mathcal{B}$, except for a number of initial conditions, whose importance is* *vanishingly small compared with that of all other initial conditions.* Here the ensemble average is defined as

$$\langle \mathcal{B} \rangle = \frac{\int d^{2f}\Omega\, \mathcal{B}(p,q)\rho(p,q)}{\int d^{2f}\Omega\, \rho(p,q)}. \tag{A2.2.20}$$

The thinking behind this was that, over a long time period, a system trajectory in Γ space passes through every configuration in the region of motion (here the energy shell), i.e. the system is ergodic, and hence the infinite time average is equal to the average in the region of motion, or the average over the microcanonical ensemble density. The ergodic hypothesis is meant to provide justification of the 'equal *a priori* probability' postulate. It is a strong condition. For a system of N particles, the infinitely long time must be much longer than $O(e^N)$, whereas the usual observation time window is $O(1)$. (When one writes $y = O(x)$ and $z = o(x)$, it implies that $\lim_{x\to\infty} y/z =$ finite $\neq 0$ and $\lim_{x\to\infty} z/x = 0$.) However, if by 'reasonable functions $\mathcal{B}$' one means large variables $O(N)$, then their values are nearly the same everywhere in the region of motion and the trajectory need not be truly ergodic for the time average to be equal to the ensemble average. Ergodicity of a trajectory is a difficult mathematical problem in mechanics.

The microcanonical ensemble is a certain model for the repetition of experiments: in every repetition, the system has 'exactly' the same energy, N and V; but otherwise there is no experimental control over its microstate. Because the microcanonical ensemble distribution depends only on the total energy, which is a constant of motion, it is time independent and mean values calculated with it are also time independent. This is as it should be for an equilibrium system. Besides the ensemble average value $\langle \mathcal{B} \rangle$, another commonly used 'average' is the most probable value, which is the value of $\mathcal{B}(p,q)$ that is possessed by the largest number of systems in the ensemble. The ensemble average and the most probable value are nearly equal if the *mean square fluctuation* is small, i.e. if

$$\frac{\langle \mathcal{B}^2 \rangle - \langle \mathcal{B} \rangle^2}{\langle \mathcal{B} \rangle^2} \ll 1. \tag{A2.2.21}$$

If this condition is not satisfied, there is no unique way of calculating the observed value of $\mathcal{B}$, and the validity of the statistical mechanics should be questioned. In all physical examples, the mean square fluctuations are of the order of $1/N$ and vanish in the thermodynamic limit.

A2.2.3.2 Mixing

In the last subsection, the microcanonical ensemble was formulated as an ensemble from which the *equilibrium* properties of a dynamical system can be determined by its energy alone. We used the postulate of equal *a priori* probability and gave a discussion of the ergodic hypothesis. The ergodicity condition, even though a strong condition, does not ensure that if one starts from a *non-equilibrium* ensemble the expectation values of dynamical functions will approach their equilibrium values as time proceeds. For this, one needs a stronger condition than ergodicity, a condition of *mixing*. Every mixing system is ergodic, but the reverse is not true.

Consider, at $t = 0$, some non-equilibrium ensemble density $\rho_{ne}(p^0, q^0)$ on the constant energy hypersurface S, such that it is normalized to one. By Liouville's theorem, at a later time t the ensemble density becomes $\rho_{ne}(\phi_{-t}(p,q))$, where $\phi_{-t}(p,q)$ is the function that takes the current phase coordinates (p,q) to their initial values time (t) ago; the function ϕ is uniquely determined by the equations of motion. The expectation value of any dynamical variable $\mathcal{B}$ at time t is therefore

$$\int_S d^{2f}\Omega\, \mathcal{B}(p,q)\rho_{ne}(\phi_{-t}(p,q)). \tag{A2.2.22}$$

As t becomes large, this should approach the equilibrium value $\langle \mathcal{B} \rangle$, which for an ergodic system is

$$\frac{\int d^{2f}\Omega\, \mathcal{B}(p,q)\rho(p,q)}{\int d^{2f}\Omega\, \rho(p,q)} = \frac{\int_S d^{2f}\Omega\, \mathcal{B}(p,q)}{\int_S d^{2f}\Omega} \tag{A2.2.23}$$

where S is the hypersurface of the energy shell for the microcanonical ensemble. This equality is satisfied if the system is *mixing*.

A system is mixing if, for every pair of functions f and g whose squares are integrable on S,

$$\lim_{t \to \pm\infty} \int_S d^{2f}\Omega \, f(p,q) g(\phi_{-t}(p,q)) = \frac{\int_S d^{2f}\Omega \, f(p,q) \int_S d^{2f}\Omega \, g(p,q)}{\int_S d^{2f}\Omega}. \tag{A2.2.24}$$

The statement of the mixing condition is equivalent to the following: if Q and R are arbitrary regions in S, and an ensemble is initially distributed uniformly over Q, then the fraction of members of the ensemble with phase points in R at time t will approach a limit as $t \to \infty$, and this limit equals the fraction of area of S occupied by R.

The ensemble density $\rho_{ne}(p_t, q_t)$ of a mixing system does not approach its equilibrium limit in the pointwise sense. It is only in a 'coarse-grained' sense that the average of $\rho_{ne}(p_t, q_t)$ over a region R in S approaches a limit to the equilibrium ensemble density as $t \to \infty$ for each fixed R.

In the condition of mixing, equation (A2.2.24), if the function g is replaced by ρ_{ne}, then the integral on the left-hand side is the expectation value of f, and at long times approaches the equilibrium value, which is the microcanonical ensemble average $\langle f \rangle$, given by the right-hand side. The condition of mixing is a sufficient condition for this result. The condition of mixing for equilibrium systems also has the implication that every equilibrium time-dependent correlation function, such as $\langle f(p,q) g(\phi_t(p,q)) \rangle$, approaches a limit of the uncorrelated product $\langle f \rangle \langle g \rangle$ as $t \to \infty$.

A2.2.3.3   Entropy

For equilibrium systems, thermodynamic entropy is related to ensemble density distribution ρ as

$$S = -k_b \langle \log \rho \rangle \tag{A2.2.25}$$

where k_b is a universal constant, the Boltzmann's constant. For equilibrium systems, ρ is a functional of constants of motion like energy and is time independent. Thus, the entropy as defined here is also invariant over time. In what follows it will be seen that this definition of entropy obeys all properties of thermodynamic entropy.

A low-density gas which is not in equilibrium, is well described by the one-particle distribution $f(\vec{v}, \vec{r}, t)$, which describes the behaviour of a particle in μ space of the particle's velocity $\vec{v}$ and position $\vec{r}$. One can obtain $f(\vec{v}, \vec{r}, t)$ from the classical density distribution $\rho(p, q, t)$ (defined earlier following the postulate of equal *a priori* probabilities) by integrating over the degrees of freedom of the remaining $(N-1)$ particles. Such a coarse-grained distribution f satisfies the Boltzmann transport equation (see section A3.1). Boltzmann used the crucial assumption of molecular chaos in deriving this equation. From f one can define an H function as $H(t) = \langle \log f \rangle_{ne}$ where the non-equilibrium average is taken over the time-dependent f. Boltzmann proved an H theorem, which states that 'if at a given instant t, the state of gas satisfies the assumption of molecular chaos, then at the instant $t + \epsilon (\epsilon \to 0)$, $dH(t)/dt \leq 0$; the equality $dH(t)/dt = 0$ is satisfied if and only if f is the equilibrium Maxwell–Boltzmann distribution'; i.e. $H(t)$ obtained from the Boltzmann transport equation is a monotonically decreasing function of t. Thus a generalization of the equilibrium definition of entropy to systems slightly away from equilibrium can be made:

$$S(t) = -k_b \langle \log f \rangle_{ne} \equiv -k_b H(t). \tag{A2.2.26}$$

Such a generalization is consistent with the Second Law of Thermodynamics, since the H theorem and the generalized definition of entropy together lead to the conclusion that the entropy of an isolated non-equilibrium system increases monotonically, as it approaches equilibrium.

A2.2.3.4   Entropy and temperature in a microcanonical ensemble

For a microcanonical ensemble, $\rho = [\Gamma(E)]^{-1}$ for each of the allowed $\Gamma(E)$ microstates. Thus for an isolated system in equilibrium, represented by a microcanonical ensemble,

$$S = k_b \log \Gamma(E). \tag{A2.2.27}$$

Consider the microstates with energy E_l such that $E_l \leq E$. The total number of such microstates is given by

$$\Sigma(E) = [h^f N!]^{-1} \int_{0 < \mathcal{H} < E} \mathrm{d}^{2f}\Omega. \tag{A2.2.28}$$

Then $\Gamma(E) = \Sigma(E + \mathrm{d}E) - \Sigma(E)$, and the density of states $\mathcal{D}(E) = \mathrm{d}\Sigma/\mathrm{d}E$. A system containing a large number of particles N, or an indefinite number of particles but with a macroscopic size volume V, *normally* has the number of states Σ, which approaches asymptotically to

$$\Sigma \sim \exp\left(N\phi\left(\frac{E}{N}, \frac{V}{N}\right)\right) \qquad \text{or} \qquad \Sigma \sim \exp\left(V\psi\left(\frac{E}{V}, \frac{N}{V}\right)\right) \tag{A2.2.29}$$

where E/N, E/V and ϕ or ψ are each $\sim O(1)$, and

$$\phi > 0 \qquad \phi' > 0 \qquad \phi'' < 0. \tag{A2.2.30}$$

Consider the three examples considered in section A2.2.2. For examples (i) and (iii), the degeneracy function $g(N, n)$ is a discrete analogue of $\Gamma(E)$. Even though $\Sigma(E)$ can be obtained from $g(N, n)$ by summing it over n from the lowest-energy state up to energy E, the largest value of n dominates the sum if N is large, so that $g(N, n)$ is also like $\Sigma(E)$. For example (ii), $\Sigma(E)$ can be obtained from the density of states $\mathcal{D}(\epsilon)$ by an integration over ϵ from zero to E. $\Sigma(E)$ so obtained conforms to the above asymptotic properties for large N, for the last two of the three examples. For the first example of 'N non-interacting spin-$\frac{1}{2}$ particles in a magnetic field', this is the case only for the energy states corresponding to $0 < m < N/2$. (The other half state space with $0 > m > (-N/2)$, corresponding to the positive magnetic potential energy in the range between zero and $\mu H N$, corresponds to the system in non-equilibrium states, which have sometimes been described using 'negative temperatures' in an equilibrium description. Such peculiarities often occur in a model system which has a finite upper bound to its energy.)

Using the asymptotic properties of $\Sigma(E)$ for large N, one can show that, within an additive constant $\sim O(\log N)$ or smaller, the three quantities $k_b \log \Gamma(E)$, $k_b \log \mathcal{D}(E)$ and $k_b \log \Sigma(E)$ are equivalent and thus any of the three can be used to obtain S for a large system. This leads to the result that $S = k_b \log \Sigma(E) = k_b N \phi$, so that *the entropy as defined is an extensive quantity*, consistent with the thermodynamic behaviour.

For an isolated system, among the independent macroscopic variables N, V and E, only V can change. Now V cannot decrease without compressing the system, and that would remove its isolation. Thus V can only increase, as for example is the case for the free expansion of a gas when one of the containing walls is suddenly removed. For such an adiabatic expansion, the number of microstates in the final state is larger; thus the entropy of the final state is larger than that of the initial state. More explicitly, note that $\Sigma(E)$ is a non-decreasing function of V, since if $V_1 > V_2$, then the integral in the defining equation, (A2.2.28), for $V = V_1$ extends over a domain of integration that includes that for $V = V_2$. Thus $S(E, V) = k_b \log \Sigma(E)$ is a non-decreasing function of V. This is *also consistent* with the Second Law of Thermodynamics.

Next, let x_i be either p_i or q_i, $(i = 1, \ldots, f)$. Consider the ensemble average $\langle x_i \, \partial\mathcal{H}/\partial x_j \rangle$:

$$\left\langle x_i \frac{\partial\mathcal{H}}{\partial x_j} \right\rangle = \frac{\int_{E < \mathcal{H} < E + \mathrm{d}E} \mathrm{d}^{2f}\Omega \, x_i(\partial\mathcal{H}/\partial x_j)}{\int_{E < \mathcal{H} < E + \mathrm{d}E} \mathrm{d}^{2f}\Omega}$$

$$= \frac{\mathrm{d}E(\partial/\partial E) \int_{\mathcal{H} < E} \mathrm{d}^{2f}\Omega \, x_i(\partial\mathcal{H}/\partial x_j)}{\mathcal{D}E \, \mathrm{d}E}.$$

Since $\partial E/\partial x_j = 0$,

$$\int_{\mathcal{H}<E} \mathrm{d}^{2f}\Omega\, x_i \frac{\partial \mathcal{H}}{\partial x_j} = \int_{\mathcal{H}<E} \mathrm{d}^{2f}\Omega\, x_i \frac{\partial(\mathcal{H}-E)}{\partial x_j}$$

$$= \int_{\mathcal{H}<E} \mathrm{d}^{2f}\Omega\, \frac{\partial}{\partial x_j}[x_i(\mathcal{H}-E)] - \delta_{ij}\int_{\mathcal{H}<E} \mathrm{d}^{2f}\Omega\,(\mathcal{H}-E).$$

The first integral on the right-hand side is zero: it becomes a surface integral over the boundary where $(\mathcal{H}-E) = 0$. Using the result in the previous equation, one obtains

$$\left\langle x_i \frac{\partial \mathcal{H}}{\partial x_j} \right\rangle = \frac{\delta_{ij}}{\mathcal{D}(E)} \frac{\partial}{\partial E}\int_{\mathcal{H}<E} \mathrm{d}^{2f}\Omega\,(E-\mathcal{H})$$

$$= \frac{\delta_{ij}}{\mathcal{D}(E)}\int_{\mathcal{H}<E} \mathrm{d}^{2f}\Omega = \frac{\delta_{ij}}{\mathcal{D}(E)}\Sigma(E)$$

$$= \delta_{ij}\frac{\Sigma(E)}{\partial\Sigma(E)/\partial E} = \delta_{ij}\left[\frac{\partial}{\partial E}\log\Sigma(E)\right]^{-1}$$

$$= \delta_{ij}\frac{k_b}{\partial S/\partial E}. \tag{A2.2.31}$$

In a microcanonical ensemble, the internal energy of a system is

$$U \equiv \langle\mathcal{H}\rangle \sim E. \tag{A2.2.32}$$

Since the temperature T relates U and S as $T = (\partial U/\partial S)_V$, it is appropriate to make the identification

$$\frac{\partial S}{\partial E} \equiv \frac{1}{T}. \tag{A2.2.33}$$

Since T is positive for systems in thermodynamic equilibrium, S and hence $\log\Sigma$ should both be monotonically increasing functions of E. This is the case as discussed above.

With this identification of T, the above result reduces to the generalized equipartition theorem:

$$\left\langle x_i \frac{\partial \mathcal{H}}{\partial x_j} \right\rangle = \delta_{ij}k_b T. \tag{A2.2.34}$$

For $i = j$ and $x_i = p_i$, one has

$$\left\langle p_i \frac{\partial \mathcal{H}}{\partial p_i} \right\rangle = k_b T \tag{A2.2.35}$$

and for $i = j$ and $x_i = q_i$,

$$\left\langle q_i \frac{\partial \mathcal{H}}{\partial q_i} \right\rangle = k_b T. \tag{A2.2.36}$$

Now since $\partial\mathcal{H}/\partial q_i = -\dot{p}_i$, one gets the virial theorem:

$$\left\langle \sum_{i=1}^{f} q_i \dot{p}_i \right\rangle = -f k_b T. \tag{A2.2.37}$$

There are many physical systems which are modelled by Hamiltonians, which can be transformed through a canonical transformation to a quadratic form:

$$\mathcal{H} = \sum_i (a_i p_i^2 + b_i q_i^2) \tag{A2.2.38}$$

where p_i and q_i are canonically conjugate variables and a_i and b_i are constants. For such a form of a Hamiltonian:

$$\sum_i \left(p_i \frac{\partial \mathcal{H}}{\partial p_i} + q_i \frac{\partial \mathcal{H}}{\partial q_i} \right) = 2\mathcal{H}. \tag{A2.2.39}$$

If f of the constants a_i and b_i are non-zero, then it follows from above that

$$\langle \mathcal{H} \rangle = \tfrac{1}{2} f k_b T. \tag{A2.2.40}$$

Each harmonic term in the Hamiltonian contributes $\tfrac{1}{2} k_b T$ to the average energy of the system, which is the theorem of the equipartition of energy. Since this is also the internal energy U of the system, one can compute the heat capacity

$$\frac{C_V}{k_b} = \frac{f}{2}. \tag{A2.2.41}$$

This is a classical result valid only at high temperatures. At low temperatures, quantum mechanical attributes of a degree of freedom can partially or fully freeze it, thereby modifying or removing its contribution to U and C_V.

A2.2.3.5 *Thermodynamics in a microcanonical ensemble: classical ideal gas*

The definition of entropy and the identification of temperature made in the last subsection provides us with a connection between the microcanonical ensemble and thermodynamics.

   A quasistatic thermodynamic process corresponds to a slow variation of E, V and N. This is performed by coupling the system to external agents. During such a process the ensemble is represented by uniformly distributed points in a region in Γ space, and this region slowly changes as the process proceeds. The change is slow enough that at every instant we have a microcanonical ensemble. Then the change in the entropy during an infinitesimal change in E, V and N during the quasistatic thermodynamic process is

$$dS(E, V, N) = \left(\frac{\partial S}{\partial E} \right)_{V,N} dE + \left(\frac{\partial S}{\partial V} \right)_{E,N} dV + \left(\frac{\partial S}{\partial N} \right)_{E,V} dN. \tag{A2.2.42}$$

The coefficient of dE is the inverse absolute temperature as identified above. We now define the pressure and chemical potential of the system as

$$P \equiv T \left(\frac{\partial S}{\partial V} \right)_{E,N} \qquad \mu \equiv -T \left(\frac{\partial S}{\partial N} \right)_{E,V}. \tag{A2.2.43}$$

Then one has

$$dS = \frac{1}{T}(dE + P\,dV - \mu\,dN) \qquad \text{or } dE = T\,dS - P\,dV + \mu\,dN. \tag{A2.2.44}$$

This is the First Law of Thermodynamics.

   The complete thermodynamics of a system can now be obtained as follows. Let the isolated system with N particles, which occupies a volume V and has an energy E within a small uncertainty dE, be modelled by a microscopic Hamiltonian $\mathcal{H}$. First, find the density of states $\mathcal{D}(E)$ from the Hamiltonian. Next, obtain the entropy as $S(E, V, N) = k_b \log \mathcal{D}(E)$ or, alternatively, by either of the other two equivalent expressions involving $\Gamma(E)$ or $\Sigma(E)$. Then, solve for E in terms of S, V and N. This is the internal energy of the system: $U(S, V, N) = E(S, V, N)$. Finally, find other thermodynamic functions as follows: the absolute temperature from $T = (\partial U / \partial S)_{V,N}$, the pressure from $P = T(\partial S / \partial V)_{E,N} = -(\partial U / \partial V)_{S,N}$, the Helmholtz free energy from $A = U - TS$, the enthalpy from $H = U + PV$, the Gibbs free energy from $G = U + PV - TS$, $\mu = G/N$ and the heat capacity at constant volume from $C_V = (\partial U/\partial T)_{V,N}$.

To illustrate, consider an ideal classical gas of N molecules occupying a volume V and each with mass M and three degrees of translational motion. The Hamiltonian is

$$\mathcal{H} = \frac{1}{2M} \sum_{i=1}^{N} p_i^2. \tag{A2.2.45}$$

Calculate the $\Sigma(E)$ first. It is

$$\Sigma(E) = [h^{3N} N!]^{-1} \int_{0 < \mathcal{H} < E} d^{6N}\Omega = \frac{1}{N!} \left(\frac{V}{h^3}\right)^N \int_{\mathcal{H} < E} d^3 p_1 \dots d^3 p_N.$$

If $P_o = (2ME)^{\frac{1}{2}}$, then the integral is the volume of a $3N$-sphere of radius P_o which is also equal to $C_{3N} P_o^{3N}$ where C_{3N} is the volume of a unit sphere in $3N$ dimensions. It can be shown that

$$C_{3N} = \frac{\pi^{3N/2}}{(3N/2)!}.$$

For large N, $N! \sim N^N e^{-N}$ and C_{3N} reduces to

$$C_{3N} = \left(\frac{2\pi}{3N}\right)^{3N/2} \exp(3N/2).$$

This gives

$$\Sigma(E) = \frac{C_{3N}}{N!} \left(\frac{V}{h^3}(2ME)^{\frac{3}{2}}\right)^N. \tag{A2.2.46}$$

Now one can use $S = k_b \log \Sigma$. Then, for large N, for entropy one obtains

$$S(E, V, N) = Nk_b \left[\frac{5}{2} + \log\left(\frac{V}{N}\left(\frac{4\pi ME}{3h^2 N}\right)^{\frac{3}{2}}\right)\right]. \tag{A2.2.47}$$

It is now easy to invert this result to obtain $E(S, V, N) \equiv U(S, V, N)$:

$$U(S, V, N) = \frac{3h^2}{4\pi} N \left(\frac{N}{V}\right)^{\frac{2}{3}} \exp\left(\frac{2}{3}\frac{S}{Nk_b} - \frac{5}{3}\right). \tag{A2.2.48}$$

As expected, S and U are extensive, i.e. are proportional to N. From U one can obtain the temperature

$$T = \left(\frac{\partial U}{\partial S}\right)_{V,N} = \frac{2}{3}\frac{U}{Nk_b}. \tag{A2.2.49}$$

From this result it follows that

$$C_V = \left(\frac{\partial U}{\partial T}\right)_{V,N} = \frac{3}{2}Nk_b. \tag{A2.2.50}$$

Finally, the equation of state is

$$P = -\left(\frac{\partial U}{\partial V}\right)_{S,N} = \frac{2}{3}\frac{U}{V} = \frac{Nk_bT}{V} \tag{A2.2.51}$$

and the chemical potential is

$$\mu = k_b T \log\left(\frac{N}{V}\left(\frac{h^2}{2\pi Mk_bT}\right)^{\frac{3}{2}}\right). \tag{A2.2.52}$$

For practical calculations, the microcanonical ensemble is not as useful as other ensembles corresponding to more commonly occurring experimental situations. Such equilibrium ensembles are considered next.

A2.2.3.6   *Interaction between systems*

Between two systems there can be a variety of interactions. Thermodynamic equilibrium of a system implies thermal, chemical and mechanical equilibria. It is therefore logical to consider, in sequence, the following interactions between two systems: *thermal contact*, which enables the two systems to share energy; *material contact*, which enables exchange of particles between them; and *pressure transmitting contact*, which allows an exchange of volume between the two systems. In each of the cases, the combined composite system is supposed to be isolated (surrounded by adiabatic walls as described in section A2.1).

In addition, there could be a mechanical or electromagnetic interaction of a system with an external entity which may do work on an otherwise isolated system. Such a contact with a work source can be represented by the Hamiltonian $\mathcal{H}(p, q, x)$ where x is the coordinate (for example, the position of a piston in a box containing a gas, or the magnetic moment if an external magnetic field is present, or the electric dipole moment in the presence of an external electric field) describing the interaction between the system and the external work source. Then the force, canonically conjugate to x, which the system exerts on the outside world is

$$X = \frac{\partial \mathcal{H}(p, q, x)}{\partial x}. \qquad (A2.2.53)$$

A thermal contact between two systems can be described in the following way. Let two systems with Hamiltonians $\mathcal{H}_\mathrm{I}$ and $\mathcal{H}_\mathrm{II}$ be in contact and interact with Hamiltonian $\mathcal{H}'$. Then the composite system (I+II) has Hamiltonian $\mathcal{H} = \mathcal{H}_\mathrm{I} + \mathcal{H}_\mathrm{II} + \mathcal{H}'$. The interaction should be weak, such that the microstate of the composite system, say l, is specified by giving the microstate l' of system I and the microstate l'' of system II, with the energy E_l of the composite system given, to a good approximation, by $E_l = E_{l'}^\mathrm{I} + E_{l''}^\mathrm{II}$ where $l = (l', l'')$. The existence of the weak interaction is supposed to allow a sufficiently frequent exchange of energy between the two systems in contact. Then, after sufficient time, one expects the composite system to reach a final state regardless of the initial states of the subsystems. In the final state, every microstate (l', l'') of the composite system will be realized with equal probability, consistent with the postulate of equal *a priori* probability. Any such final state is a state of statistical equilibrium, the corresponding ensemble of states is called a *canonical ensemble*, and corresponds to thermal equilibrium in thermodynamics. The thermal contact as described here corresponds to a diathermic wall in thermodynamics (see section A2.1).

Contacts between two systems which enable them to exchange energy (in a manner similar to thermal contact) and to exchange particles are other examples of interaction. In these cases, the microstates of the composite system can be given for the case of a weak interaction by, $(N, l) = (N', l'; N'', l'')$. The sharing of the energy and the number of particles lead to the constraints: $E_l(N) = E_{l'}^\mathrm{I}(N') + E_{l''}^\mathrm{II}(N'')$ and $N = N' + N''$. The corresponding equilibrium ensemble is called a *grand canonical ensemble*, or a T–μ ensemble.

Finally, if two systems are separated by a movable diathermic (perfectly conducting) wall, then the two systems are able to exchange energy and volume: $E_l(V) = E_{l'}^\mathrm{I}(V') + E_{l''}^\mathrm{II}(V'')$ and $V = V' + V''$. If the interaction is weak, the microstate of the composite system is $(V, l) = (V', l'; V'', l'')$, and the corresponding equilibrium ensemble is called the T–P ensemble.

A2.2.4   Canonical ensemble

Consider two systems in thermal contact as discussed above. Let the system II (with volume V_R and particles N_R) correspond to a reservoir R which is much larger than the system I (with volume V and particles N) of interest. In order to find the canonical ensemble distribution one needs to obtain the probability that the system I is in a specific microstate ν which has an energy E_ν. When the system is in this microstate, the reservoir will have the energy $E_\mathrm{R} = E_T - E_\nu$ due to the constraint that the total energy of the isolated composite system I+II is fixed and denoted by E_T; but the reservoir can be in any one of the $\Gamma_\mathrm{R}(E_T - E_\nu)$ possible states that the mechanics within the reservoir dictates. Given that the microstate of the system of interest is specified to be ν, the total number of accessible states for the composite system is clearly $\Gamma_R(E_T - E_\nu)$. Then, by the postulate of equal *a priori* probability, the probability that the system will be in state ν (denoted by P_ν) is proportional to $\Gamma_R(E_T - E_\nu)$:

$$P_\nu(E_\nu) = \frac{1}{C}\Gamma_R(E_T - E_\nu)$$

where the proportionality constant is obtained by the normalization of P_ν,

$$C = \sum_\nu \Gamma_R(E_T - E_\nu)$$

where the sum is over all microstates accessible to the system I. Thus

$$P_\nu(E_\nu) = \frac{\Gamma_R(E_T - E_\nu)}{\sum_\nu \Gamma_R(E_T - E_\nu)}$$

which can be rewritten as

$$P_\nu(E_\nu) = \frac{\exp[\log \Gamma_R(E_T - E_\nu)]}{\sum_\nu \exp[\log \Gamma_R(E_T - E_\nu)]} \equiv \frac{\exp[S_R(E_T - E_\nu)/k_b]}{\sum_\nu \exp[S_R(E_T - E_\nu)/k_b]}$$

where the following definition of statistical entropy is introduced

$$S(E, V, N) \equiv k_b \log \Gamma(E, V, N). \tag{A2.2.54}$$

Now, since the reservoir is much bigger than the system I, one expects $E_T \gg E_\nu$. Thermal equilibrium between the reservoir and the system implies that their temperatures are equal. Therefore, using the identification of T in section A2.1.4, one has

$$\frac{\partial S_R(E_R)}{\partial E_R} = \frac{\partial S_I(E_\nu)}{\partial E_\nu} = \frac{\partial S_T}{\partial E_T} = \frac{1}{T}. \tag{A2.2.55}$$

Then it is natural to use the expansion of $S_R(E_R)$ around the maximum value of the reservoir energy, E_T:

$$S_R(E_T - E_\nu) = S_R(E_T) - \frac{\partial S_R(E_T)}{\partial E_T}E_\nu + \cdots.$$

Using the leading terms in the expansion and the identification of the common temperature T, one obtains

$$S_R(E_T - E_\nu) = S_R(E_T) - E_\nu/(k_b T)$$

from which it follows that

$$P_\nu(E_\nu) = \frac{\exp(-E_\nu/(k_b T))}{\sum_\nu \exp(-E_\nu/(k_b T))}. \tag{A2.2.56}$$

Note that in this normalized probability, *the properties of the reservoir enter the result only through the common equilibrium temperature T*. The accuracy of the expansion used above can be checked by considering the next term, which is

$$\frac{1}{2}\frac{\partial^2 S}{\partial E^2}E_\nu^2.$$

Its ratio to the first term can be seen to be $(\partial T/\partial E_T)E_\nu/2T$. Since E_ν is proportional to the number of particles in the system N and E_T is proportional to the number of particles in the composite system $(N + N_R)$, the ratio of the second-order term to the first-order term is proportional to $N/(N + N_R)$. Since the reservoir is assumed to be much bigger than the system. (i.e. $N_R \gg N$) this ratio is negligible, and the truncation of the expansion is justified. The combination $1/(k_b T)$ occurs frequently and is denoted by β below.

The above derivation leads to the identification of the canonical ensemble density distribution. More generally, consider a system with volume V and N_A particles of type A, N_B particles of type B, etc., such that $N = N_A + N_B + \cdots$, and let the system be in thermal equilibrium with a much larger heat reservoir at temperature T. Then if $\mathcal{H}$ is the system Hamiltonian, the canonical distribution is (quantum mechanically)

$$\rho = \frac{\exp(-\beta \mathcal{H})}{\text{Tr}[\exp(-\beta \mathcal{H})]}. \tag{A2.2.57}$$

The corresponding classical distribution is

$$\rho(p, q) \, d^{2f}\Omega = \frac{e^{-\beta \mathcal{H}(p,q)} \, d^{2f}\Omega}{h^f N_A! N_B! \ldots Q_N} \tag{A2.2.58}$$

where f is the total number of degrees of freedom for the N-particle system and

$$Q_N(\beta, V) = \frac{1}{h^f N_A! N_B! \ldots} \int e^{-\beta \mathcal{H}(p,q)} \, d^{2f}\Omega \tag{A2.2.59}$$

which, for a one-component system, reduces to

$$Q_N(\beta, V) = \frac{1}{h^f N!} \int e^{-\beta \mathcal{H}(p,q)} \, d^{2f}\Omega \tag{A2.2.60}$$

This result is the classical analogue of

$$Q_N(\beta, V) = \sum_{\nu} \exp(-\beta E_\nu) \equiv \text{Tr}[\exp(-\beta \mathcal{H})]. \tag{A2.2.61}$$

$Q_N(\beta, V)$ is called the canonical partition function, and plays a central role in determining the thermodynamic behaviour of the system. The constants in front of the integral in (A2.2.59) and (A2.2.60) can be understood in terms of the uncertainty principle and indistinguishability of particles, as was discussed earlier in section A2.2.3.1 while obtaining (A2.2.18). Later, in section A2.2.5.5, the classical limit of an ideal quantum gas is considered, which also leads to a similar understanding of these multiplicative constants, which arise on account of overcounting of microstates in classical mechanics.

The canonical distribution corresponds to the probability density for the system to be in a specific microstate with energy $E \sim \mathcal{H}$; from it one can also obtain the probability $\mathcal{P}(E)$ that the system has an energy between E and $E + dE$ if the density of states $\mathcal{D}(E)$ is known. This is because, classically,

$$[h^f N_A! N_B! \ldots]^{-1} \int_{E < \mathcal{H} < E+dE} d^{2f}\Omega = \mathcal{D}(E) \, dE \tag{A2.2.62}$$

and, quantum mechanically, the sum over the degenerate states with $E < \mathcal{H} < E + dE$ also yields the extra factor $\mathcal{D}(E) \, dE$. The result is

$$\mathcal{P}(E) \, d(E) = [Q_N]^{-1} e^{-\beta E} \mathcal{D}(E) \, dE. \tag{A2.2.63}$$

Then, the partition function can also be rewritten, as

$$Q_N = \int e^{-\beta E} \mathcal{D}(E) \, dE. \tag{A2.2.64}$$

A2.2.4.1 Thermodynamics in a canonical ensemble

In the microcanonical ensemble, one has specified $E \sim U(S, V, N)$ and T, P and μ are among the derived quantities. In the canonical ensemble, the system is held at fixed T, and the change of a thermodynamic variable from S in a microcanonical ensemble to T in a canonical ensemble is achieved by replacing the internal energy $U(S, V, N)$ by the Helmholtz free energy $A(T, V, N) \equiv (U - TS)$. The First Law statement for dU, equation (A2.2.44) now leads to

$$dA = \mu \, dN - P \, dV - S \, dT. \tag{A2.2.65}$$

If one denotes the averages over a canonical distribution by $\langle \cdots \rangle$, then the relation $A = U - TS$ and $U = \langle \mathcal{H} \rangle$ leads to the statistical mechanical connection to the thermodynamic free energy A:

$$A = -k_\mathrm{b} T \log Q_N. \tag{A2.2.66}$$

To see this, note that $S = -k_\mathrm{b} \langle \log \rho \rangle$. Thus

$$S = -k_\mathrm{b} \langle \log(Q_N^{-1} \, e^{-\beta \mathcal{H}}) \rangle = k_\mathrm{b} \log Q_N + T^{-1} \langle \mathcal{H} \rangle = k_\mathrm{b} \log Q_N + T^{-1} U$$

which gives the result $A = U - TS = -k_\mathrm{b} T \log Q_N$. For any canonical ensemble system, its thermodynamic properties can be found once its partition function is obtained from the system Hamiltonian. The sequence can be

$$\mathcal{H} \to Q_N \to A \to (\mu, P, S) \tag{A2.2.67}$$

where the last connection is obtained from the differential relations

$$\mu = \left(\frac{\partial A}{\partial N} \right)_{V,T} \qquad P = \left(\frac{\partial A}{\partial V} \right)_{N,T} \qquad S = \left(\frac{\partial A}{\partial T} \right)_{N,V}. \tag{A2.2.68}$$

One can trivially obtain the other thermodynamic potentials U, H and G from the above. It is also interesting to note that the internal energy U and the heat capacity $C_{V,N}$ can be obtained directly from the partition function. Since $Q_N(\beta, V) = \sum_\nu \exp(-\beta E_\nu)$, one has

$$
\begin{aligned}
U \equiv \langle E_\nu \rangle &= \frac{\sum_\nu E_\nu \exp(-\beta E_\nu)}{\sum_\nu \exp(-\beta E_\nu)} \\
&= -\frac{\partial}{\partial \beta} \log Q_N(\beta, V) = \frac{\partial}{\partial \beta} (\beta A).
\end{aligned} \tag{A2.2.69}
$$

Fluctuations in energy are related to the heat capacity $C_{V,N}$, and can be obtained by twice differentiating $\log Q_N$ with respect to β, and using equation (A2.2.69):

$$
\begin{aligned}
\langle (E_\nu - \langle E \rangle)^2 \rangle &= \langle E_\nu^2 \rangle - \langle E \rangle^2 \\
&= -\frac{\partial U}{\partial \beta} = k_\mathrm{b} T^2 \frac{\partial U}{\partial T} = k_\mathrm{b} T^2 C_{V,N}.
\end{aligned} \tag{A2.2.70}
$$

Both $\langle E \rangle$ and $C_{V,N}$ are extensive quantities and proportional to N or the system size. The root mean square fluctuation in energy is therefore proportional to $N^{\frac{1}{2}}$, and the relative fluctuation in energy is

$$\frac{\langle (E_\nu - \langle E \rangle)^2 \rangle^{\frac{1}{2}}}{\langle E \rangle} \sim \frac{1}{N^{\frac{1}{2}}}. \tag{A2.2.71}$$

This behaviour is characteristic of thermodynamic fluctuations. This behaviour also implies the equivalence of various ensembles in the thermodynamic limit. Specifically, as $N \to \infty$ the energy fluctuations vanish, the partition of energy between the system and the reservoir becomes uniquely defined and the thermodynamic properties in microcanonical and canonical ensembles become identical.

A2.2.4.2   Expansion in powers of $\hbar$

In the relation (A2.2.66), one can use the partition function evaluated using either (A2.2.59) or (A2.2.61). The use of (A2.2.59) gives the first term in an expansion of the quantum mechanical A in powers of $\hbar$ in the quasi-classical limit. In this section the next non-zero term in this expansion is evaluated. For this consider the partition function (A2.2.61). The trace of $\exp(-\beta\mathcal{H})$ can be obtained using the wavefunctions of free motion of the ideal gas of N particles in volume V:

$$\psi_p = V^{-N/2}\, e^{(\mathrm{i}/\hbar)\sum_j p_j q_j} \tag{A2.2.72}$$

where q_j are the coordinates and $p_j = \hbar k_j$ are the corresponding momenta of the N particles, whose $3N$ degrees of freedom are labelled by the suffix j. The particles may be identical (with same mass M) or different. For identical particles, the wavefunctions above have to be made symmetrical or antisymmetrical in the corresponding $\{q_i\}$ depending on the statistics obeyed by the particles. This effect, however, leads to exponentially small correction in A and can be neglected. The other consequence of the indistinguishability of particles is in the manner of how the momentum sums are done. This produces a correction which is third order in $\hbar$, obtained in section A2.2.5.5, and does not affect the $O(\hbar^2)$ term that is calculated here. In each of the wavefunctions ψ, the momenta p_j are definite constants and form a dense discrete set with spacing between the neighbouring p_j proportional to V^{-1}. Thus, the summation of the matrix elements $\langle\psi_p|\exp(-\beta\mathcal{H})|\psi_p\rangle$ with respect to all p_j can be replaced by an integration:

$$Q_N(\beta, V) = \mathrm{Tr}\langle\psi_p|\exp(-\beta\mathcal{H})|\psi_p\rangle \tag{A2.2.73}$$

$$= \frac{1}{h^{3N} N_A! N_B! \ldots} \int \mathrm{d}^{3N}p\, \mathrm{d}^{3N}q\, I \tag{A2.2.74}$$

where

$$I = e^{-(\mathrm{i}/\hbar)\sum_j p_j q_j}\, e^{-\beta\mathcal{H}}\, e^{(\mathrm{i}/\hbar)\sum_j p_j q_j}.$$

When $\beta = 0$, $I = 1$. For systems in which the Hamiltonian $\mathcal{H}$ can be written as

$$\mathcal{H} = \sum_j \frac{p_j^2}{2M_j} + U = -\frac{1}{2}\hbar^2 \sum_j \frac{1}{M_j}\frac{\partial^2}{\partial q_j^2} + U \tag{A2.2.75}$$

with $U = U(\{q_j\})$ as the potential energy of interaction between N particles, the integral I can be evaluated by considering its derivative with respect to β, (note that the operator $\mathcal{H}$ will act on all factors to its right):

$$\frac{\partial I}{\partial \beta} = -e^{-(\mathrm{i}/\hbar)\sum_j p_j q_j}\mathcal{H}\{e^{(\mathrm{i}/\hbar)\sum_j p_j q_j} I\} \tag{A2.2.76}$$

$$= -E(p, q)I + \sum_j \frac{\hbar^2}{2M_j}\left(\frac{2\mathrm{i}}{\hbar}p_j\frac{\partial I}{\partial q_j} + \frac{\partial^2 I}{\partial q_j^2}\right) \tag{A2.2.77}$$

where $E(p, q) = (\sum_j (p_j^2/2M_j) + U)$ is the classical form of the energy. By using the substitution, $I = \exp(-\beta E(p, q))\chi$ and expanding $\chi = 1 + \hbar\chi_1 + \hbar^2\chi_2 + \cdots$, one can obtain the quantum corrections to the classical partition function. Since for $\beta = 0$, $I = 1$, one also has for $\beta = 0$, $\chi = 1$, and $\chi_1 = \chi_2 = 0$. With this boundary condition, one obtains the result that

$$\chi_1 = -\frac{1}{2}\mathrm{i}\beta^2 \sum_j \frac{p_j}{M_j}\frac{\partial U}{\partial q_j}$$

and

$$\chi_2 = -\frac{1}{8}\beta^4 \left(\sum_j \frac{p_j}{M_j} \frac{\partial U}{\partial q_j} \right)^2 + \frac{1}{6}\beta^3 \sum_j \sum_k \frac{p_j}{M_j} \frac{p_k}{M_k} \frac{\partial^2 U}{\partial q_j \partial q_k} + \frac{1}{6}\beta^3 \sum_j \frac{1}{M_j} \left(\frac{\partial U}{\partial q_j} \right)^2 - \frac{1}{4}\beta^2 \sum_j \frac{1}{M_j} \frac{\partial^2 U}{\partial q_j^2}.$$

For the partition function, the contribution from χ_1, which is the first-order correction in $\hbar$, vanishes identically. One obtains

$$Q_N(\beta, V) = Q_N^{\text{cl}}(1 + \hbar^2 \langle \chi_2 \rangle^{\text{cl}} + \cdots) \tag{A2.2.78}$$

where the superscript (cl) corresponds to the classical value, and $\langle \chi_2 \rangle^{\text{cl}}$ is the classical canonical ensemble average of χ_2. The free energy A can then be inferred as

$$A = A^{\text{cl}} - \beta^{-1} \log(1 + \hbar^2 \langle \chi_2 \rangle^{\text{cl}} + \cdots) \tag{A2.2.79}$$

$$\approx A^{\text{cl}} - \beta^{-1}\hbar^2 \langle \chi_2 \rangle^{\text{cl}}. \tag{A2.2.80}$$

One can formally evaluate $\langle \chi_2 \rangle^{\text{cl}}$. Since $\langle p_j p_k \rangle = M_j \beta^{-1} \delta_{jk}$, one obtains

$$\langle \chi_2 \rangle^{\text{cl}} = \frac{\beta^3}{24} \sum_j \frac{1}{M_j} \left\langle \left(\frac{\partial U}{\partial q_j} \right)^2 \right\rangle - \frac{\beta^2}{12} \sum_j \frac{1}{M_j} \left\langle \frac{\partial^2 U}{\partial q_j^2} \right\rangle. \tag{A2.2.81}$$

This can be further simplified by noting that

$$\int \frac{\partial^2 U}{\partial q_j^2} \mathrm{e}^{-\beta U} \, \mathrm{d}q_j = \frac{\partial U}{\partial q_j} \mathrm{e}^{-\beta U} + \beta \int \left(\frac{\partial U}{\partial q_j} \right)^2 \mathrm{e}^{-\beta U} \, \mathrm{d}q_j \tag{A2.2.82}$$

which implies that

$$\left\langle \frac{\partial^2 U}{\partial q_j^2} \right\rangle = \beta \left\langle \left(\frac{\partial U}{\partial q_j} \right)^2 \right\rangle. \tag{A2.2.83}$$

It follows that

$$\langle \chi_2 \rangle^{\text{cl}} = -\frac{\beta^3}{24} \sum_j \frac{1}{M_j} \left\langle \left(\frac{\partial U}{\partial q_j} \right)^2 \right\rangle \tag{A2.2.84}$$

with the end result that

$$Q_N(\beta, V) = Q_N^{\text{cl}} \left(1 - \hbar^2 \frac{\beta^3}{24} \sum_j \frac{1}{M_j} \left\langle \left(\frac{\partial U}{\partial q_j} \right)^2 \right\rangle \right) \tag{A2.2.85}$$

and

$$A = A^{\text{cl}} + \hbar^2 \frac{\beta^2}{24} \sum_j \frac{1}{M_j} \left\langle \left(\frac{\partial U}{\partial q_j} \right)^2 \right\rangle. \tag{A2.2.86}$$

The leading order quantum correction to the classical free energy is always positive, is proportional to the sum of mean square forces acting on the particles and decreases with either increasing particle mass or increasing temperature. The next term in this expansion is of order $\hbar^4$. This feature enables one to independently calculate the leading correction due to quantum statistics, which is $O(\hbar^3)$. The result calculated in section A2.2.5.5 is

$$A_3 = \pm \frac{\pi^{\frac{3}{2}}}{2\gamma} \frac{N^2 \beta^{\frac{1}{2}} \hbar^3}{V M^{\frac{3}{2}}} \tag{A2.2.87}$$

for an ideal quantum gas of N identical particles. The upper sign is for Fermi statistics, the lower is for Bose statistics and γ is the degeneracy factor due to nuclear and electron spins.

In the following three subsections, the three examples described in A2.2.2 are considered. In each case the model system is thermal equilibrium with a large reservoir at temperature $T = (k_b\beta)^{-1}$. Then the partition function for each system is evaluated and its consequences for the thermodynamic behaviour of the model system are explored.

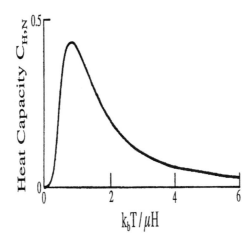

Figure A2.2.1. Heat capacity of a two-state system as a function of the dimensionless temperature, $k_b T/(\mu H)$.

A2.2.4.3   Application to ideal systems: two-state model

Let us consider first the two-state model of non-interacting spin-$\frac{1}{2}$ particles in a magnetic field. For a system with only one such particle there are two non-degenerate energy levels with energies $\pm \mu H$, and the partition function is $Q_1 = \exp(-\beta \mu H) + \exp(\beta \mu H) = 2\cosh(\beta \mu H)$. For N such indistinguishable spin-$\frac{1}{2}$ particles, the canonical partition function is

$$Q_N = \frac{(Q_1)^N}{N!} = \frac{2^N}{N!}(\cosh(\beta \mu H))^N$$

The internal energy is

$$U = -\frac{\partial \log Q_N}{\partial \beta} = -N\mu H \tanh(\beta \mu H).$$

If H is ∞ (very large) or T is zero, the system is in the lowest possible and a non-degenerate energy state and $U = -N\mu H$. If either H or β is zero, then $U = 0$, corresponding to an equal number of spins up and down. There is a symmetry between the positive and negative values of $\beta \mu H$, but negative β values do not correspond to thermodynamic equilibrium states. The heat capacity is

$$C_{H,N} = -\frac{1}{k_b T^2}\left(\frac{\partial U}{\partial \beta}\right)_{H,N} = Nk_b((\beta \mu H)\mathrm{sech}(\beta \mu H))^2.$$

Figure A2.2.1 shows $C_{H,N}$, is units of Nk_b, as a function of $(\beta \mu H)$. $C_{H,N}$ is zero in the two limits of zero and infinite values of $(\beta \mu H)$, which also implies the limits of $T = \infty$ and $T = 0$. For small $(\beta \mu H)$, it approaches zero as $\sim(\beta \mu H)^2$ and for large $(\beta \mu H)$ as $(\beta \mu H)^2 \exp(-2\beta \mu H)$. It has a maximum value of $0.439Nk_b$ around $\beta \mu H = 1.2$. This behaviour is characteristic of any two-state system, and the maximum in the heat capacity is called a Schottky anomaly.

From the partition function, one also finds the Helmholtz free energy as

$$\beta F = -\log Q_N = -N[\log(2) + \log(N!) + \log(\cosh(\beta \mu H))].$$

One can next obtain the entropy either from $S = (U - F)/T$ or from $S = -(\partial F/\partial T)_{V,N,H}$, and one can verify that the result is the same.

It is also instructive to start from the expression for entropy $S = k_b \log(g(N, m))$ for a specific energy partition between the two-state system and the reservoir. Using the result for $g(N, m)$ in section A2.2.2, and noting that $E = -(2\mu H)m$, one gets (using the Stirling approximation $N! \approx (2\pi N)^{\frac{1}{2}} N^N e^{-N}$),

$$k_b^{-1} S = -N\left[\left(\frac{1}{2} + \frac{m}{N}\right) \log\left(\frac{1}{2} + \frac{m}{N}\right) + \left(\frac{1}{2} - \frac{m}{N}\right) \log\left(\frac{1}{2} - \frac{m}{N}\right)\right].$$

Since $E = -(2\mu H)m$, a given spin excess value $2m$ implies a given energy partition. The free energy for such a specific energy partition is

$$\beta F = N\left[-(2\beta\mu H)\frac{m}{N} + \left(\frac{1}{2} + \frac{m}{N}\right) \log\left(\frac{1}{2} + \frac{m}{N}\right) + \left(\frac{1}{2} - \frac{m}{N}\right) \log\left(\frac{1}{2} - \frac{m}{N}\right)\right].$$

This has to be minimized with respect to E or equivalently m/N to obtain the thermal equilibrium result. The value of m/N that corresponds to equilibrium is found to be

$$\log\left(\frac{N + 2m}{N - 2m}\right) = 2\beta\mu H$$

which corresponds to $\langle 2m \rangle = N \tanh(\beta\mu H)$ and leads to the same U as above. It also gives the equilibrium magnetization as $\mathcal{M} = N\mu \tanh(\beta\mu H)$.

A2.2.4.4   Application to ideal systems: classical ideal gas

Consider a system of N non-interacting point particles in a three-dimensional cubical box of volume $V = L^3$. First consider one classical particle with energy $E = p^2/(2M)$. The partition function is

$$\begin{aligned}
Q_1 &= h^{-3} \int_V dV \int_{-\infty}^{+\infty} dp_x \int_{-\infty}^{+\infty} dp_y \int_{-\infty}^{+\infty} dp_z \, e^{-(p_x^2 + p_y^2 + p_z^2)/(2Mk_bT)} \\
&= \frac{V}{h^3}\left(\int_{-\infty}^{+\infty} dp_x \, e^{-p_z^2/(2Mk_bT)}\right)^3 \\
&= \frac{V}{h^3}(2\pi Mk_bT)^{\frac{3}{2}} = V\left(\frac{2\pi Mk_bT}{h^2}\right)^{\frac{3}{2}} \equiv \frac{V}{V_q}
\end{aligned} \tag{A2.2.88}$$

where the definition of the quantum volume V_q associated with the thermal deBroglie wavelength, $\lambda_T \sim h/(2\pi Mk_bT)^{\frac{1}{2}}$, is introduced. The same result is obtained using the density of states $\mathcal{D}(\epsilon)$ obtained for this case in section A2.2.2. Even though this $\mathcal{D}(\epsilon)$ was obtained using quantum considerations, the sum over n was replaced by an integral which is an approximation that is valid when k_bT is large compared to energy level spacing. This high-temperature approximation leads to a classical behaviour.

For an ideal gas of N indistinguishable point particles one has $Q_N = Q_1^N/N! = (V/V_q)^N/N!$. For large N one can again use the Stirling approximation for $N!$ and obtain the Helmholtz free energy

$$F = -k_bT \log Q_N = Nk_bT \log\left(e\frac{N}{V}V_q\right) = Nk_bT \log\left(e\frac{N}{V}\left(\frac{2\pi Mk_bT}{h^2}\right)^{-\frac{3}{2}}\right).$$

(The term $\frac{1}{2}\log(2\pi N)k_bT$ is negligible compared to terms proportional to Nk_bT.) The entropy obtained from the relation $S = -(\partial F/\partial T)_{N,V}$ agrees with the expression, equation (A2.2.47), obtained for the microcanonical ensemble, and one also obtains $U = F + TS = \frac{3}{2}Nk_bT$ consistent with the equipartition law. The ideal equation of state $P = Nk_bT/V$ is also obtained from evaluating $P = -(\partial F/\partial V)_{N,T}$. Thus one obtains the same thermodynamic behaviour from the canonical and microcanonical ensembles. This is generally the case when N is very large since the fluctuations around the average behave as $N^{-\frac{1}{2}}$. A quantum ideal gas with either Fermi or Bose statistics is treated in subsections A2.2.5.4 to A2.2.5.7.

A2.2.4.5   Ideal gas of diatomic molecules

Consider a gas of N non-interacting diatomic molecules moving in a three-dimensional system of volume V. Classically, the motion of a diatomic molecule has six degrees of freedom—three translational degrees corresponding to the centre of mass motion, two more for the rotational motion about the centre of mass and one additional degree for the vibrational motion about the centre of mass. The equipartition law gives $\langle E_{\text{trans}} \rangle = \frac{3}{2} N k_b T$. In a similar manner, since the rotational Hamiltonian has rotational kinetic energy from two orthogonal angular momentum components, in directions each perpendicular to the molecular axis, equipartition gives $\langle E_{\text{rot}} \rangle = N k_b T$. For a rigid dumb-bell model, one would then get $\langle E_{\text{total}} \rangle = \frac{5}{2} N k_b T$, since no vibration occurs in a rigid dumb-bell. The corresponding heat capacity per mole (where $N = N_a$ is the Avogadro's number and $R = N_a k_b$ is the gas constant), is $C_v = \frac{5}{2} R$ and $C_p = \frac{7}{2} R$. If one has a vibrating dumb-bell, the additional vibrational motion has two quadratic terms in the associated Hamiltonian—one for the kinetic energy of vibration and another for the potential energy as in a harmonic oscillator. The vibrational motion thus gives an additional $\langle E_{\text{vib}} \rangle = N k_b T$ from the equipartition law, which leads to $\langle E_{\text{total}} \rangle = \frac{7}{2} N k_b T$ and heat capacities per mole as $C_v = \frac{7}{2} R$ and $C_p = \frac{9}{2} R$.

These results do not agree with experimental results. At room temperature, while the translational motion of diatomic molecules may be treated classically, the rotation and vibration have quantum attributes. In addition, quantum mechanically one should also consider the electronic degrees of freedom. However, typical electronic excitation energies are very large compared to $k_b T$ (they are of the order of a few electronvolts, and 1 eV corresponds to $T \approx 10\,000$ K). Such internal degrees of freedom are considered frozen, and an electronic cloud in a diatomic molecule is assumed to be in its ground state ϵ_0 with degeneracy g_0. The two nuclei A and B, which along with the electronic cloud make up the molecule, have spins I_A and I_B, and the associated degeneracies $(2I_A + 1)$ and $(2I_B + 1)$, respectively. If the molecule is homonuclear, A and B are indistinguishable and, by interchanging the two nuclei, but keeping all else the same, one obtains the same configuration. Thus for a homonuclear molecule, the configurations can be overcounted by a factor of two if the counting scheme used is the same as that for heteronuclear molecules. Thus, the degeneracy factor in counting the internal states of a diatomic molecule is $g = g_0 (2I_A + 1)(2I_B + 1)/(1 + \delta_{AB})$ where δ_{AB} is zero for the heteronuclear case and one for the homonuclear case.

The energy of a diatomic molecule can be divided into translational and internal contributions: $\epsilon_j = (\hbar k)^2/(2M) + \epsilon_{\text{int}}$, and $\epsilon_{\text{int}} = \epsilon_0 + \epsilon_{\text{rot}} + \epsilon_{\text{vib}}$. In the canonical ensemble for an ideal gas of diatomic molecules in thermal equilibrium at temperature $T = (k_b \beta)^{-1}$ the partition function then factorizes:

$$Q_N = (N!)^{-1} [Q_{\text{trans}}]^N [Q_{\text{int}}]^N$$

where the single molecule translational partition function Q_{trans} is the same as Q_1 in equation (A2.2.88) and the single-molecule internal partition function is

$$Q_{\text{int}} = g\, e^{-\beta \epsilon_0} Q_{\text{rot}} Q_{\text{vib}}.$$

The rotational and vibrational motions of the nuclei are uncoupled, to a good approximation, on account of a mismatch in time scales, with vibrations being much faster than the rotations (electronic motions are even faster than the vibrational ones). One typically models these as a rigid rotation plus a harmonic oscillation, and obtains the energy eigenstates for such a model diatomic molecule. The resulting vibrational states are non-degenerate, are characterized by a vibrational quantum number $v = 0, 1, 2, \ldots$ and with an energy $\epsilon_{\text{vib}} \equiv \epsilon_v = (\frac{1}{2} + v)\hbar\omega_0$ where ω_0 is the characteristic vibrational frequency. Thus

$$Q_{\text{vib}} = \sum_{J=0}^{\infty} e^{-\beta\hbar\omega_0(\frac{1}{2}+v)} = [e^{\frac{1}{2}\beta\hbar\omega_0} - e^{\frac{1}{2}\beta\hbar\omega_0}]^{-1}.$$

The rotational states are characterized by a quantum number $J = 0, 1, 2, \ldots$ are degenerate with degeneracy $(2J + 1)$ and have energy $\epsilon_{\text{rot}} \equiv \epsilon_J = J(J + 1)\hbar^2/(2I_o)$ where I_o is the molecular moment of inertia. Thus

$$Q_{\text{rot}} = \sum_{J=0}^{\infty} (2J + 1)\, e^{-J(J+1)\theta_r/T}$$

where $\theta_r = \hbar^2/(2I_o k_b)$. If the spacing between the rotational levels is small compared to $k_b T$, i.e. if $T \gg \theta_r$, the sum can be replaced by an integral (this is appropriate for heavy molecules and is a good approximation for molecules other than hydrogen):

$$Q_{\text{rot}} \approx \int_{J=0}^{\infty} dJ\, (2J + 1)\, e^{-J(J+1)\theta_r/T} = \frac{T}{\theta_r}$$

which is the high-temperature, or classical, limit. A better evaluation of the sum is obtained with the use of the Euler–Maclaurin formula:

$$\sum_{J=0}^{\infty} f(J) = \int_0^{\infty} dJ\, f(J) + \frac{1}{2} f(0) - \frac{1}{12} f'(0) + \frac{1}{720} f'''(0) - \frac{1}{30240} f^v(0) + \cdots.$$

Putting $f(J) = (2J + 1)\exp(-J(J + 1)\theta_r/T)$, one obtains

$$Q_{\text{rot}} \approx \frac{T}{\theta_r} + \frac{1}{3} + \frac{1}{15}\frac{\theta_r}{T} + \frac{4}{315}\left(\frac{\theta_r}{T}\right)^2 + \cdots.$$

If $T \ll \theta_r$, then only a first few terms in the sum need to be retained:

$$Q_{\text{rot}} \approx 1 + 3\, e^{-2\theta_r/T} + 5\, e^{-6\theta_r/T} + \cdots.$$

Once the partition function is evaluated, the contributions of the internal motion to thermodynamics can be evaluated. Q_{int} depends only on T, and has no effect on the pressure. Its effect on the heat capacity C_v can be obtained from the general expression $C_v = (k_b T^2)^{-1}(\partial^2 \log Q_N/\partial \beta^2)$. Since the partition function factorizes, its logarithm and, hence, the heat capacity, reduces to additive contributions from translational, rotational and vibrational contributions: $C_v = C_v^{\text{trans}} + C_v^{\text{rot}} + C_v^{\text{vib}}$, where the translational motion (treated classically) yields $C_v^{\text{trans}} = \frac{3}{2}N k_b T$. The rotational part at high temperatures gives

$$C_v^{\text{rot}} = N k_b \left(1 + \frac{1}{45}\left(\frac{\theta_r}{T}\right)^2 + \frac{16}{945}\left(\frac{\theta_r}{T}\right)^3 + \cdots\right)$$

which shows that C_v^{rot} decreases at high T, reaching the classical equipartition value from above at $T = \infty$. At low temperatures,

$$C_v^{\text{rot}} \approx 12 N k_b \left(\frac{\theta_r}{T}\right)^2 e^{-2\theta_r/T}$$

so that as $T \to 0$, C_v^{rot} drops to zero exponentially. The vibrational contribution C_v^{vib} is given by

$$C_v^{\text{vib}} = N k_b \left(\frac{\theta_v}{T}\right)^2 \frac{e^{\theta_v/T}}{(e^{\theta_v/T} - 1)^2} \qquad \text{where } \theta_v = \frac{\hbar\omega_o}{k_b}.$$

For $T \gg \theta_v$, C_v^{vib} is very nearly $N k_b$, the equipartition value, and for $T \ll \theta_v$, C_v^{vib} tends to zero as $(\theta_v/T)^2 \exp(-\theta_v/T)$. For most diatomic molecules θ_v is of the order of 1000 K and θ_r is less than 100 K. For HCl, $\theta_r = 15$ K; for N_2, O_2 and NO it is between 2 and 3 K; for H_2, D_2 and HD it is, respectively, 85, 43 and 64 K. Thus, at room temperature, the rotational contribution could be nearly $N k_b$ and the vibrational contribution could be only a few per cent of the equipartition value. Figure A2.2.2 shows the temperature dependence of C_p for HD, HT and DT, various isotopes of the hydrogen molecule.

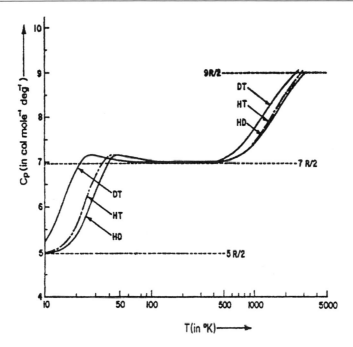

Figure A2.2.2. The rotational–vibrational specific heat, C_p, of the diatomic gases HD, HT and DT as a function of temperature. From *Statistical Mechanics* by Raj Pathria. Reprinted by permission of Butterworth Heinemann.

A2.2.4.6 Application to ideal systems: black body radiation

This subsection, and the next, deals with a system of N non-interacting harmonic oscillators.

Electromagnetic radiation in thermal equilibrium within a cavity is often approximately referred to as the black-body radiation. A classical black hole is an ideal black body. Our own star, the Sun, is pretty black! A perfect black body absorbs all radiation that falls onto it. By Kirchhoff's law, which states that 'a body must emit at the same rate as it absorbs radiation if equilibrium is to be maintained', the emissivity of a black body is highest. As shown below, the use of classical statistical mechanics leads to an infinite emissivity from a black body. Planck quantized the standing wave modes of the electromagnetic radiation within a black-body cavity and solved this anomaly. He considered the distribution of energy U among N oscillators of frequency ω. If U is viewed as divisible without limit, then an infinite number of distributions are possible. Planck considered 'U as made up of an entirely determined number of finite equal parts' of value $\hbar\omega$. This quantization of the electromagnetic radiation leads to the concept of *photons* of energy quanta $\hbar\omega$, each of which having a Hamiltonian of the form of a harmonic oscillator. A state of the free electromagnetic field is specified by the number, n, for each of such oscillators and n then corresponds to the number of photons in a state with energy $\hbar\omega$. Photons obey Bose–Einstein statistics. Denote by n_j the number of photons with energy $\epsilon_j \equiv \hbar\omega_j$. Then $n_j = 0, 1, 2, \ldots$ and the canonical partition function is

$$Q = \sum_\nu e^{-\beta E_\nu} = \sum_{n_1=0}^{\infty} \sum_{n_2=0}^{\infty} \cdots \sum_{n_j=0}^{\infty} \cdots e^{-\beta(n_1\epsilon_1 + n_2\epsilon_2 + \cdots + n_j\epsilon_j + \cdots)}.$$

Here the zero point energy is temporarily suppressed. Now the exponential is a product of independent factors.

Thus one gets

$$Q = \prod_j \left(\sum_{n_j=0}^{\infty} e^{-\beta n_j \epsilon_j} \right) = \prod_j \left(\frac{1}{1 - e^{-\beta \epsilon_j}} \right)$$

on account of the geometric nature of the series being summed. One should note that photons are massless and their total number is indeterminate. Since $\log Q = -\beta A$, one can obtain various properties of the photon gas. Specifically consider the average occupation number of the jth state:

$$
\begin{aligned}
\langle n_j \rangle &= \frac{\sum_\nu n_j e^{-\beta E_\nu}}{\sum_\nu e^{-\beta E_\nu}} = \frac{\sum_{n_1,n_2,...} n_j e^{-\beta(n_1\epsilon_1 + \cdots + n_j\epsilon_j + \cdots)}}{Q} \\
&= \frac{\partial \log Q}{\partial(-\beta\epsilon_j)} = \frac{\partial}{\partial(-\beta\epsilon_j)} \left\{ \sum_j [-\log(1 - e^{-\beta\epsilon_j})] \right\} \\
&= \frac{1}{e^{\beta\epsilon_j} - 1} \equiv \frac{1}{e^{\beta\hbar\omega_j} - 1}.
\end{aligned}
$$ (A2.2.89)

This is the Planck distribution function. The thermal average energy in the jth mode is (including the zero point energy)

$$\langle \epsilon_j \rangle = \frac{1}{2}\hbar\omega_j + \frac{\hbar\omega_j}{e^{\beta\hbar\omega_j} - 1}.$$

Since for small $\beta\hbar\omega_j = y$, $(\exp(y) - 1)^{-1} \approx [y(1 + y/2 + \cdots)]^{-1} \approx y^{-1}(1 - y/2) = (y^{-1} - 1/2)$, one obtains, when $\epsilon_j \ll k_b T$, the result for the high-temperature limit: $\langle \epsilon_j \rangle \to k_b T$. This is also the average energy for a classical harmonic oscillator with two quadratic degrees of freedom (one kinetic and one potential) in the Hamiltonian, an equipartition result. For low temperatures one has $\epsilon_j \gg k_b T$ and $\langle \epsilon_j \rangle \to (\frac{1}{2}\hbar\omega_j + \hbar\omega_j e^{-\beta\hbar\omega_j})$. The oscillator settles down in the ground state at zero temperature.

Any cavity contains an infinite number of electromagnetic modes. For radiation confined to a perfectly conducting cubical cavity of volume $V = L^3$, the modes are given by the electric field components of the form:

$$E_x = E_{x0} \sin \omega t \cos(n_x\pi x/L) \sin(n_y\pi x/L) \sin(n_z\pi x/L)$$

$$E_y = E_{y0} \sin \omega t \sin(n_x\pi x/L) \cos(n_y\pi x/L) \sin(n_z\pi x/L)$$

$$E_z = E_{z0} \sin \omega t \sin(n_x\pi x/L) \sin(n_y\pi x/L) \cos(n_z\pi x/L).$$

Within the cavity $\vec{\nabla} \cdot \vec{E} = 0$, which in Fourier space is $\vec{k} \cdot \vec{E} = 0$. Thus, only two of the three components of $\vec{E}$ are independent. The electromagnetic field in a cavity is a transversely polarized field with two independent polarization directions, which are mutually perpendicular and are each normal to the propagation direction $\vec{k}$ of the $\vec{E}$ field, which satisfies the electromagnetic wave equation, $c^2\nabla^2\vec{E} = \partial^2\vec{E}/\partial t^2$. Substituting the form of the $\vec{E}$ field above, one gets

$$c^2\pi^2 n^2 = \omega^2 L^2 \qquad \text{where } n = (n_x^2 + n_y^2 + n_z^2)^{\frac{1}{2}}$$

so that the quantized photon modes have frequencies of the form $\omega_n = n\pi c/L$. The total energy of the photons in the cavity is then

$$U = \sum_n \langle \epsilon_n \rangle = \sum_n \frac{\hbar\omega_n}{e^{\beta\hbar\omega_n} - 1}.$$

Here the zero point energy is ignored, which is appropriate at reasonably large temperatures when the average occupation number is large. In such a case one can also replace the sum over n by an integral. Each of the triplet (n_x, n_y, n_z) can take the values $0, 1, 2, \ldots, \infty$. Thus the sum over (n_x, n_y, n_z) can be replaced by an integral over the volume element $dn_x\, dn_y\, dn_z$ which is equivalent to an integral in the positive octant of the three-dimensional n-space. Since there are two independent polarizations for each triplet (n_x, n_y, n_z), one has

$$\sum_n (\cdots) = 2\frac{1}{8} \int_0^\infty 4\pi n^2\, dn (\cdots).$$

Then

$$U = \pi \int_0^\infty dn\, n^2 \frac{\hbar\omega_n}{e^{\beta\hbar\omega_n} - 1} = V \frac{\hbar}{\pi^2 c^3} \int_0^\infty d\omega \frac{\omega^3}{e^{\beta\hbar\omega} - 1} \equiv V \int d\omega\, u_\omega.$$

Since $\int_0^\infty dx\, x^3/(e^x - 1) = \pi^4/15$, one obtains the result for the energy per unit volume as

$$\frac{U}{V} = \frac{\pi^2 k_b^4}{15\hbar^3 c^3} T^4. \tag{A2.2.90}$$

This is known as the Stefan–Boltzmann law of radiation. If in this calculation of total energy U one uses the classical equipartition result $\langle \epsilon_n \rangle = k_b T$, one encounters the integral $\int_0^\infty d\omega\, \omega^2$ which is infinite. This divergence, which is the Rayleigh–Jeans result, was one of the historical results which collectively led to the inevitability of a quantum hypothesis. This divergence is also the cause of the infinite emissivity prediction for a black body according to classical mechanics.

The quantity u_ω introduced above is the spectral density defined as the energy per unit volume per unit frequency range and is

$$u_\omega = \frac{\hbar}{\pi^2 c^3} \frac{\omega^3}{e^{\beta\hbar\omega} - 1}. \tag{A2.2.91}$$

This is known as the Planck radiation law. Figure A2.2.3 shows this spectral density function. The surface temperature of a hot body such as a star can be estimated by approximating it by a black body and measuring the frequency at which the maximum emission of radiant energy occurs. It can be shown that the maximum of the Planck spectral density occurs at $\hbar\omega_{max}/(k_b T) \approx 2.82$. So a measurement of ω_{max} yields an estimate of the temperature of the hot body. From the total energy U, one can also obtain the entropy of the photon gas (black-body radiation). At a constant volume, $dS = dU/T = (4\pi^2 k_b^4 V)/(15\hbar^3 c^3)T^2\, dT$. This can be integrated with the result

$$S = \frac{4\pi^2 k_b^4 V}{45\hbar^3 c^3} T^3.$$

The constant of integration is zero: at zero temperature all the modes go to the unique non-degenerate ground state corresponding to the zero point energy. For this state $S \sim \log(g) = \log(1) = 0$, a confirmation of the Third Law of Thermodynamics for the photon gas.

A2.2.4.7 Application to ideal systems: elastic waves in a solid

The energy of an elastic wave in a solid is quantized just as the energy of an electromagnetic wave in a cavity. The quanta of the elastic wave energy are called *phonons*. The thermal average number of phonons in an elastic wave of frequency ω is given, just as in the case of photons, by

$$\langle n(\omega) \rangle = (\exp(\beta\hbar\omega) - 1)^{-1}.$$

Phonons are normal modes of vibration of a low-temperature solid, where the atomic motions around the equilibrium lattice can be approximated by harmonic vibrations. The coupled atomic vibrations can be diagonalized into uncoupled normal modes (phonons) if a harmonic approximation is made. In the simplest analysis of the contribution of phonons to the average internal energy and heat capacity one makes two assumptions: (i) the frequency of an elastic wave is independent of the strain amplitude and (ii) the velocities of all elastic waves are equal and independent of the frequency, direction of propagation and the direction of polarization. These two assumptions are used below for all the modes and leads to the famous Debye model.

There are differences between photons and phonons: while the total number of photons in a cavity is infinite, the number of elastic modes in a finite solid is finite and equals $3N$ if there are N atoms in a three-dimensional solid. Furthermore, an elastic wave has three possible polarizations, two transverse and one longitudinal, in contrast to only two transverse polarizations for photons. Thus the sum of a quantity over all phonon modes is approximated by

$$\sum_n (\cdots) = \frac{3}{8} \int_0^{n_D} 4\pi n^2 \, dn (\cdots)$$

where the maximum number n_D is obtained from the constraint that the total number of phonon modes is $3N$:

$$\frac{3}{8} \int_0^{n_D} 4\pi n^2 \, dn = 3N$$

which gives $n_D = (6N/\pi)^{\frac{1}{3}}$. Keeping in mind the differences noted above, the total thermal energy contributed by phonons can be calculated in a manner analogous to that used above for photons. In place of the velocity of light c, one has the velocity of sound v and $\omega_n = n\pi v/L$. The maximum value n_D corresponds to the highest allowed mode frequency $\omega_D = n_D \pi v/L$, and ω_D is referred to as the Debye frequency. The calculation for U then proceeds as

$$U = \sum_n \langle \epsilon_n \rangle = \sum_n \frac{\hbar \omega_n}{e^{\beta \hbar \omega_n} - 1}$$

$$= \frac{3\pi}{2} \int_0^{n_D} dn \, n^2 \frac{\hbar \omega_n}{e^{\beta \hbar \omega_n} - 1}$$

$$= \frac{3V}{2\pi^2 v^3} \int_0^{\omega_D} d\omega \, \omega^2 \frac{\hbar \omega}{e^{\beta \hbar \omega} - 1} = \frac{3V}{2\pi^2 v^3 \hbar^3 \beta^4} \int_0^{x_D} dx \frac{x^3}{e^x - 1}.$$

The upper limit of the dimensionless variable x_D is typically written in terms of the Debye temperature θ_D as $x_D = \theta_D/T$, where using $x_D = \beta \hbar \omega_D = \beta \hbar \pi v n_D/L = \beta \hbar v (6\pi^2 N/V)^{\frac{1}{3}}$, one identifies the Debye temperature as

$$\theta_D = (\hbar v/k_b)(6\pi^2 N/V)^{\frac{1}{3}}. \tag{A2.2.92}$$

Since $\omega_D^3 = 6\pi^2 v^3 N/V$, one can also write

$$U = \int_0^\infty d\omega \, g(\omega) \frac{\hbar \omega}{e^{\beta \hbar \omega} - 1} \tag{A2.2.93}$$

where $g(\omega) \, d\omega$ is the number of phonon states with a frequency between ω and $\omega + d\omega$, and is given by

$$g(\omega) = \begin{cases} \frac{9N}{\omega_D^3} \omega^2 & \text{if } \omega < \omega_D \\ 0 & \text{if } \omega > \omega_D. \end{cases} \tag{A2.2.94}$$

$g(\omega)$ is essentially the density of states and the above expression corresponds to the Debye model.

In general, the phonon density of states $g(\omega) \, d\omega$ is a complicated function which can be directly measured from experiments, or can be computed from the results from computer simulations of a crystal. The explicit analytic expression of $g(\omega)$ for the Debye model is a consequence of the two assumptions that were made

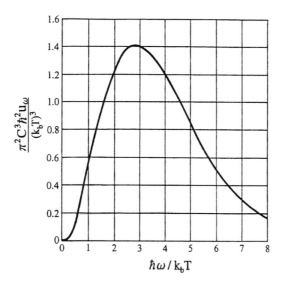

Figure A2.2.3. Planck spectral density function as a function of the dimensionless frequency $\hbar\omega/(k_b T)$.

above for the frequency and velocity of the elastic waves. An even simpler assumption about $g(\omega)$ leads to the Einstein model, which first showed how quantum effects lead to deviations from the classical equipartition result as seen experimentally. In the Einstein model, one assumes that only one level at frequency ω_E is appreciably populated by phonons so that $g(\omega) = \delta(\omega - \omega_E)$ and $U = (\hbar\omega_E)/(e^{\beta\hbar\omega_E} - 1)$, for each of the Einstein modes. $\hbar\omega_E/k_b$ is called the Einstein temperature θ_E.

High-temperature behaviour. Consider T much higher than a characteristic temperature like θ_D or θ_E. Since $\beta\hbar\omega$ is then small compared to 1, one can expand the exponential to obtain

$$\frac{\hbar\omega}{e^{\beta\hbar\omega} - 1} \approx \frac{1}{\beta}$$

and

$$U = \sum_n \frac{\hbar\omega_n}{e^{\beta\hbar\omega_n} - 1} \approx k_b T \sum_n 1 = 3Nk_b T \qquad (A2.2.95)$$

as expected by the equipartition law. This leads to a value of $3Nk_b$ for the heat capacity C_V. This is known as the Dulong and Petit's law.

Low-temperature behaviour. In the Debye model, when $T \ll \theta_D$, the upper limit, x_D, can be approximately replaced by ∞, the integral over x then has a value $\pi^4/15$ and the total phonon energy reduces to

$$U(T) \approx \frac{3V}{2\pi^2 v^3 \hbar^3 \beta^4} \frac{\pi^4}{15} = \frac{3\pi^3 Nk_b}{5\theta_D^3} T^4$$

proportional to T^4. This leads to the heat capacity, for $T \ll \theta_D$,

$$C_V = \left(\frac{\partial U}{\partial T}\right)_{V,N} = \frac{12\pi^4 Nk_b}{5\theta_D^3} T^3 \equiv A_{\mathrm{ph}} T^3. \qquad (A2.2.96)$$

This result is called the Debye T^3 law. Figure A2.2.4 compares the experimental and Debye model values for the heat capacity C_p. It also gives Debye temperatures for various solids. One can also evaluate

Debye Temperatures (K)

Substance	Pb	Tl	Hg	I	Cd	Na	KBr	Ag	Ca
	88	96	97	106	168	172	177	215	226

Substance	KCl	Zn	NaCl	Cu	Al	Fe	CaF	FeS	C
	230	235	281	315	398	453	474	645	1860

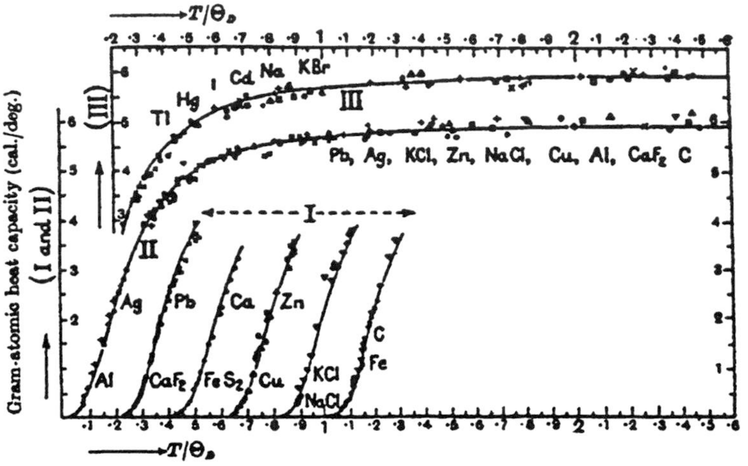

Figure A2.2.4. Experimental and Debye values for the heat capacity C_p. From Born and Huang [1] by permission of Oxford University Press.

C_V for the Einstein model: as expected it approaches the equipartition result at high temperatures but decays exponentially to zero as T goes to zero. The Debye model is more appropriate for the acoustic branches of the elastic modes of a harmonic solid. For molecular solids one has in addition optical branches in the elastic wave dispersion, and the Einstein model is more appropriate to describe the contribution to U and C_V from the optical branch. The above discussion for phonons is suitable for non-metallic solids. In metals, one has, in addition, the contribution from the electronic motion to U and C_V. This is discussed later, in section A2.2.5.6.

A2.2.5 Grand canonical ensemble

Now consider two systems that are in thermal and diffusive contact, such that there can be sharing of both energy and particles between the two. Again let I be the system and II be a much larger reservoir. Since the composite system is isolated, one has the situation in which the volume of each of the two are fixed at V' and V'', respectively, and the total energy and total number of particles are shared: $E_l = E_{l'}^{\mathrm{I}} + E_{l''}^{\mathrm{II}}$ where $l = (l', l'')$ and $N = N' + N''$. We shall use the notation $E = E' + E''$ for the former of these two constraints. For a given partition the allowed microstates of the system I is given by $\Gamma_{\mathrm{I}}(E', N')$ and that for the sytem II by $\Gamma_{\mathrm{II}}(E'', N'') \equiv \Gamma_{\mathrm{II}}(E - E', N - N')$. Then the total number of allowed microstates for the composite system, subject to the two constraints, is

$$\Gamma_{\mathrm{C}}(E, N) = \sum_{N'} \sum_{E'} \Gamma_{\mathrm{I}}(E', N')\Gamma_{\mathrm{II}}(E - E', N - N').$$

Among all possible partitions in the above expression, the equilibrium partition corresponds to the most probable partition, for which $d\Gamma_C = 0$. Evaluating this differential yields the following relation:

$$0 = \frac{d\Gamma_C}{\Gamma_I \Gamma_{II}} = \left(\frac{1}{\Gamma_I} \frac{\partial \Gamma_I}{\partial E'} - \frac{1}{\Gamma_{II}} \frac{\partial \Gamma_{II}}{\partial E''} \right) dE' + \left(\frac{1}{\Gamma_I} \frac{\partial \Gamma_I}{\partial N'} - \frac{1}{\Gamma_{II}} \frac{\partial \Gamma_{II}}{\partial N''} \right) dN'.$$

Since E' and N' are independent variables, their variations are arbitrary. Hence, for the above equality to be satisfied, each of the two bracketed expressions must vanish when the (E, N) partition is most probable. The vanishing of the coefficient of dE' implies the equality of temperatures of I and II, consistent with thermal equilibrium:

$$\beta_I \equiv \frac{\partial \log \Gamma_I}{\partial E'} = \frac{\partial \log \Gamma_{II}}{\partial E''} \equiv \beta_{II}. \tag{A2.2.97}$$

The result that the coefficient of dN' is zero for the most probable partition is the consequence of the chemical equilibrium between the system and the reservoir. It leads us to identify the chemical potential μ as

$$\frac{\partial \log \Gamma}{\partial N} \equiv -\beta u \tag{A2.2.98}$$

in analogy to the thermodynamic definition. Then, since $\beta_I = \beta_{II}$, the vanishing of the coefficient of dN' leads to the equality of chemical potentials: $\mu_I = \mu_{II}$. In a manner similar to that used to obtain the canonical distribution, one can expand

$$\Gamma_{II}(E - E', N - N') = \exp[S_{II}(E - E', N - N')/k_b]$$
$$= \Gamma_{II}(E, N) \exp\left(-\frac{1}{k_b} \left(E' \frac{\partial S}{\partial E} + N' \frac{\partial S}{\partial N} \right) \right)$$
$$\propto \exp[-\beta(E' - \mu N')].$$

With this result and arguments similar to those used in the last section, one finds the grand canonical ensemble distribution as (quantum mechanically)

$$\rho = \frac{\exp(-\beta[\mathcal{H} - \mu N])}{\sum_{N=0}^{\infty} \text{Tr}[\exp(-\beta[\mathcal{H} - \mu N])]}. \tag{A2.2.99}$$

The corresponding classical distribution is

$$\rho(p, q; N) \, d^{2f}\Omega = \frac{e^{-\beta[\mathcal{H}(p,q;N) - \mu N]} \, d^{2f}\Omega}{h^f N! \, \Xi} \tag{A2.2.100}$$

where f is the total number of degrees of freedom if the system has N particles, and the grand partition function $\Xi(\beta, \mu, V)$ is given by

$$\Xi(\beta, \mu, V) = \sum_{N=0}^{\infty} \frac{1}{h^f N!} \int e^{-\beta[\mathcal{H}(p,q) - \mu N]} \, d^{2f}\Omega \tag{A2.2.101}$$

which is the classical analogue of

$$\sum_{N=0}^{\infty} \sum_{\nu} \exp(-\beta[E_\nu - \mu N]) \equiv \sum_{N=0}^{\infty} \text{Tr}[\exp(-\beta[\mathcal{H} - \mu N])]. \tag{A2.2.102}$$

In the above, the sum over N has the upper limit of infinity. This is clearly correct in the thermodynamic limit. However, for a system with finite volume, V, depending on the 'hard core size' of its constituents, there will be a maximum number of particles, $M(V)$, that can be packed in volume V. Then, for all N such that $N > M(V)$, the value of $(-\beta \mathcal{H})$ becomes infinity and all terms in the N sum with $N > M(V)$ vanish. Thus, provided the inter-particle interactions contain a strongly repulsive part, the N sum in the above discussion can be extended to infinity.

If, in this ensemble, one wants to find only the probability that the system has N particles, one sums the distribution over the energy microstates to obtain:

$$\mathcal{P}(N) = \frac{e^{\beta \mu N} Q_N(\beta, V)}{\Xi(\beta, \mu, V)}. \tag{A2.2.103}$$

The combination $e^{\beta \mu}$ occurs frequently. It is called the fugacity and is denoted by z. The grand canonical ensemble is also known as $T-\mu$ ensemble.

A2.2.5.1 T–P ensemble

In many experiments the sample is in thermodynamic equilibrium, held at constant temperature and pressure, and various properties are measured. For such experiments, the $T-P$ ensemble is the appropriate description. In this case the system has fixed N and shares energy and volume with the reservoir: $E = E' + E''$ and $V = V' + V''$, i.e. the system and the reservoir are connected by a pressure transmitting movable diathermic membrane which enables the sharing of the energy and the volume. The most probable partition leads to the conditions for thermal (equality of temperatures) and mechanical (equality of pressures) equilibria. The later condition is obtained after identifying pressure P as

$$\frac{\partial \log \Gamma}{\partial V} \equiv \beta P. \tag{A2.2.104}$$

The $T-P$ ensemble distribution is obtained in a manner similar to the grand canonical distribution as (quantum mechanically)

$$\rho = \frac{\exp(-\beta[H + PV])}{\int_0^\infty dV \, \text{Tr}[\exp(-\beta[H + PV])]} \tag{A2.2.105}$$

and classically as

$$\rho(p, q; V) \, d^{2f}\Omega = \frac{e^{-\beta[\mathcal{H}(p,q;V)+PV]} \, d^{2f}\Omega}{h^f N! Y} \tag{A2.2.106}$$

where the $T-P$ partition function $Y(T, P, N)$ is given by

$$Y(T, P, N) = \frac{1}{h^f N!} \int_0^\infty dV \int e^{-\beta[\mathcal{H}(p,q;V)+PV]} \, d^{2f}\Omega. \tag{A2.2.107}$$

Its quantum mechanical analogue is

$$Y(T, P, N) = \int_0^\infty dV \sum_\nu \exp(-\beta[E_\nu(V) + PV]) \tag{A2.2.108}$$

$$\equiv \int_0^\infty dV \, \text{Tr}[\exp(-\beta[\mathcal{H} + PV])]. \tag{A2.2.109}$$

The $T-P$ partition function can also be written in terms of the canonical partition function Q_N as:

$$Y(T, P, N) = \int_0^\infty Q_N(\beta, V) \, e^{-\beta PV} \, dV \tag{A2.2.110}$$

and the probability that the system will have a volume between V and $V + dV$ is given by

$$P(V)\,d(V) = Q_N(\beta, V)\frac{e^{-\beta PV}\,dV}{Y(T, P, N)}. \tag{A2.2.111}$$

From the canonical ensemble where (V, T, N) are held fixed, one needs to change V to P as an independent variable in order to obtain the T–P ensemble where (P, T, N) are fixed. This change is done through a Legendre transform, $G = A + PV$, which replaces the Helmholtz free energy by the Gibbs free energy as the relevant thermodynamic potential for the T–P ensemble. Now, the internal energy U and its natural independent variables S, V, and N are all extensive quantities, so that for an arbitrary constant a,

$$U(aS, aV, aN) = aU(S, V, N). \tag{A2.2.112}$$

Differentiating both sides with respect to a and using the differential form of the First Law, $dU = T\,dS - P\,dV + \mu\,dN$, one obtains the *Gibbs–Duhem equation*:

$$U = TS - PV + \mu N \tag{A2.2.113}$$

which implies that $G = A + PV = U - TS + PV = \mu N$. The connection to thermodynamics in the T–P ensemble is made by the identification

$$G(T, P, N) = -k_b T \log Y(T, P, N). \tag{A2.2.114}$$

The average value and root mean square fluctuations in volume V of the T–P ensemble system can be computed from the partition function $Y(T, P, N)$:

$$\langle V \rangle = -\left(\frac{\partial \log Y}{\partial(\beta P)}\right)_{T,N} \tag{A2.2.115}$$

$$\langle V^2 \rangle - \langle V \rangle^2 = -\frac{1}{\beta}\left(\frac{\partial \langle V \rangle}{\partial P}\right)_{T,N}. \tag{A2.2.116}$$

The entropy S can be obtained from

$$S = -\left(\frac{\partial G}{\partial T}\right)_{P,N} = k_b\left(\log Y - \beta\frac{\partial}{\partial\beta}\log Y\right). \tag{A2.2.117}$$

A2.2.5.2 Thermodynamics in a grand canonical ensemble

In a canonical ensemble, the system is held at fixed (V, T, N). In a grand canonical ensemble the (V, T, μ) of the system are fixed. The change from N to μ as an independent variable is made by a Legendre transformation in which the dependent variable A, the Helmholtz free energy, is replaced by the grand potential

$$\Omega_G = A - \mu N = U - TS - \mu N = -PV. \tag{A2.2.118}$$

Therefore, from the differential relation, equation (A2.2.65), one obtains,

$$d\Omega_G = -S\,dT - P\,dV - N\,d\mu \tag{A2.2.119}$$

which implies

$$N = -\left(\frac{\partial\Omega_G}{\partial\mu}\right)_{V,T} \qquad P = -\left(\frac{\partial\Omega_G}{\partial V}\right)_{\mu,T} \qquad S = -\left(\frac{\partial\Omega_G}{\partial T}\right)_{\mu,V}. \tag{A2.2.120}$$

Statistical mechanics of weakly interacting systems

Using equations (A2.2.101) and (A2.2.60), one has

$$\Xi(\beta, \mu, V) = \sum_{N=0}^{\infty} e^{\beta \mu N} Q_N(\beta, V) = \sum_{N=0}^{\infty} e^{\beta(\mu N + k_b T \log Q_N)}. \tag{A2.2.121}$$

Using this expression for Ξ and the relation $A = -k_b T \log Q_N$, one can show that the average of $(\mu N - A)$ in the grand canonical ensemble is

$$\langle \mu N - A \rangle = \frac{\partial}{\partial \beta} (\log \Xi(\beta, \mu, V)). \tag{A2.2.122}$$

The connection between the grand canonical ensemble and thermodynamics of fixed (V, T, μ) systems is provided by the identification

$$\log \Xi(\beta, \mu, V) = -\beta \Omega_G = \beta P V. \tag{A2.2.123}$$

Then one has

$$k_b T \log \Xi(\beta, \mu, V) = \mu \langle N \rangle - \langle A \rangle. \tag{A2.2.124}$$

In the grand canonical ensemble, the number of particles fluctuates. By differentiating $\log \Xi$, equation (A2.2.121) with respect to $\beta \mu$ at fixed V and β, one obtains

$$\langle N \rangle = \frac{1}{\beta} \frac{\partial \log \Xi}{\partial \mu} \tag{A2.2.125}$$

and

$$\langle (N - \langle N \rangle)^2 \rangle = \langle N^2 \rangle - \langle N \rangle^2 = \frac{1}{\beta^2} \frac{\partial^2 \log \Xi}{\partial \mu^2} = \frac{1}{\beta} \frac{\partial \langle N \rangle}{\partial \mu}. \tag{A2.2.126}$$

Since $\partial \langle N \rangle / \partial \mu \sim \langle N \rangle$, the fractional root mean square fluctuation in N is

$$\frac{\langle (N - \langle N \rangle)^2 \rangle^{\frac{1}{2}}}{\langle N \rangle} \sim \frac{1}{N^{\frac{1}{2}}}. \tag{A2.2.127}$$

There are two further useful results related to $\langle (N - \langle N \rangle)^2 \rangle$. First is its connection to the isothermal compressibility $\kappa_T = -V^{-1}(\partial P / \partial V)_{\langle N \rangle, T}$, and the second to the spatial correlations of density fluctuations in a grand canonical system.

Now since $\Omega_G = -PV$, the Gibbs–Duhem equation gives $d\Omega_G = -S\,dT - P\,dV - \langle N \rangle\,d\mu = -P\,dV - V\,dp$, which implies that $d\mu = (V\,dP - S\,dT)/\langle N \rangle$. Let $v = V/\langle N \rangle$ be the specific volume, and express μ as $\mu(v, T)$. Then the result for $d\mu$ gives

$$\left(\frac{\partial \mu}{\partial v} \right)_T = v \left(\frac{\partial P}{\partial v} \right)_T.$$

Now a change in v can occur either through V or $\langle N \rangle$:

$$\left(\frac{\partial}{\partial v} \right)_{V, T} = -\frac{\langle N \rangle}{v} \left(\frac{\partial}{\partial \langle N \rangle} \right)_{V, T}$$

$$\left(\frac{\partial}{\partial v} \right)_{\langle N \rangle, T} = \langle N \rangle \left(\frac{\partial}{\partial V} \right)_{\langle N \rangle, T}$$

These two should lead to an equivalent change in v. Thus one obtains

$$\frac{\langle N \rangle}{v} \left(\frac{\partial \mu}{\partial \langle N \rangle} \right)_{V,T} = -V \left(\frac{\partial P}{\partial V} \right)_{\langle N \rangle, T}$$

the substitution of which yields, for the mean square number fluctuations, the result

$$\frac{\langle (N - \langle N \rangle)^2 \rangle}{\langle N \rangle} = \frac{\kappa_T}{\beta v}. \tag{A2.2.128}$$

For homogeneous systems, the average number density is $n_o = \langle N \rangle / V \equiv v^{-1}$. Let us define a local number density through

$$n(\vec{r}) = \sum_{i=1}^{N} \delta(\vec{r} - \vec{r}_i) \tag{A2.2.129}$$

where $\vec{r}$ is a point within the volume V of the grand ensemble system in which, at a given instant, there are N particles whose positions are given by the vectors $\vec{r}_i$, $i = 1, 2, \ldots, N$. One has $N = \int dV \, n(\vec{r})$ and, for homogeneous systems, $\langle N \rangle = \int dV \, \langle n(\vec{r}) \rangle = \int dV \, n_o = V n_o$. One can then define the fluctuations in the local number density as $\delta n = n - n_o$, and construct the spatial density–density correlation function as

$$G(\vec{r} - \vec{r}') \equiv n_o^{-2} \langle \delta n(\vec{r}) \delta n(\vec{r}') \rangle. \tag{A2.2.130}$$

$G(\vec{r})$ is also called the pair correlation function and is sometimes denoted by $h(\vec{r})$. Integration over $\vec{r}$ and $\vec{r}'$ through the domain of system volume gives, on the one hand,

$$\int_V d\vec{r}' \int_V d\vec{r} \, G(\vec{r} - \vec{r}') = V \int d\vec{r} \, G(\vec{r})$$

and, on the other,

$$\int_V d\vec{r}' \int_V d\vec{r} \, G(\vec{r} - \vec{r}') = n_o^{-2} \int_V d\vec{r}' \int_V d\vec{r} \, [\langle n(\vec{r}) n(\vec{r}') \rangle - n_o^2]$$

$$= n_o^{-2} (\langle N^2 \rangle - \langle N \rangle^2) = \langle N \rangle n_o^{-1} \frac{\kappa_T}{\beta}.$$

Comparing the two results and substituting the relation of the mean square number fluctuations to isothermal compressibility, equation (A2.2.128) one has

$$\int_V d\vec{r} \, G(\vec{r}) = k_b T \kappa_T. \tag{A2.2.131}$$

The correlation function $G(\vec{r})$ quantifies the density fluctuations in a fluid. Characteristically, density fluctuations scatter light (or any radiation, like neutrons, with which they can couple). Then, if a radiation of wavelength λ is incident on the fluid, the intensity of radiation scattered through an angle θ is proportional to the structure factor

$$S(\vec{q}) = n_o \int_V d\vec{r} \, e^{-i\vec{q} \cdot \vec{r}} G(\vec{r}) \tag{A2.2.132}$$

where $|\vec{q}| = 4\pi \sin(\theta/2)/\lambda$. The limiting value of $S(\vec{q})$ as $q \to 0$ is then proportional to κ_T. Near the critical point of a fluid, anomalous density fluctuations create a divergence of κ_T which is the cause of the phenomenon of critical opalescence: density fluctuations become correlated over a lengthscale which is long compared to a molecular lengthscale and comparable to the wavelength of the incident light. This causes the light to be strongly scattered, whereby multiple scattering becomes dominant, making the fluid medium appear turbid or opaque.

For systems in which the constituent particles interact via short-range pair potentials, $W = \sum_{i=1}^{N} \sum_{j=1}^{(i-1)} u(|(\vec{r}_i - \vec{r}_j)|)$, there are two relations, that one can prove by evaluating the average of the total energy $E = K + W$, where K is the total kinetic energy, and the average pressure P, that are valid in general. These are

$$\frac{\langle E \rangle}{\langle N \rangle} = \frac{3}{2} k_b T + \frac{1}{2} n_o \int_V d\vec{r}\; g(r) u(r) \tag{A2.2.133}$$

and the virial equation of state,

$$P = n_o k_b T \left(1 - \frac{n_o}{6 k_b T} \int_V d\vec{r}\; g(r) r \frac{du(r)}{dr} \right). \tag{A2.2.134}$$

Here $g(r) = G(r) + 1$ is called a radial distribution function, since $n_o g(r)$ is the conditional probability that a particle will be found at $\vec{r}$ if there is another at the origin. For strongly interacting systems, one can also introduce the potential of the mean force $w(r)$ through the relation $g(r) = \exp(-\beta w(r))$. Both $g(r)$ and $w(r)$ are also functions of temperature T and density n_o.

A2.2.5.3   Density expansion

For an imperfect gas, i.e. a low-density gas in which the particles are, most of the time, freely moving as in an ideal gas and only occasionally having binary collisions, the potential of the mean force is the same as the pair potential $u(r)$. Then, $g(r) \approx \exp(-\beta u(r))[1 + O(n_o)]$, and from equation (A2.2.133) the change from the ideal gas energy, $\Delta U = \langle E \rangle - \langle E \rangle_{\text{ideal}}$, to leading order in n_o, is

$$\frac{\Delta U}{N} \approx \frac{1}{2} n_o \int_V d\vec{r}\; u(r)\, e^{-\beta u(r)} = -\frac{1}{2} n_o \frac{\partial}{\partial \beta} \int_V d\vec{r}\; [e^{-\beta u(r)} - 1]$$

$$= -\frac{1}{2} n_o \frac{\partial}{\partial \beta} \int_V d\vec{r}\; f(r) \tag{A2.2.135}$$

where

$$f(r) = e^{-\beta u(r)} - 1. \tag{A2.2.136}$$

FigureA2.2.5 shows a sketch of $f(r)$ for Lennard–Jones pair potential. Now if ΔA is the excess Helmholtz free energy relative to its ideal gas value, then $(-\beta \Delta A) = \log(Q/Q_{\text{ideal}})$ and $\Delta U/N = [\partial(\beta \Delta A/N)/(\partial \beta)]$. Then, integrating with respect to β, one obtains

$$-\beta \Delta A/N = \frac{1}{2} n_o \int_V d\vec{r}\; f(r) + O(n_o^2). \tag{A2.2.137}$$

One can next obtain pressure P from the above by

$$\beta P = n_o + n_o^2 \frac{\partial(\beta \Delta A/N)}{\partial n_o} = n_o + n_o^2 B_2(T) + O(n_o^3) \tag{A2.2.138}$$

where

$$B_2(T) = -\frac{1}{2} \int d\vec{r}\; f(r). \tag{A2.2.139}$$

The same result can also be obtained directly from the virial equation of state given above and the low-density form of $g(r)$. $B_2(T)$ is called the second virial coefficient and the expansion of P in powers of n_o is known as the virial expansion, of which the leading non-ideal term is deduced above. The higher-order terms in

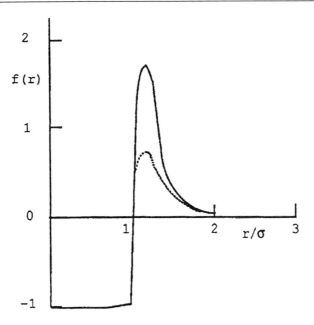

Figure A2.2.5. Sketch of $f(r)$ for the Lennard-Jones pair potential $u(r) = 4\epsilon[(\sigma/r)^{12} - (\sigma/r)^6]$; full curve $-\beta\epsilon = 1.0$ and broken curve $-\beta\epsilon = 0.5$. From Plischke and Bergensen 1985, further reading.

the virial expansion for P and in the density expansion of $g(r)$ can be obtained using the methods of cluster expansion and cumulant expansion.

For purely repulsive potentials ($u(r) > 0$), $f(r)$ is negative and $B_2(T)$ is positive. For purely attractive potentials, on the other hand, $f(r)$ is always positive leading to a negative $B_2(T)$. Realistic interatomic potentials contain both a short-range repulsive potential (due to the strong short distance overlap of electronic wavefunctions of the two atoms) and a weaker longer range van der Waals attractive potential. The temperature dependence of $B_2(T)$ can be used to qualitatively probe the nature of interatomic potential. At a certain temperature T_B, known as the Boyle temperature, the effects of attractive and repulsive potentials balance exactly, giving $B_2(T_B) = 0$. A phenomenological extension of the ideal gas equation of state was made by van der Waals more than 100 years ago. For one mole of gas,

$$pV = RT \implies \left(P + \frac{a}{v^2}\right)(v - b) = RT. \tag{A2.2.140}$$

Here b corresponds to the repulsive part of the potential, which is equivalent to the excluded volume due to the finite atomic size, and a/v^2 corresponds to the attractive part of the potential. The van der Waals equation of state is a very good qualitative description of liquids as well as imperfect gases. Historically, it is the first example of a mean field theory. It fails only in the neighbourhood of a critical point due to its improper treatment of the density fluctuations.

A2.2.5.4   Ideal quantum gases

Thermodynamics of ideal quantum gases is typically obtained using a grand canonical ensemble. In principle this can also be done using a canonical ensemble partition function, $Q = \sum_v \exp(-\beta E_v)$. For the photon and phonon gases, the canonical ensemble was used in sections A2.2.4.6 and A2.2.4.7. Photons and phonons are massless and their total number indeterminate, since they can be created or destroyed, provided the momentum and energy are conserved in the process. On the other hand, for an ideal gas consisting of particles with non-zero mass, in a canonical ensemble, the total number of particles is fixed at N. Thus, in the occupation number representation of the single-particle states j, the sum of all n_j is constrained to be N:

$$Q_N = \sum_{n_1, n_2, \dots, n_j, \dots} \delta\left(N - \sum_j n_j\right) \exp\left(-\beta \sum_j n_j \epsilon_j\right)$$

where ϵ_j is the energy of the jth single-particle state. The restriction on the sum over n_j creates a complicated combinatorial problem, which even though solvable, is non-trivial. This constraint is removed by considering the grand canonical partition function:

$$
\begin{aligned}
\Xi(\beta, \mu, V) &= \sum_{N=0}^{\infty} e^{\beta\mu N} Q_N(\beta, V) \\
&= \sum_{N=0}^{\infty} e^{\beta\mu N} \sum_{n_1, n_2, \dots, n_j, \dots} \delta\left(N - \sum_j n_j\right) \exp\left(-\beta \sum_j n_j \epsilon_j\right) \\
&= \sum_{n_1, n_2, \dots, n_j, \dots} \exp\left(-\beta \sum_j (\epsilon_j - \mu) n_j\right).
\end{aligned}
\tag{A2.2.141}
$$

Now the exponential factors for various n_j within the sum are independent, which simplifies the result as

$$\Xi = e^{\beta PV} = \sum_{n_1, n_2, \dots, n_j, \dots} \prod_j e^{-\beta(\epsilon_j - \mu)n_j} = \prod_j \sum_{n_j} e^{-\beta(\epsilon_j - \mu)n_j}.$$

The sum over n_j can now be performed, but this depends on the statistics that the particles in the ideal gas obey. Fermi particles obey the Pauli exclusion principle, which allows only two possible values: $n_j = 0, 1$. For Bose particles, n_j can be any integer between zero and infinity. Thus the grand partition function is

$$\Xi = \prod_j [1 + e^{-\beta(\epsilon_j - \mu)}] \qquad \text{for fermions} \tag{A2.2.142}$$

and

$$\Xi = \prod_j [1 - e^{-\beta(\epsilon_j - \mu)}]^{-1} \qquad \text{for bosons.} \tag{A2.2.143}$$

This leads to, using equations (A2.2.123),

$$\beta PV = -\beta\Omega_G = \log \Xi = \pm \sum_j \log[1 \pm e^{-\beta(\epsilon_j - \mu)}] \tag{A2.2.144}$$

where the upper sign corresponds to fermions and the lower sign to bosons. From equation (A2.2.141), the average occupation number $\langle n_j \rangle = \partial(\log \Xi)/\partial(\beta\mu)$. From this one obtains

$$\langle n_j \rangle = [e^{\beta(\epsilon_j - \mu)} \pm 1]^{-1} \tag{A2.2.145}$$

where again the upper sign corresponds to fermions and the lower sign to bosons. From this, one has, for the total number of particles, $\langle N \rangle$,

$$\langle N \rangle = \sum_j \langle n_j \rangle = \sum_j [e^{\beta(\epsilon_j - \mu)} \pm 1]^{-1} \tag{A2.2.146}$$

and for the total internal energy $U \equiv \langle E \rangle$

$$U = \sum_j \epsilon_j \langle n_j \rangle = \sum_j \epsilon_j [e^{\beta(\epsilon_j - \mu)} \pm 1]^{-1}. \tag{A2.2.147}$$

When the single-particle states j are densely packed within any energy interval of $k_b T$, the sum over j can be replaced by an integral over energy such that

$$\sum_j \cdots \to \int_0^\infty d\epsilon\, \mathcal{D}(\epsilon) \cdots = \frac{\gamma V}{4\pi^2} \left(\frac{2M}{\hbar^2}\right)^{\frac{3}{2}} \int_0^\infty d\epsilon\, \epsilon^{\frac{1}{2}} \cdots \tag{A2.2.148}$$

Using equation (A2.2.88), this can be rewritten as

$$\sum_j \cdots \to \frac{2\gamma V}{\pi^{\frac{1}{2}} V_q} \beta^{\frac{3}{2}} \int_0^\infty d\epsilon\, \epsilon^{\frac{1}{2}} \cdots \tag{A2.2.149}$$

Using this approximation, expressions for $\langle N \rangle$, U and P reduce to

$$n_0 \equiv \frac{\langle N \rangle}{V} = \frac{2\gamma}{\pi^{\frac{1}{2}} V_q} \beta^{\frac{3}{2}} \int_0^\infty d\epsilon\, \epsilon^{\frac{1}{2}} [e^{\beta(\epsilon_j - \mu)} \pm 1]^{-1}$$

$$= \frac{2\gamma}{\pi^{\frac{1}{2}} V_q} \int_0^\infty dy\, y^{\frac{1}{2}} [e^{y - \beta\mu} \pm 1]^{-1} \tag{A2.2.150}$$

$$\equiv \frac{2\gamma}{\pi^{\frac{1}{2}} V_q} F_{\frac{1}{2}}(\beta\mu) \tag{A2.2.151}$$

$$\frac{U}{V} = \frac{2\gamma}{\pi^{\frac{1}{2}} V_q} \beta^{\frac{3}{2}} \int_0^\infty d\epsilon\, \epsilon^{\frac{3}{2}} [e^{\beta(\epsilon_j - \mu)} \pm 1]^{-1}$$

$$= \frac{2\gamma}{\pi^{\frac{1}{2}} V_q} \beta^{-1} \int_0^\infty dy\, y^{\frac{3}{2}} [e^{y - \beta\mu} \pm 1]^{-1} \tag{A2.2.152}$$

$$\equiv \frac{2\gamma}{\pi^{\frac{1}{2}} V_q} \beta^{-1} F_{\frac{3}{2}}(\beta\mu) \tag{A2.2.153}$$

and

$$P = \pm \frac{2\gamma}{\pi^{\frac{1}{2}} V_q} \beta^{\frac{1}{2}} \int_0^\infty d\epsilon\, \epsilon^{\frac{1}{2}} \log[1 \pm e^{\beta(\epsilon - \mu)}] \tag{A2.2.154}$$

$$= \frac{2}{3} \frac{2\gamma}{\pi^{\frac{1}{2}} V_q} \beta^{\frac{3}{2}} \int_0^\infty d\epsilon\, \epsilon^{\frac{3}{2}} [e^{\beta(\epsilon_j - \mu)} \pm 1]^{-1} \tag{A2.2.155}$$

$$\equiv \frac{2}{3} \frac{2\gamma}{\pi^{\frac{1}{2}} V_q} \beta^{-1} F_{\frac{3}{2}}(\beta\mu). \tag{A2.2.156}$$

An integration by parts was used to deduce equation (A2.2.155) from equation (A2.2.154). Comparing the results for U and P, one finds that, just as for the classical gas, for ideal quantum gases, also, the relation $U = \frac{3}{2} PV$ is satisfied. In the above results it was found that $P = P(\beta\mu)$ and $\langle N \rangle / V \equiv n_0 = n_0(\beta\mu)$. In principle, one has to eliminate $(\beta\mu)$ between the two in order to deduce the equation of state, $P = P(\beta, n_0)$, for ideal quantum gases. Now $F_{\frac{3}{2}}(\beta\mu)$ is a function of a single variable. Therefore P is a homogeneous function of order $\frac{5}{2}$ in μ and $k_b T$. Similarly n_0 is a homogeneous function of order $\frac{3}{2}$ in μ and $k_b T$; and so is $S/V = (\partial P / \partial T)_{V,\mu}$. This means that $S/\langle N \rangle$ is a homogeneous function of order zero, i.e. $S/\langle N \rangle = \phi(\beta\mu)$, which in turn implies that for an adiabatic process $\beta\mu$ remains constant. Thus, from the expressions above for P and $\langle N \rangle / V$, one has for adiabatic processes, $PV^{\frac{5}{3}} = \text{constant}$, $VT^{\frac{3}{2}} = \text{constant}$ and $T^{\frac{5}{2}}/P = \text{constant}$.

A2.2.5.5   Ideal quantum gases—classical limit

When the temperature is high and the density is low, one expects to recover the classical ideal gas limit. The number of particles is still given by $N = \sum_j n_j$. Thus the average number of particles is given by equation (A2.2.146). The average density $\langle N \rangle / V = n_o$ is the thermodynamic density. At low n_o and high T one expects many more accessible single-particle states than the available particles, and $\langle N \rangle = \sum_j \langle n_j \rangle$ means that each $\langle n_j \rangle$ must be small compared to one. Thus, from equation (A2.2.145) for $\langle n_j \rangle$, the classical limit corresponds to the limit when $\exp(\beta(\epsilon_j - \mu)) \gg 1$. This has to be so for any ϵ_j, which means that the fugacity $z = \exp(-\beta\mu) \gg 1$ or $(-\beta\mu) \gg 1$ at low n_o and high T. In this classical limit, $\langle n_j \rangle = \exp(-\beta(\epsilon_j - \mu))$. The chemical potential μ is determined from $\langle N \rangle = \sum_j \langle n_j \rangle$, which leads to the result that

$$\mu = k_b T \log\left(\frac{\langle N \rangle}{\sum_j e^{-\beta\epsilon_j}} \right) \tag{A2.2.157}$$

with the final result that

$$\langle n_j \rangle = \langle N \rangle \frac{e^{-\beta\epsilon_j}}{\sum_j e^{-\beta\epsilon_j}}.$$

This is the classical Boltzmann distribution in which $\langle n_j \rangle / \langle N \rangle$, the probability of finding a particle in the single-particle state j, is proportional to the classical Boltzmann factor $e^{-\beta\epsilon_j}$.

Now $\log Q_N = -\beta A$, $A = G - PV = \mu\langle N \rangle - PV$ and $\beta PV = \log \Xi$. Thus the canonical partition function is

$$\log Q(\langle N \rangle, V, T) = -\beta\mu\langle N \rangle + \log \Xi$$

which leads to the classical limit result:

$$\log Q = -\beta\mu\langle N \rangle \pm \sum_j \log[1 \pm e^{-\beta(\epsilon_j - \mu)}]$$

$$= -\beta\mu\langle N \rangle + \sum_j e^{-\beta(\epsilon_j - \mu)}$$

$$= -\beta\mu\langle N \rangle + \sum_j \langle n_j \rangle = -\beta\mu\langle N \rangle + \langle N \rangle$$

where the approximation $\log(1 + x) \approx x$ for small x is used. Now, from the result for μ above, in equation (A2.2.157), one has

$$\beta\mu = \log\langle N \rangle - \log \sum_j e^{-\beta\epsilon_j}. \tag{A2.2.158}$$

Thus

$$\log Q = -\langle N \rangle \log\langle N \rangle + \langle N \rangle + \langle N \rangle \log \sum_j e^{-\beta\epsilon_j}.$$

For large $\langle N \rangle$, $\langle N \rangle \log\langle N \rangle - \langle N \rangle \approx \log(\langle N \rangle !)$ whereby

$$Q(\langle N \rangle, V, T) = \frac{1}{\langle N \rangle !}\left(\sum_j e^{-\beta\epsilon_j} \right)^{\langle N \rangle}.$$

This result is identical to that obtained from a canonical ensemble approach in the thermodynamic limit, where the fluctuations in N vanish and $\langle N \rangle = N$. The single-particle expression for the canonical partition function $Q_1 = \sum_j e^{-\beta\epsilon_j}$ can be evaluated using $\epsilon_j = (\hbar\pi)^2 V^{-\frac{2}{3}} n_j^2/(2M)$ for a particle in a cubical box of volume V. In the classical limit, the triplet of quantum numbers n_j can be replaced by a continuous variable through the transformation $\sum_j \rightarrow (\gamma V/\pi^3) \int_0^\infty dk_x \int_0^\infty dk_y \int_0^\infty dk_z$, and $\hbar n_j \rightarrow \hbar (V^{\frac{1}{3}}/\pi)k \equiv p$, which is the momentum of the classical particle. The transformation leads to the result that

$$Q_1 = (\gamma V/h^3) \int d^3 p \, \exp(-\beta p^2/(2M)).$$

This is the same as that in the canonical ensemble. All the thermodynamic results for a classical ideal gas then follow, as in section A2.2.4.4. In particular, since from equation (A2.2.158) the chemical potential is related to Q_1, which was obtained in equation (A2.2.88), one obtains

$$\beta\mu = \log\langle N\rangle - \log Q_1 = \log[(\langle N\rangle V_q/(\gamma V)] = \log(n_o V_q/\gamma)$$

or, equivalently, $z = n_o V_q/\gamma$. The classical limit is valid at low densities when $n_o \ll (\gamma/V_q)$, i.e. when $z = \exp(\beta\mu) \ll 1$. For $n_o \geq (\gamma/V_q)$ one has a quantum gas. Equivalently, from the definition of V_q one has a quantum gas when $k_b T$ is below $k_b T_o \equiv (2\pi\hbar^2/M)(n_o/\gamma)^{\frac{2}{3}}$.

If $z = \exp(\beta\mu) \ll 1$, one can also consider the leading order quantum correction to the classical limit. For this consider the thermodynamic potential Ω_G given in equation (A2.2.144). Using equation (A2.2.149), one can convert the sum to an integral, integrate by parts the resulting integral and obtain the result:

$$\Omega_G = -\frac{4\gamma V}{3\pi^{\frac{1}{2}} V_q \beta} \int_0^\infty dy \, \frac{y^{\frac{3}{2}}}{z^{-1}e^y \pm 1} \qquad (A2.2.159)$$

where $z = \exp(\beta\mu)$ is the fugacity, which has an ideal gas value of $n_o V_q/\gamma$. Note that the integral is the same as $F_{\frac{3}{2}}(\beta\mu)$. Since V_q is proportional to $\hbar^3$, and z is small, the expansion of the integrand in powers of z is appropriate and leads to the leading quantum correction to the classical ideal gas limit. Using

$$[z^{-1}e^y \pm 1]^{-1} = z\,e^{-y}[1 \pm z\,e^{-y}]^{-1} \qquad (A2.2.160)$$
$$= z\,e^{-y}[1 \mp z\,e^{-y} + O(z^2)] \qquad (A2.2.161)$$

Ω_G can be evaluated with the result

$$\Omega_G = -PV = -\frac{\gamma V z}{V_q \beta}\left[1 \mp \frac{z}{2^{\frac{5}{2}}} + O(z^2)\right]. \qquad (A2.2.162)$$

The first term is the classical ideal gas term and the next term is the first-order quantum correction due to Fermi or Bose statistics, so that one can write

$$\Omega_G = \Omega_G^{cl} \pm \frac{\gamma V}{V_q \beta}\frac{z^2}{2^{\frac{5}{2}}} + O(z^3). \qquad (A2.2.163)$$

The small additions to all thermodynamic potentials are the same when expressed in terms of appropriate variables. Thus the first-order correction term when expressed in terms of V and β is the correction term for the Helmholtz free energy A:

$$A = A^{cl} \pm \frac{\pi^{\frac{3}{2}}}{2\gamma}\frac{N^2 \beta^{\frac{1}{2}}}{V M^{\frac{3}{2}}}\hbar^3 + \cdots \qquad (A2.2.164)$$

where the classical limiting value $z = n_o V_q/\gamma$, and the definition in equation (A2.2.88) of V_q is used. Finally, one can obtain the correction to the ideal gas equation of state by computing $P = -(\partial A/\partial V)_{\beta,N}$. The result is

$$P = n_o k_b T\left[1 \pm \frac{n_o}{2\gamma}\left(\frac{\pi\beta}{M}\right)^{\frac{3}{2}}\hbar^3 + \cdots\right]. \qquad (A2.2.165)$$

The leading correction to the classical ideal gas pressure term due to quantum statistics is proportional to $\hbar^3$ and to n_0. The correction at constant density is larger in magnitude at lower temperatures and lighter mass. The coefficient of n_0 can be viewed as an effective second virial coefficient $B_2^{\text{eff}}(T)$. The effect of quantum statistics at this order of correction is to add to a classical ideal gas some effective interaction. The upper sign is for a Fermi gas and yields a positive $B_2^{\text{eff}}(T)$ equivalent to an effective repulsive interaction which is a consequence of the Pauli exclusion rule. The lower sign is for a Bose gas which yields a negative $B_2^{\text{eff}}(T)$ corresponding to an effective attractive interaction. This is an indicator of the tendency of Bose particles to condense in the lowest-energy state. This phenomena is treated in section A2.2.5.7.

A2.2.5.6   Ideal Fermi gas and electrons in metals

The effects of quantum statistics are relevant in many forms of matter, particularly in solids at low temperatures. The presence of strong Coulombic interaction between electrons and ions leads one to expect that the behaviour of such systems will be highly complex and nonlinear. It is then remarkable that numerous metals and semiconductors can be well described in many respect in terms of models of effectively non-interacting 'particles'. One such example is the thermal properties of conducting electrons in metals, which can be well approximated by an ideal gas of fermions. This approximation works at high densities on account of the Pauli exclusion principle. No two indistinguishable fermions occupy the same state, many single-particle states are filled and the lowest energy of unoccupied states is many times $k_b T$, so that the energetics of interactions between electrons become negligible. If the conduction electrons (mass m_e) in a metal are modelled by an ideal Fermi gas, the occupation number in the jth single-particle state is (from equation (A2.2.145))

$$\langle n_j \rangle = f(\epsilon_j) \qquad \text{where } f(\epsilon_j) = [e^{\beta(\epsilon_j - \mu)} + 1]^{-1} \tag{A2.2.166}$$

with $\epsilon_j = (\hbar k)^2/(2m_e)$ and $k = n\pi V^{-\frac{1}{3}}$ as for a free particle in a cubical box. Consider first the situation at $T = 0$, $\beta = \infty$. Since μ depends on T, it is useful to introduce the Fermi energy and Fermi temperature as $\epsilon_F \equiv k_b T_F \equiv \mu(T = 0)$. At $T = 0$, $\langle n_j \rangle$ is one for $\epsilon_j < \epsilon_F$ and is zero for $\epsilon_j > \epsilon_F$. Due to the Pauli exclusion principle, each single-particle state n_j is occupied by one spin-up electron and one spin-down electron. The total available, N, electrons fill up the single-particle states up to the Fermi energy ϵ_F which therefore depends on N. If n_F is defined via $\epsilon_F = (\hbar)^2/(2m_e)V^{-\frac{2}{3}}(\pi n_F)^2$, then $N = (2)(\frac{1}{8})(4\pi n_F^3/3) = \pi n_F^3/3$, which gives the relation between N and ϵ_F:

$$\epsilon_F \equiv k_b T_F = (\hbar)^2/(2m_e)(3\pi^2 N/V)^{\frac{2}{3}}. \tag{A2.2.167}$$

The total energy of the Fermi gas at $T = 0$ is

$$U_o = \sum_j \langle n_j \rangle \epsilon_j = \sum_{n < n_F} \epsilon_n = 2\left(\frac{1}{8}\right)4\pi \int_0^{n_F} dn \, n^2 \epsilon_n$$

$$= \frac{\pi^3 \hbar^2}{2m_e V^{\frac{2}{3}}} \int_0^{n_F} dn \, n^4 = \frac{\pi^3 \hbar^2}{10 m_e V^{\frac{2}{3}}} n_F^5 = \frac{3}{5} N \epsilon_F. \tag{A2.2.168}$$

The average kinetic energy per particle at $T = 0$, is $\frac{3}{5}$ of the Fermi energy ϵ_F. At constant N, the energy increases as the volume decreases since $\epsilon_F \sim V^{-\frac{2}{3}}$. Due to the Pauli exclusion principle, the Fermi energy gives a repulsive contribution to the binding of any material. This is balanced by the Coulombic attraction between ions and electrons in metals.

The thermal average of a physical quantity X can be computed at any temperature through

$$\langle X \rangle = \sum_n f(\epsilon_n, T, \mu) X_n.$$

This can be expressed, in terms of the density of states $\mathcal{D}(\epsilon)$, as

$$\langle X \rangle = \int d\epsilon\, \mathcal{D}(\epsilon) f(\epsilon, T, \mu) X(\epsilon).$$

For the total number of particles N and total energy U one has

$$N = \int d\epsilon\, \mathcal{D}(\epsilon) f(\epsilon, T, \mu)$$

and

$$U = \int d\epsilon\, \mathcal{D}(\epsilon)\epsilon f(\epsilon, T, \mu).$$

For an ideal gas the density of states is computed in section A2.2.2 (equation A2.2.8). Its use in evaluating N at $T = 0$ gives

$$N = \frac{\pi n_F^3}{3} = \frac{V}{3\pi^2}\left(\frac{2m_e\epsilon_F}{\hbar^2}\right)^{\frac{3}{2}} \qquad \text{and} \qquad \mathcal{D}(\epsilon_F) = \frac{V}{2\pi^2}\left(\frac{2m_e}{\hbar^2}\right)^{\frac{3}{2}}\epsilon_F^{\frac{1}{2}}. \tag{A2.2.169}$$

This gives a simple relation

$$\mathcal{D}(\epsilon_F) = 3N/(2\epsilon_F) = 3N/(2k_b T_F). \tag{A2.2.170}$$

At $T = 0$, N and U obtained above can also be found using

$$N = \int_0^{\epsilon_F} d\epsilon\, \mathcal{D}(\epsilon) \qquad \text{and} \qquad U = \int_0^{\epsilon_F} d\epsilon\, \mathcal{D}(\epsilon)\epsilon. \tag{A2.2.171}$$

If the increase in the total energy of a system of N conduction electrons when heated from zero to T is denoted by ΔU, then

$$\Delta U = \int_0^\infty d\epsilon\, \mathcal{D}(\epsilon)\epsilon f(\epsilon) - \int_0^\infty d\epsilon\, \mathcal{D}(\epsilon)\epsilon.$$

Now multiplying the identity $(N = \int_0^\infty d\epsilon\, \mathcal{D}(\epsilon) f(\epsilon) = \int_0^{\epsilon_F} d\epsilon\, \mathcal{D}(\epsilon))$ by ϵ_F, one has

$$\left(\int_0^{\epsilon_F} + \int_{\epsilon_F}^\infty\right) d\epsilon\, \epsilon_F \mathcal{D}(\epsilon) f(\epsilon) = \int_0^{\epsilon_F} d\epsilon\, \epsilon_F \mathcal{D}(\epsilon).$$

Using this, one can rewrite the expression for ΔU in physically transparent form:

$$\Delta U = \int_{\epsilon_F}^\infty d\epsilon\, (\epsilon - \epsilon_F)\mathcal{D}(\epsilon) f(\epsilon) + \int_0^{\epsilon_F} d\epsilon\, (\epsilon_F - \epsilon)\mathcal{D}(\epsilon)[1 - f(\epsilon)]. \tag{A2.2.172}$$

The first integral is the energy needed to move electrons from ϵ_F to orbitals with energy $\epsilon > \epsilon_F$, and the second integral is the energy needed to bring electrons to ϵ_F from orbitals below ϵ_F. The heat capacity of the electron gas can be found by differentiating ΔU with respect to T. The only T-dependent quantity is $f(\epsilon)$. So one obtains

$$C_{el} = \frac{\partial U}{\partial T} = \int_0^\infty d\epsilon\, (\epsilon - \epsilon_F)\mathcal{D}(\epsilon)\frac{\partial f}{\partial T}.$$

Now, typical Fermi temperatures in metals are of the order of $50\,000$ K. Thus, at room temperature, T/T_F is very small compared to one. So, one can ignore the T dependence of μ, to obtain

$$\frac{\partial f}{\partial T} = k_b\beta\frac{x\,e^x}{(e^x + 1)^2} \qquad \text{where } x = \beta(\epsilon - \epsilon_F).$$

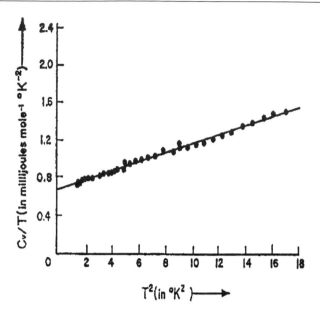

Figure A2.2.6. Electronic contribution to the heat capacity C_V of copper at low temperatures between 1 and 4 K. (From Corak *et al* [2])

This is a very sharply peaked function around ϵ_F with a width of the order of $k_b T$. (At $T = 0$, $f(\epsilon)$ is a step function and its temperature derivative is a delta function at ϵ_F.) Thus in the integral for C_{el}, one can replace $\mathcal{D}(\epsilon)$ by its value at ϵ_F, transform the integration variable from ϵ to x and replace the lower limit of x, which is $(-\beta\epsilon_F)$, by $(-\infty)$. Then one obtains

$$C_{el} = k_b^2 T \mathcal{D}(\epsilon_F) \int_{-\beta\epsilon_F}^{\infty} dx \, \frac{x^2 e^x}{(e^x + 1)^2} \approx k_b^2 T \mathcal{D}(\epsilon_F) \int_{-\infty}^{\infty} dx \, \frac{x^2 e^x}{(e^x + 1)^2}$$

$$= \frac{\pi^2}{3} \mathcal{D}(\epsilon_F) k_b^2 T = \frac{\pi^2}{2} N k_b \frac{T}{T_F} \equiv A_{el} T \qquad (A2.2.173)$$

where equation (A2.2.170) is used. This result can be physically understood as follows. For small T/T_F, the number of electrons excited at T from the $T = 0$ step-function Fermi distribution is of the order of NT/T_F and the energy of each of these electrons is increased by about $k_b T$. This gives $\Delta U \sim N k_b T^2/T_F$ and $C_{el} \sim N k_b T/T_F$.

In typical metals, both electrons and phonons contribute to the heat capacity at constant volume. The temperature-dependent expression

$$C_V = A_{el} T + A_{ph} T^3 \qquad (A2.2.174)$$

where A_{ph} is given in equation (A2.2.96) obtained from the Debye theory discussed in section A2.2.4.7, fits the low-temperature experimental measurements of C_V for many metals quite well, as shown in figure A2.2.6 for copper.

A2.2.5.7   Ideal Bose gas and Bose–Einstein condensation

In an ideal Bose gas, at a certain transition temperature a remarkable effect occurs: a macroscopic fraction of the total number of particles condenses into the lowest-energy single-particle state. This effect, which occurs when the Bose particles have non-zero mass, is called Bose–Einstein condensation, and the key to its understanding is the chemical potential. For an ideal gas of photons or phonons, which have zero mass, this effect does not occur. This is because their total number is arbitrary and the chemical potential is effectively zero for the photon or phonon gas.

From equation (A2.2.145), the average occupation number of an ideal Bose gas is

$$\langle n_j \rangle = [e^{\beta(\epsilon_j - \mu)} - 1]^{-1} \equiv [z^{-1} e^{\beta \epsilon_j} - 1]^{-1}.$$

There is clearly a possible singularity in $\langle n_j \rangle$ if $(\epsilon_j - \mu)$ vanishes. Let the energy scale be chosen such that the ground-state energy $\epsilon_0 = 0$. Then the ground-state occupancy is

$$\langle n_0 \rangle = [e^{\beta \mu} - 1]^{-1} \equiv [z^{-1} - 1]^{-1}.$$

At $T = 0$, it is expected that all the N particles will be in the ground state. Now if at low temperatures, $N \approx \langle n_0 \rangle$ is to be large, such as 10^{20}, then one must have z very close to one and $\beta \mu \ll 1$. Thus

$$N = \lim_{T \to 0} \frac{1}{e^{-\beta \mu} - 1} \approx \frac{1}{1 - \beta \mu - 1} = -\frac{k_b T}{\mu}$$

which gives the chemical potential of a Bose gas, as $T \to 0$, to be

$$\mu = -\frac{k_b T}{N} \qquad \text{and the fugacity} \qquad z = e^{\beta \mu} \approx 1 - \frac{1}{N}. \tag{A2.2.175}$$

The chemical potential for an ideal Bose gas has to be lower than the ground-state energy. Otherwise the occupancy $\langle n_j \rangle$ of some state j would become negative.

Before proceeding, an order of magnitude calculation is in order. For $N = 10^{20}$ at $T = 1$ K, one obtains $\mu = -1.4 \times 10^{-36}$ ergs. For a He^4 atom (mass M) in a cube with $V = L^3$, the two lowest states correspond to $(n_x, n_y, n_z) = (1, 1, 1)$, and $(2, 1, 1)$. The difference in these two energies is $\Delta \epsilon = \epsilon(211) - \epsilon(111) = 3\hbar^2 \pi^2 / (2ML^2)$. For a box with $L = 1$ cm containing He^4 particles, $\Delta \epsilon = 2.5 \times 10^{-30}$ ergs. This is very small compared to $k_b T$, which even at 1 mK is 1.38×10^{-19} ergs. On the other hand, $\Delta \epsilon$ is large compared to μ, which at 1 mK is -1.4×10^{-39} ergs. Thus the occupancy of the (211) orbital is $\langle n_{211} \rangle \approx [\exp(\beta \Delta \epsilon) - 1]^{-1} \approx [\beta \Delta \epsilon]^{-1} \approx 0.5 \times 10^{11}$, and $\langle n_{111} \rangle \approx N \approx 10^{20}$, so that the ratio $\langle n_{211} \rangle / \langle n_{111} \rangle \approx 10^{-9}$, a very small fraction.

For a spin-zero particle in a cubic box, the density of states is

$$\mathcal{D}(\epsilon) = \frac{V}{4\pi^2} \left(\frac{2M}{\hbar^2} \right)^{\frac{3}{2}} \epsilon^{\frac{1}{2}}.$$

The total number of particles in an ideal Bose gas at low temperatures needs to be written such that the ground-state occupancy is separated from the excited-state occupancies:

$$N = \sum_j \langle n_j \rangle = \langle n_0 \rangle + \sum_{j \neq 0} \langle n_j \rangle \equiv N_0(T) + N_e(T)$$

$$= N_0(T) + \int_0^\infty d\epsilon \, \mathcal{D}(\epsilon) [z^{-1} e^{\beta \epsilon} - 1]^{-1}$$

$$= \frac{1}{z^{-1} - 1} + \frac{V}{4\pi^2} \left(\frac{2M}{\hbar^2} \right)^{\frac{3}{2}} \int_0^\infty d\epsilon \, \frac{\epsilon^{\frac{1}{2}}}{z^{-1} e^{\beta \epsilon} - 1}. \tag{A2.2.176}$$

Since $\mathcal{D}(\epsilon)$ is zero when $\epsilon = 0$, the ground state does not contribute to the integral for N_e. At sufficiently low temperatures, N_0 will be very large compared to one, which implies z is very close to one. Then one can

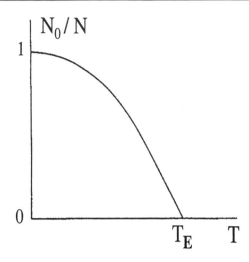

Figure A2.2.7. Fraction of Bose particles in the ground state as a function of the temperature.

approximate z by one in the integrand for N_e. Then the integral can be evaluated by using the transformation $x = \beta\epsilon$ and the known value of the integral

$$\int_0^\infty dx \, \frac{x^{\frac{1}{2}}}{e^x - 1} = 1.306\pi^{\frac{1}{2}}.$$

The result for the total number in the excited states is

$$N_e(T) = \frac{1.306V}{4}\left(\frac{2Mk_bT}{\pi\hbar^2}\right)^{\frac{3}{2}} = 2.612n_q V \tag{A2.2.177}$$

where $n_q \equiv V_q^{-1} = [(Mk_bT)/(2\pi\hbar^2)]^{\frac{3}{2}}$ is the quantum concentration. The fraction $N_e/N \approx 2.612n_q/n$ where $n = N/V$. This ratio can also be written as

$$\frac{N_e}{N} = \left(\frac{T}{T_E}\right)^{\frac{3}{2}} \qquad \text{where } T_E = \frac{2\pi\hbar^2}{Mk_b}\left(\frac{N}{2.612V}\right)^{\frac{2}{3}}. \tag{A2.2.178}$$

T_E is called the Einstein temperature; $N_e(T_E) = N$. Above T_E the ground-state occupancy is not a macroscopic number. Below T_E, however, N_0 begins to become a macroscopic fraction of the total number of particles according to the relation

$$N_0 = N - N_e = N\left[1 - \left(\frac{T}{T_E}\right)^{\frac{3}{2}}\right]. \tag{A2.2.179}$$

The function $N_0(T)$ is sketched in figure A2.2.7. At zero temperature all the Bose particles occupy the ground state. This phenomenon is called the Bose–Einstein condensation and T_E is the temperature at which the transition to the condensation occurs.

A2.2.6 Summary

In this chapter, the foundations of equilibrium statistical mechanics are introduced and applied to ideal and weakly interacting systems. The connection between statistical mechanics and thermodynamics is made by introducing ensemble methods. The role of mechanics, both quantum and classical, is described. In particular, the concept and use of the density of states is utilized. Applications are made to ideal quantum and classical gases, ideal gas of diatomic molecules, photons and the black body radiation, phonons in a harmonic solid, conduction electrons in metals and the Bose–Einstein condensation. Introductory aspects of the density expansion of the equation of state and the expansion of thermodynamic quantities in powers of $\hbar$ are also given. Other chapters deal with the applications to the strongly interacting systems, and the critical phenomena. Much of this section is restricted to equilibrium systems. Other sections discuss kinetic theory of fluids, chaotic and other dynamical systems, and other non-equilibrium phenomena.

References

[1]   Born M and Huang K 1954 *Dynamical Theory of Crystal Lattices* (Oxford: Clarendon)
[2]   Corak *et al* 1955 *Phys. Rev.* **98** 1699

Further Reading

Chandler D 1987 *Introduction to Modern Statistical Mechanics* (Oxford: Oxford University Press)
Hill T L 1960 *Introduction to Statistical Thermodynamics* (Reading, MA: Addison-Wesley)
Huang K 1987 *Statistical Mechanics* 2nd edn (New York: Wiley)
Kittel C and Kroemer H 1980 *Thermal Physics* 2nd edn (San Francisco, CA: Freeman)
Landau L D and Lifshitz E M 1980 *Statistical Physics* part 1, 3rd edn (Oxford: Pergamon)
Ma S-K 1985 *Statistical Mechanics* (Singapore: World Scientific)
Pathria R K 1972 *Statistical Mechanics* (Oxford: Pergamon)
Pauli W 1977 Statistical mechanics *Lectures on Physics* vol 4, ed C P Enz (Cambridge, MA: MIT)
Plischke M and Bergensen B 1989 *Equilibrium Statistical Physics* (Englewood Cliffs, NJ: Prentice-Hall)
Toda M, Kubo R and Saito N 1983 *Statistical Physics* I (Berlin: Springer)

A2.3
Statistical mechanics of strongly interacting systems: liquids and solids

Jayendran C Rasaiah

Had I been present at the creation, I would have given some useful hints for the better ordering of the universe.
Alphonso X, Learned King of Spain, 1252–1284

A2.3.1   Introduction

Statistical mechanics provides a link between the microscopic properties of a system at an atomic or molecular level, and the equilibrium and dynamic properties measured in a laboratory. The statistical element follows from the enormously large number of particles involved, of the order of Avogadro's number (6.023×10^{23}), and the assumption that the measured properties (e.g. the pressure) are averages over instantaneous values. The equilibrium properties are determined from the partition function, while the transport coefficients of a system, not far from its equilibrium state, are related to equilibrium time correlation functions in the so-called linear response regime.

Fluctuations of observables from their average values, unless the observables are constants of motion, are especially important, since they are related to the response functions of the system. For example, the constant volume specific heat C_v of a fluid is a response function related to the fluctuations in the energy of a system at constant N, V and T, where N is the number of particles in a volume V at temperature T. Similarly, fluctuations in the number density ($\rho = N/V$) of an open system at constant μ, V and T, where μ is the chemical potential, are related to the isothermal compressibility κ_T, which is another response function. Temperature-dependent fluctuations characterize the dynamic equilibrium of thermodynamic systems, in contrast to the equilibrium of purely mechanical bodies in which fluctuations are absent.

In this chapter we discuss the main ideas and results of the equilibrium theory of strongly interacting systems. The partition function of a weakly interacting system, such as an ideal gas, is easily calculated to provide the absolute free energy and other properties (e.g. the entropy). The determination of the partition function of a strongly interacting system, however, is much more difficult, if not impossible, except in a few special cases. The special cases include several one-dimensional systems (e.g. hard rods, the one-dimensional (1D) Ising ferromagnet), the two-dimensional (2D) Ising model for a ferromagnet at zero magnetic field and the entropy of ice. Onsager's celebrated solution of the 2D Ising model at zero field profoundly influenced our understanding of strongly interacting systems near the critical point, where the response functions diverge. Away from this region, however, the theories of practical use to most chemists, engineers and physicists are approximations based on a mean-field or average description of the prevailing interactions. Theories of fluids in which, for example, the weaker interactions due to dispersion forces or the polarity of the molecules are treated as perturbations to the harsh repulsive forces responsible for the structure of the fluid, also fall into the mean-field category.

The structure of a fluid is characterized by the spatial and orientational correlations between atoms and molecules determined through x-ray and neutron diffraction experiments. Examples are the atomic pair correlation functions (g_{oo}, g_{oh}, g_{hh}) in liquid water. An important feature of these correlation functions is that the thermodynamic properties of a system can be calculated from them. The information they contain is equivalent to that present in the partition function, and is more directly related to experimental observations. It is therefore natural to focus attention on the theory of these correlation functions, which is now well developed, especially in the region away from the critical point. Analytic and numerical approximations to the correlations functions are more readily formulated than for the corresponding partition functions from which they are derived. This has led to several useful theories, which include the scaled particle theory for hard bodies and integral equations approximations for the two body correlation functions of simple fluids. Examples are the Percus–Yevick, mean spherical and hypernetted chain approximations which are briefly described in this chapter and perturbation theories of fluids which are treated in greater detail.

We discuss classical non-ideal liquids before treating solids. The strongly interacting fluid systems of interest are hard spheres characterized by their harsh repulsions, atoms and molecules with dispersion interactions responsible for the liquid–vapour transitions of the rare gases, ionic systems including strong and weak electrolytes, simple and not quite so simple polar fluids like water. The solid phase systems discussed are ferromagnets and alloys.

A2.3.2   Classical non-ideal fluids

The main theoretical problem is to calculate the partition function given the classical Hamitonian

$$H(r^N, p^N) = K(p^N) + E_{int} + U_N(r^N, \omega^N) \tag{A2.3.1}$$

where $K(p^N)$ is the kinetic energy, E_{int} is the internal energy due to vibration, rotation and other internal degrees of freedom and

$$U_N(r^N, \omega^N) = \sum u_{ij}(r_{ij}, \omega_i, \omega_j) + \sum u_{ijk}(r_{ij}, r_{ik}, r_{kj}, \omega_i, \omega_j, \omega_k) + \cdots \tag{A2.3.2}$$

is the intermolecular potential composed of two-body, three-body and higher-order interactions. Here p^N stands for the sets of momenta $\{p_1, p_2, \ldots, p_N\}$ of the N particles, and likewise r^N and ω^N are the corresponding sets of the positions and angular coordinates of the N particles and r_{ij} is the distance between particles i and j. For an ideal gas $U_N(r^N, \omega^N) = 0$.

A2.3.2.1   Interatomic potentials

Information about interatomic potentials comes from scattering experiments as well as from model potentials fitted to the thermodynamic and transport properties of the system. We will confine our discussion mainly to systems in which the total potential energy $U(r^N, \omega^N)$ for a given configuration $\{r^N, \omega^N\}$ is pairwise additive, which implies that the three- and higher-body potentials are ignored. This is an approximation because the fluctuating electron charge distribution in atoms and molecules determines their polarizability which is not pair-wise additive. However, the total potential can be approximated as the sum of effective pair potentials.

A few of the simpler pair potentials are listed below.

(a) The potential for hard spheres of diameter σ

$$u_{ij}(r) = \begin{cases} \infty & r < a_{ij} \\ 0 & r > a_{ij}. \end{cases} \tag{A2.3.3}$$

(b) The square well or mound potential

$$u_{ij} = \begin{cases} \infty & r < a_{ij} \\ d_{ij} & a_{ij} < r < b_{ij} \\ 0 & b_{ij} < r. \end{cases} \tag{A2.3.4}$$

Table A2.3.1. Parameters for the Lennard-Jones potential.

Substance	σ (Å)	ε/k (K)
He	2.556	10.22
Ne	2.749	35.6
Ar	3.406	119.8
Kr	3.60	171
Xe	4.10	221
CH_4	3.817	148.2

(c) The Lennard-Jones potential

$$u_{ij}^{LJ}(r) = 4\varepsilon \left[\left(\frac{\sigma}{r} \right)^{12} - \left(\frac{\sigma}{r} \right)^{6} \right] \tag{A2.3.5}$$

which has two parameters representing the atomic size σ and the well depth ε of the interatomic potential. The r^{-6} dependence of the attractive part follows from the dispersive forces between the particles, while the r^{-12} dependence is a convenient representation of the repulsive forces. The potential is zero at $r = \sigma$ and $-\varepsilon$ at the minimum when $r = 2^{1/6}\sigma$. Typical values of ε and σ are displayed in table A2.3.1.

(d) The Coulomb potential between charges e_i and e_j separated by a distance r

$$u_{ij}^{Coul}(r) = \frac{e_i e_j}{r}. \tag{A2.3.6}$$

(e) The point dipole potential

$$u^{DD}(r, \omega_i, \omega_j) = -\frac{1}{r^3}[3(\mu_i \cdot r_{ij})(\mu_j \cdot r_{ij})/r^2 - (\mu_i \cdot \mu_j)] \tag{A2.3.7}$$

where μ_i is the dipole moment of particle l and $r = |r_{ij}|$ is the intermolecular separation between the point dipoles i and j.

Thermodynamic stability requires a repulsive core in the interatomic potential of atoms and molecules, which is a manifestation of the Pauli exclusion principle operating at short distances. This means that the Coulomb and dipole interaction potentials between charged and uncharged real atoms or molecules must be supplemented by a hard core or other repulsive interactions. Examples are as follows.

(f) The restricted primitive model (RPM) for ions in solution

$$u^{RPM}(r) = u_{ij}^{HS}(r) + u_{ij}^{Coul}(r) \tag{A2.3.8}$$

in which the positive or negative charges are embedded in hard spheres of the same size in a continuum solvent of dielectric constant ε. An extension of this is the primitive model (PM) electrolyte in which the restriction of equal sizes for the oppositely charged ions is relaxed.

Other linear combinations of simple potentials are also widely used to mimic the interactions in real systems. An example is the following.

(g) The Stockmayer potential for dipolar molecules:

$$u_{ij}^{S}(r, \omega_i, \omega_j) = u^{LJ}(r) + u^{DD}(r, \omega_i, \omega_j) \tag{A2.3.9}$$

which combines the Lennard-Jones and point dipole potentials.

Table A2.3.2. Halide–water, alkali metal cation–water and water–water potential parameters (SPC/E model). In the SPC/E model for water, the charges on H are at 1.000 Å from the Lennard-Jones centre at O. The negative charge is at the O site and the HOH angle is 109.47°.

Ion/water	σ_{io} (Å)	ε_{io} (kJ mol^{-1})	Charge (q)
F$^-$	3.143	0.6998	-1
Cl$^-$	3.785	0.5216	-1
Br$^-$	3.896	0.4948	-1
I$^-$	4.168	0.5216	-1
Li$^+$	2.337	0.6700	$+1$
Na$^+$	2.876	0.5216	$+1$
K$^+$	3.250	0.5216	$+1$
Rb$^+$	3.348	0.5216	$+1$
Cs$^+$	3.526	0.5216	$+1$

Water–water	σ_{oo} (A)	ε_{oo} (kJ mol^{-1})	Charge (q)
O(H$_2$O)	3.169	0.6502	-0.8476
H(H$_2$O)			$+0.4238$

Important applications of atomic potentials are models for water (TIP3, SPC/E) in which the intermolecular potential consists of atom–atom interactions between the oxygen and hydrogen atoms of distinct molecules, with the characteristic atomic geometry maintained (i.e. an HOH angle of 109° and a intramolecular OH distance of 1 Å) by imposing constraints between atoms of the same molecule. For example, the effective simple point charge model (SPC/E) for water is defined as a linear combination of Lennard-Jones interactions between the oxygen atoms of distinct molecules and Coulombic interactions between the charges adjusted for a self-polarization correction.

The SPC/E model approximates many-body effects in liquid water and corresponds to a molecular dipole moment of 2.35 Debye (D) compared to the actual dipole moment of 1.85 D for an isolated water molecule. The model reproduces the diffusion coefficient and thermodynamics properties at ambient temperatures to within a few per cent, and the critical parameters (see below) are predicted to within 15%. The same model potential has been extended to include the interactions between ions and water by fitting the parameters to the hydration energies of small ion–water clusters. The parameters for the ion–water and water–water interactions in the SPC/E model are given in table A2.3.2.

A2.3.2.2 Equations of state, the virial series and the liquid–vapour critical point

The equation of state of a fluid relates the pressure (P), density (ρ) and temperature (T),

$$P = P(\rho, T). \tag{A2.3.10}$$

It is determined experimentally; an early study was the work of Andrews on carbon dioxide [1]. The exact form of the equation of state is unknown for most substances except in rather simple cases, e.g. a 1D gas of hard rods. However, the ideal gas law $P = \rho kT$, where k is Boltzmann's constant, is obeyed even by real fluids at high temperature and low densities, and systematic deviations from this are expressed in terms of the virial series:

$$Z = P/\rho kT = 1 + B_2(T)\rho + B_3(T)\rho^2 + \cdots \tag{A2.3.11}$$

which is an expansion of the compressibility factor $Z = P/\rho kT$ in powers of the number density ρ at constant temperature. Here $B_2(T)$, $B_3(T)$, ..., $B_n(T)$ etc are the second, third, ... and nth virial coefficients

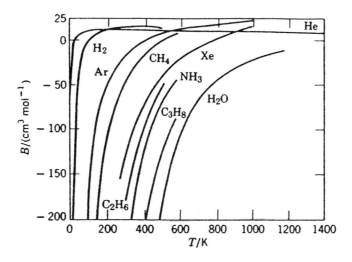

Figure A2.3.1. Second virial coefficient $B_2(T)$ of several gases as a function of temperature T. (From [10]).

determined by the intermolecular potentials as discussed later in this chapter. They can be determined experimentally, but the radius of convergence of the virial series is not known. Figure A2.3.1 shows the second virial coefficient plotted as a function of temperature for several gases.

The temperature at which $B_2(T)$ is zero is the Boyle temperature T_B. The excess Helmholtz free energy follows from the thermodynamic relation

$$\frac{\beta A^{\text{ex}}}{N} = \int_0^\rho \left(\frac{P}{\rho kT} - 1 \right) \, \mathrm{d} \ln \rho$$
$$= \sum_{n=2}^{\infty} \frac{B_n(T)}{n-1} \rho^{n-1}. \tag{A2.3.12}$$

The first seven virial coefficients of hard spheres are positive and no Boyle temperature exists for hard spheres.

Statistical mechanical theory and computer simulations provide a link between the equation of state and the interatomic potential energy functions. A fluid–solid transition at high density has been inferred from computer simulations of hard spheres. A vapour–liquid phase transition also appears when an attractive component is present in the interatomic potential (e.g. atoms interacting through a Lennard-Jones potential) provided the temperature lies below T_c, the critical temperature for this transition. This is illustrated in figure A2.3.2 where the critical point is a point of inflexion of the critical isotherm in the P–V plane.

Below T_c, liquid and vapour coexist and their densities approach each other along the coexistence curve in the T–V plane until they coincide at the critical temperature T_c. The coexisting densities in the critical region are related to $T - T_c$ by the power law

$$|\rho_l - \rho_g| \approx |T - T_c|^\beta \tag{A2.3.13}$$

where β is called a critical exponent. The pressure P approaches the critical pressure P_c along the critical isotherm like

$$P - P_c \approx |\rho - \rho_c|^\delta \tag{A2.3.14}$$

which defines another critical exponent δ. The isothermal compressibility κ_T and the constant volume specific heat C_V are response functions determined by fluctuations in the density and the energy. They diverge at the

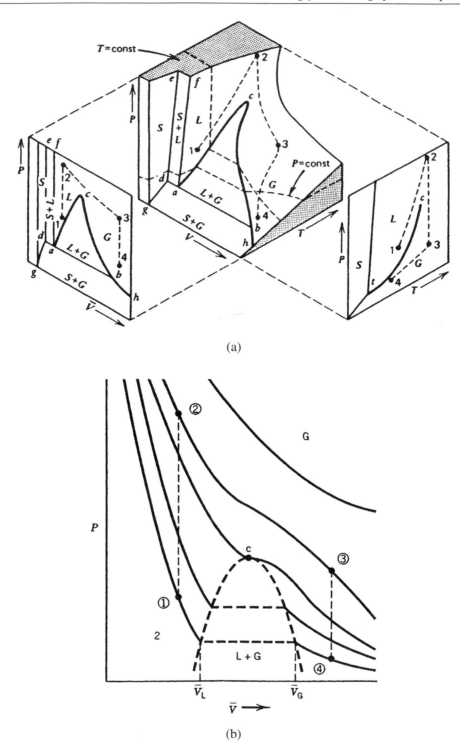

(a)

(b)

Figure A2.3.2. (a) P–V–T surface for a one-component system that contracts on freezing. (b) P–V isotherms in the region of the critical point.

critical point, and determine two other critical exponents α and γ defined, along the critical isochore, by

$$\kappa_T \approx |T - T_c|^\gamma \tag{A2.3.15}$$

and

$$C_V \approx |T - T_c|^\alpha. \tag{A2.3.16}$$

As discussed elsewhere in this encyclopaedia, the critical exponents are related by the following expressions:

$$\alpha + 2\beta + \gamma = 2$$
$$\alpha + \beta(\delta + 1) = 2$$
$$\gamma - \beta(\delta - 1) = 0. \tag{A2.3.17}$$

The individual values of the exponents are determined by the symmetry of the Hamiltonian and the dimensionality of the system.

Although the exact equations of state are known only in special cases, there are several useful approximations collectively described as mean-field theories. The most widely known is van der Waals' equation [2]

$$P = \frac{\rho k T}{1 - b\rho} - a\rho. \tag{A2.3.18}$$

The parameters a and b are characteristic of the substance, and represent corrections to the ideal gas law due to the attractive (dispersion) interactions between the atoms and the volume they occupy due to their repulsive cores. We will discuss van der Waals' equation in some detail as a typical example of a mean-field theory.

van der Waals' equation shows a liquid–gas phase transition with the critical constants $\rho_c = (1/3b)$, $P_c = a/(27b^2)$, $T_c = 8a/(27kb)$. This follows from the property that at the critical point on the P–V plane there is a point of inflexion and $(d^2 P/dV^2)_c = (dP/dV)_c = 0$. These relations determine the parameters a and b from the experimentally determined critical constants for a substance. The compressibility factor $Z_c = P_c/\rho_c k T_c$, however, is 3/8 in contrast to the experimental value of about ≈ 0.30 for the rare gases. By expanding van der Waals' equation in powers of ρ one finds that the second virial coefficient

$$B_2(T) = b - \frac{a}{kT}. \tag{A2.3.19}$$

This is qualitatively of the right form. As $T \to \infty$, $B_2(T) \to b$ and $B_2(T) = 0$ at the Boyle temperature, $T_B = a/(kb)$. This provides another route to determining the parameters a and b.

van der Waals' equation of state is a cubic equation with three distinct solutions for V at a given P and T below the critical values. Subcritical isotherms show a characteristic loop in which the middle portion corresponds to positive $(dP/dV)_T$ representing an unstable region.

The coexisting densities below T_c are determined by the equalities of the chemical potentials and pressures of the coexisting phases, which implies that the horizontal line joining the coexisting vapour and liquid phases obeys the condition

$$\mu_{\text{vapour}} - \mu_{\text{liquid}} = \int_{\text{liquid}}^{\text{vapour}} V \, dp = 0 \qquad (T \text{ constant}). \tag{A2.3.20}$$

This is the well known equal areas rule derived by Maxwell [3], who enthusiastically publicized van der Waal's equation (see figure A2.3.3). The critical exponents for van der Waals' equation are typical mean-field exponents $\alpha \approx 0$, $\beta = 1/2$, $\gamma = 1$ and $\delta = 3$. This follows from the assumption, common to van der Waals' equation and other mean-field theories, that the critical point is an analytic point about which the free energy and other thermodynamic properties can be expanded in a Taylor series.

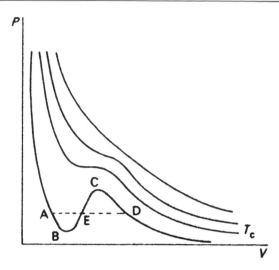

Figure A2.3.3. $P-V$ isotherms for van der Waals' equation of state. Maxwell's equal areas rule (area ABE = area ECD) determines the volumes of the coexisting phases at subcritical temperatures.

van der Waals' equation can be written in the reduced form

$$P_R = \frac{8T_R}{(3V_R - 1)} - \frac{3}{V_R^2} \qquad\qquad (A2.3.21)$$

using the reduced variables $V_R = V/V_c$, $P_R = P/P_c$ and $T_R = T/T_c$. The reduced second virial coefficient

$$B_{2,R}(T_R) = \frac{B_2(T)}{\bar{V}_c} = \left[\frac{1}{3} - \frac{9}{8T_R}\right], \qquad\qquad (A2.3.22)$$

where $\bar{V}_c = V_c/N$ and the reduced Boyle temperature $T_{R,B} = 27/8$. The reduced forms for the equation of state and second virial coefficient are examples of the law of corresponding states. The statistical mechanical basis of this law is discussed in this chapter and has wider applicability than the van der Waals form.

A2.3.3 Ensembles

An ensemble is a collection of systems with the same $r + 2$ variables, at least one of which must be extensive. Here r is the number of components. For a one-component system there are just three variables and the ensembles are characterized by given values of N, V, T (canonical), μ, V, T (grand canonical), N, V, E (microcanonical) and N, P, T (constant pressure). Our discussion of strongly interacting systems of classical fluids begins with the fundamental equations in canonical and grand canonical ensembles. The results are equivalent to each other in the thermodynamic limit. The particular choice of an ensemble is a matter of convenience in developing the theory, or in treating the fluctuations like the density or the energy.

A2.3.3.1 *Canonical ensemble* (N, V, T)

This is a collection of closed systems with the same number of particles N and volume V (constant density) for each system at temperature T. The partition function

$$Q(N, V, T) = \frac{Q_{\text{int}}}{N!\Lambda^{3N}} Z(N, V, T) \qquad\qquad (A2.3.23)$$

where Q_{int} is the internal partition function (PF) determined by the vibration, rotation, electronic states and other degrees of freedom. It can be factored into contributions q_{int} from each molecule so that $Q_{int} = q_{int}^N$. The factor $1/\Lambda^{3N}$ is the translational PF in which $\Lambda = h/(2\pi m kT)^{1/2}$ is the thermal de Broglie wavelength. The configurational PF assuming classical statistics for this contribution is

$$Z(N, V, T) = \frac{1}{\Omega^N} \int \exp(-\beta U_N(r^N, \omega^N)) \, dr^N \, d\omega^N \qquad (A2.3.24)$$

where Ω is 4π for linear molecules and $8\pi^2$ for nonlinear molecules. The classical configurational PF is independent of the momenta and the masses of the particles. In the thermodynamic limit ($N \to \infty$, $V \to \infty$, $N/V = \rho$), the Helmholtz free energy

$$A = -kT \ln Q(N, V, T). \qquad (A2.3.25)$$

Other thermodynamic properties are related to the PF through the equation

$$dA = -S \, dT - p \, dV + \sum_i \mu_i \, dN_i \qquad (A2.3.26)$$

where μ_i is the chemical potential of species i, and the summation is over the different species. The pressure

$$P = -kT (\partial \ln Z/\partial V)_{N,T} \qquad (A2.3.27)$$

and since the classical configurational PF Z is independent of the mass, so is the equation of state derived from it. Differences between the equations of state of isotopes or isotopically substituted compounds (e.g. H_2O and D_2O) are due to quantum effects.

For an ideal gas, $U(r^N, \omega^N) = 0$ and the configurational PF $Z(N, V, T) = V^N$. Making use of Sterling's approximation for $N! \approx (e/N)^N$ for large N, it follows that the Helmholtz free energy

$$A^{ideal} = -NkT \ln(q_{int}e/\Lambda^3) + NkT \ln \rho \qquad (A2.3.28)$$

and the chemical potential

$$\mu^{ideal} = (\partial A^{ideal}/\partial N)_{V,T} = kT \ln(\Lambda^3/q_{int}) + kT \ln \rho. \qquad (A2.3.29)$$

The excess Helmholtz free energy

$$A^{ex} = A - A^{ideal} = -kT \ln[Z(N, V, T)/V^N] \qquad (A2.3.30)$$

and the excess chemical potential

$$\begin{aligned} \mu^{ex} &= (\partial A^{ex}/\partial N)_{V,T} \\ &\approx A^{ex}(N+1, V, T) - A^{ex}(N, V, T) \qquad \text{for large } N \\ &= kT \ln[V Z(N, V, T)/Z(N+1, V, T)]. \end{aligned} \qquad (A2.3.31)$$

Confining our attention to angularly-independent potentials to make the argument and notation simpler,

$$Z(N+1, V, T) = \int \exp(-\beta U_{N+1}(r^{N+1})) \, dr^{N+1} = V \int \exp(-\beta U_{N+1}(r^{N+1})) \, dr^N$$

in which

$$U_{N+1}(r^{N+1}) = U_N(r^N) + \Delta U_N(r_{N+1}, r^N)$$

where $\Delta U_N\,(r_{N+1}, r^N)$ is the interaction of the $(N + 1)$th particle situated at r_{N+1} with the remaining N particles at coordinates $r^N = \{r_1, r_2, \ldots, r_N\}$. Substituting this into the expression for $Z(N + 1, V, T)$ we see that the ratio

$$\frac{Z(N + 1, V, T)}{Z(N, V, T)} = \frac{V \int \exp(-\beta U_N(r^N)) \exp(-\beta \Delta U_N(r_{N+1}, r^N))\, dr_{N+1}\, dr^N}{Z(N, V, T)}.$$

But $\exp(-\beta U_N(r^N))/Z(N, V, T)$ is just the probability that the N particle system is in the configuration $\{r^N\}$, and it follows that

$$\mu^{\text{ex}} = -kT \ln[Z(N + 1, V, T)/VZ(N, V, T)]$$
$$= -kT \ln\langle\exp(-\beta \Delta U_N(r_{N+1}, r^N))\rangle_N \qquad (\text{A2.3.32})$$

where $\langle\cdots\rangle_N$ represents a configurational average over the canonical N particle system. This expression for μ^{ex}, proposed by Widom, is the basis of the particle insertion method used in computer simulations to determine the excess chemical potential. The exponential term is either zero or one for hard core particles, and

$$\mu^{\text{ex}} = -kT \ln P(0, v) \qquad (\text{A2.3.33})$$

where $P(0, v)$ is the probability of forming a cavity of the same size and shape as the effective volume v occupied by the particle.

A2.3.3.2   Grand canonical ensemble (μ, V, T)

This is a collection of systems at constant μ, V and T in which the number of particles can fluctuate. It is of particular use in the study of open systems. The PF

$$\Xi(\mu, V, T) = \sum_{n=0}^{\infty} Q(N, V, T)\lambda^N = \sum_{n=0}^{\infty} \frac{Z(N, V, T)}{N!} z^N \qquad (\text{A2.3.34})$$

where the absolute activity $\lambda = \exp(\mu/kT)$ and the fugacity $z = q_{\text{int}}\lambda/\Lambda^3$. The first equation shows that the grand PF is a generating function for the canonical ensemble PF. The chemical potential

$$\mu = kT \ln(\Lambda^3/q_{\text{int}}) + kT \ln z. \qquad (\text{A2.3.35})$$

For an ideal gas $Z(N, V, T) = V^N$, we have seen earlier that

$$\mu^{\text{ideal}} = kT \ln(\Lambda^3/q_{\text{int}}) + kT \ln \rho. \qquad (\text{A2.3.36})$$

The excess chemical potential

$$\mu^{\text{ex}} = \mu - \mu^{\text{ideal}} = kT \ln(z/\rho) \qquad (\text{A2.3.37})$$

and $(z/\rho) \to 1$ as $\rho \to 0$. In the thermodynamic limit $(v \to \infty)$, the pressure

$$p = \frac{kT}{V} \ln \Xi(\mu, V, T). \qquad (\text{A2.3.38})$$

The characteristic thermodynamic equation for this ensemble is

$$d(pV) = S\,dT + p\,dv + \sum N_i\,d\mu_i \qquad (\text{A2.3.39})$$

from which the thermodynamic properties follow. In particular, for a one-component system,

$$\langle N \rangle = kT \{ \partial (PV)/\partial \mu \}_{T,V} = kT \{ z\partial \ln \Xi / \partial z \}_{T,V}. \tag{A2.3.40}$$

Note that the average density $\rho = \langle N \rangle / V$. Defining $\chi(z) = (1/V) \ln \Xi$, one finds that

$$P/kT = \chi(z)$$
$$\rho = z\partial \chi(z)/\partial z. \tag{A2.3.41}$$

On eliminating z between these equations, we get the equation of state $P = P(\rho, T)$ and the virial coefficients by expansion in powers of the density. For an ideal gas, $Z(N, V, T) = V^N$, $\Xi = \exp(zV)$ and $P/kT = z = \chi(z) = \rho$.

A2.3.3.3   The virial coefficients

The systematic calculation of the virial coefficients, using the grand canonical ensemble by eliminating z between the equations presented above, is a technically formidable problem. The solutions presented using graph theory to represent multidimensional integrals are elegant, but impractical beyond the first six or seven virial coefficients due to the mathematical complexity and labour involved in calculating the integrals. However, the formal theory led to the development of density expansions for the correlation functions, which have proved to be extremely useful in formulating approximations for them.

A direct and transparent derivation of the second virial coefficient follows from the canonical ensemble. To make the notation and argument simpler, we first assume pairwise additivity of the total potential with no angular contribution. The extension to angularly-independent non-pairwise additive potentials is straightforward. The total potential

$$U_N(r^N) = \sum u_{ij}(r_{ij}) \tag{A2.3.42}$$

and the configurational PF assuming classical statistics is

$$Z(N, V, T) = \int \exp\left(-\beta \sum_{i<j} u(r_{ij})\right) dr_1 \ldots dr_N = \int \prod_{i<j} \exp(-\beta u(r_{ij})) \, dr_1 \ldots dr_N. \tag{A2.3.43}$$

The Mayer function defined by

$$f_{ij}(r_{ij}) = \exp(-\beta u_{ij}(r_{ij})) - 1 \tag{A2.3.44}$$

figures prominently in the theoretical development [7]. It is a step function for hard spheres:

$$f_{ij}^{HS}(r_{ij}) = \begin{cases} -1 & r < \sigma \\ 0 & r > \sigma \end{cases} \tag{A2.3.45}$$

where σ is the sphere diameter. More generally, for potentials with a repulsive core and an attractive well, $f_{ij}(r_{ij})$ has the limiting values -1 and 0 as $r \to 0$ and ∞, respectively, and a maximum at the interatomic distance corresponding to the minimum in $u_{ij}(r_{ij})$. Substituting $(1 + f_{ij}(r_{ij}))$ for $\exp(-\beta u_{ij}(r_{ij}))$ decomposes the configurational PF into a sum of terms:

$$Z(N, V, T) = \int \prod_{i<j} (1 + f_{ij}(r_{ij})) \, dr_1 \ldots dr_N$$
$$= \int \left[1 + \sum_{i<j} f_{ij}(r_{ij}) + \cdots \right] dr_1 \ldots dr_N$$
$$= V^N + \frac{N(N-1)}{2} V^{N-2} \int f_{12}(r_{12}) \, dr_1 \, dr_2 + \cdots$$

$$= V^N \left[1 + \frac{N(N-1)}{2V^2} \int f_{12}(r_{12}) \, dr_1 \, dr_2 + \cdots \right]$$

$$= V^N \left[1 - \frac{N(N-1)}{V} I_2(T) + \cdots \right] \tag{A2.3.46}$$

where

$$I_2(T) = -(1/2V) \int\int f_{12}(r_{12}) \, dr_1 \, dr_2.$$

The third step follows after interchanging summation and integration and recognizing that the $N(N-1)/2$ terms in the sum are identical. The pressure follows from the relation

$$P = kT \left(\frac{\partial \ln Z(N, V, T)}{\partial V} \right)_{N,T}$$

$$= \frac{NkT}{V} + \frac{kT}{1 - [N(N-1)/V]I_2(T)} \frac{N(N-1)}{V^2} I_2(T) + \cdots$$

$$= \frac{NkT}{V} \left[1 + \frac{N-1}{V} I_2(T) + \cdots \right]. \tag{A2.3.47}$$

In the thermodynamic limit ($N \to \infty$, $V \to \infty$ with $N/V = \rho$), this is just the virial expansion for the pressure, with $I_2(T)$ identified as the second virial coefficient

$$B_2(T) = -1/(2V) \int\int f_{12}(r_{12}) \, dr_1 \, dr_2$$

$$= -1/(2V) \int\int f_{12}(r_{12}) \, dr_1 \, dr_{12}.$$

The second step involves a coordinate transformation to the origin centred at particle 1 with respect to which the coordinates of particle 2 are defined. Since f_{12} depends only on the distance between particles 1 and 2, integration over the position of 1 gives a factor V that cancels the V in the denominator:

$$B_2(T) = -(1/2) \int f_{12}(r_{12}) \, dr_{12}. \tag{A2.3.48}$$

Finally, the assumed spherical symmetry of the interactions implies that the volume element dr_{12} is $4\pi r_{12}^2 \, dr_{12}$. For angularly-dependent potentials, the second virial coefficient

$$B_2(T) = -1/(2\Omega^2) \int\int f_{12}(r_{12}, \omega_1, \omega_2) \, dr_{12} \, d\omega_1 \, d\omega_2 \tag{A2.3.49}$$

where $f_{12}(r_{12}, \omega_1, \omega_2)$ is the corresponding Mayer f-function for an angularly-dependent pair potential $u_{12}(r_{12}, \omega_1, \omega_2)$.

The nth virial coefficient can be written as sums of products of Mayer f-functions integrated over the coordinates and orientations of n particles. The third virial coefficient for spherically symmetric potentials is

$$B_3(T) = -1/(3\Omega^3) \int\int f_{12}(r_{12}) f_{13}(r_{13}) f_{23}(r_{23}) \, dr_{12} \, dr_{13}. \tag{A2.3.50}$$

If we represent the f-bond by a line with two open circles to denote the coordinates of the particle 1 and 2, then the first two virial coefficients can be depicted graphically as

$$B_2 \ (T) = \text{-}1/2 \ \bullet\!\!-\!\!-\!\!\circ \tag{A2.3.51}$$

and

$$B_3\,(T) = -1/3 \quad \triangle \qquad\qquad (A2.3.52)$$

where blackening a circle implies integration over the coordinates of the particle represented by the circle. The higher virial coefficients can be economically expressed in this notation by extending it to include the symmetry number [5].

For hard spheres of diameter σ, $f_{12}(r_{12}) = -1$ for $r < \sigma$ and is zero otherwise. It follows that

$$B_2^{HS} = b_0 = 2\pi\sigma^3/3 \qquad\qquad (A2.3.53)$$

where the second virial coefficient, abbreviated as b_0, is independent of temperature and is four times the volume of each sphere. This is called the excluded volume correction per molecule; the difference between the system volume V and the excluded volume of the molecules is the actual volume available for further occupancy. The factor four arises from the fact that σ is the distance of closest approach of the centers of the two spheres and the excluded volume for a pair is the volume of a sphere of radius σ. Each molecule contributes half of this to the second virial coefficient which is equal to b_0. The third and fourth virial coefficients for hard spheres have been calculated exactly and are

$$B_3^{HS}/b_0^2 = 5/8 \qquad\qquad (A2.3.54)$$
$$B_4^{HS}/b_0^3 = [2707\pi + 438\sqrt{2} - 4131\,\mathrm{arc\,cos}\,(1/3)]/4480 \qquad\qquad (A2.3.55)$$

while the fifth, sixth and seventh virial coefficients were determined numerically by Rhee and Hoover [8]:

$$B_5^{HS}/b_0^4 \approx 0.1103$$
$$B_6^{HS}/b_0^5 \approx 0.0386$$
$$B_7^{HS}/b_0^6 \approx 0.0138. \qquad\qquad (A2.3.56)$$

They are positive and independent of temperature.

The virial series in terms of the packing fraction $\eta = \pi\rho\sigma^3/3$ is then

$$P/\rho kT = 1 + 4\eta + 10\eta^2 + 18.36\eta^3 + 28.25\eta^4 + 39.5\eta^5 + \cdots \qquad\qquad (A2.3.57)$$

which, as noticed by Carnahan and Starling [9], can be approximated as

$$P/\rho kT = 1 + 4\eta + 10\eta^2 + 18\eta^3 + 28\eta^4 + 40\eta^5 + \cdots. \qquad\qquad (A2.3.58)$$

This is equivalent to approximating the first few coefficients in this series by

$$C_n^{HS} \approx (n-1)^2 + 3(n-1) \qquad\quad \text{for } n \geq 2.$$

Assuming that this holds for all n enables the series to be summed exactly when

$$\frac{P_{CS}}{\rho kT} = 1 + \sum_{n=2}^{\infty} [(n-1)^2 + 3(n-1)]\eta^{n-1}$$
$$= \frac{1 + \eta + \eta^2 - \eta^3}{(1-\eta)^3}. \qquad\qquad (A2.3.59)$$

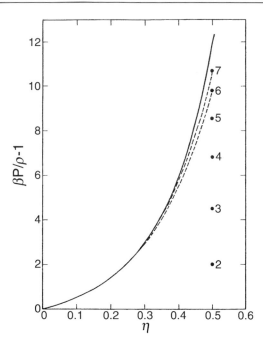

Figure A2.3.4. The equation of state $P/\rho kT - 1$, calculated from the virial series and the CS equation of state for hard spheres, as a function of $\eta = \pi\rho\sigma^3/6$ where $\rho\sigma^3$ is the reduced density.

This is Carnahan and Starling's (CS) equation of state for hard spheres; it agrees well with the computer simulations of hard spheres in the fluid region. The excess Helmholtz free energy

$$\frac{\beta A^{\text{ex}}}{N} = \int_0^\eta \left(\frac{P}{\rho kT} - 1\right) d\ln\eta = \frac{\eta(4 - 3\eta)}{(1 - \eta)^2}. \tag{A2.3.60}$$

Figure A2.3.4 compares $P/\rho kT - 1$, calculated from the CS equation of state for hard spheres, as a function of the reduced density $\rho\sigma^3$ with the virial expansion.

These equations provide a convenient and accurate representation of the thermodynamic properties of hard spheres, especially as a reference system in perturbation theories for fluids.

A2.3.3.4   Quantum corrections and path integral methods

We have so far ignored quantum corrections to the virial coefficients by assuming classical statistical mechanics in our discussion of the configurational PF. Quantum effects, when they are relatively small, can be treated as a perturbation (Friedman 1995) when the leading correction to the PF can be written as

$$Q(N, V, T) = Q_{\text{class}}(N, V, T)\left[1 - \frac{\beta^2\hbar^2}{24}\sum_{i=1}^{3N}\frac{\langle U''\rangle}{m_i} + O(\hbar^2)\right] \tag{A2.3.61}$$

where $U'' = (\partial^2 U_N/\partial r_i^2)$ is the curvature in the potential energy function for a given configuration. The curvature of the pair potential is greatest near the minimum in the potential, and is analogous to a force constant, with the corresponding angular frequency given by $(\langle U''\rangle/m_i)^{1/2}$. Expressing this as a wavenumber

ν in cm^{-1}, the leading correction to the classical Helmholtz free energy of a system with a pairwise additive potential is given by

$$\beta(A - A_{\text{classical}}) = \left(\frac{298.16}{T}\right)^2 \sum_{i=1}^{3N} \left(\frac{\nu_i}{1015.1 \text{ cm}^{-1}}\right)^2 \tag{A2.3.62}$$

which shows that the quantum correction is significant only if the mean curvature of the potential corresponds to a frequency of 1000 cm^{-1} or more [4] and the temperature is low. Thus the quantum corrections to the second virial coefficient of light atoms or molecules He^4, H$_2$, D$_2$ and Ne are significant [6, 8]. For angularly-dependent potentials such as for water, the quantum effects of the rotational contributions to the second virial coefficient also contribute to significant deviations from the classical value (10 or 20%) at low temperatures [10].

When quantum effects are large, the PF can be evaluated by path integral methods [11]. Our exposition follows a review article by Gillan [12]. Starting with the canonical PF for a system of N particles

$$Q(N, V, T) = \text{Tr } e^{-\beta H} = \sum_n \langle n | e^{-\beta H} | n \rangle \tag{A2.3.63}$$

where $\beta = 1/kT$ and the trace is taken over a complete set of orthonormal states $|n\rangle$. If the states $|n\rangle$ are the eigenstates of the Hamiltonian operator, the PF simplifies to

$$Q(N, V, T) = \sum_n \exp(-\beta E_n) \tag{A2.3.64}$$

where the E_n are the eigenvalues. The average value of an observable A is given by

$$\langle A \rangle = Q(N, V, T)^{-1} \sum_n \langle n | \hat{A} | n \rangle \exp(-\beta E_n) \tag{A2.3.65}$$

where $\hat{A}$ is the operator corresponding to the observable A. The above expression for $\langle A \rangle$ is the sum of the expectation values of A in each state weighted by the probabilities of each state at a temperature T. The difficulty in evaluating this for all except simple systems lies in (a) the enormous number of variables required to represent the state of N particles which makes the sum difficult to determine and (b) the symmetry requirements of quantum mechanics for particle exchange which must be incorporated into the sum.

To make further progress, consider first the PF of a single particle in a potential field $V(x)$ moving in one dimension. The Hamiltonian operator

$$\hat{H} = -\frac{\hbar^2}{2m} \frac{d^2}{dx^2} + V(x). \tag{A2.3.66}$$

The eigenvalues and eigenfunctions are E_n and $\phi_n(x)$ respectively. The density matrix

$$\rho(x, x'; \beta) = \sum_n \phi_n(x) \phi_n(x') \exp(-\beta E_n) \tag{A2.3.67}$$

is a sum over states; it is a function of x and x' and the temperature. The PF

$$Q(V, T) = \sum_n \exp(-\beta E_n) = \int dx \, \rho(x, x : \beta) \tag{A2.3.68}$$

which is equivalent to setting $x = x'$ in the density matrix and integrating over x. The average value of an observable A is then given by

$$\langle A \rangle = \int \hat{A} \rho(x, x')_{x=x'} \, dx \tag{A2.3.69}$$

where $\hat{A}\rho(x, x')_{x=x'}$ means $\hat{A}$ operates on $\rho(x, x'; \beta)$ and then x' is set equal to x. A differential equation for the density matrix follows from differentiating it with respect to β, when one finds

$$\frac{d}{d\beta}\rho(x, x'; \beta) = -\hat{H}\rho(x, x'; \beta) \tag{A2.3.70}$$

which is the Bloch equation. This is similar in form to the time-dependent Schrödinger equation:

$$i\hbar\frac{d\psi(x, t)}{dt} = \hat{H}\psi(x, t). \tag{A2.3.71}$$

The time evolution of the wavefunction $\psi(x, t)$ is given by

$$\psi(x, t) = \int dx'\, K(x, x'; t)\psi(x', 0) \tag{A2.3.72}$$

where $K(x, x'; t)$ is the propagator which has two important properties. The first follows from differentiating the above equation with respect to time, when it is seen that the propagator obeys the equation

$$i\hbar\frac{dK(x, x'; t)}{dt} = \hat{H}K(x, x'; t) \tag{A2.3.73}$$

with the boundary condition

$$K(x, x'; 0) = \delta(x - x'). \tag{A2.3.74}$$

Comparing this with the Bloch equation establishes a correspondence between t and $i\beta\hbar$. Putting $t = i\beta\hbar$, one finds

$$\rho(x, x'; \beta) = K(x, x'; -i\beta\hbar). \tag{A2.3.75}$$

The boundary condition is equivalent to the completeness relation

$$\sum_n \phi_n(x)\phi_n(x') = \delta(x - x'). \tag{A2.3.76}$$

The second important property of the propagator follows from the fact that time is a serial quantity, and

$$\psi(x, t_1 + t_2) = \int dx''\, K(x, x''; t_2)\psi(x'', t_1)$$
$$= \int dx'\, dx''\, K(x, x''; t_2)K(x'', x'; t_1)\psi(x', 0) \tag{A2.3.77}$$

which implies that

$$K(x, x'; t_1 + t_2) = \int dx''\, K(x, x''; t_2)K(x'', x'; t_1). \tag{A2.3.78}$$

A similar expression applies to the density matrix, from its correspondence with the propagator. For example,

$$\rho(x, x'; \beta) = \int dx''\, \rho(x, x''; \beta/2)\rho(x'', x'; \beta/2) \tag{A2.3.79}$$

and generalizing this to the product of P factors at the inverse temperature β/P,

$$\rho(x_0, x_p; \beta) = \int dx_1\, dx_2 \ldots dx_{P-1}\rho(x_0, x_1; \beta/P)\rho(x_1, x_2; \beta/P) \ldots \rho(x_{P-1}, x_P; \beta/P). \tag{A2.3.80}$$

Here P is an integer. By joining the ends x_0 and x_P and labelling this x, one sees that

$$Q(N, V, T) = \int dx\, \rho(x, x; \beta)$$

$$= \int dx_1 \ldots dx_P\, \rho(x_1, x_2; \beta/P)\rho(x_2, x_3; \beta/P)\ldots\rho(x_{P-1}, x_P; \beta/P) \quad \text{(A2.3.81)}$$

which has an obvious cyclic structure of P beads connecting P density matrices at an inverse temperature of β/P. This increase in temperature by a factor P for the density matrix corresponds to a short time approximation for the propagator reduced by the same factor P.

To evaluate the density matrix at high temperature, we return to the Bloch equation, which for a free particle ($V(x) = 0$) reads

$$\frac{d\rho(x, x'; \beta)}{d\beta} = \frac{\hbar^2}{2m}\frac{d^2\rho(x, x'; \beta)}{dx^2} \quad \text{(A2.3.82)}$$

which is similar to the diffusion equation

$$\frac{d\rho}{dt} = -D\frac{d^2\rho}{dx^2}. \quad \text{(A2.3.83)}$$

The solution to this is a Gaussian function, which spreads out in time. Hence the solution to the Bloch equation for a free particle is also a Gaussian:

$$\rho(x, x'; \beta) = \left(\frac{m}{2\pi\beta\hbar^2}\right)^{1/2}\exp\left[-\frac{m}{2\beta\hbar^2}(x - x')^2 - \beta V\right] \quad \text{(A2.3.84)}$$

where V is zero. The above solution also applies to a single-particle, 1D system in a constant potential V. For a single particle acted on by a potential $V(x)$, we can treat it as constant in the region between x and x', with a value equal to its mean, when β is small. One then has the approximation

$$\rho(x, x'; \beta/P) \approx \left(\frac{m}{2\pi\beta\hbar^2}\right)^{1/2}\exp\left[-\frac{mP}{2\beta\hbar^2}(x - x')^2 - \frac{\beta[V(x) + V(x')]}{2P}\right] \quad \text{(A2.3.85)}$$

which leads to the following expression for the PF of a single particle in a potential $V(x)$:

$$Q(V, T) \approx Q_P(V, T) = \left(\frac{mP}{2\pi\beta\hbar}\right)^{P/2}\int dx_1\, dx_2 \ldots dx_P \exp\left[-\beta\sum_{s=1}^{P}\frac{mP}{2\beta\hbar^2}(x_{s+1} - x_s)^2 + P^{-1}V_s(x)\right]. \quad \text{(A2.3.86)}$$

Feynman showed that this is exact in the limit $P \to \infty$. The expression for Q_P has the following important characteristics.

(a) It is the classical PF for a polymer chain of P beads coupled by the harmonic force constant

$$k = \frac{mP}{\beta^2\hbar^2} \quad \text{(A2.3.87)}$$

with each bead acted on by a potential $V(x)/P$. The spring constant comes from the kinetic energy operator in the Hamiltonian.

(b) The cyclic polymer chain has a mean extension, characterized by the root mean square of its radius of gyration

$$\Delta^2 = P^{-1}\left\langle\sum_{s=1}^{P}\Delta x_s^2\right\rangle \quad \text{(A2.3.88)}$$

where $\Delta x_s = x_s - \bar{x}$ and $\bar{x}$ is the instantaneous centre of mass of the chain, defined by

$$\bar{x} = P^{-1} \sum_{s=1}^{P} x_s. \tag{A2.3.89}$$

For free particles, the mean square radius of gyration is essentially the thermal wavelength to within a numerical factor, and for a 1D harmonic oscillator in the $P \to \infty$ limit,

$$\Delta^2 = (\beta m \omega_0^2)^{-1}[(\beta \hbar \omega_0/2) \coth(\beta \hbar \omega_0/2) - 1] \tag{A2.3.90}$$

where ω_0 is the frequency of the oscillator. As $T \to 0$, this tends to the mean square amplitude of vibration in the ground state.

(c) The probability distribution of finding the particle at x_1 is given by

$$P(x_1) \approx \frac{\int \ldots \int dx_2 \ldots dx_P \exp[-(k/2)(x_{s+1} - x_s)^2 + P^{-1}V(x_s)]}{\int \ldots \int dx_1 \ldots dx_P \exp[-(k/2)(x_{s+1} - x_s)^2 + P^{-1}V(x_s)]} \tag{A2.3.91}$$

which shows that it is the same as the probability of finding one particular bead at x_1 in the classical isomorph, which is the same as $1/P$ times the probability of finding any particular one of the P beads at x_1. The isomorphism between the quantum system and the classical polymer chain allows a variety of techniques, including simulations, to study these systems.

The eigenfunctions of a system of two particles are determined by their positions x and y, and the density matrix is generalized to

$$\rho(x, y; x', y'; \beta) = \sum_n \phi_n(x, y) \exp(-\beta E_n) \phi_n^*(x', y') \tag{A2.3.92}$$

with the PF given by

$$Q(2, V, T) = \int dx\, dy\, \rho(x, y; x, y; \beta). \tag{A2.3.93}$$

In the presence of a potential function $U(x, y)$, the density matrix in the high-temperature approximation has the form

$$\rho(x, y; x, y; \beta/P) \approx \left(\frac{mP}{2\pi\beta\hbar^2}\right)^3 \exp\left\{-\frac{mP}{2\beta\hbar^2}[(x - x')^2 + (y - y')^2] + \frac{\beta}{2P}[U(x, y) - U(x', y')]\right\}. \tag{A2.3.94}$$

Using this in the expression for the PF, one finds

$$Q(2, V, T) \approx \left(\frac{mP}{2\pi\beta\hbar^2}\right)^{3P} \int dx_1\, dy_1 \ldots dx_P\, dy_P$$

$$\times \exp\left(-\beta \sum_{s=1}^{P}[(1/2)K(x_{s+1} - x_s)^2 + (1/2)K(y_{s+1} - y_s)^2 + P^{-1}U(x_s, y_s)]\right). \tag{A2.3.95}$$

There is a separate cyclic polymer chain for each of the two particles, with the same force constant between adjacent beads on each chain. The potential acting on each bead in a chain is reduced, as before, by a factor $1/P$ but interacts only with the corresponding bead of the same label in the second chain. The generalization to many particles is straightforward, with one chain for each particle, each having the same number of beads coupled by harmonic springs between adjacent beads and with interactions between beads of the same label on different chains. This, however, is still an approximation as the exchange symmetry of the wavefunctions of the system is ignored.

The invariance of the Hamiltonian to particle exchange requires the eigenfunctions to be symmetric or antisymmetric, depending on whether the particles are bosons or fermions. The density matrix for bosons and fermions must then be sums over the corresponding symmetric and anti-symmetric states, respectively. Important applications of path integral simulations are to mixed classical and quantum systems, e.g. an electron in argon. For further discussion, the reader is referred to the articles by Gillan, Chandler and Wolynes, Berne and Thirumalai, Alavi and Makri in the further reading section.

We return to the study of classical systems in the remaining sections.

A2.3.3.5 1D hard rods

This is an example of a classical non-ideal system for which the PF can be deduced exactly [13]. Consider N hard rods of length d in a 1D extension of length L which takes the place of the volume in a three-dimensional (3D) system. The canonical PF

$$Q(N, L, T) = \frac{Z(N, L, T)}{N! \Lambda^N} \tag{A2.3.96}$$

where the configurational PF

$$Z(N, L, T) = \int\!\!\int \ldots \int \exp\left(-\beta \sum_{i<j} u_{ij}(r_{ij})\right) dr_1 \ldots dr_N = \int\!\!\int \ldots \int \prod_{i<j} \exp(-\beta u_{ij}(r_{ij})) \, dr_1 \ldots dr_N. \tag{A2.3.97}$$

For hard rods of length d

$$u_{ij}(r_{ij}) = \begin{cases} \infty & r_{ij} < d \\ 0 & r_{ij} > d \end{cases} \tag{A2.3.98}$$

so that an exponential factor in the integrand is zero unless $r_{ij} > d$. Another restriction in 1D systems is that since the particles are ordered in a line they cannot pass each other. Hence $d/2 < r_1 < L - Nd + d/2, 3d/2 < r2 < L - Nd + 3d/2$ etc. Changing variables to $x = r_1 - d/2, x = r_2 - 3d/2, \ldots, x_N = r - (N-1)d/2$, we have

$$Z(N, L, T) = \int \ldots \int_{0 < x_i < L - Nd} dx_1 \ldots dx_N = (L - Nd)^N. \tag{A2.3.99}$$

The pressure

$$P = kT (d \ln Z/dL)_{N,T} = \rho kT/(1 - \rho d) \tag{A2.3.100}$$

where the density of rods $\rho = N/L$. This result is exact. Expanding the denominator (when $\rho d < 1$) leads to the virial series for the pressure:

$$P/\rho kT = 1 + d\rho + d^2 \rho^2 + \cdots. \tag{A2.3.101}$$

The nth virial coefficient $B_n^{\text{HR}} = d^{n-1}$ is independent of the temperature. It is tempting to assume that the pressure of hard spheres in three dimensions is given by a similar expression, with d replaced by the excluded volume b_0, but this is clearly an approximation as shown by our previous discussion of the virial series for hard spheres. This is the excluded volume correction used in van der Waals' equation, which is discussed next. Other 1D models have been solved exactly in [14–16].

A2.3.3.6   Mean-field theory—van der Waals' equation

van der Waals' equation corrects the ideal gas equation for the attractive and repulsive interactions between molecules. The approximations employed are typical of mean-field theories. We consider a simple derivation, assuming that the total potential energy $U_N(r^N)$ of the N molecules is pairwise additive and can be divided into a reference part and a remainder in any given spatial configuration $\{r^N\}$. This corresponds roughly to repulsive and attractive contributions, and

$$U_N(r^N) = U_N^0(r^N) + U_N^{\text{attr}}(r^N) \tag{A2.3.102}$$

where $U_N^0(r^N)$ is the energy of the reference system and

$$U_N^{\text{attr}}(r^N) = \sum_{i<j} u_{ij}^{\text{attr}}(r_{ij}) \tag{A2.3.103}$$

is the attractive component. This separation assumes the cross terms are zero. In the above sum, there are $N(N-1)/2$ terms that depend on the relative separation of the particles. The total attractive interaction of each molecule with the rest is replaced by an approximate average interaction, expressed as

$$U_N^{\text{attr}}(r_N) = \frac{N(N-1)}{2V} \int u_{12}^{\text{attr}}(r_{12}) 4\pi r_{12}^2 \, dr_{12} \approx -a\frac{N^2}{V} \tag{A2.3.104}$$

where

$$a = -\frac{1}{2} \int u_{12}^{\text{attr}}(r_{12}) 4\pi r_{12}^2 \, dr_{12}. \tag{A2.3.105}$$

The PF

$$Z(N, V, T) = \int \exp(-\beta U_N(r^N) \, dr^N$$

$$\approx \exp(\beta a N^2/V) Z^0(N, V, T) \tag{A2.3.106}$$

in which $Z^0(N, V, T)$ is the configurational PF of the reference system. The Helmholtz free energy and pressure follow from the fundamental equations for the canonical ensemble

$$A = -kT \ln Q(N, V, T) = A^0 + a N^2/V \tag{A2.3.107}$$

and

$$P = kT (\ln Z(N, V, T)/T)_{N,T} = P^0 - a N^2/V^2. \tag{A2.3.108}$$

Assuming a hard sphere reference system with the pressure given by

$$P^0 = NkT/(V - Nb) \tag{A2.3.109}$$

we immediately recover van der Waals' equation of state

$$P = NkT/(V - Nb) - a N^2/V^2 \tag{A2.3.110}$$

since $\rho = N/V$. An improvement to van der Waals' equation would be to use a more accurate expression for the hard sphere reference system, such as the CS equation of state discussed in the previous section. A more complete derivation that includes the Maxwell equal area rule was given in [18].

A2.3.3.7   The law of corresponding states

van der Waals' equation is one of several two-parameter, mean-field equations of state (e.g. the Redlich–Kwong equation) that obey the law of corresponding states. This is a scaling law in which the thermodynamic properties are expressed as functions of reduced variables defined in terms of the critical parameters of the system.

A theoretical basis for the law of corresponding states can be demonstrated for substances with the same intermolecular potential energy function but with different parameters for each substance. Conversely, the experimental verification of the law implies that the underlying intermolecular potentials are essentially similar in form and can be transformed from substance to substance by scaling the potential energy parameters. The potentials are then said to be conformal. There are two main assumptions in the derivation:

(a)   quantum effects are neglected, i.e. classical statistics is assumed;

(b)   the total potential is pairwise additive

$$U_N(r^N) = \sum_{i<j} u_{ij}(r_{ij}) \tag{A2.3.111}$$

and characterized by two parameters: a well depth ε and a size σ

$$u_{ij}(r_{ij}) = \varepsilon \phi(r_{ij}/\sigma). \tag{A2.3.112}$$

The configurational PF

$$Z(N, V, T) = \int \ldots \int \exp(-\beta U_N(r^N))\, dr_1\, dr_2 \ldots dr_N$$

$$= \sigma^{3N} \int \ldots \int \exp\left(-\beta\varepsilon \sum_{i<j} \phi(r_{ij})\right) d(r_1/\sigma^3) \ldots d(r_N/\sigma^3)$$

$$= \sigma^{3N} Z^*(N, V^*, T^*) \tag{A2.3.113}$$

where $V^* = V/\sigma^{3N}$, $T^* = kT/\varepsilon$ and $Z^*(N, V, T)$ is the integral in the second line.

The excess Helmholtz free energy

$$A^{\mathrm{ex}} = -kT \ln(Z(N, V, T)/V^N) = -kT \ln(Z^*(N, V^*, T^*)/V^{*N})$$

$$= -NkT[\rho^* f(T^*, \rho^*)] \tag{A2.3.114}$$

where $\rho^* = \rho\sigma^3$, $\rho = N/V$ and

$$f(T^*, \rho^*) = [Z^*(N, V^*, T^*)]^{1/N}/N. \tag{A2.3.115}$$

It follows that atoms or molecules interacting with the same pair potential $\varepsilon\phi(r_{ij}/\sigma)$, but with different ε and σ, have the same thermodynamic properties, derived from A^{ex}/NkT, at the same scaled temperature T^* and scaled density ρ^*. They obey the same scaled equation of state, with identical coexistence curves in scaled variables below the critical point, and have the same scaled vapour pressures and second virial coefficients as a function of the scaled temperature. The critical compressibility factor $P_c V_c/RT_c$ is the same for all substances obeying this law and provides a test of the hypothesis. Table A2.3.3 lists the critical parameters and the compressibility factors of rare gases and other simple substances.

The compressibility factor $Z_c = P_c V_c/RT_c$ of the rare gases Ar, Kr and Xe at the critical point is nearly 0.291, and they are expected to obey a law of corresponding states, but one very different from the prediction of van der Waals' equation discussed earlier, for which the compressibility factor at the critical point is 0.375.

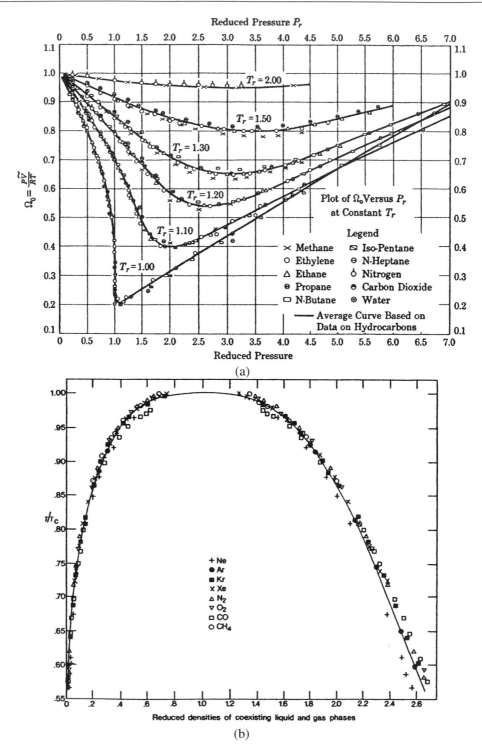

Figure A2.3.5. (a) $P/\rho kT$ as a function of the reduced variables T_R and P_R and (b) coexisting liquid and vapour densities in reduced units ρ_R as a function of T_R for several substances (after [19]).

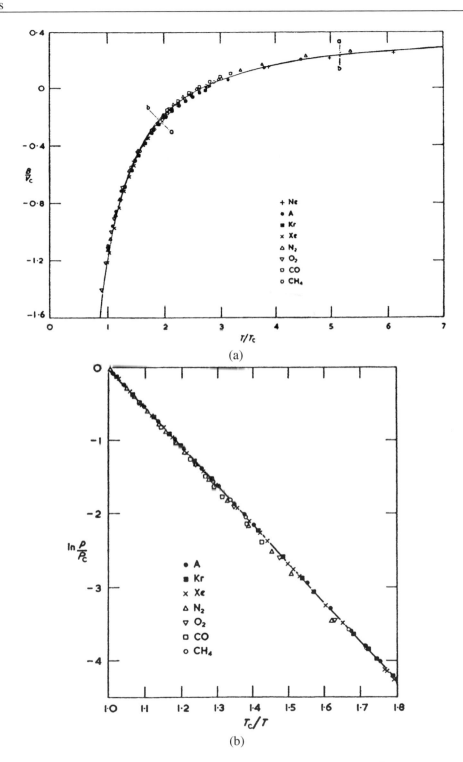

Figure A2.3.6. (a) Reduced second virial coefficient B_2/V_c as a function of T_R and (b) $\ln P_R$ versus $1/T_R$ for several substances (after [19]).

Table A2.3.3. Critical constants.

Substance	T_c (K)	P_c (atm)	V_c (cm^3 mol^{-1})	$P_c V_c / R T_c$
He	5.21	2.26	57.76	0.305
Ne	44.44	26.86	41.74	0.307
Ar	150.7	48	75.2	0.292
Kr	209.4	54.3	92.2	0.291
Xe	289.8	58.0	118.8	0.290
CH_4	190.6	45.6	98.7	0.288
H_2	33.2	12.8	65.0	0.305
N_2	126.3	33.5	126.3	0.292
O_2	154.8	50.1	78.0	0.308
CO_2	304.2	72.9	94.0	0.274
NH_3	405.5	111.3	72.5	0.242
H_2O	647.4	218.3	55.3	0.227

Deviations of Z_c from 0.291 for the other substances listed in the table are small, except for CO_2, NH_3 or H_2O, for which the molecular charge distributions contribute significantly to the intermolecular potential and lead to deviations from the law of corresponding states. The effect of hydrogen bonding in water contributes to its anomalous properties; this is mimicked by the charge distribution in the SPC/E or other models discussed in section A2.3.2. The pair potentials of all the substances listed in the table, except CO_2, NH_3 or H_2O, are fairly well represented by the Lennard-Jones potential—see table A2.3.1. The lighter substances, He, H_2 and to some extent Ne, show deviations due to quantum effects. The rotational PF of water in the vapour phase also has significant contribution from this source.

The equation of state determined by $Z^*(N, V^*, T^*)$ is not known in the sense that it cannot be written down as a simple expression. However, the critical parameters depend on ε and σ, and a test of the law of corresponding states is to use the reduced variables T_R, P_R and V_R as the scaled variables in the equation of state. Figure A2.3.5(b) illustrates this for the liquid–gas coexistence curves of several substances. As first shown by Guggenheim [19], the curvature near the critical point is consistent with a critical exponent β closer to 1/3 rather than the 1/2 predicted by van der Waals' equation. This provides additional evidence that the law of corresponding states obeyed is not the form associated with van der Waals' equation. Figure A2.3.5(b) shows that $P/\rho kT$ is approximately the same function of the reduced variables T_R and P_R

$$P/\rho kT = f(T_R, P_R) \tag{A2.3.116}$$

for several substances.

Figure A2.3.6 illustrates the corresponding states principle for the reduced vapour pressure P_R and the second virial coefficient as functions of the reduced temperature T_R showing that the law of corresponding states is obeyed approximately by the substances indicated in the figures. The usefulness of the law also lies in its predictive value.

A2.3.4 Correlation functions of simple fluids

The correlation functions provide an alternate route to the equilibrium properties of classical fluids. In particular, the two-particle correlation function of a system with a pairwise additive potential determines all of its thermodynamic properties. It also determines the compressibility of systems with even more complex three-body and higher-order interactions. The pair correlation functions are easier to approximate than the PFs to which they are related; they can also be obtained, in principle, from x-ray or neutron diffraction experiments.

This provides a useful perspective of fluid structure, and enables Hamiltonian models and approximations for the equilibrium structure of fluids and solutions to be tested by direct comparison with the experimentally determined correlation functions. We discuss the basic relations for the correlation functions in the canonical and grand canonical ensembles before considering applications to model systems.

A2.3.4.1   Canonical ensemble

The probability of observing the configuration $\{r^N\}$ in a system of given N, V and T is

$$P(r^N) = \frac{\exp(-\beta U(r^N))}{Z(N, V, T)} \qquad (A2.3.117)$$

where $Z(N, V, T)$ is the configurational PF and

$$\int P(r^N)\,dr^N = 1. \qquad (A2.3.118)$$

The probability function $P(r^N)$ cannot be factored into contributions from individual particles, since they are coupled by their interactions. However, integration over the coordinates of all but a few particles leads to reduced probability functions containing the information necessary to calculate the equilibrium properties of the system.

Integration over the coordinates of all but one particle provides the one-particle correlation function:

$$\rho_N^{(1)}(r_1) = N \int \cdots \int P(r^N)\,dr_2 \ldots dr_N = \left\langle \sum_{i=1}^{N} \delta(r_1 - r_i') \right\rangle_{CE} \qquad (A2.3.119)$$

where $\langle D\rangle_{CE}$ denotes the average value of a dynamical variable D in the canonical ensemble. Likewise, the two- and n-particle reduced correlation functions are defined by

$$\rho_N^{(2)}(r_1, r_2) = N(N-1) \int \cdots \int P(r^N)\,dr_3 \ldots dr_N = \left\langle \sum_{i=1}^{N}\sum_{\substack{i=1 \\ i \neq j}}^{N} \delta(r_1 - r_i')\delta(r_2 - r_j') \right\rangle_{CE} \qquad (A2.3.120)$$

$$\rho_N^{(n)}(r_1, \ldots, r_n) = \frac{N!}{(N-n)!} \int \cdots \int P(r^N)\,dr_{n+1} \ldots dr_N \qquad (A2.3.121)$$

where n is an integer. Integrating these functions over the coordinates $r_1, r_2, \ldots, r_N$ gives the normalization factor $N!/(N-n)!$. In particular,

$$\int \rho_N^{(1)}(r_1)dr_1 = N$$

$$\int \rho_N^{(2)}(r_1, r_2)dr_1\,dr_2 = N - 1. \qquad (A2.3.122)$$

For an isotropic fluid, the one-particle correlation function is independent of the position and

$$\rho_N^{(1)}(r_1) = N/V = \rho \qquad (A2.3.123)$$

which is the fluid density. The two-particle correlation function depends on the relative separation between particles. Assuming no angular dependence in the pair interaction,

$$\rho_N^{(2)}(r_1, r_2) = \rho_N^{(2)}(r) = \rho^2 g(r) \qquad (A2.3.124)$$

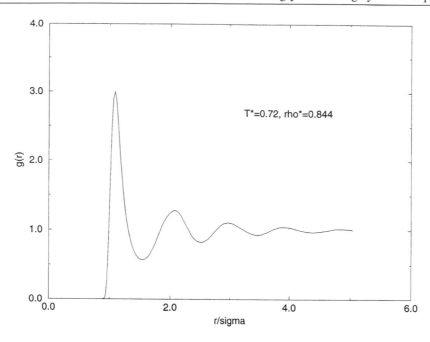

Figure A2.3.7. The radial distribution function $g(r)$ of a Lennard-Jones fluid representing argon at $T^* = 0.72$ and $\rho^* = 0.844$ determined by computer simulations using the Lennard-Jones potential.

where $r = |r_{12}|$. The second relation defines the radial distribution function $g(r)$ which measures the correlation between the particles in the fluid at a separation r. Thus, we could regard $\rho g(r)$ as the average density of particles at r_{12}, given that there is one at the origin r_1. Since the fluid is isotropic, the average local density in a shell of radius r and thickness Δr around each particle is independent of the direction, and

$$g(r) = \frac{\langle N(r, \Delta r) \rangle}{[(N-1)/V]V_{\text{shell}}} \qquad (A2.3.125)$$

where $V_{\text{shell}} = 4\pi r^2 \Delta r$ is the volume of the shell and $\langle N(r, \Delta r) \rangle$ is the average number of particles in the shell. We see that the pair distribution function $g(r)$ is the ratio of the average number $\langle N(r, \Delta r) \rangle$ in the shell of radius r and thickness Δr to the number that would be present if there were no particle interactions. At large distances, this interaction is zero and $g(r) \rightarrow 1$ as $r \rightarrow \infty$. At very small separations, the strong repulsion between real atoms (the Pauli exclusion principle working again) reduces the number of particles in the shell, and $g(r) \rightarrow 0$ as $r \rightarrow 0$. For hard spheres of diameter σ, $g(r)$ is exactly zero for $r < \sigma$. Figure A2.3.7 illustrates the radial distribution function $g(r)$ of argon, a typical monatomic fluid, determined by a molecular dynamics (MD) simulation using the Lennard-Jones potential for argon at $T^* = 0.72$ and $\rho^* = 0.84$, and figure A2.3.8 shows the corresponding atom–atom radial distribution functions $g_{\text{oo}}(r)$, $g_{\text{oh}}(r)$ and $g_{\text{hh}}(r)$ of the SPC/E model for water at 25 °C also determined by MD simulations. The correlation functions are in fair agreement with the experimental results obtained by x-ray and neutron diffraction.

Between the limits of small and large r, the pair distribution function $g(r)$ of a monatomic fluid is determined by the direct interaction between the two particles, and by the indirect interaction between the same two particles through other particles. At low densities, it is only the direct interaction that operates through the Boltzmann distribution and

$$g(r) \approx \exp(-\beta u(r)) \qquad \text{(low-density approximation).} \qquad (A2.3.126)$$

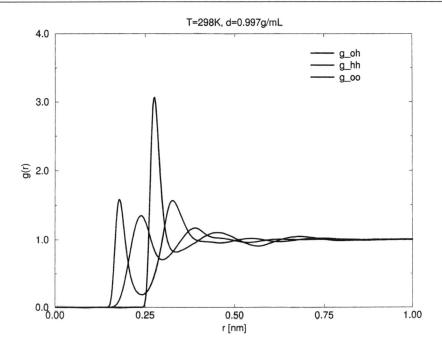

Figure A2.3.8. Atom–atom distribution functions $g_{oo}(r)$, $g_{oh}(r)$ and $g_{hh}(r)$ for liquid water at 25 °C determined by MD simulations using the SPC/E model. Curves are from the leftmost peak: g_{oh}, g_{hh}, g_{oo} are red, green, blue, respectively.

At higher densities, the effect of indirect interactions is represented by the cavity function $y(r, \rho, T)$, which multiplies the Boltzmann distribution

$$g(r) \approx \exp(-\beta u(r)) y(r, \rho, T). \tag{A2.3.127}$$

$y(r, \rho, T) \rightarrow 1$ as $\rho \rightarrow 0$, and it is a continuous function of r even for hard spheres at $r = \sigma$, the diameter of the spheres. It has a density expansion similar to the virial series:

$$y(r, \rho, T) = 1 + \sum_{n=1}^{\infty} y_n(r, T) \rho^n. \tag{A2.3.128}$$

The coefficient of ρ is the convolution integral of Mayer f-functions:

$$y_1(r, T) = \int f_{13}(r_{13}) f_{32}(r_{32}) \, d\mathbf{r}_3 = \quad . \tag{A2.3.129}$$

In the graphical representation of the integral shown above, a line represents the Mayer function $f(r_{ij})$ between two particles i and j. The coordinates are represented by open circles that are labelled, unless it is integrated over the volume of the system, when the circle representing it is blackened and the label erased. The black circle in the above graph represents an integration over the coordinates of particle 3, and is not labelled. The coefficient of ρ^2 is the sum of three terms represented graphically as

$$y_2(r, T) = \quad + 2 \quad + \quad + \quad \tag{A2.3.130}$$

In general, each graph contributing to $y_n(r, T)$ has n black circles representing the number of particles through which the indirect interaction occurs; this is weighted by the nth power of the density in the expression for $g(r)$. This observation, and the symmetry number of a graph, can be used to further simplify the graphical notation, but this is beyond the scope of this article. The calculation or accurate approximation of the cavity function are important problems in the correlation function theory of non-ideal fluids.

For hard spheres, the coefficients $y_n(r)$ are independent of temperature because the Mayer f-functions, in terms of which they can be expressed, are temperature independent. The calculation of the leading term $y_1(r)$ is simple, but the determination of the remaining terms increases in complexity for larger n. Recalling that the Mayer f-function for hard spheres of diameter σ is -1 when $r < \sigma$, and zero otherwise, it follows that $y_1(r, T)$ is zero for $r > 2\sigma$. For $r < 2\sigma$, it is just the overlap volume of two spheres of radii 2σ and a simple calculation shows that

$$y_1(r) = \begin{cases} \pi\sigma^3 \left\{ \dfrac{4}{3} - \left(\dfrac{r}{\sigma}\right) + \dfrac{1}{12}\left(\dfrac{r}{\sigma}\right)^3 \right\} & r < 2\sigma \\ 0 & r > 2\sigma. \end{cases} \tag{A2.3.131}$$

This leads to the third virial coefficient for hard spheres. In general, the nth virial coefficient of pairwise additive potentials is related to the coefficient $y_n(r, T)$ in the expansion of $g(r)$, except for Coulombic systems for which the virial coefficients diverge and special techniques are necessary to resum the series.

The pair correlation function has a simple physical interpretation as the potential of mean force between two particles separated by a distance r

$$w(r) = -kT \ln g(r) = u(r) - kT \ln y(r). \tag{A2.3.132}$$

As $\rho \to 0$, $y(r) \to 1$ and $w(r) \to u(r)$. At higher densities, however, $w(r) \neq u(r)$. To understand its significance, consider the *mean force* between *two fixed* particles at r_1 and r_2, separated by the distance $r = |r_1 - r_2|$. The mean force on particle 1 is the force averaged over the positions and orientations of all other particles, and is given by

$$\begin{aligned} \langle F_1(r) \rangle &= -\langle \nabla_1 U_N(r^N) \rangle \\ &= \frac{-\int \ldots \int \nabla_1 U_N(r^N) \exp(-\beta U_N(r^N)) \, dr_3 \ldots dr_N}{Z(N, V, T)} \\ &= kT \nabla_1 \ln \left[\int \ldots \int \exp(-\beta U_N(r^N)) \, dr_3 \ldots dr_N \right] \\ &= kT \nabla_1 \ln \left[\frac{Z(N, V, T)\rho^{(2)}(r_1, r_2)}{N(N-1)} \right] \\ &= kT \nabla_1 \ln g(r) = -\nabla_1 w(r) \end{aligned} \tag{A2.3.133}$$

where we have used the definition of the two-particle correlation function, $\rho^{(2)}(r_1, r_2)$, and its representation as $\rho^2 g(r)$ for an isotropic fluid in the last two steps. It is clear that the negative of the gradient of $w(r)$ is the force on the fixed particles, averaged over the motions of the others. This explains its characterization as the potential of mean force.

The concept of the potential of mean force can be extended to mixtures and solutions. Consider two ions in a sea of water molecules at fixed temperature T and solvent density ρ. The potential of mean force $w(r)$ is the direct interaction between the ions $u_{ij}(r) = u_{ij}^*(r) + q_i q_j/r$, plus the interaction between the ions through water molecules which is $-kT \ln y_{ij}(r)$. Here $u_{ij}^*(r)$ is the short-range potential and $q_i q_j/r$ is the Coulombic potential between ions. Thus,

$$w_{ij}(r) = u_{ij}^*(r) + q_i q_j/r - kT \ln y(r). \tag{A2.3.134}$$

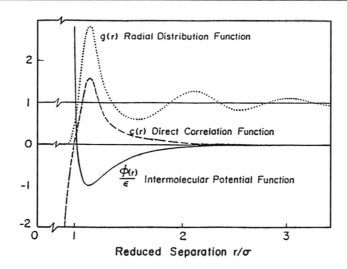

Figure A2.3.9. Plots of $g(r)$ and $c(r)$ versus r for a Lennard-Jones fluid at $T^* = 0.72$, $\rho^* = 0.84$, compared to $\beta u(r)$.

At large distances, $u_{ij}^*(r)$ 0 and $w_{ij}(r) \simeq q_i q_j / \varepsilon r$ where ε is the macroscopic dielectric constant of the solvent. This shows that the dielectric constant ε of a polar solvent is related to the cavity function for two ions at large separations. One could extend this concept to define a local dielectric constant $\varepsilon(r)$ for the interaction between two ions at small separations.

The direct correlation function $c(r)$ of a homogeneous fluid is related to the pair correlation function through the Ornstein–Zernike relation

$$h(r_{12}) = c(r_{12}) + \rho \int c(r_{13}) h(r_{32}) \, dr_3 \qquad (A2.3.135)$$

where $h(r) = g(r) - 1$ differs from the pair correlation function only by a constant term. $h(r) \to 0$ as $r \to \infty$ and is equal to -1 in the limit of $r = 0$. For hard spheres of diameter σ, $h(r) = -1$ inside the hard core, i.e. $r < \sigma$. The function $c(r)$ has the range of the intermolecular potential $u(r)$, and is generally easier to approximate. Figure A2.3.9 shows plots of $g(r)$ and $c(r)$ for a Lennard-Jones fluid at the triple point $T^* = 0.72$, $\rho^* = 0.84$, compared to $\beta u(r) = \phi(r)/\epsilon$.

The second term in the Ornstein–Zernike equation is a convolution integral. Substituting for $h(r)$ in the integrand, followed by repeated iteration, shows that $h(r)$ is the sum of convolutions of c-functions or 'bonds' containing one or more c-functions in series. Representing this graphically with $c(r) = \circ\text{-}\text{-}\text{-}\text{-}\circ$, we see that

$$h(r) = \circ\text{-}\text{-}\text{-}\circ + \rho \; \circ\text{-}\text{-}\text{-}\bullet\text{-}\text{-}\text{-}\circ + \rho^2 \; \circ\text{-}\text{-}\text{-}\bullet\text{-}\text{-}\bullet\text{-}\text{-}\text{-}\circ + \cdots \qquad (A2.3.136)$$

$h(r) - c(r)$ is the sum of series diagrams of c-bonds, with black circles signifying integration over the coordinates. It represents only part of the indirect interactions between two particles through other particles. The remaining indirect interactions cannot be represented as series diagrams and are called bridge diagrams. We now state, without proof, that the logarithm of the cavity function

$$\ln y(r) = h(r) - c(r) + B(r) \qquad (A2.3.137)$$

where the bridge diagram $B(r)$ has the f-bond density expansion

$$B(r_{12}) = \text{⧄} + \cdots .$$

$$\quad\quad\quad\quad 1 \quad\ 2 \quad\quad\quad\quad\quad\quad\quad\quad\quad\quad\quad\quad\quad\text{(A2.3.138)}$$

Only the first term in this expansion is shown. It is identical to the last term shown in the equation for $y_2(r)$, which is the coefficient of ρ^2 in the expansion of the cavity function $y(r)$.

It follows that the exact expression for the pair correlation function is

$$g(r) = h(r) + 1 = \exp(-\beta u(r) + h(r) - c(r) + B(r)). \quad\quad\quad \text{(A2.3.139)}$$

Combining this with the Ornstein–Zernike equation, we have two equations and three unknowns $h(r), c(r)$ and $B(r)$ for a given pair potential $u(r)$. The problem then is to calculate or approximate the bridge functions for which there is no simple general relation, although some progress for particular classes of systems has been made recently.

The thermodynamic properties of a fluid can be calculated from the two-, three- and higher-order correlation functions. Fortunately, only the two-body correlation functions are required for systems with pairwise additive potentials, which means that for such systems we need only a theory at the level of the two-particle correlations. The average value of the total energy

$$\langle E \rangle = \langle KE \rangle + \langle E_{\text{int}} \rangle + \langle U_N \rangle \quad\quad\quad \text{(A2.3.140)}$$

where the translational kinetic energy $\langle KE \rangle = 3/2NkT$ is determined by the equipartition theorem. The rotational, vibrational and electronic contributions to $\langle E_{\text{int}} \rangle$ are separable and determined classically or quantum mechanically. The average potential energy

$$\langle U_N(r_N) \rangle = \frac{\int \ldots \int U_N(r_N) \exp(-\beta U_N(r_N)) \, dr_N}{Z(N, V, T)}. \quad\quad\quad \text{(A2.3.141)}$$

For a pairwise additive potential, each term in the sum of pair potentials gives the same result in the above expression and there are $N(N-1)/2$ such terms. It follows that

$$\langle U_N(r_N) \rangle = \frac{N(N-1)}{2} \frac{\int \ldots \int u_{12}(r_{12}) \exp(-\beta U_N(r_N)) \, dr_1 \, dr_2 \, dr^{N-2}}{Z(N, V, T)}$$

$$= \frac{1}{2} \int \ldots \int u_{12}(r_{12}) \rho^{(2)}(r_1, r_2) \, dr_1 \, dr_2. \quad\quad\quad \text{(A2.3.142)}$$

For a fluid $\rho^{(2)}(r_1, r_2) = \rho^2 g(r_{12})$ where the number density $\rho = N/V$. Substituting this in the integral, changing to relative coordinates with respect to particle 1 as the origin, and integrating over r_1 to give V, leads to

$$\frac{\langle U_N \rangle}{N} = \frac{\rho}{2} \int \ldots \int u_{12}(r_{12}) g(r_{12}) \, dr_{12}. \quad\quad\quad \text{(A2.3.143)}$$

The pressure follows from the virial theorem, or from the characteristic thermodynamic equation and the PF. It is given by

$$\frac{PV}{\rho kT} = 1 - \frac{\rho}{6kT} \int r \frac{du_{12}(r)}{dr} g(r) \, dr_{12} \quad\quad\quad \text{(A2.3.144)}$$

which is called the virial equation for the pressure. The first term is the ideal gas contribution to the pressure and the second is the non-ideal contribution. Inserting $g(r) = \exp(-\beta u(r)) y(r)$ and the density expansion of the cavity function $y(r)$ into this equation leads to the virial expansion for the pressure. The nth virial coefficient, $B_n(T)$, given in terms of the coefficients $y_n(r, T)$ in the density expansion of the cavity function is

$$B_n(T) = -\frac{1}{6kT} \int r \frac{du_{12}(r)}{dr} \exp(-\beta u(r)) y_{n-2}(r, T) \, d\mathbf{r}_{12}. \tag{A2.3.145}$$

The virial pressure equation for hard spheres has a simple form determined by the density ρ, the hard sphere diameter σ and the distribution function at contact $g(\sigma+)$. The derivative of the hard sphere potential is discontinuous at $r = \sigma$, and

$$\phi(r) = \exp(-\beta u(r)) = \begin{cases} 1 & r < \sigma \\ 0 & r > \sigma \end{cases}$$

is a step function. The derivative of this with respect to r is a delta function

$$\frac{d\phi}{dr} = -\beta \frac{du(r)}{dr} \exp(-\beta(u(r)) = \delta(r - \sigma+)$$

and it follows that

$$\frac{du(r)}{dr} g(r) = -\frac{1}{\beta} \delta(r - \sigma+) y(r).$$

Inserting this expression in the virial pressure equation, we find that

$$\frac{P}{\rho kT} = 1 + \frac{2\pi}{3} \rho \sigma^3 g(\sigma+) \tag{A2.3.146}$$

where we have used the fact that $y(\sigma) = g(\sigma+)$ for hard spheres. The virial coefficients of hard spheres are thus also related to the contact values of the coefficients $y_n(\sigma)$ in the density expansion of the cavity function. For example, the expression $y_2(r)$ for hard spheres given earlier leads to the third virial coefficient $B_3^{HS} = 5b_0^2/8$.

We conclude this section by discussing an expression for the excess chemical potential in terms of the pair correlation function and a parameter λ, which couples the interactions of one particle with the rest. The idea of a coupling parameter was introduced by Onsager [20] and Kirkwood [21]. The choice of λ depends on the system considered. In an electrolyte solution it could be the charge, but in general it is some variable that characterizes the pair potential. The potential energy of the system

$$U_N(\mathbf{r}^N; \lambda) = U_{N-1}(\mathbf{r}^{N-1}) + \lambda \sum_{j=1}^{N} u_{1j}(\mathbf{r}_{1j}) \tag{A2.3.147}$$

where the particle at $\mathbf{r}_1$ couples with the remaining $N - 1$ particles and $0 \le \lambda \le 1$. The configurational PF for this system

$$Z(N, V, T; \lambda) = \int \cdots \int \exp(-\beta U_N(\mathbf{r}^N; \lambda) \, d\mathbf{r}^N. \tag{A2.3.148}$$

When $\lambda = 1$, we recover the PF $Z(N, V, T)$ for the fully coupled system. In the opposite limit of $\lambda = 0$, $Z(N, V, T; 0) = V Z(N - 1, V, T)$, where $Z(N - 1, V, T)$ refers to a fully coupled system of $N - 1$ particles. Our previous discussion of the chemical potential showed that the excess chemical potential is related to the logarithm of the ratio $Z(N, V, T; 0)/V Z(N - 1, V, T)$ for large N:

$$\mu^{ex} = -kT \ln[Z(N, V, T)/V Z(N - 1, V, T)]$$

$$= -kT \int_0^1 \frac{\partial \ln Z(N, V, T; \lambda)}{\partial \lambda} \, d\lambda. \tag{A2.3.149}$$

The integral is easily simplified for a pairwise additive system, and one finds

$$\frac{dZ(N, V, T; \lambda)}{d\lambda} = -\beta \int \cdots \int \sum_{j=2}^{N} u_{1j}(r_{1j}) \exp(-\beta U_N(r^N, \lambda)) \, dr^N$$

$$= \beta(N-1) \int \cdots \int u_{12}(r_{12}) \exp(-\beta U_N(r^N, \lambda)) dr^N.$$

Dividing by $Z(N, V, T; \lambda)$ and recalling the definition of the correlation function

$$\frac{d \ln Z(N, V, T)}{d\lambda} = \frac{\beta}{N} \int \cdots \int u_{12}(r_{12}) \rho^{(2)}(r_1, r_2; \lambda) \, dr_1 \, dr_1.$$

For a fluid, $\rho^{(2)}(r_1, r_2; \lambda) = \rho^2 g(r_{12}; \lambda)$. Changing to relative coordinates, integrating over the coordinates of particle 1 and inserting this in the expression for the excess chemical potential leads to the final result

$$\mu^{\text{ex}} = \rho \int_0^1 d\lambda \int u(r_{12}) g(r_{12}; \lambda) \, dr_{12}. \tag{A2.3.150}$$

This is Kirkwood's expression for the chemical potential. To use it, one needs the pair correlation function as a function of the coupling parameter λ as well as its spatial dependence. For instance, if λ is the charge on a selected ion in an electrolyte, the excess chemical potential follows from a theory that provides the dependence of $g(r_{12}; \lambda)$ on the charge and the distance r_{12}. This method of calculating the chemical potential is known as the Guntelburg charging process, after Guntelburg who applied it to electrolytes.

By analogy with the correlation function for the fully coupled system, the pair correlation function $g(r; \lambda)$ for an intermediate values of λ is given by

$$g(r_{12}; \lambda) = \exp(-\beta \lambda u_{12}(r_{12})) y(r_{12}, \rho, T; \lambda) \tag{A2.3.151}$$

where $y(r, \rho, T; \lambda)$ is the corresponding cavity function for the partially coupled system. Kirkwood derived an integral equation for $g(r; \lambda)$ in terms of a three-body correlation function approximated as the product of two-body correlation functions called the superposition approximation. The integral equation, which can be solved numerically, gives results of moderate accuracy for hard spheres and Lennard-Jones systems. A similar approximation is due to Born and Green [23, 24] and Yvon [22]. Other approximations for $g(r)$ are discussed later in this chapter.

The presence of three-body interactions in the total potential energy leads to an additional term in the internal energy and virial pressure involving the three-body potential $u_{123}(r_1, r_2, r_3)$, and the corresponding three-body correlation function $g^{(3)}(r_1, r_2, r_3)$. The expression for the energy is then

$$\frac{\langle U_N \rangle}{N} = \frac{\rho}{2} \int \cdots \int u_{12}(r_{12}) g(r_{12}) \, dr_{12} + \frac{\rho}{6} \int \cdots \int u_{123}(r_1, r_2, r_3) g^{(3)}(r_{12}, r_{13}, r_{23}) \, dr_2 \, dr_3. \tag{A2.3.152}$$

The virial equation for the pressure is also modified by the three-body and higher-order terms, and is given in general by

$$P = \rho kT - \frac{1}{DV} \left\langle \sum_{i<j} r_i \cdot \nabla_i U_N(r^N) \right\rangle \tag{A2.3.153}$$

where D is the dimensionality of the system.

A2.3.4.2   Grand canonical ensemble (μ, V, T)

The grand canonical ensemble is a collection of open systems of given chemical potential μ, volume V and temperature T, in which the number of particles or the density in each system can fluctuate. It leads to an important expression for the compressibility κ_T of a one-component fluid:

$$\rho k T \kappa_T = \frac{\langle N^2 \rangle - \langle N \rangle^2}{\langle N \rangle} \tag{A2.3.154}$$

where the compressibility can be determined experimentally from light scattering or neutron scattering experiments. Generalizations of the above expression to multi-component systems have important applications in the theory of solutions [25].

It was shown in section A2.3.3.2 that the grand canonical ensemble (GCE) PF is a generating function for the canonical ensemble PF, from which it follows that correlation functions in the GCE are just averages of the fluctuating numbers N and $N - 1$

$$\int \langle \rho^{(1)}(r_1) \rangle_{\mathrm{GCE}} \, dr_1 = \langle N \rangle$$

$$\iint \langle \rho^{(2)}(r_1, r_2) \rangle_{\mathrm{GCE}} \, dr_1 dr_2 = \langle N - 1 \rangle. \tag{A2.3.155}$$

We see that

$$\iint [\langle \rho^{(2)}(r_1, r_2) \rangle - \langle \rho^{(1)}(r_1) \rangle \langle \rho^{(1)}(r_2) \rangle] \, dr_1 dr_2 = \langle N(N-1) \rangle - \langle N \rangle^2$$

where the subscript GCE has been omitted for convenience. The right-hand side of this is just $\langle N^2 \rangle - \langle N \rangle^2 - \langle N \rangle$. The pair correlation function $g(r_1, r_2)$ is defined by

$$\langle \rho^{(2)}(r_1, r_2) \rangle = \langle \rho^{(1)}(r_1) \rangle \langle \rho^{(1)}(r_2) \rangle g(r_1, r_2) \tag{A2.3.156}$$

and it follows that

$$\iint \langle \rho_1(r_1) \rangle \langle \rho_1(r_2) \rangle [g(r_1, r_2) - 1] \, dr_1 dr_2 = \langle N^2 \rangle - \langle N \rangle^2 - \langle N \rangle.$$

For an isotropic fluid, the singlet density is the density of the fluid, i.e. $\langle \rho^{(1)}(r) \rangle = \langle N \rangle / V = \rho$, and the pair correlation function $g(r_1, r_2)$ depends on the interparticle separation $r_{12} = |r_1 - r_2|$. Using this in the above integral, changing to relative coordinates with respect to particle 1 as the origin and integrating over its coordinates, one finds

$$\rho^2 V \int [g(r) - 1] \, dr_{12} = \langle N^2 \rangle - \langle N \rangle^2 - \langle N \rangle.$$

Division by $\langle N \rangle$ and taking note of the fluctuation formula (A2.3.144) for the compressibility leads to the fundamental relation

$$\rho k T \kappa_T = 1 + \rho \int [g(r_{12}) - 1] \, dr_{12} \tag{A2.3.157}$$

called the compressibility equation which is not limited to systems with pairwise additive potentials. Integrating the compressibility with respect to the density provides an independent route to the pressure, aside from the pressure calculated from the virial equation. The exact pair correlation function for a given model system should give the same values for the pressure calculated by different routes. This serves as a test for the accuracy of an approximate $g(r)$ for a given Hamiltonian.

Table A2.3.4. The critical exponents γ, ν and η.

Exponent	MFT	Ising ($d = 2$)	Numerical ($d = 3$)
ν	1/2	1	0.630 ± 0.001
γ	1	7/4	1.239 ± 0.002
η	0	1/4	0.03

The first term in the compressibility equation is the ideal gas term and the second term, the integral of $g(r) - 1 = h(r)$, represents the non-ideal contribution due to the correlation or interaction between the particles. The correlation function $h(r)$ is zero for an ideal gas, leaving only the first term. The correlations between the particles in a fluid displaying a liquid–gas critical point are characterized by a correlation length ζ that becomes infinitely large as the critical point is approached. This causes the integral in the compressibility equation and the compressibility κ_T to diverge.

The divergence in the correlation length ζ is characterized by the critical exponent ν defined by

$$\zeta = |T - T_{\mathrm{c}}|^{-\nu} \tag{A2.3.158}$$

while the divergence in the compressibility, near the critical point, is characterized by the exponent γ as discussed earlier. The correlation function near the critical region has the asymptotic form [26]

$$h(r) \approx \frac{f(r/\zeta)}{r^{D-2+\eta}} \tag{A2.3.159}$$

where D is the dimensionality and η is a critical exponent. Substituting this in the compressibility equation, it follows with $D = 3$ that

$$kT\kappa_T \approx \xi^{2-\eta} \int f(x)x^{1-\eta}x\,\mathrm{d}x \tag{A2.3.160}$$

where $x = r/\zeta$. Inserting the expressions for the temperature dependence of the compressibility and the correlation length near the critical point, one finds that the exponents are related by

$$\gamma = \nu(2 - \eta). \tag{A2.3.161}$$

Table A2.3.4 summarizes the values of these critical exponents in two and three dimensions and the predictions of mean field theory.

The compressibility equation can also be written in terms of the direct correlation function. Taking the Fourier transform of the Ornstein–Zernike equation

$$\tilde{h}(k) = \tilde{c}(k) + \rho\tilde{c}(k)\tilde{h}(k) \tag{A2.3.162}$$

where we have used the property that the Fourier transform of a convolution integral is the product of Fourier transforms of the functions defining the convolution. Here the Fourier transform of a function $f(r)$ is defined by

$$\tilde{f}(k) = \int f(r)\exp(-ikr)\,\mathrm{d}r. \tag{A2.3.163}$$

From the Ornstein–Zernike equation in Fourier space one finds that

$$1 + \rho\tilde{h}(k) = [1 - \rho\tilde{c}(k)]^{-1}$$

when $k = 0$, $1 + \rho \tilde{h}(0)$ is just the right-hand side of the compressibility equation. Taking the inverse, it follows that

$$\beta \left(\frac{\partial P}{\partial \rho} \right)_T = [1 - \rho \tilde{c}(0)] = 1 - \rho \int c(r) \, dr. \qquad (A2.3.164)$$

At the critical point $\beta (\partial P / \partial \rho)_T = 0$, and the integral of the direct correlation function remains finite, unlike the integral of $h(r)$.

A2.3.4.3   Integral equation approximations for a fluid

The equilibrium properties of a fluid are related to the correlation functions which can also be determined experimentally from x-ray and neutron scattering experiments. Exact solutions or approximations to these correlation functions would complete the theory. Exact solutions, however, are usually confined to simple systems in one dimension. We discuss a few of the approximations currently used for 3D fluids.

Successive n and $n + 1$ particle density functions of fluids with pairwise additive potentials are related by the Yvon–Born–Green (YBG) hierarchy [6]

$$\nabla_1 \rho^{(n)}(r^n) = \beta \left(F_1^{\text{ext}} + \sum_{j=2}^{n} F_{1j} \right) \rho^{(n)}(r^{(n)}) + \beta \int F_{1,n+1} \rho^{(n+1)}(r^{n+1}) \, dr^{n+1} \qquad (A2.3.165)$$

where $F_1^{\text{ext}} = -\nabla_1 \phi$ is the external force, $F_{1j} = -\nabla_1 u(r_{1j})$ and $r^n \equiv \{r_1, r_2, r_3 \ldots r_n\}$ is the set of coordinates of n particles. The simplest of these occurs when $n = 1$, and it relates the one- and two-particle density functions of a fluid in an inhomogeneous field, e.g. a fluid near a wall:

$$kT \ln \rho(r_1, [\phi]) = -\nabla_1 \phi(r_1) - \int \rho(r_2 | r_1; [\phi]) \nabla_1 u(r_{12}) \, dr_2 \qquad (A2.3.166)$$

where $\rho(r_2 | r_1; [\phi]) = \rho^{(2)}(r_1, r_2; [\phi]) / \rho(r_1; [\phi])$ and the superscript 1 is omitted from the one-particle local density. For an homogeneous fluid in the absence of an external field, $F^{\text{ext}} = 0$ and $\rho^{(n)}(r^n) = \rho^n g^{(n)}(r^n)$ and the YBG equation leads to

$$\nabla_1 g^{(n)}(r^n) = -\beta \sum_{j=2}^{n} \nabla_1 u(r_{1j}) g^{(n)}(r^n) - \beta \int \nabla_1 u(r_{1,n+1}) g^{(n+1)}(r^{n+1}) \, dr^{n+1}. \qquad (A2.3.167)$$

Kirkwood derived an analogous equation that also relates two- and three-particle correlation functions but an approximation is necessary to uncouple them. The superposition approximation mentioned earlier is one such approximation, but unfortunately it is not very accurate. It is equivalent to the assumption that the potential of average force of three or more particles is pairwise additive, which is not the case even if the total potential is pair decomposable. The YBG equation for $n = 1$, however, is a convenient starting point for perturbation theories of inhomogeneous fluids in an external field.

We will describe integral equation approximations for the two-particle correlation functions. There is no single approximation that is equally good for all interatomic potentials in the 3D world, but the solutions for a few important models can be obtained analytically. These include the Percus–Yevick (PY) approximation [27, 28] for hard spheres and the mean spherical (MS) approximation for charged hard spheres, for hard spheres with point dipoles and for atoms interacting with a Yukawa potential. Numerical solutions for other approximations, such as the hypernetted chain (HNC) approximation for charged systems, are readily obtained by fast Fourier transform methods.

The Ornstein–Zernike equation

$$h(r_{12}) = c(r_{12}) + \rho \int c(r_{13}) h(r_{32}) \, dr_3 \qquad (A2.3.168)$$

and the exact relation for the pair correlation function

$$g(r) = h(r) + 1 = \exp[(-\beta u(r)) + h(r) - c(r) + B(r)] \tag{A2.3.169}$$

provide a convenient starting point for the discussion of these approximations. This equivalent to the exact relation

$$c(r) = \beta u(r) - \ln g(r) + h(r) + B(r) \tag{A2.3.170}$$

for the direct correlation function. As $r \to \infty$, $c(r) \to -\beta u(r)$ except at $T = T_c$. Given the pair potential $u(r)$, we have two equations for the three unknowns $h(r)$, $c(r)$ and $B(r)$; one of these is the Ornstein–Zernike relation and the other is either one of the exact relations cited above. Each of the unknown functions has a density expansion which is the sum of integrals of products of Mayer f-functions, which motivates their approximation by considering different classes of terms. In this sense, the simplest approximation is the following.

(a) Hypernetted chain approximation

This sets the bridge function

$$B(r) = 0. \tag{A2.3.171}$$

It is accurate for simple low valence electrolytes in aqueous solution at $25\,^\circ\text{C}$ and for molten salts away from the critical point. The solutions are obtained numerically. A related approximation is the following.

(b) Percus–Yevick (PY) approximation

In this case [27, 28], the function $\exp[(h(r) - c(r)]$ in the exact relation for $g(r)$ is linearized after assuming $B(r) = 0$, when

$$g(r) \simeq \exp(-\beta u(r))[1 + h(r) - c(r)] = \exp(-\beta u(r))[g(r) - c(r)]. \tag{A2.3.172}$$

Rearranging this, we have the PY approximation for the direct correlation function

$$c(r) = f(r)y(r). \tag{A2.3.173}$$

This expression is combined with the Orstein–Zernick equation to obtain the solution for $c(r)$.

For hard spheres of diameter σ, the PY approximation is equivalent to $c(r) = 0$ for $r > \sigma$ supplemented by the core condition $g(r) = 0$ for $r < \sigma$. The analytic solution to the PY approximation for hard spheres was obtained independently by Wertheim [32] and Thiele [33]. Solutions for other potentials (e.g. Lennard-Jones) are obtained numerically.

(c) Mean spherical approximation

In the MS approximation, for hard core particles of diameter σ, one approximates the direct correlation function by

$$c(r) = -\beta u(r) \qquad \text{for } r > \sigma \tag{A2.3.174}$$

and supplements this with the exact relation

$$g(r) = 0 \qquad \text{for } r < \sigma. \tag{A2.3.175}$$

The solution determines $c(r)$ inside the hard core from which $g(r)$ outside this core is obtained via the Ornstein–Zernike relation. For hard spheres, the approximation is identical to the PY approximation. Analytic solutions have been obtained for hard spheres, charged hard spheres, dipolar hard spheres and for particles interacting with the Yukawa potential. The MS approximation for point charges (charged hard spheres in the limit of zero size) yields the Debye–Huckel limiting law distribution function.

It would appear that the approximations listed above are progressively more drastic. Their accuracy, however, is unrelated to this progression and depends on the nature of the intermolecular potential. Approximations that are good for systems with strong long-range interactions are not necessarily useful when the interactions are short ranged. For example, the HNC approximation is accurate for simple low valence electrolytes in aqueous solution in the normal preparative (0–2 M) range at 25 °C, but fails near the critical region. The PY approximation, on the other hand, is poor for electrolytes, but is much better for hard spheres. The relative accuracy of these approximations is determined by the cancellation of terms in the density expansions of the correlation functions, which depends on the range of the intermolecular potential.

A2.3.5 Equilibrium properties of non-ideal fluids

A2.3.5.1 Integral equation and scaled particle theories

Theories based on the solution to integral equations for the pair correlation functions are now well developed and widely employed in numerical and analytic studies of simple fluids [6]. Further improvements for simple fluids would require better approximations for the bridge functions $B(r)$. It has been suggested that these functions can be scaled to the same functional form for different potentials. The extension of integral equation theories to molecular fluids was first accomplished by Chandler and Andersen [30] through the introduction of the site–site direct correlation function $c_{\alpha\beta}(r)$ between atoms in each molecule and a site–site Ornstein–Zernike relation called the reference interaction site model (RISM) equation [31]. Approximations, corresponding to the closures for simple monatomic fluids, enable the site–site pair correlation functions $h_{\alpha\beta}(r)$ to be obtained. The theory has been successfully applied to simple molecules and to polymers.

Integral equation approximations for the distribution functions of simple atomic fluids are discussed in the following.

(a) Hard spheres

(i) *PY and MS approximations.* The two approximations are identical for hard spheres, as noted earlier. The solution yields the direct correlation function inside the hard core as a cubic polynomial:

$$c(r) = -\lambda_1 - 6\lambda_2\eta(r/\sigma) - (1/2)\lambda_1\eta(r/\sigma)^3 \quad r < \sigma$$
$$= 0 \qquad\qquad\qquad\qquad\qquad\qquad\qquad r > \sigma. \tag{A2.3.176}$$

In this expression, the packing fraction $\eta = \pi\rho\sigma^3/6$, and the other two parameters are related to this by

$$\lambda_1 = (1 + 2\eta)^2/(1 - \eta)^4 \qquad \lambda_2 = -(1 + \eta/2)^2/(1 - \eta)^4.$$

The solution was first obtained independently by Wertheim [32] and Thiele [33] using Laplace transforms. Subsequently, Baxter [34] obtained the same solutions by a Wiener–Hopf factorization technique. This method has been generalized to charged hard spheres.

The pressure from the virial equation is calculated by noting that $h(r) - c(r)$ is continuous at $r = \sigma$, and $c(r) = 0$ for $r > \sigma$. It follows that

$$h(\sigma+) - c(\sigma+) = h(\sigma-) - c(\sigma-)$$

and since $c(\sigma+) = 0$ and $h(\sigma-) = -1$, we have $g(\sigma+) = 1 + h(\sigma+) = c(\sigma-)$. This gives an expression for the pressure of hard spheres in the PY approximation in terms of $c(\sigma-)$, equivalent to the virial pressure

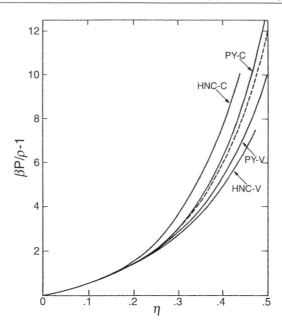

Figure A2.3.10. Equation of state for hard spheres from the PY and HNC approximations compared with the CS equation (- - -). C and V refer to the compressibility and virial routes to the pressure (after [6]).

equation (A2.3.146)

$$\frac{P}{\rho kT} = 1 - \frac{2\pi}{3}\rho\sigma^3 c(\sigma-).$$ (A2.3.177)

Setting $r = \sigma$ in the solution for $c(r)$, it follows that

$$\frac{P_V}{\rho kT} = \frac{1 + 2\eta + 3\eta^2}{(1 - \eta)^2}$$ (A2.3.178)

where the subscript V denotes the pressure calculated from the virial pressure equation. The pressure from the compressibility equation follows from the expression for $(dP/d\rho)_T$ in terms of the integral of the direct correlation function $c(r)$; the upper limit of this integral is $r = \sigma$ in the PY approximation for hard spheres since $c(r) = 0$ for $r > \sigma$. One finds that the pressure P_c from the compressibility equation is given by

$$\frac{P_c}{\rho kT} = \frac{1 + \eta + \eta^2}{(1 - \eta)^3}.$$ (A2.3.179)

The CS equation for the pressure is found to be the weighted mean of the pressure calculated from the virial and compressibility equations:

$$P_{CS} = (1/3)P_V + (2/3)P_c.$$

Figure A2.3.10 compares the virial and pressure equations for hard spheres with the pressure calculated form the CS equations and also with the pressures determined in computer simulations.

The CS pressures are close to the machine calculations in the fluid phase, and are bracketed by the pressures from the virial and compressibility equations using the PY approximation. Computer simulations show a fluid–solid phase transition that is not reproduced by any of these equations of state. The theory has been extended to mixtures of hard spheres with additive diameters by Lebowitz [35], Lebowitz and Rowlinson [35], and Baxter [36].

(ii) Scaled particle theory. The virial equation for the pressure of hard spheres is determined by the contact value $g(\sigma+)$ of the pair correlation functions, which is related to the average density of hard spheres in contact with a spherical cavity of radius σ, from which the spheres are excluded. The fixed cavity affects the fluid in the same way that a hard sphere at the centre of this cavity would influence the rest of the fluid. Reiss, Frisch and Lebowitz [37] developed an approximate method to calculate this, and found that the pressure for hard spheres is identical to the pressure from the compressibility equation in the PY approximation given in equation A2.3.178.

The method has been extended to mixtures of hard spheres, to hard convex molecules and to hard spherocylinders that model a nematic liquid crystal. For mixtures (*m* subscript) of hard convex molecules of the same shape but different sizes, Gibbons [38] has shown that the pressure is given by

$$\frac{P}{\rho kT} = \frac{1}{1 - \xi_m} + \frac{AB}{(1 - \xi_m)^2} + \frac{B^2 C}{3(1 - \xi_m)^3} \tag{A2.3.180}$$

where

$$\xi_m = \rho \sum_i x_i V_i \qquad A = \sum_i x_i \overline{R}_i$$

$$B = \sum_i x_i S_i \qquad C = \sum_i x_i \overline{R}_i^2 \tag{A2.3.181}$$

where $\overline{R}_i$ is the radius of particle i averaged over all orientations, V_i and S_i are the volume and surface area of the particle i, respectively, and x_i is its mole fraction. The pressure corresponding to the PY compressibility equation is obtained for parameters corresponding to hard sphere mixtures. We refer the reader to the review article by Reiss in the further reading section for more detailed discussions.

(iii) Gaussian statistics. Chandler [39] has discussed a model for fluids in which the probability $P(N, v)$ of observing N particles within a molecular size volume v is a Gaussian function of N. The moments of the probability distribution function are related to the n-particle correlation functions $g^{(n)}(r_1, r_2, \ldots, r_n)$, and

$$\alpha_n = \langle N(N - 1) \ldots N - n + 1) \rangle = \rho^n \int_v \ldots \int_v g^{(n)}(r_1, \ldots, r_n) \, \mathrm{d}r_1 \ldots \mathrm{d}r_n.$$

The inversion of this leads to an expression for $P(N, v)$:

$$P(N, v) = \sum_{n=N}^{\infty} (-1)^{n-N} \frac{\alpha_n}{N!(n - N)!}$$

involving all of the moments of the probability distribution function. The Gaussian approximation implies that only the first two moments $\langle N \rangle_v$ and $\langle N^2 \rangle_v$, which are determined by the density and the pair correlation function, are sufficient to determine the probability distribution $P(N, v)$. Computer simulation studies of hard spheres by Crooks and Chandler [40] and even water by Hummer *et al* [41] have shown that the Gaussian model is accurate at moderate fluid densities; deviations for hard spheres begin to occur at very low and high densities near the ideal gas limit and close to the transition to a solid phase, respectively.

The assumption of Gaussian fluctuations gives the PY approximation for hard sphere fluids and the MS approximation on addition of an attractive potential. The RISM theory for molecular fluids can also be derived from the same model.

(b) Strong electrolytes

The long-range interactions between ions lie at the opposite extreme to the harsh repulsive interactions between hard spheres. The methods used to calculate the thermodynamic properties through the virial expansion cannot be directly applied to Coulombic systems since the integrals entering into the virial coefficients diverge. The correct asymptotic form of the thermodynamic properties at low concentrations was first obtained by Debye and Hückel in their classic study of charged hard spheres [42] by linearizing the Poisson–Boltzmann equation as discussed below. This immediately excludes serious consideration of ion pairing, but this defect, especially in low dielectric solvents, was taken into account by Bjerrum [43], who assumed that all oppositely charged ions within a distance $e_+e_-/2\varepsilon kT$ were paired, while the rest were free. The free ions were treated in the Debye–Hückel approximation.

The Debye treatment is not easily extended to higher concentrations and special methods are required to incorporate these improvements. One method, due to Mayer [44], resums the virial expansion to cancel out the divergences of the integrals. Mayer obtained the Debye–Hückel limiting law and the first correction to this as a convergent renormalized second virial coefficient that automatically incorporates the effect of ion pairing. Improvements due to Outhwaite, Bhuyian and others, involve modifications of Debye and Hückel's original treatment of the Poisson–Boltzmann equation to yield a modified Poisson–Boltzmann (MPB) equation for the average electrostatic potential $\psi_i(r)$ of an ion. We refer to the review article by Outhwaite (1974) in the further reading section for a detailed discussion.

Two widely used theories of electrolytes at room temperature are the MS and HNC approximations for the pair correlation functions. The approximations fail or are less successful near the critical point. The solutions to the HNC approximation in the usual laboratory concentration range are obtained numerically, where fast Fourier transform methods are especially useful [45]. They are accurate for low valence electrolytes in aqueous solution at room temperature up to 1 or 2 M. However, the HNC approximation does not give a numerical solution near the critical point. The MS approximation of charged hard spheres can be solved analytically, as first shown by Waisman and Lebowitz [46]. This is very convenient and useful in mapping out the properties of electrolytes of varying charges over a wide range of concentrations. The solution has been extended recently to charged spheres of unequal size [47] and to sticky charged hard spheres [48, 49]. Ebeling [50] extended Bjerrum's theory of association by using the law of mass action to determine the number of ion pairs while treating the free ions in the MS approximation supplemented with the second ionic virial coefficient. Ebeling and Grigoro [51] located a critical point from this theory. The critical region of electrolytes is known to be characterized by pairing and clustering of ions and it has been observed experimentally that dimers are abundant in the vapour phase of ionic fluids. The nature of the critical exponents in this region, whether they are classical or non-classical, and the possibilities of a crossover from one to the other are currently under study [52–56]. Computer simulation studies of this region are also under active investigation [57–59]. Koneshan and Rasaiah [60] have observed clusters of sodium and chloride ions in simulations of aqueous sodium chloride solutions under supercritical conditions.

Strong electrolytes are dissociated into ions that are also paired to some extent when the charges are high or the dielectric constant of the medium is low. We discuss their properties assuming that the ionized gas or solution is electrically neutral, i.e.

$$\sum_{i=1}^{\sigma} c_i e_i = 0 \tag{A2.3.182}$$

where c_i is the concentration of the free ion i with charge e_i and σ is the number of ionic species. The local charge density at a distance r from the ion i is related to the ion concentrations c_j and pair distribution functions $g_{ij}(r)$ by

$$\rho_i(r) = \sum_{j=1}^{\sigma} c_j e_j g_{ij}(r). \tag{A2.3.183}$$

The electroneutrality condition can be expressed in terms of the integral of the charge density by recognizing the obvious fact that the total charge around an ion is equal in magnitude and opposite in sign to the charge on the central ion. This leads to the zeroth moment condition

$$-e_i = \int \rho_i(r)\,\mathrm{d}\boldsymbol{r}. \tag{A2.3.184}$$

The distribution functions also satisfy a second moment condition, as first shown by Stillinger and Lovett [61]:

$$-\frac{3\varepsilon_0 kT}{2\pi} = \sum_{i=1}^{\sigma} c_i e_i \int \rho_i(r) r^2 \,\mathrm{d}\boldsymbol{r} \tag{A2.3.185}$$

where ε_0 is the dielectric constant of the medium in which the ions are immersed. The Debye–Hückel limiting law and the HNC and MS approximations satisfy the zeroth and second moment conditions.

The thermodynamic properties are calculated from the ion–ion pair correlation functions by generalizing the expressions derived earlier for one-component systems to multicomponent ionic mixtures. For ionic solutions it is also necessary to note that the interionic potentials are solvent averaged ionic potentials of average force:

$$u_{ij}(r; T.P) = u_{ij}^*(r; T, P) + \frac{e_i e_j}{\varepsilon_0 r}. \tag{A2.3.186}$$

Here $u_{ij}^*(r, T, P)$ is the short-range potential for ions, and ε_0 is the dielectric constant of the solvent. The solvent averaged potentials are thus actually free energies that are functions of temperature and pressure. The thermodynamic properties calculated from the pair correlation functions are summarized below.

(i) The virial equation provides the osmotic coefficient measured in isopiestic experiments:

$$\phi_v = 1 - \frac{1}{6ckT} \sum_{i=1}^{\sigma} \sum_{j=1}^{\sigma} c_i c_j \int r \frac{\partial u_{ij}}{\partial r} g_{ij}(r)\,\mathrm{d}\boldsymbol{r}. \tag{A2.3.187}$$

(ii) The generalization of the compressibility equation, taking into account electroneutrality,

$$\frac{\partial \ln \gamma_\pm}{\partial \ln c} = \frac{1}{cG_\pm} - 1 \tag{A2.3.188}$$

where

$$G_\pm = \int (g_\pm - 1)\,\mathrm{d}\boldsymbol{r} \tag{A2.3.189}$$

provides the concentration dependence of the mean activity coefficient γ determined experimentally from cell EMFs.

(iii) The energy equation is related to the heat of dilution determined from calorimetric measurements

$$E^{\mathrm{ex}} = \frac{1}{2} \sum_{i=1}^{\sigma} \sum_{j=1}^{\sigma} c_i c_j \int \frac{\partial[\beta u_{ij}(r)]}{\partial \beta} g_{ij}(r)\,\mathrm{d}\boldsymbol{r}. \tag{A2.3.190}$$

For an ionic solution

$$\frac{\partial[\beta u_{ij}(r)]}{\partial r} = \frac{e_i e_j}{\varepsilon_0 r}\left[1 + \frac{\partial \ln \varepsilon_0}{\partial \ln T}\right] + \frac{\partial[\beta u_{ij}^*(r)]}{\partial r} \tag{A2.3.191}$$

and $\mathrm{d}\ln \varepsilon_0/\mathrm{d}\ln T = -1.3679$ for water at $25\,^{\circ}\mathrm{C}$.

(iv) The equation for the excess volume is related to the partial molar volumes of the solute determined from density measurements

$$V^{\mathrm{ex}} = \frac{1}{2} \sum_{i=1}^{\sigma} \sum_{j=1}^{\sigma} c_i c_j \int \frac{\partial[\beta u_{ij}(r)]}{\partial P} g_{ij}(r)\,\mathrm{d}\boldsymbol{r}. \tag{A2.3.192}$$

In an ionic solution

$$\frac{\partial[\beta u_{ij}(r)]}{\partial P} = \frac{e_i e_j}{\varepsilon_0 r}\left[\frac{\partial \ln \varepsilon_0}{\partial \ln P}\right] + \frac{\partial[\beta u_{ij}^*(r)]}{\partial P}$$

(A2.3.193)

where $d \ln \varepsilon_0/d \ln P_0 = 47.1 \times 10^{-6}$ for water at $25\,°C$.

The theory of strong electrolytes due to Debye and Hückel derives the exact limiting laws for low valence electrolytes and introduces the idea that the Coulomb interactions between ions are screened at finite ion concentrations.

(c) The Debye–Hückel theory

The model used is the RPM. The average electrostatic potential $\psi_i(r)$ at a distance r away from an ion i is related to the charge density $\rho_i(r)$ by Poisson's equation

$$\nabla^2\psi_i(r) = -\frac{4\pi\rho_i(r)}{\varepsilon_0} = -\frac{4\pi}{\varepsilon_0}\sum_{j=1}^{\sigma} c_j e_j g_{ij}.$$

(A2.3.194)

Debye and Hückel [42] assumed that the ion distribution functions are related to $\psi_i(r)$ by

$$g_{ij}(r) = \exp(-\beta e_j \psi_i(r))$$

which is an approximation. This leads to the PB equation

$$\nabla^2\psi_i(r) = \begin{cases} -\frac{4\pi}{\varepsilon_0}\sum_{j=1}^{\sigma} c_j e_j \exp(-\beta e_j \psi_i(r)) & r > \sigma \\ 0 & r < \sigma. \end{cases}$$

(A2.3.195)

Linearizing the exponential,

$$g_{ij}(r) = 1 - \beta e_j \psi_i(r)$$

(A2.3.196)

in the PB equation leads to the Debye–Hückel differential equation:

$$\nabla^2\psi_i(r) = \begin{cases} \kappa^2(\psi_i(r)) & r > \sigma \\ 0 & r < \sigma \end{cases}$$

(A2.3.197)

where κ is defined by

$$\kappa^2 = \frac{4\pi}{\varepsilon_0 kT}\sum_{j=1}^{\sigma} c_j e_j^2.$$

(A2.3.198)

The solution to this differential equation is

$$g_{ij}(r) = \begin{cases} 1 - \frac{e_i e_j}{\varepsilon_0 kTr}\frac{\exp(-\kappa(r-\sigma))}{(1+\kappa\sigma)} & r > \sigma \\ 0 & r < \sigma \end{cases}$$

(A2.3.199)

which obeys the zeroth moment or electroneutrality condition, but not the second moment condition.

The mean activity coefficient γ_{+-} of a single electrolyte in this approximation is given by

$$\ln\gamma_\pm = -\frac{A|z_+z_-|\sqrt{I}}{1+Ba\sqrt{I}}$$

(A2.3.200)

where a is the effective distance of closest approach of the ions, and A and B are constants determined by the temperature T and the dielectric constant of the solvent ε_0. This expression is widely used to calculate the activity coefficients of simple electrolytes in the usual preparative range. The contributions of the hard cores to non-ideal behaviour are ignored in this approximation.

When $\kappa\sigma \ll 1$ (i.e. at very low concentrations), we have the Debye–Hückel limiting law distribution function:

$$g_{ij}(r) = 1 - \beta e_i e_j \exp(-\kappa r)/\varepsilon_0 r \quad (r > \sigma)$$
$$= 0 \quad (r < \sigma) \qquad (A2.3.201)$$

which satisfies both the zeroth and second moment conditions. It also has an interesting physical interpretation. The total charge $P_i(r)\,dr$ in a shell of radius r and thickness dr around an ion is

$$P_i(r)\,dr = \rho_i(r)4\pi r^2\,dr = -\kappa^2 e_i r \exp(-\kappa r)\,dr \qquad (A2.3.202)$$

which has a maximum at a distance $r = 1/\kappa$, which is called the Debye length or the radius of the 'ionic atmosphere'. Each ion is pictured as surrounded by a cloud or 'ionic atmosphere' whose net charge is opposite in sign to the central ion. The cloud charge $P_i(r)$ has a maximum at $r = 1/\kappa$. The limiting law distribution function implies that the electrostatic potential

$$\psi_i(r) = e_i \exp(-\kappa r)/\varepsilon_0 r. \qquad (A2.3.203)$$

Expanding the exponential, one finds for small κr that

$$\psi_i(r) = \frac{e_i}{\varepsilon_0 r} - \frac{e_i}{\varepsilon_0(1/\kappa)}. \qquad (A2.3.204)$$

The first term is the Coulomb field of the ion, and the second is the potential due to the ion atmosphere at an effective distance equal to $1/\kappa$. For a univalent aqueous electrolyte at 298 K,

$$1/\kappa = 3.043/\sqrt{C}\ \text{Å}$$

where C is the total electrolyte concentration in moles per litre.

The thermodynamic properties derived from the limiting law distribution functions are

$$\frac{E^{\text{ex}}}{NkT} = \frac{\kappa^3}{8\pi c}\left[1 + \frac{\partial \ln \varepsilon_0}{\partial \ln T}\right] \qquad (A2.3.205)$$

$$\ln \gamma_{\pm} = \ln \gamma^{\text{HS}} - \frac{\kappa^3}{8\pi c} \qquad (A2.3.206)$$

$$\phi = \phi^{\text{HS}} - \frac{\kappa^3}{24\pi c} \qquad (A2.3.207)$$

$$\frac{A^{\text{ex}}}{NkT} = \frac{A^{\text{ex,HS}}}{NkT} - \frac{\kappa^3}{24\pi c} \qquad (A2.3.208)$$

where $c = \sum c_i$ is the total ionic concentration and the superscript HS refers to the properties of the corresponding uncharged hard sphere system. Debye and Hückel assumed ideal behaviour for the uncharged system ($\phi^{\text{HS}} = \gamma^{\text{HS}} = 1$ and $A^{\text{ex,HS}} = 0$).

The Debye–Hückel limiting law predicts a square-root dependence on the ionic strength $I = 1/2\sum c_i z_i^2$ of the logarithm of the mean activity coefficient ($\log \gamma_{\pm}$), the heat of dilution (E^{ex}/VI) and the excess volume

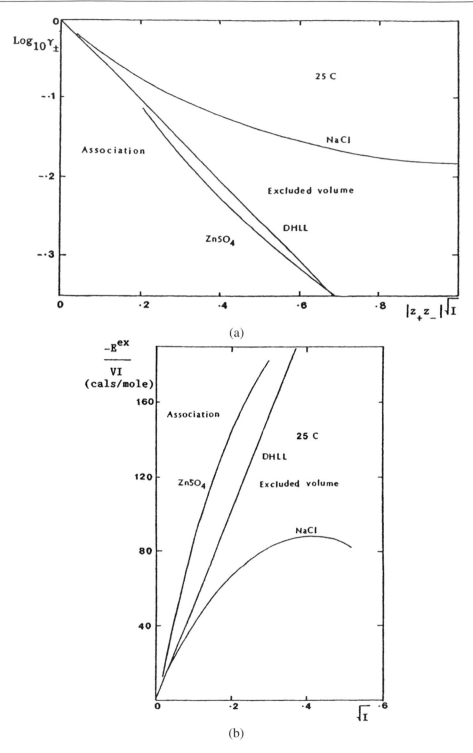

Figure A2.3.11. The mean activity coefficients and heats of dilution of NaCl and ZnSO$_4$ in aqueous solution at 25 °C as a function of $|z_+z_-|\sqrt{I}$, where I is the ionic strength. DHLL = Debye–Hückel limiting law.

(V^{ex}); it is considered to be an exact expression for the behaviour of an electrolyte at infinite dilution. Some experimental results for the activity coefficients and heats of dilution are shown in figure A2.3.11 for aqueous solutions of NaCl and ZnSO$_4$ at 25 °C; the results are typical of the observations for 1–1 (e.g. NaCl) and 2–2 (e.g. ZnSO$_4$) aqueous electrolyte solutions at this temperature.

The thermodynamic properties approach the limiting law at infinite dilution, but deviate from it at low concentrations, in different ways for the two charge types. Evidence from the ionic conductivity of 2–2 electrolyte solutions suggests that the negative deviations from the limiting law observed for these solutions are due to ion pairing or association. The opposite behaviour found for aqueous 1–1 electrolytes, for which ion pairing is negligble at room temperature, is caused by the finite size of the ions and is the excluded volume effect. The Debye–Hückel theory ignores ion association and treats the effect of the sizes of the ions incompletely. The limiting law slopes and deviations from them depend strongly on the temperature and dielectric constant of the solvent and on the charges on the ions. An aqueous solution of sodium chloride, for instance, behaves like a weak electrolyte near the critical temperature of water because the dielectric constant of the solvent decreases rapidly with increasing temperature.

As pointed out earlier, the contributions of the hard cores to the thermodynamic properties of the solution at high concentrations are not negligible. Using the CS equation of state, the osmotic coefficient of an uncharged hard sphere solute (in a continuum solvent) is given by

$$\phi^{\text{HS}} = 1 + \frac{4\eta - 2\eta^2}{(1 - \eta^3)} \tag{A2.3.209}$$

where $\eta = c\sigma^3/6$. For a 1 M solution this contributes 0.03 to the deviation of the osmotic coefficient from ideal behaviour.

(d) Mayer's theory

The problem with the virial expansion when applied to ionic solutions is that the virial coefficients diverge. This difficulty was resolved by Mayer who showed how the series could be resumed to cancel the divergencies and yield a new expansion for a charged system. The terms in the new series are ordered differently from those in the original expansion, and the Debye–Hückel limiting law follows as the leading correction due to the non-ideal behaviour of the corresponding uncharged system. In principle, the theory enables systematic corrections to the limiting law to be obtained as at higher electrolyte concentrations. The results are quite general and are applicable to any electrolyte with a well defined short-range potential $u_{ij}^*(r)$, besides the RPM electrolyte.

The principle ideas and main results of the theory at the level of the second virial coefficient are presented below. The Mayer f-function for the solute pair potential can be written as the sum of terms:

$$f_{ij}(r) = f_{ij}^*(r) + (1 + f_{ij}^*(r)) \sum_{n=1}^{\infty} \frac{1}{n!} (-\beta e_i e_j / \varepsilon_0 r)^n \tag{A2.3.210}$$

where $f_{ij}^*(r)$ is the corresponding Mayer f-function for the short-range potential $u_{ij}^*(r)$ which we represent graphically as $_i\text{O}----\text{O}_j$ and $\beta = 1/kT$. Then the above expansion can be represented graphically as

$$\tag{A2.3.211}$$

$_i$o———o$_j$ represents $f_{ij}(r)$, the Mayer f-function for the pair potential $u_{ij}(r)$, and $\text{o}\sim\sim\sim\text{o} = -\beta e_i e_j / \varepsilon_0 r$ represents the Coulomb potential multiplied by $-\beta$. The graphical representation of the virial coefficients in terms of Mayer f-bonds can now be replaced by an expansion in terms of f^* bonds (o–––o) and Coulomb bonds (o$\sim\sim\sim$o). Each f-bond is replaced by an f^*-bond and the sum of one or more Coulomb bonds in parallel with or without an f^*-bond in parallel. The virial coefficients then have the following graphical representation:

$$i \bullet\!\!-\!\!-\!\!-\!\!\bullet j = i \bullet\!-\!-\!-\!\bullet j + i \bullet\!\sim\!\sim\!\bullet j +_i \langle\!\sim\!\sim\!\rangle j + \cdots$$

$$\triangle = \triangle + \triangle +$$

$$\square = \square + \square +   \text{(A2.3.212)}$$

$$\vdots$$

$$\frac{A^{\text{ex}}}{NkT} = \frac{A^{\text{ex.o}}}{NkT} - \frac{\kappa^3}{12\pi c} + \cdots \text{HT}$$

where HT stands for higher-order terms. There is a symmetry number associated with each graph which we do not need to consider explicitly in this discussion. Each black circle denotes summation over the concentration c_i and integration over the coordinates of species i. The sum over all graphs in which the f-bond is replaced by an f^*-bond gives the free energy $A^{\text{ex}*}$ of the corresponding uncharged system. The effect of the Coulomb potential on the expansion is more complicated because of its long range. The second term in the expansion of the second virial coefficient is the bare Coulomb bond multiplied by $-\beta$. If we multiply this by a screening function and carry out the integration the result is finite, but it contributes nothing to the overall free energy because of electroneutrality. This is because the contribution of the charge e_i from the single Coulomb bond at a vertex when multiplied by c_i and summed over i is zero. The result for a graph with a cycle of Coulomb bonds, however, is finite. Each vertex in these graphs has two Coulomb bonds leading into it and instead of $c_i e_i$ we have $\sum c_i e_i^2$ (which appears as a factor in the definition of κ^2). This is not zero unless the ion concentration is also zero. Mayer summed all graphs with cycles of Coulomb bonds and found that this leads to the Debye–Hückel limiting law expression for the excess free energy! The essential mechanism behind this astonishing result is that the long-range nature of the Coulomb interaction requires that the ions be considered collectively rather than in pairs, triplets etc, which is implied by the conventional virial expansion. The same mechanism is also responsible for the modification of the interaction between two charges by the presence of others, which is called 'screening'. The sum of all chains of Coulomb bonds between two ions represents the direct interaction as well as the sum of indirect interactions (of the longest range) through other ions. The latter is a subset of the graphs which contribute to the correlation function $h_{ij}(r) = g_{ij}(r) - 1$ and has the

graphical representation

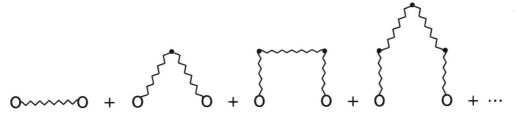

Explicit calculation of this sum shows that it is the Debye screened potential

$$q_{ij}(r) = = -\beta e_i e_j \exp(-\kappa r)/\varepsilon_0 r. \tag{A2.3.213}$$

Going beyond the limiting law it is found that the modified (or renormalized) virial coefficients in Mayer's theory of electrolytes are functions of the concentration through their dependence on κ. The ionic second virial coefficient $B_2(\kappa)$ is given by [62]

$$B_2(\kappa) = -\frac{1}{2} \sum_{i=1}^{\sigma} \sum_{j=1}^{\sigma} c_i c_j \int [(\exp(-\beta u_{ij}^*(r) + q_{ij}(r)) - 1 - q_{ij}(r) - q_{ij}(r)^2/2] \, d\mathbf{r}. \tag{A2.3.214}$$

This expression contains the contribution of the short-range potential included earlier in A^{ex^*}, so that the excess free energy, to this level of approximation, is

$$\left(\frac{A^{ex}}{NkT}\right)_{DHLL+B_2} = -\frac{\kappa^3}{12\pi c} + \frac{B_2(\kappa)}{c}. \tag{A2.3.215}$$

This is called the DHLL + B_2 approximation. On carrying out the integrations over $q_{ij}(r)$ and $q_{ij}(r)^2/2$ and using the electroneutrality condition, this can be rewritten as [63]

$$\left(\frac{A^{ex}}{NkT}\right)_{DHLL+B_2} = -\frac{5\kappa^3}{96\pi c} + \frac{S_2(\kappa)}{c} \tag{A2.3.216}$$

where

$$S_2(\kappa) = -\frac{1}{2} \sum_{i=1}^{\sigma} \sum_{j=1}^{\sigma} c_i c_j \int [(\exp(-\beta u_{ij}^*(r) + q_{ij}(r)) - 1] \, d\mathbf{r}. \tag{A2.3.217}$$

This has the form of a second virial coefficient in which the Debye screened potential has replaced the Coulomb potential. Expressions for the other excess thermodynamic properties are easily derived.

Mayer's theory is formally exact within the radius of convergence of the virial series and it predicts the properties characteristic of all charge types without the need to introduce any additional assumptions. Unfortunately, the difficulty in calculating the higher virial coefficients limits the range of concentrations to which the theory can be applied with precision. The DHLL + B_2 approximation is qualitatively correct in reproducing the association effects observed at low concentrations for higher valence electrolytes and the excluded volume effects observed for all electrolytes at higher concentrations.

(e) The MS approximation

The MS approximation for the RPM, i.e. charged hard spheres of the same size in a continuum dielectric, was solved by Waisman and Lebowitz [46] using Laplace transforms. The solutions can also be obtained [47] by an extension of Baxter's method to solve the PY approximation for hard spheres and sticky hard spheres. The method can be further extended to solve the MS approximation for unsymmetrical electrolytes (with hard cores of unequal size) and weak electrolytes, in which chemical bonding is mimicked by a delta function interaction. We discuss the solution to the MS approximation for the symmetrically charged RPM electrolyte.

For the RPM of an electrolyte the MS approximation is

$$c_{ij}(r) = -\beta u_{ij}(r) = -e_i e_j / \varepsilon_0 r \qquad \text{for } r > \sigma \qquad (A2.3.218)$$

with the exact relation

$$h_{ij}(r) = g_{ij}(r) - 1 = -1 \qquad \text{for } r < \sigma. \qquad (A2.3.219)$$

The generalization of the Orstein–Zernike equation to a mixture is

$$h_{ij}(r_{12}) = c_{ij}(r_{12}) + \sum_{k=1}^{\sigma} \rho_k \int c_{ik}(r_{13}) h_{kj}(r_{32}) \, \mathrm{d}r_3 \qquad (A2.3.220)$$

where i and j refer to the ionic species (positive and negative ions), ρ_i is the concentration (or number density) of the ith species and σ is the number of ionic species. Taking Fourier transforms and using the convolution theorem puts this in matrix form

$$\tilde{\mathbf{H}} = \tilde{\mathbf{C}} + \tilde{\mathbf{H}} \mathbf{P} \tilde{\mathbf{C}} \qquad (A2.3.221)$$

where $\mathbf{H}$ and $\mathbf{C}$ are matrices whose elements are the Fourier transforms of h_{ij} and c_{ij}, and $\mathbf{P}$ is a diagonal matrix whose elements are the concentrations ρ_i of the ions. The correlation function matrix is symmetric since $c_{+-} = c_{-+}$ and $h_{+-} = h_{-+}$. The RPM symmetrically charged electrolyte has the additional simplification

$$|e_+| = |e_-| = e$$

$$\rho_+ = \rho_- = \rho/2 \qquad (A2.3.222)$$

and

$$c_{++} = c_{--} \qquad h_{++} = h_{--} \qquad (A2.3.223)$$

where e is the magnitude of the charge and ρ is the total ion concentration. Defining the sum and difference functions

$$F_s = (F_+ + F_-)/2 \qquad \text{and} \qquad F_D = (F_+ - F_-)/2 \qquad (A2.3.224)$$

of the direct and indirect correlation functions c_{ij} and h_{ij}, the Ornstein–Zernike equation separates into two equations

$$h_s = c_s + \rho c_s * h_s \qquad (A2.3.225)$$

$$h_D = c_D - \rho c_D * h_D \qquad (A2.3.226)$$

where * stands for a convolution integral and the core condition is replaced by

$$h_s = -1 \qquad h_D = 0 \qquad \text{for } 0 < r < \sigma. \qquad (A2.3.227)$$

The MS solution for c_s turns out to be identical to the MS (or PY) approximation for hard spheres of diameter σ; it is a cubic polynomial in r/σ. The solution for c_D is given by

$$c_D = \begin{cases} \dfrac{\beta e^2}{\varepsilon_0 k T \sigma} \left[2B - B^2 \left(\dfrac{r}{\sigma} \right) \right] & 0 < r < \sigma \\[3mm] \dfrac{\beta e^2}{\varepsilon_0 k T r} & r > \sigma \end{cases} \qquad (A2.3.228)$$

where

$$B = \frac{[(1+x) - (1+2x)^{1/2}]}{x} \tag{A2.3.229}$$

and $x = \kappa\sigma$. The excess energy of a fully dissociated strong electrolyte in the MSA approximation is

$$\frac{E^{\text{ex}}}{NkT} = \frac{-x[(1+x) - (1+2x)^{1/2}]}{4\pi c\sigma^3}. \tag{A2.3.230}$$

Integration with respect to β, from $\beta = 0$ to finite β, leads to the excess Helmholtz free energy:

$$\frac{A^{\text{ex}} - A^{\text{ex,HS}}}{NkT} = -\frac{[6x + 3x^2 + 2 - 2(1+2x)^{3/2}]}{12\pi\rho\sigma^3} \tag{A2.3.231}$$

where $A^{\text{ex,HS}}$ is the excess free energy of hard spheres. The osmotic coefficient follows from this and is given by

$$\phi^{\text{E}} = \phi^{\text{HS}} + \frac{[3x + 3x(1+2x)^{1/2} - 2(1+2x)^{3/2} + 2]}{12\pi\rho\sigma^3} \tag{A2.3.232}$$

where ϕ^{HS} is the osmotic coefficient of the uncharged hard spheres of diameter σ in the MS or PY approximation. The excess Helmholtz free energy is related to the mean activity coefficient $\gamma_\pm$ by

$$A^{\text{ex}} = NkT[\ln \gamma_\pm + (1 - \phi)] \tag{A2.3.233}$$

and the activity coefficient from the energy equation, calculated from ϕ^{E} and $A^{\text{ex,E}}$ is given by

$$\ln \gamma_\pm^{\text{E}} = \ln \gamma^{\text{HS}} + \frac{-x[(1+x) - (1+2x)^{1/2}]}{4\pi c\sigma^3}. \tag{A2.3.234}$$

The second term on the right is $\beta E^{\text{ex}}/NkT$. This is true for any theory that predicts $\beta(A^{\text{ex}} - A^{\text{HS}})$ as a function of $x = \kappa\sigma$ only, which is the case for the MS approximation.

The thermodynamic properties calculated by different routes are different, since the MS solution is an approximation. The osmotic coefficient from the virial pressure, compressibility and energy equations are not the same. Of these, the energy equation is the most accurate by comparison with computer simulations of Card and Valleau [63]. The osmotic coefficients from the virial and compressibility equations are

$$\phi_{\text{V}} = \phi^{\text{HS}} + \frac{x^2 B}{12\pi\rho\sigma^3} \tag{A2.3.235}$$

$$\phi_{\text{C}} = \phi^{\text{HS}}. \tag{A2.3.236}$$

In the limit of zero ion size, i.e. as $\sigma \to 0$, the distribution functions and thermodynamic functions in the MS approximation become identical to the Debye–Hückel limiting law.

(f) The HNC approximation

The solutions to this approximation are obtained numerically. Fast Fourier transform methods and a reformulation of the HNC (and other integral equation approximations) in terms of the screened Coulomb potential by Allnatt [64] are especially useful in the numerical solution. Figure A2.3.12 compares the osmotic coefficient of a 1–1 RPM electrolyte at 25 °C with each of the available Monte Carlo calculations of Card and Valleau [63].

The agreement is excellent up to a 1 molar concentration. The excess energies for 1–1, 2–1, 2–2 and 3–1 charge types calculated from the MS and HNC approximations are shown in figure A2.3.13. The Monte

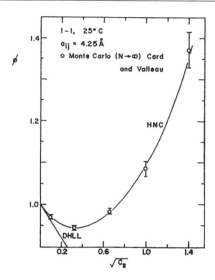

Figure A2.3.12. The osmotic coefficient of a 1–1 RPM electrolyte compared with the Monte Carlo results of [63].

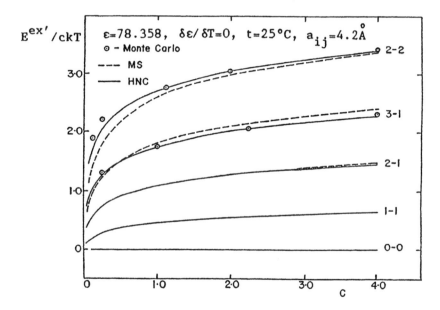

Figure A2.3.13. The excess energy of 1–1, 2–1, 3–1 and 2–2 RPM electrolytes in water at 25 °C. The full and dashed curves are from the HNC and MS approximations, respectively. The Monte Carlo results of Card and Valleau [63] for the 1–3 and 2–2 charge types are also shown.

Carlo results for 2–2 and 3–1 electrolytes are also shown in the same figure. The agreement is good, even for the energies of the higher valence electrolytes. However, as illustrated in figure A2.3.14, the HNC and MS approximations deteriorate in accuracy as the charges on the ions are increased [67].

The osmotic coefficients from the HNC approximation were calculated from the virial and compressibility equations; the discrepancy between ϕ_V and ϕ_C is a measure of the accuracy of the approximation. The osmotic coefficients calculated via the energy equation in the MS approximation are comparable in accuracy to the HNC approximation for low valence electrolytes. Figure A2.3.15 shows deviations from the Debye–Hückel

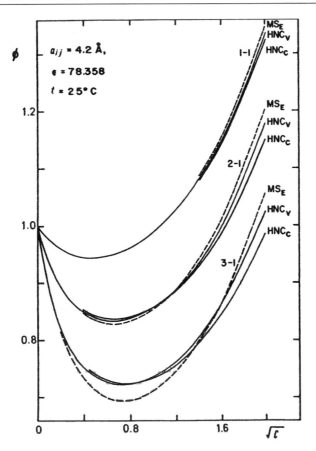

Figure A2.3.14. Osmotic coefficients for 1–1, 2–1 and 3–1 RPM electrolytes according to the MS and HNC approximations.

limiting law for the energy and osmotic coefficient of a 2–2 RPM electrolyte according to several theories. The negative deviations from the limiting law are reproduced by the HNC and DHLL + B_2 equations but not by the MS approximation.

In figure A2.3.16, the theoretical HNC osmotic coefficients for a range of ion size parameters in the primitive model are compared with experimental data for the osmotic coefficients of several 1–1 electrolytes at 25 °C. Choosing $a_{+-} = r_+ + r_-$ to fit the data at low concentrations, it is found that the calculated osmotic coefficients are too large at the higher concentrations. On choosing a_{+-} to be the sum of the Pauling radii of the ions, and a short-range potential given by a square well or mound d_{ij} equal to the width of a water molecule (2.76 Å), it is found that the osmotic coefficients can be fitted to the accuracy shown in figure A2.3.17 [65]. There are other models for the short-range potential which produce comparable fits for the osmotic coefficients showing that the square well approximation is by no means unique [66].

A2.3.5.2   Weak electrolytes

In a weak electrolyte (e.g. an aqueous solution of acetic acid) the solute molecules AB are incompletely dissociated into ions A^+ and B^- according to the familiar chemical equation

$$AB = A^+ + B^-. \qquad (A2.3.237)$$

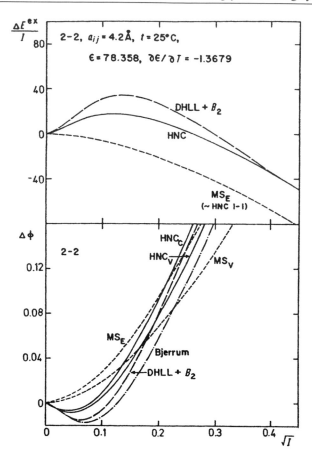

Figure A2.3.15. Deviations (Δ) of the heat of dilution E^{ex}/I and the osmotic coefficient ϕ from the Debye–Hückel limiting law for 1–1 and 2–2 RPM electrolytes according to the DHLL + B_2, HNC and MS approximations.

The forces binding the atoms in AB are chemical in nature and must be introduced, at least approximately, in the Hamiltonian in a theoretical treatment of this problem. The binding between A and B in the dimer AB is quite distinct from the formation of ion pairs in higher valence electrolytes (e.g. aqueous solutions of $ZnSO_4$ at room temperature) where the Coulomb interactions between the ions lead to ion pairs which account for the anomalous conductance and activity coefficients at low concentration. The greater shielding of the ion charges with increasing electrolyte concentration would induce the ion pairs to dissociate as the concentration rises, whereas the dimer population produced by the chemical bonding represented in the above chemical reaction would increase with the concentration of the solution.

Weak electrolytes in which dimerization (as opposed to ion pairing) is the result of chemical bonding between oppositely charged ions have been studied using a sticky electrolyte model (SEM). In this model, a delta function interaction is introduced in the Mayer f-function for the oppositely charged ions at a distance $L = \sigma$, where σ is the hard sphere diameter. The delta function mimics bonding and the Mayer f-function

$$f_{+-} = -1 + L\zeta\delta(r - L)/12 \qquad r \leq \sigma \qquad (A2.3.238)$$

where ζ is the sticking coefficient. This induces a delta function in the correlation function $h_{+-}(r)$ for

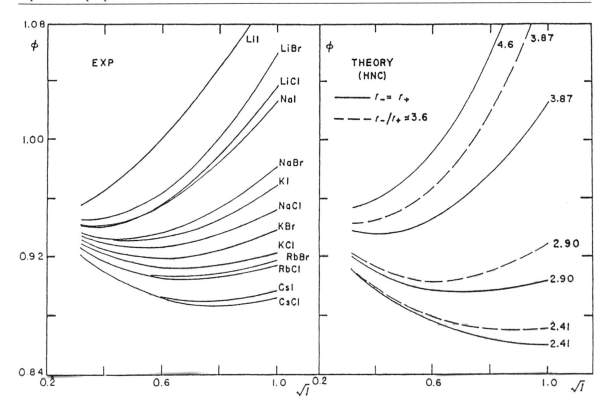

Figure A2.3.16. Theoretical HNC osmotic coefficients for a range of ion size parameters in the primitive model compared with experimental data for the osmotic coefficients of several 1–1 electrolytes at 25 °C. The curves are labelled according to the assumed value of $a_{+-} = r_+ + r_-$.

oppositely charged ions with a different coefficient λ:

$$h_{+-} = -1 + L\lambda\delta(r - L)/12 \qquad r \leq \sigma. \qquad (A2.3.239)$$

The interaction between ions of the same sign is assumed to be a pure hard sphere repulsion for $r \leq \sigma$. It follows from simple steric considerations that an exact solution will predict dimerization only if $L < \sigma/2$, but polymerization may occur for $\sigma/2 < L = \sigma$. However, an approximate solution may not reveal the full extent of polymerization that occurs in a more accurate or exact theory. Cummings and Stell [69] used the model to study chemical association of uncharged atoms. It is closely related to the model for adhesive hard spheres studied by Baxter [70].

The association 'constant' K defined by $K = \rho_{AB}/\rho_+\rho_-$ is

$$K = \frac{\pi\lambda(L/\sigma^3)}{3(1 - \langle N \rangle)^2} \qquad (A2.3.240)$$

where the average number of dimers $\langle N \rangle = \eta\lambda(L/\sigma)^3$ and $\eta = \pi\rho\sigma^3/6$, in which ρ is the total ionic density. We can now distinguish three different cases:

$\lambda = 0$	no dimers	strong electrolyte (RPM)
$\lambda = (\sigma/L)^3/\eta$	all dimers if $L < \sigma/2$	dipolar dumb-bells
$0 < \lambda < (\sigma/L)^3/\eta$	ions + dimers	weak electrolyte (SEM).

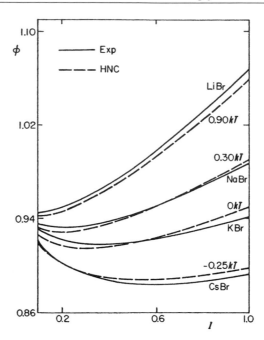

Figure A2.3.17. Theoretical (HNC) calculations of the osmotic coefficients for the square well model of an electrolyte compared with experimental data for aqueous solutions at 25 °C. The parameters for this model are $a_{+-} = r_+$(Pauling) $+ r_-$(Pauling), $d_{++} = d_{--} = 0$ and d_{+-} as indicated in the figure.

Either the same or different approximations may be used to treat the binding at $r = L$ and the remaining electrical interactions between the ions. The excess energy of the sticky electrolyte is given by

$$\frac{E^{ex}}{NkT} = \frac{\langle N \rangle}{2} \frac{\partial \ln \zeta}{\partial \beta} - \left(1 + \frac{\partial \ln \varepsilon_0}{\partial \ln T}\right) \frac{\kappa H}{2} \qquad (A2.3.241)$$

where

$$H = \kappa \int_\sigma^\infty h_D(r) r \, dr \qquad (A2.3.242)$$

and $h_D(r) = [h_{+-}(r) - h_{++}(r)]/2$. The first term is the binding energy and the second is the energy due to the interactions between the charges which can be determined analytically in the MS approximation and numerically in the HNC approximation. For any integer $n = \sigma/L$, $H' = H/\sigma$ in the MS approximation has the form [48]

$$H' = \frac{(a_1 + a_2 x) - (a_1^2 + 2xa_3)^{1/2}}{24 a_4 \eta} \qquad (A2.3.243)$$

where a_i $(i = 1$ to $4)$ are functions of the reduced ion concentration η, the association parameter λ and n. When $\lambda = 0$, $a_i = 1$, the average number of dimers $\langle N \rangle = 0$ and the energy of the RPM strong electrolyte in the MS approximation discussed earlier is recovered. The effect of a hard sphere solvent on the degree of dissociation of a weak electrolyte enhances the association parameter λ due to the packing effect of the solvent, while adding a dipole to the solvent has the opposite effect [71].

The PY approximation for the binding leads to negative results for λ; the HNC approximation for this is satisfactory. Figure A2.3.18 shows the excess energy as a function of the weak electrolyte concentration for the RPM and SEM for a 2–2 electrolyte.

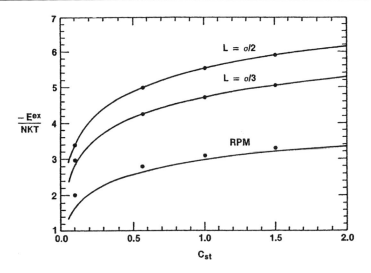

Figure A2.3.18. The excess energy E^{ex} in units of NkT as a function of the concentration c_{st} for the RPM and SEM 2–2 electrolyte. The curves and points are results of the HNC/MS and HNC approximations, respectively, for the binding and the electrical interactions. The ion parameters are $\sigma = 4.2$ Å, and $\varepsilon = 73.4$. The sticking coefficients $\zeta = 1.6 \times 10^6$ and 2.44×10^6 for $L = \sigma/2$ and $\sigma/3$, respectively.

In the limit $\lambda = (\sigma/L)^3/\eta$ with $L < \sigma/2$, the system should consist of dipolar dumb-bells. The asymptotic form of the direct correlation function (defined through the Ornstein–Zernike equation) for this system (in the absence of a solvent) is given by

$$c_{ij}(r) = -\beta A e_i e_j/r \qquad (A2.3.244)$$

where $A = \varepsilon/(\varepsilon - 1)$ and ε is the dielectric constant of the system of dipolar dumb-bells. The energy of dipolar dumb-bells, excluding the binding energy, in the MS approximation is [48]

$$\frac{E^{ex}}{N_D kT} = \frac{-x[c_1 + c_2 x'] - (c_1^2 + 2c_3 x')^{1/2}}{24\eta} \qquad (A2.3.245)$$

where x' is the reduced dipole moment defined by

$$x' = \kappa\sigma = 2n(A\pi\rho/kT)^{1/2}\mu \qquad (A2.3.246)$$

dipole moment $\mu = eL = e\sigma/n$, $N_D = N/2$ is the number of dipoles, the coefficients c_i ($i = 1$ to 3) depend on the dipole elongation and n is an integer. This provides an analytic solution for the energy of dipolar dumb-bells in the MSA approximation; it suffers from the defect that it tends to a small but finite constant in the limit of zero density and should strictly be applicable only for $L < \sigma/2$.

A2.3.6 Perturbation theory

The attractive dispersive forces between the atoms of a simple homogeneous fluid increase their cohesive energy, but their vector sums nearly cancel, producing little alteration in the structure determined primarily by the repulsive part of the interatomic potential. Charges, dipoles and hydrogen bonding, as in water molecules, increase the cohesive energy of molecules and produce structural changes. Despite this, the harsh interatomic repulsions dominate the structure of simple fluids. This observation forms the physical basis of perturbation theory. van der Waals implicitly used this idea in his equation of state in which the attractive part of the interaction is treated as a perturbation to the repulsive part in a mean-field approximation.

In perturbation theories of fluids, the pair total potential is divided into a reference part and a perturbation

$$u(1, 2) = u^0(1, 2) + w(1, 2) \tag{A2.3.247}$$

where $u^0(1, 2)$ is the pair potential of the reference system which usually has the features that determine the size and shape of the molecules, while the perturbation $w(1, 2)$ contains dispersive and attractive components which provide the cohesive energy of the fluid. The equilibrium properties of the system are calculated by expansion in a suitable parameter about the reference system, whose properties are assumed known to the extent that is necessary. The reference system may be anisotropic, i.e. with $u^0(1, 2) = u^0(r_{12}, \Omega_1, \Omega_2)$, where (Ω_1, Ω_2) represent the angular coordinates of atoms 1 and 2, or it may be isotropic when $u^0(1, 2) = u^0(r_{12})$.

The most common choice for a reference system is one with hard cores (e.g. hard spheres or hard spheroidal particles) whose equilibrium properties are necessarily independent of temperature. Although exact results are lacking in three dimensions, excellent approximations for the free energy and pair correlation functions of hard spheres are now available to make the calculations feasible.

The two principal methods of expansion used in perturbation theories are the high-temperature λ expansion of Zwanzig [72], and the γ expansion introduced by Hemmer [73]. In the λ-expansion, the perturbation $w(1, 2)l$ is modulated by the switching parameter λ which varies between 0 and 1, thereby turning on the perturbation. The free energy is expanded in powers of λ, and reduces to that of the reference system when $\lambda = 0$. In the γ expansion, the perturbation is long ranged of the form $w(r) = -\gamma^3 \phi(\gamma r)$, and the free energy is expanded in powers of γ about $\gamma = 0$. In the limit as $\gamma \to 0$, the free energy reduces to a mean-field van der Waals-like equation. The γ expansion is especially useful in understanding long-range perturbations, such as Coulomb and dipolar interactions, but difficulties in its practical implementation lie in the calculation of higher-order terms in the expansion. Another perturbation approach is the mode expansion of Andersen and Chandler [74], in which the configurational integral is expanded in terms of collective coordinates that are the Fourier transforms of the particle densities. The expansion is especially useful for electrolytes and has been optimized and improved by adding the correct second virial coefficient. Combinations of the λ and γ expansions, the union of the λ and virial expansions and other improvements have also been discussed in the literature. Our discussion will be mainly confined to the λ expansion and to applications of perturbation theory to determining free energy differences by computer simulation. We conclude the section with a brief discussion of perturbation theory of inhomogeneous fluids.

A2.3.6.1   The λ expansion

The first step is to divide the total potential into two parts: a reference part and the remainder treated as a perturbation. A coupling parameter λ is introduced to serve as a switch which turns the perturbation on or off. The total potential energy of N particles in a given configuration $(r_1, \ldots, r_N)$ is

$$U_N(r_1, \ldots, r_N; \lambda) = U_N^0(r_1, \ldots, r_N) + \lambda W_N(r, \ldots, r_N) \tag{A2.3.248}$$

where $0 = \lambda = 1$, $U_N^0(r_1, \ldots, r_N)$ is the reference potential and $W_N(r_1, \ldots, r_N)$ is the perturbation. When λ is zero the perturbation is turned off, and it is on when $\lambda = 1$.

The configurational PF

$$
\begin{aligned}
Z(N, V, T; \lambda) &= \int \exp(-\beta U_N(r^N; \lambda)) \, dr^N \\
&= \int \exp(-\beta U_N^0(r^N)) \exp(-\beta \lambda W_N(r^N)) \, dr^N. \tag{A2.3.249}
\end{aligned}
$$

Multiplying and dividing by the configurational PF of the reference system

$$Z^0(N, V, T) = \int \exp(-\beta U_N^0(r^N)) \, dr^N \tag{A2.3.250}$$

one finds that

$$Z(N, V, T; \lambda) = Z^0(N, V, T)\langle \exp(-\lambda W_N(r^N))\rangle_0 \tag{A2.3.251}$$

where $\langle \ldots \rangle_0$ is an average over the reference system. The Helmholtz free energy

$$A(N, V, T; \lambda) = -kT \ln Q(N, V, T; \lambda) = -kT \ln \frac{Z(N, V, T; \lambda)}{N! \Lambda^{3N}}. \tag{A2.3.252}$$

It follows that the change in Helmholtz free energy due to the perturbation is

$$A(N, V, T; \lambda) - A^0(N, V, T; \lambda) = -kT \ln\langle \exp(-\beta\lambda W_N(r^N))\rangle_0. \tag{A2.3.253}$$

This equation was first derived by Zwanzig [72]. Note that β and λ always occur together. Expanding about $\lambda = 0$ at constant β (or equivalently about $\beta = 0$ at constant λ) one finds

$$-\beta \Lambda A(\lambda) = \ln\langle \exp(-\beta\lambda W_N(r^N))\rangle_0$$

$$= \ln\left\langle \sum_{n=0}^{\infty} \frac{(-\beta\lambda)^n}{n!} W_N(r^N) \right\rangle_0 = \sum_{n=1}^{\infty} (-\beta\lambda)^n a_n \tag{A2.3.254}$$

which defines the coefficients a_n. By comparing the coefficients of $(-\beta\lambda)^n$ for different n, one finds

$$a_1 = \langle W_N \rangle_0$$

$$a_2 = \frac{1}{2}[\langle W_N^2 \rangle_0 - \langle W_N \rangle_0^2]$$

$$a_3 = \frac{1}{3!}[\langle W_N^3 \rangle_0 - 3\langle W_N \rangle_0 \langle W_N^2 \rangle_0 + 2\langle W_N \rangle_0^3] \text{ etc} \tag{A2.3.255}$$

where the averages are over the reference system whose properties are assumed to be known.

The first term in the high-temperature expansion, a_1, is essentially the mean value of the perturbation averaged over the reference system. It provides a strict upper bound for the free energy called the Gibbs–Bogoliubov inequality. It follows from the observation that $\exp(-x)1 - x$ which implies that $\ln\langle \exp(-x)\rangle \ln\langle 1 - x \rangle - \langle x \rangle$. Hence

$$\ln\langle \exp(-\beta\lambda W_N(r^N))\rangle_0 \geq -\beta\lambda\langle W_N(r^N)\rangle_0.$$

Multiplying by -1 reverses this inequality, and we see that

$$\beta \Delta A(\lambda) \leq \beta\lambda\langle W_N(r^N)\rangle_0 = \beta\lambda a_1 \tag{A2.3.256}$$

which proves the result. The higher-order terms in the high-temperature expansion represent fluctuations about the mean.

Assuming the perturbing potential is pairwise additive,

$$W_N(r^N) = \sum_{i<j} w_{ij}(r_{ij}) \tag{A2.3.257}$$

we have

$$
\begin{aligned}
a_1 = \langle W_N \rangle_0 &= \frac{\int \dots \int \sum_{i<j} w_{ij}(r_{ij}) \exp(-\beta U_N^0(r^N)) \, dr^N}{Z^0(N, V, T)} \\
&= \frac{N(N-1)}{2} \frac{\int \dots \int \exp(-\beta U_N^0(r^N)) \, dr_3 . dr_N \, dr_1 \, dr_2}{Z^0(N, V, T)} \\
&= \int \dots \int \rho_N^{0,(2)}(r_1, r_2) \, dr_1 r_2
\end{aligned}
$$

where translational and rotational invariance of the reference fluid system implies that $\rho^{0,(2)}(r_1, r_2) = \rho^2 g_N^0(r_{12})$. Using this in the above expression, changing to relative coordinates and integrating over coordinates of 1, one has

$$
a_1 = \frac{\rho N}{2} \int w_{12}(r_{12}) g_N^0(r_{12}) \, dr_{12} \tag{A2.3.258}
$$

which was first obtained by Zwanzig. As discussed above, this provides an upper bound for the free energy, so that

$$
\frac{\Delta A(\lambda)}{N} \leq \frac{\rho}{2} \int \lambda w_{12}(r_{12}) g_N^0(r_{12}) \, dr_{12}. \tag{A2.3.259}
$$

The high-temperature expansion, truncated at first order, reduces to van der Waals' equation, when the reference system is a fluid of hard spheres.

The second-order term, a_2, was also obtained by Zwanzig, and involves two-, three- and four-body correlation functions for an N-particle system. Before passage to the thermodynamic limit,

$$
\begin{aligned}
a_2 = \frac{1}{2} \int \dots \int \rho_N^{(2)}(r_1, r_2) w_{12}(r_{12}) \, dr_1 \, dr_2 \\
+ \int \int \int \rho_N^{(3)}(r_1, r_2, r_3) w_{12}(r_{12}) w_{23}(r_{23}) \, dr_1 \, dr_2 \, dr_3 \\
+ \frac{1}{4} \int \int \int \int [\rho_N^{(4)}(r_1, r_2, r_3, r_4) - \rho_N^{(2)}(r_1, r_2)\rho_N^{(2)}(r_3, r_4)] w_{12}(r_{12}) w_{23}(r_{23}) \, dr_1 \, dr_2 \, dr_3 \, dr_4.
\end{aligned}
$$
$$\tag{A2.3.260}$$

Evaluating its contribution to the free energy of the system requires taking the thermodynamic limit ($N \to \infty$) for the four-particle distribution function. Lebowitz and Percus [75] and Hiroike [76] showed that the asymptotic behaviour of $\rho_N^{(4)}$ in the canonical ensemble, when the 1,2 and 3,4 pairs are widely separated, is given by

$$
\rho_N^{(4)}(r_1, r_2, r_3, r_4) = \rho_N^{(2)}(r_1, r_2)\rho_N^{(2)}(r_3, r_4) + x(r_1, r_2, r_3, r_4)/N + O(N^{-2}) \tag{A2.3.261}
$$

where the O($1/N$) term makes a finite contribution to the last term in a_2. This correction can also be evaluated in the grand canonical ensemble.

The high-temperature expansion could also be derived as a Taylor expansion of the free energy in powers of λ about $\lambda = 0$:

$$
A(\lambda) = A_0 + \lambda(\delta A/\delta \lambda)_{\lambda=0} + (\lambda^2/2)(\delta^2 A/\delta \lambda^2)_{\lambda=0} + \cdots \tag{A2.3.262}
$$

so that the coefficients of the various powers of λ are related to a_n, with

$$
\begin{aligned}
\frac{\partial A}{\partial \lambda} &= \frac{1}{Z(N, V, T; \lambda)} \int \dots \int W_N(r^N) \exp(-\beta U(r^N)) \, dr^N \\
&= \langle W_N(r^N) \rangle_\lambda.
\end{aligned} \tag{A2.3.263}
$$

It follows that

$$A(\lambda) = A(0) + \int_0^\lambda \langle W_N(r^N) \rangle_\lambda \, d\lambda. \qquad (A2.3.264)$$

Assuming the perturbing potential is pairwise additive, an argument virtually identical to the calculation of $a_1 = \langle W_N(r^N) \rangle_0$ shows that

$$\langle W_N(r^N) \rangle_\lambda = \frac{1}{2} \int \cdots \int \rho_N^{(2)}(r_1, r_2; \lambda) w_{12}(r_{12}) \, dr_1 \, dr_2$$

where $\rho_N^{(2)}(r_1, r_2; \lambda)$ is the two-particle density correlation function in an N-particle system with potential $U_N(r_1, \ldots, r_N; \lambda)$. Substituting this in the expression for $A(\lambda)$, we have

$$\Delta A(\lambda) = \frac{1}{2} \int_0^\lambda d\lambda \int \cdots \int \rho_N^{(2)}(r_1, r_2; \lambda) w_{12}(r_{12}) \, dr_1 \, dr_2 \qquad (A2.3.265)$$

where $\Delta A(\lambda) = A(\lambda) - A(0)$. Expanding the two-particle density correlation function in powers of λ,

$$\rho_N^{(2)}(r_1, r_2; \lambda) = \rho_N^{0,(2)}(r_1, r_2) + \lambda \left(\frac{\partial \rho_N^{(2)}(r_1, r_2; \lambda)}{\partial \lambda} \right)_{\lambda=0} + \cdots \qquad (A2.3.266)$$

we see that the zeroth order term in λ yields the first-order term a_1 in the high-temperature expansion for the free energy, and the first-order term in λ gives the second-order term a_2 in this expansion. As is usual for a fluid, $\rho_N^{(2)}(r_1, r_2; \lambda) = \rho^2 g_N^0(r_{12}; \lambda)$ and

$$g_N^{(2)}(r_{12}; \lambda) = g_N^{0,(2)}(r_{12}) + \lambda \left(\frac{\partial g(r_{12}; \lambda)}{\partial \lambda} \right)_{\lambda=0} + \cdots . \qquad (A2.3.267)$$

But by definition

$$g_N^{(2)}(r_{12}; \lambda) = \exp[-\beta(u_{12}^0(r_{12}) + \lambda w_{12}(r_{12}))] y_N(r_{12}; \lambda) \qquad (A2.3.268)$$

where $y(r_{12}, \lambda)$ is the cavity function, and $u_{12}^0(r_{12})$ is the pair potential of the reference system, from which it follows that

$$g_N^{(2)}(r_{12}; \lambda) \approx [1 - \beta \lambda w_{12}(r_{12})] g_N^{0,(2)}(r_{12}) \qquad (A2.3.269)$$

which suggests $g_N^{(2)}(r_{12}; \lambda) \simeq g_N^0(r_{12})$ when $\beta w_{12}(r_{12}) \beta \varepsilon \ll 1$, where ε is the depth of the potential well. It also suggests an improved approximation

$$g_N^{(2)}(r_{12}; \lambda) \approx \exp(-\beta u_{12}(r_{12}; \lambda)) y_N^0(r_{12}) \qquad (A2.3.270)$$

where $u_{12}(r_{12}; \lambda) = u_{12}^0(r_{12}) + \lambda w_{12}(r_{12})$ is the pair potential. The calculation of the second-order term in the high-temperature expansion involves the three- and four-body correlation functions which are generally not known even for a hard sphere reference system. The situation becomes worse for the higher-order terms in the perturbation expansion. However, determination of the first-order term in this expansion requires only the pair correlation function of the reference system, for which a convenient choice is a fluid of hard spheres whose equilibrium properties are known. Barker and Henderson [77] suggested a hard sphere diameter defined by

$$d = \int_0^\infty [1 - \exp(-\beta u_{12}(r))] \, dr \qquad (A2.3.271)$$

where $u_{12}(r)$ is the pair potential. This diameter is temperature dependent and the free energy needs to be calculated to second order to obtain the best results.

Truncation at the first-order term is justified when the higher-order terms can be neglected. When $\beta\varepsilon \ll 1$, a judicious choice of the reference and perturbed components of the potential could make the higher-order terms small. One choice exploits the fact that a_1, which is the mean value of the perturbation over the reference system, provides a strict upper bound for the free energy. This is the basis of a variational approach [78, 79] in which the reference system is approximated as hard spheres, whose diameters are chosen to minimize the upper bound for the free energy. The diameter depends on the temperature as well as the density. The method was applied successfully to Lennard-Jones fluids, and a small correction for the softness of the repulsive part of the interaction, which differs from hard spheres, was added to improve the results.

A very successful first-order perturbation theory is due to Weeks, Chandler and Andersen [80], in which the pair potential $u(r)$ is divided into a reference part $u^0(r)$ and a perturbation $w(r)$

$$u(r) = u^0(r) + w(r) \tag{A2.3.272}$$

in which

$$u^0(r) = \begin{cases} u(r) + \varepsilon & (r < R_{\min}) \\ 0 & (r > R_{\min}) \end{cases} \tag{A2.3.273}$$

and

$$w(r) = \begin{cases} -\varepsilon & (r < R_{\min}) \\ u(r) & (r > R_{\min}) \end{cases} \tag{A2.3.274}$$

where ε is the depth of the potential well which occurs at $r = R_{\min}$. This division into reference and perturbed parts is very fortuitous. The second step is to relate the reference system to an equivalent hard sphere fluid with a pair potential $u^{HS}(r)$. This is done by defining $v(r)$ by

$$\exp(-\beta v(r)) = \exp(-\beta u^{HS}(r)) + \alpha[\exp(-\beta u^0(r)) - \exp(-\beta u^{HS}(r))] \tag{A2.3.275}$$

in which α is an expansion parameter with $0 \le \alpha \le 1$. The free energy of the system with the pair potential $v(r)$ is expanded in powers of α to $O(\alpha^2)$ to yield

$$A' = A^{HS} - \frac{N\rho\alpha}{2} \int [\exp(-\beta u^0(r)) - \exp(-\beta u^{HS}(r))] y^{HS}(r; d) \, dr + O(\alpha^2) \tag{A2.3.276}$$

where $y^{HS}(r, d)$ is the cavity function of hard spheres of diameter d determined by annihilating the term of order α by requiring the integral to be zero. The diameter d is temperature and density dependent as in the variational theory. The free energy of the fluid with the pair potential $u(r)$ is now calculated to first order using the approximation

$$g^0(r) \exp(-\beta u(r)) y^{HS}(r, d). \tag{A2.3.277}$$

This implies, with the indicated choice of hard sphere diameter d, that the compressibilities of the reference system and the equivalent of the hard sphere system are the same.

Another important application of perturbation theory is to molecules with anisotropic interactions. Examples are dipolar hard spheres, in which the anisotropy is due to the polarity of the molecule, and liquid crystals in which the anisotropy is due also to the shape of the molecules. The use of an anisotropic reference system is more natural in accounting for molecular shape, but presents difficulties. Hence, we will consider only isotropic reference systems, in which the reference potential $u^0(r_{12})$ is usually chosen in one of two ways. In the first choice, $u^0(r_{12})$ is defined by

$$\exp[-\beta u^0(r_{12})] = \Omega^{-2} \int\int \exp[-\beta u(r_{12}, \Omega_1, \Omega_2)] \, d\Omega_1 \, d\Omega_2 \tag{A2.3.278}$$

which can be applied even to hard non-spherical molecules. The ensuing reference potential, first introduced by Rushbrooke [81], is temperature dependent and was applied by Cook and Rowlinson [82] to spheroidal molecules. It is more complicated to use than the temperature-independent reference potential defined by the simple averaging

$$u^0(r_{12}) = \Omega^{-2} \iint u(r_{12}, \Omega_1, \Omega_2)\, d\Omega_1\, d\Omega_2. \tag{A2.3.279}$$

This choice was introduced independently by Pople [83] and Zwanzig [84].

We assume that the anisotropic pair interaction can be written as

$$u(r_{12}, \Omega_1, \Omega_2; \lambda) = u^0(r_{12}) + \lambda w(r_{12}, \Omega_1, \Omega_2) \tag{A2.3.280}$$

where the switching function λ lies between zero and one. The perturbation is fully turned on when λ is one and is switched off when λ is zero. In the λ expansion for the free energy of the fluid,

$$\Delta A(\lambda) = \lambda A_1 + \lambda^2 A_2 + \lambda^3 A_3 + \cdots \tag{A2.3.281}$$

one finds that the leading term of order λ

$$A_1 = \frac{\rho N}{2\Omega^2} \iiint g^0(r_{12}) w(r_{12}, \Omega_1, \Omega_2)\, dr_{12}\, d\Omega_1\, d\Omega_2 \tag{A2.3.282}$$

vanishes on carrying out the angular integration due to the spherical symmetry of the reference potential. The expressions for the higher-order terms are

$$A_2 = -\frac{\beta\rho}{4\Omega^2} \iiint g^0(r_{12}) w(r_{12}, \Omega_1, \Omega_2)^2\, dr_{12}\, d\Omega_1\, d\Omega_2 \tag{A2.3.283}$$

$$A_3 = \frac{\beta^2\rho}{12\Omega^2} \iiint g^0(r_{12}) w(r_{12}, \Omega_1, \Omega_2)^3\, dr_{12}\, d\Omega_1\, d\Omega_2 + \frac{\beta^2\rho^2}{6\Omega^3} \iiint g^{0,(3)}(r_{12}, r_{23}, r_{31})$$
$$\times w(r_{12}, \Omega_1, \Omega_2) w(r_{23}, \Omega_1, \Omega_2) w(r_{13}, \Omega_1, \Omega_2)\, dr_{12}\, dr_{13}\, d\Omega_1\, d\Omega_2\, d\Omega_3. \tag{A2.3.284}$$

The expansion of the perturbation $w(r_{12}, \Omega_1, \Omega_2)$ in terms of multipole potentials (e.g. dipole–dipole, dipole–quadrupole, quadrupole–quadrupole) using spherical harmonics

$$w(r_{12}, \Omega_1, \Omega_2) = \sum\sum X^{l_1 l_2; m}(r) Y^{l_1}_{m_1}(\Omega_1) Y^{l_2}_{m_2}(\Omega_2) \tag{A2.3.285}$$

leads to additional simplifications due to symmetry. For example, for molecules with only dipole and quadrupolar interactions, all terms in which the dipole moment appears an odd number of times at an integration vertex vanish. In particular, for a pure dipolar fluid, the two-body integral contributing to A_3 vanishes. Angular integration of the three-body integral leads to the Axelrod–Teller three-body potential:

$$u(r_1, r_2, r_3) = (3\cos\theta_1 \cos\theta_2 \cos\theta_3 + 1)/(r_{12}r_{13}r_{23})^3 \tag{A2.3.286}$$

where $\theta_1, \theta_2, \theta_3$ and r_{12}, r_{13}, r_{23} are the angles and sides of the triangle formed by the three dipoles so that only the spatial integration remains to evaluate A_3. This is accomplished by invoking the superposition approximation

$$g^{0,(3)}(r_{12}, r_{13}, r_{23}) = g^0(r_{12}) g^0(r_{13}) g^0(r_{23}) \tag{A2.3.287}$$

which makes only a small error when it contributes to the third-order perturbation term. Tables of the relevant integrals for dipoles and multipoles associated with a hard sphere reference system are available [85].

For many molecules the reduced dipole moment $\mu* = (\mu^2/(\varepsilon\sigma^3))^{1/2}$ is greater than 1 and the terms in the successive terms in the λ expansion oscillate widely. Stell, Rasaiah and Narang [85] suggested taming this by replacing the truncated expansion by the Padé approximant

$$\Delta A = A_2 \left(1 - \frac{A_2}{A_3} \right)^{-1} \tag{A2.3.288}$$

which reproduces the expected behaviour that as μ becomes large the free energy A increases as μ^2. The Padé approximant is quite successful in reproducing the thermodynamic behaviour of polar fluids. However, the critical exponents, as in all mean-field theories, are the classical exponents.

The generalization of the λ expansion to multicomponent systems is straightforward but requires knowledge of the reference system pair correlation functions of all the different species. Application to electrically neutral Coulomb systems is complicated by the divergence of the leading term of order λ in the expansion, but this difficulty can be circumvented by exploiting the electroneutrality condition and using a screened Coulomb potential

$$u_{ij}(r) = u^0_{ij}(r) + \frac{e_i e_j}{\varepsilon_0 r} \exp(-\alpha r) \tag{A2.3.289}$$

where α is the screening parameter and $u^0_{ij}(r)$ is the reference potential. The term of order λ, generalized to the multicomponent electrolyte, is

$$A'_1 = \frac{V}{2\varepsilon_0} \sum_{i=1}^{\sigma} \sum_{j=1}^{\sigma} e_i e_j c_i c_j \int_0^{\infty} g^0_{ij}(r) \frac{\exp(-\alpha r)}{r} \, \mathrm{d}r. \tag{A2.3.290}$$

The integrals are divergent in the limit $\alpha = 0$. However, substituting $g^0_{ij}(r) = h^0_{ij}(r) + 1$, and making use of the electroneutrality condition in the form

$$\sum_i^{\sigma} \sum_j^{\sigma} e_i e_j c_i c_j = \left(\sum_{i=1}^{\sigma} e_i c_i \right)^2 = 0 \tag{A2.3.291}$$

one finds, on taking the limit $\alpha \to 0$, that

$$A_1 = \frac{V}{2\varepsilon_0} \sum_{i=1}^{\sigma} \sum_{j=1}^{\sigma} e_i e_j c_i c_j \int_0^{\infty} \frac{h^0_{ij}(r)}{r} \, \mathrm{d}r \tag{A2.3.292}$$

in which the divergences have been subtracted out. It follows from the Gibbs–Bogoliubov inequality that the first two terms form an upper bound, and

$$A \leq A^0 + A_1. \tag{A2.3.293}$$

For a symmetrical system in which the reference species are identical (e.g. hard spheres of the same size), the integral can be taken outside the summation, which then adds up to zero due to the electroneutrality condition, to yield

$$A_1 = 0. \tag{A2.3.294}$$

The reference free energy in this case is an upper bound for the free energy of the electrolyte. A lower bound for the free energy difference ΔA between the charged and uncharged RPM system was derived by Onsager [86]; this states that $\Delta A/N > -e^2/\varepsilon\sigma$. Improved upper and lower bounds for the free energy have been discussed by Gillan [87].

The expression for κ^2 shows that it is the product of the ionic concentration c and $e^2/\varepsilon_0 kT$, which is called the Bjerrum parameter. The virial series is an expansion in the total ionic concentration c at a fixed

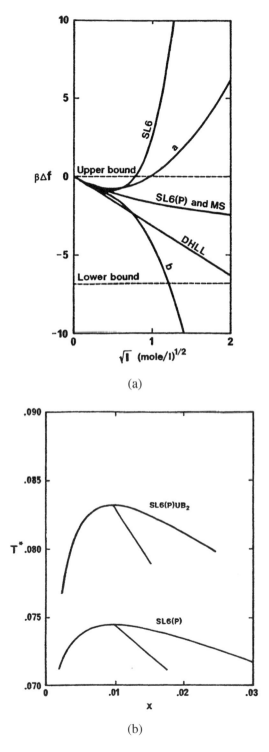

(a)

(b)

Figure A2.3.19. Coexistence curve for the RPM predicted by SLR(P) and SL6(P) ∪ B$_2$. The reduced temperature $T^* = \varepsilon k T \sigma / e^2$ and $x = \rho \sigma^3$ (after [85]).

value of $e^2/\varepsilon_0 kT$. A theory due to Stell and Lebowitz (SL) [88], on the other hand, is an expansion in the Bjerrum parameter at constant c. The leading terms in this expansion are

$$\frac{A^{\text{ex}}}{NkT} = \frac{A^0}{NkT} + \frac{A_1}{NkT} - \frac{\kappa_1^3}{12\pi c} \tag{A2.3.295}$$

where we have already seen that the first two terms form an upper bound. Here κ_1^{-1} is a modified Debye length defined by

$$\kappa_1^2 = \kappa^2 + \frac{4\pi}{\varepsilon_0 kT} \sum_{i=1}^{\sigma} \sum_{j=1}^{\sigma} e_i e_j c_i c_j \int_0^\infty h_{ij}^0 \, d\boldsymbol{r}. \tag{A2.3.296}$$

By the same argument as before, the integral may be taken outside the summations for a symmetrical reference system (e.g. the RPM electrolyte) and applying the electroneutrality condition one sees that in this case $\kappa_1 = \kappa$. Since the first two terms in the SL expansion form an upper bound for the free energy, the limiting law as $T \to \infty$ at constant c must always be approached from one side, unlike the Debye–Hückel limiting law which can be approached from above or below as $c \to 0$ at fixed temperature T (e.g. $ZnSO_4$ and HCl in aqueous solutions).

Examination of the terms to $O(\kappa^6)$ in the SL expansion for the free energy show that the convergence is extremely slow for a RPM 2–2 electrolyte in 'aqueous solution' at room temperature. Nevertheless, the series can be summed using a Padé approximant similar to that for dipolar fluids which gives results that are comparable in accuracy to the MS approximation as shown in figure A2.3.19(a). However, unlike the DHLL + B_2 approximation, neither of these approximations produces the negative deviations in the osmotic and activity coefficients from the DHLL observed for higher valence electrolytes at low concentrations. This can be traced to the absence of the complete renormalized second virial coefficient in these theories; it is present only in a linearized form. The union of the Pade approximant (SL6(P)), derived from the SL theory to $O(\kappa^6)$, and the Mayer expansion carried as far as DHLL + B_2

$$\text{SL6(P)} \cup B_2 = \text{SL6(P)} + B_2 - \text{SL6(P)} \cap B_2 \tag{A2.3.297}$$

produces the right behaviour at low concentrations and has an accuracy comparable to the MS approximation at high concentrations. Figure A2.3.19(b) shows the coexistence curves for charged hard spheres predicted by SL6(P) and the SL6(P) $\cup B_2$.

By integrating over the hard cores in the SL expansion and collecting terms it is easily shown this expansion may be viewed as a correction to the MS approximation which still lacks the complete second virial coefficient. Since the MS approximation has a simple analytic form within an accuracy comparable to the Pade (SL6(P)) approximation it may be more convenient to consider the union of the MS approximation with Mayer theory. Systematic improvements to the MS approximation for the free energy were used to determine the critical point and coexistence curves of charged, hard spheres by Stell, Wu and Larsen [89], and are discussed by Haksjold and Stell [90].

A2.3.6.2 Computational alchemy

Perturbation theory is also used to calculate free energy differences between distinct systems by computer simulation. This computational alchemy is accomplished by the use of a switching parameter λ, ranging from zero to one, that transforms the Hamiltonian of one system to the other. The linear relation

$$U(\lambda) = \lambda U_{\text{C}} + (1 - \lambda)U_{\text{B}} \tag{A2.3.298}$$

interpolates between the energies $U_{\text{B}} \, (\boldsymbol{r}^N, \boldsymbol{\omega}^N)$ and $U_{\text{C}} \, (\boldsymbol{r}^N, \boldsymbol{\omega}^N)$ of the initial and final states of molecules C and B and allows for fictitious intermediate states also to be sampled. The switching parameter could be the dihedral angle in a peptide or polymer chain, the charge on an atom or one of the parameters ε or σ defining the size or well depth of its pair interaction with the environment.

It follows from our previous discussion of perturbation theory that

$$\Delta A(\mathrm{B} \to \mathrm{C}) = A_{\mathrm{C}} - A_{\mathrm{B}} = -kT \ln \langle \exp[-\beta(U_{\mathrm{C}} - U_{\mathrm{B}})] \rangle_B \tag{A2.3.299}$$

which is the free energy difference between the two states as a function of their energy difference sampled over the equilibrium configurations of one of the states. In the above expression, the averaging is over the equilibrium states of B, and C is treated as a perturbation of B. One could equally well sample over C and treat B as a perturbation of C. The averages are calculated using Monte Carlo or molecular dynamics discussed elsewhere; convergence is rapid when the initial and final states are similar. Since free energy differences are additive, the change in free energy between widely different states can also be determined through multiple simulations via closely-spaced intermediates determined by the switching function λ which gradually mutates B into C. The total free energy difference is the sum of these changes, so that

$$\Delta A(\mathrm{B} \to \mathrm{C}) = \int_0^1 \frac{\delta \Delta A(\lambda)}{\delta \lambda} \, d\lambda \tag{A2.3.300}$$

and one calculates the derivative by using perturbation theory for small increments in λ. The accuracy can be improved by calculating incremental changes in both directions.

A closely-related method for determining free energy differences is characterized as thermodynamic integration. The configurational free energy of an intermediate state

$$A(\lambda) = -kT \ln Z(\lambda) \tag{A2.3.301}$$

from which it follows that

$$\Delta A(\mathrm{B} \to \mathrm{C}) = \int_0^1 \left\langle \frac{\delta U(\lambda)}{\delta \lambda} \right\rangle_\lambda d\lambda \tag{A2.3.302}$$

where the derivative pertains to equilibrated intermediate states. This forms the basis of the 'slow growth' method, in which the perturbation is applied linearly over a finite number of time steps and the free energy difference computed as the sum of energy differences. This method, however, samples over non-equilibrium intermediate states and the choice of the number of time steps over which the perturbation is applied and the corresponding accuracy of the calculation must be determined empirically.

Free energy perturbation (FEP) theory is now widely used as a tool in computational chemistry and biochemistry [91]. It has been applied to determine differences in the free energies of solvation of two solutes, free energy differences in conformational or tautomeric forms of the same solute by mutating one molecule or form into the other. Figure A2.3.20 illustrates this for the mutation of $CH_3OH \to CH_3CH_3$ [92].

There are many other applications. They include determination of the ratios of the partition coefficients (P_B/P_C) of solutes B and C in two different solvents by using the thermodynamic cycle:

$$
\begin{array}{ccc}
 & \Delta G_t(\mathrm{B}) & \\
\mathrm{B(solvent\ 1)} & \longrightarrow & \mathrm{B(solvent\ 2)} \\
\Delta G_{\mathrm{sol}}(\mathrm{B} \to \mathrm{C})_1 \downarrow & & \downarrow \Delta G_{\mathrm{sol}}(\mathrm{B} \to \mathrm{C})_2 \\
\mathrm{C(solvent\ 1)} & \longrightarrow & \mathrm{C(solvent\ 2)} \\
 & \Delta G_t(\mathrm{C}). &
\end{array}
\tag{A2.3.303}
$$

It follows that

$$
\begin{aligned}
2.3RT \ln(P_B/P_C) &= \Delta G_t(\mathrm{C}) - \Delta G_t(\mathrm{B}) \\
&= \Delta G_{\mathrm{sol}}(\mathrm{B} \to \mathrm{C})_2 - \Delta G_{\mathrm{sol}}(\mathrm{B} \to \mathrm{C})_1
\end{aligned}
\tag{A2.3.304}
$$

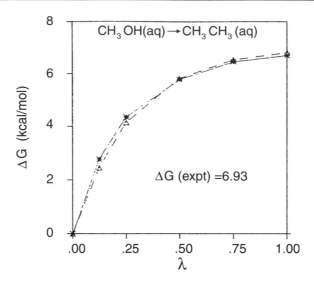

Figure A2.3.20. Free energy change in the transformation of $CH_3OH \rightarrow CH_3CH_3$ (after [92]).

where $\Delta G_t(B)$ and $\Delta G_t(C)$ are the free energies of transfer of solute B and C, respectively, from solvent 1 to 2 and $\Delta G_{sol}(B \rightarrow C)_1$ and $\Delta G_{sol}(B \rightarrow C)_2$ are differences in the solvation energies of B and C in the respective solvents. Likewise, the relative pK_as of two acids HA and HB in the same solvent can be calculated from the cycle depicted below for the acid HA

$$\Delta G_{diss}^{gas}(HA)$$
$$HA(gas) \longrightarrow H^+(gas) + A^-(gas)$$
$$\Delta G_{hyd}(HA) \downarrow \qquad \downarrow \Delta G_{hyd}(H^+) \qquad \downarrow \Delta G_{hyd}(A^-)$$
$$HA(soln) \longrightarrow H^+(soln) + A^-(soln)$$
$$\Delta G_{diss}^{soln}(HA) \tag{A2.3.305}$$

with a corresponding cycle for HB. From the cycle for HA, we see that

$$2.3RT pK_a(HA) = \Delta G_{diss}^{soln}(HA)$$
$$= +\Delta G_{hyd}(H^+) + \Delta G_{hyd}(A^-) - \Delta G_{hyd}(HA) \tag{A2.3.306}$$

with a similar expression for $pK_a(HB)$. The difference in the two pK_as is related to the differences in the gas phase acidities (free energies of dissociations of the acids), the free energies of hydration of B^- and A^- and the corresponding free energies of the undissociated acids:

$$2.3RT[pK_a(HA) - pK_a(HB)] = \Delta G_{diss}^{gas}(HA) - \Delta G_{diss}^{gas}(HB) + \Delta G_{hyd}(HA) - \Delta G_{hyd}(HB)$$
$$+ \Delta G_{hyd}(A^-) - \Delta G_{hyd}(B^-).$$

The relative acidities in the gas phase can be determined from *ab initio* or molecular orbital calculations while differences in the free energies of hydration of the acids and the cations are obtained from FEP simulations in which HA and A^- are mutated into HB and B^-, respectively.

Another important application of FEP is in molecular recognition and host–guest binding with its dependence on structural alterations. The calculations parallel our discussion of acid dissociation constants and have been used to determine the free energies of binding of A–T and C–G base pairs in solution from the corresponding binding energies in the gas phase. The relative free energies of binding two different ligands L_1 and L_2 to the same host are obtained from the following cycle:

$$
\begin{array}{ccc}
& \Delta G_1 & \\
E \quad + \quad L_1 & \longrightarrow & EL_1 \\
\downarrow \qquad \quad \downarrow \Delta G_3 & \quad \downarrow \Delta G_4 & \\
E \quad + \quad L_2 & \longrightarrow & EL_2 \\
& \Delta G_2. &
\end{array}
\tag{A2.3.307}
$$

The difference in the free energy change when L_1 is replaced by L_2 is

$$
\Delta G_2 - \Delta G_1 = \Delta G_4 - \Delta G_3
\tag{A2.3.308}
$$

which is determined in FEP simulations by mutating L_1 to L_2 and EL_1 to EL_2.

FEP theory has also been applied to modelling the free energy profiles of reactions in solution. An important example is the solvent effect on the SN2 reaction

$$
Cl^- + CH_3Cl \rightarrow [Cl - -CH_3 - -Cl]^- \rightarrow ClCH_3 + Cl^-
\tag{A2.3.309}
$$

as illustrated from the work of Jorgenson [93] in figure A2.3.21.

The gas phase reaction shows a double minimum and a small barrier along the reaction coordinate which is the difference between the two C–CL distances. The minima disappear in aqueous solution and this is accompanied by an increase in the height of the barrier. The behaviour in dimethyl formamide is intermediate between these two.

A2.3.6.3 Inhomogeneous fluids

An inhomogeneous fluid is characterized by a non-uniform singlet density $\rho(r_1, [\phi])$ that changes with distance over a range determined by an external field. Examples of an external field are gravity, the walls enclosing a system or charges on an electrode. They are important in studies of interfacial phenomena such as wetting and the electrical double layer. The attractive interatomic forces in such systems do not effectively cancel due to the presence of the external field and perturbation theories applied to homogeneous systems are not very useful. Integral equation methods that ignore the bridge diagrams are also not very successful.

As discussed earlier, the singlet density $\rho(r_1, [\phi])$ in an external field ϕ due to a wall is given by

$$
kT \ln \rho(r_1, [\phi]) = -\nabla_1 \phi(r_1) - \int \rho(r_2|r_1; [\phi]) \nabla_1 u(r_{12}) \, dr_2
\tag{A2.3.310}
$$

where the conditional density $\rho(r_2|r_1; [\phi]) = \rho^{(2)}(r_1, r_2; [\phi])/\rho(r_1; [\phi])$. Weeks, Selinger and Broughton (WSB) [94] use this as a starting point of a perturbation theory in which the potential is separated into two parts,

$$
u(r_{12}) = u^R(r_{12}) + w(r_{12})
\tag{A2.3.311}
$$

where $u^R(r_{12})$ is the reference potential and $w(r_{12})$ the perturbation. An effective field ϕ^R for the reference system is chosen so that the singlet density is unchanged from that of the complete system, implying the same average force on the atoms at r_1:

$$
\rho^R(r_1, [\phi^R]) = \rho(r_1, [\phi]).
\tag{A2.3.312}
$$

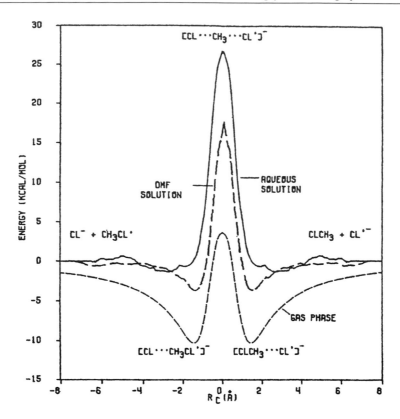

Figure A2.3.21. Free energy profile of the SN2 reaction $Cl^-+CH_3Cl\rightarrow[Cl-CH_3-Cl]\rightarrow ClCH_3+Cl^-$ in the gas phase, dimethyl formamide and in water (from [93]).

The effective field is determined by assuming that the conditional probabilities are the same, i.e.

$$\rho^R(r_2|r_1;[\phi^R]) = \rho(r_2|r_1;[\phi]) \qquad (A2.3.313)$$

when it follows that

$$\nabla_1[\phi^R(r_1) - \phi(r_1)] = \int \rho^R(r_2|r_1;[\phi^R])\nabla_1 w(r_{12})\,dr_2. \qquad (A2.3.314)$$

The conditional probability $\rho^R(r_2|r_1;[\phi^R])$ differs from the singlet density $\rho^R(r_2;[\phi^R])$ mainly when 1 and 2 are close and the gradient of the perturbation $\nabla_1 w(r_{12})$ is small. Replacing $\rho^R(r_2|r_1;[\phi^R])$ by $\rho^R(r_2,[\phi^R])$, taking the gradient outside the integral and integrating,

$$[\phi^R(r_1) - \phi(r_1)] = \int \rho^R(r_2;[\phi^R] - \rho^B)w(r_{12})\,dr_2$$

$$= \rho^B \int g^R(r_2;[\phi^R]) - 1)w(r_{12})\,dr_2 \qquad (A2.3.315)$$

where ρ^B is the bulk density far from the wall and $\rho^R(r_2,[\phi^R]) = \rho^B g^R(r_2,[\phi^R])$. This equation can be solved by standard methods provided the reference fluid distribution functions in the external field ϕ^R are known, for example through computer simulations. Other approximate methods to do this have also been devised by WSB [95].

A2.3.7   Solids and alloys

A2.3.7.1   *Introduction*

Our discussion of solids and alloys is mainly confined to the Ising model and to systems that are isomorphic to it. This model considers a periodic lattice of N sites of any given symmetry in which a spin variable $s_i = 1$ is associated with each site and interactions between sites are confined only to those between nearest neighbours. The total potential energy of interaction

$$U_N(\{s_k\}) = -J \sum_{\langle ij \rangle} s_i s_j - H \sum_i s_i \qquad (A2.3.316)$$

where $\{s_k\}$ denotes the spin variable $\{s_1, s_2, \ldots, s_N\}$, J is the coupling constant between neighbouring sites and H is the external field which acts on the spins s_i at each site. The notation $\langle ij \rangle$ denotes summation over the nearest-neighbour sites; there are $Nq/2$ terms in this sum where q is the coordination number of each site. Ferromagnetic systems correspond to $J > 0$, for which the spins are aligned in domains either up ↑↑↑↑↑↑ or down ↓↓↓↓↓↓ at temperatures below the critical point, while in an antiferromagnet $J < 0$, and alternating spins ↑↓↑↓↑↓↑ on the lattice sites dominate at the lowest temperatures. The main theoretical problem in these systems is to predict the critical temperature and the phase diagram. Of added interest is the isomorphism to the lattice gas and to a two-component alloy so that the phase diagram for an Ising ferromagnetic can be mapped on to those for these systems as well. This analogy is further strengthened by the universality hypothesis, which states that the critical exponents and properties near the critical point are identical, to the extent that they depend only on the dimensionality of the system and the symmetry of the Hamiltonian. The details of the intermolecular interactions are thus of less importance.

A2.3.7.2   *Ising model*

The partition function (PF) for the Ising [96] model for a system of given N, H and T is

$$Z(N, H, T) = \exp(-\beta G) = \sum_{\{s_k\}} \exp[-\beta U_N(\{s_k\})]$$

$$= \sum_{s_1 = \pm 1} \sum_{s_2 = \pm 1} \cdots \sum_{s_N = \pm 1} \exp \beta \left[J \sum_{\langle ij \rangle} s_i s_j + H \sum_i s_i \right]. \qquad (A2.3.317)$$

There are 2^N terms in the sum since each site has two configurations with spin either up or down. Since the number of sites N is finite, the PF is analytic and the critical exponents are classical, unless the thermodynamic limit ($N \to \infty$) is considered. This allows for the possibility of non-classical exponents and ensures that the results for different ensembles are equivalent. The characteristic thermodynamic equation for the variables N, H and T is

$$dG = -S\,dT - M\,dH + \mu\,dN \qquad (A2.3.318)$$

where M is the total magnetization and μ is the chemical potential. Since the sites are identical, the average magnetization per site is independent of its location, $m(H, T) = \langle s_i \rangle$ for all sites i. The total magnetization

$$M = Nm(H, T) = N\langle s_0 \rangle. \qquad (A2.3.319)$$

The magnetization per site

$$\langle s_0 \rangle = \frac{1}{Z} \sum_{\{s_k\}} s_0 \exp[-\beta(U_N(\{s_k\}))] = \frac{1}{N\beta} \left\{ \frac{\partial \ln Z(N, H, T)}{\partial H} \right\} \qquad (A2.3.320)$$

follows from the PF. As $H \to 0$

$$m(H, T) = \langle s_0 \rangle = 0 \qquad (T > T_c) \qquad (A2.3.321)$$

unless the temperature is less than the critical temperature T_c when the magnetization lies between -1 and $+1$,

$$-1 \leq m(H, T) \leq 1 \qquad (T < T_c) \qquad (A2.3.322)$$

and $m(H, T)$ versus H is a symmetrical odd function as shown in the figure A2.3.22.

At $T = 0$, all the spins are either aligned up or down. The magnetization per site is an order parameter which vanishes at the critical point. Along the coexistence curve at zero field

$$m(H, T) \approx (T - T_c)^\beta \qquad (A2.3.323)$$

where β here is a critical exponent identical to that for a fluid in the same dimension due to the universality. Since $s_i = \pm 1$, at all temperatures $\langle s_i^2 \rangle = 1$.

To calculate the spin correlation functions $\langle s_i s_j \rangle$ between any two sites, multiply the expression for $\langle s_0 \rangle$ by Z when

$$\langle s_0 \rangle Z = \sum_{\{s_k\}} s_0 \exp \beta \left[J \sum_{\langle ij \rangle} s_i s_j + H \sum_i s_i \right].$$

Differentiating with respect to H and dividing by Z, we have

$$\left\{ \frac{\partial \langle s_0 \rangle}{\partial H} \right\}_T + N\beta \langle s_0 \rangle \langle s_i \rangle = \beta \sum_i \langle s_0 s_i \rangle$$

where the factor N comes from the fact that there are N identical sites. The magnetic susceptibility per site

$$\chi_T(H) = \left(\frac{\partial \langle s_0 \rangle}{\partial H} \right)_T = \beta \sum_i [\langle s_0 s_i \rangle - \langle s_0 \rangle \langle s_i \rangle]. \qquad (A2.3.324)$$

For $T > T_c$, $\langle s_0 \rangle = \langle s_i \rangle = 0$ when $H = 0$. Separating out the term $i = 0$ from the rest and noting that $\langle s_0^2 \rangle = 1$, we have

$$\chi_T(0) = \beta \left[1 + \sum_{i \neq 0} \langle s_0 s_i \rangle \right] \qquad \text{for } T > T_c \qquad (A2.3.325)$$

which relates the susceptibility at zero field to the sum of the pair correlation function over different sites. This equation is analogous to the compressibility equation for fluids and diverges with the same exponent γ as the critical temperature is approached from above:

$$\chi_T(0) \simeq |T - T_c|^{-\gamma}. \qquad (A2.3.326)$$

The correlation length $\zeta = |T - T_c|^{-\nu}$ diverges with the exponent ν. Assuming that when $T > T_c$ the site correlation function decays as [26]

$$\langle s_i s_0 \rangle = \frac{f(r/\xi)}{r^{D-2+\eta}} \qquad (A2.3.327)$$

where r is the distance between sites, D is the dimensionality and η is another critical exponent, one finds, as for fluids, that $(2 - \eta)\nu = \gamma$.

An alternative formulation of the nearest-neighbour Ising model is to consider the number of up [↑] and down [↓] spins, the numbers of nearest-neighbour pairs of spins [↑↑], [↓↓], [↑↓] and their distribution over the lattice sites. Not all of the spin densities are independent since

$$N = [\uparrow] + [\downarrow] \qquad (A2.3.328)$$

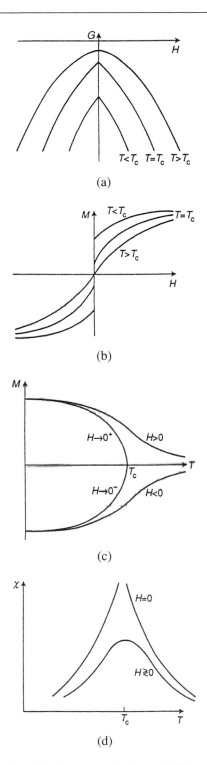

Figure A2.3.22. (a) The free energy G and (b) the magnetization $m(H, T)$ as a function of the magnetic field H at different temperatures. (c) The magnetization $m(H, T)$ and (d) the susceptibility χ as a function of temperature.

and, if q is the coordination number of each site,

$$q[\uparrow] = 2[\uparrow\uparrow] + [\uparrow\downarrow] \tag{A2.3.329}$$

$$q[\downarrow] = 2[\downarrow\downarrow] + [\uparrow\downarrow]. \tag{A2.3.330}$$

Thus, only two of the five quantities $[\uparrow], [\downarrow], [\uparrow\uparrow], [\downarrow\downarrow], [\uparrow\downarrow]$ are independent. We choose the number of down spins $[\downarrow]$ and nearest-neighbour pairs of down spins $[\downarrow\downarrow]$ as the independent variables. Adding and subtracting the above two equations,

$$qN = 2([\uparrow\uparrow] + [\downarrow\downarrow] + [\uparrow\downarrow]) \tag{A2.3.331}$$

$$q([\uparrow] - [\downarrow]) = 2([\uparrow\uparrow] - [\downarrow\downarrow]) \tag{A2.3.332}$$

and

$$[\uparrow\uparrow] + [\downarrow\downarrow] = qN/2 - [\uparrow\downarrow]. \tag{A2.3.333}$$

Defining the magnetization per site as the average number of up spins minus down spins,

$$\langle s_i \rangle = m(H, T) = \{[\uparrow] - [\downarrow]\}/N = 2([\uparrow\uparrow] - [\downarrow\downarrow])/N \tag{A2.3.334}$$

where the last relation follows because we consider only nearest-neighbour interactions between sites. The lattice Hamiltonian

$$U_N(\{s_k\}) = -J \sum_{\langle ij \rangle} s_i s_j - H \sum_i s_i$$

$$= -J([\downarrow\downarrow] + [\uparrow\uparrow] - [\uparrow\downarrow]) - H([\uparrow] - [\downarrow]). \tag{A2.3.335}$$

Making use of the relations between the spin densities, the energy of a given spin configuration can be written in terms of the numbers of down spins $[\downarrow]$ and nearest-neighbour down spins $[\downarrow\downarrow]$:

$$U_N(\{s_k\}) = -J\left(\frac{qN}{2} - 2[\uparrow\downarrow]\right) - H(N - 2[\downarrow])$$

$$= -N\left(\frac{qJ}{2} + H\right) + 2(qJ + H)[\downarrow] - -4J[\downarrow\downarrow]). \tag{A2.3.336}$$

For given J, H, q and N, the PF is determined by the numbers $[\downarrow]$ and $[\downarrow\downarrow]$ and their distribution over the sites, and is given by

$$Z(N, H, T) = e^{[(\beta Jq/2 + H)N]} \sum_{[\downarrow]} e^{-2\beta(qJ+H)[\downarrow]} \sum_{[\downarrow\downarrow]} g_N([\downarrow], [\downarrow\downarrow]) e^{4\beta J[\downarrow\downarrow]} \tag{A2.3.337}$$

where the sum over the number of nearest-neighbour down spins $[\downarrow\downarrow]$ is for a given number of down spins $[\downarrow]$, and $g_N([\downarrow], [\downarrow\downarrow])$ is the number of ways of distributing $[\downarrow]$ and $[\downarrow\downarrow]$ over N sites. Summing this over all $[\downarrow\downarrow]$ for fixed $[\downarrow]$ just gives the number of ways of distributing $[\downarrow]$ down spins over N sites, so that

$$\sum_{[\downarrow\downarrow]} g_N([\downarrow], [\downarrow\downarrow]) = \frac{N!}{([\downarrow])!(N - [\downarrow])!}. \tag{A2.3.338}$$

In this formulation a central problem is the calculation of $g_N([\downarrow], [\downarrow\downarrow])$.

The Ising model is isomorphic with the lattice gas and with the nearest-neighbour model for a binary alloy, enabling the solution for one to be transcribed into solutions for the others. The three problems are thus essentially one and the same problem, which emphasizes the importance of the Ising model in developing our understanding not only of ferromagnets but other systems as well.

A2.3.7.3   Lattice gas

This model for a fluid was introduced by Lee and Yang [97]. The system is divided into cells with occupation numbers

$$n_i = \begin{cases} 1 & \text{cell } i \text{ is occupied} \\ 0 & \text{cell } i \text{ is not occupied.} \end{cases} \tag{A2.3.339}$$

No more than one particle may occupy a cell, and only nearest-neighbour cells that are both occupied interact with energy $-\varepsilon$. Otherwise the energy of interactions between cells is zero. The total energy for a given set of occupation numbers $\{n\} = (n_1, n_2, \ldots, n_N)$ of the cells is then

$$U_N(\{n\}) = -\varepsilon \sum_{\langle ij \rangle} n_i n_j \tag{A2.3.340}$$

where the sum is over nearest-neighbour cells. The grand PF for this system is

$$\Xi(\mu, N, T) = \exp(\beta p N) = \sum_{n_1 = 0.1} \cdots \sum_{n_N = 0.1} \exp \beta \left[\varepsilon \sum_{\langle ij \rangle} n_i n_j + \mu \sum_i n_i \right]. \tag{A2.3.341}$$

The relationship between the lattice gas and the Ising model follows from the observation that the cell occupation number

$$n_i = \frac{(1 + s_i)}{2} = \begin{cases} 1 & s_i = 1 \\ 0 & s_i = -1 \end{cases} \tag{A2.3.342}$$

which associates the spin variable $s_i = \pm 1$ of the Ising model with the cell occupation number of the lattice gas. To calculate the energy, note that

$$\sum_{\langle ij \rangle} n_i n_j = \frac{1}{4} \sum_{\langle ij \rangle} (1 + s_i)(1 + s_j) = \frac{Nq}{8} + \frac{q}{4} \sum_i s_i + \frac{1}{4} \sum_{\langle ij \rangle} s_i s_j \tag{A2.3.343}$$

where the second equality follows from

$$\sum_{\langle ij \rangle} 1 = \frac{Nq}{2} \qquad \sum_{\langle ij \rangle} s_i = \frac{q}{2} \sum_i s_i \tag{A2.3.344}$$

Also

$$\sum_i n_i = \sum_i \frac{(1 + s_i)}{2} = \frac{N}{2} + \frac{1}{2} \sum_i s_i. \tag{A2.3.345}$$

It follows that the exponent appearing in the PF for the lattice gas,

$$\varepsilon \sum_{\langle ij \rangle} n_i n_j + \mu \sum_i n_i = \frac{\varepsilon Nq}{8} + \frac{\mu N}{2} + \frac{\varepsilon}{4} \sum_{\langle ij \rangle} s_i s_j + \left(\frac{\varepsilon q}{4} + \frac{\mu}{2} \right) \sum_i s_i. \tag{A2.3.346}$$

Using this in the lattice gas grand PF,

$$\exp(\beta p N) = \exp \left[\beta \left(\frac{\varepsilon Nq}{8} + \frac{\mu N}{2} \right) \right] \sum_{\{s_k\}} \exp \left[\frac{\beta \varepsilon}{4} \sum_{\langle ij \rangle} s_i s_j + \beta \left(\frac{\varepsilon q}{4} + \frac{\mu}{2} \right) \sum_i s_i \right]. \tag{A2.3.347}$$

Comparing with the PF of the Ising model

$$\exp(-\beta G) = \sum_{\{s_k\}} \exp \left[\beta J \sum_{\langle ij \rangle} s_i s_j + \beta H \sum_i s_i \right] \tag{A2.3.348}$$

one sees that they are of the same form, with solutions related by the following transcription table:

Ising	Lattice gas
$4J$	ε
H	$\left(\dfrac{\varepsilon q}{4}+\dfrac{\mu}{2}\right)=Jq+\mu/2$
$-G$	$pN-\left(\dfrac{\varepsilon Nq}{8}+\dfrac{\mu N}{2}\right)$

It follows from this that

$$\rho(\text{lattice gas}) = \langle n_i\rangle = (1+\langle s_i\rangle)/2 = (1+m)/2$$
$$\mu(\text{lattice gas}) = 2H - 2Jq$$
$$P(\text{lattice gas}) = -\frac{G}{N} + \frac{\varepsilon q}{8} + \frac{\mu}{2} = -\frac{G}{N} - \frac{Jq}{2} + H. \tag{A2.3.349}$$

At $H = 0$, $\mu(\text{lattice gas}) = -2Jq$ and the chemical potential is analytic even at $T = T_c$. From the thermodynamic relation,

$$d\mu = -S_M\,dT + V_M\,dp \tag{A2.3.350}$$

where S_M and V_M are the molar entropy and volume, it follows that

$$\rho\left(\frac{\partial^2\mu}{\partial T^2}\right)_\rho = -\frac{1}{T}\frac{C}{V_M} + \left(\frac{\partial^2 P}{\partial T^2}\right)_\rho. \tag{A2.3.351}$$

The specific heat along the critical isochore hence has the same singularity as $(\partial^2 P/\partial T^2)_\rho$ for a lattice gas.

The relationship between the lattice gas and the Ising model is also transparent in the alternative formulation of the problem, in terms of the number of down spins $[\downarrow]$ and pairs of nearest-neighbour down spins $[\downarrow\downarrow]$. For a given degree of site occupation $[\downarrow]$,

$$U_{[\downarrow]} = -\varepsilon[\downarrow\downarrow] \tag{A2.3.352}$$

and the lattice gas canonical ensemble PF

$$Q([\downarrow], N, T) = \sum_{[\downarrow\downarrow]} g_N([\downarrow],[\downarrow\downarrow])\exp(\beta\varepsilon[\downarrow\downarrow]). \tag{A2.3.353}$$

Removing the restriction on fixed $[\downarrow]$, by considering the grand ensemble which sums over $[\downarrow]$, one has

$$\exp(\beta pN) = \Xi(z, N, T) = \sum_{[\downarrow]} z^{[\downarrow]}\sum_{[\downarrow\downarrow]} g_N([\downarrow],[\downarrow\downarrow])\exp(\beta\varepsilon[\downarrow\downarrow]) \tag{A2.3.354}$$

where the fugacity $z = \exp(\beta\mu)$. Comparing this with the PF for the Ising model in this formulation, the entries in the transcription table given above are readily derived. Note that

$$m(T, H) \iff (1 - 2\rho) \tag{A2.3.355}$$

and

$$2H \iff -kT\ln(z/\sigma) \tag{A2.3.356}$$

where $\sigma = \exp(-2\beta qJ) = \exp(-\beta q\varepsilon/2)$. Since m is an odd function of H, for the Ising ferromagnet $(1 - 2r)$ must be an odd function of $kT\ln(z/\sigma)$ for a lattice gas and $m = 0$ corresponds to $\rho = 1/2$ for

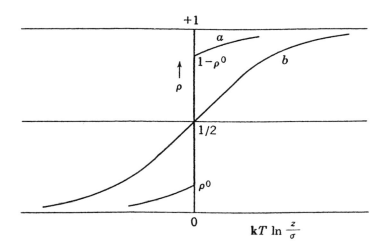

Figure A2.3.23. The phase diagram for the lattice gas.

a lattice gas. The liquid and vapour branches of the lattice gas are completely symmetrical about $\rho = 1/2$ when $T < T_c$. The phase diagram on the $\rho - \mu$ is illustrated in figure A2.3.23.

For two symmetrically placed points A and B on the isotherm, i.e. conjugate phases,

$$\rho(z) = 1 - \rho(z')$$
$$\ln(z/\sigma) = -\ln(z'/\sigma), \text{ i.e. } zz' = \sigma^2$$
$$\beta p(z) - (1/2)\ln z = \beta p(z') - (1/2)\ln z'$$
$$E/N - \rho(z)q\varepsilon/2 = E'/N - \rho(z')q\varepsilon/2 \tag{A2.3.357}$$

from which it follows that for the conjugate phases

$$\rho(z) + \rho(z') = 1$$
$$\mu(z, T) + \mu(z', T) = \varepsilon$$
$$p(z, T) - p(z', T) = (1/2)[\mu(z, T) - \mu(z', T)]$$
$$E(z, T) - E(z', T) = (Nq\varepsilon/2)[\rho(z, T) - \rho(z', T)]. \tag{A2.3.358}$$

A2.3.7.4 Binary alloy

A binary alloy of two components A and B with nearest-neighbour interactions ε_{AA}, ε_{BB} and ε_{AB}, respectively, is also isomorphic with the Ising model. This is easily seen on associating spin up with atom A and spin down with atom B. There are no vacant sites, and the occupation numbers of the site i are defined by

$$n_{i,A} = (1/2)(1 - s_i) \qquad n_{i,B} = (1/2)(1 + s_i). \tag{A2.3.359}$$

Summing over the sites

$$\sum_i (n_{i,A} + n_{i,B}) = N_A + N_B = N \tag{A2.3.360}$$

where N_A and N_B are the number of atoms of A and B, respectively, distributed over N sites. For an open system,

$$\langle N_A \rangle = \left\langle (1/2) \left[1 - \sum_i s_i \right] \right\rangle = (N/2)[1 - m(H, T)]$$

$$\langle N_B \rangle = \left\langle (1/2) \left[1 + \sum_i s_i \right] \right\rangle = (N/2)[1 + m(H, T)]. \tag{A2.3.361}$$

The coordination number of each site is q, and

$$q N_A = 2 N_{AA} + N_{AB} \tag{A2.3.362}$$

$$q N_B = 2 N_{BB} + N_{AB} \tag{A2.3.363}$$

$$N = N_A + N_A = (2/q)[N_{AA} + N_{BB} + N_{AB}]. \tag{A2.3.364}$$

On a given lattice of N sites, one number from the set $\{N_A, N_B\}$ and another from the set $\{N_{AA}, N_{BB}, N_{AB}\}$ determine the rest. We choose N_A and N_{AA} as the independent variables. Assuming only nearest-neighbour interactions, the energy of a given configuration

$$U_{[N_A]}(N_A, N_{AA}) = \varepsilon_{AA} N_{AA} + \varepsilon_{BB} N_{BB} + \varepsilon_{AB} N_{AB}$$

$$= \frac{q N \varepsilon_B}{2} + q N_A (\varepsilon_{AB} - \varepsilon_{AA}) + N_{AA} (\varepsilon_{AA} + \varepsilon_{BB} - 2\varepsilon_{AB}) \tag{A2.3.365}$$

which should be compared with the corresponding expressions for the lattice gas and the Ising model. The grand PF for the binary alloy is

$$\Xi(N, z_A, z_B, T) = \sum_{N_A=0}^{N} z_A^{N_A} z_B^{N_B} \sum_{N_{AA}, \text{fixed } N_A} q_N(N_A, N_{AA}) \exp[-\beta U_{N_A}(N_A, N_{AA})]$$

$$= (z_B^{N_B} e^{\beta q \varepsilon_{BB}}) \sum_{N_A=0}^{N} [(z_A/z_B) e^{-\beta q (\varepsilon_{AB} - \varepsilon_{AA})}] \sum_{N_{AA}(\text{fixed} N_A)} g_N(N_A, N_{AA}) e^{-\beta q (\varepsilon_{AA} + \varepsilon_{BB} - 2\varepsilon_{AB}) N_{AA}}$$

$$\tag{A2.3.366}$$

where $g_N(N_A, N_{AA})$ is the number of ways of distributing N_A and N_{AA} over N lattice sites and $z_A = \exp(\beta \mu_A)$ and $z_B = \exp(\beta \mu_B)$ are the fugacities of A and B, respectively. Comparing the grand PF for the binary alloy with the corresponding PFs for the lattice gas and Ising model leads to the following transcription table:

Ising model	Lattice gas	Binary alloy
$-4J$	ε	$\varepsilon_{AA} + \varepsilon_{BB} - 2\varepsilon_{AB}$
$-2(qJ + H)$	μ	$\mu_A + \mu_B + q(\varepsilon_{AA} - \varepsilon_{AB})$

When $2\varepsilon_{AB} > (\varepsilon_{AA} + \varepsilon_{BB})$, the binary alloy corresponds to an Ising ferromagnet ($J > 0$) and the system splits into two phases: one rich in A and the other rich in component B below the critical temperature T_c. On the other hand, when $2\varepsilon_{AB} < (\varepsilon_{AA} + \varepsilon_{BB})$, the system corresponds to an antiferromagnet: the ordered phase below the critical temperature has A and B atoms occupying alternate sites.

A2.3.8   Mean-field theory and extensions

Our discussion shows that the Ising model, lattice gas and binary alloy are related and present one and the same statistical mechanical problem. The solution to one provides, by means of the transcription tables, the solution to the others. Historically, however, they were developed independently before the analogy between the models was recognized.

We now turn to a mean-field description of these models, which in the language of the binary alloy is the Bragg–Williams approximation and is equivalent to the Curie–Weiss approximation for the Ising model. Both these approximations are closely related to the van der Waals description of a one-component fluid, and lead to the same classical critical exponents $\alpha = 0$, $\beta = 1/2$, $\delta = 3$ and $\gamma = 1$.

As a prelude to discussing mean-field theory, we review the solution for non-interacting magnets by setting $J = 0$ in the Ising Hamiltonian. The PF

$$Z(N, H, T) = \sum_{s_i=\pm1} \cdots \sum_{s_N=\pm1} e^{\beta H \sum_{i=1}^{N} s_i} = \sum_{s_i=\pm1} \cdots \sum_{s_N=\pm1} \prod_{i=1}^{N} e^{\beta H s_i}$$

$$= \left(\sum_{s_i=\pm1} e^{-\beta H s_i} \right)^N = (e^{\beta H s_i} + e^{-\beta H s_i})^N = 2^N \cosh N(\beta H) \qquad (A2.3.367)$$

where the third step follows from the identity of all the N lattice sites. The magnetization per site

$$m(H, T) = \frac{1}{N\beta} \left(\frac{\partial \ln Z}{\partial H} \right)_T = \tanh(\beta H) \qquad (A2.3.368)$$

and the graph of $m(H, T)$ versus H is a symmetrical sigmoid curve through the origin with no residual magnetization at any temperature when $H = 0$. This is because there are no interactions between the sites. We will see that a modification of the local field at a site that includes, even approximately, the effect of interactions between the sites leads to a critical temperature and residual magnetization below this temperature.

The local field at site i in a given configuration is

$$H_i = J \sum_{\langle j \rangle} s_i + H = qJ\langle s_i \rangle + H - J \sum_{\langle j \rangle} (s_i - \langle s_i \rangle) \qquad (A2.3.369)$$

where the last term represents a fluctuation from the average value of the spin $\langle s_i \rangle$ at site i which is the magnetization $m(H, T)$ per site. In the mean-field theory, this fluctuation is ignored and the effective mean field at all sites is

$$H_{\text{eff}} = qJm(H, T) + H. \qquad (A2.3.370)$$

Substituting this in the expressions for the PF for non-interacting magnets with the external field replaced by the effective field H_{eff}, we have

$$Z_{\text{eff}}(N, H, T) = 2^N \cosh^N \beta[qJM(H, T) + H] \qquad (A2.3.371)$$

and by differentiation with respect to H,

$$m(H, T) = \tanh[\beta(qJm(H, T) + H)] \qquad (A2.3.372)$$

from which it follows that

$$\left(\frac{1 + m}{1 - m} \right) = \exp[2\beta(qJM + H)]. \qquad (A2.3.373)$$

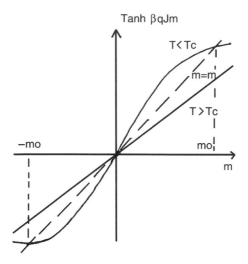

Figure A2.3.24. Plot of tanh[$\beta q J m(0, T)$] versus $m(0, T)$ at different temperatures.

Since $dG = -S\,dT - M\,dH$, integration with respect to H yields the free energy per site:

$$
\begin{aligned}
\frac{G}{N} &= \frac{kT}{2} \ln\left[\frac{(1 - m^2)}{4}\right] + \frac{q J m^2}{2} \\
&= -\left(\frac{J q m^2}{2} + m H\right) + \frac{kT}{2}\left[(1 + m) \ln\left(1 + \frac{m}{2}\right) + (1 - m) \ln\left(1 - \frac{m}{2}\right)\right]
\end{aligned}
\tag{A2.3.374}
$$

where the first two terms represent the energy contribution, and the last term is the negative of the temperature times the contribution of the entropy to the free energy. It is apparent that this entropy contribution corresponds to ideal mixing.

At zero field ($H = 0$),

$$
m(0, T) = \tanh[\beta q J m(0, T)]
\tag{A2.3.375}
$$

which can be solved graphically by plotting tanh[$\beta q J m(0, T)$] versus $m(0, T)$ and finding where this cuts the line through the origin with a slope of one, see figure A2.3.24.

Since

$$
\tanh(\beta q J m) = \beta q J m - (1/3)(\beta q J m)^3 + \cdots
\tag{A2.3.376}
$$

the slope as $m \to 0$ is $\beta q J$. A solution exists for the residual magnetization when the slope is greater than 1. This implies that the critical temperature $T_c = q J / k$, which depends on the coordination number q and is independent of the dimensionality. Table A2.3.5 compares the critical temperatures predicted by mean-field theory with the 'exact' results. Mean-field theory is seriously in error for 1D systems but its accuracy improves with the dimensionality. For $D \geq 4$ it is believed to be exact.

It follows from our equation (A2.3.373) that

$$
H = \frac{1}{2\beta} \ln\left(\frac{1 + m}{1 - m}\right) - q J m.
\tag{A2.3.377}
$$

The magnetization is plotted as a function of the field in figure A2.3.25.

When $T > T_c$, $m = 0$ and the susceptibility at zero field $\chi_T(0) > 0$. At $T = T_c$, $(dH/dm)_{T,H=0} = 0$ which implies that the susceptibility diverges, i.e. $\chi_T(0) = \infty$.

Table A2.3.5. Critical temperatures predicted by mean-field theory (MFT) and the quasi-chemical (QC) approximation compared with the exact results.

		kT_c/J		
D	q	MFT	QC	Exact
1	2	2	0	0
2	4(sq)	4	2.88	2.27
3	6(sc)	6	4.93	4.07

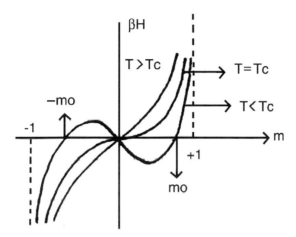

Figure A2.3.25. The magnetic field versus the magnetization $m(H, T)$ at different temperatures.

When $T < T_c$, the graph of H versus m shows a van der Waals like loop, with an unstable region where the susceptibility $\chi_T(0) < 0$. In the limit $H \to 0$, there are three solutions for the residual magnetization $m = (-m_0, 0, m_0)$, of which the solution $m = 0$ is rejected as unphysical since it lies in the unstable region. The symmetrically disposed acceptable solutions for the residual magnetizations are solutions to

$$m_0 = \tanh[\beta q J m_0]$$
$$= \tanh[(T_c/T)m_0] = (T_c/T)m_0 + (1/3)[(T_c/T)m_0]^3 + \cdots$$

from which it follows that as $T \to T_c$

$$m_0 \simeq 3^{1/2}(T/T_c)[1 - T/T_c]^{1/2}. \tag{A2.3.378}$$

This shows that the critical exponent $\beta = 1/2$.

The susceptibility at finite field H is given by

$$\chi_T(H) = (\partial m/\partial H)_T = \frac{\beta(1 - m^2)}{1 - \beta q J(1 - m^2)}. \tag{A2.3.379}$$

Recalling that $\beta q J = T_c/T$, the susceptibility at zero field ($H \to 0$)

$$\chi_T(0) = \frac{(1 - m_0^2)}{k[(T - T_c) + T_c m_0^2]}. \tag{A2.3.380}$$

For $T > T_c$, $m_0 = 0$ and

$$\chi_T(0) \simeq A_1(T - T_c)^{-\gamma} = (1/k)(T - T_c)^{-1} \tag{A2.3.381}$$

which shows that the critical exponent $\gamma = 1$ and the amplitude A_1 of the divergence of $\chi_T(0)$ is N/k, when the critical point is approached from above T_c. When $T < T_c$, $m_0^2 \simeq 3(T_c - T)/T_c$ and

$$\chi_T(0) \simeq A_2(T - T_c)^{-\gamma} = (1/2k)(T - T_c)^{-1} \tag{A2.3.382}$$

which shows that the critical exponent γ remains the same but the amplitude A_2 of the divergence is $1/2k$ when the critical point is approached from below T_c. This is half the amplitude when T_c is approached from above.

Along the critical isotherm, $T = T_c$,

$$H = kT_c(1/2)\ln[(1 + m_0)/(1 - m_0) - m_0]$$
$$\approx kT_c[m_0 + m_0^3/3 + m_0^5/5 + \cdots - m_0] \approx kT_c m_0^3/3 + \cdots . \tag{A2.3.383}$$

It follows that the critical exponent δ defined by $H \approx m_0^\delta$ is 3.

Fluctuations in the magnetization are ignored by mean-field theory and there is no correlation between neighbouring sites, so that

$$\langle s_i s_j \rangle = \langle s_i \rangle \langle s_j \rangle \tag{A2.3.384}$$

and the spins are randomly distributed over the sites. As seen earlier, the entropy contribution to the free energy is that of ideal mixing of up and down spins. The average energy

$$\langle U_N \rangle = -J \left\langle \sum_{\langle ij \rangle} s_i s_j \right\rangle - H \left\langle \sum_i s_i \right\rangle$$
$$= J \sum_{\langle ij \rangle} \langle s_i \rangle \langle s_j \rangle - H \sum_i \langle s_i \rangle$$
$$= -J(Nq/2)m^2 - HNm \tag{A2.3.385}$$

which is in accord with our interpretation of the terms contributing to the free energy in the mean-field approximation. Since $m = 0$ at zero field for $T > T_c$ and $m = \pm m_0$ at zero field when $T < T_c$, the configurational energy at zero field ($H = 0$) is given by

$$\langle U_N(H = 0) \rangle = \begin{cases} 0 & T > T_c \\ J(Nq/2)m_0^2 & T < T_c. \end{cases} \tag{A2.3.386}$$

This shows very clearly that the specific heat has a jump discontinuity at $T = T_c$:

$$C_{H=0}(T) = (1/N)\langle \partial U_N(H = 0)/\partial T \rangle$$
$$= \begin{cases} 0 & T > T_c \\ -qJm_0(\mathrm{d}m_0/\mathrm{d}T) & T < T_c. \end{cases} \tag{A2.3.387}$$

The neglect of fluctuations in mean-field theory implies that

$$[\downarrow\downarrow] \propto [\downarrow]^2 \qquad [\uparrow\uparrow] \propto [\uparrow]^2 \qquad [\uparrow\downarrow] \propto [\uparrow][\downarrow] \tag{A2.3.388}$$

and it follows that

$$\frac{[\downarrow\downarrow][\uparrow\uparrow]}{[\uparrow\downarrow]^2} = \frac{1}{4}. \tag{A2.3.389}$$

This is the equilibrium constant for the 'reaction'

$$[\downarrow\downarrow] + [\uparrow\uparrow] = 2[\uparrow\downarrow] \qquad (A2.3.390)$$

assuming the energy change is zero. An obvious improvement is to use the correct energy change $(4J)$ for the 'reaction', when

$$\frac{[\downarrow\downarrow][\uparrow\uparrow]}{[\uparrow\downarrow]^2} = \frac{1}{4}\exp(4J/kT). \qquad (A2.3.391)$$

This is the quasi-chemical approximation introduced by Fowler and Guggenheim [98] which treats the nearest-neighbour pairs of sites, and not the sites themselves, as independent. It is exact in one dimension. The critical temperature in this approximation is

$$T_c = (2J/k)[1/\ln(q/(q-2))] \qquad (A2.3.392)$$

which predicts the correct result of $T_c = 0$ for the 1D Ising model, and better estimates than mean-field theory, as seen in table A2.3.5, for the same model in two and three dimensions ($d = 2$ and 3). Bethé [99] obtained equivalent results by a different method. Mean-field theory now emerges as an approximation to the quasi-chemical approximation, but the critical exponents in the quasi-chemical approximation are still the classical values. Figure A2.3.26 shows mean-field and quasi-chemical approximations for the specific heat and residual magnetization of a square lattice ($d = 2$) compared to the exact results.

A2.3.8.1   Landau's generalized mean-field theory

An essential feature of mean-field theories is that the free energy is an analytical function at the critical point. Landau [100] used this assumption, and the up–down symmetry of magnetic systems at zero field, to analyse their phase behaviour and determine the mean-field critical exponents. It also suggests a way in which mean-field theory might be modified to conform with experiment near the critical point, leading to a scaling law, first proposed by Widom [101], which has been experimentally verified.

Assume that the free energy can be expanded in powers of the magnetization m which is the order parameter. At zero field, only even powers of m appear in the expansion, due to the up–down symmetry of the system, and

$$G = G_0 + a_2 m^2 + a_4 m^4 + \cdots \qquad (A2.3.393)$$

where the coefficients a_i are temperature dependent and $a_4 > 0$ but a_2 may be positive, negative or zero as illustrated in figure A2.3.27.

A finite residual magnetization of $\pm m_0$ is obtained only if $a_2 < 0$; and the critical temperature corresponds to $a_2 = 0$. Assume that a_2 is linear in $t = (T - T_c)/T_c$, near T_c, when

$$a_2 \approx a_2^* t. \qquad (A2.3.394)$$

The free energy expansion reads

$$G = G_0 + a_2^* t m^2 + a_4 m^4 + \cdots. \qquad (A2.3.395)$$

The equilibrium magnetization corresponds to a minimum free energy which implies that

$$(dG/dm) = 0 = 2a_2^* t m + 4a_4 m^3 + \cdots.$$

It follows that

$$m_0^2 = [a_2^*/(2a_4)](-t) \qquad (A2.3.396)$$

which implies that m_0 is real and finite if $t < 0$ (i.e. $T < T_c$) and the critical exponent $\beta = 1/2$. For t the only solution is $m_0 = 0$. Hence m_0 changes continuously with t and is zero for $t \geq 0$.

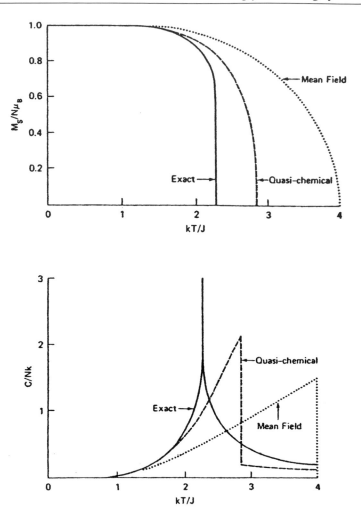

Figure A2.3.26. Mean-field and quasi-chemical approximations for the specific heat and residual magnetization of a square lattice $(d = 2)$ compared to the exact results.

Along the coexistence curve, $t < 0$,

$$G = G_0 + a_2^* t m_0^2 + a_4 m_0^4$$
$$= G_0 - a_2^* t^2/(2a_4) + O(t^4) \tag{A2.3.397}$$

and the specific heat

$$C_{H=0} = \frac{1}{T}\left(\frac{\partial^2 G}{\partial T^2}\right) = \begin{cases} 0 & t > 0 \\ \dfrac{1}{T_c}\left(\dfrac{a_2^{*2}}{a_4}\right) & t < 0. \end{cases} \tag{A2.3.398}$$

There is jump discontinuity in the specific heat as the temperature passes from below to above the critical temperature.

To determine the critical exponents γ and δ, a magnetic interaction term $-hm$ is added to the free energy and

$$G = G_0 - hm + a_2^* t m^2 + a_4 m^4 + \cdots. \tag{A2.3.399}$$

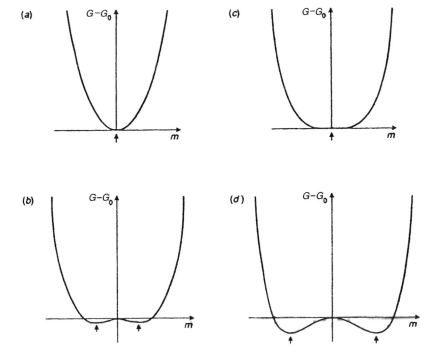

Figure A2.3.27. The free energy as a function of m in the Landau theory for (a) $a_2 > 0$, (b) $a_2 = 0$, (c) $a_2 < 0$ and (d) $a_2 < 0$ with $a_4 > 0$.

Minimizing the free energy with respect to m, one finds

$$h = 2a_2^* tm + 4a_4 m^3 + \cdots. \tag{A2.3.400}$$

Along the critical isochore, $t = 0$ and

$$h \approx 4a_4 m^3 \tag{A2.3.401}$$

which implies that $\delta = 3$.

For $t \neq 0$, the inverse susceptibility

$$\chi_T(0)^{-1} = (dh/dm)_t = 2a_2^* t + 12a_4 m^3 \tag{A2.3.402}$$

which, as $h \rightarrow 0$, leads to

$$\chi_T(0)^{-1} = (dh/dm)_t = 2a_2^* t + 12a_4 m_0^3. \tag{A2.3.403}$$

When $t > 0$, $m_0 = 0$ and

$$\chi_T(0)_{H=0} = (1/2a_2^*)t^{-1} \tag{A2.3.404}$$

while for $t < 0$, $m_0^2 = [a_2^*/(2a_4)](-t)$ and

$$\chi_T(0)_{H=0} = -(1/4a_2^*)t^{-1}. \tag{A2.3.405}$$

This implies that the critical exponent $\gamma = 1$, whether the critical temperature is approached from above or below, but the amplitudes are different by a factor of 2, as seen in our earlier discussion of mean-field theory. The critical exponents are the classical values $\alpha = 0$, $\beta = 1/2$, $\delta = 3$ and $\gamma = 1$.

The assumption that the free energy is analytic at the critical point leads to classical exponents. Deviations from this require that this assumption be abandoned. In mean-field theory,

$$h = am(t + bm^3) + \cdots \tag{A2.3.406}$$

near the critical point, which implies $\beta = 1/2$ and $\gamma = 1$. Modifying this to

$$h = am(t + bm^{1+\gamma/\beta}) \tag{A2.3.407}$$

implies that $\delta = 1 + \gamma/\beta$, which is correct. Widom postulated that

$$h = m\phi(t, m^{1/\beta}) \tag{A2.3.408}$$

where ϕ is a generalized homogeneous function of degree γ

$$\phi(\lambda t, (\lambda m)^{1/\beta}) = \lambda^\gamma \phi(t, m^{1/\beta}). \tag{A2.3.409}$$

This is Widom's scaling assumption. It predicts a scaled equation of state, like the law of corresponding states, that has been verified for fluids and magnets [102].

A2.3.9   High- and low-temperature expansions

Information about the behaviour of the 3D Ising ferromagnet near the critical point was first obtained from high- and low-temperature expansions. The expansion parameter in the high-temperature series is $\tanh K$, and the corresponding parameter in the low-temperature expansion is $\exp(-2K)$. A 2D square lattice is self-dual in the sense that the bisectors of the line joining the lattice points also form a square lattice and the coefficients of the two expansions, for the 2D square lattice system, are identical to within a factor of two. The singularity occurs when

$$\tanh K = \exp(-2K). \tag{A2.3.410}$$

Kramers and Wannier [103] used this to locate the critical temperature $T_c = 2.27\,J/k$.

A2.3.9.1   The high-temperature expansion

The PF at zero field

$$Z(N, 0, T) = \sum_{\{s\}} \exp \sum_{\langle ij \rangle} K s_i s_j = \sum_{\{s\}} \prod_{\langle ij \rangle} K s_i s_j \tag{A2.3.411}$$

where $K = \beta J$ and $\{s\}$ implies summation over the spins on the lattice sites. Since $s_i s_j = \pm 1$,

$$
\begin{aligned}
\exp(K s_i s_j) &= \exp(\pm K) = \cosh K \pm \sinh K \\
&= \cosh K + s_i s_j \sinh K \\
&= \cosh K[1 + s_i s_j \tanh K]
\end{aligned}
\tag{A2.3.412}
$$

from which it follows that

$$
\begin{aligned}
Z(N, 0, T) &= (\cosh K)^{qN/2} \sum_{\{s\}} \prod_{\langle ij \rangle} (1 + s_i s_j \tanh K) \\
&= (\cosh K)^{qN/2} \sum_{\{s\}} \left[1 + \tanh K \sum_l s_i s_j + \tanh^2 K \sum_l (s_i s_j)(s_k s_l) + \cdots \right]
\end{aligned}
\tag{A2.3.413}
$$

where $\sum_l$ is the sum over all possible sets of l pairs of nearest-neighbour spins. The expansion parameter is $\tanh K$ which $\to 0$ as $T \to \infty$ and becomes 1 as $T \to 0$. The expansion coefficients can be expressed graphically. A coefficient $(s_i s_j)(s_k s_l) \ldots (s_p s_q)$ of $\tanh^r K$ is the product or sum of products of graphs with r bonds in which each bond is depicted as a line joining two sites. Note also that

$$\sum_{s_i=\pm1} s_i^n = \begin{cases} 2 & n \text{ even} \\ 0 & n \text{ odd}. \end{cases} \tag{A2.3.414}$$

Hence, on summing over the graphs, the only non-zero terms are closed polygons with an even number of bonds at each site, i.e. s_i must appear an even number of times at a lattice site in a graph that does not add up to zero on summing over the spins on the sites.

Each lattice point extraneous to the sites connected by graphs also contributes a factor of two on summing over spin states. Hence all lattice points contribute a factor of 2^N whether they are connected or not, and

$$Z(N,0,T) = (\cosh K)^{qN/2} 2^N \sum_{r=0}^{qN/2} n(r,N) \tanh^r K \tag{A2.3.415}$$

where (for $r \neq 0$), $n(r,N)$ is the number of distinct-side polygons (closed graphs) drawn on N sites such that there are an even number of bonds on each site. For $r=0$, no lattice site is connected, but define $n(N)=1$. Also, since closed polygons cannot be connected on one or two sites, $n(1,N)=n(2,N)=0$. The problem then is to count $n(r,N)$ for all r. On an infinite lattice of identical sites, $n(r,N)=Np(r)$ where $p(r)$ is the number of r-side polygons that can be constructed on a given lattice site. This number is closely connected to the structure of the lattice.

A2.3.9.2 The low-temperature expansion
At zero field (see equation (A2.3.335)),

$$\sum_{\langle ij \rangle} K s_i s_j = [\uparrow\uparrow] + [\downarrow\downarrow] - [\uparrow\downarrow] = qN/2 - 2[\uparrow\downarrow] \tag{A2.3.416}$$

and

$$Z(N,0,T) = \exp(qN/2)\sum_{[\uparrow\downarrow]} \exp(-2K[\uparrow\downarrow])$$
$$= \exp(qN/2)\sum_r m(r,N)(-2Kr) \tag{A2.3.417}$$

where $m(r,N)$ is the number of configurations with $[\uparrow\downarrow]=r$. The high- and low-temperature expansions are complementary.

The dual lattice is obtained by drawing the bisectors of lines connecting neighbouring lattice points. Examples of lattices in two dimensions and their duals are shown in figure A2.3.28. A square lattice is self-dual.

Consider a closed polygon which appears in the high-temperature expansion. Put up spins $[\uparrow]$ on the sites of the lattice inside the polygon and down spins $[\downarrow]$ on the lattice sites outside. The spins across the sides of the closed polygons are oppositely paired. The PF can be calculated equally well by counting closed polygons on the original lattice (high-temperature expansion) or oppositely paired spins on the dual lattice (low-temperature expansion). Both expansions are exact.

For a 2D square lattice $q=4$, and the high- and low-temperature expansions are related in a simple way

$$2n(r,N) = m(r,N)^{\text{dual}} \tag{A2.3.418}$$

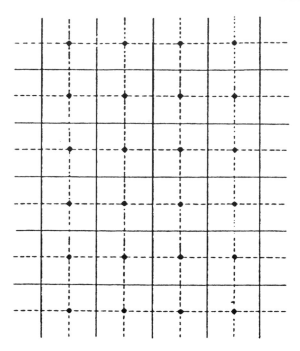

Square lattice and its dual

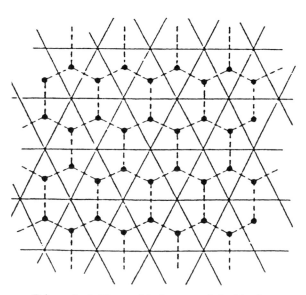

Triangular lattice and its hexagonal dual lattice

Figure A2.3.28. Square and triangular lattices and their duals. The square lattice is self-dual.

where the factor two comes from the fact that reversing all the spins does not change the number of oppositely paired spins. The dual of the lattice is a square lattice. Hence the PFs in the two expansions have the same coefficients except for irrelevant constant factors:

$$Z(N, 0, T) = 2^N (\cosh K)^{2N} \sum_{r=0}^{2N} n(r, N) \tanh K$$

$$= 2 \exp(2KN) \sum_{r=0}^{2N} n(r, N) \exp(-2Kr). \qquad (A2.3.419)$$

Both expansions are exact and assuming there is only one singularity, identified with the critical point, this must occur when

$$\tanh K = \exp(-2K). \qquad (A2.3.420)$$

With $x = \exp(-2K)$, this implies that

$$x = (1 - x)/(1 + x)$$

which leads to a quadratic equation

$$x^2 + 2x - 1 = 0. \qquad (A2.3.421)$$

The solutions are $x = -1 \pm \sqrt{2}$. Since $K = \beta J$ is necessarily not negative, the only acceptable solution is $x = -1 + \sqrt{2}$. Identifying the singularity with the critical point, the solution $x = \exp(2K_c) = -1 + \sqrt{2}$ is equivalent to the condition

$$\sinh(2K_c) = \sinh(2J/kT_c) = 1 \qquad (A2.3.422)$$

from which it follows that the critical temperature $T_c = 2.27J/k$. This result was known before Onsager's solution to the 2D Ising model at zero field.

More generally, for other lattices and dimensions, numerical analysis of the high-temperature expansion provides information on the critical exponents and temperature. The high-temperature expansion of the susceptibility may be written in powers of $K = \beta J$ as

$$\chi_T(0) = \sum_{n=0}^{\infty} a_n(\beta J)^n. \qquad (A2.3.423)$$

Suppose the first $n + 1$ coefficients are known, where $n \simeq 15$. The susceptibility diverges as $(1 - \beta/\beta_c)^{-\gamma}$ as $\beta \to \beta_{c-}$ and we have

$$\chi_T(0) \approx A(1 - \beta/\beta_c)^{-\gamma}$$

$$= A \left[1 + \gamma(\beta/\beta_c) + \cdots + \frac{\gamma(\gamma + 1) \ldots (\gamma + n - 1)}{n!} (\beta/\beta_c)^n + \cdots \right]. \qquad (A2.3.424)$$

For large n

$$a_n J^n \approx A \frac{\gamma(\gamma + 1) \ldots (\gamma + n - 1)}{n! \beta_c^n}. \qquad (A2.3.425)$$

Taking the ratio of successive terms and dividing by the coordination number q

$$r_n = \frac{a_n}{q a_{n-1}} = \frac{kT_c}{qJ} \left(1 + \frac{\gamma - 1}{n} \right). \qquad (A2.3.426)$$

Plotting r versus $1/n$ gives kT_c/qJ as the intercept and $(kT_c/qJ)(1 - \gamma)$ as the slope from which T_c and γ can be determined. Figure A2.3.29 illustrates the method for lattices in one, two and three dimensions and compares it with mean-field theory which is independent of the dimensionality.

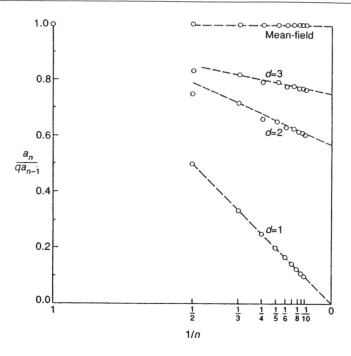

Figure A2.3.29. Calculation of the critical temperature T_c and the critical exponent γ for the magnetic susceptibility of Ising lattices in different dimensions from high-temperature expansions.

A2.3.10   Exact solutions to the Ising model

The Ising model has been solved exactly in one and two dimensions; Onsager's solution of the model in two dimensions is only at zero field. Information about the Ising model in three dimensions comes from high- and low-temperature expansions pioneered by Domb and Sykes [104] and others. We will discuss the solution to the 1D Ising model in the presence of a magnetic field and the results of the solution to the 2D Ising model at zero field.

A2.3.10.1   One dimension

We will describe two cases: open and closed chains of N sites. For an open chain of N sites, the energy of a spin configuration $\{s_k\}$ is

$$U_N(\{s_k\}) = -J \sum_{i=1}^{N-1} s_i s_{i+1} - H \sum_{i=1}^{N} s_i \qquad (A2.3.427)$$

and for a closed chain of N sites with periodic boundary conditions $s_{N+1} = s_1$

$$U_N(\{s_k\}) = -J \sum_{i=1}^{N} s_i s_{i+1} - \frac{H}{2} \sum_{i=1}^{N} (s_i + s_{i+1}). \qquad (A2.3.428)$$

Both systems give the same results in the thermodynamic limit. We discuss the solution for the open chain at zero field and the closed chain for the more general case of $H \neq 0$.

(a) Open chain at zero field, i.e. $H = 0$

The PF

$$Z(N, 0, T) = \sum_{s_i = \pm 1} \cdots \sum_{s_N = \pm 1} \exp\left(\beta J \sum_{i=1}^{N-1} s_i s_{i+1}\right)$$

$$= \sum_{s_i = \pm 1} \cdots \sum_{s_N = \pm 1} \exp\left(\beta J \sum_{i=1}^{N-2} s_i s_{i+1}\right) \sum_{s_N = \pm 1} \exp(\beta J s_{N-1} s_N). \qquad \text{(A2.3.429)}$$

Doing the last sum

$$Z(N, 0, T) = Z(N - 1, 0, T)[\exp(\beta J s_{N-1}) + \exp(\beta J s_{N-1})]$$

$$= Z(N - 1, 0, T) 2 \cosh(\beta J) \qquad \text{(A2.3.430)}$$

since $s_{N-1} = \pm 1$. Proceeding by iteration, starting from $N = 1$, which has just two states with the spin up or down

$$Z(1, 0, T) = 2$$

$$Z(2, 0, T) = Z(1, 0, T) 2 \cosh(\beta J) = 2^2 \cosh(\beta J)$$

$$Z(3, 0, T) = 2^3 \cosh^2(\beta J)$$

$$\cdots$$

$$Z(N, 0, T) = 2^N \cosh^{N-1}(\beta J). \qquad \text{(A2.3.431)}$$

The free energy G in the thermodynamic limit ($N \to \infty$) follows from

$$-\frac{\beta G}{N} = \lim_{N-\infty} \frac{1}{N} \ln Z(N, 0, T)$$

$$= \ln 2 + \lim_{N-\infty} \left(\frac{N-1}{N}\right) \ln \cosh(\beta J) = \ln[2 \cosh(\beta J)]. \qquad \text{(A2.3.432)}$$

(b) Closed chain, $H \neq 0$

The PF in this case is

$$Z(N, H, T) = \sum_{s_1 = \pm 1} \cdots \sum_{s_N = \pm 1} \exp\left[\left(\beta J \sum_{k=1}^{N} s_k s_{k+1}\right) + \frac{\beta H}{2} \sum_{k=1}^{N} (s_k + s_{k+1})\right]$$

$$= \sum_{s_1 = \pm 1} \cdots \sum_{s_N = \pm 1} \prod_{k=1}^{N} \exp \beta\left[J s_k s_{k+1} + \frac{H}{2}(s_k + s_{k+1})\right]$$

$$= \sum_{s_1 = \pm 1} \cdots \sum_{s_N = \pm 1} P_{s_1 s_2} P_{s_2 s_3} \cdots P_{s_N s_1} \qquad \text{(A2.3.433)}$$

where $P_{s_1 s_2}$ are the elements of a 2×2 matrix called the transfer matrix

$$\mathbf{P} = \begin{pmatrix} P_{11} & P_{1-1} \\ P_{-11} & P_{-1-1} \end{pmatrix} = \begin{pmatrix} \exp \beta(J + H) & \exp(-\beta J) \\ \exp(-\beta J) & \exp \beta(J - H) \end{pmatrix} \qquad \text{(A2.3.434)}$$

with the property that $\sum_{s_2} P_{s_1 s_2} P_{s_2 s_3} = (\mathbf{P}^2)_{s_1 s_3}$. It follows for the closed chain that

$$Z(N, H, T) = \sum_{s_1 = \pm 1} (\mathbf{P}^N)_{s_1 s_1} = \mathrm{Tr}\mathbf{P}^N \qquad (A2.3.435)$$

where $\mathbf{P}^N$ is also a 2×2 matrix.

The trace is evaluated by diagonalizing the matrix $\mathbf{P}$ using a similarity transformation $\mathbf{S}$:

$$\mathbf{P}' = \mathbf{S}^{-1}\mathbf{P}\mathbf{S} = \begin{pmatrix} \lambda_+ & 0 \\ 0 & \lambda_- \end{pmatrix} \qquad (A2.3.436)$$

where the diagonal elements of the matrix $\mathbf{P}'$ are the eigenvalues of $\mathbf{P}$, and

$$\mathbf{P}'^N = \begin{pmatrix} \lambda_+^N & 0 \\ 0 & \lambda_-^N \end{pmatrix}. \qquad (A2.3.437)$$

Noting that

$$\mathbf{P}'^N = \mathbf{S}^{-1}\mathbf{P}\mathbf{S}\mathbf{S}^{-1}\mathbf{P}\mathbf{S} \dots \mathbf{S}^{-1}\mathbf{P}\mathbf{S} = \mathbf{S}^{-1}\mathbf{P}^N\mathbf{S}$$

by virtue of the property that $\mathbf{S}\mathbf{S}^{-1} = \mathbf{I}$, where $\mathbf{I}$ is the identity matrix, we see that

$$\mathrm{Tr}[\mathbf{P}'^N] = \mathrm{Tr}[\mathbf{S}^{-1}\mathbf{P}^N\mathbf{S}] = \mathrm{Tr}[\mathbf{S}^{-1}\mathbf{S}\mathbf{P}^N] = \mathrm{Tr}[\mathbf{P}^N]$$

which leads to

$$Z(N, H, T) = \lambda_+^N + \lambda_-^N. \qquad (A2.3.438)$$

Assuming the eigenvalues are not degenerate and $\lambda_+ > \lambda_-$,

$$Z(N, H, T) = \lambda_+^N [1 + (\lambda_-/\lambda_+)^N].$$

In the thermodynamic limit of $N \to \infty$,

$$\frac{-\beta G}{N} = \lim_{N \to \infty} \frac{1}{N} \ln Z(N, H, T) = \ln \lambda_+. \qquad (A2.3.439)$$

This is an important general result which relates the free energy per particle to the largest eigenvalue of the transfer matrix, and the problem reduces to determining this eigenvalue.

The eigenvalues of the transfer matrix are the solutions to

$$\det |\mathbf{P} - \lambda\mathbf{I}| = 0.$$

This leads to a quadratic equation whose solutions are

$$\lambda_\pm = \exp(\beta J)\{\cosh(\beta H) \pm [\sinh^2(\beta H) + \exp(-4\beta J)]^{1/2}\} \qquad (A2.3.440)$$

which confirms that the eigenvalues are not degenerate. The free energy per particle

$$-\frac{\beta G}{N} = \beta J + \ln\{\cosh(\beta H) + [\sinh^2(\beta H) + \exp(-4\beta J)]^{1/2}\}. \qquad (A2.3.441)$$

This reduces to the results for the free energy at zero field ($H = 0$)

$$\frac{-\beta G}{N} = \ln[2\cosh(\beta J)] \qquad (A2.3.442)$$

and the free energy of non-interacting magnets in an external field

$$\frac{-\beta G}{N} = \ln[2\cosh(\beta H)] \tag{A2.3.443}$$

which were derived earlier. At finite T (i.e. $T > 0$), λ_+ is analytic and there is no phase transition. However, as $T \to 0$,

$$\lambda_+ \to \exp(K)[\cosh(h) + (\sinh^2(h))^{1/2}(1 + O(\exp(-4K)))]$$
$$= \exp(K)[\cosh(h) + |\sinh(h)|(1 + O(\exp(-4K)))]$$

where $K = \beta J$ and $h = \beta H$. But $\cosh(h) + |\sinh(h)| = \exp|h|$, and it follows that

$$\lambda_+ \to \exp(K + |h|)$$

as $T \to 0$. We see from this that as $T \to 0$

$$-\frac{G}{N} = kT \ln \lambda_+ = kT[K + |h|] = J + |H| \tag{A2.3.444}$$

and

$$m = \frac{1}{N}\left(\frac{\partial G}{\partial H}\right)_T = \begin{cases} +1 & H > 0 \\ -1 & H < 0 \end{cases} \tag{A2.3.445}$$

which implies a residual magnetization $m_0 = \pm 1$ at zero field and a first-order phase transition at $T = 0$. For $T \neq 0$, there is no discontinuity in m as H passes through zero from positive to negative values or *vice versa*, and differentiation of G with respect to H at constant T provides the magnetization per site

$$m(H, T) = \frac{\sinh(\beta H)}{[\sinh^2(\beta H) + \exp(-4\beta J)]^{1/2}} \tag{A2.3.446}$$

which is an odd function of H with $m \to 0$ as $H \to 0$. Note that this reduces to the result

$$m(H, T) = \tanh(\beta H) \tag{A2.3.447}$$

for non-interacting magnets.

As $H \to 0$, $\sinh(\beta J) \to \beta J$, $m(H, T) \to \beta H \exp(2\beta J)$ and

$$\chi_T(0) = (dm/dH)_T = \beta \exp(2\beta J) \tag{A2.3.448}$$

which diverges exponentially as $T \to 0$, which is also characteristic of a phase transition at $T = 0$.

The average energy $\langle E \rangle$ follows from the relation

$$\langle E \rangle / N = -(1/N)(d \ln Z/d\beta)_{H,J} = -(d \ln \lambda_+/d\beta)_{H,J} \tag{A2.3.449}$$

and at zero field

$$\langle E \rangle_{H=0}/N = -J \tanh(\beta J). \tag{A2.3.450}$$

The specific heat at zero field follows easily,

$$C_{H=0} = -\frac{N}{kT^2}\left(\frac{\partial \langle E \rangle_{H=0}}{\partial \beta}\right) = Nk(\beta J)^2 \operatorname{sech}^2(\beta J) \tag{A2.3.451}$$

and we note that it passes through a maximum as a function of T.

The spin correlation functions and their dependence on the distance between sites and the coupling between adjacent sites are of great interest in understanding the range of these correlations. In general, for a closed chain

$$\langle s_i s_{i+n} \rangle = Z(N, H, T)^{-1} \sum_{s_1 = \pm 1} \cdots \sum_{s_N = \pm} s_i s_{i+n} \exp \left(\sum_{j=1}^{N} K s_i s_{i+1} + h s_i \right). \tag{A2.3.452}$$

For nearest-neighbour spins

$$\langle s_j s_{j+1} \rangle = [NZ(N, H, T)]^{-1}[dZ(N, H, T)/dK] \tag{A2.3.453}$$

and making use of $Z(N, H, T) = \lambda_+^N [1 + (\lambda_- / \lambda_+)^N]$ in the thermodynamic limit $(N \to \infty)$

$$\langle s_i s_{i+1} \rangle = (\partial \ln \lambda_+ / \partial K)$$
$$= 1 - \frac{2 \exp(-4K)[\sinh^2 h + \exp(-4K)]^{-1/2}}{\cosh h + [\sinh^2 h + \exp(-4K)]^{1/2}}. \tag{A2.3.454}$$

At zero field $(H = 0)$, $h = 0$ and

$$\langle s_i s_{i+1} \rangle = \tanh K \tag{A2.3.455}$$

which shows that the correlation between neighbouring sites approaches 1 as $T \to 0$. The correlation between non-nearest neighbours is easily calculated by assuming that the couplings $(K_1, K_2, K_3, \ldots, K_N)$ between the sites are different, in which case a simple generalization of the results for equal couplings leads to the PF at zero field

$$Z(N, 0, T) = 2^N \prod_{j=1}^{N-1} \cosh K_j. \tag{A2.3.456}$$

Repeating the earlier steps one finds, as expected, that the coupling K_i between the spins at the sites i and $i + 1$ determines their correlation:

$$\langle s_i s_{i+1} \rangle = Z^{-1}(dZ(N, H, T)/dK_i) = \tanh K_i. \tag{A2.3.457}$$

Now notice that since $s_{i+1}^2 = 1$,

$$\langle s_i s_{i+1} s_{i+1} s_{i+1} \rangle = \langle s_i s_{1+2} \rangle = Z^{-1} \left(\frac{\partial^2 Z}{\partial K_i \partial K_{i+1}} \right)$$
$$= \tanh K_i \tanh K_{i+1}. \tag{A2.3.458}$$

In the limit $K_i = K_{i+1} = K$,

$$\langle s_i s_{i+2} \rangle = \tanh^2 K \tag{A2.3.459}$$

and repeating this argument serially for the spin correlations between i and $i + n$ sites

$$\langle s_i s_{i+n} \rangle = \tanh^n K \tag{A2.3.460}$$

so the correlation between non-neighbouring sites approaches 1 as $T \to 0$ since the spins are all aligned in this limit.

The correlation length ζ follows from the above relation, since

$$\langle s_i s_{i+j} \rangle = \exp(j \ln \tanh K) = \exp(-j \ln \coth K) = \exp(-j/\zeta) \tag{A2.3.461}$$

from which it follows that

$$\zeta = 1/\ln \coth(K). \tag{A2.3.462}$$

As expected, as $T \to 0$, $K \to \infty$ and the correlation length $\zeta \approx \exp(\beta J)/2 \to \infty$, while in the opposite limit, as $T \to \infty$, $\zeta \to 0$.

A2.3.10.2   Two dimensions

Onsager's solution to the 2D Ising model in zero field ($H = 0$) is one of the most celebrated results in theoretical chemistry [105]; it is the first example of critical exponents. Also, the solution for the Ising model can be mapped onto the lattice gas, binary alloy and a host of other systems that have Hamiltonians that are isomorphic to the Ising model Hamiltonian.

By a deft application of the transfer matrix technique, Onsager showed that the free energy is given by

$$-\frac{\beta G}{N} = \ln \cosh(2\beta J) + \frac{1}{2\pi} \int_0^\pi d\phi \ln \frac{[1 + (1 - \kappa^2 \sin^2 \varphi)]^{1/2}}{2} \tag{A2.3.463}$$

where

$$\kappa = \frac{2 \sinh(2\beta J)}{\cosh^2(2\beta J)} \tag{A2.3.464}$$

which is zero at $T = 0$ and $T = \infty$ and passes through a maximum of 1 when $\beta J_c = 0.440\,69$. This corresponds to a critical temperature $T_c = 2.269 J/k$ when a singularity occurs in the Gibbs free energy, since $[1 + (1 - \kappa^2 \sin^2 \phi)^{1/2}] \to 0$ as $T \to T_c$ and $\phi \to \pi/2$. As $T \to T_c$,

$$C_{H=0} \approx \frac{8k}{\pi} \frac{J}{kT_c} \ln|T - T_c|^{-1} \tag{A2.3.465}$$

so that the critical exponent $\alpha = 0_{\log}$. The spontaneous magnetization

$$m_0 = \begin{cases} 0 & T > T_c \\ [1 - \sinh^{-4}(2\beta J)]^{1/8} & T < T_c \end{cases} \tag{A2.3.466}$$

and the critical exponent $\beta = 1/8$. This result was first written down by Onsager during a discussion at a scientific meeting, but the details of his derivation were never published. Yang [107] gave the first published proof of this remarkably simple result. The spin correlation functions at $T = T_c$ decay in a simple way as shown by Kaufman and Onsager [106],

$$\langle s_i s_{i+j} \rangle 1/r^{1/4} \tag{A2.3.467}$$

where r is the distance between the sites.

A2.3.11   Summary

We have described the statistical mechanics of strongly interacting systems. In particular those of non-ideal fluids, solids and alloys. For fluids, the virial coefficients, the law of corresponding states, integral equation approximations for the correlation functions and perturbation theories are treated in some detail, along with applications to hard spheres, polar fluids, strong and weak electrolytes and inhomogeneous fluids. The use of perturbation theory in computational studies of the free energy of ligand binding and other reactions of biochemical interest is discussed. In treating solids and alloys, the Ising model and its equivalence to the lattice gas model and a simple model of binary alloys, is emphasized. Mean-field approximations to this model and the use of high- and low-temperature approximations are described. Solutions to the 1D Ising model with and without a magnetic field are derived and Onsager's solution to the 2D case is briefly discussed.

References

[1] Andrews T 1869 On the continuity of the gaseous and liquid states *Phil. Trans. R. Soc.* **159** 575
[2] van der Waals J H 1873 Over de continuiteit van den gas-en vloeistof toestand *Thesis* University of Leiden (English transl. 1988 *Studies in Statistical Mechanics* ed J S Rowlinson (Amsterdam: North-Holland))
[3] Maxwell J C 1874 Van der Waals on the continuity of the gaseous and liquid states *Nature* **10** 477
    Maxwell J C 1875 On the dynamical evidence of the molecular constitution of bodies *Nature* **11** 357
[4] Bett K E, Rowlinson J S and Saville G 1975 *Thermodynamics for Chemical Engineers* (Cambridge, MA: MIT Press)
[5] Stell G 1964 Cluster expansions for classical systems in equilibrium *The Equilibrium Theory of Classical Fluids* ed H L Frisch and J L Lebowitz (New York: Benjamin)
[6] Hansen J P and McDonald I 1976 *Theories of Simple Liquids* (New York: Academic)
[7] Mayer J G and Mayer M G 1940 *Statistical Mechanics* (New York: Wiley)
[8] Rhee F H and Hoover W G 1964 Fifth and sixth virial coefficients for hard spheres and hard disks *J. Chem. Phys.* **40** 939
    Rhee F H and Hoover W G 1967 Seventh virial coefficients for hard spheres and hard disks *J. Chem. Phys.* **46** 4181
[9] Carnahan N F and Starling K E 1969 Equation of state for nonattracting rigid spheres *J. Chem. Phys.* **51** 635
[10] Harvey A N 1999 Applications of first-principles calculations to the correlation of water's second virial coefficient *Proc. 13th Int. Conf. of the Properties of Water and Steam (Toronto, 12–16 September 1999)*
[11] Feynman R P 1972 *Statistical Mechanics, a Set of Lectures* (New York: Benjamin/Cummings)
[12] Gillan M J 1990 Path integral simulations of quantum systems *Computer Modeling of Fluids and Polymers* ed C R A Catlow *et al* (Dordrecht: Kluwer)
[13] Tonks L 1936 The complete equation of state of one, two and three dimensional gases of hard elastic spheres *Phys. Rev.* **50** 955
[14] Takahashi H 1942 *Proc. Phys. Math. Soc. Japan* **24** 60
[15] Lieb E H and Mattis D C 1966 *Mathematical Physics in One Dimension* (New York: Academic)
[16] Cho C H, Singh S and Robinson G W 1996 An explanation of the density maximum in water *Phys. Rev. Lett.* **76** 1651
[17] Kac M, Uhlenbeck G E and Hemmer P 1963 On van der Waals theory of vapor–liquid equilibrium. I. Discussion of a one-dimensional model *J. Math. Phys.* **4** 216
[18] van Kampen N G 1964 Condensation of a classical gas with long-range attraction *Phys. Rev.* A **135** 362
[19] Guggenheim E A 1945 The principle of corresponding states *J. Chem. Phys.* **13** 253
[20] Onsager L 1933 Theories of concentrated electrolytes *Chem. Rev.* **13** 73
[21] Kirkwood J G 1935 Statistical mechanics of fluid mixtures *J. Chem. Phys.* **3** 300
    Kirkwood J G 1936 Statistical mechanics of liquid solutions *Chem. Rev.* **19** 275
[22] Yvon J 1935 *Actualitiés Scientifiques et Industriel* (Paris: Herman et Cie)
[23] Born M and Green H S 1946 A general kinetic theory of liquids: I. The molecular distribution functions *Proc. R. Soc.* A **188** 10
[24] Born M and Green H S 1949 *A General Kinetic Theory of Liquids* (Cambridge: Cambridge University Press)
[25] Kirkwood J G and Buff F P 1951 Statistical mechanical theory of solutions I *J. Chem. Phys.* **19** 774
[26] Fisher M 1964 Correlation functions and the critical region of simple fluids *J. Math. Phys.* **5** 944
[27] Percus J K 1982 Non uniform fluids *The Liquid State of Matter: Fluids, Simple and Complex* ed E W Montroll and J L Lebowitz (Amsterdam: North-Holland)
[28] Percus J K and Yevick G J 1958 Analysis of classical statistical mechanics by means of collective coordinates *Phys. Rev.* **110** 1
[29] Stell G 1977 Fluids with long-range forces: towards a simple analytic theory *Statistical Mechanics part A, Equilibrium Techniques* ed B Berne (New York: Plenum)
[30] Chandler D and Andersen H C 1972 Optimized cluster expansions for classical fluids II. Theory of molecular liquids *J. Chem. Phys.* **57** 1930
[31] Chandler D 1982 Equilibrium theory of polyatomic fluids *The Liquid State of Matter: Fluids, Simple and Complex* ed E W Montroll and J L Lebowitz (Amsterdam: North-Holland)
[32] Wertheim M S 1963 Exact solution of the Percus–Yevick equation for hard spheres *Phys. Rev. Lett.* **10** 321
    Wertheim M S 1964 Analytic solution of the Percus–Yevick equation *J. Math. Phys.* **5** 643
[33] Thiele E 1963 Equation of state for hard spheres *J. Chem. Phys.* **39** 474
[34] Baxter R J 1968 Ornstein Zernike relation for a disordered fluid *Aust. J. Phys.* **21** 563
[35] Lebowitz J L 1964 Exact solution of the generalized Percus–Yevick equation for a mixture of hard spheres *Phys. Rev.* **133** A895
    Lebowitz J L and Rowlinson J S 1964 Thermodynamic properties of hard sphere mixtures *J. Chem. Phys.* **41** 133
[36] Baxter R J 1970 Ornstein Zernike relation and Percus–Yevick approximation for fluid mixtures *J. Chem. Phys.* **52** 4559
[37] Reiss H, Frisch H l and Lebowitz J L 1959 Statistical mechanics of rigid spheres *J. Chem. Phys.* **31** 361
    Reiss H 1977 Scaled particle theory of hard sphere fluids *Statistical Mechanics and Statistical Methods in Theory and Application* ed U Landman (New York: Plenum) pp 99–140
[38] Gibbons R M 1969 Scaled particle theory for particles of arbitrary shape *Mol. Phys.* **17** 81
[39] Chandler D 1993 Gaussian field model of fluids with an application to polymeric fluid *Phys. Rev.* E **48** 2989
[40] Crooks G E and Chandler D 1997 Gaussian statistics of the hard sphere fluid *Phys. Rev.* E **56** 4217
[41] Hummer G, Garde S, Garcîa A E, Pohorille A and Pratt L R 1996 An information theory model of hydrophobic interactions *Proc. Natl Acad. Sci.* **93** 8951
[42] Debye P and Huckel E 1923 *Phys. Z.* **24** 305

[43] Bjerrum N 1926 *Kgl. Dansk Videnskab, Selskab* **7** No 9

[44] Mayer J 1950 Theory of ionic solutions *J. Chem. Phys.* **18** 1426

[45] Rasaiah J C and Friedman H L 1968 Integral equation methods in computations of equilibrium properties of ionic solutions *J. Chem. Phys.* **48** 2742

Rasaiah J C and Friedman H L 1969 Integral equation computations for 1-1 electrolytes. Accuracy of the method *J. Chem. Phys.* **50** 3965

[46] Waisman E and Lebowitz J K 1972 Mean spherical model integral equation for charged hard spheres I. Method of solution *J. Chem. Phys.* **56** 3086

Waisman E and Lebowitz J K 1972 Mean spherical model integral equation for charged hard spheres II. Results *J. Chem. Phys.* **56** 3093

[47] Blum L 1980 Primitive electrolytes in the mean spherical model *Theoretical Chemistry: Advances and Perspectives* vol 5 (New York: Academic)

[48] Rasaiah J C 1990 A model for weak electrolytes *Int. J. Thermophys.* **11** 1

[49] Zhou Y and Stell G 1993 Analytic approach to molecular liquids V. Symmetric dissociative dipolar dumb-bells with the bonding length $\sigma/3 = L = \sigma/2$ and related systems *J. Chem. Phys.* **98** 5777

[50] Ebeling W 1968 Zur Theorie der Bjerrumschen Ionenassoziation in Electrolyten *Z. Phys. Chem. (Leipzig)* **238** 400

[51] Ebeling W and Grigoro M 1980 Analytical calculation of the equation of state and the critical point in a dense classical fluid of charged hard spheres *Phys. (Leipzig)* **37** 21

[52] Pitzer K S 1995 Ionic fluids: near-critical and related properties *J. Phys. Chem.* **99** 13 070

[53] Fisher M and Levin Y 1993 Criticality in ionic fluids: Debye Huckel Theory, Bjerrum and beyond *Phys. Rev. Lett.* **71** 3826

Fisher M 1996 The nature of criticality in ionic fluids *J. Phys.: Condens. Matter* **8** 9103

[54] Stell G 1995 Criticality and phase transitions in ionic fluids *J. Stat. Phys.* **78** 197

Stell G 1999 New results on some ionic fluid problems, new approaches to problems in liquid state theory *Proc. NATO Advanced Study Institute (Patte Marina, Messina, Italy 1998)* ed C Caccamo, J P Hansen and G Stell (Dordrecht: Kluwer)

[55] Anisimov M A, Povodyrev A A, Sengers J V and Levelt-Sengers J M H 1997 Vapor–liquid equilibria, scaling and crossover in aqueous solutions of sodium chloride near the critical line *Physica* A **244** 298

[56] Jacob J, Kumar A, Anisimov M A, Povodyrev A A. and Sengers J V 1998 Crossover from Ising to mean-field critical behavior in an aqueous electrolyte solution *Phys. Rev.* E **58** 2188

[57] Orkoulas G and Panagiotopoulos A Z 1999 Phase behavior of the restricted primitive model and square-well fluids from Monte Carlo simulations in the grand canonical ensemble *J. Chem. Phys.* **110** 1581

[58] Valleau J P and Torrie G M 1998 Heat capacity of the restricted primitive model *J. Chem. Phys.* **108** 5169

[59] Camp P J and Patey G N 1999 Ion association in model ionic fluids *Phys. Rev.* E **60** 1063

Camp P J and Patey G N 1999 Ion association and condensation in primitive models of electrolytes *J. Chem. Phys.*

[60] Koneshan S and Rasaiah J C 2000 Computer simulation studies of aqueous sodium chloride solutions at 298K and 683K *J. Chem. Phys.* **113** 8125

[61] Stillinger F H and Lovett R 1968 General restriction on the distribution of ions in electrolytes *J. Chem. Phys.* **48** 1991

[62] Friedman H L 1962 *Ionic Solution Theory* (New York: Interscience)

[63] Card D N and Valleau J 1970 Monte Carlo study of the thermodynamics of electrolyte solutions *J. Chem. Phys.* **52** 6232

Rasaiah J C, Card D N and Valleau J 1972 Calculations on the 'restricted primitive model' for 1-1-electrolyte solutions *J. Chem. Phys.* **56** 248

[64] Allnatt A 1964 Integral equations in ionic solution theory *Mol. Phys.* **8** 533

[65] Rasaiah J C 1970 Equilibrium properties of ionic solutions; the primitive model and its modification for aqueous solutions of the alkali halides at 25 °C *J. Chem. Phys.* **52** 704

[66] Ramanathan P S and Friedman H L 1971 Study of a refined model for aqueous 1-1 electrolytes *J. Chem. Phys.* **54** 1086

[67] Rasaiah J C 1972 Computations for higher valence electrolytes in the restricted primitive model *J. Chem. Phys.* **56** 3071

[68] Valleau J P and and Cohen L K 1980 Primitive model electrolytes. I. Grand canonical Monte Carlo computations *J. Chem. Phys.* **72** 5932

Valleau J P, Cohen L K and Card D N 1980 Primitive model electrolytes. II. The symmetrical electrolyte *J. Chem. Phys.* **72** 5942

[69] Cummings P T and Stell G 1984 Statistical mechanical models of chemical reactions analytic solution of models of $A + B \Leftrightarrow AB$ in the Percus–Yevick approximation *Mol. Phys.* **51** 253

Cummings P T and Stell G 1984 Statistical mechanical models of chemical reactions II. Analytic solutions of the Percus–Yevick approximation for a model of homogeneous association *Mol. Phys.* **51** 253

[70] Baxter R J 1968 Percus–Yevick equation for hard spheres with surface adhesion *J. Chem. Phys.* **49** 2770

[71] Zhu J and Rasaiah J C 1989 Solvent effects in weak electrolytes II. Dipolar hard sphere solvent an the sticky electrolyte model with $L = \sigma$ *J. Chem. Phys.* **91** 505

[72] Zwanzig R 1954 High temperature equation of state by a perturbation method I. Nonpolar Gases *J. Chem. Phys.* **22** 1420

[73] Hemmer P C 1964 On van der Waals theory of vapor–liquid equilibrium IV. The pair correlation function and equation of state for long-range forces *J. Math. Phys.* **5** 75

[74] Andersen H C and Chandler D 1970 Mode expansion in equilibrium statistical mechanics I. General theory and application to electron gas *J. Chem. Phys.* **53** 547

Chandler D and Andersen H C 1971 Mode expansion in equilibrium statistical mechanics II. A rapidly convergent theory of ionic solutions *J. Chem. Phys.* **54** 26

Andersen H C and Chandler D 1971 Mode expansion in equilibrium statistical mechanics III. Optimized convergence and application to ionic solution theory *J. Chem. Phys.* **55** 1497

[75] Lebowitz J L and Percus J 1961 Long range correlations in a closed system with applications to nonuniform fluids *Phys. Rev.* **122** 1675

[76] Hiroike K 1972 Long-range correlations of the distribution functions in the canonical ensemble *J. Phys. Soc. Japan* **32** 904

[77] Barker J and Henderson D 1967 Perturbation theory and equation of state for a fluids II. A successful theory of liquids *J. Chem. Phys.* **47** 4714

[78] Mansoori G A and Canfield F B 1969 Variational approach to the equilibrium properties of simple liquids I *J. Chem. Phys.* **51** 4958

Mansoori G A and Canfield F B 1970 *J. Chem. Phys.* **53** 1618

[79] Rasaiah J C and Stell G 1970 Upper bounds on free energies in terms of hard sphere results *Mol. Phys.* **18** 249

[80] Weeks J, Chandler D and Anderson H C 1971 Role of repulsive forces in determining the equilibrium structure of simple liquids *J. Chem. Phys.* **54** 5237

Chandler D, Weeks J D and Andersen H C 1983 The van der Waals picture of liquids, solids and phase transformations *Science* **220** 787

[81] Rushbrooke G 1940 On the statistical mechanics of assemblies whose energy-levels depend on temperature *Trans. Faraday Soc.* **36** 1055

[82] Cook and Rowlinson J S 1953 Deviations form the principles of corresponding states *Proc. R. Soc.* A **219** 405

[83] Pople J 1954 Statistical mechanics of assemblies of axially symmetric molecules I. General theory *Proc. R. Soc.* A **221** 498

Pople J 1954 Statistical mechanics of assemblies of axially symmetric molecules II. Second virial coefficients *Proc. R. Soc.* A **221** 508

[84] Zwanzig R 1955 High temperature equation of state by a perturbation method II. Polar gases *J. Chem. Phys.* **23** 1915

[85] Larsen B, Rasaiah J C and Stell G 1977 Thermodynamic perturbation theory for multipolar and ionic fluids *Mol. Phys.* **33** 987

Stell G, Rasaiah J C and Narang H 1974 *Mol. Phys.* **27** 1393

[86] Onsager L 1939 Electrostatic interaction of molecules *J. Phys. Chem.* **43** 189

[87] Gillan M 1980 Upper bound on the free energy of the restricted primitive model for ionic liquids *Mol. Phys.* **41** 75

[88] Stell G and Lebowitz J 1968 Equilibrium properties of a system of charged particles *J. Chem. Phys.* **49** 3706

[89] Stell G, Wu K C and Larsen B 1976 Critical point in a fluid of charged hard spheres *Phys. Rev. Lett.* **211** 369

[90] Haksjold B and Stell G 1982 The equilibrium studies of simple ionic liquids *The Liquid State of Matter: Fluids, Simple and Complex* ed E W Montroll and J L Lebowitz (Amsterdam: North-Holland)

[91] Straatsmaa T P and McCammon J A 1992 Computational alchemy *Ann. Rev. Phys. Chem.* **43** 407

[92] Jorgenson W L and Ravimohan C 1985 Monte Carlo simulation of the differences in free energy of hydration *J. Chem. Phys.* **83** 3050

[93] Jorgenson W 1989 Free energy calculations: a breakthrough in modeling organic chemistry in solution *Accounts Chem. Res.* **22** 184

[94] Weeks J D, Selinger R L B and Broughton J Q 1995 Self consistent treatment of attractive forces in non uniform liquids *Phys. Rev. Lett.* **75** 2694

[95] Weeks J D, Vollmayr K and Katsov K 1997 Intermolecular forces and the structure of uniform and non uniform fluids *Physica* A **244** 461

Weeks J D, Katsov K and Vollmayr K 1998 Roles of repulsive and attractive forces in determining the structure of non uniform liquids: generalized mean field theory *Phys. Rev. Lett.* **81** 4400

[96] Ising E 1925 *Z. Phys.* 31 253

[97] Lee T D and Yang C N 1952 Statistical theory of equations of state and phase transitions II. Lattice gas and Ising models *Phys. Rev.* **87** 410

[98] Fowler R H and Guggenheim E A 1940 Statistical thermodynamics of super-lattices *Proc. R. Soc.* A **174** 189

[99] Bethé H 1935 Statistical theory of superlattices *Proc. R. Soc.* A **150** 552

[100] Landau L D 1935 quoted in Landau L D and Lifshitz E M 1958 *Statistical Physics* ch XIV, section 135 (Oxford: Pergamon)

[101] Widom B 1965 Equation of state near the critical point *J. Chem. Phys.* **43** 3898

[102] Neece G A and Widom B 1969 Theories of liquids *Ann. Rev. Phys. Chem.* **20** 167

[103] Kramers H A and Wannier G H 1941 Statistics of the two-dimensional ferromagnet part I *Phys. Rev.* **60** 252

Kramers H A and Wannier G H 1941 Statistics of the two-dimensional ferromagnet part II *Phys. Rev.* **60** 263

[104] Domb C and Sykes M F 1957 On the susceptibility of a ferromagnetic above the Curie point *Proc. R. Soc.* A **240** 214

Domb C and Sykes M F 1957 Specific heat of a ferromagnetic Substance above the Curie point *Phys. Rev.* **129** 567

[105] Onsager L 1944 Crystal statistics I. A two-dimensional model with an order–disorder transition *Phys. Rev.* **65** 117

[106] Kaufman B 1949 Crystal statistics II. Partition function evaluated by Spinor analysis *Phys. Rev.* **65** 1232

Onsager L and Kaufman B 1949 Crystal statistics III. Short range order in a binary Ising lattice *Phys. Rev.* **65** 1244

[107] Yang C N 1952 The spontaneous magnetization of a two-dimensional Ising lattice *Phys. Rev.* **85** 809 (**87** 404)

Further Reading

Alavi A 1996 Path integrals and *ab initio* molecular dynamics *Monte Carlo and Molecular Dynamics of Condensed Matter Systems* ed K Binder and G Ciccotti (Bologna: SIF)

Anisimov M A and Sengers J V 1999 Crossover critical phenomena in aqueous solutions *Proc. 13th Int. Conf. on the Properties of Water and Steam (Toronto, September 12–16 1999)*

Barker J A and Henderson D 1976 What is a liquid? Understanding the states of matter *Rev. Mod. Phys.* **48** 587

Berne B J and Thirumalai D 1986 On the simulation of quantum systems: path integral methods *Ann. Rev. Phys. Chem.* **37** 401

Chandler D 1987 *Introduction to Modern Statistical Mechanics* (Oxford: Oxford University Press)

Chandler D and Wolynes P 1979 Exploiting the isomorphism between quantum theory and classical statistical mechanics of polyatomic fluids *J. Chem. Phys.* **70** 2914

Debendetti P G 1996 *Metastable Liquids, Concepts and Principles* (Princeton, NJ: Princeton University Press)

Domb C 1996 *The Critical Point. A Historical Introduction to the Modern Theory of Critical Phenomena* (London: Taylor and Francis)

Eyring H, Henderson D, Stover B J and Eyring E 1982 *Statistical Mechanics and Dynamics* (New York: Wiley)

Fisher M 1983 Scaling, universality and renormalization group theory *Critical Phenomena (Lecture Notes in Physics vol 186)* (Berlin: Springer)

Friedman H 1985 *A Course in Statistical Mechanics* (Englewood Cliffs, NJ: Prentice-Hall)

Friedman H L and Dale W T 1977 Electrolyte solutions at equilibrium *Statistical Mechanics part A, Equilibrium Techniques* ed B J Berne (New York: Plenum)

Goldenfeld L 1992 *Lectures in Phase Transitions and Renormalization Group* (New York: Addison-Wesley)

Goodstein D L 1974 *States of Matter* (Englewood Cliffs, NJ: Prentice-Hall and Dover)

Guggenheim E A 1967 *Thermodynamics and Advanced Treatment* 5th edn (Amsterdam: North-Holland)

McQuarrie D 1976 *Statistical Mechanics* (New York: Harper and Row)

Rice S A and Gray P 1965 *The Statistical Mechanics of Simple Liquids* (New York: Interscience)

Wilde R E and Singh S 1998 *Statistical Mechanics* (New York: Wiley)

Hirschfelder J O, Curtiss C F and Bird R B 1954 *Molecular Theory of Gases and Liquids* (New York: Wiley)

Rowlinson J and Swinton J 1983 *Liquids and Liquid Mixtures* 3rd edn (London: Butterworth)

Voth G 1996 Path integral centroid methods *Advances in Chemical Physics, New methods in Computational Quantum Mechanics* vol XCIII, ed I Prigogine and S A Rice

Makri N 1999 Time dependent quantum methods for large systems *Ann. Rev. Phys. Chem.* **50** 167

Reiss H and Hammerich A D S 1986 Hard spheres: scaled particle theory and exact relations on the existence and structure of the fluid/solid phase transition *J. Phys. Chem.* **90** 6252

Stillinger F 1973 Structure in aqueous solutions from the standpoint of scaled particle theory *J. Solution Chem.* **2** 141

Widom B 1967 Intermolecular forces and the nature of the liquid state *Science* **375** 157

Longuet-Higgins H C and Widom B 1964 A rigid sphere model for the melting of argon *Mol. Phys.* **8** 549

Smith W R 1972 Perturbation theory in the classical statistical mechanics of fluids *Specialist Periodical Report* vol 1 (London: Chemical Society)

Watts R O 1972 Integral equation approximations in the theory of fluids *Specialist Periodical Report* vol 1 (London: Chemical Society)

Mitchell D J, McQuarrie D A, Szabo A and Groeneveld J 1977 On the second-moment condition of Stillinger and Lovett *J. Stat. Phys.* **17** 1977

Vlachy V 1999 Ionic effects beyond Poisson–Boltzmann theory *Ann. Rev. Phys. Chem.* **50** 145

Outhwaite C W 1974 Equilibrium theories of electrolyte solutions *Specialist Periodical Report* (London: Chemical Society)

Rasaiah J C 1987 Theories of electrolyte solutions *The Liquid State and its Electrical Properties (NATO Advanced Science Institute Series Vol 193)* ed E E Kunhardt, L G Christophous and L H Luessen (New York: Plenum)

Rasaiah J C 1973 A view of electrolyte solutions *J. Solution Chem.* **2** 301

Stell G, Patey G N and Høye J S 1981 Dielectric constant of fluid models: statistical mechanical theory and its quantitative implementation *Adv. Chem. Phys.* **48** 183

Wertheim M 1979 Equilibrium statistical mechanics of polar fluids *Ann. Rev. Phys. Chem.* **30** 471

Reynolds C, King P M and Richards W G 1992 Free energy calculations in molecular biophysics *Mol. Phys.* **76** 251

Pratt L 1997 Molecular theory of hydrophobic effects *Encyclopedia of Computational Chemistry*

Lynden-Bell R M and Rasaiah J C 1997 From hydrophobic to hydrophilic behavior: a simulation study of solvation entropy and free energy of simple solutes *J. Chem. Phys.* **107** 1981

Hummer G, Garde S, Garcia A E, Paulitis M E and Pratt L R 1998 Hydrophobic effects on a molecular scale *J. Phys. Chem.* **102** 10 469

Lum K, Chandler D and Weeks J D 1999 Hydrophobicity at small and large length scales *J. Phys. Chem.* B **103** 4570

Pratt L R and Hummer G (eds) 1999 Simulation and theory of electrostatic interactions in solution; computational chemistry, biophysics and aqueous solutions *AIP Conf. Proc. (Sante Fe, NM, 1999)* vol 492 (New York: American Institute of Physics)

Stanley H E 1971 *Introduction to Phase Transitions and Critical Phenomena* (Oxford: Oxford University Press)

Ziman J M 1979 *Models of Disorder* (Cambridge: Cambridge University Press)

Yeomans Y M 1992 *Statistical Mechanics of Phase Transitions* (Oxford: Oxford University Press)

Stanley H E 1999 Scaling, universality and renormalization: three pillars of modern critical phenomena *Rev. Mod. Phys.* **71** S358

Kadanoff L P 1999 *Statistical Physics: Statics, Dynamics and Renormalization* (Singapore: World Scientific)

A2.4
Fundamentals of electrochemistry

Andrew Hamnett

Electrochemistry is concerned with the study of the interface between an electronic and an ionic conductor and, traditionally, has concentrated on: (i) the nature of the ionic conductor, which is usually an aqueous or (more rarely) a non-aqueous solution, polymer or superionic solid containing mobile ions; (ii) the structure of the electrified interface that forms on immersion of an electronic conductor into an ionic conductor; and (iii) the electron-transfer processes that can take place at this interface and the limitations on the rates of such processes.

Ionic conductors arise whenever there are mobile ions present. In electrolyte solutions, such ions are normally formed by the dissolution of an ionic solid. Provided the dissolution leads to the complete separation of the ionic components to form essentially independent anions and cations, the electrolyte is termed *strong*. By contrast, *weak* electrolytes, such as organic carboxylic acids, are present mainly in the undissociated form in solution, with the total *ionic* concentration orders of magnitude lower than the formal concentration of the solute. Ionic conductivity will be treated in some detail below, but we initially concentrate on the equilibrium structure of liquids and ionic solutions.

A2.4.1   The elementary theory of liquids

Modern-day approaches to ionic solutions need to be able to contend with the following problems:

(1) the nature of the solvent itself, and the interactions taking place in that solvent;

(2) the changes taking place on the dissolution of an ionic electrolyte in the solvent;

(3) macroscopic and microscopic studies of the properties of electrolyte solutions.

Even the description of the solvent itself presents major theoretical problems: the partition function for a liquid can be written in the classical limit [1, 2] as

$$Q(T, V, N) = \frac{q^N}{(8\pi^2)^N N! \Lambda^{3N}} \int \cdots \int d\boldsymbol{X}^N \exp[-\beta U_N(\boldsymbol{X}^N)] \qquad (A2.4.1)$$

where $\beta = 1/kT$ and the integral in (1) is over both spatial and orientation coordinates (i.e. both Cartesian and Eulerian coordinates) of each of the N molecules, and is termed the *configurational integral*, Z_N. In this equation, q is the partition coefficient for the internal degrees of freedom in each molecule (rotational, vibrational and electronic), Λ is the translational partition function $h/(2\pi mkT)^{1/2}$ and $U_N(\boldsymbol{X}^N)$ is the energy associated with the instantaneous configuration of the N molecules defined by $\boldsymbol{X}^N$. Clearly, the direct evaluation of (1) for all but the simplest cases is quite impossible, and modern theories have made an indirect attack on (1) by defining *distribution functions*. If we consider, as an example, two particles, which we fix with total coordinates X_1 and X_2, then the joint probability of finding particle 1 in volume dX_1 and particle 2 in volume dX_2 is

$$P^{(2)}(X_1, X_2) \, dX_1 \, dX_2 = dX_1 \, dX_2 \int \ldots \int dX_3, \ldots, dX_N P(X_1, X_2, X_3, \ldots, X_N) \qquad \text{(A2.4.2)}$$

where $P(X_1, \ldots, X_N)$ is the Boltzmann factor:

$$P(X^N) = \frac{\exp[-\beta U_N(X^N)]}{\int \ldots \int dX^N \exp[-\beta U_N(X^N)]}.$$

In fact, given that there are $N(N-1)$ ways of choosing a pair of particles, the pair distribution function, $\rho^{(2)}(X_1, X_2) \, dX_1 \, dX_2 = N(N-1)P^{(2)}(X_1, X_2) \, dX_1 \, dX_2$ is the probability of finding any particle at X_1 in volume dX_1 and a different particle at X_2 in volume dX_2. A little reflection will show that for an isotropic liquid, the value of $\rho^{(1)}(X_1)$ is just the number density of the liquid, $\rho = N/V$, since we can integrate over all orientations and, if the liquid is isotropic, its density is everywhere constant.

A2.4.1.1   Correlation functions

The pair distribution function clearly has dimensions (density)^2, and it is normal to introduce the pair correlation function $g(X_1, X_2)$ defined by

$$g(X_1, X_2) = \frac{\rho^{(2)}(X_1, X_2)}{\rho^{(1)}(X_1)\rho^{(2)}(X_2)} \qquad \text{(A2.4.3)}$$

and we can average over the orientational parts of both molecules, to give

$$g(R_1, R_2) = \frac{1}{(8\pi^2)^2} \int\int d\Omega_1 \, d\Omega_2 \, g(X_1, X_2). \qquad \text{(A2.4.4)}$$

Given that R_1 can be arbitrarily chosen as anywhere in the sample volume of an isotropic liquid in the absence of any external field, we can transform the variables R_1, R_2 into the variables R_1, R_{12}, where R_{12} is the vector separation of molecules 1 and 2. This allows us to perform one of the two integrations in equation (A2.4.2) above, allowing us to write the probability of a second molecule being found at a distance $r(= |r_{12}|)$ from a central molecule as $\rho g(r) r \, dr \sin\theta \, d\theta \, d\phi$. Integration over the angular variables gives the number of molecules found in a spherical shell at a distance r from a central molecule as

$$N(r) \, dr = 4\pi r^2 \rho g(r) \, dr. \qquad \text{(A2.4.5)}$$

The function $g(r)$ is central to the modern theory of liquids, since it can be measured experimentally using neutron or x-ray diffraction and can be related to the interparticle potential energy. Experimental data [1] for two liquids, water and argon (iso-electronic with water) are shown in figure A2.4.1, plotted as a function of $R^* = R/\sigma$, where σ is the effective diameter of the species, and is roughly the position of the first maximum in $g(R)$. For water, $\sigma = 2.82$ Å, very close to the intermolecular distance in the normal tetrahedrally bonded form of ice, and for argon, $\sigma = 3.4$ Å. The second peak for argon is at $R^* = 2$, as expected for a spherical molecular system consisting roughly of concentric spheres. However, for water, the second peak in $g(R)$ is found at $R^* = 1.6$, which corresponds closely to the second-nearest-neighbour distance in ice, strongly supporting the model for the structure of water that is ice-like over short distances. This strongly structured model for water in fact dictates many of its anomalous properties.

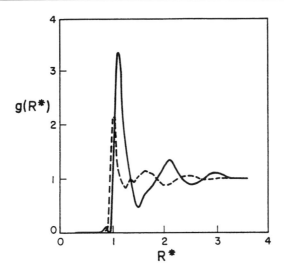

Figure A2.4.1. Radial distribution function $g(R^*)$ for water (dashed curve) at $4\,°\text{C}$ and 1 atm and for liquid argon (full curve) at 84.25 K and 0.71 atm as functions of the reduced distance $R^* = R/\sigma$, where σ is the molecular diameter; from [1].

The relationship between $g(R)$ and the interparticle potential energy is most easily seen if we assume that the interparticle energy can be factorized into pairwise additive potentials as

$$U_N(\mathbf{X}^N) \equiv U_N(\mathbf{X}_1, \mathbf{X}_2, \dots, \mathbf{X}_N) = \sum_{i<j}^{N} u(r_{ij}) \tag{A2.4.6}$$

where the summation is over all pairs i, j. From equation A2.4.1 we can calculate the total internal energy U as

$$U = -\left(\frac{\partial \ln Q}{\partial \beta}\right)_{V,N} = -\frac{1}{Q}\left\{\frac{1}{N!\Lambda^{3N}}\int d\mathbf{R}^N\left[-\sum_{i<j}^{N} u(r_{ij})\right]\exp(-U_N) - \frac{3NQ}{\Lambda}\frac{\partial\Lambda}{\partial\beta}\right\} \tag{A2.4.7}$$

and where, for simplicity, we have ignored internal rotational and orientational effects. For an isotropic liquid, the summation in (A2.4.6) over pairs of molecules yields $N(N-1)/2$ equal terms, which can be written as the product of a two-particle integral over the $u(r_{12})$ and integrals of the type shown in (A2.4.2) above. After some algebra, we find

$$U = \frac{3}{2}NkT + \frac{1}{2}\int_V d\mathbf{r}_1\int_V d\mathbf{r}_2\, u(r_{12})\rho^{(2)}(\mathbf{r}_1, \mathbf{r}_2) = \frac{3}{2}NkT + \frac{\rho N}{2}\int_0^{\infty} dr\, 4\pi r^2 g(r)u(r). \tag{A2.4.8}$$

To the same order of approximation, the pressure P can be written as

$$\frac{\beta P}{\rho} = 1 - \frac{\beta\rho}{6}\int_0^{\infty} dr\, 4\pi r^3 \frac{du(r)}{dr} g(r) \tag{A2.4.9}$$

where in both (A2.4.8) and (A2.4.9) the potential $u(r)$ is assumed to be sufficiently short range for the integrals to converge. Other thermodynamic functions can be calculated once $g(r)$ is known, and it is of considerable importance that $g(r)$ can be obtained from $u(r)$ and, ideally, that the inverse process can also be carried out.

Unfortunately, this latter process is much more difficult to do in such a way as to distinguish different possible $u(r)$s with any precision.

Clearly, the assumption of pairwise additivity is unlikely to be a good one for water; indeed, it will break down for any fluid at high density. Nonetheless, $g(r)$ remains a good starting point for any liquid, and we need to explore ways in which it can be calculated. There are two distinct methods: (a) solving equations relating $g(r)$ to $u(r)$ by choosing a specific $u(r)$; (b) by simulation methods using molecular dynamic or Monte Carlo methods.

There are two approaches commonly used to derive an analytical connection between $g(r)$ and $u(r)$: the Percus–Yevick (PY) equation and the hypernetted chain (HNC) equation. Both are derived from attempts to form functional Taylor expansions of different correlation functions. These auxiliary correlation functions include:

(i)   the *total correlation function*,

$$h(r_1, r_2) = g(r_1, r_2) - 1;$$ (A2.4.10)

(ii) the *background correlation function*,

$$y(r_1, r_2) = g(r_1, r_2) \exp[\beta u(r_1, r_2)];$$ (A2.4.11)

(iii) the *direct correlation function*, $C(r_1, r_2)$, defined through the Ornstein–Zernike relation:

$$h(r_1, r_2) - C(r_1, r_2) \equiv \rho \int dr_3 h(r_1, r_3) C(r_3, r_2).$$ (A2.4.12)

The singlet *direct correlation function* $C^{(1)}(r)$ is defined through the relationship

$$C^{(1)}(r) \equiv \ln[\rho^{(1)}(r)\Lambda^3] + \beta[w(r) - \mu]$$ (A2.4.13)

where $\rho^{(1)}$ is as defined above, μ is the chemical potential and $w(r)$ the local one-body potential in an inhomogeneous system.

The PY equation is derived from a Taylor expansion of the direct correlation function, and has the form

$$y(r_1, r_2) \approx 1 + \rho \int dr_3\, C(r_2, r_3) h(r_3, r_1)$$ (A2.4.14)

and comparison with the Ornstein–Zernike equation shows that $C(r_1, r_2) \approx g(r_1, r_2) - y(r_1, r_2) \approx y(r_1, r_2) f(r_1, r_2)$, where $f(r_1, r_2) \equiv \exp[-\beta u(r_1, r_2)] - 1$. Substitution of this expression into (A2.4.14) finally gives us, in terms of the pair correlation coefficient alone

$$g(r_1, r_2) \exp[\beta u(r_1, r_2)] = 1 + \rho \int dr_3 g(r_2, r_3)\, e^{\beta u(r_2, r_3)}[e^{-\beta u(r_2, r_3)} - 1][g(r_3, r_1) - 1].$$ (A2.4.15)

This integral equation can be solved by expansion of the integrand in bipolar coordinates [2, 3]. Further improvement to the PY equation can be obtained by analytical fit to simulation studies as described below.

The HNC equation uses, instead of the expression for $C(r_1, r_2)$ from (A2.4.14) above, an expression $C(r_1, r_2) \approx h(r_1, r_2) - \ln(y(r_1, r_2))$, which leads to the first-order HNC equation:

$$\ln(y(r_1, r_2)) \approx \rho \int dr_3\, C(r_1, r_3) h(r_3 r_2).$$ (A2.4.16)

Comparison with the PY equation shows that the HNC equation is nonlinear, and this does present problems in numerical work, as well as preventing any analytical solutions being developed even in the simplest of cases.

In the limit of low densities, (A2.4.15) shows that the zeroth-order approximation for $g(r)$ has the form

$$g(r_1, r_2) \approx e^{-\beta u(r_1, r_2)}$$ (A2.4.17)

a form that will be useful in our consideration of the electrolyte solutions described below.

A2.4.1.2 Simulation methods

Simulation methods for calculating $g(r)$ have come into their own in the past 20 years as the cost of computing has fallen. The Monte Carlo method is the simplest in concept: this depends essentially on identifying a statistical or Monte Carlo approach to the solution of equation (A2.4.2). As with all Monte Carlo integrations, a series of random values of the coordinates $X_1, \ldots, X_N$ is generated, and the integrand evaluated. The essential art in the technique is to pick predominantly configurations of high probability, or at least to eliminate the wasteful evaluation of the integrand for configurations of high energy. This is achieved by moving one particle randomly from the previous configuration, i, and checking the energy difference $\Delta U = U_{i+1} - U_i$. If $\Delta U < 0$ the configuration is accepted, and if $\Delta U > 0$, the value of $\exp(-\beta \Delta U)$ is compared to a second random number ξ, where $0 < \xi < 1$. If $\exp(-\beta \Delta U) > \xi$ the configuration is accepted, otherwise it is rejected and a new single-particle movement generated. A second difficulty is that the total number of particles that can be treated by Monte Carlo techniques is relatively small unless a huge computing resource is available: given this, boundary effects would be expected to be dominant, and so periodic boundary conditions are imposed, in which any particle leaving through one surface re-enters the system through the opposite surface. Detailed treatments of the Monte Carlo technique were first described by Metropolis *et al* [4]; the method has proved valuable not only in the simulation of realistic interparticle potentials, but also in the simulation of model potentials for comparison with the integral equation approaches above.

The alternative simulation approaches are based on molecular dynamics calculations. This is conceptually simpler that the Monte Carlo method: the equations of motion are solved for a system of N molecules, and periodic boundary conditions are again imposed. This method permits both the equilibrium and transport properties of the system to be evaluated, essentially by numerically solving the equations of motion

$$m \frac{d^2 R_k}{dt^2} = \sum_{j=1, j \neq k}^{N} \mathbf{F}(R_{kj}) = - \sum_{j=1, j \neq k}^{N} \nabla_k U(R_{kj}) \tag{A2.4.18}$$

by integrating over discrete time intervals δt. Details are given elsewhere [2].

A2.4.2 Ionic solutions

There is, in essence, no limitation, other than the computing time, to the accuracy and predictive capacity of molecular dynamic and Monte Carlo methods, and, although the derivation of realistic potentials for water is a formidable task in its own right, we can anticipate that accurate simulations of water will have been made relatively soon. However, there remain major theoretical problems in deriving any analytical theory for water, and indeed any other highly-polar solvent of the sort encountered in normal electrochemistry. It might be felt, therefore, that the extension of the theory to analytical descriptions of ionic solutions was a well-nigh hopeless task. However, a major simplification of our problem is allowed by the possibility, at least in more dilute solutions, of smoothing out the influence of the solvent molecules and reducing their influence to such average quantities as the dielectric permittivity, ε_m, of the medium. Such a viewpoint is developed within the McMillan–Mayer theory of solutions [1, 2], which essentially seeks to partition the interaction potential into three parts: that due to the interaction between the solvent molecules themselves, that due to the interaction between the solvent and the solute and that due to the interaction between the solute molecules dispersed within the solvent. The main difference from the dilute fluid results presented above is that the potential energy $u(r_{ij})$ is replaced by the potential of mean force $W(r_{ij})$ for two particles and, for N_a particles of solute in the solvent, by the expression

$$W(X^{N_a}; z_a \to 0) = -kT \ln g^{(N_a)}(X^{N_a}; z_a \to 0)$$

where z_a is the so-called *activity* defined as $z_a = q_a e^{-\beta \mu_a} / \Lambda_a^3$ (cf equation (A2.4.1)); it has units of number density.

The McMillan–Mayer theory allows us to develop a formalism similar to that of a dilute interacting fluid for solute dispersed in the solvent provided that a sensible description of W can be given. At the limit of dilution, when intersolute interactions can be neglected, we know that the chemical potential of a can be written as $\mu_a = W(a|s) + kT \ln(\rho_a \Lambda_a^3 q_a^{-1})$, where $W(a|s)$ is the potential of mean force for the interaction of a solute molecule with the solvent. If we define $\gamma_a^0 = \underset{\rho_a \to 0}{\mathrm{Lt}} (z_a 8\pi^2/\rho_a) = e^{\beta W(a|s)}$ then the grand canonical partition function can be written in the form:

$$\Xi(T, V, \lambda_a, \lambda_s) = \sum_{N_a \geq 0} \frac{(z_a/\gamma_a^0)^{N_a}}{N_a!} \Xi(T, V, \lambda_s) \int e^{-\beta W(X^{N_a}; z_a \to 0)} \, dX^{N_a} \tag{A2.4.19}$$

where we have successfully partitioned the solute–solute interactions into a modified configuration integral, the solute–solvent interactions into γ_a^0 and the solvent–solvent interactions into the partition coefficient $\Xi(T, V, \lambda_s)$.

A2.4.2.1 The structure of water and other polar solvents

In terms of these three types of interactions, we should first consider the problems of water and other polar solvents in more detail. Of the various components of the interaction between water molecules, we may consider the following.

(1) At very short distances, less than about 2.5 Å, a reasonable description of the interaction will be strongly repulsive, to prevent excessive interpenetration; a Lennard–Jones function will be adequate:

$$U_{\mathrm{LJ}}(R) = 4\varepsilon \left[\left(\frac{\sigma}{R} \right)^{12} - \left(\frac{\sigma}{R} \right)^{6} \right]. \tag{A2.4.20}$$

(2) At distances of a few molecular diameters, the interaction will be dominated by electric multipole interactions; for dipolar molecules, such as water, the dominant term will be the dipole–dipole interaction:

$$U_{\mathrm{DD}}(X_1, X_2) = R_{12}^{-3}[\vec{\mu}_1 \cdot \vec{\mu}_2 - 3(\vec{\mu}_1 \cdot u_{12})(\vec{\mu}_2 \cdot u_{12})] \tag{A2.4.21}$$

where u_{12} is a unit vector in the direction of the vector $R_2 - R_1$.

(3) At intermediate distances, 2.4 Å $\leq R \leq$ 4 Å, there is a severe analytical difficulty for water and other hydrogen-bonded solvents; in that the hydrogen-bond energy is quite large, but is extremely orientation dependent. If the water molecule is treated as tetrahedral, with two O–H vectors h_{i1}, h_{i2} and two lone-pair vectors l_{i1}, l_{i2} for the ith molecule, then the hydrogen-bond energy has the form

$$U_{\mathrm{HB}}(X_1, X_2) = \varepsilon_{\mathrm{HB}} G(X_1, X_2)$$
$$= \varepsilon_{\mathrm{HB}} G_\sigma(R_{ij} - R_{\mathrm{H}}) \left\{ \sum_{\alpha,\beta=1}^{2} G_\sigma[(h_{i\alpha} \cdot u_{ij}) - 1]G_\sigma[(l_{j\beta} \cdot u_{ij}) + 1] \right.$$
$$\left. + G_\sigma[(l_{i\alpha} \cdot u_{ij}) - 1]G_\sigma[(h_{j\beta} \cdot u_{ij}) + 1] \right\} \tag{A2.4.22}$$

an expression that looks unwieldy but is quite straightforward to apply numerically. The function $G_\sigma(x)$ is defined either as unity for $|x| < \sigma$ and zero for $|x| \leq \sigma$ or in a Gaussian form: $G_\sigma = \exp(-x^2/2\sigma^2)$.

The form of the hydrogen-bonded potential leads to a strongly structured model for water, as discussed above. In principle, this structure can be defined in terms of the average number of hydrogen bonds formed by a single water molecule with its neighbours. In normal ice this is four, and we expect a value close to this

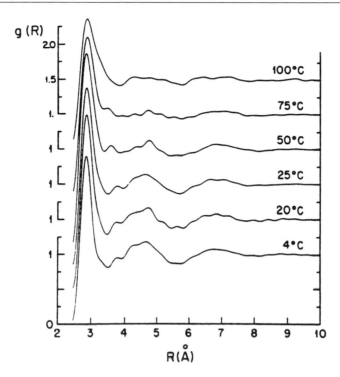

Figure A2.4.2. The temperature dependence of $g(R)$ of water. From [1].

in water close to the freezing point. We also intuitively expect that this number will decrease with increasing temperature, an expectation confirmed by the temperature dependence of $g(R)$ for water in figure A2.4.2 [1]. The picture should be seen as highly dynamic, with these hydrogen bonds forming and breaking continuously, with the result that the clusters of water molecules characterizing this picture are themselves in a continuous state of flux.

A2.4.2.2 *Hydration and solvation of ions*

The solute–solvent interaction in equation (A2.4.19) is a measure of the solvation energy of the solute species at infinite dilution. The basic model for ionic hydration is shown in figure A2.4.3 [5]: there is an inner hydration sheath of water molecules whose orientation is essentially determined entirely by the field due to the central ion. The number of water molecules in this inner sheath depends on the size and chemistry of the central ion; being, for example, four for Be^{2+}, but six for Mg^{2+}, Al^{3+} and most of the first-row transition ions. Outside this primary shell, there is a secondary sheath of more loosely bound water molecules oriented essentially by hydrogen bonding, the evidence for which was initially indirect and derived from ion mobility measurements. More recent evidence for this secondary shell has now come from x-ray diffraction and scattering studies and infrared (IR) measurements. A further highly diffuse region, the tertiary region, is probably present, marking a transition to the hydrogen-bonded structure of water described above. The ion, as it moves, will drag at least part of this solvation sheath with it, but the picture should be seen as essentially dynamic, with the well defined inner sheath structure of figure A2.4.3 being mainly found in highly-charged ions of high electronic stability, such as Cr^{3+}. The enthalpy of solvation of *cations* primarily depends on the charge on the central ion and the effective ionic radius, the latter begin the sum of the normal Pauling ionic

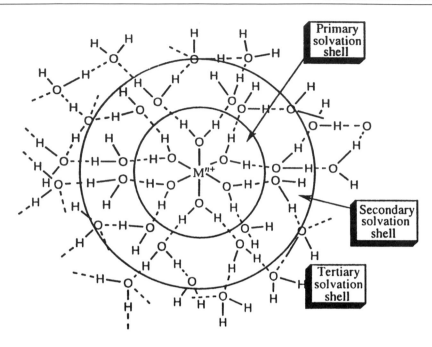

Figure A2.4.3. The localized structure of a hydrated metal cation in aqueous solution (the metal ion being assumed to have a primary hydration number of six). From [5].

radius and the radius of the oxygen atom in water (0.85 Å). A reasonable approximate formula has

$$\Delta H^0_{\text{hyd}} = -695Z^2/(r_+ + 0.85) \text{ [kJ mol}^{-1}]. \tag{A2.4.23}$$

In general, anions are less strongly hydrated than cations, but recent neutron diffraction data have indicated that even around the halide ions there is a well defined primary hydration shell of water molecules, which, in the case of Cl^-, varies from four to six in constitution; the exact number being a sensitive function of concentration and the nature of the accompanying cation.

(a) Methods for determining the structure of the solvation sheath

Structural investigations of metal–ion hydration have been carried out by spectroscopic, scattering and diffraction techniques, but these techniques do not always give identical results since they measure in different timescales. There are three distinct types of measurement:

(1) those giving an average structure, such as neutron and x-ray scattering and diffraction studies;

(2) those revealing dynamic properties of coordinated water molecules, such as nuclear magnetic resonance (NMR) and quasi-elastic scattering methods;

(3) those based on energetic discrimination between water in the different hydration sheaths and bulk water, such as IR, Raman and thermodynamic studies.

First-order difference neutron scattering methods for the analysis of concentrated solutions of anions and cations were pioneered by Enderby [6] and co-workers and some results for Ni^{2+} plotted as $\Delta g(r)$ are shown in figure A2.4.4 [5]. The sharp M–O and M–D pair correlations are typical of long-lived inner hydration sheaths, with the broader structure showing the second hydration sheath being clearly present, but more

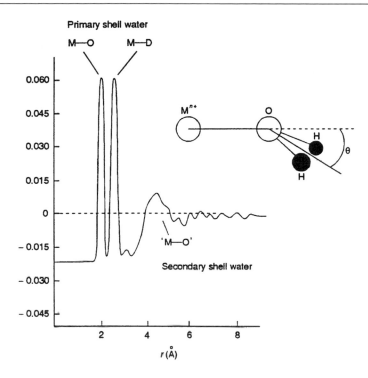

Figure A2.4.4. Plot of the radial distribution difference function $\Delta g(r)$ against distance r (pm) for a 1.46 M solution of NiCl$_2$ in D$_2$O. From [5].

diffuse. Note that the water molecule is tilted so that the D–O–D$\cdots$M atoms are not coplanar. This tilt appears to be concentration dependent, and decreases to zero below 0.1 M NiCl$_2$. It is almost certainly caused by interaction between the hydrogen-bonded secondary sheaths around the cations, a fact that will complicate the nature of the potential of the mean force, discussed in more detail below. The secondary hydration sheaths have been studied by large-angle x-ray scattering (LAXS). For Cr^{3+}, a well defined secondary sheath containing 13 ± 1 molecules of water could be identified some 4.02 ± 0.2 Å distant from the central ion. The extended x-ray absorption-edge fine structure (EXAFS) technique has also been used to study the local environment around anions and cations: in principle the technique is ideally suited for this, since it has high selectivity for the central ion and can be used in solutions more dilute than those accessible to neutron or x-ray scattering. However, the technique also depends on the capacity of the data to resolve different structural models that may actually give rise to rather similar EXAFS spectra. The sensitivity of the technique also falls away for internuclear distances in excess of 2.5 Å.

The secondary hydration sheath has also been studied using vibrational spectroscopy. In the presence of highly-charged cations, such as Al^{3+}, Cr^{3+} and Rh^{3+}, frequency shifts can be seen due to the entire primary and secondary hydration structure, although the *number* of water molecules hydrating the cation is somewhat lower than that expected on the basis of neutron data or LAXS data. By contrast, comparison of the Raman and neutron diffraction data for Sc^{3+} indicates the presence of [Sc(H$_2$O)$_7$]$^{3+}$ in solution, a result supported by the observation of pentagonal bipyramidal coordination in the x-ray structure of the aqua di-μ-hydroxo dimer [Sc$_2$(OH)$_2$]$^{4+}$.

The hydration of more inert ions has been studied by ^{18}O labelling mass spectrometry. ^{18}O-enriched water is used, and an equilibrium between the solvent and the hydration around the central ion is first attained, after which the cation is extracted rapidly and analysed. The method essentially reveals the number of oxygen atoms that exchange slowly on the timescale of the extraction, and has been used to establish the existence of the stable $[Mo_3O_4]^{4+}$ cluster in aqueous solution.

One of the most powerful methods for the investigation of hydration is NMR, and both ^1H and ^{17}O nuclei have been used. By using paramagnetic chemical shift reagents such as Co^{2+} and Dy^{3+}, which essentially shift the peak position of bulk water, hydration measurements have been carried out using ^1H NMR on a number of tripositive ions. ^{17}O NMR measurements have also been carried out and, by varying the temperature, the dynamics of water exchange can also be studied. The hydration numbers measured by this technique are those for the inner hydration sheath and, again, values of four are found for Be^{2+} and six for many other di- and tri-positive cations. The hydration numbers for the alkali metals' singly-positive cations have also been determined by this method, with values of around three being found.

Hydration and solvation have also been studied by conductivity measurements; these measurements give rise to an effective radius for the ion, from which a hydration number can be calculated. These effective radii are reviewed in the next section.

A2.4.3   Ionic conductivity

A2.4.3.1   The Boltzmann transport equation

The motion of particles in a fluid is best approached through the Boltzmann transport equation, provided that the combination of internal and external perturbations does not substantially disturb the equilibrium. In other words, our starting point will be the statistical thermodynamic treatment above, and we will consider the effect of both the internal and external fields. Let the chemical species in our fluid be distinguished by the Greek subscripts $\alpha, \beta, \ldots$ and let $f_\alpha(r, c, t) \, dV \, dc_x \, dc_y \, dc_z$ be the number of molecules of type α located in volume dV at r and having velocities between c_x and $c_x + dc_x$ etc. Note that we expect c and r are independent. Let the external force on molecules of type α be F_α. At any space point, r, the rate of increase of f_α, $(\partial f_\alpha/\partial t)$, will be determined by:

(1)   the nett flow of molecules of type α with velocity c into dV, $-\nabla \cdot (c f_\alpha) = -c \cdot \text{grad}(f_\alpha)$;

(2)   acceleration of molecules in dV into and out of the range dc by F_α, $-(1/m_\alpha)\nabla_c \cdot (F_\alpha f_\alpha)$;

(3)   accelerations, de-excitations, etc of local molecules by intermolecular collisions. This is the most troublesome part analytically: it will be composed of terms corresponding to *gain* of molecules in dV at c and a *loss* by collisions. We will not write down an explicit expression for the nett collision effect, but rather write $(\partial f_\alpha/\partial t)_{\text{coll}}$.

The nett result is

$$\left(\frac{\partial f_\alpha}{\partial t}\right) + c \cdot \vec{\nabla} f_\alpha + \frac{1}{m_\alpha}\vec{\nabla}_c \cdot F_\alpha f_\alpha = \left(\frac{\partial f_\alpha}{\partial t}\right)_{\text{coll}} \tag{A2.4.24}$$

which is Boltzmann's transport equation. To make progress, we make the assumption now that in first order, f_α on the left-hand side of the equation is the equilibrium value, f_α^0. We further make the so-called relaxation-time approximation that $(\partial f_\alpha/\partial t)_{\text{coll}} = ((f_\alpha^0 - f_\alpha)/\tau)$, where τ is, in principle, a function of c, or at least of $|c|$. We then have, from (A2.4.24) $(z_\alpha e_0/m_\alpha)E \cdot \vec{\nabla}_c(f_\alpha) = ((f_\alpha^0 - f_\alpha)/\tau)$, where the charge on ions of type α is $z_\alpha e_0$ and the applied electric field is E. Given that the current density, J, in dV is

$$\iiint z_\alpha e_0 c f_\alpha \, dc \equiv \iiint z_\alpha e_0 c (f_\alpha - f_\alpha^0) \, dc \tag{A2.4.25}$$

substituting (A2.4.25) from the Boltzmann equation and evaluating the conductivity, κ_α, from ions of type α, we have, after carrying out the spatial integrations

$$\kappa_\alpha = -\frac{z_\alpha^2 e_0^2}{m_\alpha} \iiint \tau c \cdot \vec{\nabla}_c f_\alpha \, dc = \frac{z_\alpha^2 e_0^2}{m_\alpha} \iiint f_\alpha \nabla_c(\tau c) \, dc \approx \frac{z_\alpha^2 e_0^2 N \tau}{m_\alpha} \tag{A2.4.26}$$

where N is the number of ions per unit volume. From elementary analysis, if we define a mean ionic drift velocity v in the direction of the applied electric field, E, the conductivity contribution from ions of type α will be $N z_\alpha e_0 v / |E| \equiv N z_\alpha e_0 u$, where u is termed the *mobility*; from which we can see that $u = z_\alpha e_0 \tau / m_\alpha$.

A2.4.3.2 *The elementary theory of ionic conductivity [7]*

An alternative approach is to consider ions of charge $z_\alpha e_0$ accelerated by the electric field strength, E, being subject to a frictional force, K_R, that increases with velocity, v, and is given, for simple spherical ions of radius r_α, by the Stokes formula, $K_R = 6\pi \eta r_\alpha v$, where η is the viscosity of the medium. After a short induction period, the velocity attains a limiting value, v_{max}, corresponding to the exact balance between the electrical and frictional forces:

$$z_\alpha e_0 E = 6\pi \eta r_\alpha v_{max} \qquad (A2.4.27)$$

and the terminal velocity is given by

$$v_{max} = z_\alpha e_0 E / (6\pi \eta r_\alpha) \qquad (A2.4.28)$$

and it is evident that $\tau = m_\alpha / (6\pi \eta r_\alpha)$. It follows that, for given values of η and E, each type of ion will have a transport velocity dependent on the charge and the radius of the solvated ion and a direction of migration dependent on the sign of the charge.

For an electrolyte solution containing both anions and cations, with the terminal velocity of the cations being v_{max}^+, and the number of ions of charge $z^+ e_0$ per unit volume being N^+, the product $A N^+ v_{max}^+$ corresponds just to that quantity of positive ions that passes per unit time through a surface of area A normal to the direction of flow. The product $A N^- v_{max}^-$ can be defined analogously, and the amount of charge carried through this surface per unit time, or the current per area A, is given by

$$I = I^+ + I^- = A e_0 (N^+ z^+ v_{max}^+ + N^- z^- v_{max}^-) = A e_0 (N^+ z^+ u^+ + N^- z^- u^-) \times |E| \qquad (A2.4.29)$$

where the u are the mobilities defined above. If the potential difference between the electrodes is ΔV, and the distance apart of the electrodes is l, then the magnitude of the electric field $|E| = \Delta V / l$. Since $I = G \Delta V$, where G is the conductance, G is given by

$$G = (A/l) e_0 (N^+ z^+ u^+ + N^- z^- u^-). \qquad (A2.4.30)$$

The conductivity is obtained from this by division by the geometric factor (A/l), giving

$$\kappa = e_0 (N^+ z^+ u^+ + N^- z^- u^-). \qquad (A2.4.31)$$

It is important to recognize the approximations made here: the electric field is supposed to be sufficiently small so that the equilibrium distribution of velocities of the ions is essentially undisturbed. We are also assuming that the we can use the relaxation approximation, and that the relaxation time τ is independent of the ionic concentration and velocity. We shall see below that these approximations break down at higher ionic concentrations: a primary reason for this is that ion–ion interactions begin to affect both τ and F_α, as we shall see in more detail below. However, in very dilute solutions, the ion scattering will be dominated by solvent molecules, and in this limiting region (A2.4.31) will be an adequate description.

Measurement of the conductivity can be carried out to high precision with specially designed cells. In practice, these cells are calibrated by first measuring the conductance of an accurately known standard, and then introducing the sample under study. Conductances are usually measured at about 1 kHz AC rather than with DC voltages in order to avoid complications arising from electrolysis at anode and cathode [8].

The conductivity of solutions depends, from (A2.4.31), on both the concentration of ions and their mobility. Typically, for 1 M NaCl in water at $18\,^\circ\text{C}$, a value of 7.44 Ω^{-1} m^{-1} is found: by contrast, 1 M

Table A2.4.1. Typical transport numbers for aqueous solutions.

Electrolyte	t_0^+	$t_0^-\,(=1-t_0^+)$
KCl	0.4906	0.5094
NH_4Cl	0.4909	0.5091
HCl	0.821	0.179
KOH	0.274	0.726
NaCl	0.3962	0.6038
$NaOOCCH_3$	0.5507	0.4493
$KOOCCH_3$	0.6427	0.3573
$CuSO_4$	0.375	0.625

H_2SO_4 has a conductivity of $36.6\ \Omega^{-1}\ m^{-1}$ at the same temperature and acetic acid, a *weak* electrolyte, has a conductivity of only $0.13\ \Omega^{-1}\ m^{-1}$.

In principle, the effects of the concentration of ions can be removed by dividing (A2.4.31) by the concentration. Taking Avagadro's constant as L and assuming a concentration of solute c mol m^{-3}, then from the electroneutrality principle we have $N^+z^+ = N^-z^- = v_\pm z_\pm cL$ and clearly

$$\Lambda \equiv \frac{\kappa}{c} = v_\pm z_\pm Le_0(u^+ + u^-) \equiv v_\pm z_\pm F(u^+ + u^-) \tag{A2.4.32}$$

where Λ is termed the *molar conductivity* and F is the Faraday, which has the numerical value $96\,485$ C mol^{-1}.

In principle, Λ should be independent of the concentration according to (A2.4.31), but this is not found experimentally. At very low concentrations Λ is roughly constant, but at higher concentrations substantial changes in the mobilities of the ions are found, reflecting increasing ion–ion interactions. Even at low concentrations the mobilities are not constant and, empirically, for strong electrolytes, Kohlrausch observed that Λ decreased with concentration according to the expression

$$\Lambda = \Lambda_0 - k\sqrt{c/c^0} \tag{A2.4.33}$$

where Λ_0 is the molar conductivity extrapolated to zero concentration and c^0 is the standard concentration (usually taken as 1 M). Λ_0 plays an important part in the theory of ionic conductivity since at high dilution the ions should be able to move completely independently, and as a result equation (A2.4.32) expressed in the form

$$\Lambda_0 = v_\pm z_\pm F(u_0^+ + u_0^-) \equiv v_+\lambda_0^+ + v_-\lambda_0^- \tag{A2.4.34}$$

is exactly true.

The fraction of current carried by the cations is clearly $I^+/(I^+ + I^-)$; this fraction is termed the *transport number* of the cations, t^+, and evidently

$$t^+ = \frac{u^+}{u^+ + u^-}. \tag{A2.4.35}$$

In general, since the mobilities are functions of the concentration, so are the transport numbers, but limiting transport numbers can be defined by analogy to (A2.4.34). The measurement of transport numbers can be carried out straightforwardly, allowing an unambiguous partition of the conductivity and assessment of the individual ionic mobilities at any concentration. Some typical transport numbers are given in table A2.4.1 [7], for aqueous solutions at $25\,^\circ$C and some limiting single-ion molar conductivities are given in table A2.4.2 [7].

Table A2.4.2. Limiting single-ion conductivities.

Ion	λ_0^+, λ_0^- $(\Omega^{-1}\,mol^{-1}\,cm^2)$	Ion	λ_0^+, λ_0^- $(\Omega^{-1}\,mol^{-1}\,cm^2)$
		Ag^+	62.2
H^+	349.8	Na^+	50.11
OH^-	197	Li^+	38.68
K^+	73.5	$[Fe(CN)_6]^{4-}$	440
NH_4^+	73.7	$[Fe(CN)_6]^{3-}$	303
Rb^+	77.5	$[CrO_4]^{2-}$	166
Cs^+	77	$[SO_4]^{2-}$	161.6
		I^-	76.5
Ba^{2+}	126.4	Cl^-	76.4
Ca^{2+}	119.6	NO_3^-	71.5
Mg^{2+}	106	CH_3COO^-	40.9
		$C_6H_5COO^-$	32.4

A2.4.3.3 *The solvation of ions from conductivity measurements*

We know from equations (A2.4.32) and (A2.4.34) that the limiting ionic conductivities are directly proportional to the limiting ionic mobilities: in fact

$$\lambda_0^+ = z^+ F u_0^+ \tag{A2.4.36}$$

$$\lambda_0^- = z^- F u_0^-. \tag{A2.4.37}$$

At infinite dilution, the assumption of a constant relaxation time is reasonable and, using Stokes law as well, we have

$$u_0 = z e_0 / 6\pi \eta r. \tag{A2.4.38}$$

At first sight, we would expect that the mobilities of more highly-charged ions would be larger, but it is apparent from table A2.4.2 that this is not the case; the *mobilities* of Na^+ and Ca^{2+} are comparable, even though equation (A2.4.38) would imply that the latter should be about a factor of two larger. The explanation lies in the fact that r also increases with charge, which, in turn, can be traced to the increased size of the hydration sheath in the doubly-charged species, since there is an increased attraction of the water dipoles to the more highly-charged cations.

It is also possible to explain, from hydration models, the differences between equally-charged cations, such as the alkali metals ($\lambda_0^{K^+} = 73.5$, $\lambda_0^{Na^+} = 50.11$ and $\lambda_0^{Li^+} = 38.68$, all in units of $\Omega^{-1}\,mol^{-1}\,cm^2$). From atomic physics it is known that the radii of the bare ions is in the order $Li^+ < Na^+ < K^+$. The attraction of the water dipoles to the cation increases strongly as the distance between the charge centres of the cation and water molecule decreases, with the result that the total radius of the ion and bound water molecules actually increases in the order $K^+ < Na^+ < Li^+$, and this accounts for the otherwise rather strange order of mobilities.

The differing extent of hydration shown by the different types of ion can be determined experimentally from the amount of water carried over with each type of ion. A simple measurement can be carried out by adding an electrolyte such as LiCl to an aqueous solution of sucrose in a Hittorf cell. Such a cell consists of two compartments separated by a narrow neck [7]; on passage of charge the strongly hydrated Li^+ ions will migrate from the anode to the cathode compartment, whilst the more weakly hydrated Cl^- ions migrate towards the anode compartment; the result is a slight increase in the concentration of sucrose in the anode compartment, since the sucrose itself is essentially electrically neutral and does not migrate in the electric field. The change in concentration of the sucrose can either be determined analytically or by measuring the change in rotation of plane polarized light transmitted through the compartment. Measurements carried out in this way lead to hydration numbers for ions, these being the number of water molecules that migrate with each cation or anion. Values of 10–12 for Mg^{2+}, 5.4 for K^+, 8.4 for Na^+ and 14 for Li^+ are clearly in reasonable agreement with the values inferred from the Stokes law arguments above. They are also in agreement with the measurements carried out using large organic cations to calibrate the experiment, since these are assumed not to be hydrated at all.

Anions are usually less strongly hydrated, as indicated above, and from equation (A2.4.38) this would suggest that increasing the charge on the anion should lead unequivocally to an increase in mobility and hence to an increase in limiting ionic conductivity. An inspection of table A2.4.2 shows this to be borne out to some extent by the limited data available. The rather low conductivities exhibited by organic anions is a result of their considerably larger size; even taking hydration into account, their total diameter normally exceeds that of the simple anions.

One anomaly immediately obvious from table A2.4.2 is the much higher mobilities of the proton and hydroxide ions than expected from even the most approximate estimates of their ionic radii. The origin of this behaviour lies in the way in which these ions can be accommodated into the water structure described above. Free protons cannot exist as such in aqueous solution: the very small radius of the proton would lead to an enormous electric field that would polarize any molecule, and in an aqueous solution the proton immediately attaches itself to the oxygen atom of a water molecule, giving rise to an H_3O^+ ion. In this ion, however, the positive charge does not simply reside on a single hydrogen atom; NMR spectra show that all three hydrogen atoms are equivalent, giving a structure similar to that of the NH_3 molecule. The formation of a water cluster around the H_3O^+ ion and its subsequent fragmentation may then lead to the positive charge being transmitted across the cluster without physical migration of the proton, and the limiting factor in proton motion becomes hydrogen-bonded cluster formation and not conventional migration. It is clear that this model can be applied to the anomalous conductivity of the hydroxide ion without any further modification. Hydrogen-atom tunnelling from a water molecule to an OH^- ion will leave behind an OH^- ion, and the migration of OH^- ions is, in fact, traceable to the migration of H^+ in the opposite direction. This type of mechanism is supported by the observation of the effect of temperature. It is found that the mobility of the proton goes through a maximum at a temperature of 150 °C (where, of course, the measurements are carried out under pressure). This arises because as the temperature in increased from ambient, the main initial effect is to loosen the hydrogen-bonded local structure that inhibits reorientation. However, at higher temperatures, the thermal motion of the water molecules becomes so marked that cluster formation becomes inhibited.

The complete hydration shell of the proton consists of both the central H_3O^+ unit and further associated water molecules; mass spectrometric evidence would suggest that a total of four water molecules form the actual $H_9O_4^+$ unit, giving a hydration number of four for the proton. Of course, the measurement of this number by the Hittorf method is not possible since the transport of protons takes place by a mechanism that does not involve the actual movement of this unit. By examining concentration changes and using large organic cations as calibrants, a hydration number of one is obtained, as would be expected.

From equations (A2.4.36) and (A2.4.37), we can calculate the magnitudes of the mobilities for cations and anions. As an example, from table A2.4.2, the limiting ionic conductivity for the Na^+ ion is 50.11×10^{-4} $m^2\,\Omega^{-1}\,mol^{-1}$. From this we obtain a value of $u_0^+ = \lambda_0^+/F \approx 5.19 \times 10^{-8}$ $m^2\,V^{-1}\,s^{-1}$, which implies that in a field of 100 V m^{-1}, the sodium ion would move a distance of about 2 cm in 1 h. The mobilities of other ions have about the same magnitude $(4–8) \times 10^{-8}$ $m^2\,V^{-1}\,s^{-1}$), with the marked exception of the proton. This has an apparent mobility of 3.63×10^{-7} $m^2\,V^{-1}\,s^{-1}$, almost an order of magnitude higher, reflecting the different conduction mechanism described above. These mobilities give rise to velocities that are small compared to thermal velocities, at least for the small electric fields normally used, confirming the validity of the analysis carried out above.

With the knowledge now of the magnitude of the mobility, we can use equation (A2.4.38) to calculate the radii of the ions; thus for lithium, using the value of 0.000 89 kg $m^{-1}\,s^{-1}$ for the viscosity of pure water (since we are using the conductivity at infinite dilution), the radius is calculated to be 2.38×10^{-10} m ($\equiv 2.38$ Å). This can be contrasted with the crystalline ionic radius of Li$^+$, which has the value 0.78 Å. The difference between these values reflects the presence of the hydration sheath of water molecules; as we showed above, the transport measurements suggest that Li$^+$ has a hydration number of 14.

From equation (A2.4.38) we can, finally, deduce Walden's rule, which states that the product of the ionic mobility at infinite dilution and the viscosity of the pure solvent is a constant. In fact

$$u_0\eta = ze_0/6\pi r = \text{constant} \tag{A2.4.39}$$

whereby $\lambda_0\eta = \text{constant}$ and $\Lambda_0\eta = \text{constant}$. This rule permits us to make an estimate of the change of u_0 and λ_0 with a change in temperature and the alteration of the solvent, simply by incorporating the changes in viscosity.

A2.4.4   Ionic interactions

The McMillan–Mayer theory offers the most useful starting point for an elementary theory of ionic interactions, since at high dilution we can incorporate all ion–solvent interactions into a limiting chemical potential, and deviations from solution ideality can then be explicitly connected with ion–ion interactions only. Furthermore, we may assume that, at high dilution, the interaction energy between two ions (assuming only two are present in the solution) will be of the form

$$u(r_{12}) = \begin{cases} +\infty & r \leq d \\ \dfrac{z_1 z_2 e_0^2}{4\pi\varepsilon\varepsilon_0 r_{12}} & r > d \end{cases} \tag{A2.4.40}$$

where in the limiting dilution law, first calculated by Debye and Hückel (DH), d is taken as zero. It should be emphasized that $u(r)$ is not the potential of mean force, $W(r)$, defined in the McMillan–Mayer theory above; this latter needs to be worked out by calculating the *average electrostatic potential* (AEP), $\psi_i(r)$ surrounding a given ion, i, with charge $z_i e_0$. This is because although the interaction between any ion j and this central ion is given by (A2.4.40), the work required to bring the ion j from infinity to a distance r from i is influenced by other ions surrounding i. Oppositely charged ions will tend to congregate around the central ion, giving rise to an ionic 'atmosphere' or *cosphere*, which intervenes between ions i and j, *screening* the interaction represented in (A2.4.40). The resulting AEP is the sum of the central interaction and the interaction with the ionic cosphere, and it can be calculated by utilizing the Poisson equation:

$$\nabla^2\psi_i(r) = -\frac{q_{(i)}(r)}{\varepsilon\varepsilon_0} \tag{A2.4.41}$$

where $q_{(i)}(r)$ is the charge density (i.e. the number of charges per unit volume) at a distance r from the centre i. In terms of the pair correlation coefficient defined above:

$$q_{(i)}(r) = e_0 \sum_j z_j \rho_j g_{ji}(r) \tag{A2.4.42}$$

where ρ_j is the number density of ions of type j. From (A2.4.20) above, we have $g_{ji} = \mathrm{e}^{-\beta W_{ji}(r)}$, and it is the potential of the mean force W that is related to the AEP. The first approximation in the DH theory is then to write

$$W_{ji}(r) \approx e_0 z_j \psi_i(r) \tag{A2.4.43}$$

whence $g_{ji}(r) \approx e^{-\beta z_j e_0 \psi_i(r)}$, which was originally given by DH as a consequence of the Boltzmann law, but clearly has a deep connection with statistical thermodynamics of fluids. From (A2.4.41) and (A2.4.42), we have

$$\nabla^2 \psi_i(r) = -\frac{e_0}{\varepsilon \varepsilon_0} \sum_j z_j \rho_j \, e^{-\beta z_j e_0 \psi}. \qquad (A2.4.44)$$

The major deficiency of the equation as written is that there is no excluded volume, a deficiency DH could rectify for the central ion, but not for all ions around the central ion. This deficiency has been addressed within the DH framework by Outhwaite [9].

To solve (A2.4.44), the assumption is made that $\beta z_j e_0 \psi_i(r) \ll 1$, so the exponential term can be expanded. Furthermore, we must have $\sum_j z_j \rho_j = 0$ since the overall solution is electroneutral. Finally we end up with

$$\nabla^2 \psi_i(r) = -\frac{e_0}{\varepsilon \varepsilon_0} \sum_j z_j \rho_j \, e^{-\beta z_j e_0 \psi_i(r)} \approx \frac{e_0^2 \psi_i(r)}{\varepsilon \varepsilon_0 kT} \sum_j \rho_j z_j^2 \equiv \kappa^2 \psi_i(r) \qquad (A2.4.45)$$

where κ has the units of inverse length. The *ionic strength*, I, is defined as

$$I = \frac{1}{2} \sum_j \rho_j (z_j e_0)^2 \qquad (A2.4.46)$$

so $\kappa^2 = 2I/\varepsilon \varepsilon_0 kT$. For aqueous solutions at $25\,^\circ$C, κ^{-1} m$^{-1} = 3.046 \times 10^{-10}/(I)^{1/2}$. Equation (A2.4.45) can be solved straightforwardly providing the assumption is made that the mean cosphere around each ion is spherical. On this basis (A2.4.45) reduces to

$$\frac{1}{r^2}\frac{d}{dr}\left(r^2 \frac{d\psi_i}{dr}\right) = \kappa^2 \psi \qquad (A2.4.47)$$

which solves to give

$$\psi_i(r) = \frac{A}{r} e^{-\kappa r} + \frac{B}{r} e^{+\kappa r} \qquad (A2.4.48)$$

where A and B are constants of integration. B is clearly zero and, in the original DH model with no core repulsion term, A was fixed by the requirement that as $r \to 0$, $\psi_i(r)$ must behave as $z_i e_0/4\pi\varepsilon\varepsilon_0 r$. In the extended model, equation (A2.4.47) is also solved for the central ion, and the integration constants determined by matching ψ and its derivative at the ionic radius boundary. We finally obtain, for the limiting DH model:

$$\psi_i(r) = \frac{z_i e_0}{4\pi\varepsilon\varepsilon_0 r} e^{-\kappa r}. \qquad (A2.4.49)$$

Given that $W_{ji}(r) \approx e_0 z_j \psi_i(r)$, we finally obtain for the pair correlation coefficient

$$g_{ji}(r) = \exp[-\beta z_j e_0 \psi_i(r)] \approx 1 - \frac{z_i z_j e_0^2}{4\pi\varepsilon\varepsilon_0 kT r} e^{-\kappa r}. \qquad (A2.4.50)$$

An alternative derivation, which uses the Ornstein–Zernicke equation (equation (A2.4.12)), was given by Lee [2]. The Ornstein–Zernicke equation can be written as

$$h_{ij}(rr') - C_{ij}(rr') = \sum_l \rho_l \int ds \, C_{il}(rs) h_{ij}(sr'). \qquad (A2.4.51)$$

Given $y(r, r') \approx 1$ for very dilute solutions, the PY condition leads to

$$C_{ij} \approx f_{ij} = e^{-\beta u_{ij}} - 1 \approx -u_{ij}/kT \qquad \text{and} \qquad h_{ij} = e^{-\beta W_{ij}} - 1 \approx -W_{ij}/kT \qquad (A2.4.52)$$

whence W_{ij} satisfies the integral equation

$$-W_{ij}(r) = -\frac{z_i z_j e_0^2}{4\pi\varepsilon\varepsilon_0 r} + \sum_l \beta\rho_l \int ds \left(\frac{z_i z_j e_0^2}{4\pi\varepsilon\varepsilon_0 s}\right) W_{lj}(|r - s|). \qquad (A2.4.53)$$

This can be solved by standard techniques to yield

$$W_{ij}(r) = \frac{z_i z_j e_0^2}{4\pi\varepsilon\varepsilon_0} \frac{e^{-\kappa r}}{r}. \qquad (A2.4.54)$$

A result identical to that above.

From these results, the thermodynamic properties of the solutions may be obtained within the McMillan–Mayer approximation; i.e. treating the dilute solution as a quasi-ideal gas, and looking at deviations from this model solely in terms of ion–ion interactions, we have

$$\frac{U - U^{\text{ideal}}}{V} = \frac{1}{2}\sum_i\sum_j \rho_i\rho_j \int dr\, 4\pi r^2 u_{ij}(r) g_{ij}(r) = \frac{1}{2}\frac{e_0^2}{4\pi\varepsilon\varepsilon_0}\sum_i\sum_j \rho_i\rho_j z_i z_j \int dr\, \frac{1}{r}[4\pi r^2 g_{ij}(r)]$$

$$(A2.4.55)$$

using (A2.4.50) and $\sum_j z_j\rho_j = 0$, this can be evaluated to give

$$\frac{U - U^{\text{ideal}}}{V} = -\frac{\kappa^3 kT}{8\pi}. \qquad (A2.4.56)$$

The chemical potential may be calculated from the expression

$$\frac{\mu_j}{kT} = \ln(\rho_j\Lambda_j^3) + \sum_i \frac{\rho_i}{kT}\int_0^1 d\xi \int_0^\infty dr\, 4\pi r^2 u_{ij}(r) g_{ij}(r;\xi) \qquad (A2.4.57)$$

where ξ is a coupling parameter which determines the extent to which ion j is coupled to the remaining ions in the solution. This is closely related to the work of charging the jth ion in the potential of all the other ions and, for the simple expression in (A2.4.57), the charging can be represented by writing the charge on the jth ion in the equation for g_{ij} as $z_j\xi$, with ξ increasing from zero to one as in (A2.4.57). Again, using (A2.4.50) and $\sum_j z_j\rho_j = 0$, we find

$$\frac{\mu_j}{kT} = \ln(\rho_j\Lambda_j^3) - \frac{z_j^2 e_0^2\kappa}{8\pi\varepsilon\varepsilon_0 kT}. \qquad (A2.4.58)$$

This is, in fact, the main result of the DH analysis. The activity coefficient is clearly given by

$$\ln\gamma_j = -\frac{z_j^2 e_0^2\kappa}{8\pi\varepsilon\varepsilon_0 kT} \qquad (A2.4.59)$$

where again the activity coefficient is referred to as a dilute non-interacting 'gas' of solvated ions in the solvent. From equation (A2.4.46), we can express (A2.4.59) in terms of the ionic strength I, and we find:

$$\ln\gamma_j = -Az_j^2\sqrt{I} \equiv -1.172z_j^2\sqrt{I} \qquad (A2.4.60)$$

where I is the ionic strength at a standard concentration of 1 mol kg^{-1} and A is a constant that depends solely on the properties of the solvent, and the equivalence refers to water at 25 °C. It should be realized that separate ionic activity coefficients are not, in fact, accessible experimentally, and only the *mean* activity coefficient, defined for a binary electrolyte $A_{\nu+}B_{\nu-}$ by $\gamma_\pm = (\gamma_+^{\nu+}\gamma_-^{\nu-})^{1/(\nu_+ + \nu_-)}$ is accessible. Straightforward algebra gives

$$\ln\gamma_\pm = -A|z_+ z_-|\sqrt{I} \equiv -1.172|z_+ z_-|\sqrt{I}. \qquad (A2.4.61)$$

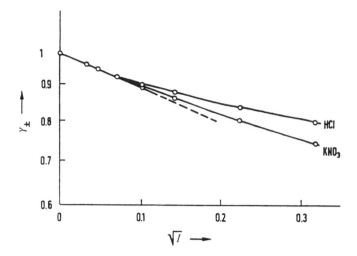

Figure A2.4.5. Theoretical variation of the activity coefficient $\gamma_\pm$ with $\sqrt{I}$ from equation (A2.4.61) and experimental results for 1–1 electrolytes at 25 °C. From [7].

We have seen that the DH theory in the limiting case neglects excluded volume effects; in fact the excluded volume of the *central* ion can be introduced into the theory as explained after (A2.4.48). If the radius of the ions is taken as a_0 for all ions, we have, in first order,

$$\ln \gamma_\pm = -\frac{|z_+ z_-| e_0^2 \kappa}{8\pi \varepsilon \varepsilon_0 kT (1 + a_0 \kappa)}. \qquad (A2.4.62)$$

For different electrolytes of the same charge, expression (A2.4.61) predicts the same values for $\gamma_\pm$; in other words, the limiting law does not make allowance for any differences in size or other ionic properties. For 1–1 electrolytes, this is experimentally found to be the case for concentrations below 10^{-2} M, although for multi-charged electrolytes, the agreement is less good, even for 10^{-3} mol kg^{-1}. In table A2.4.3 [7] some values of $\gamma_\pm$ calculated from (A2.4.61) are collected and compared to some measured activity coefficients for a few simple electrolytes. Figure A2.4.5 [7] shows these properties graphed for 1–1 electrolytes to emphasize the nature of the deviations from the limiting law. It is apparent from the data in both the table and the figure that deviations from the limiting law are far more serious for 2–1 electrolytes, such as H_2SO_4 and Na_2SO_4. In the latter case, for example, the limiting law is in serious error even at 0.005 M.

A2.4.4.1 *Beyond the limiting law*

At concentrations greater than 0.001 mol kg^{-1}, equation (A2.4.61) becomes progressively less and less accurate, particularly for unsymmetrical electrolytes. It is also clear, from table A2.4.3, that even the properties of electrolytes of the same charge type are no longer independent of the chemical identity of the electrolyte itself, and our neglect of the factor κa_0 in the derivation of (A2.4.61) is also not valid. As indicated above, a partial improvement in the DH theory may be made by including the effect of finite size of the central ion alone. This leads to the expression

$$\ln \gamma_\pm = -\frac{|z_+ z_-| e_0^2 \kappa}{8\pi \varepsilon \varepsilon_0 kT (1 + a_0 \kappa)} = -\frac{A |z_+ z_-| \sqrt{I}}{(1 + B a_0 \sqrt{I})} \qquad (A2.4.63)$$

where the parameter B also depends only on the properties of the solvent, and has the value 3.28×10^9 m^{-1} for water at 25 °C. The parameter a_0 is adjustable in the theory, and usually the product $B a_0$ is close to unity.

Table A2.4.3. $\gamma_\pm$ values of various electrolytes at different concentration.

m (mol kg^{-1})	I	Equation (A2.4.60)	Electrolyte		
			HCl	KNO$_3$	LiF
1–1 electrolytes					
0.001	0.001	0.9636	0.9656	0.9649	0.965
0.002	0.002	0.9489	0.9521	0.9514	0.951
0.005	0.005	0.9205	0.9285	0.9256	0.922
0.010	0.010	0.8894	0.9043	0.8982	0.889
0.020	0.020	0.8472	0.8755	0.8623	0.850
0.050	0.050		0.8304	0.7991	
0.100	0.100		0.7964	0.7380	
			H$_2$SO$_4$	Na$_2$SO$_4$	
1–2 or 2–1 electrolytes					
0.001	0.003	0.8795	0.837	0.887	
0.002	0.006	0.8339	0.767	0.847	
0.005	0.015	0.7504	0.646	0.778	
0.010	0.030	0.6662	0.543	0.714	
0.020	0.060		0.444	0.641	
0.050	0.150			0.536	
0.100	0.300		0.379	0.453	
			CdSO$_4$	CuSO$_4$	
2–2 electrolytes					
0.001	0.004	0.7433	0.754	0.74	
0.002	0.008	0.6674	0.671		
0.005	0.020	0.5152	0.540	0.53	
0.010	0.040		0.432	0.41	
0.020	0.080		0.336	0.315	
0.050	0.200		0.277	0.209	
0.100	0.400		0.166	0.149	

Even (A2.4.63) fails at concentrations above about 0.1 M, and the mean activity coefficient for NaCl shown in figure A2.4.6 [2] demonstrates that in more concentrated solutions the activity coefficients begin to rise, often exceeding the value of unity. This rise can be traced to more than one effect. As we shall see below, the inclusion of ion-exclusion effects for all the ions gives rise to this phenomenon. In addition, the ion–ion interactions at higher concentrations cannot really be treated by a hard-sphere model anyway, and models taking into account the true ion–ion potential for solvated ions at close distances are required. Furthermore, the number of solvent molecules essentially immobilized in the solvent sheath about each ion becomes a significant fraction of the total amount of solvent present. This can be exemplified by the case of sulphuric acid: given that each proton requires four water molecules for solvation and the sulphate ion can be estimated to require one, each mole of H$_2$SO$_4$ will require 9 mol of water. One kilogram of water contains approximately 55 mol, so that a 1 mol kg^{-1} solution of H$_2$SO$_4$ will only leave 46 mol of 'free' water. The effective concentration of an electrolyte will, therefore, be appreciably higher than its analytical value, and this effect becomes more marked the higher the concentration. A further effect also becomes important at higher concentrations: implicit in our whole approach is the assumption that the free energy of the solvation of the ions is independent of concentration. However, if we look again at our example of sulphuric acid,

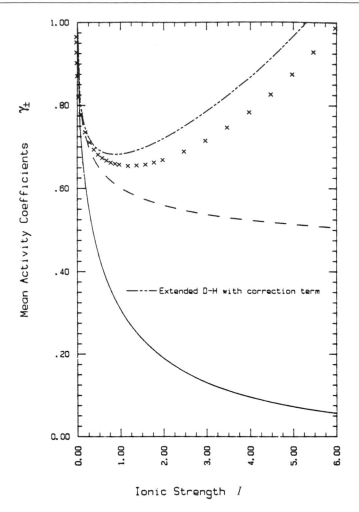

Figure A2.4.6. Mean activity coefficient for NaCl solution at 25 °C as a function of the concentration: full curve from (A2.4.61); dashed curve from (A2.4.63); dot-dashed curve from (A2.4.64). The crosses denote experimental data. From [2].

it is clear that for $m > 6$ mol kg^{-1}, apparently all the water is present in the solvation sheaths of the ions! Of course what actually occurs is that the extent of solvation of the ions changes, in effect decreasing the stability of the ions. However, this process essentially invalidates the McMillan–Mayer approach, or at the least requires the potential of mean force to be chosen in such a way as to reproduce the change in solvation energy.

Within the general DH approach, equation (A2.4.63) may be further modified by adding a linear term, as suggested by Hitchcock [8]:

$$\ln \gamma_{\pm} = -\frac{A|z_+z_-|\sqrt{I}}{(1 + Ba_0\sqrt{I})} + bI \tag{A2.4.64}$$

where b is a parameter to be fitted to the data. As can be seen from figure A2.4.6 this accounts for the behaviour quite well, but the empirical nature of the parameter b and the lack of agreement on its interpretation mean that (A2.4.64) can only be used empirically.

The simplest extension to the DH equation that does at least allow the qualitative trends at higher concentrations to be examined is to treat the excluded volume rationally. This model, in which the ion of charge $z_i e_0$ is given an ionic radius d_i is termed the *primitive model*. If we assume an essentially spherical equation for the u_{ij}:

$$u_{ij} = \begin{cases} +\infty & r \le d_{ij} \\ \dfrac{z_i z_j e_0^2}{4\pi \varepsilon \varepsilon_0 r_{ij}} & r > d_{ij}. \end{cases} \tag{A2.4.65}$$

This can be treated analytically within the *mean spherical approximation* for which

$$g_{ij} = 0 \qquad r < d_{ij}$$

$$C_{ij}(r) \approx -\frac{z_i z_j e_0^2}{4\pi \varepsilon \varepsilon_0 kT r} \qquad r > d_{ij}. \tag{A2.4.66}$$

These equations were solved by Blum [10], and a characteristic inverse length, 2Γ, appears in the theory. This length is implicitly given by the equation

$$2\Gamma = \alpha \left\{ \sum_{l=1}^{n} \rho_i \left[\frac{z_i - (\pi/2\Delta)d_i^2 P_n}{1 + \Gamma d_i} \right]^2 \right\}^{1/2} \tag{A2.4.67}$$

where

$$P_n \equiv \frac{1}{\Omega} \sum_k \frac{\rho_k d_k z_k}{1 + \Gamma d_k}$$

$$\Omega \equiv 1 + \frac{\pi}{2\Delta} \sum_k \frac{\rho_k d_k^3}{1 + \Gamma d_k}$$

$$\zeta_n \equiv \sum_k \rho_k (d_k)^n$$

$$\Delta \equiv 1 - \frac{\pi \zeta_3}{6}$$

$$\alpha^2 \equiv \frac{e_0^2}{\varepsilon \varepsilon_0 kT}.$$

In this formalism, which is already far from transparent, the internal energy is given by

$$-\frac{U - U^{\text{HS}}}{VkT} = \frac{e_0^2}{4\pi \varepsilon \varepsilon_0 kT} \left\{ \Gamma \sum_{i=1}^{n} \frac{\rho_i z_i^2}{1 + \Gamma d_i} + \frac{\pi}{2\Delta} \Omega P_n^2 \right\} \tag{A2.4.68}$$

and the mean activity coefficient by

$$\ln \gamma_\pm - \ln \gamma_\pm^{\text{HS}} = \frac{U - U^{\text{HS}}}{NkT} - \frac{\alpha^2}{8\rho} \left(\frac{P_n}{\Delta} \right)^2 \tag{A2.4.69}$$

where the superscript HS refers to solutions of the pure hard-sphere model as given by Lee [2].

The integral equation approach has also been explored in detail for electrolyte solutions, with the PY equation proving less useful than the HNC equation. This is partly because the latter model reduces cleanly to the MSA model for small $h(12)$ since

$$C(12) = h(12) - \ln y(12) = h(12) - \ln[1 + h(12)] - \beta u(12) \approx -\beta u(12) + \tfrac{1}{2}(h(12))^2 + \cdots.$$

Table A2.4.4. Osmotic coefficients obtained by various methods.

Concentration (mol dm^{-3})	Monte Carlo	MSA	PY	HNC
0.00911	0.97	0.969	0.97	0.97
0.10376	0.945	0.931	0.946	0.946
0.425	0.977	0.945	0.984	0.980
1.00	1.094	1.039	1.108	1.091
1.968	1.346	1.276	1.386	1.340

Using the Ornstein–Zernicke equation, numerical solutions for the restricted primitive model can be obtained.

In principle, simulation techniques can be used, and Monte Carlo simulations of the primitive model of electrolyte solutions have appeared since the 1960s. Results for the osmotic coefficients are given for comparison in table A2.4.4 together with results from the MSA, PY and HNC approaches. The primitive model is clearly deficient for values of r_{ij} close to the closest distance of approach of the ions. Many years ago, Gurney [11] noted that when two ions are close enough together for their solvation sheaths to overlap, some solvent molecules become freed from ionic attraction and are effectively returned to the bulk [12].

The potential model for this approach has the form

$$u_{ij} = \begin{cases} +\infty & r_{ij} \leq d_{ij} \\ \dfrac{z_i z_j e_0^2}{4\pi \varepsilon \varepsilon_0 r_{ij}} & r_{ij} > d_{ij} + 2r_w \\ (A_g)_{ij} + \dfrac{z_i z_j e_0^2}{4\pi \varepsilon \varepsilon_0 r_{ij}} & d_{ij} < r_{ij} < d_{ij} + 2r_w. \end{cases} \tag{A2.4.70}$$

The A_g are essentially adjustable parameters and, clearly, unless some of the parameters in (A2.4.70) are fixed by physical argument, then calculations using this model will show an improved fit for purely algebraic reasons. In principle, the radii can be fixed by using tables of ionic radii; calculations of this type, in which just the A_g are adjustable, have been carried out by Friedman and co-workers using the HNC approach [12]. Further refinements were also discussed by Friedman [13], who pointed out that an additional term u_{cavity} is required to account for the fact that each ion is actually in a cavity of low dielectric constant, ε_c, compared to that of the bulk solvent, ε. A real difficulty discussed by Friedman is that of making the potential continuous, since the discontinuous potentials above may lead to artefacts. Friedman [13] addressed this issue and derived formulae that use repulsion terms of the form $B_{ij}[(r_i + r_j)/r_{ij}]^n$, rather than the hard-sphere model presented above.

A quite different approach was adopted by Robinson and Stokes [8], who emphasized, as above, that if the solute dissociated into v ions, and a total of h molecules of water are required to solvate these ions, then the real concentration of the ions should be corrected to reflect only the bulk solvent. Robinson and Stokes derive, with these ideas, the following expression for the activity coefficient:

$$\ln \gamma_\pm = -\frac{A|z_+ z_-|\sqrt{I}}{(1 + Ba_0\sqrt{I})} - \frac{h}{v}\ln a_A - \ln[1 + 0.001 W_A(v - h)m] \tag{A2.4.71}$$

where a_A is the activity of the solvent, W_A is its molar mass and m is the molality of the solution. Equation (A2.4.71) has been extensively tested for electrolytes and, provided h is treated as a parameter, fits remarkably well for a large range of electrolytes up to molalities in excess of two. Unfortunately, the values of h so derived, whilst showing some sensible trends, also show some rather counter-intuitive effects, such as an increase from Cl^- to I^-. Furthermore, the values of h are not additive between cations and anions in solution, leading to significant doubts about the interpretation of the equation. Although considerable effort has gone into finding alternative, more accurate expressions, there remains considerable doubt about the overall physical framework.

A2.4.4.2 Interionic interactions and the conductivity

Equation (A2.4.24) determines the mobility of a single ion in solution, and contains no correction terms for the interaction of the ions themselves. However, in solution, the application of an electric field causes positive and negative ions to move in opposite directions and the symmetrical spherical charge distribution of equation (A2.4.49) becomes distorted. Each migrating ion will attempt to rebuild its atmosphere during its motion, but this rebuilding process will require a certain time, termed the relaxation time, so that the central ion, on its progress through the solution, will always be a little displaced from the centre of charge of its ionic cloud. The result of this is that each central ion will experience a retarding force arising from its associated ionic cloud, which is migrating in the opposite direction, an effect termed the relaxation or asymmetry effect. Obviously this effect will be larger the nearer, on average, the ions are in solution; in other words, the effect will increase at higher ionic concentrations.

In addition to the relaxation effect, the theory of Debye, Hückel and Onsager also takes into account a second effect, which arises from the Stokes law discussed above. We saw that each ion travelling through the solution will experience a frictional effect owing to the viscosity of the liquid. However, this frictional effect itself depends on concentration, since, with increasing concentration, encounters between the solvent sheaths of oppositely charged ions will become more frequent. The solvent molecules in the solvation sheaths are moving with the ions, and therefore an individual ion will experience an additional drag associated with the solvent molecules in the solvation sheaths of oppositely charged ions; this is termed the electrophoretic effect.

The quantitative calculation of the dependence of the electrolyte conductivity on concentration begins from expression (A2.4.49) for the potential exerted by a central ion and its associated ionic cloud. As soon as this ionic motion begins, the ion will experience an effective electric field E_{rel} in a direction opposite to that of the applied electric field, whose magnitude will depend on the ionic mobility. In addition, there is a second effect identified by Onsager due to the movement of solvent sheaths associated with the oppositely charged ions encountered during its own migration through the solution. This second term, the electrophoresis term, will depend on the viscosity of the liquid, and combining this with the reduction in conductivity due to relaxation terms we finally emerge with the Debye–Hückel–Onsager equation [8]:

$$\Lambda = \Lambda_0 - \Lambda_0 \frac{z_+ z_- e_0^2}{24\pi \varepsilon \varepsilon_0 kT} \frac{2q\kappa}{1 + \sqrt{q}} - \frac{L e_0^2 (z_+ + |z_-|)\kappa}{6\pi \eta} \tag{A2.4.72}$$

where

$$q = \frac{z_+ z_-}{z_+ + |z_-|} \frac{\lambda_0^+ + \lambda_0^-}{|z_-|\lambda_0^+ + z_+\lambda_0^+} \tag{A2.4.73}$$

L is Avagadro's constant and κ is defined above. It can be seen that there are indeed two corrections to the conductivity at infinite dilution: the first corresponds to the relaxation effect, and is correct in (A2.4.72) only under the assumption of a zero ionic radius. For a finite ionic radius, a_0, the first term needs to be modified: Falkenhagen [8] originally showed that simply dividing by a term $(1 + \kappa a_0)$ gives a first-order correction, and more complex corrections have been reviewed by Pitts et al [14], who show that, to a second order, the relaxation term in (A2.4.72) should be divided by $(1 + \kappa a_0)(1 + \kappa a_0 \sqrt{q})$. The electrophoretic effect should also be corrected in more concentrated solutions for ionic size; again to a first order, it is sufficient to divide by the correction factor $(1 + \kappa a_0)$. Note that for a completely dissociated 1–1 electrolyte $q = 0.5$, and expression

Table A2.4.5. Experimental and theoretical values of $B_1\Lambda_0 + B_2$ for various salts in aqueous solution at 291 K.

Salt	Observed value of $B_1\Lambda_0 + B_2$ (m^2 Ω^{-1} mol^{-1})	Calculated value of $B_1\Lambda_0 + B_2$ (m^2 Ω^{-1} mol^{-1})
LiCl	7.422×10^{-3}	7.343×10^{-3}
NaCl	7.459×10^{-3}	7.569×10^{-3}
KCl	8.054×10^{-3}	8.045×10^{-3}
LiNO$_3$	7.236×10^{-3}	7.258×10^{-3}

(A2.4.72) can be re-written in terms of the molarity, c/c^0, remembering that κ can be expressed either in terms of molalities or molarities; in the latter case we have

$$\kappa^2 = \frac{e_0^2 L c^0}{\varepsilon\varepsilon_0 kT}\frac{\sum_i z_i^2 c_i}{c^0} = \frac{e_0^2 L c^0}{\varepsilon\varepsilon_0 kT}\sum_i z_i^2 \nu_i (c/c^0) \tag{A2.4.74}$$

where c^0 is the standard concentration of 1 M. Finally, we see

$$\Lambda = \Lambda_0 - (B_1\Lambda_0 + B_2)\sqrt{c/c^0} \tag{A2.4.75}$$

in which B_1 and B_2 are independent of concentration. This is evidently identical in form to Kohlrausch's empirical law already discussed earlier (equation (A2.4.33)). Equation (A2.4.75) is valid in the same concentration range as the DH limiting law, i.e. for molalities below 0.01 M for symmetrical electrolytes, and for unsymmetrical electrolytes to even lower values. In fact, for symmetrical singly-charged 1–1 electrolytes, useful estimations of the behaviour can be obtained, with a few per cent error, for up to 0.1 mol kg^{-1} concentrations, but symmetrical multi-charged electrolytes ($z^+ = z^- \neq 1$) usually show deviations, even at 0.01 M.

At higher concentrations, division by the factor $(1 + \kappa a_0)$ gives rise to an expression of the form

$$\Lambda = \Lambda_0 - \frac{(B_1\Lambda_0 + B_2)}{1 + \kappa a_0}\sqrt{c/c^0} \tag{A2.4.76}$$

which is valid for concentrations up to about 0.1 M

In aqueous solution, the values of B_1 and B_2 can be calculated straightforwardly. If Λ is expressed in m^2 Ω^{-1} mol^{-1} and c^0 is taken as 1 mol dm^{-3}, then, for water at 298 K and $z^+ = |z^-| = 1$, $B_1 = 0.229$ and $B_2 = 6.027 \times 10^{-3}$ m^2 Ω^{-1} mol^{-1}. At 291 K the corresponding values are 0.229 and 5.15×10^{-3} m^2 Ω^{-1} mol^{-1} respectively. Some data for selected 1–1 electrolytes is given in table A2.4.5 [7]; it can be seen that Onsager's formula is very well obeyed.

A2.4.5 The electrified double layer

Once an electrode, which for our purposes may initially be treated as a conducting plane, is introduced into an electrolyte solution, several things change. There is a substantial loss of symmetry, the potential experienced by an ion will now be not only the screened potential of the other ions but will contain a term arising from the field due to the electrode and a term due to the image charge in the electrode. The structure of the solvent is also perturbed: next to the electrode the orientation of the molecules of solvent will be affected by the electric field at the electrode surface and the nett orientation will derive from both the interaction with the electrode and with neighbouring molecules and ions. Finally, there may be a sufficiently strong interaction between the

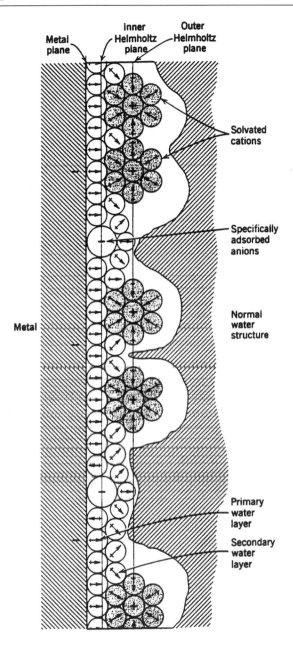

Figure A2.4.7. Hypothetical structure of the electrolyte double layer. From [15].

ions and the electrode surface such that the ions lose at least some of their inner solvation sheath and adsorb onto the electrode surface.

The classical model of the electrified interface is shown in figure A2.4.7 [15], and the following features are apparent.

(1) There is an ordered layer of solvent dipoles next to the electrode surface, the extent of whose orientation is expected to depend on the charge on the electrode.

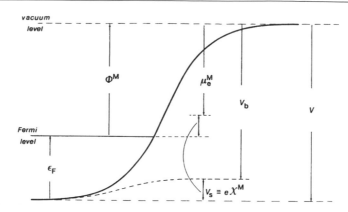

Figure A2.4.8. Potential energy profile at the metal–vacuum interface. Bulk and surface contributions to V are shown separately. From [16].

(2) There is, or may be, an inner layer of specifically adsorbed *anions* on the surface; these anions have displaced one or more solvent molecules and have lost part of their inner solvation sheath. An imaginary plane can be drawn through the centres of these anions to form the *inner Helmholtz plane* (IHP).

(3) The layer of solvent molecules not directly adjacent to the metal is the closest distance of approach of solvated *cations*. Since the enthalpy of solvation of cations is normally substantially larger than that of anions, it is normally expected that there will be insufficient energy to strip the cations of their inner solvation sheaths, and a second imaginary plane can be drawn through the centres of the solvated cations. This second plane is termed the *outer Helmholtz plane* (OHP).

(4) Outside the OHP, there may still be an electric field and hence an imbalance of anions and cations extending in the form of a *diffuse layer* into the solution.

Owing to the various uncompensated charges at the interface there will be associated changes in the potential, but there are subtleties about what can actually be measured that need some attention.

A2.4.5.1 The electrode potential

Any measurement of potential must describe a reference point, and we will take as this point the potential of an electron well separated from the metal and at rest *in vacuo*. By reference to figure A2.4.8 [16], we can define the following quantities.

(1) The Fermi energy ε_F which is the difference in energy between the bottom of the conduction band and the Fermi level; it is positive and in the simple Sommerfeld theory of metals [17], $\varepsilon_F = h^2 k_F^2 / 2m = \hbar^2 (3\pi^2 n_e)^{2/3} / 2m$ where n_e is the number density of electrons.

(2) The work function Φ^M, which is the energy required to remove an electron from the inner Fermi level to vacuum.

(3) The surface potential of the phase, χ^M, due to the presence of surface dipoles. At the metal–vacuum interface these dipoles arise from the fact that the electrons in the metal can relax at the surface to some degree, extending outwards by a distance of the order of 1 Å, and giving rise to a spatial imbalance of charge at the surface.

(4) The chemical potential of the electrons in the metal, μ_e^M, a *negative* quantity.

(5) The potential energy of the electrons, V, which is a *negative* quantity that can be partitioned into bulk and surface contributions, as shown.

Of the quantities shown in figure A2.4.8 Φ^M is measurable, as is ε_F, but the remainder are not and must be calculated. Values of 1–2 V have been obtained for χ^M, although smaller values are found for the alkali metals.

If two metals with different work functions are placed in contact there will be a flow of electrons from the metal with the *lower* work function to that with the higher work function. This will continue until the *electrochemical* potentials of the electrons in the two phases are equal. This change gives rise to a *measurable* potential difference between the two metals, termed the contact potential or Volta potential difference. Clearly $\Delta_{M_2}^{M_1}\Phi = e_0\Delta_{M_1}^{M_2}\psi$, where $\Delta_{M_1}^{M_2}\psi$ is the Volta potential difference between a point close to the surface of M_1 and that close to the surface of M_2. The actual number of electrons transferred is very small, so the Fermi energies of the two phases will be unaltered, and only the value of the potential V will have changed. If we assume that the χ^M are unaltered as well, and we define the potential *inside* the metal as ϕ, then the equality of electrochemical potentials also leads to

$$-\mu_e^{M_1} + e_0\Delta_{M_2}^{M_1}\phi + \mu_e^{M_2} = 0. \tag{A2.4.77}$$

This internal potential, ϕ, is *not* directly measurable; it is termed the Galvani potential, and is the target of most of the modelling discussed below. Clearly we have $\Delta_{M_2}^{M_1}\phi = \Delta_{M_2}^{M_1}\chi + \Delta_{M_2}^{M_1}\psi$.

Once a metal is immersed in a solvent, a second dipolar layer will form at the metal surface due to the alignment of the solvent dipoles. Again, this contribution to the potential is not directly measurable; in addition, the metal dipole contribution itself will change since the distribution of the electron cloud will be modified by the presence of the solvent. Finally, there will be a contribution from free charges both on the metal and in the electrolyte. The overall contribution to the Galvani potential difference between the metal and solution then consists of these four quantities, as shown in figure A2.4.9 [16]. If the potential due to dipoles at the metal–vacuum interface for the metal is χ^M and for the solvent–vacuum interface is χ^S, then the Galvani potential difference between metal and solvent can be written either as

$$\Delta_S^M\phi = (\chi_M + \delta\chi_M) - (\chi_S + \delta\chi_S) + g(\text{ion}) \equiv g_S^M(\text{dip}) + g(\text{ion}) \tag{A2.4.78}$$

or as

$$\Delta_S^M\phi = \delta_S^M\chi + \Delta_S^M\psi \tag{A2.4.79}$$

where $\delta\chi^M, \delta\chi^S$ are the changes in surface dipole for metal and solvent on forming the interface and the g values are local to the interface. In (A2.4.78) we pass across the interface, and in (A2.4.79) we pass into the vacuum from both the metal and the solvent. As before, the value of $\Delta_S^M\psi$, the Volta potential difference, is measurable experimentally, but it is evident that we cannot associate this potential difference with that due to free charges at the interface, since there are changes in the dipole contribution on both sides as well. Even if there are no free charges at the interface (at the point of zero charge, or PZC), the Volta potential difference is not zero unless $\delta\chi_M = \delta\chi_S$; i.e. the free surfaces of the two phases will still be charged unless the changes in surface dipole of solvent and metal exactly balance. In practice, this is not the case: careful measurements [18] show that $\Delta_{H_2O}^{Hg}\psi = -0.26$ V at the PZC; showing that the dipole changes do not, in fact, compensate. Historically, this discussion is of considerable interest, since a bitter dispute between Galvani and Volta over the origin of the EMF when two different metals are immersed in the same solution could, in principle, be due just to the Volta potential difference between the metals. In fact, it is easy to see that if conditions are such that there are no free charges on either metal, the difference in potential between them, again a measurable quantity, is given by

$$\Delta E_{\sigma=0} = \Delta\Phi + (\Delta_S^{M_1}\psi)_{\sigma=0} - (\Delta_S^{M_2}\psi)_{\sigma=0} \tag{A2.4.80}$$

showing that the difference in work functions would only account for the difference in the electrode potentials if the two Volta terms were actually zero.

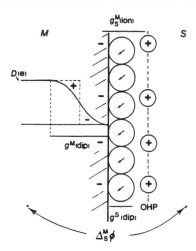

Figure A2.4.9. Components of the Galvani potential difference at a metal–solution interface. From [16].

A2.4.5.2 *Interfacial thermodynamics of the diffuse layer*

Unlike the situation embodied in section A2.4.1, in which the theory was developed in an essentially isotropic manner, the presence of an electrode introduces an essentially non-isotropic element into the equations. Neglecting rotational-dependent interactions, we see that the overall partition function can be written

$$Q(T, V, N) = \frac{q^N}{N!\Lambda^{3N}} \int \cdots \int dr^N \exp\left[-\beta\left\{ \sum_{i<j}^N u(r_{ij}) + \sum_k^N w(x_k) \right\} \right] \tag{A2.4.81}$$

where $w(x_k)$ is the contribution to the potential energy deriving from the electrode itself, and x_k is the distance between the kth particle and the electrode surface. Clearly, if $w(x_k) \to 0$, then we recover the partition function for the isotropic fluid. We will work within the McMillan–Mayer theory of solutions, so we will evaluate (A2.4.81) for ions in a continuum, recognizing the need to replace $u(r_{ij})$ by a potential of mean force. In a similar way, an exact analytical form for $w(x_k)$ is also expected to prove difficult to derive. A complete account of w must include the following contributions.

(1) A short-range contribution, $w^s(x_k)$, which takes into account the nearest distance of approach of the ion to the electrode surface. For ions that do not specifically adsorb this will be the OHP, distance h from the electrode. For ions that do specifically adsorb $w^s(x_k)$ will be more complex, having contributions both from short-range attractive forces and from the energy of de-solvation.

(2) A contribution from the charge on the surface, $w^{(Q_e)}(x_k)$. If this charge density is written Q_e then elementary electrostatic theory shows that $w^{(Q_e)}(x_k)$ will have the unscreened form

$$w^{(Q_e)}(x_k) = \text{constant} + \frac{z_k e_0 Q_e}{\varepsilon\varepsilon_0} x_k. \tag{A2.4.82}$$

(3) An energy of attraction of the ion to its intrinsic image, $w^{(im)}(x_k)$, of unscreened form

$$w^{(im)}(x_k) = \frac{z_k^2 e_0^2}{16\pi\varepsilon\varepsilon_0 x_k}. \tag{A2.4.83}$$

In addition, the energy of interaction between any two ions will contain a contribution from the mirror potential of the second ion; $u(r_{ij})$ is now given by a short-range term and a term of the form

$$u^{(el)}(r_{ij}) = \frac{z_i z_j e_0^2}{4\pi\varepsilon\varepsilon_0}\left(\frac{1}{r_{ij}} - \frac{1}{r_{ij}^*}\right) \tag{A2.4.84}$$

where r_{ij}^* is the distance between ion i and the image of ion j.

Note that there are several implicit approximations made in this model: the most important is that we have neglected the effects of the electrode on orientating the solvent molecules at the surface. This is highly significant: image forces arise whenever there is a discontinuity in the dielectric function and the simple model above would suggest that, at least the layer of solvent next to the electrode should have a dielectric function rather different, especially at low frequencies, from the bulk dielectric function. Implicit in (A2.4.82)–(A2.4.84) is also the fact that ε is assumed independent of x, an assumption again at variance with the simple model presented in A2.4.5.1. In principle, these deficiencies could be overcome by modifying the form of the short-range potentials, but it is not obvious that this will be satisfactory for the description of the image forces, which are intrinsically long range.

The most straightforward development of the above equations has been given by Martynov and Salem [19], who show that to a reasonable approximation in dilute electrolytes:

$$kT \ln(\rho_\alpha^{(1)}(x_\alpha)/\rho_{\alpha0}^{(1)}) + w_\alpha^s(x_\alpha) + z_\alpha e_0\phi(x_\alpha) + z_\alpha^2 e_0^2\delta\varphi(x_\alpha) = 0 \tag{A2.4.85}$$

$$kT \ln(g(r_1, r_j)) + w_{\alpha\beta}^s(R_{ij}) + z_\alpha e_0\psi_\alpha(r_i, r_j) = 0 \tag{A2.4.86}$$

where $\phi(x)$ is the Galvani potential in the electrolyte at distance x from the electrode, $\delta\varphi(x_\alpha)$ is the change in the single-ion mirror-plane potential on moving the ion from infinity to point x_α, $\rho_{\alpha0}^{(1)}$ is the number density of ions of type α at a distance remote from the electrode, and $z_\alpha e_0\psi_\alpha(r_i, r_j)$ is the binary electrostatic potential determined by solution of the relevant Poisson equation:

$$\nabla_j^2\psi_\alpha(r_i, r_j) = -\frac{1}{\varepsilon\varepsilon_0}\left[z_\alpha e_0\delta(R_{ij}) + \sum_\beta z_\beta e_0\rho_\beta^{(1)}(r_j)\left(\frac{g_{\alpha\beta}^{(2)}(r_i, r_j)}{\rho_\alpha^{(1)}(r_i)/\rho_{\alpha0}^{(1)}} - \frac{\rho_\beta^{(1)}(r_j)}{\rho_{\beta0}^{(1)}}\right)\right]. \tag{A2.4.87}$$

The physical meaning of the second term in (A2.4.87) is that the bracket gives the excess concentration of ions β at point r_j given an ion α at point r_i. Finally, we need the Poisson equation for the Galvani potential at distance x from the electrode, which is given by

$$\frac{d^2\phi}{dx^2} = -\frac{\sum_\alpha z_\alpha e_0\rho_\alpha^{(1)}(x)}{\varepsilon\varepsilon_0}. \tag{A2.4.88}$$

By using the expressions for $\rho_\alpha^{(1)}$ and $g_{\alpha\beta}^{(2)}$ from (A2.4.85) and (A2.4.96) in (A2.4.87) and (A2.4.88), solutions may in principle be obtained for the various potentials and charge distributions in the system. The final equations for a dilute electrolyte are

$$\frac{d^2\phi}{dx^2} = -\frac{\sum_\alpha z_\alpha e_0 n_\alpha^0 \exp\{-\beta[w^s(x) + z_\alpha e_0\phi(x) + z_\alpha^2 e_0^2\delta\varphi(x)]\}}{\varepsilon\varepsilon_0} \tag{A2.4.89}$$

where $n_\alpha^0 \equiv \rho_{\alpha 0}^{(1)}$, and which is to be solved under the boundary conditions

$$x = h \qquad \frac{\mathrm{d}\phi}{\mathrm{d}x} = -\frac{Q_\mathrm{e}}{\varepsilon\varepsilon_0}$$

$$x \to \infty \qquad \phi \to 0$$

and

$$\nabla_j^2 \psi_\alpha(\mathbf{r}_i, \mathbf{r}_j) = -\frac{1}{\varepsilon\varepsilon_0}\left\{ z_\alpha e_0 \delta(R_{ij}) + \sum_\beta z_\beta e_0 \rho_\beta^{(1)}(x)[\exp[-(w_{\alpha\beta}^\mathrm{s}(R_{ij}) + z_\alpha e_0 \psi_\alpha(\mathbf{r}_i, \mathbf{r}_j))/kT] - 1] \right\}$$

(A2.4.90)

with the boundary conditions

$$x_j = 0 \qquad \frac{\partial \psi_\alpha}{\partial x_j} = 0$$

$$R_{ij} \to \infty \qquad \psi_\alpha \to 0.$$

Equations (A2.4.89) and (A2.4.90) are the most general equations governing the behaviour of an electrolyte near an electrode, and solving them would, in principle, give a combined DH ionic atmosphere and a description of the ionic distribution around each electrode.

The zeroth-order solution to the above equations is the Goüy–Chapman theory dating from the early part of the 20th century [20]. In this solution, the ionic atmosphere is ignored, as is the mirror image potential for the ion. Equation (A2.4.90) can therefore be ignored and equation (A2.4.89) reduces to

$$\frac{\mathrm{d}^2\phi}{\mathrm{d}x^2} = -\frac{\sum_\alpha z_\alpha e_0 n_\alpha^0 \exp\{-\beta[z_\alpha e_0 \phi(x)]\}}{\varepsilon\varepsilon_0}$$

(A2.4.91)

where we have built in the further assumption that $w^\mathrm{s}(x) = 0$ for $x > h$ and $w^\mathrm{s}(x) = \infty$ for $x < h$. This corresponds to the hard-sphere model introduced above. Whilst (A2.4.91) can be solved for general electrolyte solutions, a solution in closed form can be most easily obtained for a 1–1 electrolyte with ionic charges $\pm z$. Under these circumstances, (A2.4.91) reduces to

$$\frac{\mathrm{d}^2\phi}{\mathrm{d}x^2} = \frac{2z e_0 n^0}{\varepsilon\varepsilon_0} \sinh\left(\frac{z e_0}{kT}\phi\right)$$

(A2.4.92)

where we have assumed that $n_+^0 = n_-^0 = n^0$. Integration under the boundary conditions above gives:

$$Q_\mathrm{e} = (8kT\varepsilon\varepsilon_0 n^0)^{1/2} \sinh\left(\frac{z e_0 \phi(h)}{2kT}\right)$$

(A2.4.93)

$$\phi(h) = \frac{2kT}{z e_0} \sinh^{-1}\left(\frac{Q_\mathrm{e}}{(8kT\varepsilon\varepsilon_0 n^0)^{1/2}}\right)$$

(A2.4.94)

$$\phi(x) = \frac{4kT}{z e_0} \tanh^{-1}\left\{ e^{-\kappa(x-h)} \tanh\left(\frac{z e_0 \phi(h)}{4kT}\right)\right\}$$

(A2.4.95)

and κ has the same meaning as in the DH theory, $\kappa^2 = 2z^2 e_0^2 n^0/\varepsilon\varepsilon_0 kT$. These are the central results of the Goüy–Chapman theory. Clearly, if $z e_0 \phi(h)/4kT$ is small then $\phi(x) \sim \phi(h) e^{-\kappa(x-h)}$ and the potential decays exponentially in the bulk of the electrolyte. The basic physics is similar to the DH analysis, in that the actual field due to the electrode becomes screened by the ionic charges in the electrolyte.

A better approximation may be obtained by expansion of ϕ and ψ_α in powers of the dimensionless variable $\overline{q} = (ze_0^2/\varepsilon\varepsilon_0 kT)\kappa$. If $\phi \approx \phi_0 + \overline{q}\phi_1$ and $\psi \approx \overline{q}\psi_1$, then it is possible to show that

$$\nabla_j^2 \psi_1(r_i, r_j) - \kappa^2 \psi_1(r_i, r_j) = -\frac{ze_0}{\varepsilon\varepsilon_0}\delta(R_{ij}) \tag{A2.4.96}$$

and

$$\frac{d^2\phi_1}{dx^2} - \kappa^2\phi_1 = -\frac{e_0}{kT}\phi_0\delta\varphi \tag{A2.4.97}$$

where $\delta\varphi = kT e^{-2\kappa z}/4e_0\kappa z$ is the screened image potential. Solutions to equations (A2.4.96 and A2.4.97) have been obtained, and it is found that, for a given Q_e, the value of $\phi(h)$ is always smaller than that predicted by the Goüy–Chapman theory. This theory, both in the zeroth-order and first-order analyses given here, has been the subject of considerable analytical investigation [15], but there has been relatively little progress in devizing more accurate theories, since the approximations made even in these simple derivations are very difficult to correct accurately.

A2.4.5.3 Specific ionic adsorption and the inner layer

Interaction of the water molecules with the electrode surface can be developed through simple statistical models. Clearly for water molecules close to the electrode surface, there will be several opposing effects: the hydrogen bonding, tending to align the water molecules with those in solution; the electric field, tending to align the water molecules with their dipole moments perpendicular to the electrode surface; and dipole–dipole interactions, tending to orient the nearest-neighbour dipoles in opposite directions. Simple estimates [21] based on 20 kJ mol^{-1} for each hydrogen bond suggest that the orientation energy pE becomes comparable to this for $E \sim 5 \times 10^9 \text{ V m}^{-1}$; such field strengths will be associated with surface charges of the order of 0.2–0.3 C m^{-2} or 20–$30 \text{ }\mu\text{C cm}^{-2}$ assuming $p = 6.17 \times 10^{-30} \text{ C m}$ and the dielectric function for water at the electrode surface of about six. This corresponds to all molecules being strongly oriented. These are comparable to the fields expected at reasonably high electrode potentials. Similarly, the energy of interaction of two dipoles lying antiparallel to each other is $-p^2/(4\pi\varepsilon_0 E^3)$; for $R \sim 4 \text{ Å}$ the orientational field needs to be in excess of 10^9 V m^{-1}, a comparable number.

The simplest model for water at the electrode surface has just two possible orientations of the water molecules at the surface, and was initially described by Watts-Tobin [22]. The associated potential drop is given by

$$g(\text{dip}) = -\frac{(N_+ - N_-)p}{\varepsilon\varepsilon_0} \tag{A2.4.98}$$

and if the total potential drop across the inner region of dimension h_i is $\Delta\phi$:

$$N_+/N_- = \exp[-(U_0 - 2p\Delta\phi/h_i)/kT] \tag{A2.4.99}$$

where U_0 is the energy of interaction between neighbouring dipoles. A somewhat more sophisticated model is to assume that water is present in the form of both monomers and dimers, with the dimers so oriented as not to give any nett contribution to the value of $g(\text{dip})$.

A further refinement has come with the work of Parsons [23], building on an analysis by Damaskhin and Frumkin [24]. Parsons suggested that the solvent molecules at the interface could be thought of as being either free or associated as clusters. In a second case the nett dipole moment would be reduced from the value found for perpendicular alignment since the clusters would impose their own alignment. The difficulty with such models is that the structure of the clusters themselves is likely to be a function of the electric field, and simulation methods show, see below, that this is indeed the case.

The experimental data and arguments by Trassatti [25] show that at the PZC, the water dipole contribution to the potential drop across the interface is relatively small, varying from about 0 V for Au to about 0.2 V for In and Cd. For transition metals, values as high as 0.4 V are suggested. The basic idea of water clusters on the electrode surface dissociating as the electric field is increased has also been supported by *in situ* Fourier transform infrared (FTIR) studies [26], and this model also underlies more recent statistical mechanical studies [27].

The model of the inner layer suggests that the interaction energy of water molecules with the metal will be at a minimum somewhere close to the PZC, a result strongly supported by the fact that adsorption of less polar organic molecules often shows a maximum at this same point [18]. However, particularly at anodic potentials, there is now strong evidence that simple anions may lose part of their hydration or solvation sheath and migrate from the OHP to the IHP. There is also evidence that some larger cations, such as $[R_4N]^+$, Tl^+ and Cs^+ also undergo specific adsorption at sufficiently negative potentials. The evidence for specific adsorption comes not only from classical experiments in which the surface tension of mercury is studied as a function of the potential (electrocapillarity), and the coverage derived from rather indirect reasoning [28], but also more direct methods, such as the measurement of the amount of material removed from solution, using radioactive tracers and ellipsometry. A critical problem is much of this work, particularly in those data derived from electrocapillarity, is that the validity of the Goüy–Chapman model must be assumed, an assumption that has been queried. The calculation of the free energy change associated with this process is not simple, and the following effects need to be considered.

(1)   The energy gained on moving from the OHP to the IHP. The electrostatic part of this will have the form $(z_k e_0 Q_e / \varepsilon \varepsilon_0)(x_{OHP} - x_{IHP})$, but the de-solvation part is much more difficult to estimate.

(2)   The fact that more than one molecule of water may be displaced for each anion adsorbed, and that the adsorption energy of these water molecules will show a complex dependence on the electrode potential.

(3)   The fact that a chemical bond may form between a metal and an anion, leading to, at least, a partial discharge of the ion.

(4)   The necessity to calculate the electrostatic contribution to both the ion–electrode attraction and the ion–ion repulsion energies, bearing in mind that there are at least two dielectric function discontinuities in the simple double-layer model above.

(5)   That short-range contributions to both the ion–ion and ion–electrode interactions must be included.

These calculations have, as their aim, the generation of an *adsorption isotherm*, relating the concentration of ions in the solution to the coverage in the IHP and the potential (or more usually the *charge*) on the electrode. No complete calculations have been carried out incorporating all the above terms. In general, the analytical form for the isotherm is

$$\ln(f(\theta)) = \text{constant} + \ln a_\pm + AQ_e + g(\theta) \qquad \text{(A2.4.100)}$$

where $f(\theta)$ and $g(\theta)$ are functions of the coverage. For models where lateral interactions are dominant, $g(\theta)$ will have a $\theta^{1/2}$ dependence: if multiple electrostatic imaging is important, a term linear in θ will be found. Whereas, if dispersion interactions between ions on the surface are important, then a term in θ^3 becomes significant. The form of $f(\theta)$ is normally taken as $\theta/(1 - \theta)^p$ where p is the number of water molecules displaced by one adsorbed ion. Details of the various isotherms are given elsewhere [28], but modern simulation methods, as reviewed below, are needed to make further progress.

A2.4.5.4   Simulation techniques

The theoretical complexity of the models discussed above and the relative difficulty of establishing unequivocal models from the experimental data available has led to an increasing interest in Monte Carlo and, particularly,

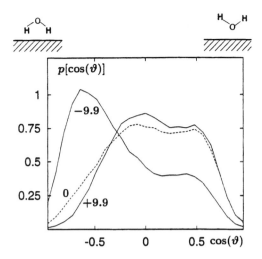

Figure A2.4.10. Orientational distribution of the water dipole moment in the adsorbate layer for three simulations with different surface charge densities (in units of $\mu C\ cm^{-2}$ as indicated). In the figure $\cos\theta$ is the angle between the water dipole vector and the surface normal that points into the aqueous phase. From [30].

molecular dynamics approaches. Such studies have proved extremely valuable in establishing quite independent models of the interface against which the theories described above can be tested. In particular, these simulation techniques allow a more realistic explicit treatment of the solvent and ions on an equal footing. Typically, the solvent is treated within a rigid multipole model, in which the electrical distribution is modelled by a rigid distribution of charges on the various atoms in the solvent molecule. Dispersion and short-range interactions are modelled using Lennard-Jones or similar model potentials, and the interaction of water with the metal surface is generally modelled with a corrugated potential term to take account of the atomic structure of the metal. Such potentials are particularly marked for metal–oxygen interactions. In the absence of an electrical charge on the electrode the Pt–O interaction energy is usually given by an expression of the form

$$U_{Pt-O} = [A\,e^{-\alpha r} - B\,e^{-\beta r}]f(x, y) + C\,e^{-\gamma r}[1 - f(x, y)] \qquad (A2.4.101)$$

where $f(x, y) = e^{-\lambda(x^2+y^2)}$ and the Pt–H interaction is weakly repulsive, of the form

$$U_{Pt-H} = D\,e^{-\mu r}. \qquad (A2.4.102)$$

This potential will lead to a single water molecule adsorbing at the PZC on Pt with the dipole pointing *away* from the surface and the oxygen atom pointing directly at a Pt-atom site (on-top configuration).

The main difficulty in these simulations is the long-range nature of the Coulomb interactions, since both mirror-plane images and real charges must be included, and the finite nature of the simulated volume must also be included. A more detailed discussion is given by Benjamin [29], and the following conclusions have been reached.

(1) Only at extremely high electric fields are the water molecules fully aligned at the electrode surface. For electric fields of the size normally encountered, a distribution of dipole directions is found, whose half-width is strongly dependent on whether specific adsorption of ions takes place. In the absence of such adsorption the distribution function steadily narrows, but in the presence of adsorption the distribution may show little change from that found at the PZC; an example is shown in figure A2.4.10 [30].

(2) The pair correlation functions g_{OO}, g_{OH} and g_{HH} have been obtained for water on an uncharged electrode surface. For Pt(100), the results are shown in figure A2.4.11 [29], and compared to the correlation

functions for the second, much more liquid-like layer. It is clear that the first solvation peak is enhanced by comparison to the liquid, but is in the same position and emphasizing the importance of hydrogen bonding in determining nearest O–O distances: however, beyond the first peak there are new peaks in the pair correlation function for the water layer immediately adjacent to the electrode that are absent in the liquid, and result from the periodicity of the Pt surface. By contrast, these peaks have disappeared in the second layer, which is very similar to normal liquid water.

(3) Simulation results for turning on the electric field at the interface in a system consisting of a water layer between two Pt electrodes 3 nm apart show that the dipole density initially increases fairly slowly, but that between 10 and 20 V nm^{-1} there is an apparent phase transition from a moderately ordered structure, in which the ordering is confined close to the electrodes only, to a substantially ordered layer over the entire 3 nm thickness. Effectively, at this field, which corresponds to the energy of about four hydrogen bonds, the system loses all the ordering imposed by the hydrogen bonds and reverts to a purely linear array of dipoles.

(4) For higher concentrations of aqueous electrolyte, the simulations suggest that the ionic densities do not change monotonically near the electrode surface, as might be expected from the Goüy–Chapman analysis above, but oscillate in the region $x < 10$ Å. This oscillation is, in part, associated with the oscillation in the oxygen atom density caused by the layering effect occurring in liquids near a surface.

(5) At finite positive and negative charge densities on the electrode, the counterion density profiles often exhibit significantly higher maxima, i.e. there is an overshoot, and the derived potential actually shows oscillations itself close to the electrode surface at concentrations above about 1 M.

(6) Whether the potentials are derived from quantum mechanical calculations or classical image forces, it is quite generally found that there is a stronger barrier to the adsorption of cations at the surface than anions, in agreement with that generally found experimentally.

A2.4.6 Thermodynamics of electrified interfaces

If a metal, such as copper, is placed in contact with a solution containing the ions of that metal, such as from aqueous copper sulphate, then we expect an equilibrium to be set up of the following form:

$$Cu^0 \leftrightarrow Cu^{2+} + 2e_M^- \tag{A2.4.103}$$

where the subscript M refers to the metal. As indicated above, there will be a potential difference across the interface, between the Galvani potential in the interior of the copper and that in the interior of the electrolyte. The effects of this potential difference must be incorporated into the normal thermodynamic equations describing the interface, which is done, as above, by defining the *electrochemical potential* of a species with charge $z_i e_0$ per ion or $z_i F$ per mole. If one mole of z-valent ions is brought from a remote position to the interior of the solution, in which there exists a potential ϕ, then the work done will be $zF\phi$; this work term must be added to or subtracted from the free energy per mole, μ, depending on the relative signs of the charge and ϕ, and the condition for equilibrium for component i partitioned between two phases with potentials $\phi(I)$ and $\phi(II)$ is

$$\mu_i(I) + z_i F\phi(I) = \mu_i(II) + z_i F\phi(II) \tag{A2.4.104}$$

where $\phi(I)$ and $\phi(II)$ are the Galvani or inner potentials in the interior of phases (I) and (II). The expression $\mu_i + z_i F$ is referred to as the electrochemical potential, $\tilde{\mu}_i$. We have

$$\tilde{\mu}_i = \mu_i + z_i F\phi = \mu_i^0 + RT \ln a_i + z_i F\phi \tag{A2.4.105}$$

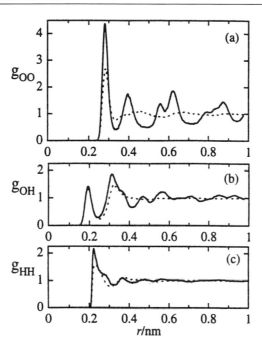

Figure A2.4.11. Water pair correlation functions near the Pt(100) surface. In each panel, the full curve is for water molecules in the first layer, and the broken curve is for water molecules in the second layer. From [30].

and the condition for electrochemical equilibrium can be written for our copper system:

$$\tilde{\mu}_{Cu}(M) = \tilde{\mu}_{Cu^{2+}}(S) + 2\tilde{\mu}_{e^-}(M) \tag{A2.4.106}$$

where the labels M and S refer to the metal and to the solution respectively. Assuming the copper atoms in the metal to be neutral, so that $\tilde{\mu}_{Cu^0} = \mu_{Cu^0}$, we then have

$$\mu_{Cu^0}^0(M) + RT \ln(a_{Cu}(M)) = \mu_{Cu^{2+}}^0(S) + RT \ln(a_{Cu^{2+}}) + 2F\phi_S + \mu_{e^-}^0(M) + 2RT \ln(a_{e^-}) - 2F\phi_M \tag{A2.4.107}$$

Given that the concentration of both the copper atoms and the electrons in the copper metal will be effectively constant, so that two of the activity terms can be neglected, we finally have, on rearranging (A2.4.107),

$$\Delta\phi \equiv \phi_M - \phi_S = \frac{\mu_{Cu^{2+}}^0(S) + \mu_{e^-}^0(M) - \mu_{Cu^0}^0(M)}{2F} + \frac{RT}{2F} \ln(a_{Cu^{2+}}) \equiv \Delta\phi_0 + \left(\frac{RT}{2F}\right) \ln(a_{Cu^{2+}}) \tag{A2.4.108}$$

where $\Delta\phi_0$ is the Galvani potential difference at equilibrium between the electrode and the solution in the case where $a_{Cu^{2+}}(aq) = 1$, and is referred to as the standard Galvani potential difference. It can be seen, in general, that the Galvani potential difference will alter by a factor of $(RT/zF) \ln 10 \equiv 0.059/z$ V at 298 K, for every order of magnitude change in activity of the metal ion, where z is the valence of the metal ion in solution.

A2.4.6.1 The Nernst equation for redox electrodes

In addition to the case of a metal in contact with its ions in solution there are other cases in which a Galvani potential difference between two phases may be found. One case is the immersion of an inert electrode, such as platinum metal, into an electrolyte solution containing a substance 'S' that can exist in either an oxidized or reduced form through the loss or gain of electrons from the electrode. In the simplest case, we have

$$S_{ox} + e^- \leftrightarrow S_{red} \qquad (A2.4.109)$$

an example being

$$Fe^{3+} + e^- \leftrightarrow Fe^{2+} \qquad (A2.4.110)$$

where the physical process described is the exchange of *electrons* (not ions) between the electrolyte and the electrode: at no point is the electron conceived as being free in the solution. The equilibrium properties of the redox reaction (A2.4.109) can, in principle, be treated in the same way as above. At equilibrium, once a double layer has formed and a Galvani potential difference set up, we can write

$$\tilde{\mu}_{S_{ox}} + n\tilde{\mu}_{e^-}^M = \mu_{S_{red}} \qquad (A2.4.111)$$

and, bearing in mind that the positive charge on 'ox' must exceed 'red' by $|ne^-|$ if we are to have electroneutrality, then (A2.4.111) becomes

$$\mu_{S_{ox}}^0 + RT \ln(a_{S_{ox}}) + nF\phi_S + n\mu_{e^-}^0 - nF\phi_M = \mu_{S_{red}}^0 + RT \ln(a_{S_{red}}) \qquad (A2.4.112)$$

whence

$$\Delta\phi = \phi_M - \phi_S = \frac{\mu_{S_{ox}}^0 + n\mu_{e^-}^0 - \mu_{S_{red}}^0}{nF} + RT \ln\left(\frac{a_{S_{ox}}}{a_{S_{red}}}\right) \equiv \Delta\phi^0 + RT \ln\left(\frac{a_{S_{ox}}}{a_{S_{red}}}\right) \qquad (A2.4.113)$$

where the standard Galvani potential difference is now defined as that for which the activities of S_{ox} and S_{red} are equal. As can be seen, an alteration of this ratio by a factor of ten leads to a change of $0.059/n$ V in $\Delta\phi$ at equilibrium. It can also be seen that $\Delta\phi$ will be independent of the magnitudes of the activities provided that their ratio is a constant.

For more complicated redox reactions, a general form of the Nernst equation may be derived by analogy with (A2.4.113). If we consider a stoichiometric reaction of the following type:

$$\nu_1 S_1 + \nu_2 S_2 + \cdots + \nu_i S_i + ne^- \leftrightarrow \nu_j S_j + \cdots \nu_p S_p \qquad (A2.4.114)$$

which can be written in the abbreviated form

$$\sum_{ox} \nu_{ox} S_{ox} + ne^- \leftrightarrow \sum_{red} S_{red} \qquad (A2.4.115)$$

then straightforward manipulation leads to the generalized Nernst equation:

$$\Delta\phi = \Delta\phi^0 + \frac{RT}{nF} \ln\left(\frac{\prod_{ox} a_{ox}^{\nu_{ox}}}{\prod_{red} a_{red}^{\nu_{red}}}\right) \qquad (A2.4.116)$$

where the notation

$$\prod_{ox} a_{ox}^{\nu_{ox}} = a_{S_{ox_1}}^{\nu_{ox_1}} a_{S_{ox_2}}^{\nu_{ox_2}} \cdots a_{S_{ox_i}}^{\nu_{ox_i}}. \qquad (A2.4.117)$$

As an example, the reduction of permanganate in acid solution follows the equation

$$MnO_4^- + 8H_3O^+ + 5e^- \leftrightarrow Mn^{2+} + 12H_2O \qquad (A2.4.118)$$

and the potential of a platinum electrode immersed in a solution containing both permanganate and Mn^{2+} is given by

$$\Delta\phi = \Delta\phi^0 + \frac{RT}{nF} \ln\left(\frac{a_{MnO_4^-} a_{H_3O^+}^8}{a_{Mn^{2+}}}\right) \qquad (A2.4.119)$$

assuming that the activity of neutral H_2O can be put equal to unity.

A2.4.6.2   The Nernst equation for gas electrodes

The Nernst equation above for the dependence of the equilibrium potential of redox electrodes on the activity of solution species is also valid for uncharged species in the gas phase that take part in electron exchange reactions at the electrode–electrolyte interface. For the specific equilibrium process involved in the reduction of chlorine:

$$Cl_2 + 2e^- \leftrightarrow 2Cl^- \tag{A2.4.120}$$

the corresponding Nernst equation can easily be shown to be

$$\Delta\phi = \Delta\phi^0 + \frac{RT}{2F} \ln\left(\frac{a_{Cl_2}(aq)}{a_{Cl^-}^2}\right) \tag{A2.4.121}$$

where a_{Cl_2} (aq) is the activity of the chlorine gas dissolved in water. If the Cl_2 solution is in equilibrium with chlorine at pressure p_{Cl_2} in the gas phase, then

$$\mu_{Cl_2}(gas) = \mu_{Cl_2}(aq). \tag{A2.4.122}$$

Given that $\mu_{Cl_2}(gas) = \mu_{Cl_2}^0(gas) + RT \ln(p_{Cl_2}/p^0)$ and $\mu_{Cl_2}(aq) = \mu_{Cl_2}^0(aq) + RT \ln(a_{Cl_2}(aq))$, where p^0 is the standard pressure of 1 atm ($\equiv 101\,325$ Pa), then it is clear that

$$a_{Cl_2}(aq) = \left(\frac{p_{Cl_2}}{p^0}\right) \exp\left(\frac{\mu_{Cl_2}^0(gas) - \mu_{Cl_2}^0(aq)}{RT}\right) \tag{A2.4.123}$$

and we can write

$$\Delta\phi = \Delta\phi^{0'} + \left(\frac{RT}{2F}\right) \ln\left(\frac{p_{Cl_2}}{p^0 a_{Cl^-}^2}\right) \tag{A2.4.124}$$

where $\Delta\phi^{0'}$ is the Galvani potential difference under the standard conditions of $p_{Cl_2} = p^0$ and $a_{Cl^-} = 1$.

A2.4.6.3   The measurement of electrode potentials and cell voltages

Although the results quoted above are given in terms of the Galvani potential difference between a metal electrode and an electrolyte solution, direct measurement of this Galvani potential difference between an electrode and an electrolyte is not possible, since any voltmeter or similar device will incorporate unknowable surface potentials into the measurement. In particular, any contact of a measurement probe with the solution phase will have to involve a second phase boundary between the metal and the electrolyte somewhere; at this boundary an electrochemical equilibrium will be set up and with it a second equilibrium Galvani potential difference, and the overall potential difference measured by this instrument will in fact be the difference of two Galvani voltages at the two interfaces. In other words, even at zero current, the actual EMF measured for a galvanic cell will be the difference between the two Galvani voltages $\Delta\phi(I)$ and $\Delta\phi(II)$ for the two interfaces, as shown in figure A2.4.12 [7].

Figure A2.4.12 shows the two possibilities that can exist, in which the Galvani potential of the solution, ϕ_S, lies between $\phi(I)$ and $\phi(II)$ and in which it lies below (or, equivalently, above) the Galvani potentials of the metals. It should be emphasized that figure A2.4.12 is highly schematic: in reality the potential near the phase boundary in the solution changes initially linearly and then exponentially with distance away from the electrode surface, as we saw above. The other point is that we have assumed that ϕ_S is a constant in the region between the two electrodes. This will only be true provided the two electrodes are immersed in the same solution and that no current is passing.

It is clear from figure A2.4.12 that the EMF or potential difference, E, between the two metals is given by

$$E = \Delta\phi(II) - \Delta\phi(I) = \phi(II) - \phi(I) \tag{A2.4.125}$$

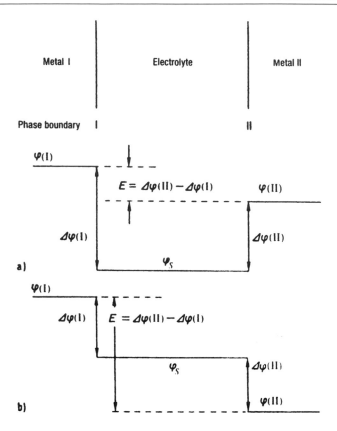

Figure A2.4.12. The EMF of a galvanic cell as the difference between the equilibrium Galvani potentials at the two electrodes: (a) $\Delta\phi(\mathrm{I}) > 0$, $\Delta\phi(\mathrm{II}) > 0$ and (b) $\Delta\phi(\mathrm{I}) > 0$, $\Delta\phi(\mathrm{II}) < 0$. From [7].

where we adopt the normal electrochemical convention that the EMF is always equal to the potential on the metal on the right of the figure minus the potential of the metal on the left. It follows that once the Galvani potential of any one electrode is known it should be possible, at least in principle, to determine the potentials for all other electrodes. In practice, since the Galvani potential of no single electrode is known the method adopted is to arbitrarily choose one *reference* electrode and assign a value for its Galvani potential. The choice actually made is that of the hydrogen electrode, in which hydrogen gas at one atmosphere pressure is bubbled over a platinized platinum electrode immersed in a solution of unit H_3O^+ activity. From the discussion in section A2.4.6.2, it will be clear that provided an equilibrium can be established rapidly for such an electrode, its Galvani potential difference will be a constant, and changes in the measured EMF of the complete cell as conditions are altered at the other electrode will actually reflect the changes in the Galvani potential difference of that electrode.

Cells need not necessarily contain a reference electrode to obtain meaningful results; as an example, if the two electrodes in figure A2.4.12 are made from the same metal, M, but these are now in contact with two solutions of the same metal ions, M^{z+} but with differing ionic activities, which are separated from each other by a glass frit that permits contact, but impedes diffusion, then the EMF of such a cell, termed a concentration cell, is given by

$$E = \Delta\phi(\mathrm{II}) - \Delta\phi(\mathrm{I}) = \frac{RT}{zF} \ln\left(\frac{a_{M^{z+}}(\mathrm{II})}{a_{M^{z+}}(\mathrm{I})}\right). \qquad (A2.4.126)$$

Equation (A2.4.126) shows that the EMF increases by $0.059/z$ V for each decade change in the activity ratio in the two solutions.

A2.4.6.4 Conventions in the description of cells

In order to describe any electrochemical cell a convention is required for writing down the cells, such as the concentration cell described above. This convention should establish clearly where the boundaries between the different phases exist and, also, what the overall cell reaction is. It is now standard to use vertical lines to delineate phase boundaries, such as those between a solid and a liquid or between two immiscible liquids. The junction between two miscible liquids, which might be maintained by the use of a porous glass frit, is represented by a single vertical dashed line, ¦, and two dashed lines, ¦¦, are used to indicate two liquid phases joined by an appropriate electrolyte bridge adjusted to minimize potentials arising from the different diffusion coefficients of the anion and cation (so-called 'junction potentials').

The cell is written such that the cathode is to the right when the cell is acting in the galvanic mode, and electrical energy is being generated from the electrochemical reactions at the two electrodes. From the point of view of external connections, the cathode will appear to be the positive terminal, since electrons will travel in the external circuit from the anode, where they pass from electrolyte to metal, to the cathode, where they pass back into the electrolyte. The EMF of such a cell will then be the difference in the Galvani potentials of the metal electrodes on the right-hand side and the left-hand side. Thus, the concentration cell of section A2.4.6.3 would be represented by $M|M^{z+}(I)|M^{z+}(II)|M$.

In fact, some care is needed with regard to this type of concentration cell, since the assumption implicit in the derivation of (A2.4.126) that the potential in the solution is constant between the two electrodes, cannot be entirely correct. At the phase boundary between the two solutions, which is here a semi-permeable membrane permitting the passage of water molecules but not ions between the two solutions, there will be a potential jump. This so-called liquid-junction potential will increase or decrease the measured EMF of the cell depending on its sign. Potential jumps at liquid–liquid junctions are in general rather small compared to normal cell voltages, and can be minimized further by suitable experimental modifications to the cell.

If two redox electrodes both use an inert electrode material such as platinum, the cell EMF can be written down immediately. Thus, for the hydrogen/chlorine fuel cell, which we represent by the cell $H_2(g)|Pt|HCl(m)|Pt|Cl_2(g)$ and for which it is clear that the cathodic reaction is the reduction of Cl_2 as considered in section A2.4.6.2:

$$E = \Delta\phi(Cl_2/Cl^-) - \Delta\phi(H_2/H_3O^+(aq))$$
$$= E^0 - \left(\frac{RT}{2F}\right)\ln(a_{H_3O}^2 a_{Cl^-}^2) + \left(\frac{RT}{2F}\right)\ln\{(p_{H_2}/p^0)/(p_{Cl_2}/p^0)\} \qquad (A2.4.127)$$

where E^0 is the standard EMF of the fuel cell, or the EMF at which the activities of H_3O^+ and Cl^- are unity and the pressures of H_2 and Cl_2 are both equal to the standard pressure, p^0.

A2.4.7 Electrical potentials and electrical current

The discussion in earlier sections has focussed, by and large, on understanding the equilibrium structures in solution and at the electrode–electrolyte interface. In this last section, some introductory discussion will be given of the situation in which we depart from equilibrium by permitting the flow of electrical current through the cell. Such current flow leads not only to a potential drop across the electrolyte, which affects the cell voltage by virtue of an ohmic drop IR_i (where R_i is the internal resistance of the electrolyte between the electrodes), but each electrode exhibits a characteristic current–voltage behaviour, and the overall cell voltage will, in general, reflect both these effects.

A2.4.7.1   The concept of overpotential

Once current passes through the interface, the Galvani potential difference will differ from that expected from the Nernst equation above; the magnitude of the difference is termed the *overpotential*, which is defined heuristically as

$$\eta = \Delta\phi - \Delta\phi_r = E - E_r \tag{A2.4.128}$$

where the subscript r refers to the 'rest' situation, i.e. to the potential measured in the absence of any current passing. Provided equilibrium can be established, this rest potential will correspond to that predicted by the Nernst equation. Obviously, the sign of η is determined by whether E is greater than or less than E_r.

At low currents, the rate of change of the electrode potential with current is associated with the limiting rate of electron transfer across the phase boundary between the electronically conducting electrode and the ionically conducting solution, and is termed the electron transfer overpotential. The electron transfer rate at a given overpotential has been found to depend on the nature of the species participating in the reaction, and the properties of the electrolyte and the electrode itself (such as, for example, the chemical nature of the metal). At higher current densities, the primary electron transfer rate is usually no longer limiting; instead, limitations arise through the slow transport of reactants from the solution to the electrode surface or, conversely, the slow transport of the product away from the electrode (diffusion overpotential) or through the inability of chemical reactions coupled to the electron transfer step to keep pace (reaction overpotential).

Examples of the latter include the adsorption or desorption of species participating in the reaction or the participation of chemical reactions before or after the electron transfer step itself. One such process occurs in the evolution of hydrogen from a solution of a weak acid, HA: in this case, the electron transfer from the electrode to the proton in solution must be preceded by the acid dissociation reaction taking place in solution.

A2.4.7.2   The theory of electron transfer

The rate of simple chemical reactions can now be calculated with some confidence either within the framework of activated-complex theory or directly from quantum mechanical first principles, and theories that might lead to analogous predictions for simple electron transfer reactions at the electrode–electrolyte interface have been the subject of much recent investigation. Such theories have hitherto been concerned primarily with greatly simplified models for the interaction of an ion in solution with an inert electrode surface. The specific adsorption of electroactive species has been excluded and electron transfer is envisaged only as taking place between the strongly solvated ion in the outer Helmholtz layer and the metal electrode. The electron transfer process itself can only be understood through the formalism of quantum mechanics, since the transfer itself is a tunnelling phenomenon that has no simple analogue in classical mechanics.

Within this framework, by considering the physical situation of the electrode double layer, the free energy of activation of an electron transfer reaction can be identified with the reorganization energy of the solvation sheath around the ion. This idea will be carried through in detail for the simple case of the strongly solvated Fe^{3+}/Fe^{2+} couple, following the change in the ligand–ion distance as the critical reaction variable during the transfer process.

In aqueous solution, the oxidation of Fe^{2+} can be conceived as a reaction of two aquo-complexes of the form

$$[Fe(H_2O)_6]^{2+} \leftrightarrow [Fe(H_2O)_6]^{3+} + e^-. \tag{A2.4.129}$$

The H_2O molecules of these aquo-complexes constitute the inner solvation shell of the ions, which are, in turn, surrounded by an external solvation shell of more or less uncoordinated water molecules forming part of the water continuum, as described in section A2.4.2 above. Owing to the difference in the solvation energies, the radius of the Fe^{3+} aquo-complex is smaller than that of Fe^{2+}, which implies that the mean distance of the vibrating water molecules at their normal equilibrium point must change during the electron transfer. Similarly, changes must take place in the outer solvation shell during electron transfer, all of which implies

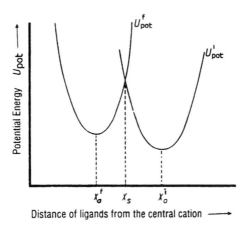

Figure A2.4.13. Potential energy of a redox system as a function of ligand–metal separation. From [7].

that the solvation shells themselves inhibit electron transfer. This inhibition by the surrounding solvent molecules in the inner and outer solvation shells can be characterized by an activation free energy $\Delta G^{\ddagger}$.

Given that the tunnelling process itself requires no activation energy, and that tunnelling will take place at some particular configuration of solvent molecules around the ion, the entire activation energy referred to above must be associated with ligand/solvent movement. Furthermore, from the Franck–Condon principle, the electron tunnelling process will take place on a rapid time scale compared to nuclear motion, so that the ligand and solvent molecules will be essentially stationary during the actual process of electron transfer.

Consider now the aquo-complexes above, and let x be the distance of the centre of mass of the water molecules constituting the inner solvation shell from the central ion. The binding interaction of these molecules leads to vibrations of frequency $f = \omega/2\pi$ taking place about an equilibrium point x_0 and, if the harmonic approximation is valid, the potential energy change U_{pot} associated with the ligand vibration can be written in parabolic form as

$$U_{\mathrm{pot}} = \tfrac{1}{2}M\omega^2(x - x_0)^2 + B + U_{\mathrm{el}} \qquad (A2.4.130)$$

where M is the mass of the ligands, B is the binding energy of the ligands and U_{el} is the electrical energy of the ion-electrode system. The total energy of the system will also contain the kinetic energy of the ligands, written in the form $p^2/2M$, where p is the momentum of the molecules during vibrations:

$$U_{\mathrm{tot}} = \frac{p^2}{2M} + \frac{1}{2}M\omega^2(x - x_0)^2 + B + U_{\mathrm{el}}. \qquad (A2.4.131)$$

It is possible to write two such equations for the initial state, i, (corresponding to the reduced aquo-complex $[\mathrm{Fe(H_2O)_6}]^{2+}$) and the final state, f, corresponding to the oxidized aquo-complex and the electron now present in the electrode. Clearly

$$U_{\mathrm{tot}}^{i} = \frac{p^2}{2M} + \frac{1}{2}M\omega^2(x - x_0^i)^2 + B^i + U_{\mathrm{el}}^{i} \qquad (A2.4.132)$$

with a corresponding equation for state f, and with the assumption that the frequency of vibration does not alter between the initial and final states of the aquo-complex. During electron transfer, the system moves, as shown in figure A2.4.13 ([7]), from an equilibrium situation centred at x_0 along the parabolic curve labelled U_{pot}^{i} to the point x_s where electron transfer takes place; following this, the system will move along the curve labelled U_{pot}^{f} to the new equilibrium situation centred on x_0^f.

The point at which electron transfer takes place clearly corresponds to the condition $U^i_{pot} = U^f_{pot}$; equating equations (A2.4.132) for the states i and f we find that

$$x_s = \frac{B^f + U^f_{el} - B^i - U^i_{el} + (M\omega^2/2)([x^0_f]^2 - [x^0_i]^2)}{M\omega^2(x^f_0 - x^i_0)}. \tag{A2.4.133}$$

The activation energy, U_{act}, is defined as the minimum additional energy above the zero-point energy that is needed for a system to pass from the initial to the final state in a chemical reaction. In terms of equation (A2.4.132), the energy of the initial reactants at $x = x_s$ is given by

$$U^i = \frac{p^2}{2M} + \frac{1}{2}M\omega^2(x_s - x^i_0)^2 + B^i + U^i_{el} \tag{A2.4.134}$$

where $B^i + U^i_{el}$ is the zero-point energy of the initial state. The minimum energy required to reach the point x_s is clearly that corresponding to the momentum $p = 0$. By substituting for x_s from equation (A2.4.133), we find

$$U_{act} = \frac{M\omega^2}{2}(x_s - x^i_0)^2 = \frac{(U_s + U^f_{el} - U^i_{el} + B^f - B^i)^2}{4U_s}. \tag{A2.4.135}$$

where U_s has the value $(M\omega^2/2)(x^f_0 - x^i_0)^2$. U_s is termed the *reorganization energy* since it is the additional energy required to deform the complex from initial to final value of x. It is common to find the symbol λ for U_s, and model calculations suggest that U_s normally has values in the neighbourhood of 1 eV (10^5 J mol^{-1}) for the simplest redox processes.

In our simple model, the expression in (A2.4.135) corresponds to the activation energy for a redox process in which only the interaction between the central ion and the ligands in the primary solvation shell is considered, and this only in the form of the totally symmetrical vibration. In reality, the rate of the electron transfer reaction is also influenced by the motion of molecules in the outer solvation shell, as well as by other vibrational modes of the inner shell. These can be incorporated into the model provided that each type of motion can be treated within the simple harmonic approximation. The total energy of the system will then consist of the kinetic energy of all the atoms in motion together with the potential energy arising from each vibrational degree of freedom. It is no longer possible to picture the motion, as in figure A2.4.13, as a one-dimensional translation over an energy barrier, since the total energy is a function of a large number of normal coordinates describing the motion of the entire system. Instead, we have two potential energy surfaces for the initial and final states of the redox system, whose intersection described the reaction hypersurface. The reaction pathway will proceed now *via* the saddle point, which is the minimum of the total potential energy subject to the condition $U^i_{pot} = U^f_{pot}$ as above.

This is a standard problem [31] and essentially the same result is found as in equation (A2.4.135), save that the B^i and B^f now become the sum over all the binding energies of the central ion in the initial and final states and U_s is now given by

$$U_s = \sum_j \frac{M_j\omega^2_j}{2}(x^f_{j,0} - x^i_{j,0})^2 \tag{A2.4.136}$$

where M_j is the effective mass of the jth mode and ω_j is the corresponding frequency; we still retain the approximation that these frequencies are all the same for the initial and final states.

With the help of U_s, an expression for the rate constant for the reaction

$$[Fe(H_2O)_6]^{2+} \leftrightarrow [Fe(H_2O)_6]^{3+} + e^-_M \tag{A2.4.137}$$

can be written

$$k_f = k^0_f \exp\left(-\frac{U_{act}}{kT}\right) = A\exp\left(-\frac{(U_s + U^f_{el} - U^i_{el} + B^f - B^i)^2}{4U_skT}\right) \tag{A2.4.138}$$

where A is the so-called frequency factor and e_M^- refers to an electron in the metal electrode. The rate constant for the back reaction is obtained by interchanging the indices i and f in equation (A2.4.74). It will be observed that under these circumstances U_s remains the same and we obtain

$$k_b = A \exp\left(-\frac{(U_s + U_{el}^i - U_{el}^f + B^i - B^f)^2}{4 U_s kT}\right) \tag{A2.4.139}$$

A2.4.7.3   The exchange current

It is now possible to derive an expression for the actual current density from (A2.4.138) and (A2.4.139), assuming reaction (A2.4.137), and, for simplicity, assuming that the concentrations, c, of Fe^{2+} and Fe^{3+} are equal. The potential difference between the electrode and the outer Helmholtz layer, $\Delta\phi$, is incorporated into the electronic energy of the $Fe^{3+} + e_M^-$ system through a potential-dependent term of the form

$$U_{el}^f = U_{el,0}^f - e_0\Delta\phi \tag{A2.4.140}$$

where the minus sign in (A2.4.140) arises through the negative charge on the electron. Inserting this into (A2.4.138) and (A2.4.139) and multiplying by concentration and the Faraday to convert from rate constants to current densities, we have

$$j^+ = FAc \exp\left(-\frac{(U_s + U_{el,0}^f - U_{el}^i + B^f - B^i - e_0\Delta\phi)^2}{4 U_s kT}\right) \tag{A2.4.141}$$

$$j^- = -FAc \exp\left(-\frac{(U_s + U_{el}^i - U_{el,0}^f + B^i - B^f + e_0\Delta\phi)^2}{4 U_s kT}\right) \tag{A2.4.142}$$

where we adopt the convention that positive current involves *oxidation*. At the rest potential, $\Delta\phi_r$, which is actually the same as the standard Nernst potential $\Delta\phi_0$ when assuming that the activity coefficients of the ions are also equal, the rates of these two reactions are equal, which implies that the terms in brackets in the two equations must also be equal when $\Delta\phi = \Delta\phi_0$. From this it is clear that

$$e_0\Delta\phi_0 = U_{el,0}^f - U_{el}^i + B^f - B^i \tag{A2.4.143}$$

and if we introduce the overpotential, $\eta = \Delta\phi - \Delta\phi_0$, evidently

$$j^+ = FAc \exp\left[-\frac{(U_s - e_0\eta)^2}{4 U_s kT}\right] \tag{A2.4.144}$$

$$j^- = -FAc \exp\left[-\frac{(U_s + e_0\eta)^2}{4 U_s kT}\right] \tag{A2.4.145}$$

from which we obtain the exchange current density as the current at $\eta = 0$:

$$j_0 = FAc \exp\left[-\frac{U_s}{4kT}\right] \tag{A2.4.146}$$

and the activation energy of the exchange current density can be seen to be $U_s/4$. If the overpotential is small, such that $e_0\eta \ll U_s$ (and recalling that U_s lies in the region of about 1 eV), the quadratic form of (A2.4.144) and (A2.4.145) can be expanded with neglect of terms in η^2. Recalling, also, that $e_0/k_B = F/R$, we then finally obtain

$$j = j^+ + j^- = FAc \exp\left(-\frac{U_s}{4kT}\right)\left\{\exp\left(\frac{F\eta}{2RT}\right) - \exp\left(-\frac{F\eta}{2RT}\right)\right\} \tag{A2.4.147}$$

Table A2.4.6.

System	Electrolyte	Temperature (°C)	Electrode	j_0 (A cm^{-2})	β
Fe^{3+}/Fe^{2+} (0.005 M)	1 M H_2SO_4	25	Pt	2×10^{-3}	0.58
$K_3Fe(CN)_6/K_4Fe(CN)_6$ (0.02 M)	0.5 M K_2SO_4	25	Pt	5×10^{-2}	0.49
$Ag/10^{-3}$ M Ag^+	1 M $HClO_4$	25	Ag	1.5×10^{-1}	0.65
$Cd/10^{-2}$ M Cd^{2+}	0.4 M K_2SO_4	25	Cd	1.5×10^{-3}	0.55
$Cd(Hg)/1.4 \times 10^{-3}$ M Cd^{2+}	0.5 M Na_2SO_4	25	Cd(Hg)	2.5×10^{-2}	0.8
$Zn(Hg)/2 \times 10^{-2}$ M Zn^{2+}	1 M $HClO_4$	0	Zn(Hg)	5.5×10^{-3}	0.75
Ti^{4+}/Ti^{3+} (10^{-3} M)	1 M acetic acid	25	Pt	9×10^{-4}	0.55
H_2/OH^-	1 M KOH	25	Pt	10^{-3}	0.5
H_2/H^+	1 M H_2SO_4	25	Hg	10^{-12}	0.5
H_2/H^+	1 M H_2SO_4	25	Pt	10^{-3}	0.5
O_2/H^+	1 M H_2SO_4	25	Pt	10^{-6}	0.25
O_2/OH^-	1 M KOH	25	Pt	10^{-6}	0.3

which is the simplest form of the familiar Butler–Volmer equation with a symmetry factor $\beta = \frac{1}{2}$. This result arises from the strongly simplified molecular model that we have used above and, in particular, the assumption that the values of ω_j are the same for all normal modes. Relaxation of this assumption leads to a more general equation:

$$j = j^+ + j^- = FAc \exp\left(-\frac{U_s}{4kT}\right)\left\{\exp\left(\frac{\beta F\eta}{RT}\right) - \exp\left(-\frac{(1-\beta)F\eta}{RT}\right)\right\} \quad \text{(A2.4.148)}$$

$$= j^0 \left\{\exp\left(\frac{\beta F\eta}{RT}\right) - \exp\left(-\frac{(1-\beta)F\eta}{RT}\right)\right\}. \quad \text{(A2.4.149)}$$

For a more general reaction of the form Ox+ne^- $\leftrightarrow$ Red, with differing concentrations of Ox and Red, the exchange current density is given by

$$j^0 = nFA(c_{Ox}^{\beta}, c_{Red}^{1-\beta}) \exp\left(-\frac{U_{act}}{kT}\right). \quad \text{(A2.4.150)}$$

Some values for j^0 and β for electrochemical reactions of importance are given in table A2.4.6, and it can be seen that the exchange currents can be extremely dependent on the electrode material, particularly for more complex processes such as hydrogen oxidation. Many modern electrochemical studies are concerned with understanding the origin of these differences in electrode performance.

References

[1] Ben-Naim A 1992 *Statistical Thermodynamics for Chemists and Biologists* (London: Plenum)
[2] Lee L L 1988 *Molecular Thermodynamics of Nonideal Fluids* (Boston: Butterworths)
[3] Kihara T 1953 *Rev. Mod. Phys.* **25** 831
[4] Metropolis N A, Rosenbluth A W, Rosenbluth M N, Teller A H and Teller E 1953 *J. Chem. Phys.* **21** 1087
[5] Richens D T 1997 *The Chemistry of Aqua-ions* (Chichester: Wiley)
[6] Enderby J E 1983 *Ann. Rev. Phys. Chem.* **34** 155
    Enderby J E 1983 *Contemp. Phys.* **24** 561
[7] Hamann C H, Hamnett A and Vielstich W 1998 *Electrochemistry* (Weinheim: Wiley)

[8] Robinson R A and Stokes R H 1959 *Electrolyte Solutions* (London: Butterworth)

[9] Outhwaite C W 1975 *Statistical Mechanics* ed K Singer (London: Chemistry Society)

[10] Blum L 1975 *Mol. Phys.* **30** 1529

[11] Gurney R W 1954 *Ionic Processes in Solution* (New York: Dover)

[12] Desnoyers J E and Jolicoeur C 1969 *Modern Aspects of Electrochemistry* vol 5, ed B E Conway and J O'M Bockris (New York: Plenum)

[13] Friedman H L 1969 *Modern Aspects of Electrochemistry* vol 6, ed B E Conway and J O'M Bockris (New York: Plenum)

[14] Pitts E, Tabor B E and Daly J 1969 *Trans. Farad. Soc.* **65** 849

[15] Barlow C A and MacDonald J R 1967 *Adv. Electrochem. Electrochem. Eng.* **6** 1

[16] Trassatti S 1980 *Comprehensive Treatise of Electrochemistry* ed J O'M Bockris, B E Conway and E Yeager (New York: Plenum)

[17] Ashcroft N W and Mermin N D 1976 *Solid-State Physics* (New York: Holt, Rinehart and Winston)

[18] Frumkin A N, Petrii O A and Damaskin B B 1980 *Comprehensive Treatise of Electrochemistry* ed J O'M Bockris, B E Conway and E Yeager (New York: Plenum)

[19] Martynov G A and Salem R R 1983 *Electrical Double Layer at a Metal–Dilute Electrolyte Solution Interface* (Berlin: Springer)

[20] Goüy G 1910 *J. Phys.* **9** 457
     Chapman D L 1913 *Phil. Mag.* **25** 475

[21] Bockris J O'M and Khan S U 1993 *Surface Electrochemistry* (New York: Plenum)

[22] Watts-Tobin R J 1961 *Phil. Mag.* **6** 133

[23] Parsons R 1975 *J. Electroanal. Chem.* **53** 229

[24] Damaskhin B B and Frumkin A N 1974 *Electrochim. Acta* **19** 173

[25] Trassatti S 1986 *Trends in Interfacial Electrochemistry (NATO ASI Series 179)* ed A Fernando Silva (Dordrecht: Reidel)

[26] Bewick A, Kunimatsu K, Robinson J and Russell J W 1981 *J. Electroanal. Chem.* **276** 175
     Habib M A and Bockris J O'M 1986 *Langmuir* **2** 388

[27] Guidelli R 1986 *Trends in Interfacial Electrochemistry (NATO ASI Series 179)* ed A Fernando Silva (Dordrecht: Reidel)

[28] Habib M A and Bockris J O'M 1980 *Comprehensive Treatise of Electrochemistry* ed J O'M Bockris, B E Conway and E Yeager (New York: Plenum)

[29] Benjamin I 1997 *Mod. Aspects Electrochem.* **31** 115

[30] Spohr E 1999 *Electrochim. Acta* **44** 1697

[31] Schmickler W 1996 *Interfacial Electrochemistry* (Oxford: Oxford University Press)

A2.5
Phase transitions and critical phenomena

Robert L Scott

A2.5.1 One-component first-order transitions

The thermodynamic treatment of simple phase transitions is straightforward and is discussed in A2.1.6 and therefore need not be repeated here. In a one-component two-phase system, the phase rule yields one degree of freedom, so the transition between the two phases can be maintained along a pressure–temperature line. Figure A2.5.1 shows a typical p, T diagram with lines for fusion (solid–liquid), sublimation (solid–gas), and vaporization (liquid–gas) meeting at a triple point (solid–liquid–gas). Each of these lines can, at least in principle, be extended as a metastable line (shown as a dashed line) beyond the triple point. (Supercooling of gases below the condensation point, supercooling of liquids below the freezing point and superheating of liquids above the boiling point are well known; superheating of solids above the melting point is more problematic.) The vaporization line (i.e. the vapour pressure curve) ends at a critical point, with a unique pressure, temperature, and density, features that will be discussed in detail in subsequent sections. Because this line ends it is possible for a system to go around it and move continuously from gas to liquid without a phase transition; above the critical temperature the phase should probably just be called a 'fluid'.

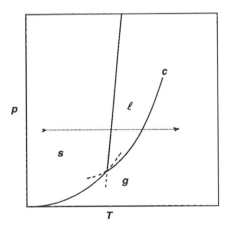

Figure A2.5.1. Schematic phase diagram (pressure p versus temperature T) for a typical one-component substance. The full lines mark the transitions from one phase to another (g, gas; ℓ, liquid; s, solid). The liquid–gas line (the vapour pressure curve) ends at a critical point (c). The dotted line is a constant pressure line. The dashed lines represent metastable extensions of the stable phases.

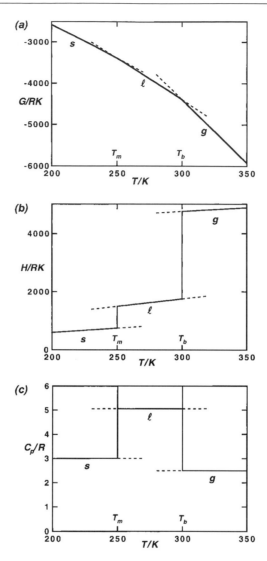

Figure A2.5.2. Schematic representation of the behaviour of several thermodynamic functions as a function of temperature T at constant pressure for the one-component substance shown in figure A2.5.1. (The constant-pressure path is shown as a dotted line in figure A2.5.1.) (a) The molar Gibbs free energy $\overline{G}$, (b) the molar enthalpy $\overline{H}$, and (c) the molar heat capacity at constant pressure $\overline{C}_p$. The functions shown are dimensionless (R is the gas constant per mole, while K is the temperature unit Kelvin). The dashed lines represent metastable extensions of the stable phases beyond the transition temperatures.

Figure A2.5.2 shows schematically the behaviour of several thermodynamic functions along a constant-pressure line (shown as a dotted line in figure A2.5.1)—the molar Gibbs free energy $\overline{G}$ (for a one-component system the same as the chemical potential μ), the molar enthalpy $\overline{H}$ and the molar heat capacity at constant pressure $\overline{C}_p$. Again, at least in principle, each of the phases can be extended into a metastable region beyond the equilibrium transition.

It will be noted that the free energy (figure A2.5.2(a)) is necessarily continuous through the phase transitions, although its first temperature derivative (the negative of the entropy S) is not. The enthalpy $H = G + TS$ (shown in figure A2.5.2(b)) is similarly discontinuous; the vertical discontinuities are of course the enthalpies of transition. The graph for the molar heat capacity $\overline{C}_p$ (figure A2.5.2(c)) looks superficially like that for the enthalpy, but represents something quite different at the transition. The vertical line with an arrow at a transition temperature is a mathematical delta function, representing an ordinate that is infinite and an abscissa (ΔT) that is zero, but whose product is nonzero, an 'area' that is equal to the molar enthalpy of transition $\Delta \overline{H}$.

Phase transitions at which the entropy and enthalpy are discontinuous are called 'first-order transitions' because it is the first derivatives of the free energy that are discontinuous. (The molar volume $\overline{V} = (\partial \overline{G}/\partial p)_T$ is also discontinuous.) Phase transitions at which these derivatives are continuous but second derivatives of G are discontinuous (e.g. the heat capacity, the isothermal compressibility, the thermal expansivity etc) are called 'second order'.

The initial classification of phase transitions made by Ehrenfest (1933) was extended and clarified by Pippard [1], who illustrated the distinctions with schematic heat capacity curves. Pippard distinguished different kinds of second- and third-order transitions and examples of some of his second-order transitions will appear in subsequent sections; some of his types are unknown experimentally. Theoretical models exist for third-order transitions, but whether these have ever been found is unclear.

A2.5.2 Phase transitions in two-component systems

Phase transitions in binary systems, normally measured at constant pressure and composition, usually do not take place entirely at a single temperature, but rather extend over a finite but nonzero temperature range. Figure A2.5.3 shows a temperature–mole fraction (T, x) phase diagram for one of the simplest of such examples, vaporization of an ideal liquid mixture to an ideal gas mixture, all at a fixed pressure, (e.g. 1 atm). Because there is an additional composition variable, the sample path shown in the figure is not only at constant pressure, but also at a constant total mole fraction, here chosen to be $x = 1/2$.

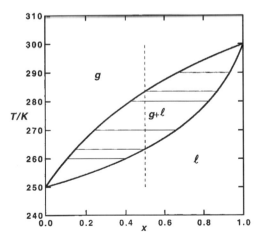

Figure A2.5.3. Typical liquid–gas phase diagram (temperature T versus mole fraction x at constant pressure) for a two-component system in which both the liquid and the gas are ideal mixtures. Note the extent of the two-phase liquid–gas region. The dashed vertical line is the direction ($x = 1/2$) along which the functions in figure A2.5.5 are determined.

As the temperature of the liquid phase is increased, the system ultimately reaches a phase boundary, the 'bubble point' at which the gas phase (vapour) begins to appear, with the composition shown at the left end of the horizontal two-phase 'tie-line'. As the temperature rises more gas appears and the relative amounts of the two phases are determined by applying a lever-arm principle to the tie-line: the ratio of the fraction f_g of molecules in the gas phase to that f_ℓ in the liquid phase is given by the inverse of the ratio of the distances from the phase boundary to the position of the overall mole fraction x_0 of the system,

$$f_g/f_\ell = (x_\ell - x_0)/(x_0 - x_g).$$

With a further increase in the temperature the gas composition moves to the right until it reaches $x = 1/2$ at the phase boundary, at which point all the liquid is gone. (This is called the 'dew point' because, when the gas is cooled, this is the first point at which drops of liquid appear.) An important feature of this behaviour is that the transition from liquid to gas occurs gradually over a nonzero range of temperature, unlike the situation shown for a one-component system in figure A2.5.1. Thus the two-phase region is bounded by a dew-point curve and a bubble-point curve.

Figure A2.5.4 shows for this two-component system the same thermodynamic functions as in figure A2.5.2, the molar Gibbs free energy $\overline{G} = x_1\mu_1 + x_2\mu_2$, the molar enthalpy $\overline{H}$ and the molar heat capacity $\overline{C}_p$, again all at constant pressure, but now also at constant composition, $x = 1/2$. Now the enthalpy is continuous because the vaporization extends over an appreciable temperature range. Moreover, the heat capacity, while discontinuous at the beginning and at the end of the transition, is *not* a delta function. Indeed the graph appears to satisfy the definition of a second-order transition (or rather two, since there are two discontinuities).

However, this behaviour is not restricted to mixtures; it is also found in a one-component fluid system observed along a constant-volume path rather than the constant-pressure path illustrated in figure A2.5.2. Clearly it would be confusing to classify the same transition as first- or second-order depending on the path. Pippard described such one-component constant-volume behaviour (discussed by Gorter) as a 'simulated second-order transition' and elected to restrict the term 'second-order' to a path along which two phases became more and more nearly alike until at the transition they became identical. As we shall see, that is what is seen when the system is observed along a path through a critical point. Further clarification of this point will be found in subsequent sections.

It is important to note that, in this example, as in 'real' second-order transitions, the curves for the two-phase region cannot be extended beyond the transition; to do so would imply that one had more than 100% of one phase and less than 0% of the other phase. Indeed it seems to be a quite general feature of all known second-order transitions (although it does not seem to be a thermodynamic requirement) that some aspect of the system changes gradually until it becomes complete at the transition point.

Three other examples of liquid–gas phase diagrams for a two-component system are illustrated in figure A2.5.5, all a result of deviations from ideal behaviour. Such deviations in the liquid mixture can sometimes produce azeotropic behaviour, in which there are maximum or minimum boiling mixtures (shown in figure A2.5.5(a) and figure A2.5.5(b)). Except at the azeotropic composition (that of the maximum or minimum), a constant-composition path through the vaporization yields the same kind of qualitative behaviour shown in figure A2.5.4. Behavior like that shown in figure A2.5.2 is found only on the special path through the maximum or minimum, where the entire vaporization process occurs at a unique temperature.

A third kind of phase diagram in a two-component system (as shown in figure A2.5.5(c)) is one showing liquid–liquid phase separation below a critical-solution point, again at a fixed pressure. (On a T, x diagram, the critical point is always an extremum of the two-phase coexistence curve, but not always a maximum. Some binary systems show a minimum at a lower critical-solution temperature; a few systems show closed-loop two-phase regions with a maximum and a minimum.) As the temperature is increased at any composition other than the critical composition $x = x_c$, the compositions of the two coexisting phases adjust themselves

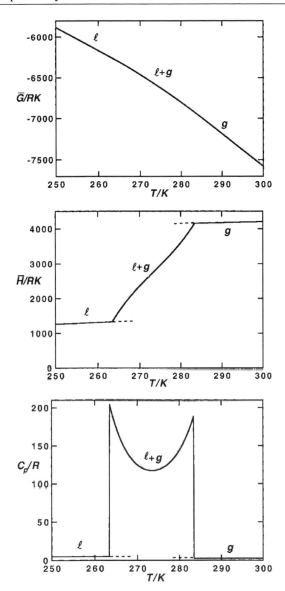

Figure A2.5.4. Thermodynamic functions $\overline{G}$, $\overline{H}$, and $\overline{C}_p$ as a function of temperature T at constant pressure and composition ($x = 1/2$) for the two-component system shown in figure A2.5.3. Note the difference between these and those shown for the one-component system shown in figure A2.5.2. The functions shown are dimensionless as in figure A2.5.2. The dashed lines represent metastable extensions (superheating or supercooling) of the one-phase systems.

to keep the total mole fraction unchanged until the coexistence curve is reached, above which only one phase persists. Again, the behaviour of the thermodynamic functions agrees qualitatively with that shown in figure A2.5.4, except that there is now only one transition line, not two. However, along any special path leading through the critical point, there are special features in the thermodynamic functions that will be discussed in subsequent sections. First, however, we return to the one-component fluid to consider the features of its critical point.

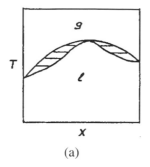 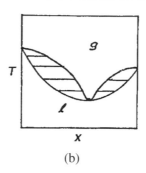 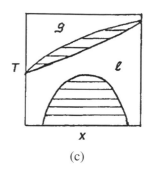

(a)   (b)   (c)

Figure A2.5.5. Phase diagrams for two-component systems with deviations from ideal behaviour (temperature T versus mole fraction x at constant pressure). Liquid–gas phase diagrams with maximum (a) and minimum (b) boiling mixtures (azeotropes). (c) Liquid–liquid phase separation, with a coexistence curve and a critical point.

A2.5.3   Analytic treatment of critical phenomena in fluid systems. The van der Waals equation

All simple critical phenomena have similar characteristics, although all the analogies were not always recognized in the beginning. The liquid–vapour transition, the separation of a binary mixture into two phases, the order–disorder transition in binary alloys, and the transition from ferromagnetism to paramagnetism all show striking common features. At a low temperature one has a highly ordered situation (separation into two phases, organization into a superlattice, highly ordered magnetic domains, etc). At a sufficiently high temperature all long-range order is lost, and for all such cases one can construct a phase diagram (not always recognized as such) in which the long-range order is lost gradually until it vanishes at a critical point.

A2.5.3.1   The van der Waals fluid

Although later models for other kinds of systems are symmetrical and thus easier to deal with, the first analytic treatment of critical phenomena is that of van der Waals (1873) for coexisting liquid and gas [2]. The familiar van der Waals equation gives the pressure p as a function of temperature T and molar volume $\overline{V}$,

$$p = RT/(\overline{V} - b) - a/\overline{V}^2 \tag{A2.5.1}$$

where R is the gas constant per mole and a and b are constants characteristic of the particular fluid. The constant a is a measure of the strength of molecular attraction, while b is the volume excluded by a mole of molecules considered as hard spheres.

Figure A2.5.6 shows a series of typical p, V isotherms calculated using equation (A2.5.1). (The temperature, pressure and volume are in reduced units to be explained below.) At sufficiently high temperatures the pressure decreases monotonically with increasing volume, but below a critical temperature the isotherm shows a maximum and a minimum.

The coexistence lines are determined from the requirement that the potentials (which will subsequently be called 'fields'), i.e. the pressure, the temperature, and the chemical potential μ, be the same in the conjugate (coexisting) phases, liquid and gas, at opposite ends of the tie-line. The equality of chemical potentials is equivalent to the requirement—for these variables (p, V), not necessarily for other choices—that the two areas between the horizontal coexistence line and the smooth cubic curve be equal. The full two-phase line is the stable situation, but the continuation of the smooth curve represents metastable liquid or gas phases (shown as a dashed curve); these are sometimes accessible experimentally. The part of the curve with a positive slope (shown as a dotted curve) represents a completely unstable situation, never realizable in practice because a negative compressibility implies instant separation into two phases. In analytic treatments like this of

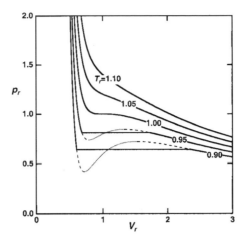

Figure A2.5.6. Constant temperature isotherms of reduced pressure p_r *versus* reduced volume V_r for a van der Waals fluid. Full curves (including the horizontal two-phase tie-lines) represent stable situations. The dashed parts of the smooth curve are metastable extensions. The dotted curves are unstable regions.

van der Waals, the maxima and minima in the isotherms (i.e. the boundary between metastable and unstable regions) define a 'spinodal' curve; in principle this line distinguishes between nucleation with an activation energy and 'spinodal decomposition'. (see A3.3.) It should be noted that, if the free energy is nonanalytic as we shall find necessary in subsequent sections, the concept of the spinodal becomes unclear and can only be maintained by invoking special internal constraints. However, the usefulness of the distinction between activated nucleation and spinodal decomposition can be preserved.

With increasing temperature the two-phase tie-line becomes shorter and shorter until the two conjugate phases become identical at a critical point where the maximum and minimum in the isotherm have coalesced. Thus the critical point is defined as that point at which $(\partial p/\partial \overline{V})_T$ and $(\partial^2 p/\partial \overline{V}^2)_T$ are simultaneously zero, or where the equivalent quantities $(\partial^2 \overline{A}/\partial \overline{V})_T$ and $(\partial^3 \overline{A}/\partial \overline{V}^3)_T$ are simultaneously zero. These requirements yield the critical constants in terms of the constants R, a and b,

$$p_c = a/(27b^2) \qquad \overline{V}_c = 3b \qquad T_c = 8a/(27Rb).$$

Equation (A2.5.1) can then be rewritten in terms of the reduced quantities $p_r = p/p_c$, $T_r = T/T_c$, and $V_r = \overline{V}/\overline{V}_c$

$$p_r = 8T_r/(3V_r - 1) - 3/V_r^2. \tag{A2.5.2}$$

It is this equation with the reduced quantities that appears in figure A2.5.6.

Since the pressure $p = -(\partial \overline{A}/\partial \overline{V})_T$, integration of equation (A2.5.1) yields $\overline{A}(T, \overline{V})$

$$\overline{A}(T, \overline{V}) - \overline{A}^\circ(T, \overline{V}^\circ) = -RT \ln[(\overline{V} - b)/\overline{V}^\circ] - a/\overline{V}$$

where $\overline{A}(T, \overline{V}^\circ)$ is the molar free energy of the ideal gas at T and $\overline{V}^\circ$. (It is interesting to note that the van der Waals equation involves a simple separation of the free energy into entropy and energy; the first term on the right is just $-T(\overline{S} - \overline{S}^\circ)$, while the second is just $\overline{U} - \overline{U}^\circ$.)

The phase separation shown in figures A2.5.6 can also be illustrated by the entirely equivalent procedure of plotting the molar Helmholtz free energy $\overline{A}(T, \overline{V})$ as a function of the molar volume $\overline{V}$ for a series of constant temperatures, shown in figure A2.5.7. At constant temperature and volume, thermodynamic equilibrium requires that the Helmholtz free energy must be minimized. It is evident for temperatures below

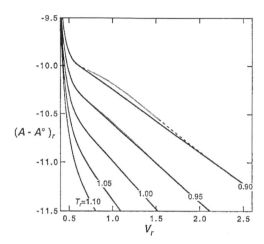

Figure A2.5.7. Constant temperature isotherms of reduced Helmholtz free energy A_r versus reduced volume V_r. The two-phase region is defined by the line simultaneously tangent to two points on the curve. The dashed parts of the smooth curve are metastable one-phase extensions while the dotted curves are unstable regions. (The isotherms are calculated for an unphysical $V_r^\circ = 0.1$, the only effect of which is to separate the isotherms better.)

the critical point that for certain values of the molar volume the molar free energy $\overline{A}(T, \overline{V})$ can be lowered by separation into two phases. The exact position of the phase separation is found by finding a straight line that is simultaneously tangent to the curve at two points; the slope at any point of the curve is $(\partial \overline{A}/\partial \overline{V}) = -p$, so the pressures are equal at the two tangent points. Similarly the chemical potential $\mu = \overline{A} + p\overline{V}$ is the same at the two points. That the dashed and dotted parts of the curve are metastable or unstable is clear because they represent higher values of $\overline{A}$ than the corresponding points on the two-phase line. (The metastable region is separated from the completely unstable region by a point of inflection on the curve.)

The problem with figures A2.5.6 and A2.5.7 is that, because it extends to infinity, volume is not a convenient variable for a graph. A more useful variable is the molar density $\rho = 1/\overline{V}$ or the reduced density $\rho_r = 1/V_r$ which have finite ranges, and the familiar van der Waals equation can be transformed into an alternative although relatively unfamiliar form by choosing as independent variables the chemical potential μ and the density ρ.

Unlike the pressure where $p = 0$ has physical meaning, the zero of free energy is arbitrary, so, instead of the ideal gas volume, we can use as a reference the molar volume of the real fluid at its critical point. A reduced Helmholtz free energy A_r in terms of the reduced variables T_r and V_r can be obtained by replacing a and b by their values in terms of the critical constants

$$A_r = [\overline{A}(T, \overline{V}) - \overline{A}(T, \overline{V}_c)]/(p_c \overline{V}_c) = -(8/3)T_r \ln[(3V_r - 1)/2] - 3(1 - V_r)/V_r.$$

Then, since the chemical potential for a one-component system is just $\mu = \overline{G} = \overline{A} + p\overline{V}$, a reduced chemical potential can be written in terms of a reduced density $\rho_r = \rho/\rho_c = \overline{V}_c/\overline{V}$

$$\mu_r = [\mu(T, \rho) - \mu(T, \rho_c)](\rho_c/p_c)$$
$$= -(8/3)T_r[\ln((3 - \rho_r)/(2\rho_r)) + (3/2)(\rho_r - 1)/(3 - \rho_r)] - 6(\rho_r - 1). \qquad (A2.5.3)$$

Equation (A2.5.3) is a μ_r, ρ_r equation of state, an alternative to the p_r, V_r equation (A2.5.2).

The van der Waals μ_r, ρ_r isotherms, calculated using equation (A2.5.3), are shown in figure A2.5.8. It is immediately obvious that these are much more nearly antisymmetric around the critical point than are the

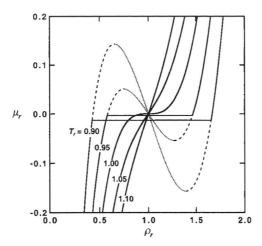

Figure A2.5.8. Constant temperature isotherms of reduced chemical potential μ_r versus reduced density ρ_r for a van der Waals fluid. Full curves (including the horizontal two-phase tie-lines) represent stable situations. The dashed parts of the smooth curve are metastable extensions, while the dotted curves are unstable regions.

corresponding p_r, V_r isotherms in figure A2.5.6 (of course, this is mainly due to the finite range of ρ_r from 0 to 3). The symmetry is not exact, however, as a careful examination of the figure will show. This choice of variables also satisfies the equal-area condition for coexistent phases; here the horizontal tie-line makes the chemical potentials equal and the equal-area construction makes the pressures equal.

For a system in which the total volume remains constant, the same minimization condition that applies to $\overline{A}$ also applies to $\overline{A}/\overline{V} = \overline{A}\rho$, or to $(A\rho)_r$, a quantity that can easily be expressed in terms of T_r and ρ_r,

$$(A\rho)_r = -(8/3)T_r\rho_r \ln[(3 - \rho_r)/(2\rho_r)] - 3\rho_r(\rho_r - 1) - (4T_r - 3)(\rho_r - 1).$$

It is evident that, for the system shown in figure A2.5.9, $\overline{A}/\overline{V} = \overline{A}\rho$ or $(A\rho)_r$ can be minimized for certain values of ρ on the low-temperature isotherms if the system separates into two phases rather than remaining as a single phase. As in figure A2.5.7 the exact position of the phase separation is found by finding a straight line that is simultaneously tangent to the curve at two points; the slope at any point on the curve is the chemical potential μ, as is easily established by differentiating $\overline{A}\rho$ with respect to ρ,

$$[\partial(\overline{A}\rho)/\partial\rho]_T = \overline{A} + \rho(\partial\overline{A}/\partial\rho)_T = \overline{A} - \overline{V}(\partial\overline{A}/\partial\overline{V})_T = \overline{A} + p\overline{V} = \mu.$$

(The last term in the equation for $(A\rho)_r$ has been added to avoid adding a constant to μ_r; doing so does not affect the principle, but makes figure A2.5.9 clearer.) Thus the common tangent satisfies the condition of equal chemical potentials, $\mu_\ell = \mu_g$. (The common tangent also satisfies the condition that $p_\ell = p_g$ because $p = \mu\rho - \rho\overline{A}$.) The two points of tangency determine the densities of liquid and gas, ρ_ℓ and ρ_g, and the relative volumes of the two phases are determined by the lever-arm rule.

The T_r, ρ_r coexistence curve can be calculated numerically to any desired precision and is shown in figure A2.5.10. The spinodal curve (shown dotted) satisfies the equation

$$(\partial^2 A_r/\partial V_r^2)_{T_r} = -(\partial p_r/\partial V_r)_{T_r} = 6\rho_r^2[4T_r/(3\rho_r - 1)^2 - \rho_r] = 0.$$

Alternatively, expansion of equations (A2.5.1), (A2.5.2) or (A2.5.3) into Taylor series leads ultimately to series expressions for the densities of liquid and gas, ρ_ℓ and ρ_g, in terms of their sum (called the 'diameter')

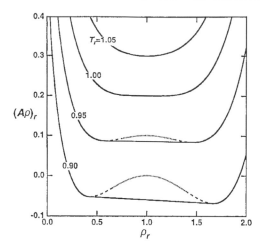

Figure A2.5.9. $(A\rho)_r$, the Helmholtz free energy per unit volume in reduced units, of a van der Waals fluid as a function of the reduced density ρ_r for several constant temperatures above and below the critical temperature. As in the previous figures the full curves (including the tangent two-phase tie-lines) represent stable situations, the dashed parts of the smooth curve are metastable extensions, and the dotted curves are unstable regions. See text for details.

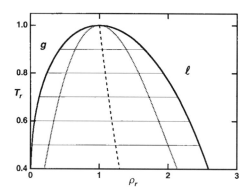

Figure A2.5.10. Phase diagram for the van der Waals fluid, shown as reduced temperature T_r versus reduced density ρ_r. The region under the smooth coexistence curve is a two-phase liquid–gas region as indicated by the horizontal tie-lines. The critical point at the top of the curve has the coordinates (1,1). The dashed line is the diameter, and the dotted curve is the spinodal curve.

and their difference:

$$(\rho_\ell + \rho_g)/(2\rho_c) = 1 + (2/5)(1 - T_r) + (128/875)(1 - T_r)^2 + \cdots \tag{A2.5.4}$$

$$(\rho_\ell - \rho_g)/(2\rho_c) = 2(1 - T_r)^{1/2} - (13/25)(1 - T_r)^{3/2} + \cdots$$

$$\text{or } [(\rho_\ell - \rho_g)/(2\rho_c)]^2 = 4(1 - T_r) - (52/25)(1 - T_r)^2 + \cdots. \tag{A2.5.5}$$

Note that equation (A2.5.5), like equation (A2.5.4), is just a power series in $(1 - T_r) = (T_c - T)/T_c$, a variable that will appear often and will henceforth be represented by t. All simple equations of state (e.g. the Dieterici and Berthelot equations) yield equations of the same form as (A2.5.4) and (A2.5.5); only the coefficients differ. There are better expressions for the contribution of the hard-sphere fluid to the pressure than the van der Waals $RT/(\overline{V} - b)$, but the results are similar. Indeed it can be shown that any analytic equation of state, however complex, must necessarily yield similar power series.

If the small terms in t^2 and higher are ignored, equation (A2.5.4) is the 'law of the rectilinear diameter' as evidenced by the straight line that extends to the critical point in figure A2.5.10; this prediction is in good qualitative agreement with most experiments. However, equation (A2.5.5), which predicts a parabolic shape for the top of the coexistence curve, is unsatisfactory as we shall see in subsequent sections.

The van der Waals energy $\overline{U} = -a/\overline{V} = -a\rho$. On a path at constant total critical density $\rho = \rho_c$, which is a constant-volume path, the energy of the system will be the sum of contributions from the two conjugate phases, the densities and amounts of which are changing with temperature. With proper attention to the amounts of material in the two phases that maintain the constant volume, this energy can be written relative to the one-phase energy $\overline{U}_c$ at the critical point,

$$(\overline{U} - \overline{U}_c)/(a\rho_c) = (\overline{U} - \overline{U}_c)/(9RT_c/8) = -(\rho_\ell + \rho_g)/\rho_c + \rho_\ell\rho_g/\rho_c^2 + 1$$
$$= -(\rho_\ell + \rho_g)/\rho_c + (\rho_\ell + \rho_g)^2/(2\rho_c)^2 - (\rho_\ell - \rho_g)^2/(2\rho_c)^2 + 1$$

or, substituting from equations (A2.5.4) and (A2.5.5),

$$(\overline{U} - \overline{U}_c)/(RT_c) = (9/2)[-t + (14/25)t^2 + \cdots].$$

Differentiating this with respect to $T = -T_c t$ yields a heat capacity at constant volume,

$$\overline{C}_V = (\partial\overline{U}/\partial T)_{\overline{V}} = (9/2)R[1 - (28/25)t + \cdots]. \qquad (A2.5.6)$$

This is of course an excess heat capacity, an amount in addition to the contributions of the heat capacities $\overline{C}_V$ of the liquid and vapour separately. Note that this excess (as well as the total heat capacity shown later in figure A2.5.26) is always finite, and increases only linearly with temperature in the vicinity of the critical point; at the critical point there is no further change in the excess energy, so this part of the heat capacity drops to zero. This behaviour looks very similar to that of the simple binary system in section A2.5.2. However, unlike that system, in which there is no critical point, the experimental heat capacity $\overline{C}_V$ along the critical isochore (constant volume) appears to diverge at the critical temperature, contrary to the prediction of equation (A2.5.6).

Finally, we consider the isothermal compressibility $\kappa_T = -(\partial \ln V/\partial p)_T = (\partial \ln \rho/\partial p)_T$ along the coexistence curve. A consideration of figure A2.5.6 shows that the compressibility is finite and positive at every point in the one-phase region except at the critical point. Differentiation of equation (A2.5.2) yields the compressibility along the critical isochore:

$$\rho_c\kappa_T = 1/[6(T_r - 1)] = 1/(-6t) \qquad (\rho = \rho_c, T_r \geq 1).$$

At the critical point (and anywhere in the two-phase region because of the horizontal tie-line) the compressibility is infinite. However the compressibility of each conjugate phase can be obtained as a series expansion by evaluating the derivative (as a function of ρ_r) for a particular value of T_r, and then substituting the values of ρ_r for the ends of the coexistence curve. The final result is

$$\rho_c\kappa_T = 1/[12t \pm (216/5)t^{3/2} + \cdots] = [1/(12t)][1 \mp (12/5)t^{1/2} + \cdots] \qquad (\text{coex}, T_r \leq 1) \qquad (A2.5.7)$$

where in the $\pm$ and the $\mp$, the upper sign applies to the liquid phase and the lower sign to the gas phase. It is to be noted that although the compressibility becomes infinite as one approaches T_c from either direction, its value at a small δT below T_c is only half that at the same δT above T_c; this means that there is a discontinuity at the critical point.

Although the previous paragraphs hint at the serious failure of the van der Waals equation to fit the shape of the coexistence curve or the heat capacity, failures to be discussed explicitly in later sections, it is important to recognize that many of the other predictions of analytic theories are reasonably accurate. For example, analytic equations of state, even ones as approximate as that of van der Waals, yield reasonable values (or at least 'ball park estimates') of the critical constants p_c, T_c, and $\overline{V}_c$. Moreover, in two-component systems where the critical point expands to a critical line in p, T space, or in three-component systems where there are critical surfaces, simple models yield many useful predictions. It is only in the vicinity of critical points that analytic theories fail.

A2.5.3.2 The van der Waals fluid mixture

Van der Waals (1890) extended his theory to mixtures of components A and B by introducing mole-fraction-dependent parameters a_m and b_m defined as quadratic averages

$$a_m = (1-x)^2 a_{AA} + 2x(1-x)a_{AB} + x^2 a_{BB} \tag{A2.5.8}$$

$$b_m = (1-x)^2 b_{AA} + 2x(1-x)b_{AB} + x^2 b_{BB} \tag{A2.5.9}$$

where the as and bs extend the meanings defined in section A2.5.3.1 for the one-component fluid to the three kinds of pairs in the binary mixture; x is the mole fraction of the second component (B). With these definitions of a_m and b_m, equation (A2.5.1) for the pressure remains unchanged, but an entropy of mixing must be added to the equation for the Helmholtz free energy.

$$\overline{A}(T,\overline{V}) - (1-x)\overline{A}_A(T,\overline{V}^\circ) - x\overline{A}_B(T,\overline{V}^\circ)$$
$$= RT[(1-x)\ln(1-x) + x\ln x] + RT\ln[(\overline{V}-b_m)/\overline{V}^\circ] - a_m/\overline{V}.$$

Van der Waals and especially van Laar simplified these expressions by assuming a geometric mean for a_{AB} and an arithmetic mean for b_{AB},

$$a_{12} = (a_{11}a_{22})^{1/2} \quad \text{and} \quad b_{12} = (b_{11}+b_{22})/2.$$

Then equations (A2.5.8) and (A2.5.9) for a_m and b_m become

$$a_m = [(1-x)a_{AA}^{1/2} + xa_{BB}^{1/2}]^2 \tag{A2.5.10}$$

$$b_m = (1-x)b_{AA} + xb_{BB}. \tag{A2.5.11}$$

With these simplifications, and with various values of the as and bs, van Laar (1906–1910) calculated a wide variety of phase diagrams, determining critical lines, some of which passed continuously from liquid–liquid critical points to liquid–gas critical points. Unfortunately, he could only solve the difficult coupled equations by hand and he restricted his calculations to the geometric mean assumption for a_{12} (i.e. to equation (A2.5.10)). For a variety of reasons, partly due to the eclipse of the van der Waals equation, this extensive work was largely ignored for decades.

A2.5.3.3 Global phase diagrams

Half a century later Van Konynenburg and Scott (1970, 1980) [3] used the van der Waals equation to derive detailed phase diagrams for two-component systems with various parameters. Unlike van Laar they did not restrict their treatment to the geometric mean for a_{AB}, and for the special case of $b_{AA} = b_{BB} = b_{AB}$ (equal-sized molecules), they defined two reduced variables,

$$\zeta = (a_{BB} - a_{AA})/(a_{BB} + a_{AA}) \tag{A2.5.12}$$

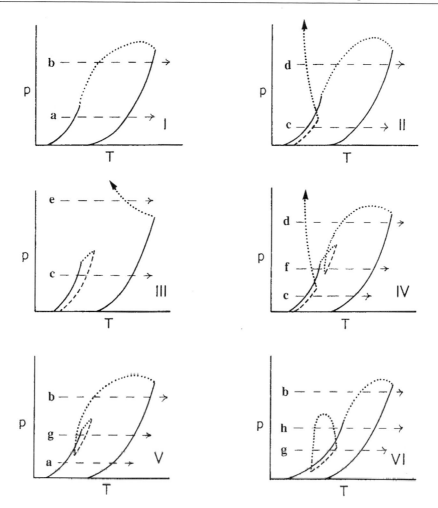

Figure A2.5.11. Typical pressure–temperature phase diagrams for a two-component fluid system. The full curves are vapour pressure lines for the pure fluids, ending at critical points. The dotted curves are critical lines, while the dashed curves are three-phase lines. The dashed horizontal lines are not part of the phase diagram, but indicate constant-pressure paths for the (T, x) diagrams in figure A2.5.12. All but the type VI diagrams are predicted by the van der Waals equation for binary mixtures. Adapted from figures in [3].

$$\lambda = (a_{BB} - 2a_{AB} + a_{BB})/(a_{BB} + a_{AA}). \qquad (A2.5.13)$$

Physically, ζ is a measure of the difference in the energies of vaporization of the two species (roughly a difference in normal boiling point), and λ is a measure of the energy of mixing. With these definitions equation (A2.5.8) can be rewritten as

$$a_m/a_{AA} = [(1 - \zeta) + 2x(\zeta - \lambda) + x^2\lambda]/(1 - \zeta).$$

If a_{AB} is weak in comparison to a_{AA} and a_{BB}, λ is positive and separation into two phases may occur.

With this formulation a large number of very different phase diagrams were calculated using computers that did not exist in 1910. Six principal types of binary fluid phase diagrams can be distinguished by considering

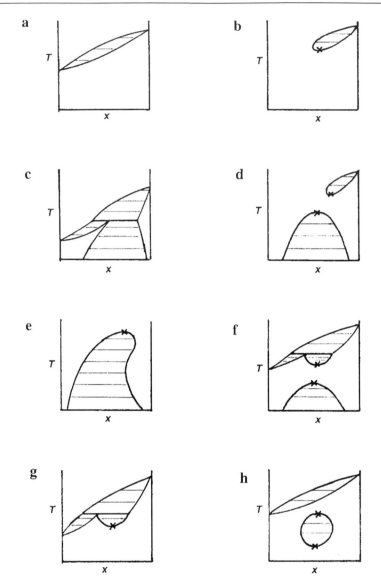

Figure A2.5.12. Typical temperature T versus mole fraction x diagrams for the constant-pressure paths shown in figure A2.5.11. Note the critical points ($\times$) and the horizontal three-phase lines.

where critical lines begin and end. These are presented in figure A2.5.11 as the p, T projections of p, T, x diagrams, which show the vapour pressure curves for the pure substances, critical lines and three-phase lines. To facilitate understanding of these projections, a number of diagrams showing T versus x for a fixed pressure, identified by horizontal lines in figure A2.5.11, are shown in figure A2.5.12. Note that neither of these figures shows any solid state, since the van der Waals equation applies only to fluids. The simple van der Waals equation for mixtures yields five of these six types of phase diagrams. Type VI, with its low-pressure (i.e. below 1 atm) closed loop between a lower critical-solution temperature (LCST) and an upper critical-solution temperature (UCST), cannot be obtained from the van der Waals equation without making λ temperature dependent.

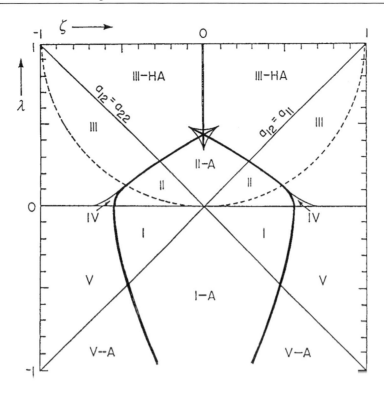

Figure A2.5.13. Global phase diagram for a van der Waals binary mixture for which $b_{AA} = b_{BB}$. The coordinates λ and ζ are explained in the text. The curves separate regions of the different types shown in figure A2.5.11. The heavier curves are tricritical lines (explained in section A2.5.9). The 'shield region' in the centre of the diagram where three tricritical lines intersect consists of especially complex phase diagrams not yet found experimentally. Adapted from figures in [3].

The boundaries separating these principal types of phase behaviour are shown on a λ, ζ diagram (for equal-sized molecules) in figure A2.5.13. For molecules of different size, but with the approximation of equation (A2.5.10), more global phase diagrams were calculated using a third parameter,

$$\xi = (b_{BB} - b_{AA})/(b_{BB} + b_{AA})$$

and appropriately revised definitions for ζ and λ. For different-sized molecules ($\xi \neq 0$), the global phase diagram is no longer symmetrical, but the topology is the same.

In recent years global phase diagrams have been calculated for other equations of state, not only van der Waals-like ones, but others with complex temperature dependences. Some of these have managed to find type VI regions in the overall diagram. Some of the recent work was brought together at a 1999 conference [4].

A2.5.4   Analytic treatments of other critical phenomena

A2.5.4.1   Liquid–liquid phase separation in a simple binary mixture

The previous section showed how the van der Waals equation was extended to binary mixtures. However, much of the early theoretical treatment of binary mixtures ignored equation-of-state effects (i.e. the contributions of the expansion beyond the volume of a close-packed liquid) and implicitly avoided the distinction between

```
B  A  B  B  A  A  A  B  B  B        A  A  A  A  A  A  A  A  A  A
A  B  A  B  B  B  B  B  B  A        A  A  A  A  A  A  A  A  A  A
B  B  B  B  A  B  A  A  A  A        A  A  A  A  A  A  A  A  A  A
A  B  A  A  B  A  B  B  A  A        A  A  A  A  A  A  A  A  A  A
A  A  B  B  B  A  A  A  B  A        A  A  A  A  A  A  A  A  A  A
A  B  B  B  A  B  A  A  A  A        B  B  B  B  B  B  B  B  B  B
B  A  B  A  A  B  B  B  A  B        B  B  B  B  B  B  B  B  B  B
B  B  A  B  B  B  B  B  A  A        B  B  B  B  B  B  B  B  B  B
A  B  B  A  B  A  A  A  A  A        B  B  B  B  B  B  B  B  B  B
A  B  B  A  B  A  B  A  A  A        B  B  B  B  B  B  B  B  B  B
           (a)                                (b)
```

Figure A2.5.14. Quasi-lattice representation of an equimolar binary mixture of A and B (a) randomly mixed at high temperature, and (b) phase separated at low temperature.

constant pressure and constant volume by putting the molecules, assumed to be equal in size, into a kind of pseudo-lattice. Figure A2.5.14 shows schematically an equimolar mixture of A and B, at a high temperature where the distribution is essentially random, and at a low temperature where the mixture has separated into two virtually one-component phases.

The molar Helmholtz free energy of mixing (appropriate at constant volume) for such a symmetrical system of molecules of equal size, usually called a 'simple mixture', is written as a function of the mole fraction x of the component B

$$\Delta \overline{A}^M = \overline{A}_m - (1-x)\mu_A^\circ - x\mu_B^\circ$$
$$= RT[x \ln x + (1-x)\ln(1-x)] + Kx(1-x) \qquad (A2.5.14)$$

where the μ°'s are the chemical potentials of the pure components. The Gibbs free energy of mixing $\Delta \overline{G}^M$ is (at constant pressure) $\Delta \overline{A}^M + p\Delta \overline{V}^M$ and many theoretical treatments of such a system ignore the volume change on mixing and use the equation above for $\Delta \overline{G}^M$, which is the quantity of interest for experimental measurements at constant pressure. Equation (A2.5.14) is used to plot $\Delta \overline{A}^M$ or equivalently $\Delta \overline{G}^M$ versus x for several temperatures in figure A2.5.15. As in the case of the van der Waals fluid a tangent line determines phase separation, but here the special symmetry requires that it be horizontal and that the mole fractions of the conjugate phases x' and x'' satisfy the condition $x'' = 1 - x'$. The critical-solution point occurs where $(\partial^2 \Delta \overline{G}^M / \partial x^2)_{T,p}$ and $(\partial^3 \Delta \overline{G}^M / \partial x^3)_{T,p}$ are simultaneously zero; for this special case, this point is at $x_c = 1/2$ and $T_c = K/2R$. The reduced temperatures that appear on the isotherms in figure A2.5.15 are then defined as $T_r = T/T_c = 2RT/K$.

In the simplest model the coefficient K depends only on the differences of the attractive energies $-\varepsilon$ of the nearest-neighbour pairs (these energies are negative relative to those of the isolated atoms, but here their magnitudes ε are expressed as positive numbers)

$$K = \overline{N}(z/2)(\varepsilon_{AA} - 2\varepsilon_{AB} + \varepsilon_{BB}) = \overline{N}w.$$

Here $\overline{N}$ is the number of molecules in one mole and z is the coordination number of the pseudo-lattice (4 in figure A2.5.14); so $\overline{N}(z/2)$ is the total number of nearest-neighbour pairs in one mole of the mixture. The quantity w is the *interchange energy*, the energy involved in the single exchange of molecules between the two pure components. (Compare the parameter λ for the van der Waals mixture in the section above

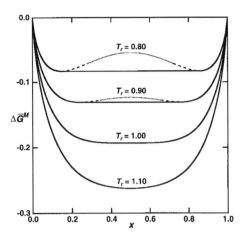

Figure A2.5.15. The molar Gibbs free energy of mixing $\Delta \overline{G}^M$ versus mole fraction x for a simple mixture at several temperatures. Because of the symmetry of equation (A2.5.15) the tangent lines indicating two-phase equilibrium are horizontal. The dashed and dotted curves have the same significance as in previous figures.

(equation (A2.5.13))). If w is a constant, independent of temperature, then K is temperature independent and $Kx(1-x)$ is simply the molar energy of mixing $\Delta \overline{U}^M$ (frequently called the molar enthalpy of mixing $\Delta \overline{H}^M$ when the volume change is assumed to be zero). If the chemical potentials μ_1 and μ_2 are derived from the free energy of mixing, μ, x isotherms are obtained that are qualitatively similar to the μ, ρ isotherms shown in figure A2.5.8, an unsurprising result in view of the similarity of the free energy curves for the two cases.

For such a symmetrical system one may define a 'degree of order' or 'order parameter' $s = 2x - 1$ such that s varies from -1 at $x = 0$ to $+1$ at $x = 1$. Then $x = (1 + s)/2$, and $1 - x = (1 - s)/2$ and equation (A2.5.14) can be rewritten as:

$$\Delta \overline{G}^M \cong \Delta \overline{A}^M = RT \left[\frac{(1+s)}{2} \ln \frac{(1+s)}{2} + \frac{(1-s)}{2} \ln \frac{(1-s)}{2} \right] + K \frac{(1-s^2)}{4}. \tag{A2.5.15}$$

It is easy to derive the coexistence curve. Because of the symmetry, the double tangent is horizontal and the coexistent phases occur at values of s where $(\partial \Delta \overline{G}^M / \partial s)_T$ equals zero.

$$RT \ln[(1+s)/(1-s)] = 2RT \tanh^{-1} s = Ks.$$

Even if K is temperature dependent, the coexistence curve can still be defined in terms of a reduced temperature $T_r = 2RT/K(T)$, although the reduced temperature is then no longer directly proportional to the temperature T.

$$(T_r/2) \ln[(1+s)/(1-s)] = T_r \tanh^{-1} s = s. \tag{A2.5.16}$$

Figure A2.5.16 shows the coexistence curve obtained from equation (A2.5.16). The logarithms (or the hyperbolic tangent) can be expanded in a power series, yielding

$$(T_r - 1)s + (T_r/3)s^3 + (T_r/5)s^5 + \cdots = 0.$$

This series can be reverted and the resulting equation is very simple:

$$s^2 = (x' - x'')^2 = 3t - (12/5)t^2 + \cdots. \tag{A2.5.17}$$

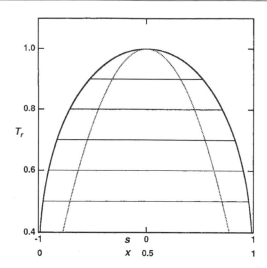

Figure A2.5.16. The coexistence curve, $T_r = K/(2R)$ versus mole fraction x for a simple mixture. Also shown as an abscissa is the order parameter s, which makes the diagram equally applicable to order–disorder phenomena in solids and to ferromagnetism. The dotted curve is the spinodal.

The leading term in equation (A2.5.17) is the same kind of parabolic coexistence curve found in section A2.5.3.1 from the van der Waals equation. The similarity between equations (A2.5.5) and (A2.5.17) should be obvious; the form is the same even though the coefficients are different.

The derivative $(\partial \Delta \overline{G}^M / \partial x)_T = \mu_B - \mu_A$, where μ_B and μ_A are the chemical potentials of the two species, is the analogue of the pressure in the one-component system. From equation (A2.5.15) one obtains

$$(\partial \Delta \overline{G}^M / \partial x)_T = 2(\partial \Delta \overline{G}^M / \partial s)_T = RT[\ln(1+s) - \ln(1-s)] - Ks.$$

At $s = 0$ this derivative obviously vanishes for all temperatures, but this is simply a result of the symmetry. The second derivative is another matter:

$$(\partial^2 \Delta \overline{G}^M / \partial x^2)_T = 4(\partial^2 \Delta \overline{G}^M / \partial s^2)_T = 2[2RT/(1-s^2) - K] = 4RT_c[T_r/(1-s^2) - 1].$$

This vanishes at the critical-solution point as does $(\partial p/\partial \overline{V})_T$ at the one-component fluid critical point. Thus an 'osmotic compressibility' or 'osmotic susceptibility' can be defined by analogy with the compressibility κ_T of the one-component fluid. Its value along the simple-mixture coexistence curve can be obtained using equation (A2.5.17) and is found to be proportional to t^{-1}. The osmotic compressibility of a binary mixture diverges at the critical point just like the compressibility of a one-component fluid (compare this to equation (A2.5.7)).

For a temperature-independent K, the molar enthalpy of mixing is

$$\Delta \overline{H}^M = Kx(1-x) = 2RT_c(1-s^2)/4 = (RT_c/2)[1 - 3t + (12/5)t^2 - \cdots]$$

and the excess mixing contribution to the heat capacity (now at constant pressure) is

$$\Delta \overline{C}^M_{p,x=x_c} = (R/2)[3 - (24/5)t + \cdots].$$

Again, as in the case of $\overline{C}_V$ for the van der Waals fluid, there is a linear increase up to a finite value at the critical point and then a sudden drop to the heat capacity of the one-phase system because the liquids are now completely mixed.

Few if any binary mixtures are exactly symmetrical around $x = 1/2$, and phase diagrams like that sketched in figure A2.5.5(c) are typical. In particular one can write for mixtures of molecules of different size (different molar volumes $\overline{V}_A^{\circ}$ and $\overline{V}_B^{\circ}$) the approximate equation

$$\Delta \overline{G}^M = RT[x \ln \phi + (1 - x) \ln(1 - \phi)] + [(1 - x)\overline{V}_A^{\circ} + x\overline{V}_B^{\circ}]K'\phi(1 - \phi)$$

which is a combination of the Flory entropy of mixing for polymer solutions with an enthalpy of mixing due to Scatchard and Hildebrand. The variable ϕ is the volume fraction of component B, $\phi = x_B\overline{V}_B^{\circ}/(x_A\overline{V}_A^{\circ} + x_B\overline{V}_B^{\circ})$, and the parameter K' now has the dimensions of energy per unit volume. The condition for a critical-solution point, that the two derivatives cited above must simultaneously equal zero, yields the results

$$K'(T_c) = (RT_c/2)[(1/\overline{V}_A^{\circ})^{1/2} + (1/\overline{V}_B^{\circ})^{1/2}]^2$$

$$\phi_c = (\overline{V}_A^{\circ})^{1/2}/[(\overline{V}_A^{\circ})^{1/2} + (\overline{V}_B^{\circ})^{1/2}].$$

This simple model continues to ignore the possibility of volume changes on mixing, so for simplicity the molar volumes $\overline{V}_A^{\circ}$ and $\overline{V}_B^{\circ}$ are taken as those of the pure components. It should come as no surprise that in this unsymmetrical system both $\phi' + \phi''$ and $\phi' - \phi''$ must be considered and that the resulting equations have extra terms that look like those in equations (A2.5.4) and (A2.5.5) for the van der Waals mixture; as in that case, however, the top of the coexistence curve is still parabolic. Moreover the parameter K' is now surely temperature dependent (especially so for polymer solutions), and the calculation of a coexistence curve will depend on such details.

As in the one-fluid case, the experimental sums are in good agreement with the law of the rectilinear diameter, but the experimental differences fail to give a parabolic shape to the coexistence curve.

It should be noted that a strongly temperature-dependent K (or K') can yield more than one solution to the equation $T_c = K/2R$. Figure A2.5.17 shows three possible examples of a temperature-dependent K for the simple mixture: (a) a constant K as assumed in the discussion above, (b) a K that slowly decreases with T, the most common experimental situation, and (c) a K that is so sharply curved that it produces not only an upper critical-solution temperature (UCST), but also a lower critical-solution temperature (LCST) below which the fluids are completely miscible (i.e. the type VI closed-loop binary diagram of section A2.5.3.3). The position of the curves can be altered by changing the pressure; if the two-phase region shrinks until the LCST and UCST merge, one has a 'double critical point' where the curve just grazes the critical line. A fourth possibility (known experimentally but not shown in figure A2.5.17) is an opposite curvature producing a low-temperature UCST and a high-temperature LCST with a one-phase region at intermediate temperatures; if these two critical-solution temperatures coalesce, one has a 'critical double point'.

A2.5.4.2 Order–disorder in solid mixtures

In a liquid mixture with a negative K (negative interchange energy w), the formation of unlike pairs is favoured and there is no phase separation. However, in a crystal there is long-range order and at low temperatures, although there is no physical phase separation, a phase transition from a disordered arrangement to a regular arrangement of alternating atoms is possible The classic example is that of β-brass (CuZn) which crystallizes in a body-centred cubic lattice. At high temperature the two kinds of atoms are distributed at random, but at low temperature they are arranged on two interpenetrating *superlattices* such that each Cu has eight Zn nearest neighbours, and each Zn eight Cu nearest neighbours, as shown in figure A2.5.18; this is like the arrangement of ions in a crystal of CsCl.

The treatment of such order–disorder phenomena was initiated by Gorsky (1928) and generalized by Bragg and Williams (1934) [5]. For simplicity we restrict the discussion to the symmetrical situation where there are equal amounts of each component ($x = 1/2$). The lattice is divided into two superlattices α and

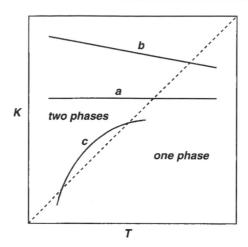

Figure A2.5.17. The coefficient K as a function of temperature T. The line $K = 2RT$ (shown as dashed line) defines the critical point and separates the two-phase region from the one-phase region. (a) A constant K as assumed in the simplest example; (b) a slowly decreasing K, found frequently in experimental systems, and (c) a sharply curved $K(T)$ that produces two critical-solution temperatures with a two-phase region in between.

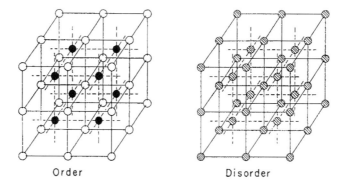

Figure A2.5.18. Body-centred cubic arrangement of β-brass (CuZn) at low temperature showing two interpenetrating simple cubic superlattices, one all Cu, the other all Zn, and a single lattice of randomly distributed atoms at high temperature. Reproduced from Hildebrand J H and Scott R L 1950 *The Solubility of Nonelectrolytes* 3rd edn (New York: Reinhold) p 342.

β, like those in the figure, and a degree of order s is defined such that the mole fraction of component B on superlattice β is $(1 + s)/4$ while that on superlattice α is $(1 - s)/4$. Conservation conditions then yield the mole fraction of A on the two superlattices

$$x_A^\alpha = x_B^\beta = (1 + s)/4 \qquad \text{and} \qquad x_A^\beta = x_B^\alpha = (1 - s)/4.$$

If the entropy and the enthalpy for the separate mixing in each of the half-mole superlattices are calculated and then combined, the following equation is obtained:

$$\Delta \overline{G}^M = RT \left[\frac{(1 + s)}{2} \ln \frac{(1 + s)}{2} + \frac{(1 - s)}{2} \ln \frac{(1 - s)}{2} \right] - \overline{N} w \frac{(1 + s)^2}{4}. \tag{A2.5.18}$$

Note that equation (A2.5.18) is almost identical with equation (A2.5.15). Only the final term differs and then only by the sign preceding s^2. Now, however, the interchange energy can be negative if the unlike attraction is stronger than the like attractions; then of course $K = \overline{N}w$ is also negative. If a reduced temperature T_r is defined as $-2RT/K$, a plot of $\Delta\overline{G}^M$ versus s for various T_rs is identical to that in figure A2.5.15. For all values of $K/(2RT)$ above -1 (i.e. $T_r > 1$), the minimum occurs at $s = 0$, corresponding to complete disorder when each superlattice is filled with equal amounts of A and B. However, for values below -1, i.e. $T_r < 1$, the minimum occurs at nonzero values of s, values that increase with decreasing temperature. Recall that $K/(2RT) = +1$ defined the critical temperature for phase separation in a symmetrical binary mixture; here a value of -1 defines the limit of long-range ordering. Thus for order–disorder behaviour $T_c = -K/2R$ defines a kind of critical temperature, although, by analogy with magnetic phenomena in solids, it is more often called the Curie point.

The free energy minimum is found by differentiating equation (A2.5.18) with respect to s at constant T and setting the derivative equal to zero. In its simplest form the resultant equation is

$$RT \ln[(1 + s)/(1 - s)] = 2RT \tanh^{-1} s = -Ks$$

exactly the same as equation (A2.5.13) for phase separation in simple mixtures except that this has $-Ks$ instead of $+Ks$. However, since it is a negative K that produces superlattice separation, the effect is identical, and figures A2.5.15 and A2.5.16 apply to both situations. The physical models are different, but the mathematics are just the same. This 'disordering curve', like the coexistence curve, is given by equation (A2.5.15) and is parabolic, and, for a temperature-independent K, the molar heat capacity $\overline{C}_p$ for the equimolar alloy will be exactly the same as that for the simple mixture.

Other examples of order–disorder second-order transitions are found in the alloys CuPd and Fe$_3$Al. However, not all ordered alloys pass through second-order transitions; frequently the partially ordered structure changes to a disordered structure at a first-order transition.

Nix and Shockley [6] gave a detailed review of the status of order–disorder theory and experiment up to 1938, with emphasis on analytic improvements to the original Bragg–Williams theory, some of which will be discussed later in section A2.5.4.4.

A2.5.4.3 Magnetism

The magnetic case also turns out to be similar to that of fluids, as Curie and Weiss recognized early on, but later for a long period this similarity was overlooked by those working on fluids (mainly chemists) and by those working on magnetism (mainly physicists). In a ferromagnetic material such as iron, the magnetic interactions between adjacent atomic magnetic dipoles causes them to be aligned so that a region (a 'domain') has a substantial magnetic dipole. Ordinarily the individual domains are aligned at random, and there is no overall magnetization. However, if the sample is placed in a strong external magnetic field, the domains can be aligned and, if the temperature is sufficiently low, a 'permanent magnet' is made, permanent in the sense that the magnetization is retained even though the field is turned off. Above a certain temperature, the Curie temperature T_C, long-range ordering in domains is no longer possible and the material is no longer ferromagnetic, but only paramagnetic. Individual atoms can be aligned in a magnetic field, but all ordering is lost if the field is turned off. (The use of a subscript C for the Curie temperature should pose no serious confusion, since it is a kind of critical temperature too.)

The little atomic magnets are of course quantum mechanical, but Weiss's original theory of paramagnetism and ferromagnetism (1907) [7] predated even the Bohr atom. He assumed that in addition to the external magnetic field B_0, there was an additional internal 'molecular field' B_i proportional to the overall magnetization M of the sample,

$$B = B_0 + B_i = B_0 + \lambda M.$$

If this field is then substituted into the Curie law appropriate for independent dipoles one obtains

$$M/B = M/(B_0 + \lambda M) = C/T \tag{A2.5.19}$$

where C is the Curie constant. The experimental magnetic susceptibility χ is defined as just M/B_0, since the internal field cannot be measured. Rearrangement of equation (A2.5.19) leads to the result

$$\chi = M/B_0 = C/(T - C\lambda) = C/(T - T_C). \tag{A2.5.20}$$

Equation (A2.5.20) is the Curie–Weiss law, and T_C, the temperature at which the magnetic susceptibility becomes infinite, is the Curie temperature. Below this temperature the substance shows spontaneous magnetization and is ferromagnetic. Normally the Curie temperature lies between 1 and 10 K. However, typical ferromagnetic materials like iron have very much larger values for quantum-mechanical reasons that will not be pursued here.

Equations (A2.5.19) and (A2.5.20) are valid only for small values of B_0 and further modelling is really not possible without some assumption, usually quantum mechanical, about the magnitude and orientation of the molecular magnets. This was not known to Weiss, but in the simplest case (half-integral spins), the magnetic dipole has the value of the Bohr magneton β_e, and the maximum possible magnetization M_{max} when all the dipoles are aligned with the field is $N\beta_e/V$, where N/V is the number of dipoles per unit volume.

If an order parameter s is defined as M/M_{max}, it can be shown that

$$s = \tanh[(s + \beta_e B_0/kT_C)(T_C/T)] = \tanh[(s + B_r)/T_r]. \tag{A2.5.21}$$

Isotherms of $\beta_e B_0/kT_C$, which might be called a reduced variable B_r, versus s are shown in figure A2.5.19 and look rather similar to the p_r, V_r plots for a fluid (figure A2.5.6). There are some differences, however, principally the symmetry that the fluid plots lack. At values of $T > T_C$, the curves are smooth and monotonic, but at T_C, as required, the magnetic susceptibility become infinite (i.e. the slope of B_r versus s becomes horizontal).

For $T < T_C$ ($T_r < 1$), however, the isotherms are S-shaped curves, reminiscent of the p_r, V_r isotherms that the van der Waals equation yields at temperatures below the critical (figure A2.5.6). As in the van der Waals case, the dashed and dotted portions represent metastable and unstable regions. For zero external field,

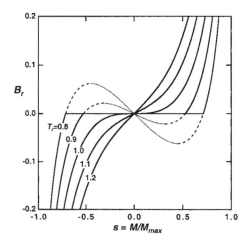

Figure A2.5.19. Isotherms showing the reduced external magnetic field $B_r = \beta_e B_0/kT_C$ versus the order parameter $s = M/M_{max}$, for various reduced temperatures $T_r = T/T_C$.

there are two solutions, corresponding to two spontaneous magnetizations. In effect, these represent two 'phases' and the horizontal line is a 'tie-line'. Note, however, that unlike the fluid case, even as shown in μ_r, ρ_r form (figure A2.5.8), the symmetry causes all the 'tie-lines' to lie on top of one another at $B_r = 0$ ($B_0 = 0$).

For $B_0 = 0$, equation (A2.5.21) reduces to

$$s = \tanh(s/T_r)$$

which, while it looks somewhat different, is exactly the same as equation (A2.5.16) and yields exactly the same parabolic 'coexistence curve' as that from equation (A2.5.17). Experimentally, as we shall see in the next section, the curve is not parabolic, but more nearly cubic. More generally, equation (A2.5.21) may be used to plot T_r versus s for fixed values of B_r as shown in figure A2.5.20. The similarity of this to a typical phase diagram $(T, \rho$ or $T, x)$ is obvious. Note that for nonzero values of the external field B_r the curves always lie outside the 'two-phase' region.

Related to these ferromagnetic materials, but different, are antiferromagnetic substances like certain transition-metal oxides. In these crystals, there is a complicated three-dimensional structure of two inter-penetrating superlattices not unlike those in CuZn. Here, at low temperatures, the two superlattices consist primarily of magnetic dipoles of opposite orientation, but above a kind of critical temperature, the Néel temperature T_N, all long-range order is lost and the two superlattices are equivalent. For $B_0 = 0$ the behaviour of an antiferromagnet is exactly analogous to that of a ferromagnet with a similar 'coexistence curve' $s(T_r)$, but for nonzero magnetic fields they are different. Unlike a ferromagnet at its Curie temperature, the susceptibility of an antiferromagnet does not diverge at the Néel temperature; extrapolation using the Curie–Weiss law yields a negative Curie temperature. Below the Néel temperature the antiferromagnetic crystal is anisotropic because there is a preferred axis of orientation. The magnetic susceptibility is finite, but varies with the angle between the crystal axis and the external field.

A related phenomenon with electric dipoles is 'ferroelectricity' where there is long-range ordering (nonzero values of the polarization P even at zero electric field E) below a second-order transition at a kind of critical temperature.

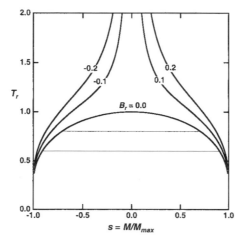

Figure A2.5.20. The reduced temperature $T_r = T/T_C$ versus the order parameter $s = M/M_{max}$ for various values of the reduced magnetic field B_r. Note that for all nonzero values of the field the curves lie outside the 'two-phase' region.

A2.5.4.4   Mean field versus 'molecular field'

Apparently Weiss believed (although van der Waals did not) that the interactions between molecules were long-range and extended over the entire system; under such conditions, it was reasonable to assume that the energies could be represented as proportional to the populations of the various species. With the development of theories of intermolecular forces in the 1920s that showed that intermolecular interactions were usually very short-range, this view was clearly unrealistic. In the discussions of liquid and solid mixtures in the preceding sections it has been assumed that the principal interactions, or perhaps even the only ones, are between nearest neighbours; this led to energies proportional to the interchange energy w. It was therefore necessary to introduce what is clearly only an approximation, that the probability of finding a particular molecular species in the nearest-neighbour shell (or indeed any more distant shell) around a given molecule is simply the probability of finding that species in the entire system. This is the 'mean-field' approximation that underlies many of the early analytic theories.

However, one can proceed beyond this zeroth approximation, and this was done independently by Guggenheim (1935) with his 'quasi-chemical' approximation for simple mixtures and by Bethe (1935) for the order–disorder solid. These two approximations, which turned out to be identical, yield some enhancement to the probability of finding like or unlike pairs, depending on the sign of w and on the coordination number z of the lattice. (For the unphysical limit of z equal to infinity, they reduce to the mean-field results.)

The integral under the heat capacity curve is an energy (or enthalpy as the case may be) and is more or less independent of the details of the model. The quasi-chemical treatment improved the heat capacity curve, making it sharper and narrower than the mean-field result, but it still remained finite at the critical point. Further improvements were made by Bethe with a second approximation, and by Kirkwood (1938). Figure A2.5.21 compares the various theoretical calculations [6]. These modifications lead to somewhat lower values of the critical temperature, which could be related to a flattening of the coexistence curve. Moreover, and perhaps more important, they show that a short-range order persists to higher temperatures, as it must because of the preference for unlike pairs; the excess heat capacity shows a discontinuity, but it does not drop to zero as mean-field theories predict. Unfortunately these improvements are still analytic and in the vicinity of the critical point still yield a parabolic coexistence curve and a finite heat capacity just as the mean-field treatments do.

Figure A2.5.22 shows [6] the experimental heat capacity of β-brass (CuZn) measured by Moser in 1934. Note that the experimental curve is sharper and goes much higher than any of the theoretical curves in figure A2.5.21; however, at that time it was still believed to have a finite limit.

A2.5.4.5   The critical exponents

It has become customary to characterize various theories of critical phenomena and the experiments with which they are compared by means of the exponents occurring in certain relations that apply in the limit as the critical point is approached. In general these may be defined by the equation

$$E = \lim_{Y \to Y_c} \left[\frac{\partial \ln |X - X_c|}{\partial \ln |Y - Y_c|} \right]_{\text{path}}$$

where E is an exponent, X and Y are properties of the system, and the path along which the derivative is evaluated must be specified.

Exponents derived from the analytic theories are frequently called 'classical' as distinct from 'modern' or 'nonclassical' although this has nothing to do with 'classical' versus 'quantum' mechanics or 'classical' versus 'statistical' thermodynamics. The important thermodynamic exponents are defined here, and their classical values noted; the values of the more general nonclassical exponents, determined from experiment

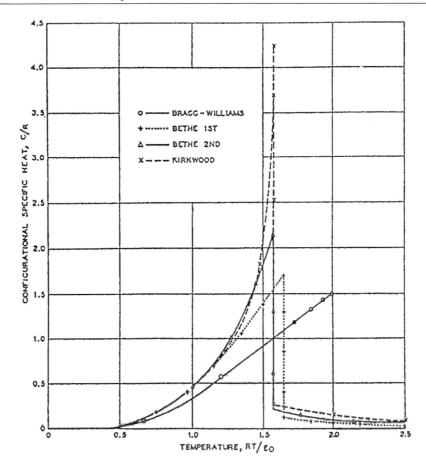

Figure A2.5.21. The heat capacity of an order–disorder alloy like β-brass calculated from various analytic treatments. Bragg–Williams (mean-field or zeroth approximation); Bethe-1 (first approximation also Guggenheim); Bethe-2 (second approximation); Kirkwood. Each approximation makes the heat capacity sharper and higher, but still finite. Reproduced from [6] Nix F C and Shockley W 1938 *Rev. Mod. Phys.* **10** 14, figure 13. Copyright (1938) by the American Physical Society.

and theory, will appear in later sections. The equations are expressed in reduced units in order to compare the amplitude coefficients in subsequent sections.

(a) The heat-capacity exponent α.

An exponent α governs the limiting slope of the molar heat capacity, variously $\overline{C}_V$, $\overline{C}_{p,x}$, or $\overline{C}_M$, along a line through the critical point,

$$\overline{C}(\rho_c T_c / p_c) = A^{\pm} t^{-\alpha} + \cdots \qquad (A2.5.22)$$

where the $\pm$ recognizes that the coefficient A^+ for the function above the critical point will differ from the A^- below the critical point. A similar quantity is the thermal expansivity $\alpha_p = (\partial \ln V / \partial T)_p$. For all these analytic theories, as we have seen on pages 533 and 539, the heat capacity remains finite, so $\alpha = 0$. As we shall see, these properties actually diverge with exponents slightly greater than zero. Such divergences are called 'weak'.

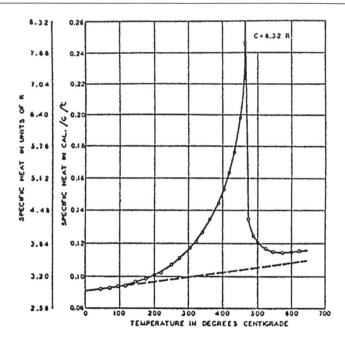

Figure A2.5.22. The experimental heat capacity of a β-brass (CuZn) alloy containing 48.9 atomic percent Zn as measured by Moser (1934). The dashed line is calculated from the specific heats of Cu and Zn assuming an ideal mixture. Reproduced from [6] Nix F C and Shockley W 1938 *Rev. Mod. Phys.* **10** 4, figure 4. Copyright (1938) by the American Physical Society.

(b) The coexistence-curve exponent β.

In general the width of the coexistence line ($\Delta\rho$, Δx, or ΔM) is proportional to an order parameter s, and its absolute value may be written as

$$|(\rho - \rho_c)/\rho_c| = |s| = Bt^\beta + \cdots. \tag{A2.5.23}$$

As we have seen, all the analytic coexistence curves are quadratic in the limit, so for all these analytic theories, the exponent $\beta = 1/2$.

(c) The susceptibility exponent γ.

A third exponent γ, usually called the 'susceptibility exponent' from its application to the magnetic susceptibility χ in magnetic systems, governs what in pure-fluid systems is the isothermal compressibility κ_T, and what in mixtures is the osmotic compressibility, and determines how fast these quantities diverge as the critical point is approached (i.e. as $T_r \to 1$).

$$p_c\kappa_T = p_c(\partial \ln V/\partial p)_T = \Gamma^{\pm}t^{-\gamma} + \cdots. \tag{A2.5.24}$$

For analytic theories, γ is simply 1, and we have seen that for the van der Waals fluid Γ^+/Γ^- equals 2. Divergences with exponents of the order of magnitude of unity are called 'strong'.

(d) The critical-isotherm exponent δ.

Finally the fourth exponent δ governs the limiting form of the critical isotherm, in the fluid case, simply

$$|[p(T_c) - p_c]/p_c| = D|[V(T_c) - V_c]/V_c|^\delta + \cdots. \qquad (A2.5.25)$$

Since all the analytic treatments gave cubic curves, their δ is obviously 3.

Exponent values derived from experiments on fluids, binary alloys, and certain magnets differ substantially from all those derived from analytic (mean-field) theories. However it is surprising that the experimental values appear to be the same from all these experiments, not only for different fluids and fluid mixtures, but indeed the same for the magnets and alloys as well (see section A2.5.5).

(e) Thermodynamic inequalities.

Without assuming analyticity, but by applying thermodynamics, Rushbrooke (1963) and Griffiths (1964) derived general constraints relating the values of the exponents.

$$\alpha_2^- + 2\beta + \gamma_1^- \geq 2$$

$$\alpha_2^- + \beta(1 + \delta) \geq 2.$$

Here α_2^- is the exponent for the heat capacity measured along the critical isochore (i.e. in the two-phase region) below the critical temperature, while γ_1^- is the exponent for the isothermal compressibility measured in the one-phase region at the edge of the coexistence curve. These inequalities say nothing about the exponents α^+ and γ^+ in the one-phase region above the critical temperature.

Substitution of the classical values of the exponents into these equations shows that they satisfy these conditions as equalities.

A2.5.5   The experimental failure of the analytic treatment

Nearly all experimental 'coexistence' curves, whether from liquid–gas equilibrium, liquid mixtures, order–disorder in alloys, or in ferromagnetic materials, are far from parabolic, and more nearly cubic, even far below the critical temperature. This was known for fluid systems, at least to some experimentalists, more than one hundred years ago. Verschaffelt (1900), from a careful analysis of data (pressure–volume and densities) on isopentane, concluded that the best fit was with $\beta = 0.34$ and $\delta = 4.26$, far from the classical values. Van Laar apparently rejected this conclusion, believing that, at least very close to the critical temperature, the coexistence curve must become parabolic. Even earlier, van der Waals, who had derived a classical theory of capillarity with a surface-tension exponent of $3/2$, found (1893) that experimental results on three liquids yielded lower exponents (1.23–1.27); he too apparently expected that the discrepancy would disappear closer to the critical point. Goldhammer (1920) formulated a law of corresponding states for a dozen fluids assuming that the exponent β was $1/3$. For reasons that are not entirely clear, this problem seems to have attracted little attention for decades after it was first pointed out. (This interesting history has been detailed by Levelt Sengers [8, 9].)

In 1945 Guggenheim [10], as part of an extensive discussion of the law of corresponding states, showed that, when plotted as reduced temperature T_r versus reduced density ρ_r, all the coexistence-curve measurements on three inert gases (Ar, Kr, Xe) fell on a single curve, and that Ne, N_2, O_2, CO and CH_4 also fit the same curve very closely. Moreover he either rediscovered or re-emphasized the fact that the curve was unequivocally cubic (i.e. $\beta = 1/3$) over the entire range of experimental temperatures, writing for ρ_r

$$\rho_r = 1 + (3/4)t \pm (7/4)t^{1/3}. \qquad (A2.5.26)$$

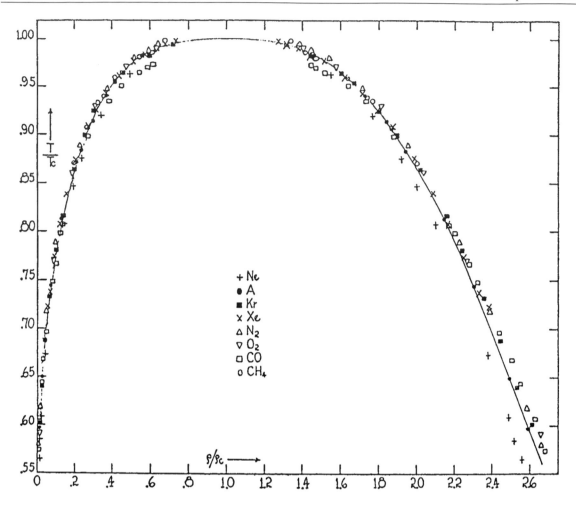

Figure A2.5.23. Reduced temperature $T_r = T/T_c$ versus reduced density $\rho_r = \rho/\rho_c$ for Ne, Ar, Kr, Xe, N$_2$, O$_2$, CO, and CH$_4$. The full curve is the cubic equation (A2.5.26). Reproduced from [10], p 257 by permission of the American Institute of Physics.

Figure A2.5.23 reproduces Guggenheim's figure, with experimental results and the fit to equation (A2.5.25). It is curious that he never commented on the failure to fit the analytic theory even though that treatment—with the quadratic form of the coexistence curve—was presented in great detail in *Statistical Thermodynamics* (Fowler and Guggenheim, 1939). The paper does not discuss any of the other critical exponents, except to fit the vanishing of the surface tension σ at the critical point to an equation

$$\sigma = \sigma_0 t^{11/9}.$$

This exponent $11/9$, now called μ, is almost identical with that found by van der Waals in 1893.

In 1953 Scott [11] pointed out that, if the coexistence curve exponent was $1/3$, the usual conclusion that the corresponding heat capacity remained finite was invalid. As a result the heat capacity might diverge and he suggested an exponent $\alpha = 1/3$. Although it is now known that the heat capacity does diverge, this suggestion attracted little attention at the time.

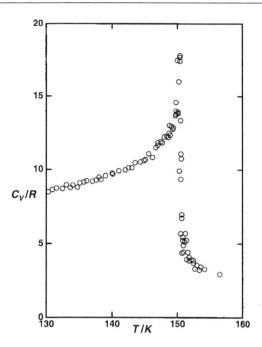

Figure A2.5.24. The heat capacity of argon in the vicinity of the critical point, as measured by Voronel and coworkers. Adapted from figure 1 of [12].

However, the discovery in 1962 by Voronel and coworkers [12] that the constant-volume heat capacity of argon showed a weak divergence at the critical point, had a major impact on uniting fluid criticality with that of other systems. They thought the divergence was logarithmic, but it is not quite that weak, satisfying equation (A2.5.21) with an exponent α now known to be about 0.11. The equation applies both above and below the critical point, but with different coefficients; A^- is larger than A^+. Thus the heat capacity (figure A2.5.24) is quite asymmetrical around T_c and appears like a sharp discontinuity.

In 1962 Heller and Benedek made accurate measurements of the zero-field magnetization of the antiferromagnet MnF_2 as a function of temperature and reported a β of 0.335 ± 0.005, a result supporting an experimental parallelism between fluids and magnets.

By 1966 the experimental evidence that the classical exponents were wrong was overwhelming and some significant theoretical advances had been made. In that year an important conference on critical phenomena [13] was held at the US National Bureau of Standards, which brought together physicists and chemists, experimentalists and theoreticians. Much progress had already been made in the preceding several years, and finally the similarity between the various kinds of critical phenomena was clearly recognized. The next decade brought near resolution to the problems.

A2.5.6   The Ising model and the gradual solution of the problem

A2.5.6.1   *The Ising model*

In 1925 Ising [14] suggested (but solved only for the relatively trivial case of one dimension) a lattice model for magnetism in solids that has proved to have applicability to a wide variety of other, but similar, situations. The mathematical solutions, or rather attempts at solution, have made the Ising model one of the most famous problems in classical statistical mechanics.

The model is based on a classical Hamiltonian $\mathcal{H}$ (here shown in script to distinguish it from the enthalpy H)

$$\mathcal{H} = -\sum_{i<j} J_{ij}\sigma_i\sigma_j - h\sum_i \sigma_i$$

where σ_i and σ_j are scalar numbers ($+1$ or -1) associated with occupancy of the lattice sites. In the magnetic case these are obviously the two orientations of the spin $s = 1/2$, but without any vector significance. The same Hamiltonian can be used for the lattice-solid mixture, where $+1$ signifies occupancy by molecule A, while -1 signifies a site occupied by molecule B (essentially the model used for the order–disorder transition in section A2.5.4.2). For the 'lattice gas', $+1$ signifies a site occupied by a molecule, while -1 signifies an unoccupied site (a 'hole').

The parameter J_{ij} is a measure of the energy of interaction between sites i and j while h is an external potential or field common to the whole system. The term $h\sum_i \sigma_i$ is a generalized work term (i.e. $-pV$, μN, VB_0M, etc), so $\mathcal{H}$ is a kind of generalized enthalpy. If the interactions J are zero for all but nearest-neighbour sites, there is a single nonzero value for J, and then

$$\mathcal{H} = -J \sum_{nn,i<j} \sigma_i\sigma_j - h\sum_i \sigma_i.$$

Thus any nearest-neighbour pair with the same signs for σ (spins parallel) contributes a term $-J$ to the energy and hence to $\mathcal{H}$. Conversely any nearest-neighbour pair with opposite signs for σ (spins opposed) contributes $+J$ to the energy. (If this doesn't seem right when extended to the lattice gas or to the lattice solid, it should be noted that a shift of the zero of energy resolves this problem and yields exactly the same equation. Thus, in the lattice mixture, there is only one relevant energy parameter, the interchange energy w.) What remained to be done was to derive the various thermodynamic functions from this simple Hamiltonian.

The standard analytic treatment of the Ising model is due to Landau (1937). Here we follow the presentation by Landau and Lifschitz [15], which casts the problem in terms of the order–disorder solid, but this is substantially the same as the magnetic problem if the vectors are replaced by scalars (as the Ising model assumes). The thermodynamic potential, in this case $G(T, p, s)$, is expanded as a Taylor series in even powers of the order parameter s (because of the symmetry of the problem there are no odd powers)

$$G(T, p, s) = G_0 + G_2 s^2 + G_4 s^4 + G_6 s^6 + \cdots.$$

Here the coefficients G_2, G_4, and so on, are functions of p and T, presumably expandable in Taylor series around $p - p_c$ and $T - T_c$. However, it is frequently overlooked that the derivation is accompanied by the comment that 'since ... the second-order transition point must be some singular point of the thermodynamic potential, there is every reason to suppose that such an expansion cannot be carried out up to terms of arbitrary order', but that 'there are grounds to suppose that its singularity is of higher order than that of the terms of the expansion used'. The theory developed below was based on this assumption.

For the kind of transition above which the order parameter is zero and below which other values are stable, the coefficient A_2 must change sign at the transition point and A_4 must remain positive. As we have seen, the dependence of s on temperature is determined by requiring the free energy to be a minimum (i.e. by setting its derivative with respect to s equal to zero). Thus

$$(\partial G/\partial s)_{T,p} = 2G_2 s + 4G_4 s^3 + 6G_6 s^5 + \cdots = 0.$$

If the G coefficients are expanded (at constant pressure p_c) in powers of t, this can be rewritten as

$$(-g_{21}t + g_{22}t^2 + \cdots) + (g_{40} - g_{41}t + \cdots)s^2 + (g_{60} + \cdots)s^4 + \cdots = 0.$$

Reverting this series and simplifying yields the final result in powers of t

$$s^2 = (g_{22}/g_{40})t - [(g_{22}g_{44}^2 - g_{21}g_{41}g_{40} + g_{41}^2 g_{60})/(g_{40}^2 g_{60})]t^2 + \cdots \qquad (A2.5.27)$$

and we see that, like all the previous cases considered, this curve too is quadratic in the limit. (The derivation here has been carried to higher powers than shown in [15].) These results are more general than the analytic results in previous sections (in the sense that the coefficients are more general), but the basic conclusion is the same; moreover other properties like the heat capacity are also described in the analytic forms discussed in earlier sections. There is no way of explaining the discrepancies without abandoning the assumption of analyticity. (It is an interesting historical note that many Russian scientists were among the last to accept this failure; they were sure that Landau had to have been right, and ignored his stated reservations.)

That analyticity was the source of the problem should have been obvious from the work of Onsager (1944) [16] who obtained an exact solution for the two-dimensional Ising model in zero field and found that the heat capacity goes to infinity at the transition, a logarithmic singularity that yields $\alpha = 0$, but not the $\alpha = 0$ of the analytic theory, which corresponds to a finite discontinuity. (While diverging at the critical point, the heat capacity is symmetrical without an actual discontinuity, so perhaps should be called third-order.) Subsequently Onsager (1948) reported other exponents, and Yang (1952) completed the derivation. The exponents are rational numbers, but not the classical ones. The 'coexistence curve' is nearly flat at its top, with an exponent $\beta = 1/8$, instead of the mean-field value of $1/2$. The critical isotherm is also nearly flat at T_C; the exponent δ (determined later) is 15 rather than the 3 of the analytic theories. The susceptibility diverges with an exponent $\gamma = 7/4$, a much stronger divergence than that predicted by the mean-field value of 1.

The classical treatment of the Ising model makes no distinction between systems of different dimensionality, so, if it fails so badly for $d = 2$, one might have expected that it would also fail for $d = 3$. Landau and Lifschitz [15] discussed the Onsager and Yang results, but continued to emphasize the analytic conclusions for $d = 3$.

A2.5.6.2   *The assumption of homogeneity. The 'scaling' laws*

The first clear step away from analyticity was made in 1965 by Widom [17] who suggested that the assumption of analytic functions be replaced by the less severe assumption that the singular part of the appropriate thermodynamic function was a homogeneous function of two variables, $(\rho_r - 1)$ and $(1 - T_r)$. A homogeneous function $f(u, v)$ of two variables is one that satisfies the condition

$$f(\lambda^{a_u} u, \lambda^{a_v} v) = \lambda f(u, v).$$

If one assumes that the singular part A^* of the Helmholtz free energy is such a function

$$A^*[\lambda^{a_\rho} (\rho_r - 1), \lambda^{a_T} (1 - T_r)] = \lambda[(\rho_r - 1), (1 - T_r)]$$

then a great deal follows. In particular, the reduced chemical potential $\mu_r = [\mu(\rho, T) - \mu(\rho_c, T)](\rho_c/p_c)$ of a fluid can be written as

$$\mu_r[\lambda^{a_\rho/(1-a_\rho)} (\rho_r - 1), \lambda^{a_T/(1-a_\rho)} (1 - T_r)] = \lambda \mu_r[(\rho_r - 1), (1 - T_r)].$$

(The brackets symbolize 'function of', not multiplication.) Since there are only two parameters, a_ρ and a_T, in this expression, the homogeneity assumption means that all four exponents α, β, γ and δ must be functions of these two; hence the inequalities in section A2.5.4.5(e) must be equalities. Equations for the various other thermodynamic quantities, in particular the singular part of the heat capacity C_V and the isothermal compressibility κ_T, may be derived from this equation for μ_r. The behaviour of these quantities as the critical point is approached can be satisfied only if

$$a_\rho = 1/(\delta + 1) = \beta/(2 - \alpha) \qquad \text{and} \qquad a_T = a_\rho/\beta.$$

This implies that μ_r may be written in a scaled form

$$\mu_r = [\mu(\rho, T) - \mu(\rho_c, T)](\rho_c/p_c) = (\rho_r - 1)|\rho_r - 1|^{\delta-1} Dh(x/x_0) \qquad \text{(A2.5.28)}$$

where $h(x/x_0)$ is an analytic function of $x = (T_r - 1)/|\rho_r - 1|^{1/\beta}$ and x_0, the value of x at the critical point, $x_0 = B^{-1/\beta}$. The curve $x = -x_0$ is the coexistence curve, the curve $x = 0$ is the critical isotherm, and the curve $x = \infty$ is the critical isochore. All the rest of the thermodynamic behaviour in the critical region can be derived from this equation, with the appropriate exponents as functions of β and δ. Note that there are now not only just two independent exponents, but also only two independent amplitudes, B and D, the amplitudes in equations (A2.5.23) and (A2.5.25). This homogeneity assumption is now known as the 'principle of two-scale-factor universality'. This principle, proposed as an approximation, seems to have stood the test of time; no further generalization seems to be needed. (We shall return to discuss exponents and amplitudes in section A2.5.7.1.)

An unexpected conclusion from this formulation, shown in various degrees of generality in 1970–71, is that for systems that lack the symmetry of simple lattice models the slope of the diameter of the coexistence curve should have a weak divergence proportional to $t^{-\alpha}$. This is very hard to detect experimentally because it usually produces only a small addition to the classical linear term in the equation for the diameter

$$(\rho_l + \rho_g)/(2\rho_c) = \rho_d = 1 + A_{1-\alpha} t^{1-\alpha} + A_1 t + \cdots.$$

However this effect was shown convincingly first [18] by Jüngst, Knuth and Hensel (1985) for the fluid metals caesium and rubidium (where the effect is surprisingly large) and then by Pestak *et al* (1987) for a series of simple fluids; figure A2.5.25 shows the latter results [19]. Not only is it clear that there is curvature very close to the critical point, but it is also evident that for this reason critical densities determined by extrapolating a linear diameter may be significantly too high. The magnitude of the effect (i.e. the value of the coefficient $A_{1-\alpha}$), seems to increase with the polarizability of the fluid.

A2.5.6.3 *The 'reason' for the nonanalyticity: fluctuations*

No system is exactly uniform; even a crystal lattice will have fluctuations in density, and even the Ising model must permit fluctuations in the configuration of spins around a given spin. Moreover, even the classical treatment allows for fluctuations; the statistical mechanics of the grand canonical ensemble yields an exact relation between the isothermal compressibility κ_T and the number of molecules N in volume V:

$$\sigma_N^2 = \langle N^2 \rangle - \langle N \rangle^2 = kT\kappa_T(N^2/V)$$

where σ is the standard deviation of the distribution of N's, and the brackets indicate averages over the distribution.

If the finite size of the system is ignored (after all, N is probably 10^{20} or greater), the compressibility is essentially infinite at the critical point, and then so are the fluctuations. In reality, however, the compressibility diverges more sharply than classical theory allows (the exponent γ is significantly greater than 1), and thus so do the fluctuations.

Microscopic theory yields an exact relation between the integral of the radial distribution function $g(r)$ and the compressibility

$$RT\rho\kappa_T = 1 + \rho \int (g(r) - 1)\, dr = 1 + \rho \int h(r)\, dr$$

where $g(r)$ is the radial distribution function which is the probability density for finding a molecule a distance r from the centre of a specified molecule, and $h(r)$ is the pair correlation function. At sufficiently long

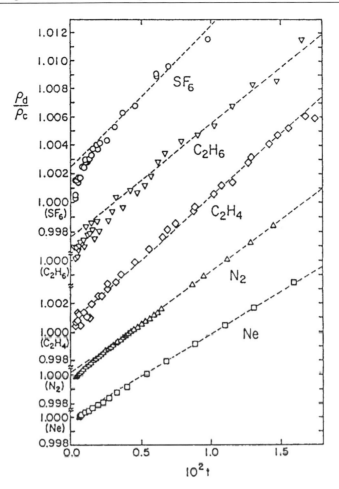

Figure A2.5.25. Coexistence-curve diameters as functions of reduced temperature for Ne, N_2, C_2H_4, C_2H_6, and SF_6. Dashed lines indicate linear fits to the data far from the critical point. Reproduced from [19] Pestak M W, Goldstein R E, Chan M H W, de Bruyn J R, Balzarini D A and Ashcroft N W 1987 *Phys. Rev.* B **36** 599, figure 3. Copyright (1987) by the American Physical Society.

distances $g(r)$ must become unity while $h(r)$ must become zero. Since κ_T diverges at the critical point, so also must the integral. The only way the integral can diverge is for the integrand to develop a very long tail. The range of the fluctuations is measured by the correlation length ξ. Near, but not exactly at, the critical point, the behaviour of $h(r)$ can be represented by the Ornstein–Zernike (1914, 1916) equation

$$h(r) = \exp(-r/\xi)/r$$

while, at the critical point, $h(r) \propto 1/r^{d-2+\eta}$, where d is the dimensionality of the system and η is a very small number (zero classically). The correlation length ξ increases as the critical point is approached and it will ultimately diverge. On the critical isochore, $\rho = \rho_c$, one finds

$$\xi = \xi_0|t|^{-\nu}$$

where classically $\nu = \gamma/2 = 1/2$. If the hypothesis of homogeneity is extended to the correlation length, what has become known as hyperscaling yields relations between the exponents ν and η and the thermodynamic exponents:

$$\nu = (2 - \alpha)/d \qquad \text{and} \qquad 2 - \eta = d(1 + \delta)/(1 - \delta).$$

Here d is the dimensionality of the system. (One recovers the analytic values with $d = 4$.)

Fluctuations in density and composition produce opalescence, a recognized feature of the critical region. Since systems very close to a critical point become visibly opaque, the fluctuations must extend over ranges comparable to the wavelength of light (i.e. to distances very much greater than molecular dimensions). Measurements of light scattering can yield quantitative information about the compressibility and thus about the magnitude of the fluctuations. Such measurements in the critical region showed the failure of the analytic predictions and yielded the first good experimental determinations of the exponent γ. As predicted even by classical theory the light scattering (i.e. the compressibility) on the critical isochore at a small temperature δT above the critical temperature is larger than that at the same δT below the critical temperature along the coexistence curve.

What this means is that mean-field (analytic) treatments fail whenever the range of correlations greatly exceeds the range of intermolecular forces. It follows that under these circumstances there should be no difference between the limiting behaviour of an Ising lattice and the nonlattice fluids; they should have the same exponents. Nearly a century after the introduction of the van der Waals equation for fluids, Kac, Uhlenbeck and Hemmer (1963) [20] proved that, in a one-dimensional system, it is exact for an intermolecular interaction that is infinite in range and infinitesimal in magnitude. (It is interesting to note that, in disagreement with van der Waals, Boltzmann insisted that the equation could only be correct if the range of the interactions were infinite.)

Moreover, well away from the critical point, the range of correlations is much smaller, and when this range is of the order of the range of the intermolecular forces, analytic treatments should be appropriate, and the exponents should be 'classical'. The need to reconcile the nonanalytic region with the classical region has led to attempts to solve the 'crossover' problem, to be discussed in section A2.5.7.2.

A2.5.6.4 A uniform geometric view of critical phenomena: 'fields' and 'densities'

While there was a general recognition of the similarity of various types of critical phenomena, the situation was greatly clarified in 1970 by a seminal paper by Griffiths and Wheeler [21]. In particular the difference between variables that are 'fields' and those that are 'densities' was stressed. A 'field' is any variable that is the same in two phases at equilibrium, (e.g. pressure, temperature, chemical potential, magnetic field). Conversely a 'density' is a variable that is different in the two phases (e.g. molar volume or density, a composition variable like mole fraction or magnetization). The similarity between different kinds of critical phenomena is seen more clearly when the phase diagram is shown exclusively with field variables. (Examples of this are figures A2.5.1 and A2.5.11.)

The field-density concept is especially useful in recognizing the parallelism of path in different physical situations. The criterion is the number of densities held constant; the number of fields is irrelevant. A path to the critical point that holds only fields constant produces a strong divergence; a path with one density held constant yields a weak divergence; a path with two or more densities held constant is nondivergent. Thus the compressibility κ_T of a one-component fluid shows a strong divergence, while C_V in the one-component fluid is comparable to C_{px} (constant pressure and composition) in the two-component fluid and shows a weak divergence.

The divergences of the heat capacity C_V and of the compressibility κ_T for a one-component fluid are usually defined as along the critical isochore, but if the phase diagram is shown in field space (p versus T as in figures A2.5.1 or A2.5.11), it is evident that this is a 'special' direction along the vapour pressure curve. Indeed any direction that lies within the coexistence curve (e.g. constant enthalpy etc) and intersects that curve at the critical point will yield the same exponents. Conversely any path that intersects this special direction, such as the critical isobar, will yield different exponents. These other directions are not unique; there is no such thing as orthogonality in thermodynamics. Along the critical isobar, the compressibility divergence is still strong, but the exponent is reduced by renormalization from γ to $\gamma/\beta\delta$, nearly a 40% reduction. The weak divergence of C_V is reduced by a similar amount from α to $\alpha/\beta\delta$.

Another feature arising from field-density considerations concerns the coexistence curves. For one-component fluids, they are usually shown as temperature T versus density ρ, and for two-component systems, as temperature versus composition (e.g. the mole fraction x); in both cases one field is plotted against one density. However in three-component systems, the usual phase diagram is a triangular one at constant temperature; this involves two densities as independent variables. In such situations exponents may be 'renormalized' to higher values; thus the coexistence curve exponent may rise to $\beta/(1-\alpha)$. (This 'renormalization' has nothing to do with the 'renormalization group' to be discussed in the next section.)

Finally the concept of fields permits clarification of the definition of the order of transitions [22]. If one considers a space of all fields (e.g. figure A2.5.1, but not figure A2.5.3), a first-order transition occurs where there is a discontinuity in the first derivative of one of the fields with respect to another (e.g. $(\partial\mu/\partial T)_p = -\overline{S}$ and $(\partial\mu/\partial p)_T = \overline{V}$), while a second-order transition occurs when the corresponding first derivative is continuous but the second is not and so on. Thus the Ehrenfest–Pippard definitions are preserved if the paths are not defined in terms of any densities. A feature of a critical point, line, or surface is that it is located where divergences of various properties, in particular correlation lengths, occur. Moreover it is reasonable to assume that at such a point there is always an order parameter that is zero on one side of the transition and that becomes nonzero on the other side. Nothing of this sort occurs at a first-order transition, even the gradual liquid–gas transition shown in figures A2.5.3 and A2.5.4.

A2.5.6.5   The calculation of exponents

From 1965 on there was an extensive effort to calculate, or rather to estimate, the exponents for the Ising model. Initially this usually took the form of trying to obtain a low-temperature expansion (i.e. in powers of T) or a high-temperature expansion (i.e. in powers of $1/T$) of the partition function, in the hope of obtaining information about the ultimate form of the series, and hence to learn about the singularities at the critical point. Frequently this effort took the form of converting the finite series (sometimes with as many as 25 terms) into a Padé approximant, the ratio of two finite series. From this procedure, estimates of the various critical exponents (normally as the ratio of two integers) could be obtained. For the two-dimensional Ising model these estimates agreed with the values deduced by Onsager and Yang, which encouraged the belief that those for the three-dimensional model might be nearly correct. Indeed the $d = 3$ exponents estimated from theory were in reasonable agreement with those deduced from experiments close to the critical point. In this period much of the theoretical progress was made by Domb, Fisher, Kadanoff, and their coworkers.

In 1971 Wilson [23] recognized the analogy between quantum-field theory and the statistical mechanics of critical phenomena and developed a renormalization-group (RG) procedure that was quickly recognized as a better approach for dealing with the singularities at the critical point. New calculation methods were developed, one of which, expansion in powers of $\varepsilon = 4 - d$, where d is the dimension taken as a continuous variable, was first proposed by Wilson and Fisher (1972). These new procedures led to theoretical values of the critical exponents with much smaller estimates of uncertainty. The best current values are shown in table A2.5.1 in section A2.5.7.1. The RG method does assume, without proof, the homogeneity hypothesis and thus that the exponent inequalities are equalities. Some might wish that these singularities and exponents could be derived from a truly molecular statistical-mechanical theory; however, since the singular behaviour arises from the approach of the correlations to infinite distance, this does not seem likely in the foreseeable future. This history, including a final chapter on the renormalization group, is discussed in detail in a recent (1996) book by Domb [23].

A2.5.6.6 *Extended scaling. Wegner corrections*

In 1972 Wegner [25] derived a power-series expansion for the free energy of a spin system represented by a Hamiltonian roughly equivalent to the scaled equation (A2.5.28), and from this he obtained power-series expansions of various thermodynamic quantities around the critical point. For example the compressibility can be written as

$$\kappa_T = \kappa_T^0 + \Gamma_0 t^{-\gamma} + \Gamma_1 t^{-\gamma + \Delta_1} + \Gamma_2 t^{-\gamma + \Delta_2} + \cdots.$$

The new parameters in the exponents, Δ_1 and Δ_2, are exactly or very nearly 0.50 and 1.00 respectively. Similar equations apply to the 'extended scaling' of the heat capacity and the coexistence curve for the determination of α and β.

The Wegner corrections have been useful in analysing experimental results in the critical region. The 'correct' exponents are the limiting values as T_r approaches unity, not the average values over a range of temperatures. Unfortunately the Wegner expansions do not converge very quickly (if they converge at all), so the procedure does not help in handling a crossover to the mean-field behaviour at lower temperatures where the correlation length is of the same order of magnitude as the range of intermolecular forces. A consistent method of handling crossover is discussed in section A2.5.7.2.

A2.5.6.7 *Some experimental problems*

The scientific studies of the early 1970s are full of concern whether the critical exponents determined experimentally, particularly those for fluids, could be reconciled with the calculated values, and at times it appeared that they could not be. However, not only were the theoretical values more uncertain (before RG calculations) than first believed, but also there were serious problems with the analysis of the experiments, in addition to those associated with the Wegner corrections outlined above. Scott [26] has discussed in detail experimental difficulties with binary fluid mixtures, but some of the problems he cited apply to one-component fluids as well.

An experiment in the real world has to deal with gravitational effects. There will be gravity-induced density gradients and concentration gradients such that only at one height in an experimental cell will the system be truly at the critical point. To make matters worse, equilibration in the critical region is very slow. These problems will lead to errors of uncertain magnitude in the determination of all the critical exponents. For example, the observed heat capacity will not display an actual divergence because the total enthalpy is averaged over the whole cell and only one layer is at the critical point.

Another problem can be the choice of an order parameter for the determination of β and of the departure from linearity of the diameter, which should be proportional to $t^{1-\alpha}$. In the symmetrical systems, the choice of the order parameter s is usually obvious, and the symmetry enforces a rectilinear diameter. Moreover, in the one-component fluid, the choice of the reduced density ρ/ρ_c has always seemed the reasonable choice. However, for the two-component fluid, there are two order parameters, density and composition. It is not the density ρ that drives the phase separation, but should the composition order parameter be mole fraction x, volume fraction ϕ, or what? For the coexistence exponent β the choice is ultimately immaterial if one gets close enough to the critical temperature, although some choices are better than others in yielding an essentially cubic curve over a greater range of reduced temperature. (Try plotting the van der Waals coexistence curve against molar volume $\overline{V}$ instead of density ρ.) However this ambiguity can have a very serious effect on any attempt to look for experimental evidence for departures from the rectilinear diameter in binary mixtures; an unwise choice for the order parameter can yield an exponent 2β rather than the theoretical $1 - \alpha$ (previously discussed in section A2.5.6.2) thus causing a much greater apparent departure from linearity.

A2.5.7 The current status of the Ising model; theory and experiment

Before reviewing the current knowledge about Ising systems, it is important to recognize that there are non-Ising systems as well. A basic feature of the Ising model is that the order parameter is a scalar, even in the magnetic system of spin $1/2$. If the order parameter is treated as a vector, it has a dimensionality n, such that $n = 1$ signifies a scalar (the Ising model), $n = 2$ signifies a vector with two components (the XY model), $n = 3$ signifies a three-component vector (the Heisenberg model), $n = \infty$ is an unphysical limit to the vector concept (the so-called spherical model), and $n \rightarrow 0$ is a curious mathematical construct that seems to fit critical phenomena in some polymer equilibria. Some of these models will be discussed in subsequent sections, but first we limit ourselves to the Ising model.

A2.5.7.1 The Ising exponents and amplitudes

There is now consensus on some questions about which there had been lingering doubts.

(a) There is now agreement between experiment and theory on the Ising exponents. Indeed it is now reasonable to assume that the theoretical values are better, since their range of uncertainty is less.

(b) There is no reason to doubt that the inequalities of section A2.5.4.5(e) are other than equalities. The equalities are assumed in most of the theoretical calculations of exponents, but they are confirmed (within experimental error) by the experiments.

(c) The exponents apply not only to solid systems (e.g. order–disorder phenomena and simple magnetic systems), but also to fluid systems, regardless of the number of components. (As we have seen in section A2.5.6.4 it is necessary in multicomponent systems to choose carefully the variable to which the exponent is appropriate.)

(d) There is no distinction between the exponents above and below the critical temperature. Thus $\gamma^+ = \gamma^- = \gamma$ and $\alpha^+ = \alpha_2^- = \alpha_1^- = \alpha$. However, there is usually a significant difference in the coefficients above and below (e.g. A^+ and A^-); this produces the discontinuities at the critical point.

Many of the earlier uncertainties arose from apparent disagreements between the theoretical values and experimental determinations of the critical exponents. These were resolved in part by better calculations, but mainly by measurements closer and closer to the critical point. The analysis of earlier measurements assumed incorrectly that the measurements were close enough. (Van der Waals and van Laar were right that one needed to get closer to the critical point, but were wrong in expecting that the classical exponents would then appear.) As was shown in section A2.5.6.7, there are additional contributions from 'extended' scaling.

Moreover, some uncertainty was expressed about the applicability to fluids of exponents obtained for the Ising lattice. Here there seemed to be a serious discrepancy between theory and experiment, only cleared up by later and better experiments. By hindsight one should have realized that long-range fluctuations should be independent of the presence or absence of a lattice.

Table A2.5.1 shows the Ising exponents for two and three dimensions, as well as the classical exponents. The uncertainties are those reported by Guida and Zinn-Justin [27]. These exponent values satisfy the equalities (as they must, considering the scaling assumption) which are here reprised as functions of β and γ:

$$\alpha = 2 - 2\beta - \gamma$$
$$\delta = (\beta + \gamma)/\beta$$
$$\nu = (2\beta + \gamma)/d$$
$$\eta = 2 - d\beta\gamma/(2\beta + \gamma) = 2 - \gamma/\nu.$$

The small uncertainties in the calculated exponents seem to preclude the possibility that the $d = 3$ exponents are rational numbers (i.e. the ratio of integers). (At an earlier stage this possibility had been

Table A2.5.1. Ising model exponents.

	Exponent	$d = 2$	$d = 3$	Classical ($d \geq 4$)
α	Heat capacity, $\overline{C}_V, \overline{C}_{p,x}, \overline{C}_M$	0 (log)	0.109 ± 0.004	0 (finite jump)
β	Coexistence, $\Delta\rho, \Delta x, \Delta M$	1/8	0.3258 ± 0.0014	1/2
γ	Compressibility, $\kappa_T, \alpha_p, \kappa_{T,\mu}, \chi_T$	7/4	1.2396 ± 0.0013	1
δ	Critical isotherm, $p(V), \mu(x), B_0(M)$	15	4.8047 ± 0.0044	3
ν	Correlation length, ξ	1	0.6304 ± 0.0013	1/2
η	Critical correlation function	1/4	0.0335 ± 0.0025	0

suggested, since not only the classical exponents, but also the $d = 2$ exponents are rational numbers; pre-RG calculations had suggested $\beta = 5/16$ and $\gamma = 5/4$.)

As noted earlier in section A2.5.6.2, the assumption of homogeneity and the resulting principle of two-scale-factor universality requires the amplitude coefficients to be related. In particular the following relations can be derived:

$$\alpha A^+ \Gamma^+ / B^2 = 0.0574 \pm 0.0020$$
$$\Gamma^+ D B^{\delta-1} = 1.669 \pm 0.018$$
$$A^+ / A^- = 0.537 \pm 0.019$$
$$\Gamma^+ / \Gamma^- = 4.79 \pm 0.10.$$

These numerical values come from theory [26] and are in good agreement with recent experiments.

A2.5.7.2 *Crossover from mean-field to the critical region*

At temperatures well below the critical region one expects a mean-field treatment (or at least a fully analytic one) to be applicable, since the correlations will be short range. In the critical region, as we have seen, when the correlation length becomes far greater than the range of intermolecular forces, the mean-field treatment fails. Somewhere between these two limits the treatment of the problem has to 'cross over'. Early attempts to bridge the gap between the two regimes used switching functions, and various other solutions have been proposed. A reasonably successful treatment has been developed during the past few years by Anisimov and Sengers and their collaborators. (Detailed references will be found in a recent review chapter [28].)

As a result of long-range fluctuations, the local density will vary with position; in the classical Landau–Ginzburg theory of fluctuations this introduces a gradient term. A Ginzburg number N_G is defined (for a three-dimensional Ising system) as proportional to a dimensionless parameter ξ_0^6 / v_0^2 which may be regarded as the inverse sixth power of a normalized interaction range. (ξ_0 is the coefficient of the correlation length equation in section A2.5.6.3 and v_0 is a molecular volume.) The behaviour of the fluid will be nonanalytic (Ising-like) when $\tau = (T_c - T)/T = t/(1-t)$ is much smaller than N_G, while it is analytic (van der Waals-like) when τ is much greater than N_G. A significant result of this recent research is that the free energy can be rescaled to produce a continuous function over the whole range of temperatures.

For simple fluids N_G is estimated to be about 0.01, and Kostrowicka Wyczalkowska *et al* [29] have used this to apply crossover theory to the van der Waals equation with interesting results. The critical temperature T_c is reduced by 11% and the coexistence curve is of course flattened to a cubic. The critical density ρ_c is almost unchanged (by 2%), but the critical pressure p_c is reduced greatly by 38%. These changes reduce the

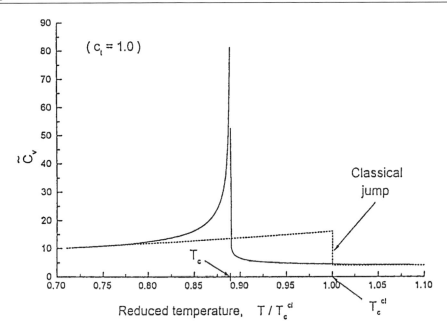

Figure A2.5.26. Molar heat capacity $\overline{C}_V$ of a van der Waals fluid as a function of temperature: from mean-field theory (dotted line); from crossover theory (full curve). Reproduced from [29] Kostrowicka Wyczalkowska A, Anisimov M A and Sengers J V 1999 Global crossover equation of state of a van der Waals fluid *Fluid Phase Equilibria* **158–160** 532, figure 4, by permission of Elsevier Science.

critical compression factor $(p\overline{V}/RT)_c$ from 3.75 to 2.6; the experimental value for argon is 2.9. The molar heat capacity $\overline{C}_V$ for the classical van der Waals fluid and the crossover van der Waals fluid are compared in figure A2.5.26.

Povodyrev *et al* [30] have applied crossover theory to the Flory equation (section A2.5.4.1) for polymer solutions for various values of N, the number of monomer units in the polymer chain, obtaining the coexistence curve and values of the coefficient β_{eff} from the slope of that curve. Figure A2.5.27 shows their comparison between classical and crossover values of β_{eff} for $N = 1$, which is of course just the simple mixture. As seen in this figure, the crossover to classical behaviour is not complete until far below the critical temperature.

Sengers and coworkers (1999) have made calculations for the coexistence curve and the heat capacity of the real fluid SF_6 and the real mixture 3-methylpentane + nitroethane and the agreement with experiment is excellent; their comparison for the mixture [28] is shown in figure A2.5.28.

However, for more complex fluids such as high-polymer solutions and concentrated ionic solutions, where the range of intermolecular forces is much longer than that for simple fluids and N_G is much smaller, mean-field behaviour is observed much closer to the critical point. Thus the crossover is sharper, and it can also be nonmonotonic.

A2.5.8   Other examples of second-order transitions

There are many other examples of second-order transitions involving critical phenomena. Only a few can be mentioned here.

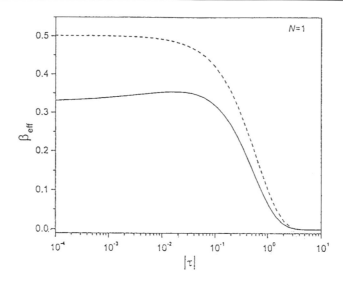

Figure A2.5.27. The effective coexistence curve exponent $\beta_{\text{eff}} = \mathrm{d}\ln x/\mathrm{d}\ln\tau$ for a simple mixture ($N = 1$) as a function of the temperature parameter $\tau = t/(1-t)$ calculated from crossover theory and compared with the corresponding curve from mean-field theory (i.e. from figure A2.5.15). Reproduced from [30], Povodyrev A A, Anisimov M A and Sengers J V 1999 Crossover Flory model for phase separation in polymer solutions *Physica* A **264** 358, figure 3, by permission of Elsevier Science.

A2.5.8.1   Two-dimensional Ising systems

No truly two-dimensional systems exist in a three-dimensional world. However monolayers absorbed on crystalline or fluid surfaces offer an approximation to two-dimensional behaviour. Chan and coworkers [31] have measured the coexistence curve for methane adsorbed on graphite by an ingenious method of determining the maximum in the heat capacity at various coverages. The coexistence curve (figure A2.5.29) is fitted to $\beta = 0.127$, very close to the theoretical $1/8$. A 1992 review [32] summarizes the properties of rare gases on graphite.

A2.5.8.2   The XY model ($n = 2$)

If the scalar order parameter of the Ising model is replaced by a two-component vector ($n = 2$), the *XY* model results. An important example that satisfies this model is the λ-transition in helium, from superfluid helium-II to ordinary liquid helium, occurring for the isotope ^4He and for mixtures of ^4He with ^3He. (This is the transition at 1.1 K, not the liquid–gas critical point at 5.2 K, which is Ising.) Calculations indicate that at the $n = 2$ transition, the heat capacity exponent α is very small, but negative. If so, the heat capacity does not diverge, but rather reaches a maximum just at the λ-point, as shown in the following equation:

$$\overline{C}(n = 2, d = 3) = \overline{C}_{\text{max}} - At^{-\alpha}$$

where $\overline{C}_{\text{max}}$ is the value at the λ-transition. At first this prediction was hard to distinguish experimentally from a logarithmic divergence but experiments in space under conditions of microgravity by Lipa and coworkers (1996) have confirmed it [33] with an $\alpha = -0.01285$, a value within the limits of uncertainty of the theoretical calculations. The results above and below the transition were fitted to the same value of $\overline{C}_{\text{max}}$ and α but with $A^+/A^- = 1.054$. Since the heat capacity is finite and there is no discontinuity, this should perhaps be called a third-order transition.

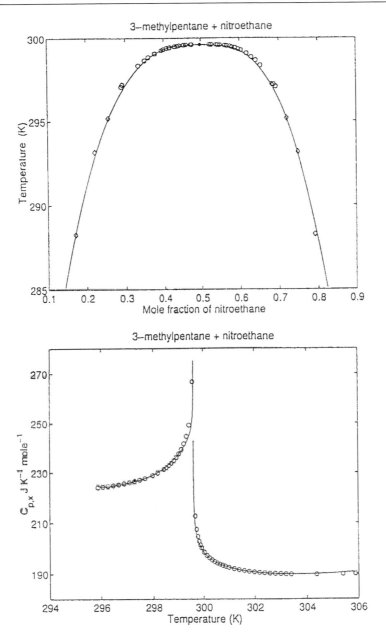

Figure A2.5.28. The coexistence curve and the heat capacity of the binary mixture 3-methylpentane + nitroethane. The circles are the experimental points, and the lines are calculated from the two-term crossover model. Reproduced from [28], 2000 *Supercritical Fluids—Fundamentals and Applications* ed E Kiran, P G Debenedetti and C J Peters (Dordrecht: Kluwer) Anisimov M A and Sengers J V Critical and crossover phenomena in fluids and fluid mixtures, p 16, figure 3, by kind permission from Kluwer Academic Publishers.

The liquid-crystal transition between smectic-A and nematic for some systems is an *XY* transition. Depending on the value of the MacMillan ratio, the ratio of the temperature of the smectic-A-nematic transition

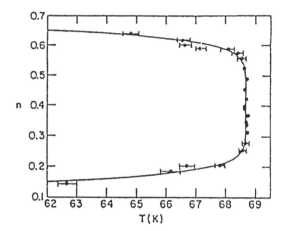

Figure A2.5.29. Peak positions of the liquid–vapour heat capacity as a function of methane coverages on graphite. These points trace out the liquid–vapour coexistence curve. The full curve is drawn for $\beta = 0.127$. Reproduced from [31] Kim H K and Chan M H W *Phys. Rev. Lett.* **53** 171 (1984) figure 2. Copyright (1984) by the American Physical Society.

to that of the nematic-isotropic transition (which is Ising), the behaviour of such systems varies continuously from a λ-type transition to a tricritical one (see section A2.5.9). Garland and Nounesis [34] reviewed these systems in 1994.

A2.5.8.3 The Heisenberg model (n = 3)

While the behaviour of some magnetic systems is Ising-like, others require a three-dimensional vector. In the limit of where the value of the quantum number J goes to infinity (i.e. where all values of the magnetic quantum number M are possible), the Heisenberg model ($n = 3$) applies. The exponents β and γ are somewhat larger than the Ising or XY values; the exponent α is substantially negative (about -0.12).

A2.5.8.4 Polymerization systems (n → 0)

Some equilibrium polymerizations are such that over a range of temperatures only the monomer exists in any significant quantity, but below or above a unique temperature polymers start to form in increasing number. Such a polymerization temperature is a critical point, another kind of second-order transition. The classic example is that of the ring-chain transition in sulfur, but more recently similar behaviour has been found in a number of 'living polymers'. Wheeler and coworkers [35] have shown that these systems can best be treated as examples of the mathematical limit of the n-vector model with $n \to 0$. The heat capacity in such a system diverges more strongly than that of an Ising system ($\alpha = 0.235$ [27]); the heat capacity of sulfur fits the model qualitatively, but there are chemical complications.

   Mixtures of such polymeric substances with solvents show a line of critical points that in theory end at a tricritical point. (See section A2.5.9 for further discussion of tricritical phenomena.)

A2.5.8.5 Superconductivity

Alone among all known physical phenomena, the transition in low-temperature ($T_c < 25$ K) superconducting materials (mainly metals and alloys) retains its classical behaviour right up to the critical point; thus the exponents are the analytic ones. Unlike the situation in other systems, such superconducting interactions are truly long range and thus mean field.

For the newer high-temperature superconducting materials, the situation is different. These substances crystallize in structures that require a two-component order parameter and show XY behaviour, usually three dimensional (i.e. $n = 2$, $d = 3$). Pasler *et al* [36] have measured the thermal expansivity of $YBa_2Cu_3O_{7-\delta}$ and have found the exponent α to be 0 ± 0.018, which is consistent with the small negative value calculated for the XY model and found for the λ-transition in helium.

A2.5.9   Multicritical points

An ordinary critical point such as those discussed in earlier sections occurs when two phases become more and more nearly alike and finally become one. Because this involves two phases, it is occasionally called a 'bicritical point'. A point where three phases simultaneously become one is a 'tricritical point'. There are two kinds of tricritical points, symmetrical and unsymmetrical; there is a mathematical similarity between the two, but the physical situation is so different that they need to be discussed quite separately. One feature that both kinds have in common is that the dimension at and above which modern theory yields agreement between 'classical' and 'nonclassical' treatments is $d = 3$, so that analytic treatments (e.g. mean-field theories) are applicable to paths leading to tricritical points, unlike the situation with ordinary critical points where the corresponding dimension is $d = 4$. (In principle there are logarithmic corrections to these analytic predictions for $d = 3$, but they have never been observed directly in experiments.)

A 1984 volume reviews in detail theories and experiments [37] on multicritical points; some important papers have appeared since that time.

A2.5.9.1   Symmetrical tricritical points

In the absence of special symmetry, the phase rule requires a minimum of three components for a tricritical point to occur. Symmetrical tricritical points do have such symmetry, but it is easiest to illustrate such phenomena with a true ternary system with the necessary symmetry. A ternary system comprised of a pair of enantiomers (optically active d- and l-isomers) together with a third optically inert substance could satisfy this condition. While liquid–liquid phase separation between enantiomers has not yet been found, ternary phase diagrams like those shown in figure A2.5.30 can be imagined; in these diagrams there is a necessary symmetry around a horizontal axis that represents equal amounts of the two enantiomers.

Now consider such a symmetrical system, that of a racemic mixture of the enantiomers plus the inert third component. A pair of mirror-image conjugate phases will not physically separate or even become turbid, since they have exactly the same density and the same refractive index. Unless we find evidence to the contrary, we might conclude that this is a binary mixture with a T, x phase diagram like one of those on the right-hand side of figure A2.5.30. In particular any symmetrical three-phase region will have to shrink symmetrically, so it may disappear at a tricritical point, as shown in two of the four 'pseudobinary' diagrams. The dashed lines in these diagrams are two-phase critical points, and will show the properties of a second-order transition. Indeed, a feature of these diagrams is that with increasing temperature, a first-order transition ends at a tricritical point that is followed by a second-order transition line. (This is even more striking if the phase diagram is shown in field space as a p, T or μ, T diagram.)

These unusual 'pseudobinary' phase diagrams were derived initially by Meijering (1950) from a 'simple mixture' model for ternary mixtures. Much later, Blume, Emery and Griffiths (1971) deduced the same diagrams from a three-spin model of helium mixtures. The third diagram on the right of figure A2.5.30 is essentially that found experimentally for the fluid mixture $^4He + {}^3He$; the dashed line (second-order transition) is that of the λ-transition.

Symmetrical tricritical points are predicted for fluid mixtures of sulfur or living polymers in certain solvents. Scott (1965) in a mean-field treatment [38] of sulfur solutions found that a second-order transition

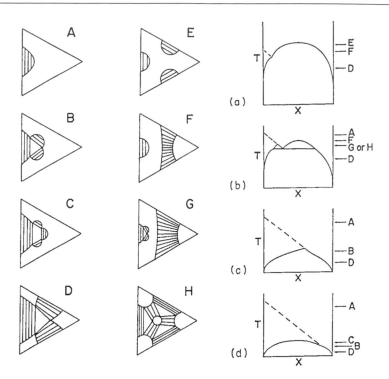

Figure A2.5.30. Left-hand side: Eight hypothetical phase diagrams (A through H) for ternary mixtures of d- and l-enantiomers with an optically inactive third component. Note the symmetry about a line corresponding to a racemic mixture. Right-hand side: Four T, x diagrams ((a) through (d)) for 'pseudobinary' mixtures of a racemic mixture of enantiomers with an optically inactive third component. Reproduced from [37] 1984 *Phase Transitions and Critical Phenomena* ed C Domb and J Lebowitz, vol 9, ch 2, Knobler C M and Scott R L Multicritical points in fluid mixtures. Experimental studies pp 213–14, (Copyright 1984) by permission of the publisher Academic Press.

line (the critical polymerization line) ended where two-phase separation of the polymer and the solvent begins; the theory yields a tricritical point at that point. Later Wheeler and Pfeuty [39] extended their $n \to 0$ treatment of equilibrium polymerization to sulfur solutions; in mean field their theory reduces to that of Scott, and the predictions from the nonclassical formulation are qualitatively similar. The production of impurities by slow reaction between sulfur and the solvent introduces complications; it can eliminate the predicted three-phase equilibrium, flatten the coexistence curve and even introduce an unsymmetrical tricritical point.

Symmetrical tricritical points are also found in the phase diagrams of some systems forming liquid crystals.

A2.5.9.2 *Unsymmetrical tricritical points*

While, in principle, a tricritical point is one where three phases simultaneously coalesce into one, that is not what would be observed in the laboratory if the temperature of a closed system is increased along a path that passes exactly through a tricritical point. Although such a difficult experiment is yet to be performed, it is clear from theory (Kaufman and Griffiths 1982, Pegg *et al* 1990) and from experiments in the vicinity of tricritical points that below the tricritical temperature T_t only two phases coexist and that the volume of one shrinks precipitously to zero at T_t.

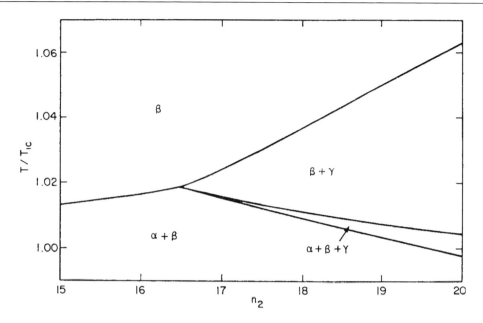

Figure A2.5.31. Calculated T/T_{1c}, n_2 phase diagram in the vicinity of the tricritical point for binary mixtures of ethane ($n_1 = 2$) with a higher hydrocarbon of continuous n_2. The system is in a sealed tube at fixed tricritical density and composition. The tricritical point is at the confluence of the four lines. Because of the fixing of the density and the composition, the system does not pass through critical end points; if the critical end-point lines were shown, the three-phase region would be larger. An experiment increasing the temperature in a closed tube would be represented by a vertical line on this diagram. Reproduced from [40], figure 8, by permission of the American Institute of Physics.

While the phase rule requires three components for an unsymmetrical tricritical point, theory can reduce this requirement to two components with a continuous variation of the interaction parameters. Lindh *et al* (1984) calculated a phase diagram from the van der Waals equation for binary mixtures and found (in accord with figure A2.5.13) that a tricritical point occurred at sufficiently large values of the parameter ζ (a measure of the difference between the two components).

One can effectively reduce the three components to two with 'quasibinary' mixtures in which the second component is a mixture of very similar higher hydrocarbons. Figure A2.5.31 shows a phase diagram [40] calculated from a generalized van der Waals equation for mixtures of ethane ($n_1 = 2$) with normal hydrocarbons of different carbon number n_2 (treated as continuous). It is evident that, for some values of the parameter n_2, those to the left of the tricritical point at $n_2 = 16.48$, all that will be observed with increasing temperature is a two-phase region ($\alpha + \beta$) above which only the β phase exists. Conversely, for larger values of n_2, those to the right of the tricritical point, increasing the temperature takes the system from the two-phase region ($\alpha + \beta$) through a narrow three-phase region ($\alpha + \beta + \gamma$) to a different two-phase region ($\beta + \gamma$).

Most of the theoretical predictions have now been substantially verified by a large series of experiments in a number of laboratories. Knobler and Scott and their coworkers (1977–1991) have studied a number of quasibinary mixtures, in particular ethane + (hexadecane + octadecane) for which the experimental $n_2 = 17.6$. Their experimental results essentially confirm the theoretical predictions shown in figure A2.5.31.

A2.5.9.3 Higher-order critical points

Little is known about higher order critical points. Tetracritical points, at least unsymmetrical ones, require four components. However for tetracritical points, the crossover dimension $d = 2$, so any treatment can surely be mean-field, or at least analytic.

A2.5.10 Higher-order phase transitions

We have seen in previous sections that the two-dimensional Ising model yields a symmetrical heat capacity curve that is divergent, but with no discontinuity, and that the experimental heat capacity at the λ-transition of helium is finite without a discontinuity. Thus, according to the Ehrenfest–Pippard criterion these transitions might be called third-order.

It has long been known from statistical mechanical theory that a Bose–Einstein ideal gas, which at low temperatures would show condensation of molecules into the ground translational state (a condensation in momentum space rather than in position space), should show a third-order phase transition at the temperature at which this condensation starts. Normal helium (^{4}He) is a Bose–Einstein substance, but is far from ideal at low temperatures, and the very real forces between molecules make the λ-transition to He II very different from that predicted for a Bose–Einstein gas.

Recent research (1995–) has produced at very low temperatures (nanokelvins) a Bose–Einstein condensation of magnetically trapped alkali metal atoms. Measurements [41] of the fraction of molecules in the ground state of ^{87}Rb as a function of temperature show good agreement with the predictions for a finite number of noninteracting bosons in the three-dimensional harmonic potential produced by the magnets; indeed the difference in this occupancy differs only slightly from that predicted for translation in a 3D box. However the variation of the energy as a function of temperature is significantly different from that predicted for a 3D box; the harmonic potential predicts a discontinuity in the heat capacity which is confirmed by experiment; thus this transition is second-order rather than third-order.

Acknowledgments

I want to thank Anneke and Jan Sengers for supplying me with much information and for critical reading of parts of the manuscript. However any errors, omissions or misplaced emphases are entirely my own.

References

[1] Pippard A B 1957 *The Elements of Classical Thermodynamics* (Cambridge: Cambridge University Press) pp 136–59

[2] Van der Waals J D 1873 Over de continuiteit van den gas- en vloeistoftoestand *PhD Thesis* Sijthoff, Leiden (Engl. Transl. 1988 *J. D. van der Waals: On the Continuity of the Gaseous and Liquid States* ed J S Rowlinson, vol. XIV of *Studies in Statistical Mechanics* ed J L Lebowitz (Amsterdam: North-Holland))

[3] Van Konynenburg P H and Scott R L 1980 Critical lines and phase equilibria in van der Waals mixtures *Phil. Trans. R. Soc.* **298** 495–540

[4] *Workshop on Global Phase Diagrams* 1999 *PCCP* **1** 4225–326 (16 papers from the workshop held at Walberberg, Germany 21–24 March 1999)

[5] Bragg W L and Williams E J 1934 The effect of thermal agitation on atomic arrangement in alloys *Proc. R. Soc.* A **145** 699–730

[6] Nix F C and Shockley W 1938 Order and disorder in alloys *Rev. Mod. Phys.* **10** 1–69

[7] Weiss P 1907 L'Hypothèse du champ moleculaire et la propriété ferromagnétique *J. Phys. Radium Paris* **6** 661–90

[8] Levelt Sengers J M H 1976 Critical exponents at the turn of the century *Physica* A **82** 319–51

[9] Levelt Sengers J M H 1999 Mean-field theories, their weaknesses and strength *Fluid Phase Equilibria* **158–160** 3–17

[10] Guggenheim E A 1945 The principle of corresponding states *J. Chem. Phys.* **13** 253–61

[11] Scott R L 1953 Second-order transitions and critical phenomena *J. Chem. Phys.* **21** 209–11

[12] Bagatskii M I, Voronel A V and Gusak V G 1962 Determination of heat capacity C_v of argon in the immediate vicinity of the critical point *Zh. Eksp. Teor. Fiz.* **43** 728–9

[13] Green M S and Sengers J V (eds) 1966 *Critical Phenomena, Proc. Conf. (April, 1965)* (Washington: National Bureau of Standards Miscellaneous Publication 273)

[14] Ising E 1925 Beitrag sur theorie des ferromagnetismus *Z. Phys.* **31** 253–8

[15] Landau L D and Lifschitz E M 1969 *Statistical Physics* 2nd English edn (Oxford: Pergamon) chapter XIV. The quotation is from the first English edition (1959). The corresponding statement in the second English edition is slightly more cautious. However even the first edition briefly reports the results of Onsager and Yang for two dimensions, but leaves the reader with the belief (or the hope?) that somehow three dimensions is different.

[16] Onsager L 1944 Crystal Statistics. I. A two-dimensional model with an order–disorder transition *Phys. Rev.* **65** 117–49

[17] Widom B 1965 Equation of state in the neighborhood of the critical point *J. Chem. Phys.* **43** 3898–905

[18] Jüngst S, Knuth B and Hensel F 1985 Observation of singular diameters in the coexistence curves of metals *Phys. Rev. Lett.* **55** 2160–3

[19] Pestak M W, Goldstein R E, Chan M H W, de Bruyn J R, Balzarini D A and Ashcroft N W 1987 Three-body interactions, scaling variables, and singular diameters in the coexistence curves of fluids *Phys. Rev.* B **36** 599–614

[20] Kac M, Uhlenbeck G E and Hemmer P C 1963 On the van der Waals theory of the vapor-liquid equilibrium. I. Discussion of a one-dimensional model *J. Math. Phys.* **4** 216–28

[21] Griffiths R B and Wheeler J C 1970 Critical points in multicomponent systems *Phys. Rev.* A **2** 1047–64

[22] Wheeler J C 2000 Personal communication

[23] Wilson K G 1971 Renormalization group and critical phenomena. I. Renormalization group and the Kadanoff scaling picture *Phys. Rev.* B **4** 3174–83
      Wilson K G 1971 Renormalization group and critical phenomena. II. Phase space cell analysis of critical behaviour *Phys. Rev.* B **4** 3184–205

[24] Domb C 1996 *The Critical Point. A Historical Introduction to the Modern Theory of Critical Phenomena* (London and Bristol, PA: Taylor and Francis)

[25] Wegner F J 1972 Corrections to scaling laws *Phys. Rev.* B **5** 4529–36

[26] Scott R L 1978 Critical exponents for binary fluid mixtures *Specialist Periodical Reports, Chem. Thermodynam.* **2** 238–74

[27] Guida R and Zinn-Justin J 1998 Critical exponents of the N-vector model *J. Phys. A Mathematical and General* **31** 8103–21

[28] Anisimov M A and Sengers J V 2000 Critical and crossover phenomena in fluids and fluid mixtures *Supercritical Fluids- Fundamentals and Applications* ed E Kiran, P G Debenedetti and C J Peters (Dordrecht: Kluwer) pp 1–33

[29] Kostrowicka Wyczalkowska A, Anisimov M A and Sengers J V 1999 Global crossover equation of state of a van der Waals fluid *Fluid Phase Equilibria* **158–160** 523–35

[30] Povodyrev A A, Anisimov M A and Sengers J V 1999 Crossover Flory model for phase separation in polymer solutions *Physica* A **264** 345–69

[31] Kim H K and Chan M H W 1984 Experimental determination of a two-dimensional liquid-vapor critical exponent *Phys. Rev. Lett.* **53** 170–3

[32] Shrimpton N D, Cole M W, Steele W A and Chan M H W 1992 Rare gases on graphite *Surface Properties of Layered Structures* ed G Benedek, (Dordrecht: Kluwer) pp 219–69

[33] Lipa J A, Swanson D R, Nissen J A, Chui T C P and Israelsson U E 1996 Heat capacity and thermal relaxation of bulk helium very near the lambda point *Phys. Rev. Lett.* **76** 944–7

[34] Garland C W and Nounesis G 1994 Critical behavior at nematic-smectic-A phase transitions *Phys. Rev.* E **49** 2964–71

[35] Wheeler J C, Kennedy S J and Pfeuty P 1980 Equilibrium polymerization as a critical phenomenon *Phys. Rev. Lett.* **45** 1748–52

[36] Pasler V, Schweiss P, Meingast C, Obst B, Wühl H, Rykov A I and Tajima S 1998 3D-XY critical fluctuations of the thermal expansivity in detwinned YBa$_2$Cu$_3$O$_{7-\delta}$ single crystals near optimal doping *Phys. Rev. Lett.* **81** 1094–7

[37] Domb C and Lebowitz J (eds) 1984 *Phase Transitions and Critical Phenomena* vol 9 (London, New York: Academic) ch 1. Lawrie I D and Sarbach S: Theory of tricritical points; ch 2. Knobler C M and Scott R L Multicritical points in fluid mixtures. Experimental studies.

[38] Scott R L 1965 Phase equilibria in solutions of liquid sulfur. I. Theory *J. Phys. Chem.* **69** 261–70

[39] Wheeler J C and Pfeuty P 1981 Critical points and tricritical points in liquid sulfur solutions *J. Chem. Phys.* **74** 6415–30

[40] Pegg I L, Knobler C M and Scott R L 1990 Tricritical phenomena in quasibinary mixtures. VIII. Calculations from the van der Waals equation for binary mixtures *J. Chem. Phys.* **92** 5442–53

[41] Ensher J R, Jin D S, Mathews M R, Wieman C E and Cornell E A 1996 Bose–Einstein condensation in a dilute gas: measurement of energy and ground-state occupation *Phys. Rev. Lett.* **77** 4984–7

Further Reading

Domb C 1996 *The Critical Point. A Historical Introduction to the Modern Theory of Critical Phenomena* (London and Bristol, PA: Taylor and Francis)

A historical survey. Details of classical treatment of fluids, magnets, light scattering and correlations. The 'Onsager revolution'. Reconciliation of the 'classical' and 'modern' approach. Discussion of calculations and renormalization group theory. Extensive bibliography.

Domb C, Green M S and Lebowitz J (eds) 1972–1987 *Phase Transitions and Critical Phenomena* 19 volumes (London: Academic)

Stanley H E 1971 *Introduction to Phase Transitions and Critical Phenomena* (New York: Oxford University Press)

Reprinted in paperback, 1987. Details of many of the kinds of transitions and critical points are discussed in this book. Unfortunately not revised since the development of renormalization group theory.

Van der Waals J D 1988 *On the Continuity of the Gaseous and Liquid States* vol XIV of (*Studies in Statistical Mechanics* vol XIV) ed J L Lebowitz (Amsterdam: North-Holland)

English translations of the 1873 thesis and the 1889 paper on liquid mixtures. The introductory essay by Rowlinson J S (119 pages) covers many of the features discussed in this article.

PART A3

DYNAMICAL PROCESSES

A3.1
Kinetic theory: transport and fluctuations

J R Dorfman

A3.1.1 Introduction

The kinetic theory of gases has a long history, extending over a period of a century and a half, and is responsible for many central insights into, and results for, the properties of gases, both in and out of thermodynamic equilibrium [1]. Strictly speaking, there are two familiar versions of kinetic theory, an informal version and a formal version. The informal version is based upon very elementary considerations of the collisions suffered by molecules in a gas, and upon elementary probabilistic notions regarding the velocity and free path distributions of the molecules. In the hands of Maxwell, Boltzmann and others, the informal version of kinetic theory led to such important predictions as the independence of the viscosity of a gas on its density at low densities, and to qualitative results for the equilibrium thermodynamic properties, the transport coefficients, and the structure of microscopic boundary layers in a dilute gas. The more formal theory is also due to Maxwell and Boltzmann, and may be said to have had its beginning with the development of the Boltzmann transport equation in 1872 [2]. At that time Boltzmann obtained, by heuristic arguments, an equation for the time dependence of the spatial and velocity distribution function for particles in the gas. This equation provided a formal foundation for the informal methods of kinetic theory. It leads directly to the Maxwell–Boltzmann velocity distribution for the gas in equilibrium. For non-equilibrium systems, the Boltzmann equation leads to a version of the second law of thermodynamics (the Boltzmann H-theorem), as well as to the Navier–Stokes equations of fluid dynamics, with explicit expressions for the transport coefficients in terms of the intermolecular potentials governing the interactions between the particles in the gas [3]. It is not an exaggeration to state that the kinetic theory of gases was one of the great successes of nineteenth century physics. Even now, the Boltzmann equation remains one of the main cornerstones of our understanding of non-equilibrium processes in fluid as well as solid systems, both classical and quantum mechanical. It continues to be a subject of investigation in both the mathematical and physical literature and its predictions often serve as a way of distinguishing different molecular models employed to calculate gas properties. Kinetic theory is typically used to describe the non-equilibrium properties of dilute to moderately dense gases composed of atoms, or diatomic or polyatomic molecules. Such properties include the coefficients of shear and bulk viscosity, thermal conductivity, diffusion, as well as gas phase chemical reaction rates, and other, similar properties.

In this section we will survey both the informal and formal versions of the kinetic theory of gases, starting with the simpler informal version. Here the basic idea is to combine both probabilistic and mechanical arguments to calculate quantities such as the equilibrium pressure of a gas, the mean free distance between collisions for a typical gas particle, and the transport properties of the gas, such as its viscosity and thermal conductivity. The formal version again uses both probabilistic and mechanical arguments to obtain an equation, the Boltzmann transport equation, that determines the distribution function, $f(\mathbf{r}, \mathbf{v}, t)$, that describes the number of gas particles in a small spatial region, $\delta\mathbf{r}$, about a point $\mathbf{r}$, and in a small region of velocities, $\delta\mathbf{v}$, about a given velocity **v**, at some time t. The formal theory forms the basis for almost all applications of kinetic theory to realistic systems.

We will almost always treat the case of a dilute gas, and almost always consider the approximation that the gas particles obey classical, Hamiltonian mechanics. The effects of quantum properties and/or of higher densities will be briefly commented upon. A number of books have been devoted to the kinetic theory of gases. Here we note that some of the interesting and easily accessible ones are those of Boltzmann [2], Chapman and Cowling [3], Hirshfelder *et al* [4], Hanley [5], Fertziger and Kaper [6], Resibois and de Leener [7], Liboff [8] and Present [9]. Most textbooks on the subject of statistical thermodynamics have one or more chapters on kinetic theory [10–13].

A3.1.2   The informal kinetic theory for the dilute gas

We begin by considering a gas composed of N particles in a container of volume V. We suppose, first, that the particles are single atoms, interacting with forces of finite range denoted by a. Polyatomic molecules can be incorporated into this informal discussion, to some extent, but atoms and molecules interacting with long-range forces require a separate treatment based upon the Boltzmann transport equation. This equation is capable of treating particles that interact with infinite-range forces, at least if the forces approach zero sufficiently rapidly as the separation of the particles becomes infinite. Typical potential energies describing the interactions between particles in the gas are illustrated in figure A3.1.1, where we describe Lennard–Jones (LJ) and Weeks–Chandler–Anderson (WCA) potentials. The range parameter, a, is usually taken to be a value close to the first point where the potential energy becomes negligible for all greater separations. While choice of the location of this point is largely subjective, it will not be a serious issue in what follows, since the results to be described below are largely qualitative order-of-magnitude results. However we may usefully take the distance a to represent the effective *diameter* of a particle.

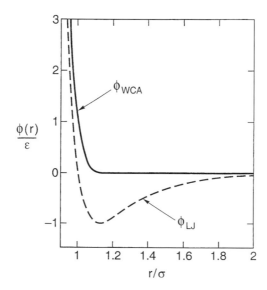

Figure A3.1.1. Typical pair potentials. Illustrated here are the Lennard–Jones potential, ϕ_{LJ}, and the Weeks–Chandler–Anderson potential, ρ_{WCA}, which gives the same repulsive force as the Lennard–Jones potential. The relative separation is scaled by σ, the distance at which the Lennard–Jones first passes through zero. The energy is scaled by the well depth, ε.

The dilute gas condition can be stated as the condition that the available volume per particle in the container is much larger that the volume of the particle itself. In other words

$$\frac{V}{N} \gg a^3 \quad \text{or} \quad na^3 \ll 1 \tag{A3.1.1}$$

where $n = N/V$ is the average number of particles per unit volume. We will see below that this condition is equivalent to the requirement that the mean free path between collisions, which we denote by λ, is much greater than the size of a particle, a. Next we suppose that the state of the gas can be described by a distribution function $f(\mathbf{r}, \mathbf{v}, t)$, such that $f(\mathbf{r}, \mathbf{v}, t)\, d\mathbf{r}\, d\mathbf{v}$ is the number of gas particles in $d\mathbf{r}$ about $\mathbf{r}$, and in $d\mathbf{v}$ about $\mathbf{v}$ at time t. To describe the state of a gas of polyatomic molecules, or of any mixture of different particles, we would need to include additional variables in the argument of f to describe the internal states of the molecules and the various components of the mixture. To keep the discussion simple, we will consider gases of monoatomic particles, for the time being.

At this point it is important to make some clarifying remarks: (1) clearly one cannot regard $d\mathbf{r}$ in the above expression, strictly, as a mathematical differential. It cannot be infinitesimally small, since $d\mathbf{r}$ much be large enough to contain some particles of the gas. We suppose instead that $d\mathbf{r}$ is large enough to contain some particles of the gas but small compared with any important physical length in the problem under consideration, such as a mean free path, or the length scale over which a physical quantity, such as a temperature, might vary. (2) The distribution function $f(\mathbf{r}, \mathbf{v}, t)$ typically does not describe the *exact* state of the gas in the sense that it tells us exactly how many particles are in the designated regions at the given time t. To obtain and use such an exact distribution function one would need to follow the motion of the individual particles in the gas, that is, solve the mechanical equations for the system, and then do the proper counting. Since this is clearly impossible for even a small number of particles in the container, we have to suppose that f is an ensemble average of the microscopic distribution functions for a very large number of identically prepared systems. This, of course, implies that kinetic theory is a branch of the more general area of statistical mechanics. As a result of these two remarks, we should regard any distribution function we use as an ensemble average rather than an exact expression for our particular system, and we should be careful when examining the variation of the distribution with space and time, to make sure that we are not too concerned with variations on spatial scales that are of the order or less than the size of a molecule, or on time scales that are of the order of the duration of a collision of a particle with a wall or of two or more particles with each other.

A3.1.2.1   Equilibrium properties from kinetic theory

The equilibrium state for a gas of monoatomic particles is described by a spatially uniform, time independent distribution function whose velocity dependence has the form of the Maxwell–Boltzmann distribution, obtained from equilibrium statistical mechanics. That is, $f(\mathbf{r}, \mathbf{v}, t)$ has the form $f_{\mathrm{eq}}(\mathbf{v})$ given by

$$f_{\mathrm{eq}}(\mathbf{v}) = n\varphi(v) \tag{A3.1.2}$$

where

$$\varphi(v) = \left(\frac{\beta m}{2\pi}\right)^{3/2} e^{-\beta \mathbf{v}^2 / 2m} \tag{A3.1.3}$$

is the usual Maxwell–Boltzmann velocity distribution function. Here m is the mass of the particle, and the quantity $\beta = (k_B T)^{-1}$, where T is the equilibrium thermodynamic temperature of the gas and k_B is Boltzmann's constant, $k_B = 1.380 \times 10^{-23}$ J K^{-1}.

We are now going to use this distribution function, together with some elementary notions from mechanics and probability theory, to calculate some properties of a dilute gas in equilibrium. We will calculate the pressure that the gas exerts on the walls of the container as well as the rate of effusion of particles from a very small hole in the wall of the container. As a last example, we will calculate the mean free path of a molecule between collisions with other molecules in the gas.

(a) The pressure

To calculate the pressure, we need to know the force per unit area that the gas exerts on the walls of the vessel. We calculate the force as the negative of the rate of change of the vector momentum of the gas particles as they strike the container. We consider then some small area, A, on the wall of the vessel and look at particles with a particular velocity $\mathbf{v}$, chosen so that it is physically possible for particles with this velocity to strike the designated area from within the container. We consider a small time interval δt, and look for all particles with velocity $\mathbf{v}$ that will strike this area, A over time interval δt. As illustrated in figure A3.1.2, all such particles must lie in a small 'cylinder' of base area A, and height, $|\mathbf{v} \cdot \hat{n}| \delta t$, where $\hat{n}$ is a unit normal to the surface of the container at the small area A, and directed toward the interior of the vessel. We will assume that the gas is very dilute and that we can ignore the collisions between particles, and take only collisions of particles with the wall into account. Every time such a particle hits our small area of the wall, its momentum changes, since its momentum after a collision differs from its momentum prior to the collision. Let us suppose that the particles make elastic, specular collisions with the surface, so that the momentum change per particle at each collision is $\Delta \mathbf{p} = -2(\mathbf{p} \cdot \hat{n})\hat{n} = -2m(\mathbf{v} \cdot \hat{n})\hat{n}$. This vector is directed in toward the container. Now to calculate the total change in the momentum of the gas in time δt due to collisions with the wall at the point of interest, we have to know how many particles with velocity $\mathbf{v}$ collide with the wall, multiply the number of collisions by the change in momentum per collision, and then integrate over all possible values of the velocity $\mathbf{v}$ than can lead to such a collision. To calculate the number of particles striking the small area, A, in time interval δt, we have to invoke probabilistic arguments, since we do not know the actual locations and the velocities of all the particles at the beginning of the time interval. We do know that if we ignore possible collisions amongst the particles themselves, all of the particles with velocity $\mathbf{v}$ colliding with A in time δt will have to reside in the small cylinder illustrated in figure A3.1.2, with volume $A|\mathbf{v} \cdot \hat{n}| \delta t$. Now, using the distribution function f given by equation (A3.1.2), we find that the number, $\delta \mathcal{N}(\mathbf{v})$, of particles with velocity $\mathbf{v}$ in the range $d\mathbf{v}$, in the collision cylinder is

$$\delta \mathcal{N}(\mathbf{v}) = n\varphi(v)|\mathbf{v} \cdot \hat{n}|A\delta t \, d\mathbf{v}. \tag{A3.1.4}$$

Now each such particle adds its change in momentum, as given above, to the total change of momentum of the gas in time δt. The total change in momentum of the gas is obtained by multiplying $\delta \mathcal{N}$ by the change in momentum per particle and integrating over all allowed values of the velocity vector, namely, those for which $\mathbf{v} \cdot \hat{n} \leq 0$. That is

$$\Delta \mathbf{p}_{\text{total}} = -2A\hat{n}\delta t n \int_{\mathbf{v} \cdot n \leq 0} d\mathbf{v}|\mathbf{v} \cdot \hat{n}|(\mathbf{v} \cdot \hat{n})\varphi(v). \tag{A3.1.5}$$

Finally the pressure, P, exerted by the gas on the container, is the negative of the force per unit area that the wall exerts on the gas. This force is measured by the change in momentum of the gas per unit time. Thus we are led to

$$P = 2n \int_{\mathbf{v} \cdot \hat{n} \leq 0} d\mathbf{v}|\mathbf{v} \cdot \hat{n}|^2 \varphi(v)$$

$$= \frac{n}{\beta} = nk_B T. \tag{A3.1.6}$$

Here we have carried out the velocity integral over the required half-space and used the explicit form of the Maxwell–Boltzmann distribution function, given by equation (A3.1.3).

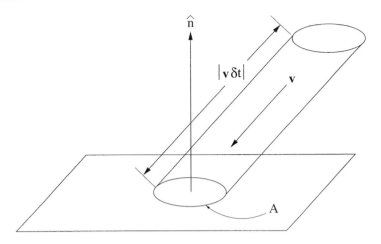

Figure A3.1.2. A collision cylinder for particles with velocity **v** striking a small region of area A on the surface of a container within a small time interval δt. Here $\hat{n}$ is a unit normal to the surface at the small region, and points into the gas.

(b) The rate of effusion through a small hole

It is a simple matter now to calculate number of particles per unit area, per unit time, that pass through a small hole in the wall of the vessel. This quantity is called the rate of effusion, denoted by n_e, and it governs the loss of particles in a container when there is a small hole in the wall separating the gas from a vacuum, say. This number is in fact obtained by integrating the quantity, $\delta \mathcal{N}(\mathbf{v})$ over all possible velocities having the proper direction, and then dividing this number by $A\delta t$. Thus we find

$$n_e = n \int_{\mathbf{v} \cdot \hat{n} \leq 0} |\mathbf{v} \cdot \hat{n}| \varphi(v)$$

$$= \frac{n\bar{v}}{4} \tag{A3.1.7}$$

where $\bar{v}$ is the average speed of a particle in a gas in equilibrium, given by

$$\bar{v} = \left(\frac{8}{m\pi\beta} \right)^{1/2}. \tag{A3.1.8}$$

The result, (A3.1.7), can be viewed also as the number of particles per unit area per unit time colliding from one side of any small area in the gas, whether real or fictitious. We will use this result in the next section when we consider an elementary kinetic theory for transport coefficients in a gas with some kind of flow taking place.

(c) The mean free path

The previous calculations, while not altogether trivial, are among the simplest uses one can make of kinetic theory arguments. Next we turn to a somewhat more sophisticated calculation, that for the mean free path of a particle between collisions with other particles in the gas. We will use the general form of the distribution function at first, before restricting ourselves to the equilibrium case, so as to set the stage for discussions in later sections where we describe the formal kinetic theory. Our approach will be first to compute the average frequency with which a particle collides with other particles. The inverse of this frequency is the mean time between collisions. If we then multiply the mean time between collisions by the mean speed, given by equation (A3.1.8), we will obtain the desired result for the mean free path between collisions. It is important to point out that one might choose to define the mean free path somewhat differently, by using the root mean square velocity instead of $\bar{v}$, for example. The only change will be in a numerical coefficient. The important issue will be to obtain the dependence of the mean free path upon the density and temperature of the gas and on the size of the particles. The numerical factors are not that important.

Let us focus our attention for the moment on a small volume in space, $\mathbf{dr}$, and on particles in the volume with a given velocity $\mathbf{v}$. Let us sit on such a particle and ask if it might collide in time δt with another particle whose velocity is $\mathbf{v}_1$, say. Taking the effective diameter of each particle to be a, as described above, we see that our particle with velocity $\mathbf{v}$ presents a cross sectional area of size πa^2 for collisions with other particles. If we focus on collisions with another particle with velocity $\mathbf{v}_1$, then, as illustrated in figure A3.1.3, a useful coordinate system to describe this collision is one in which the particle with velocity $\mathbf{v}$ is located at the origin and the z-axis is aligned along the direction of the vector $\mathbf{g} = \mathbf{v}_1 - \mathbf{v}$. In this coordinate system, the centre of the particle with velocity $\mathbf{v}_1$ must be somewhere in the collision cylinder of volume $\pi a^2 |\mathbf{g}| \delta t$ in order that a collision between the two particles takes place in the time interval δt. Now in the small volume $\mathbf{dr}$ there are $f(\mathbf{r}, \mathbf{v}, t)\, \mathbf{dr}$ particles with velocity $\mathbf{v}$ at time t, each one with a collision cylinder of the above type attached to it. Thus the total volume, $\delta \mathcal{V}(\mathbf{v}, \mathbf{v}_1)$ of these $(\mathbf{v}, \mathbf{v}_1)$ collision cylinders is

$$\delta \mathcal{V}(\mathbf{v}, \mathbf{v}_1) = \pi a^2 |\mathbf{g}| \delta t f(\mathbf{r}, \mathbf{v}, t). \tag{A3.1.9}$$

Now, again, we use a probabilistic argument to say that the number of particles with velocity $\mathbf{v}_1$ in this total volume is given by the product of the total volume and the number of particles per unit volume with velocity $\mathbf{v}_1$, that is, $\delta \mathcal{V}(\mathbf{v}, v_1) f(\mathbf{r}, \mathbf{v}_1, t)$. To complete the calculation, we suppose that the gas is so dilute that each of the collision cylinders has either zero or one particle with velocity $\mathbf{v}_1$ in it, and that each such particle actually collides with the particle with velocity $\mathbf{v}$. Thus the total number of collisions suffered by particles with velocity $\mathbf{v}$ in time δt is

$$\pi a^2 \delta t f(\mathbf{r}, \mathbf{v}, t) \int \mathbf{dv}_1 |\mathbf{v}_1 - \mathbf{v}| f(\mathbf{r}, \mathbf{v}_1, t).$$

Then it follows that the total number of collisions per unit time suffered by particles with *all* velocities is

$$\pi a^2 \int \mathbf{dv} \int \mathbf{dv}_1 |\mathbf{v}_1 - \mathbf{v}| f(\mathbf{r}, \mathbf{v}, t) f(\mathbf{r}, \mathbf{v}_1, t). \tag{A3.1.10}$$

Notice that each collision is counted twice, once for the particle with velocity $\mathbf{v}$ and once for the particle with velocity $\mathbf{v}_1$. We also note that we have assumed that the distribution functions f do not vary over distances which are the lengths of the collision cylinders, as the interval δt approaches some small value, but still large compared with the duration of a binary collision.

Our first result is now the average collision frequency obtained from the expression, (A3.1.10), by dividing it by the average number of particles per unit volume. Here it is convenient to consider the equilibrium case, and to use (A3.1.2) for f. Then we find that the average collision frequency, v, for the particles is

$$v = n\pi a^2 \int \mathbf{dv}\, \mathbf{dv}_1 |\mathbf{v}_1 - \mathbf{v}| \varphi(v) \varphi(v_1)$$

$$= n\pi a^2 \left(\frac{16}{\beta \pi m} \right)^{1/2}. \tag{A3.1.11}$$

The average time between collisions is then v^{-1}, and in this time the particle will typically travel a distance λ, the mean free path, where

$$\lambda = \bar{v} v^{-1} = \frac{1}{2^{1/2} \pi n a^2}. \tag{A3.1.12}$$

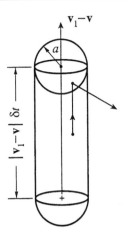

Figure A3.1.3. The collision cylinder for collisions between particles with velocities $\mathbf{v}$ and $\mathbf{v}_1$. The origin is placed at the centre of the particle with velocity $\mathbf{v}$ and the z-axis is in the direction of $\mathbf{v}_1 - \mathbf{v}$. The spheres indicate the range, a, of the intermolecular forces.

This is the desired result. It shows that the mean free path is inversely proportional to the density and the collision cross section. This is a physically sensible result, and could have been obtained by dimensional arguments alone, except for the unimportant numerical factor.

A3.1.2.2 The mean free path expressions for transport coefficients

One of the most useful applications of the mean free path concept occurs in the theory of transport processes in systems where there exist gradients of average but local density, local temperature, and/or local velocity. The existence of such gradients causes a transfer of particles, energy or momentum, respectively, from one region of the system to another.

The kinetic theory of transport processes in gases rests upon three basic assumptions.

(i)   The gas is dense enough that the mean free path is small compared with the characteristic size of the container. Consequently, the particles collide with each other much more often than they collide with the walls of the vessel.

(ii) As stated above, the gas is sufficiently dilute that the mean free path is much larger than the diameter of a particle.

(iii) The local density, temperature and density vary slowly over distances of the order of a mean free path.

If these assumptions are satisfied then the ideas developed earlier about the mean free path can be used to provide qualitative but useful estimates of the transport properties of a dilute gas. While many varied and complicated processes can take place in fluid systems, such as turbulent flow, pattern formation, and so on, the principles on which these flows are analysed are remarkably simple. The description of both simple and complicated flows in fluids is based on five hydrodynamic equations, the Navier–Stokes equations. These equations, in turn, are based upon the mechanical laws of conservation of particles, momentum and energy in a fluid, together with a set of phenomenological equations, such as Fourier's law of thermal conduction and Newton's law of fluid friction. When these phenomenological laws are used in combination with the conservation equations, one obtains the Navier–Stokes equations. Our goal here is to derive the phenomenological laws from elementary mean free path considerations, and to obtain estimates of the associated transport coefficients. Here we will consider thermal conduction and viscous flow as examples.

(a) Thermal conduction

We can obtain an understanding of Fourier's law of thermal conduction by considering a very simple situation, frequently encountered in the laboratory. Imagine a layer of gas, as illustrated in figure A3.1.4, which is small enough to exclude convection, but many orders of magnitude larger than a mean free path. Imagine further that the temperature is maintained at constant values, T_1 and T_2, $T_2 > T_1$, along two planes separated by a distance L, as illustrated. We suppose that the system has reached a stationary state so that the local temperature at any point in the fluid is constant in time and depends only upon the z-component of the location of the point. Now consider some imaginary plane in the fluid, away from the boundaries, and look at the flow of particles across the plane. We make a major simplification and assume that all particles crossing the plane carry with them the local properties of the system a mean free path above and below the plane. That is, suppose we examine the flow of particles through the plane, coming from above it. Then we can say that the number of particles crossing the plane per unit area and per unit time from above, i.e. the particle current density, j_n^- heading down, is given by

$$j_n^-(z) = \tfrac{1}{4} n(z + \lambda) \bar{v}(z + \lambda) \tag{A3.1.13}$$

where z is the height of the plane we consider, λ is the mean free path, and we use (A3.1.7) for this current density. Similarly, the upward flux is

$$j_n^+(z) = \tfrac{1}{4} n(z - \lambda) \bar{v}(z - \lambda). \tag{A3.1.14}$$

In a steady state, with no convection, the two currents must be equal, $j_n^+(z) = j_n^-(z) \equiv j_n(z)$. Now we assume that each particle crossing the place carries the energy per particle characteristic of the location at a mean free path above or below the plane. Thus the upward and downward energy current densities, $j_e^\pm$, are

$$j_e^\pm(z) = j_n(z) e(z \mp \lambda) \tag{A3.1.15}$$

where $e(z \mp \lambda)$ is the local energy per particle at a distance λ below and above the plane. The net amount of energy transferred per unit area per unit time in the positive z direction, $q(z)$, is then

$$
\begin{aligned}
q(z) = j_e^+ - j_e^- &= j_n(z)[e(z - \lambda) - e(z + \lambda)] \\
&= -2 j_n(z) \lambda \frac{\partial e(z)}{\partial z} + O\left(\frac{\partial^2 e}{\partial^3 z}\right).
\end{aligned}
\tag{A3.1.16}
$$

Neglecting derivatives of the third order and higher, we obtain Fourier's law of thermal conduction

$$q(z) = -k \frac{\partial T}{\partial z} \tag{A3.1.17}$$

where the coefficient of thermal conductivity, k, is given by

$$k = \frac{1}{2} n \bar{v} \lambda \frac{\partial e}{\partial T}. \tag{A3.1.18}$$

The result is, of course, a case of the more general expression of Fourier's law, namely

$$\mathbf{q} = -k \nabla T \tag{A3.1.19}$$

adjusted to the special situation that the temperature gradient is in the z-direction. Since k is obviously positive, our result is in accord with the second law of thermodynamics, which requires heat to flow from hotter to colder regions.

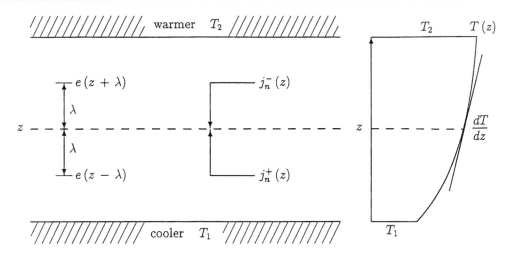

Figure A3.1.4. Steady state heat conduction, illustrating the flow of energy across a plane at a height z.

We can easily obtain an expression for k by using the explicit forms for $\bar{v}$ and λ, given in (A3.1.8) and (A3.1.12). Thus, in this approximation

$$k = \frac{c_v (k_B T)^{1/2}}{a^2 m^{1/2} \pi^{3/2}}. \tag{A3.1.20}$$

where c_v is the specific heat per particle. We have assumed that the gradients are sufficiently small that the local average speed and mean free path can be estimated by their (local) equilibrium values. The most important consequences of this result for the thermal conductivity are its independence of the gas density and its variation with temperature as $T^{1/2}$. The independence of density is well verified at low gas pressures, but the square-root temperature dependence is only verified at high temperatures. Better results for the temperature dependence of κ can be obtained by use of the Boltzmann transport equation, which we discuss in the next section. The temperature dependence turns out to be a useful test of the functional form of the intermolecular potential energy.

(b) The shear viscosity

A distribution of velocities in a fluid gives rise to a transport of momentum in the fluid in complete analogy with the transport of energy which results from a distribution of temperatures. To analyse this transport of momentum in a fluid with a gradient in the average local velocity, we use the same method as employed in the case of thermal conduction. That is, we consider a layer of fluid contained between two parallel planes, moving with velocities in the x-direction with values U_1 and U_2, $U_2 > U_1$, as illustrated in figure A3.1.5. We suppose that the width of the layer is very large compared with a mean free path, and that the fluid adjacent to the moving planes moves with the velocity of the adjacent plane. If the velocities are not so large as to develop a turbulent flow, then a steady state can be maintained with an average local velocity, $\mathbf{u}(x, y, z)$, in the fluid of the form, $\mathbf{u}(x, y, z) = u_x(z)\hat{\mathbf{x}}$, where $\hat{\mathbf{x}}$ is a unit vector in the x-direction.

The molecules of the gas are in constant motion, of course, and there is a transport of particles in all directions in the fluid. If we consider a fictitious plane in the fluid, far from the moving walls, at a height z, then there will be a flow of particles from above and below the plane. The particles coming from above will carry a momentum with them typical of the average flow at a height $z + \lambda$, while those coming from below will carry the typical momentum at height $z - \lambda$, where λ is the mean free path length. Due to the velocity gradient in the fluid there will be a net transport of momentum across the plane, tending to slow down the

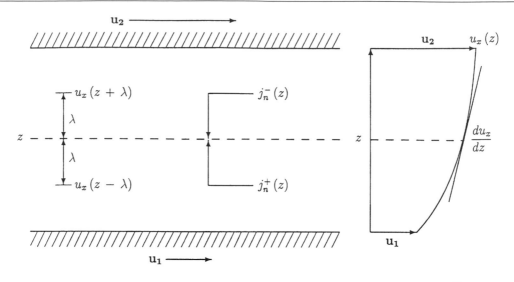

Figure A3.1.5. Steady state shear flow, illustrating the flow of momentum across a plane at a height z.

faster regions and to accelerate the slower regions. This transport of momentum leads to viscous forces (or stresses if measured per unit area) in the fluid, which in our case will be in the x-direction. The analysis of this viscous stress is almost identical to that for thermal conduction.

Following the method used above, we see that there will be an upward flux of momentum in the x-direction, $j_{p_z}^+(z)$, across the plane at z given by

$$j_{p_x}^+(z) = j_n(z)mu_x(z - \lambda) \tag{A3.1.21}$$

and a downward flux

$$j_{p_x}^-(z) = j_n(z)mu_x(z + \lambda). \tag{A3.1.22}$$

The net upward flow in the x-component of the momentum is called the shear stress, σ_{zx}, and by combining (A3.1.21) and (A3.1.22), we see that

$$\sigma_{zx} = j_{p_x}^+(z) - j_{p_x}^-(z) = j_n(z)mu_x(z - \lambda) - j_n(z)mu_x(z + \lambda)$$
$$= -\frac{1}{2}m\lambda n(z)\bar{v}(z)\frac{\partial u_x(z)}{\partial z} + \cdots . \tag{A3.1.23}$$

Here we have neglected derivatives of the local velocity of third and higher orders. Equation (A3.1.23) has the form of the phenomenological Newton's law of friction

$$\sigma_{zx} = -\eta\frac{\partial u_x(z)}{\partial z} \tag{A3.1.24}$$

if we identify the coefficient of shear viscosity η with the quantity

$$\eta = \tfrac{1}{2}m\lambda n(z)\bar{v}(z). \tag{A3.1.25}$$

An explicit expression for the coefficient of shear viscosity can be obtained by assuming the system is in local thermodynamic equilibrium and using the previously derived expression for λ and $\bar{v}$. Thus we obtain

$$\eta = \frac{(mk_B T)^{1/2}}{\pi^{3/2}a^2}. \tag{A3.1.26}$$

As in the case of thermal conductivity, we see that the viscosity is independent of the density at low densities, and grows with the square root of the gas temperature. This latter prediction is modified by a more systematic calculation based upon the Boltzmann equation, but the independence of viscosity on density remains valid in the Boltzmann equation approach as well.

(c) The Euken factor

We notice, using (A3.1.20) and (A3.1.26), that this method leads to a simple relation between the coefficients of shear viscosity and thermal conductivity, given by

$$\frac{k}{m\eta c_v} = 1. \tag{A3.1.27}$$

That is, this ratio should be a universal constant, valid for all dilute gases. A more exact calculation based upon the Boltzmann equation shows that the right-hand side of equation (A3.1.27) should be replaced by 2.5 instead of 1, plus a correction that varies slightly from gas to gas. The value of 2.5 holds with a very high degree of accuracy for dilute monatomic gases [5]. However, when this ratio is computed for diatomic and polyatomic gases, the value of 2.5 is no longer recovered.

Euken advanced a very simple argument which allowed him to extend the Boltzmann equation formula for $k/(m\eta c_v)$ to diatomic and polyatomic gases. His argument is that when energy is transported in a fluid by particles, the energy associated with each of the internal degrees of freedom of a molecule is transported. However, the internal degrees of freedom play no role in the transport of momentum. Thus we should modify (A3.1.20) to include these internal degrees of freedom. If we also modify it to correct for the factor of 2.5 predicted by the Boltzmann equation, we obtain

$$k = \tfrac{1}{2}n\bar{v}\lambda(Cc_v^{tr} + c_v^i) \tag{A3.1.28}$$

where $C = 2.5$, c_v^{tr} is the translational specific heat per molecule, and c_v^i is the specific heat per molecule associated with the internal degrees of freedom. We can easily obtain a better value for the ratio, $k/(m\eta c_v)$ in terms of the ratio of specific heat at constant pressure per molecule to the specific heat at constant volume, $\gamma = c_p/c_v$, as

$$k/(m\eta c_v) = \tfrac{1}{4}(9\gamma - 5). \tag{A3.1.29}$$

The right-hand side of (A3.1.29), called the Euken factor, provides a reasonably good estimate for this ratio [11].

A3.1.3 The Boltzmann transport equation

In 1872, Boltzmann introduced the basic equation of transport theory for dilute gases. His equation determines the time-dependent position and velocity distribution function for the molecules in a dilute gas, which we have denoted by $f(\mathbf{r}, \mathbf{v}, t)$. Here we present his derivation and some of its major consequences, particularly the so-called H-theorem, which shows the consistency of the Boltzmann equation with the irreversible form of the second law of thermodynamics. We also briefly discuss some of the famous debates surrounding the mechanical foundations of this equation.

We consider a large vessel of volume V, containing N molecules which interact with central, pairwise additive, repulsive forces. The latter requirement allows us to avoid the complications of long-lived 'bound' states of two molecules which, though interesting, are not central to our discussion here. We suppose that the pair potential has a strong repulsive core and a finite range a, such as the WCA potential illustrated in figure A3.1.1. Now, as before, we define a distribution function, $f(\mathbf{r}, \mathbf{v}, t)$, for the gas over a six-dimensional position and velocity space, $(\mathbf{r}, \mathbf{v})$, such that

$$f(\mathbf{r}, \mathbf{v}, t)\delta\mathbf{r}\delta\mathbf{v} \equiv \text{the number of particles in } \delta\mathbf{r}\delta\mathbf{v} \text{ around } \mathbf{r} \text{ and } \mathbf{v} \text{ at time } t. \tag{A3.1.30}$$

To get an equation for $f(\mathbf{r}, \mathbf{v}, t)$, we take a region $\delta\mathbf{r}\delta\mathbf{v}$ about a point $(\mathbf{r}, \mathbf{v})$, that is large enough to contain a lot of particles, but small compared with the range of variation of f.

There are four mechanisms that change the number of particles in this region. The particles can:

(i)   flow into or out of $\delta\mathbf{r}$, the *free-streaming term*,
(ii) leave the $\delta\mathbf{v}$ region as a result of a direct collision, the *loss term*,
(iii) enter the $\delta\mathbf{v}$ region after a restituting collision, the *gain term*, and
(iv) collide with the wall of the container (if the region contains part of the walls), the *wall term*.

We again assume that there is a time interval δt which is long compared with the duration of a binary collision but is too short for particles to cross a cell of size $\delta\mathbf{r}$. Then the change in the number of particles in $\delta\mathbf{r}\delta\mathbf{v}$ in time δt can be written as

$$[f(\mathbf{r}, \mathbf{v}, t + \delta t) - f(\mathbf{r}, \mathbf{v}, t)]\delta\mathbf{r}\delta\mathbf{v} = \Gamma_f - \Gamma_- + \Gamma_+ + \Gamma_w \tag{A3.1.31}$$

where Γ_f, Γ_-, Γ_+, and Γ_w represent the changes in f due to the four mechanisms listed above, respectively. We suppose that each particle in the small region suffers at most one collision during the time interval δt, and calculate the change in f.

The computation of Γ_f is relatively straightforward. We simply consider the free flow of particles into and out of the region in time δt. An expression for this flow in the x-direction, for example, can be obtained by considering two thin layers of size $v_x\delta t\delta r_y\delta r_z$ that contain particles that move into or out of a cell with its centre at (x, y, z) in time δt (see figure A3.1.6).

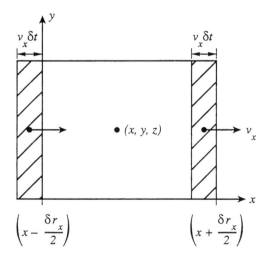

Figure A3.1.6. A schematic illustration of flow into and out of a small region. The hatched areas represent regions where particles enter and leave the region in time δt.

The free streaming term can be written as the difference between the number of particles entering and leaving the small region in time δt. Consider, for example, a cubic cell and look at the faces perpendicular to the x-axis. The flow of particles across the faces at $x - \frac{1}{2}\delta r_x$ and at $x + \frac{1}{2}\delta r_x$ is

$$\Gamma_f^{(x)} = v_x\delta t\delta r_y\delta r_z\delta\mathbf{v}[f(x - \tfrac{1}{2}\delta r_x, y, z, \mathbf{v}, t) - f(x + \tfrac{1}{2}\delta r_x, y, z, \mathbf{v}, t)] \tag{A3.1.32}$$

and similar expressions exist for the y- and z-directions. The function f is supposed to be sufficiently smooth that it can be expanded in a Taylor series around (x, y, z). The zeroth-order terms between the parentheses cancel and the first-order terms add up. Neglecting terms of order δ^2 and higher and summing over all directions then yields

$$\Gamma_{\mathrm{f}} = -\delta\mathbf{v}\delta t\delta\mathbf{r}(\mathbf{v} \cdot \nabla)f(\mathbf{r}, \mathbf{v}, t). \tag{A3.1.33}$$

Next we consider the computation of the loss term, Γ_-. As in the calculation of the mean free path, we need to calculate the number of collisions suffered by particles with velocity $\mathbf{v}$ in the region $\delta\mathbf{r}\delta\mathbf{v}$ in time δt, assuming that each such collision results in a change of the velocity of the particle. We carry out the calculation in several steps. First, we focus our attention on a particular particle with velocity $\mathbf{v}$, and suppose that it is going to collide sometime during the interval $[t, t + \delta t]$ with a particle with velocity $\mathbf{v}_1$. Now examine again the coordinate system with origin at the center of the particle with velocity $\mathbf{v}$, and with the z-axis directed along the vector $\mathbf{g} = \mathbf{v}_1 - \mathbf{v}$. By examining figure A3.1.3, one can easily see that if the particle with velocity $\mathbf{v}_1$ is somewhere at time t within the *collision cylinder* illustrated there, with volume $|\mathbf{v}_1 - \mathbf{v}|\pi a^2\delta t$, this particle will collide sometime during the interval $[t, t + \delta t]$ with the particle with velocity $\mathbf{v}$, if no other particles interfere, which we assume to be the case. These collision cylinders will be referred to as $(\mathbf{v}_1, \mathbf{v})$-collision cylinders. We also ignore the possibility that the particle with velocity $\mathbf{v}_1$ might, at time t, be somewhere within the action sphere of radius a about the centre of the velocity-$\mathbf{v}$ particle, since such events lead to terms that are of higher order in the density than those we are considering here, and such terms do not even exist if the duration of a binary collision is strictly zero, as would be the case for hard spheres, for example.

We now compute Γ_- by noting again the steps involved in calculating the mean free path, but applying them now to the derivation of an expression for Γ_-.

- The number of $(\mathbf{v}_1, \mathbf{v})$-collision cylinders in the region $\delta\mathbf{r}\delta\mathbf{v}$ is equal to the number of particles with velocity $\mathbf{v}$ in this region, $f(\mathbf{r}, \mathbf{v}, t)\delta\mathbf{r}\delta\mathbf{v}$.
- Each $(\mathbf{v}_1, \mathbf{v})$-collision cylinder has the volume given above, and the total volume of these cylinders is equal to the product of the volume of each such cylinder with the number of these cylinders, that is $f(\mathbf{r}, \mathbf{v}, t)|\mathbf{v}_1 - \mathbf{v}|\pi a^2\delta\mathbf{r}\delta\mathbf{v}\delta t$.
- If we wish to know the number of $(\mathbf{v}_1, \mathbf{v})$-collisions that actually take place in this small time interval, we need to know exactly where each particle is located and then follow the motion of *all* the particles from time t to time $t + \delta t$. In fact, this is what is done in computer simulated molecular dynamics. We wish to avoid this exact specification of the particle trajectories, and instead carry out a plausible argument for the computation of Γ_-. To do this, Boltzmann made the following assumption, called the *Stosszahlansatz*, which we encountered already in the calculation of the mean free path:

Stosszahlansatz. *The total number of $(\mathbf{v}_1, \mathbf{v})$-collisions taking place in δt equals the total volume of the $(\mathbf{v}_1, \mathbf{v})$-collision cylinders times the number of particles with velocity $\mathbf{v}_1$ per unit volume.*

After integration over $\mathbf{v}_1$, we obtain

$$\Gamma_- = \delta\mathbf{r}\delta\mathbf{v}f(\mathbf{r}, \mathbf{v}, t) \int d\mathbf{v}_1\, \delta t\pi a^2|\mathbf{v}_1 - \mathbf{v}|f(\mathbf{r}, \mathbf{v}_1, t). \tag{A3.1.34}$$

The gas has to be dilute because the collision cylinders are assumed not to overlap, and also because collisions between more than two particles are neglected. Also it is assumed that f hardly changes over $\delta\mathbf{r}$ so that the distribution functions for both colliding particles can be taken at the same position $\mathbf{r}$.

The assumptions that go into the calculation of Γ_- are referred to collectively as the *assumption of molecular chaos*. In this context, this assumption says that the probability that a pair of particles with given velocities will collide can be calculated by considering each particle separately and ignoring any correlation between the probability for finding one particle with velocity $\mathbf{v}$ and the probability for finding another with velocity $\mathbf{v}_1$ in the region $\delta\mathbf{r}$.

For the construction of Γ_+, we need to know how two particles can collide in such a way that one of them has velocity $\mathbf{v}$ after the collision. The answer to this question can be found by a more careful examination of the 'direct' collisions which we have just discussed. To proceed with this examination, we note that the factor πa^2 appearing in (A3.1.34) can also be written as an integral over the impact parameters and azimuthal angles of the $(\mathbf{v}_1, \mathbf{v})$ collisions. That is, $\pi a^2 = \int b \, db \int d\epsilon$, where b, the impact parameter, is the initial distance between the centre of the incoming $\mathbf{v}_1$-particle and the axis of the collision cylinder (z-axis), and ϵ is the angle between the x-axis and the position of particle 2 in the x–y plane. Here $0 \le b \le a$, and $0 \le \epsilon \le 2\pi$. The laws of conservation of linear momentum, angular momentum, and energy require that both the impact parameter b, and $|\mathbf{g}| = |\mathbf{v}_1 - \mathbf{v}|$, the magnitude of the relative velocity, be the same before and after the collision. To see what this means let us follow the two particles through and beyond a direct collision. We denote all quantities after the collision by primes. The conservation of momentum

$$\mathbf{v}_1 + \mathbf{v} = \mathbf{v}_1' + \mathbf{v}'$$

implies, after squaring and using conservation of energy

$$v_1^2 + v^2 = v_1'^2 + v'^2$$

that

$$\mathbf{v}_1 \cdot \mathbf{v} = \mathbf{v}_1' \cdot \mathbf{v}'.$$

By multiplying this result by a factor of -2, and adding the result to the conservation of energy equation, one easily finds $|\mathbf{g}| = |\mathbf{g}'| = |\mathbf{v}_1' - \mathbf{v}'|$. This result, taken together with conservation of angular momentum, $\mu g b = \mu g' b'$, where $\mu = \frac{1}{2} m$ is the reduced mass of the two-particle system, shows that b is also conserved, $b = b'$. This is illustrated in figure A3.1.7.

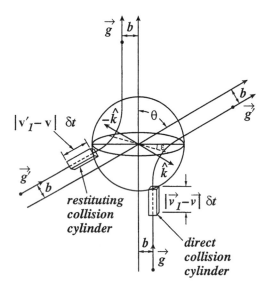

Figure A3.1.7. Direct and restituting collisions in the relative coordinate frame. The collision cylinders as well as the appropriate scattering and azimuthal angles are illustrated.

Next, we denote the line between the centres of the two particles at the point of closest approach by the unit vector $\hat{\mathbf{k}}$. In figure A3.1.7, it can also be seen that the vectors $-\mathbf{g}$ and $\mathbf{g}'$ are each other's mirror images in the direction of $\hat{\mathbf{k}}$ in the plane of the trajectory of particles:

$$\mathbf{g}' = \mathbf{g} - 2(\mathbf{g} \cdot \hat{\mathbf{k}})\hat{\mathbf{k}} \qquad (A3.1.35)$$

and thus $(\mathbf{g} \cdot \hat{\mathbf{k}}) = -(\mathbf{g}' \cdot \hat{\mathbf{k}})$. Together with conservation of momentum this gives

$$\mathbf{v}_1' = \mathbf{v}_1 - (\mathbf{g} \cdot \hat{\mathbf{k}})\hat{\mathbf{k}}$$
$$\mathbf{v}' = \mathbf{v} + (\mathbf{g} \cdot \hat{\mathbf{k}})\hat{\mathbf{k}}. \qquad (A3.1.36)$$

The main point of this argument is to show that if particles with velocities $\mathbf{v}'$ and $\mathbf{v}_1'$ collide in the right geometric configuration with impact parameter b, such a collision will result in one of the particles having the velocity of interest, $\mathbf{v}$, after the collision. These kinds of collisions which produce particles with velocity $\mathbf{v}$, contribute to Γ_+, and are referred to as 'restituting' collisions. This is illustrated in figure A3.1.8, where particles having velocities $\mathbf{v}'$ and $\mathbf{v}_1'$ are arranged to collide in such a way that the unit vector of closest approach, $\hat{\mathbf{k}}$, is replaced by $-\hat{\mathbf{k}}$. Consider, then, a collision with initial velocities $\mathbf{v}_1'$ and $\mathbf{v}'$ and the same impact parameter as in the direct collision, but with $\hat{\mathbf{k}}$ replaced by $-\hat{\mathbf{k}}$. The final velocities are now $\mathbf{v}_1''$ and $\mathbf{v}''$, which are equal to $\mathbf{v}_1$ and $\mathbf{v}$, respectively, because

$$\mathbf{v}_1'' = \mathbf{v}_1' - (\mathbf{g}' \cdot \hat{\mathbf{k}})\hat{\mathbf{k}} = \mathbf{v}_1' + (\mathbf{g} \cdot \hat{\mathbf{k}})\hat{\mathbf{k}} = \mathbf{v}_1 \qquad (A3.1.37)$$

and

$$\mathbf{v}'' = \mathbf{v}' - (\mathbf{g} \cdot \hat{\mathbf{k}})\hat{\mathbf{k}} = \mathbf{v}. \qquad (A3.1.38)$$

Thus the *increase* of particles in our region due to restituting collisions with an impact parameter between b and $b + db$ and azimuthal angle between ϵ and $\epsilon + d\epsilon$ (see figure A3.1.7) can be obtained by adjusting the expression for the *decrease* of particles due to a 'small' collision cylinder:

Loss: $\delta t b\, db\, d\epsilon |\mathbf{g}| f(\mathbf{r}, \mathbf{v}, t) f(\mathbf{r}, \mathbf{v}_1, t)\delta\mathbf{v}\delta\mathbf{v}_1\delta\mathbf{r}$

Gain: $\delta t b\, db\, d\epsilon |\mathbf{g}'| f(\mathbf{r}, \mathbf{v}', t) f(\mathbf{r}, \mathbf{v}_1', t)\delta\mathbf{v}_1'\delta\mathbf{v}'\delta\mathbf{r}$

where b has to be integrated from 0 to a, and ϵ from 0 to 2π. Also, by considering the Jacobian for the transformation to relative and centre-of-mass velocities, one easily finds that $d\mathbf{v}_1\, d\mathbf{v} = d\mathbf{v}\, d\mathbf{g}$, where $\mathbf{v}$ is the velocity of the centre-of-mass of the two colliding particles with respect to the container. After a collision, $\mathbf{g}$ is rotated in the centre-of-mass frame, so the Jacobian of the transformation $(\mathbf{v}, \mathbf{g}) \rightarrow (\mathbf{v}', \mathbf{g}')$ is unity and $d\mathbf{v}\, d\mathbf{g} = d\mathbf{v}'\, d\mathbf{g}'$. So

$$d\mathbf{v}_1\, d\mathbf{v} = d\mathbf{v}\, d\mathbf{g} = d\mathbf{v}'\, d\mathbf{g}' = d\mathbf{v}_1'\, d\mathbf{v}' \qquad (A3.1.39)$$

Now we are in the correct position to compute Γ_+, using exactly the same kinds of arguments as in the computation of Γ_-, namely, the construction of collision cylinders, computing the total volume of the relevant cylinders and again making the *Stosszahlansatz*. Thus, we find that

$$\Gamma_+ = \iiint d\mathbf{v}_1'\, b\, db\, d\epsilon |\mathbf{v}_1' - \mathbf{v}'| f(\mathbf{r}, \mathbf{v}', t) f(\mathbf{r}, \mathbf{v}_1', t)\delta\mathbf{r}\delta\mathbf{v}'\delta t. \qquad (A3.1.40)$$

For every value of the velocity $\mathbf{v}$, the velocity ranges $d\mathbf{v}_1'\, \delta\mathbf{v}'$ in the above expression are only over that range of velocities $\mathbf{v}', \mathbf{v}_1'$ such that particles with velocity in the range $\delta\mathbf{v}$ about $\mathbf{v}$ are produced in the $(\mathbf{v}', \mathbf{v}_1')$-collisions. If we now use the equalities, equation (A3.1.39), as well as the fact that $|\mathbf{g}| = |\mathbf{g}'|$, we can write

$$\Gamma_+ = \iiint d\mathbf{v}_1\, b\, db\, d\epsilon |\mathbf{v}_1 - \mathbf{v}| f(\mathbf{r}, \mathbf{v}', t) f(\mathbf{r}, \mathbf{v}_1', t)\delta\mathbf{r}\delta\mathbf{v}\delta t. \qquad (A3.1.41)$$

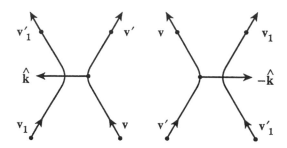

Figure A3.1.8. Schematic illustration of the direct and restituting collisions.

The term describing the interaction with the walls, Γ_w, is discussed in a paper by Dorfman and van Beijeren [14].

Finally, all of the Γ-terms can be inserted in (A3.1.31), and dividing by $\delta t \delta \mathbf{r} \delta \mathbf{v}$ gives the Boltzmann transport equation

$$\frac{\partial f(\mathbf{r}, \mathbf{v}, t)}{\partial t} + \mathbf{v} \cdot \nabla f(\mathbf{r}, \mathbf{v}, t) = J(f, f) + T_w \tag{A3.1.42}$$

where $J(f, f) = \int \int \int \, d\mathbf{v}_1 \, b \, db \, d\epsilon \, |\mathbf{v}_1 - \mathbf{v}| [f' f'_1 - f_1 f]$.

The primes and subscripts on the fs refer to their velocity arguments, and the primed velocities in the gain term should be regarded as functions of the unprimed quantities according to (A3.1.36). It is often convenient to rewrite the integral over the impact parameter and the azimuthal angle as an integral over the unit vector $\hat{\mathbf{k}}$ as

$$gb \, db \, d\epsilon = B(\mathbf{g}, \hat{\mathbf{k}}) \, d\hat{\mathbf{k}} \tag{A3.1.43}$$

where

$$d\hat{\mathbf{k}} = \sin(\pi - \psi) \, d(\pi - \psi) \, d\epsilon \tag{A3.1.44}$$

and ψ is the angle between $\mathbf{g}$ and $\hat{\mathbf{k}}$. Then $d\hat{\mathbf{k}} = |\sin \psi \, d\psi \, d\epsilon|$, so that

$$B(\mathbf{g}, \hat{\mathbf{k}}) = |\mathbf{v}_1 - \mathbf{v}| \left| \frac{b}{\sin \psi} \right| \left| \frac{db}{d\psi} \right| \tag{A3.1.45}$$

with the restriction for purely repulsive potentials that $\mathbf{g} \cdot \hat{\mathbf{k}} < 0$. As can be seen in figure A3.1.7,

$$B(\mathbf{g}', \hat{\mathbf{k}}) = B(\mathbf{g}, -\hat{\mathbf{k}}). \tag{A3.1.46}$$

Let us apply this to the situation where the molecules are hard spheres of *diameter a*. We have $db/d\psi = d(a \sin \psi)/d\psi = a \cos \psi$ (see figure A3.1.9), and $B(\mathbf{g}, \hat{\mathbf{k}}) = ga^2 \cos \psi = a^2 |(\mathbf{g} \cdot \hat{\mathbf{k}})|$.

The Boltzmann equation for hard spheres is given then as

$$\frac{\partial f}{\partial t} + (\mathbf{v} \cdot \nabla f) = a^2 \int \, d\mathbf{v}_1 \int_{\mathbf{g} \cdot \hat{\mathbf{k}} < 0} d\hat{\mathbf{k}} |\mathbf{g} \cdot \hat{\mathbf{k}}| [f'_1 f' - f_1 f]. \tag{A3.1.47}$$

This completes the heuristic derivation of the Boltzmann transport equation. Now we turn to Boltzmann's argument that his equation implies the Clausius form of the second law of thermodynamics, namely, that the entropy of an isolated system will increase as the result of any irreversible process taking place in the system. This result is referred to as *Boltzmann's H-theorem*.

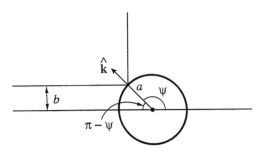

Figure A3.1.9. Hard sphere collision geometry in the plane of the collision. Here a is the diameter of the spheres.

A3.1.3.1 Boltzmann's H-theorem

Boltzmann showed that under very general circumstances, there exists a time-dependent quantity, $H(t)$, that never increases in the course of time. This quantity is given by

$$H(t) \equiv \int d\mathbf{r}_1 \int d\mathbf{v}_1 \, f(\mathbf{r}_1, \mathbf{v}_1, t)[\ln f(\mathbf{r}_1, \mathbf{v}_1, t) - 1]. \qquad (A3.1.48)$$

Here, the spatial integral is to be carried out over the entire volume of the vessel containing the gas, and for convenience we have changed the notation slightly. Now we differentiate H with time,

$$\frac{dH(t)}{dt} = \int d\mathbf{r}_1 \int d\mathbf{v}_1 \, \frac{\partial f_1}{\partial t} \ln f_1 \qquad (A3.1.49)$$

and use the Boltzmann equation to find that

$$\frac{dH}{dt} = \int\!\!\int d\mathbf{r}_1 \, d\mathbf{v}_1 [-(\mathbf{v}_1 \cdot \nabla f_1) + J(f_1, f_1) + T_w] \ln f_1. \qquad (A3.1.50)$$

We are going to carry out some spatial integrations here. We suppose that the distribution function vanishes at the surface of the container and that there is no flow of energy or momentum into or out of the container. (We mention in passing that it is possible to relax this latter condition and thereby obtain a more general form of the second law than we discuss here. This requires a careful analysis of the wall-collision term T_w. The interested reader is referred to the article by Dorfman and van Beijeren [14]. Here, we will drop the wall operator since for the purposes of this discussion it merely ensures that the distribution function vanishes at the surface of the container.) The first term can be written as

$$-\int\!\!\int d\mathbf{r}_1 \, d\mathbf{v}_1 \, \mathbf{v}_1 \cdot \nabla[f_1(\ln f_1 - 1)]. \qquad (A3.1.51)$$

This can be evaluated easily in terms of the distribution function at the walls of the closed container and therefore it is zero. The second term of (A3.1.50) is based on the *Stosszahlansatz*, and is

$$\frac{dH(t)}{dt} = \int\!\!\int\!\!\int\!\!\int d\mathbf{r}_1 \, d\mathbf{v}_1 \, d\mathbf{v}_2 \, d\hat{\mathbf{k}} \, B(\mathbf{g}, \hat{\mathbf{k}}) \Psi(\mathbf{v}_1)(f_1' f_2' - f_1 f_2) \qquad (A3.1.52)$$

with $\Psi(\mathbf{v}_1) = \ln f_1$. The integrand may be symmetrized in $\mathbf{v}_1$ and $\mathbf{v}_2$ to give

$$\frac{dH(t)}{dt} = \frac{1}{2} \int\!\!\int\!\!\int\!\!\int d\mathbf{r}_1 \, d\mathbf{v}_1 \, d\mathbf{v}_2 \, d\hat{\mathbf{k}} \, B(\mathbf{g}, \hat{\mathbf{k}})[\Psi(\mathbf{v}_1) + \Psi(\mathbf{v}_2)](f_1' f_2' - f_1 f_2).$$

For each collision there is an inverse one, so we can also express the time derivative of the H-function in terms of the inverse collisions as

$$\frac{\mathrm{d}H(t)}{\mathrm{d}t} = +\frac{1}{2} \int\int\int\int \mathrm{d}\mathbf{r}_1 \, \mathrm{d}\mathbf{v}_1' \, \mathrm{d}\mathbf{v}_2' \, \mathrm{d}\hat{\mathbf{k}} \, B(\mathbf{g}', -\hat{\mathbf{k}})[\Psi(\mathbf{v}_1') + \Psi(\mathbf{v}_2')](f_1 f_2 - f_1' f_2')$$

$$= -\frac{1}{2} \int\int\int\int \mathrm{d}\mathbf{r}_1 \, \mathrm{d}\mathbf{v}_1 \, \mathrm{d}\mathbf{v}_2 \, \mathrm{d}\hat{\mathbf{k}} \, B(\mathbf{g}, \hat{\mathbf{k}})[\Psi(\mathbf{v}_1') + \Psi(\mathbf{v}_2')](f_1' f_2' - f_1 f_2).$$

We obtain the H-theorem by adding these expressions and dividing by two,

$$\frac{\mathrm{d}H(t)}{\mathrm{d}t} = \frac{1}{4} \int\int\int\int \mathrm{d}\mathbf{r}_1 \, \mathrm{d}\mathbf{v}_1 \, \mathrm{d}\mathbf{v}_2 \, \mathrm{d}\hat{\mathbf{k}} \, B(\mathbf{g}, \hat{\mathbf{k}})[\Psi_1 + \Psi_2 - \Psi_1' - \Psi_2'](f_1' f_2' - f_1 f_2).$$

Now, using $\Psi(\mathbf{v}_1) = \ln f_1$, we obtain

$$\frac{\mathrm{d}H}{\mathrm{d}t} = \frac{1}{4} \int\int\int\int \mathrm{d}\mathbf{r}_1 \, \mathrm{d}\mathbf{v}_1 \, \mathrm{d}\mathbf{v}_2 \, \mathrm{d}\hat{\mathbf{k}} \, B(\mathbf{g}, \hat{\mathbf{k}}) \ln\left(\frac{f_1 f_2}{f_1' f_2'}\right)(f_1' f_2' - f_1 f_2).$$

If $f_1 f_2 \neq f_1' f_2'$, the integrand is negative;

$$f_1 f_2 < f_1' f_2' \qquad \text{the second factor is positive, the first is negative;}$$
$$f_1 f_2 > f_1' f_2' \qquad \text{the second is negative, and the first is positive.}$$

Both cases give a decreasing $H(t)$. That is

$$\frac{\mathrm{d}H(t)}{\mathrm{d}t} \leq 0. \tag{A3.1.53}$$

The integral is zero only if for all $\mathbf{v}_1$ and $\mathbf{v}_2$

$$f_1 f_2 = f_1' f_2'. \tag{A3.1.54}$$

This is Boltzmann's H-theorem

We now show that when H is constant in time, the gas is in equilibrium. The existence of an equilibrium state requires the rates of the restituting and direct collisions to be equal; that is, that there is a detailed balance of gain and loss processes taking place in the gas.

Taking the natural logarithm of (A3.1.54), we see that $\ln f_1 + \ln f_2$ has to be conserved for an equilibrium solution of the Boltzmann equation. Therefore, $\ln f_1$ can generally be expressed as a linear combination with constant coefficients of the $(d + 2)$ quantities conserved by binary collisions, i.e. (i) the number of particles, (ii) the d components of the linear momentum, where d is the number of dimensions, and (iii) the kinetic energy: $\ln f_1 = A + \mathbf{B} \cdot \mathbf{v}_1 + C v_1^2$. (Adding an angular momentum term to $\ln(f_1 f_2)$ is not independent of conservation of momentum, because the positions of the particles are the same.) The particles are assumed to have no internal degrees of freedom. Then

$$f_1 \propto \exp(\mathbf{B} \cdot \mathbf{v}_1 + C v_1^2) = A \exp[-\tfrac{1}{2}\beta m(\mathbf{v}_1 - \mathbf{u})^2]. \tag{A3.1.55}$$

When H has reached its minimum value this is the well known Maxwell–Boltzmann distribution for a gas in thermal equilibrium with a uniform motion $\mathbf{u}$. So, argues Boltzmann, solutions of his equation for an isolated system approach an equilibrium state, just as real gases seem to do.

Up to a negative factor ($-k_B$, in fact), differences in H are the same as differences in the thermodynamic entropy between initial and final equilibrium states. Boltzmann thought that his H-theorem gave a foundation of the increase in entropy as a result of the collision integral, whose derivation was based on the *Stosszahlansatz*.

(a) The reversibility and the recurrence paradoxes

Boltzmann's H-theorem raises a number of questions, particularly the central one: how can a gas that is described exactly by the reversible laws of mechanics be characterized by a quantity that always decreases? Perhaps a *non-mechanical* assumption was introduced here. If so, this would suggest, although not imply, that Boltzmann's equation might not be a useful description of nature. In fact, though, this equation is so useful and accurate a predictor of the properties of dilute gases, that it is now often used as a test of intermolecular potential models.

   The question stated above was formulated in two ways, each using an exact result from classical mechanics. One way, associated with the physicist Loschmidt, is fairly obvious. If classical mechanics provides a correct description of the gas, then associated with any physical motion of a gas, there is a time-reversed motion, which is also a solution of Newton's equations. Therefore if H decreases in one of these motions, there ought to be a physical motion of the gas where H increases. This is contrary to the H-theorem. The other objection is based on the recurrence theorem of Poincare [15], and is associated with the mathematician Zermelo. Poincare's theorem states that in a bounded mechanical system with finite energy, any initial state of the gas will eventually recur as a state of the gas, to within any preassigned accuracy. Thus, if H decreases during part of the motion, it must eventually increase so as to approach, arbitrarily closely, its initial value.

   The recurrence paradox is easy to refute and was done so by Boltzmann. He pointed out that the recurrence time even for a system of a several particles, much less a system of 10^{23} particles, is so enormously long (orders of magnitude larger than the age of the universe) that one will never live long enough to observe a recurrence. The usual response to Loschmidt is to argue that while the gas is indeed a mechanical system, almost all initial states of the gas one is likely to encounter in the laboratory will show an approach to equilibrium as described by the H-theorem. That is, the Boltzmann equation describes the most typical behaviour of a gas. While an anti-Boltzmann-like behaviour is not ruled out by mechanics, it is very unlikely, in a statistical sense, since such a motion would require a very careful (to put it mildly) preparation of the initial state. Thus, the reversibility paradox is more subtle, and the analysis of it eventually led Boltzmann to the very fruitful idea of an ergodic system [16]. In any case, there is no reason to doubt the validity of the Boltzmann equation for the description of irreversible processes in dilute gases. It describes the typical behaviour of a laboratory system, while any individual system may have small fluctuations about this typical behaviour.

A3.1.3.2   The Chapman–Enskog normal solutions of the Boltzmann equation

The practical value of the Boltzmann equation resides in the utility of the predictions that one can obtain from it. The form of the Boltzmann is such that it can be used to treat systems with long range forces, such as Lennard–Jones particles, as well as systems with finite-range forces. Given a potential energy function, one can calculate the necessary collision cross sections as well as the various restituting velocities well enough to derive practical expressions for transport coefficients from the Boltzmann equation. The method for obtaining solutions of the equation used for fluid dynamics is due to Enskog and Chapman, and proceeds by finding solutions that can be expanded in a series whose first term is a Maxwell–Boltzmann distribution of local equilibrium form. That is, the first takes the form given by (A3.1.55), with the quantities A, β and $\mathbf{u}$ being functions of $\mathbf{r}$ and t. One then assumes that the local temperature, $(k_B\beta)^{-1}$, mean velocity, $\mathbf{u}$, and local density, n, are slowly varying in space and time, where the distance over which they change, L, say is large compared with a mean free path, λ. The higher terms in the Chapman–Enskog solution are then expressed in a power series in gradients of the five variables, n, β and $\mathbf{u}$, which can be shown to be an expansion in powers of $l/L \ll 1$. Explicit results are then obtained for the first, and higher, order solution in l/L, which in turn lead to Navier–Stokes as well as higher order hydrodynamic equations. Explicit expressions are obtained for the various transport coefficients, which can then be compared with experimental data. The agreement is sufficiently close that the theoretical results provide a useful way for checking the accuracy of various trial potential energy functions. A complete account of the Chapman–Enskog solution method can be found in the book by Chapman and Cowling [3], and comparisons with experiments, the extension to polyatomic molecules, and to quantum gases, are discussed at some length in the books of Hirshfelder *et al* [4], of Hanley [5] and of Kestin [17] as well as in an enormous literature.

A3.1.3.3 *Extension of the Boltzmann equation to higher densities*

It took well over three quarters of a century for kinetic theory to develop to the point that a systematic extension of the Boltzmann equation to higher densities could be found. This was due to the work of Bogoliubov, Green and Cohen, who borrowed methods from equilibrium statistical mechanics, particularly the use of cluster expansion techniques, to obtain a virial expansion of the right-hand side of the Boltzmann equation to include the effects of collisions among three, four and higher numbers of particles, successively. However, this virial expansion was soon found to diverge term-by-term in the density, beyond the three-body term in three dimensions, and beyond the two-body term in two dimensions. In order to obtain a well behaved generalized Boltzmann equation one has to sum these divergent terms. This has been accomplished using various methods. One finds that the transport coefficients for moderately dense gases cannot be expressed strictly in a power series of the gas density, but small logarithmic terms in the density also appear. Moreover, one finds that long-range correlations exist in a non-equilibrium gas, that make themselves felt in the effects of light scattering by a dense fluid with gradients in temperature or local velocity. Reviews of the theory of non-equilibrium processes in dense gases can be found in articles by Dorfman and van Beijeren [14], by Cohen [18] and by Ernst [19], as well as in the book of Resibois and de Leener [7].

A3.1.4 Fluctuations in gases

Statistical mechanics and kinetic theory, as we have seen, are typically concerned with the average behaviour of an ensemble of similarly prepared systems. One usually hopes, and occasionally can demonstrate, that the variations of these properties from one system to another in the ensemble, or that the variation with time of the properties of any one system, are very small. There is a well developed theory for equilibrium fluctuations of these types. In this theory one can relate, for example, the specific heat at constant volume of a system in contact with a thermal reservoir to the mean square fluctuations of the energy of the system. It is also well known that the scattering of light by a fluid system is determined by the density fluctuations in the system, caused by the motion of the particles in the system. These thermal fluctuations in density, temperature and other equilibrium properties of the system typically scale to zero as the number of degrees of freedom in the system becomes infinite. A good account of these equilibrium fluctuations can be found in the text by Landau and Lifshitz [20].

When a system is not in equilibrium, the mathematical description of fluctuations about some time-dependent ensemble average can become much more complicated than in the equilibrium case. However, starting with the pioneering work of Einstein on Brownian motion in 1905, considerable progress has been made in understanding time-dependent fluctuation phenomena in fluids. Modern treatments of this topic may be found in the texts by Keizer [21] and by van Kampen [22]. Nevertheless, the non-equilibrium theory is not yet at the same level of rigour or development as the equilibrium theory. Here we will discuss the theory of Brownian motion since it illustrates a number of important issues that appear in more general theories.

We consider the motion of a large particle in a fluid composed of lighter, smaller particles. We also suppose that the mean free path of the particles in the fluid, λ, is much smaller than a characteristic size, R, of the large particle. The analysis of the motion of the large particle is based upon a method due to Langevin. Consider the equation of motion of the large particle. We write it in the form

$$M \frac{d\mathbf{v}(t)}{dt} = -\zeta \mathbf{v}(t) + \mathbf{F}(t) \qquad (A3.1.56)$$

where M is the mass of the large particle, $\mathbf{v}(t)$ is its velocity at time t, the quantity $-\zeta\mathbf{v}(t)$ represents the hydrodynamic friction exerted by the fluid on the particle, while the term $\mathbf{F}(t)$ represents the fluctuations in the force on the particle produced by the discontinuous nature of the collisions with the fluid particles. If the Brownian particle is spherical, with radius R, then ζ is usually taken to have the form provided by Stokes' law of friction on a slowly moving particle by a continuum fluid,

$$\zeta = 6\pi\eta R \qquad (A3.1.57)$$

where η is the shear viscosity of the fluid. The fact that the fluid is not a continuum is incorporated in the fluctuating force $\mathbf{F}(t)$. This fluctuating force is taken to have the following properties

$$\langle \mathbf{F}(t) \rangle = 0 \qquad (A3.1.58)$$

$$\langle \mathbf{F}_i(t_1)\mathbf{F}_j(t_2) \rangle = A\delta(t_1 - t_2)\delta_{\mathrm{Kr}}(i, j) \qquad (A3.1.59)$$

where A is some constant yet to be determined, $\delta(t_1 - t_2)$ is a Dirac delta function in the time interval between t_1 and t_2, and $\delta_{\mathrm{Kr}}(i, j)$ is a Kronecker delta function in the components of the fluctuating force in the directions denoted by i, j. The angular brackets denote an average, but the averaging process is somewhat subtle, and discussed in some detail in the book of van Kampen [22]. For our purposes, we will take the average to be over an ensemble constructed by following a long trajectory of one Brownian particle, cutting the trajectory into a large number of smaller, disjoint pieces, all of the same length, and then taking averages over all of the pieces. Thus the pieces of the long trajectory define the ensemble over which we average. The delta function correlation in the random force assumes that the collisions of the Brownian particle with the particles of the fluid are instantaneous and uncorrelated with each other.

If we now average the Langevin equation, (A3.1.56), we obtain a very simple equation for $\langle \mathbf{v}(t) \rangle$, whose solution is clearly

$$\langle \mathbf{v}(t) \rangle = \langle \mathbf{v}(0) \rangle \, e^{-\zeta t/M}. \qquad (A3.1.60)$$

Thus the average velocity decays exponentially to zero on a time scale determined by the friction coefficient and the mass of the particle. This average behaviour is not very interesting, because it corresponds to the average of a quantity that may take values in all directions, due to the noise and friction, and so the decay of the average value tells us little about the details of the motion of the Brownian particle. A more interesting quantity is the mean square velocity, $\langle \mathbf{v}(t)^2 \rangle$, obtained by solving (A3.1.56), squaring the solution and then averaging. Here the correlation function of the random force plays a central role, for we find

$$\langle \mathbf{v}^2(t) \rangle = \langle \mathbf{v}^2(0) \rangle \, e^{-2\zeta t/M} + \frac{3A}{2\zeta}(1 - e^{-2\zeta t/M}). \qquad (A3.1.61)$$

Notice that this quantity does not decay to zero as time becomes long, but rather it reaches the value

$$\langle \mathbf{v}^2(t) \rangle \to \frac{3A}{2\zeta} \qquad (A3.1.62)$$

as $t \to \infty$. Here we have the first appearance of A, the coefficient of the correlation of the random force. It is reasonable to suppose that in the infinite-time limit, the Brownian particle becomes equilibrated with the surrounding fluid. This means that the average kinetic energy of the Brownian particle should approach the value $3k_\mathrm{B}T/2$ as time gets large. This is consistent with (A3.1.62), if the coefficient A has the value

$$A = \frac{2\zeta k_\mathrm{B}T}{M}. \qquad (A3.1.63)$$

Thus, the requirement that the Brownian particle becomes equilibrated with the surrounding fluid fixes the unknown value of A, and provides an expression for it in terms of the friction coefficient, the thermodynamic temperature of the fluid, and the mass of the Brownian particle. Equation (A3.1.63) is the simplest and best known example of a *fluctuation–dissipation theorem*, obtained by using an equilibrium condition to relate the strength of the fluctuations to the frictional forces acting on the particle [22].

Two more important ideas can be illustrated by means of the Langevin approach to Brownian motion. The first result comes from a further integration of the velocity equation to find an expression for the fluctuating displacement of the moving particle, and for the mean square displacement as a function of time. By carrying out the relevant integrals and using the fluctuation–dissipation theorem, we can readily see that the mean square displacement, $\langle (\mathbf{r}(t) - \mathbf{r}(0))^2 \rangle$, grows linearly in time t, for large times, as

$$\langle (\mathbf{r}(t) - \mathbf{r}(0))^2 \rangle = \frac{6 k_B T M}{\zeta} t. \tag{A3.1.64}$$

Now for a particle undergoing diffusion, it is also known that its mean square displacement grows linearly in time, for long times, as

$$\langle (\mathbf{r}(t) - \mathbf{r}(0))^2 \rangle = 6Dt \tag{A3.1.65}$$

where D is the diffusion coefficient of the particle. By comparing (A3.1.64) and (A3.1.65) we see that

$$D = \frac{k_B T M}{\zeta}. \tag{A3.1.66}$$

This result is often called the *Stokes–Einstein formula* for the diffusion of a Brownian particle, and the Stokes' law friction coefficient $6\pi \eta R$ is used for ζ.

The final result that we wish to present in this connection is an example of the Green–Kubo time-correlation expressions for transport coefficients. These expressions relate the transport coefficients of a fluid, such as viscosity, thermal conductivity, etc, in terms of time integrals of some average time-correlation of an appropriate microscopic variable. For example, if we were to compute the time correlation function of one component of the velocity of the Brownian particle, $\langle v_x(t_1) v_x(t_2) \rangle$, we would obtain

$$\langle \mathbf{v}_x(t_1) \mathbf{v}_x(t_2) \rangle = k_B T e^{-\zeta |t_1 - t_2|/M} \tag{A3.1.67}$$

for large times, neglecting factors that decay exponentially in both t_1 and t_2. The Green–Kubo formula for diffusion relates the diffusion coefficient, D to the time integral of the time-correlation of the velocity through

$$D = \int_0^\infty \mathrm{d}t \langle v_x(0) v_x(t) \rangle \tag{A3.1.68}$$

a result which clearly reproduces (A3.1.66). The Green–Kubo formulae are of great interest in kinetic theory and non-equilibrium statistical mechanisms since they provide a new set of functions, the time-correlation functions, that tell us more about the microscopic properties of the fluid than do the transport coefficients themselves, and that are very useful for analysing fluid behaviour when making computer simulations of the fluid.

References

[1] Brush S 1966–1972 *Kinetic Theory* vols 1–3 (New York; Pergamon)
[2] Boltzmann L 1995 *Lectures on Gas Theory* translator S Brush (New York: Dover)
[3] Chapman S and Cowling T G 1970 *The Mathematical Theory of Non-Uniform Gases* 3rd edn (Cambridge: Cambridge University Press)
[4] Hirschfelder J O, Curtiss C F and Bird R B 1954 *Molecular Theory of Gases and Liquids* (New York: Wiley)
[5] Hanley H J M 1970 *Transport Phenomena in Fluids* (New York: Marcel Dekker)
[6] Fertziger J H and Kaper H G 1972 *Mathematical Theory of Transport Processes in Gases* (Amsterdam: North Holland)
[7] Resibois P and de Leener M 1977 *Classical Kinetic Theory of Fluids* (New York: Wiley)

[8]   Liboff R L 1998 *Kinetic Theory: Classical, Quantum, and Relativistic Descriptions* 2nd edn (New York: Wiley)
[9]   Present R D 1958 *Kinetic Theory of Gases* (New York: McGraw-Hill)
[10] McQuarrie D A 1976 *Statistical Mechanics* (New York: Harper and Row)
[11] Kestin J and Dorfman J R 1970 *A Course in Statistical Thermodynamics* (New York: Academic)
[12] Huang K 1990 *Statistical Mechanics* 2nd edn (New York: Wiley)
[13] Wannier G 1987 *Statistical Mechanics* (New York: Dover)
[14] Dorfman J R and van Beijeren H 1977 The kinetic theory of gases *Statistical Mechanics, Part B: Time-Dependent Processes* ed B J Berne (New York: Plenum)
[15] Uhlenbeck G E and Ford G W 1963 *Lectures in Statistical Mechanics* (Providence, RI: American Mathematical Society)
[16] Arnold V I and Avez A 1968 *Ergodic Problems of Classical Mechanics* (New York: Benjamin)
[17] Kestin J (ed) 1973 *Transport Phenomena, AIP Conference Proceedings No. 11* (New York: American Institute of Physics)
[18] Cohen E G D 1993 Fifty years of kinetic theory *Physica* A **194** 229
[19] Ernst M H 1998 Bogoliubov–Choh–Uhlenbeck theory: cradle of modern kinetic theory *Progress in Statistical Physics* ed W Sung *et al* (Singapore: World Scientific)
[20] Landau L D and Lifshitz E M 1980 *Statistical Physics, Part I* 3rd edn, translators J B Sykes and M J Kearney (Oxford: Pergamon)
[21] Keizer J 1987 *Statistical Thermodynamics of Nonequilibrium Processes* (New York: Springer)
[22] van Kampen N G 1992 *Stochastic Processes in Physics and Chemistry* (Amsterdam: North Holland)

Further Reading

Cohen E G D 1993 Kinetic theory: understanding nature through collisions *Am. J. Phys.* **61** 524

Balescu R 1997 *Statistical Dynamics: Matter Out of Equilibrium* (London: Imperial College Press)

McLennan J A 1989 *Introduction to Nonequilibrium Statistical Mechanics* (Englewood Cliffs; Prentice-Hall)

Lifshitz E M and Pitaevskii L P 1981 *Physical Kinetics* translators J B Sykes and R N Franklin (London: Pergamon)

A3.2
Non-equilibrium thermodynamics

Ronald F Fox

A3.2.1 Introduction

Equilibrium thermodynamics may be developed as an autonomous macroscopic theory or it may be derived from microscopic statistical mechanics. The intrinsic beauty of the macroscopic approach is partially lost with the second treatment. Its beauty lies in its internal consistency. The advantage of the second treatment is that certain quantities are given explicit formulae in terms of fundamental constants, whereas the purely macroscopic approach must use measurements to determine these same quantities. The Stefan–Boltzmann constant is a prime example of this dichotomy. Using purely macroscopic thermodynamic arguments, Boltzmann showed that the energy density emitted per second from a unit surface of a black body is σT^4 where T is the temperature and σ is the Stefan–Boltzmann constant, but it takes statistical mechanics to produce the formula

$$\sigma = \frac{2\pi^5 k^4}{15c^2 h^3}$$

in which k is Boltzmann's constant, c is the speed of light and h is Planck's constant. This beautiful formula depends on three fundamental constants, an exhibition of the power of the microscopic viewpoint. Likewise, non-equilibrium thermodynamics may be developed as a purely autonomous macroscopic theory or it may be derived from microscopic kinetic theory, either classically or quantum mechanically. The separation between the macroscopic and microscopic approaches is a little less marked than for the equilibrium theory because the existence of the microscopic underpinning leads to the existence of fluctuations in the macroscopic picture, as well as to the celebrated Onsager reciprocal relations. On purely macroscopic grounds, the fluctuation–dissipation relation that connects the relaxation rates to the strengths of the fluctuations may be established, but it takes the full microscopic theory to compute their quantitative values, at least in principle. In practice, these computations are very difficult. This presentation is primarily about the macroscopic approach, although at the end the microscopic approach, based on linear response theory, is also reviewed.

The foundations for the macroscopic approach to non-equilibrium thermodynamics are found in Einstein's theory of Brownian movement [1] of 1905 and in the equation of Langevin [2] of 1908. Uhlenbeck and Ornstein [3] generalized these ideas in 1930 and Onsager [4, 5] presented his theory of irreversible processes in 1931. Onsager's theory [4] was initially deterministic with little mention of fluctuations. His second paper [5] used fluctuation theory to establish the reciprocal relations. The fundamental role of fluctuations was generalized by the works of Chandrasekhar [6] in 1943, of Wang and Uhlenbeck [7] in 1945, of Casimir [8] in 1945, of Prigogine [9] in 1947 and of Onsager and Machlup [10, 11] in 1953. In 1962 de Groot and Mazur [12] published their definitive treatise *Non-Equilibrium Thermodynamics* which greatly extended the applicability of the theory as well as deepening its foundations. By this time, it was clearly recognized that the mathematical setting for non-equilibrium thermodynamics is the theory of stationary, Gaussian–Markov processes. The Onsager reciprocal relations may be most easily understood within this context. Nevertheless, the issue of the most general form for stationary, Gaussian–Markov processes, although broached by de Groot and Mazur [12], was not solved until the work of Fox and Uhlenbeck [13–15] in 1969. For example, this work made it possible to rigorously extend the Onsager theory to hydrodynamics and to the Boltzmann equation.

A3.2.2   General stationary Gaussian–Markov processes

A3.2.2.1   Characterization of random processes

Let $a(t)$ denote a time dependent random process. $a(t)$ is a random process because at time t the value of $a(t)$ is not definitely known but is given instead by a probability distribution function $W_1(a, t)$ where a is the value $a(t)$ can have at time t with probability determined by $W_1(a, t)$. $W_1(a, t)$ is the first of an infinite collection of distribution functions describing the process $a(t)$ [7, 13]. The first two are defined by

$W_1(a, t) \, da$ = probability at time t that the value of $a(t)$ is between a and $a + da$

$W_2(a_1, t_1; a_2, t_2) \, da_1 \, da_2$ = probability that at time t_1 the value of $a(t)$ is between a_1 and $a_1 + da_1$ and that at time t_2 the value of $a(t)$ is between a_2 and $a_2 + da_2$.

The higher order distributions are defined analogously. W_2 contains W_1 through the identity

$$W_1(a_1, t_1) = \int da_2 \, W_2(a_1, t_1; a_2, t_2).$$

Similar relations hold for the higher distributions. The Gaussian property of the process implies that all statistical information is contained in just W_1 and W_2.

The condition that the process $a(t)$ is a *stationary* process is equivalent to the requirement that all the distribution functions for $a(t)$ are invariant under time translations. This has as a consequence that $W_1(a, t)$ is independent of t and that $W_2(a_1, t_1; a_2, t_2)$ only depends on $t = t_2 - t_1$. An *even* stationary process [4] has the additional requirement that its distribution functions are invariant under time reflection. For W_2, this implies $W_2(a_1; a_2, t) = W_2(a_2; a_1, t)$. This is called *microscopic reversibility*. It means that the quantities are even functions of the particle velocities [12]. It is also possible that the variables are odd functions of the particle velocities [8], say, b_1 and b_2 for which $W_2(b_1; b_2, t) = W_2(-b_2; -b_1, t)$. In the general case considered later, the thermodynamic quantities are a mixture of even and odd [14, 15]. For the odd case, the presence of a magnetic field, B, or a coriolis force depending on angular velocity ω, requires that B and ω also change sign during time reversal and microscopic reversibility reads as $W_2(b_1; b_2, B, \omega t) = W_2(-b_2; -b_1, -B, -\omega, t)$. Examples of even processes include heat conduction, electrical conduction, diffusion and chemical reactions [4]. Examples of odd processes include the Hall effect [12] and rotating frames of reference [4]. Examples of the general setting that lacks even or odd symmetry include hydrodynamics [14] and the Boltzmann equation [15].

Before defining a *Markov* process $a(t)$, it is necessary to introduce *conditional* probability distribution functions $P_2(a_1, t_1 \mid a_2, t_2)$ and $P_3(a_1, t_1; a_2, t_2 \mid a_3, t_3)$ defined by

$P_2(a_1, t_1 \mid a_2, t_2) \, da_2$ = probability at time t_2 that the value of $a(t)$ is between a_2 and $a_2 + da_2$ *given* that at time $t_1 < t_2 \, a(t)$ had the definite value a_1.

$P_3(a_1, t_1; a_2, t_2 \mid a_3, t_3) \, da_3$ = probability at time t_3 that the value of $a(t)$ is between a_3 and $a_3 + da_3$ *given* that at time $t_2 < t_3 \, a(t)$ had the definite value a_2 *and* at time $t_1 < t_2 \, a(t)$ had the definite value a_1.

These conditional distributions are related to the W_n by

$$W_2(a_1, t_1; a_2, t_2) = W_1(a_1, t_1) P_2(a_1, t_1 \mid a_2, t_2)$$
$$W_2(a_1, t_1; a_2, t_2; a_3, t_3) = W_2(a_1, t_1; a_2, t_2) P_3(a_1, t_1; a_2, t_2 \mid a_3, t_3)$$

and so forth for the higher order distributions. The *Markov* property of $a(t)$ is defined by

$$P_2(a_2, t_2 \mid a_3, t_3) = P_3(a_1, t_1; a_2, t_2 \mid a_3, t_3) \qquad (A3.2.1)$$

which means that knowledge of the value of $a(t)$ at time t_1 does not influence the distribution of values of $a(t)$ at time $t_3 > t_1$ if there is also information giving the value of $a(t)$ at the intermediate time t_2. Therefore, a Markov process is completely characterized by its W_1 and P_2 or equivalently by only W_2. A stationary Markov process has distributions satisfying the Smoluchowski equation (also called the Chapman–Kolmogorov equation)

$$W_2(a_1; a_2 t) = \int da\, W_2(a_1; a, t - s) P_2(a \mid a_2, s) \qquad \text{for all } s \in [0, t]. \qquad (A3.2.2)$$

Proof. For $t_1 < t_3 < t_2$ and using (A3.2.1)

$$W_2(a_1; a_2, t_2 - t_1) = \int da_3\, W_3(a_1, t_i; a_3, t_3; a_2, t_2) = \int da_3\, W_2(a_1, t_1; a_3, t_3) P_3(a_1, t_i; a_3, t_3 \mid a_2, t_2)$$

$$= \int da_3\, W_2(a_1, t_1; a_3, t_3) P_2(a_3, t_3 \mid a_2 t_2).$$

Setting $s = t_2 - t_3$ and $t = t_2 - t_1$ and $a_3 = a$ gives (A3.2.2) for a stationary process $a(t)$. QED

While the Smoluchowski equation is necessary for a Markov process, in general it is not sufficient, but known counter-examples are always non-Gaussian as well.

A3.2.2.2 The Langevin equation

The prototype for all physical applications of stationary Gaussian–Markov processes is the treatment of Brownian movement using the Langevin equation [2, 3, 7]. The Langevin equation describes the time change of the velocity of a slowly moving colloidal particle in a fluid. The effect of the interactions between the particle and the fluid molecules produces two forces. One force is an average effect, the *frictional drag* that is proportional to the velocity, whereas the other force is a fluctuating force, $\tilde{F}(t)$, that has mean value zero. Therefore, a particle of mass M obeys the Langevin equation

$$M \frac{du}{dt} = -\alpha u + \tilde{F}(t) \qquad (A3.2.3)$$

where α is the frictional drag coefficient and u is the particle's velocity. It is the fluctuating force, $\tilde{F}(t)$, that makes u a random process. For a sphere of radius R in a fluid of viscosity η, $\alpha = 6\pi \eta R$, a result obtained from hydrodynamics by Stokes in 1854. To characterize this process it is necessary to make assumptions about $\tilde{F}(t)$. $\tilde{F}(t)$ is taken to be a stationary Gaussian process that is called *white noise*. This is defined by the correlation formula

$$\langle \tilde{F}(t) \tilde{F}(s) \rangle = 2\lambda \delta(t - s) \qquad (A3.2.4)$$

where $\langle \ldots \rangle$ denotes averaging over $\tilde{F}(t)$, λ is a constant and the Dirac delta function of time expresses the quality of whiteness for the noise. The linearity of (A3.2.3) is sufficient to guarantee that u is also a stationary Gaussian process, although this claim requires some care as is shown below.

Equation (A3.2.3) must be solved with respect to some initial value for the velocity, $u(0)$. In the conditional distribution for the process $u(t)$, the initial value, $u(0)$, is denoted by u_0 giving $P_2(u_0 \mid u, t)$. Because $u(t)$ is a Gaussian process, $P_2(u_0 \mid u, t)$ is completely determined by the mean value of $u(t)$ and by its mean square. Using (A3.2.4) and recalling that $\langle \tilde{F}(t) \rangle = 0$ it is easy to prove that

$$\langle u(t) \rangle = u(0) \exp\left[-\frac{\alpha}{M} t \right] \tag{A3.2.5}$$

$$\langle u^2(t) \rangle = u^2(0) \exp\left[-\frac{2\alpha}{M} t \right] + \frac{\lambda}{\alpha M} \left(1 - \exp\left[-\frac{2\alpha}{M} t \right] \right) \tag{A3.2.6}$$

$$\langle u(t)u(s) \rangle = u^2(0) \exp\left[-\frac{\alpha}{M}(t+s) \right] + \frac{\lambda}{\alpha M} \left(\exp\left[-\frac{\alpha}{M}|t-s| \right] - \exp\left[-\frac{\alpha}{M}(t+s) \right] \right). \tag{A3.2.7}$$

Using $\sigma^2 = \lambda/\alpha M$ and $\rho(t) = \exp[-(\alpha/M)t]$, $P_2(u_0 \mid u, t)$ is given by

$$P_2(u_0 \mid u, t) = \frac{\exp[-(u - u_0\rho(t))^2/2\sigma^2(1 - \rho^2(t))]}{\sqrt{2\pi\sigma^2(1 - \rho^2(t))}} \tag{A3.2.8}$$

which is checked by seeing that it reproduces (A3.2.5) and (A3.2.6). From (A3.2.8), it is easily seen that the $t \to \infty$ limit eliminates any influence of u_0

$$\lim_{t \to \infty} P_2(u_0 \mid u, t) = W_1(u) = \frac{\exp[-\alpha M u^2/2\lambda]}{\sqrt{2\pi\lambda/\alpha M}}.$$

However, $W_1(u)$ should also be given by the equilibrium Maxwell distribution

$$W_1(u) = \frac{\exp[-Mu^2/2kT]}{\sqrt{2\pi kT/M}} \tag{A3.2.9}$$

in which k is Boltzmann's constant and T is the equilibrium temperature of the fluid. The equality of these two expressions for $W_1(u)$ results in Einstein's relation

$$\lambda = kT\alpha. \tag{A3.2.10}$$

Putting (A3.2.10) into (A3.2.4) gives the prototype example of what is called the *fluctuation–dissipation relation*

$$\langle \tilde{F}(t)\tilde{F}(s) \rangle = 2kT\alpha\delta(t-s).$$

Looking back at (A3.2.7), we see that a second average over $u(0)$ can be performed using $W_1(u_0)$. This second type of averaging is denoted by $\{\ldots\}$. Using (A3.2.9) we obtain

$$\{u^2(0)\} = \frac{kT}{M}$$

which with (A3.2.7) and (A3.2.10), the Einstein relation, implies

$$\{\langle u(t)u(s) \rangle\} = \frac{kT}{M} \exp\left[-\frac{\alpha}{M}|t-s| \right].$$

This result clearly manifests the stationarity of the process that is not yet evident in (A3.2.7).

Using $W_2 = W_1 P_2$, (A3.2.8) and (A3.2.9) may be used to satisfy the Smoluchowski equation, (A3.2.2), another necessary property for a stationary process. Thus $u(t)$ is an example of a stationary Gaussian–Markov process. In the form given by (A3.2.3), the process $u(t)$ is also called an Ornstein–Uhlenbeck process ('OU process').

Consider an ensemble of Brownian particles. The approach of P_2 to W_1 as $t \to \infty$ represents a kind of diffusion process in velocity space. The description of Brownian movement in these terms is known as the *Fokker–Planck* method [16]. For the present example, this equation can be shown to be

$$\frac{\partial}{\partial t} P_2(u, t) = \frac{\alpha}{M} \frac{\partial}{\partial u} (u P_2(u, t)) + \frac{kT\alpha}{M^2} \frac{\partial^2}{\partial u^2} P_2(u, t) \tag{A3.2.11}$$

subject to the initial condition $P_2(u, 0) = \delta(u - u_0)$. The solution to (A3.2.11) is given by (A3.2.8). The Langevin equation and the Fokker–Planck equation provide equivalent complementary descriptions [17].

A3.2.3 Onsager's theory of non-equilibrium thermodynamics

A3.2.3.1 Regression equations and fluctuations

For a system which is close to equilibrium, it is assumed that its state is described by a set of extensive thermodynamic variables, $a_1(t), a_2(t), \ldots, a_n(t)$ where n is very much less than the total number of degrees of freedom for all of the molecules in the system. The latter may be of order 10^{24} while the former may be fewer than 10. In equilibrium, the a_i are taken to have value zero so that the non-equilibrium entropy is given by

$$S = S_0 - \tfrac{1}{2} k a_i E_{ij} a_j \tag{A3.2.12}$$

where S_0 is the maximum equilibrium value of the entropy, E_{ij} is a symmetric positive definite time independent entropy matrix and repeated indices are to be summed, a convention used throughout this presentation. *Thermodynamic forces* are defined by

$$X_l = \frac{\partial S}{\partial a_l} = -k E_{lj} a_j. \tag{A3.2.13}$$

Onsager postulates [4, 5] the phenomenological equations for irreversible processes given by

$$R_{ij} \frac{\mathrm{d}}{\mathrm{d}t} a_j \equiv R_{ij} J_j = X_i = -k E_{ij} a_j \tag{A3.2.14}$$

in which the J_j are called the *thermodynamic fluxes*, and which is a natural generalization of the linear phenomenological laws such as Fourier's law of heat conduction, Newton's law of internal friction etc. The matrix R_{ij} is real with eigenvalues having positive real parts and it is invertible. These equations are *regression* equations whose solutions approach equilibrium asymptotically in time.

Since the a_i are thermodynamic quantities, their values fluctuate with time. Thus, (A3.2.14) is properly interpreted as the averaged regression equation for a random process that is actually driven by random thermodynamic forces, $\tilde{e}_i(t)$. The completed equations are coupled Langevin-like equations

$$R_{ij} \frac{\mathrm{d}}{\mathrm{d}t} a_j = R_{ij} J_j = X_i + \tilde{e}_i = -k E_{ij} a_j + \tilde{e}_i. \tag{A3.2.15}$$

The mean values of the $\tilde{e}_i(t)$ are zero and each is assumed to be stationary Gaussian white noise. The linearity of these equations guarantees that the random process described by the a_i is also a stationary Gaussian–Markov process [12]. Denoting the inverse of R_{ij} by L_{ij} and using the definition

$$\tilde{F}_i = L_{ij} \tilde{e}_i$$

(A3.2.15) may be rewritten as

$$\frac{\mathrm{d}}{\mathrm{d}t} a_i = J_i = L_{ij} X_j + \tilde{F}_i = -k L_{ij} E_{jk} a_k + \tilde{F}_i. \tag{A3.2.16}$$

Since the $\tilde{F}_i$ are linearly related to the $\tilde{e}_i(t)$, they are also stationary Gaussian white noises. This property is explicitly expressed by

$$\langle \tilde{F}_i(t)\tilde{F}_j(t)\rangle = 2Q_{ij}\delta(t-s) \qquad\qquad (A3.2.17)$$

in which Q_{ij}, the force–force correlation matrix, is necessarily symmetric and positive definite. While (A3.2.16) suggests that the fluxes may be coupled to any force, symmetry properties may be applied to show that this is not so. By establishing the tensor character of the different flux and force components, it can be shown that only fluxes and forces of like character can be coupled. This result is called Curie's principle [9, 12].

Let $G_{ij} = kL_{ik}E_{kj}$. The solution to (A3.2.16) is

$$a_i(t) = (\exp[-\mathbf{G}t])_{ij}a_j(0) + \int_0^t \mathrm{d}s\,(\exp[-\mathbf{G}(t-s)])_{ij}\tilde{F}_j(s). \qquad (A3.2.18)$$

The statistics for the initial conditions, $a_j(0)$, are determined by the equilibrium distribution obtained from the entropy in (A.3.2.12) and in accordance with the Einstein–Boltzmann–Planck formula

$$W_1(a_1, a_2, \ldots, a_n) = \left(\frac{\|\mathbf{E}\|}{(2\pi)^n}\right)^{1/2} \exp\left[-\frac{1}{2}a_i E_{ij}a_j\right] \qquad (A3.2.19)$$

where $\|\mathbf{E}\|$ is the determinant of E_{ij}. $\{\ldots\}$ will again denote averaging with respect to W_1 while $\langle\ldots\rangle$ continues to denote averaging over the $\tilde{F}_i$. Notice in (A3.2.19) that W_1 now depends on n variables at a single time, a natural generalization of the situation reviewed above for the one-dimensional Langevin equation. This simply means the process is n dimensional.

A3.2.3.2   *Two time correlations and the Onsager reciprocal relations*

From (A3.2.18) it follows that

$$\langle a_i(t)\rangle = (\exp[-\mathbf{G}t])_{ij}a_j(0)$$

and

$$\{\langle a(t)\rangle\} = 0.$$

All the other information needed for this process is contained in the two time correlation matrix because the process is Gaussian. A somewhat involved calculation [18] results (for $t_2 > t_1$) in

$$\chi_{ij}(t_2, t_1) \equiv \{\langle a_i(t_2)a_j(t_1)\rangle\} = (\exp[-\mathbf{G}(t_2 - t_1)])_{ik}(\exp[-\mathbf{G}t_1]\mathbf{E}^{-1}\exp[-\mathbf{G}^\dagger t_1])_{kj}$$

$$+ 2(\exp[-\mathbf{G}(t_2 - t_1)])_{ik}\int_0^{t_1} \mathrm{d}s\,(\exp[-\mathbf{G}(t_1 - s)]\mathbf{Q}\exp[-\mathbf{G}^\dagger(t_1 - s)])_{kj}. \quad (A3.2.20)$$

If we set $t_2 = t_1 = t$, stationarity requires that $\chi_{ij}(t, t) = (\mathbf{E}^{-1})_{ij}$ because (A3.2.19) implies

$$\{a_i, a_j\} = \int \mathrm{d}^n a\, W_1(a_1, a_2, \ldots, a_n)a_i a_j = (\mathbf{E}^{-1})_{ij}.$$

By looking at $\mathbf{G}\chi(t, t) + \chi(t, t)\mathbf{G}^\dagger$, the resulting integral implied by (A3.2.20) for $t_2 = t_1 = t$ contains an exact differential [18] and one obtains

$$\mathbf{GE}^{-1} + \mathbf{E}^{-1}\mathbf{G}^\dagger = \mathbf{G}\chi(t, t) + \chi(t, t)\mathbf{G}^\dagger = 2\mathbf{Q} + \exp[-\mathbf{G}t](\mathbf{GE}^{-1} + \mathbf{E}^{-1}\mathbf{G}^\dagger - 2\mathbf{Q})\exp[-\mathbf{G}^\dagger t].$$

This is compatible with stationarity if and only if

$$\mathbf{GE}^{-1} + \mathbf{E}^{-1}\mathbf{G}^\dagger = 2\mathbf{Q} \qquad\qquad (A3.2.21)$$

which is the general case fluctuation–dissipation relation [12]. Inserting this identity into (A3.2.17) and using similar techniques reduces $\chi_{ij}(t_2, t_1)$ to the manifestly stationary form

$$\chi_{ij}(t_2 - t_1) = (\exp[-\mathbf{G}|t_2 - t_1|])_{ik}(\mathbf{E}^{-1})_{kj}. \tag{A3.2.22}$$

In Onsager's treatment, the additional restriction that the a_i are even functions of the particle velocities is made. As indicated above, this implies microscopic reversibility which in the present n-dimensional case means (for $t = t_2 - t_1 > 0$) $W_2(a_1, a_2, \ldots, a_n; a'_1, a'_2, \ldots, a'_n, t) = W_2(a'_1, a'_2, \ldots, a'_n; a_1, a_2, \ldots, a_n, t)$. This implies

$$\chi_{ij}(t_2 - t_1) \equiv \int d^n a \, d^n a' \, W_2(a_1, a_2, \ldots, a_n t_1; a'_1, a'_2, \ldots, a'_n, t_2) a'_i a_j$$

$$= \int d^n a \, d^n a' \, W_2(a'_1, a'_2, \ldots, a'_n, t_1; a_1, a_2, \ldots, a_n, t_2) a_j a'_i \equiv \chi_{ji}(t_2 - t_1). \tag{A3.2.23}$$

Take the t_2-derivative of this equation by using (A3.2.22) and set $t_2 = t_1$

$$\frac{d}{dt_2} \chi_{ij}(t_2 - t_1)_{t_2=t_1} = -G_{ik}(\mathbf{E}^{-1})_{kj} = \frac{d}{dt_2} \chi_{ji}(t_2 - t_1)_{t_2=t_1} = -G_{jk}(\mathbf{E}^{-1})_{ki}. \tag{A3.2.24}$$

Inserting the definition of $\mathbf{G}$ gives the celebrated Onsager reciprocal relations [4, 5]

$$L_{ij} = L_{ji}. \tag{A3.2.25}$$

If odd variables, the b, are also included, then a generalization by Casimir [8] results in the Onsager–Casimir relations

$$L_{ij}(\boldsymbol{B}, \omega) = L_{ji}(-\boldsymbol{B}, -\omega)$$
$$L_{im}(\boldsymbol{B}, \omega) = -L_{mi}(-\boldsymbol{B}, -\omega)$$
$$L_{nm}(\boldsymbol{B}, \omega) = L_{mn}(-\boldsymbol{B}, -\omega) \tag{A3.2.26}$$

wherein the indices i and j are for variables a and the indices m and n are for variables b. These are proved in a similar fashion. For example, when there are mixtures of variables a and b, microscopic reversibility becomes $W_2(a_1, \ldots, a_p, b_{p+1}, \ldots, b_n; a'_1, \ldots, a'_p, b'_{p+1}, \ldots, b'_n, \boldsymbol{B}, \omega, t) = W_2(a'_1, \ldots, a'_p, -b'_{p+1}, \ldots, -b'_n; a_1, \ldots, a_p, -b_{p+1}, \ldots, -b_n, -\boldsymbol{B}, -\omega, t)$. A cross-correlation between an even variable and an odd variable is given by

$$\chi_{im}(\boldsymbol{B}, \omega, t_2 - t_1) \equiv \int d^p a \, d^{n-p} b \, d^p a' \, d^{n-p} b' \, W_2(a_1, \ldots, a_p, b_{p+1}, \ldots, b_n, t_1; a'_1, \ldots a'_p,$$

$$b'_{p+1}, \ldots, b'_n, \boldsymbol{B}, \omega, t_2) a'_i b_m$$

$$= \int d^p a \, d^{n-p} b \, d^p a' \, d^{n-p} b' \, W_2(a'_1, \ldots, a'_p, -b'_{p+1}, \ldots, -b'_n, t_1; a_1, \ldots, a_p,$$

$$- b_{p+1}, \ldots, -b_n, -\boldsymbol{B}, -\omega, t_2) b_m a'_i$$

$$\equiv -\chi_{mi}(-\boldsymbol{B}, -\omega, t_2 - t_1)$$

in which the last identity follows from replacing all b by their negatives. Differentiation of this expression with respect to t_2 followed by setting $t_2 = t_1$ results in the middle result of (A3.2.26).

In the general case, (A3.2.23) cannot hold because it leads to (A3.2.24) which requires $\mathbf{GE}^{-1} = (\mathbf{GE}^{-1})^{\dagger}$ which is in general not true. Indeed, the simple example of the Brownian motion of a harmonic oscillator suffices to make the point [7, 14, 18]. In this case the equations of motion are [3, 7]

$$M \frac{dx}{dt} = p \quad \text{and} \quad \frac{dp}{dt} + M\omega^2 x = -\frac{\alpha}{M} p + \tilde{F} \tag{A3.2.27}$$

where M is the oscillator mass, ω is the oscillator frequency and α is the friction coefficient. The fluctuating force, $\tilde{F}$, is Gaussian white noise with zero mean and correlation formula

$$\langle \tilde{F}(t)\tilde{F}(s)\rangle = 2D\delta(t - s).$$

Define y by $y = M\omega x$. The identifications

$$a_i = \begin{pmatrix} y \\ p \end{pmatrix} \qquad A_{ij} \begin{pmatrix} 0 & -\omega \\ \omega & 0 \end{pmatrix} \qquad S_{ij} = \begin{pmatrix} 0 & 0 \\ 0 & \alpha/M \end{pmatrix} \qquad \tilde{F}_i = \begin{pmatrix} 0 \\ \tilde{F} \end{pmatrix}$$

permit writing (A3.2.27) as

$$\frac{\mathrm{d}}{\mathrm{d}t}a_i = -A_{ij}a_j - S_{ij}a_j + \tilde{F}_i.$$

Clearly, $G = A + S$ in this example. The entropy matrix can be obtained from the Maxwell–Boltzmann distribution

$$W_1(y, p) = W_0 \exp\left[-\frac{p^2 + y^2}{2MkT}\right]$$

which implies that

$$E_{ij} = \frac{1}{MkT}\begin{pmatrix} 1 & 0 \\ 0 & 1 \end{pmatrix}.$$

The Q of (A3.2.17) is clearly given by

$$Q_{ij} = \begin{pmatrix} 0 & 0 \\ 0 & D \end{pmatrix}.$$

In this case the fluctuation–dissipation relation, (A3.2.21), reduces to $D = kT\alpha$. It is also clear that $\mathbf{GE}^{-1} = (\mathbf{A} + \mathbf{S})/MkT$ which is not self-adjoint.

A3.2.3.3   *Least dissipation variational principles*

Onsager [4, 5] generalized Lord Rayleigh's 'principle of the least dissipation of energy' [19]. In homage to Lord Rayleigh, Onsager retained the name of the principle (i.e. the word *energy*) although he clearly stated [4] that the role of the potential in this principle is played by the *rate of increase of the entropy* [9]. This idea is an attempt to extend the highly fruitful concept of an underlying variational principle for dynamics, such as Hamilton's principle of least action for classical mechanics, to irreversible processes. Because the regression equations are linear, the parallel with Lord Rayleigh's principle for linear friction in mechanics is easy to make. It has also been extended to velocity dependent forces in electrodynamics [20].

From (A3.2.12), it is seen that the rate of entropy production is given by

$$\frac{\mathrm{d}}{\mathrm{d}t}S = -\left(\frac{\mathrm{d}}{\mathrm{d}t}a_i\right)kE_{ij}a_j = J_iX_i.$$

Wherein the definition of the thermodynamic fluxes and forces of (A3.2.13) and (A3.2.14) have been used. Onsager defined [5] the analogue of the Rayleigh dissipation function by

$$\Phi = \tfrac{1}{2}R_{ij}J_iJ_i.$$

When the reciprocal relations are valid in accord with (A3.2.25) then R is also symmetric. The variational principle in this case may be stated as

$$0 = \delta[J_iX_i - \Phi] = [X_i - R_{ij}J_j]\delta J_i$$

wherein the variation is with respect to the fluxes for fixed forces. The second variation is given by $-R$, which has negative eigenvalues (for symmetric R the eigenvalues are real and positive), which implies that the difference between the entropy production rate and the dissipation function is a maximum for the averaged irreversible process, hence *least dissipation* [5]. By multiplying this by the temperature, the free energy is generated from the entropy and, hence, Onsager's terminology of least dissipation of *energy*. Thus, the principle of least dissipation of entropy for near equilibrium dynamics is already found in the early work of Onsager [5, 9].

Related variational principles for non-equilibrium phenomena have been developed by Chandrasekhar [21]. How far these ideas can be taken remains an open question. Glansdorff and Prigogine [22] attempted to extend Onsager's principle of the least dissipation of entropy [4, 5, 9] to non-linear phenomena far away from equilibrium. Their proposal was ultimately shown to be overly ambitious [23]. A promising approach for the non-linear steady state regime has been proposed by Keizer [23, 24]. This approach focuses on the covariance of the fluctuations rather than on the entropy, although in the linear regime around full equilibrium, the two quantities yield identical principles. In the non-linear regime the distinction between them leads to a novel thermodynamics of steady states that parallels the near equilibrium theory. An experiment for non-equilibrium electromotive force [25] has confirmed this alternative approach and strongly suggests that a fruitful avenue of investigation for far from equilibrium thermodynamics has been opened.

A3.2.4 Applications

A3.2.4.1 *Soret effect and Dufour effect [12]*

Consider an isotropic fluid in which viscous phenomena are neglected. Concentrations and temperature are non-uniform in this system. The rate of entropy production may be written

$$\frac{dS}{dt} = J_q \cdot \nabla \frac{1}{T} + \sum_{i=1}^{n} J_i \cdot \nabla \left(-\frac{\mu_i}{T} \right) \tag{A3.2.28}$$

in which J_q is the heat flux vector and J_i is the mass flux vector for species i which has chemical potential μ_i. This is over-simplified for the present discussion because the n mass fluxes, J_i, are not linearly independent [12]. This fact may be readily accommodated by eliminating one of the fluxes and using the Gibbs–Duhem relation [12]. It is straightforward to identify the thermodynamic forces, X_i, using the generic form for the entropy production in (A3.2.27). The fluxes may be expressed in terms of the forces by

$$J_q = L_{qq} \nabla \frac{1}{T} - \sum_{i=1}^{n} L_{qi} \nabla \frac{\mu_i}{T}$$

$$J_i = L_{iq} \nabla \frac{1}{T} - \sum_{j=1}^{n} L_{ij} \nabla \frac{\mu_j}{T}.$$

The Onsager relations in this case are

$$L_{qi} = L_{iq} \quad \text{and} \quad L_{ij} = L_{ji}.$$

The coefficients, L_{iq}, are characteristic of the phenomenon of *thermal diffusion*, i.e. the flow of matter caused by a temperature gradient. In liquids, this is called the Soret effect [12]. A reciprocal effect associated with the coefficient L_{qi} is called the *Dufour effect* [12] and describes heat flow caused by concentration gradients. The Onsager relation implies that measurement of one of these effects is sufficient to determine the coupling for both. The coefficient L_{qq} is proportional to the heat conductivity coefficient and is a single scalar quantity in an isotropic fluid even though its associated flux is a vector. This fact is closely related to the Curie principle. The remaining coefficients, L_{ij}, are proportional to the mutual diffusion coefficients (except for the diagonal ones which are proportional to self-diffusion coefficients).

Chemical reactions may be added to the situation giving an entropy production of

$$\frac{\mathrm{d}S}{\mathrm{d}t} = J_q \cdot \nabla \frac{1}{T} + \sum_{i=1}^{n} J_i \cdot \nabla \left(-\frac{\mu_i}{T} \right) - \frac{1}{T} \sum_{j=1}^{r} J_j A_j$$

in which there are r reactions with variable progress rates related to the J_j and with chemical affinities A_j. Once again, these fluxes are not all independent and some care must be taken to rewrite everything so that symmetry is preserved [12]. When this is done, the Curie principle decouples the vectorial forces from the scalar fluxes and *vice versa* [9]. Nevertheless, the reaction terms lead to additional reciprocal relations because

$$J_j = - \sum_{k=1}^{r} L_{jk} A_k$$

implies that $L_{jk} = L_{kj}$.

These are just a few of the standard examples of explicit applications of the Onsager theory to concrete cases. There are many more involving acoustical, electrical, gravitational, magnetic, osmotic, thermal and other processes in various combinations. An excellent source for details is the book by DeGroot and Mazur [12], which was published in a Dover edition in 1984, making it readily accessible and inexpensive. There, one will find many specific accounts. For example, in the case of thermal and electric or thermal and electromagnetic couplings: (1) the *Hall effect* is encountered where the Hall coefficient is related to Onsager's relation through the resistivity tensor; (2) the *Peltier effect* is encountered where Onsager's relation implies the *Thompson relation* between the thermo-electric power and the Peltier heat and (3) *galvanomagnetic* and *thermomagnetic* effects are met along with the *Ettinghausen effect*, the *Nernst effect* and the *Bridgman relation*. In the case of so-called *discontinuous systems*, the *thermomolecular pressure effect, thermal effusion* and the *mechanocaloric effect* are encountered as well as *electro-osmosis*. Throughout, the entropy production equation plays a central role [12].

A3.2.4.2   The fluctuating diffusion equation

A byproduct of the preceding analysis is that the Onsager theory immediately determines the form of the fluctuations that should be added to the diffusion equation. Suppose that a solute is dissolved in a solvent with concentration c. The diffusion equation for this is

$$\frac{\partial}{\partial t} c + \nabla \cdot J = 0 \qquad \text{where } J = -D\nabla c \tag{A3.2.29}$$

in which D is the diffusion constant. This is called *Fick's law* of diffusion [12]. From (A3.2.28), the thermodynamic force is seen to be (at constant T)

$$X = -\nabla \frac{\mu}{T} = -\frac{1}{T} \frac{\partial \mu}{\partial c} \nabla c = -\frac{k}{c_0} \nabla c$$

wherein it has been assumed that the solute is a nonelectrolyte exhibiting ideal behaviour with $\mu = kT \ln(c)$ and c_0 is the equilibrium concentration. Since this is a continuum system, the general results developed above need to be continuously extended as follows. The entropy production in a volume V may be written as

$$\frac{\mathrm{d}}{\mathrm{d}t} S = \int \mathrm{d}^3 r (D\nabla c) \cdot \left(\frac{k}{c_0} \nabla c \right) = \frac{Dk}{c_0} \int \mathrm{d}^3 r (\nabla c) \cdot (\nabla c) = \frac{Dk}{c_0} \int \mathrm{d}^3 r \int \mathrm{d}^3 r' (\nabla c(r)) \cdot \delta(r - r')(\nabla c(r))$$

$$= \frac{Dk}{c_0} \int \mathrm{d}^3 r \int \mathrm{d}^3 r' c(r)(\nabla \cdot \nabla' \delta(r - r'))c(r'). \tag{A3.2.30}$$

The continuous extension of (A3.2.12) becomes

$$S(t) = S_0 - \frac{k}{2} \int d^3r \int d^3r' \, c(r) E(r - r') c(r').$$

The time derivative of this expression together with (A3.2.29) implies

$$\frac{d}{dt} S = -\frac{k}{2} \int d^3r \int d^3r' [(D\nabla^2 c(r)) E(r - r') + E(r - r')(D\nabla'^2 c(r'))]$$

$$= -\frac{kD}{2} \int d^3r \int d^3r' [c(r)(\nabla^2 E(r - r') + \nabla'^2 E(r - r')) c(r')]. \tag{A3.2.31}$$

Equations (A3.2.30) and (A3.2.31) imply the identity for the entropy matrix

$$E(r - r') = \frac{1}{c_0} \delta(r - r').$$

Equation (A3.2.29) also implies that the extension of G is now

$$G(r - r') = -D\nabla^2 \delta(r - r').$$

The extension of the fluctuation–dissipation relation of (A3.2.21) becomes

$$2Q(r - r') = \int d^3r'' [G(r - r'') E^{-1}(r'' - r') + E^{-1}(r - r'') G(r'' - r')] = -2Dc_0 \nabla^2 \delta(r - r').$$

This means that the fluctuating force can be written as

$$\tilde{F}(r, t) = \nabla \cdot \tilde{g}(r, t) \qquad \text{where } \langle \tilde{g}_\alpha(r, t) \tilde{g}_\beta(r', s) \rangle = 2Dc_0 \delta_{\alpha\beta} \delta(r - r') \delta(t - s).$$

The resulting fluctuating diffusion equation is

$$\frac{\partial}{\partial t} c = D\nabla^2 c + \tilde{F}.$$

The quantity $\tilde{g}$ can be thought of as a fluctuating mass flux.

Two applications of the fluctuating diffusion equation are made here to illustrate the additional information the fluctuations provide over and beyond the deterministic behaviour. Consider an infinite volume with an initial concentration, c, that is constant, c_0, everywhere. The solution to the averaged diffusion equation is then simply $\langle c \rangle = c_0$ for all t. However, the two-time correlation function may be shown [26] to be

$$\chi_{cc}(r, t; r', s) = \langle (c(r, t) - c_0)(c(r', s) - c_0) \rangle = c_0 \left[\delta(r - r') - \frac{1}{(8\pi D|t - s|)^{3/2}} \exp\left[-\frac{|r - r'|^2}{8D|t - s|} \right] \right].$$

As the time separation $|t - s|$ approaches ∞ the second term in this correlation vanishes and the remaining term is the equilibrium density–density correlation formula for an ideal solution. The second possibility is to consider a non-equilibrium initial state, $c(r, t) = c_0 \delta(r)$. The averaged solution is [26]

$$\langle c(r, t) \rangle = c_0 \frac{1}{(4\pi Dt)^{3/2}} \exp\left[-\frac{r^2}{4Dt} \right]$$

whereas the two-time correlation function may be shown after extensive computation [26] to be

$$\chi_{cc}(r, t; r', s) = \langle (c(r, t) - c_0)(c(r, s) - c_0) \rangle = c_0 \left[\delta(r - r') \frac{1}{(4\pi D|t - s|)^{3/2}} \exp \left[-\frac{|r + r'|^2}{4D|t - s|} \right] \right.$$
$$\left. - \frac{1}{(8\pi D|t - s|)^3} \exp \left[-\frac{|r - r'|^2}{8D|t - s|} \right] \exp \left[-\frac{|r - r'|^2}{8D|t - s|} \right] \right].$$

This covariance function vanishes as $|t - s|$ approaches ∞ because the initial density profile has a finite integral, that creates a vanishing density when it spreads out over the infinite volume.

This example illustrates how the Onsager theory may be applied at the macroscopic level in a self-consistent manner. The ingredients are the averaged regression equations and the entropy. Together, these quantities permit the calculation of the fluctuating force correlation matrix, $\mathbf{Q}$. Diffusion is used here to illustrate the procedure in detail because diffusion is the simplest known case exhibiting continuous variables.

A3.2.4.3 *Fluctuating hydrodynamics*

A proposal based on Onsager's theory was made by Landau and Lifshitz [27] for the fluctuations that should be added to the Navier–Stokes hydrodynamic equations. Fluctuating stress tensor and heat flux terms were postulated in analogy with the Onsager theory. However, since this is a case where the variables are of mixed time reversal character, the 'derivation' was not fully rigorous. This situation was remedied by the derivation by Fox and Uhlenbeck [13, 14, 18] based on general stationary Gaussian–Markov processes [12]. The precise form of the Landau proposal is confirmed by this approach [14].

Let $\Delta\rho$, Δu and ΔT denote the deviations of the mass density, ρ, the velocity field, u, and the temperature, T, from their full equilibrium values. The fluctuating, linearized Navier–Stokes equations are

$$\frac{\partial}{\partial t} \Delta\rho + \rho_{eq} \nabla \cdot \Delta u = 0$$

$$\rho_{eq} \frac{\partial}{\partial t} \Delta u_\alpha + A_{eq} \frac{\partial}{\partial x_\alpha} \Delta\rho + B_{eq} \frac{\partial}{\partial x_\alpha} \Delta T = \frac{\partial}{\partial x_\beta} \left[2\eta \Delta D_{\alpha\beta} + (\xi - \tfrac{2}{3}\eta) \Delta D_{\gamma\gamma} \delta_{\alpha\beta} \right] + \frac{\partial}{\partial x_\beta} \tilde{S}_{\alpha\beta}$$

$$\rho_{eq} C_{eq} \frac{\partial}{\partial t} \Delta T = K \nabla^2 \Delta T - T_{eq} B_{eq} \nabla \cdot \Delta u + \nabla \cdot \tilde{g} \qquad (A3.2.32)$$

in which η is the shear viscosity, ξ is the bulk viscosity, K is the heat conductivity, the subscript 'eq' denotes equilibrium values and A_{eq}, B_{eq} and C_{eq} are defined [14] by

$$A_{eq} = \left(\frac{\partial p}{\partial \rho} \right)_{eq} \qquad B_{eq} = \left(\frac{\partial p}{\partial T} \right)_{eq} \qquad C_{eq} = \left(\frac{\partial \varepsilon}{\partial T} \right)_{eq}$$

in which p is the pressure and ε is the energy per unit mass. $D_{\alpha\beta}$ is the strain tensor and $\tilde{S}_{\alpha\beta}$ is the fluctuating stress tensor while $\tilde{g}_\alpha$ is the fluctuating heat flux vector. These fluctuating terms are Gaussian white noises with zero mean and correlations given by

$$\langle \tilde{S}_{\alpha\beta}(r, t) \tilde{S}_{\mu\nu}(r', t') \rangle = 2kT_{eq}\delta(r - r')\delta(t - t') \left[\eta(\delta_{\alpha\mu}\delta_{\beta\nu} + \delta_{\alpha\nu}\delta_{\beta\mu}) + \left(\xi - \frac{2}{3}\eta \right) \delta_{\alpha\beta}\delta_{\mu\nu} \right]$$

$$\langle \tilde{g}_\alpha(r, t) \tilde{g}_\beta(r', t') \rangle = 2kT_{eq}^2\delta(r - r')\delta(t - t')K\delta_{\alpha\beta} \qquad (A3.2.33)$$

$$\langle \tilde{S}_{\alpha\beta}(r, t) \tilde{g}_\mu(r', t') \rangle = 0.$$

The lack of correlation between the fluctuating stress tensor and the fluctuating heat flux in the third expression is an example of the Curie principle for the fluctuations. These equations for fluctuating hydrodynamics are arrived at by a procedure very similar to that exhibited in the preceding section for diffusion. A crucial ingredient is the equation for entropy production in a fluid

$$\frac{d}{dt}S(t) = \int d^3r \left[\frac{K}{T^2}(\nabla T)\cdot(\nabla T) + \frac{1}{T}\left(2\eta D_{\alpha\beta}D_{\alpha\beta} + \left(\xi - \frac{2}{3}\eta\right)(D_{\alpha\alpha})^2 \right) \right].$$

This expression determines the entropy matrix needed for the fluctuation–dissipation relation [14] used to obtain (A3.2.33).

Three interesting applications of these equations are made here. The first is one of perspective. A fluid in full equilibrium will exhibit fluctuations. In fact, these fluctuations are responsible for Rayleigh–Brillouin light scattering in fluids [28]. From the light scattering profile of an equilibrium fluid, the viscosities, heat conductivity, speed of sound and sound attenuation coefficient can be determined. This is a remarkable exhibition of how non-equilibrium properties of the fluid reside in the equilibrium fluctuations. Jerry Gollub once posed to the author the question: 'how does a fluid know to make the transition from steady state conduction to steady state convection at the threshold of instability in the Rayleigh–Benard system [21]?' The answer is that the fluid fluctuations are incessantly testing the stability and nucleate the transition when threshold conditions exist. Critical opalescence [28] is a manifestation of this macroscopic influence of the fluctuations.

The second application is to temperature fluctuations in an equilibrium fluid [18]. Using (A3.2.32) and (A3.2.33) the correlation function for temperature deviations is found to be

$$\langle \Delta T(r,t)\Delta T(r',t') \rangle = \frac{kT_{eq}^2}{\rho_{eq}C_{eq}}\left(\frac{\rho_{eq}C_{eq}}{4\pi K|t-t'|}\right)^{3/2}\exp\left[-\frac{\rho_{eq}C_{eq}|r-r'|^2}{4K|t-t'|}\right]. \tag{A3.2.34}$$

When the two times are identical, the formula simplifies to

$$\langle \Delta T(r)\Delta T(r') \rangle = \frac{kT_{eq}^2}{\rho_{eq}C_{eq}}\delta(r-r').$$

Define the temperature fluctuations in a volume V by

$$\Delta T_V = \frac{1}{V}\int d^3r\,\Delta T(r).$$

This leads to the well known formula

$$\langle \Delta T_V \Delta T_V \rangle = \frac{kT_{eq}^2}{\rho_{eq}VC_{eq}} = \frac{kT_{eq}^2}{C_V}$$

in which C_V is the ordinary heat capacity since C_{eq} is the heat capacity per unit mass. This formula can be obtained by purely macroscopic thermodynamic arguments [29]. However, the dynamical information in (A3.2.34) cannot be obtained from equilibrium thermodynamics alone.

The third application is to velocity field fluctuations. For an equilibrium fluid the velocity field is, on average, zero everywhere but it does fluctuate. The correlations turn out to be

$$\langle u_\alpha(r,t)u_\beta(r',t') \rangle = \frac{kT_{eq}}{\rho_{eq}}\left[(4\pi\nu|t-t'|^2)^{-3/2}\exp\left[-\frac{|r-r'|^2}{4\nu|t-t'|}\right] + \frac{\partial^2}{\partial x_\alpha \partial x_\beta}\left[(4\pi|r-r'|)^{-1}\Phi\left(\frac{|r-r'|}{2\nu^{1/2}|t-t'|^{1/2}}\right) \right] \right] \tag{A3.2.35}$$

in which $\nu = \eta/\rho_{eq}$, the kinematic viscosity and $\Phi(x)$ is defined by

$$\Phi(x) = \frac{2}{\sqrt{\pi}}\int_0^x dy\,\exp(-y^2).$$

When $r = r'$ or for $\nu^{1/2}|t - t'|^{1/2} \gg |r - r'|$, (A3.2.35) simplifies greatly, yielding

$$\langle u_\alpha(r, t) u_\beta(r, t')\rangle = \frac{3}{2}\frac{kT_{eq}}{\rho_{eq}}(4\pi\nu|t - t'|^2)^{-3/2}\delta_{\alpha\beta}$$

which is indistinguishable from the famous long-time-tail result [30].

A3.2.4.4 *Fluctuating Boltzmann equation*

Onsager's theory can also be used to determine the form of the fluctuations for the Boltzmann equation [15]. Since hydrodynamics can be derived from the Boltzmann equation as a *contracted description*, a contraction of the fluctuating Boltzmann equation determines fluctuations for hydrodynamics. In general, a contraction of the description creates a new description which is non-Markovian, i.e. has memory. The Markov approximation to the contraction of the fluctuating Boltzmann equation is identical with fluctuating hydrodynamics [15]. This is an example of the internal consistency of the Onsager approach. Similarly, it is possible to consider the hydrodynamic problem of the motion of a sphere in a fluctuating fluid described by fluctuating hydrodynamics (with appropriate boundary conditions). A contraction of this description [14] produces Langevin's equation for Brownian movement. Thus, three levels of description exist in this hierarchy: fluctuating Boltzmann equation, fluctuating hydrodynamic equations and Langevin's equation. The general theory for such hierarchies of description and their contractions can be found in the book by Keizer [31].

A3.2.5 Linear response theory

Linear response theory is an example of a microscopic approach to the foundations of non-equilibrium thermodynamics. It requires knowledge of the Hamiltonian for the underlying microscopic description. In principle, it produces explicit formulae for the relaxation parameters that make up the Onsager coefficients. In reality, these expressions are extremely difficult to evaluate and approximation methods are necessary. Nevertheless, they provide a deeper insight into the physics.

The linear response of a system is determined by the lowest order effect of a perturbation on a dynamical system. Formally, this effect can be computed either classically or quantum mechanically in essentially the same way. The connection is made by converting quantum mechanical commutators into classical Poisson brackets, or *vice versa*. Suppose that the system is described by Hamiltonian $H + H_{ex}$ where H_{ex} denotes an external perturbation that may depend on time and generally does not commute with H. The density matrix equation for this situation is given by the Bloch equation [32]

$$\frac{\partial}{\partial t}\rho = -\frac{i}{\hbar}[H + H_{ex}, \rho] \tag{A3.2.36}$$

where ρ denotes the density matrix and the square brackets containing two quantities separated by a comma denotes a commutator. In the classical limit, the density matrix becomes a phase space distribution, f, of the coordinates and conjugate momenta and the Bloch equation becomes Liouville's equation [32]

$$\frac{\partial}{\partial t}f = \sum_i\left(\frac{\partial(H + H_{ex})}{\partial q_i}\frac{\partial f}{\partial p_i} - \frac{\partial(H + H_{ex})}{\partial p_i}\frac{\partial f}{\partial q_i}\right) \equiv \{H + H_{ex}, f\} \tag{A3.2.37}$$

in which the index i labels the different degrees of freedom and the second equality defines the Poisson bracket. Both of these equations may be expressed in terms of Liouville operators in the form

$$\frac{\partial}{\partial t}\rho = i(L + L_{ex})\rho \tag{A3.2.38}$$

where quantum mechanically these operators are defined by (A3.2.36), and classically ρ means f and the operators are defined by (A3.2.37) [32].

Assuming explicit time dependence in L_{ex}, (A3.2.38) is equivalent to the integral equation

$$\rho(t) = \exp[i(t - t_0)L]\rho(t_0) + \int_{t_0}^{t} ds \exp[i(t - t_0)L]iL_{ex}(s)\rho(s)$$

as is easily proved by t-differentiation. Note that the exponential of the quantum mechanical Liouville operator may be shown to have the action

$$\exp[itL]A = \exp\left[-\frac{i}{\hbar}H\right] A \exp\left[\frac{i}{\hbar}H\right]$$

in which A denotes an arbitrary operator. This identity is also easily proved by t-differentiation.

The usual context for linear response theory is that the system is prepared in the infinite past, $t_0 \rightarrow -\infty$, to be in equilibrium with Hamiltonian H and then H_{ex} is turned on. This means that $\rho(t_0)$ is given by the canonical density matrix

$$\rho(t_0) = \rho_{eq} = \frac{1}{Z} \exp[-\beta H]$$

where $\beta = 1/kT$ and $Z = \text{Trace} \exp[-\beta H]$. Clearly

$$\exp[itL]\rho_{eq} = \rho_{eq}.$$

Thus, to first order in L_{ex}, $\rho(t)$ is given by

$$\rho(t) = \rho_{eq} - \frac{i}{\hbar} \int_{-\infty}^{t} ds \exp\left[-\frac{i}{\hbar}(t - s)H\right] [H_{ex}(s), \rho_{eq}] \exp\left[\frac{i}{\hbar}(t - s)H\right] + \cdots. \qquad \text{(A3.2.39)}$$

The classical analogue is

$$f(t) = f_{eq} + \int_{-\infty}^{t} ds \exp[i(t - s)L]\{H_{ex}(s), f_{eq}\} + \cdots.$$

Let B denote an observable value. Its expectation value at time t is given by

$$\langle B \rangle \equiv \text{Ex}(B) \equiv \text{Trace}(B\rho(t)).$$

Denote the deviation of B from its equilibrium expectation value by $\Delta B = B - \text{Trace}(B\rho_{eq})$. From (A3.2.39), the deviation of the expectation value of B from its equilibrium expectation value, $\delta\langle B \rangle$, is

$$\delta\langle B \rangle = -\frac{i}{\hbar} \int_{-\infty}^{t} ds \, \text{Trace}\left(B \exp\left[-\frac{i}{\hbar}(t - s)H\right] [H_{ex}(s), \rho_{eq}] \exp\left[\frac{i}{\hbar}(t - s)H\right] \right)$$

$$= -\frac{i}{\hbar} \int_{-\infty}^{t} ds \, \text{Trace}\left(\Delta B \exp\left[-\frac{i}{\hbar}(t - s)H\right] [H_{ex}(s), \rho_{eq}] \exp\left[\frac{i}{\hbar}(t - s)H\right] \right)$$

$$= -\frac{i}{\hbar} \int_{-\infty}^{t} ds \, \text{Trace}\left(\exp\left[\frac{i}{\hbar}(t - s)H\right] \Delta B \exp\left[-\frac{i}{\hbar}(t - s)H\right] [H_{ex}(s), \rho_{eq}] \right)$$

$$= -\frac{i}{\hbar} \int_{-\infty}^{t} ds \, \text{Trace}(\Delta B(t - s)[H_{ex}(s), \rho_{eq}])$$

where $B(t - s)$ is the Heisenberg operator solution to the Heisenberg equation of motion

$$\frac{d}{dt} \Delta B = \frac{i}{\hbar} [H, \Delta B]. \tag{A3.2.40}$$

The second equality follows from the fact that going from B to ΔB involves subtracting a c-number from B and that c-number can be taken outside the trace. The resulting trace is the trace of a commutator, which vanishes. The invariance of the trace to cyclic permutations of the order of the operators is used in the third equality. It is straightforward to write

$$\text{Trace}(\Delta B(t - s)[H_{ex}(s), \rho_{eq}]) = \text{Trace}([\Delta B(t - s), H_{ex}(s)]\rho_{eq}).$$

The transition from H_{ex} to ΔH_{ex} inside the commutator is allowed since $\text{Trace}(H_{ex}\rho_{eq})$ is a c-number and commutes with any operator. Thus, the final expression is [32]

$$\delta \langle B(t) \rangle = -\frac{i}{\hbar} \int_{-\infty}^{t} ds \, \text{Trace}([\Delta B(t - s), \Delta H_{ex}(s)]\rho_{eq}).$$

If the external perturbation is turned on with a time dependent function $F(t)$ and H_{ex} takes the form $AF(t)$ where A is a time independent operator (or H_{ex} is the sum of such terms), then

$$\delta \langle B(t) \rangle = -\frac{i}{\hbar} \int_{-\infty}^{t} ds \, \text{Trace}([\Delta B(t - s), \Delta A]\rho_{eq}) F(s) \equiv \int_{-\infty}^{t} ds \, \Phi_{BA}(t - s) F(s)$$

which defines the linear response function $\Phi_{BA}(t - s)$. This quantity may be written compactly as

$$\Phi_{BA}(t) = -\frac{i}{\hbar} \langle [\Delta B(t), \Delta A] \rangle_{eq}.$$

An identical expression holds classically [32] if $-i/\hbar$ times the commutator is replaced by the classical Poisson bracket.

The Heisenberg equation of motion, (A3.2.40), may be recast for imaginary times $t = -i\hbar\lambda$ as

$$\frac{d}{d\lambda} A = [H, A]$$

with the solution

$$A(\lambda) = \exp[\lambda H] A \exp[-\lambda H].$$

Therefore, the Kubo identity [32] follows

$$\exp[-\beta H] \int_{0}^{\beta} d\lambda \exp[\lambda H][A, H] \exp[-\lambda H]$$

$$= \exp[-\beta H] \int_{0}^{\beta} d\lambda \exp[\lambda H] \left(-\frac{d}{d\lambda} A \right) \exp[-\lambda H]$$

$$= \exp[-\beta H] \int_{0}^{\beta} d\lambda (H \exp[\lambda H] A \exp[-\lambda H] - \exp[\lambda H] A \exp[-\lambda H] H)$$

$$= \exp[-\beta H] \int_{0}^{\beta} d\lambda [H, \exp[\lambda H] A \exp[-\lambda H]]$$

$$= \exp[-\beta H] \int_{0}^{\beta} d\lambda \frac{d}{d\lambda} (\exp[\lambda H] A \exp[-\lambda H])$$

$$= \exp[-\beta H](\exp[\beta H] A \exp[-\beta H] - A)$$

$$= A \exp[-\beta H] - \exp[-\beta H] A = [A, \exp[-\beta H]].$$

Therefore,

$$
\begin{aligned}
\Phi_{BA}(t) &= -\frac{i}{\hbar}\mathrm{Trace}([\Delta B(t), \Delta A]\rho_{eq}) = -\frac{i}{\hbar}\mathrm{Trace}\frac{1}{Z}([\Delta B(t), \Delta A]\exp[-\beta H]) \\
&= -\frac{i}{\hbar}\mathrm{Trace}\frac{1}{Z}(\Delta B(t)[\Delta A, \exp[-\beta H]]) \\
&\quad - \frac{i}{\hbar}\mathrm{Trace}\frac{1}{Z}\left(\Delta B(t)\exp[-\beta H]\int_0^\beta d\lambda \exp[\lambda H]\left(-\frac{d}{d\lambda}\Delta A\right)\exp[-\lambda H]\right) \\
&= -\frac{i}{\hbar}\mathrm{Trace}\left(\Delta B(t)\rho_{eq}\int_0^\beta d\lambda \exp[\lambda H]\left(-\frac{d}{d\lambda}\Delta A\right)\exp[-\lambda H]\right) \\
&= -\frac{i}{\hbar}\left\langle \Delta B(t)\int_0^\beta d\lambda \exp[\lambda H]\left(-\frac{d}{d\lambda}\Delta A\right)\exp[-\lambda H]\right\rangle_{eq} \\
&= -\frac{i}{\hbar}\left\langle \Delta B(t)\int_0^\beta d\lambda \exp[\lambda H][\Delta A, H]\exp[-\lambda H]\right\rangle_{eq}.
\end{aligned}
\tag{A3.2.41}
$$

Using the Heisenberg equation of motion, (A3.2.40), the commutator in the last expression may be replaced by the time-derivative operator

$$
i\hbar\frac{d}{dt}\Delta A = [\Delta A, H].
$$

This converts (A3.2.41) into

$$
\Phi_{BA}(t) = \left\langle \Delta B(t)\int_0^\beta d\lambda \exp[\lambda H]\frac{d}{dt}\Delta A \exp[-\lambda H]\right\rangle_{eq}
$$

where the time derivative of ΔA is evaluated at $t = 0$. The quantity $kT\Phi_{BA}(t)$ is called the *canonical correlation* of $d/dt\,\Delta A$ and ΔB [32]. It is invariant under time translation by $\exp[-iH\tau/\hbar]$ because both ρ_{eq} and $\exp[\pm\lambda H]$ commute with this time evolution operator and the trace operation is invariant to cyclic permutations of the product of operators upon which it acts. Thus

$$
\begin{aligned}
\Phi_{BA}(t+\tau) &= \mathrm{Trace}\left(\rho_{eq}\exp\left[-\frac{i}{\hbar}H\tau\right]\Delta B(t)\exp\left[\frac{i}{\hbar}H\tau\right]\int_0^\beta d\lambda \exp\left[H\left(\lambda - \frac{i}{\hbar}t\right)\right]\left(\frac{d}{dt}\Delta A\right)\right. \\
&\quad \left. \times \exp\left[H\left(-\lambda + \frac{i}{\hbar}t\right)\right]\right) \\
&= \mathrm{Trace}\left(\exp\left[\frac{i}{\hbar}H\tau\right]\rho_{eq}\exp\left[-\frac{i}{\hbar}H\tau\right]\Delta B(t)\int_0^\beta d\lambda \exp\left[H\frac{i}{\hbar}t\right]\exp\left[H\left(\lambda - \frac{i}{\hbar}t\right)\right]\right. \\
&\quad \left. \times \left(\frac{d}{dt}\Delta A\right)\exp[-H\lambda]\right) = \Phi_{BA}(t).
\end{aligned}
\tag{A3.2.42}
$$

Consider the canonical correlation of ΔA and ΔB, $C(\Delta A, \Delta B)$, defined by

$$
C(\Delta A(0), \Delta B(t)) = kT\langle \Delta B(t)\int_0^\beta d\lambda \exp[\lambda H]\Delta A(0)\exp[-\lambda H]\rangle_{eq}.
$$

The analysis used for (A3.2.42) implies

$$
C(\Delta A(\tau), \Delta B(t+\tau)) = C(\Delta A(0), \Delta B(t))
$$

which means that this correlation is independent of τ, i.e. *stationary*. Taking the τ-derivative implies

$$C\left(\frac{\mathrm{d}}{\mathrm{d}t}\Delta A(0), \Delta B(t)\right) + C\left(\Delta A(0), \frac{\mathrm{d}}{\mathrm{d}t}\Delta B(t)\right) = 0.$$

This is equivalent to

$$
\begin{aligned}
\Phi_{BA}(t) &= \left\langle \Delta B(t)\int_0^\beta \mathrm{d}\lambda\, \exp[\lambda H]\left(\frac{\mathrm{d}}{\mathrm{d}t}\Delta A(0)\right)\exp[-\lambda H]\right\rangle_{\mathrm{eq}} \\
&= -\left\langle \left(\frac{\mathrm{d}}{\mathrm{d}t}\Delta B(t)\right)\int_0^\beta \mathrm{d}\lambda\, \exp[\lambda H]\Delta A(0)\exp[-\lambda H]\right\rangle_{\mathrm{eq}}.
\end{aligned}
\tag{A3.2.43}
$$

In different applications, one or the other of these two equivalent expressions may prove useful.

As an example, let B be the current J_i corresponding to the displacement A_i appearing in H_{ex}. Clearly

$$J_i(t) = \frac{\mathrm{d}}{\mathrm{d}t}A_i(t) = \frac{\mathrm{i}}{\hbar}[H, A_i].$$

Because this current is given by a commutator, its equilibrium expectation value is zero. Using the first expression in (A3.2.43), the response function is given by

$$\Phi_{ij}(t) = \left\langle J_i(t)\int_0^\beta \mathrm{d}\lambda\, \exp[\lambda H]J_j(0)\exp[-\lambda H]\right\rangle_{\mathrm{eq}}.
\tag{A3.2.44}$$

For a periodic perturbation, $\delta\langle\Delta B(t)\rangle$ is also periodic. The *complex admittance* [30] is given by

$$\chi_{BA}(t) = \int_0^\infty \mathrm{d}t\, \Phi_{BA}(t)\,\mathrm{e}^{\mathrm{i}\omega t}.$$

For the case of a current as in (A3.2.44) the result is the Kubo formula [32] for the complex conductivity

$$\sigma_{ij}(\omega) = \int_0^\infty \mathrm{d}t\, \mathrm{e}^{\mathrm{i}\omega t}\left\langle J_i(t)\int_0^\beta \mathrm{d}\lambda\, \exp[\lambda H]J_j(0)\exp[-\lambda H]\right\rangle_{\mathrm{eq}}.$$

Several explicit applications of these relations may be found in the books by Kubo *et al* [32] and by McLennan [33].

There are other techniques leading to results closely related to Kubo's formula for the conductivity coefficient. Notable among them is the Mori–Zwanzig theory [34, 35] based on projection operator techniques and yielding the generalized Langevin equation [18]. The formula for the conductivity coefficient is an example of the general formula for relaxation parameters, the Green–Kubo formula [36, 37]. The examples of Green–Kubo formulae for viscosity, thermal conduction and diffusion are in the book by McLennan [33].

A3.2.6 Prospects

The current frontiers for the subject of non-equilibrium thermodynamics are rich and active. Two areas dominate interest: non-linear effects and molecular bioenergetics. The linearization step used in the near equilibrium regime is inappropriate far from equilibrium. Progress with a microscopic kinetic theory [38] for non-linear fluctuation phenomena has been made. Careful experiments [39] confirm this theory. Non-equilibrium long range correlations play an important role in some of the light scattering effects in fluids in far from equilibrium states [38, 39].

The role of non-equilibrium thermodynamics in molecular bioenergetics has experienced an experimental revolution during the last 35 years. Membrane energetics is now understood in terms of chemiosmosis [40]. In chemiosmosis, a trans-membrane electrochemical potential energetically couples the oxidation–reduction energy generated during catabolism to the adenosine triphosphate (ATP) energy needed for chemosynthesis during anabolism. Numerous advances in experimental technology have opened up whole new areas of exploration [41]. Quantitative analysis using non-equilibrium thermodynamics to account for the free energy and entropy changes works accurately in a variety of settings. There is a rich diversity of problems to be worked on in this area. Another biological application brings the subject back to its foundations. Rectified Brownian movement (involving a Brownian ratchet) is being invoked as the mechanism behind many macromolecular processes [42]. It may even explain the dynamics of actin and myosin interactions in muscle fibres [43]. In rectified Brownian movement, metabolic free energy generated during catabolism is used to bias boundary conditions for ordinary diffusion, thereby producing a non-zero flux. In this way, thermal fluctuations give the molecular mechanisms of cellular processes their vitality [44].

References

[1]   Einstein A 1956 *Investigations on the Theory of Brownian Movement* (New York: Dover). This book is based on a series of papers Einstein published from 1905 until 1908
[2]   Langevin P 1908 Sur la theorie du mouvement brownien *C. R. Acad. Sci. Paris* **146** 530
[3]   Uhlenbeck G E and Ornstein L S 1930 On the theory of the Brownian motion *Phys. Rev.* **36** 823
[4]   Onsager L 1931 Reciprocal relations in irreversible processes. I *Phys. Rev.* **37** 405
[5]   Onsager L 1931 Reciprocal relations in irreversible processes. II *Phys. Rev.* **38** 2265
[6]   Chandrasekhar S 1943 Stochastic problems in physics and astronomy *Rev. Mod. Phys.* **15** 1
[7]   Wang M C and Uhlenbeck G E 1945 On the theory of Brownian motion II *Rev. Mod. Phys.* **17** 323
[8]   Casimir H B G On Onsager's principle of microscopic reversibility *Rev. Mod. Phys.* **17** 343
[9]   Prigogine I 1947 *Etude Thermodynamique des Phenomenes Irreversibles* (Liege: Desoer)
[10] Onsager L and Machlup S 1953 Fluctuations and irreversible processes *Phys. Rev.* **91** 1505
[11] Machlup S and Onsager L 1953 Fluctuations and irreversible processes. II. Systems with kinetic energy *Phys. Rev.* **91** 1512
[12] de Groot S R and Mazur P 1962 *Non-Equilibrium Thermodynamics* (Amsterdam: North-Holland)
[13] Fox R F 1969 Contributions to the theory of non-equilibrium thermodynamics *PhD Thesis* Rockefeller University, New York
[14] Fox R F and Uhlenbeck G E 1970 Contributions to non-equilibrium thermodynamics. I. Theory of hydrodynamical fluctuations *Phys. Fluids* **13** 1893
[15] Fox R F and Uhlenbeck G E 1970 Contributions to non-equilibrium thermodynamics. II. Fluctuation theory for the Boltzmann equation *Phys. Fluids* **13** 2881
[16] Risken H 1984 *The Fokker–Planck Equation, Methods of Solution and Application* (Berlin: Springer)
[17] Arnold L 1974 *Stochastic Differential Equations* (New York: Wiley–Interscience)
[18] Fox R F 1978 Gaussian stochastic processes in physics *Phys. Rev.* **48** 179
[19] Rayleigh J W S 1945 *The Theory of Sound* vol 1 (New York: Dover) ch 4
[20] Goldstein H 1980 *Classical Mechanics* 2nd edn (Reading, MA: Addison-Wesley) ch 1
[21] Chandrasekhar S 1961 *Hydrodynamic and Hydromagnetic Stability* (London: Oxford University Press)
[22] Glansdorff P and Prigogine I 1971 *Thermodynamic Theory of Structure, Stability and Fluctuations* (London: Wiley–Interscience)
[23] Lavenda B H 1985 *Nonequilibrium Statistical Thermodynamics* (New York: Wiley) ch 3
[24] Keizer J E 1987 *Statistical Thermodynamics of Nonequilibrium Processes* (New York: Springer) ch 8
[25] Keizer J and Chang O K 1987 *J. Chem. Phys.* **87** 4064
[26] Keizer J E 1987 *Statistical Thermodynamics of Nonequilibrium Processes* (New York: Springer) ch 6
[27] Landau L D and Lifshitz E M 1959 *Fluid Mechanics* (London: Pergamon) ch 17
[28] Berne B J and Pecora R 1976 *Dynamic Light Scattering* (New York: Wiley) ch 10
[29] Landau L D and Lifshitz E M 1958 *Statistical Physics* (London: Pergamon) ch 12, equation (111.6)
[30] Fox R F 1983 Long-time tails and diffusion *Phys. Rev.* A **27** 3216
[31] Keizer J E 1987 *Statistical Thermodynamics of Nonequilibrium Processes* (New York: Springer) ch 9
[32] Kubo R, Toda M and Hashitsume N 1985 *Statistical Physics II* (Berlin: Springer) ch 4
[33] McLennan J A 1989 *Introduction to Non-Equilibrium Statistical Mechanics* (Englewood Cliffs, NJ: Prentice-Hall) ch 9
[34] Zwanzig R 1961 Memory effects in irreversible thermodynamics *Phys. Rev.* **124** 983
[35] Mori H 1965 Transport, collective motion and Brownian motion *Prog. Theor. Phys.* **33** 423
[36] Green M S 1954 Markov random processes and the statistical mechanics of time-dependent phenomena. II. Irreversible processes in fluids *J. Chem. Phys.* **22** 398

[37] Kubo R, Yokota M and Nakajima S 1957 Statistical–mechanical theory of irreversible processes. II. Response to thermal disturbance
      J. Phys. Soc. Japan **12** 1203
[38] Kirkpatrick T R, Cohen E G D and Dorfman J R 1982 Light scattering by a fluid in a nonequilibrium steady state. II. Large gradients
      Phys. Rev. A **26** 995
[39] Segre P N, Gammon R W, Sengers J V and Law B M 1992 Rayleigh scattering in a liquid far from thermal equilibrium *Phys. Rev.*
      A **45** 714
[40] Harold F M 1986 *The Vital Force: A Study of Bioenergetics* (New York: Freeman)
[41] de Duve C 1984 *A Guided Tour of the Living Cell* vols 1 and 2 (New York: Scientific American)
[42] Peskin C S, Odell G M and Oster G F 1993 Cellular motions and thermal fluctuations: the Brownian ratchet *Biophys. J.* **65** 316
[43] Huxley A F 1957 Muscle structure and theories of contraction *Prog. Biophys. Biophys. Chem.* **7** 255
[44] Fox R F 1998 Rectified Brownian movement in molecular and cell biology *Phys. Rev.* E **57** 2177

Further Reading

Wax N (ed) 1954 *Selected Papers on Noise and Stochastic Processes* (New York: Dover)
van Kampen N G 1981 *Stochastic Processes in Physics and Chemistry* (Amsterdam: North-Holland)
Katchalsky A and Curran P F 1965 *Nonequilibrium Thermodynamics in Biophysics* (Cambridge, MA: Harvard University
      Press)
Lavenda B H 1985 *Nonequilibrium Statistical Thermodynamics* (New York: Wiley)
Keizer J E 1987 *Statistical Thermodynamics of Nonequilibrium Processes* (New York: Springer)
McLennan J A 1989 *Introduction to Non-equilibrium Statistical Mechanics* (Englewood Cliffs, NJ: Prentice-Hall)

A3.3
Dynamics in condensed phase (including nucleation)

Rashmi C Desai

A3.3.1 Introduction

Radiation probes such as neutrons, x-rays and visible light are used to 'see' the structure of physical systems through elastic scattering experiments. Inelastic scattering experiments measure both the structural and dynamical correlations that exist in a physical system. For a system which is in thermodynamic equilibrium, the molecular dynamics create spatio-temporal correlations which are the manifestation of thermal fluctuations around the equilibrium state. For a condensed phase system, dynamical correlations are intimately linked to its structure. For systems in equilibrium, linear response theory is an appropriate framework to use to inquire on the spatio-temporal correlations resulting from thermodynamic fluctuations. Appropriate response and correlation functions emerge naturally in this framework, and the role of theory is to understand these correlation functions from first principles. This is the subject of section A3.3.2.

A system of interest may be macroscopically homogeneous or inhomogeneous. The inhomogeneity may arise on account of interfaces between coexisting phases in a system or due to the system's finite size and proximity to its external surface. Near the surfaces and interfaces, the system's translational symmetry is broken; this has important consequences. The spatial structure of an inhomogeneous system is its average equilibrium property and has to be incorporated in the overall theoretical structure, in order to study spatio-temporal correlations due to thermal fluctuations around an inhomogeneous spatial profile. This is also illustrated in section A3.3.2.

Another possibility is that a system may be held in a constrained equilibrium by external forces and thus be in a non-equilibrium steady state (NESS). In this case, the spatio-temporal correlations contain new ingredients, which are also exemplified in section A3.3.2.

There are also important instances when the system in neither in equilibrium nor in a steady state, but is actually evolving in time. This happens, for example, when a binary homogeneous mixture at high temperature is suddenly quenched to a low-temperature non-equilibrium state in the middle of the coexistence region of the mixture. Following the quench, the mixture may be in a metastable state or in an unstable state, as defined within a mean field description. The subsequent dynamical evolution of the system, which follows an initial thermodynamic metastability or instability, and the associated kinetics, is a rich subject involving many fundamental questions, some of which are yet to be fully answered. The kinetics of thermodynamically unstable systems and phenomena like spinodal decomposition are treated in section A3.3.3, after some introductory remarks. The late-stage kinetics of domain growth is discussed in section A3.3.4. The discussion in this section is applicable to late-stage growth regardless of whether the initial post-quench state was thermodynamically unstable or metastable. The study of metastable states is connected with the subject of nucleation and subsequent growth kinetics (treated in section A3.3.4). Homogeneous nucleation is the subject of section A3.3.5. As will be clear from section A3.3.3.1, the distinction between the spinodal decomposition and nucleation is not sharp. Growth morphology with apparent nucleation characteristics can occur when post-quench states are within the classical spinodal, except when the binary mixture is symmetric.

The specific examples chosen in this section, to illustrate the dynamics in condensed phases for the variety of system-specific situations outlined above, correspond to long-wavelength and low-frequency phenomena. In such cases, conservation laws and broken symmetry play important roles in the dynamics, and a macroscopic hydrodynamic description is either adequate or is amenable to an appropriate generalization. There are other examples where short-wavelength and/or high-frequency behaviour is evident. If this is the case, one would require a more microscopic description. For fluid systems which are the focus of this section, such descriptions may involve a kinetic theory of dense fluids or generalized hydrodynamics which may be linear or may involve nonlinear mode coupling. Such microscopic descriptions are not considered in this section.

A3.3.2   Equilibrium systems: thermal fluctuations and spatio-temporal correlations

In this section, we consider systems in thermodynamic equilibrium. Even though the system is in equilibrium, molecular constituents are in constant motion. We inquire into the nature of the thermodynamic fluctuations which have at their root the molecular dynamics. The space–time correlations that occur on account of thermodynamic fluctuations can be probed through inelastic scattering experiments, and the range of space and time scales explored depends on the wavenumber and frequency of the probe radiation. We illustrate this by using inelastic light scattering from dense fluids. Electromagnetic radiation couples to matter through its dielectric fluctuations. Consider a non-magnetic, non-conducting and non-absorbing medium with the average dielectric constant ϵ_0. Let the incident electric field be a plane wave of the form

$$\vec{E}_i(\vec{r}, t) = \vec{n}_i E_0 \exp[i(\vec{k}_i \cdot \vec{r} - w_i t)]$$

where $\vec{n}_i$ is a unit vector in the direction of the incident field, E_0 is the field amplitude, $\vec{k}_i$ is the incident wavevector and w_i is the incident angular frequency. The plane wave is incident upon a medium with local dielectric function $\epsilon(\vec{r}, t) = \epsilon_0 \mathsf{I} + \delta\epsilon(\vec{r}, t)$, where $\delta\epsilon(\vec{r}, t)$ is the dielectric tensor fluctuation at position $\vec{r}$ and time t, and I is a unit second-rank tensor. Basic light scattering theory can be used to find the inelastically scattered light spectrum. If the scattered field at the detector is also in the direction $\vec{n}_i$ (i.e. $\vec{n}_f = \vec{n}_i$ for a polarized light scattering experiment), the scattered wavevector is $\vec{k}_f = \vec{k}_i - \vec{k}$ and the scattered frequency is $w_f = w_i - w$, then, apart from some known constant factors that depend on the geometry of the experiment and the incident field, the inelastically scattered light intensity is proportional to the spectral density of the local dielectric fluctuations. If the medium is isotropic and made up of spherically symmetrical molecules, then the dielectric tensor is proportional to the unit tensor I: $\epsilon(\vec{r}, t) = [\epsilon_0 + \delta\epsilon(\vec{r}, t)]\mathsf{I}$. From the dielectric equation of state $\epsilon_0 = \epsilon(\rho_0, T_0)$, one can proceed to obtain the local dielectric fluctuation as $\delta\epsilon(\vec{r}, t) = (\partial\epsilon/\partial\rho)_T \delta\rho(\vec{r}, t) + (\partial\epsilon/\partial T)_\rho \delta T(\vec{r}, t)$. In many simple fluids, it is experimentally found that the thermodynamic derivative $(\partial\epsilon/\partial T)_\rho$ is approximately zero. One then has a simple result that

$$I_\epsilon(\vec{k}, w) \sim \left(\frac{\partial\epsilon}{\partial\rho}\right)_T^2 S_{\rho\rho}(\vec{k}, w) \tag{A3.3.1}$$

where $S_{\rho\rho}(\vec{k}, w)$ is the spectrum of density fluctuations in the simple fluid system. $S_{\rho\rho}(\vec{k}, w)$ is the space–time Fourier transform of the density–density correlation function $S_{\rho\rho}(\vec{r}, t; \vec{r}', t') = \langle \delta\rho(\vec{r}, t)\delta\rho(\vec{r}', t') \rangle$.

Depending on the type of scattering probe and the scattering geometry, other experiments can probe other similar correlation functions. Elastic scattering experiments effectively measure frequency integrated spectra and, hence, probe only the space-dependent static structure of a system. Electron scattering experiments probe charge density correlations, and magnetic neutron scattering experiments the spin density correlations. Inelastic thermal neutron scattering from a non-magnetic system is a sharper probe of density–density correlations in a system but, due to the shorter wavelengths and higher frequencies involved, these results are complementary to those obtained from inelastic polarized light scattering experiments. The latter provide space–time correlations in the long-wavelength hydrodynamic regime.

In order to analyse results from such experiments, it is appropriate to consider a general framework, linear response theory, which is useful whenever the probe radiation weakly couples to the system. The linear response framework is also convenient for utilizing various symmetry and analyticity properties of correlation functions and response functions, thereby reducing the general problem to determining quantities which are amenable to approximations in such a way that the symmetry and analyticity properties are left intact. Such approximations are necessary in order to avoid the full complexity of many-body dynamics. The central quantity in the linear response theory is the response function. It is related to the corresponding correlation function (typically obtained from experimental measurements) through a fluctuation dissipation theorem. In the next section, section A3.3.2.1, we discuss only the subset of necessary results from the linear response theory, which is described in detail in the book by Forster (see Further Reading).

A3.3.2.1 Linear response theory

Consider a set of physical observables $\{A_i(\vec{r}, t)\}$. If a small external field $\delta a_i^{\text{ext}}(\vec{r}, t)$ couples to the observable A_i, then in presence of a set of small external fields $\{\delta a_l\}$, the Hamiltonian H of a system is perturbed to

$$\mathcal{H}(t) = H - \sum_i \int d\vec{r}\, A_i(\vec{r})\delta a_i^{\text{ext}}(\vec{r}, t) \tag{A3.3.2}$$

in a Schrödinger representation. One can use time-dependent perturbation theory to find the linear response of the system to the small external fields. If the system is in equilibrium at time $t = -\infty$, and is evolved under $\mathcal{H}(t)$, the effect on $A_i(\vec{r}, t)$ which is $\delta\langle A_i \rangle = \langle A_i \rangle_{\text{noneq}} - \langle A_i \rangle_{\text{eq}}$ can be calculated to first order in external fields. The result is (causality dictates the upper limit of time integration to t)

$$\delta\langle A_i(\vec{r}, t)\rangle = \sum_j \int_{-\infty}^{t} dt' \int d\vec{r}'\, 2i\chi_{ij}''(\vec{r}, t; \vec{r}', t')\delta a_j^{\text{ext}}(\vec{r}', t') \tag{A3.3.3}$$

where the response function (matrix) is given by

$$\chi_{ij}''(\vec{r}, t; \vec{r}', t') = \chi_{ij}''(\vec{r} - \vec{r}', t - t') = \left\langle \frac{1}{2\hbar}[A_i(\vec{r}, t), A_j(\vec{r}', t')] \right\rangle \tag{A3.3.4}$$

in a translationally invariant system. Note that the response function is an equilibrium property of the system with Hamiltonian H, independent of the small external fields $\{\delta a_i\}$. In the classical limit (see section A2.2.3) the quantum mechanical commutator becomes the classical Poisson bracket and the response function reduces to $\langle (i/2)[A_i(\vec{r}, t), A_j(\vec{r}', t')]_{\text{P.B.}}\rangle$.

Since typical scattering experiments probe the system fluctuations in the frequency–wavenumber space, the Fourier transform $\chi_{ij}''(\vec{k}, w)$ is closer to measurements, which is in fact the imaginary (dissipative) part of the response function (matrix) defined as

$$\chi_{ij}(\vec{k}, z) = \int \frac{dw}{\pi} \frac{\chi_{ij}''(\vec{k}, w)}{w - z} \quad (\text{Im } z \neq 0). \tag{A3.3.5}$$

The real part of $\chi_{ij}(\vec{k}, w)$, χ_{ij}' is the dispersive (reactive) part of χ_{ij}, and the definition of χ_{ij} implies a relation between χ_{ij}' and χ_{ij}'' which is known as the Kramers–Kronig relation.

The response function $\chi_{ij}''(\vec{r}, t)$, which is defined in equation (A3.3.4), is related to the corresponding correlation function, $S_{ij}(\vec{r}, t)$ through the fluctuation dissipation theorem:

$$\chi_{ij}''(\vec{k}, w) = \frac{1}{2\hbar}(1 - e^{-\beta\hbar w})S_{ij}(\vec{k}, w). \tag{A3.3.6}$$

The fluctuation dissipation theorem relates the dissipative part of the response function (χ'') to the correlation of fluctuations (A_i), for any system in thermal equilibrium. The left-hand side describes the dissipative behaviour of a many-body system: all or part of the work done by the external forces is irreversibly distributed into the infinitely many degrees of freedom of the thermal system. The correlation function on the right-hand side describes the manner in which a fluctuation arising spontaneously in a system in thermal equilibrium, even in the absence of external forces, may dissipate in time. In the classical limit, the fluctuation dissipation theorem becomes $\chi''_{ij}(\vec{k}, w) = (\beta/2)w S_{ij}(\vec{k}, w)$.

There are two generic types of external fields that are of general interest. In one of these, which relates to the scattering experiments, the external fields are to be taken as periodic perturbations

$$\delta a_i^{\text{ext}}(\vec{r}, t) = \delta a_i(\vec{r}) \exp[-(\eta - iw)t]$$

where η is an infinitesimally small negative constant, and $\delta a_i(\vec{r})$ can also be a periodic variation in $\vec{r}$, as in the case for incident plane wave electromagnetic radiation considered earlier.

In the other class of experiments, the system, in equilibrium at $t = -\infty$, is adiabatically perturbed to a non-equilibrium state which gets fully switched on by $t = 0$, through the field, $\delta a_i^{\text{ext}}(\vec{r}, t) = \delta a_i(\vec{r}) e^{\epsilon t}, t \leq 0$, with ϵ an infinitesimally small positive constant. At $t = 0$ the external field is turned off, and the system so prepared in a non-equilibrium state will, if left to itself, relax back to equilibrium. This is the generic relaxation experiment during which the decay of the initial ($t = 0$) value is measured. Such an external field will produce, at $t = 0$, spatially varying initial values $\delta\langle A_i(\vec{r}, t = 0)\rangle$ whose spatial Fourier transforms are given by

$$\delta\langle A_i(\vec{k}, t = 0)\rangle = \sum \chi_{ij}(\vec{k})\delta a_j(\vec{k}) \tag{A3.3.7}$$

where

$$\chi_{ij}(\vec{k}) = \int \frac{dw}{\pi} \frac{\chi''_{ij}(\vec{k}, w)}{w}. \tag{A3.3.8}$$

If $\delta a_i(\vec{r})$ is slowly varying in space, the long-wavelength limit $\chi_{ij}(\vec{k} \to 0)$ reduces to a set of static susceptibilities or thermodynamic derivatives. Now, since for $t > 0$ the external fields are zero, it is useful to evaluate the one-sided transform

$$\delta\langle A_i\rangle(\vec{k}, z) = \int_0^\infty dt\, e^{izt} \delta\langle A_i(\vec{k}, t)\rangle$$

$$= \sum_j \int \frac{dw}{i\pi} \frac{\chi''_{ij}(\vec{k}, w)}{w(w - z)} \cdot \delta a_j(\vec{k})$$

$$= \frac{1}{iz} \sum_j [\chi(\vec{k}, z)\chi^{-1}(\vec{k}) - 1]_{ij} \cdot \delta\langle A_j(\vec{k}, 0)\rangle. \tag{A3.3.9}$$

The second equality is obtained using the form of the external field $\delta a_i^{\text{ext}}(\vec{r}, t)$ specific to the relaxation experiments. The last equality is to be read as a matrix equation. The system stability leads to the positivity of all susceptibilities $\chi(\vec{k})$, so that its inverse exists. This last equality is superior to the second one, since the external fields δa_i have been eliminated in favour of the initial values $\delta\langle A_j(\vec{k}, t = 0)\rangle$, which are directly measurable in a relaxation experiment. It is then possible to analyse the relaxation experiments by obtaining the measurements for positive times and comparing them to $\delta\langle A_i(\vec{k}, t)\rangle$ as evaluated in terms of the initial values using some approximate model for the dynamics of the system's evolution. One such model is a linear hydrodynamic description.

A3.3.2.2   Fluctuations in the hydrodynamic domain

We start with a simple example: the decay of concentration fluctuations in a binary mixture which is in equilibrium. Let $\delta C(\vec{r}, t) = C(\vec{r}, t) - C_0$ be the concentration fluctuation field in the system where C_0 is the mean concentration. C is a conserved variable and thus satisfies a continuity equation:

$$\frac{\partial \delta C}{\partial t} = -\vec{\nabla} \cdot \vec{j}_c(\vec{r}, t) \qquad \text{(A3.3.10)}$$

where a phenomenological linear constitutive relation relates the concentration flux $\vec{j}_c$ to the gradient of the local chemical potential $\mu(\vec{r}, t)$ as follows:

$$\vec{j}_c(\vec{r}, t) = -L\vec{\nabla}\mu(\vec{r}, t). \qquad \text{(A3.3.11)}$$

Here L is the Onsager coefficient and the minus sign $(-)$ indicates that the concentration flow occurs from regions of high μ to low μ in order that the system irreversibly flows towards the equilibrium state of a uniform chemical potential. In a system slightly away from equilibrium, the dependence of μ on the thermodynamic state variables, concentration, pressure and temperature (C, p, T), would, in general, relate changes in the chemical potential like $\vec{\nabla}\mu$ to $\vec{\nabla}C$, $\vec{\nabla}p$ and $\vec{\nabla}T$. However, for most systems the thermodynamic derivatives $\partial\mu/\partial p$ and $\partial\mu/\partial T$ are small, and one has, to a good approximation, $\vec{\nabla}\mu = (\partial\mu/\partial C)_{p,T}\vec{\nabla}\delta C$. This linear approximation is not always valid; however, it is valid for the thermodynamic fluctuations in a binary mixture at equilibrium. It enables us to thus construct a closed linear equation for the concentration fluctuations, the diffusion equation:

$$\frac{\partial \delta C}{\partial t} = D\nabla^2\delta C \qquad \text{(A3.3.12)}$$

where $D = L(\partial\mu/\partial C)_{p,T}$ is the diffusion coefficient, which we assume to be a constant. The diffusion equation is an example of a hydrodynamic equation. The characteristic ingredients of a hydrodynamic equation are a conservation law and a linear transport law.

The solutions of such partial differential equations require information on the spatial boundary conditions and initial conditions. Suppose we have an infinite system in which the concentration fluctuations vanish at the infinite boundary. If, at $t = 0$ we have a fluctuation at origin $\delta C(\vec{r}, 0) = \Delta C_0 \delta(\vec{r})$, then the diffusion equation can be solved using the spatial Fourier transforms. The solution in Fourier space is $C(k, t) = \exp(-Dk^2 t)\Delta C_0$, which can be inverted analytically since the Fourier transform of a Gaussian is a Gaussian. In real space, the initial fluctuation decays in time in a manner such that the initial delta function fluctuation broadens to a Gaussian whose width increases in time as $(Dt)^{d/2}$ for a d-dimensional system, while the area under the Gaussian remains equal to ΔC_0 due to the conservation law. Linear hydrodynamics are not always valid. For this example, near the consolute (critical) point of the mixture, the concentration fluctuations nonlinearly couple to transverse velocity modes and qualitatively change the result. Away from the critical point, however, the above, simple analysis illustrates the manner in which the thermodynamic fluctuations decay in the hydrodynamic (i.e. long-wavelength) regime. The diffusion equation and its solutions constitute a rich subject with deep connections to brownian motion theory [1]: both form a paradigm for many other models of dynamics in which diffusion-like decay and damping play important roles.

In dense systems like liquids, the molecular description has a large number of degrees of freedom. There are, however, a few collective degrees of freedom, collective modes, which when perturbed through a fluctuation, relax to equilibrium very slowly, i.e. with a characteristic decay time that is long compared to the molecular interaction time. These modes involve a large number of particles and their relaxation time is proportional to the square of their characteristic wavelength, which is large compared to the intermolecular separation. Hydrodynamics is suitable to describe the dynamics of such long-wavelength, slowly-relaxing modes.

In a hydrodynamic description, the fluid is considered as a continuous medium which is locally homogeneous and isotropic, with dissipation occurring through viscous friction and thermal conduction. For a one-component system, the hydrodynamic (collective) variables are deduced from conservation laws and broken symmetry. We first consider (section A3.3.2.3) the example of a Rayleigh–Brillouin spectrum of a one-component monatomic fluid. Here conservation laws play the important role. In the next example (section A3.3.2.4), we use a fluctuating hydrodynamic description for capillary waves at a liquid–vapour interface where broken symmetry plays an important role. A significant understanding of underlying phenomena for each of these examples has been obtained using linear hydrodynamics [2], even though, in principle, nonlinear dynamical aspects are within the exact dynamics of these systems.

In the next section we discuss linear hydrodynamics and its role in understanding the inelastic light scattering experiments from liquids, by calculating the density–density correlation function, $S_{\rho\rho}$.

A3.3.2.3   Rayleigh–Brillouin spectrum

The three conservation laws of mass, momentum and energy play a central role in the hydrodynamic description. For a one-component system, these are the only hydrodynamic variables. The mass density has an interesting feature in the associated continuity equation: the mass current (flux) is the momentum density and thus itself is conserved, in the absence of external forces. The mass density $\rho(\vec{r}, t)$ satisfies a continuity equation which can be expressed in the form (see, for example, the book on fluid mechanics by Landau and Lifshitz, cited in the Further Reading)

$$\left(\frac{\partial}{\partial t} + \vec{v} \cdot \vec{\nabla}\right) \rho = -\rho \vec{\nabla} \cdot \vec{v}. \tag{A3.3.13}$$

The equation of momentum conservation, along with the linear transport law due to Newton, which relates the dissipative stress tensor to the rate of strain tensor $e_{ik} = \frac{1}{2}(\nabla_i v_k + \nabla_k v_i)$, and which introduces two transport coefficients, shear viscosity η and bulk viscosity η_b, lead to the equation of motion for a Newtonian fluid:

$$\left(\frac{\partial}{\partial t} + \vec{v} \cdot \vec{\nabla}\right) \vec{v} = -\frac{1}{\rho} \vec{\nabla} p + \nu \nabla^2 \vec{v} + (\nu_l - \nu) \vec{\nabla}(\vec{\nabla} \cdot \vec{v}) \tag{A3.3.14}$$

where the kinematic viscosity $\nu = \eta/\rho$ and the kinematic longitudinal viscosity $\nu_l = (\frac{4}{3}\eta + \eta_b)/\rho$.

The energy conservation law also leads to an associated continuity equation for the total energy density. The total energy density contains both the kinetic energy density per unit volume and the internal energy density. The energy flux is made up of four terms: a kinematic term, the rates of work done by reversible pressure and dissipative viscous stress, and a dissipative heat flux. It is the dissipative heat flux that is assumed to be proportional to the temperature gradient and this linear transport law, Fourier's law, introduces as a proportionality coefficient, the coefficient of thermal conductivity, κ. From the resulting energy equation, one can obtain the equation for the rate of entropy balance in the system, which on account of the irreversibility and the arrow of time implied by the Second Law of Thermodynamics leads to the result that each of the transport coefficients η, η_b and κ is a positive definite quantity. Using the mass conservation equation (A3.3.13), and thermodynamic relations which relate entropy change to changes in density and temperature, the entropy balance equation can be transformed to the hydrodynamic equation for the local temperature $T(\vec{r}, t)$:

$$\left(\frac{\partial}{\partial t} + \vec{v} \cdot \vec{\nabla}\right) T = -\alpha^{-1}(\gamma - 1)\vec{\nabla} \cdot \vec{v} + (\rho C_V)^{-1}[\vec{\nabla} \cdot (\kappa \vec{\nabla} T) + 2\eta[e_{ik} - \frac{1}{3}e_{jj}\delta_{ik}]^2 + \eta_b[e_{jj}]^2] \tag{A3.3.15}$$

where α is the thermal expansion coefficient, $\gamma = C_p/C_V$, C_p the heat capacity per unit mass at constant pressure and C_V the same at constant volume. e_{ik} is the rate of the strain tensor defined above, and a repeated subscript implies a summation over that subscript, here and below.

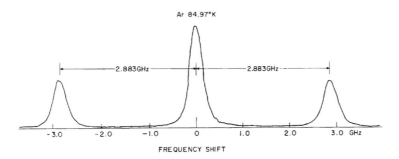

Figure A3.3.1. Rayleigh–Brillouin spectrum from liquid argon, taken from [4].

The three equations (A3.3.13), (A3.3.14) and (A3.3.15) are a useful starting point in many hydrodynamic problems. We now apply them to compute the density–density correlation function

$$S_{\rho\rho}(\vec{r}, t; \vec{r}', t') = \langle \delta\rho(\vec{r}, t)\delta\rho(\vec{r}', t')\rangle. \tag{A3.3.16}$$

Since the fluctuations are small, it is appropriate to linearize the three equations in $\delta\rho(\vec{r}, t) = \rho - \rho_0$, $\delta T(\vec{r}, t) = T - T_0$ and $\vec{v}(\vec{r}, t) = \vec{v}$, by expanding around their respective equilibrium values ρ_0, T_0 and zero, where we assume that the mean fluid velocity is zero. The linearization eliminates the advective term $\vec{v} \cdot \vec{\nabla}$ etc from each of the three equations, and also removes the bilinear viscous dissipation terms from the temperature equation (A3.3.15). The $\vec{\nabla}p$ term in the velocity equation (A3.3.14) can be expressed in terms of density and temperature gradients using thermodynamic derivative identities:

$$\vec{\nabla}p = \left(\frac{\partial p}{\partial \rho}\right)_T \left[\vec{\nabla}\delta\rho - \left(\frac{\partial \rho}{\partial T}\right)_p \vec{\nabla}\delta T\right] = \frac{c^2}{\gamma}\left[\vec{\nabla}\delta\rho + \rho_0\alpha\vec{\nabla}\delta T\right]$$

where $(\partial p/\partial \rho)_T = c_T^2 = c^2/\gamma$ with c_T and c the isothermal and adiabatic speeds of sound, respectively. The momentum equation then linearizes to

$$\frac{\partial \vec{v}}{\partial t} + \frac{c^2}{\gamma\rho_0}[\vec{\nabla}\delta\rho + \rho_0\alpha\vec{\nabla}\delta T] - \nu\nabla^2\vec{v} - (\nu_l - \nu)\vec{\nabla}(\vec{\nabla} \cdot \vec{v}) = 0. \tag{A3.3.17}$$

The linearized equations for density and temperature are:

$$\frac{\partial \delta\rho}{\partial t} + \rho_0\vec{\nabla} \cdot \vec{v} = 0 \tag{A3.3.18}$$

and

$$\frac{\partial \delta T}{\partial t} + \alpha^{-1}(\gamma - 1)\vec{\nabla} \cdot \vec{v} - \gamma D_T\nabla^2\delta T = 0. \tag{A3.3.19}$$

Here the thermal diffusivity $D_T \equiv \kappa/(\rho_0 C_p)$. These two equations couple only to the longitudinal part $\Psi \equiv \vec{\nabla} \cdot \vec{v}$ of the fluid velocity. From equation (A3.3.17) it is easy to see that Ψ satisfies

$$\frac{\partial \Psi}{\partial t} + \frac{c^2}{\gamma\rho_0}[\nabla^2\delta\rho + \rho_0\alpha\nabla^2\delta T] - \nu_l\nabla^2\Psi. \tag{A3.3.20}$$

Out of the five hydrodynamic modes, the polarized inelastic light scattering experiment can probe only the three modes represented by equations (A3.3.18)–(A3.3.20). The other two modes, which are in equation (A3.3.17), decouple from the density fluctuations; these are due to transverse velocity components which is the vorticity $\vec{\omega} \equiv \vec{\nabla} \times \vec{v}$. Vorticity fluctuations decay in a manner analogous to that of the concentration fluctuations discussed in section A3.3.2.2, if one considers the vorticity fluctuation Fourier mode of the wavevector $\vec{k}$. Then the correlations of the kth Fourier mode of vorticity also decays in an exponential manner with the form $\exp(-\nu k^2 t)$.

The density fluctuation spectrum can be obtained by taking a spatial Fourier transform and a temporal Laplace transform of the three coupled equations (A3.3.18)–(A3.3.20), and then solving the resulting linear coupled algebraic set for the density fluctuation spectrum. (See details in the books by Berne–Pecora and Boon–Yip.) The result for $S_{\rho\rho}(\vec{k}, w)$ given below is proportional to its frequency integral $S_{\rho\rho}(\vec{k})$ which is the liquid structure factor discussed earlier in section A2.2.5.2. The density fluctuation spectrum is

$$
\begin{aligned}
\frac{S_{\rho\rho}(k, w)}{S_{\rho\rho}(k)} &= 2 \, \mathrm{Re} \lim_{\epsilon \to 0} \frac{\langle \delta\rho^*(k, t=0)\delta\rho(k, s=\epsilon+iw)\rangle}{\langle \delta\rho^*(k, t=0)\delta\rho(k, t=0)\rangle} \\
&= \frac{\gamma - 1}{\gamma} \frac{2 D_T k^2}{w^2 + (D_T k^2)^2} + \frac{1}{\gamma}\left[\frac{\Gamma k^2}{(w + ck)^2 + (\Gamma k^2)^2} + \frac{\Gamma k^2}{(w - ck)^2 + (\Gamma k^2)^2} \right] \\
&\quad + \frac{1}{\gamma}[\Gamma + (\gamma - 1)D_T]\frac{k}{c}\left[\frac{(w + ck)}{(w + ck)^2 + (\Gamma k^2)^2} - \frac{(w - ck)}{(w - ck)^2 + (\Gamma k^2)^2} \right] \quad \text{(A3.3.21)}
\end{aligned}
$$

where $\Gamma = \frac{1}{2}[\nu_l + (\gamma - 1)D_T]$.

This is the result for monatomic fluids and is well approximated by a sum of three Lorentzians, as given by the first three terms on the right-hand side. The physics of these three Lorentzians can be understood by thinking about a local density fluctuation as made up of thermodynamically independent entropy and pressure fluctuations: $\rho = \rho(s, p)$. The first term is a consequence of the thermal processes quantified by the entropy fluctuations at constant pressure, which lead to the decaying mode $[(\gamma - 1)/\gamma] \exp[-D_T k^2 |t|]$ and the associated Lorentzian known as the *Rayleigh* peak is centred at zero frequency with a half-width at half-maximum of $D_T k^2$. The next two terms (Lorentzians) arise from the mechanical part of the density fluctuations, the pressure fluctuations at constant entropy. These are the adiabatic sound modes $(1/\gamma) \exp[-\Gamma k^2 |t|] \cos[\omega(k)|t|]$ with $\omega(k) = \pm ck$, and lead to the two spectral lines (Lorentzians) which are shifted in frequency by $-ck$ (Stokes line) and $+ck$ (anti-Stokes line). These are known as the *Brillouin–Mandelstam* doublet. The half-width at half-maximum of this pair is Γk^2 which gives the attenuation of acoustic modes. In dense liquids, the last two terms in the density fluctuation spectrum above are smaller by orders of magnitude compared to the three Lorentzians, and lead to s-shaped curves centred at $w = \pm ck$. They cause a weak asymmetry in the Brillouin peaks which induces a slight pulling of their position towards the central Rayleigh peak. The Rayleigh–Brillouin spectrum from liquid argon, as measured by an inelastic polarized light scattering experiment, is shown in figure A3.3.1. An accurate measurement of the Rayleigh–Brillouin lineshape can be used to measure many of the thermodynamic and transport properties of a fluid. The ratio of the integrated intensity of the Rayleigh peak to those of the Brillouin peaks, known as the Landau–Placzek ratio, is $(I_R)/(2I_B) = (\gamma - 1)$, and directly measures the ratio of specific heats γ. From the position of the Brillouin peaks one can obtain the adiabatic speed of sound c, and knowing γ and c one can infer isothermal compressibility. From the width of the Rayleigh peak, one can obtain thermal diffusivity (and if C_p is known, the thermal conductivity κ). Then from the width of the Brillouin peaks, one can obtain the longitudinal viscosity (and, if shear viscosity is known, the bulk viscosity).

A large variety of scattering experiments (inelastic light scattering using polarized and depolarized set ups, Raman scattering, inelastic neutron scattering) have been used over the past four decades to probe and understand the spatio–temporal correlations and molecular dynamics in monatomic and polyatomic fluids in equilibrium, spanning the density range from low-density gases to dense liquids. In the same fashion, concentration fluctuations in binary mixtures have also been probed. See [3–8] for further reading for these topics.

In the next section, we consider thermal fluctuations in an inhomogeneous system.

A3.3.2.4 Capillary waves

In this section we discuss the frequency spectrum of excitations on a liquid surface. While we used linearized equations of hydrodynamics in the last section to obtain the density fluctuation spectrum in the bulk of a homogeneous fluid, here we use linear *fluctuating* hydrodynamics to derive an equation of motion for the instantaneous position of the interface. We then use this equation to analyse the fluctuations in such an inhomogeneous system, around equilibrium and around a NESS characterized by a small temperature gradient. More details can be found in [9, 10].

Surface waves at an interface between two immiscible fluids involve effects due to gravity (g) and surface tension (σ) forces. (In this section, σ denotes surface tension and σ_{ik} denotes the stress tensor. The two should not be confused with one another.) In a hydrodynamic approach, the interface is treated as a sharp boundary and the two bulk phases as incompressible. The Navier–Stokes equations for the two bulk phases (balance of macroscopic forces is the ingredient) along with the boundary condition at the interface (surface tension σ enters here) are solved for possible harmonic oscillations of the interface of the form, $\exp[-(iw+\epsilon)t+i\vec{q}\cdot\vec{x}]$, where w is the frequency, ϵ is the damping coefficient, $\vec{q}$ is the 2–d wavevector of the periodic oscillation and $\vec{x}$ a 2–d vector parallel to the surface. For a liquid–vapour interface which we consider, away from the critical point, the vapour density is negligible compared to the liquid density and one obtains the hydrodynamic dispersion relation for surface waves $w_s^2 = (\sigma/\rho_0)q^3 + gq$. The term gq in the dispersion relation arises from the gravity waves, and dominates for macroscopic wavelengths, but becomes negligible for wavelengths shorter than the capillary constant $(2\sigma/g\rho_0)^{1/2}$, which is of the order of a few millimetres for water. In what follows we discuss phenomena at a planar interface (for which g is essential), but restrict ourselves to the capillary waves regime and set $g = 0^+$. Capillary wave dispersion is then $w_c(q) = (\frac{\sigma}{\rho_0})^{1/2}q^{3/2}$, and the damping coefficient $\epsilon(q) = (2\eta/\rho_0)q^2$.

Consider a system of coexisting liquid and vapour contained in a cubical box of volume L^3. An external, infinitesimal gravitational field locates the liquid of density ρ_1 in the region $z < -\xi$, while the vapour of lower density ρ_v is in the region $z > \xi$. A flat surface, of thickness 2ξ, is located about $z = 0$ in the $\vec{x} = (x, y)$ plane. The origin of the z axis is defined in accord with Gibbs' prescription:

$$\int_{-L/2}^{0} [\rho(z) - \rho_1]\,dz + \int_{0}^{L/2} [\rho(z) - \rho_v]\,dz = 0 \qquad (A3.3.22)$$

where $\rho(z)$ is the equilibrium density profile of the inhomogeneous system. Let us first consider the system in equilibrium. Let it also be away from the critical point. Then $\rho_v \ll \rho_1$ and the interface thickness is only a few nanometres, and a model with zero interfacial width and a step function profile (Fowler model) is appropriate. Also, since the speed of sound is much larger than the capillary wave speed we can assume the liquid to be incompressible, which implies a constant ρ in the liquid and, due to the mass continuity equation (equation (A3.3.13)), also implies $\vec{\nabla} \cdot \vec{v} = 0$. Furthermore, if the amplitude of the capillary waves is small, the nonlinear convective (advective) term $(\vec{v}.\vec{\nabla}\vec{v})$ can also be ignored in (A3.3.14). The approach of fluctuating hydrodynamics corresponds to having additional Gaussian random stress-tensor fluctuations in the Newtonian transport law and analogous heat flux fluctuations in the Fourier transport law. These fluctuations arise from those short lifetime degrees of freedom that are not included in a hydrodynamic description, a description based only on long-lifetime conserved hydrodynamic variables. The equations of motion for the bulk fluid for $z < 0$ are:

$$\frac{\partial v_x}{\partial x} + \frac{\partial v_z}{\partial z} = 0 \qquad (A3.3.23)$$

which is the continuity equation, and

$$\frac{\partial v_x}{\partial t} = -\frac{\partial}{\partial x}\frac{p}{\rho} + \frac{\eta}{\rho}\left(\frac{\partial^2}{\partial x^2} + \frac{\partial^2}{\partial z^2}\right)v_x + \frac{\partial}{\partial x}\left(\frac{s_{xx}}{\rho}\right) + \frac{\partial}{\partial z}\left(\frac{s_{xz}}{\rho}\right)$$

$$\frac{\partial v_z}{\partial t} = -\frac{\partial}{\partial z}\frac{p}{\rho} + \frac{\eta}{\rho}\left(\frac{\partial^2}{\partial x^2} + \frac{\partial^2}{\partial z^2}\right)v_z + \frac{\partial}{\partial z}\left(\frac{s_{zz}}{\rho}\right) + \frac{\partial}{\partial x}\left(\frac{s_{xz}}{\rho}\right).$$

These are the two components of the Navier–Stokes equation including fluctuations s_{ij}, which obey the fluctuation dissipation theorem, valid for incompressible, classical fluids:

$$\langle s_{ik}(\vec{x}, z, t)s_{lm}(\vec{x}', z', t')\rangle_{eq} = 2kT\eta[\delta_{il}\delta_{km} + \delta_{im}\delta_{kl} - \tfrac{2}{3}\delta_{ik}\delta_{lm}]\delta(\vec{x} - \vec{x}')\delta(z - z')\delta(t - t'). \qquad (A3.3.24)$$

This second moment of the fluctuations around equilibrium also defines the form of ensemble $\langle\cdots\rangle_{eq}$ for the equilibrium average at temperature T.

Surface properties enter through the Young–Laplace equation of state for the 'surface pressure' P_{sur}:

$$P_{sur} = -\sigma\frac{\partial^2}{\partial x^2}\zeta \qquad \text{at } z = 0. \qquad (A3.3.25)$$

The *non-conserved* variable $\zeta(\vec{x}, t)$ is a *broken symmetry variable*; it is the instantaneous position of the Gibbs' surface, and it is the translational symmetry in z direction that is broken by the inhomogeneity due to the liquid–vapour interface. In a more microscopic statistical mechanical approach [9], it is related to the number density fluctuation $\delta\rho(\vec{x}, z, t)$ as

$$\zeta(\vec{x}, t) \simeq (\rho_l - \rho_v)^{-1}\int_{-\xi}^{\xi} dz\, \delta\rho(\vec{x}, z, t) \qquad (A3.3.26)$$

but in the present hydrodynamic approach it is defined by

$$\frac{\partial\zeta}{\partial t} = v_z \qquad \text{at } z = 0. \qquad (A3.3.27)$$

The boundary conditions at the $z = 0$ surface arise from the mechanical equilibrium, which implies that both the normal and tangential forces are balanced there. This leads to

$$p = -\sigma\frac{\partial^2\zeta}{\partial x^2} + 2\eta\frac{\partial v_z}{\partial z} + s_{zz} \qquad \text{at } z = 0 \qquad (A3.3.28)$$

$$\eta\left(\frac{\partial v_x}{\partial z} + \frac{\partial v_z}{\partial x}\right) = -s_{xz} \qquad \text{at } z = 0. \qquad (A3.3.29)$$

If the surface tension is a function of position, then there is an additional term, $\partial\sigma/\partial x$, to the right-hand side in the last equation. From the above description it can be shown that the equation of motion for the Fourier component $\zeta(\vec{q}, t)$ of the broken symmetry variable ζ is

$$\frac{\partial^2\zeta(\vec{q}, t)}{\partial t^2} + 2\epsilon(q)\frac{\partial\zeta(\vec{q}, t)}{\partial t} + w_c^2(q)\zeta(\vec{q}, t) = -\frac{q^2}{\rho}\int_{-\infty}^{0} dz\, e^{qz}[s_{zz}(\vec{q}, z, t) - s_{xx}(\vec{q}, z, t) - 2is_{xz}(\vec{q}, z, t)]$$

$$(A3.3.30)$$

where $\epsilon(q)$ and $w_c(q)$ are the damping coefficient and dispersion relation for the capillary waves defined earlier. This damped driven harmonic oscillator equation is driven by spontaneous thermal fluctuations and is valid in the small viscosity limit. It does not have any special capillary wave fluctuations. The thermal random force fluctuations s_{ij} are in the bulk and are coupled to the surface by the e^{qz} factor. This surface–bulk coupling is an essential ingredient of any hydrodynamic theory of the liquid surface: the surface is not a separable phase.

We now evaluate the spectrum of interfacial fluctuations $S(\vec{q}, w)$. It is the space–time Fourier transform of the correlation function $\langle\zeta(\vec{x}, t)\zeta(\vec{x}', t')\rangle$. It is convenient to do this calculation first for the fluctuations around a NESS which has a small constant temperature gradient, no convection and constant pressure. The corresponding results for the system in equilibrium are obtained by setting the temperature gradient to zero. There are three steps in the calculation: first, solve the full nonlinear set of hydrodynamic equations in the steady state, where the time derivatives of all quantities are zero; second, linearize about the steady-state solutions; third, postulate a non-equilibrium ensemble through a generalized fluctuation dissipation relation.

A steady-state solution of the full nonlinear hydrodynamic equations is $\vec{v} = 0$, $p = $ constant and $\partial T/\partial x = $ constant, where the yz walls perpendicular to the xy plane of the interface are kept at different temperatures. This steady-state solution for a *small* temperature gradient means that the characteristic length scale of the temperature gradient $(\partial \ln T/\partial x)^{-1} \ll L$. The solution also implicitly means that the thermal expansion coefficient and surface excess entropy are negligible, i.e. $(\partial \ln \rho/\partial \ln T)$ and $(\partial \ln \sigma/\partial \ln T)$ are both approximately zero, which in turn ensures that there is no convection in the bulk or at the surface. We again assume that the fluid is incompressible and away from the critical point. Then, linearizing around the steady-state solution once again leads, for ζ, to the equation of motion identical to (A3.3.30), which in Fourier space $(\vec{q}, w)$ can be written as (assuming $\epsilon \ll w_c$)

$$[w^2 + 2\epsilon(q)w + w_c^2(q)]\zeta(\vec{q}, w) = -\frac{q^2}{\rho}\int_{-\infty}^{0} dz\, e^{qz}[s_{zz}(\vec{q}, z, w) - s_{xx}(\vec{q}, z, w) - 2is_{xz}(\vec{q}, z, w)]. \quad (A3.3.31)$$

The shear viscosity η is, to a good approximation, independent of T. So the only way the temperature gradient can enter the analysis is through the form of the non-equilibrium ensemble, i.e. through the random forces s_{ik}. Now we assume that the short-ranged random forces have the same form of the real-space correlation as in the thermal equilibrium case above (equation (A3.3.24)), but with T replaced by $T(\vec{x}) = T_0 + (\partial T/\partial \vec{x})_0 \cdot \vec{x}$. Thus the generalized fluctuation dissipation relation for a NESS, which determines the NESS ensemble, is

$$\langle s_{ik}(\vec{q}, z, w)s_{lm}(\vec{q}', z', w')\rangle_{\text{NESS}} = 2kT\eta\left[\delta_{il}\delta_{km} + \delta_{im}\delta_{kl} - \frac{2}{3}\delta_{ik}\delta_{lm}\right]\delta(z - z')\delta(w - w')(2\pi)^3$$

$$\times \left[\delta(\vec{q} - \vec{q}') - \left(\frac{\partial \ln T}{\partial \vec{x}}\right)_0 \cdot \left(\frac{\partial}{\partial i(\vec{q} - \vec{q}')}\right)\delta(\vec{q} - \vec{q}')\right]. \quad (A3.3.32)$$

Then, from equations (A3.3.31) and (A3.3.32), we obtain the spectrum of interfacial fluctuations:

$$\langle \zeta(\vec{x}, w)\zeta(\vec{x}', w')\rangle = 2\pi\delta(w - w')\int \frac{d^2q}{(2\pi)^2}\, e^{i\vec{q}\cdot(\vec{x}-\vec{x}')}S(\vec{q}, w). \quad (A3.3.33)$$

In the absence of a temperature gradient, i.e. in thermal equilibrium, the dynamic structure factor $S(\vec{q}, w)$ is

$$S(\vec{q}, w) = \frac{8kT\eta q^3/\rho^2}{[w^2 - w_c^2(q)]^2 + 4\epsilon^2(q)w^2} \quad (A3.3.34)$$

which is sharply and symmetrically peaked at the capillary wave frequencies $w_c(q) = \pm(\sigma q^3/\rho)^{1/2}$. In the NESS, the result has asymmetry and is given by

$$S_{\text{NESS}}(\vec{q}, w) = S(\vec{q}, w)(1 - \tfrac{1}{2}\Delta(\vec{q}, w)) \quad (A3.3.35)$$

where

$$\Delta(\vec{q}, w) = \frac{[2w^2 + w_c(q)^2]4w(\eta/\rho)\vec{q} \cdot (\partial \ln T/\partial \vec{x})_0}{[w^2 - w_c^2(q)]^2 + 4\epsilon^2(q)w^2}. \quad (A3.3.36)$$

Since ζ is the 'surface-averaged' part of $\delta\rho$ from equation (A3.3.36), $S(\vec{q}, w)$ is the appropriately 'surface-averaged' density fluctuation spectrum near an interface, and is thus experimentally accessible. The correction term $\Delta(\vec{q}, w)$ is an odd function of frequency which creates an asymmetry in the heights of the two ripplon peaks. This is on account of the small temperature gradient breaking the time reversal symmetry: there are more ripples travelling from the hot side to the cold side than from cold to hot. One can also calculate the zeroth and first frequency moments of $S(\vec{q}, w)$:

$$S^0(q) = \frac{kT_0}{\sigma q^2} \tag{A3.3.37}$$

and

$$S^1(q) = -\left[\frac{3}{8\eta}\hat{q} \cdot \left(\frac{\partial \ln T}{\partial \vec{x}}\right)_0\right]\frac{kT_0}{q^2}. \tag{A3.3.38}$$

These are both long ranged in the long-wavelength limit $q \to 0$: $S^0(q)$ due to broken translational symmetry and $S^1(q)$ due to broken time reversal symmetry. $S^1(q)$ vanishes for fluctuations around equilibrium, and $S^0(q)$ is the same for both NESS and equilibrium. The results above are valid only for $L_\nabla \gg L \gg l_c$ where the two bounding length scales are respectively characteristic of the temperature gradient

$$L_\nabla = \left[\hat{q} \cdot \left(\frac{\partial \ln T}{\partial \vec{x}}\right)_0\right]^{-1}$$

and of the capillary wave mean free path

$$l_c \equiv \frac{w_c(q)}{q\epsilon(q)}.$$

The correction due to the temperature gradient in the capillary wave peak heights is the corresponding fractional difference, which can be obtained by evaluating $\Delta(\vec{q}, w = w_c)$. The result is simple:

$$\Delta(\vec{q}, w = w_c) = 3\frac{l_c(q)}{L_\nabla}.$$

For the system in thermal equilibrium, one can compute the time-dependent mean square displacement $\langle |\zeta|^2\rangle(q, t)$, from the damped forced harmonic oscillator equation for ζ, equation (A3.3.30). The result is

$$\langle |\zeta|^2\rangle(q, t) = \frac{kT_0}{\sigma q^2}\left[1 - \exp\left(-\frac{4\eta}{\rho}q^2 t\right)\right] \tag{A3.3.39}$$

which goes to $S^0(q)$ as $t \to \infty$ as required. By integrating it over the two-dimensional wavevector $\vec{q}$, one can find the mean square displacement of the interface:

$$\langle \zeta^2\rangle(t) = \frac{kT_0}{4\sigma}[\ln \tau + E_1(\tau) + C]$$

where $\tau = (4\eta q_{max}^2/\rho)t$ with $q_{max} = 2\pi/a$, with a being a typical molecular size (diameter), E_1 is Euler's integral and $C \sim 0.577$ is Euler's number. Thus, as $t \to \infty$, $\langle \zeta^2\rangle(t)$ diverges as $\ln t$, which is the dynamic analogue of the well known infrared divergence of the interfacial thickness. Numerically, the effect is small: at $t \sim 10^{18}$ s we find using typical values of T, σ, η and ρ such that $[\langle \zeta^2\rangle(t)]^{1/2} \sim 9$ Å.

From equation (A3.3.39) one can also proceed by first taking the $t \to \infty$ limit and then integrating over $\vec{q}$, with the result

$$\langle \zeta^2\rangle = \frac{kT_0}{2\pi\sigma}\ln\frac{q_{max}}{q_{min}}$$

where $q_{min} = 2\pi/L$. Again $\langle \zeta^2\rangle$ shows a logarithmic infrared divergence as $2\pi/L \to 0$ which is the conventional result obtained from equilibrium statistical mechanics. This method hides the fact, which is transparent in the dynamic treatment, that the source of the divergence is the spontaneous random force fluctuations in the bulk, which drive the oscillations of the surface ζ. The equilibrium description of capillary wave excitations of a surface are often introduced in the framework of the so-called capillary wave model: the true free energy is functionally expanded around a 'bare' free energy in terms of the suppressed density fluctuations $\delta\rho$, and these 'capillary wave fluctuations' are assumed to be of the form

$$\delta\rho_{cwf} = -\zeta(x)\frac{d\rho(z)}{dz}. \tag{A3.3.40}$$

It can be shown that this form leads to an unphysical dispersion relation for capillary waves: $w_c^2 \sim q^4$, rather than $\sim q^3$. This is precisely because of the neglect of the surface–bulk coupling in the above assumed form. One can show that a fluctuation *consistent* with capillary wave dispersion is

$$\delta\rho(\vec{x}, z, t) = -\zeta(x, t)\left[\frac{d\rho(z)}{dz} - \frac{q^2\sigma}{c^2}e^{qz}\right] \tag{A3.3.41}$$

for $z < \xi$, where one neglects the vapour density, and where c is the speed of sound in the bulk phase coupled to the surface. Thus, if one wants to introduce density fluctuations into the description, the entire fluid has to be self-consistently treated as compressible. Physically, the first term $\zeta(d\rho(z)/dz)$ corresponds to a perturbation, or kick, of the interface, and the second term self-consistently accounts for the pressure fluctuations in the bulk due to that kick. Neglecting the second term amounts to violating momentum conservation, resulting in an incorrect 'energy–momentum relation' for the capillary wave excitations.

A3.3.3   Non-equilibrium time-evolving systems

There are many examples in nature where a system is not in equilibrium and is evolving in time towards a thermodynamic equilibrium state. (There are also instances where non-equilibrium and time variation appear to be a persistent feature. These include chaos, oscillations and strange attractors. Such phenomena are not considered here.)

A pervasive natural phenomenon is the growth of order from disorder which occurs in a variety of systems. As a result, an interdisciplinary area rich in problems involving the formation and evolution of spatial structures has developed, which combines non-equilibrium dynamics and nonlinear analysis. An important class of such problems deals with the kinetics of phase ordering and phase separation, which are characteristics of any first-order phase transition. Examples of such growth processes occur in many diverse systems, such as chemically reacting systems, biological structures, simple and binary fluids, crystals, polymer melts and metallic alloys. It is interesting that such a variety of systems, which display growth processes, have common characteristics. In the remainder of chapter A3.3 we focus our attention on such common features of kinetics, and on the models which attempt to explain them. Substantial progress has occurred over the past few decades in our understanding of the kinetics of domain growth during first-order phase transitions.

Consider an example of phase separation. It is typically initiated by a rapid change (quench) in a thermodynamic variable (often temperature, and sometimes pressure) which places a disordered system in a post-quench initial non-equilibrium state. The system then evolves towards an inhomogeneous ordered state of coexisting phases, which is its final equilibrium state. Depending on the nature of the quench, the system can be placed in a post-quench state which is thermodynamically unstable or metastable (see figure A3.3.2). In the former case, the onset of separation is spontaneous, and the kinetics that follows is known as *spinodal decomposition*. For the metastable case, the nonlinear fluctuations are required to initiate the separation process; the system is said to undergo phase separation through *homogeneous nucleation* if the system is pure and through *heterogeneous nucleation* if system has impurities or surfaces which help initiate nucleation. The phase transformation kinetics of supercooled substances via homogeneous nucleation is a

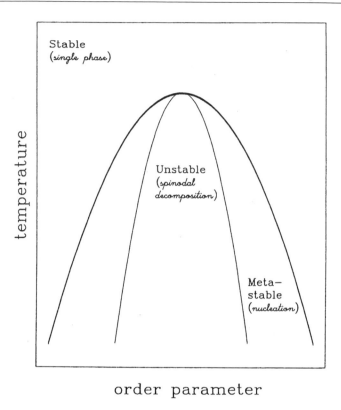

Figure A3.3.2. A schematic phase diagram for a typical binary mixture showing stable, unstable and metastable regions according to a van der Waals mean field description. The coexistence curve (outer curve) and the spinodal curve (inner curve) meet at the (upper) critical point. A critical quench corresponds to a sudden decrease in temperature along a constant order parameter (concentration) path passing through the critical point. Other constant order parameter paths ending within the coexistence curve are called off-critical quenches.

fundamental topic. It is also important in science and technology: gases can be compressed way beyond their equilibrium pressures without forming liquids, and liquids can be supercooled several decades below their freezing temperature without crystallizing.

In both cases the late stages of kinetics show power law domain growth, the nature of which does not depend on the initial state; it depends on the nature of the fluctuating variable(s) which is (are) driving the phase separation process. Such a fluctuating variable is called the order parameter; for a binary mixture, the order parameter $\phi(\vec{r}, t)$ is the relative concentration of one of the two species and its fluctuation around the mean value is $\delta c(\vec{r}, t) = c(\vec{r}, t) - c_0$. In the disordered phase, the system's concentration is homogeneous and the order parameter fluctuations are microscopic. In the ordered phase, the inhomogeneity created by two coexisting phases leads to a macroscopic spatial variation in the order parameter field near the interfacial region. In a magnetic system, the average magnetization characterises the para–ferro magnetic transition and is the order parameter. Depending on the system and the nature of the phase transition, the order parameter may be scalar, vector or complex, and may be conserved or non-conserved.

Here we shall consider two simple cases: one in which the order parameter is a non-conserved scalar variable and another in which it is a conserved scalar variable. The latter is exemplified by the binary mixture phase separation, and is treated here at much greater length. The former occurs in a variety of examples, including some order–disorder transitions and antiferromagnets. The example of the para–ferro transition is one in which the magnetization is a conserved quantity in the absence of an external magnetic field, but becomes non-conserved in its presence.

For a one-component fluid, the vapour–liquid transition is characterized by density fluctuations; here the order parameter, mass density ρ, is also conserved. The equilibrium structure factor $S(\vec{k})$ of a one component fluid is discussed in section A2.2.5.2 and is the Fourier transform of the density–density correlation function. For each of the examples above one can construct the analogous order parameter correlation function. Its spatial Fourier transform (often also denoted by $S(\vec{k})$) is, in most instances, measurable through an appropriate elastic scattering experiment. In a quench experiment which monitors the kinetics of phase transition, the relevant structure evolves in time. That is, the *equal-time* correlation function of the order parameter fluctuations $\langle \delta\phi(\vec{r}, t)\delta\phi(0, t)\rangle_{noneq}$, which would be time independent in equilibrium, acquires time dependence associated with the growth of order in the non-equilibrium system. Its spatial Fourier transform, $S(\vec{k}, t)$ is called the time-dependent structure factor and is experimentally measured.

The evolution of the system following the quench contains different stages. The early stage involves the emergence of macroscopic domains from the initial post-quench state, and is characterized by the formation of interfaces (domain walls) separating regions of space where the system approaches one of its final coexisting states (domains). Late stages are dominated by the motion of these interfaces as the system acts to minimize its surface free energy. During this stage the mean size of the domains grows with time while the total amount of interface decreases. Substantial progress in the understanding of late stage domain growth kinetics has been inspired by the discovery of *dynamical scaling*, which arises when a single length dominates the time evolution. Then various measures of the morphology depend on time only through this length (an instantaneous snapshot of the order parameter's space dependence is referred to as the system's morphology at that time). The evolution of the system then acquires self-similarity in the sense that the spatial patterns formed by the domains at two different times are statistically identical apart from a global change of the length scale.

The time-dependent structure factor $S(\vec{k}, t)$, which is proportional to the intensity $I(k, t)$ measured in an elastic scattering experiment, is a measure of the strength of the spatial correlations in the ordering system with wavenumber k at time t. It exhibits a peak whose position is inversely proportional to the average domain size. As the system phase separates (orders) the peak moves towards increasingly smaller wavenumbers (see figure A3.3.3).

A signature of the dynamical scaling is evidenced by the collapse of the experimental data to a scaled form, for a d-dimensional system:

$$S(k, t) = (R(t))^d S_0(kR(t)) \tag{A3.3.42}$$

where S_0 is a time-independent function and $R(t)$ is a characteristic length (such as the average domain size) (see figure A3.3.4). To the extent that other lengths in the system, such as the interfacial width, play important roles in the kinetics, the dynamical scaling may be valid only asymptotically at very late times.

Another important characteristic of the late stages of phase separation kinetics, for asymmetric mixtures, is the cluster size distribution function of the minority phase clusters: $n(R, \tau)dR$ is the number of clusters of minority phase per unit volume with radii between R and $R + dR$. Its zeroth moment gives the mean number of clusters at time τ and the first moment is proportional to the mean cluster size.

A3.3.3.1 *Langevin models for phase transition kinetics*

Considerable amount of research effort has been devoted, especially over the last three decades, on various issues in domain growth and dynamical scaling. See the reviews [13–17].

Although in principle the microscopic Hamiltonian contains the information necessary to describe the phase separation kinetics, in practice the large number of degrees of freedom in the system makes it necessary

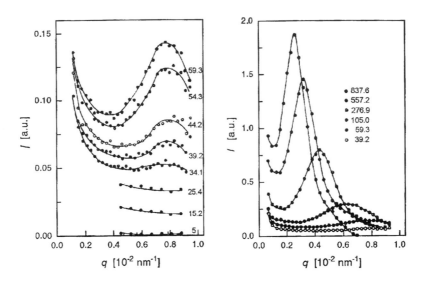

Figure A3.3.3. Time-dependent structure factor as measured through light scattering experiments from a phase separating mixture of polystyrene (PS) ($M = 1.5 \times 10^5$) and poly(vinylmethylether) (PVME) ($M = 4.6 \times 10^4$) following a fast quench from a homogeneous state to $T = 101\,°C$ located in the two-phase region. The time in seconds following the quench is indicated for each structure factor curve. Taken from [11].

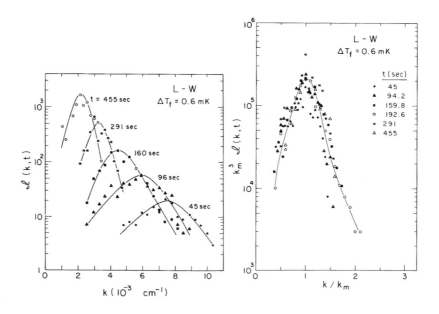

Figure A3.3.4. Time-dependent structure factor as measured through light scattering experiments from a phase separating mixture of 2,6-lutidine and water, following a fast quench from a homogeneous state through the critical point to a temperature 0.6 mK below the critical temperature. The time in seconds, following the quench is indicated for each structure factor curve. In the figure on the right-hand side the data collapse indicates dynamic scaling. Taken from [12].

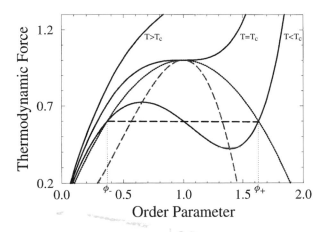

Figure A3.3.5. Thermodynamic force as a function of the order parameter. Three equilibrium isotherms (full curves) are shown according to a mean field description. For $T < T_c$, the isotherm has a van der Waals loop, from which the use of the Maxwell equal area construction leads to the horizontal dashed line for the equilibrium isotherm. Associated coexistence curve (dotted curve) and spinodal curve (dashed line) are also shown. The spinodal curve is the locus of extrema of the various van der Waals loops for $T < T_c$. The states within the spinodal curve are thermodynamically unstable, and those between the spinodal and coexistence curves are metastable according to the mean field description.

to construct a reduced description. Generally, a subset of slowly varying macrovariables, such as the hydro-dynamic modes, is a useful starting point. The equation of motion of the macrovariables can, in principle, be derived from the microscopic Hamiltonian, but in practice one often begins with a phenomenological set of equations. The set of macrovariables are chosen to include the order parameter and all other slow variables to which it couples. Such slow variables are typically obtained from the consideration of the conservation laws and broken symmetries of the system. The remaining degrees of freedom are assumed to vary on a much faster timescale and enter the phenomenological description as random thermal noise. The resulting coupled nonlinear stochastic differential equations for such a chosen 'relevant' set of macrovariables are collectively referred to as the Langevin field theory description.

In two of the simplest Langevin models, the order parameter ϕ is the only relevant macrovariable; in model A it is non-conserved and in model B it is conserved. (The labels A, B, etc have historical origin from the Langevin models of critical dynamics; the scheme is often referred to as the Hohenberg–Halperin classification scheme.) For model A, the Langevin description assumes that, on average, the time rate of change of the order parameter is proportional to (the negative of) the thermodynamic force that drives the phase transition. For this single variable case, the thermodynamic force is canonically conjugate to the order parameter: i.e. in a thermodynamic description, if ϕ is a state variable, then its canonically conjugate force is $\partial f/\partial\phi$ (see figure A3.3.5), where f is the free energy.

In a field theory description, the thermodynamic free energy f is generalized to a free energy functional $\mathcal{F}[\phi(\vec{r}, t)]$, leading to the thermodynamic force as the analogous functional derivative. The Langevin equation for model A is then

$$\frac{\partial \phi}{\partial t} = -M\frac{\delta \mathcal{F}}{\delta \phi} + \eta(\vec{r}, t) \tag{A3.3.43}$$

where the proportionality coefficient M is the mobility coefficient, which is related to the random thermal noise η through the fluctuation dissipation theorem:

$$\langle \eta(\vec{r}, t)\eta(\vec{r}', t')\rangle = kTM\delta(\vec{r} - \vec{r}')\delta(t - t'). \tag{A3.3.44}$$

The phenomenology of model B, where ϕ is conserved, can also be outlined simply. Since ϕ is conserved, it obeys a conservation law (continuity equation):

$$\frac{\partial \phi}{\partial t} = -\vec{\nabla} \cdot \vec{j}(\vec{r}, t) \tag{A3.3.45}$$

where (provided $\vec{j}$ itself is not a conserved variable) one can write the transport law

$$\vec{j}(\vec{r}, t) = -[M\vec{\nabla}\mu(\vec{r}, t) + \vec{\zeta}^*] \tag{A3.3.46}$$

with $\vec{\zeta}^*$ being the order parameter current arising from thermal noise, and $\mu(\vec{r}, t)$, which is the local chemical potential, being synonymous with the thermodynamic force discussed above. It is related to the free energy functional as

$$\mu(\vec{r}, t) = \frac{\delta \mathcal{F}}{\delta \phi} \tag{A3.3.47}$$

Putting it all together, one has the Langevin equation for model B:

$$\frac{\partial \phi}{\partial t} = +M\nabla^2 \left(\frac{\delta \mathcal{F}}{\delta \phi} \right) + \zeta \tag{A3.3.48}$$

where $\zeta = \vec{\nabla} \cdot \vec{\zeta}^*$ is the random thermal noise which satisfies the fluctuation dissipation theorem:

$$\langle \zeta(\vec{r}, t)\zeta(\vec{r}', t') \rangle = -2kTM\nabla^2\delta(\vec{r} - \vec{r}')\delta(t - t'). \tag{A3.3.49}$$

As is evident, the free energy functional $\mathcal{F}$ plays a crucial role in the model A/B kinetics. It contains a number of terms. One of these is the local free energy term $f(\phi)$ which can be thought of as a straightforward generalization of the thermodynamic free energy function in which the global thermodynamic variable ϕ is replaced by its local field value $\phi(\vec{r}, t)$. Many universal features of kinetics are insensitive to the detailed shape of $f(\phi)$. Following Landau, one often uses for it a form obtained by expanding around the value of ϕ at the critical point, ϕ_c. If the mean value of ϕ is $\overline{\phi}$, then

$$\delta\phi \equiv (\phi - \overline{\phi}) = (\phi - \phi_c) + (\phi_c - \overline{\phi}) \equiv \phi^* + \phi_o \tag{A3.3.50}$$

with $\phi^* = (\phi - \phi_c)$ and $\phi_o = (\phi_c - \overline{\phi})$. The Landau expansion is written in terms of ϕ^* as

$$f(\phi) = \tfrac{1}{2}a_o(T - T_c)\phi^{*2} + \tfrac{1}{4}u\phi^{*4} - \mathcal{H}\phi \tag{A3.3.51}$$

where an external field $\mathcal{H}$ is assumed to couple linearly to ϕ. In the absence of $\mathcal{H}$, $f(\phi)$ has a single minimum for temperatures above T_c at $\phi^* = 0$, and two minima below T_c at $\phi^* = \pm[a_o(T_c - T)/u]^{1/2} \equiv \pm\phi_{min}^*$ corresponding to the two coexisting ordered phases in equilibrium (see figure A3.3.6).

The free energy functional also contains a square gradient term which is the cost of the inhomogeneity of the order parameter at each point in the system. Such a surface energy cost term occurs in many different contexts; it was made explicit for binary mixtures by Cahn and Hilliard [18], for superconductors by Ginzburg, and is now commonplace. It is often referred to as the Ginzburg term. This Landau–Ginzburg free energy functional is

$$\mathcal{F}[\phi(\vec{r}, t)] = \int d^d r \left[f(\phi) + \frac{\kappa}{2}(\nabla\phi)^2 \right] \tag{A3.3.52}$$

where the coefficient κ of the square gradient term is related to the interfacial (domain wall) tension σ through the mean field expression

$$\sigma = \kappa \int_{-\infty}^{\infty} dz \left(\frac{d\phi}{dz} \right)^2 \tag{A3.3.53}$$

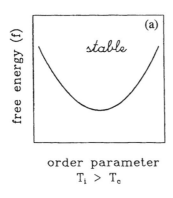
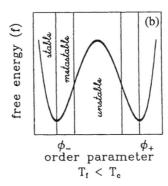

Figure A3.3.6. Free energy as a function of the order parameter ϕ^* for the homogeneous single phase (a) and for the two-phase regions (b), $\mathcal{H} = 0$.

where z is the direction of the local normal to the interface, and $\phi(z)$ is the equilibrium order parameter profile.

There is a class of systems where, in addition to the above two terms, there is a non-local free energy functional term in $\mathcal{F}$. This can arise due to elastic fields, or due to other long-range repulsive interactions (LRRI) originating from the coherent action of molecular dipoles (electric or magnetic). The square gradient (Ginzburg) term is a short-range attractive term which then competes with the LRRI, resulting in a rich variety of structures often termed supercrystals. For such systems, which include Langmuir monolayers and uniaxial magnetic garnet films, the kinetics is much richer. It will not be considered here. See [19, 20] for reviews.

With the form of free energy functional prescribed in equation (A3.3.52), equations (A3.3.43) and (A3.3.48) respectively define the problem of kinetics in models A and B. The Langevin equation for model A is also referred to as the time-dependent Ginzburg–Landau equation (if the noise term is ignored); the model B equation is often referred to as the Cahn–Hilliard–Cook equation, and as the Cahn–Hilliard equation in the absence of the noise term.

For deep quenches, where the post-quench T is far below T_c, the equations are conveniently written in terms of scaled (dimensionless) variables: $\psi(\vec{x}, \tau) = \phi^*/\phi^*_{min}$ and $\vec{x} = \vec{r}/\xi$, where the correlation length $\xi = (\kappa/(a_o|T_c - T|))^{1/2}$, and the dimensionless time τ is defined as equal to $[(2Ma_o|T_c - T|)t]$ for model A and equal to $[(2Ma_o^2(T_c - T)^2/\kappa)t]$ for model B. In terms of these variables the model B Langevin equation can be written as

$$\frac{\partial \psi}{\partial \tau} = -\frac{1}{2}\nabla_x^2(\nabla_x^2\psi + \psi - \psi^3) + \epsilon^{1/2}\mu(\vec{x}, \tau) \tag{A3.3.54}$$

where

$$\epsilon = \frac{kTu}{a_o^2(T_c - T)^2}\left(\frac{a_o|T_c - T|}{\kappa}\right)^{d/2} \tag{A3.3.55}$$

is the strength of the random thermal noise μ which satisfies

$$\langle\mu(\vec{x}, \tau)\mu(\vec{x}', \tau')\rangle = -\nabla_x^2\delta(\vec{x} - \vec{x}')\delta(\tau - \tau'). \tag{A3.3.56}$$

Similarly, the dimensionless model A Langevin equation can also be obtained. The result is recovered by replacing the outermost ∇_x^2 by (-1) in equation (A3.3.54) and by $(-\frac{1}{2})$ in equation (A3.3.56).

Using the renormalization group techniques, it has been shown, by Bray [16], that the thermal noise is irrelevant in the deep-quench kinetics. This is because the free energy has two stable fixed points to which the system can flow: for $T > T_c$ it is the infinite temperature fixed point, and for $T < T_c$ it is the zero-temperature strong coupling fixed point. Since at $T = 0$, the strength of the noise ϵ vanishes, the thermal noise term μ

can be neglected in model B phase separation kinetics during which $T < T_c$. The same conclusion was also obtained, earlier, [21] from a numerical simulation of equations (A3.3.54) and (A3.3.56). In what follows, we ignore the thermal noise term. One must note, however, that there are many examples of kinetics where the thermal noise can play an important role. See for example a recent monograph [22].

For critical quench experiments there is a symmetry $\phi_o = 0$ and from equation (A3.3.50) $\delta\phi = \phi^*$, leading to a symmetric local free energy (figure A3.3.6) and a scaled order parameter whose average is zero, $\delta\psi = \psi$. For off-critical quenches this symmetry is lost. One has $\delta\phi = \phi^* + \phi_o$ which scales to $\delta\psi = \psi + \psi_o$ with $\psi_o = \phi_o/\phi^*_{\min}$. ψ_o is a measure of how far off-critical the system is. For $\psi_o = \pm 1$ the system will be quenched to the coexistence curve and $\psi_o = 0$ corresponds to a quench through the critical point. In general, one has to interpret the dimensionless order parameter ψ in (A3.3.54) as a mean value plus the fluctuations. If one replaces ψ in (A3.3.54) by $\psi + \psi_o$, the mean value of the order parameter ψ_o becomes explicit and the average of such a replaced ψ becomes zero, so that now ψ is the order parameter fluctuation. The conservation law dictates that the average value of the order parameter remains equal to ψ_o throughout the time evolution. Since the final equilibrium phase corresponds to $\psi_\pm = \pm 1$, non-zero ψ_o reflects an asymmetry in the spatial extent of these two phases. The degree of asymmetry is given by the lever rule. A substitution of ψ by $\psi + \psi_o$ in (A3.3.54) yields the following nonlinear partial differential equation (we ignore the noise term):

$$\frac{\partial\psi}{\partial\tau} = -\frac{1}{2}\nabla^2_x([q^2_c + \nabla^2_x]\psi - 3\psi_o\psi^2 - \psi^3) \qquad (A3.3.57)$$

where $q^2_c = (1 - 3\psi^2_o)$. For a critical quench, when $\psi_o = 0$, the bilinear term vanishes and q^2_c becomes one, so that the equation reduces to the symmetric equation (A3.3.54). In terms of the scaled variables, it can be shown that the equation of the classical spinodal, shown in figures A3.3.2 and A3.3.5, is $q^2_c = 0$ or $|\psi_o| = 1/\sqrt{3}$. For states within the classical mean field spinodal, $q^2_c > 0$.

Equation (A3.3.57) must be supplied with appropriate initial conditions describing the system prior to the onset of phase separation. The initial post-quench state is characterized by the order parameter fluctuations characteristic of the pre-quench initial temperature T_i. The role of these fluctuations has been described in detail in [23]. However, again using the renormalization group arguments, any initial short-range correlations should be irrelevant, and one can take the initial conditions to represent a completely disordered state at $T = \infty$. For example, one can choose the white noise form $\langle \psi(\vec{x}, 0)\psi(\vec{x}', 0)\rangle = \epsilon_o\delta(\vec{x} - \vec{x}')$, where $\langle\cdots\rangle$ represents an average over an ensemble of initial conditions, and ϵ_o controls the size of the initial fluctuations in ψ; $\epsilon_o \ll 1$.

The fundamental problem of understanding phase separation kinetics is then posed as finding the nature of late-time solutions of deterministic equations such as (A3.3.57) subject to random initial conditions.

A linear stability analysis of (A3.3.57) can provide some insight into the structure of solutions to model B. The linear approximation to (A3.3.57) can be easily solved by taking a spatial Fourier transform. The result for the kth Fourier mode is

$$\psi(\vec{k}, \tau) = e^{\gamma_k \tau}\psi(\vec{k}, 0) \qquad (A3.3.58)$$

where the exponential growth exponent γ_k is given by

$$\gamma_k = \tfrac{1}{2}k^2(q^2_c - k^2). \qquad (A3.3.59)$$

For $0 < k < q_c$, γ_k is positive, and the corresponding Fourier mode fluctuations grow in time, i.e. these are the *linearly* unstable modes of the system. The maximally unstable mode occurs at $k_m = q_c/\sqrt{2}$ and overwhelms all other growing modes due to exponential growth in the linear approximation. The structure factor can also be computed analytically in this linear approximation, and has a time invariant maximum at $k = k_m$.

In binary polymer mixtures (polymer melts), the early time experimental observations can be fitted to a structure factor form obtained from a linear theory on account of its slow dynamics.

The limitations and range of validity of the linear theory have been discussed in [17, 23, 24]. The linear approximation to equations (A3.3.54) and (A3.3.57) assumes that the nonlinear terms are small compared to the linear terms. As $\psi(\vec{k}, \tau)$ increases with time, at some crossover time t_{cr} the linear approximation becomes invalid. This occurs roughly when $\langle \psi^2 \rangle$ becomes comparable to $(\psi_{sp} - \psi_o)^2 = \psi_{sp}^2 q_c^2$. One can obtain t_{cr} using equation (A3.3.58), in which k can be replaced by k_m, since the maximally unstable mode grows exponentially faster than other modes. Then the dimensionless crossover time $\tau_{cr} \equiv t_{cr}(2M\kappa/\xi_o^4)$ is obtained from

$$(\psi_{sp} - \psi_o)^2 = \langle |\psi(\vec{k}_m, \tau_{cr})|^2 \rangle = e^{2\gamma_{k_m} \tau_{cr}} \langle |\psi(\vec{k}_m, 0)|^2 \rangle$$

where the initial fluctuation spectrum is to be determined from the Ornstein–Zernicke theory, at the pre-quench temperature T_o:

$$\langle |\psi(\vec{k}, 0)|^2 \rangle = \frac{\epsilon_o}{(k_m^2 + q_c^2)}.$$

Here ϵ_o is given by equation (A3.3.55) evaluated at T_o, and can be written as $\epsilon_o = kT_o u \kappa^{-2} \xi_o^{4-d/2}$. Using the values $k_m^2 = q_c^2/2$, $\psi_{sp}^2 = 1/3$, and $\xi_o^2 = \kappa/[a_o(T_o - T_c)]$, one obtains

$$2\gamma_{k_m} \tau_{cr} = \frac{d}{4} \ln(\kappa) + \ln(q_c^4) + \ln\left(\frac{[a_o(T_o - T_c)]^{(2-d/4)}}{2kT_o u} \right).$$

As is evident from the form of the square gradient term in the free energy functional, equation (A3.3.52), κ is like the square of the effective range of interaction. Thus, the dimensionless crossover time depends only weakly on the range of interaction as $\ln(\kappa)$. For polymer chains of length N, $\kappa \sim N$. Thus for practical purposes, the dimensionless crossover time τ_{cr} is not very different for polymeric systems as compared to the small molecule case. On the other hand, the scaling of t_{cr} to τ_{cr} is through a characteristic time which itself increases linearly with κ, and one has

$$t_{cr} = \frac{2\kappa[(d/4)\ln(\kappa) + \ln(q_c^4) + \ln(([a_o(T_o - T_c)]^{(2-\frac{d}{4})})/2kT_o u)]}{q_c^4 M a_o^2 (T_o - T_c)^2}$$

which behaves like $\kappa \ln(\kappa) \sim N \ln(N)$ for polymeric systems. It is clear that the longer time for the validity of linear theory for polymer systems is essentially a longer characteristic time phenomenon.

For initial post-quench states in the metastable region between the classical spinodal and coexistence curves, q_c^2 is negative and so is γ_k for all values of k. Linear stability analysis is not adequate for the metastable region, since it predicts that all modes are stable. Nonlinear terms are important and cannot be ignored in the kinetics leading to either nucleation or spinodal decomposition. The transition from spinodal decomposition to nucleation is also not well defined because nonlinear instabilities play an increasingly more important role as the 'classical spinodal' is approached from within.

A3.3.3.2   Unstable states and kinetics of spinodal decomposition

Equation (A3.3.57) is an interesting nonlinear partial differential equation, but it is mathematically intractable. It contains quadratic and cubic nonlinear terms. The cubic term treats both phases in a symmetric manner; for a symmetric binary mixture only this term survives and leads to a labyrinthian morphology in which both phases have an equal share of the system volume. This term is the source of spinodal decomposition. For the symmetric case the partial differential equation is parameter free and there is no convenient small expansion parameter, especially during early times (the linear approximation loses its validity around $\tau \sim 10$). At late times, the ratio of the interfacial width to the time-dependent domain size $\xi/R(\tau)$ was used as a small parameter by Pego [25] in a matched asymptotic expansion method. It leads to useful connections of this nonlinear problem to the Mullins–Sekerka instability for the slowest timescale and to the classic Stefan problem on a faster timescale.

The quadratic term treats the two phases in an asymmetric manner and is the source of nucleation-like morphology. As the off criticality ψ_o increases, the quadratic term gradually assumes a greater role compared to the cubic nonlinear term. Nucleation-like features in the kinetics occur even for a 49–51 mixture in principle, and are evident at long enough times, since the minority phase will form clusters within the majority background phase for any asymmetric mixture.

While approximate analytical methods have played a role in advancing our understanding of the model B kinetics, complimentary information from laboratory experiments and numerical simulations have also played an important role. Figures A3.3.3 and A3.3.4 show the time-dependent structure factors from laboratory experiments on a binary polymer melt and a small molecule binary mixture, respectively. Compared to the conceptual model B Langevin equation discussed above, real binary mixtures have additional physical effects: for a binary polymer melt, hydrodynamic interactions play a role at late times [17]; for a small molecule binary fluid mixture, hydrodynamic flow effects become important at late times [26]; and for a binary alloy, the elastic effects play a subsidiary, but important, role [37]. In each of these systems, however, there is a broad range of times when model B kinetics are applicable. Comparing the approximate theory of model B kinetics with the experimental results from such systems may not be very revealing, since the differences may be due to effects not contained in model B. Comparing an approximate theory to computer simulation results provides a good test for the theory, provided a good estimate of the numerical errors in the simulation can be made.

In the literature there are numerical simulations of equation (A3.3.57) for both two- and three-dimensional systems [21, 23, 28–31]. For a two-dimensional system, morphology snapshots of the order parameter field ψ are shown in figure A3.3.7 for late times, as obtained from the numerical simulations of (A3.3.57). The light regions correspond to positive values of ψ and the dark regions to negative values. For the critical quench case (figures A3.3.7(a) and (b)), the (statistical) symmetry of ψ between the two phases is apparent. The topological difference between the critical and off-critical quench evolutions at late times is also clear: bicontinuous for critical quench and isolated closed cluster topology for asymmetric off-critical quench. Domain coarsening is also evident from these snapshots for each of the two topologies. For the off-critical quench, from such snapshots one can obtain the time evolution of the cluster size distribution.

From a snapshot at time τ, the spatial correlation function $G(x, \tau) \equiv \langle \psi(\vec{x}, \tau) \psi(0, \tau) \rangle$ can be computed, where $\langle \cdots \rangle$ includes the angular average assuming the system to be spatially isotropic. Repeating this for various snapshots yields the full space-and-time-dependent correlation function $G(x, \tau)$. Its spatial Fourier transform is essentially the time-dependent structure factor $S(k, t)$ measured in light scattering experiments (see figures A3.3.3 and A3.3.4). There are a number of ways to obtain the time-dependent domain size, $R(\tau)$: (i) first zero of $G(x, \tau)$, (ii) first moment of $S(k, t)$, (iii) value k_m where $S(k, t)$ is a maximum. The result that is now firmly established from experiments and simulations, is that

$$R(\tau) \sim \tau^{1/3} \tag{A3.3.60}$$

independent of the system dimensionality d. In the next section (section A3.3.4) we describe the classic theory of Lifshitz, Slyozov and Wagner, which is one of the cornerstone for understanding the $\tau^{1/3}$ growth law and asymptotic cluster size distribution for quenches to the coexistence curve.

As in the experiments, the simulation results also show dynamic scaling at late times. The scaling function $S_o(kR(\tau))$ at late times has the large k behaviour $S_o(y) \sim y^{-(d+1)}$ known as Porod's law [13, 16]. This result is understood to be the consequence of the sharp interfaces at late times. The small k behaviour, $S_o(y) \sim y^4$ was independently predicted in [32, 33], and was put on a firm basis in [34].

Interfaces play a central role in phase transition kinetics of both models A and B. Figure A3.3.8 shows the interfacial structure corresponding to figure A3.3.7(b). One can see the relationship between the interfacial width and the domain size for a late-stage configuration. The upper part of the figure demarks the interfacial regions of the system where $0.75\psi_- < \psi < 0.75\psi_+$. The lower plot gives a cross sectional variation of ψ as

Figure A3.3.7. The order parameter field morphology, for $\psi_0 = 0.0$ at (a) $\tau = 500$ and (b) $\tau = 5000$; and for $\psi_0 = 0.4$ at (c) $\tau = 500$ and (d) $\tau = 5000$. The dark regions have $\psi < 0$. From [28].

the system is traversed. The steep gradients in ψ in the lower plot clearly indicates the sharpness of interfaces at late times.

In figure A3.3.9, the early-time results of the interface formation are shown for $\psi_0 = 0.48$. The classical spinodal corresponds to $\psi_0 \sim 0.58$. Interface motion can be simply monitored by defining the domain boundary as the location where $\psi = 0$. Surface tension smooths the domain boundaries as time increases. Large interconnected clusters begin to break apart into small circular droplets around $\tau = 160$. This is because the quadratic nonlinearity eventually outpaces the cubic one when off-criticality is large, as is the case here.

Some features of late-stage interface dynamics are understood for model B and also for model A. We now proceed to discuss essential aspects of this interface dynamics. Consider the Langevin equations without noise. Equation (A3.3.57) can be written in a more general form:

$$\frac{\partial \psi}{\partial \tau} = -\nabla_x^2 (\xi^2 \nabla_x^2 \psi - f'(\psi)) \tag{A3.3.61}$$

where we have absorbed the factor $\frac{1}{2}$ in the time units of τ, introduced ξ even though it is one, in order to keep track of characteristic lengths, and denoted the thermodynamic force (chemical potential) by f'. At late times the domain size $R(\tau)$ is much bigger than the interfacial width ξ. Locally, therefore, the interface appears to be planar. Let its normal be in direction u, and let $u = u_0$ at the point within the interface where $\psi = 0$. Then $\psi = \psi(u, \vec{s}, \tau)$ where $\vec{s}$ refers to the $(d-1)$ coordinates parallel to the interface at point $\vec{x}$. In essence, we have used the interface specific coordinates: $\vec{x} = \vec{R}(\vec{s}) + u\hat{n}(s)$, where $\hat{n}(s)$ is a unit normal at the interface, pointing from the ψ_- phase into the ψ_+ phase. The stationary solution of (A3.3.61), when

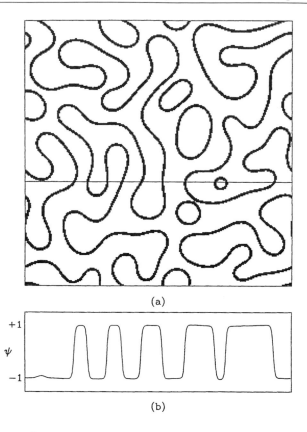

(a)

(b)

Figure A3.3.8. Interface structure for $\tau = 5000$, $\psi_o = 0$. In (a) the shaded regions correspond to interfaces separating the domains. In (b) a cross sectional view of the order parameter ψ is given. The location of the cross section is denoted by the horizontal line in (a). From [35].

$\xi/R(\tau)$ is very small, satisfies

$$\xi^2 \frac{d^2\psi_o}{du^2} = \frac{\partial f}{\partial \psi_0} \tag{A3.3.62}$$

which has a kink profile solution $\psi_o(u) = \pm\tanh((u - u_o)/(\sqrt{2}\xi)$ for the double well free energy $f(\psi_o) = \psi_o^4/4 - \psi_o^2/2$. By linearizing around such a kink solution, one obtains a linear eigenvalue problem. Its lowest eigenmode is $\zeta_o(u) \equiv d\psi_o(u)/du$ and the corresponding eigenvalue is zero. It is localized within the interface and is called the Goldstone mode arising out of the broken translational symmetry at the interface. The higher eigenmodes, which are constructed to be orthogonal to the Goldstone mode, are the capillary wave fluctuation modes with the dispersion relation that is a generalised version [35] of that discussed in section A3.3.2.4. The orthogonality to the Goldstone mode leads to a constraint which is used to show an important relation between the local interface velocity $v(\vec{x})$ and local curvature $K(\vec{s})$. It is, to the lowest order in ξK,

$$\sigma\xi^2 K(s) = \int du\, d\vec{x}'\, G(\vec{x}, \vec{x}')\zeta_o(u)\zeta_o(u')v(\vec{x}') \tag{A3.3.63}$$

where $G(\vec{x}, \vec{x}')$ is the diffusion Green's function satisfying $\nabla^2 G(\vec{x}, \vec{x}') = \delta(\vec{x} - \vec{x}')$. The mean field surface tension σ, defined in equation (A3.3.53), is the driving force for the interface dynamics. The diffusion Green's function couples the interface motion, at two points $(\vec{s}, \vec{s}')$ on the interface, inextricably to the bulk

Figure A3.3.9. Time dependence of the domain boundary morphology for $\psi_o = 0.48$. Here the domain boundary is the location where $\psi = 0$. The evolution is shown for early-time τ values of (a) 50, (b) 100, (c) 150, (d) 200, (e) 250 and (f) 300. From [29].

dynamics. For a conserved order parameter, the interface dynamics and late-stage domain growth involve the *evaporation–diffusion–condensation mechanism* whereby large droplets (small curvature) grow at the expense of small droplets (large curvature). This is also the basis for the Lifshitz–Slyozov analysis which is discussed in section A3.3.4.

If the order parameter is not conserved, the results are much simpler and were discussed by Lifshitz and by Allen and Cahn [36]. For model A, equation (A3.3.61) is to be replaced by the time-dependent Ginzburg–Landau equation which is obtained by removing the overall factor of $(-\nabla_x^2)$ from the right-hand side. This has the consequence that, in the constraint, equation (A3.3.63), the diffusion Green's function is replaced by $\delta(\vec{x} - \vec{x}')$ and the integrals can be performed with the right-hand side reducing to $-\sigma v(s)$. The surface tension then cancels from both sides and one gets the Allen–Cahn result:

$$-\xi^2 K(s) = v(s). \tag{A3.3.64}$$

For model A, the interfaces decouple from the bulk dynamics and their motion is driven entirely by the local curvature, and the surface tension plays only a background, but still an important, role. From this model A interface dynamics result, one can also simply deduce that the domains grow as $R(\tau) \sim \tau^{1/2}$: at some late time, a spherical cluster of radius R grows; since $K \sim (d-1)/R$ and $v \sim -dR/d\tau$, one has $R^2 \sim (d-1)\tau$.

A3.3.4 Late-stage growth kinetics and Ostwald ripening

Late stages of model B dynamics for asymmetric quenches may be described by the Lifshitz–Slyozov–Wagner (LSW) theory of coarsening. When the scalar order parameter is conserved, the late-stage coarsening is referred to as Ostwald ripening. The LSW analysis is valid for late-stage domain growth following either spinodal decomposition or nucleation. A recent paper [37] has combined the steady-state homogeneous nucleation theory described in the next section, A3.3.5, with the LSW analysis in a new model for the entire process of phase separation.

If the initial condition places the post-quench system just inside and quite near the coexistence curve, the conservation law dictates that one (minority) phase will occupy a much smaller 'volume' fraction than the other (majority) phase in the final equilibrium state.

The dynamics is governed by interactions between different domains of the minority phase. At late times these will have attained spherical (circular) shape for a three (two)-dimensional system. For model B systems, the classic work of Lifshitz and Slyozov [38] and the independent work by Wagner [39] form the theoretical cornerstone. The late-stage dynamics is mapped onto a diffusion equation with sources and sinks (i.e. domains) whose boundaries are time dependent. The Lifshitz–Slyozov (LS) treatment of coarsening is based on a mean field treatment of the diffusive interaction between the domains and on the assumption of an infinitely sharp interface with well defined boundary conditions. The analysis predicts the onset of dynamical scaling. As in section A3.3.3.1 we shall denote the extent of the off-criticality by $\psi_0 > 0$. The majority phase equilibrates at $\psi_+ = +1$ and the minority phase at $\psi_- = -1$. At late times, the minority clusters have radius $R(\tau)$ which is much larger than the interface width ξ. An important coupling exists between the interface and the majority phase through the surface tension σ. This coupling is manifested through a Gibbs–Thomson boundary condition, which is given later.

The LS analysis is based on the premise that the clusters of the minority phase compete for growth through an evaporation–condensation mechanism, whereby larger clusters grow at the expense of smaller ones. (Material (of the minority phase) evaporates from a smaller cluster, diffuses through the majority phase background matrix and condenses on a larger cluster.) That is, the dominant growth mechanism is the transport of the order parameter from interfaces of high curvature to regions of low curvature by diffusion through the intervening bulk phases. The basic model B equations, (A3.3.48) and (A3.3.52), can be linearized around the majority phase bulk equilibrium value of the order parameter, $\psi_+ = 1$ (which corresponds to the off-criticality $\psi_0 = -1$), by using $\psi = 1 + \delta\psi$ and keeping only up to first-order terms in $\delta\psi$. The result in dimensionless form is

$$\frac{\partial}{\partial \tau}\delta\psi = -\xi^2 \nabla^4 \delta\psi + \left(\frac{\delta^2 f}{\delta\psi^2}\right)_{\psi=1} \nabla^2 \delta\psi \qquad (A3.3.65)$$

where we have kept the interfacial width ξ as a parameter to be thought of as one; we retain it in order to keep track of the length scales in the problem. Since at late times the characteristic length scales are large compared to ξ, the ∇^4 term is negligible and $\delta\psi$ satisfies a diffusion equation,

$$\frac{\partial}{\partial \tau}\delta\psi = f''(1)\nabla^2 \delta\psi. \qquad (A3.3.66)$$

Due to the conservation law, the diffusion field $\delta\psi$ relaxes in a time much shorter than the time taken by significant interface motion. If the domain size is $R(\tau)$, the diffusion field relaxes over a time scale $\tau_D \sim R^2$.

However a typical interface velocity is shown below to be $\sim R^{-2}$. Thus in time τ_D, interfaces move a distance of about one, much smaller compared to R. This implies that the diffusion field $\delta\psi$ is essentially always in equilibrium with the interfaces and, thus, obeys Laplace's equation

$$\nabla^2 \delta\psi = 0 \qquad (A3.3.67)$$

in the bulk.

A3.3.4.1   Gibbs–Thomson boundary condition

To derive the boundary condition, it is better to work with the chemical potential instead of the diffusion field. We have

$$\frac{\partial\psi}{\partial\tau} = -\nabla \cdot \vec{j} \qquad (A3.3.68)$$

$$\vec{j} = -\nabla\mu \qquad (A3.3.69)$$

and

$$\mu = f'(\psi) - \xi^2\nabla^2\psi. \qquad (A3.3.70)$$

In the bulk, linearizing μ leads to $\mu = f''(\psi_+)\delta\psi - \xi^2\nabla^2\delta\psi$, where the ∇^2 term is again negligible, so that μ is proportional to $\delta\psi$. Thus μ also obeys Laplace's equation

$$\nabla^2\mu = 0. \qquad (A3.3.71)$$

Let us analyse μ near an interface. The Laplacian in the curvilinear coordinates $(u, \vec{s})$ can be written such that (A3.3.71) becomes (near the interface)

$$\mu = f'(\psi) - \xi^2\left(\frac{\partial\psi}{\partial u}\right)_\tau K - \xi^2\left(\frac{\partial^2\psi}{\partial u^2}\right)_\tau \qquad (A3.3.72)$$

where $K = \vec{\nabla} \cdot \hat{n}$ is the total curvature. The value of μ at the interface can be obtained from (A3.3.72) by multiplying it with $(\partial\psi/\partial u)_\tau$ (which is sharply peaked at the interface) and integrating over u across the interface. Since μ and K vary smoothly through the interface, one obtains a general result that, at the interface,

$$\mu\Delta\psi = \Delta f - \xi^2\sigma K \qquad (A3.3.73)$$

where $\Delta\psi$ is the change in ψ across the interface and Δf is the difference in the minima of the free energy f for the two bulk phases. For the symmetric double well, $\Delta f = 0$ and $\Delta\psi = 2$. Thus

$$\mu = -\tfrac{1}{2}\xi^2\sigma K. \qquad (A3.3.74)$$

We make two side remarks.

(i)   If the free energy minima have unequal depths, then this calculation can also be done. See [14].

(ii)   Far away from the interface, $\mu = f''(\psi_+)\delta\psi = 2\delta\psi$ for the ψ^4 form. Then one also has for the supersaturation

$$\delta\psi(\infty) = -\lim_{u\to\infty}\delta\psi(u) = -\mu/2 = +\xi^2\sigma K/4. \qquad (A3.3.75)$$

The supersaturation $\epsilon \equiv \delta\psi(\infty)$ is the mean value of $\delta\psi$, which reflects the presence of other subcritical clusters in the system.

Equation (A3.3.73) is referred to as the Gibbs–Thomson boundary condition. Equation (A3.3.74) determines μ on the interfaces in terms of the curvature, and between the interfaces μ satisfies Laplace's equation,

equation (A3.3.71). Now, since $\vec{j} = -\vec{\nabla}\mu$, an interface moves due to the imbalance between the current flowing into and out of it. The interface velocity is therefore given by

$$j_{\text{out}} - j_{\text{in}} = v\Delta\psi \tag{A3.3.76}$$

and also from equation (A3.3.69),

$$j_{\text{out}} - j_{\text{in}} = -\left[\frac{\partial\mu}{\partial u}\right] = -[\hat{n} \cdot \nabla\mu]. \tag{A3.3.77}$$

Here $[\cdots]$ denotes the discontinuity in $\cdots$ across the interface. Equations (A3.3.71), (A3.3.74), (A3.3.76) and (A3.3.77) together determine the interface motion.

Consider a single spherical domain of minority phase ($\psi_- = -1$) in an infinite sea of majority phase ($\psi_+ = +1$). From the definition of μ in (A3.3.70), $\mu = 0$ at infinity. Let $R(\tau)$ be the domain radius. The solution of Laplace's equation, (A3.3.71), for $d > 2$, with a boundary condition at ∞ and equation (A3.3.74) at $r = R$, is spherically symmetric and is, using $K = (d-1)/R$,

$$\mu = -\frac{(d-1)\sigma\xi^2}{2r} \qquad \text{for } r \geq R \tag{A3.3.78}$$

and

$$\mu = -\frac{(d-1)\sigma\xi^2}{2R} \qquad \text{for } r \leq R. \tag{A3.3.79}$$

Then, using equations (A3.3.76) and (A3.3.77), we obtain, since $\Delta\psi = 2$,

$$\frac{dR}{d\tau} = v = -\frac{1}{2}\left[\frac{\partial\mu}{\partial r}\right]_{R-\epsilon}^{R+\epsilon} = -\frac{(d-1)\xi^2\sigma}{4R^2}. \tag{A3.3.80}$$

Integrating equation (A3.3.80), we get (setting $\xi = 1$)

$$R^3(\tau) = R^3(0) - \tfrac{3}{4}(d-1)\sigma\tau \tag{A3.3.81}$$

which leads to a 'R^3 proportional to τ' time dependence of the evaporating domain: the domain evaporates in time τ proportional to $R^3(0)$.

A3.3.4.2   LS analysis for growing droplets

Again consider a single spherical droplet of minority phase ($\psi_- = -1$) of radius R immersed in a sea of majority phase. But now let the majority phase have an order parameter at infinity that is (slightly) smaller than +1, i.e. $\psi(\infty) \equiv \psi_0 < 1$. The majority phase is now 'supersaturated' with the dissolved minority species, and if the minority droplet is large enough it will grow by absorbing material from the majority phase. Otherwise it will evaporate as above. The two regimes are separated by a critical radius R_c.

Let $f(\pm 1) = 0$ by convention, then the Gibbs–Thomson boundary condition, equation (A3.3.73), becomes at $r = R$,

$$(1 + \psi_0)\mu = f(\psi_0) - \frac{(d-1)\sigma}{R}. \tag{A3.3.82}$$

At $r = \infty$, from equation (A3.3.70),

$$\mu = f'(\psi_0). \tag{A3.3.83}$$

The solution of Laplace's equation, (A3.3.71), with these boundary conditions is, for $d = 3$,

$$\mu = \begin{cases} f'(\psi_o) + \left(\dfrac{f(\psi_o)}{1+\psi_o} - f'(\psi_o)\right)\dfrac{R}{r} - \dfrac{2\sigma}{(1+\psi_o)}\dfrac{1}{r} & r \geq R \qquad\text{(A3.3.84)} \\[2ex] \dfrac{f(\psi_o)}{1+\psi_o} - \dfrac{2\sigma}{(1+\psi_o)}\dfrac{1}{R} & r \leq R. \qquad\text{(A3.3.85)} \end{cases}$$

Using equations (A3.3.76), (A3.3.77) and (A.3.3.84), one finds the interface velocity $v \equiv dR/d\tau$ as

$$\frac{dR}{d\tau} = \left(\frac{f(\psi_o)}{(1+\psi_o)^2} - \frac{f'(\psi_o)}{(1+\psi_o)}\right)\frac{1}{R} - \frac{2\sigma}{(1+\psi_o)^2}\frac{1}{R^2}. \qquad\text{(A3.3.86)}$$

For a small supersaturation, $\psi_o = 1 - \epsilon$ with $\epsilon \ll 1$. To leading (non-trivial) order in ϵ, equation (A3.3.86) reduces to

$$v(R) \equiv \frac{dR}{d\tau} = \frac{\sigma}{2R}\left(\frac{1}{R_c} - \frac{1}{R}\right) \qquad\text{(A3.3.87)}$$

with $R_c = \sigma/(f''(1)\epsilon)$ as the critical radius.

The form of $v(R)$ in (A3.3.87) is valid only for $d = 3$. If we write it as

$$\frac{dR}{d\tau} = \frac{\alpha_d}{R}\left(\frac{1}{R_c} - \frac{1}{R}\right) \qquad\text{(A3.3.88)}$$

then the general expression (see [40]) for α_d is $\alpha_d = (d-1)(d-2)\sigma/4$. For $d = 2$, α_d vanishes due to the singular nature of the Laplacian in two-dimensional systems. For $d = 2$ and in the limit of a small (zero) volume fraction of the minority phase, equation (A3.3.87) is modified to (see the appendix of [28]),

$$\frac{dR}{d\tau} = \frac{\sigma}{4R\ln(4\tau)}\left(\frac{1}{R_c} - \frac{1}{R}\right) \qquad\text{(A3.3.89)}$$

with $R_c = \sigma/(2f''(1)\epsilon)$. A change of variable $\tau^* = \tau/\ln(4\tau)$ converts (A3.3.89) into the same form as (A3.3.87), but now the time-like variable has a logarithmic modification.

In the LS analysis, an assembly of drops is considered. Growth proceeds by evaporation from drops with $R < R_c$ and condensation onto drops $R > R_c$. The supersaturation ϵ changes in time, so that $\epsilon(\tau)$ becomes a sort of mean field due to all the other droplets and also implies a time-dependent critical radius $R_c(\tau) = \sigma/[f''(1)\epsilon(\tau)]$. One of the starting equations in the LS analysis is equation (A3.3.87) with $R_c(\tau)$.

For a general dimension d, the cluster size distribution function $n(R, \tau)$ is defined such that $n(R, \tau)dR$ equals the number of clusters per unit 'volume' with a radius between R and $R+dR$. Assuming no nucleation of new clusters and no coalescence, $n(R, \tau)$ satisfies a continuity equation

$$\frac{\partial n}{\partial \tau} + \frac{\partial}{\partial R}(vn) = 0 \qquad\text{(A3.3.90)}$$

where $v \equiv dR/d\tau$ is given by equation (A3.3.87). Finally, the conservation law is imposed on the entire system as follows. Let the spatial average of the conserved order parameter be $(1 - \epsilon_o)$. At late times the supersaturation $\epsilon(\tau)$ tends to zero giving the constraint

$$\epsilon_o = \epsilon(\tau) + V_d \int_0^\infty dR\, R^d n(R, \tau) \sim V_d \int_0^\infty dR\, R^d n(R, \tau) \qquad\text{(A3.3.91)}$$

where V_d is the volume of the d-dimensional unit sphere. Equations (A3.3.88), (A3.3.90) and (A3.3.91) constitute the LS problem for the cluster size distribution function $n(R, \tau)$. The LS analysis of these equations starts by introducing a scaling distribution of droplet sizes. For a d-dimensional system, one writes

$$n(R, \tau) = R_c^{-(d+1)} f\left(\frac{R}{R_c}\right). \qquad\text{(A3.3.92)}$$

Equation (A3.3.91) becomes, denoting R/R_c by x,

$$\epsilon_0 = 2V_d \int_0^\infty \mathrm{d}x \, x^d f(x) \tag{A3.3.93}$$

and fixes the normalization of $f(x)$. If equation (A3.3.92) is substituted into equation (A3.3.90) we obtain, using the velocity equation (A3.3.88),

$$\frac{\dot{R}_c}{R_c^{d+2}}\left[(d+1)f(x) + x\frac{\mathrm{d}f}{\mathrm{d}x}\right] = \frac{\alpha_d}{R_c^{d+4}}\left[\left(\frac{2}{x^3} - \frac{1}{x^2}\right)f(x) + \left(\frac{1}{x} - \frac{1}{x^2}\right)\frac{\mathrm{d}f}{\mathrm{d}x}\right]. \tag{A3.3.94}$$

For the consistency of the scaling form, equation (A3.3.92), R_c dependence should drop out from equation (A3.3.94); i.e.

$$R_c^2 \dot{R}_c = \alpha_d \gamma \tag{A3.3.95}$$

which integrates to

$$R_c(\tau) = (3\alpha_d \gamma \tau)^{1/3}. \tag{A3.3.96}$$

Equation (A3.3.94) simplifies to

$$\left[\frac{2}{x^3} - \frac{1}{x^2} - \gamma(d+1)\right]f(x) = \left[\gamma x - \frac{1}{x} + \frac{1}{x^2}\right]\frac{\mathrm{d}f}{\mathrm{d}x} \tag{A3.3.97}$$

which integrates to

$$\ln f(x) = \int^x \frac{\mathrm{d}y}{y}\frac{(2 - y - \gamma(d+1)y^3)}{(\gamma y^3 - y + 1)}. \tag{A3.3.98}$$

Due to the normalization integral, equation (A3.3.93), $f(x)$ cannot be non-zero for arbitrarily large x; $f(x)$ must vanish for x greater than some cut-off value x_0, which must be a pole of the integrand in equation (A3.3.98) on the positive real axis. For this to occur $\gamma \le \frac{4}{27} \equiv \gamma_0$. Equations (A3.3.88) and (A3.3.96) together yield an equation for $x = R/R_c$:

$$\frac{\mathrm{d}x}{\mathrm{d}\tau} = \frac{1}{3\gamma\tau}\left(\frac{1}{x} - \frac{1}{x^2} - \gamma x\right) \tag{A3.3.99}$$

$$\equiv \frac{1}{3\gamma\tau}g(x). \tag{A3.3.100}$$

The form of $g(x)$ is shown in figures A3.3.10. For $\gamma < \gamma_0$, all drops with $x > x_1$ will asymptotically approach the size $x_2 R_c(\tau)$, which tends to infinity with τ as $\tau^{1/3}$ from equation (A3.3.96). For $\gamma > \gamma_0$, all points move to the origin and the conservation condition again cannot be satisfied. The only allowed solution consistent with conservation condition, equation (A3.3.93), is that γ asymptotically approaches γ_0 from above. In doing this it takes an infinite time. (If it reaches γ_0 in finite time, all drops with $x > \frac{3}{2}$ would eventually arrive at $x = \frac{3}{2}$ and become stuck and one has a repeat of the $\gamma < \gamma_0$ case.) $\gamma = \gamma_0 = \frac{4}{27}$ then corresponds to a double pole in the integrand in (A3.3.98). LS show that $\gamma(\tau) = \gamma_0[1 - \tilde{\epsilon}^2(\tau)]$ with $\tilde{\epsilon}(\tau) \to 0$ as $\tau \to \infty$. For a asymptotic scaled distribution, one uses $\gamma = \gamma_0 = \frac{4}{27}$ and evaluates the integral in (A3.3.98) to obtain

$$f(x) = \begin{cases} \text{constant } x^2 \, (3+x)^{-(1+\frac{4d}{9})}\left(\frac{3}{2} - x\right)^{-(2+\frac{5d}{9})} \exp\left(-\frac{d}{3-2x}\right) & \text{for } x < \frac{3}{2} \\ 0 & \text{for } x \ge \frac{3}{2} \end{cases} \tag{A3.3.101}$$

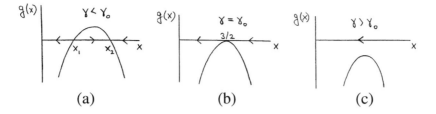

Figure A3.3.10. $g(x)$ as a function of x for the three possible classes of γ.

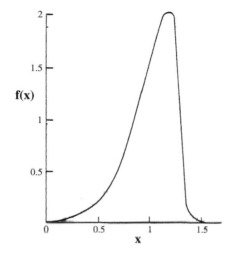

Figure A3.3.11. The asymptotic cluster size distribution $f(x)$ from LS analysis for $d = 3$.

where the normalization constraint, equation A3.3.93, can be used to determine the constant. $f(x)$ is the scaled LS cluster distribution and is shown in figure A3.3.11 for $d = 3$.

In section A3.3.3 the Langevin models that were introduced for phase transition kinetics utilized the Landau mean field expansion of the free energy, equations (A3.3.50) and (A3.3.51). In spite of this, many of the subsequent results based on (A3.3.57) as a starting point are more broadly valid and are dependent only on the existence of a double-well nature of the free energy functional as shown in figure A3.3.6. Also, as the renormalization group analysis shows, the role of thermal noise is irrelevant for the evolution of the initially unstable state. Thus, apart from the random fluctuations in the initial state, which are essential for the subsequent growth of unstable modes, the mean field description is a good theoretical starting point in understanding spinodal decomposition and the ensuing growth. The late-stage growth analysis given in this section is a qualitatively valid starting point for quenches with sufficient off-criticality, and becomes correct asymptotically as the off-criticality ψ_0 increases, bringing the initial post-quench state closer to the coexistence curve, where it is one. In general, one has to add to the LS analysis the cluster–cluster interactions. The current states of such extensions (which are non-trivial) are reviewed in [37, 40, 41]. The main results are that the universal scaling form of the LS cluster distribution function $f(x)$ given above acquires a dependence on $\delta\psi_0 \equiv (1 - \psi_0)$ which measures the proximity to the coexistence curve (it is essentially the volume fraction for the vapour–liquid nucleation); also, the $\tau^{1/3}$ growth law for the domain size has the form: $R(\tau) = [K(\delta\psi_0)\tau]^{1/3}$, where $K(\delta\psi_0)$ is a monotonically increasing function of $\delta\psi_0$.

A3.3.5 Nucleation kinetics—metastable systems

In this section, we restrict our discussion to homogeneous nucleation, which has a illustrious history spanning at least six decades (see [37, 42] and references therein). Heterogeneous nucleation occurs more commonly in nature, since suspended impurities or imperfectly wetted surfaces provide the interface on which the growth of the new phase is initiated. Heterogeneous nucleation is treated in a recent book by Debenedetti (see section 3.4 of this book, which is listed in Further Reading); interesting phenomena of breath figures and dew formation are related to heterogeneous nucleation, and are discussed in [43–45].

In contrast to spinodal decomposition, where small-amplitude, long-wavelength fluctuations initiate the growth, the kinetics following an initial metastable state requires large-amplitude (nonlinear) fluctuations for the stable phase to nucleate. A qualitative picture for the nucleation event is as follows. For the initial metastable state, the two minima of the local free energy functional are not degenerate, in contrast to the initially unstable case shown in figure A3.3.6. The metastable system is initially in the higher of the two minimum energy states and has to overcome a free energy barrier (provided by the third extremum which is a maximum) in order to go over to the absolute minimum, which is the system's ground state. For this, it requires an activation energy which is obtained through rarely occurring large-amplitude fluctuations of the order parameter, in the form of a critical droplet. Physically, the rarity of the nonlinear fluctuation introduces large characteristic times for nucleation to occur.

The central quantity of interest in homogeneous nucleation is the nucleation rate J, which gives the number of droplets nucleated per unit volume per unit time for a given supersaturation. The free energy barrier is the dominant factor in determining J; J depends on it exponentially. Thus, a small difference in the different model predictions for the barrier can lead to orders of magnitude differences in J. Similarly, experimental measurements of J are sensitive to the purity of the sample and to experimental conditions such as temperature. In modern field theories, J has a general form

$$J = \frac{1}{\tau^*}\Omega\, e^{(E_c/kT)} \tag{A3.3.102}$$

where τ^* is the time scale for the macroscopic fluctuations and Ω is the volume of phase space accessible for fluctuations. The barrier height to nucleation E_c is described below.

A homogeneous metastable phase is always stable with respect to the formation of infinitesimal droplets, provided the surface tension σ is positive. Between this extreme and the other thermodynamic equilibrium state, which is inhomogeneous and consists of two coexisting phases, a critical size droplet state exists, which is in unstable equilibrium. In the 'classical' theory, one makes the capillarity approximation: the critical droplet is assumed homogeneous up to the boundary separating it from the metastable background and is assumed to be the same as the new phase in the bulk. Then the work of formation $W(R)$ of such a droplet of arbitrary radius R is the sum of the free energy gain of the new stable phase droplet and the free energy cost due to the new interface that is formed:

$$W(R) = 4\pi R^2 \sigma - \tfrac{4}{3}\pi R^3 \Delta f \tag{A3.3.103}$$

where Δf is the positive bulk free energy difference per unit volume between the stable and metastable phases. From this, by maximizing $W(R)$ with respect to R, one obtains the barrier height to nucleation $W(R_c) \equiv E_c$ and the critical radius R_c: $R_c = 2\sigma/(\Delta f)$ and $E_c = (16\pi/3)\sigma^3/(\Delta f)^2$. For a supercooled vapour nucleating into liquid drops, Δf is given by $kT\rho_l \ln(\epsilon)$, where ρ_l is the bulk liquid density and $\epsilon = P/P_e$ is the supersaturation ratio, which is the ratio of the actual pressure P to the equilibrium vapour pressure P_e of the liquid at the same T. For the case of the nucleation of a crystal from a supercooled liquid, $\Delta f = \Delta\mu/v$, where v is the volume per particle of the solid and $\Delta\mu$ is the chemical potential difference between the bulk solid and the bulk liquid. These results, given here for three-dimensional systems, can be easily generalized to an arbitrary d-dimensional case [37]. Often, it is useful to use the capillary length $l_c = (2\sigma v)/(kT)$ as the unit of length: the critical radius is $R_c = l_c/\epsilon(t)$, where the supersaturation $\epsilon(t) = \Delta\mu/(kT)$, and the nucleation barrier is $(E_c/kT) = (\epsilon_0/\epsilon(t))^2$, where the dimensionless quantity $\epsilon_0 = l_c[(4\pi\sigma)/(3kT)]^{1/2}$.

Early (classical) theories of homogeneous nucleation are based on a microscopic description of cluster dynamics (see reference (1) in [37]). A kinetic equation for the droplet number density $n_i(t)$ of a given size i at time t is written, in which its time rate of change is the difference between J_{i-1} and J_i, where J_i is the rate at which droplets of size i grow to size $i + 1$ by gaining a single molecule [13]. By providing a model for the forward and backward rates at which a cluster gains or loses a particle, J_i is related to $\{n_i(t)\}$, and a set of coupled rate equations for $\{n_i(t)\}$ is obtained. The nucleation rate is obtained from the steady-state solution in which $J_i = J$ for large i. The result is in the form of equation (A3.3.102), with specific expressions for $J_0 \equiv \Omega/\tau^*$ obtained for various cases such as vapour–liquid and liquid–solid transitions. Classical theories give nucleation rates that are low compared to experimental measurements. Considerable effort has gone into attempts to understand 'classical' theories, and compare their results to experiments (see references in [42]).

In two classic papers [18, 46], Cahn and Hilliard developed a field theoretic extension of early theories of nucleation by considering a spatially inhomogeneous system. Their free energy functional, equations (A3.3.52), has already been discussed at length in section A3.3.3. They considered a two-component incompressible fluid. The square gradient approximation implied a slow variation of the concentration on the coarse-graining length scale ξ (i.e. a diffuse interface). In their 1959 paper [46], they determined the saddle point of this free energy functional and analysed the properties of a critical nucleus of the minority phase within the metastable binary mixture. While the results agree with those of the early theories for low supersaturation, the properties of the critical droplet change as the supersaturation is increased: (i) the work required to form a critical droplet becomes progressively less compared to 'classical' theory result, and approaches zero continuously as spinodal is approached; (ii) the interface with the exterior phase becomes more diffuse and the interior of the droplet becomes inhomogeneous in its entirety; (iii) the concentration at the droplet centre approaches that of the exterior phase; and (iv) the radius and excess concentration in the droplet at first decrease, pass through a minimum and become infinite at the spinodal. These papers provide a description of the spatially inhomogeneous critical droplet, which is not restricted to planar interfaces, and yields, for $W(R)$, an expression that goes to zero at the mean field spinodal. The Cahn–Hilliard theory has been a useful starting point in the development of modern nucleation theories.

A full theory of nucleation requires a dynamical description. In the late 1960s, the early theories of homogeneous nucleation were generalized and made rigorous by Langer [47]. Here one starts with an appropriate Fokker–Planck (or its equivalent Langevin) equation for the probability distribution function $P(\{\psi_i\}, t)$ for the set of relevant field variables $\{\psi_i\}$ which are semi-macroscopic and slowly varying:

$$\frac{\partial P}{\partial t} = -\sum_i \frac{\delta J_i}{\delta \psi_i} \tag{A3.3.104}$$

where the probability current J_i is given by

$$J_i = \sum_j M_{ij} \left[\frac{\delta \mathcal{F}}{\delta \psi_j} + kT \frac{\delta P}{\delta \psi_j} \right]. \tag{A3.3.105}$$

$\mathcal{F}$ is the free energy functional, for which one can use equation (A3.3.52). The summation above corresponds to both the sum over the semi-macroscopic variables and an integration over the spatial variable $\vec{r}$. The mobility matrix M_{ij} consists of a symmetric dissipative part and an antisymmetric non-dissipative part. The symmetric part corresponds to a set of generalized Onsager coefficients.

The decay of a metastable state corresponds to passing from a local minimum of $\mathcal{F}$ to another minimum of lower free energy which occurs only through improbable free energy fluctuations. The most probable path for this passage to occur when the nucleation barrier is high is via the saddle point. The saddle point corresponds to a critical droplet of the stable phase in a metastable phase background. The nucleation rate is given by the steady-state solution of the Fokker–Planck equation that describes a finite probability current across the saddle point. The result is of the form given in (A3.3.102). The quantity $1/\tau^*$ is also referred to as a dynamical prefactor and Ω as a statistical prefactor.

Within this general framework there have been many different systems modelled and the dynamical, statistical prefactors have been calculated. These are detailed in [42]. For a binary mixture, phase separating from an initially metastable state, the work of Langer and Schwartz [48] using the Langer theory [47] gives the nucleation rate as

$$J(t) = (l_c^3 t_c)^{-1} \frac{3\epsilon_o^6}{4\pi} \left(\frac{\epsilon(t)}{\epsilon_o} \right)^{2/3} \left(1 + \frac{\epsilon(t)}{\epsilon_o} \right)^{3.55} \exp\left[-\left(\frac{\epsilon_o}{\epsilon(t)} \right)^2 \right] \qquad (A3.3.106)$$

where l_c is the capillary length defined above and the characteristic time $t_c = l_c^2/[DvC_{eq}(\infty)]$ with D as the diffusion coefficient and $C_{eq}(\infty)$ the solute concentration in the background matrix at a planar interface in the phase separated system [37].

One can introduce a distributed nucleation rate $j(R,t)\,dR$ for nucleating clusters of radius between R and $R + dR$. Its integral over R is the total nucleation rate $J(t)$. Equation (A3.3.103) can be viewed as a radius-dependent droplet energy which has a maximum at $R = R_c$. If one assumes $j(R,t)$ to be a Gaussian function, then

$$j(R,t) = \frac{J(t)}{\sqrt{2\pi}(\delta R)} \exp\left[-\frac{(R - R_c)^2}{2(\delta R)^2} \right] \qquad (A3.3.107)$$

where $(\delta R)^2 = 2[E_c - W(R)]/|E_c''|$, with E_c and E_c'' being, respectively, the values of $W(R)$ and its second derivative evaluated at $R = R_c$. Langer [47] showed that the droplet energy is not only a function of R but can also depend on the capillary wavelength fluctuations w; i.e. $W(R) \to E(R, w)$. Then, the droplets appear at the saddle point in the surface of $E(R, w)$. The $2-d$ surface area of the droplet is given by $4\pi(R^2 + w^2)$, which gives the change in the droplet energy due to non-zero w, as $\Delta E(R) = 4\pi\sigma w^2$. Both approaches lead to the same Gaussian form of the distributed nucleation rate with $w \equiv (\delta R)$ estimated from an uncertainty in the required activation energy of the order of $kT/2$.

Just as is the case for the LSW theory of Ostwald ripening, the Langer–Schwartz theory is also valid for quenches close to the coexistence curve. Its extension to non-zero volume fractions requires that such a theory take into account cluster–cluster correlations. A framework for such a theory has been developed [37] using a multi-droplet diffusion equation for the concentration field. This equation has been solved analytically using (i) a truncated multipole expansion and (ii) a mean field Thomas–Fermi approximation. The equation has also been numerically simulated. Such studies are among the first attempts to construct a unified model for the entire process of phase separation that combines steady-state homogeneous nucleation theory with the LSW mechanism for ripening, modified to account for the inter-cluster correlations.

A3.3.6   Summary

In this brief review of dynamics in condensed phases, we have considered dense systems in various situations. First, we considered systems in equilibrium and gave an overview of how the space–time correlations, arising from the thermal fluctuations of slowly varying physical variables like density, can be computed and experimentally probed. We also considered capillary waves in an inhomogeneous system with a planar interface for two cases: an equilibrium system and a NESS system under a small temperature gradient. Finally, we considered time evolving non-equilibrium systems in which a quench brings a homogeneous system to an initially unstable (spinodal decomposition) or metastable state (nucleation) from which it evolves to a final inhomogeneous state of two coexisting equilibrium phases. The kinetics of the associated processes provides rich physics involving nonlinearities and inhomogeneities. The early-stage kinetics associated with the formation of interfaces and the late-stage interface dynamics in such systems continues to provide challenging unsolved problems that have emerged from the experimental observations on real systems and from the numerical simulations of model systems.

References

[1] Chandrasekhar S 1943 *Rev. Mod. Phys.* **15** 1
[2] Koch S W, Desai R C and Abraham F F 1982 *Phys. Rev.* A **26** 1015
[3] Mountain R D 1966 *Rev. Mod. Phys.* **38** 205
[4] Fleury P A and Boon J P 1969 *Phys. Rev.* **186** 244
    Fleury P A and Boon J P 1973 *Adv. Chem. Phys.* **24** 1
[5] Tong E and Desai R C 1970 *Phys. Rev.* A **2** 2129
[6] Desai R C and Kapral R 1972 *Phys. Rev.* A **6** 2377
[7] Weinberg M, Kapral R and Desai R C 1973 *Phys. Rev.* A **7** 1413
[8] Kapral R and Desai R C 1974 *Chem. Phys.* **3** 141
[9] Grant M and Desai R C 1983 *Phys. Rev.* A **27** 2577
[10] Jhon M, Dahler J S and Desai R C 1981 *Adv. Chem. Phys.* **46** 279
[11] Hashimoto T, Kumaki J and Kawai H 1983 *Macromolecules* **16** 641
[12] Chou Y C and Goldburg W I 1981 *Phys. Rev.* A **23** 858
    see also Wong N-C and Knobler C M 1978 *J. Chem. Phys.* **69** 725
[13] Gunton J D, San Miguel M and Sahni P S 1983 *Phase Transitions and Critical Phenomena* vol 8, ed C Domb and J L Lebowitz (New York: Academic)
[14] Furukawa H 1985 *Adv. Phys.* **34** 703
[15] Binder K 1987 *Rep. Prog. Phys.* **50** 783
[16] Bray A J 1994 *Adv. Phys.* **43** 357
[17] Glotzer S C 1995 *Ann. Rev. Comput. Phys.* **II** 1–46
[18] Cahn J W and Hilllard J E 1958 *J. Chem. Phys.* **28** 258
[19] Seul M and Andelman D 1995 *Science* **267** 476
[20] Desai R C 1997 *Phys. Can.* **53** 210
[21] Rogers T M, Elder K R and Desai R C 1988 *Phys. Rev.* B **37** 9638
[22] Garcia-Ojalvo J and Sancho J M 1999 *Noise in Spatially Extended Systems* (Berlin: Springer)
[23] Elder K R, Rogers T M and Desai R C 1988 *Phys. Rev.* B **38** 4725
[24] Binder K 1984 *Phys. Rev.* A **29** 341
[25] Pego R L 1989 *Proc. R. Soc.* A **422** 261
[26] Siggia E D 1979 *Phys. Rev.* A **20** 595
[27] Cahn J W 1961 *Acta Metall.* **9** 795
    Cahn J W 1966 *Acta Metall.* **14** 1685
    Cahn J W 1968 *Trans. Metall. Soc. AIME* **242** 166
[28] Rogers T M and Desai R C 1989 *Phys. Rev.* B **39** 11 956
[29] Elder K R and Desai R C 1989 *Phys. Rev.* B **40** 243
[30] Toral R, Chakrabarti A and Gunton J D 1989 *Phys. Rev.* B **39** 901
[31] Shinozaki A and Oono Y 1993 *Phys. Rev.* E **48** 2622
    Shinozaki A and Oono Y 1991 *Phys. Rev.* A **66** 173
[32] Yeung C 1988 *Phys. Rev. Lett.* **61** 1135
[33] Furukawa H 1989 *Phys. Rev.* B **40** 2341
[34] Fratzl P, Lebowitz J L, Penrose O and Amar J 1991 *Phys. Rev.* B **44**, 4794, see appendix B
[35] Rogers T M 1989 *PhD Thesis* University of Toronto
[36] Allen S M and Cahn J W 1979 *Acta Metall.* **27** 1085
    see also Ohta T, Jasnow D and Kawasaki K 1982 *Phys. Rev. Lett.* **49** 1223 for the model A scaled structure factor
[37] Sagui C and Grant M 1999 *Phys. Rev.* E **59** 4175 and references therein
[38] Lifshitz I M and Slyozov V V 1961 *J. Phys. Chem. Solids* **19** 35
[39] Wagner C 1961 *Z. Elektrochem.* **65** 581
[40] Yao J H, Elder K R, Guo H and Grant M 1993 *Phys. Rev.* B **47** 1410
[41] Akaiwa N and Voorhees P W 1994 *Phys. Rev.* E **49** 3860
    Akaiwa N and Voorhees P W 1996 *Phys. Rev.* E **54** R13
[42] Gunton J D 1999 *J. Stat. Phys.* **95**, 903 and references therein
[43] Fritter D, Knobler C M and Beysens D 1991 *Phys. Rev.* A **43** 2858
[44] Beysens D, Steyer A, Guenoun P, Fritter D and Knobler C M 1991 *Phase Trans.* **31** 219
[45] Rogers T M, Elder K R and Desai R C 1988 *Phys. Rev.* B **38** 5303

[46] Cahn J W and Hilliard J E 1959 *J. Chem. Phys.* **31** 688
[47] Langer J S 1967 *Ann. Phys., NY* **41** 108
Langer J S 1969 *Ann. Phys., NY* **54** 258
[48] Langer J S and Schwartz A J 1980 *Phys. Rev.* A **21** 948

Further Reading

Balucani U and Zoppi M 1994 *Dynamics of the Liquid State* (Oxford: Oxford University Press)
Berne B J and Pecora R 1976 *Dynamic Light Scattering* (New York: Wiley)
Boon J P and Yip S 1980 *Molecular Hydrodynamics* (New York: McGraw-Hill)
Forster D 1975 *Hydrodynamic Fluctuations, Broken Symmetry, and Correlation Functions* (New York: Benjamin)
Debenedetti P G 1996 *Metastable Liquids* (Princeton, NJ: Princeton University Press)
Gunton J D and Droz M 1983 *Introduction to the Theory of Metastable and Unstable States* (Berlin: Springer)
Landau L D and Lifshitz E M 1959 *Fluid Mechanics* (Reading, MA: Addison-Wesley) ch 2, 7, 16, 17. (More recent editions do not have chapter 17.)
Rowlinson J S and Widom B 1982 *Molecular Theory of Capillarity* (Oxford: Clarendon)
Strobl G 1996 *The Physics of Polymers* (Berlin: Springer) especially ch 3 and 4.

A3.4
Gas-phase kinetics

David Luckhaus and Martin Quack

A3.4.1   Introduction

Gas-phase reactions play a fundamental role in nature, for example atmospheric chemistry [1–5] and interstellar chemistry [6], as well as in many technical processes, for example combustion and exhaust fume cleansing [7–9]. Apart from such practical aspects the study of gas-phase reactions has provided the basis for our understanding of chemical reaction mechanisms on a microscopic level. The typically small particle densities in the gas phase mean that reactions occur in well defined elementary steps, usually not involving more than three particles.

At the limit of extremely low particle densities, for example under the conditions prevalent in interstellar space, ion–molecule reactions become important (see chapter A3.5). At very high pressures gas-phase kinetics approach the limit of condensed phase kinetics where elementary reactions are less clearly defined due to the large number of particles involved (see chapter A3.6).

Here, we mainly discuss homogeneous gas-phase reactions at intermediate densities where ideal gas behaviour can frequently be assumed to be a good approximation and diffusion is sufficiently fast that transport processes are not rate determining. The focus is on thermally activated reactions induced by collisions at well defined temperatures, although laser induced processes are widely used for the experimental study of such gas-phase reactions (see chapter B2.1). The aim of the present chapter is to introduce the basic concepts at our current level of understanding. It is not our goal to cover the vast original literature on the general topic of gas reactions. We refer to the books and reviews cited as well as to chapter B2.1 for specific applications.

Photochemical reactions (chapter A3.13) and heterogeneous reactions on surfaces (chapter A3.10) are discussed in separate chapters.

A3.4.2   Definitions of the reaction rate

The are many ways to define the rate of a chemical reaction. The most general definition uses the rate of change of a thermodynamic state function. Following the second law of thermodynamics, for example, the change of entropy S with time t would be an appropriate definition under reaction conditions at constant energy U and volume V:

$$v_S(t) = \left(\frac{\partial S}{\partial t}\right)_{U,V} \geq 0. \qquad (A3.4.1)$$

An alternative rate quantity under conditions of constant temperature T and volume, frequently realized in gas kinetics, would be

$$v_A(t) = -\left(\frac{\partial A}{\partial t}\right)_{T,V} \geq 0, \qquad (A3.4.2)$$

where A is the Helmholtz free energy.

For non-zero v_S and v_A the problem of defining the thermodynamic state functions under non-equilibrium conditions arises (see chapter A3.2). The definition of rate of change implied by equations (A3.4.1) and (A3.4.2) includes changes that are not due to chemical reactions.

In *reaction kinetics* it is conventional to define *reaction rates* in the context of chemical reactions with a well defined stoichiometric equation

$$0 = \sum_i v_i B_i \tag{A3.4.3}$$

where v_i are the stoichiometric coefficients of species B_i ($v_i < 0$ for reactants and $v_i > 0$ for products, by convention). This leads to the conventional definition of the 'rate of conversion':

$$v_\xi(t) = \frac{d\xi}{dt} = v_i^{-1} \frac{dn_i}{dt}. \tag{A3.4.4}$$

The 'extent of reaction' ξ is defined in terms of the amount n_i of species B_i (i.e. the amount of substance or enplethy n_i, usually expressed in moles [10]):

$$\xi(t) = \frac{n_i(t) - n_i(t = 0)}{v_i}; \tag{A3.4.5}$$

v_ξ is an extensive quantity, i.e. for two independent subsystems I and II we have $v_\xi(I + II) = v_\xi(I) + v_\xi(II)$. For homogeneous reactions we obtain the conventional definition of the 'reaction rate' v_i as rate of conversion per volume

$$v_c(t) = V^{-1} v_\xi(t) = v_i^{-1} \frac{dc_i}{dt} \tag{A3.4.6}$$

where c_i is the concentration of species B_i, for which we shall equivalently use the notation $[B_i]$ (with the common unit mol dm^{-3} and the unit of v_c being mol dm^{-3} s^{-1}). In gas kinetics it is particularly common to use the quantity particle density for concentration, for which we shall use C_i (capital letter) with

$$v_C(t) = N_A V^{-1} v_\xi(t) = v_i^{-1} \frac{dC_i}{dt}. \tag{A3.4.7}$$

N_A is Avogadro's constant. The most commonly used unit then is cm^{-3} s^{-1}, sometimes inconsistently written (molecule cm^{-3} s^{-1}). v_c is an intensive quantity. Table A3.4.1 summarizes the definitions.

Figure A3.4.1 shows as an example the time dependent concentrations and entropy for the simple decomposition reaction of chloroethane:

$$C_2H_5Cl = C_2H_4 + HCl. \tag{A3.4.8}$$

The slopes of the functions shown provide the reaction rates according to the various definitions under the reaction conditions specified in the figure caption. These slopes are similar, but not identical (nor exactly proportional), in this simple case. In more complex cases, such as oscillatory reactions (chapters A3.14 and C3.6), the simple definition of an overall rate law through equation (A3.4.6) loses its usefulness, whereas equation (A3.4.1) could still be used for an isolated system.

A3.4.3 Empirical rate laws and reaction order

A general form of the 'rate law', i.e. the differential equation for the concentrations is given by

$$v_c(t) = v_i^{-1} \frac{dc_i}{dt} = f(c_1, c_2, \ldots). \tag{A3.4.9}$$

Table A3.4.1. Definitions of the reaction rate.

Constraint	Extensive quantity	Intensive quantity	Reaction rate
U, V = constant adiabatic	Entropy S (thermodynamics of irreversible processes)	Local entropy $S_V = \dfrac{\delta S}{\delta V}$	$v_S = \dfrac{\mathrm{d}S}{\mathrm{d}t} \geq 0$ unit: $\mathrm{J\,K^{-1}\,s^{-1}}$ $v_{S_V} = \dfrac{\mathrm{d}S_V}{\mathrm{d}t} \geq 0$ unit: $\mathrm{J\,K^{-1}\,s^{-1}\,cm^{-3}}$
T, V = constant isothermal	Helmholtz energy $A = U - TS$	Local A $A_V = \dfrac{\delta A}{\delta V}$	$v_A = -\dfrac{\mathrm{d}A}{\mathrm{d}t} \geq 0$ unit: $\mathrm{J\,s^{-1}}$ $v_{A_V} = \dfrac{\mathrm{d}A_V}{\mathrm{d}t} \geq 0$ unit: $\mathrm{J\,s^{-1}\,cm^{-3}}$
V = constant isothermal or adiabatic, fixed stoichiometry	Amount of substance n_i, number of particles N_i $0 = \sum_l \nu_i B_i$ extent of reaction ξ $\mathrm{d}\xi = \nu_i^{-1}\mathrm{d}n_i$	Concentration $c_i = n_i/V$ $\dfrac{\delta \xi}{\delta V} \simeq \dfrac{\xi}{V}$ $c_i = \dfrac{N_i}{V}$	$\dfrac{\mathrm{d}n_i}{\mathrm{d}t}$ or $\dfrac{\mathrm{d}c_i}{\mathrm{d}t}$ unit: $\mathrm{mol\,s^{-1}}$ or $\mathrm{mol\,cm^{-3}\,s^{-1}}$ $\dfrac{\mathrm{d}\xi}{\mathrm{d}t}$ or $\dfrac{1}{V}\dfrac{\mathrm{d}\xi}{\mathrm{d}t} = \dfrac{1}{\nu_i}\dfrac{\mathrm{d}c_i}{\mathrm{d}t}$ unit: $\mathrm{mol\,s^{-1}}$, $\mathrm{mol\,cm^{-3}\,s^{-1}}$ or $\mathrm{molecule\,cm^{-3}\,s^{-1}}$ $v_c = \dfrac{1}{\nu_i}\dfrac{\mathrm{d}c_i}{\mathrm{d}t}$

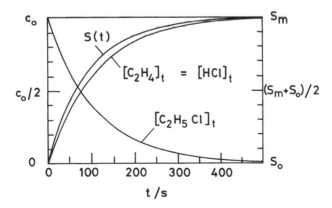

Figure A3.4.1. Concentration and entropy as functions of time for reaction equation (A3.4.8). S_m is the maximum value of the entropy (after [20]).

The functional dependence of the reaction rate on concentrations may be arbitrarily complicated and include species not appearing in the stoichiometric equation, for example, catalysts, inhibitors, etc. Sometimes, however, it takes a particularly simple form, for example, under certain conditions for elementary reactions

Table A3.4.2. Rate laws, reaction order, and rate constants.

Reaction	Rate law	Reaction order	Dimension of rate constant $[k]$
Isomerization $CH_3NC = CH_3CN$ excess of inert gas	$-\dfrac{d[CH_3NC]}{dt} = k[CH_3NC]$	First order in CH_3NC. First-order total	$[s^{-1}]$
Atom transfer $F + CHF_3 = HF + CF_3$	$-\dfrac{d[F]}{dt} = k[F][CHF_3]$	First order in F. First order in CHF_3. Second-order total	$[cm^3 mol^{-1}s^{-1}]$ or $[cm^3\ s^{-1}]$
Radical recombination $2CH_3 = C_2H_6$	$-\dfrac{1}{2}\dfrac{d[CH_3]}{dt} = k[CH_3]^2$	Second order in CH_3. Second-order total	$[cm^3 mol^{-1}s^{-1}]$ or $[cm^3\ s^{-1}]$
Decomposition $CH_3CHO = CH_4 + CO$	$-\dfrac{d[CH_3CHO]}{dt} = k[CH_3CHO]^{3/2}$	Order 3/2 in CH_3CHO and total	$[cm^{3/2}mol^{-1/2}s^{-1}]$

and for other relatively simple reactions:

$$v_c(t) = k \prod_i c_i^{m_i} \tag{A3.4.10}$$

with a concentration-independent and frequently time-independent 'rate coefficient' or 'rate constant' k. m_i is the order of the reaction with respect to the species B_i and the total order of the reaction m is given by

$$m = \sum_i m_i \tag{A3.4.11}$$

where m and m_i are real numbers. Table A3.4.2 summarizes a few examples of such rate laws. In general, one may allow for rate coefficients that depend on time (but not on concentration) [11].

If certain species are present in large excess, their concentration stays approximately constant during the course of a reaction. In this case the dependence of the reaction rate on the concentration of these species can be included in an effective rate constant k_{eff}. The dependence on the concentrations of the remaining species then defines the *apparent order of the reaction*. Take for example equation (A3.4.10) with $c_{i>1} \gg c_1$. The result would be a *pseudo m_1th order* effective rate law:

$$v_c(t) = k_{eff} c_1^{m_1} \tag{A3.4.12}$$

$$k_{eff} = k \prod_{i>1} c_i^{m_i}. \tag{A3.4.13}$$

This is the situation exploited by the so-called *isolation* method to determine the order of the reaction with respect to each species (see chapter B2.1). It should be stressed that the rate coefficient k in (A3.4.10) depends upon the definition of the v_i in the stoichiometric equation. It is a conventionally defined quantity to within multiplication of the stoichiometric equation by an arbitrary factor (similar to reaction enthalpy).

The definitions of the empirical rate laws given above do not exclude empirical rate laws of another form. Examples are reactions, where a reverse reaction is important, such as in the *cis–trans* isomerization of 1,2-dichloroethene:

$$cis - C_2H_2Cl_2 = trans - C_2H_2Cl_2 \tag{A3.4.14}$$

$$-\frac{d[cis]}{dt} = k_a[cis] - k_b[trans] \tag{A3.4.15}$$

or the classic example of hydrogen bromide formation:

$$\frac{1}{2}H_2 + \frac{1}{2}Br_2 = HBr \tag{A3.4.16}$$

$$-\frac{d[HBr]}{dt} = k_a[H_2][Br_2]^{1/2}\left(1 + k_b\frac{[HBr]}{[Br_2]}\right)^{-1}. \tag{A3.4.17}$$

Neither (A3.4.15) nor (A3.4.17) is of the form (A3.4.10) and thus neither reaction order nor a unique rate coefficient can be defined. Indeed, the number of possible rate laws that are *not* of the form of (A3.4.10) greatly exceeds those cases following (A3.4.10). However, certain particularly simple reactions necessarily follow a law of type of (A3.4.10). They are particularly important from a mechanistic point of view and are discussed in the next section.

A3.4.4   Elementary reactions and molecularity

Sometimes the reaction orders m_i take on integer values. This is generally the case, if a chemical reaction

$$A + B \rightarrow \text{ products} \tag{A3.4.18}$$

or

$$2A + B \rightarrow \text{ products} \tag{A3.4.19}$$

takes place on a microscopic scale through direct interactions between particles as implied by equation (A3.4.18) or equation (A3.4.19). Thus, the coefficients of the substances in (A3.4.18) and (A3.4.19) represent the actual number of particles involved in the reaction, rather than just the stoichiometric coefficients. To keep the distinction clear we shall reserve the reaction arrow '$\rightarrow$' for such *elementary reactions*. Sometimes the inclusion of the reverse elementary reaction will be signified by a double arrow '$\rightleftharpoons$'. Other, *compound reactions* can always be decomposed into a set of—not necessarily consecutive—elementary steps representing the *reaction mechanism*.

Elementary reactions are characterized by their *molecularity*, to be clearly distinguished from the reaction order. We distinguish *uni-* (or *mono-*), *bi-*, and *trimolecular* reactions depending on the number of particles involved in the 'essential' step of the reaction. There is some looseness in what is to be considered 'essential', but in gas kinetics the definitions usually are clearcut through the number of particles involved in a reactive collision; plus, perhaps, an additional convention as is customary in unimolecular reactions.

A3.4.4.1   *Unimolecular reactions*

Strictly unimolecular processes—sometimes also called *monomolecular*—involve only a single particle:

$$A \rightarrow \text{ products.} \tag{A3.4.20}$$

Classic examples are the spontaneous emission of light or spontaneous radioactive decay. In chemistry, an important class of monomolecular reactions is the predissociation of metastable (excited) species. An example is the formation of oxygen atoms in the upper atmosphere by predissociation of electronically excited O_2 molecules [12–14]:

$$O_2^* \rightarrow 2O. \tag{A3.4.21}$$

Excited O_2^* molecules are formed by UV light absorption. Monomolecular reactions (e.g., $c = [O_2^*]$) show a first-order rate law:

$$-\frac{dc}{dt} = kc(t). \tag{A3.4.22}$$

Integration of the differential equation with time-independent k leads to the familiar exponential decay:

$$c(t) = c(0) \exp\{-kt\}. \qquad (A3.4.23)$$

The rate constant in this case is of the order of 10^{11} s^{-1} depending on the rovibronic level considered.

Another example of current interest is the *vibrational predissociation* of hydrogen bonded complexes such as $(HF)_2$:

$$\begin{matrix} H \\ \diagdown \\ \quad (*) \quad \overset{k}{\rightarrow} 2HF. \\ F \cdots H - F \end{matrix} \qquad (A3.4.24)$$

With one quantum of non-bonded (HF)-stretching excitation (∗) the internal energy (~ 50 kJ mol^{-1}) is about four times in excess of the hydrogen bond dissociation energy (12.7 kJ mol^{-1}). At this energy the rate constant is about $k \approx 5 \times 10^7$ s^{-1} [15]. With two quanta of (HF)-stretching (at about seven times the dissociation energy) the rate constant is $k \approx 7.5 \times 10^8$ s^{-1} in all cases, depending on the rovibrational level considered [16, 17].

While monomolecular collision-free predissociation excludes the preparation process from explicit consideration, thermal unimolecular reactions involve collisional excitation as part of the unimolecular mechanism. The simple mechanism for a thermal chemical reaction may be formally decomposed into three (possibly reversible) steps (with rovibronically excited $(CH_3NC)^*$):

$$CH_3NC + M \rightleftharpoons (CH_3NC)^* + M \qquad (A3.4.25)$$

$$(CH_3NC)^* \rightleftharpoons (CH_3CN)^* \qquad (A3.4.26)$$

$$(CH_3CN)^* + M \rightleftharpoons CH_3CN + M. \qquad (A3.4.27)$$

The inert collision partner M is assumed to be present in large excess:

$$[M] \gg [CH_3NC] \qquad (A3.4.28)$$

$$[M] \approx \text{constant}. \qquad (A3.4.29)$$

This mechanism as a whole is called 'unimolecular' since the essential isomerization step equation (A3.4.26) only involves a single particle, *viz.* CH_3NC. Therefore it is often simply written as follows:

$$CH_3NC \overset{[M]}{\rightarrow} CH_3CN. \qquad (A3.4.30)$$

Experimentally, one finds the same first-order rate law as for monomolecular reactions, but with an effective rate constant k that now depends on [M].

$$-\frac{dc}{dt} = k([M])c(t). \qquad (A3.4.31)$$

The correct treatment of the mechanism (equations (A3.4.25)–(A3.4.27)), which goes back to Lindemann [18] and Hinshelwood [19], also describes the pressure dependence of the effective rate constant in the low-pressure limit ($[M] \leq [CH_3NC]$, see section A3.4.8.2).

The unimolecular rate law can be justified by a probabilistic argument. The number ($N_A V dc \propto dc$) of particles which react in a time dt is proportional both to this same time interval dt and to the number of particles present ($N_A V c \propto c$). However, this probabilistic argument need not always be valid, as illustrated in figure A3.4.2 for a simple model [20]:

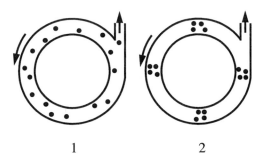

1   2

Figure A3.4.2. A simple illustration of limiting dynamical behaviour: case 1 statistical, case 2 coherent (after [20]).

A number of particles perform periodic rotations in a ring-shaped container with a small opening, through which some particles can escape. Two situations can now be distinguished.

Case 1. The particles are statistically distributed around the ring. Then, the number of escaping particles will be proportional both to the time interval (opening time) dt and to the total number of particles in the container. The result is a first-order rate law.

Case 2. The particles rotate in small packets ('coherently' or 'in phase'). Obviously, the first-order rate law no longer holds. In chapter B2.1 we shall see that this simple consideration has found a deeper meaning in some of the most recent kinetic investigations [21].

A3.4.4.2   Bimolecular reactions

Bimolecular reactions involve two particles in their essential step. In the so-called *self-reactions* they are of the same species:

$$A + A \rightarrow \text{ products} \tag{A3.4.32}$$

with the stoichiometric equation

$$2A = \text{ products.} \tag{A3.4.33}$$

Typical examples are radical recombinations:

$$CH_3 + CH_3 \rightarrow (C_2H_6)^* \tag{A3.4.34}$$

$$(C_2H_6)^* + M \rightarrow C_2H_6. \tag{A3.4.35}$$

Here the initially formed excited species $(C_2H_6)^*$ is sufficiently long lived that the deactivation step (equation (A3.4.35)) is not essential and one writes

$$CH_3 + CH_3 \xrightarrow{[M]} C_2H_6. \tag{A3.4.36}$$

The rate is given by the second-order law ($c \equiv [CH_3]$ or $c \equiv [A]$)

$$-\frac{1}{2}\frac{dc}{dt} = kc^2. \tag{A3.4.37}$$

Integration leads to

$$\frac{1}{c(t)} = 2kt + \frac{1}{c(0)}. \tag{A3.4.38}$$

Bimolecular reactions between different species

$$A + B \rightarrow \text{products} \tag{A3.4.39}$$

lead to the second-order rate law

$$-\frac{dc_A}{dt} = kc_A c_B. \tag{A3.4.40}$$

For $c_B(0) \neq c_A(0)$ the solution of this differential equation is

$$\ln\left(\frac{c_B(t)}{c_A(t)}\right) - \ln\left(\frac{c_B(0)}{c_A(0)}\right) = (c_B(0) - c_A(0))kt. \tag{A3.4.41}$$

The case of equal concentrations, $c_B = c_A = c(t)$, is similar to the case A+A in equation (A3.4.37), except for the stoichiometric factor of two. The result thus is

$$\frac{1}{c(t)} = kt + \frac{1}{c(0)}. \tag{A3.4.42}$$

If one of the reactants is present in large excess $c_B \gg c_A$ its concentration will essentially remain constant throughout the reaction. Equation (A3.4.41) then simplifies to

$$c_A(t) = c_A(0) \exp\{-k_{eff}t\} \tag{A3.4.43}$$

with the effective *pseudo first-order* rate constant $k_{eff} = kc_B$.

One may justify the differential equations (A3.4.37) and (A3.4.40) again by a probability argument. The number of reacting particles $N_A V dc \propto dc$ is proportional to the frequency of encounters between two particles and to the time interval dt. Since not every encounter leads to reaction, an additional reaction probability P_R has to be introduced. The frequency of encounters is obtained by the following simple argument. Assuming a statistical distribution of particles, the probability for a given particle to occupy a volume element δV is proportional to the concentration c. If the particles move independently from each other (ideal behaviour) the same is true for a second particle. Therefore the probability for two particles to occupy the same volume element (an encounter) is proportional to c^2. This leads to the number of particles reacting in the time interval dt:

$$N_A V dc \propto P_R c^2 dt. \tag{A3.4.44}$$

In the case of bimolecular gas-phase reactions, 'encounters' are simply collisions between two molecules in the framework of the general collision theory of gas-phase reactions (section A3.4.5.2). For a random thermal distribution of positions and momenta in an ideal gas reaction, the probabilistic reasoning has an exact foundation. However, as noted in the case of unimolecular reactions, in principle one must allow for deviations from this ideal behaviour and, thus, from the simple rate law, although in practice such deviations are rarely taken into account theoretically or established empirically.

The second-order rate law for bimolecular reactions is empirically well confirmed. Figure A3.4.3 shows the example of methyl radical recombination (equation (A3.4.36)) in a graphical representation following equation (A3.4.38) [22–24]. For this example the bimolecular rate constant is

$$k = 4.4 \times 10^{-11} \text{ cm}^3 \text{ s}^{-1} \tag{A3.4.45}$$

or

$$k = 2.6 \times 10^{13} \text{ cm}^3 \text{ mol}^{-1} \text{ s}^{-1}. \tag{A3.4.46}$$

It is clear from figure A3.4.3 that the second-order law is well followed. However, in particular for recombination reactions at low pressures, a transition to a third-order rate law (second order in the recombining species and first order in some collision partner) must be considered. If the non-reactive collision partner M is present in excess and its concentration [M] is time-independent, the rate law still is pseudo-second order with an effective second-order rate coefficient proportional to [M].

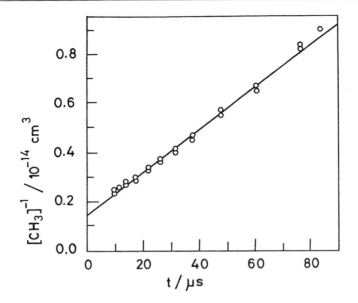

Figure A3.4.3. Methyl radical recombination as a second-order reaction (after [22, 23]).

A3.4.4.3 Trimolecular reactions

Trimolecular reactions require the simultaneous encounter of three particles. At the usually low particle densities of gas phase reactions they are relatively unlikely. Examples for trimolecular reactions are atom recombination reactions

$$A + B + M \rightarrow AB + M \tag{A3.4.47}$$

with the stoichiometric equation

$$A + B = AB. \tag{A3.4.48}$$

In contrast to the bimolecular recombination of polyatomic radicals (equation (A3.4.34)) there is no long-lived intermediate AB* since there are no extra intramolecular vibrational degrees of freedom to accommodate the excess energy. Therefore, the formation of the bond and the deactivation through collision with the inert collision partner M have to occur simultaneously (within 10–100 fs). The rate law for trimolecular recombination reactions of the type in equation (A3.4.47) is given by

$$-\frac{dc_A}{dt} = k[M]c_A c_B \tag{A3.4.49}$$

as can be derived by a probability argument similar to bimolecular reactions (and with similar limitations). Generally, collisions with different collision partners M_i may have quite different efficiencies. The rate law actually observed is therefore given by

$$-\frac{dc_A}{dt} = \sum_i k_i[M_i]c_A c_B. \tag{A3.4.50}$$

If the dominant contributions $k_i[M_i]$ are approximately constant, this leads to pseudo second-order kinetics with an effective rate constant

$$k_{\text{eff}} = \sum_i k_i[M_i]. \tag{A3.4.51}$$

The recombination of oxygen atoms affords an instructive example:

$$O + O + O \xrightarrow{k_O} O_2 + O \tag{A3.4.52}$$

$$O + O + O_2 \xrightarrow{k_{O_2}} O_2 + O_2 \tag{A3.4.53}$$

with the common stoichiometric equation

$$2O = O_2. \tag{A3.4.54}$$

Here $k_O \gg k_{O_2}$ because (A3.4.52) proceeds through a highly-excited molecular complex O_3^* with particularly efficient redistribution pathways for the excess energy. As long as $[O] \geq [O_2]$ the rate law for this trimolecular reaction is given by ($c(t) = [O]$, $k = k_O$):

$$-\frac{1}{2}\frac{dc}{dt} = kc^3. \tag{A3.4.55}$$

Integration leads to

$$\frac{1}{c(t)^2} = 4kt + \frac{1}{c(0)}. \tag{A3.4.56}$$

Trimolecular reactions have also been discussed for molecular reactions postulating concerted reactions via cyclic intermediate complexes, for example

$$2NO + O_2 \longrightarrow \quad\quad\quad \longrightarrow 2NO_2. \tag{A3.4.57}$$

Empirically, one indeed finds a third-order rate law

$$-\frac{1}{2}\frac{d[NO]}{dt} = k[NO]^2[O_2]. \tag{A3.4.58}$$

However, the postulated trimolecular mechanism is highly questionable. The third-order rate law would also be consistent with mechanisms arising from consecutive bimolecular elementary reactions, such as

$$NO + NO \rightleftharpoons (NO)_2 \tag{A3.4.59}$$

$$(NO)_2 + O_2 \rightarrow 2NO_2 \tag{A3.4.60}$$

or

$$NO + O_2 \rightleftharpoons NO_3 \tag{A3.4.61}$$

$$NO_3 + NO \rightarrow 2NO_2. \tag{A3.4.62}$$

In fact, the bimolecular mechanisms are generally more likely. Even the atom recombination reactions sometimes follow a mechanism consisting of a sequence of bimolecular reactions

$$A + M \rightleftharpoons AM \tag{A3.4.63}$$

$$AM + A \rightarrow A_2 + M. \tag{A3.4.64}$$

This so-called complex mechanism has occasionally been proven to apply [25, 26].

A3.4.5 Theory of elementary gas-phase reactions

A3.4.5.1 General theory

The foundations of the modern theory of elementary gas-phase reactions lie in the time-dependent molecular quantum dynamics and molecular scattering theory, which provides the link between time-dependent quantum dynamics and chemical kinetics (see also chapter A3.11). A brief outline of the steps in the development is as follows [27].

We start from the time-dependent Schrödinger equation for the state function (wavefunction $\Psi(t)$) of the reactive molecular system with Hamiltonian operator $\hat{H}$:

$$i\hbar \frac{\partial \Psi(t)}{\partial t} = \hat{H}\Psi(t). \tag{A3.4.65}$$

Its solution can be written in terms of the time evolution operator $\hat{U}$

$$\Psi(t) = \hat{U}(t, t_0)\Psi(t_0) \tag{A3.4.66}$$

which satisfies a similar differential equation

$$i\hbar \frac{\partial \hat{U}}{\partial t} = \hat{H}\hat{U}. \tag{A3.4.67}$$

For time-independent Hamiltonians we have

$$\hat{U}(t, t_0) = \exp[-i\hat{H}(t - t_0)/\hbar]. \tag{A3.4.68}$$

For strictly monomolecular processes the general theory would now proceed by analysing the time-dependent wavefunction as a function of space (and perhaps spin) coordinates $\{q_i\}$ of the particles in terms of time-dependent probability densities.

$$P(\{q_i\}, t) = |\Psi(\{q_i\}, t)|^2 \tag{A3.4.69}$$

which are integrated over appropriate regions of coordinate space assigned to reactants and products. These time-dependent probabilities can be associated with time-dependent concentrations, reaction rates and, if applicable, rate coefficients.

For thermal unimolecular reactions with bimolecular collisional activation steps and for bimolecular reactions, more specifically one takes the limit of the time evolution operator for $t_0 \to -\infty$ and $t \to +\infty$ to describe isolated binary collision events. The corresponding matrix representation of $\hat{U}$ is called the scattering matrix or **S**-matrix with matrix elements

$$S_{fi} = U_{fi}(t \to +\infty, t_0 \to -\infty). \tag{A3.4.70}$$

The physical interpretation of the scattering matrix elements is best understood in terms of its square modulus

$$P_{fi} = |S_{fi}|^2 \tag{A3.4.71}$$

which is the transition probability between an initial fully specified quantum state $|i\rangle$ before the collision and a final quantum state $|f\rangle$ after the collision.

In a third step the **S**-matrix is related to state-selected reaction cross sections σ_{fi}, in principle observable in beam scattering experiments [28–35], by the fundamental equation of scattering theory

$$\sigma_{fi} = \frac{\pi}{k_i^2}|\delta_{fi} - S_{fi}|^2. \tag{A3.4.72}$$

Here $\delta_{fi} = 1(0)$ is the Kronecker delta for $f = i (f \neq i)$ and k_i is the wavenumber for the collision, related to the initial relative centre of mass translational energy $E_{t,i}$ before the collision

$$k_i = \hbar^{-1}\sqrt{2\mu E_{t,i}} \tag{A3.4.73}$$

with reduced mass μ for the collision partners of mass m_A and m_B:

$$\mu = \frac{m_A m_B}{m_A + m_B}. \tag{A3.4.74}$$

Actually equation (A3.4.72) for σ_{fi} is still formal, as practically observable cross sections, even at the highest quantum state resolution usually available in molecular scattering, correspond to certain sums and averages of the individual σ_{fi}. We use capital indices for such coarse-grained state-selected cross sections

$$\sigma_{FI} = \langle\sigma_{fi}\rangle. \tag{A3.4.75}$$

In a fourth step the cross section is related to a state-selected specific bimolecular rate coefficient

$$k_{FI}(E_t) = \sigma_{FI}(E_{t,I})\sqrt{2E_{t,I}/\mu}. \tag{A3.4.76}$$

This rate coefficient can be averaged in a fifth step over a translational energy distribution $P(E_t)$ appropriate for the bulk experiment. In principle, any distribution $P(E_t)$ as applicable in the experiment can be introduced at this point. If this distribution is a thermal Maxwell–Boltzmann distribution one obtains a partially state-selected thermal rate coefficient

$$k_{FI}(T) = \left(\frac{8k_B T}{\pi\mu}\right)^{1/2} \int_0^\infty \left(\frac{E_{t,I}}{k_B T}\right)\sigma_{FI}(E_{t,I})\exp\left\{-\frac{E_{t,I}}{k_B T}\right\}\left(\frac{dE_{t,I}}{k_B T}\right). \tag{A3.4.77}$$

In a final, sixth step one may also average (sum) over a thermal (or other) quantum state distribution I (and F) and obtain the usual thermal rate coefficient

$$k(T) = \langle k_{FI}(T)\rangle. \tag{A3.4.78}$$

Figure A3.4.4 summarizes these steps in one scheme. Different theories of elementary reactions represent different degrees of approximations to certain averages, which are observed in experiments.

There are two different aspects to these approximations. One consists in the approximate treatment of the underlying many-body quantum dynamics; the other, in the statistical approach to observable average quantities. An exhaustive discussion of different approaches would go beyond the scope of this introduction. Some of the most important aspects are discussed in separate chapters (see chapters A3.7, A3.11, A3.12, A3.13).

Here, we shall concentrate on basic approaches which lie at the foundations of the most widely used models. Simplified collision theories for bimolecular reactions are frequently used for the interpretation of experimental gas-phase kinetic data. The general transition state theory of elementary reactions forms the starting point of many more elaborate versions of quasi-equilibrium theories of chemical reaction kinetics [27, 36–38].

In practice, one of the most important aspects of interpreting experimental kinetic data in terms of model parameters concerns the temperature dependence of rate constants. It can often be described phenomenologically by the Arrhenius equation [39–41]

$$k(T) = A(T)\exp\{-E_A(T)/RT\} \tag{A3.4.79}$$

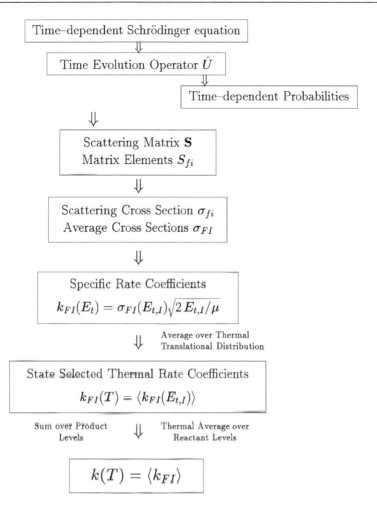

Figure A3.4.4. Steps in the general theory of chemical reactions.

where the *pre-exponential Arrhenius factor* A and the *Arrhenius activation energy* E_A generally depend on the temperature. R is the gas constant. This leads to the *definition* of the Arrhenius parameters:

$$E_A(T) \overset{\text{def}}{=} RT^2 \frac{\mathrm{d}\ln(k(T))}{\mathrm{d}T} \tag{A3.4.80}$$

$$A(T) \overset{\text{def}}{=} k(T)\exp\left\{\frac{E_A(T)}{RT}\right\}. \tag{A3.4.81}$$

The usefulness of these definitions is related to the usually weak temperature dependence of E_A and A. In the simplest models they are constant, whereas $k(T)$ shows a very strong temperature dependence.

A3.4.5.2   Simple collision theories of bimolecular reactions

A bimolecular reaction can be regarded as a reactive collision with a reaction cross section σ that depends on the relative translational energy E_t of the reactant molecules A and B (masses m_A and m_B). The *specific rate*

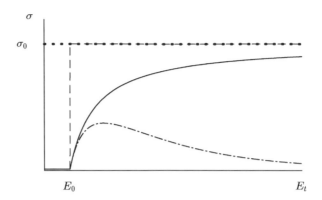

Figure A3.4.5. Simple models for effective collision cross sections σ: hard sphere without threshold (dotted line) hard sphere with threshold (dashed line) and hyperbolic threshold (full curve). E_t is the (translational) collision energy and E_0 is the threshold energy. σ_0 is the hard sphere collision cross section. The dashed–dotted curve is of the generalized type $\sigma_R(E_t > E_0) = \sigma_0(1 - E_0/E_t)\exp[(1 - E_t/E_0)/(a/E_0)]$ with the parameter $a = 3E_0$.

constant $k(E_t)$ can thus formally be written in terms of an effective reaction cross section σ, multiplied by the relative centre of mass velocity v_{rel}

$$k(E_t) = \sigma(E_t)v_{rel} = \sigma(E_t)\sqrt{2E_t/\mu}. \tag{A3.4.82}$$

Simple collision theories neglect the internal quantum state dependence of σ. The rate constant as a function of temperature T results as a thermal average over the Maxwell–Boltzmann velocity distribution $p(E_t)$:

$$k(T) = \int_0^\infty p(E_t)k(E_t)\,dE_t = \langle v_{rel}\rangle\langle\sigma\rangle. \tag{A3.4.83}$$

Here one has the thermal average centre of mass velocity

$$\langle v_{rel}\rangle = \sqrt{\frac{8k_B T}{\pi\mu}} \tag{A3.4.84}$$

and the thermally averaged reaction cross section

$$\langle\sigma\rangle \stackrel{\text{def}}{=} \int_0^\infty \left(\frac{E_t}{k_B T}\right)\sigma(E_t)\exp\left\{-\frac{E_t}{k_B T}\right\}\left(\frac{dE_t}{k_B T}\right). \tag{A3.4.85}$$

We use the symbol k_B for Boltzmann's constant to distinguish it from the rate constant k. Equation (A3.4.85) *defines* the thermal average reaction cross section $\langle\sigma\rangle$.

In principle, the reaction cross section not only depends on the relative translational energy, but also on individual reactant and product quantum states. Its sole dependence on E_t in the simplified effective expression (equation (A3.4.82)) already implies unspecified averages over reactant states and sums over product states. For practical purposes it is therefore appropriate to consider simplified models for the energy dependence of the effective reaction cross section. They often form the basis for the interpretation of the temperature dependence of thermal cross sections. Figure A3.4.5 illustrates several cross section models.

(a) Hard sphere collisions

The reactants are considered as hard spheres with radii r_A and r_B, respectively. A (reactive) collision occurs on contact yielding a constant cross section σ_0 independent of the energy:

$$\sigma_0 = \pi (r_A + r_B)^2 \tag{A3.4.86}$$

$$k(T) = \sigma_0 \sqrt{\frac{8k_B T}{\pi \mu}}. \tag{A3.4.87}$$

(b) Constant cross section with a threshold

The reaction can only occur once the collision energy reaches at least a value E_0. The reaction cross section remains constant above this threshold:

$$\sigma = \begin{cases} 0 & \text{for } E_t < E_0 \\ \sigma_0 & \text{for } E_t \geq E_0 \end{cases} \tag{A3.4.88}$$

$$k(T) = \sigma_0 \sqrt{\frac{8k_B T}{\pi \mu}} \left(1 + \frac{E_0}{k_B T}\right) \exp\left\{-\frac{E_0}{k_B T}\right\}. \tag{A3.4.89}$$

(c) Cross section with a hyperbolic threshold

Again, the reaction requires a minimum collision energy E_0, but increases only gradually above the threshold towards a finite, high-energy limit σ_0:

$$\sigma = \begin{cases} 0 & \text{for } E_t < E_0 \\ \sigma_0 \left(1 - \dfrac{E_0}{E_t}\right) & \text{for } E_t \geq E_0 \end{cases} \tag{A3.4.90}$$

$$k(T) = \sigma_0 \sqrt{\frac{8k_B T}{\pi \mu}} \exp\left\{-\frac{E_0}{k_B T}\right\}. \tag{A3.4.91}$$

(d) Generalized collision model

The hyperbolic cross section model can be generalized further by introducing a function $f(\Delta E)$ ($\Delta E = E_t - E_0$) to describe the reaction cross section above a threshold:

$$\sigma = \begin{cases} 0 & \text{for } E_t < E_0 \\ \sigma_0 \left(1 - \dfrac{E_0}{E_t}\right) f(\Delta E) & \text{for } E_t \geq E_0 \end{cases} \tag{A3.4.92}$$

$$k(T) = \sigma_0 \sqrt{\frac{8k_B T}{\pi \mu}} g(T) \exp\left\{-\frac{E_0}{k_B T}\right\} \tag{A3.4.93}$$

$$g(T) = \int_0^\infty \frac{\Delta E}{k_B T} f(\Delta E) \exp\left\{-\frac{\Delta E}{k_B T}\right\} d\left(\frac{\Delta E}{k_B T}\right). \tag{A3.4.94}$$

A3.4.6 Transition state theory

Transition state theory or 'activated complex theory' has been one of the most fruitful approximations in reaction kinetics and has had a long and complex history [42–44]. Transition state theory is originally based on the idea that reactant molecules have to pass a bottleneck to reach the product side and that they stay in quasi-equilibrium on the reactant side until this bottleneck is reached. The progress of a chemical reaction can often be described by the motion along a *reaction path* in the multidimensional molecular configuration space. Figure A3.4.6 shows typical potential energy profiles along such paths for uni- and bimolecular reactions. The effective potential energy $V(r_q)$ includes the zero point energy due to the motion orthogonal to the reaction coordinate r_q. The bottleneck is located at $r_q^{\neq}$, usually coinciding with an effective potential barrier, i.e. a first-order saddle point of the multidimensional potential hypersurface. Its height with respect to the reactants' zero point level is E_0. In its canonical form the transition state theory assumes a thermal equilibrium between the reactant molecules A and molecules X moving in some infinitesimal range δ over the barrier towards the product side. For the unimolecular case this yields the equilibrium concentration:

$$[X] = \frac{q_X}{q_A} \exp\{-E_0/k_B T\}[A] \tag{A3.4.95}$$

$$q_X = \frac{1}{2} q^{\neq} \delta \sqrt{2\pi \mu k_B T / h^2} \tag{A3.4.96}$$

where h is Planck's constant. q stands for molecular partition functions referred to the corresponding zero point level. Thus q_A is the partition function for the reactant A. $q^{\neq}$ is the restricted partition function for fixed reaction coordinate $r = r^{\neq}$ referring to the top of the effective (i.e. zero point corrected) barrier. It is often called the 'partition function of the transition state' bearing in mind that—in contrast to the X molecules— it does not correspond to any observable species. Rather, it defines the meaning of the purely technical term 'transition state'. Classically it corresponds to a $(3N - 7)$-dimensional hypersurface in the $(3N - 6)$-dimensional internal coordinate space of an N atomic system. The remainder of (A3.4.96) derives from the classical partition function for the motion in a one-dimensional box of length δ with an associated reduced mass μ. The factor of one half accounts for the fact that, in equilibrium, only half of the molecules located within $r^{\neq} \pm \delta/2$ move towards the product side.

Assuming a thermal one-dimensional velocity (Maxwell–Boltzmann) distribution with average velocity $\sqrt{2k_B T / \pi \mu}$ the reaction rate is given by the equilibrium flux if (1) the flux from the product side is neglected and (2) the thermal equilibrium is retained throughout the reaction:

$$-\frac{d[A]}{dt} = [X] \frac{\sqrt{2k_B T / \pi \mu}}{\delta}. \tag{A3.4.97}$$

Combining equations (A3.4.95)–(A3.4.97) one obtains the *first Eyring equation* for unimolecular rate constants:

$$k_{\text{uni}}(T) = \frac{k_B T}{h} \frac{q^{\neq}}{q_A} \exp\{-E_0/k_B T\}. \tag{A3.4.98}$$

A completely analogous derivation leads to the rate coefficient for bimolecular reactions, where $\tilde{q}$ are partition functions *per unit volume*:

$$k_{\text{bi}}(T) = \frac{k_B T}{h} \frac{\tilde{q}^{\neq}}{\tilde{q}_A \tilde{q}_B} \exp\{-E_0/k_B T\}. \tag{A3.4.99}$$

In the high barrier limit, $E_0 \gg k_B T$, E_0 is approximately equal to the Arrhenius activation energy. The ratio of the partition functions is sometimes called the 'statistical' or 'entropic' factor. Its product with the 'universal frequency factor' $k_B T / h$ corresponds approximately to Arrhenius' pre-exponential factor $A(T)$.

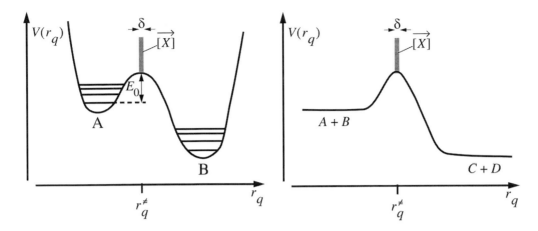

Figure A3.4.6. Potential energy along the reaction coordinate r_q for an unimolecular isomerization (left) and a bimolecular reaction (right). $r_q^{\neq}$ is the location of the transition state at the saddle point. [X] is the concentration of molecules located within δ of $r_q^{\neq}$ moving from the reactant side to the product side (indicated by the arrow, which is omitted in the text).

The quasi-equilibrium assumption in the above canonical form of the transition state theory usually gives an upper bound to the real rate constant. This is sometimes corrected for by multiplying (A3.4.98) and (A3.4.99) with a *transmission coefficient* $0 \leq \kappa \leq 1$.

In a *formal* analogy to the expressions for the thermodynamical quantities one can now *define* the standard enthalpy $\Delta^{\neq}H^{\ominus}$ and entropy $\Delta^{\neq}S^{\ominus}$ of activation. This leads to the *second Eyring equation*:

$$k(T) = \frac{k_B T}{h} \exp\{\Delta^{\neq}S^{\ominus}/R\} \exp\{-\Delta^{\neq}H^{\ominus}/RT\} \left(\frac{k_B T}{p^{\ominus}}\right)^j \qquad (A3.4.100)$$

where $p^{\ominus}$ is the standard pressure of the ideal gas ($j = 0$ for unimolecular and $j = 1$ for bimolecular reactions). As a definition (A3.4.100) is strictly identical to (A3.4.98) and (A3.4.99) if considered as a theoretical equation. Since neither $\Delta^{\neq}S^{\ominus}$ nor $\Delta^{\neq}H^{\ominus}$ are connected to observable species, equation (A3.4.100) may also be taken as an empirical equation, *viz.* an alternative representation of Arrhenius' equation (equation (A3.4.79)). In the field of thermochemical kinetics [43] one tries, however, to estimate $\Delta^{\neq}H^{\ominus}$ and $\Delta^{\neq}S^{\ominus}$ on the basis of molecular properties.

There is an immediate connection to the collision theory of bimolecular reactions. Introducing internal partition functions q_{int}, excluding the (separable) degrees of freedom for overall translation,

$$q = q_{int} q_{trans} \qquad (A3.4.101)$$

with

$$q_{trans} = V \left(\frac{2\pi M k_B T}{h^2}\right)^{3/2} \qquad (A3.4.102)$$

and comparing with equation (A3.4.83) the transition state theory expression for the effective thermal cross section of reaction becomes

$$\langle \sigma \rangle = \frac{h^2}{8\pi \mu_{AB} k_B T} \frac{q_{int}^{\neq}}{q_{int,A} q_{int,B}} \exp\{-E_0/k_B T\} \qquad (A3.4.103)$$

where V is the volume, $M = m_A + m_B$ is the total mass, and μ_{AB} is the reduced mass for the relative translation of A and B. One may interpret equation (A3.4.103) as the transition state version of the collision theory of bimolecular reactions: Transition state theory is used to calculate the thermally averaged reaction cross section to be inserted into equation (A3.4.83).

A3.4.7 Statistical theories beyond canonical transition state theory

Transition state theory may be embedded in the more general framework of statistical theories of chemical reactions, as summarized in figure A3.4.7 [27, 36]. Such theories have aimed at going beyond canonical transition state theory in several ways. The first extension concerns reaction systems with potential energy schemes depicted in figure A3.4.8 (in analogy to figure A3.4.6), where one cannot identify a saddle point on the potential hypersurface to be related to a transition state. The left-hand diagram corresponds to a complex forming bimolecular reaction, and the right-hand to a direct barrierless bimolecular reaction. The individual sections (the left- and right-hand parts) of the left diagram correspond to the two unimolecular dissociation channels for the intermediate characterized by the potential minimum. These unimolecular dissociation channels correspond to simple bond fissions. The general types of reactions shown in figure A3.4.8 are quite abundant in gas kinetics. Most ion molecule reactions as well as many radical–radical reactions are of this type. Thus, most of the very fast reactions in interstellar chemistry, atmospheric and combustion chemistry belong to this class of reaction, where standard canonical transition state theory cannot be applied and extension is clearly necessary. A second extension of interest would apply the fundamental ideas of transition state theory to state-selected reaction cross sections (see section A3.4.5.1). This theoretical program is carried out in the framework of phase space theory [45, 46] and of the statistical adiabatic channel model [27, 47], the latter being more general and containing phase space theory as a special case. In essence, the statistical adiabatic channel model is a completely state-selected version of the transition state theory. Here, the starting point is the **S**-matrix element (equation (A3.4.104)), which in the statistical limit takes the statistically averaged form

$$\langle |S_{fi}|^2 \rangle_{F,I,\Delta E} = \begin{cases} W(E, J)^{-1} & \text{for strongly coupled channels} \\ \delta_{f,i} & \text{for weakly coupled channels} \end{cases} \tag{A3.4.104}$$

where $W(E, J)$ is the total number of adiabatically open reaction channels for a given total angular momentum quantum number J (or any other good quantum number). $\langle \rangle_{F,I,\Delta E}$ refers to the averaging over groups of final and initial states ('coarse graining') and over suitably chosen collision energy intervals ΔE. Following the lines of the general theory of reaction cross sections, section A3.4.5.1, and starting from equation (A3.4.104) one can derive all the relevant kinetic specific reaction cross sections, specific rate constants and lifetimes in unimolecular reactions and the thermal rate constants analogous to transition state theory.

We summarize here only the main results of the theory and refer to a recent review [27] for details. The total number of adiabatically open channels is computed by searching for channel potential maxima $V_{a,\max}$. The channel potentials $V_a(r_q)$ are obtained by following the quantum energy levels of the reaction system along the reaction path r_q. An individual adiabatic channel connects an asymptotic scattering channel (corresponding to a reactant or to a product quantum level) with the reaction complex. One has the total number of open channels as a function of energy E, angular momentum J and other good quantum numbers:

$$W(E, J, \ldots) = \sum_{a(J\ldots)} h(E - V_{a,\max}). \tag{A3.4.105}$$

Here $h(x)$ is the Heaviside step function with $h(x > 0) = 1$ and $h(x \leq 0) = 0$ (not to be confused with Planck's constant). The limit $a(J \ldots)$ indicates that the summation is restricted to channel potentials with a given set of good quantum numbers $(J \ldots)$.

A state-to-state integral reaction cross section from reactant level a to product level b takes the form

$$\sigma_{ba} = \frac{\pi}{g_a k_a^2} \sum_{J=0}^{\infty} (2J + 1) \frac{W(E, J, a)W(E, J, b)}{W(E, J)}. \tag{A3.4.106}$$

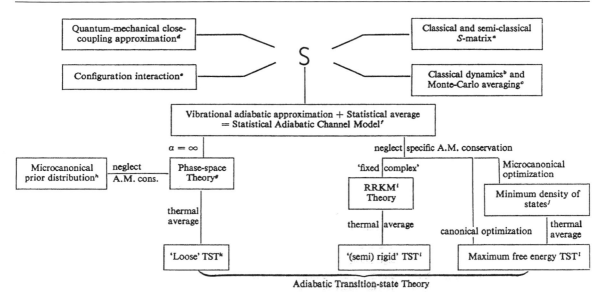

Figure A3.4.7. Summary of statistical theories of gas kinetics with emphasis on complex forming reactions (in the figure A.M. is the angular momentum, after Quack and Troe [27, 36, 74]). The indices refer to the following references: (*a*) [75–77]; (*b*) [78]; (*c*) [79–81]; (*d*) [82–85]; (*e*) [86–88]; (*f*) [36, 37, 47, 89, 90]; (*g*) [45, 46, 91]; (*h*) [92–95]; (*i*) [96–105]; (*j*) [106–109]; (*k*) [88, 94, 98, 99]; and (*l*) [94, 106–112].

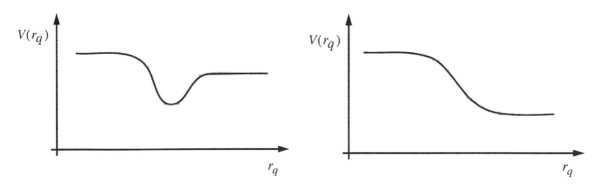

Figure A3.4.8. Potential energy profiles for reactions without barrier. Complex forming bimolecular reaction (left) and direct barrierless bimolecular reaction (right).

Here the levels consist of several states. g_a is the reactant level degeneracy and k_a is the collision wavenumber (see equation (A3.4.73)).

A specific unimolecular rate constant for the decay of a highly excited molecule at energy E and angular momentum J takes the form

$$k(E, J, \ldots) = \gamma \frac{W(E, J, \ldots)}{h\rho(E, J, \ldots)} \tag{A3.4.107}$$

where γ is a dimensionless transmission coefficient (usually $0 \leq \gamma \leq 1$) and $\rho(E, J, \ldots)$ is the density of molecular states. These expressions are relevant in the theory of thermal and non-thermal unimolecular reactions and are generalizations of the Rice–Ramsperger–Kassel–Marcus (RRKM) theory (see chapter A3.12).

Finally, the generalization of the partition function $q^{\neq}$ in transition state theory (equation (A3.4.96)) is given by

$$Q_{\text{int}}^* = \sum_a \exp(-V_{a,\text{max}}/k_B T) = \int_0^\infty W(E) \exp(-E/k_B T) \left(\frac{dE}{k_B T} \right) \qquad (A3.4.108)$$

with the total number of open channels

$$W(E) = \sum_a \sum_{J=0}^\infty (2J+1) W(E, J, a). \qquad (A3.4.109)$$

These equations lead to forms for the thermal rate constants that are perfectly similar to transition state theory, although the computations of the partition functions are different in detail. As described in figure A3.4.7 various levels of the theory can be derived by successive approximations in this general state-selected form of the transition state theory in the framework of the statistical adiabatic channel model. We refer to the literature cited in the diagram for details.

It may be useful to mention here one currently widely applied approximation for barrierless reactions, which is now frequently called microcanonical and canonical variational transition state theory (equivalent to the 'minimum density of states' and 'maximum free energy' transition state theory in figure A3.4.7). This type of theory can be understood by considering the partition functions $Q(r_q)$ as functions of r_q similar to equation (A3.4.108) but with $V_a(r_q)$ instead of $V_{a,\text{max}}$. Obviously $Q(r_q) \geq Q^*$ so that the best possible choice for a transition state results from minimizing the partition function along the reaction coordinate r_q:

$$Q^{\neq}(T) = \min_{r_q} Q(r_q, T) = Q(r_q^{\neq}, T). \qquad (A3.4.110)$$

Equation (A3.4.110) represents the canonical form ($T = $ constant) of the 'variational' theory. Minimization at constant energy yields the analogous microcanonical version. It is clear that, in general, this is only an approximation to the general theory, although this point has sometimes been overlooked. One may also define a free energy

$$A(r_q) = -k_B T \ln Q(r_q) \qquad (A3.4.111)$$

which leads to a maximum free energy condition

$$A^{\neq}(T) = \max_{r_q} A(r_q, T) = A(r_q^{\neq}, T). \qquad (A3.4.112)$$

The free energy as a function of reaction coordinates has been explicitly represented by Quack and Troe [36, 112] for the reaction

$$C_2H_6 \longrightarrow 2CH_3 \qquad (A3.4.113)$$

but the general concept goes back to Eyring (see [27, 36]).

A3.4.8 Gas-phase reaction mechanisms

The kinetics of a system of elementary reactions forming a reaction mechanism are described by a system of coupled differential equations. Disregarding transport processes there is one differential equation for each species involved. Few examples for these systems of coupled differential equations can be solved exactly in closed form. The accurate solution more generally requires integration by numerical methods. In the simplest case of reversible elementary reactions the stoichiometry is sufficient to decouple the differential equations leading to simple rate laws. For more complicated compound reaction mechanisms this can only be achieved with more or less far reaching approximations, usually concerning reactive intermediates. The most important are *quasi-equilibrium* (or partial equilibrium) and the *quasi-stationarity* (or quasi-steady-state), whose practical importance goes far beyond gas-phase kinetics.

A3.4.8.1   Elementary reactions with back-reaction

The simplest possible gas-phase reaction mechanisms consist of an elementary reaction and its back reaction. Here we consider uni- and bimolecular reactions yielding three different combinations. The resulting rate laws can all be integrated in closed form.

(a) Unimolecular reactions with unimolecular back reaction

The equation

$$A \underset{k_{-1}}{\overset{k_1}{\rightleftharpoons}} C$$

is the elementary mechanism of reversible isomerization reactions, for example

$$CH_3NC \underset{[M]}{\overset{[M]}{\rightleftharpoons}} CH_3CN. \tag{A3.4.114}$$

The rate law is given by

$$-\frac{dc_A}{dt} = k_1 c_A - k_{-1} c_C. \tag{A3.4.115}$$

Exploiting the stoichiometric equation one can eliminate c_C. Integration yields the simple relaxation of the initial concentrations into the equilibrium, $c_A^{eq} = c_A(\infty)$, with a relaxation time τ:

$$c_A(t) - c_A^{eq} = (c_A(t) - c_A^{eq}) \exp\{-t/\tau\} \tag{A3.4.116}$$

$$\tau = \frac{1}{k_1 + k_{-1}}. \tag{A3.4.117}$$

(b) Bimolecular reactions with unimolecular back reaction

For example

$$2A \underset{k_{-1}}{\overset{k_2}{\rightleftharpoons}} A_2.$$

The rate law is given by

$$-\frac{1}{2}\frac{dc_A}{dt} = k_2 c_A^2 - k_{-1} c_{A_2}. \tag{A3.4.118}$$

After transformation to the turnover variable $x = (c_A(0) - c_A(t))/2 = c_{A_2}(t) - c_{A_2}(0)$ integration yields

$$\ln\left(\frac{x + x_e - c_A(0) - K/4}{x - x_e}\right) - \ln\left(\frac{c_A(0) + K/4 - x_e}{x_e}\right) = k_2(4c_A(0) + K - 8x_e)t \tag{A3.4.119}$$

$$x_e = \frac{c_A(0)}{2} + \frac{K}{8} - \left(\frac{K}{4}\left(c_{A_2}(0) + \frac{c_A(0)}{2}\right) + \left(\frac{K}{8}\right)^2\right)^{1/2} \tag{A3.4.120}$$

where $K = k_2/k_{-1}$ is the equilibrium constant.

(c) Bimolecular reactions with bimolecular back-reaction

For example

$$A + B \underset{k_{-2}}{\overset{k_2}{\rightleftarrows}} C + D.$$

The rate law is given by

$$-\frac{dc_A}{dt} = k_2 c_A c_B - k_{-2} c_C c_D \tag{A3.4.121}$$

After transformation to the turnover variable $x = c_A(0) - c_A(t)$, integration yields

$$\ln\left(\frac{1 - [x/(a + b)]}{1 - [x/(a - b)]}\right) = 2k_2(1 - K^{-1})bt \tag{A3.4.122}$$

$$a = \frac{c_A(0) + c_B(0) + K^{-1}[c_C(0) + c_D(0)]}{2(1 - K^{-1})} \tag{A3.4.123}$$

$$b = \left(a^2 - \frac{c_A(0)c_B(0) - K^{-1}c_C(0)c_D(0)}{1 - K^{-1}}\right)^{1/2} \tag{A3.4.124}$$

where $K = k_2/k_{-2}$ is the equilibrium constant.

Bimolecular steps involving identical species yield correspondingly simpler expressions.

A3.4.8.2   *The Lindemann–Hinshelwood mechanism for unimolecular reactions*

The system of coupled differential equations that result from a compound reaction mechanism consists of several different (reversible) elementary steps. The kinetics are described by a system of coupled differential equations rather than a single rate law. This system can sometimes be decoupled by assuming that the concentrations of the intermediate species are small and quasi-stationary. The *Lindemann mechanism of thermal unimolecular reactions* [18, 19] affords an instructive example for the application of such approximations. This mechanism is based on the idea that a molecule A has to pick up sufficient energy before it can undergo a monomolecular reaction, for example, bond breaking or isomerization. In thermal reactions this energy is provided by collisions with other molecules M in the gas to produce excited species A*:

$$A + M \underset{k_d}{\overset{k_a}{\rightleftarrows}} A^* + M \tag{A3.4.125}$$

$$A^* \overset{k_r}{\rightarrow} \text{products.} \tag{A3.4.126}$$

Two important points must be noted here.

(1)   The collision partners may be any molecule present in the reaction mixture, i.e., inert bath gas molecules, but also reactant or product species. The activation (k_a) and deactivation (k_d) rate constants in equation (A3.4.125) therefore represent the effective average rate constants.

(2)   The collision (k_a, k_d) and reaction (k_r) efficiencies may significantly differ between different excited reactant states. This is essentially neglected in the Lindemann–Hinshelwood mechanism. In particular, the *strong collision* assumption implies that so much energy is transferred in a collision that the collision efficiency can be regarded as effectively independent of the energy.

With $k_1 = k_a[M]$ and $k_{-1} = k_d[M]$ the resulting system of differential equations is

$$-\frac{d[A]}{dt} = k_1[A] - k_{-1}[A^*] \tag{A3.4.127}$$

$$-\frac{d[A^*]}{dt} = -k_1[A] + k_r[A^*] + k_{-1}[A^*] \tag{A3.4.128}$$

$$\frac{d[\text{products}]}{dt} = k_r[A^*]. \tag{A3.4.129}$$

If the excitation energy required to form activated species A^* is much larger than $k_B T$ its concentration will remain small. This is fulfilled if $k_a \ll k_d$. Following Bodenstein, $[A^*]$ is then assumed to be quasi-stationary, i.e. after some initialization phase the concentration of activated species remains approximately constant (strictly speaking the ratio $[A^*]/[A]$ remains approximately constant (see section A3.4.8.3)):

$$-\frac{d[A^*]}{dt} \approx 0 \tag{A3.4.130}$$

$$\Rightarrow \quad [A^*]_{QS} = \frac{k_1[A]}{k_{-1} + k_r} \tag{A3.4.131}$$

This yields the quasi-stationary reaction rate with an effective unimolecular rate constant

$$v_c = \frac{d[\text{products}]}{dt} = k_{\text{eff}}[A] = k_r[A^*]_{QS} \tag{A3.4.132}$$

$$k_{\text{eff}} = \frac{k_1 k_r}{k_{-1} + k_r} = \frac{k_a[M]k_r}{k_d[M] + k_r}. \tag{A3.4.133}$$

The effective rate law correctly describes the pressure dependence of unimolecular reaction rates at least qualitatively. This is illustrated in figure A3.4.9. In the limit of high pressures, i.e. large $[M]$, k_{eff} becomes independent of $[M]$ yielding the high-pressure rate constant k_∞ of an effective first-order rate law. At very low pressures, product formation becomes much faster than deactivation. k_{eff} now depends linearly on $[M]$. This corresponds to an effective second-order rate law with the pseudo first-order rate constant k_0:

$$k_\infty = \frac{k_a}{k_d} k_r \tag{A3.4.134}$$

$$k_0 = k_a[M]. \tag{A3.4.135}$$

In addition to $[A^*]$ being quasi-stationary the *quasi-equilibrium* approximation assumes a virtually unperturbed equilibrium between activation and deactivation (equation (A3.4.125)):

$$\frac{[A^*]}{[A]} = \frac{k_a}{k_d}. \tag{A3.4.136}$$

This approximation is generally valid if $k_r \ll k_{-1}$. For the Lindemann mechanism of unimolecular reactions this corresponds to the high-pressure limit $k_{\text{eff}} = k_\infty$.

The approximate results can be compared with the long time limit of the exact stationary state solution derived in section A3.4.8.3:

$$k_{\text{eff}} = \tfrac{1}{2}\left\{ k_1 + k_{-1} + k_r - [(k_1 + k_{-1} + k_r)^2 - 4k_1 k_r]^{1/2} \right\}. \tag{A3.4.137}$$

This leads to the quasi-stationary rate constant of equation (A3.4.133) if $4k_1 k_r \ll (k_1 + k_{-1} + k_r)^2$, which is more general than the Bodenstein condition $k_a \ll k_d$.

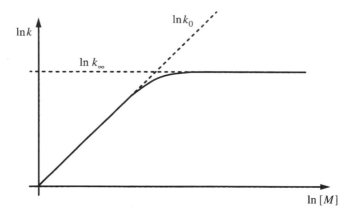

Figure A3.4.9. Pressure dependence of the effective unimolecular rate constant. Schematic fall-off curve for the Linde-mann–Hinshelwood mechanism. k_∞ is the (constant) high-pressure limit of the effective rate constant k_{eff} and k_0 is the low-pressure limit, which depends linearly on the concentration of the inert collision partner [M].

A3.4.8.3   Generalized first-order kinetics

The Lindemann mechanism for thermally activated unimolecular reactions is a simple example of a particular class of compound reaction mechanisms. They are mechanisms whose constituent reactions individually follow first-order rate laws [11, 20, 36, 48–56]:

$$A_i \overset{K_{ji}}{\rightarrow} A_j \qquad i, j = 1, \dots, N \tag{A3.4.138}$$

where N is the number of different species involved. With $c_i = [A_i]$ this leads to the following system of N coupled differential equations called *generalized first-order kinetics*:

$$-\frac{dc_i}{dt} = \sum_{j=1}^{N} K_{ij} c_j \qquad i = 1, \dots, N. \tag{A3.4.139}$$

The individual reactions need not be unimolecular. It can be shown that the relaxation kinetics after small perturbations of the equilibrium can always be reduced to the form of (A3.4.138) in terms of extension variables from equilibrium, even if the underlying reaction system is not of first order [51, 52, 57, 58].

Generalized first-order kinetics have been extensively reviewed in relation to technical chemical appli-cations [59] and have been discussed in the context of copolymerization [53]. From a theoretical point of view, the general class of coupled kinetic equations (A3.4.138) and (A3.4.139) is important, because it allows for a general closed-form solution (in matrix form) [49]. Important applications include the Pauli master equation for statistical mechanical systems (in particular gas-phase statistical mechanical kinetics) [48] and the investigation of certain simple reaction systems [49, 50, 55]. It is the basis of the many-level treatment of thermal unimolecular reactions in terms of the appropriate master equations for energy transfer [36, 55, 60–63]. Generalized first-order kinetics also form the basis for certain statistical limiting cases of multiphoton induced chemical reactions and laser chemistry [54, 56].

Written in matrix notation, the system of first-order differential equations, (A3.4.139) takes the form

$$-\frac{d\mathbf{c}(t)}{dt} = \mathbf{K}\mathbf{c}(t) \tag{A3.4.140}$$

With time independent matrix $\mathbf{K}$ it has the general solution

$$\mathbf{c}(t) = \exp\{-\mathbf{K}t\}\mathbf{c}(0). \tag{A3.4.141}$$

The exponential function of the matrix can be evaluated through the power series expansion of exp(). $\mathbf{c}$ is the column vector whose elements are the concentrations c_i. The matrix elements of the *rate coefficient matrix* $\mathbf{K}$ are the first-order rate constants K_{ij}. The system is called *closed* if all reactions and back reactions are included. Then $\mathbf{K}$ is of rank $N - 1$ with positive eigenvalues, of which exactly one is zero. It corresponds to the equilibrium state, with concentrations c_i^{eq} determined by the principle of microscopic reversibility:

$$\frac{K_{ij}}{K_{ji}} = \frac{c_i^{eq}}{c_j^{eq}}. \tag{A3.4.142}$$

In this case $\mathbf{K}$ is similar to a real symmetric matrix and equation (A3.4.141) can easily be solved by diagonalization of $\mathbf{K}$.

If some of the reactions of (A3.4.138) are neglected in (A3.4.139), the system is called open. This generally complicates the solution of (A3.4.141). In particular, the system no longer has a well defined equilibrium. However, as long as the eigenvalues of $\mathbf{K}$ remain positive, the kinetics at long times will be dominated by the smallest eigenvalue. This corresponds to a stationary state solution.

As an example we take again the Lindemann mechanism of unimolecular reactions. The system of differential equations is given by equations (A3.4.127)–(A3.4.129). The rate coefficient matrix is

$$\mathbf{K} = \begin{pmatrix} k_1 & -k_{-1} & \vdots & 0 \\ -k_1 & k_{-1} + k_r & \vdots & 0 \\ \cdots\cdots\cdots\cdots\cdots\cdots\cdots & & \\ 0 & -k_r & & 0 \end{pmatrix}. \tag{A3.4.143}$$

Since the back reaction, products $\to$ A*, has been neglected this is an open system. Still $\mathbf{K}$ has a trivial zero eigenvalue corresponding to complete reaction, i.e. pure products. Therefore we only need to consider (A3.4.127) and (A3.4.128) and the corresponding (2×2) submatrix indicated in equation (A3.4.143).

The eigenvalues $\lambda_1 < \lambda_2$ of $\mathbf{K}$ are both positive

$$\lambda_{1,2} = \tfrac{1}{2} \left\{ k_1 + k_{-1} + k_r \pm [(k_1 + k_{-1} + k_r)^2 - 4k_1 k_r]^{1/2} \right\} > 0. \tag{A3.4.144}$$

For long times, the smaller eigenvalue λ_1 will dominate (A3.4.141), yielding the stationary solution

$$\begin{pmatrix} c_A(t) \\ c_{A^*}(t) \end{pmatrix} = \exp\{-\lambda_1 t\} \begin{pmatrix} a \\ b \end{pmatrix} \tag{A3.4.145}$$

where a and b are time-independent functions of the initial concentrations. With the condition $\lambda_1 \ll \lambda_2$ one obtains the effective unimolecular rate constant

$$k_{eff} = -\frac{d \ln(c_A + c_{A^*})}{dt} = \lambda_1 \overset{\lambda_1 \ll \lambda_2}{\approx} \frac{k_1 k_r}{k_1 + k_{-1} + k_r}. \tag{A3.4.146}$$

For $k_a \ll k_d$ this is identical to the quasi-stationary result, equation (A3.4.133), although only the ratio [A*]/[A] = b/a (equation (A3.4.145)) is stationary and not [A*] itself. This suggests d[A*]/dt $\ll$ d[A]/dt as a more appropriate formulation of quasi-stationarity. Furthermore, the general stationary state solution (equation (A3.4.144)) for the Lindemann mechanism contains cases that are not usually retained in the Bodenstein quasi-steady-state solution.

An important example for the application of general first-order kinetics in gas-phase reactions is the master equation treatment of the fall-off range of thermal unimolecular reactions to describe non-equilibrium effects in the weak collision limit when activation and deactivation cross sections (equation (A3.4.125)) are to be retained in detail [60].

General first-order kinetics also play an important role for the so-called *local eigenvalue analysis* of more complicated reaction mechanisms, which are usually described by nonlinear systems of differential equations. Linearization leads to effective general first-order kinetics whose analysis reveals information on the time scales of chemical reactions, species in steady states (quasi-stationarity), or partial equilibria (quasi-equilibrium) [64–66].

A3.4.8.4 General compound reaction mechanisms

More general compound reaction mechanisms lead to systems of differential equations of different orders. They can sometimes be treated by applying a quasi-stationarity or a quasi-equilibrium approximation. Often, this may even work for simple chain reactions. Chain reactions generally consist of four types of reaction steps: In the *chain initiation* steps, reactive species (radicals) are produced from stable species (reactants or catalysts). They react with stable species to form other reactive species in the *chain propagation*. Reactive species recovered in the chain propagation steps are called *chain carriers*. Propagation steps where one reactive species is replaced by another less-reactive species are sometimes termed *inhibiting*. *Chain branching* occurs if more than one reactive species are formed. Finally, the chain is *terminated* by reactions of reactive species, which yield stable species, for example through recombination in the gas phase or at the surface of the reaction vessel.

The assumption of quasi-stationarity can sometimes be justified if there is no significant chain branching, for example in HBr formation at 200–300 °C:

$$
\begin{array}{llll}
(1) & \text{initialization} & \text{Br}_2 \xrightarrow{[M]} 2\text{Br}^* \\
(2) & \text{propagation} & \text{Br}^* + \text{H}_2 \rightarrow \text{HBr} + \text{H}^* \\
(3) & & \text{H}^* + \text{Br}_2 \rightarrow \text{HBr} + \text{Br}^* \\
(4) & \text{inhibition} & \text{H}^* + \text{HBr} \rightarrow \text{H}_2 + \text{Br}^* \\
(5) & \text{termination} & \text{Br}^* + \text{Br}^* \xrightarrow{[M]} \text{Br}_2 \\
(6) & & \text{H}^* + \text{H}^* \xrightarrow{[M]} \text{H}_2 \\
(7) & & \text{H}^* + \text{Br}^* \xrightarrow{[M]} \text{HBr}.
\end{array}
\qquad (\text{A3.4.147})
$$

Chain carriers are indicated by an asterisk. Assuming quasi-stationarity for $[\text{H}^*]$ and $[\text{Br}^*]$ and neglecting (6) and (7) (because $[\text{H}^*] \ll [\text{Br}^*]$) yields

$$
\frac{d[\text{HBr}]}{dt} = \frac{2k_2(k_1/k_5)^{1/2}[\text{H}_2][\text{Br}_2]^{1/2}}{1 + (k_4[\text{HBr}]/k_3[\text{Br}_2])}.
\qquad (\text{A3.4.148})
$$

The resulting rate law agrees with the form found experimentally. Of course the postulated mechanism can only be proven by measuring the rate constants of the individual elementary steps separately and comparing calculated rates of equation (A3.4.148) with observed rates of HBr formation.

In general, the assumption of quasi-stationarity is difficult to justify *a priori*. There may be several possible choices of intermediates for which the assumption of quasi-stationary concentrations appears justified. It is possible to check for consistency, for example $[\text{A}^*]_{QS} \ll [\text{A}]$, but the final justification can only come from a comparison with the exact solution. These have usually to be obtained with numerical solvers for systems of differential equations [7–9]. In particular, if transport phenomena with complicated boundary conditions must be taken into account this is the only viable solution. Modern fields of application include atmospheric chemistry and combustion chemistry [67, 68]. A classic example is the H_2/O_2 reaction. The mechanism includes more than 60 elementary steps and has been discussed in great detail [69]. A recent analysis of the explosion limits of this system in the range of 0.3–15.7 atm and 850–1040 K included 19 reversible elementary

Table A3.4.3. Rate constants for the reaction of H_2 with O_2 [73]. The rate constants are given in terms of the following expression: $k(T) = A(T/K)^b \exp(-E/RT)$.

Reaction	A $(cm^3 \, mol^{-1} \, s^{-1})$	b	E $(kJ \, mol^{-1})$
$OH + H_2 \rightarrow H_2O + H$	1.2×10^9	1.3	15.2
$H + H_2O \rightarrow OH + H_2$	4.5×10^9	1.3	78.7
$H + O_2 \rightarrow OH + O$	2.2×10^{14}	0	70.4
$OH + O \rightarrow H + O_2$	1.0×10^{13}	0	0
$O + H_2 \rightarrow OH + H$	1.8×10^{10}	1.0	37.3
$OH + H \rightarrow O + H_2$	8.3×10^9	1.0	29.1
$OH + OH \rightarrow H_2O + O$	1.5×10^9	1.14	0
$O + H_2O \rightarrow OH + OH$	1.6×10^{10}	1.14	72.4
$H + HO_2 \rightarrow OH + OH$	1.5×10^{14}	0	4.2
$H + HO_2 \rightarrow H_2 + O_2$	2.5×10^{13}	0	2.9
$OH + HO_2 \rightarrow H_2O + O$	1.5×10^{13}	0	0
$O + HO_2 \rightarrow OH + O_2$	2.0×10^{13}	0	0
	$cm^6 \, mol^{-2} \, s^{-1}$		
$H + H + M \rightarrow H_2 + M$	9.0×10^{16}	−0.6	0
$O + OH + M \rightarrow HO_2 + M$	2.2×10^{22}	−2.0	0
$H + O_2 + M \rightarrow HO_2 + M$	2.3×10^{18}	−0.8	0

reactions [67]. Table A3.4.3 summarizes some of the major reactions for the hydrogen–oxygen reaction. A simplified mechanism involves only six reactions:

$$
\begin{array}{llll}
(1) & \text{initiation} & H_2 + O_2 & \rightarrow & 2OH \\
(2) & \text{propagation} & OH + H_2 & \rightarrow & H_2O + H \\
(3) & \text{branching} & H + O_2 & \rightarrow & OH + O \\
(4) & \text{branching} & O + H_2 & \rightarrow & OH + H \\
(5) & \text{termination} & 2H & \xrightarrow{[M],\text{wall}} & H_2 \\
(6) & \text{termination} & H + O_2 & \xrightarrow{[M]} & HO_2
\end{array}
$$
(A3.4.149)

$$3H_2 + O_2 = 2H + 2H_2O \ (2 + 2 + 3 + 4).$$

Reaction (5) proceeds mostly heterogeneously, reaction (6) mostly homogeneously. This mechanism can be integrated with simplifying assumptions to demonstrate the main features of gas-phase explosion kinetics [8]. The importance of numerical treatments, however, cannot be overemphasized in this context. Over the decades enormous progress has been made in the numerical treatment of differential equations of complex gas-phase reactions [8, 70, 71]. Complex reaction systems can also be seen in the context of nonlinear and self-organizing reactions, which are separate subjects in this encyclopedia (see chapters A3.14, C3.6).

A3.4.9 Summarizing overview

Although the field of gas-phase kinetics remains full of challenges it has reached a certain degree of maturity. Many of the fundamental concepts of kinetics, in general take a particularly clear and rigorous form in gas-phase kinetics. The relation between fundamental quantum dynamical theory, empirical kinetic treatments, and experimental measurements, for example of combustion processes [72], is most clearly established in gas-phase kinetics. It is the aim of this article to review some of these most basic aspects. Details can be found in the sections on applications as well as in the literature cited.

References

[1] Molina M J and Rowland F S 1974 *Nature* **249** 810

[2] Barker J R (ed) 1995 *Progress and Problems in Atmospheric Chemistry (Advanced Series in Physical Chemistry)* vol 3 (Singapore: World Scientific)

[3] Crutzen P J 1995 Overview of tropospheric chemistry: developments during the past quarter century and a look ahead *Faraday Discuss.* **100** 1–21

[4] Molina M J, Molina L T and Golden D M 1996 Environmental chemistry (gas and gas–solid interactions): the role of physical chemistry *J. Phys. Chem.* **100** 12 888

[5] Crutzen P J 1996 Mein leben mit O_3, NO_x, etc. *Angew. Chem.* **108** 1878–98 (*Angew. Chem. Int. Ed. Engl.* **35** 1758–77)

[6] Herbst E 1987 Gas phase chemical processes in molecular clouds *Interstellar Processes* ed D J Hollenbach and H A Tronson (Dordrecht: Reidel) pp 611–29

[7] Gardiner W C Jr (ed) 1984 *Combustion Chemistry* (New York: Springer)

[8] Warnatz J, Maas U and Dibble R W 1999 *Combustion: Physical and Chemical Fundamentals, Modelling and Simulation, Experiments, Polutant Formation* 2nd edn (Heidelberg: Springer)

[9] Gardiner W C Jr (ed) 2000 *Gas-Phase Combustion Chemistry* 2nd edn (Heidelberg: Springer)

[10] Mills I, Cvitaš T, Homann K, Kallay N and Kuchitsu K 1993 *Quantities, Units and Symbols in Physical Chemistry* 2nd edn (Oxford: Blackwell) (3rd edn in preparation)

[11] Quack M 1984 On the mechanism of reversible unimolecular reactions and the canonical ('high pressure') limit of the rate coefficient at low pressures *Ber. Bunsenges. Phys. Chem.* **88** 94–100

[12] Herzberg G 1989 *Molecular Spectra and Molecular Structure. I. Spectra of Diatomic Molecules* (Malabar, FL: Krieger)

[13] Ackermann M and Biaume F 1979 *J. Mol. Spectrosc.* **35** 73

[14] Cheung A S C, Yoshino K, Freeman D E, Friedman R S, Dalgarno A and Parkinson W H 1989 The Schumann–Runge absorption-bands of $^{16}O^{18}O$ in the wavelength region 175–205 nm and spectroscopic constants of isotopic oxygen molecules *J. Mol. Spectrosc.* **134** 362–89

[15] Pine A S, Lafferty W J and Howard B J 1984 Vibrational predissociation, tunneling, and rotational saturation in the HF and DF dimers *J. Chem. Phys.* **81** 2939–50

[16] Quack M and Suhm M A 1998 Spectroscopy and quantum dynamics of hydrogen fluoride clusters *Adv. in Mol. Vibr. Coll. Dyn.* vol 3 (JAI) pp 205–48

[17] He Y, Müller H B, Quack M and Suhm M A 2000 *J. Chem. Phys.*

[18] Lindemann F A 1922 Discussion on 'the radiation theory of chemical reactions' *Trans. Faraday Soc.* **17** 598–9

[19] Hinshelwood C N 1933 *The Kinetics of Chemical Change in Gaseous Systems* 3rd edn (Oxford: Clarendon)

[20] Quack M and Jans-Bürli S 1986 *Molekulare Thermodynamik und Kinetik. Teil 1: Chemische Reaktionskinetik* (Zürich: Fachvereine) (New English edition in preparation by D Luckhaus and M Quack)

[21] Rosker M J, Rose T S and Zewail A 1988 Femtosecond real-time dynamics of photofragment–trapping resonances on dissociative potential-energy surfaces *Chem. Phys. Lett.* **146** 175–9

[22] van den Bergh H E, Callear A B and Norström R J 1969 *Chem. Phys. Lett.* **4** 101–2

[23] Callear A B and Metcalfe M P 1976 *Chem. Phys.* **14** 275

[24] Glänzer K, Quack M and Troe J 1977 High temperature UV absorption and recombination of methyl radicals in shock waves *16th Int. Symp. on Combustion* (Pittsburgh: The Combustion Institute) pp 949–60

[25] Hippler H, Luther K and Troe J 1974 On the role of complexes in the recombination of halogen atoms *Ber. Bunsenges. Phys. Chem.* **78** 178–9

[26] van den Bergh H and Troe J 1975 NO-catalyzed recombination of iodine atoms. Elementary steps of the complex mechanism *Chem. Phys. Lett.* **31** 351–4

[27] Quack M and Troe J 1998 Statistical adiabatic channel models *Encyclopedia of Computational Chemistry* ed P v R Schleyer *et al* (New York: Wiley) pp 2708–26

[28] Faubel M and Toennies J P 1978 *Adv. Atom. Mol. Phys.* **13** 229

[29] Zare R N 1979 Kinetics of state selected species *Faraday Discuss. Chem. Soc.* **67** 7–15

[30] Bernstein R B and Zare R N 1980 State to state reaction dynamics *Phys. Today* **33** 43

[31] Bernstein R B (ed) 1982 *Chemical Dynamics via Molecular Beam and Laser Techniques (The Hinshelwood Lectures, Oxford, 1980)* (Oxford: Oxford University Press)

[32] Faubel M 1983 *Adv. Atom. Mol. Phys.* **19** 345

[33] Polanyi J C 1987 *Science* **236** 680

[34] Lee Y T 1987 *Science* **236** 793

[35] Herschbach D R 1987 *Angew. Chem. Int. Ed. Engl.* **26** 1221

[36] Quack M and Troe J 1977 Unimolecular reactions and energy transfer of highly excited molecules *Gas Kinetics and Energy Transfer* vol 2 (London: The Chemical Society)

[37] Quack M and Troe J 1981 Statistical methods in scattering *Theoretical Chemistry: Advances and Perspectives* vol 6B (New York: Academic) pp 199–276

[38] Truhlar D G, Garrett B C and Klippenstein S J 1996 Current status of transition-state theory *J. Phys. Chem.* **100** 12 771–800

[39] Arrhenius S 1889 Über die Reaktionsgeschwindigkeit bei der Inversion von Rohrzucker durch Säuren *Z. Physik. Chem.* **4** 226–48
[40] Arrhenius S 1899 Zur Theorie der chemischen Reaktionsgeschwindigkeiten *Z. Physik. Chem.* **28** 7–35
[41] van't Hoff J H 1884 *Études de Dynamique Chimique* (Amsterdam: Müller)
[42] Denbigh K 1981 *Principles of Chemical Equilibrium* 4th edn (London: Cambridge University Press)
[43] Benson S W 1976 *Thermochemical Kinetics* 2nd edn (New York: Wiley)
[44] Laidler K J and King M C 1983 The development of transition state theory *J. Phys. Chem.* **87** 2657–64
[45] Pechukas P and Light J C 1965 *J. Chem. Phys.* **42** 3281–91
[46] Nikitin E E 1965 *Theor. Exp. Chem.* **1** 144 (Engl. Transl.)
[47] Quack M and Troe J 1974 Specific rate constants of unimolecular processes ii. Adiabatic channel model *Ber. Bunsenges. Phys. Chem.* **78** 240–52
[48] Pauli W Jr 1928 Über das H–Theorem vom Anwachsen der Entropie vom Standpunkt der neuen Quantenmechanik *Probleme der modernen Physik* ed P Debye (Leipzig: Hirzel) pp 30–45
[49] Jost W 1947 *Z. Naturf.* a **2** 159
[50] Jost W 1950 *Z. Phys. Chem.* **195** 317
[51] Eigen M 1954 *Faraday Discuss. Chem. Soc.* **17** 194
[52] Eigen M and Schoen J 1955 *Ber. Bunsenges. Phys. Chem.* **59** 483
[53] Horn F 1971 General first order kinetics *Ber. Bunsenges. Phys. Chem.* **75** 1191–201
[54] Quack M 1979 Master equations for photochemistry with intense infrared light *Ber. Bunsenges. Phys. Chem.* **83** 757–75
[55] Quack M and Troe J 1981 Current aspects of unimolecular reactions *Int. Rev. Phys. Chem.* **1** 97–147
[56] Quack M 1998 Multiphoton excitation *Encyclopedia of Computational Chemistry* vol 3, ed P v R Schleyer *et al* (New York: Wiley) pp 1775–91
[57] Castellan G W 1963 *Ber. Bunsenges. Phys. Chem.* **67** 898
[58] Bernasconi C F (ed) 1976 *Relaxation Kinetics* (New York: Academic)
[59] Wei J and Prater C D 1962 The structure and analysis of complex reaction systems *Advances in Catalysis* (New York: Academic) pp 203–392
[60] Gilbert R G, Luther K and Troe J 1983 Theory of thermal unimolecular reactions in the fall-off range. II. Weak collision rate constants *Ber. Bunsenges. Phys. Chem.* **87** 169–77
[61] Pilling M J 1996 *Ann. Rev. Phys. Chem.* **47** 81
[62] Venkatesh P K, Dean A M, Cohen M H and Carr R W 1997 *J. Chem. Phys.* **107** 8904
[63] Venkatesh P K, Dean A M, Cohen M H and Carr R W 1999 *J. Chem. Phys.* **111** 8313–29
[64] Lam S H and Goussis D A 1988 Understanding complex chemical kinetics with computational singular perturbation *22nd Int. Symp. on Combustion* ed M C Salamony (Pittsburgh, PA: The Combustion Institute) pp 931–41
[65] Maas U and Pope S B 1992 Simplifying chemical kinetics: intrinsic low-dimensional manifolds in composition space *Comb. Flame* **88** 239
[66] Warnatz J, Maas U and Dibble R W 1999 *Combustion: Physical and Chemical Fundamentals, Modelling and Simulation, Experiments, Polutant Formation* (Heidelberg: Springer)
[67] Mueller M A, Yetter R A and Dryer F L 1999 Flow reactor studies and kinetic modelling of the $H_2/O_2/NO_x$ reaction *Int. J. Chem. Kinet.* **31** 113–25
[68] Mueller M A, Yetter R A and Dryer D L 1999 Flow reactor studies and kinetic modelling of the $H_2/O_2/NO_x$ and $CO/H_2O/O_2/NO_x$ reactions *Int. J. Chem. Kinet.* **31** 705–24
[69] Dougherty E P and Rabitz H 1980 Computational kinetics and sensitivity analysis of hydrogen–oxygen combustion *J. Chem. Phys.* **72** 6571
[70] Gear C W 1971 *Numerical Initial Value Problems in Ordinary Differential Equations* (Englewood Cliffs, NJ: Prentice-Hall)
[71] Deuflhard P and Wulkow M 1989 Computational treatment of polyreaction kinetics by orthogonal polynomials of a discrete variable *Impact of Computing in Science and Engineering* vol 1
[72] Ebert V, Schulz C, Volpp H R, Wolfrum J and Monkhouse P 1999 Laser diagnostics of combustion processes: from chemical dynamics to technical devices *Israel J. Chem.* **39** 1–24
[73] Warnatz J 1979 *Ber. Bunsenges. Phys. Chem.* **83** 950
[74] Quack M 1977 Detailed symmetry selection rules for reactive collisions *Mol. Phys.* **34** 477–504
[75] Miller W H 1970 *J. Chem. Phys.* **53** 1949
[76] Marcus R A 1971 *J. Chem. Phys.* **54** 3965
[77] Miller W H 1971 *Adv. Chem. Phys.* **30** 77–136
[78] Slater N B 1959 *Theory of Unimolecular Reactions* (Ithaca, NY: Cornell University Press)
[79] Bunker D L 1971 *Methods in Computational Physics* vol 10 (New York: Academic) pp 287–325
[80] Porter R N 1974 *Ann. Rev. Phys. Chem.* **25** 317–55
[81] Polanyi J C and Schreiber J L 1973 *Physical Chemistry—An Advanced Treatise* vol 6 (New York: Academic)
[82] Gordon R G 1971 *Methods in Computational Physics* vol 10 (London: Academic) p 82
[83] Redmon M J and Micha D A 1974 *Chem. Phys. Lett.* **28** 341
[84] Shapiro M 1972 *J. Chem. Phys.* **56** 2582
[85] Micha D A 1973 *Acc. Chem. Res.* **6** 138–44

[86]   Mies F H and Krauss M 1966 *J. Chem. Phys.* **45** 4455–68
[87]   Mies F H 1968 *Phys. Rev.* **175** 164–75
[88]   Mies F H 1969 *J. Chem. Phys.* **51** 787–97
        Mies F H 1969 *J. Chem. Phys.* **51** 798–807
[89]   Quack M and Troe J 1975 *Ber. Bunsenges. Phys. Chem.* **79** 170–83
[90]   Quack M and Troe J 1975 *Ber. Bunsenges. Phys. Chem.* **79** 469–75
[91]   White R A and Light J C 1971 Statistical theory of bimolecular exchange reactions: angular distribution *J. Chem. Phys.* **55** 379–87
[92]   Kinsey J L 1971 *J. Chem. Phys.* **54** 1206
[93]   Ben-Shaul A, Levine R D and Bernstein R B 1974 *J. Chem. Phys.* **61** 4937
[94]   Quack M and Troe J 1976 *Ber. Bunsenges. Phys. Chem.* **80** 1140
[95]   Levine R D and Bernstein R B (eds) 1989 *Molecular Reaction Dynamics and Chemical Reactivity* (Oxford: Oxford University Press)
[96]   Robinson P J and Holbrook K A 1972 *Unimolecular Reactions* (London: Wiley)
[97]   Forst W 1973 *Theory of Unimolecular Reactions* (New York: Academic)
[98]   Marcus R A and Rice O K J 1951 *Phys. Colloid Chem.* **55** 894–908
[99]   Marcus R A 1952 *J. Chem. Phys.* **20** 359–64
[100]   Rosenstock H M, Wallenstein M B, Wahrhaftig A L and Eyring H 1952 *Proc. Natl Acad. Sci. USA* **38** 667–78
[101]   Marcus R A 1965 *J. Chem. Phys.* **43** 2658
[102]   Pilling M J and Smith I W M (eds) 1987 *Modern Gas Kinetics. Theory, Experiment and Application* (Oxford: Blackwell)
[103]   Gilbert R G and Smith S C (eds) 1990 *Theory of Unimolecular and Recombination Reactions* (Oxford: Blackwell)
[104]   Holbrook K A, Pilling M J and Robertson S H (eds) 1996 *Unimolecular Reactions* 2nd edn (Chichester: Wiley)
[105]   Baer T and Hase W L (eds) 1996 *Unimolecular Reaction Dynamics* (Oxford: Oxford University Press)
[106]   Bunker D L and Pattengill M 1968 *J. Chem. Phys.* **48** 772–6
[107]   Wong W A and Marcus R A 1971 *J. Chem. Phys.* **55** 5625–9
[108]   Gaedtke H and Troe J 1973 *Ber. Bunsenges. Phys. Chem.* **77** 24–9
[109]   Garret D C and Truhlar D G 1979 *J. Chem. Phys.* **70** 1592–8
[110]   Glasstone S, Laidler K J and Eyring H 1941 *The Theory of Rate Processes* (New York: McGraw-Hill)
[111]   Wardlaw D M and Marcus R A 1988 *Adv. Chem. Phys.* **70** 231–63
[112]   Quack M and Troe J 1977 *Ber. Bunsenges. Phys. Chem.* **81** 329–37

Further Reading

Johnston H S 1966 *Gas Phase Reaction Rate Theory* (Ronald)
Nikitin E E 1974 *Theory of Elementary Atomic and Molecular Processes in Gases* (Oxford: Clarendon)
Quack M and Jans-Bürli S 1986 *Molekulare Thermodynamik und Kinetik. Teil 1. Chemische Reaktionskinetik* (Zürich: Fachvereine)
Pilling M J and Smith I W M (eds) 1987 *Modern Gas Kinetics. Theory, Experiment and Application* (Oxford: Blackwell)
Laidler K J 1987 *Chemical Kinetics* (New York: Harper Collins)
Gilbert R G and Smith S C 1990 *Theory of Unimolecular and Recombination Reactions* (Oxford: Blackwell)
Baer T and Hase W L 1996 *Unimolecular Reaction Dynamics* (Oxford: Oxford University Press)
Holbrook K A, Pilling M J and Robertson S H 1996 *Unimolecular Reactions* 2nd edn (Chichester: Wiley)
Warnatz J, Maas U and Dibble R W 1999 *Combustion: Physical and Chemical Fundamentals, Modelling and Simulation, Experiments, Pollutant Formation* (Heidelberg: Springer)

A3.5
Ion chemistry

A A Viggiano and Thomas M Miller

A3.5.1 Introduction

Ion chemistry is a product of the 20th century. J J Thomson discovered the electron in 1897 and identified it as a constituent of all matter. Free positive ions (as distinct from ions deduced to exist in solids or electrolytes) were first produced by Thomson just before the turn of the century. He produced beams of light ions, and measured their mass-to-charge ratios, in the early 1900s, culminating in the discovery of two isotopes of neon in 1912 [1]. This year also marked Thomson's discovery of H_3^+, which turns out to be the single most important astrophysical ion and which may be said to mark the beginning of the study of the *chemistry* of ions. Thomson noted that 'the existence of this substance is interesting from a chemical point of view', and the problem of its structure soon attracted the distinguished theorist Niels Bohr [2]. (In 1925, the specific reaction producing H_3^+ was recognized [2].) The mobilities of electrons and ions drifting in weak electric fields were first measured by Thomson, Rutherford and Townsend at the Cavendish Laboratory of Cambridge University in the closing years of the 19th century. The average mobility of the negative charge carrier was observed to increase dramatically in some gases, while the positive charge carrier mobility was unchanged— the *anomalous mobility problem*—which led to the hypothesis of electron attachment to molecules to form negative ions [3]. In 1936, Eyring, Hirschfelder and Taylor calculated the rate constant for an ion–molecule reaction (the production of H_3^+!), showing it to be 100 times greater than for a typical neutral reaction, but it was not until 20 years later that any ion–molecule rate constant was measured experimentally [4]. Negative ion–molecule reactions were not studied at all until 1957 [5].

In this section, the wide diversity of techniques used to explore ion chemistry and ion structure will be outlined and a sampling of the applications of ion chemistry will be given in studies of lamps, lasers, plasma processing, ionospheres and interstellar clouds.

Note that chemists tend to refer to positive ions as *cations* (attracted to the cathode in electrolysis) and negative ions as *anions* (attracted to an anode). In this section of the encyclopedia, the terms *positive ion* and *negative ion* will be used for the sake of clarity.

A3.5.2 Methodologies

A3.5.2.1 Spectroscopy

(a) Action spectroscopy

The term *action spectroscopy* refers to how a particular 'action', or process, depends on photon energy. For example, the photodissociation of O_2^- with UV light leads to energetic $O^- + O$ fragments; the kinetic energy released has been studied as a function of photon energy by Lavrich *et al* [6, 7]. Many of the processes discussed in this section may yield such an action spectrum and we will deal with the processes individually.

(b) Laser induced fluorescence

Laser induced fluorescence (LIF) detection of molecules has served as a valuable tool in the study of gas-phase processes in combustion, in the atmosphere, and in plasmas [8–12]. In the LIF technique, laser light is used to excite a particular level of an atom or molecule which then radiates (fluoresces) to some lower excited state or back to the ground state. It is the fluorescence photon which signifies detection of the target. Detection may be by measurement of the total fluorescence signal or by resolved fluorescence, in which the various rovibrational populations are separated. LIF is highly selective and may be used to detect molecules with densities as low as 10^5 cm^{-3} in low pressure situations (<0.1 Pa) where collisional quenching is negligible. In the presence of an atmosphere of air, the detection limit is about 10^{10} cm^{-3}. The use of LIF for ions is more difficult than for neutrals because a typical ion number density may be orders of magnitude lower than for neutrals. Nevertheless, important LIF work with ions has been reported.

LIF has been used to study state-selected ion–atom and ion–molecule collisions in gas cells. Ar$^+$ reactions with N_2 and CO were investigated by Leone and colleagues in the 1980s [13, 14] and that group has continued to contribute new understanding of the drifting and reaction of ions in gases, including studies of velocity distributions and rotational alignment [15–19]. The vibrational state dependence of the charge transfer reaction

$$N_2^+(v = 0, 1, 2) + X \rightarrow N_2 + X^+$$

where X = Ar or O_2 and the collisional deactivation reaction

$$N_2^+(v = 1, 2) + X \rightarrow N_2^+(v' < v) + X$$

were studied by this group using LIF [20]. They showed that charge transfer is enhanced by vibrational excitation and that vibrational deactivation is much more likely with O_2 than with Ar.

We also consider here a reaction that displays both electron-transfer and proton-transfer channels,

$$DBr^+(^2\Pi_i, v^+, J^+) + HBr \rightarrow HBr^+(^2\Pi_{i'}, v'^+, J'^+) + DBr$$

and

$$HBr^+(^2\Pi_i, v^+ = 0, J^+) + HBr \rightarrow H_2Br^+ + Br$$

studied via LIF of the $A^2\Sigma$–$X^2\Pi_{1/2,3/2}$ (0, 0) bands of HBr$^+$, using photons in the range 358–378 nm [21, 22]. For the electron transfer reaction, it was found that any excess energy in the process was statistically partitioned among all degrees of freedom of the complex and was manifested in the LIF spectra as rotational heating. Flow tube experiments tuned to different Br isotopes also showed a hydrogen-atom transfer channel in the HBr$^+$ + HBr reaction.

LIF is also used with liquid and solid samples. For example, LIF is used to detect UO_2^{2+} ions in minerals; the uranyl ion is responsible for the bright green fluorescence given off by minerals such as autunite and opal upon exposure to UV light [23].

(c) Photodissociation of ions

Photodissociation of molecular ions occurs when a photon is absorbed by the ion and the energy is released (at least partly) by the breaking of one of the molecular bonds. The photodissociation of a molecular ion is conceptually similar to that for neutral molecules, but the experimental techniques differ. Photodissociation events are divided into two categories: *direct dissociation*, in which the photoexcitation is from a bound state to a repulsive state and *predissociation*, in which a quasi-bound state is accessed in the excitation. Direct dissociation takes place rapidly (fs to ps timescale). The shape of the direct dissociation cross section curve against photon energy is governed by the (Franck–Condon) overlap of wavefunctions of the initial state (usually the ground state) and the final, repulsive state. It will normally consist of peaks corresponding to vibrational

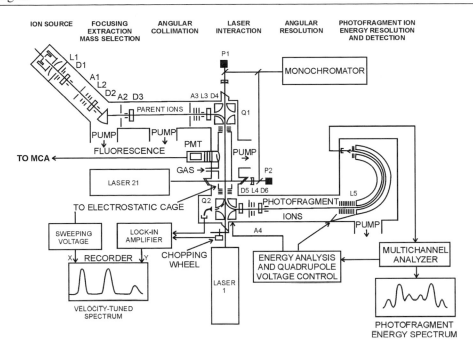

Figure A3.5.1. The fast ion beam photofragment spectrometer at SRI International. 'L' labels electrostatic lenses, 'D' labels deflectors and 'A' labels apertures.

structure in the initial level of the target ion with shapes skewed by the overlap with the repulsive state. One can model these shapes to obtain the potential curves of both the initial and repulsive states. Predissociation, in contrast, may take place over a much longer timescale; the lifetime of a particular predissociating state may be determined from the width of the resonance observed. Measurements of the lifetime for a series of predissociating states gives a picture of the predissociation mechanism. Photodissociation cross sections tend to peak in the 10^{-18} cm^2 range and hence are often given in Mb (megabarn) units.

There are many experimental methods by which photodissociation of ions have been studied. The earliest were crossed-beams experiments on H$_2^+$ beginning in the late 1960s [24–26] and experiments on a variety of ions in the 1970s using drift tubes [27–29]. Later techniques allowed more detailed information to be obtained on state symmetries and kinetic energy releases [30–32]. Figure A3.5.1 shows the fast ion beam photofragment spectrometer at SRI International; similar apparatus is in use at other institutions [33]. The apparatus consists of an ion source and mass selector, two electrostatic quadrupole benders that allow a laser beam to interact coaxially with the ion beam, a product-ion (photofragment) energy analyser and a particle detector. An interesting feature of the coaxial beam technique, aside from the long interaction region, is that sub-Doppler line widths can be obtained because of a thousandfold or more narrowing of the ion velocity distribution in the centre-of-mass reference frame for typical keV ion energies. By the same token, photofragment ions that differ in energy by a tenth of an electron volt in the centre-of-mass frame will be separated by typically 10–20 eV in the laboratory frame. This simplifies the job of the photofragment energy analyser. An example of a photofragment kinetic energy spectrum is shown in figure A3.5.2. If the laser beam is sent at right angles to the ion beam (instead of coaxially), the optical polarization vector can be rotated to map out angular distributions of photofragments. (In the coaxial arrangement, the optical polarization is necessarily always perpendicular to the ion beam direction.)

In the past decade there has been photodissociation work on doubly charged positive ions, e.g., N$_2^{2+}$, NO$_2^{2+}$, CF$_3^{2+}$, CCl$_3^{2+}$, SiF$_2^{2+}$ and SiF$_3^{2+}$ [34]. Interestingly, the result of photoexcitation of the latter four molecular

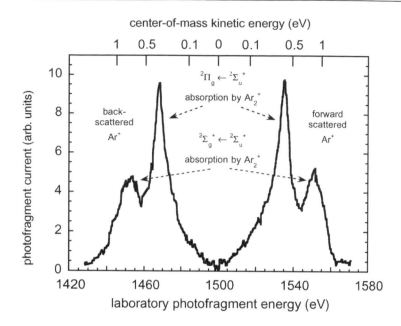

Figure A3.5.2. The Ar^+ photofragment energy spectrum for the dissociation of 3 keV Ar_2^+ ions at 752.5 nm. The upper scale gives the kinetic energy release in the centre-of-mass reference frame, both parallel and antiparallel to the ion beam velocity vector in the laboratory.

ions is the loss of neutrals, as a consequence of the electronic structures. There are two possible scenarios, illustrated by

$$NO^{2+} + h\nu \rightarrow N^+ + O^+ \text{ and } SiF_3^{2+} + h\nu \rightarrow SiF_2^{2+} + F.$$

There has been much activity in the study of photodissociation of cluster ions, dating back to the 1970s when it was realized that most ions in the earth's lower atmosphere were heavily clustered [7, 35, 36].

(d) Photoelectron spectroscopy

Photoelectron spectroscopy (PES) of negative ions involves irradiation of an ion beam with laser light and energy analysis of the electrons liberated when the photon energy exceeds the binding energy of the electron. The kinetic energy of the detached electron is the difference between the photon energy and the binding energy of the electron [37–47]. Analysis of the electron energy thus gives a direct measurement of the electron affinity of the corresponding neutral atom or molecule, a very important thermochemical quantity. Generally speaking, PES yields more information about the *neutral* atom or molecule than the corresponding negative ion, because the target ion is ideally in its ground state and the electron kinetic energy is then dependent on the final state of the neutral product. The energy resolution of a PES experiment is usually adequate (often 5–10 meV) to resolve vibrational structure due to the neutral molecule, certainly for low-mass systems of few atoms and likewise electronic structure, including singlet–triplet splittings and fine structure separations. In a few cases, rotational energy levels have been resolved. Features may appear in a photoelectron spectrum due to excited levels of the target negative ion and give valuable information about the structure of the negative ion, but at the cost of complicating the spectrum.

An example of a PES apparatus is shown in figure A3.5.3. A PES apparatus consists of (a) an ion source, (b) a fixed-frequency laser, (c) an interaction region, (d) an electron energy analyser and (e) an electron

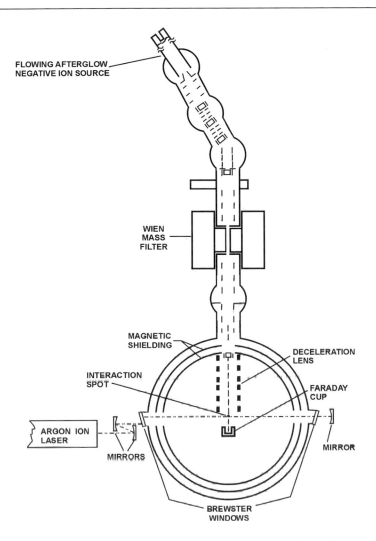

Figure A3.5.3. The negative ion photoelectron spectrometer used at the University of Colorado. The apparatus now contains a UV-buildup cavity inside the vacuum system (not shown in this sketch).

detector. Ion sources include gas discharge, sputtering, electron-impact and flowing afterglow. The laser may be cw (the argon-ion laser operated at 488 nm is common) or pulsed (which allows frequency doubling etc.). Recent trends have been toward UV laser light because the negative ions of importance in practical chemistry (e.g., atmospheric chemistry and biochemistry) tend to be strongly bound and because the more energetic light allows one to access more electronic and vibrational states. The interaction region may include a magnetic field that routes detached electrons toward the energy analyser. The energy analyser is either a hemispherical electrostatic device or a time-of-flight energy analyser; the latter is especially suited to a pulsed-laser system.

Data obtained with a PES apparatus are shown in figure A3.5.4 [48]. Interpretation of the spectrum for a diatomic molecule is particularly straightforward. Peaks to the left of the origin band (each band containing unresolved rotational structure) are spaced by the vibrational separation in the neutral molecule, and their relative intensities are determined by the amount of spatial overlap between wavefunctions for the negative ion and the neutral molecule (Franck–Condon factors). Peaks to the right of the origin band are spaced by the vibrational separation in the negative ion and their relative intensities give the effective temperature of the

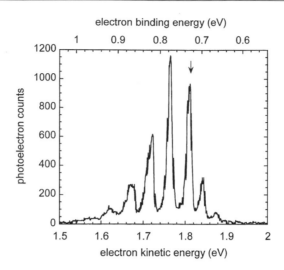

Figure A3.5.4. The 488 nm photoelectron spectrum of NaCl$^-$. The arrow marks the origin band, for transitions from NaCl$^-$ ($v = 0$) to NaCl ($v = 0$), from which the electron affinity of NaCl is obtained.

ion source. Subtracting the electron kinetic energy corresponding to the origin band from the photon energy yields the electron affinity of the molecule. The energy of the maximum of the envelope of neutral molecule peaks is referred to as the *vertical detachment energy*, i.e., the energy required from the ground vibrational state of the negative ion to the neutral molecule with no change in nuclear geometry. Photoelectron spectra are often far more complicated than the example shown, especially for polyatomic molecules. The origin band may have zero intensity, making the electron affinity difficult to determine directly. Fine structure at least doubles the number of peaks in the spectrum. PES experiments have been carried out for doubly charged negative ions [49] and using multiphoton detachment [37].

It is advantageous if the laser system permits rotation of the optical polarization. Detached electrons correlated with different final electronic states of the neutral molecule will generally be emitted with different angular distributions about the direction of polarization. Measurement of the angular distribution helps in the interpretation of complex photoelectron spectra. The angular distribution $f(\theta)$ of photoelectrons is [50]

$$f(\theta) = [1 + \beta P_2(\cos \theta)]/4\pi$$

where θ is the angle between the optical polarization vector and the direction of emission of a photoelectron, β is the asymmetry parameter (β is in the range -1 to 2 and in general depends on photon energy) and $P_2(\cos \theta)$ is the second-order Legendre polynomial, $(3 \cos^2 \theta - 1)/2$. $P_2(\cos \theta)$ is zero if $\theta = 54.7°$, in which case the detected electron signal is proportional to the angle-averaged detachment cross section. The distribution function is independent of the azimuthal angle.

ZEKE (zero kinetic energy) photoelectron spectroscopy has also been applied to negative ions [51]. In ZEKE work, the laser wavelength is swept through photodetachment thresholds and only electrons with near-zero kinetic energy are allowed into a detector, resulting in narrow threshold peaks. The resulting resolution (2–3 cm^{-1}) is superior to that commonly encountered with PES (40–300 cm^{-1}).

PES of neutral molecules to give positive ions is a much older field [52]. The information is valuable to chemists because it tells one about unoccupied orbitals in the neutral that may become occupied in chemical reactions. Since UV light is needed to ionize neutrals, UV lamps and synchrotron radiation have been used as well as UV laser light. With suitable electron-energy resolution, vibrational states of the positive ions can be resolved, as with the negative-ion PES described above. The angular distribution of photoelectrons can be also determined as described above.

(e) Absorption

In absorption spectroscopy, the attenuation of light as it passes through a sample is measured as a function of wavelength. The attenuation is due to rovibrational or electronic transitions occurring in the sample. Mapping out the attenuation versus photon frequency gives a description of the molecule or molecules responsible for the absorption. The attenuation at a particular frequency follows the Beer–Lambert law,

$$I = I_0 \exp(-\sigma n L)$$

where I and I_0 are the attenuated and unattenuated intensities, σ is the cross section, n is the number density of target molecules and L is the path length. Broadening of spectral lines may be observed, and is classed as *homogeneous broadening* (e.g., collisions with other molecules and laser power effects) or *inhomogeneous broadening* (e.g., Doppler broadening) [53, 54]. So-called 'UV/vis' absorption spectroscopy is a standard tool for analysis of chemical samples. In organic samples, absorption by functional groups in the sample aids in identification of the species because it is strongly dependent upon the relative number of single, double, and triple bonds.

Microwave spectra (giving pure rotational spectra) are especially useful for the detection of interstellar molecular ions (in some cases the microwave spectrum has first been observed in interstellar spectra!). Typically a DC glow discharge tube is used to produce the target ion (e.g., HCO^+) [55, 56]. If the photons travel parallel or antiparallel to the electric field direction, there is a small but measurable Doppler shift in frequencies. This is due to the drift velocity of ions in the electric field, which may aid in distinguishing ion spectra from neutral spectra, but in any case must be accounted for [55, 58]. Infrared spectroscopy has also been carried out on ions in a glow discharge tube using the beat frequency between a fixed-frequency visible laser and a tunable dye laser. The difference frequency laser, in the IR, irradiates a long discharge tube. The method was first used to study the important astrophysical ion H_3^+ [59]. *Velocity modulation spectroscopy* utilizes an audio frequency glow discharge coupled with phase synchronous demodulation of the absorbed IR laser radiation to take advantage of the Doppler shift occurring for ions drifting in glow discharge tubes. Many important positive ions, such as H_3O^+, NH_4^+ and H_3^+, have been studied with this technique with the high precision common to IR spectroscopy [58, 60].

Far-infrared spectra of great sensitivity may be obtained with *laser magnetic resonance* (LMR). The sensitivity comes about because the gas sample is located inside the long cavity of the laser where the circulating power is typically 100 times that used in extracavity work. A discharge in the gas cell produces the radical species to be studied, and an axial magnetic field is varied to bring energy levels into resonance with one of many laser lines. HBr^+ was the first ion to be observed with LMR, in 1979. OH^- was one of the first negative ions to have been detected by direct absorption spectroscopy [61].

Strictly speaking, the term absorption spectroscopy refers to measurements of light intensity. In practice, the absorption may be deduced from the detection of electrons or ions produced in the process, such as in absorption of light leading to photodetachment or photodissociation, i.e., action spectra. In the absorption spectroscopy of ions, this is a natural tack to take as the charged-particle production can be detected with greater precision than is possible for a measurement of a small change in the light intensity. The most important of such experiment types is the coaxial beams spectrometer, one of which is described in detail in the section on dissociation of ions. In these experiments, the ions are identified by mass and collimated, and interact with the laser beam over a long distance (0.25–1 m). The method was first used with ions in 1976 for HD^+, with the absorption events detected via enhanced production of buffer gas ions as a result of charge transfer reactions [62]. Since this time many small ions have been studied in great detail, notably H_3^+, O_2^+, CH^+, H_2O^+ and CO^+ [31, 32, 63]. A few negative ions have been studied using coaxial fast-ion/laser beams. The high-resolution IR spectrum of NH^-, for example, was studied in this manner. The negative ion was excited to an autodetaching state with a photon energy greater than the electron affinity of NH. Detection of autodetached electrons signified an absorption event. Aside from determination of spectroscopic constants for NH^-, information on autodetachment dynamics was obtained [64].

A3.5.2.2   Kinetics and dynamics

In principle the study of ion–molecule kinetics and dynamics is no different from studies of the same processes in neutral species; however, there are additional forces that govern reactivity, often leading to behaviour that is fundamentally different from neutral processes. An important factor in determining ion–molecule rate constants and cross sections is the rate at which the reactants collide, i.e. the collision rate. In contrast to neutral kinetics, the collision rate at low energies or temperatures is determined not by the size of the molecule but by electrical forces. The ion–molecule collision rate is determined by the classical capture cross section for a point charge interacting with a structureless multipole. This was first described analytically for a point charge interacting with a polarizable species with no other multipole. In this case, the collisional value of the rate constant is independent of temperature [65]. The only other force of any significance is from the ion–permanent dipole interaction. Other forces, such as those between the ion and the quadrupole moment of the neutral, and between the neutral dipole and the induced dipole of the ion, have been shown to be of minor importance [58–63]. If the physical size of the reactants is greater than the capture radius, e.g. at translational energies of several tenths of an electron volt and greater, more conventional notions apply. Except for species with very small polarizabilities and systems of large mass, ion–molecule collision rates are above 10^{-9} molecules cm^{-3} s^{-1}, or about a factor of ten larger than neutral collision rates.

Several processes are unique to ions. A common reaction type in which no chemical rearrangement occurs but rather an electron is transferred to a positive ion or from a negative ion is termed charge transfer or electron transfer. Proton transfer is also common in both positive and negative ion reactions. Many proton- and electron-transfer reactions occur at or near the collision rate [72]. A reaction pertaining only to negative ions is associative detachment [73, 74],

$$A^- + B \rightarrow AB + e.$$

Associative detachment reactions are important in controlling the electron concentration in the earth's mesosphere [75]. Reactions in which more than one neutral product are formed also occur and are sometimes referred to as reactive detachment [76].

Several reactivity trends are worth noting. Reactions that are rapid frequently stay rapid as the temperature or centre-of-mass kinetic energy of the reactants is varied. Slow exothermic reactions almost always show behaviour such that the rate constant decreases with increasing temperature at low temperature or kinetic energies and then increases at higher temperature. As an example, figure A3.5.5 shows rate constants for the charge-transfer reaction of Ar^+ with O_2 as a function of temperature. The data are from five separate experiments and four experimental techniques [77–81] and cover the extremely wide temperature range of 0.8 K to 1400 K. The extremely low temperature data are relatively flat. At approximately 20 K, the rate constants decrease. The decrease is described by a power law. A minimum is found at 800–900 K and a steep increase is found above 1000 K. The position of the minimum varies considerably for other reactions.

The decrease in reactivity with increasing temperature is due to the fact that many low-energy ion–molecule reactions proceed through a double-well potential with the following mechanism [82]:

$$A^\pm + B \underset{k_r}{\overset{k_f}{\rightleftharpoons}} (AB)^{*\pm}$$

$$(AB)^{*\pm} \overset{k_{prod}}{\longrightarrow} C^\pm + D.$$

The minimum energy pathway for the reaction of Cl^- with CH_3Br is shown in figure A3.5.6 [83]. As the reactants approach they are attracted by the ion–dipole and ion–induced-dipole forces and enter the entrance channel complex. As the reaction proceeds along the minimum energy path, the potential energy increases due to the forces necessary for rearrangement. The species then enter the product well and finally separate into products. The two wells are separated by a barrier that is often below the energy of the reactants but

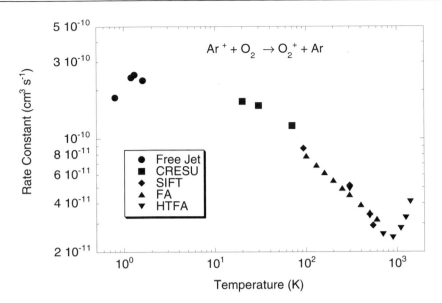

Figure A3.5.5. Rate constants for the reaction of Ar$^+$ with O$_2$ as a function of temperature. CRESU stands for the French translation of reaction kinetics at supersonic conditions, SIFT is selected ion flow tube, FA is flowing afterglow and HTFA is high temperature flowing afterglow.

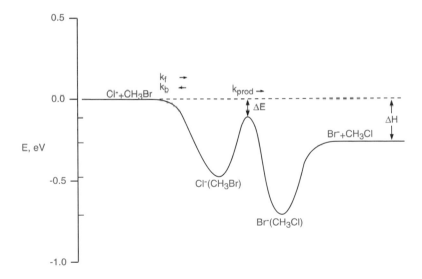

Figure A3.5.6. Minimum energy pathway for the reaction of Cl$^-$ with CH$_3$Br.

still plays an important role in controlling reactivity. The decrease in the overall rate constant with increasing temperature is due to the rate constant for collision complex dissociation to reactants, k_b, increasing more rapidly with temperature than the rate constant for the complex going to products, k_{prod} [68]. The increase at higher energies and temperatures is often due to new channels opening, including new vibrational and electronic states as well as new chemical channels.

Rotational and translational energy have been shown to be equivalent in controlling most reactions [84]. Vibrational energy often increases reactivity; however, sometimes it does not affect reactivity, or occasionally decreases reactivity. The following sections describe several of the more common techniques used to measure ion–molecule rate constants or cross sections.

(a) Flow tubes

Flow tube studies of ion–molecule reactions date back to the early 1960s, when the flowing afterglow was adapted to study ion kinetics [85]. This represented a major advance since the flowing afterglow is a thermal device under most situations and previous instruments were not. Since that time, many iterations of the ion–molecule flow tube have been developed and it is an extremely flexible method for studying ion–molecule reactions [86–92].

The basic flow system is conceptually straightforward. A carrier gas, often helium, flows into the upstream end of a tube approximately 1 m long with a radius of several centimetres. This buffer gas pressure is approximately 100 Pa. Ions are created either in the flow tube or injected from an external source at the upstream end of the pipe. The carrier gas transports the ions downstream at approximately 100 m s^{-1}. Part way down the tube, a neutral reactant is added and the ions created in the source region are transformed into products. Conditions are chosen so that all ion chemistry leading to the reactant ion is complete and the ions are thermalized before they encounter the neutral reagent gas. The rate constant is determined by sampling a small portion of the gas with a quadrupole mass spectrometer and monitoring the disappearance of the primary ion and the appearance of product ions. For the reaction of $A^+ + B \rightarrow$ products, the rate constant is given by

$$k = 1/[B]\tau \ln([A_0^+]/[A^+]) \tag{A3.5.1}$$

where [B] is the reactant neutral concentration in the flow tube, τ is the reaction time and $[A_0^+]$ and $[A^+]$ are the ion concentrations with and without reactant neutrals in the flow tube. This equation assumes that the concentration of B is much greater than that of A^+, i.e. first order kinetics apply. This situation applies to all ion kinetics since it is difficult to make large quantities of ions. Fortunately, the derivation of the rate constant depends only on the relative ion concentration, which is much easier to measure than the absolute concentration. The reaction time is determined from the flow velocity of the carrier gas. The average ion flow velocity is approximately 1.6 times the average neutral flow velocity [87], a result of ion diffusion, ions being neutralized on the flow tube walls and the carrier gas having a parabolic flow profile characteristic of laminar flow.

The basic system described above can be easily modified to study many processes. Figure A3.5.7 shows an example of a modern ion–molecule flow tube [93] with a number of interesting features. First, ions are created external to the flow tube. Any suitable ion source can be used, including high- and low-pressure electron-impact ion sources, a supersonic-expansion source (shown) or a flow-tube source. Once created the ions are injected into a quadrupole mass spectrometer and only ions with the proper mass are injected into the flow tube through a Venturi inlet. Under favourable circumstances, only one ion species enters the flow tube. This configuration is called the selected-ion flow tube (SIFT) [89–91]. Alternatively, ions can be created in the carrier-gas flow by a filament or discharge. Neutral reagents are added through a variety of inlets. Unstable species such as O, H and N atoms, molecular radicals and vibrationally excited diatomics can be injected by passing the appropriate gas through a microwave discharge. In a SIFT, the chemistry is usually straightforward since there is only one reactant ion and one neutral present in the flow tube.

Flowing afterglows and SIFTs have been operated between 80 K and 1800 K. In addition, the ion kinetic energy can be varied by adding a drift tube either at room temperature [94–97] or with temperature variability [84]. A drift tube is a series of rings electrically connected by uniform resistors. Applying a voltage to the resistor chain forms a uniform electric field in the flow tube. Ions are accelerated by the electric field and

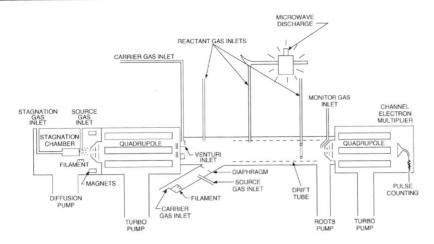

Figure A3.5.7. Schematic diagram of a selected ion flow drift tube with supersonic expansion ion source.

decelerated by collisions so that a steady-state velocity results. The ion kinetic energy in the centre of mass, $\langle\mathrm{KE}\rangle_{\mathrm{cm}}$, is given by the Wannier expression [98]

$$\langle\mathrm{KE}\rangle_{\mathrm{cm}} = \frac{(m_i + m_b)m_n}{2(m_i + m_n)}v_d^2 + 3/2kT \tag{A3.5.2}$$

where m_i, m_n and m_b are the mass of the ion, neutral and buffer respectively, v_d is the velocity of the ion due to the electric field and T is the temperature. Equation (A3.5.2) shows that the kinetic energy in a drift tube is the sum of a thermal component ($\frac{3}{2}kT$) and a drift field component. At a fixed kinetic energy, varying the contribution of each term yields information on rotational and vibrational effects [84]. Excited-state effects can be studied in a number of other ways. Electronic excitation often occurs in SIFT studies of atomic ions. Vibrational effects can be studied by exciting neutral diatomics in a microwave discharge. Ion vibrations can be excited in the source and monitored by LIF or judicious choice of a reactant neutral, i.e. one that reacts differently with excited states than for ground states. Often one looks for a reaction that is endothermic with respect to the ground state and energetically allowed for the excited state. Product-state information can be obtained by the monitor method or through optical spectroscopy. This list of possibilities is not exhaustive but it does give a sample of the type of information that can be obtained.

(b) Traps

Another powerful class of instrumentation used to study ion–molecule reactivity is trapping devices. Traps use electric and magnetic fields to store ions for an appreciable length of time, ranging from milliseconds to thousands of seconds. Generally, these devices run at low pressure and thus can be used to obtain data at pressures well below the range in which flow tubes operate.

The most widely used type of trap for the study of ion–molecule reactivity is the ion-cyclotron-resonance (ICR) [99] mass spectrometer and its successor, the Fourier-transform mass spectrometer (FTMS) [100, 101]. Figure A3.5.8 shows the cubic trapping cell used in many FTMS instruments [101]. Ions are created in or injected into a cubic cell in a vacuum of 10^{-2} Pa or lower. A magnetic field, B, confines the motion in the $x–y$ plane through ion-cyclotron motion. The frequency of motion, ω, is given as $1.537 \times 10^7 B\,e/m$ where B is the magnetic field in tesla (typically 1–7 T), e is the charge of an electron and m the mass in atomic mass units. To trap ions in the z direction, a potential is placed on the two end electrodes. The ions oscillate in the z direction until their motion is damped by collisions. The magnetic field adds little energy to the motion, and

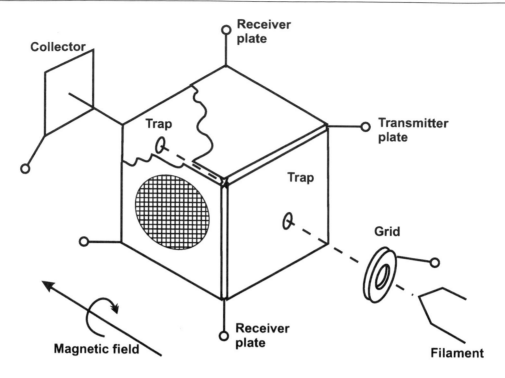

Figure A3.5.8. Schematic diagram of the cell used in a Fourier transform mass spectrometer.

the ions can be described as thermal. Ions are detected by applying a radio frequency (RF) pulse (a chirp) to the transmitter plates. The RF pulse causes ions with the matching cyclotron frequency to absorb energy. The ions are not only energized but quickly move coherently. Image currents on the receiver plates are detected. By putting a rapidly varying RF pulse (0–2 MHz) on the transmitter plate one obtains image currents as a function of time. A fast Fourier transform yields the frequency spectrum that is directly related to the mass spectrum by the equation described above.

Kinetics measurements are made by adding a known concentration of reactant gas to the cell and monitoring the time evolution of the ion intensities. Rate constants are derived from equation (A3.5.1). Many useful tricks can be employed. The most useful is to chirp the cell so that all ions except those with the correct mass are excited out of the cell. In this manner the kinetics are simplified; only one ion and one neutral exist at time zero, and product information is easily obtained. A mass-specific excitation pulse adds energy such that ions may acquire enough energy to dissociate upon collision with background gas. The pattern of the dissociation often yields structural information. The ICR is particularly suited to the study of radiative association [101] and radiative cooling [102] of ions since the pressure is low and the trapping time can be long.

Another class of trapping device that is gaining importance is the radiofrequency trap [103]. Quadrupole ion traps (also called Paul traps) are three-dimensional traps with rotationally symmetric ring and endcap electrodes. An RF voltage of opposite phase is applied to the ring and endcaps, respectively, to create a quadrupolar RF field. This type of trap suffers from electric field heating of the ions and can be classified as a nonthermal device. More innovative traps in limited use are the ring electrode trap and 22-pole trap. Both of these devices trap ions in a large field-free region and produce thermal ions. Reactions at very low temperatures have been studied with these types of trap [81, 103].

(c) Beams

The guided-ion beam has become the instrument of choice for studying ion–molecule reactions at elevated kinetic energies [103]. In many guided-ion beam systems the lowest energy obtainable is slightly above thermal energy ($\sim$0.1–0.2 eV), although it can be as low as the thermal energy of the target gas. The upper range varies but is generally in the tens of electron volts.

In essence, a guided-ion beam is a double mass spectrometer. Figure A3.5.9 shows a schematic diagram of a guided-ion beam apparatus [104]. Ions are created and extracted from an ion source. Many types of source have been used and the choice depends upon the application. Combining a flow tube such as that described in this chapter has proven to be versatile and it ensures the ions are thermalized [105]. After extraction, the ions are mass selected. Many types of mass spectrometer can be used; a Wien ExB filter is shown. The ions are then injected into an octopole ion trap. The octopole consists of eight parallel rods arranged on a circle. An RF voltage is applied to alternating sets of rods to trap ions in the centre of the octopole in an approximately square well potential. Little energy is transferred to the ions. The surrounding part of the octopole is a chamber where reactive gas is added. Typical pressures of added gas are of the order of 10^{-2} Pa. Pressure is kept low so single-collision conditions apply; the primary ions collide at most once with the reactant gas. The collision cell is generally run at room temperature although cooled and heated versions have been used. The main advantage of the octopole collision cell arrangement, over the arrangement used in early beam apparatuses, is greater collection efficiency of the product ions since products scattering in all directions are collected. The primary ions react with the reactant neutral and the resulting mix of ions exits the octopole to be mass analysed and detected. A quadrupole mass filter is often used for mass analysis although other mass spectrometers can be used. The reaction cross section is derived from the Beer–Lambert law, $I = I_0 \exp(-\sigma n L)$ where I and I_0 are the reactant ion signals with and without the reactant gas, n is the number density in the collision cell and L is the length of the collision cell. In the single-collision limit, I is taken as the product ion signal and I_0 as the primary ion signal.

As with most methods for studying ion–molecule kinetics and dynamics, numerous variations exist. For low-energy processes, the collision cell can be replaced with a molecular beam perpendicular to the ion beam [106]. This greatly reduces the thermal energy spread of the reactant neutral. Another approach for low energies is to use a merged beam [103]. In this system the supersonic expansion is aimed at the throat of the octopole, and the ions are passed through a quadrupole bender and merged with the neutral beam. Exceedingly low collision energies can be obtained with this arrangement. Another important modification is obtained by adding a second octopole between the collision cell and the mass analyser. This allows the product-ion flight times to be measured, thus yielding the kinetic-energy release in the reaction to provide important dynamical information. Laser radiation can be introduced into the octopole to measure product distributions or to study dissociative processes. One valuable use of guided-ion beams has been the study of thresholds for endothermic processes in order to measure bond strengths and other thermodynamic quantities.

Two techniques exist for measuring the angular distribution of products. In the crossed-beam setup, the octopole/collision cell is replaced with an interaction zone defined by the overlap of an ion beam and a supersonic neutral beam [106]. The angular distribution is measured by moving a mass spectrometer to detect ions at various angles. A simpler approach is to measure the product transmission as a function of trapping potential on the octopole [103]. The derivative of the signal yields the angular information but with limited resolution. Angular distributions are often used to determine the extent of collision complex formation.

(d) Other techniques

While the techniques described above are the most common and versatile for measuring ion–molecule kinetics, several other techniques are worth mentioning. An important technique for measuring ion energetics is the

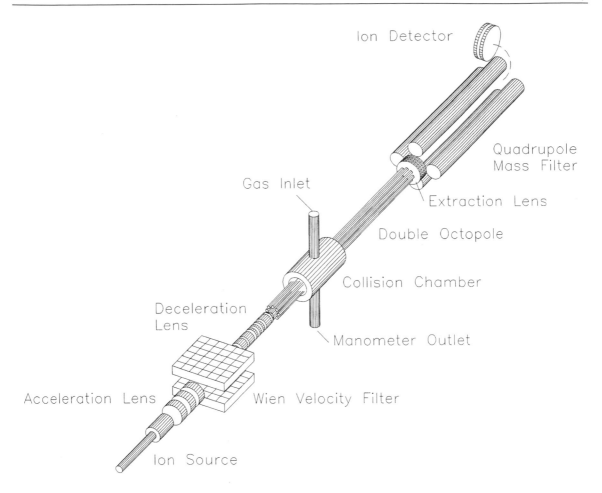

Figure A3.5.9. Schematic diagram of a guided-ion beam.

pulsed, high-pressure mass spectrometer (PHPMS) [107]. In PHPMS, a pulsed beam of 2 keV electrons enters a small chamber containing reactants and a buffer gas. The chamber is maintained at ~500 Pa. Ion signals are then recorded as a function of time until equilibrium is established. Knowledge of the ion signals and partial pressures of the reactant neutral(s) yields the equilibrium constant. Temperature variation allows the enthalpy and entropy of reaction to be derived. Important thermodynamic information obtained by this technique includes ligand bond strengths, proton affinities, gas phase basicities, electron affinities and ionization energies. Information on kinetics can also be obtained.

Several instruments have been developed for measuring kinetics at temperatures below that of liquid nitrogen [81]. Liquid helium cooled drift tubes and ion traps have been employed, but this apparatus is of limited use since most gases freeze at temperatures below about 80 K. Molecules can be maintained in the gas phase at low temperatures in a free jet expansion. The CRESU apparatus (acronym for the French translation of reaction kinetics at supersonic conditions) uses a Laval nozzle expansion to obtain temperatures of 8–160 K. The merged ion beam and molecular beam apparatus are described above. These techniques have provided important information on reactions pertinent to interstellar-cloud chemistry as well as the temperature dependence of reactions in a regime not otherwise accessible. In particular, information on ion–molecule collision rates as a function of temperature has proven valuable in refining theoretical calculations.

Most ion–molecule techniques study reactivity at pressures below 1000 Pa; however, several techniques now exist for studying reactions above this pressure range. These include time-resolved, atmospheric-pressure, mass spectrometry; optical spectroscopy in a pulsed discharge; ion-mobility spectrometry [108] and the turbulent flow reactor [109].

A3.5.3   Applications

A3.5.3.1   Ion structure and energetics

The molecular constants that describe the structure of a molecule can be measured using many optical techniques described in section A3.5.1 as long as the resolution is sufficient to separate the rovibrational states [110–112]. Absorption spectroscopy is difficult with ions in the gas phase, hence many ion species have been first studied by matrix isolation methods [113], in which the IR spectrum is observed for ions trapped within a frozen noble gas on a liquid-helium cooled surface. The measured frequencies may be shifted as much as 1% from gas phase values because of the weak interaction with the matrix.

These days, remarkably high-resolution spectra are obtained for positive and negative ions using coaxial-beam spectrometers and various microwave and IR absorption techniques as described earlier. Information on molecular bond strengths, isomeric forms and energetics may also be obtained from the techniques discussed earlier. The kinetics of cluster-ion formation, as studied in a selected-ion flow tube (SIFT) or by high-pressure mass spectrometry, may be interpreted in terms of cluster bond strengths [114]. In addition, the chemistry of ions may be used to identify the structure. For example, the ionic product of reaction between O_2^+ and CH_4 at 300 K has been identified as CH_2OOH^+ from its chemistry; the reaction mechanism is insertion [115]. Collision-induced dissociation (in a SIFT apparatus, a triple-quadrupole apparatus, a guided-ion beam apparatus, an ICR or a beam-gas collision apparatus) may be used to determine ligand-bond energies, isomeric forms of ions and gas-phase acidities.

Photoelectron spectra of cluster ions yields cluster-bond strengths, because each added ligand increases the binding energy of the extra electron in the negative ion by the amount of the ligand bond strength (provided the bond is electrostatic and does not appreciably affect the chromophore ion) [116].

One example of the determination of molecular constants can be taken from the photoelectron spectrum for $NaCl^-$ shown in figure A3.5.4 [48]. The peak spacing to the left of the origin band is 45 meV: the nominal vibrational frequency ω_e in neutral NaCl. The spectral resolution is not good enough to specify the small anharmonic correction, $\omega_e x_e$, or the rotational constant, B_e, but these, along with the equilibrium separation, r_e (=2.361 Å), are accurately known from optical spectra. The spectrum in figure A3.5.4 also provides new information about the negative ion: the peaks to the right of the origin band are spaced by 33 meV, the nominal vibrational spacing in $NaCl^-$. The distribution of peak heights everywhere is determined by the Franck–Condon overlap of wavefunctions for NaCl and $NaCl^-$ vibrational states, so the data give the ion temperature and the magnitude of the change in r_e between the neutral and negative ion (0.136 Å in this case). Vibrational frequencies and bond-energy considerations imply that $r_e(NaCl^-) > r_e(NaCl)$. Therefore, $r_e(NaCl^-) = 2.497$ Å, and $B_e = 0.195$ cm^{-1}. Finally, the position of the origin peak gives the electron binding energy (the electron affinity of NaCl, 0.727 eV) and a thermochemical cycle allows one to calculate the bond energy of $NaCl^-$ (all other quantities being known):

$$D_0(\text{Na–Cl}^-) = D_0(\text{Na–Cl}) + EA(\text{NaCl}) - EA(\text{Cl})$$

yielding $D_0(\text{Na–Cl}^-) = 1.34$ eV. Admittedly, this is a simple spectrum to interpret. Had the system involved a Π state instead of Σ states, additional peaks would have complicated the spectrum (but yielded additional information if resolved). Had the molecules been polyatomic, where vibrations may be bending modes or stretches and combinations, the spectra and interpretation become more complex—the systems must be described in terms of normal modes instead of the more intuitive, but coupled, stretches and bends.

A3.5.3.2   Thermochemistry

The principles of ion thermochemistry are the same as those for neutral systems; however, there are several important quantities pertinent only to ions. For positive ions, the most fundamental quantity is the adiabatic ionization potential (IP), defined as the energy required at 0 K to remove an electron from a neutral molecule [117–119].

Positive ions also form readily by adding a proton to a neutral atom or molecule [120]

$$M + H^+ \rightarrow MH^+.$$

The proton affinity, PA, is defined (at 298 K) as [117]

$$PA = \Delta H_f^0(M) + \Delta H_f(H^+) - \Delta H_f(MH^+).$$

Negative ions also have two unique thermodynamic quantities associated with them: the electron affinity, EA, defined as the negative of the enthalpy change for addition of an electron to a molecule at 0 K [117, 121, 122]

$$M + e^- \rightarrow M^-$$

and the gas-phase acidity of a molecule, defined as the Gibbs energy change at 298 K, $\Delta G_{acid}(AH)$, for the process [117, 121]

$$AH \rightarrow H^+ + A^-.$$

The enthalpy for this process is the proton affinity of the negative ion.

Much effort has gone into determining these quantities since they are fundamental to ionic reactivity. Examples include thermodynamic equilibrium measurements for all quantities and photoelectron studies for determination of EAs and IPs. The most up-to-date tabulation on ion thermochemistry is the *NIST Chemistry WebBook* (webbook.nist.gov/chemistry) [123].

Neutral thermochemistry can be determined by studying ion thermochemistry. For example, the following cycle can be used to determine a neutral bond strength,

$$
\begin{array}{ll}
A^- \rightarrow A + e^- & EA(A) \\
AH \rightarrow A^- + H^+ & \Delta H_{acid}(AH) \\
\underline{H^+ + e^- \rightarrow H} & \underline{-\ IP(H)} \\
AH \rightarrow A + H & D(A\text{–}H)
\end{array}
$$

where $D(A\text{–}H)$ is the bond dissociation energy or enthalpy for dissociating a hydrogen atom from AH. Often it is easier to determine the EA and anionic proton affinity than it is to determine the bond strength directly, especially when AH is a radical. The IP of hydrogen is well known. As an example, this technique has been used to determine all bond strengths in ethylene and acetylene [124].

A3.5.3.3   Cluster properties

A gas phase ionic cluster can be described as a core ion solvated by one or more neutral atoms or molecules and it is often represented as $A^{\pm}(B)_n$ or $A^{\pm} \cdot B_n$, where $A^{\pm}$ is the core ion and B are the ligand molecules. Of course, the core and the ligand can be the same species, e.g. the hydrated electron. The interactions governing the properties of these species are often similar to those governing liquid-phase ionic solvation. Modern techniques allow clusters with a specific number of neutral molecules to be studied, providing information on the evolution of properties as a function of solvation number. This leads to insights into the fundamental properties of solutions and has made this field an active area of research.

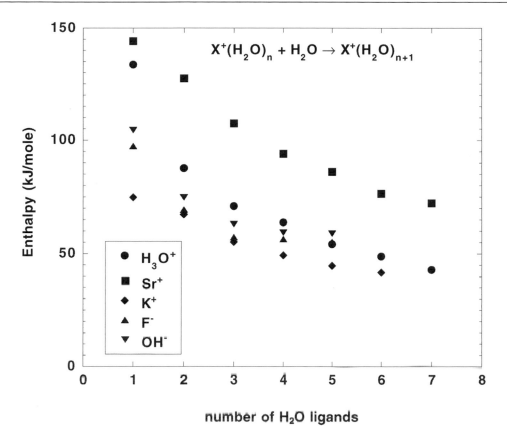

Figure A3.5.10. Bond strengths of water clustering to various core ions as a function of the number of water molecules.

The most fundamental of cluster properties are the bond strengths and entropy changes for the process [125]

$$A^{\pm} \cdot B_n + B + M = A^{\pm} \cdot B_{n=1} + M.$$

The thermodynamic quantities are derived from equilibrium measurements as a function of temperature. The measurements are frequently made in a high-pressure mass spectrometer [107]. The pertinent equation is $\ln(K_{n,n+1}) = -\Delta G^0/RT = -\Delta H^0/RT + \Delta S^0/T$. Another important method to determine bond strengths is from threshold measurements in collisional dissociation experiments [126]. Typically, ΔH^0 changes for $n = 0$ are 1.5 to several times the solution value [127]. The value usually drops monotonically with increasing n. The step size can have discontinuities as solvent shells are filled. The discontinuities in the thermodynamic properties appear as magic numbers in the mass spectra, i.e. ions of particular stability such as those with a closed solvation shell tend to be more abundant than those with one more or less ligand. A graph of bond strengths for H_2O bonding to several ions against cluster size is shown in figure A3.5.10 [125].

Clusters can undergo a variety of chemical reactions, some relevant only to clusters. The simplest reaction involving clusters is association, namely the sticking of ligands to an ionic core. For association to occur, the ion–neutral complex must release energy either by radiating or by collision with an inert third body. The latter is an important process in the earth's atmosphere [128–131] and the former in interstellar clouds [101]. Cluster ions can be formed by photon or electron interaction with a neutral cluster produced in a supersonic expansion [132]. Another process restricted to clusters is ligand-switching or the replacement of one ligand for another. Often exothermic ligand-switching reactions take place at rates near the gas kinetic limit, especially for small values of n [72, 133]. Chemical-reactivity studies as a function of cluster size show a variety of trends [93, 127, 133]. Proton-transfer reactions are often unaffected by solvation, while nucleophilic-displacement reactions are often shut down by as few as one or two solvent molecules. Increasing solvation number can also change the type of reactivity. A good example is the reaction of $NO^+(H_2O)_n$ with H_2O. These associate for small n but react to form $H_3O^+(H_2O)_n$ ions for $n = 3$. This is an important process in much of the earth's atmosphere. Neutral reactions have been shown to proceed up to 30 orders of magnitude faster when clustered to inert alkali ions than in the absence of the ionic clustering [134].

Caging is an important property in solution and insight into this phenomenon has been obtained by studying photodestruction of $Br_2^-(M)_n$ and $I_2^-(M)_n$ clusters, where M is a ligand such as Ar or CO_2. When the X_2^- core is photoexcited above the dissociation threshold of X_2^-, the competition between the two processes forming $X^-(M)_m$ and $X_2^-(M)_m$ indicates when caging is occurring. For $I_2^-(CO_2)_n$ the caging is complete at $n = 16$ [127, 135].

An important class of molecule often described as clusters may better be referred to as micro-particles. This class includes metal, semiconductor and carbon clusters. Particularly interesting are the carbon clusters, C_n^+. Mass spectra from a carbon cluster ion source show strong magic numbers at C_{60}^+ and C_{70}^+ [136]. This led to the discovery of the class of molecules called buckminsterfullerenes. Since that time, other polyhedra have been discovered, most notably metallocarbohedrenes [137]. The first species of this type discovered was $Ti_8C_{12}^+$. Much of the work done on metal clusters has been focused on the transition from cluster properties to bulk properties as the clusters become larger, e.g. the transition from quantum chemistry to band theory [127].

A3.5.3.4 *Atmospheric chemistry*

Atmospheric ions are important in controlling atmospheric electrical properties and communications and, in certain circumstances, aerosol formation [128, 130, 131, 138–145]. In addition, ion composition measurements can be used to derive trace neutral concentrations of the species involved in the chemistry. Figure A3.5.11 shows the total-charged-particle concentration as a function of altitude [146]. The total density varies between 10^3 and 10^6 ions cm^{-3}. The highest densities occur above 100 km. Below 100 km the total ion density is roughly constant even though the neutral density changes by a factor of approximately 4×10^6. Most negative charge is in the form of electrons above 80 km, while negative ions dominate below this altitude.

Above approximately 80 km, the prominent bulge in electron concentration is called the ionosphere. In this region ions are created from UV photoionization of the major constituents—O, NO, N_2 and O_2. The ionosphere has a profound effect on radio communications since electrons reflect radio waves with the same frequency as the plasma frequency, $f = 8.98 \times 10^3 n_e^{1/3}$, where n_e is the electron density in cm^{-3} [147]. The large gradient in electron density ensures that a wide variety of frequencies are absorbed. It is this phenomenon that allows one to hear distant radio signals. Ion chemistry plays a major role in determining the electron density. Diatomic ions recombine rapidly with electrons while monatomic ions do not. Monatomic positive ions do not destroy electrons until they are converted to diatomic ions. The most important reaction in the ionosphere is the reaction of O^+ with N_2,

$$O^+ + N_2 \rightarrow NO^+ + N.$$

Although this reaction is exothermic, the reaction has a small rate constant. This is one of the most studied ion–molecule reactions, and dependences on many parameters have been measured [148].

More complex ions are created lower in the atmosphere. Almost all ions below 70–80 km are cluster ions. Below this altitude range free electrons disappear and negative ions form. Three-body reactions become important. Even though the complexity of the ions increases, the determination of the final species follows

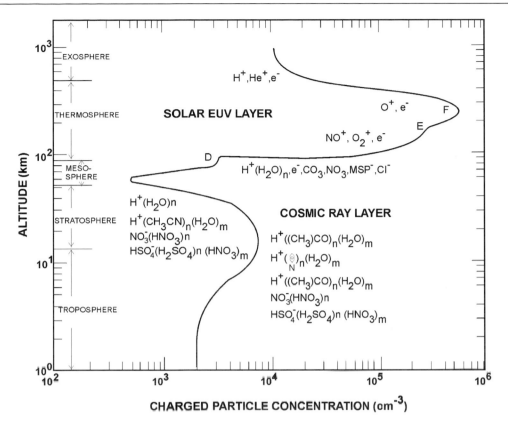

Figure A3.5.11. Charged particle concentrations in the atmosphere.

a rather simple scheme. For positive ions, formation of $H^+(H_2O)_n$ is rapid, occurring in times of the order of milliseconds or shorter in the stratosphere and troposphere. After formation of $H^+(H_2O)_n$, the chemistry involves reaction with species that have a higher proton affinity than that of H_2O. The resulting species can be written as $H^+(X)_m(H_2O)_n$. The main chemical processes include ligand exchange and proton transfer as well as association and dissociation of H_2O ligands. Examples of species X include NH_3 [149], CH_3COCH_3 [150] and CH_3CN [151]. The rate constants are large, so the proton hydrates are transformed even when the concentration of X is low.

The negative ion chemistry is equally clear. $NO_3^-(HNO_3)_m(H_2O)_n$ ions are formed rapidly. Only acids, HX, stronger than HNO_3 react with this series of ions producing $X^-(HX)_m(H_2O)_n$. Most regions of the atmosphere have low concentrations of such acids. The two exceptions are a layer of H_2SO_4 in the 30–40 km region [152, 153] and H_2SO_4 and CH_3SO_3H which play an important role near the ground under some circumstances [154].

Ion-composition measurements can be used to derive the concentrations of X and HX involved in the chemistry. This remains the only practical method of monitoring abundances of H_2SO_4 and CH_3CN in the upper atmosphere. Concentrations as low as 10^4 molecules cm^{-3} have been measured.

A3.5.3.5 Astrochemistry

The astrochemistry of ions may be divided into topics of interstellar clouds, stellar atmospheres, planetary atmospheres and comets. There are many areas of astrophysics (stars, planetary nebulae, novae, supernovae) where highly ionized species are important, but beyond the scope of 'ion chemistry'. (Still, molecules, including H_2O, are observed in solar spectra [155] and a surprise in the study of Supernova 1987A was the identification of molecular species, CO, SiO and possibly H_3^+ [156, 157].) In the early universe, after expansion had cooled matter to the point that molecules could form, the small fraction of positive and negative ions that remained was crucial to the formation of molecules, for example [156]

$$H^- + H \rightarrow H_2 + e^- \text{ and } H^+ + H \rightarrow H_2^+ + h\nu.$$

The formation of molecules was the first step toward local gravitational collapses which led to, among other things, the production of this encyclopedia.

Interstellar clouds of gases contain mostly H, H_2 and He, but the minority species are responsible for the interesting chemistry that takes place, just as in the earth's atmosphere. Interstellar clouds are divided into two types: *diffuse*, with atomic or molecular concentrations in the neighbourhood of 100 cm^{-3} and temperatures of 100–200 K, in which ionization is accomplished primarily by stellar UV light, and *dense* (or dark) clouds, with densities of 10^4–10^6 cm^{-3} and temperatures of 10–20 K, in which ionization is a result of galactic cosmic rays since visible and UV light cannot penetrate the dense clouds [156, 158]. The dense clouds also contain particulate matter (referred to as dust or grains). Close to 100 molecular species, as large as 13-atomic, have been detected in interstellar clouds by RF and MM spectroscopy; among these are nine types of molecular positive ion. It is assumed that the neutral molecular species (except H_2) are mainly synthesized through ion–molecule reactions, followed by electron–ion recombination, since neutral–neutral chemical reactions proceed very slowly in the cold temperatures of the clouds, except on grain surfaces. Ion–molecule reactions are typically even faster at lower temperatures. Extensive laboratory studies of ion–molecule reactions, including work at very low temperatures, have mapped out the reaction schemes that take place in interstellar clouds. In dense clouds the reactions

$$H_2^+ + H_2 \rightarrow H_3^+ + H \text{ and } He^+ + CO \rightarrow C^+ + O + He$$

are of paramount importance. These reactions are followed by reactions with C and H_2 to produce CH_3^+, that subsequently undergoes reaction with many neutral molecules to give ion products such as CH_5^+, $C_2H_5OH_2^+$ and CH_3CNH^+. Many of the reactions involve radiative association. Dissociative electron–ion recombination then yields neutrals such as CH_4 (methane), C_2H_5OH (ethanol) and CH_3CN (acetonitrile) [158]. It is often joked that diffuse interstellar clouds contain enough grain alcohol to keep space travellers happy on their long journeys. In diffuse clouds, the reaction scheme is more varied and leads to smaller molecules in general.

The above mentioned H_3^+ plays a central role in the lower ionospheres of the Jovian planets due to the dominance of H_2 in these atmospheres (>99% H_2 and He) [2, 157, 159–161]. Hydrocarbon ions are formed as a result of proton-transfer collisions with trace gases such as methane, ethane, ethylene and acetylene. Subsequent reactions build up larger hydrocarbon ions. Positive ions are destroyed in these ionospheres principally by neutralization through electron–ion recombination. The ionospheric chemistry of the inner planets is more complex due to the different concentrations and reactivity of the N_2, O_2, CO and CO_2 mixtures on these planets. In the lower ionospheres of the inner planets, at altitudes where molecular ions are stable, O_2^+ and NO^+ dominate, along with CO_2^+ on Mars.

Comets contain an *ion tail* that is tens of thousands of kilometres long. It results from solar UV photoionization of gases in the coma surrounding the comet nucleus [162]. A number of ions have been identified in comet comas including OH^+, H_2O^+, CH^+, CO_2^+, CO^+, N_2^+ and NH_3^+. Significantly, NH_4^+ and H_3O^+ have also been detected and are clearly the result of ion–molecule reactions. It is useful to model the ion processes as a way of learning about the concentrations of neutrals that must be present. Electron–ion recombination is the major loss mechanism.

A3.5.3.6   Plasma chemistry

The chemistry of plasmas is often one of neutral chemistry, since the concentration of neutral atoms and molecules usually exceeds the charged particle density by many orders of magnitude. Clearly, intense plasmas, such as in stellar interiors or in fusion plasmas, fall outside the scope of chemistry. Even in the weakly coupled plasmas it is the charged particles that are creating the neutral radicals and otherwise control the situation—the plasma decays when the source of ionization is turned off. An example is plasma etching of silicon surfaces: neutral F atoms produced in a glow discharge in a buffer gas seeded with SF_6, for example, react with surface atoms to produce gaseous SiF_4 and leave behind an etched surface.

The most important parameter describing a weakly coupled plasma is the Debye shielding length, or *Debye length* [36]. In the plasma, charges tend to cluster around a charge of opposite sign, even while undergoing random thermal motion. The cluster of charges shields the central charge so that the net radial potential function for the cluster is an exponential decay, with a decay constant λ_d, the Debye length. For an electron/positive-ion plasma, with a uniform spatial distribution of ions and a Boltzmann distribution of electron energies, the Debye length for each species is given by

$$\lambda_d = (\eta k T / 4\pi e^2 n)^{1/2}$$

where η is the dielectric constant of the medium, k is Boltzmann's constant, T and n are the temperature and number density of electrons or positive ions and e is the electron charge. The Debye length may be different for electrons and ions if the respective temperatures are not the same. The number densities of positive and negative charges in a plasma are always equal. The defining distinction between a plasma and a simple collection of free charges is that the *extent of a plasma must be much greater than* λ_d. A second useful notion is that of the *plasma sheath* that forms between the plasma and any physical boundary. Inside the sheath, charged particles are diffusing toward the boundary, where they are neutralized. The plasma sheath is of the order of a Debye length. Strong electric fields may exist across the plasma sheath. A good example is found in a glow discharge tube (e.g., a fluorescent lamp) where most of the change in the electric potential across the tube is in the plasma sheaths around the electrodes, not across the plasma itself.

Depending on the electron and ion temperatures in a plasma, all of the processes mentioned in this section of the encyclopedia may be taking place simultaneously in the plasma [163]. Understanding, or modelling, the plasma may be quite complicated [164]. Flame chemistry involves charged particles [165]. Most of the early investigations and classifications of electron and ion interactions came about in attempts to understand electric discharges, and these continue today in regard to electric power devices, such as switches and high-intensity lamps [166]. Often the goal is to prevent discharges in the face of high voltages. Military applications involving the earth's ionosphere funded refined work during and following the Second World War. Newer applications such as gas discharge lasers have driven recent studies of plasma chemistry. The rare-gas halide excimer laser is a marvellous example of plasma chemistry, because the lasing molecule may be formed in recombination between a positive and a negative ion, for example [167–169]

$$Ar^+ + F^- \rightarrow ArF^* \text{ (excimer state)} \rightarrow ArF \text{ (ground state)} + h\nu \text{ (193 nm)}$$

or

$$Ar_2^+ + F^- \rightarrow ArF^* \text{ (excimer state)} + Ar \rightarrow ArF \text{ (ground state)} + h\nu \text{ (193 nm)} + Ar.$$

The Ar_2^+ is formed from $Ar^* + Ar$, where the metastable Ar^* is a product of electron-impact or charge-transfer collisions. The F^- is formed by dissociative electron attachment to F_2 or NF_3. The population inversion required for light amplification is simple to obtain in the ArF laser since the ground state of the lasing molecule is not bound, except by van der Waals forces and quickly dissociates upon emission of the laser light.

References

[1] Evans R D 1955 *The Atomic Nucleus* (New York: McGraw-Hill)
[2] Oka T 1992 The infrared spectrum of H_3^+ in laboratory and space plasmas *Rev. Mod. Phys.* **64** 1141–9
[3] Loeb L B 1960 *Basic Processes of Gaseous Electronics* 2nd edn (Berkeley, CA: University of California)
[4] Franklin J L (ed) 1979 *Ion–Molecule Reactions, Part I, Kinetics and Dynamics* (Stroudsburg, PA: Dowden, Hutchinson and Ross)
[5] Muschlitz E E 1957 Formation of negative ions in gases by secondary collision processes *J. Appl. Phys.* **28** 1414–18
[6] Lavrich D J, Buntine M A, Serxner D and Johnson M A 1993 Excess energy-dependent photodissociation probabilities for O_2^- in water clusters: $O_2^- (H_2O)_n$, $1 \leq n \leq 33$ *J. Chem. Phys.* **99** 5910–16
[7] Lavrich D J, Buntine M A, Serxner D and Johnson M A 1995 Excess energy-dependent photodissociation probabilities for O_2^- in water clusters: $O_2^-.(H_2O)_n$, $1 \leq n \leq 33$ *J. Phys. C: Solid State Phys.* **99** 8453–7
[8] Zare R N and Dagdigian P J 1974 Tunable laser fluorescence method for product state analysis *Science* **185** 739–46
[9] Greene C H and Zare R N 1983 Determination of product population and alignment using laser-induced fluorescence *J. Chem. Phys.* **78** 6741–53
[10] Crosley D R 1981 Collisional effects on laser-induced fluorescence *Opt. Eng.* **20** 511–21
[11] Crosley D R 1996 Applications to combustion *Atomic, Molecular, and Optical Physics Handbook* ed G W F Drake (Woodbury, NY: AIP)
[12] Altkorn R and Zare R N 1984 Effects of saturation on laser-induced fluorescence measurements of population and polarization *Annual Review of Physical Chemistry* ed B S Rabinovitch, J M Schurr and H L Strauss (Palo Alto, CA: Annual Reviews)
[13] Hamilton C E, Bierbaum V M and Leone S R 1985 Product vibrational state distributions of thermal energy charge transfer reactions determined by laser-induced fluorescence in a flowing afterglow: $Ar^+ + CO \rightarrow CO^+(v = 0–6) + Ar$ *J. Chem. Phys.* **83** 2284–92
[14] Sonnenfroh D M and Leone S R 1989 A laser-induced fluorescence study of product rotational state distributions in the charge transfer reaction: $Ar^+(^2P_{3/2}) + N_2 \rightarrow Ar + N_2^+(X)$ at 0.28 and 0.40 eV *J. Chem. Phys.* **90** 1677–85
[15] Dressler R A, Meyer H and Leone S R 1987 Laser probing of the rotational alignment of N_2^+ drifted in He *J. Chem. Phys.* **87** 6029–39
[16] Anthony E B, Schade W, Bastian M J, Bierbaum V M and Leone S R 1997 Laser probing of velocity-subgroup dependent rotational alignment of N_2^+ drifted in He *J. Chem. Phys.* **106** 5413–22
[17] Duncan M A, Bierbaum V M, Ellison G B and Leone S R 1983 Laser-induced fluorescence studies of ion collisional excitation in a drift field: rotational excitation of N_2^+ in He *J. Chem. Phys.* **79** 5448–56
[18] Leone S R 1989 Laser probing of ion collisions in drift fields: state excitation, velocity distributions, and alignment effects *Gas Phase Bimolecular Collisions* ed M N R Ashford and J E Baggott (London: Royal Society of Chemistry)
[19] de Gouw J A, Krishnamurthy M and Leone S R 1997 The mobilities of ions and cluster ions drifting in polar gases *J. Chem. Phys.* **106** 5937–42
[20] Kato S, Frost M J, Bierbaum V M and Leone S R 1994 Vibrational specificity for charge transfer versus deactivation in N_2^+ $(v = 0, 1, 2) + Ar$ and O_2 reactions *Can. J. Chem.* **72** 625–36
[21] Xie J and Zare R N 1992 Determination of the absolute thermal rate constants for the charge-transfer reaction $DBr^+(^2\Pi, v^+)+HBr \rightarrow HBr^+(^2\Pi', v'^+) + DBr$ *J. Chem. Phys.* **96** 4293–302
[22] Green R J, Xie J, Zare R N, Viggiano A A and Morris R A 1997 Rate constants and products for the reaction of HBr^+ with HBr and DBr *Chem. Phys. Lett.* **277** 1–5
[23] deNeufville J P, Kasden A and Chimenti R J L 1981 Selective detection of uranium by laser-induced fluorescence: a potential remote-sensing technique. 1: Optical characteristics of uranyl geologic targets *Appl. Opt.* **20** 1279–96
[24] von Busch F and Dunn G H 1972 Photodissociation of H_2^+ and D_2^+: experimental *Phys. Rev.* A **5** 1726–43
[25] Ozenne J-B, Pham D and Durup J 1972 Photodissociation of H_2^+ by monochromatic light with energy analysis of the ejected H^+ ions *Chem. Phys. Lett.* **17** 422–4
[26] van Asselt N P F B, Maas J G and Los L 1974 Laser induced photodissociation of H_2^+ ions *Chem. Phys. Lett.* **24** 555–8
[27] Lee L C, Smith G P, Miller T M and Cosby P C 1978 Photodissociation cross sections of Ar_2^+, Kr_2^+, and Xe_2^+ from 6200 to 8600 Å *Phys. Rev.* A **17** 2005–11
[28] Lee L C and Smith G P 1979 Photodissociation cross sections of Ne_2^+, Ar_2^+, Kr_2^+, and Xe_2^+, from 3500 to 5400 Å *Phys. Rev.* A **19** 2329–34
[29] Lee L C, Smith G P, Moseley J T, Cosby P C and Guest J A 1979 Photodissociation and photodetachment of Cl_2^-, ClO^-, Cl_3^- and $BrCl_2^-$ *J. Chem. Phys.* **70** 3237–46
[30] Moseley J T 1984 Determination of ion molecular potential curves using photodissociative processes *Applied Atomic Collision Physics* ed H S W Massey, E W McDaniel and B Bederson (New York: Academic)
[31] Moseley J and Durup J 1981 Fast ion beam photofragment spectroscopy *Annual Review of Physical Chemistry* ed B S Rabinovitch, J M Schurr and H L Strauss (Palo Alto, CA: Annual Reviews)
[32] Moseley J T 1985 Ion photofragment spectroscopy *Photodissociation and Photoionization* ed K P Lawley (New York: Wiley)
[33] Huber B A, Miller T M, Cosby P C, Zeman H D, Leon R L, Moseley J T and Peterson J R 1977 Laser-ion coaxial beams spectrometer *Rev. Sci. Instrum.* **48** 1306–13
[34] Lee Y-Y, Leone S R, Champkin P, Kaltoyannis N and Price S D 1997 Laser photofragmentation and collision-induced reactions of SiF_2^{2+} and SiF_3^{2+} *J. Chem. Phys.* **106** 7981–94

[35] Farrar J M 1993 Electronic photodissociation spectroscopy of mass-selected clusters: solvation and the approach to the bulk *Cluster Ions* ed C Y Ng and I Provis (New York: Wiley)

[36] McDaniel E W 1989 *Atomic Collisions: Electron and Photon Projectiles* (New York: Wiley)

[37] Pegg D J 1996 Photodetachment *Atomic, Molecular, and Optical Physics Handbook* ed G W F Drake (Woodbury, NY: AIP)

[38] Dessent C E H and Johnson M A 1998 Fundamentals of negative ion photoelectron spectroscopy *Fundamentals and Applications of Gas Phase Ion Chemistry* ed K R Jennings (Berlin: Kluwer)

[39] Lineberger W C 1982 Negative ion photoelectron spectroscopy *Applied Atomic Collision Physics, Vol 5, Special Topics* ed H S W Massey, E W McDaniel and B Bederson (New York: Academic)

[40] Mead R D, Stevens A E and Lineberger W C 1984 Photodetachment in negative ion beams *Gas Phase Ion Chemistry* ed M T Bowers (New York: Academic)

[41] Drzaic P S, Marks J and Brauman J I 1984 Electron photodetachment from gas phase molecular anions *Gas Phase Ion Chemistry: Ions and Light* ed M T Bowers (New York: Academic)

[42] Cordermann R R and Lineberger W C 1979 Negative ion spectroscopy *Annual Review of Physical Chemistry* ed B S Rabinovitch, J M Schurr and H L Strauss (Palo Alto, CA: Annual Reviews)

[43] Miller T M 1981 Photodetachment and photodissociation of ions *Advances in Electronics and Electron Physics* ed L Marton and C Marton (New York: Academic)

[44] Esaulov A V 1986 Electron detachment from atomic negative ions *Ann. Phys., Paris* **11** 493–592

[45] Dunbar R C 1979 Ion photodissociation *Gas Phase Ion Chemistry* ed M T Bowers (New York: Academic)

[46] Wang L, Lee Y T and Shirley D A 1987 Molecular beam photoelectron spectroscopy of SO_2: geometry, spectroscopy, and dynamics of SO_2^+ *J. Chem. Phys.* **87** 2489–97

[47] Pollard J E, Trevor D J, Lee Y T and Shirley D A 1981 Photoelectron spectroscopy of supersonic molecular beams *Rev. Sci. Instrum.* **52** 1837–46

[48] Miller T M, Leopold D G, Murray K K and Lineberger W C 1986 Electron affinities of the alkali halides and the structure of their negative ions *J. Chem. Phys.* **85** 2368–75

[49] Wang L-S, Ding C-F, Wang X-B and Nicholas J B 1998 Probing the potential barriers in intramolecular electrostatic interactions in free doubly charged anions *Phys. Rev. Lett.* at press

[50] Cooper J and Zare R N 1968 Angular distributions of photoelectrons *J. Chem. Phys.* **48** 942–3

[51] Yourshaw I, Zhao Y and Neumark D M 1996 Many-body effects in weakly bound anion and neutral clusters: zero electron kinetic energy spectroscopy and threshold photodetachment spectroscopy of Ar_nBr^- ($n = 2–9$) and Ar_nI^- ($n = 2–19$) *J. Chem. Phys.* **105** 351–73

[52] Wang K and McKoy V 1995 High-resolution photoelectron spectroscopy of molecules *Annual Review of Physical Chemistry* ed H L Strauss, G T Babcock and S R Leone (Palo Alto, CA: Annual Reviews)

[53] Stenholm S 1996 Absorption and gain spectra *Atomic, Molecular, and Optical Physics Handbook* ed G F W Drake (New York: AIP)

[54] Miller T A 1982 Light and radical ions *Annual Review of Physical Chemistry* ed B S Rabinovitch, J M Schurr and H L Strauss (Palo Alto, CA: Annual Reviews)

[55] Woods R C, Saykally R J, Anderson T G, Dixon T A and Szanto P G 1981 The molecular structure of HCO^+ by the microwave substitution method *J. Chem. Phys.* **75** 4256–60

[56] Saykally R J and Woods R C 1981 High resolution spectroscopy of molecular ions *Annual Reviews of Physical Chemistry* ed B S Rabinovitch, J M Schurr and H L Strauss (Palo Alto, CA: Annual Reviews)

[57] Haese N N, Pan F-S and Oka T 1983 Doppler shift and ion mobility measurements of ArH^+ in a He DC glow discharge by infrared laser spectroscopy *Phys. Rev. Lett.* **50** 1575–8

[58] Gudeman C S and Saykally R J 1984 Velocity modulation infrared laser spectroscopy of molecular ions *Annual Review of Physical Chemistry* ed B S Rabinovitch, J M Schurr and H L Strauss (Palo Alto, CA: Annual Reviews)

[59] Huet T R, Kabbadj Y, Gabrys C M and Oka T 1994 The $\nu_2 + \nu_3 - \nu_2$ band of NH_3^+ *J. Mol. Spectrosc.* **163** 206–13

[60] Kabbadj Y, Huet T R, Uy D and Oka T 1996 Infrared spectroscopy of the amidogen ion, NH_2^+ *J. Mol. Spectrosc.* **175** 277–88

[61] Rosenbaum N H, Owrutsky J C, Tack L M and Saykally R J 1986 Velocity modulation laser spectroscopy of negative ions: the infrared spectrum of hydroxide (OH^-) *J. Chem. Phys.* **84** 5308–13

[62] Wing W H, Ruff G A, Lamb W E and Spezeski J J 1976 Observation of the infrared spectrum of the hydrogen molecular ion HD^+ *Phys. Rev. Lett.* **36** 1488–91

[63] Carrington A and McNab I R 1989 The infrared predissociation spectrum of H_3^+ *Accounts Chem. Res.* **22** 218–22

[64] Neumark D M, Lykke K R, Andersen T and Lineberger W C 1985 Infrared spectrum and autodetachment dynamics of NH^- *J. Chem. Phys.* **83** 4364–73

[65] Gioumousis G and Stevenson D P 1958 Reactions of gaseous molecule ions with gaseous molecules. V. Theory *J. Chem. Phys.* **29** 294–9

[66] Su T and Chesnavich W J 1982 Parametrization of the ion–polar molecule collision rate constant by trajectory calculations *J. Chem. Phys.* **76** 5183–5

[67] Su T 1985 Kinetic energy dependences of ion polar molecule collision rate constants by trajectory calculations *J. Chem. Phys.* **82** 2164–6

[68] Troe J 1992 Statistical aspects of ion–molecule reactions *State-Selected and State-to-State Ion–Molecule Reaction Dynamics: Theory* ed M Baer M and C-Y Ng (New York: Wiley)

[69] Clary D C, Smith D and Adams N G 1985 Temperature dependence of rate coefficients for reactions of ions with dipolar molecules *Chem. Phys. Lett.* **119** 320–6

[70] Bhowmik P K and Su T 1986 Trajectory calculations of ion–quadrupolar molecule collision rate constants *J. Chem. Phys.* **84** 1432–4

[71] Su T, Viggiano A A and Paulson J F 1992 The effect of the dipole-induced dipole potential on ion–polar molecule collision rate constants *J. Chem. Phys.* **96** 5550–1

[72] Ikezoe Y, Matsuoka S, Takebe M and Viggiano A A 1987 *Gas Phase Ion–Molecule Reaction Rate Constants Through 1986* (Tokyo: Maruzen)

[73] Fehsenfeld F C 1975 Associative Detachment *Interactions Between Ions and Molecules* ed P Ausloos (New York: Plenum)

[74] Viggiano A A and Paulson J F 1983 Temperature dependence of associative detachment reactions *J. Chem. Phys.* **79** 2241–5

[75] Ferguson E E 1972 Review of laboratory measurements of aeronomic ion–neutral reactions *Ann. Geophys.* **28** 389

[76] Van Doren J M, Miller T M, Miller A E S, Viggiano A A, Morris R A and Paulson J F 1993 Reactivity of the radical anion OCC^- *J. Am. Chem. Soc.* **115** 7407–14

[77] Adams N G, Bohme D K, Dunkin D B and Fehsenfeld F C 1970 Temperature dependence of the rate coefficients for the reactions of Ar with O_2, H_2, and D_2 *J. Chem. Phys.* **52** 1951

[78] Rebrion C, Rowe B R and Marquette J B 1989 Reactions of Ar^+ with $H_2 N_2 O_2$ and CO at 20, 30, and 70 K *J. Chem. Phys.* **91** 6142–7

[79] Midey A J and Viggiano A A 1998 Rate constants for the reaction of Ar^+ with O_2 and CO as a function of temperature from 300 to 1400 K: derivation of rotational and vibrational energy effects *J. Chem. Phys.* at press

[80] Dotan I and Viggiano A A 1993 Temperature, kinetic energy, and rotational temperature dependences for the reactions of $Ar^+ (^2P_{3/2})$ with O_2 and CO *Chem. Phys. Lett.* **209** 67–71

[81] Smith M A 1994 Ion–molecule reaction dynamics at very low temperatures *Unimolecular and Bimolecular Ion–Molecule Reaction Dynamics* ed C-Y Ng, T Baer and I Powis (New York: Wiley)

[82] Olmstead W N and Brauman J I 1977 Gas phase nucleophilic displacement reactions *J. Am. Chem. Soc.* **99** 4219–28

[83] Seeley J V, Morris R A, Viggiano A A, Wang H and Hase W L 1997 Temperature dependencies of the rate constants and branching ratios for the reactions of $Cl^- (H_2O)_{0-3}$ with $CH_3 Br$ and thermal dissociation rates for $Cl^- (CH_3 Br)$ *J. Am. Chem. Soc.* **119** 577–84

[84] Viggiano A A and Morris R A 1996 Rotational and vibrational energy effects on ion–molecule reactivity as studied by the VT-SIFDT technique *J. Phys. Chem.* **100** 19 227–40

[85] Ferguson E E, Fehsenfeld F C, Dunkin D B, Schmeltekopf A L and Schiff H I 1964 Laboratory studies of helium ion loss processes of interest in the ionosphere *Planet. Space Sci.* **12** 1169–71

[86] Graul S T and Squires R R 1988 Advances in flow reactor techniques for the study of gas-phase ion chemistry *Mass Spectrom. Rev.* **7** 263–358

[87] Ferguson E E, Fehsenfeld F C and Schmeltekopf A L 1969 Flowing afterglow measurements of ion–neutral reactions *Adv. At. Mol. Phys.* **5** 1–56

[88] Ferguson E E 1992 A personal history of the early development of the flowing afterglow technique for ion molecule reactions studies *J. Am. Soc. Mass Spectrom.* **3** 479–86

[89] Adams N G and Smith D 1976 The selected ion flow tube (SIFT); a technique for studying ion–neutral reactions *Int. J. Mass Spectrom. Ion Phys.* **21** 349

[90] Smith D and Adams N G 1988 The selected ion flow tube (SIFT): studies of ion–neutral reactions *Adv. At. Mol. Phys.* **24** 1–49

[91] Adams N G and Smith D 1988 Flowing afterglow and SIFT *Techniques for the Study of Ion–Molecule Reactions* ed J M Farrar and W H Saunders Jr (New York: Wiley)

[92] Hierl P M *et al* 1996 Flowing afterglow apparatus for the study of ion–molecule reactions at high temperatures *Rev. Sci. Instrum.* **67** 2142–8

[93] Viggiano A A, Arnold S T and Morris R A 1998 Reactions of mass selected cluster ions in a thermal bath gas *Int. Rev. Phys. Chem.* **17** 147–84

[94] McFarland M, Albritton D L, Fehsenfeld F C, Ferguson E E and Schmeltekopf A L 1973 Flow-drift technique for ion mobility and ion–molecule reaction rate constant measurements. I. Apparatus and mobility measurements *J. Chem. Phys.* **59** 6610–19

[95] McFarland M, Albritton D L, Fehsenfeld F C, Ferguson E E and Schmeltekopf A L 1973 Flow-drift technique for ion mobility and ion–molecule reaction rate constant measurements. II. Positive ion reactions of N^+, O^+, and N_2^+ with O_2 and O^+ with N_2 from thermal to ≈ 2 eV *J. Chem. Phys.* **59** 6620–8

[96] McFarland M, Albritton D L, Fehsenfeld F C, Ferguson E E and Schmeltekopf A L 1973 Flow-drift technique for ion mobility and ion–molecule reaction rate constant measurements. III. Negative ion reactions of O^- + CO, NO, H_2, and D_2 *J. Chem. Phys.* **59** 6629–35

[97] Lindinger W and Smith D 1983 Influence of translational and internal energy on ion–neutral reactions *Reactions of Small Transient Species* ed A Fontijn and M A A Clyne (New York: Academic)

[98] Viehland L A and Robson R E 1989 Mean energies of ion swarms drifting and diffusing through neutral gases *Int. J. Mass Spectrom. Ion Processes* **90** 167–86

[99]   Kemper P R and Bowers M T 1988 Ion cyclotron resonance spectrometry *Techniques for the Study of Ion–Molecule Reactions* ed J M Farrar and W H Saunders Jr (New York: Wiley)
[100] Freiser B S 1988 Fourier Transform Mass Spectrometry *Techniques for the Study of Ion–Molecule Reactions* ed J M Farrar and W H Saunders Jr (New York: Wiley)
[101] Dunbar R C 1994 Ion–molecule radiative association *Unimolecular and Bimolecular Ion–Molecule Reaction Dynamic* ed C-Y Ng, T Baer and I Powis (New York: Wiley)
[102] Heninger M, Fenistein S, Durup-Ferguson M, Ferguson E E, Marx R and Mauclaire G 1986 Radiative lifetime for $v = 1$ and $v = 2$ ground state NO$^+$ ions *Chem. Phys. Lett.* **131** 439–43
[103] Gerlich D 1992 Inhomogeneous RF fields: a versatile tool for the study of processes with slow ions *State-Selected and State-to-State Ion–Molecule Reaction Dynamics: Part 1. Experiment* ed C Ng and M Baer (New York: Wiley)
[104] Watson L R, Thiem T L, Dressler R A, Salter R H and Murad E 1993 High temperature mass spectrometric studies of the bond energies of gas-phase ZnO, NiO, and CuO *J. Phys. Chem.* **97** 5577–80
[105] Schultz R H and Armentrout P B 1991 Reactions of N$_4^+$ with rare gases from thermal to 10 eV center of mass energy: collision induced dissociation, charge transfer, and ligand exchange *Int. J. Mass Spectrom. Ion Proc.* **107** 29–48
[106] Futrell J H 1992 Crossed-molecular beam studies of state-to-state reaction dynamics *State-Selected and State-to-State Ion–Molecule Reaction Dynamics: Part 1. Experiment* ed C Ng and M Baer (New York: Wiley)
[107] Kebarle P 1988 Pulsed electron high pressure mass spectrometer *Techniques for the Study of Ion–Molecule Reactions* ed J M Farrar and W H Saunders Jr (New York: Wiley)
[108] Knighton W B and Grimsurd E P 1996 Gas phase ion chemistry under conditions of very high pressure *Advances in Gas Phase Ion Chemistry* ed N G Adams and L M Babcock (JAI)
[109] Seeley J V, Jayne J T and Molina M J 1993 High pressure fast-flow technique for gas phase kinetics studies *Int. J. Chem. Kinet.* **25** 571–94
[110] Huber K P and Herzberg G 1979 *Molecular Spectra and Molecular Structure. IV. Constants of Diatomic Molecules* (New York: Van Nostrand Reinhold)
[111] Mallard W G and Linstrom P J (eds) 1988 *NIST Standard Reference Database* 69:http://webbook.nist.gov/chemistry/
[112] Bates D R 1991 Negative ions: structure and spectra *Advances in Atomic, Molecular and Optical Physics* ed D R Bates and B Bederson (New York: Academic)
[113] Shida T 1991 Photochemistry and spectroscopy of organic ions and radicals *Annual Review of Physical Chemistry* ed H L Strauss, G T Babcock and S R Leone (Palo Alto, CA: Annual Reviews)
[114] Castleman A W and Märk T D 1986 Cluster ions: their formation, properties, and role in elucidating the properties of matter in the condensed state *Gaseous Ion Chemistry and Mass Spectrometry* ed J H Futrell (New York: Wiley)
[115] Van Doren J M, Barlow S E, DePuy C H, Bierbaum V M, Dotan I and Ferguson E E 1986 Chemistry and structure of the CH$_3$O$_2^+$ product of the O$_2^+$ + CH$_4$ reaction *J. Phys. Chem.* **90** 2772–7
[116] Papanikolas J M, Gord J R, Levinger N E, Ray D, Vorsa V and Lineberger W C 1991 Photodissociation and geminate recombination dynamics of I$_2^-$ in mass-selected I$_2^-$(CO$_2$)$_n$ cluster ions *J. Phys. Chem.* **90** 8028–40
[117] Lias S G, Bartmess J E, Liebman J F, Holmes J L, Levin R D and Mallard W G 1988 Gas-phase ion and neutral thermochemistry *J. Phys. Chem. Ref. Data* **17, Supplement 1** 1–861
[118] Lias S G 1998 Ionization energy evaluation *NIST Chemistry WebBook, NIST Standard Reference Database Number 69* ed W G Mallard and P J Linstrom (Gaithersburg, MD: National Institute of Standards and Technology)
[119] Lias S G 1997 Ionization energies of gas phase molecules *Handbook of Chemistry and Physics* ed D R Lide (Boca Raton, FL: CRC Press)
[120] Hunter E P and Lias S G 1998 Proton affinity evaluation (WebBook) *NIST Standard Reference Database Number 69* ed W G Mallard and P J Linstrom (Gaithersburg, MD: National Institute of Standards and Technology)
[121] Bartmess J E 1998 Negative ion energetics data *NIST Chemistry WebBook, NIST Standard Reference Database Number 69* ed W G Mallard and P J Linstrom (Gaithersburg, MD: National Institute of Standards and Technology)
[122] Miller T M 1997 Electron affinities *Handbook of Chemistry and Physics* ed D R Lide (Boca Raton, FL: CRC)
[123] Mallard W G and Linstrom P J (eds) 1998 *NIST Standard Reference Database No 69 (Gaithersburg)* http://webbook.nist.gov
[124] Ervin K M, Gronert S, Barlow S E, Gilles M K, Harrison A G, Bierbaum V M, DePuy C H, Lineberger W C and Ellison G B 1990 Bond strengths of ethylene and acetylene 1990 *J. Am. Chem. Soc.* **112** 5750–9
[125] Keesee R G and Castleman A W Jr 1986 Thermochemical data on gas-phase ion–molecule association and clustering reactions *J. Phys. Chem. Ref. Data* **15** 1011
[126] Armentrout P B and Baer T 1996 Gas phase ion dynamics and chemistry *J. Phys. Chem.* **100** 12 866–77
[127] Castleman A W Jr and Bowen K H Jr 1996 Clusters: structure, energetics, and dynamics of intermediate states of matter *J. Phys. Chem.* **100** 12 911–44
[128] Viggiano A A and Arnold F 1995 Ion chemistry and composition of the atmosphere *Atmospheric Electrodynamics* ed H Volland (Boca Raton, FL: CRC Press)
[129] Viggiano A A 1993 *In-situ* mass spectrometry and ion chemistry in the stratosphere and troposphere *Mass Spectrom. Rev.* **12** 115–37
[130] Ferguson E E, Fehsenfeld F C and Albritton D L 1979 Ion chemistry of the earth's atmosphere *Gas Phase Ion Chemistry* ed M T Bowers (San Diego, CA: Academic)

[131] Ferguson E E 1979 Ion–molecule reactions in the atmosphere *Kinetics of Ion–Molecule Reactions* ed P Ausloos (New York: Plenum)

[132] Johnson M A and Lineberger W C 1988 Pulsed methods for cluster ion spectroscopy *Techniques for the Study of Ion–Molecule Reactions* ed J M Farrar and W H Saunders Jr (New York: Wiley)

[133] Bohme D K and Raksit A B 1984 Gas phase measurements of the influence of stepwise solvation on the kinetics of nucleophilic displacement reactions with CH_3Cl and CH_3Br at room temperature *J. Am. Chem. Soc.* **106** 3447–52

[134] Viggiano A A, Deakyne C A, Dale F and Paulson J F 1987 Neutral reactions in the presence of alkali ions *J. Chem. Phys.* **87** 6544–52

[135] Vorsa V, Campagnola P J, Nandi S, Larsson M and Lineberger W C 1996 Protofragments of $I_2^-.Ar_n$ clusters: observation of metastable isomeric ionic fragments *J. Chem. Phys.* **105** 2298–308

[136] Kroto H W, Heath J R, O'Brian S C, Curl R F and Smalley R E 1985 C_{60}: Buckminsterfullerene *Nature* **318** 162–3

[137] Guo B C, Kerns K P and Castleman A W Jr 1992 $Ti_8C_{12}^+$-metallo-carbohedrenes: a new class of molecular clusters? *Science* **255** 1411–13

[138] Reid G C 1976 Ion chemistry of the D-region *Advances in Atomic and Molecular Physics* ed D R Bates and B Bederson (Orlando, FL: Academic)

[139] Smith D and Adams N G 1980 Elementary plasma reactions of environmental interest *Topics in Current Chemistry* ed F L Boschke (Berlin: Springer)

[140] Ferguson E E and Arnold F 1981 Ion chemistry of the stratosphere *Accounts. of Chem. Res.* **14** 327–34

[141] Thomas L 1983 Modelling of the ion composition of the middle atmosphere *Ann. Geophys.* **1** 61–73

[142] Arnold F and Viggiano A A 1986 Review of rocket-borne ion mass spectrometry in the middle atmosphere *Middle Atmosphere Program Handbook, Vol. 19* ed R A Goldberg (Urbana, IL: SCOSTEP)

[143] Brasseur G and Solomon S 1986 *Aeronomy of the Middle Atmosphere* 2nd edn (Boston, MA: D Reidel)

[144] Brasseur G and De Baets P 1986 Ions in the mesosphere and lower thermosphere: a two-dimensional model *J. Geophys. Res.* **91** 4025–46

[145] Viggiano A A, Morris R A and Paulson J F 1994 Effects of O_2^- and SF_6 vibrational energy on the rate constant for charge transfer between O_2^- and SF_6 *Int. J. Mass Spectrom. Ion Processes* **135** 31–7

[146] Arnold F 1980 The middle atmosphere ionized component *Vth ESA-PAC Symposium on European Rocket and Balloon Programmes and Related Research* (Bournemouth, UK: ESA) pp 479–95

[147] Book D L 1987 *NRL Plasma Formulary* (Washington, DC: Naval Research Laboratory)

[148] Hierl P M, Dotan I, Seeley J V, Van Doren J M, Morris R A and Viggiano A A 1997 Rate constants for the reactions of O^+ with N_2 and O_2 as a function of temperature (300–1800 K) *J. Chem. Phys.* **106** 3540–4

[149] Eisele F L 1986 Identification of tropospheric ions *J. Geophys. Res.* **91** 7897–906

[150] Hauck G and Arnold F 1984 Improved positive-ion composition measurements in the upper troposphere and lower stratosphere and the detection of acetone *Nature* **311** 547–50

[151] Schlager H and Arnold F 1985 Balloon-borne fragment ion mass spectrometry studies of stratospheric positive ions: unambiguous detection of $H^+(CH_3CN)_1(H_2O)$-clusters *Planet. Space Sci.* **33** 1363–6

[152] Viggiano A A and Arnold F 1981 The first height measurements of the negative ion composition of the stratosphere *Planet. Space Sci.* **29** 895–906

[153] Arnold F and Henschen G 1978 First mass analysis of stratospheric negative ions *Nature* **257** 521–2

[154] Eisele F L 1989 Natural and anthropogenic negative ions in the troposphere *J. Geophys. Res.* **94** 2183–96

[155] Oka T 1997 Water on the sun—molecules everywhere *Science* **277** 328–9

[156] Dalgarno A and Lepp S 1996 Applications of atomic and molecular physics to astrophysics *Atomic, Molecular, and Optical Physics Handbook* ed G W F Drake (Woodbury, NY: AIP)

[157] Dalgarno A and Fox J 1994 Ion chemistry in atmospheric and astrophysical plasmas *Unimolecular and Bimolecular Ion–Molecule Reaction Dynamics* ed C-Y Ng, T Baer and I Powis (New York: Wiley)

[158] Smith D and Spanel P 1995 Ions in the terrestrial atmosphere and in interstellar clouds *Mass Spectrom. Rev.* **14** 255–78

[159] Dalgarno A 1994 Terrestrial and extraterrestrial H_3^+ *Advances in Atomic, Molecular and Optical Physics* ed B Bederson and A Dalgarno (New York: Academic)

[160] Fox J L 1996 Aeronomy *Atomic, Molecular, and Optical Physics Handbook* ed G W F Drake (Woodbury, NY: AIP)

[161] Geballe T R and Oka T 1996 Detection of H3+ in interstellar space *Nature* **384** 334–5

[162] Haeberli R M, Altwegg K, Balsiger H and Geiss J 1995 Physics and chemistry of ions in the pile-up region of comet P/Halley *Astron. Astrophys.* **297** 881–91

[163] Capitelli M, Celiberto R and Cacciatore M 1994 Needs for cross sections in plasma chemistry *Advances in Atomic, Molecular and Optical Physics* ed B Bederson, H Walther and M Inokuti (New York: Academic)

[164] Garscadden A 1996 Conduction of electricity in gases *Atomic, Molecular, and Optical Handbook* ed G W F Drake (Woodbury, NY: AIP)

[165] Fontijn A 1982 Combustion and flames *Applied Atomic Collision Physics, Vol 5, Special Topics* ed H S W Massey, E W McDaniel and B Bederson (New York: Academic)

[166] Weymouth J F 1982 Collision phenomena in electrical discharge lamps *Applied Atomic Collision Physics, Vol 5, Special Topics* ed H S W Massey, E W McDaniel and B Bederson (New York: Academic)

[167] Huestis D L 1982 Introduction and overview *Applied Atomic Collision Physics, Vol 3, Gas Lasers* ed H S W Massey, E W McDaniel, B Bederson and W L Nighan (New York: Academic)

[168] Chantry P J 1982 Negative ion formation in gas lasers *Applied Atomic Collision Physics Vol 3, Gas Lasers* ed H S W Massey, E W McDaniel, B Bederson and W L Nighan (New York: Academic)

[169] Rokni M and Jacob J H 1982 Rare-gas halide lasers *Applied Atomic Collision Physics, Vol 3, Gas Lasers* ed H S W Massey, E W McDaniel, B Bederson and W L Nighan (New York: Academic)

Further Reading

Farrar J M and Saunders W H Jr (eds) 1988 *Techniques for the Study of Ion–Molecule Reactions* (New York: Wiley)

The best place to start for a detailed look at the instrumentation for the study of ion–molecule chemistry.

Ng C-Y, Baer T and Powis I (eds) 1994 *Unimolecular and Bimolecular Ion–Molecule Reaction Dynamics* (New York: Wiley)

An excellent reference for recent work on ion chemistry.

Ng C-Y and Baer T (eds) 1992 *State-Selected and State-to-State Ion–Molecule Reaction Dynamics* vols 1 and 2 (New York: Wiley)

A comprehensive look at the effect of state selection on ion–molecule reactions from both experimental and theoretical viewpoints.

Bowers M T (ed) 1979 and 1984 *Gas Phase Ion Chemistry* vols 1–3 (New York: Academic)

An older look at the field of ion chemistry. Most of the concepts are still valid and this series is a good foundation for a beginner in the field.

Adams N G and Babcock L M (eds) 1992, 1996 and 1998 *Advances in Gas Phase Ion Chemistry* vols 1–3 (Greenwich, CT: JAI)

On ongoing series about current topics in ion chemistry.

Drake G W F (ed) 1996 *Atomic, Molecular, and Optical Physics Handbook* (Woodbury, NY: AIP)

Fundamental chemical physics descriptions of both ion and neutral processes.

A3.6
Chemical kinetics in condensed phases

Jorg Schroeder

A3.6.1   Introduction

The transition from the low-pressure gas to the condensed phase is accompanied by qualitative changes in the kinetics of many reactions caused by the presence of a dense solvent environment permanently interacting with reactants (also during their motion along the reaction path). Though this solvent influence in general may be a complex phenomenon as contributions of different origin tend to overlap, it is convenient to single out aspects that dominate the kinetics of certain types of reactions under different physical conditions.

Basic features of solvent effects can be illustrated by considering the variation of the rate constant k_{uni} of a unimolecular reaction as one gradually passes from the low-pressure gas phase into the regime of liquid-like densities [1] (see figure A3.6.1). At low pressures, where the rate is controlled by thermal activation in isolated binary collisions with bath gas molecules, k_{uni} is proportional to pressure, i.e. it is in the low-pressure limit k_0. Raising the pressure further, one reaches the fall-off region where the pressure dependence of k_{uni} becomes increasingly weaker until, eventually, it attains the constant so-called high-pressure limit k_∞. At this stage, collisions with bath gas molecules, which can still be considered as isolated binary events, are sufficiently frequent to sustain an equilibrium distribution over rotational and vibrational degrees of freedom of the reactant molecule, and k_∞ is determined entirely by the intramolecular motion along the reaction path. k_∞ may be calculated by statistical theories (see chapter A3.4) if the potential-energy (hyper)surface (PES) for the reaction is known. What kind of additional effects can be expected, if the density of the compressed bath gas approaches that of a dense fluid? Ideally, there will be little further change, as equilibration becomes even more effective because of permanent energy exchange with the dense heat bath. So, even with more confidence than in the gas phase, one could predict the rate constant using statistical reaction rate theories such as, for example, transition state theory (TST). However, this ideal picture may break down if (i) there is an appreciable change in charge distribution or molar volume as the system moves along the reaction path from reactant to product state, (ii) the reaction entails large-amplitude structural changes that are subject to solvent frictional forces retarding the motion along the reaction path or (iii) motion along the reaction path is sufficiently fast that thermal equilibrium over all degrees of freedom of the solute and the bath cannot be maintained.

(i)   This situation can still be handled by quasi-equilibrium models such as TST, because the solvent only influences the equilibrium energetics of the system. The ensuing phenomena may be loosely referred to as 'static' solvent effects. These may be caused by electronic solute–solvent interactions that change the effective PES by shifting intersection regions of different electronic states or by lowering or raising potential-energy barriers, but also by solvent structural effects that influence the free-energy change along the reaction path associated with variations in molar volume.

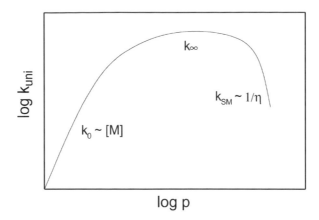

Figure A3.6.1. Pressure dependence of unimolecular rate constant k_{uni}.

(ii) The decrease of the rate constant due to the viscous drag exerted by the solvent medium requires an extension of statistical rate models to include diffusive barrier crossing, because the no-recrossing postulate of TST is obviously violated. In the so-called Smoluchowski limit, one would expect an inverse dependence of k_{uni} on solvent viscosity η at sufficiently high pressure. A reaction rate constant is still well defined and kinetic rate equations may be used to describe the course of the reaction.

This is no longer the case when (iii) motion along the reaction path occurs on a time scale comparable to other relaxation times of the solute or the solvent, i.e. the system is partially non-relaxed. In this situation dynamic effects have to be taken into account explicitly, such as solvent-assisted intramolecular vibrational energy redistribution (IVR) in the solute, solvent-induced electronic surface hopping, dephasing, solute–solvent energy transfer, dynamic caging, rotational relaxation, or solvent dielectric and momentum relaxation.

The introductory remarks about unimolecular reactions apply equivalently to bimolecular reactions in condensed phase. An essential additional phenomenon is the effect the solvent has on the rate of approach of reactants and the lifetime of the collision complex. In a dense fluid the rate of approach evidently is determined by the mutual diffusion coefficient of reactants under the given physical conditions. Once reactants have met, they are temporarily trapped in a solvent cage until they either diffusively separate again or react. It is common to refer to the pair of reactants trapped in the solvent cage as an encounter complex. If the 'unimolecular' reaction of this encounter complex is much faster than diffusive separation: i.e., if the effective reaction barrier is sufficiently small or negligible, the rate of the overall bimolecular reaction is diffusion controlled.

As it has appeared in recent years that many fundamental aspects of elementary chemical reactions in solution can be understood on the basis of the dependence of reaction rate coefficients on solvent density [2–5], increasing attention is paid to reaction kinetics in the gas-to-liquid transition range and supercritical fluids under varying pressure. In this way, the essential differences between the regime of binary collisions in the low-pressure gas phase and that of a dense environment with typical many-body interactions become apparent. An extremely useful approach in this respect is the investigation of rate coefficients, reaction yields and concentration–time profiles of some typical model reactions over as wide a pressure range as possible, which permits the continuous and well controlled variation of the physical properties of the solvent. Among these the most important are density, polarity and viscosity in a continuum description or collision frequency, local solvent shell structure and various relaxation time scales in a microscopic picture.

Progress in the theoretical description of reaction rates in solution of course correlates strongly with that in other theoretical disciplines, in particular those which have profited most from the enormous advances in computing power such as quantum chemistry and equilibrium as well as non-equilibrium statistical mechanics of liquid solutions where Monte Carlo and molecular dynamics simulations in many cases have taken on the traditional role of experiments, as they allow the detailed investigation of the influence of intra- and intermolecular potential parameters on the microscopic dynamics not accessible to measurements in the laboratory. No attempt, however, will be made here to address these areas in more than a cursory way, and the interested reader is referred to the corresponding chapters of the encyclopedia.

In the sections below a brief overview of static solvent influences is given in A3.6.2, while in A3.6.3 the focus is on the effect of transport phenomena on reaction rates, i.e. diffusion control and the influence of friction on intramolecular motion. In A3.6.4 some special topics are addressed that involve the superposition of static and transport contributions as well as some aspects of dynamic solvent effects that seem relevant to understanding the solvent influence on reaction rate coefficients observed in homologous solvent series and compressed solution. More comprehensive accounts of dynamics of condensed-phase reactions can be found in chapters A3.8, A3.13, B3.3, C3.1, C3.2 and C3.5.

A3.6.2   Static solvent effects

The treatment of equilibrium solvation effects in condensed-phase kinetics on the basis of TST has a long history and the literature on this topic is extensive. As the basic ideas can be found in most physical chemistry textbooks and excellent reviews and monographs on more advanced aspects are available (see, for example, the recent review article by Truhlar *et al* [6] and references therein), the following presentation will be brief and far from providing a complete picture.

A3.6.2.1   Separation of time scales

A reactive species in liquid solution is subject to permanent random collisions with solvent molecules that lead to statistical fluctuations of position, momentum and internal energy of the solute. The situation can be described by a reaction coordinate X coupled to a huge number of solvent bath modes. If there is a reaction barrier $E_0^\pm$ ('+' refers to the forward direction and '−' to the reverse reaction), in a way similar to what is common in gas phase reaction kinetics, one may separate the reaction into the elementary steps of activation of A or B, barrier crossing, and equilibration of B or A, respectively (see figure A3.6.2). The time scale $\tau_r^\pm$ for mounting and crossing the barrier is determined by the magnitude of statistical fluctuations $X(t) = \langle X(t) \rangle$ at temperature T, where $\langle \rangle$ indicates ensemble average. In a canonical ensemble this is mainly the Boltzmann factor $\tau_r^\pm \sim e^{E_0^\pm/kT}$, where k denotes Boltzmann's constant. Obviously, the reaction is a rare event if the barrier is large. On the other hand, the time scale for energy relaxation τ_s in a potential well is inversely proportional to the curvature of the potential V along X,

$$\tau_s \sim \sqrt{\mu \left/ \left.\frac{\partial^2 V}{\partial X^2}\right|_{X=\mathrm{A,B}}\right.}$$

where μ denotes reduced mass. So the overall time scale for the reaction

$$\tau_r^\pm \approx \tau_{s(\mathrm{A,B})} \exp(E_0^\pm/kT) \gg \tau_{s(\mathrm{A,B})}$$

for $E_0^\pm \gg kT$. If at the same time $\tau_r^\pm$ is also significantly larger than all other relevant time constants of the solute–bath system (correlation time of the bath, energy and momentum relaxation time, barrier passage time), X may be considered to be a random variable and the motion of the reacting species along this reaction coordinate a stochastic Markov process under the influence of a statistically fluctuating force. This simply means that before, during and after the reaction all degrees of freedom of the solute–solvent system but X are in thermodynamic equilibrium. In this case quasi-equilibrium models of reaction rates are applicable. If the additional requirements are met that (i) each trajectory crossing the transition state at the barrier top never

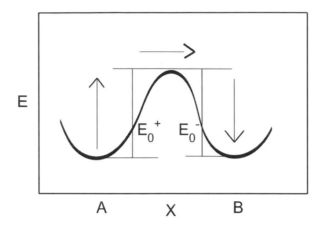

Figure A3.6.2. Activation and barrier crossing.

recrosses and (ii) the Born–Oppenheimer approximation is fulfilled, TST can be used to calculate the reaction rate and provide an upper limit to the real rate coefficient (see chapter A3.12).

A3.6.2.2   Thermodynamic formulation of TST and reference states

For analysing equilibrium solvent effects on reaction rates it is common to use the thermodynamic formulation of TST and to relate observed solvent-induced changes in the rate coefficient to variations in Gibbs free-energy differences between solvated reactant and transition states with respect to some reference state. Starting from the simple one-dimensional expression for the TST rate coefficient of a unimolecular reaction $A \xrightarrow{k_{TST}} P$

$$k_{TST} = \frac{kT}{h} \frac{Q^{\ddagger}}{Q_A} \exp(-E_0/kT) = \frac{kT}{h} \frac{[A^{\ddagger}]}{[A]} \tag{A3.6.1}$$

where Q_A and $Q^{\dagger}$ denote the partition functions of reactant and transition state per volume, respectively, E_0 is the barrier height, and [A], [A‡] stand for equilibrium concentration of reactant and, in a formal not physical sense, the transition state, respectively. Defining an equilibrium constant in terms of activities a

$$K_a^{\ddagger} \equiv \frac{a^{\ddagger}}{a_A} = \frac{\gamma^{\ddagger} [A^{\dagger}]}{\gamma_A [A]} = \frac{Q^{\ddagger}}{Q_A} \exp(-E_0/kT) \tag{A3.6.2}$$

with corresponding activity coefficients denoted by γ, one obtains for the rate coefficient from equations (A3.6.1) and (A3.6.2)

$$k_{TST} = \frac{kT}{h} \frac{[A^{\ddagger}]}{[A]} = \frac{kT}{h} \frac{Q^{\ddagger}}{Q_A} \exp(-E_0/kT) \frac{\gamma_A}{\gamma^{\ddagger}} \equiv k_{TST}^0 \frac{\gamma_A}{\gamma^{\ddagger}} \tag{A3.6.3}$$

where k_{TST}^0 is a standard rate coefficient which depends on the reference state chosen. If one uses the dilute-gas phase as reference, i.e. $k_{TST}^0 = k_{gas}$, all equilibrium solvation effects according to equation (A3.6.3) are included in the ratio of activity coefficients $\gamma_A/\gamma^{\ddagger}$ which is related to the Gibbs free energy of activation for the reaction in the gas $\Delta G_{gas}^{\dagger}$ and in solution $\Delta G_{solution}^{\ddagger}$:

$$kT \ln\left(\frac{k_{solution}}{k_{gas}}\right) = kT \ln\left(\frac{\gamma_A}{\gamma^{\ddagger}}\right) = \Delta G_{gas}^{\ddagger} - \Delta G_{solution}^{\ddagger}. \tag{A3.6.4}$$

Since $\Delta G^{\ddagger}_{gas} - \Delta G^{\ddagger}_{solution}$ in equation (A3.6.4) is equal to the difference between the Gibbs free energy of solvation of reactant and transition state, $\Delta G_{sol}(A) - \Delta G_{sol}(A^{\ddagger})$, one has a direct correlation between equilibrium solvation free enthalpies and rate coefficient ratios. It is common practice in physical organic chemistry to use as a reference state not the gas phase, but a suitable reference solvent M, such that one correlates measured rate coefficient ratios of equation (A3.6.4) to relative changes in Gibbs free energy of solvation

$$[\Delta G_{sol,S}(A) - \Delta G_{sol,S}(A^{\ddagger})] - [\Delta G_{sol,M}(A) - \Delta G_{sol,M}(A^{\ddagger})] \equiv \delta_M \Delta G^{\ddagger}. \qquad (A3.6.5)$$

The shorthand notation in the rhs of equation (A3.6.5) is frequently referred to as the Leffler–Grunwald operator [7].

Considering a bimolecular reaction $A+B \xrightarrow{k_{TST}} P$, one correspondingly obtains for the rate constant ratio

$$k_{solution}/k_{gas} = \gamma_A \gamma_B / \gamma^{\ddagger}. \qquad (A3.6.6)$$

In the TST limit, the remaining task strictly speaking does not belong to the field of reaction kinetics: it is a matter of obtaining sufficiently accurate reactant and transition state structures and charge distributions from quantum chemical calculations, constructing sufficiently realistic models of the solvent and the solute–solvent interaction potential, and calculating from these ingredients values of Gibbs free energies of solvation and activity coefficients. In many cases, a microscopic description may prove a task too complex, and one rather has to use simplifying approximations to characterize influences of different solvents on the kinetics of a reaction in terms of some macroscopic physical or empirical solvent parameters. In many cases, however, this approach is sufficient to capture the kinetically significant contribution of the solvent–solute interactions.

A3.6.2.3 Equilibrium solvation—macroscopic description

(a) Naïve view of solvent cavity effects

Considering equation (A3.6.3), if activity coefficients of reactant and transition state are approximately equal, for a unimolecular reaction one should observe $k_{solution} \approx k_{gas}$. This in fact is observed for many unimolecular reactions where the reactant is very similar to the transition state, i.e. only a few bond lengths and angles change by a small amount and there is an essentially constant charge distribution. There are, however, also large deviations from this simplistic prediction, in particular for dissociation reactions that require separation of fragments initially formed inside a common solvent cavity into individually solvated products.

For a bimolecular reaction in such a case one obtains from equation (A3.6.6) $k_{solution} \approx \gamma \cdot k_{gas}$, so one has to estimate the activity coefficient of a reactant to qualitatively predict the solvent effect. Using ad hoc models of solvation based on the free-volume theory of liquids or the cohesive energy density of a solvent cavity, purely thermodynamic arguments yield $\gamma \sim 10^2$–10^3 [8–10].

The reason for this enhancement is intuitively obvious: once the two reactants have met, they temporarily are trapped in a common solvent shell and form a short-lived so-called encounter complex. During the lifetime of the encounter complex they can undergo multiple collisions, which give them a much bigger chance to react before they separate again, than in the gas phase. So this effect is due to the microscopic solvent structure in the vicinity of the reactant pair. Its description in the framework of equilibrium statistical mechanics requires the specification of an appropriate interaction potential.

(b) Electrostatic effects–Onsager and Born models

If the charge distribution changes appreciably during the reaction, solvent polarity effects become dominant and in liquid solution often mask the structural influences mentioned above. The calculation of solvation energy differences between reactant and transition state mainly consists of estimating the Gibbs free energies of solvation ΔG_{sol} of charges, dipoles, quadrupoles etc in a polarizable medium. If the solute itself is considered non-polarizable and the solvent a continuous linear dielectric medium without internal structure, then $\Delta G_{sol} = \frac{1}{2}E_{int}$, where E_{int} is the solute–solvent interaction energy [11]. Reactant and transition state are modelled as point charges or point dipoles situated at the centre of a spherical solvent cavity. The point charge or the point dipole will polarize the surrounding dielectric continuum giving rise to an electric field which in turn will act on the charge distribution inside the cavity. The energy of the solute in this so-called reaction field may be calculated by a method originally developed by Onsager. Using his reaction field theory [12, 13], one obtains the molar Gibbs free energy of solvation (with respect to vacuum) of an electric point dipole μ_{el} in a spherical cavity of radius r embedded in a homogeneous dielectric of dielectric constant ε as

$$\Delta G_{sol,dip} = -N_A \frac{\varepsilon - 1}{2\varepsilon + 1} \frac{\mu_{el}^2}{4\pi \varepsilon_0 r^3} \qquad (A3.6.7)$$

with ε_0 and N_A denoting vacuum permittivity and Avogadro's constant, respectively. The dielectric constant inside the cavity in this approximation is assumed to be unity. Applying this expression to a solvent series study of a reaction involving large charge separation, such as the Menshutkin reaction of triethylamine with ethyliodide

$$Et_3N \quad + \quad EtI \quad \rightarrow \quad (Et_3^+NEt^-N)^{\ddagger} \rightarrow Et_4N^+ + I^-$$
$$\mu_a, r_a \qquad \mu_b, r_b \qquad \mu_{ab}^{\dagger}, r_{ab}^{\dagger}$$

one obtains

$$\delta_M \Delta G^{\ddagger} = -N_A \frac{\varepsilon - 1}{2\varepsilon + 1} \frac{1}{4\pi \varepsilon_0} \left(\frac{(\mu_{ab}^{\ddagger})^2}{(r_{ab}^{\ddagger})^3} - \frac{\mu_a^2}{r_a^3} - \frac{\mu_b^2}{r_b^3} \right)$$

predicting a linear relationship between $\ln(k_{solvent}/k_{reference})$ and $(\varepsilon - 1)/(2\varepsilon + 1)$ which is only approximately reflected in the experimental data covering a wide range of solvents [14] (see figure A3.6.3). This is not surprising, in view of the approximate character of the model and, also, because a change of solvent does not only lead to a variation in the dielectric constant, but at the same time may be accompanied by a change in other kinetically relevant properties of the medium, demonstrating a general weakness of this type of experimental approach.

Within the framework of the same dielectric continuum model for the solvent, the Gibbs free energy of solvation of an ion of radius r_{ion} and charge $z_{ion}e$ may be estimated by calculating the electrostatic work done when hypothetically charging a sphere at constant radius r_{ion} from $q = 0 \rightarrow q = z_{ion}e$. This yields the Born equation [13]

$$\Delta G_{sol,ion} = -N_A \frac{z_{ion}^2 e^2}{8\pi \varepsilon_0 r_{ion}} \left(1 - \frac{1}{\varepsilon} \right) \qquad (A3.6.8)$$

such that for a reaction of the type

$$A^{z+} \quad + \quad B \quad \rightarrow \quad (AB^{z+})^{\ddagger} \quad \rightarrow \quad P$$
$$z, r_A \qquad \mu_B, r_B \qquad z, r^{\ddagger}$$

the change in effective barrier height (difference of Gibbs free energy of solvation changes between transition state and reactants) according to equations (A3.6.7) and (A3.6.8) equals

$$\delta_M \Delta G^{\ddagger} = -\frac{N_A}{8\pi \varepsilon_0} \left[(ze)^2 \left(1 - \frac{1}{\varepsilon} \right) \left(\frac{1}{r^{\dagger}} - \frac{1}{r_A} \right) - \frac{\mu_B^2}{r_B^3} \frac{2(\varepsilon - 1)}{2\varepsilon + 1} \right].$$

This formula does not include the charge–dipole interaction between reactants A and B. The correlation between measured rate constants in different solvents and their dielectric parameters in general is of a similar quality as illustrated for neutral reactants. This is not, however, due to the approximate nature of the Born model itself which, in spite of its simplicity, leads to remarkably accurate values of ion solvation energies, if the ionic radii can be reliably estimated [15].

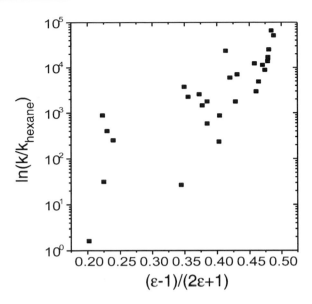

Figure A3.6.3. Solvent polarity dependence of the rate constant for the Menshutkin reaction (data from [14]).

Onsager's reaction field model in its original form offers a description of major aspects of equilibrium solvation effects on reaction rates in solution that includes the basic physical ideas, but the inherent simplifications seriously limit its practical use for quantitative predictions. It since has been extended along several lines, some of which are briefly summarized in the next section.

(c) Improved dielectric continuum models

Onsager's original reaction field method imposes some serious limitations: the description of the solute as a point dipole located at the centre of a cavity, the spherical form of the cavity and the assumption that cavity size and solute dipole moment are independent of the solvent dielectric constant.

Kirkwood generalized the Onsager reaction field method to arbitrary charge distributions and, for a spherical cavity, obtained the Gibbs free energy of solvation in terms of a multipole expansion of the electrostatic field generated by the charge distribution [12, 13]

$$\Delta G_{sol,K} = \frac{N_A e^2}{8\pi \varepsilon_0} \sum_{n=0}^{\infty} \frac{(n+1)(1-\varepsilon)}{(n+1)\varepsilon + n} \frac{1}{r^{2n+1}} \sum_{k=1}^{N} \sum_{l=1}^{N} z_k z_l \mathbf{r}_k^n \mathbf{r}_l^n P_n(\cos \vartheta_{kl}) \tag{A3.6.9}$$

where N is the number of point charges in the cavity, vectors $\mathbf{r}_i$ denote their position, ϑ_{ij} the angle between respective vectors, and P_n are the Legendre polynomials. This expression reduces to equations (A3.6.8) and (A3.6.7) for $n = 0$ and $n = 1$, respectively. It turns out that usually it is sufficient to consider terms up to $n \approx 4$ to achieve convergence of the expansion. The absolute value of the solvation energy calculated from equation (A3.6.9), however, critically depends on the size and the shape of the cavity. Even when the charge distribution of reactants and transition state can be calculated to sufficient accuracy by advanced quantum chemical methods, this approach only will give useful quantitative information about the solvent dependence of reaction rates, if the cavity does not change appreciably along the reaction path from reactants to transition state and if it is largely solvent independent.

As this condition usually is not met, considerable effort has gone into developing methods to calculate solvation energies for cavities of arbitrary shape that as closely as possible mimic the topology of the interface between solute molecule and solvent continuum. Among these are various implementations of boundary element methods [16], in which a cavity surface of arbitrary shape is divided into surface elements carrying a specified surface charge. In one of the more simple variants, a virtual charge scheme as proposed by Miertuš [17], the charge distribution of the solute $\rho^0(\mathbf{r}_j)$ reflects itself in corresponding polarization surface charge densities at the cavity interface $\sigma(\mathbf{s}_i)$ that are assigned to each of m surface elements $\mathbf{s}_i$ and assumed to be constant across the respective surface areas ΔS_i. The electric potential generated by these virtual charges is

$$V_\sigma = \sum_{i=1}^{m} \Delta S_i \frac{\sigma(\mathbf{s}_i)}{4\pi\varepsilon_0 r_i}. \tag{A3.6.10}$$

The surface charge density on each surface element is determined by the boundary condition

$$\sigma(\mathbf{s}_i) = \frac{\varepsilon_0(\varepsilon-1)}{\varepsilon} \left(\frac{dV}{d\mathbf{n}_i}\right)_{\mathbf{s}_i} \tag{A3.6.11}$$

where ε denotes the static dielectric constant of the solvent, and the derivative of the total electrical potential V at the interface is taken with respect to the normal vector $\mathbf{n}_i$ of each surface element $\mathbf{s}_i$. V is the sum of contributions from the solute charges ρ^0 and the induced polarization surface charges $\sigma(\mathbf{s}_i)$. Using equations (A3.6.10) and (A3.6.11), the virtual surface charge densities $\sigma(\mathbf{s}_i)$ can be calculated iteratively, and the Gibbs free energy of solvation is then half the electrostatic interaction energy of the solute charge distribution in the electric potential generated by the induced polarization surface charges

$$\Delta G_{\text{sol}} = \frac{1}{2} \int V_\sigma(\mathbf{r})\rho^0(\mathbf{r})\,d\mathbf{r}. \tag{A3.6.12}$$

Of course, one has to fix the actual shape and size of the cavity, before one can apply equation (A3.6.12). Since taking simply ionic or van der Waals radii is too crude an approximation, one often uses basis-set-dependent *ab initio* atomic radii and constructs the cavity from a set of intersecting spheres centred on the atoms [18, 19]. An alternative approach, which is comparatively easy to implement, consists of using an electrical equipotential surface to define the solute–solvent interface shape [20].

The most serious limitation remaining after modifying the reaction field method as mentioned above is the neglect of solute polarizability. The reaction field that acts back on the solute will affect its charge distribution as well as the cavity shape as the equipotential surface changes. To solve this problem while still using the polarizable continuum model (PCM) for the solvent, one has to calculate the surface charges on the solute by quantum chemical methods and represent their interaction with the solvent continuum as in classical electrostatics. The Hamiltonian of the system thus is written as the sum of the Hamilton operator for the isolated solute molecule and its interaction with the macroscopic electrostatic reaction field. The coupled equations of the solute subject to the reaction field induced in the solvent are then solved self-consistently to obtain the electron density of the solute in the presence of the polarizable dielectric—the basis of self-consistent reaction field (SCRF) models [21]. Whether this is done in the framework of, for example, Hartree–Fock theory or density functional theory, is a question of optimizing quantum chemical techniques outside the topics addressed here.

If reliable quantum mechanical calculations of reactant and transition state structures in vacuum are feasible, treating electrostatic solvent effects on the basis of SRCF-PCM using cavity shapes derived from methods mentioned above is now sufficiently accurate to predict variations of Gibbs free energies of activation $\delta\Delta G^\ddagger$ with solvent polarity reliably, at least in the absence of specific solute–solvent interactions. For instance, considering again a Menshutkin reaction, in this case of pyridine with methylbromide, Pyr+MeBr→MePyr$^+$+Br$^-$, in cyclohexane and di-n-butyl ether, the difference between calculated and experimental values of $\Delta G^\ddagger$ is only about 2% and 4%, respectively [22, 23].

As with SCRF-PCM only macroscopic electrostatic contributions to the Gibbs free energy of solvation are taken into account, short-range effects which are limited predominantly to the first solvation shell have to be considered by adding additional terms. These correct for the neglect of effects caused by solute–solvent electron correlation including dispersion forces, hydrophobic interactions, dielectric saturation in the case of multiply charged ions and solvent structural influences on cavitation. In many cases, however, the electrostatic contribution dominates and dielectric continuum models provide a satisfactory description.

A3.6.2.4   Equilibrium solvent effects—microscopic view

Specific solute–solvent interactions involving the first solvation shell only can be treated in detail by discrete solvent models. The various approaches like point charge models, supermolecular calculations, quantum theories of reactions in solution, and their implementations in Monte Carlo methods and molecular dynamics simulations like the Car–Parrinello method are discussed elsewhere in this encyclopedia. Here only some points will be briefly mentioned that seem of relevance for later sections.

(a) Point charge distribution model [11]

Considering, for simplicity, only electrostatic interactions, one may write the solute–solvent interaction term of the Hamiltonian for a solute molecule surrounded by S solvent molecules as

$$\hat{H}_{elstat} = \sum_{s=1}^{S}\left[\sum_{\lambda=1}^{N}\sum_{\alpha=1}^{N_S}\frac{1}{r_{\lambda\alpha s}} - \sum_{\lambda=1}^{N}\sum_{a=1}^{M_S}\frac{Z_a}{r_{\lambda a s}} + \sum_{l=1}^{M}\sum_{a=1}^{M_S}\frac{Z_l Z_a}{r_{las}} - \sum_{l=1}^{M}\sum_{\alpha=1}^{N_S}\frac{Z_l}{r_{l\alpha s}}\right] \tag{A3.6.13}$$

where the solute contains N electrons and M nuclei with charges Z_l and the solvent molecules N_s electrons and M_s nuclei with charge Z_a. In the point charge method equation (A3.6.13) reduces to

$$\hat{H}_{pc} = \sum_{p=1}^{P}\sum_{\lambda=1}^{N}\frac{q_p}{r_{\lambda p}} - \sum_{p=1}^{P}\sum_{l=1}^{M}\frac{q_p Z_l}{r_{lp}}. \tag{A3.6.14}$$

Here the position r_{ip} of the point charges located on the solvent molecules q_p is determined by the structure of the solvent shell and the electron density distribution within the solvent molecule. In this type of model, the latter is assumed to be fixed, i.e. the solvent molecules are considered non-polarizable while solving the Schrödinger equation for the coupled system.

Instead of using point charges one may also approximate the interaction Hamiltonian in terms of solute electrons and nuclei interacting with solvent point dipoles μ_d

$$\hat{H}_{pd} = \sum_{d=1}^{D}\sum_{\lambda=1}^{N}\frac{\mu_d r_{d\lambda}}{r_{d\lambda}^3} - \sum_{d=1}^{D}\sum_{l=1}^{M}\frac{\mu_d r_{dl} Z_l}{r_{dl}^3}. \tag{A3.6.15}$$

In either case, the structure of the solvation shell has to be calculated by other methods supplied or introduced *ad hoc* by some further model assumptions, while charge distributions of the solute and within solvent molecules are obtained from quantum chemistry.

(b) Solvation shell structure

The quality of the results that can be obtained with point charge or dipole models depends critically on the input solvation shell structure. In view of the computer power available today, taking the most rigorous route is feasible in many cases, i.e. using statistical methods to calculate distribution functions in solution. In this way the average structure of solvation shells is accessible, that is, to be used in equilibrium solvation calculations required to obtain, for example, TST rate constants.

Assuming that additive pair potentials are sufficient to describe the inter-particle interactions in solution, the local equilibrium solvent shell structure can be described using the pair correlation function $g^{(2)}(\mathbf{r}_1, \mathbf{r}_2)$. If the potential only depends on inter-particle distance, $g^{(2)}(\mathbf{r}_1, \mathbf{r}_2)$ reduces to the radial distribution function $g(r) \equiv g^{(2)}(|\mathbf{r}_1 - \mathbf{r}_2|)$ such that $\rho \cdot 4\pi r^2 dr g(r)$ gives the number of particles in a spherical shell of thickness dr at distance r from a reference particle (ρ denotes average particle density). The local particle density is then simply $\rho \cdot g(r)$. The radial distribution function can be obtained experimentally in neutron scattering experiments by measuring the angular dependence of the scattering amplitude, or by numerical simulation using Monte Carlo methods.

(c) Potential of mean force

At low solvent density, where isolated binary collisions prevail, the radial distribution function $g(r)$ is simply related to the pair potential $u(r)$ via $g_0(r) = \exp[-u(r)/kT]$. Correspondingly, at higher density one defines a function $w(r) \equiv -kT \ln[g(r)]$. It can be shown that the gradient of this function is equivalent to the mean force between two particles obtained by holding them at fixed distance r and averaging over the remaining $N - 2$ particles of the system. Hence $w(r)$ is called the potential of mean force. Choosing the low-density system as a reference state one has the relation

$$\lim_{\rho \to 0} g(r) = g_0(r) \Rightarrow \lim_{\rho \to 0} w(r) = u(r)$$

and $\Delta w(r) \equiv w(r) - u(r)$ describes the average many-body contribution such as, for example, effects due to solvation shell structure. In the language of the thermodynamic formulation of TST, the ratio of rate constants in solution and dilute-gas phase consequently may be written as

$$kT \ln \frac{k_{solution}}{k_{gas}} = -\delta \Delta w^{\ddagger} \equiv -[\Delta w(r^{\ddagger}) - \Delta w(r_{react})]. \tag{A3.6.16}$$

A3.6.2.5   Pressure effects

The inherent difficulties in interpreting the effects observed in solvent series studies of chemical reaction rates, which offer little control over the multitude of parameters that may influence the reaction, suggest rather using a single liquid solvent and varying the pressure instead, thereby changing solvent density and polarity in a well known way. One also may have to consider, of course, variations in the local solvent shell structure with increasing pressure.

(a) Activation volume

In the thermodynamic formulation of TST the pressure dependence of the reaction rate coefficient defines a volume of activation [24–26]

$$\left(\frac{\partial \ln k}{\partial p} \right)_T = -\frac{1}{RT} \left(\frac{\partial \Delta G^{\ddagger}}{\partial p} \right)_T \equiv -\frac{\Delta V^{\ddagger}}{RT} \tag{A3.6.17}$$

with $\Delta V^{\ddagger} = \bar{V}^{\ddagger} - \sum_i^{reacts} \bar{V}_i$, the difference of the molar volume of transition state and the sum over molar volumes of reactants. Experimental evidence shows that $|\Delta V^{\ddagger}|$ is of the order of 10^0–10^1 cm^3 mol^{-1} and usually pressure dependent [27]. It is common practice to interpret it using geometric arguments considering reactant and transition state structures and by differences in solvation effects between reactant and transition state. If one uses a molar concentration scale (standard state 1 mol dm^{-3}), an additional term $+\kappa_{\text{solv}}\Delta v^{\ddagger}$ appears in the rhs of equation (A3.6.16), the product of isothermal solvent compressibility and change in sum over stoichiometric coefficients between reactants and transition state.

There is one important *caveat* to consider before one starts to interpret activation volumes in terms of changes of structure and solvation during the reaction: the pressure dependence of the rate coefficient may also be caused by transport or dynamic effects, as solvent viscosity, diffusion coefficients and relaxation times may also change with pressure [2]. Examples will be given in subsequent sections.

(b) Activation volume in a dielectric continuum

If, in analogy to equation (A3.6.5), one denotes the change of activation volume with respect to some reference solvent as $\delta_{\text{M}}\Delta V^{\ddagger}$ and considers only electrostatic interactions of reactant and transition state with a dielectric continuum solvent, one can calculate it directly from

$$\delta_{\text{M}}\Delta V^{\ddagger} = \left(\frac{\partial(\delta_{\text{M}}\Delta G^{\ddagger})}{\partial p}\right)_{\text{T}} \tag{A3.6.18}$$

by using any of the models mentioned above. If the amount of charge redistribution is significant and the solvent is polar, the dielectric contribution to $\Delta V^{\ddagger}$ by far dominates any so-called intrinsic effects connected with structural changes between reactant and transition state. For the Menshutkin reaction, for example, equation (A3.6.17) gives

$$\delta_{\text{M}}\Delta V^{\ddagger} = \frac{-N_{\text{A}}}{4\pi\varepsilon_0}\left[\frac{3}{(2\varepsilon(p)+1)^2}\left(\frac{\partial\varepsilon}{\partial p}\right)_{\text{T}}\right]\left(\frac{(\mu_{\text{ab}}^{\ddagger})^2}{(r_{\text{ab}}^{\ddagger})^3} - \frac{\mu_{\text{a}}^2}{r_{\text{a}}^3} - \frac{\mu_{\text{b}}^2}{r_{\text{b}}^3}\right)$$

which includes a positive term resulting from the pressure dependence of the dielectric constant (in square brackets) and represents the experimentally observed pressure dependence of the activation volume quite satisfactorily [25]. For the Menshutkin reaction, only the large dipole moment of the transition state needs to be considered, resulting in a negative activation volume, a typical example of electrostriction. If one assumes that the neglect of solute polarizability is justified and, in addition, the cavity radius is constant, one may use this kind of expression to estimate transition state dipole moments. Improved continuum models as outlined in the preceding sections may, of course, also be applied to analyse activation volumes.

(c) Activation volume and local solvent structure

In a microscopic equilibrium description the pressure-dependent local solvent shell structure enters through variations of the potential of mean force, $(\partial\delta\Delta w^{\ddagger}/\partial p)_{\text{T}}$, such that the volume of activation contains a contribution related to the pressure dependence of radial distribution functions for reactants and transition state, i.e.

$$\Delta V_{\text{local}}^{\ddagger} = -kT\left\{\frac{\partial}{\partial p}\left[\ln\left(\frac{g(r^{\ddagger})g_0(r_{\text{react.}})}{g_0(r^{\ddagger})g(r_{\text{react.}})}\right)\right]\right\}_{\text{T}}.$$

This contribution of local solvent structure to $\Delta V^{\dagger}$ may be quite significant and, even in nonpolar solvents, in many cases outweigh the intrinsic part. It essentially describes a caging phenomenon, as with increasing pressure the local solvent density or packing fraction of solvent molecules around reactants and transition state increases, thereby enhancing the stability of the solvent cage. This constitutes an equilibrium view of caging in contrast to descriptions of the cage effect in, for example, photodissociation where solvent friction is assumed to play a central role.

How large the magnitude of this packing effect can be was demonstrated in simple calculations for the atom transfer reaction $CH_3+CH_4 \rightarrow CH_4+CH_3$ using a binary solution of hard spheres at infinite dilution as the model system [28]. Allowing spheres to partially overlap in the transition state, i.e. assuming a common cavity, reaction rates were calculated by variational TST for different solute-to-solvent hard-sphere ratios $r_\sigma = \sigma_M/\sigma_s$ and solvent densities $\rho_s\sigma_s^3$. Increasing the latter from 0.70 to 0.95 led to an enhancement of the relative rate constant $k_{\text{solution}}/k_{\text{gas}}$ by factors of 8.5, 15.5 and 53 for r_σ equal to 0.93, 1.07 and 1.41, respectively, thus clearly showing the effect of local packing density. With respect to the calculated gas phase value the rate constants at the highest density were 95, 280 and 2670 times larger, respectively. This behaviour is typical for 'tight' transition states, whereas for loose transition states as they appear, for example, in isomerization reactions, this caging effect is orders of magnitude smaller.

A3.6.3 Transport effects

If reactant motion along the reaction path in the condensed phase involves significant displacement with respect to the surrounding solvent medium and there is non-negligible solute–solvent coupling, frictional forces arise that oppose the reactive motion. The overall rate of intrinsically fast reactions for which, for example, the TST rate constant is sufficiently large, therefore, may be influenced by the viscous drag that the molecules experience on their way from reactants to products. As mentioned in the introduction, dynamic effects due to other partially non-relaxed degrees of freedom will not be considered in this section.

For a bimolecular reaction, this situation is easily illustrated by simply writing the reaction as a sequence of two steps as

$$A + B \underset{k_{\text{sep}}}{\overset{k_{\text{diff}}}{\rightleftharpoons}} (A \cdots B) \overset{k_{\text{mol}}}{\longrightarrow} (\text{products}) \tag{A3.6.19}$$

where brackets denote common solvent cage (encounter complex), k_{diff} is the rate constant of diffusive approach of reactants, sometimes called the 'encounter rate', k_{sep} is that of diffusive separation of the unreacted encounter pair and k_{mol} that of the reactive step in the encounter complex. If $k_{\text{mol}} \gg k_{\text{sep}}$, the overall reaction rate constant k essentially equals k_{diff}, and the reaction is said to be diffusion controlled. One important implicit assumption of this phenomenological description is that diffusive approach and separation are statistically independent processes, i.e. the lifetime of the encounter pair is sufficiently long to erase any memory about its formation history. Examples of processes that often become diffusion controlled in solution are atom and radical recombination, electron and proton transfer, fluorescence quenching and electronic energy transfer.

In a similar phenomenological approach to unimolecular reactions involving large-amplitude motion, the effect of friction on the rate constant can be described by a simple transition formula between the high-pressure limit k_∞ of the rate constant at negligible solvent viscosity and the so-called Smoluchowski limit of the rate constant, k_{SM}, approached in the high-damping regime at large solvent viscosity [2]:

$$\frac{1}{k} = \frac{1}{k_\infty} + \frac{1}{k_{\text{SM}}}. \tag{A3.6.20}$$

As k_{SM} is inversely proportional to solvent viscosity, in sufficiently viscous solvents the rate constant k becomes equal to k_{SM}. This concerns, for example, reactions such as isomerizations involving significant rotation around single or double bonds, or dissociations requiring separation of fragments, although it may be difficult to experimentally distinguish between effects due to local solvent structure and solvent friction.

Systematic experimental investigations of these transport effects on reaction rates can either be done by varying solvents in a homologous series to change viscosity without affecting other physicochemical or chemical properties (or as little as possible) or, much more elegantly and experimentally demanding, by varying pressure and temperature in a single solvent, maintaining control over viscosity, polarity and density at the same time. As detailed physical insight is gained by the latter approach, the few examples shown all will be from pressure-dependent experimental studies. Computer experiments involving stochastic trajectory simulations or classical molecular dynamics simulations have also been extremely useful for understanding details of transport effects on chemical reaction rates, though they have mostly addressed dynamic effects and been less successful in actually providing a quantitative connection with experimentally determined solvent or pressure dependences of rate constants or quantum yields of reactions.

A3.6.3.1 Diffusion and bimolecular reactions

(a) Diffusion-controlled rate constant

Smoluchowski theory [29, 30] and its modifications form the basis of most approaches used to interpret bimolecular rate constants obtained from chemical kinetics experiments in terms of diffusion effects [31]. The Smoluchowski model is based on Brownian motion theory underlying the phenomenological diffusion equation in the absence of external forces. In the standard picture, one considers a dilute fluid solution of reactants A and B with $[A] \ll [B]$ and asks for the time evolution of $[B]$ in the vicinity of A, i.e. of the density distribution $\rho(r, t) \equiv [B](r, t)/[B]_{t=0} \simeq [B](r(t))/[B]_{t=0}$ ($[B]$ is assumed not to change appreciably during the reaction). The initial distribution and the outer and inner boundary conditions are chosen, respectively, as

$$\rho(r, 0) = \begin{cases} 0 & \text{for } r \leq R \\ 1 & \text{for } r > R \end{cases}$$

$$\rho(r \to \infty, t) = 1 \qquad \text{for } t \geq 0$$

$$k_{\text{mol}}\rho(R) = 4\pi R^2 D_{\text{AB}} \left. \frac{\partial \rho}{\partial r} \right|_R \tag{A3.6.21}$$

where R is the encounter radius and D_{AB} the mutual diffusion coefficient of reactants. The reflecting boundary condition [32] at the encounter distance R ensures that, once a stationary concentration of encounter pairs is established, the intrinsic reaction rate in the encounter pair, $k_{\text{mol}}\rho(R)$, equals the rate of diffusive formation of encounter pairs. In this formulation k_{mol} is a second-order rate constant. Solving the diffusion equation

$$\frac{\partial \rho}{\partial t} = D_{\text{AB}} \left[\frac{\partial^2 \rho}{\partial r^2} + \frac{2}{r} \frac{\partial \rho}{\partial r} \right] \tag{A3.6.22}$$

subject to conditions (A3.6.21) and realizing that the observed reaction rate coefficient $k(t)$ equals $k_{\text{mol}}\rho(R, t)$, one obtains

$$k(t) = \frac{k_{\text{mol}}}{1 + x} \{1 + x \exp[y^2(1 + x)^2 t] \text{erfc}[y(1 + x)\sqrt{t}]\} \tag{A3.6.23}$$

using the abbreviations $x \equiv k_{\text{mol}}/4\pi R D_{\text{AB}}$ and $y \equiv \sqrt{D_{\text{AB}}} R$. The time-dependent terms reflect the transition from the initial to the stationary distribution. After this transient term has decayed to zero, the reaction rate attains its stationary value

$$k = \frac{k_{\text{mol}}}{1 + x} = \frac{4\pi R D_{\text{AB}} k_{\text{mol}}}{4\pi R D_{\text{AB}} + k_{\text{mol}}} = \frac{k_{\text{diff}} k_{\text{mol}}}{k_{\text{diff}} + k_{\text{mol}}} \tag{A3.6.24}$$

such that for $k_{\text{mol}} \gg k_{\text{diff}}$ one reaches the diffusion limit $k \simeq k_{\text{diff}}$. Comparing equation (A3.6.24) with the simple kinetic scheme (A3.6.19), one realizes that at this level of Smoluchowski theory one has $k_{\text{sep}} = k_{\text{diff}}\rho(R)$, i.e. there is no effect due to caging of the encounter complex in the common solvation shell.

There exist numerous modifications and extensions of this basic theory that not only involve different initial and boundary conditions, but also the inclusion of microscopic structural aspects [31]. Among these are hydrodynamic repulsion at short distances that may be modelled, for example, by a distance-dependent diffusion coefficient

$$D_{\text{AB}}(r) \simeq D_{\text{AB}} \left[1 - \frac{1}{2} \exp\left(1 - \frac{r}{R} \right) \right]$$

or the potential of mean force *via* the radial distribution function $g(r)$, which leads to a significant reduction of the steady-state rate constant by about one-third with respect to the Smoluchowski value [33, 34]:

$$k = k_{\text{mol}} g(R) \left[1 + k_{\text{mol}} g(R) \int \frac{dr}{4\pi r^2 D_{\text{AB}}(r) g(r)} \right]^{-1}.$$

Diffusion-controlled reactions between ions in solution are strongly influenced by the Coulomb interaction accelerating or retarding ion diffusion. In this case, the diffusion equation for ρ concerning motion of one reactant about the other stationary reactant, the Debye–Smoluchowski equation,

$$\frac{\partial \rho}{\partial t} = D_{\text{AB}} \nabla \cdot \left[\nabla \rho + \frac{\rho}{kT} \nabla V(r) \right] \tag{A3.6.25}$$

includes the gradient of the potential energy $V(r)$ of the ions in the Coulomb field. Using boundary conditions equivalent to equation (A3.6.21) and an initial condition corresponding to a Boltzmann distribution of interionic distances

$$\rho(r, 0) = e^{-V(r)/kT} = e^{-R_C/r} \qquad R_C = \frac{z_A z_B e^2}{4\pi \varepsilon \varepsilon_0 kT}$$

and solving equation (A3.6.25), one obtains the steady-state solution

$$k = 4\pi R_C D_{\text{AB}} \left[\left(1 + \frac{4\pi R_C D_{\text{AB}}}{k_{\text{mol}}} \right) e^{R_C/R} - 1 \right]^{-1}.$$

Many additional refinements have been made, primarily to take into account more aspects of the microscopic solvent structure, within the framework of diffusion models of bimolecular chemical reactions that encompass also many-body and dynamic effects, such as, for example, treatments based on kinetic theory [35]. One should keep in mind, however, that in many cases the practical value of these advanced theoretical models for a quantitative analysis or prediction of reaction rate data in solution may be limited.

(b) Transition from gaseous to liquid solvent—onset of diffusion control

Instead of concentrating on the diffusion limit of reaction rates in liquid solution, it can be instructive to consider the dependence of bimolecular rate coefficients of elementary chemical reactions on pressure over a wide solvent density range covering gas and liquid phase alike. Particularly amenable to such studies are atom recombination reactions whose rate coefficients can be easily investigated over a wide range of physical conditions from the dilute-gas phase to compressed liquid solution [3, 4].

As discussed above, one may try to represent the density dependence of atom recombination rate coefficients k in the spirit of equation (A3.6.24) as

$$\frac{1}{k} \approx \frac{1}{k_{\text{rec}}^g} + \frac{1}{k_{\text{diff}}} \tag{A3.6.26}$$

where k_{rec}^g denotes the low-pressure second-order rate coefficient proportional to bath gas density, and k_{diff} is the second-order rate coefficient of diffusion-controlled atom recombination as discussed in the previous section. In order to apply equation (A3.6.26), a number of items require answers specific to the reaction under study: (i) the density dependence of the diffusion coefficient D_{AA}, (ii) the magnitude of the encounter radius R, (iii) the possible participation of excited electronic states and (iv) the density dependence of k_{rec}^g. After these have been dealt with adequately, it can be shown that for many solvent bath gases, the phenomenon of the turnover from a molecular reaction into a diffusion-controlled recombination follows equation (A3.6.26) without any apparent discontinuity in the rate coefficient k at the gas–liquid phase transition, as illustrated for iodine atom recombination in argon [36, 37]. For this particular case, D_{AA} is based on and extrapolated from experimental data, R is taken to be one-half the sum of the Lennard-Jones radii of iodine atom and solvent molecule, and the density-dependent contribution of excited electronic states is implicitly considered by making the transition from the measured k_{rec}^g in dilute ethane gas to k_{diff} in dense liquid ethane.

A more subtle point concerns scaling of k_{rec}^g with density. Among the various possibilities that exist, either employing local densities obtained from numerically calculated radial distribution functions [38]

$$k_{rec}^g(\rho) = k_{rec}^g(\rho_0) \frac{\rho g(r; \rho)}{\rho_0 g(r; \rho_0)}$$

or taking into account that in the gas phase the reaction is controlled to a large extent by the energy transfer mechanism, such that $k_{rec}^g \approx \beta_c Z_{LJ}$ where β_c is a collision efficiency and Z_{LJ} the Lennard-Jones collision frequency, are probably the most practical. As $Z_{LJ} \sim 1/D_{AM}$ throughout the whole density range, $k_{rec}^g(\rho)$ in the latter case may be estimated by scaling with the diffusion coefficient [37]

$$k_{rec}^g(\rho) = k_{rec}^g(\rho_0) \frac{D_{AM}(\rho_0)}{D_{AM}(\rho)}.$$

Although the transition to diffusion control is satisfactorily described in such an approach, even for these apparently simple elementary reactions the situation in reality appears to be more complex due to the participation of weakly bonding or repulsive electronic states which may become increasingly coupled as the bath gas density increases. These processes manifest themselves in iodine atom and bromine atom recombination in some bath gases at high densities where marked deviations from 'normal' behaviour are observed [3, 4]. In particular, it is found that the transition from k_{rec}^g to k_{diff} is significantly broader than predicted by equation (A3.6.26), the reaction order of iodine recombination in propane is higher than 3, and S-shaped curves are observed with He as a bath gas [36] (see figure A3.6.4). This is in contrast to the recombination of the methyl radicals in Ar which can be satisfactorily described by a theory of particle encounter kinetics using appropriate interaction potentials and a modified friction for relative motion [39]. The only phenomena that cannot be reproduced by such treatments were observed at moderate gas pressures between 1 and 100 bar. This indicates that the kinetics of the reaction in this density regime may be influenced to a large extent by reactant–solute clustering or even chemical association of atoms or radicals with solvent molecules.

This problem is related to the question of appropriate electronic degeneracy factors in chemical kinetics. Whereas the general belief is that, at very low gas pressures, only the electronic ground state participates in atom recombination and that, in the liquid phase, at least most of the accessible states are coupled somewhere 'far out' on the reaction coordinate, the transition between these two limits as a function of solvent density is by no means understood. Direct evidence for the participation of different electronic states in iodine geminate recombination in the liquid phase comes from picosecond time-resolved transient absorption experiments in solution [40, 41] that demonstrate the participation of the low-lying, weakly bound iodine A and A′ states, which is also taken into account in recent mixed classical–quantum molecular dynamics simulations [42, 43].

A3.6.3.2 Unimolecular reactions and friction

So far the influence of the dense solvent environment on the barrier crossing process in a chemical reaction has been ignored. It is evident from the typical pressure dependence of the rate coefficient k of a unimolecular reaction from the low-pressure gas phase to the compressed-liquid phase that the prerequisites of TST are only met, if at all, in a narrow density regime corresponding to the plateau region of the curve. At low pressures, where the rate is controlled by thermal activation in binary collisions with the solvent molecules, k is proportional to pressure. This regime is followed by a plateau region where k is pressure independent and controlled by intramolecular motion along the reaction coordinate. Here k attains the so-called high-pressure

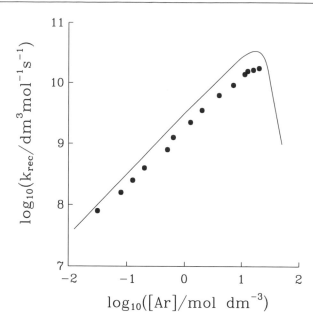

Figure A3.6.4. Pressure dependence of atom recombination rate constant of iodine in argon: experiment (points) [36] and theory (full line) [120].

limit k_∞ which can be calculated by statistical theories if the PES for the reaction is known. If the reaction entails large-amplitude structural changes, further increasing the pressure can lead to a decrease of k as a result of frictional forces retarding the barrier crossing process. In the simplest approach, k eventually approaches an inverse dependence on solvent friction, the so-called Smoluchowski limit k_{SM} of the reaction rate.

The transition from k_0 to k_∞ on the low-pressure side can be constructed using multidimensional unimolecular rate theory [1, 44], if one knows the barrier height for the reaction and the vibrational frequencies of the reactant and transition state. The transition from k_∞ to k_{SM} can be described in terms of Kramers' theory [45] which, in addition, requires knowledge of the pressure dependence of the solvent friction acting on the molecule during the particular barrier crossing process. The result can be compared with rate coefficients measured over a wide pressure range in selected solvents to test the theoretical models that are used to describe this so-called Kramers' turnover of the rate coefficient.

(a) Kramers' theory

Kramers' solution of the barrier crossing problem [45] is discussed at length in chapter A3.8 dealing with condensed-phase reaction dynamics. As the starting point to derive its simplest version one may use the Langevin equation, a stochastic differential equation for the time evolution of a slow variable, the reaction coordinate $\mathbf{r}$, subject to a rapidly statistically fluctuating force $\mathbf{F}$ caused by microscopic solute–solvent interactions under the influence of an external force field generated by the PES V for the reaction

$$M\dot{\mathbf{u}} = -\gamma\mathbf{u} + \mathbf{F}(t) - \nabla_{\mathbf{r}}V \qquad (A3.6.27)$$

where dots denote time derivative, M is the mass moving with velocity $\mathbf{u}$ along the reaction path and γ is the constant friction coefficient for motion along that path. The assumption is that there are no memory effects in the solvent bath, i.e. one considers a Markov process such that for the ensemble average $\langle\mathbf{F}(t)\cdot\mathbf{F}(t')\rangle \sim \delta(t-t')$. The corresponding two-dimensional Fokker–Planck equation for the probability distribution in phase space can be solved for the potential-barrier problem involving a harmonic well and a parabolic barrier in the limit of low and large friction. Since the low-friction limit, corresponding to the reaction in the gas phase, is correctly described by multidimensional unimolecular rate theory, only the solution in the large-friction limit is of interest in this context. One obtains a correction factor F_{Kr} to the high-pressure limit of the reaction rate constant k_∞

$$F_{Kr} = \left[\left(\left(\frac{\gamma/M}{2\omega_B} \right)^2 - 1 \right)^{1/2} - \frac{\gamma/M}{2\omega_B} \right] \qquad (A3.6.28)$$

which contains as an additional parameter the curvature of the parabolic barrier top, the so-called imaginary barrier frequency ω_B. F_{Kr} is less than unity and represents the dynamic effect of trajectories recrossing the barrier top, in contrast to the central assumption of canonical and microcanonical statistical theories, like TST or RRKM theory. In the high-damping limit, when $\gamma/M \gg \omega_B$, F_{Kr} reduces to $\omega_B M/\gamma$ which simply represents the Smoluchowski limit where velocities relax much faster than the barrier is crossed. As γ approaches zero, F_{Kr} goes to unity and the rate coefficient becomes equal to the high-pressure limit k_∞. In contrast to the situation in the Smoluchowski limit, the velocities do not obey a Maxwell–Boltzmann distribution.

(b) Pressure dependence of reaction rates

If other fall-off broadening factors arising in unimolecular rate theory can be neglected, the overall dependence of the rate coefficient on pressure or, equivalently, solvent density may be represented by the expression [1, 2]

$$k(\rho) = \frac{k_0 \rho k_\infty}{k_0 \rho + k_\infty} F_{Kr}(\rho). \qquad (A3.6.29)$$

This ensures the correct connection between the one-dimensional Kramers model in the regime of large friction and multidimensional unimolecular rate theory in that of low friction, where Kramers' model is known to be incorrect as it is restricted to the energy diffusion limit. For low damping, equation (A3.6.29) reduces to the Lindemann–Hinshelwood expression, while in the case of very large damping, it attains the Smoluchowski limit

$$k_{SM} = k_\infty \frac{\omega_B}{\gamma/M}. \qquad (A3.6.30)$$

Sometimes it may be convenient to use an even simpler interpolation formula that connects the different rate coefficient limits [4]

$$\frac{1}{k} \approx \frac{1}{k_0 \rho} + \frac{1}{k_\infty} + \frac{1}{k_{SM}} \Rightarrow k \approx \frac{k_0 \rho k_\infty}{k_\infty + k_0(1 + \gamma/M\omega_B)} \qquad (A3.6.31)$$

for which numerical simulations have shown that it is accurate to within 10–20%.

Predicting the solvent or density dependence of rate constants by equations (A3.6.29) or (A3.6.31) requires the same ingredients as the calculation of TST rate constants plus an estimate of ω_B and a suitable model for the friction coefficient γ and its density dependence. While in the framework of molecular dynamics simulations it may be worthwhile to numerically calculate friction coefficients from the average of the relevant time correlation functions, for practical purposes in the analysis of kinetic data it is much more convenient and instructive to use experimentally determined macroscopic solvent parameters.

As in the case of atom recombination, a convenient 'pressure scale' to use across the entire range is the inverse of the binary diffusion coefficient, D_{AM}^{-1}, of reactant A in solvent M, as compared to density ρ in the low-pressure gas and the inverse of solvent viscosity η^{-1} in liquid solution [46]. According to kinetic theory the diffusion coefficient in a dilute Lennard-Jones gas is given by

$$D_{AM} = \frac{3}{2\sqrt{2}} \frac{kT}{\mu_{AM}} \frac{\Omega^{(2,2)*}}{\Omega^{(1,1)*}} \frac{1}{Z_{LJ}\rho} \equiv \frac{A_D}{Z_{LJ}\rho}$$

with reduced collision integrals $\Omega^{(l,j)*}$ for Lennard-Jones well depths $\varepsilon_{AM} = \sqrt{\varepsilon_A \varepsilon_M}$ and reduced mass μ_{AM}, such that the low-pressure rate coefficient is

$$k_0 = \frac{A_D}{D_{AM}} \int_{E_0}^{\infty} f(E)\, dE \equiv \frac{A_D}{D_{AM} k_{00}}.$$

In liquid solution, Brownian motion theory provides the relation between diffusion and friction coefficient $D_{AM} = kT/\gamma$. Substituting correspondingly in equation (A3.6.31), one arrives at an expression representing the pressure dependence of the rate constant in terms of the pressure-dependent diffusion coefficient:

$$k \approx \frac{k_{00} A_D k_\infty D_{AM}}{k_\infty + k_{00} A_D (D_{AM} + kT/\mu_{AM}\omega_B)}. \tag{A3.6.32}$$

As data of the binary diffusion coefficient $D_{AM}(\rho, T)$ are not available in many cases, one has to resort to taking the solvent self-diffusion coefficient $D_M(\rho, T)$ which requires rescaling in the low-pressure regime according to

$$\frac{D_M}{D_{AM}} = \left[\frac{2\mu_{AM}}{M}\right]^{1/2} \left[\frac{\sigma_M}{\sigma_{AM}}\right]^2 \frac{\Omega_M^{(1,1)*}}{\Omega_{AM}^{(1,1)*}}.$$

In the Smoluchowski limit, one usually assumes that the Stokes–Einstein relation $(D\eta/kT)\sigma = C$ holds, which forms the basis of taking the solvent viscosity as a measure for the zero-frequency friction coefficient appearing in Kramers' expressions. Here C is a constant whose exact value depends on the type of boundary conditions used in deriving Stokes' law. It follows that the diffusion coefficient ratio is given by $D_M/D_{AM} = C_M\sigma_{AM}/C_{AM}\sigma_M$, which may be considered as approximately pressure independent.

(c) Extensions of Kramers' basic model

As extensions of the Kramers theory [47] are essentially a topic of condensed-phase reaction dynamics, only a few remarks are in place here. These concern the barrier shape and the dimensionality in the high-damping regime. The curvature at the parabolic barrier top obviously determines the magnitude of the friction coefficient at which the rate constant starts to decrease below the upper limit defined by the high-pressure limit: for relatively sharp barriers this 'turnover' will occur at comparatively high solvent density corresponding almost to liquid phase densities, whereas reactions involving flat barriers will show this phenomenon in the moderately dense gas, maybe even in the unimolecular fall-off regime before they reach k_∞.

Non-parabolic barrier tops cause the prefactor to become temperature dependent [48]. In the Smoluchowski limit, $k_{SM} \propto T^n$, $|n| \sim 1$, with $n > 0$ and $n < 0$ for curvatures smaller and larger than parabolic, respectively. For a cusp-shaped barrier top, i.e. in the limit $\omega_B \to \infty$ as might be applicable to electron transfer reactions, one obtains [45]

$$k_{SM} = k_\infty \frac{\omega_A M \sqrt{\pi}}{\gamma} \sqrt{\frac{E_0}{kT}}$$

where ω_A is the harmonic frequency of the potential well in this one-dimensional model. In the other limit, for an almost completely flat barrier top, the transition curve is extremely broad and the maximum of k is far below k_∞ [49]. A qualitatively different situation arises when reactant and product well are no longer separated by a barrier, but one considers escape out of a Lennard-Jones potential well. In this case, dynamics inside the well and outside on top of the 'barrier' plateau are no longer separable and, in a strict sense, the Smoluchowski limit is not reached any more. The stationary rate coefficient in the high-damping limit turns out to be [50]

$$\left(\frac{k}{k_\infty}\right)^{LJ} = \frac{1}{\gamma\sigma_{LJ}} \sqrt{\frac{\pi MkT}{2}} \frac{1}{L/\sigma_{LJ} - 2^{1/6}} \left[1 - \frac{2\varepsilon_{LJ}M}{(L\gamma)^2} + \cdots\right].$$

The original Kramers model is restricted to one-dimensional barriers and cannot describe effects due to the multidimensional barrier topology that may become important in cases where the system does not follow the minimum energy path on the PES but takes a detour across a higher effective potential energy barrier which is compensated by a gain in entropy. Considering a two-dimensional circular reaction path, the Smoluchowski limit of the rate coefficient obtained by solving the two-dimensional Fokker–Planck equation in coordinate space was shown to be [51]

$$k_{SM}^{2D} = k_{SM} \left[1 + \frac{kT}{M(\omega_\perp r_c)^2} \right]$$

where $\omega_\perp$ is the harmonic frequency of the transverse potential well and r_c the radius of curvature of the reaction path. This result is in good agreement with corresponding Langevin simulations [52]. A related concept is based on the picture that with increasing excitation of modes transverse to the reaction path the effective barrier curvature may increase according to $\omega_B^{eff}(E_\perp) \propto (E_\perp/b)^a$, where a and b are dimensionless parameters [53]. Approximating the topology of the saddle point region by a combination of a parabolic barrier top and a transverse parabolic potential, one arrives at a rate constant in the Smoluchowski limit given by

$$k_{SM}^{2D} = \frac{k_\infty}{\gamma/M} \omega_B(T) \qquad \text{with } \omega_B(T) \propto \left(\frac{T}{a}\right)^b \Gamma\left(b + \frac{1}{2}\right).$$

Multidimensionality may also manifest itself in the rate coefficient as a consequence of anisotropy of the friction coefficient [54]. Weak friction transverse to the minimum energy reaction path causes a significant reduction of the effective friction and leads to a much weaker dependence of the rate constant on solvent viscosity. These conclusions based on two-dimensional models also have been shown to hold for the general multidimensional case [55–61].

To conclude this section it should be pointed out again that the friction coefficient has been considered to be frequency independent as implied in assuming a Markov process, and that zero-frequency friction as represented by solvent viscosity is an adequate parameter to describe the effect of friction on observed reaction rates.

(d) Frequency-dependent friction

For very fast reactions, as they are accessible to investigation by pico- and femtosecond laser spectroscopy, the separation of time scales into slow motion along the reaction path and fast relaxation of other degrees of freedom in most cases is no longer possible and it is necessary to consider dynamical models, which are not the topic of this section. But often the temperature, solvent or pressure dependence of reaction rate coefficients determined in chemical kinetics studies exhibit a signature of underlying dynamic effects, which may justify the inclusion of some remarks at this point.

The key quantity in barrier crossing processes in this respect is the barrier curvature ω_B which sets the time window for possible influences of the dynamic solvent response. A sharp barrier entails short barrier passage times during which the memory of the solvent environment may be partially maintained. This non-Markov situation may be expressed by a generalized Langevin equation including a time-dependent friction kernel $\gamma(t)$ [62]

$$M\ddot{u} = -\int_0^t \gamma(t - \tau)\mathbf{u}(\tau)\,d\tau + \mathbf{F}(t) = \nabla_\mathbf{r} V$$

in which case the autocorrelation function of the randomly fluctuating force is no longer a δ-function but obeys $\langle \mathbf{F}(t) \cdot \mathbf{F}(t') \rangle = kT\gamma(t - t')$. This ensures that a Maxwell–Boltzmann distribution is re-established after decay of the solvent response. Adding the assumption of a Gaussian friction kernel, a generalized Fokker–Planck equation with time-dependent friction may be set up, and for a piecewise parabolic potential one obtains an expression for the rate coefficient, the so-called Grote–Hynes formula [63]:

$$k_{GH} = \frac{k_\infty}{\omega_B}\lambda_r. \tag{A3.6.33}$$

λ_r is the reactive frequency or unstable mode which is related to the friction coefficient by the implicit equation

$$\lambda_r = \frac{\omega_B^2}{\lambda_r + (\gamma(\lambda_r)/M)} \tag{A3.6.34}$$

with $\gamma(\lambda_r)$ being the Laplace transform of the time-dependent friction, $\hat{\gamma}(\lambda_r) = \int_0^\infty \exp(-\lambda_r t) \gamma(t) \, dt$. It is obvious that calculation of k_{GH} requires knowledge of potential barrier parameters and the complete viscoelastic response of the solvent, demonstrating the fundamental intimate link between condensed-phase reaction dynamics and solvation dynamics. This kind of description may be equivalently transferred to the dielectric response of the solvent causing dielectric friction effects in reactions with significant and fast charge rearrangement [64–66].

In the Smoluchowski limit the reaction is by definition the slow coordinate, such that $\hat{\gamma}(\lambda_r) \approx \hat{\gamma}(0) = \int_0^\infty \gamma(t) \, dt$, $\hat{\gamma}(0) \gg \lambda_r$ and $k_{GH} \approx k_{SM} = k_\infty \omega_B M / \hat{\gamma}(0)$. Though the time-dependent friction in principle is accessible *via* molecular dynamics simulations, for practical purposes in chemical kinetics in most cases analytical friction models have to be used including a short-time Gaussian 'inertial' component and a hydrodynamic tail at longer times. In the Grote–Hynes description the latter term only comes into play when the barrier top is sufficiently flat. As has been pointed out, the reactive mode frequency λ_r can be interpreted as an effective barrier curvature such that coupling of the reaction coordinate to the solvent changes position and shape of the barrier in phase space.

Because of the general difficulty encountered in generating reliable potentials energy surfaces and estimating reasonable friction kernels, it still remains an open question whether by analysis of experimental rate constants one can decide whether non-Markovian bath effects or other influences cause a particular solvent or pressure dependence of reaction rate coefficients in condensed phase. From that point of view, a purely empirical friction model might be a viable alternative, in which the frequency-dependent friction is replaced by a state-dependent friction $\hat{\gamma}(\omega) \rightarrow K^2 = \omega_B^2/(A/\eta + B)$ that is described in terms of properties of PES and solute–solvent interaction, depicting the reaction as occurring in a frozen environment of fixed microscopic viscosity [67, 68].

(e) Microscopic friction

The relation between the microscopic friction acting on a molecule during its motion in a solvent environment and macroscopic bulk solvent viscosity is a key problem affecting the rates of many reactions in condensed phase. The sequence of steps leading from friction to diffusion coefficient to viscosity is based on the general validity of the Stokes–Einstein relation and the concept of describing friction by hydrodynamic as opposed to microscopic models involving local solvent structure. In the hydrodynamic limit the effect of solvent friction on, for example, rotational relaxation times of a solute molecule is [69]

$$\tau_{rot} = 1/6 D_{rot} = (V_h/kT) \eta f_{bc} C + \tau_0 \tag{A3.6.35}$$

where V_h is the hydrodynamic volume of the solute in the particular solvent, whereas f_{bc} and C are parameters describing hydrodynamic boundary conditions and correcting for aspherical shape, respectively. τ_0 in turn may be related to the relaxation time of the free rotor. Though in many cases this equation correctly reproduces the viscosity dependence of τ_{rot}, in particular when solute and solvent molecules are comparable in size there are quite a number of significant deviations. One may incorporate this size effect by explicitly considering the first solvation shell on the solute surface which, under the assumption of slip boundary conditions, gives for the correction factor C in equation (A3.6.35):

$$C_{size} = \frac{f_{slip} V_h}{f_{slip} V_h + BkT \kappa_T \eta (4/\sigma_r^2 + 1)} \tag{A3.6.36}$$

with isothermal compressibility κ_T, ratio of radii of solvent to solute σ_r and a temperature-dependent parameter B. If one compares equation (A3.6.36) with the empirical friction model mentioned above, one realizes that both contain a factor of the form $C = 1/1 + a\eta$, suggesting that these models might be physically related.

Another, purely experimental possibility to obtain a better estimate of the friction coefficient for rotational motion γ_{rot} in chemical reactions consists of measuring rotational relaxation times τ_{rot} of reactants and calculating it according to equation (A3.6.35) as $\gamma_{rot} = 6kT\tau_{rot}$.

A3.6.4   Selected reactions

A3.6.4.1   Photoisomerization

According to Kramers' model, for flat barrier tops associated with predominantly small barriers, the transition from the low- to the high-damping regime is expected to occur in low-density fluids. This expectation is borne out by an extensively studied model reaction, the photoisomerization of *trans*-stilbene and similar compounds [70, 71] involving a small energy barrier in the first excited singlet state whose decay after photoexcitation is directly related to the rate coefficient of *trans-cis*-photoisomerization and can be conveniently measured by ultrafast laser spectroscopic techniques.

(a) Pressure dependence of photoisomerization rate constants

The results of pressure-dependent measurements for *trans*-stilbene in supercritical *n*-pentane [46] (figure A3.6.5) and the prediction from the model described by equation (A3.6.29), using experimentally determined microcanonical rate coefficients in jet-cooled *trans*-stilbene to calculate k_∞, show two marked discrepancies between model calculation and measurement: (1) experimental values of k are an order of magnitude higher already at low pressure and (2) the decrease of k due to friction is much less pronounced than predicted. As interpretations for the first observation, several ideas have been put forward that will not be further discussed here, such as a decrease of the effective potential barrier height due to electrostatic solute–solvent interactions enhanced by cluster formation at relatively low pressures [72, 73], or incomplete intramolecular vibrational energy redistribution in the isolated molecule [74–80], or Franck–Condon cooling in the excitation process [79, 80]. The second effect, the weak viscosity dependence, which was first observed in solvent series experiments in liquid solution [81–83], has also led to controversial interpretations: (i) the macroscopic solvent viscosity is an inadequate measure for microscopic friction acting along the reaction path [84, 85], (ii) the multidimensional character of the barrier crossing process leads to a fractional power dependence of k on $1/\eta$ [54, 81, 86, 87], (iii) as the reaction is very fast, one has to take into account the finite response time of the solvent, i.e. consider frequency-dependent friction [81, 87] and (iv) the effective barrier height decreases further with increasing electronic polarizability and polarity of the solvent, and the observed phenomenon is a manifestation of the superposition of a static solvent effect and hydrodynamic solvent friction correctly described by η [88]. One may test these hypotheses by studying molecular rotational motion and reaction independently in compressed sample solutions. A few examples will serve here to illustrate the main conclusions one can draw from the experimental results.

(b) Microscopic and frequency-dependent friction

Rotational relaxation times τ_{rot} of *trans*-stilbene and E,E-diphenylbutadiene (DPB) in liquid solvents like subcritical ethane and *n*-octane show a perfectly linear viscosity dependence with a slope that depends on the solvent [89] (figure A3.6.6), showing that microscopic friction acting during molecular rotational diffusion is proportional to the macroscopic solvent viscosity and that the relevant solute–solvent coupling changes with solvent. It seems reasonable to assume, therefore, that a corresponding relation also holds for microscopic friction governing diffusive motion along the reaction path.

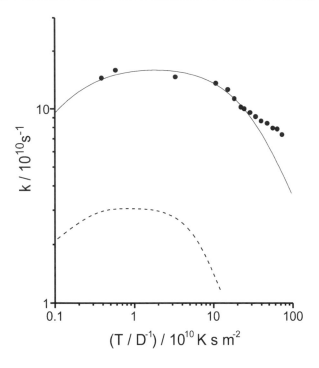

Figure A3.6.5. Photoisomerization rate constant of *trans*-stilbene in *n*-pentane *versus* inverse of the self-diffusion coefficient. Points represent experimental data, the dashed curve is a model calculation based on an RRKM fit to microcanonical rate constants of isolated *trans*-stilbene and the solid curve a fit that uses a reaction barrier height reduced by solute–solvent interaction [46].

The validity of this assumption is apparent in the viscosity dependence of rate coefficients for S_1-photoisomerization reactions in a number of related molecules such as *cis*-stilbene [90] (see figure A3.6.7), tetraphenylethylene (TPE) [91], DPB [92] and 'stiff' *trans*-stilbene [93] (where the phenyl ring is fixed by a five-membered ring to the ethylenic carbon atom). In all these cases a study of the pressure dependence reveals a linear correlation between k and $1/\eta$ in *n*-alkane and *n*-alkanol solvents, again with a solvent-dependent slope. The time scale for motion along the reaction path extends from several hundred picoseconds in DPB to a couple of hundred femtoseconds in *cis*-stilbene. There is no evidence for a frequency dependence of the friction coefficient in these reactions. As the time scale for the similar reaction in *trans*-stilbene is between 30 and 300 ps, one may conclude that also in this case the dynamics is mainly controlled by the zero-frequency friction which, in turn, is adequately represented by the macroscopic solvent viscosity. Therefore, the discrepancy between experiment and model calculation observed for *trans*-stilbene in compressed-liquid *n*-alkanes does not indicate a breakdown of the simple friction model in the Kramers–Smoluchowski theory. This result is in contrast to the analysis of solvent series study in linear alkanes, in which a solvent size effect of the microviscosity was made responsible for weak viscosity dependence [94]. Surprisingly, in a different type of non-polar solvent like methylcyclohexane, an equally weak viscosity dependence was found when the pressure was varied [95]. So the details of the viscosity influence are still posing puzzling questions.

(c) Effective barrier height

Measuring the pressure dependence of k at different temperatures shows that the apparent activation energy at constant viscosity decreases with increasing viscosity [46, 89] (figure A3.6.8). From a detailed analysis one

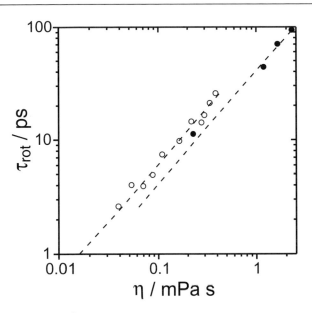

Figure A3.6.6. Viscosity dependence of rotational relaxation times of *trans*-stilbene in ethane (open circles) and *n*-octane (full circles) [89].

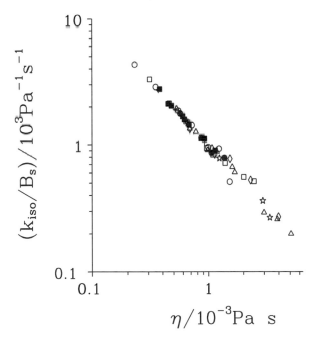

Figure A3.6.7. Viscosity dependence of reduced S_1-decay rate constants of *cis*-stilbene in various solvents [90]. The rate constants are divided by the slope of a linear regression to the measured rate constants in the respective solvent.

can extract an effective barrier height E_0 along the reaction path that decreases linearly with increasing density of the solvent. The magnitude of this barrier shift effect is more than a factor of two in nonpolar solvents

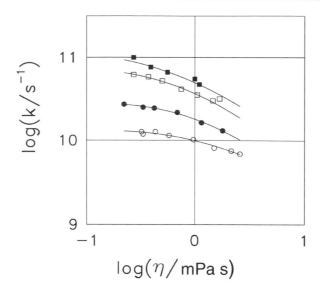

Figure A3.6.8. Isotherms of $k(\eta)$ for *trans*-stilbene photoisomerization in *n*-hexane at temperatures between 300 K (bottom) and 480 K (top). The curvature of the isotherms is interpreted as a temperature-dependent barrier shape [89].

like *n*-hexane or *n*-pentane [46]. It is interesting to note that in compressed-liquid *n*-propanol one almost reaches the regime of barrierless dynamics [96]. This is also evident in the room-temperature $k(\eta)$ isotherm measured in *n*-butanol (figure A3.6.9) which turns into linear k *versus* $1/\eta$ dependence at higher pressures, indicating that there is no further decrease of the effective barrier height. Thus the unexpected dependence of the reaction rate on solvent viscosity is connected with specific properties of the PES of *trans*-stilbene in its first excited singlet state, because corresponding measurements for, for example, DPB or TPE in *n*-alkanes and *n*-alkanols do not show any evidence for deviations from standard Kramers–Smoluchowski behaviour.

As a multidimensional PES for the reaction from quantum chemical calculations is not available at present, one does not know the reason for the surprising barrier effect in excited *trans*-stilbene. One could suspect that *trans*-stilbene possesses already a significant amount of zwitterionic character in the conformation at the barrier top, implying a fairly 'late' barrier along the reaction path towards the twisted perpendicular structure. On the other hand, it could also be possible that the effective barrier changes with viscosity as a result of a multidimensional barrier crossing process along a curved reaction path.

(d) Solvation dynamics

The dependence of k on viscosity becomes even more puzzling when the time scale of motion along the reaction coordinate becomes comparable to that of solvent dipole reorientation around the changing charge distribution within the reacting molecule—in addition to mechanical, one also has to consider dielectric friction. For *trans*-stilbene in ethanol, the $k(\eta)$ curve exhibits a turning point which is caused by a crossover of competing solvation and reaction time scales [97] (figure A3.6.10): as the viscosity increases the dielectric relaxation time of the solvent increases more rapidly than the typical time necessary for barrier crossing. Gradually, the solvation dynamics starts to freeze out on the time scale of reactive motion, the polar barrier is no longer decreased by solvent dipole reorientation and the rate coefficient drops more rapidly with increasing viscosity. As soon as the solvent dipoles are completely 'frozen', one has the same situation as in a non-polar solvent: i.e. only the electronic polarizability of the solvent causes further decrease of the barrier height.

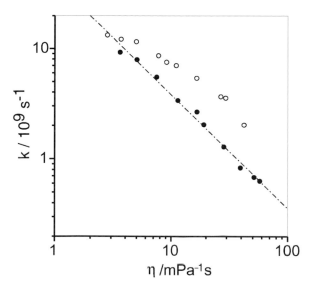

Figure A3.6.9. Viscosity dependence of photoisomerization rate constants of *trans*-stilbene (open circles) and E,E-diphenylbutadiene (full circles) in *n*-butanol. The broken line indicates a η^{-1}-dependence of k [96].

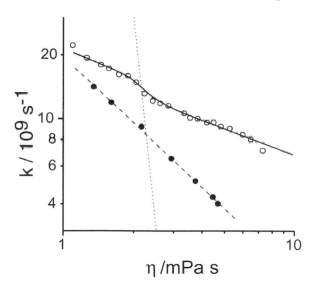

Figure A3.6.10. Viscosity dependence of photoisomerization rate constants of *trans*-stilbene (open circles) and E,E-diphenylbutadiene (full circles) in ethanol. The dashed line indicates a η^{-1}-dependence of k, the dotted line indicates the viscosity dependence of the dielectric relaxation time of ethanol and the solid curve is the result of a kinetic model describing the parallel processes of reaction and solvent relaxation [97].

A3.6.4.2 *Chair-boat inversion of cyclohexane*

As mentioned above, in liquid solution most reactions are expected to have passed beyond the low-damping regime where the dynamics is dominated by activating and deactivating collisions between reactants' solvent molecules. In general, this expectation is met, as long as there is a sufficiently strong intramolecular coupling of the reaction coordinate to a large number of the remaining modes of the reactant at the transition state which leads to fast IVR within the reactant. In this case, the high-pressure limit of unimolecular rate theory is reached, and additional coupling to the liquid solvent environment leads to a decrease of the rate coefficient through the factor F_{Kr}. From this point of view, the observation of rate coefficient maxima in liquid solution would appear to signal a breakdown of RRKM theory. In particular it has been argued that, for the case of weak *intra*molecular coupling, a strong coupling of the reaction coordinate to the solvent could effectively decrease the volume of phase space accessible to the reactant in the liquid with respect to the gas phase [98, 99]. As the relative strength of *intra*- and *inter*molecular coupling may change with solvent properties, the breakdown of the RRKM model might be accompanied by the appearance of a rate coefficient maximum in liquid solution as a function of solvent friction.

Among the few reactions for which an increase of a reaction rate coefficient in liquid solution with increasing reactant–solvent coupling strength has been observed, the most notable is the thermal chair-boat isomerization reaction of cyclohexane (figure A3.6.11) and 1,1-difluorocyclohexane [100–103]. The observed pressure dependence of the rate coefficients along different isotherms was analysed in terms of one-dimensional transition state theory by introducing a transmission coefficient κ describing the effect of solvent friction $k_{obs} = \kappa k_{TST}$. In the intermediate- to high-damping regime, κ can be identified with the Kramers term F_{Kr}. The observed pressure-dependent activation volumes $\Delta V_{OBS}^{\neq}$ were considered to represent the sum of a pressure-independent intrinsic activation volume $\Delta V_{TST}^{\neq}$ and a pressure-dependent formal collisional activation volume $\Delta V_{COLL}^{\neq}$ arising from the increase of that reactant–solvent coupling with pressure which corresponds to viscous effects

$$RT \left(\frac{\partial \ln k_{TST}}{\partial p} \right)_T = -\Delta V_{TST}^{\neq}$$

$$RT \left(\frac{\partial \ln \kappa}{\partial p} \right)_T = -\Delta V_{COLL}^{\neq}.$$

The intrinsic volume of activation was estimated to correspond to the molar volume difference between cyclohexene and cyclohexane, adding the molar volume difference between ethane and ethene to account for the two missing protons and shortened double bond in cyclohexane. This yields a value of $\Delta V_{TST}^{\neq} = -1.5$ cm^3 mol^{-1}. Then, knowing the pressure dependence of the solvent viscosity, the viscosity dependence of the relative transmission coefficient κ was estimated from

$$\frac{\kappa(\eta)}{\kappa(1.5 \text{ cP})} = \frac{k_{obs}(\eta)}{k_{obs}(1.5 \text{ cP})} \exp \left[\frac{p \Delta V_{TST}^{\neq}}{RT} \right].$$

The experimental values of $\kappa(\eta)$ have a maximum at a viscosity close to 3 cP and varies by about 15% over the entire viscosity range studied. As discussed above, this unexpected dependence of κ on solvent friction in liquid CS$_2$ is thought to be caused by a relatively weak intramolecular coupling of the reaction coordinate to the remaining modes in cyclohexane. At viscosities below the maximum, motion along the reaction coordinate due to the reduction of the accessible phase space region is fast. The barrier passage is still in the inertial regime, and the strong coupling to the solvent leads to increasingly rapid stabilization in the product well. With increasing solvent friction, the barrier crossing enters the diffusive regime and begins to show a slowdown with further increasing solvent viscosity.

This interpretation of the experimentally determined pressure dependence of the isomerization rate rests on the assumptions that (i) the barrier height for the reaction is independent of pressure and (ii) the estimate of the intrinsic volume of activation is reliable to within a factor of two and $\Delta V_{TST}^{\neq}$ does not change with pressure. As pointed out previously, due to the differences in the pressure dependences of solvent viscosity and density, a change of the barrier height with solvent density can give rise also to an apparent maximum of the rate coefficient as a function of viscosity. In particular, a decrease of E_0 with pressure by about

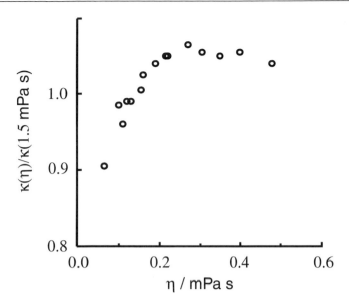

Figure A3.6.11. Viscosity dependence of transmission coefficient of the rate of cyclohexane chair-boat inversion in liquid solution (data from [100]).

1 kJ mol^{-1} could explain the observed non-monotonic viscosity dependence. Therefore, the constancy to within 0.05 kJ mol^{-1} of the isoviscous activation energy over a limited viscosity range from 1.34 to 2.0 cP lends some support to the first assumption.

From stochastic molecular dynamics calculations on the same system, in the viscosity regime covered by the experiment, it appears that *intra*- and *inter*molecular energy flow occur on comparable time scales, which leads to the conclusion that cyclohexane isomerization in liquid CS$_2$ is an activated process [99]. Classical molecular dynamics calculations [104] also reproduce the observed non-monotonic viscosity dependence of κ. Furthermore, they also yield a solvent contribution to the free energy of activation for the isomerization reaction which in liquid CS$_2$ *increases* by about 0.4 kJ mol^{-1}, when the solvent density is increased from 1.3 to 1.5 g cm^{-3}. Thus the molecular dynamics calculations support the conclusion that the high-pressure limit of this unimolecular reaction is not attained in liquid solution at ambient pressure. It has to be remembered, though, that the analysis of the measured isomerization rates depends critically on the estimated value of $\Delta V_{\text{TST}^{\neq}}$. What is still needed is a reliable calculation of this quantity in CS$_2$.

A3.6.4.3 *Photolytic cage effect and geminate recombination*

For very fast reactions, the competition between geminate recombination of a pair of initially formed reactants and its escape from the common solvent cage is an important phenomenon in condensed-phase kinetics that has received considerable attention both theoretically and experimentally. An extremely well studied example is the photodissociation of iodine for which the quantum yield Φ_d decreases from unity in the dilute-gas phase by up to a factor of ten or more in compressed-liquid solution. An intuitively appealing interpretation of this so-called photolytic cage effect, predicted by Franck and Rabinovitch in the 1930s [105], is based on models describing it as diffusive escape of the pair [106], formed instantaneously at t_0 with initial separation $\mathbf{r}_0$, from the solvent cage under the influence of Stokes friction subject to inner boundary conditions similar

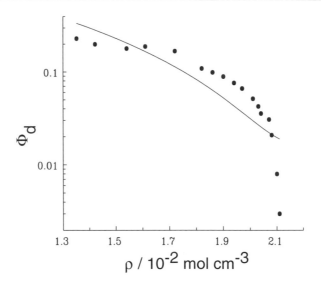

Figure A3.6.12. Photolytic cage effect of iodine in supercritical ethane. Points represent measured photodissociation quantum yields [37] and the solid curve is the result of a numerical simulation [111].

to equation (A3.6.21) [31],

$$\frac{\partial \rho}{\partial t} = D_{AA}\nabla^2\rho + \delta(\mathbf{r} - r_0)\delta(t - t_0) \qquad k_{\text{mol}}\rho(\mathbf{r}, t) = 4\pi R D_{AA}\frac{\partial \rho}{\partial r}\bigg|_R.$$

Solving this diffusion problem yields an analytical expression for the time-dependent escape probability $q(t)$:

$$q(t) = 1 - \frac{x}{z(1 + x)}\left\{\text{erfc}\left(\frac{z - 1}{2y\sqrt{t}}\right) - \exp[(z - 1)(1 + x) + y^2 t(1 + x)^2]\text{erfc}\left[y\sqrt{t}(1 + x) + \frac{z - 1}{2y\sqrt{t}}\right]\right\}$$

where x and y are as defined above and $z = r_0/R$. This equation can be compared with time-resolved measurements of geminate recombination dynamics in liquid solution [107, 108] if the parameters r_0, R, D_{AA} and k_{mol} are known or can be reliably estimated. This simple diffusion model, however, does not satisfactorily represent the observed dynamics, which is in part due to the participation of different electronic states. Direct evidence for this comes from picosecond time-resolved transient absorption experiments in solution that demonstrate the involvement of the low-lying, weakly bound iodine A and A' states. In these experiments it was possible to separate geminate pair dynamics and vibrational energy relaxation of the initially formed hot iodine molecules [40, 41, 109]. The details of the complex steps of recombination dynamics are still only partially understood and the subject of mixed quantum–classical molecular dynamics simulations [110].

In order to probe the importance of van der Waals interactions between reactants and solvent, experiments in the gas–liquid transition range appear to be mandatory. Time-resolved studies of the density dependence of the cage and cluster dynamics in halogen photodissociation are needed to extend earlier quantum yield studies which clearly demonstrated the importance of van der Waals clustering at moderate gas densities [37, 111] (see figure A3.6.12). The pressure dependence of the quantum yield established the existence of two different regimes for the cage effect: (i) at low solvent densities, excitation of solvent-clustered halogen molecules leads to predissociation of the van der Waals bond and thereby to stabilization of the halogen molecule, whereas (ii), at high liquid phase densities, the hard-sphere repulsive caging takes over which leads to a strong reduction in the photodissociation quantum yield.

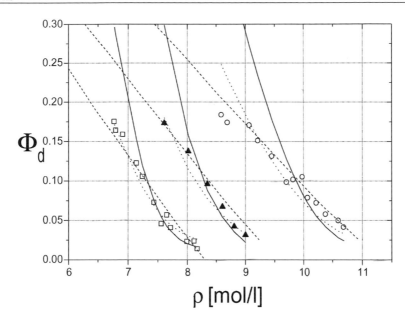

Figure A3.6.13. Density dependence of the photolytic cage effect of iodine in compressed liquid n-pentane (circles), n-hexane (triangles), and n-heptane (squares) [38]. The solid curves represent calculations using the diffusion model [37], the dotted and dashed curves are from 'static' caging models using Carnahan–Starling packing fractions and calculated radial distribution functions, respectively [38].

Attractive long-range and repulsive short-range forces both play a role in the cage effect, though each type dominates in a different density range. Whereas the second component has traditionally been recognized as being responsible for caging in liquid solution and solids, theoretical models and molecular dynamics calculations [112, 113] have confirmed the idea that complex formation between halogen and solvent molecules in supercritical solvents is important in photodissociation dynamics and responsible for the lowering of quantum yields at moderate gas densities [114, 115].

The traditional diffusion model permits estimation of the magnitude of the cage effect in solution according to [37]

$$\lim_{t \to \infty} q(t) = 1 - \frac{x}{z(1 + x)}$$

which should directly represent overall photodissociation quantum yields measured in dense solvents, as in this quantity dynamical effects are averaged out as a consequence of multiple collisions in the cage and effective collision-induced hopping between different electronic states at large interatomic distances. The initial separation of the iodine atom pair in the solvent cage may be calculated by assuming that immediately after excitation, the atoms are spherical particles subject to Stokes friction undergoing a damped motion on a repulsive potential represented by a parabolic branch. This leads to an excitation energy dependence of the initial separation [37]

$$z - 1 = \frac{\sqrt{1 - c^2}}{2c} \exp\left(-\frac{\pi c}{\sqrt{1 - c^2}}\right) \qquad \text{with } c = \frac{6\pi \eta \sigma_I R}{\sqrt{m_I(h\nu - D_0)}} \text{ for } c < 1$$

where σ_I and m_I are radius and mass of the iodine atom, respectively, $h\nu$ is the photon energy and D_0 the dissociation energy of iodine molecules. Obviously, $c \geq 1$ corresponds to the overdamped case for which $r_0 = R$ irrespective of initial energy. As in experiments a fairly weak dependence of Φ_d on excitation wavelength was found, it seems that, at least at liquid phase densities, separation of the iodine pair is overdamped, a finding corroborated by recent classical molecular dynamics simulations using simple model potentials [38].

The simple diffusion model of the cage effect again can be improved by taking effects of the local solvent structure, i.e. hydrodynamic repulsion, into account in the same way as discussed above for bimolecular reactions. The consequence is that the potential of mean force tends to favour escape at larger distances ($r_0 > 1.5R$) more than it enhances caging at small distances, leading to larger overall photodissociation quantum yields [116, 117].

The analysis of recent measurements of the density dependence of Φ_d has shown, however, that considering only the variation of solvent structure in the vicinity of the atom pair as a function of density is entirely sufficient to understand the observed changes in Φ_d with pressure and also with size of the solvent molecules [38]. Assuming that iodine atoms colliding with a solvent molecule of the first solvation shell under an angle α less than α_{max} (the value of α_{max} is solvent dependent and has to be found by simulations) are reflected back onto each other in the solvent cage, Φ_d is given by

$$\Phi_d = 1 - 4\pi\rho(1 - \cos\alpha_{max}) \int_0^{r_{shell}} r^2 g(r) \, dr$$

where the solvation shell radius shell is obtained from Lennard-Jones radii (figure A3.6.13).

As these examples have demonstrated, in particular for fast reactions, chemical kinetics can only be appropriately described if one takes into account dynamic effects, though in practice it may prove extremely difficult to separate and identify different phenomena. It seems that more experiments under systematically controlled variation of solvent environment parameters are needed, in conjunction with numerical simulations that as closely as possible mimic the experimental conditions to improve our understanding of condensed-phase reaction kinetics. The theoretical tools that are available to do so are covered in more depth in other chapters of this encyclopedia and also in comprehensive reviews [6, 118, 119].

References

[1] Troe J 1975 Unimolecular reactions: experiments and theories *Kinetics of Gas Reactions* ed W Jost (New York: Academic) p 835
[2] Troe J 1978 Kinetic phenomena in gases at high pressure *High Pressure Chemistry* ed H Kelm (Amsterdam: Reidel) pp 489–520
[3] Schroeder J and Troe J 1987 Elementary reactions in the gas–liquid transition range *Ann. Rev. Phys. Chem.* **38** 163
[4] Schroeder J and Troe J 1993 Solvent effects in the dynamics of dissociation, recombination and isomerization reactions *Activated Barrier Crossing* ed G R Fleming and P Hänggi (Singapore: World Scientific) p 206
[5] Kajimoto O 1999 Solvation in supercritical fluids: its effects on energy transfer and chemical reactions *Chem. Rev.* **99** 355–89
[6] Truhlar D G, Garrett B C and Klippenstein S J 1996 Current status of transition state theory *J. Phys. Chem.* A **100** 12 771–800
[7] Leffler J E and Grunwald E 1963 *Rates and Equilibria in Organic Reactions* (New York: Wiley)
[8] Hildebrand J H, Prausnitz J M and Scott R L 1970 *Regular and Related Solutions* (New York: Van Nostrand)
[9] Reichardt C 1988 *Solvents and Solvent Effects in Organic Chemistry* (Weinheim: VCH)
[10] Steinfeld J I, Francisco J S and Hase W L 1989 *Chemical Kinetics and Dynamics* (Englewood Cliffs, NJ: Prentice-Hall)
[11] Simkin B Ya and Sheikhet I I 1995 *Quantum Chemical and Statistical Theory of Solutions* (London: Ellis Horwood)
[12] Fröhlich H 1958 *Theory of Dielectrics* (New York: Plenum)
[13] Böttcher C J F 1973 *Theory of Dielectric Polarization* (Amsterdam: Elsevier)
[14] Abraham M H 1974 Solvent effects on transition states and reaction rates *Prog. Phys. Org. Chem.* **11** 1–87
[15] Popvych O and Tomkins R P T 1981 *Nonaqueous Solution Chemistry* (New York: Wiley)
[16] Brebbia C A and Walker S 1980 *Boundary Element Technique in Engineering* (London: Newnes-Butterworth)
[17] Miertuš S, Scrocco E and Tomasi J 1981 Electrostatic interactions of a solute with a continuum. A direct utilization of *ab initio* molecular potentials for the provision of solvent effects *Chem. Phys.* **55** 117–25
[18] Aguilar M A and Olivares del Valle F J 1989 Solute–solvent interactions. A simple procedure for constructing the solvent capacity for retaining a molecular solute *Chem. Phys.* **129** 439–50
[19] Aguilar M A and Olivares del Valle F J 1989 A computation procedure for the dispersion component of the interaction energy in continuum solute solvent models *Chem. Phys.* **138** 327–36
[20] Rivail J L 1989 *New Theoretical Concepts for Understanding Organic Reactions* ed J Bertran and I G Cizmadia (Amsterdam: Kluwer) p 219
[21] Cramer C J and Truhlar D G 1996 Continuum solvation models *Solvent Effects and Chemical Reactivity* ed O Tapia and J Bertran (Dordrecht: Kluwer) pp 1–80

[22] Mineva T, Russo N and Sicilia E 1998 Solvation effects on reaction profiles by the polarizable continuum model coupled with Gaussian density functional method *J. Comp. Chem.* **19** 290–9

[23] Castejon H and Wiberg K B 1999 Solvent effects on methyl transfer reactions. 1. The Menshutkin reaction *J. Am. Chem. Soc.* **121** 2139–46

[24] Evans M G and Polanyi M 1935 Some applications of the transition state method to the calculation of reaction velocities, especially in solution *Trans. Faraday Soc.* **31** 875–94

[25] Isaacs N S 1981 *Liquid Phase High Pressure Chemistry* (Chichester: Wiley-Interscience)

[26] Schmidt R 1998 Interpretation of reaction and activation volumes in solution *J. Phys. Chem.* A **102** 9082–6

[27] Basilevsky M V, Weinberg N N and Zhulin V M 1985 Pressure dependence of activation and reaction volumes *J. Chem. Soc. Faraday Trans.* 1 **81** 875–84

[28] Ladanyi B M and Hynes J T 1986 Transition state solvent effects on atom transfer rates in solution *J. Am. Chem. Soc.* **108** 585–93

[29] Smoluchowski Mv 1918 Versuch einer mathematischen Theorie der Koagulationskinetik kolloider Lösungen *Z. Phys. Chem.* **92** 129–39

[30] Smoluchowski Mv 1915 Über Brownsche Molekularbewegung unter Einwirkung äußerer Kräfte und deren Zusammenhang mit der verallgemeinerten Diffusionsgleichung *Ann. Phys.* **48** 1103–12

[31] Rice S A 1985 Diffusion-limited reactions *Comprehensive Chemical Kinetics* vol 25, ed C H Bamford, C F H Tipper and R G Compton (Amsterdam: Elsevier)

[32] Collins F C and Kimball G E 1949 Diffusion-controlled rate processes *J. Colloid Sci.* **4** 425

[33] Northrup S H and Hynes J T 1978 On the description of reactions in solution *Chem. Phys. Lett.* **54** 244

[34] Northrup S H and Hynes J T 1980 The stable states picture of chemical reactions. I. Formulation for rate constants and initial condition effects *J. Chem. Phys.* **73** 2700–14

[35] Kapral R 1981 Kinetic theory of chemical reactions in liquids *Adv. Chem. Phys.* **48** 71

[36] Hippler H, Luther K and Troe J 1973 Untersuchung der Rekombination von Jodatomen in stark komprimierten Gasen und in Fluessigkeiten *Ber. Bunsenges Phys. Chem.* **77** 1104–14

[37] Otto B, Schroeder J and Troe J 1984 Photolytic cage effect and atom recombination of iodine in compressed gases and liquids: experiments and simple models *J. Chem. Phys.* **81** 202

[38] Schwarzer D, Schroeder J and Schröder Ch 2000 Quantum yields for the photodissociation of iodine in compressed liquids and supercritical fluids *Z. Phys. Chem.* **214**

[39] Sceats M G 1988 *Chem. Phys. Lett.* **143** 123

[40] Harris A L, Berg M and Harris C B 1986 Studies of chemical reactivity in the condensed phase. I. The dynamics of iodine photodissociation and recombination on a picosecond time scale and comparison to theories for chemical reactions in solution *J. Chem. Phys.* **84** 788

[41] Paige M E, Russell D J and Harris C B 1986 Studies of chemical reactivity in the condensed phase. II. Vibrational relaxation of iodine in liquid xenon following geminate recombination *J. Chem. Phys.* **85** 3699–700

[42] Wang W, Nelson K A, Xiao L and Coker D F 1994 Molecular dynamics simulation studies of solvent cage effects on photodissociation in condensed phases *J. Chem. Phys.* **101** 9663–71

[43] Batista V S and Coker D F 1996 Nonadiabatic molecular dynamics simulation of photodissociation and geminate recombination of I_2 liquid xenon *J. Chem. Phys.* **105** 4033–54

[44] Gilbert R G and Smith S C 1990 *Theory of Unimolecular and Recombination Reactions* (Oxford: Blackwell)

[45] Kramers H A 1940 Brownian motion in a field of force and the diffusion model of chemical reactions *Physica* **7** 284–304

[46] Schroeder J, Troe J and Vöhringer P 1995 Photoisomerization of *trans*-stilbene in compressed solvents: Kramers turnover and solvent induced barrier shift *Z. Phys. Chem.* **188** 287

[47] Hänggi P, Talkner P and Borkovec M 1990 Reaction-rate theory: fifty years after Kramers *Rev. Mod. Phys.* **62** 251–341

[48] Brinkman H C 1956 Brownian motion in a field of force and the diffusion theory of chemical reactions *Physica* **12** 149–55

[49] Garrity D K and Skinner J L 1983 Effect of potential shape on isomerization rate constants for the BGK model *Chem. Phys. Lett.* **95** 46–51

[50] Larson R S and Lightfoot E J 1988 Thermally activated escape from a Lennard-Jones potential well *Physica* A **149** 296–312

[51] Larson R S and Kostin M D 1982 Kramers' theory of chemical kinetics: curvilinear reaction coordinates *J. Chem. Phys.* **77** 5017–25

[52] Larson R S 1986 Simulation of two-dimensional diffusive barrier crossing with a curved reaction path *Physica* A **137** 295–305

[53] Gehrke C, Schroeder J, Schwarzer D, Troe J and Voss F 1990 Photoisomerization of diphenylbutadiene in low-viscosity nonpolar solvents: experimental manifestations of multidimensional Kramers behavior and cluster effects *J. Chem. Phys.* **92** 4805–16

[54] Agmon N and Kosloff R 1987 Dynamics of two-dimensional diffusional barrier crossing *J. Phys. Chem.* **91** 1988–96

[55] Berezhkovskii A M, Berezhkovskii L M and Zitserman V Yu 1989 The rate constant in the Kramers multidimensional theory and th *Chem. Phys.* **130** 55–63

[56] Berezhkovskii A M and Zitzerman V Yu 1990 Activated rate processes in a multidimensional case *Physica* A **166** 585–621

[57] Berezhkovskii A M and Zitserman V Yu 1991 Activated rate processes in the multidimensional case. Consideration of recrossings in the multidimensional Kramers problem with anisotropic friction *Chem. Phys.* **157** 141–55

[58]   Berezhkovskii A M and Zitserman V Yu 1991 Comment on: diffusion theory of multidimensional activated rate processes: the role of anisotropy *J. Chem. Phys.* **95** 1424

[59]   Berezhkovskii A M and Zitserman V Yu 1992 Generalization of the Kramers–Langer theory: decay of the metastable state in the case of strongly anisotropic friction *J. Phys. A: Math. Gen.* **25** 2077–92

[60]   Berezhkovskii A M and Zitserman V Yu 1992 Multidimensional activated rate processes with slowly relaxing mode *Physica* A **187** 519–50

[61]   Berezhkovskii A M and Zitserman V Yu 1993 Multi-dimensional Kramers theory of the reaction rate with highly anisotropic friction. Energy diffusion for the fast coordinate versus overdamped regime for the slow coordinate *Chem. Phys. Lett.* **212** 413–19

[62]   Zwanzig R 1973 Nonlinear generalized langevin equations *J. Stat. Phys.* **9** 215–20

[63]   Grote R F and Hynes J T 1980 The stable states picture of chemical reactions. II. Rate constants for condensed and gas phase reaction models *J. Chem. Phys.* **73** 2715–32

[64]   Van der Zwan G and Hynes J T 1982 Dynamical polar solvent effects on solution reactions: A simple continuum model *J. Chem. Phys.* **76** 2993–3001

[65]   Van der Zwan G and Hynes J T 1983 Nonequilibrium solvation dynamics in solution reaction *J. Chem. Phys.* **78** 4174–85

[66]   Van der Zwan G and Hynes J T 1984 A simple dipole isomerization model for non-equilibrium solvation dynamics in reactions in polar solvents *Chem. Phys.* **90** 21–35

[67]   Zhu S-B, Lee J, Robinson G W and Lin S H 1988 A microscopic form of the extended Kramers equation. A simple friction model for cis-trans isomerization reactions *Chem. Phys. Lett.* **148** 164–8

[68]   Zhu S-B, Lee J, Robinson G W and Lin S H 1989 Theoretical study of memory kernel and velocity correlation function for condensed phase isomerization. I. Memory kernel *J. Chem. Phys.* **90** 6335–9

[69]   Dote J L, Kivelson D and Schwartz R N 1981 A molecular quasi-hydrodynamic free-space model for molecular rotational relaxation *J. Phys. Chem.* **85** 2169–80

[70]   Waldeck D H 1991 *Chem. Rev.* **91** 415

[71]   Waldeck D H 1993 Photoisomerization dynamics of stilbenes in polar solvents *J. Mol. Liq.* **57** 127–48

[72]   Schroeder J, Schwarzer D, Troe J and Voss F 1990 Cluster and barrier effects in the temperature and pressure dependence of the photoisomerization of trans.stilbene *J. Chem. Phys.* **93** 2393–404

[73]   Meyer A, Schroeder J and Troe J 1999 Photoisomerization of *trans*-stilbene in moderately compressed gases: pressure-dependent effective barriers *J. Phys. Chem.* A **103** 10 528–39

[74]   Khundkar L R, Marcus R A and Zewail A H 1983 Unimolecular reactions at low energies and RRKM-behaviour: isomerization and dissociation *J. Phys. Chem.* **87** 2473–6

[75]   Syage J A, Felker P M and Zewail A H 1984 Picosecond dynamics and photoisomerization of stilbene in supersonic beams. I. Spectra and mode assignments *J. Chem. Phys.* **81** 4685–705

[76]   Syage J A, Felker P M and Zewail A H 1984 Picosecond dynamics and photoisomerization of stilbene in supersonic beams. II. Reaction rates and potential energy surface *J. Chem. Phys.* **81** 4706–23

[77]   Nordholm S 1989 Photoisomerization of stilbene—a theoretical study of deuteration shifts and limited internal vibrational redistribution *Chem. Phys.* **137** 109–20

[78]   Bolton K and Nordholm S 1996 A classical molecular dynamics study of the intramolecular energy transfer of model trans-stilbene *Chem. Phys.* **203** 101–26

[79]   Leitner D M and Wolynes P G 1997 Quantum energy flow during molecular isomerization *Chem. Phys. Lett.* **280** 411–18

[80]   Leitner D M 1999 Influence of quantum energy flow and localization on molecular isomerization in gas and condensed phases *Int. J. Quant. Chem.* **75** 523–31

[81]   Rothenberger G, Negus D K and Hochstrasser R M 1983 Solvent influence on photoisomerization dynamics *J. Chem. Phys.* **79** 5360–7

[82]   Sundström V and Gillbro T 1984 Dynamics of the isomerization of trans-stilbene in n-alcohols studied by ultraviolet picosecond absorption recovery *Chem. Phys. Lett.* **109** 538–43

[83]   Sundström V and Gillbro T 1985 Dynamics of *trans-cis* photoisomerization of stilbene in hydrocarbon solutions *Ber. Bunsenges Phys. Chem.* **89** 222–6

[84]   Courtney S H, Kim S K, Canonica S and Fleming G R 1986 Rotational diffusion of stilbene in alkane and alcohol solutions *J. Chem. Soc. Faraday Trans. 2* **82** 2065–72

[85]   Lee M, Haseltine J N, Smith A B III and Hochstrasser R M 1989 Isomerization processes of electronically excited stilbene and diphenylbutadiene in liquids: Are they one-dimensional? *J. Am. Chem. Soc.* **111** 5044–51

[86]   Park N S and Waldeck D H 1989 Implications for multidimensional effects on isomerization dynamics: photoisomerization study of 4,4′-dimethylstilbene in *n*-alkane solvents *J. Chem. Phys.* **91** 943–52

       Park N S and Waldeck D H 1990 On the dimensionality of stilbene isomerization *Chem. Phys. Lett.* **168** 379–84

[87]   Velsko S P and Fleming G R 1982 Photochemical isomerization in solution. Photophysics of diphenylbutadiene *J. Chem. Phys.* **76** 3553–62

[88]   Schroeder J and Troe J 1985 Solvent shift and transport contributions in reactions in dense media *Chem. Phys. Lett.* **116** 453

[89]   Schroeder J 1997 Picosecond kinetics of *trans-cis* photoisomerisations: from jet-cooled molecules to compressed solutions *Ber. Bunsenges Phys. Chem.* **101** 643

[90] Nikowa L, Schwarzer D, Troe J and Schroeder J 1992 Viscosity and solvent dependence of low barrier processes: photoisomerization of *cis*-stilbene in compressed liquid solvents *J. Chem. Phys.* **97** 4827

[91] Schroeder J 1996 The role of solute-solvent interactions in the dynamics of unimolecular reactions in compressed solvents *J. Phys.: Condens. Matter* **8** 9379

[92] Gehrke C, Mohrschladt R, Schroeder J, Troe J and Vöhringer P 1991 Photoisomerization dynamics of diphenylbutadiene in compressed liquid alkanes and in solid environment *Chem. Phys.* **152** 45

[93] Mohrschladt R, Schroeder J, Troe J, Vöhringer P and Votsmeier M 1994 Solvent influence on barrier crossing in the S_1-state of *cis*- and *trans*-'stiff' stilbene *Ultrafast Phenomena IX* ed P F Barbara *et al* (New York: Springer) pp 499–503

[94] Saltiel J and Sun Y-P 1989 Intrinsic potential energy barrier for twisting in the *trans*-stilbene S1 State in hydrocarbon solvents *J. Phys. Chem.* **93** 6246–50

   Sun Y-P and Saltiel J 1989 Application of the Kramers equation to stilbene photoisomerization in *n*-alkanes using translational diffusion coefficients to define microviscosity *J. Phys. Chem.* **93** 8310–16

[95] Vöhringer P 1993 Photoisomerisierung in komprimierten Lösungen. Dissertation, Göttingen University

[96] Schroeder J, Schwarzer D, Troe J and Vöhringer P 1994 From barrier crossing to barrierless relaxation dynamics: photoisomerization of *trans*-stilbene in compressed alkanols *Chem. Phys. Lett.* **218** 43

[97] Mohrschladt R, Schroeder J, Schwarzer D, Troe J and Vöhringer P 1994 Barrier crossing and solvation dynamics in polar solvents: photoisomerzation of *trans*-stilbene and E,E-diphenylbutadiene in compressed alkanols *J. Chem. Phys.* **101** 7566

[98] Borkovec M and Berne B J 1985 Reaction dynamics in the low pressure regime: the Kramers model and collision models of molecules with many degrees of freedom *J. Chem. Phys.* **82** 794–9

   Borkovec M, Straub J E and Berne B J 1986 The influence of intramolecular vibrational relaxation on the pressure dependence of unimolecular rate constants *J. Chem. Phys.* **85** 146–9

   Straub J E and Berne B J 1986 Energy diffusion in many-dimensional Markovian systems: the consequences of competition between inter- and intramolecular vibrational energy transfer *J. Chem. Phys.* **85** 2999–3006

[99] Kuharski R A, Chandler D, Montgomery J, Rabii F and Singer S J 1988 Stochastic molecular dynamics study of cyclohexane isomerization *J. Phys. Chem.* **92** 3261

[100] Hasha D L, Eguchi T and Jonas J 1982 High pressure NMR study of dynamical effects on conformational Isomerization of cyclohehane *J. Am. Chem. Soc.* **104** 2290

[101] Ashcroft J, Besnard M, Aquada V and Jonas J 1984 *Chem. Phys. Lett.* **110** 420

[102] Ashcroft J and Xie C-L 1989 *J. Chem. Phys.* **90** 5386

[103] Campbell D M, Mackowiak M and Jonas J 1992 *J. Chem. Phys.* **96** 2717

[104] Wilson M and Chandler D 1990 *Chem. Phys.* **149** 11

[105] Franck J and Rabinowitch E 1934 Some remarks about free radicals and photochemistry of solutions *Trans. Faraday Soc.* **30** 120

[106] Noyes R M 1961 Effects of diffusion on reaction rates *Prog. React. Kinet.* **1** 129

[107] Chuang T J, Hoffman G W and Eisenthal K B 1974 Picosecond studies of the cage effect and collision induced predissociation of iodine in liquids *Chem. Phys. Lett.* **25** 201

[108] Langhoff C A, Moore B and DeMeuse M 1983 Diffusion theory and picosecond atom recombination *J. Chem. Phys.* **78** 1191

[109] Kelley D F, Abul-Haj N A and Jang D J 1984 *J. Chem. Phys.* **80** 4105

   Harris A L, Brown J K and Harris C B 1988 *Ann. Rev. Phys. Chem.* **39** 341

[110] Wang W, Nelson K A, Xiao L and Coker D F 1994 Molecular dynamics simulation studies of solvent cage effects on photodissociation in condensed phases *J. Chem. Phys.* **101** 9663–71

   Batista V S and Coker D F 1996 Nonadiabatic molecular dynamics simulation of photodissociation and geminate recombination of I_2 liquid xenon *J. Chem. Phys.* **105** 4033–54

[111] Luther K and Troe J 1974 Photolytic cage effect of iodine in gases at high pressure *Chem. Phys. Lett.* **24** 85–90

   Dutoit J C, Zellweger J M and van den Bergh H 1990 *J. Chem. Phys.* **93** 242

[112] Bunker D L and Davidson B S 1972 Photolytic cage effect. Monte Carlo experiments *J. Am. Chem. Soc.* **94** 1843

[113] Murrell J N, Stace A J and Dammel R 1978 Computer simulation of the cage effect in the photodissociation of iodine *J. Chem. Soc. Faraday Trans.* II **74** 1532

[114] Dardi P S and Dahler J S 1990 Microscopic models for iodine photodissociation quantum yields in dense fluids *J. Chem. Phys.* **93** 242–56

[115] Dardi P S and Dahler J S 1993 A model for nonadiabatic coupling in the photodissociation of I_2-solvent complexes *J. Chem. Phys.* **98** 363–72

[116] Northrup S H and Hynes J T 1979 Short range caging effects for reactions in solution. I. Reaction rate constants and short range caging picture *J. Chem. Phys.* **71** 871–83

[117] Northrup S H and Hynes J T 1979 Short range caging effects for reactions in solution. II. Escape probability and time dependent reactivity *J. Chem. Phys.* **71** 884

[118] Hynes J T 1985 The theory of reactions in solution *Theory of Chemical Reaction Dynamics* ed M Baer (Boca Raton, FL: CRC Press) pp 171–234

[119] Tapia O and Bertran J (eds) 1996 Solvent effects and chemical reactivity *Understanding Chemical Reactivity* vol 17 (Dordrecht: Kluwer)

[120] Zawadski A G and Hynes J T 1989 Radical recombination rate constants from gas to liquid phase *J. Phys. Chem.* **93** 7031–6

A3.7
Molecular reaction dynamics in the gas phase

Daniel M Neumark

A3.7.1 Introduction

The field of gas phase reaction dynamics is primarily concerned with understanding how the microscopic forces between atoms and molecules govern chemical reactivity. This goal is targeted by performing exacting experiments which yield measurements of detailed attributes of chemical reactions, and by developing state-of-the-art theoretical techniques in order to calculate accurate potential energy surfaces for reactions and determine the molecular dynamics that occur on these surfaces. It has recently become possible to compare experimental results with theoretical predictions on a series of benchmark reactions. This convergence of experiment and theory is leading to significant breakthroughs in our understanding of how the peaks and valleys on a potential energy surface can profoundly affect the measurable properties of a chemical reaction.

In most of gas phase reaction dynamics, the fundamental reactions of interest are bimolecular reactions,

$$A + BC \rightarrow AB + C \tag{A3.7.1}$$

and unimolecular photodissociation reactions,

$$ABC \xrightarrow{h\nu} AB + C. \tag{A3.7.2}$$

There are significant differences between these two types of reactions as far as how they are treated experimentally and theoretically. Photodissociation typically involves excitation to an excited electronic state, whereas bimolecular reactions often occur on the ground-state potential energy surface for a reaction. In addition, the initial conditions are very different. In bimolecular collisions one has no control over the reactant orbital angular momentum (impact parameter), whereas in photodissociation one can start with cold molecules with total angular momentum $J \approx 0$. Nonetheless, many theoretical constructs and experimental methods can be applied to both types of reactions, and from the point of view of this chapter their similarities are more important than their differences.

The field of gas phase reaction dynamics has been extensively reviewed elsewhere [1–3] in considerably greater detail than is appropriate for this chapter. Here, we begin by summarizing the key theoretical concepts and experimental techniques used in reaction dynamics, followed by a 'case study', the reaction F + H$_2$ → HF + H, which serves as an illustrative example of these ideas.

A3.7.2 Theoretical background: the potential energy surface

Experimental and theoretical studies of chemical reactions are aimed at obtaining a detailed picture of the potential energy surface on which these reactions occur. The potential energy surface represents the single most important theoretical construct in reaction dynamics. For N particles, this is a $3N - 6$ dimensional function $V(q_1 \ldots q_{3N-6})$ that gives the potential energy as a function of nuclear internal coordinates. The potential energy surface for any reaction can, in principle, be found by solving the electronic Schrödinger equation at many different nuclear configurations and then fitting the results to various functional forms, in order to obtain a smoothly varying surface in multiple dimensions. In practice, this is extremely demanding from a computational perspective. Thus, much of theoretical reaction dynamics as recently as a few years ago was performed on highly approximate model surfaces for chemical reactions which were generated using simple empirical functions (the London–Eyring–Polanyi–Sato potential, for example [4]). The $H + H_2$ reaction was the first for which an accurate surface fitted to *ab initio* points was generated [5, 6]. However, recent conceptual and computational advances have made it possible to construct accurate surfaces for a small number of benchmark systems, including the $F + H_2$, $Cl + H_2$ and $OH + H_2$ reactions [7–9]. Even in these systems, one must be concerned with the possibility that a single Born–Oppenheimer potential energy surface is insufficient to describe the full dynamics [10].

Let us consider the general properties of a potential energy surface for a bimolecular reaction involving three atoms, i.e. equation (A3.7.1) with A, B and C all atomic species. A three-atom reaction requires a three-dimensional function. It is more convenient to plot two-dimensional surfaces in which all coordinates but two are allowed to vary. Figure A3.7.1 shows a typical example of a potential energy surface contour plot for a collinear three-atom reaction. The dotted curve represents the minimum energy path, or reaction coordinate, that leads from reactants on the lower right to products on the upper left. The reactant and product valleys (often referred to as the entrance and exit valleys, respectively) are connected by the transition-state region, where the transformation from reactants to products occurs, and ends in the product valley at the upper left. The potential energy surface shown in figure A3.7.1 is characteristic of a 'direct' reaction, in that there is a single barrier (marked by ‡ in figure A3.7.1) along the minimum energy path in the transition-state region. In the other general class of bimolecular reaction, a 'complex' reaction, one finds a well rather than a barrier in the transition-state region.

The barrier on the surface in figure A3.7.1 is actually a saddle point; the potential is a maximum along the reaction coordinate but a minimum along the direction perpendicular to the reaction coordinate. The classical transition state is defined by a slice through the top of the barrier perpendicular to the reaction coordinate. This definition holds for multiple dimensions as well; for N particles, the classical transition state is a saddle point that is unbound along the reaction coordinate but bound along the $3N-7$ remaining coordinates. A cut through the surface at the transition state perpendicular to the reaction coordinate represents a $3N-7$ dimensional dividing surface that acts as a 'bottleneck' between reactants and products. The nature of the transition state and, more generally, the region of the potential energy in the vicinity of the transition state (referred to above as the transition-state region) therefore plays a major role in determining many of the experimental observables of a reaction such as the rate constant and the product energy and angular distributions. For this reason, the transition-state region is the most important part of the potential energy surface from a computational (and experimental) perspective.

Once such an *ab initio* potential energy surface for a reaction is known, then all properties of the reaction can, in principle, be determined by carrying out multidimensional quantum scattering calculations. This is again computationally very demanding, and for many years it was more useful to perform classical and quasi-classical trajectory calculations to explore dynamics on potential energy surfaces [11]. The simpler calculations led to very valuable generalizations about reaction dynamics, showing, for example, that for an exothermic reaction with an entrance channel barrier, reactant translation was far more effective than vibration in surmounting the barrier and thus forming products, and are still very useful, since quantum effects in chemical reactions are often relatively small. However, recent conceptual and computational advances [12–14] have now made it possible to carry out exact quantum scattering calculations on multidimensional potential energy surfaces, including the benchmark surfaces mentioned above. Comparison of such calculations with experimental observables provides a rigorous test of the potential energy surface.

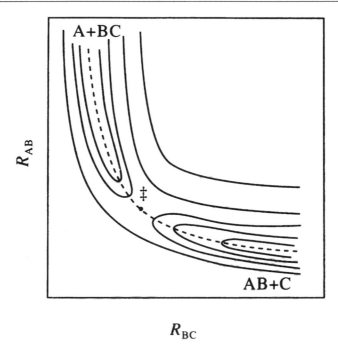

Figure A3.7.1. Two-dimensional contour plot for direct collinear reaction A+BC → AB+C. Transition state is indicated by ‡.

A3.7.3 Experimental techniques in reaction dynamics

We now shift our focus to a general discussion of experimental chemical reaction dynamics. Given that the goal of these experiments is to gain an understanding of the reaction potential energy surface, it is important to perform experiments that can be interpreted as cleanly as possible in terms of the underlying surface. Hence, bimolecular and unimolecular reactions are often studied under 'single-collision' conditions, meaning that the number density in the experiment is sufficiently low that each reactant atom or molecule undergoes at most one collision with another reactant or a photon during the course of the experiment, and the products are detected before they experience any collisions with bath gases, walls, etc. One can therefore examine the results of single-scattering events without concern for the secondary collisions and reactions that often complicate the interpretation of more standard chemical kinetics experiments. Moreover, the widespread use of supersonic beams in reaction dynamics experiments [15, 16] allows one to perform reactions under well defined initial conditions; typically the reactants are rotationally and vibrationally very cold, and the spread in collision energies (for bimolecular reactions) is narrow. The study of photodissociation reactions [2, 17] has been greatly facilitated by recent developments in laser technology, which now permit one to investigate photodissociation at virtually any wavelength over a spectral range extending from the infrared to vacuum ultraviolet (VUV).

What attributes of bimolecular and unimolecular reactions are of interest? Most important is the identity of the products, without which any further characterization is impossible. Once this is established, more detailed issues can be addressed. For example, in any exothermic reaction, one would like to determine how the excess energy is partitioned among the translational, rotational, vibrational and electronic degrees of freedom of the products. Under the ideal of 'single-collision' conditions, one can measure the 'nascent' internal energy

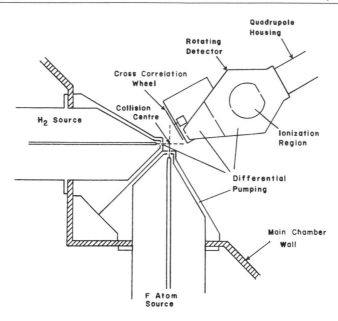

Figure A3.7.2. Schematic illustration of crossed molecular beams experiment for F + H + 2 reaction.

distribution of the products, i.e. the distribution resulting from the reaction before any relaxation (collisional or otherwise) has occurred. Measurements of the product angular distribution provide considerable insight into the topology and symmetry of the potential energy surface(s) on which the reaction occurs. More recently, the measurement of product alignment and orientation has become an area of intense interest; in photodissociation reactions, for example, one can determine if the rotational angular momentum of a molecular fragment is randomly oriented or if it tends to be parallel or perpendicular to the product velocity vector.

An incredible variety of experimental techniques have been developed over the years to address these issues. One of the most general is the crossed molecular beams method with mass spectrometric detection of the products, an experiment developed by Lee, Herschbach and co-workers [18, 19]. A schematic illustration of one version of the experiment is shown in figure A3.7.2. Two collimated beams of reactants cross in a vacuum chamber under single-collision conditions. The scattered products are detected by a rotatable mass spectrometer, in which the products are ionized by electron impact and mass selected by a quadrupole mass spectrometer. By measuring mass spectra as a function of scattering angle, one obtains angular distributions for all reaction products. In addition, by chopping either the products or one of the reactant beams with a rapidly spinning slotted wheel, one can determine the time of flight of each product from the interaction region, where the two beams cross, to the ionizer, and from this the product translational energy E_T can be determined at each scattering angle. The resulting product translational energy distributions $P(E_T)$ also contain information on the internal energy distribution of the products via conservation of energy, so long as the reactant collision energy is well defined.

In an important variation of this experiment, one of the reactant beams is replaced by a pulsed laser which photodissociates molecules in the remaining reactant beam. Use of a pulsed laser makes it straightforward to determine the product translational energy distribution by time of flight. This experiment, photofragment translational spectroscopy, was first demonstrated by Wilson [20, 21] in and is now used in many laboratories [17].

Mass spectrometry, the primary detection method in the above crossed beams experiments, is a particularly general means of analysing reaction products, since no knowledge of the optical spectroscopy of the products is required. On the other hand, electron impact ionization often leads to extensive fragmentation, thereby complicating identification of the primary products. Very recently, tunable VUV radiation from synchrotrons has been used to ionize scattered products from both photodissociation [22] and bimolecular reactions [23]; other than the ionization mechanism, the instrument is similar in principle to that shown in figure A3.7.2. By choosing the VUV wavelength to lie above the ionization potential of the product of interest but below the lowest dissociative ionization threshold (i.e. the minimum energy for $AB + h\nu \rightarrow A^+ + B + e^-$) one can eliminate fragmentation and thus simplify interpretation of the experiments.

A complementary approach to reaction dynamics centres on probing reaction products by optical spectroscopy. Optical spectroscopy often provides higher resolution on the product internal energy distribution than the measurement of translational energy distributions, but is less universally applicable than mass spectrometry as a detection scheme. If products are formed in electronically excited states, their emission spectra (electronic chemiluminescence) can be observed, but ground-state products are more problematic. Polanyi [24] made a seminal contribution in this field by showing that vibrationally excited products in their ground electronic state could be detected by spectrally resolving their spontaneous emission in the infrared; this method of 'infrared chemiluminescence' has proved of great utility in determining product vibrational and, less frequently, rotational distributions.

However, with the advent of lasers, the technique of 'laser-induced fluorescence' (LIF) has probably become the single most popular means of determining product-state distributions; an early example is the work by Zare and co-workers on $Ba + HX$ ($X =$ F, Cl, Br, I) reactions [25]. Here, a tunable laser excites an electronic transition of one of the products (the BaX product in this example), and the total fluorescence is detected as a function of excitation frequency. This is an excellent means of characterizing molecular products with bound–bound electronic transitions and a high fluorescence quantum yield; in such cases the LIF spectra are often rotationally resolved, yielding rotational, vibrational and, for open shell species, fine-structure distributions. LIF has been used primarily for diatomic products since larger species often have efficient non-radiative decay pathways that deplete fluorescence, but there are several examples in which LIF has been used to detect polyatomic species as well.

LIF can provide more detail than the determination of the product internal energy distribution. By measuring the shape LIF profile for individual rotational lines, one can obtain Doppler profiles which yield information on the translational energy distribution of the product as well [26, 27]. In photodissociation experiments where the photolysis and probe laser are polarized, the Doppler profiles yield information on product alignment, i.e. the distribution of m_J levels for products in a particular rotational state J [28]. Experiments of this type have shown, for example, that the rotational angular momentum of the OH product from H_2O photodissociation tends to be perpendicular to v [29], the vector describing the relative velocity of the products, whereas for H_2O_2 photodissociation [30] one finds J tends to be parallel to v. These 'vector correlation' measurements [31–33] are proving very useful in unravelling the detailed dynamics of photodissociation and, less frequently, bimolecular reactions.

The above measurements are 'asymptotic', in that they involve looking at the products of reaction long after the collision has taken place. These very valuable experiments are now complemented by 'transition-state spectroscopy' experiments, in which one uses frequency- or time-domain experiments to probe the very short-lived complex formed when two reactants collide [34]. For example, in our laboratory, we have implemented a transition-state spectroscopy experiment based on negative-ion photodetachment [35]. The principle of the experiment, in which a stable negative ion serves as a precursor for a neutral transition state, is illustrated in figure A3.7.3. If the anion geometry is similar to that of the transition state, then photodetachment of the anion will access the transition-state region on the neutral surface. The resulting photoelectron spectrum can give a vibrationally resolved picture of the transition-state dynamics, yielding the frequencies of the bound vibrational modes of the transition state (i.e. those perpendicular to the reaction coordinate) and thereby realizing the goal of transition-state spectroscopy. An example of the successful application of this technique is given below in the discussion of the $F + H_2$ reaction.

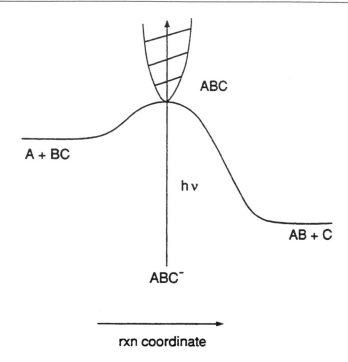

Figure A3.7.3. Principle of transition-state spectroscopy via negative-ion photodetachment.

Alternatively, one can take advantage of the developments in ultrafast laser technology and use femtosecond lasers to follow the course of a reaction in real time. In this approach, pioneered by Zewail [36], a unimolecular or bimolecular reaction is initiated by a femtosecond pump pulse, and a femtosecond probe pulse monitors some aspect of the reaction as a function of pump–probe delay time. The first example of such an experiment was the photodissociation of ICN [37]. Here the pump pulse excited ICN to a repulsive electronic state correlating to ground state $I + CN(X^2\Sigma^+)$ products. The probe pulse excited the dissociating ICN to a second repulsive state correlating to excited $I + CN(B^2\Sigma^+)$ products, and progress of the dissociation was monitored via LIF. If the probe pulse is tuned to be resonant with the CN $B \leftarrow X$ transition, the LIF signal rises monotonically on a 200 fs time scale, attributed to the time delay for the formation of CN product. On the other hand, at slightly redder probe wavelengths, the LIF signal rises then falls, indicative of the transient ICN* species formed by the pump pulse. This experiment thus represented the first observation of a molecule in the act of falling apart.

In an elegant application of this method to bimolecular reactions, the reaction $H + CO_2 \rightarrow OH + CO$ was studied by forming the $CO_2 \cdot HI$ van der Waals complex, dissociating the HI moiety with the pump pulse, allowing the resulting H atom to react with the CO_2, and then using LIF to probe the OH signal as a function of time [38]. This experiment represents the 'real-time clocking' of a chemical reaction, as it monitors the time interval between initiation of a bimolecular reaction and its completion.

The above discussion represents a necessarily brief summary of the aspects of chemical reaction dynamics. The theoretical focus of this field is concerned with the development of accurate potential energy surfaces and the calculation of scattering dynamics on these surfaces. Experimentally, much effort has been devoted to developing complementary asymptotic techniques for product characterization and frequency- and time-resolved techniques to study transition-state spectroscopy and dynamics. It is instructive to see what can be accomplished with all of these capabilities. Of all the benchmark reactions mentioned in section A3.7.2, the reaction $F + H_2 \rightarrow HF + H$ represents the best example of how theory and experiment can converge to yield a fairly complete picture of the dynamics of a chemical reaction. Thus, the remainder of this chapter focuses on this reaction as a case study in reaction dynamics.

A3.7.4   Case study: the F + H₂ reaction

The energetics for the $F + H_2$ reaction is shown in figure A3.7.4. The reaction is exothermic by 32.07 kcal mol^{-1}, so that at collision energies above 0.5 kcal mol^{-1}, enough energy is available to populate HF vibrational levels up to and including $v = 3$. Hence the determination of the HF vibration–rotation distribution from this reaction has been of considerable interest. How might one go about this? Since HF does not have an easily accessible bound excited state, LIF is not an appropriate probe technique. On the other hand, the HF vibrational transitions in the infrared are exceedingly strong, and this is the spectral region where characterization of the HF internal energy distribution has been carried out.

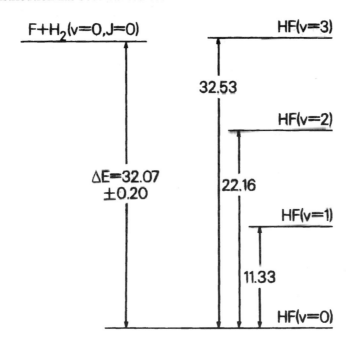

Figure A3.7.4. Energetics of the F+H₂ reaction. All energies in kcal mol^{-1}.

The first information on the HF vibrational distribution was obtained in two landmark studies by Pimentel [39] and Polanyi [24] in 1969; both studies showed extensive vibrational excitation of the HF product. Pimental found that the $F + H_2$ reaction could pump an infrared chemical laser, i.e. the vibrational distribution was inverted, with the HF($v = 2$) population higher than that for the HF($v = 1$) level. A more complete picture was obtained by Polanyi by measuring and spectrally analysing the spontaneous emission from vibrationally excited HF produced by the reaction. This 'infrared chemiluminescence' experiment yielded relative populations of 0.29, 1 and 0.47 for the HF($v = 1, 2$ and 3) vibrational levels, respectively. While improvements in these measurements were made in subsequent years, the numbers describing the vibrational populations have stayed approximately constant. The highly inverted vibrational distributions are characteristic of a potential energy surface for an exothermic reaction with a barrier in the entrance channel.

Spectroscopic determination of the HF rotational distribution is another story. In both the chemical laser and infrared chemiluminescence experiments, rotational relaxation due to collisions is faster or at least comparable to the time scale of the measurements, so that accurate determination of the nascent rotational distribution was not feasible. However, Nesbitt [40, 41] has recently carried out direct infrared absorption experiments on the HF product under single-collision conditions, thereby obtaining a full vibration–rotation distribution for the nascent products.

These spectroscopic probes have been complemented by studies using the crossed molecular beams technique. In these experiments, two well collimated and nearly monoenergetic beams of H_2 and F atoms cross in a large vacuum chamber. The scattered products are detected by a rotatable mass spectrometer, yielding the angular distribution of the reaction products. The experiment measures the transitional energy of the products via time of flight. Thus, one obtains the full transitional energy and angular distribution, $P(E_T, \theta)$, for the HF products. The first experiments of this type on the $F + D_2$ reaction were carried out by Lee [42] in 1970. Subsequent work by the Lee [43, 44] and Toennies [45, 46] groups on the $F + H_2$, D_2 and HD reactions has yielded a very complete characterization of the $P(E, \theta)$ distribution.

As an example, figure A3.7.5 shows a polar contour plot of the HF product velocity distribution at a reactant collision energy of $E_{coll} = 1.84$ kcal mol^{-1} [43]. p-H_2 refers to *para*-hydrogen, for which most of the rotational population is in the $J = 0$ level under the experimental conditions used here. This plot is in the centre-of-mass (CM) frame of reference. F atoms are coming from the right, and H_2 from the left, and the scattering angle θ is reference to the H_2 beam. The dashed circles ('Newton circles') represent the maximum speed of the HF product in a particular vibrational state, given by

$$v_{max} = \frac{m_H}{M} \sqrt{\frac{2(\Delta E + E_{coll} - E_v)}{\mu}}$$

(A3.7.3)

where ΔE is the exothermicity, E_v the vibrational energy, M is the total mass ($M = m_H + m_{HF}$) and $\mu = m_F m_H / M$ the reduced mass of the products. Thus, all the signal inside the $v = 3$ circle is from HF($v = 3$), all the signal inside the $v = 2$ circle is from HF($v = 2$) or HF($v = 3$), etc.

An important feature of figure A3.7.5 is that the contributions from different HF vibrational levels, particularly the $v = 2$ and 3 levels, are very distinct, a result of relatively little rotational excitation of the HF($v = 2$) products (i.e. if these products had sufficient rotational excitation, they would have the same translational energy as HF($v = 3$) product in low J levels). As a consequence, from figure A3.7.5 one can infer the angular distribution for each HF vibrational state, in other words, vibrationally state-resolved differential cross sections. These are quite different depending on the vibrational level. The HF($v = 2$) and ($v = 1$) products are primarily back-scattered with their angular distributions peaking at $\theta = \pi$, while the HF($v = 3$) products are predominantly forward-scattered, peaking sharply at $\theta = 0°$. In general, backward-scattered products result from low impact parameter, head-on collisions, while forward-scattered products are a signature of higher impact parameter, glancing collisions. To understand the significance of these results, it is useful to move away from experimental results and consider the development of potential energy surfaces for this reaction.

Many potential energy surfaces have been proposed for the $F + H_2$ reaction. It is one of the first reactions for which a surface was generated by a high-level *ab initio* calculation including electron correlation [47]. The resulting surface (restricted to collinear geometries) was imperfect, but it had a low barrier (1.66 kcal mol^{-1}) lying in the entrance channel, as expected for an exothermic reaction with a low activation energy ($\sim$1.0 kcal mol^{-1}). In the 1970s, several empirical surfaces were developed which were optimized so that classical trajectory calculations performed on these surfaces reproduced experimental results, primarily the rate constant and HF vibrational energy distribution. One of these, the Muckerman V surface [48], was used in many classical and quantum mechanical scattering calculations up until the mid-1980s and provided a generally accepted theoretical foundation for the $F + H_2$ reaction. However, one notable feature of this surface was its rather stiff bend potential near the transition state. With such a potential, only near-collinear collisions were likely to lead to reaction. As a consequence, the HF product angular distribution found by

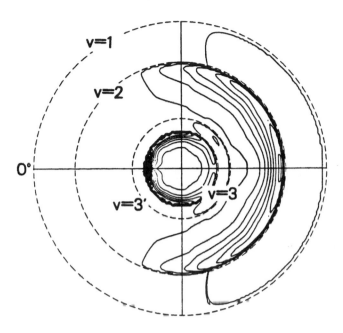

$F+p-H_2 \rightarrow HF+H$, 1.84 kcal/mole

Figure A3.7.5. Velocity–flux contour plot for HF product from the reaction $F + para\text{-}H_2 \rightarrow HF + H$ at a reactant collision energy of 1.84 kcal mol^{-1}.

scattering calculations on this surface was strongly back-scattered for all vibrational states. This is in marked disagreement with the experimental results in figure A3.7.5, which show the HF($v = 3$) distribution to be strongly forward-scattered.

At the time the experiments were performed (1984), this discrepancy between theory and experiment was attributed to quantum mechanical resonances that led to enhanced reaction probability in the HF($v = 3$) channel for high impact parameter collisions. However, since 1984, several new potential energy surfaces using a combination of *ab initio* calculations and empirical corrections were developed in which the bend potential near the barrier was found to be very flat or even non-collinear [49, 51], in contrast to the Muckerman V surface. In 1988, Sato [52] showed that classical trajectory calculations on a surface with a bent transition-state geometry produced angular distributions in which the HF($v = 3$) product was peaked at $\theta = 0°$, while the HF($v = 2$) product was predominantly scattered into the backward hemisphere ($\theta \geq 90°$), thereby qualitatively reproducing the most important features in figure A3.7.5.

At this point it is reasonable to ask whether comparing classical or quantum mechanical scattering calculations on model surfaces to asymptotic experimental observables such as the product energy and angular distributions is the best way to find the 'true' potential energy surface for the $F + H_2$ (or any other) reaction. From an experimental perspective, it would be desirable to probe the transition-state region of the $F + H_2$ reaction in order to obtain a more direct characterization of the bending potential, since this appears to be the key feature of the surface. From a theoretical perspective, it would seem that, with the vastly increased computational power at one's disposal compared to 10 years ago, it should be possible to construct a chemically accurate potential energy surface based entirely on *ab initio* calculations, with no reliance upon empirical corrections. Quite recently, both developments have come to pass and have been applied to the $F + H_2$ reaction.

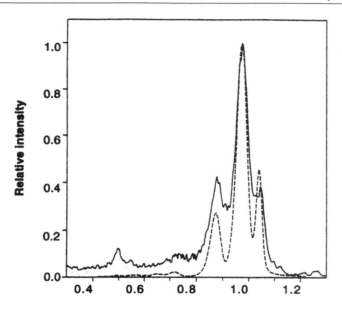

Figure A3.7.6. Photoelectron spectrum of FH_2^-. Here the F^- is complexed to *para*-H_2. Solid curve: experimental results. Dashed curve: simulated spectrum from scattering calculation on *ab initio* surface.

The transition-state spectroscopy experiment based on negative-ion photodetachment described above is well suited to the study of the $F + H_2$ reaction. The experiment is carried out through measurement of the photoelectron spectrum of the anion FH_2^-. This species is calculated to be stable with a binding energy of about 0.20 eV with respect to $F^- + H_2$ [53]. Its calculated equilibrium geometry is linear and the internuclear distances are such that good overlap with the entrance barrier transition state is expected.

The photoelectron spectrum of FH_2^- is shown in figure A3.7.6 [54]. The spectrum is highly structured, showing a group of closely spaced peaks centred around 1 eV, and a smaller peak at 0.5 eV. We expect to see vibrational structure corresponding to the bound modes of the transition state perpendicular to the reaction coordinate. For this reaction with its entrance channel barrier, the reaction coordinate at the transition state is the $F \cdots H_2$ distance, and the perpendicular modes are the F–H–H bend and H–H stretch. The bend frequency should be considerably lower than the stretch. We therefore assign the closely spaced peaks to a progression in the F–H–H bend and the small peak at 0.5 eV to a transition-state level with one quantum of vibrational excitation in the H_2 stretch.

The observation of a bend progression is particularly significant. In photoelectron spectroscopy, just as in electronic absorption or emission spectroscopy, the extent of vibrational progressions is governed by Franck–Condon factors between the initial and final states, i.e. the transition between the anion vibrational level v'' and neutral level v' is given by

$$I_{v''-v'} \propto |\langle \psi_{v''} | \psi_{v'} \rangle|^2 \tag{A3.7.4}$$

where $\psi_{v'}$ and $\psi_{v''}$ are the neutral and anion vibrational wavefunctions, respectively. Since the anion is linear, a progression in a bending mode of the neutral species can only occur if the latter is bent. Hence the FH_2^- photoelectron spectrum implies that the $FH_2^\ddagger$ transition state is bent.

While this experimental work was being carried out, an intensive theoretical effort was being undertaken by Werner and co-workers to calculate an accurate $F + H_2$ potential energy surface using purely *ab initio* methods. The many previous unsuccessful attempts indicated that an accurate calculation of the barrier height

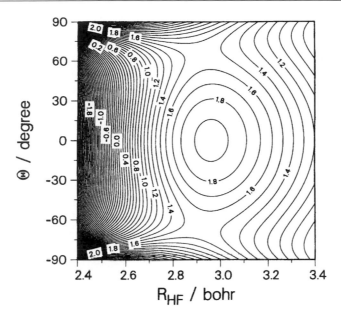

Figure A3.7.7. Two-dimensional contour plot of the Stark–Werner potential energy surface for the F + H₂ reaction near the transition state. Θ is the F–H–H bend angle.

and transition-state properties requires both very large basis sets and a high degree of electron correlation; Werner incorporated both elements in his calculation. The resulting Stark–Werner (SW) surface [7] has a bent geometry at the transition state and a barrier of 1.45 ± 0.25 kcal mol^{-1}. A two-dimensional contour plot of this potential near the transition state is shown in figure A3.7.7. The reason for the bent transition state is illuminating. The F atom has one half-filled p orbital and one might expect this to react most readily with H₂ by collinear approach of the reactants with the half-filled p orbital lined up with the internuclear axis of the H₂ molecule. On the other hand, at longer F $\cdots$ H₂ distances, where electrostatic forces dominate, there is a minimum in the potential energy surface at a T-shaped geometry with the half-filled orbital perpendicular to the H–H bond. (This arises from the quadrupole–quadrupole interaction between the F and H₂.) The interplay between favourable reactivity at a collinear geometry and electrostatic forces favouring a T-shaped geometry leads to a bent geometry at the transition state.

How good is this surface? The first test was to simulate the FH_2^- photoelectron spectrum. This calculation was carried out by Manolopoulos [54] and the result is shown as a dashed curve in figure A3.7.6. The agreement with experiment is excellent considering that no adjustable parameters are used in the calculation. In addition, Castillo *et al* [55, 56] and Aoiz *et al* [57] have performed quasi-classical and quantum scattering calculations on the SW surface to generate angular distributions for each HF product vibrational state for direct comparison to the molecular beam scattering results in figure A3.7.5. The state-specific forward-scattering of the HF($v = 3$) product is indeed reproduced in the quantum calculations and, to a somewhat lesser extent, in the quasi-classical calculations. The experimental product vibrational populations are also reproduced by the calculations. It therefore appears that scattering calculations on the SW surface agree with the key experimental results for the F + H₂ reaction.

What is left to understand about this reaction? One key remaining issue is the possible role of other electronic surfaces. The discussion so far has assumed that the entire reaction takes place on a single Born–Oppenheimer potential energy surface. However, three potential energy surfaces result from the interaction between an F atom and H₂. The spin–orbit splitting between the $^2P_{3/2}$ and $^2P_{1/2}$ states of a free F atom is 404 cm^{-1}. When an F atom interacts with H_2, the $^2P_{3/2}$ state splits into two states with A' and A'' symmetry ($^2\Sigma^+$ and $^2\Pi_{3/2}$, respectively, for collinear geometry) while the higher-lying $^2P_{1/2}$ state becomes an A' state ($^2\Pi_{1/2}$ for collinear geometry). Only the lower A' state correlates adiabatically to ground-state HF + H products; the other two states correlate to highly excited products and are therefore non-reactive in the adiabatic limit. In this limit, the excited $F(^2P_{1/2})$ state is completely unreactive.

Since this state is so low in energy, it is likely to be populated in the F atom beams typically used in scattering experiments (where pyrolysis or microwave/electrical discharges are used to generate F atoms), so the issue of its reactivity is important. The molecular beam experiments of Lee [43] and Toennies [45] showed no evidence for reaction from the $F(^2P_{1/2})$ state. However, the recent work of Nesbitt [40, 41], in which the vibrational and rotational HF distribution was obtained by very high-resolution IR spectroscopy, shows more rotational excitation of the HF($v = 3$) product than should be energetically possible from reaction with the $F(^2P_{3/2})$ state. They therefore suggested that this rotationally excited product comes from the $F(^2P_{1/2})$ state. This work prompted an intensive theoretical study of spin–orbit effects on the potential energy surface and reaction dynamics. A recent study by Alexander and co-workers [58] does predict a small amount of reaction from the $F(^2P_{1/2})$ state but concludes that the adiabatic picture is largely correct. The issue of whether a reaction can be described by a single Born–Oppenheimer surface is of considerable interest in chemical dynamics [10], and it appears that the effect of multiple surfaces must be considered to gain a complete picture of a reaction even for as simple a model system as the F + H_2 reaction.

A3.7.5 Conclusions and perspectives

This chapter has summarized some of the important concepts and results from what has become an exceedingly rich area of chemical physics. On the other hand, the very size of the field means that the vast majority of experimental and theoretical advances have been left out; the books referenced in the introduction provide a much more complete picture of the field.

Looking toward the future, two trends are apparent. First, the continued study of benchmark bimolecular and photodissociation reactions with increasing levels of detail is likely to continue and be extremely productive. Although many would claim that the 'three-body problem' is essentially solved from the perspective of chemical reaction dynamics, the possibility of multiple potential surfaces playing a role in the dynamics adds a new level of complexity even for well studied model systems such as the F + H_2 reaction considered here. Slightly more complicated benchmark systems such as the OH + H_2 and OH + CO reactions present even more of a challenge to both experiment and theory, although considerable progress has been achieved in both cases.

However, in order to deliver on its promise and maximize its impact on the broader field of chemistry, the methodology of reaction dynamics must be extended toward more complex reactions involving polyatomic molecules and radicals for which even the primary products may not be known. There certainly have been examples of this: notably the crossed molecular beams work by Lee [59] on the reactions of O atoms with a series of hydrocarbons. In such cases the spectroscopy of the products is often too complicated to investigate using laser-based techniques, but the recent marriage of intense synchrotron radiation light sources with state-of-the-art scattering instruments holds considerable promise for the elucidation of the bimolecular and photodissociation dynamics of these more complex species.

References

[1]   Levine R D and Bernstein R B 1987 *Molecular Reaction Dynamics and Chemical Reactivity* (New York: Oxford University Press)
[2]   Schinke R 1993 *Photodissociation Dynamics* (Cambridge: Cambridge University Press)
[3]   Scoles G (ed) 1988 *Atomic and Molecular Beam Methods* vols 1 and 2 (New York: Oxford University Press)
[4]   Sato S 1955 *J. Chem. Phys.* **23** 592
[5]   Siegbahn P and Liu B 1978 *J. Chem. Phys.* **68** 2457

[6] Truhlar D G and Horowitz C J 1978 *J. Chem. Phys.* **68** 2466
[7] Stark K and Werner H J 1996 *J. Chem. Phys.* **104** 6515
[8] Alagia M *et al* 1996 *Science* **273** 1519
[9] Alagia M, Balucani N, Casavecchia P, Stranges D, Volpi G G, Clary D C, Kliesch A and Werner H J 1996 *Chem. Phys.* **207** 389
[10] Butler L J 1998 *Annu. Rev. Phys. Chem.* **49** 125
[11] Polanyi J C 1972 *Acc. Chem. Res.* **5** 161
[12] Miller W H 1990 *Annu. Rev. Phys. Chem.* **41** 245
[13] Bowman J M and Schatz G C 1995 *Annu. Rev. Phys. Chem.* **46** 169
[14] Schatz G C 1996 *J. Phys. Chem.* **100** 12 839
[15] Anderson J B, Andres R P and Fenn J B 1966 *Adv. Chem. Phys.* **10** 275
[16] Miller D R 1988 *Atomic and Molecular Beam Methods* vol 1, ed G Scoles (New York: Oxford University Press) p 14
[17] Butler L J and Neumark D M 1996 *J. Phys. Chem.* **100** 12 801
[18] Lee Y T, McDonald J D, LeBreton P R and Herschbach D R 1969 *Rev. Sci. Instrum.* **40** 1402
[19] McDonald J D, LeBreton P R, Lee Y T and Herschbach D R 1972 *J. Chem. Phys.* **56** 769
[20] Busch G E and Wilson K R 1972 *J. Chem. Phys.* **56** 3626
[21] Busch G E and Wilson K R 1972 *J. Chem. Phys.* **56** 3638
[22] Sun W Z, Yokoyama K, Robinson J C, Suits A G and Neumark D M 1999 *J. Chem. Phys.* **110** 4363
[23] Blank D A, Hemmi N, Suits A G and Lee Y T 1998 *Chem. Phys.* **231** 261
[24] Polanyi J C and Tardy D C 1969 *J. Chem. Phys.* **51** 5717
[25] Cruse H W, Dagdigian P J and Zare R N 1973 *Discuss. Faraday* **55** 277
[26] Ondrey G, van Veen N and Bersohn R 1983 *J. Chem. Phys.* **78** 3732
[27] Vasudev R, Zare R N and Dixon R N 1984 *J. Chem. Phys.* **80** 4863
[28] Greene C H and Zare R N 1983 *J. Chem. Phys.* **78** 6741
[29] David D, Bar I and Rosenwaks S 1993 *J. Phys. Chem.* **97** 11 571
[30] Gericke K-H, Klee S, Comes F J and Dixon R N 1986 *J. Chem. Phys.* **85** 4463
[31] Dixon R N 1986 *J. Chem. Phys.* **85** 1866
[32] Simons J P 1987 *J. Phys. Chem.* **91** 5378
[33] Hall G E and Houston P L 1989 *Annu. Rev. Phys. Chem.* **40** 375
[34] Polanyi J C and Zewail A H 1995 *Acc. Chem. Res.* **28** 119
[35] Neumark D M 1993 *Acc. Chem. Res.* **26** 33
[36] Khundkar L R and Zewail A H 1990 *Annu. Rev. Phys. Chem.* **41** 15
[37] Dantus M, Rosker M J and Zewail A H 1987 *J. Chem. Phys.* **87** 2395
[38] Scherer N F, Khundkar L R, Bernstein R B and Zewail A H 1987 *J. Chem. Phys.* **87** 1451
[39] Parker J H and Pimentel G C 1969 *J. Chem. Phys.* **51** 91
[40] Chapman W B, Blackmon B W and Nesbitt D J 1997 *J. Chem. Phys.* **107** 8193
[41] Chapman W B, Blackmon B W, Nizkorodov S and Nesbitt D J 1998 *J. Chem. Phys.* **109** 9306
[42] Schafer T P, Siska P E, Parson J M, Tully F P, Wong Y C and Lee Y T 1970 *J. Chem. Phys.* **53** 3385
[43] Neumark D M, Wodtke A M, Robinson G N, Hayden C C and Lee Y T 1985 *J. Chem. Phys.* **92** 3045
[44] Neumark D M, Wodtke A M, Robinson G N, Hayden C C, Shobotake K, Sparks R K, Schafer T P and Lee Y T 1985 *J. Chem. Phys.* **82** 3067
[45] Faubel M, Martinezhaya B, Rusin L Y, Tappe U, Toennies J P, Aoiz F J and Banares L 1996 *Chem. Phys.* **207** 227
[46] Baer M, Faubel M, Martinez-Haya B, Rusin L, Tappe U and Toennies J P 1999 *J. Chem. Phys.* **110** 10 231
[47] Bender C F, Pearson P K, O'Neill S V and Schaefer H F 1972 *Science* **176** 1412
[48] Muckerman J T 1971 *Theoretical Chemistry—Advances and Perspectives* vol 6A, ed H Eyring and D Henderson (New York: Academic) p 1
[49] Brown F B, Steckler R, Schwenke D W, Truhlar D G and Garrett B C 1985 *J. Chem. Phys.* **82** 188
[50] Lynch G C, Steckler R, Schwenke D W, Varandas A J C, Truhlar D G and Garrett B C 1991 *J. Chem. Phys.* **94** 7136
[51] Mielke S L, Lynch G C and Truhlar D G and Schwenke D W 1993 *Chem. Phys. Lett.* **213** 10
[52] Takayanagi T and Sato S 1988 *Chem. Phys. Lett.* **144** 191
[53] Nichols J A, Kendall R A and Cole S J and Simons J 1991 *J. Phys. Chem.* **95** 1074
[54] Manolopoulos D E, Stark K, Werner H J, Arnold D W, Bradforth S E and Neumark D M 1993 *Science* **262** 1852
[55] Castillo J F, Manolopoulos D E, Stark K and Werner H J 1996 *J. Chem. Phys.* **104** 6531
[56] Castillo J F, Hartke B, Werner H J, Aoiz F J, Banares L and MartinezHaya B 1998 *J. Chem. Phys.* **109** 7224
[57] Aoiz F J, Banares L, MartinezHaya B, Castillo J F, Manolopoulos D E, Stark K and Werner H J 1997 *J. Phys. Chem.* A **101** 6403
[58] Alexander M H, Werner H J and Manolopoulos D E 1998 *J. Chem. Phys.* **109** 5710
[59] Lee Y T 1987 *Science* **236** 793

A3.8
Molecular reaction dynamics in condensed phases

Gregory A Voth

A3.8.0 Introduction

The effect of the condensed phase environment on chemical reaction rates has been extensively studied over the past few decades. The central framework for understanding these effects is provided by the transition state theory (TST) [1, 2] developed in the 1930s, the Kramers theory [3] of 1940, the Grote–Hynes [4] and related theories [5] of the 1980s and 1990s and the Yamamoto reactive flux correlation function formalism [6] as extended and further developed by a number of workers [7, 8]. Each of these seminal theoretical breakthroughs has, in turn, generated an enormous amount of research in its own right. There are many good reviews of this body of literature, some of which are cited in [5, 9–12]. It therefore serves no useful purpose to review the field again in the present chapter. Instead, the key issues involving condensed phase effects on chemical reactions will be organized around the primary theoretical concepts as they stand at the present time. Even more importantly, the gaps in our understanding and prediction of these effects will be highlighted. From this discussion it will become evident that, despite the large body of theoretical work in this field, there are significant questions that remain unanswered, as well as a need for greater contact between theory and experiment. The discussion here is by no means intended to be exhaustive, nor is the reference list comprehensive.

A3.8.1 The reactive flux

To begin, consider a system which is at equilibrium and undergoing a forward and reverse chemical reaction. For simplicity, one can focus on an isomerization reaction, but the discussion also applies to other forms of unimolecular reactions as well as to bimolecular reactions that are not diffusion limited. The equilibrium of the reaction is characterized by the mole fractions x_R and x_P of reactants and products, respectively, and an equilibrium constant K_{eq}. For gas phase reactions, it is commonplace to introduce the concept of the *minimum energy path* along some reaction coordinate, particularly if one is interested in microcanonical reaction rates. In condensed phase chemical dynamics, however, this concept is not useful. In fact, a search for the minimum energy path in a liquid phase reaction would lead one to the solid state! Instead, one considers a *free energy path* along the reaction coordinate q, and the dominant effect of a condensed phase environment is to change the nature of this path (i.e. its barriers and reactant and product wells, or minima). To illustrate this point, the free energy function along the reaction coordinate of an isomerizing molecule in the gas phase is shown by the full curve in figure A3.8.1. In the condensed phase, the free energy function will almost always be modified by the interaction with the solvent, as shown by the broken curve in figure A3.8.1. (It should be noted that, in the spirit of TST, the definition of the optimal reaction coordinate should probably be redefined for the condensed phase reaction, but for simplicity it can be taken to be the same coordinate as in the gas phase.) As can be seen from figure A3.8.1, the solvent can modify the barrier height for the reaction, the location of the

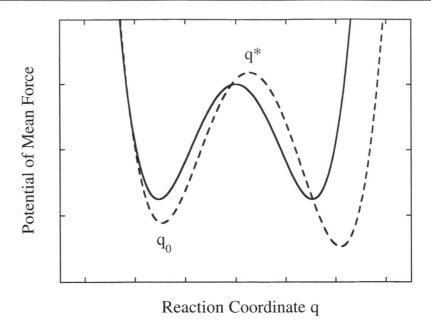

Figure A3.8.1. A schematic diagram of the PMF along the reaction coordinate for an isomerizing solute in the gas phase (full curve) and in solution (broken curve). Note the modification of the barrier height, the well positions, and the reaction free energy due to the interaction with the solvent.

barrier along q, and the reaction free energy (i.e. the difference between the reactant and product minima). It may also introduce dynamical effects that are not apparent from the curve, and it is noted here that a classical framework has been implicitly used—the generalization to the quantum regime will be addressed in a later section.

It is worth discussing the fact that a free energy can be directly relevant to the rate of a *dynamical* process such as a chemical reaction. After all, a free energy function generally arises from an ensemble average over configurations. On the other hand, most condensed phase chemical rate constants are indeed thermally averaged quantities, so this fact may not be so surprising after all, although it should be quantified in a rigorous fashion. Interestingly, the free energy curve for a condensed phase chemical reaction (cf figure A3.8.1) can be viewed, in effect, as a natural consequence of Onsager's linear regression hypothesis as it is applied to condensed phase chemical reactions, along with some additional analysis and simplifications [7].

In the spirit of Onsager, if one imagines a relatively *small* perturbation of the populations of reactants and products away from their equilibrium values, then the regression hypothesis states that the decay of these populations back to their equilibrium values will follow the same time-dependent behaviour as the decay of correlations of *spontaneous* fluctuations of the reactant and product populations in the equilibrium system. In the condensed phase, it is this powerful principle that connects a macroscopic dynamical quantity such as a kinetic rate constant with equilibrium quantities such as a free energy function along a reaction pathway and, in turn, the underlying microscopic interactions which determine this free energy function. The effect of the condensed phase environment can therefore be largely understood in the equilibrium, or quasi-equilibrium, context in terms of the modifications of the free energy curve as shown in figure A3.8.1. As will be shown later, the remaining condensed phase effects which are not included in the equilibrium picture may be defined as being 'dynamical'.

The Onsager regression hypothesis, stated mathematically for the chemically reacting system just described, is given in the classical limit by

$$\frac{\Delta N_R(t)}{\Delta N_R(0)} = \frac{\langle \delta N_R(0) \delta N_R(t) \rangle}{\langle \delta N_R(0)^2 \rangle} \tag{A3.8.1}$$

where $\Delta N_R(t) = \overline{N}_R(t) - \langle N_R \rangle$ is the time-dependent difference between the number of reactant molecules $\overline{N}_R(t)$ arising from an initial non-equilibrium (perturbed) distribution and the final equilibrium number of the reactants $\langle N_R \rangle$. On the right-hand side of the equation, $\langle \delta N_R(t) = N_R(t) - \langle N_R \rangle$ is the instantaneous fluctuation in the number of reactant molecules away from its equilibrium value in the canonical ensemble, and the notation $\langle \cdots \rangle$ denotes the ensemble average over initial conditions.

The solution to the usual macroscopic kinetic rate equations for the reactant and product concentrations yields an expression for the left-hand side of (A3.8.1) that is equal to $\Delta N_R(t) = \Delta N_R(0) \exp(-t/\tau_{rxn})$, where τ_{rxn}^{-1} is the sum of the forward and reverse rate constants, k_f and k_r, respectively. The connection with the microscopic dynamics of the reactant molecule comes about from the right-hand side of (A3.8.1). In particular, in the dilute solute limit, the reactant and product states of the reacting molecule can be identified by the reactant and product population of functions $h_R[q(t)] = 1 - h_P[q(t)]$ and $h_P[q(t)]$, respectively, where $h_P[q(t)] \equiv h[q^* - q(t)]$ and $h(x)$ is the Heaviside step function. The product population function abruptly switches from a value of zero to one as the reaction coordinate trajectory $q(t)$ passes through the barrier maximum at q^* (cf. figure A3.8.1). The important connection between the macroscopic (exponential) rate law and the decay of spontaneous fluctuations in the reactant populations, as specified by the function $h_R[q(t)] = 1 - h_P[q(t)]$ and in terms of the microscopic reaction coordinate q, is valid in a 'coarse-grained' sense in time, i.e. after a period of molecular-scale transients usually of the order of a few tens of femtoseconds. From the theoretical point of view, the importance of the connection outlined above cannot be overstated because it provides a link between the macroscopic (experimentally observed) kinetic phenomena and the molecular scale dynamics of the reaction coordinate in the equilibrium ensemble.

However, further analysis of the linear regression expression in (A3.8.1) is required to achieve a useful expression for the rate constant both from a computational and a conceptual points of view. Such an expression was first provided by Yamamoto [6], but others have extended, validated, and expounded upon his analysis in considerable detail [7, 8]. The work of Chandler [7] in this regard is followed most closely here in order to demonstrate the places in which condensed phase effects can appear in the theory, and hence in the value of the thermal rate constant. The key mathematical step is to differentiate both sides of the linear regression formula in (A3.8.1) and then carefully analyse its expected behaviour for systems having a barrier height of at least several times $k_B T$. The resulting expression for the classical forward rate constant in terms of the so-called 'reactive flux' time correlation function is given by [6–8]

$$\begin{aligned} k_f &= x_R^{-1} \langle \dot{h}_P[q(0)] h_P[q(t_{pl})] \rangle \\ &= x_R^{-1} \langle \dot{q}(0) \delta[q^* - q(0)] h_P[q(t_{pl})] \rangle \end{aligned} \tag{A3.8.2}$$

where x_R is the equilibrium mole fraction of the reactant. The classical rate constant is obtained from (A3.8.2) when the correlation function reaches a 'plateau' value at the time $t = t_{pl}$ after the molecular-scale transients have ended [7]. Upon inspection of the above expression, it becomes apparent that the classical rate constant can be calculated by averaging over trajectories initiated at the barrier top with a velocity Boltzmann distribution for the reaction coordinate and an equilibrium distribution in all other degrees of freedom of the system. Those trajectories are then weighted by their initial velocity and the initial flux over the barrier is correlated with the product state population function $h_P[q(t)]$. The time dependence of the correlation function is computed until the plateau value is reached, at which point it becomes essentially constant and the numerical value of the thermal rate constant can be evaluated. An example of such a correlation function obtained through molecular dynamics simulations is shown in figure A3.8.2.

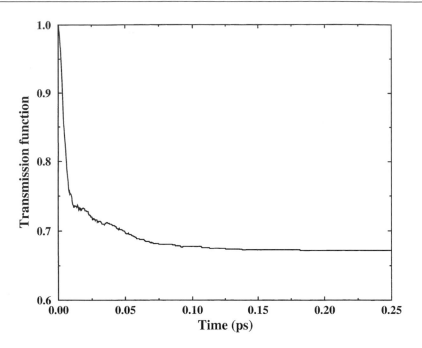

Figure A3.8.2. The correlation function $\kappa(t)$ for particular case of the reaction of methyl vinyl ketone with cyclopentadiene in water. The leveling-off of this function to reach a constant value at the plateau time t_{pl} is clearly seen.

It is important to recognize that the time-dependent behaviour of the correlation function during the molecular transient time seen in figure A3.8.2 has an important origin [7, 8]. This behaviour is due to trajectories that recross the transition state and, hence, it can be proven [7] that the classical TST approximation to the rate constant is obtained from (A3.8.2) in the $t \to 0^+$ limit:

$$
\begin{aligned}
k_f^{TST} &= x_R^{-1} \lim_{t \to 0^+} \langle \dot{q}(0)\delta[q^* - q(0)]h_P[q(t)]\rangle \\
&= x_R^{-1} \langle h[\dot{q}(0)]\dot{q}(0)\delta[q^* - q(0)]\rangle.
\end{aligned}
\tag{A3.8.3}
$$

It is, of course, widely considered that the classical TST provides the central framework for the understanding of thermal rate constants (see the review article by Truhlar *et al* [13]) and also for quantifying the dominant effects of the considered phase in chemical reactions (see below).

In order to segregate the theoretical issues of condensed phase effects in chemical reaction dynamics, it is useful to rewrite the exact classical rate constant in (A3.8.2) as [5–11]

$$
k_f = \kappa k_f^{TST}
\tag{A3.8.4}
$$

where κ is the dynamical correction factor (or 'transmission coefficient') which is given by

$$
\begin{aligned}
\kappa &= \frac{\langle \dot{q}(0)\delta[q^* - q(0)]h_P[q(t_{pl})]\rangle}{\langle h[\dot{q}(0)]\dot{q}(0)\delta[q^* - q(0)]\rangle} \\
&= \langle h_P[q(t_{pl})]\rangle_+ - \langle h_P[q(t_{pl})]\rangle_-.
\end{aligned}
\tag{A3.8.5}
$$

Here, the symbol $\langle \cdots \rangle_\pm$ denotes an averaging over the flux-weighted distribution [7, 8] for positive or negative initial velocities of the reaction coordinate. In figure A3.8.2 is shown the correlation function $\kappa(t)$ for the particular case of the reaction of methyl vinyl ketone with cyclopentadiene in water. The leveling-off of this function to reach a constant value at the plateau time t_{pl} is clearly seen.

The effect of the condensed phase environment in a thermal rate constant thus appears *both* in the value of the TST rate constant and in the value of the dynamical correction factor in (A3.8.4). These effects will be described separately in the following sections, but it should be noted that the two quantities are not independent of each other in that they both depend on the choice of the reaction coordinate q. The 'variational' choice of q amounts to finding a definition of that coordinate that causes the value of κ to be as close to unity as possible, i.e. to minimize the number of recrossing trajectories. It seems clear that an important area of research for the future will be to define theoretically the 'best' reaction coordinate in a condensed phase chemical reaction—one in which the solvent is explicitly taken into account. In charge transfer reactions, for example, a collective solvent polarization coordinate can be treated as being coupled to a solute coordinate (see, for example, [14]), but a more detailed and rigorous microscopic treatment of the full solution—phase reaction coordinate is clearly desirable for the future (see, for example, [15] for progress in this regard).

Before describing the effects of a solvent on the thermal rate constants, it is worthwhile to first reconsider the above analysis in light of current experimental work on condensed phase dynamics and chemical reactions. The formalism outlined above, while exceptionally powerful in that it provides a link between microscopic dynamics and macroscopic chemical kinetics, is intended to help us calculate and analyse *only* thermal rate constants in *equilibrium* systems. The linear regression hypothesis provides the key line of analysis for this problem. To the extent that the thermal rate constant is the quantity of interest—and many times it *is* the primary quantity of interest—this theoretical approach would appear to be the best. However, in many experiments, for example nonlinear optical experiments involving intense laser pulses and/or photoinitiated chemical reactions, the system may initially be far from equilibrium and the above theoretical analysis may not be completely applicable. Furthermore, experimentally measured quantities such as vibrational or phase relaxation rates are often only indirectly related to the thermal rate constant. It would therefore appear that more theoretical effort will be required in the future to relate experimental measurements to the particular microscopic dynamics in the liquid phase that influence the outcome of such measurements and, in turn, to the more standard quantities such as the thermal rate constant.

A3.8.2   The activation free energy and condensed phase effects

Having separated the dynamical from equilibrium (or, more accurately, quasi-equilibrium) effects, one can readily discover the origin of the activation free energy and define the concept of the potential of mean force by analysis of the expression for the TST rate constant, k_f^{TST} in (A3.8.3). The latter can be written as [7]

$$k_f^{TST} = \frac{(2\pi m\beta)^{-1/2}}{\int_{-\infty}^{q^*} dq \exp[-\beta V_{eq}(q)]} \exp[-\beta V_{eq}(q^*)] \tag{A3.8.6}$$

where $\beta = 1/k_B T$ and $V_{eq}(q)$ is the *potential of mean force* (PMF) along the reaction coordinate q. The latter quantity is all important for quantifying and understanding the effect of the condensed phase on the value of the thermal rate constant. It is defined as

$$V_{eq}(q) = -k_B T \ln \left[\int dq' \, d\boldsymbol{x} \, \delta(q - q') \exp[-\beta V'(q', \boldsymbol{x})] \right] + \text{constant} \tag{A3.8.7}$$

where $\boldsymbol{x}$ are all coordinates of the condensed phase system other than the reaction coordinate, and $V(q, \boldsymbol{x})$ is the total potential energy function. The additive constant in (A3.8.7) is irrelevant to the value of the thermal rate constant in (A3.8.6). If the PMF around its minimum in the reactant state (cf. figure A3.8.1) is expanded quadratically, i.e. $V_{eq}(q) \approx V_{eq}(q_0) + (1/2)m\omega_0^2(q - q_0)^2$, then (A3.8.6) simplifies to [5, 7]

$$k_f^{TST} = \frac{\omega_0}{2\pi} \exp(-\beta \Delta F_{cl}^*) \tag{A3.8.8}$$

where the *activation free energy* of the system is defined as $\Delta F_{cl}^* = V_{eq}(q^*) - V_{eq}(q_0)$. The PMF is often decomposed as $V_{eq}(q) = \upsilon(q) + W_{eq}(q)$, where $\upsilon(q)$ is the intrinsic contribution to the PMF from the solute potential energy function and, therefore, by definition, $W_{eq}(q)$ is the contribution arising from the solute–solvent coupling. Figure A3.8.1 illustrates how the latter coupling is responsible for the condensed phase-induced change in the activation free energy, the reaction free energy, and the position of the reactant and product wells. Thus, within the context of the TST, one can conclude that the condensed phase enters into the picture in a 'simple' way through the aforementioned modifications of the reaction coordinate free energy profile in figure A3.8.1.

In principle, nothing more is necessary to understand the influence of the solvent on the TST rate constant than the modification of the PMF, and the resulting changes in the free energy barrier height should be viewed as the dominant effect on the rate since these changes appear in an exponential form. As an example, an error in calculating the solvent contribution to the barrier of 1 kcal mol^{-1} will translate into an error of a factor of four in the rate constant—a factor which is often larger than any dynamical and/or quantum effects such as those described in later sections. This is a compelling fact for the theorist, so it is therefore no accident that the accurate calculation of the solvent contribution to the activation free energy has become the primary focus of many theoretical and computational chemists. The successful completion of such an effort requires four things: (1) an accurate representation of the solute potential, usually from highly demanding *ab initio* electronic structure calculations; (2) an accurate representation of both the solvent potential and the solute–solvent coupling; (3) an accurate computational method to compute, with good statistics, the activation free energy in condensed phase systems; and (4) improved theoretical techniques, both analytical and computational, to identify the *microscopic origin* of the dominant contributions to the activation free energy and the relationship of these effects to experimental parameters such as pressure, temperature, solvent viscosity and polarity, etc. Each of these areas has in turn generated a significant number of theoretical papers over the past few decades—too many in fact to fairly cite them here—and many of these efforts have been major steps forward. There seems to be little dispute, however, that much work remains to be done in all of these areas. Indeed, one of the computational 'grand challenges' facing theoretical chemistry over the coming decades will surely be the *quantitative prediction* (better than a factor of two) of chemical reaction rates in highly complex systems. Some of this effort may, in fact, be driven by the needs of industry and government in, for example, the environmental fate prediction of pollutants.

A3.8.3 The dynamical correction and solvent effects

While the TST estimate of the thermal rate constant is usually a good approximated to the true rate constant and contains most of the dominant solvent effects, the dynamical corrections to the rate can be important as well. In the classical limit, these corrections are responsible for a value of the dynamical correction factor κ in (A3.8.4) that drops below unity. A considerable theoretical effort has been underway over the past 50 years to develop a general theory for the dynamical correction factor (see, for example, [5–12]). One approach to the problem is a direct calculation of κ using molecular dynamics simulation and the reactive flux correlation function formalism [7, 8, 16]. This approach obviously requires the numerically exact integration of Newton's equations for the many-body potential energy surface and a good microscopic model of the condensed phase interactions. However, another approach [5, 9–12] has been to employ a *model* for the reaction coordinate dynamics around the barrier top, for example, the generalized Langevin equation (GLE) given by

$$m\ddot{q}(t) = -\frac{\mathrm{d}V_{eq}(q)}{\mathrm{d}q} - \int_0^t \mathrm{d}t'\, \eta(t-t'; q^*)\dot{q}(t') + \delta F(t). \tag{A3.8.9}$$

In this equation, m is the effective mass of the reaction coordinate, $\eta(t-t'; q^*)$ is the friction kernel calculated with the reaction coordinate 'clamped' at the barrier top, and $\delta F(t)$ is the fluctuating force from all other degrees of freedom with the reaction coordinate so configured. The friction kernel and force fluctuations are related by the fluctuation–dissipation relation

$$\eta(t; q^*) = \beta \langle \delta F(0) \delta F(t) \rangle_{q^*}. \tag{A3.8.10}$$

In the limit of a very rapidly fluctuating force, the above equation can sometimes be approximated by the simpler Langevin equation

$$m\ddot{q}(t) = -\frac{dV_{eq}(q)}{dq} - \hat{\eta}(0)\dot{q}(t') + \delta F(t) \tag{A3.8.11}$$

where $\hat{\eta}(0)$ is the so-called 'static' friction, $\hat{\eta}(0) = \int_0^\infty dt\, \eta(t'; q^*)$.

The GLE can be derived by invoking the linear response approximation for the response of the solvent modes coupled to the motion of the reaction coordinate.

It should be noted that the friction kernel is not in general independent of the reaction coordinate motion [17], i.e. a nonlinear response, so the GLE may have a limited range of validity [18–20]. Furthermore, even if the equation is valid, the strength of the friction might be so great that the second and third terms on the right-hand side of (A3.8.9) could dominate the dynamics much more so than the force generated by the PMF. It should also be noted that, even though the friction in (A3.8.9) may be adequately approximated to be dynamically independent of the value of the reaction coordinate, the equation is still in general nonlinear, depending on the nature of the PMF. For non-quadratic forms of the PMF, $V_{eq}(q)$, even the solution of the reactive dynamics from the model perspective of the GLE becomes a non-trivial problem.

Two central results have arisen from the GLE-based perspective on the dynamical correction factor. The first is the Kramers theory of 1940 [3], based on the simpler Langevin equation, while the second is the Grote–Hynes theory of 1980 [4]. Both have been extensively discussed and reviewed in the literature [5, 9–12]. The important insight of the Kramers theory is that the transmission coefficient for an isomerization or metastable escape reaction undergoes a 'turnover' as one increases the static friction from zero to large values. For weak damping (friction), the transmission coefficient is proportional to the friction, i.e. $\kappa \propto \hat{\eta}(0)$. This dependence arises because the barrier recrossings are caused by the slow energy diffusion (equilibration) in the reaction coordinate motion as it leaves the barrier region. For strong damping, on the other hand, the transmission coefficient is inversely proportional to the friction, i.e. $\kappa \propto 1/\hat{\eta}(0)$, because the barrier crossings are caused by the diffusive spatial motion of the reaction coordinate in the barrier region. For systems such as atom exchange reactions that do not involve a bound reactant state, only the spatial diffusion regime is predicted. The basic phenomenology of condensed phase activated rate processes, as mapped out by Kramers, captures the essential physics of the problem and remains the seminal work to this day.

The second key insight into the dynamical corrections to the TST was provided by the Grote–Hynes theory [4]. This theory highlights the importance of the time dependence of the friction and demonstrates how it may be taken into account at the leading order. In the overdamped regime this is done so through the insightful and compact Grote–Hynes (subscript GH) formula for the transmission coefficient [4]

$$\kappa_{GH} = \frac{\lambda_0^\ddagger}{\omega_{b,eq}} \qquad \lambda_0^\ddagger = \frac{\omega_{b,eq}^2}{\lambda_0^\ddagger + \hat{\eta}(\lambda_0^\ddagger)/m} \tag{A3.8.12}$$

where $\hat{\eta}(z)$ is the Laplace transform of the friction kernel, i.e. $\hat{\eta}(z) = \int_0^\infty dt\, e^{-zt} \eta(t)$, and $\omega_{b,eq}$ is the magnitude of the unstable PMF barrier frequency. Importantly, the derivation of this formula assumes a quadratic approximation to the barrier $V_{eq}(q) \approx V_{eq}(q^*) - (1/2)m\omega_{b,eq}^2(q - q^*)^2$ that may not always be a good one.

Research over the past decade has demonstrated that a multidimensional TST approach can also be used to calculate an even more accurate transmission coefficient than κ_{GH} for systems that can be described by the full GLE with a non-quadratic PMF. This approach has allowed for variational TST improvements [21] of the

Grote–Hynes theory in cases where the nonlinearity of the PMF is important and/or for systems which have general nonlinear couplings between the reaction coordinate and the bath force fluctuations. The Kramers turnover problem has also been successfully treated within the context of the GLE and the multidimensional TST picture [22]. A multidimensional TST approach has even been applied [15] to a realistic model of an S_N2 reaction and may prove to be a promising way to elaborate the explicit microscopic origins of solvent friction.

While there has been great progress toward an understanding and quantification of the dynamical corrections to the TST rate constant in the condensed phase, there are several quite significant issues that remain largely open at the present time. For example, even if the GLE were a valid model for calculating the dynamical corrections, it remains unclear how an accurate and predictive microscopic theory can be developed for the friction kernel $\eta(t)$ so that one does not have to resort to a molecular dynamics simulation [17] to calculate this quantity. Indeed, if one could compute the solvent friction along the reaction coordinate in such a manner, one could instead just calculate the exact rate constant using the reactive-flux formalism. A microscopic theory for the friction is therefore needed to relate the friction along the reaction coordinate to the parameters varied by experimentalists such as pressure or solvent viscosity. No complete test of Kramers theory will ever be possible until such a theoretical effort is completed. Two possible candidates in the latter vein are the instantaneous normal mode theory of liquids [23] and the damped normal mode theory [24] for liquid state dynamics.

Another key issue remaining to be resolved is whether a one-dimensional GLE as in (A3.8.11) is the optimal choice of a dynamical model in the case of strong damping, or whether a two- or multi-dimensional GLE that explicitly includes coupling to solvation and/or intramolecular modes is more accurate and/or more insightful. Such an approach might, for example, allow better contact with nonlinear optical experiments that could measure the dynamics of such additional modes. It is also entirely possible that the GLE may not even be a good approximation to the true dynamics in many cases because, for example, the friction strongly depends on the position of the reaction coordinate. In fact, a strong solvent modification of the PMF usually ensures that the friction will be spatially dependent [25]. Several analytical studies have dealt with this issue (see, for example, [26–28] and literature cited therein). Spatially-dependent friction is found to have an important effect on the dynamical correction in some instances, but in others the Grote–Hynes estimate is predicted to be robust [29]. Nevertheless, the question of the nonlinearity and the accurate modelling of real activated rate processes by the GLE remains an open one.

Another important issue has been identified by several authors [30, 31] which involves the participation of intramolecular solute modes in defining the range of the energy diffusion-limited regime of condensed phase activated dynamics. In particular, if the coupling between the reaction coordinate and such modes is strong, then the Kramers turnover behaviour as a function of the solvent friction occurs at a significantly lower value of the friction than for the simple case of the reaction coordinate coupled to the solvent bath alone. In fact, the issue of whether the turnover can be experimentally observed at all in the condensed phase hinges on this issue. To date, it has remained a challenge to calculate the effective number of intramolecular modes that are strongly coupled to the reaction coordinate; no general theory yet exists to accomplish this important goal.

As a final point, it should again be emphasized that many of the quantities that are measured experimentally, such as relaxation rates, coherences and time-dependent spectral features, are complementary to the thermal rate constant. Their information content in terms of the underlying microscopic interactions may only be indirectly related to the value of the rate constant. A better theoretical link is clearly needed between experimentally measured properties and the common set of microscopic interactions, if any, that also affect the more traditional solution phase chemical kinetics.

A3.8.4 Quantum activated rate processes and solvent effects

The discussion thus far in this chapter has been centred on classical mechanics. However, in many systems, an explicit quantum treatment is required (not to mention the fact that it is the correct law of physics). This statement is particularly true for proton and electron transfer reactions in chemistry, as well as for reactions involving high-frequency vibrations.

The exact quantum expression for the activated rate constant was first derived by Yamamoto [6]. The resulting quantum reactive flux correlation function expression is given by

$$k_f = \frac{1}{x_R \hbar \beta} \int_0^{\hbar \beta} d\tau \, \langle \dot{h}_P(-i\tau) h_P(t_{pl}) \rangle \tag{A3.8.13}$$

where $h_P(t)$ is the Heisenberg product state population operator. As opposed to the classical case, however, the $t \to 0^+$ limit of this expression is always equal to zero [32] which ensures that an *entirely different* approach from the classical analysis must be adopted in order to formulate a quantum TST (QTST), as well as a theory for its dynamical corrections. An article by Truhlar *et al* [13] describes many of the efforts over the past 60 years to develop quantum versions of the TST, and many, if not most, of these efforts have been applicable to primarily low-dimensional gas phase systems. A QTST that is useful for condensed phase reactions is an extremely important theoretical goal since a direct numerical attack on the time-dependent Schrödinger equation for many-body systems is computationally prohibitive, if not impossible. (The latter fact seems to be true in the fundamental sense, i.e. there is an exponential scaling of the numerical effort with system size for the exact solution.) In this section, some of the leading candidates for a viable condensed phase QTST will now be briefly described. The discussion should by no means be considered complete.

As a result of several complementary theoretical efforts, primarily the path integral centroid perspective [33–35], the periodic orbit [36] or instanton [37] approach and the 'above crossover' quantum activated rate theory [38], one possible candidate for a unifying perspective on QTST has emerged [39] from the ideas from [39–42]. In this theory, the QTST expression for the forward rate constant is expressed as [39]

$$k_f \approx \nu \frac{\text{Im} \, Q_b}{Q_R} \tag{A3.8.14}$$

where ν is a simple frequency factor, Q_R is the reactant partition function, and Q_b is the barrier 'partition function' which is to be interpreted in the appropriate asymptotic limit [39–42]. The frequency factor has the piecewise continuous form [39]

$$\nu = \begin{cases} \lambda_0^{\ddagger}/2\pi & \hbar \beta \lambda_0^{\ddagger} < 2\pi \\ (\hbar \beta)^{-1} & \hbar \beta \lambda_0^{\ddagger} \geq 2\pi \end{cases} \tag{A3.8.15}$$

while the barrier partition function is defined under most conditions as [39]

$$Q_b = \lim_{q_c \to iq_c} \int dq_c \, \rho_c(q_c). \tag{A3.8.16}$$

The quantity $\rho_c(q_c)$ is the Feynman path integral centroid density [43] that is understood to be expressed asymptotically as

$$\rho_c(q_c) \approx \rho_c(q^*) \exp[-\beta V_c''(q^*)(q_c - q^*)^2/2] \tag{A3.8.17}$$

where the quantum centroid potential of mean force is given by $V_c(q_c) = -k_B T \ln[\rho_c(q_c)] + \text{constant}$ and q^* is *defined* to be the value of the reaction coordinate that gives the maximum value of $V_c(q)$ in the barrier region (i.e. it may differ [33, 35] from the maximum of the classical PMF along q). The path integral centroid density along the reaction coordinate is given by the Feynman path integral expression

$$\rho_c(q_c) = \int \cdots \int Dq(\tau) Dx(\tau) \, \delta(q_c - \tilde{q}_0) \exp\{-S[q(\tau), x(\tau)]/\hbar\} \tag{A3.8.18}$$

which is a functional integral over all possible cyclic paths of the system coordinates weighted by the imaginary time action function [43]:

$$S[q(\tau), x(\tau)] = \int_0^{\hbar\beta} d\tau \left\{ \frac{m}{2}\dot{q}(\tau)^2 + \sum_{i=1}^N \frac{m_i}{2}\dot{x}_i(\tau)^2 + V[q(\tau), x(\tau)] \right\}. \tag{A3.8.19}$$

The key feature of (A3.8.18) is that the centroids of the reaction coordinate Feynman paths are constrained to be at the position q_c. The centroid $\tilde{q}_0$ of a particular reaction coordinate path $q(\tau)$ is given by the zero-frequency Fourier mode, i.e.

$$\tilde{q}_0 = \frac{1}{\hbar\beta} \int_0^{\hbar\beta} d\tau \, q(\tau) \tag{A3.8.20}$$

Under most conditions, the sign of $V_c''(q^*)$ in (A3.8.17) is negative. In such cases, the centroid variable *naturally* appears in the theory [39], and the equation for the quantum thermal rate constant from (A3.8.14)–(A3.8.17) is then given by [39]

$$k_f \approx \nu \frac{\sqrt{2\pi/\beta|V_c''(q^*)|}}{Q_R} \exp[-\beta V_c(q^*)]. \tag{A3.8.21}$$

It should be noted that in the cases where $V_c''(q^*) > 0$, the centroid variable becomes irrelevant to the quantum activated dynamics as defined by (A3.8.14) and the instanton approach [37] to evaluate Q_b based on the steepest descent approximation to the path integral becomes the approach one may take. Alternatively, one may seek a more generalized saddle point coordinate about which to evaluate A3.8.14. This approach has also been used to provide a unified solution for the thermal rate constant in systems influenced by non-adiabatic effects, i.e. to bridge the adiabatic and non-adiabatic (Golden Rule) limits of such reactions.

In the limit of reasonably high temperatures (above the so-called 'crossover' temperature), i.e. $\hbar\beta\lambda_0^{\ddagger} < 2\pi$, the above formula in (A3.8.21) is best simplified further and approximately written as

$$k_f \approx k_{GH} \frac{(2\pi m\beta)^{-1/2}}{\int_{-\infty}^{q^*} dq_c \exp[-\beta V_c(q_c)]} \exp[-\beta V_c(q^*)]. \tag{A3.8.22}$$

This formula, aside from the prefactor κ_{GH}, is often referred to as the path integral quantum transition state theory (PI–QTST) formula [33]. One clear strength of this formula is its clear analogy with and generalization of the classical TST formula in (A3.8.6). In turn, this allows for an interpretation of solvent effects on quantum activated rate constants in terms of the quantum centroid potential of mean force in a fashion analogous to the classical case. The quantum activation free energy for highly-non-trivial systems can also be directly calculated with imaginary time path integral Monte Carlo techniques [44]. Many such studies have now been carried out, but a single example will be described in the following section.

The preceding discussion has focused on the path integral centroid picture of condensed phase quantum activated dynamics, primarily because of its strong analogy with the classical case, the PMF, etc, as well as its computational utility for realistic problems. However, several recent complementary developments must be mentioned. The first is due to Pollak, Liao and Shao [45] who have significantly extended an earlier idea [30] in which the exact Heisenberg population operator in $h_P(t)$ in (A3.8.13) is replaced by one for a parabolic barrier (plus some other important manipulations, such as symmetrization of the flux operator, that were not done in [30]). The dynamical population operator then has an analytic form which in turn leads one to a purely analytic 'quantum transition state theory' approximation to (A3.8.13). This approach, which in principle can be systematically improved upon through perturbation theory, has been demonstrated to be as accurate as the path integral centroid-based formulae in (A3.8.21) and (A3.8.22) above the crossover temperature.

A second recent development has been the application [46] of the initial value representation [47] to semiclassically calculate (A3.8.13) (and/or the equivalent time integral of the 'flux–flux' correlation function).

While this approach has to date only been applied to problems with simplified harmonic baths, it shows considerable promise for applications to realistic systems, particularly those in which the real solvent 'bath' may be adequately treated by a further classical or quasiclassical approximation.

A3.8.5   Solvent effects in quantum charge transfer processes

In this section, the results of a computational study [48] will be used to illustrate the effects of the solvent—and the significant *complexity* of these effects—in quantum charge transfer processes. The particular example described here is for a 'simple' modelistic proton transfer reaction in a polar solvent. This study, while useful in its own right, also illustrates the level of detail and theoretical formalism that is likely to be necessary in the future to accurately study solvent effects in condensed phase charge transfer reactions, even at the equilibrium (quantum PMF) level.

Some obvious targets for quantum activated rate studies are proton, hydride, and hydrogen transfer reactions because they are of central importance in the solution phase and acid–base chemistry, as well as in biochemistry. These reactions are particularly interesting because they can involve large quantum mechanical effects and, since there is usually a redistribution of solute electronic charge density during the reaction, a substantial contribution to the activation free energy may have its origin from the solvent reorganization process. It is thought that intramolecular vibrations may also play a crucial role in modulating the reactive process by lowering the intrinsic barrier for the reaction.

Many of the condensed phase effects mentioned above have been studied computationally using the PI–QTST approach outlined in the first part of the last section. One such study [48] has focused on the model symmetric three-body proton transfer reaction

$$A^{-0.5} - H^{+0.5} \cdots A \rightarrow [A^{-0.25} \cdots H^{+0.5} \cdots A^{-0.25}]^{\ddagger} \rightarrow A \cdots H^{+0.5} - A^{-0.5}$$

in a polar fluid with its dipole moment chosen to model methanol. The molecular group 'A' represents a generic proton donor/acceptor group.

After some straightforward manipulations of (A3.8.22), the PI–QTST estimate of the proton transfer rate constant can be shown to be given by [48]

$$k_f^{\text{PI–QTST}} = \frac{\omega_{c,0}}{2\pi} \exp(-\beta \Delta F_c^*) \tag{A3.8.23}$$

where $\omega_{c,0} = [V_c''(q_0)/m]^{1/2}$ and the quantum activation free energy is given by [48]

$$\begin{aligned}
\Delta F_c^* &= -k_B T \ln[\rho_c(q^*)/\rho_c(q_0)] \\
&= -k_B T \ln[P_c(q_0 \rightarrow q^*)]. \tag{A3.8.24}
\end{aligned}$$

The probability $P_c(q_0 \rightarrow q^*)$ to move the reaction coordinate centroid variable from the reactant configuration to the transition state is calculated [48] by path integral Monte Carlo techniques [44] combined with umbrella sampling [48, 49]. From the calculations on the model proton transfer system above, the quantum activation free energy curves are shown in figure A3.8.3 for both a rigid and non-rigid (vibrating) intra-complex A–A (donor/acceptor) distance. Shown are both the activation curves for the complex in isolation and in the solvent. The effect of the solvent in the total activation free energy is immediately obvious, contributing 2–4 kcal mol^{-1} to its overall value. One effect of the A–A distance fluctuations is a lowering of the quantum activation free energy (i.e. increased tunnelling) both when the solvent is present and when it is not. A second interesting effect becomes evident from a comparison of the curves for the systems with the rigid versus flexible A–A distance. The contribution to the quantum activation free energy from the solvent is reduced when the A–A distance can fluctuate, resulting in a rate that is 20 times higher than in the rigid case. This novel behaviour was

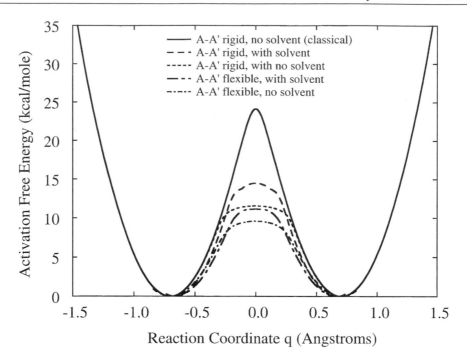

Figure A3.8.3. Quantum activation free energy curves calculated for the model A–H–A proton transfer reaction described [45]. The full line is for the classical limit of the proton transfer solute in isolation, while the other curves are for different fully quantized cases. The rigid curves were calculated by keeping the A–A distance fixed. An important feature here is the direct effect of the solvent activation process on both the solvated rigid and flexible solute curves. Another feature is the effect of a fluctuating A–A distance which both lowers the activation free energy and reduces the influence of the solvent. The latter feature enhances the rate by a factor of 20 over the rigid case.

found to arise from a *nonlinear* coupling between the intra-complex fluctuations and the solvent activation, resulting in a reduced dipole moment of the solute when there is an inward fluctuation of the A–A distance.

From the above PI–QTST studies, it was found that, in order to fully quantify the solvent effects for even a 'simple' model proton transfer reaction, one must deal with a number of complex, nonlinear interactions. Examples of other such interactions include the nonlinear dependence of the solute dipole on the position of the proton and the intrinsically nonlinear interactions arising from both solute and solvent polarizability effects [48]. In the latter context, it was found that the solvent electronic polarizability modes *must* be treated quantum mechanically when studying their influence on the proton transfer activation free energy [48]. (In general, the adequate treatment of electronic polarizability in a variety of condensed phase contexts is emerging as an extremely important problem in many contexts; condensed phase reactions may never be properly described until this problem is addressed.) The detailed calculations described above, while only for a model proton transfer system, clearly illustrate the significant challenge that lies ahead for those who hope to *quantitatively* predict the rates of computation phase chemical reactions through computer simulation.

A3.8.6   Concluding remarks

In this chapter many of the basic elements of condensed phase chemical reactions have been outlined. Clearly, the material presented here represents just an overview of the most important features of the problem. There is an extensive literature on all of the issues described herein and, more importantly, there is still much work to be done before a complete understanding of the effects of condensed phase environments on chemical reactions can be achieved. The theorist and experimentalist alike can therefore look forward to many more years of exciting and challenging research in this important area of physical chemistry.

References

[1] Eyring H 1934 The activated complex in chemical reactions *J. Chem. Phys.* **3** 107

[2] Wigner E 1937 Calculation of the rate of elementary associated reactions *J. Chem. Phys.* **5** 720

[3] Kramers H A 1940 Brownian motion in field of force and diffusion model of chemical reactions *Physica* **7** 284

[4] Grote R F and Hynes J T 1980 The stable states picture of chemical reactions. II. Rate constants for condensed and gas phase reaction models *J. Chem. Phys.* **73** 2715
Grote R F and Hynes J T 1981 Reactive modes in condensed phase reactions *J. Chem. Phys.* **74** 4465

[5] Hänggi P, Talkner P and Borkovec M 1990 Reaction-rate theory: fifty years after Kramers *Rev. Mod. Phys.* **62** 251

[6] Yamamoto T 1960 Quantum statistical mechanical theory of the rate of exchange chemical reactions in the gas phase *J. Chem. Phys.* **33** 281

[7] Chandler D 1978 Statistical mechanics of isomerization dynamics in liquids and the transition state approximation *J. Chem. Phys.* **68** 2959
Montgomery J A Jr, Chandler D and Berne B J 1979 Trajectory analysis of a kinetic theory for isomerization dynamics in condensed phases *J. Chem. Phys.* **70** 4056
Rosenberg R O, Berne B J and Chandler D 1980 Isomerization dynamics in liquids by molecular dynamics *Chem. Phys. Lett.* **75** 162

[8] Keck J 1960 Variational theory of chemical reaction rates applied to three-body recombinations *J. Chem. Phys.* **32** 1035
Anderson J B 1973 Statistical theories of chemical reactions. Distributions in the transition region *J. Chem. Phys.* **58** 4684
Bennett C H 1977 Molecular dynamics and transition state theory: the simulation of infrequent events *Algorithms for Chemical Computation (ACS Symposium Series No 46)* ed R E Christofferson (Washington, DC: American Chemical Society)
Hynes J T 1985 The theory of reactions in solution *The Theory of Chemical Reaction Dynamics* vol IV, ed M Baer (Boca Raton, FL: CRC Press)
Berne B J 1985 Molecular dynamics and Monte Carlo simulations of rare events *Multiple Timescales* ed J V Brackbill and B I Cohen (New York: Academic Press)

[9] For reviews of theoretical work on the corrections to classical TST, see
Hynes J T 1985 *Ann. Rev. Phys. Chem.* **36** 573
Berne B J, Borkovec M and Straub J E 1988 Classical and modern methods in reaction rate theory *J. Phys. Chem.* **92** 3711
Nitzan A 1988 Activated rate processes in condensed phases: the Kramers theory revisited *Adv. Chem. Phys.* **70** 489
Onuchic J N and Wolynes P G 1988 Classical and quantum pictures of reaction dynamics in condensed matter: resonances, dephasing and all that *J. Phys. Chem.* **92** 6495

[10] Fleming G and Hänggi P (eds) 1993 *Activated Barrier Crossing* (New Jersey: World Scientific)

[11] Talkner P and Hänggi P (eds) 1995 *New Trends in Kramers' Reaction Rate Theory* (Dordrecht: Kluwer)

[12] Warshel A 1991 *Computer Modeling of Chemical Reactions in Enzymes and Solutions* (New York: Wiley)

[13] Truhlar D G, Garrett B C and Klippenstein S J 1996 Current status of transition-state theory *J. Phys. Chem.* **100** 12771

[14] van der Zwan G and Hynes J T 1982 Dynamical polar solvent effects on solution reactions: a simple continuum model *J. Chem. Phys.* **76** 2993
van der Zwan G and Hynes J T 1983 Nonequilibrium solvation dynamics in solution reactions *J. Chem. Phys.* **78** 4174
van der Zwan G and Hynes J T 1984 A simple dipole isomerization model for non-equilibrium solvation dynamics in reaction in polar solvents *Chem. Phys.* **90** 21

[15] Pollak E 1993 Variational transition state theory for dissipative systems *Activated Barrier Crossing* ed G Fleming and P Hänggi (New Jersey: World Scientific) p 5
Gershinsky G and Pollak E 1995 Variational transition state theory: application to a symmetric exchange in water *J. Chem. Phys.* **103** 8501

[16] Whitnell R M and Wilson K R 1993 *Reviews of Computational Chemistry* ed K B Lipkowitz and D B Boyd (New York: VCH)

[17] Straub J E, Borkovec M and Berne B J 1987 On the calculation of dynamical friction on intramolecular degrees of freedom *J. Phys. Chem.* **91** 4995
Straub J E, Borkovec M and Berne B J 1988 Molecular dynamics study of an isomerizing diatomic Lennard-Jones fluid *J. Chem. Phys.* **89** 4833
Straub J E, Berne B J and Roux B 1990 Spatial dependence of time-dependent friction for pair diffusion in a simple fluid *J. Chem. Phys.* **93** 6804

[18] Singh S, Krishnan R and Robinson G W 1990 Theory of activated rate processes with space-dependent friction *Chem. Phys. Lett.* **175** 338

[19] Straus J B and Voth G A 1992 Studies on the influence of nonlinearity in classical activated rate processes *J. Chem. Phys.* **96** 5460

[20] Straus J B, Gomez-Llorente J M and Voth G A 1993 Manifestations of spatially-dependent friction in classical activated rate processes *J. Chem. Phys.* **98** 4082

[21] See, for example, Pollak E 1986 Theory of activated rate processes: a new derivation of Kramers' expression *J. Chem. Phys.* **85** 865

Pollak E 1987 Transition state theory for photoisomerization rates of *trans*-stilbene in the gas and liquid phases *J. Chem. Phys.* **86** 3944

Pollak E, Tucker S C and Berne B J 1990 Variational transition state theory for reaction rates in dissipative systems *Phys. Rev. Lett.* **65** 1399

Pollak E 1990 Variational transition state theory for activated rate processes *J. Chem. Phys.* **93** 1116

Pollak E 1991 Variational transition state theory for reactions in condensed phases *J. Phys. Chem.* **95** 533

Frishman A and Pollak E 1992 Canonical variational transition state theory for dissipative systems: application to generalized Langevin equations *J. Chem. Phys.* **96** 8877

Berezhkovskii A M, Pollak E and Zitserman V Y 1992 Activated rate processes: generalization of the Kramers–Grote–Hynes and Langer theories *J. Chem. Phys.* **97** 2422

[22] Pollak E, Grabert H and Hänggi P 1989 Theory of activated rate processes for arbitrary frequency dependent friction: solution of the turnover problem *J. Chem. Phys.* **91** 4073

[23] Stratt R M and Maroncelli M 1996 Nonreactive dynamics in solution: the emerging molecular view of solvation dynamics and vibrational relaxation *J. Phys. Chem.* **100** 12 981

[24] Cao J and Voth G A 1995 A theory for time correlation functions in liquids *J. Chem. Phys.* **103** 4211

[25] Haynes G R and Voth G A 1993 The dependence of the potential of mean force on the solvent friction: consequences for condensed phase activated rate theories *J. Chem. Phys.* **99** 8005

[26] Voth G A 1992 A theory for treating spatially-dependent friction in classical activated rate processes *J. Chem. Phys.* **97** 5908

[27] Haynes G R, Voth G A and Pollak E 1993 A theory for the thermally activated rate constant in systems with spatially dependent friction *Chem. Phys. Lett.* **207** 309

[28] Haynes G R, Voth G A and Pollak E 1994 A theory for the activated barrier crossing rate constant in systems influenced by space and time dependent friction *J. Chem. Phys.* **101** 7811

[29] Haynes G R and Voth G A 1995 Reaction coordinate dependent friction in classical activated barrier crossing dynamics: when it matters and when it doesn't *J. Chem. Phys.* **103** 10 176

[30] Borkovec M, Straub J E and Berne B J The influence of intramolecular vibrational relaxation on the pressure dependence of unimolecular rate constants *J. Chem. Phys.* **85** 146

Straub J E and Berne B J 1986 Energy diffusion in many dimensional Markovian systems: the consequences of the competition between inter- and intra-molecular vibrational energy transfer *J. Chem. Phys.* **85** 2999

Straub J E, Borkovec M and Berne B J 1987 Numerical simulation of rate constants for a two degree of freedom system in the weak collision limit *J. Chem. Phys.* **86** 4296

Borkovec M and Berne B J 1987 Activated barrier crossing for many degrees of freedom: corrections to the low friction result *J. Chem. Phys.* **86** 2444

Gershinsky G and Berne B J 1999 The rate constant for activated barrier crossing: the competition between IVR and energy transfer to the bath *J. Chem. Phys.* **110** 1053

[31] Hershkovitz E and Pollak E 1997 Multidimensional generalization of the PGH turnover theory for activated rate processes *J. Chem. Phys.* **106** 7678

[32] Voth G A, Chandler D and Miller W H 1989 Time correlation function and path integral analysis of quantum rate constants *J. Phys. Chem.* **93** 7009

[33] Voth G A, Chandler D and Miller W H 1989 Rigorous formulation of quantum transition state theory and its dynamical corrections *J. Chem. Phys.* **91** 7749

Voth G A 1990 Analytic expression for the transmission coefficient in quantum mechanical transition state theory *Chem. Phys. Lett.* **170** 289

[34] Gillan M J 1987 Quantum simulation of hydrogen in metals *Phys. Rev. Lett.* **58** 563

Gillam M J 1987 Quantum-classical crossover of the transition rate in the damped double well *J. Phys. C: Solid State Phys.* **20** 3621

[35] Voth G A 1993 Feynman path integral formulation of quantum mechanical transition state theory *J. Phys. Chem.* **97** 8365

[36] Miller W H 1975 Semiclassical limit of quantum mechanical transition state theory for nonseparable systems *J. Chem. Phys.* **62** 1899

[37] See, for example, Coleman S 1979 *The Whys of Subnuclear Physics* ed A Zichichi (New York: Plenum)

[38] Wolynes P G 1981 Quantum theory of activated events in condensed phases *Phys. Rev. Lett.* **47** 968

[39] Cao J and Voth G A 1996 A unified framework for quantum activated rate processes: I. General theory *J. Chem. Phys.* **105** 6856

Cao J and Voth G A 1997 A unified framework for quantum activated rate processes: II. The nonadiabatic limit *J. Chem. Phys.* **106** 1769

[40] Makarov D E and Topaler M 1995 Quantum transition-state theory below the crossover temperature *Phys. Rev. E* **52** 178

[41] Stuchebrukhov A A 1991 Green's functions in quantum transition state theory *J. Chem. Phys.* **95** 4258

[42] Zhu J J and Cukier R I 1995 An imaginary energy method-based formulation of a quantum rate theory *J. Chem. Phys.* **102** 4123

[43] Feynman R P and Hibbs A R 1965 *Quantum Mechanics and Path Integrals* (New York: McGraw-Hill)
Feynman R P 1972 *Statistical Mechanics* (Reading, MA: Addison-Wesley)

[44] For reviews of numerical path integral techniques, see
Berne B J and Thirumalai D 1987 *Ann. Rev. Phys. Chem.* **37** 401
Doll J D, Freeman D L and Beck T L 1990 *Adv. Chem. Phys.* **78** 61
Doll J D and Gubernatis J E (eds) 1990 *Quantum Simulations of Condensed Matter Phenomena* (Singapore: World Scientific)

[45] Pollak E and Liao J-L 1998 A new quantum transition state theory *J. Chem. Phys.* **108** 2733
Shao J, Liao J-L and Pollak E 1998 Quantum transition state theory—perturbation expansion *J. Chem. Phys.* **108** 9711
Liao J-L and Pollak E 1999 A test of quantum transition state theory for a system with two degrees of freedom *J. Chem. Phys.* **110** 80

[46] Wang H, Sun X and Miller W H 1998 Semiclassical approximations for the calculation of thermal rate constants for chemical reactions in complex molecular systems *J. Chem. Phys.* **108** 9726
Sun X, Wang H and Miller W H 1998 On the semiclassical description of quantum coherence in thermal rate constants *J. Chem. Phys.* **109** 4190

[47] Miller W H 1998 Quantum and semiclassical theory of chemical reaction rates *Faraday Disc. Chem. Soc.* **110** 1

[48] Lobaugh J and Voth G A 1994 A path integral study of electronic polarization and nonlinear coupling effects in condensed phase proton transfer reactions *J. Chem. Phys.* **100** 3039

[49] Valleau J P and Torrie G M 1977 *Statistical Mechanics, Part A* ed B J Berne (New York: Plenum)

A3.9
Molecular reaction dynamics: surfaces

George R Darling, Stephen Holloway and Charles Rettner

A3.9.1 Introduction

Molecular reaction dynamics is concerned with understanding elementary chemical reactions in terms of the individual atomic and molecular forces and the motions that occur during the process of chemical change. In gas phase and condensed phase reactions (discussed in sections A3.7 and A3.8) the reactants, products and all intermediates are in the same phase. This 'reduces' the complexity of such systems such that we need 'only' develop experimental and theoretical tools to treat one medium. In a surface reaction, the reactants derive from the gas phase, to which the products may or may not return, but the surface is a condensed phase exchanging energy with reactants and products and any intermediates in a nontrivial fashion. The electronic states of the surface may also play a role by changing the bonding within and between the various species, affecting the reaction as a heterogeneous catalyst (see section A3.10). Of course, the surface itself may be one of the reactants, as in the etching of silicon surfaces by halide molecules. Indeed, it might be argued that if the reactants achieve thermal equilibrium with the surface, they have become part of a new surface, with properties differing from those of the clean surface.

An individual surface reaction may be the result of several steps occurring on very different timescales. For example, a simple bimolecular reaction $A(gas) + B(gas) \rightarrow AB(gas)$ might proceed as follows: A strikes the surface, losing enough energy to stick (i.e. adsorb), B also adsorbs, A and B diffuse across the surface and meet to form AB, after some time AB acquires enough energy to escape (i.e. desorb) from the surface. Each part of this schematic process is itself complicated. In the initial collisions with the surface, the molecules can lose or gain energy, and this can be translational energy (i.e. from the centre-of-mass motion) or internal (rotational, vibrational etc) energy, or both. Internal energy can be exchanged for translational energy, or *vice versa*, or the molecule can simply fragment on impact. Thermalization (i.e. the attainment of thermal equilibrium with the surface) is a slower process, requiring possibly tens of bounces of the molecule on the surface. The subsequent diffusion of A and B towards each other is even slower, while the desorption of the product AB might occur as soon as it is formed, leaving the molecule with some of the energy released in the association step.

Why should we be interested in the dynamics of such complex systems? Apart from the intellectual rewards offered by this field, understanding reactions at surfaces can have great practical and economic value. Gas–surface chemical reactions are employed in numerous processes throughout the chemical and electronic industries. Heterogeneous catalysis lies at the heart of many synthetic cycles, and etching and deposition are key steps in the fabrication of microelectronic components. Gas–surface reactions also play an important role in the environment, from acid rain to the chemistry of the ozone hole. Energy transfer at the gas–surface interface influences flight, controls spacecraft drag, and determines the altitude of a slider above a computer hard disk. Any detailed understanding of such processes needs to be built on fundamental knowledge of the dynamics and kinetics at the molecular level.

For any given gas–surface reaction, the various elementary steps of energy transfer, adsorption, diffusion, reaction and desorption are inextricably linked. Rather than trying to study all together in a single system where they cannot easily be untangled, most progress has been made by probing the individual steps in carefully chosen systems [1–3]. For example, energy transfer in molecule–surface collisions is best studied in nonreactive systems, such as the scattering and trapping of rare-gas atoms or simple molecules at metal surfaces. We follow a similar approach below, discussing the dynamics of the different elementary processes separately. The surface must also be 'simplified' compared to technologically relevant systems. To develop a detailed understanding, we must know exactly what the surface looks like and of what it is composed. This requires the use of surface science tools (section B1.19–26) to prepare very well-characterized, atomically clean and ordered substrates on which reactions can be studied under ultrahigh vacuum conditions. The most accurate and specific experiments also employ molecular beam techniques, discussed in section B2.3.

A3.9.2   Reaction mechanisms

The basic paradigms of surface reaction dynamics originate in the pioneering studies of heterogeneous catalysis by Langmuir [4–6]. Returning to our model bimolecular reaction $A(gas) + B(gas) \rightarrow AB(gas)$, let us assume first that A adsorbs on, and comes into thermal equilibrium with, the surface. We categorize the reaction according to the behaviour of molecule B. For most surface reactions, B adsorbs and thermalizes on the surface before meeting and reacting with A, by way of a Langmuir–Hinshelwood mechanism. However, in some systems, AB can only be formed as a result of a direct collision of the incoming B with the adsorbed A. Such reactions, which are discussed in further detail in section A3.9.6, are said to occur by an Eley–Rideal mechanism. A schematic illustration of these processes is shown in figure A3.9.1.

For a Langmuir–Hinshelwood reaction, we can expect the surface temperature to be an important variable determining overall reactivity because it determines how fast A and B diffuse across the surface. If the product AB molecules thermalize before desorption, the distribution of internal and translational energies in the gas phase will also reflect the surface temperature (yielding Boltzmann distributions that are modified by a dynamical factor related by the principle of detailed balance to the energetics of adsorption [7]). The main factor discriminating between the reaction schemes is that for a Langmuir–Hinshelwood reaction, the AB molecule can have no memory of the initial state and motion of the B molecule, but these should be evident in the AB products if the mechanism is of the Eley–Rideal type. These simple divisions are of course too black and white. In all probability, the two paradigms are actually extremes, with real systems reflecting aspects of both mechanisms [8].

A3.9.3   Collision dynamics and trapping in nonreactive systems

As with any collision process, to understand the dynamics of collisions we need an appreciation of the relevant forces and masses. Far from the surface, the incoming atom or molecule will experience the van der Waals attraction of the form

$$V(z) = -C/z^3 \tag{A3.9.1}$$

where z is the distance from the surface, and C is a constant dependent on the polarizability of the particle and the dielectric properties of the solid [9]. Close to the surface, where z is 0.1 nm for a nonreactive system, this attractive interaction is overwhelmed by repulsive forces (Pauli repulsion) due to the energy cost of orthogonalizing the overlapping electronic orbitals of the incoming molecule and the surface. The net result of van der Waals attraction and Pauli repulsion is a potential with a shallow well, the physisorption well, illustrated in figure A3.9.2. The depth of this well ranges from a few meV for He adsorption to $\sim$30 meV for H_2 molecules on noble metal surfaces, and to $\sim$100 meV for Ar or Xe on metal surfaces.

The van der Waals attraction arises from the interaction between instantaneous charge fluctuations in the molecule and surface. The molecule interacts with the surface as a whole. In contrast the repulsive forces

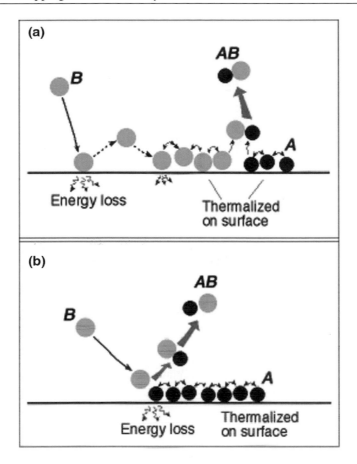

Figure A3.9.1. Schematic illustrations of (a) the Langmuir–Hinshelwood and (b) Eley–Rideal mechanisms in gas–surface dynamics.

are more short-range, localized to just a few surface atoms. The repulsion is, therefore, not homogeneous but depends on the point of impact in the surface plane, that is, the surface is corrugated.

A3.9.3.1 Binary collision (hard-cube) model

We can obtain an approximate description of the molecule–surface encounter using a binary collision model, with the projectile of mass m as one collision partner and a 'cube' having an effective mass, M, related to that of a surface atom, as the other partner. M depends on how close the projectile approaches the atoms of the surface, on the stiffness of the surface, and on the degree of corrugation of the repulsive potential. For rare–gas atoms interacting with metal surfaces, the surface electronic orbitals are delocalized and the repulsive interaction is effectively with a large cluster. The effective mass is correspondingly large, some 3–9 times the mass of a surface atom [10]. In other cases, such as O_2 colliding with Ag(111), the degree of corrugation and the effective mass may be closer to that expected for one atom [11].

Approximating the real potential by a square well and infinitely hard repulsive wall, as shown in figure A3.9.2, we obtain the hard cube model. For a well depth of W, conservation of energy and momentum

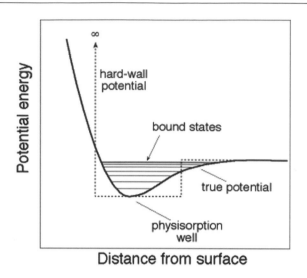

Figure A3.9.2. Interaction potential for an atom or molecule physisorbed on a surface. A convenient model is obtained by 'squaring off' the potential, which facilitates solution of the Schrödinger equation for the scattering of a quantum particle.

lead [11, 12] to the very useful Baule formula for the translational energy loss, δE, to the substrate

$$\delta E = \frac{4\mu}{(1+\mu)^2}(E+W) \tag{A3.9.2}$$

where E is the initial translational energy of the projectile and μ is the ratio m/M. This formula shows us that the energy transfer increases with the mass of the projectile, reaching a maximum when projectile and cube masses are equal. Of course the real projectile–surface interaction potential is not infinitely hard (cf figure A3.9.2). As E increases, the projectile can penetrate deeper into the surface, so that at its turning point (where it momentarily stops before reversing direction to return to the gas phase), an energetic projectile interacts with fewer surface atoms, thus making the effective cube mass smaller. Thus, we expect $\delta E/E$ to increase with E (and also with W since the well accelerates the projectile towards the surface).

The effect of surface temperature, T_s, can be included in this model by allowing the cube to move [12]. E becomes the translational energy in the frame of the centre-of-mass of projectile and cube; then we average the results over E, weighting with a Boltzmann distribution at T_s. This causes δE to decrease with increasing T_s, and when the thermal energy of the cube, kT_s, substantially exceeds E, the projectile actually gains energy in the collision! This is qualitatively consistent with experimental observations of the scattering of beams of rare-gas atoms from metal surfaces [14, 15].

A3.9.3.2   Scattering and trapping–desorption distributions

Projectiles leaving the surface promptly after an inelastic collision have exchanged energy with the surface, yet their direction of motion and translational and internal energies are clearly related to their initial values. This is called direct-inelastic (DI) scattering. At low E, the projectile sees a surface in thermal motion. This motion dominates the final energy and angular distributions of the scattering, and so this is referred to as thermal scattering. As E becomes large, the projectile penetrates the surface more deeply, seeing more of the detailed atomic structure, and the interaction comes to be dominated by scattering from individual atoms.

Eventually E becomes so large that the surface thermal motion becomes negligible, and the energy and angular distributions depend only on the atomic structure of the surface. This is known as the structure-scattering regime. Comparing experimental results with those of detailed classical molecular dynamics modelling of these phenomena can allow one to construct good empirical potentials to describe the projectile-surface interaction, as has been demonstrated for the Xe/Pt(111) [16] and Ar/Ag(111) [17] systems.

From equation (A3.9.2), we can see that at low E, the acceleration into the well dominates the energy loss, that is, δE does not reduce to zero with decreasing E. Below a critical translational energy, given by

$$E_c = \frac{4\mu W}{(1-\mu)^2} \tag{A3.9.3}$$

the projectile has insufficient energy remaining to escape from the well and it traps at the surface. Inclusion of surface temperature (cube motion) leads to a blurring of this cut-off energy so that trapping versus energy curves are predicted to be smoothed step functions. In fact, true trapping versus energy curves are closer to exponential in form, due to the combined effects of additional averaging over variations of W with surface site and with the orientation of the incident molecule. Additionally, transfer of motion normal to the surface to motion parallel to the surface, or into internal motions (rotations) can also lead to trapping, as we shall discuss below.

Trapped molecules can return to the gas phase once the thermal energy, kT_s, becomes comparable to the well depth. Having equilibrated with the surface, they have velocity, angular distribution and internal energies determined by T_s. This is visible in experiment as a scattering component (the trapping–desorption (TD) scattering component) with a very different appearance to the DI component, being peaked at and symmetrical about the surface normal, independently of the incidence conditions of the beam of projectiles. Such behaviour has been seen in many systems, for example in the scattering of Ar from Pt(111) [10] as illustrated in figure A3.9.3.

A3.9.3.3 Selective adsorption

Light projectiles impinging on a cold surface exhibit strong quantum behaviour in the scattering and trapping dynamics. Motion in the physisorption well is quantized normal to the surface, as indicated in figure A3.9.2. Although in the gas phase the projectile can have any parallel momentum, when interacting with a perfect surface, the parallel momentum can only change by whole numbers of reciprocal lattice vectors (the wavevectors corresponding to wavelengths fitting within the surface lattice) [9]. The scattering is thus into special directions, forming a diffraction pattern, which is evident even for quite massive particles such as Ar [18]. These quantizations couple to yield maxima in the trapping probability when, to accommodate the gain in parallel momentum, the projectile must drop into one of the bound states in the z-direction. In other words, the quantized gain in parallel motion leaves the projectile with more translational energy than it had initially, but the excess is cancelled by the negative energy of the bound state [19]. This is an entirely elastic phenomenon, no energy loss to the substrate is required, simply a conversion of normal for parallel motion. The trapping is undone if the parallel momentum gain is reversed.

The energies of the selective adsorption resonances are very sensitive to the details of the physisorption potential. Accurate measurement allied to computation of bound state energies can be used to obtain a very accurate quantitative form for the physisorption potential, as has been demonstrated for helium atom scattering. For molecules, we have the additional possibility of exchanging normal translations for rotational motion (the vibrational energies of light molecules are much larger than typical physisorption energies). Parallel momentum changes are effected by the surface corrugation, giving rise to corrugation mediated selective adsorption (CMSA). By analogy, rotational excitations produce rotation mediated selective adsorption (RMSA). Together these yield the acronym CRMSA. All such processes have been identified in the scattering of H_2 and its isotopomers from noble and simple metal surfaces [20]. Typical results are shown in figure A3.9.4.

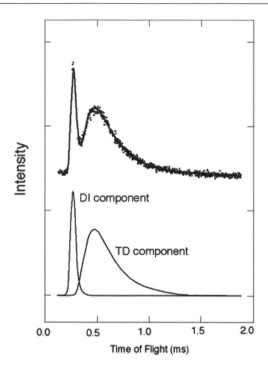

Figure A3.9.3. Time-of-flight spectra for Ar scattered from Pt(111) at a surface temperature of 100 K [10]. Points in the upper plot are actual experimental data. Curve through points is a fit to a model in which the bimodal distribution is composed of a sharp, fast moving (hence short flight time), direct-inelastic (DI) component and a broad, slower moving, trapping–desorption (TD) component. These components are shown separately in the lower curves. Parameters: $E = 12.5 \, \text{kJ mol}^{-1}$; $\theta_i = 60°$; $\theta_f = 40°$; $T_s = 100 \, \text{K}$.

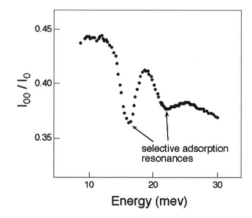

Figure A3.9.4. The ratio of specular reflectivity to incident beam intensity ratio for D_2 molecules scattering from a Cu(100) surface at 30 K [21].

The selective adsorption resonances show up as peaks in the trapping (minima in the reflectivity) because the long residence time at the surface increases the amount of energy lost to the substrate, resulting in sticking [21].

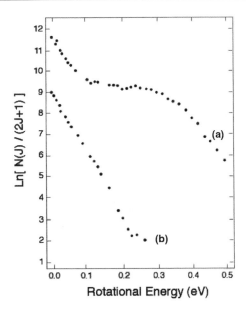

Figure A3.9.5. Population of rotational states versus rotational energy for NO molecules scattered from an Ag(111) surface at two different incidence energies and at $T_s = 520$ K [25]: (a) $E = 0.85$ eV, $\theta_i = 15°$ and (b) $E = 0.09$ eV, $\theta_i = 15°$. Results at $E = 0.85$ eV show a pronounced rotational rainbow.

A3.9.4 Molecular chemisorption and scattering

Unlike physisorption, chemisorption results from strong attractive forces mediated by chemical bonding between projectile and surface [9]. There is often significant charge transfer between surface and molecule, as in the adsorption of O_2 on metal surfaces [22]. The characteristics, well depth, distance of minimum above the surface etc. can vary greatly with surface site [23]. The degree of charge transfer can also differ, such that in many systems we can speak of there being more than one chemisorbed species [24].

The chemisorption interaction is also very strongly dependent on the molecular orientation, especially for heteronuclear molecules. This behaviour is exemplified by NO adsorption on metal surfaces, where the N end is the more strongly bound. These anisotropic interactions lead to strong steric effects and consequent rotational excitation in the scattering dynamics. Rainbows are evident in the rotational distributions [25], as can be seen in figure A3.9.5. These steric effects show up particularly strongly when the incident molecules are aligned prior to scattering (by magnetic fields) [26, 27].

The change of charge state of an adsorbed molecule leads to a change in the intramolecular bonding, usually a lengthening of the bond, which can result in vibrational excitation of the scattered molecule. Once again, this shows up in the scattering of NO from the Ag(111) surface [28], as shown in figure A3.9.6. In this case, the vibrational excitation probability is dependent on both the translational energy and the surface temperature. The translational energy dependence is probably due to the fact that the closer the molecule is to a surface, the more extended the molecular bond becomes, that is, in the language of section A3.9.5, the NO is trying to get round the elbow (see figure A3.9.8) to dissociate. It fails, and returns to the gas phase with increased vibrational energy. Surface temperature can enhance this process by supplying energy from the thermal motion of a surface atom towards the molecule [29, 30], but interaction with electronic excitations in the metal has also been demonstrated to be an efficient and likely source of energy transfer to the molecular vibrations [28, 30].

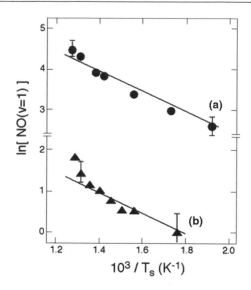

Figure A3.9.6. Population of the first excited vibrational state ($v = 1$) versus inverse of surface temperature for NO scattering from an Ag(111) surface [28]. Curves: (*a*) $E = 102$ kJ mol^{-1} and (*b*) $E = 9$ kJ mol^{-1}.

A3.9.4.1   Chemisorption and precursor states

The chemisorption of a molecule is often a precursor [31] to further reactions such as dissociation (see section A3.9.5.2), that is, the molecule must reside in the precursor state exploring many configurations until finding that leading to a reaction. Where there is more than one distinct chemisorption state, one can act as a precursor to the other [32]. The physisorption state can also act as a precursor to chemisorption, as is observed for the O_2/Ag(110) system [33].

The presence of a precursor breaks the dynamical motion into three parts [34]. First, there is the dynamics of trapping into the precursor state; secondly, there is (at least partial) thermalization in the precursor state; and, thirdly, the reaction to produce the desired species (possibly a more tightly bound chemisorbed molecule). The first two of these we can readily approach with the knowledge gained from the studies of trapping and sticking of rare-gas atoms, but the long timescales involved in the third process may perhaps more usefully be addressed by kinetics and transition state theory [35].

A3.9.5   Dynamics of dissociation reactions

A3.9.5.1   Direct dissociation of diatomics

The direct dissociation of diatomic molecules is the most well studied process in gas–surface dynamics, the one for which the combination of surface science and molecular beam techniques allied to the computation of total energies and detailed and painstaking solution of the molecular dynamics has been most successful. The result is a substantial body of knowledge concerning the importance of the various degrees of freedom (e.g. molecular rotation) to the reaction dynamics, the details of which are contained in a number of review articles [2, 36–41].

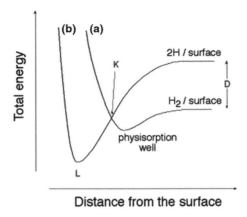

Figure A3.9.7. A representation of the Lennard-Jones model for dissociative adsorption of H_2. Curves: (a) interaction of intact molecule with surface; (b) interaction of two separately chemisorbed atoms with surface.

(a) Lennard-Jones model of hydrogen dissociation

In the 1930s Lennard-Jones [42] introduced a model that is still in use today in discussions of the dissociation of molecules at surfaces. He proposed a description based on two potential energy curves. The first, describing the interaction of the intact molecule with the surface as a physisorption potential, is shown as curve (a) in figure A3.9.7. Coupled with this, there is a second potential describing the interaction of the two separately chemisorbed atoms with the surface (curve (b) in figure A3.9.7). In equilibrium the adsorbed atoms are located at the minimum, L, of curve (b). The difference between (a) and (b) far from the surface is the gas-phase molecular dissociation energy, D. A dissociation event occurs if a molecule approaches the surface until K, where it makes a radiationless transition from (a) to (b) becoming adsorbed as atoms.

There is an inconsistency in the model in that when changing from (a) to (b) the molecular bond is instantaneously elongated. Lennard-Jones noted that although one-dimensional potential energy curves (such as shown in figure A3.9.7) can prove of great value in discussions, 'they do not lend themselves to generalization when more than one coordinate is necessary to specify a configuration'. In a quantitative theory there should be a number of additional curves between (a) and (b) corresponding to rotational and vibrational states of the molecule. In modern terms, we try to describe each degree-of-freedom relevant to the problem with a separate dimension. The potential energy curves of figure A3.9.7 then become a multidimensional surface, the potential energy surface (PES).

The model illustrated in figure A3.9.7 is primarily diabatic, the molecule jumps suddenly from one type of bonding, represented by a potential energy curve, to another. However, much of the understanding in gas–surface dynamics derives from descriptions based on motion on a single adiabatic PES, usually the ground-state PES. In the Lennard-Jones model, this would approximately correspond to whichever of (a) and (b) has the lower energy. Although this approach is successful in describing H_2 dissociation, it will not be adequate for reactions involving very sudden changes of electronic state [43]. These may occur, for example, in the O_2 reaction with simple metal surfaces [44]; they are so energetic that they can lead to light or electron emission during reaction [45].

(b) Influence of molecular vibration on reaction

Dissociation involves extension of a molecular bond until it breaks and so it might seem obvious that the more energy we can put into molecular vibration, the greater the reactivity. However, this is not always so: the

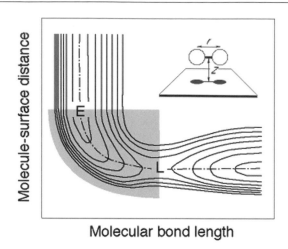

Figure A3.9.8. An elbow potential energy surface representing the dissociation of a diatomic in two dimensions—the molecular bond length and the distance from the molecule to the surface.

existence of a vibrational enhancement of dissociation reveals something about the shape, or topography of the PES itself. This is illustrated in figure A3.9.8, which shows a generic elbow PES [37]. This two-dimensional PES describes the dynamics in the molecule–surface and intramolecular bond length coordinates only. Far from the surface, it describes the intramolecular bonding of the projectile by, for example, a Morse potential. Close to the surface at large bond length, the PES describes the chemisorption of the two atoms to the surface in similar fashion to curve (b) in figure A3.9.7. The curved region linking these two extremes is the interaction region (shaded in figure A3.9.8), where the bonding is changing from one type to another. It corresponds roughly to the curve crossing point, K, in the Lennard-Jones model.

For vibrational effects in the dynamics, the location of the dissociation barrier within the curved interaction region of figure A3.9.8 is crucial. If the barrier occurs largely before the curved region, it is an 'early' barrier at point E, then vibration will not promote reaction as it occurs largely at right angles to the barrier. In contrast, if the barrier occurs when the bond is already extended, say at L (a 'late' barrier) in the figure, the vibration is now clearly helping the molecule to attack this barrier, and can substantially enhance reaction.

The consequences of these effects have been fully worked out, and agree with the Polanyi rules used in gas-phase scattering [46, 47]. Experimental observations of both the presence and absence of vibrational enhancement have been made, most clearly in hydrogen dissociation on metal surfaces. For instance, H_2 dissociation on Ni surfaces shows no vibrational enhancement [48, 49]. On Cu surfaces, however, vibrational enhancement of dissociation has been clearly demonstrated by using molecular beam techniques (section B2.6) to vary the internal and translational energies independently [50], and by examining the energy and state distributions of molecules undergoing the reverse of dissociation, the associative desorption reaction [51]. Figure A3.9.9 shows typical results presented in the form of dissociation versus translational energy curves backed out from the desorption data [52]. The curves corresponding to the vibrationally excited states clearly lie at lower energy than those for the vibrational ground-state, implying that some of the energy for the reaction comes from the H_2 vibration.

An important further consequence of curvature of the interaction region and a late barrier is that molecules that fail to dissociate can return to the gas-phase in vibrational states different from the initial, as has been observed experimentally in the H_2/Cu system [53, 55]. To undergo vibrational (de-)excitation, the molecules must round the elbow part way, but fail to go over the barrier, either because it is too high, or because the combination of vibrational and translational motions is such that the molecule moves across rather than over

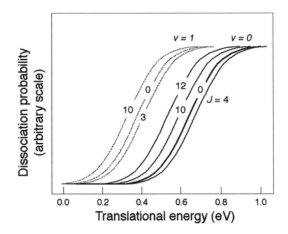

Figure A3.9.9. Dissociation probability versus incident energy for D_2 molecules incident on a Cu(111) surface for the initial quantum states indicated (v indicates the initial vibrational state and J the initial rotational state) [100]. For clarity, the saturation values have been scaled to the same value irrespective of the initial state, although in reality the saturation value is higher for the $v = 1$ state.

the barrier. Such vibrational excitation and de-excitation constrains the PES in that we require the elbow to have high curvature. Dissociation is not necessary, however, for as we have pointed out, vibrational excitation is observed in the scattering of NO from Ag(111) [55].

(c) Rotational effects: steric hindrance and centrifugal enhancement

Molecular rotation has two competing influences on the dissociation of diatomics [56–58]. A molecule will only be able to dissociate if its bond is oriented correctly with respect to the plane of the surface. If the bond is parallel to the plane, then dissociation will take place, whereas if the molecule is end-on to the surface, dissociation requires one atom to be ejected into the gas phase. In most cases, this 'reverse Eley–Rideal' process is energetically very unfavourable (although it does occur for very energetic reactions such as halide adsorption on Si surfaces, and possibly for O_2 adsorbing on reactive metals [59]). In general, molecules cannot dissociate when oriented end-on. The PES is, thus, highly corrugated in the molecular orientation coordinate. In consequence, increasing the rapidity of motion in this coordinate (i.e. increasing the rotational state) will make it more likely for the molecule to race past the small dissociation window at the parallel orientation, strike-off a more repulsive region of the PES and return to the gas phase. Therefore, dissociation is inhibited by increasing the rotational energy of the molecule. In opposition to this effect, the rotational motion can enhance reactivity when the dissociation barrier is late (i.e. occurs at extended bond length). As the molecule progresses through the interaction region, its bond begins to extend. This increases the moment of inertia and thus reduces the rotational energy. The rotational energy thus 'lost' feeds into the reaction coordinate, further stretching the molecular bond and enhancing the reaction. The combination of these two competing effects has been demonstrated in the H_2/Cu(111) system. For the first few rotational states, increases in rotation reduce the dissociation (i.e. shift the dissociation curve to higher energy) as can be seen in figure A3.9.9. Eventually, however, centrifugal enhancement wins out, and for the higher rotational states the dissociation curves are pushed to lower translational energies.

The strong dependence of the PES on molecular orientation also leads to strong coupling between rotational states, and hence rotational excitation/de-excitation in the scattering. This has been observed experimentally for H_2 scattering from Cu surfaces. Recent work has shown that for H_2 the changes in rotational state occur almost exclusively when the molecular bond is extended, that is, longer than the gas-phase equilibrium value [60].

(d) Surface corrugation and site specificity of reaction

The idea that certain sites on a surface are especially active is common in the field of heterogeneous catalysis [61]. Often these sites are defects such as dislocations or steps. But surface site specificity for dissociation reactions also occurs on perfect surfaces, arising from slight differences in the molecule–surface bonding at different locations. This is so not only of insulator and semiconductor surfaces where there is strongly directional bonding, but also of metal surfaces where the electronic orbitals are delocalized. The site dependence of the reactivity manifests itself as a strong corrugation in the PES, which has been shown to exist by *ab initio* computation of the interaction PES for H_2 dissociation on some simple and noble metal surfaces [62–64].

The dynamical implications of this corrugation appear straightforward: surface sites where the dissociation barrier is high (unfavourable reaction sites) should shadow those sites where the barrier is low (the favoured reaction sites) if the reacting molecule is incident at an angle to the surface plane. If we assume that the motion normal to the surface is important in traversing the dissociation barrier, then those molecules approaching at an angle should have lower dissociation probability than those approaching at normal incidence. This has indeed been observed in a number of dissociation systems [37], but a far more common observation is that the dissociation scales with the 'normal energy', $E_p = E \cos^2 \theta$, where E is the translational energy, and θ the angle of incidence of the beam with respect to the surface normal. Normal energy scaling, shown in figure A3.9.10, implies that the motion parallel to the surface does not affect dissociation, and the surface appears flat.

This difficulty has been resolved with the realization that the surface corrugation is not merely of the barrier energy, but of the distance of the barrier above the surface [65]. We then distinguish between *energetic corrugation* (the variation of the energetic height of the barrier) and *geometric corrugation* (a simple variation of the barrier location or shape). The two cases are indicated in figure A3.9.10. For energetic corrugation, the shadowing does lead to lower dissociation at off-normal incidence, but this can be counterbalanced by geometric corrugation, for which the parallel motion helps the molecule to attack the facing edge of the PES [65].

The site specificity of reaction can also be a state-dependent site specificity, that is, molecules incident in different quantum states react more readily at different sites. This has recently been demonstrated by Kroes and co-workers for the H_2/Cu(100) system [66]. Additionally, we can find reactivity dominated by certain sites, while inelastic collisions leading to changes in the rotational or vibrational states of the scattering molecules occur primarily at other sites. This spatial separation of the active site according to the change of state occurring (dissociation, vibrational excitation etc) is a very surface specific phenomenon.

(e) Steering dominated reaction

A very extreme version of surface corrugation has been found in the nonactivated dissociation reactions of H_2 on W [67, 68], Pd and Rh systems. In these cases, the very strong chemisorption bond of the H atoms gives rise to a very large energy release when the molecule dissociates. In consequence, at certain sites on the surface, the molecule accelerates rapidly downhill into the dissociation state. At the unfavourable sites, there are usually small dissociation barriers and, of course, molecules oriented end-on to the surface cannot dissociate. When we examine the dynamics of motion on such PESs, we find that the molecules are steered into the attractive downhill regions [69], away from the end-on orientation and away from the unfavourable reaction sites.

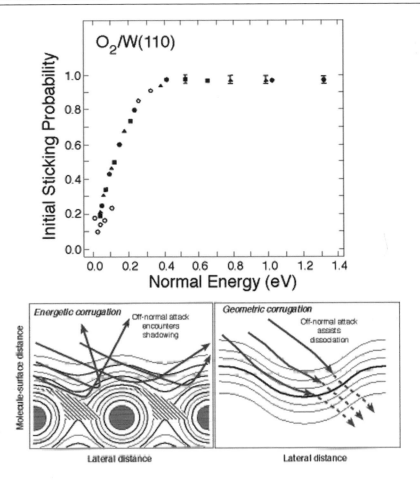

Figure A3.9.10. The dissociation probability for O_2 on W(110) [101] as a function of the normal energy, (upper). $T_g = 800$ K; θ: ($\bullet$) 0°, ($\blacktriangle$) 30°, ($\blacksquare$) 45° and ($\bigcirc$) 60°. The normal energy scaling observed can be explained by combining the two surface corrugations indicated schematically (lower diagrams).

Steering is a very general phenomenon, caused by gradients in the PES, occurring in every gas–surface system [36]. However, for these nonactivated systems showing extreme variations in the PES, the steering dominates the dissociation dynamics. At the very lowest energies, most molecules have enough time to steer into the most favourable geometry for dissociation hence the dissociation probability is high. At higher E, there is less time for steering to be effective and the dissociation decreases. The general signature of a steering dominated reaction is, therefore, a dissociation probability that falls with increasing E [49, 70, 71], as shown in figure A3.9.11. This can be contrasted with the curve usually expected for direct dissociation, figure A3.9.10, one which increases with E because, as E increases, it is easier to overcome the barriers in unfavourable geometries.

(f) Surface temperature dependence

Direct dissociation reactions are affected by surface temperature largely through the motion of the substrate atoms [72]. Motion of the surface atom towards the incoming molecule increases the likelihood of (activated)

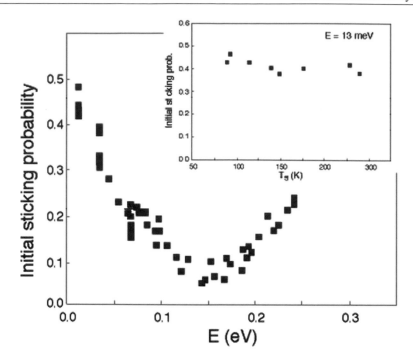

Figure A3.9.11. Dissociation of H_2 on the W(100)–c(2 × 2)–Cu surface as a function of incident energy [71]. The steering dominated reaction [102] is evident at low energy, confirmed by the absence of a significant surface temperature dependence.

dissociation, while motion away decreases the dissociation probability. For low dissociation probabilities, the net effect is an enhancement of the dissociation by increasing surface temperature, as observed in the system O_2/Pt{100}-hex-R0.7° [73].

This interpretation is largely based on the results of cube models for the surface motion. It may also be that the thermal disorder of the surface leads to slightly different bonding and hence different barrier heights. Increasing temperature also changes the populations of the excited electronic states in the surface, which may affect bonding. The contribution of these effects to the overall surface temperature dependence of reaction is presently not clear.

A3.9.5.2   Direct dissociation of polyatomic molecules

Although understanding the dissociation dynamics of diatomic molecules has come a long way, that of polyatomic molecules is much less well-developed. Quite simply, this is due to the difficulty of computing adequate PESs on which to perform suitable dynamics, when there are many atoms. Quantum dynamics also becomes prohibitively expensive as the dimensionality of the problem increases. The dissociation of CH_4 (to H + CH_3) on metal surfaces is the most studied to date [74]. This shows dependences on molecular translational energy and internal state, as well as a strong surface temperature dependence, which has been interpreted in terms of thermally assisted quantum tunnelling through the dissociation barrier. More recent experimental work has shown complicated behaviour at low E, with the possible involvement of steering or trapping [75].

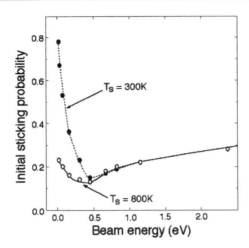

Figure A3.9.12. The dissociation probability of N_2 on the W(100) surface as a function of the energy of the molecular beam. The falling trend and pronounced surface temperature dependence are indicative of a precursor mediated reaction at low energies.

A3.9.5.3 *Precursor-mediated dissociation*

Precursor-mediated dissociation involves trapping in a molecularly chemisorbed state (or possibly several states) prior to dissociation. If the molecule thermalizes before dissociation, we can expect to observe the signature of trapping in the dissociation dynamics, that is, we expect increasing E and increasing surface temperature to decrease the likelihood of trapping, and hence of dissociation. This is exemplified by the dissociation of N_2 on W(100) in the low energy regime [76], shown in figure A3.9.12.

The thermalization stage of this dissociation reaction is not amenable to modelling at the molecular dynamics level because of the long timescales required. For some systems, such as O_2/Pt(111), a kinetic treatment is very successful [77]. However, in others, thermalization is not complete, and the internal energy of the molecule can still enhance reaction, as observed for N_2/Fe(111) [78, 79] and in the dissociation of some small hydrocarbons on metal surfaces [80]. A detailed explanation of these systems is presently not available.

A3.9.6 Eley–Rideal dynamics

The idea that reactions can occur directly when an incident reagent strikes an adsorbate is strongly supported by detailed theoretical calculations. From early classical simulations on model PESs (e.g. for H on H/tungsten [81] and O on C/platinum [82]), to more recent theoretical studies of ER reactions employing quantum-mechanical models, it has been clearly established that product molecules should show a high degree of internal excitation. As noted in section A3.9.2, certain highly facile gas–surface reactions can occur directly at the point of impact between an incident gas-phase reagent and an adsorbate; however, it is more likely that the incident reagent will 'bounce' a few times before reaction, this being to some degree accommodated in the process. A more useful working definition of an ER reaction is that it should occur before the reagents have become equilibrated at the surface. With this definition, we encompass hot-atom dynamics and what Harris and Kasemo have termed *precursor dynamics* [8]. Since the heat of adsorption of the incident reagent is not fully accommodated, ER reactions are far more exothermic than their LH counterparts.

Until relatively recently, the experimental evidence for the ER mechanism came largely from kinetic measurements, relating the rate of reaction to the incident flux and to the surface coverage and temperature [83]. For example, it has been found that the abstraction of halogens from Si(100) by incident H proceeds with a very small activation barrier, consistent with an ER mechanism. To *prove* that a reaction can occur on essentially a single gas–surface collision, however, dynamical measurements are required. The first definitive evidence for an ER mechanism was obtained in 1991, in a study showing that hyperthermal $N(C_2H_4)_3N$ can pick up a proton from a H/Pt(111) surface to give an ion with translational energy dependent on that of the incident molecule [84]. A year later, a study was reported of the formation of HD from H atoms incident on D/Cu(111), and from D incident on H/Cu(111) [85]. The angular distribution of the HD was found to be asymmetrical about the surface normal and peaked on the opposite side of the normal to that of the incident atom. This behaviour proved that the reaction must occur before the incident atom reaches equilibrium with the surface. Moreover, the angular distribution was found to depend on the translational energy of the incident atom and on which isotope was incident, firmly establishing the operation of an ER mechanism for this elementary reaction.

Conceptually similar studies have since been carried out for the reaction of H atoms with Cl/Au(111). More recently, quantum-state distributions have been obtained for both the H + Cl/Au(111) [86–88] and H(D)+D(H)/Cu(111) systems. The results of these studies are in good qualitative agreement with calculations. Even for the H(D)+D(H)/Cu(111) system [89], where we know that the incident atom cannot be significantly accommodated prior to reaction, reaction may not be direct. Detailed calculations yield much smaller cross sections for direct reaction than the overall experimental cross section, indicating that reaction may occur only after trapping of the incident atom [90].

Finally, it should also be clear that ER reactions do not necessarily yield a gas-phase product. The new molecule may be trapped on the surface. There is evidence for an ER mechanism in the addition of incident H atoms to ethylene and benzene on Cu(111) [91], and in the abstraction of H atoms from cyclohexane by incident D atoms [92], and the direct addition of H atoms to CO on Ru(001) [93].

A3.9.7 Photochemistry

The interaction of light with both clean surfaces and those having adsorbed species has been a popular research topic over the past 10 years [94]. Our understanding of processes such as photodesorption, photodissociation and photoreaction is still at a very early stage and modelling has been largely performed on a system-by-system basis rather than any general theories being applicable. One of the most important aspects of performing photochemical reactions on surfaces, which has been well documented by Polanyi and co-workers is that it is possible to align species before triggering reactions that cannot be done in the gas phase. This is frequently referred to as surface aligned photochemistry [95]. One of the key issues when light, such as that from a picosecond laser, impinges a surface covered with an adsorbate is where the actual absorption takes place. Broadly speaking there are two possible choices either in the adsorbate molecule or the surface itself. Unfortunately, although it may seem that unravelling microscopic reaction mechanisms might be quite distinct depending on what was absorbing, this is not the case and considerable effort has been spent on deciding what the dynamical consequences are for absorption into either localized or extended electronic states [96].

Of lesser interest here for a laser beam incident upon a surface are the processes that occur due to surface heating. Of greater interest are those occasions when an electronic transition is initiated and a process occurs, for in these circumstances it becomes possible to 'tune' reactivity by an external agent. A good example of this is the UV photodissociation of a range of carbonyls on Si surfaces [97]. Here it was shown explicitly that 257 nm light can selectively excite the adsorbate and then dissociation ensues. An alternative story unfolds when NO is photodesorbed from Pt surfaces. Detailed experiment and modelling shows that, in this case, the initial excitation (absorption) event occurs in the metal substrate. Following this, the excess energy is transferred to the adsorbate by a hot electron which resides for about 10–12 fs before returning to the substrate. During this time, it is possible for the NO to gain sufficient energy to overcome the adsorption bond [98].

Finally, and most recently, femtosecond lasers have been employed to investigate reactions on surfaces, one good example is the oxidation of CO on a Ru surface [99]. One of the long outstanding problems in surface dynamics is to determine the energy pathways that are responsible for irreversible processes at the surface. Both phonons and electrons are capable of taking energy from a prethermalized adsorbate and because of the time required for converting electronic motion to nuclear motion, there is the possibility that measurements employing ultrashort-pulsed lasers might be able to distinguish the dominant pathway.

A3.9.8   Outlook

Despite the considerable progress over the 1990s, the field of gas–surface reaction dynamics is still very much in its infancy. We have a relatively good understanding of hydrogen dissociation on noble metals but our knowledge of other gas–surface systems is far from complete. Even for other diatomic reagents such as N_2 or O_2 a great deal yet remains to be learned. Nevertheless, we believe that progress will take place even if in a slightly different fashion to that which is described here.

In parallel with the remarkable increase in computing power, particularly in desktop workstations, there have been significant advances also in the algorithmic development of codes that can calculate the potential energy (hyper-)surfaces that have been mentioned in this article. Most of the theoretical work discussed here has relied to a greater or lesser extent on potential energy surfaces being available from some secondary agency and this, we believe, will not be the case in the future. Software is now available which will allow the dynamicist to calculate new potentials and then deploy them to evaluate state-to-state cross sections and reaction probabilities. Although new, detailed experimental data will provide guidance, a more general understanding of gas–surface chemistry will develop further as computational power continues to increase.

References

[1]   Barker J A and Auerbach D J 1985 Gas–surface interactions and dynamics; thermal energy atomic and molecular beam studies *Surf. Sci. Rep.* **4** 1
[2]   Rettner C T and Ashfold M N R 1991 *Dynamics of Gas–Surface Interactions* (London: Royal Society of Chemistry)
[3]   Rettner C T, Auerbach D J, Tully J C and Kleyn A W 1996 Chemical dynamics at the gas–surface interface *J. Phys. Chem.* **100** 13 201
[4]   Langmuir I 1922 Chemical reactions on surfaces *Trans. Faraday Soc.* **17** 607
[5]   Holloway S 1993 Dynamics of gas–surface interactions *Surf. Sci.* **299/300** 656
[6]   Ertl G 1993 Reactions at well-defined surfaces *Surf. Sci.* **299/300** 742
[7]   Rettner C T, Michelsen H A and Auerbach D J 1993 From quantum-state-specific dynamics to reaction-rates—the dominant role of translational energy in promoting the dissociation of D_2 on Cu(111) under equilibrium conditions *Faraday Discuss.* **96** 17
[8]   Harris J and Kasemo B 1981 On precursor mechanisms for surface reactions *Surf. Sci.* **105** L281
[9]   Zangwill A 1988 *Physics at Surfaces* (Cambridge: Cambridge University Press)
[10] Head-Gordon M, Tully J C, Rettner C T, Mullins C B and Auerbach D J 1991 On the nature of trapping and desorption at high surface temperatures: theory and experiments for the Ar–Pt(111) system *J. Chem. Phys.* **94** 1516
[11] Spruit M E M, van den Hoek P J, Kuipers E W, Geuzebroek F and Kleyn A W 1989 Direct inelastic scattering of superthermal Ar, CO, NO and O2 from Ag(111) *Surf. Sci.* **214** 591
[12] Harris J 1987 Notes on the theory of atom–surface scattering *Phys. Scr.* **36** 156
[13] Harris J 1991 Mechanical energy transfer in particle–surface collisions *Dynamics of Gas–Surface Interactions* ed C T Rettner and M N R Ashfold (London: Royal Society of Chemistry) p 1
[14] Hurst J E, Becker C A, Cowin J P, Janda K C, Auerbach D J and Wharton L 1979 Observation of direct inelastic scattering in the presence of trapping-desorption scattering: Xe on Pt(111) *Phys. Rev. Lett.* **43** 1175
[15] Janda K C, Hurst J E, Cowin J P, Warton L and Auerbach D J 1983 Direct-inelastic and trapping-desorption scattering of N_2 and CH_4 from Pt(111) *Surf. Sci.* **130** 395
[16] Barker J A and Rettner C T 1992 Accurate potential energy surface for Xe/Pt(111): a benchmark gas–surface interaction potential *J. Chem. Phys.* **97** 5844
[17] Kirchner E J J, Kleyn A W and Baerends E J 1994 A comparative study of Ar/Ag(111) potentials *J. Chem. Phys.* **101** 9155
[18] Schweizer E K and Rettner C T 1989 Quantum effects in the scattering of argon from 2H-W(100) *Phys. Rev. Lett.* **62** 3085
[19] Lennard-Jones J E and Devonshire A F 1936 Diffraction and selective adsorption of atoms at crystal surfaces *Nature* **137** 1069
[20] Andersson S, Wilzén L, Persson M and Harris J 1989 Sticking in the quantum regime: H_2 and D_2 on Cu(100) *Phys. Rev. B* **40** 8146
[21] Persson M, Wilzén L and Andersson S 1990 Mean free path of a trapped physisorbed hydrogen molecule *Phys. Rev. B* **42** 5331

[22] Backx C, de Groot C P M and Biloen P 1981 Adsorption of oxygen on Ag(110) studied by high resolution ELS and TPD *Surf.*
       Sci. **104** 300
[23] Gravil P A and Bird D M 1996 Chemisorption of O_2 on Ag(110) *Surf. Sci.* **352** 248
[24] Campbell C T 1985 Atomic and molecular oxygen adsorption on Ag(111) *Surf. Sci.* **157** 43
[25] Rettner C T 1991 Inelastic scattering of NO from Ag(111): Internal state, angle, and velocity resolved measurements *J. Chem.*
       Phys. **94** 734
[26] Lahaye R J W E, Stolte S, Holloway S and Kleyn A W 1996 NO/Pt(111) orientation and energy dependence of scattering *J. Chem.*
       Phys. **104** 8301
[27] Heinzmann U, Holloway S, Kleyn A W, Palmer R E and Snowdon K J 1996 Orientation in molecule–surface interactions *J. Phys.:*
       Condens. Matter **8** 3245
[28] Rettner C T, Kimman J, Fabre F, Auerbach D J and Morawitz H 1987 Direct vibrational excitation in gas–surface collisions of
       NO with Ag(111) *Surf. Sci.* **192** 107
[29] Gates G A, Darling G R and Holloway S 1994 A theoretical study of the vibrational excitation of NO/Ag(111) *J. Chem. Phys.*
       101 6281
[30] Groß A and Brenig W 1993 Vibrational excitation of NO in NO Ag scattering revisited *Surf. Sci.* **289** 335
[31] Auerbach D J and Rettner C T 1987 Precursor states, myth or reality: a perspective from molecular beam studies *Kinetics of*
       Interface Reactions ed M Grunze and H J Kreuzer (Berlin: Springer) p 125
[32] Luntz A C, Grimblot J and Fowler D 1989 Sequential precursors in dissociative chemisorption—O_2 on Pt(111) *Phys. Rev.* B **39**
       12 903
[33] Vattuone L, Boragno C, Pupo M, Restelli P, Rocca M and Valbusa U 1994 Azimuthal dependence of sticking probability of O_2
       on Ag(110) *Phys. Rev. Lett.* **72** 510
[34] Doren D J and Tully J C 1991 Dynamics of precursor-mediated chemisorption *J. Chem. Phys.* **94** 8428
[35] Kang H C and Weinberg W H 1994 Kinetic modelling of surface rate processes *Surf. Sci.* **299/300** 755
[36] DePristo A E and Kara A 1990 Molecule–surface scattering and reaction dynamics *Adv. Chem. Phys.* **77** 163
[37] Darling G R and Holloway S 1995 The dissociation of diatomic molecules *Rep. Prog. Phys.* **58** 1595
[38] Jacobs D C 1995 The role of internal energy and approach geometry in molecule–surface reactive scattering *J. Phys.: Condens.*
       Matter **7** 1023
[39] Groß A 1996 Dynamical quantum processes of molecular beams at surfaces—hydrogen on metals *Surf. Sci.* **363** 1
[40] Groß A 1998 Reactions at surfaces studied by *ab initio* dynamics calculations *Surf. Sci. Rep.* **32** 291
[41] Kroes G J 1999 Six-dimensional quantum dynamics of dissociative chemisorption of H_2 on metal surfaces *Prog. Surf. Sci.* **60** 1
[42] Lennard-Jones J E 1932 Processes of adsorption and diffusion on solid surfaces *Trans. Faraday Soc.* **28** 333
[43] Kasemo B 1996 Charge transfer, electronic quantum processes, and dissociation dynamics in molecule–surface collisions *Surf.*
       Sci. **363** 22
[44] Katz G, Zeiri Y and Kosloff R 1999 Non-adiabatic charge transfer process of oxygen on metal surfaces *Surf. Sci.* **425** 1
[45] Böttcher A, Imbeck R, Morgante A and Ertl G 1990 Nonadiabatic surface reaction: Mechanism of electron emission in the Cs+O_2
       system *Phys. Rev. Lett.* **65** 2035
[46] Polanyi J C and Wong W H 1969 Location of energy barriers. I. Effect on the dynamics of reactions A + BC *J. Chem. Phys.* **51**
       1439
[47] Polanyi J C 1987 Some concepts in reaction dynamics *Science* **236** 680
[48] Robota H J, Vielhaber W, Lin M C, Segner J and Ertl G 1985 Dynamics of the interaction of H_2 and D_2 with Ni(110) and Ni(111)
       surfaces *Surf. Sci.* **155** 101
[49] Rendulic K D, Anger G and Winkler A 1989 Wide-range nozzle beam adsorption data for the systems H_2/Ni and H_2/Pd(100)
       Surf. Sci. **208** 404
[50] Hayden B E and Lamont C L A 1989 Coupled translational–vibrational activation in dissociative hydrogen adsorption on Cu(110)
       Phys. Rev. Lett. **63** 1823
[51] Michelsen H A, Rettner C T and Auerbach D J 1993 The adsorption of hydrogen at copper surfaces: A model system for the
       study of activated adsorption *Surface Reactions* ed R J Madix (Berlin: Springer) p 123
[52] Rettner C T, Michelsen H A and Auerbach D J 1995 Quantum-state-specific dynamics of the dissociative adsorption and associative
       desorption of H_2 at a Cu(111) surface *J. Chem. Phys.* **102** 4625
[53] Rettner C T, Auerbach D J and Michelsen H A 1992 Observation of direct vibrational-excitation in collisions of H_2 and D_2 with
       a Cu(111) surface *Phys. Rev. Lett.* **68** 2547
[54] Hodgson A, Moryl J, Traversaro P and Zhao H 1992 Energy transfer and vibrational effects in the dissociation and scattering of
       D_2 from Cu(111) *Nature* **356** 501
[55] Rettner C T, Fabre F, Kimman J and Auerbach D J 1985 Observation of direct vibrational-excitation in gas–surface collisions—NO
       on Ag(111) *Phys. Rev. Lett.* **55** 1904
[56] Beauregard J N and Mayne H R 1993 The role of reactant rotation and rotational alignment in the dissociative chemisorption of
       hydrogen on Ni(100) *Chem. Phys. Lett.* **205** 515
[57] Darling G R and Holloway S 1993 Rotational effects in the dissociative adsorption of H_2 on Cu(111) *Faraday Discuss. Chem.*
       Soc. **96** 43
[58] Darling G R and Holloway S 1994 Rotational motion and the dissociation of H_2 on Cu(111) *J. Chem. Phys.* **101** 3268

[59] Wahnström G, Lee A B and Strömquist J 1996 Motion of 'hot' oxygen adatoms on corrugated metal surfaces *J. Chem. Phys.* **105** 326

[60] Wang Z S, Darling G R and Holloway S 2000 Translation-to-rotational energy transfer in scattering of H_2 molecules from Cu(111) surfaces *Surf. Sci.* **458** 63

[61] Somorjai G A 1994 The surface science of heterogeneous catalysis *Surf. Sci.* **299/300** 849

[62] Bird D M, Clarke L J, Payne M C and Stich I 1993 Dissociation of H_2 on Mg(0001) *Chem. Phys. Lett.* **212** 518

[63] White J A, Bird D M, Payne M and Stich I 1994 Surface corrugation in the dissociative adsorption of H_2 on Cu(100) *Phys. Rev. Lett.* **73** 1404

[64] Hammer B, Scheffler M, Jacobsen K W and Nørskov J K 1994 Multidimensional potential energy surface for H_2 dissociation over Cu(111) *Phys. Rev. Lett.* **73** 1400

[65] Darling G R and Holloway S 1994 The role of parallel momentum in the dissociative adsorption of H_2 at highly corrugated surfaces *Surf. Sci.* **304** L461

[66] McCormack D A and Kroes G J 1999 A classical study of rotational effects in dissociation of H_2 on Cu(100) *Phys. Chem. Chem. Phys.* **1** 1359

[67] White J A, Bird D M and Payne M C 1995 Dissociation of H_2 on W(100) *Phys. Rev. B* **53** 1667

[68] Groß A, Wilke S and Scheffler M 1995 6-dimensional quantum dynamics of adsorption and desorption of H_2 at Pd(100)—steering and steric effects *Phys. Rev. Lett.* **75** 2718

[69] Kay M, Darling G R, Holloway S, White J A and Bird D M 1995 Steering effects in non-activated adsorption *Chem. Phys. Lett.* **245** 311

[70] Butler D A, Hayden B E and Jones J D 1994 Precursor dynamics in dissociative hydrogen adsorption on W(100) *Chem. Phys. Lett.* **217** 423

[71] Butler D A and Hayden B E 1995 The indirect channel to hydrogen dissociation on W(100)c(2 × 2)Cu—evidence for a dynamical precursor *Chem. Phys. Lett.* **232** 542

[72] Hand M R and Harris J 1990 Recoil effects in surface dissociation *J. Chem. Phys.* **92** 7610

[73] Guo X-C, Bradley J M, Hopkinson A and King D A 1994 O_2 interaction with Pt{100}-hexR0.7°—scattering, sticking and saturating *Surf. Sci.* **310** 163

[74] Luntz A C and Harris J 1991 CH_4 dissociation on metals—a quantum dynamics model *Surf. Sci.* **258** 397

[75] Walker A V and King D A 1999 Dynamics of the dissociative adsorption of methans on Pt(110)–(1 × 2) *Phys. Rev. Lett.* **82** 5156

[76] Rettner C T, Schweizer E K and Stein H 1990 Dynamics of chemisorption of N_2 on W(100): Precursor-mediated and activated dissociation *J. Chem. Phys.* **93** 1442

[77] Rettner C T and Mullins C B 1991 Dynamics of the chemisorption of O_2 on Pt(111): Dissociation via direct population of a molecularly chemisorbed precursor at high incidence kinetic energy *J. Chem. Phys.* **94** 1626

[78] Rettner C T and Stein H 1987 Effect of the translational energy on the chemisorption of N_2 on Fe(111): activated dissociation via a precursor state *Phys. Rev. Lett.* **59** 2768

[79] Rettner C T and Stein H 1987 Effect of the vibrational energy on the dissociative chemisorption of N_2 on Fe(111) *J. Chem. Phys.* **87** 770

[80] Luntz A C and Harris J 1992 The role of tunneling in precursor mediated dissociation: Alkanes on metal surfaces *J. Chem. Phys.* **96** 7054

[81] Elkowitz A B, McCreery J H and Wolken G 1976 Dynamics of atom-adsorbed atom collisions: Hydrogen on tungsten *Chem. Phys.* **17** 423

[82] Tully J C 1980 Dynamics of gas–surface interactions: reactions of atomic oxygen with adsorbed carbon on platinum *J. Chem. Phys.* **73** 6333

[83] Weinberg W H 1991 Kinetics of surface reactions *Dynamics of Gas–Surface Interactions* ed C T Rettner and M N R Ashfold (London: Royal Society of Chemistry)

[84] Kuipers E W, Vardi A, Danon A and Amirav A 1991 Surface-molecule proton transfer—a demonstration of the Eley–Ridel mechanism *Phys. Rev. Lett.* **66** 116

[85] Rettner C T 1992 Dynamics of the direct reaction of hydrogen atoms adsorbed on Cu(111) with hydrogen atoms incident from the gas phase *Phys. Rev. Lett.* **69** 383

[86] Lykke K R and Kay B D 1990 State-to-state inelastic and reactive molecular beam scattering from surfaces *Laser Photoionization and Desorption Surface Analysis Techniques* vol 1208, ed N S Nogar (Bellingham, WA: SPIE) p 1218

[87] Rettner C T and Auerbach D J 1994 Distinguishing the direct and indirect products of a gas–surface reaction *Science* **263** 365

[88] Rettner C T 1994 Reaction of an H-atom beam with Cl/Au(111)—dynamics of concurrent Eley–Rideal and Langmuir–Hinshelwood mechanisms *J. Chem. Phys.* **101** 1529

[89] Rettner C T and Auerbach D J 1996 Quantum-state distributions for the HD product of the direct reaction of H(D)/Cu(111) with D(H) incident from the gas phase *J. Chem. Phys.* **104** 2732

[90] Shalashilin D V, Jackson B and Persson M 1999 Eley–Rideal and hot atom reactions of H(D) atoms with D(H)-covered Cu(111) surfaces; quasiclassical studies *J. Chem. Phys.* **110** 11 038

[91] Xi M and Bent B E 1992 Evidence for an Eley–Rideal mechanism in the addition of hydrogen atoms to unsaturated hydrocarbons on Cu(111) *J. Vac. Sci. Technol. B* **10** 2440

[92]   Xi M and Bent B E 1993 Reaction of deuterium atoms with cyclohexane on Cu(111)—hydrogen abstraction reactions by Eley–Rideal mechanisms *J. Phys. Chem.* **97** 4167

[93]   Xie J, Mitchell W J, Lyons K J and Weinberg W H 1994 Atomic hydrogen induced decomposition of chemisorbed formate at 100 K on the Ru(001) surface *J. Chem. Phys.* **101** 9195

[94]   Dai E H L and Ho W 1995 *Laser Spectroscopy and Photochemistry at Metal Surfaces* (Singapore: World Scientific)

[95]   Polanyi J C and Rieley H 1991 Photochemistry in the adsorbed state *Dynamics of Gas–Surface Interactions* ed C T Rettner and M N R Ashfold (London: Royal Society of Chemistry) p 329

[96]   Hasselbrink E 1994 State-resolved probes of molecular desorption dynamics induced by short-lived electronic excitations *Laser Spectroscopy and Photochemistry at Metal Surfaces* ed E H L Dai and W Ho (Hong Kong: World Scientific) p 685

[97]   Ho W 1994 Surface photochemistry *Surf. Sci.* **299/300** 996

[98]   Cavanagh R R, King D S, Stephenson J C and Heinz T F 1993 Dynamics of nonthermal reactions—femtosecond surface chemistry *J. Phys. Chem.* **97** 786

[99]   Bonn M, Funk S, Hess C, Denzler D N, Stampfl C, Scheffler M, Wolf M and Ertl G 1999 Phonon versus electron-mediated desorption and oxidation of CO on Ru(001) *Science* **285** 1042

[100] Michelsen H A, Rettner C T, Auerbach D J and Zare R N 1993 Effect of rotation on the translational and vibrational energy dependence of the dissociative adsorption of D_2 on Cu(111) *J. Chem. Phys.* **98** 8294

[101] Rettner C T, DeLouise L A and Auerbach D J 1986 Effect of incidence kinetic energy and surface coverage on the dissociative chemisorption of oxygen on W(110) *J. Chem. Phys.* **85** 1131

[102] Darling G R, Kay M and Holloway S 1998 The steering of molecules in simple dissociation reactions *Surf. Sci.* **400** 314

Further Reading

Rettner C T and Ashfold M N R 1991 *Dynamics of Gas–Surface Interactions* (London: Royal Society of Chemistry)

Darling G R and Holloway S 1995 The dissociation of diatomic molecules *Rep. Prog. Phys.* **58** 1595

Rettner C T, Auerbach D J, Tully J C and Kleyn A W 1996 Chemical dynamics at the gas–surface interface *J. Phys. Chem.* **100** 13 201

A3.10
Reactions on surfaces: corrosion, growth, etching and catalysis

Todd P St Clair and D Wayne Goodman

A3.10.1 Introduction

The impact of surface reactions on society is often overlooked. How many of us pause to appreciate integrated circuitry before checking email? Yet, without growth and etching reactions, the manufacturing of integrated circuits would be quite impractical. Or consider that in 1996, the United States alone consumed 123 billion gallons of gasoline [1]. The production of this gasoline from crude petroleum is accomplished by the petroleum industry using heterogeneous catalytic reactions. Even the control of automobile exhaust emissions, an obvious environmental concern, is achieved *via* catalytic reactions using 'three-way catalysts' that eliminate hydrocarbons, CO and NO_x. The study of these types of surface reactions and others is an exciting and rapidly changing field. Nevertheless, much remains to be understood at the atomic level regarding the interaction of gases and liquids with solid surfaces.

Surface science has thrived in recent years primarily because of its success at providing answers to fundamental questions. One objective of such studies is to elucidate the basic mechanisms that control surface reactions. For example, a goal could be to determine if CO dissociation occurs prior to oxidation over Pt catalysts. A second objective is then to extrapolate this microscopic view of surface reactions to the corresponding macroscopic phenomena.

How are fundamental aspects of surface reactions studied? The surface science approach uses a simplified system to model the more complicated 'real-world' systems. At the heart of this simplified system is the use of well defined surfaces, typically in the form of oriented single crystals. A thorough description of these surfaces should include composition, electronic structure and geometric structure measurements, as well as an evaluation of reactivity towards different adsorbates. Furthermore, the system should be constructed such that it can be made increasingly more complex to more closely mimic macroscopic systems. However, relating surface science results to the corresponding real-world problems often proves to be a stumbling block because of the sheer complexity of these real-world systems.

Essential to modern surface science techniques is the attainment and maintenance of ultrahigh vacuum (UHV), which corresponds to pressures of the order of 10^{-10} Torr ($\sim 10^{-13}$ atm). At these pressures, the number of collisions between gas phase molecules and a surface are such that a surface can remain relatively contaminant-free for a period of hours. For example, in air at 760 Torr and 298 K the collision frequency is 3×10^{23} collisions cm^{-2} s^{-1}. Assuming a typical surface has 10^{15} atoms cm^{-2}, then each surface atom undergoes $\sim 10^{8}$ collisions per second. Clearly, a surface at 760 Torr has little chance of remaining clean. However, by lowering the pressure to 10^{-10} Torr, the collision frequency decreases to approximately 10^{10} collisions cm^{-2} s^{-1}, corresponding to a collision with a surface atom about every 10^{5} s. Decreasing the pressure is obviously a solution to maintaining a clean sample, which itself is crucial to sustaining well characterized surfaces during the course of an experiment.

Modern UHV chambers are constructed from stainless steel. The principal seals are metal-on-metal, thus the use of greases is avoided. A combination of pumps is normally used, including ion pumps, turbomolecular pumps, cryopumps and mechanical (roughing) pumps. The entire system is generally heatable to $\sim$500 K. This 'bakeout' for a period of 10–20 h increases gas desorption rates from the internal surfaces, ultimately resulting in lower pressure. For further reading on vacuum technology, including vacuum and pump theory, see [2, 3].

The importance of low pressures has already been stressed as a criterion for surface science studies. However, it is also a limitation because real-world phenomena do not occur in a controlled vacuum. Instead, they occur at atmospheric pressures or higher, often at elevated temperatures, and in conditions of humidity or even contamination. Hence, a major thrust in surface science has been to modify existing techniques and equipment to permit detailed surface analysis under conditions that are less than ideal. The scanning tunnelling microscope (STM) is a recent addition to the surface science arsenal and has the capability of providing atomic-scale information at ambient pressures and elevated temperatures. Incredible insight into the nature of surface reactions has been achieved by means of the STM and other *in situ* techniques.

This chapter will explore surface reactions at the atomic level. A brief discussion of corrosion reactions is followed by a more detailed look at growth and etching reactions. Finally, catalytic reactions will be considered, with a strong emphasis on the surface science approach to catalysis.

A3.10.2   Corrosion

A3.10.2.1   Introduction

Corrosion is a frequently encountered phenomenon in which a surface undergoes changes associated with exposure to a reactive environment. While materials such as plastics and cement can undergo corrosion, the term corrosion more commonly applies to metal surfaces. Rust is perhaps the most widely recognized form of corrosion, resulting from the surface oxidation of an iron-containing material such as steel. Economically, corrosion is extremely important. It has been estimated that annual costs associated with combating and preventing corrosion are 2–3% of the gross national product for industrialized countries. Equipment damage is a major component of the costs associated with corrosion. There are also costs related to corrosion prevention, such as implementation of anti-corrosive paints or other protective measures. Finally, there are indirect losses, such as plant shutdowns, when equipment or facilities need repair or replacement.

Most metals tend to corrode in an environment of air and/or water, forming metal oxides or hydrated oxides. Whether or not such a reaction is possible is dictated by the thermodynamics of the corrosion reaction. If the reaction has a negative Gibbs free energy of formation, then the reaction is thermodynamically favoured. While thermodynamics determines whether a particular reaction can occur or not, the *rate* of the corrosion reaction is determined by kinetic factors. A number of variables can affect the corrosion rate, including temperature, pH and passivation, which is the formation of a thin protective film on a metal surface. Passivation can have a tremendous influence on the corrosion rate, often reducing it to a negligible amount.

Since metals have very high conductivities, metal corrosion is usually electrochemical in nature. The term electrochemical is meant to imply the presence of an electrode process, i.e. a reaction in which free electrons participate. For metals, electrochemical corrosion can occur by loss of metal atoms through anodic dissolution, one of the fundamental corrosion reactions. As an example, consider a piece of zinc, hereafter referred to as an electrode, immersed in water. Zinc tends to dissolve in water, setting up a concentration of

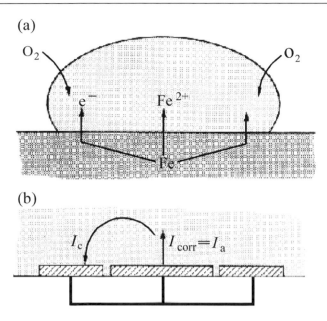

Figure A3.10.1. (a) A schematic illustration of the corrosion process for an oxygen-rich water droplet on an iron surface. (b) The process can be viewed as a short-circuited electrochemical cell [4].

Zn^{2+} ions very near the electrode surface. The term *anodic dissolution* arises because the area of the surface where zinc is *dissolving* to form Zn^{2+} is called the *anode*, as it is the source of positive current in the system. Because zinc is oxidized, a concentration of electrons builds up on the electrode surface, giving it a negative charge. This combination of negatively charged surface region with positively charged near-surface region is called an electrochemical double layer. The potential across the layer, called the electrode potential, can be as much as ± 1 V.

In moist environments, water is present either at the metal interface in the form of a thin film (perhaps due to condensation) or as a bulk phase. Figure A3.10.1 schematically illustrates another example of anodic dissolution where a droplet of slightly acidic water (for instance, due to H_2SO_4) is in contact with an Fe surface in air [4]. Because Fe is a conductor, electrons are available to reduce O_2 at the edges of the droplets. The electrons are then replaced by the oxidation reaction of Fe to Fe^{2+} (forming $FeSO_4$ if H_2SO_4 is the acid), and the rate of corrosion is simply the current induced by metal ions leaving the surface.

Corrosion protection of metals can take many forms, one of which is passivation. As mentioned above, passivation is the formation of a thin protective film (most commonly oxide or hydrated oxide) on a metallic surface. Certain metals that are prone to passivation will form a thin oxide film that displaces the electrode potential of the metal by +0.5–2.0 V. The film severely hinders the diffusion rate of metal ions from the electrode to the solid–gas or solid–liquid interface, thus providing corrosion resistance. This decreased corrosion rate is best illustrated by anodic polarization curves, which are constructed by measuring the net current from an electrode into solution (the corrosion current) under an applied voltage. For passivable metals, the current will increase steadily with increasing voltage in the so-called active region until the passivating film forms, at which point the current will rapidly decrease. This behaviour is characteristic of metals that are susceptible to passivation.

Another method by which metals can be protected from corrosion is called alloying. An alloy is a multi-component solid solution whose physical and chemical properties can be tailored by varying the alloy composition. For example, copper has relatively good corrosion resistance under non-oxidizing conditions.

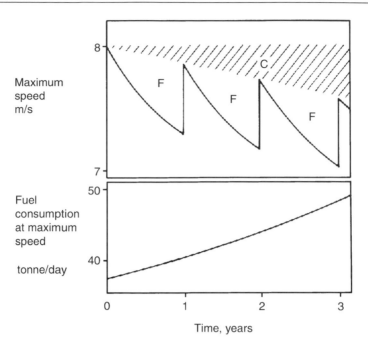

Figure A3.10.2. The influence of corrosion (C) and marine fouling (F) on the performance of a steel ship drydocked annually for cleaning and painting [5].

It can be alloyed with zinc to yield a stronger material (brass), but with lowered corrosion resistance. However, by alloying copper with a passivating metal such as nickel, both mechanical and corrosion properties are improved. Another important alloy is steel, which is an alloy between iron (>50%) and other alloying elements such as carbon.

Although alloying can improve corrosion resistance, brass and steel are not completely resistant to attack and often undergo a form of corrosion known as selective corrosion (also called de-alloying or leaching). De-alloying consists of the segregation of one alloy component to the surface, followed by the removal of this surface component through a corrosion reaction. De-zincification is the selective leaching of zinc from brasses in an aqueous solution. The consequences of leaching are that mechanical and chemical properties change with compositional changes in the alloy.

As an example of the effect that corrosion can have on commercial industries, consider the corrosive effects of salt water on a seagoing vessel. Corrosion can drastically affect a ship's performance and fuel consumption over a period of time. As the hull of a steel boat becomes corroded and fouled by marine growths, the performance of the ship declines because of increased frictional drag. Therefore, ships are drydocked periodically to restore the smoothness of the hull. Figure A3.10.2 shows the loss of speed due to corrosion and marine fouling between annual drydockings for a ship with a steel hull [5]. As corrosion effects progressively deteriorate the hull and as marine growth accumulated, the ship experienced an overall loss of speed even after drydocking and an increased fuel consumption over time. It is clear that there is strong economic motivation to implement corrosion protection.

Surface science studies of corrosion phenomena are excellent examples of *in situ* characterization of surface reactions. In particular, the investigation of corrosion reactions with STM is promising because not only can it be used to study solid–gas interfaces, but also solid–liquid interfaces.

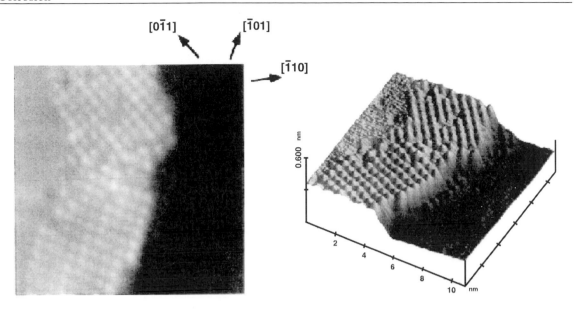

Figure A3.10.3. STM images of the early stages of sulfur segregation on Ni(111). Sulfur atoms are seen to preferentially nucleate at step edges [8].

A3.10.2.2 Surface science of corrosion

(a) The role of sulfur in corrosion

STM has been used to study adsorption on surfaces as it relates to corrosion phenomena [6, 7]. Sulfur is a well known corrosion agent and is often found in air (SO_2, H_2S) and in aqueous solution as dissolved anions (HSO_3^-) or dissolved gas (H_2S). By studying the interaction of sulfur with surfaces, insights can be gained into the fundamental processes governing corrosion phenomena. A Ni(111) sample with 10 ppm sulfur bulk impurity was used to study sulfur adsorption by annealing the crystal to segregate the sulfur to the surface [8]. Figure A3.10.3 shows a STM image of a S-covered Ni(111) surface. It was found that sulfur formed islands preferentially near step edges, and that the Ni surface reconstructed under the influence of sulfur adsorption. This reconstruction results in surface sites that have fourfold symmetry rather than threefold symmetry as on the unreconstructed (111) surface. Furthermore, the fourfold symmetry sites are similar to those found on unreconstructed Ni(100), demonstrating the strong influence that sulfur adsorption has on this surface. The mechanism by which sulfur leads to corrosion of nickel surfaces is clearly linked to the ability of sulfur to weaken Ni–Ni bonds.

(b) Anodic dissolution in alloys

This weakening of Ni–Ni surface bonds by adsorbed sulfur might lead one to expect that the corrosion rate should increase in this case. In fact, an increased anodic dissolution rate was observed for Ni_3Fe (100) in 0.05 M H_2SO_4 [9]. Figure A3.10.4 shows the anodic polarization curves for clean and S-covered single-crystal alloy surfaces. While both surfaces show the expected current increase with potential increase, the sulfur-covered surface clearly has an increased rate of dissolution. In addition, the sulfur coverage (measured using radioactive sulfur, ^{35}S) does not decrease even at the maximum dissolution rate, indicating that adsorbed sulfur is not consumed by the dissolution reaction. Instead, surface sulfur simply enhances the rate of dissolution, as expected based on the observation above that Ni–Ni bonds are significantly weakened by surface sulfur.

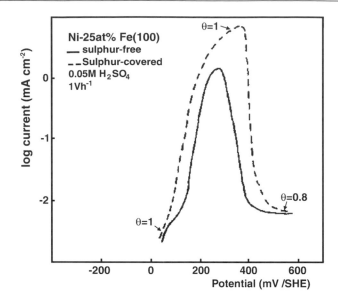

Figure A3.10.4. The effect of sulfur on the anodic polarization curves from a $Ni_{0.25}Fe(100)$ alloy in 0.05 M H_2SO_4. θ is the sulfur (^{35}S) coverage [6].

The nature of copper dissolution from CuAu alloys has also been studied. CuAu alloys have been shown to have a surface Au enrichment that actually forms a protective Au layer on the surface. The anodic polarization curve for CuAu alloys is characterized by a critical potential, E_c, above which extensive Cu dissolution is observed [10]. Below E_c, a smaller dissolution current arises that is approximately potential-independent. This critical potential depends not only on the alloy composition, but also on the solution composition. STM was used to investigate the mechanism by which copper is selectively dissoluted from a $CuAu_3$ electrode in solution [11], both above and below the critical potential. At potentials below E_c, it was found that, as copper dissolutes, vacancies agglomerate on the surface to form voids one atom deep. These voids grow two-dimensionally with increasing Cu dissolution while the second atomic layer remains undisturbed. The fact that the second atomic layer is unchanged suggests that Au atoms from the first layer are filling in holes left by Cu dissolution. In sharp contrast, for potentials above E_c, massive Cu dissolution results in a rough surface with voids that grow both parallel and perpendicular to the surface, suggesting a very fast dissolution process. These *in situ* STM observations lend insight into the mechanism by which Cu dissolution occurs in $CuAu_3$ alloys.

The characterization of surfaces undergoing corrosion phenomena at liquid–solid and gas–solid interfaces remains a challenging task. The use of STM for *in situ* studies of corrosion reactions will continue to shape the atomic-level understanding of such surface reactions.

A3.10.3   Growth

A3.10.3.1   Introduction

Thin crystalline films, or overlayers, deposited onto crystalline substrates can grow in such a way that the substrate lattice influences the overlayer lattice. This phenomenon is known as *epitaxy*; if the deposited material is different from (the same as) the substrate, the process is referred to as heteroepitaxy (homoepitaxy). Epitaxial growth is of interest for several reasons. First, it is used prevalently in the semiconductor industry for the manufacture of III/V and II/VI semiconductor devices. Second, novel phases have been grown

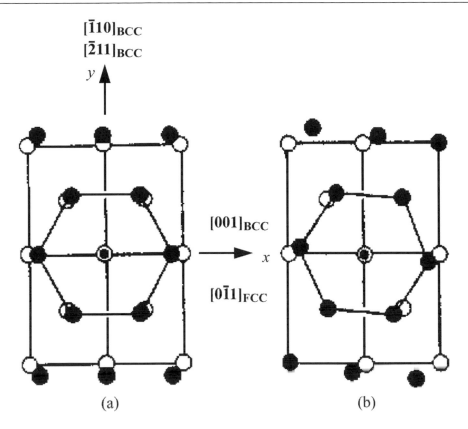

Figure A3.10.5. An fcc(111) monolayer (full circles) overlaid onto a bcc(110) substrate (open circles). (a) fcc[011] parallel to bcc[001]. (b) 5.26° rotation relative to (a). The lattice constants were chosen to produce row-matching in (b) [12].

epitaxially by exploiting such phenomena as lattice mismatch and strain. These new phases have physical and chemical properties of interest to science and engineering. Finally, fundamental catalytic studies often focus on modelling oxide-supported metal particles by depositing metal films on oxide single crystals and thin films and, in many cases, these oxide and metal films grow epitaxially.

When considering whether growth will occur epitaxially or not, arguments can be made based on geometrical considerations, or row matching. This concept is based on the idea that the overlayer must sit on minima of the substrate corrugation potential to minimize the interaction energy. For example, consider the illustration of epitaxial growth in figure A3.10.5, where an fcc(111) monolayer has been overlaid on a bcc(110) surface [12]. Figure A3.10.5(a) shows that the overlayer must be expanded or contracted in two directions to obtain row matching. Figure A3.10.5(b) shows, however, that rotation of the overlayer by 5.26° results in row matching along the most close-packed row of the lattices. Epitaxial growth clearly provides a pathway to energetically favourable atomic arrangements.

The influence of the substrate lattice makes it energetically favourable for two materials to align lattices. On the other hand, if two lattices are misaligned or mismatched in some other way, then lattice strain may result. This lattice strain can lead to a metastable atomic arrangement of the deposited material. In other words, an overlayer can respond to lattice strain by adopting a crystal structure that differs from its normal bulk structure in order to row-match the substrate lattice. This phenomenon is known as pseudomorphy.

For example, Cu (fcc) deposited on a Pd(100) surface will grow epitaxially to yield a pseudomorphic fcc overlayer [13]. However, upon increasing the copper film thickness, a body-centred tetragonal (bct) metastable phase, one not normally encountered for bulk copper, was observed. This phase transformation is due to a high degree of strain in the fcc overlayer.

Another example of epitaxy is tin growth on the (100) surfaces of InSb or CdTe ($a = 6.49$ Å) [14]. At room temperature, elemental tin is metallic and adopts a bct crystal structure ('white tin') with a lattice constant of 5.83 Å. However, upon deposition on either of the two above-mentioned surfaces, tin is transformed into the diamond structure ('grey tin') with $a = 6.49$ Å and essentially no misfit at the interface. Furthermore, since grey tin is a semiconductor, then a novel heterojunction material can be fabricated. It is evident that epitaxial growth can be exploited to synthesize materials with novel physical and chemical properties.

A3.10.3.2 Film growth techniques

There are several design parameters which distinguish film growth techniques from one another, namely generation of the source atom/molecule, delivery to the surface and the surface condition. The source molecule can be generated in a number of ways including vapour produced thermally from solid and liquid sources, decomposition of organometallic compounds and precipitation from the liquid phase. Depending on the pressures used, gas phase atoms and molecules impinging on the surface may be in viscous flow or molecular flow. This parameter is important to determining whether atom–atom (molecule–molecule) collisions, which occur in large numbers at pressures higher than UHV, can affect the integrity of the atom (molecule) to be deposited. The condition of the substrate surface may also be a concern: elevating the surface temperature may alter the growth kinetics, or the surface may have to be nearly free of defects and/or contamination to promote the proper growth mode. Two film growth techniques, molecular beam epitaxy (MBE) and vapour phase epitaxy (VPE) will be briefly summarized below. These particular techniques were chosen because of their relevance to UHV studies. The reader is referred elsewhere for more detailed discussions of the various growth techniques [15–17].

MBE is accomplished under UHV conditions with pressures of the order of $\sim 10^{-10}$ Torr. By using such low pressures, the substrate surface and deposited thin films can be kept nearly free of contamination. In MBE, the material being deposited is usually generated in UHV by heating the source material to the point of evaporation or sublimation. The gas phase species is then focused in a molecular beam onto the substrate surface, which itself may be at an elevated temperature. The species flux emanating from the source can be controlled by varying the source temperature and the species flux arriving at the surface can be controlled by the use of mechanical shutters. Precise control of the arrival of species at the surface is a very important characteristic of MBE because it allows the growth of epitaxial films with very abrupt interfaces. Several sources can be incorporated into a single vacuum chamber, allowing doped semiconductors, compounds or alloys to be grown. For instance, MBE is used prevalently in the semiconductor industry to grow GaAs/Al$_x$Ga$_{1-x}$As layers and, in such a situation, a growth chamber would be outfitted with Ga, As and Al deposition sources. Because of the compatibility of MBE with UHV surface science techniques, it is often the choice of researchers studying fundamentals of thin-film growth.

A second technique, VPE, is also used for surface science studies of overlayer growth. In VPE, the species being deposited can be generated in several ways, including vaporization of a liquid precursor into a flowing gas stream or sublimation of a solid precursor. VPE generates an unfocused vapour or cloud of the deposited material, rather than a collimated beam as in MBE. Historically, VPE played a major role in the development of III/V semiconductors. Currently, VPE is used as a tool for studying metal growth on oxides, an issue of importance to the catalysis community.

The following two sections will focus on epitaxial growth from a surface science perspective with the aim of revealing the fundamentals of thin-film growth. As will be discussed below, surface science studies of thin-film deposition have contributed greatly to an atomic-level understanding of nucleation and growth.

A3.10.3.3   Thermodynamics

The number of factors affecting thin-film growth is largely dependent upon the choice of growth technique. The overall growth mechanism may be strongly influenced by three factors: mass transport, thermodynamics and kinetics. For instance, for an exothermic (endothermic) process, increasing (decreasing) the surface temperature will decrease (increase) the growth rate for a thermodynamically limited process. On the other hand, if temperature has no effect on the growth rate, then the process may be limited by mass transport, which has very little dependence on the substrate temperature. Another test of mass transport limitations is to increase the total flow rate to the surface while keeping the partial pressures constant—if the growth rate is influenced, then mass transport limitations should be considered. Alternatively, if the substrate orientation is found to influence the growth rates, then the process is very likely kinetically limited. Thus, through a relatively straightforward analysis of the parameters affecting macroscopic quantities, such as growth rate, a qualitative description of the growth mechanism can be obtained.

The growth of epitaxial thin films by vapour deposition in UHV is a non-equilibrium kinetic phenomenon. At thermodynamic equilibrium, atomic processes are required to proceed in opposite directions at equal rates. Hence, a system at equilibrium must have equal adsorption and desorption rates, as well as equal cluster growth and cluster decay rates. If growth were occurring under equilibrium conditions, then there would be no net change in the amount of deposited material on the surface. Typical growth conditions result in systems far from equilibrium, so film growth is usually limited by kinetics considerations. Thermodynamics does play an important role, however, as will be discussed next.

Thermodynamics can lend insight into the expected growth mode by examination of energetics considerations. The energies of importance are the surface free energy of the overlayer, the interfacial energy between the substrate and the overlayer, and the surface free energy of the substrate. Generally, if the free energy of the overlayer plus the interface energy is greater than the free energy of the substrate, then Frank–van der Merwe (FM) growth will occur [18]. FM growth, also known as layer-by-layer growth, is characterized by the completion of a surface overlayer before the second layer begins forming. However, if the free energy of the overlayer plus the interface energy is less than the free energy of the substrate then the growth mode is Volmer–Weber (VW) [18]. VW, or three-dimensional (3D), growth yields 3D islands or clusters that coexist with bare patches of substrate. There is also a third growth mode, called Stranski–Krastanov (SK), which can be described as one or two monolayers of growth across the entire surface subsequently followed by the growth of 3D islands [18]. In SK growth, the sum of the surface free energy of the overlayer plus interface energy is initially greater than that of the substrate, resulting in the completion of the first monolayer, after which the surface free energy of the overlayer plus interface energy becomes greater than that of the substrate, resulting in 3D growth. It should be stressed that the energetic arguments for these growth modes are only valid for equilibrium processes. However, these descriptions provide good models for the growth modes experimentally observed even under non-equilibrium conditions.

A3.10.3.4   Nucleation and growth

The process of thin-film growth from an atomic point of view consists of the following stages: adsorption, diffusion, nucleation, growth and coarsening. Adsorption is initiated by exposing the substrate surface to the deposition source. As described above, this is a non-equilibrium process, and the system attempts to restore equilibrium by forming aggregates. The adatoms randomly walk during the diffusion process until two or more collide and subsequently nucleate to form a small cluster. A rate-limiting step is the formation of some critical cluster size, at which point cluster growth becomes more probable than cluster decay. The clusters increase in size during the growth stage, with the further addition of adatoms leading to island formation. Growth proceeds at this stage according to whichever growth mode is favoured. Once deposition has ceased, further island morphological changes occur during the coarsening stage, whereby atoms in small islands evaporate and add to other islands or adsorb onto available high-energy adsorption sites such as step edge sites. For an excellent review on the atomic view of epitaxial metal growth, see [19].

Experimentally, the variable-temperature STM has enabled great strides to be made towards understanding nucleation and growth kinetics on surfaces. The evolution of overlayer growth can be followed using STM from the first stages of adatom nucleation through the final stages of island formation. The variable-temperature STM has also been crucial to obtaining surface diffusion rates. In such cases, however, the importance of tip–sample interactions must be considered. Typically, low tunnelling currents are best because under these conditions the tip is further from the surface, thereby reducing the risk of tip–sample interactions.

Much effort in recent years has been aimed at modelling nucleation at surfaces and several excellent reviews exist [20–22]. Mean-field nucleation theory is one of these models and has a simple picture at its core. In the nucleation stage, an atom arriving at the surface from the gas phase adsorbs and then diffuses at a particular rate until it collides with another surface adatom to form a dimer. If the dimers are assumed to be stable (so that no decay occurs) and immobile (so that no diffusion occurs) then, as deposition proceeds, the concentration of dimers will increase approximately linearly until it is roughly equal to the concentration of monomers. At this point, the probability of an atom colliding with a dimer is comparable to the probability of an adatom colliding with another adatom, hence growth and nucleation compete. Once the island density has saturated, i.e. no more clusters are being formed, then the adatom mean free path is equal to the mean island separation and further deposition results in island growth. At coverages near 0.5 monolayers (ML), islands begin to coalesce and the island density decreases.

This simple and idealistic picture of nucleation and growth from mean field nucleation theory was found to be highly descriptive of the Ag/Pt(111) system at 75 K (figure A3.10.6) [23]. Figure A3.10.6 shows a series of STM images of increasing Ag coverage on Pt(111) and demonstrates the transition from nucleation to growth. At very low coverages ((a) and (b)), the average cluster size is 2.4 and 2.6 atoms, respectively, indicating that dimers and trimers are the predominant surface species. However, when the coverage was more than doubled from (a) to (b), the mean island size remained relatively constant. This result clearly indicates that deposition at these low coverages is occurring in the nucleation regime. By increasing the coverage to 0.03 ML, the Ag mean island size doubled to 6.4 atoms and the island density increased, indicating that nucleation and growth were competing. Finally, after increasing the coverage even further (d), the mean island size doubled again, while the island density saturated, suggesting that a pure growth regime dominated, with little or no nucleation occurring.

Growth reactions at surfaces will certainly continue to be the focus of much research. In particular, the synthesis of novel materials is an exciting field that holds much promise for the nanoscale engineering of materials. Undoubtedly, the advent of STM as a means of investigating growth reactions on the atomic scale will influence the future of nanoscale technology.

A3.10.4 Etching

A3.10.4.1 Introduction

Etching is a process by which material is removed from a surface. The general idea behind etching is that by interaction of an etch atom or molecule with a surface, a surface species can be formed that is easily removed. The use of a liquid to etch a surface is known as *wet etching*, while the use of a gas to etch a surface is known as *dry etching*. Wet etching has been employed since the late Middle Ages. The process then was rather simple and could be typified as follows. The metal to be etched was first coated with a wax, or in modern vernacular, a mask. Next, a pattern was cut into the wax to reveal the metal surface beneath. Then, an acid was used to etch the exposed metal, resulting in a patterned surface. Finally, the mask was removed to reveal

a) $\Theta = 0.0024$ ML   b) $\Theta = 0.006$ ML

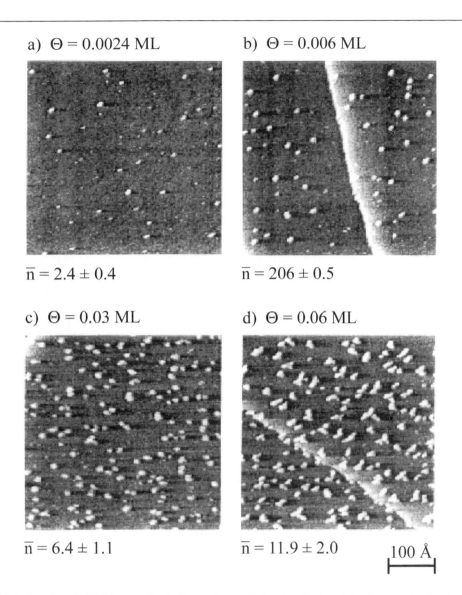

$\bar{n} = 2.4 \pm 0.4$ $\bar{n} = 206 \pm 0.5$

c) $\Theta = 0.03$ ML   d) $\Theta = 0.06$ ML

$\bar{n} = 6.4 \pm 1.1$ $\bar{n} = 11.9 \pm 2.0$ $\underset{\longmapsto}{100 \text{ Å}}$

Figure A3.10.6. A series of STM images for Ag/Pt(111) at 75 K showing the transition from nucleation to growth [23]. Coverages (Θ) and mean island sizes ($\bar{n}$) are indicated.

the finished product. Modern methods are considerably more technologically advanced, although the general principles behind etching remain unchanged.

Both wet and dry etching are used extensively in the semiconductor processing industry. However, wet etching has limitations that prevent it being used to generate micron or submicron pattern sizes for GaAs etching. The most serious of these limitations is called substrate undercutting, which is a phenomenon where etch rates parallel and perpendicular to the surface are approximately equal (isotropic etching). Substrate undercutting is much less prevalent for silicon surfaces than GaAs surfaces, thus wet etching is more commonly used to etch silicon surfaces. Generally, when patterning surfaces, anisotropic etching is preferred, where etch rates perpendicular to the surface exceed etch rates parallel to the surface. Hence, in cases of undercutting, an ill defined pattern typically results. In the early 1970s, dry etching (with CF_4/O_2, for example) became widely used for patterning. Dry methods have a distinct advantage over wet methods, namely anisotropic etching.

A form of anisotropic etching that is of some importance is that of orientation-dependent etching, where one particular crystal face is etched at a faster rate than another crystal face. A commonly used orientation-dependent wet etch for silicon surfaces is a mixture of KOH in water and isopropanol. At approximately 350 K, this etchant has an etch rate of 0.6 μm min^{-1} for the Si(100) plane, 0.1 μm min^{-1} for the Si(110) plane and 0.006 μm min^{-1} for the Si(111) plane [24]. These different etch rates can be exploited to yield anisotropically etched surfaces.

Semiconductor processing consists of a number of complex steps, of which etching is an integral step. Figure A3.10.7 shows an example of the use of etching [25] in which the goal of this particular process is to remove certain parts of a film, while leaving the rest in a surface pattern to serve as, for example, interconnection paths. This figure illustrates schematically how etching paired with a technique called photolithography can be used to manufacture a semiconductor device. In this example, the substrate enters the manufacturing stream covered with a film (for example, a SiO_2 film on a Si wafer). A liquid thin-film called a photoresist (denoted 'positive resist' or 'negative resist', as explained below) is first placed on the wafer, which is then spun at several thousand rotations per minute to spread out the film and achieve a uniform coating. Next, the wafer is exposed through a mask plate to an ultraviolet (UV) light source. The UV photons soften certain resists (positive resists) and harden others (negative resists). Next, a developer solution is used to remove the susceptible area, leaving behind the remainder according to the mask pattern. Then, the wafer is etched to remove all of the surface film not protected by the photoresist. Finally, the remaining photoresist is removed, revealing a surface with a patterned film. Thus the role of etching in semiconductor processing is vital and it is evident that motivation exists to explore etching reactions on a fundamental level.

A3.10.4.2   Dry etching techniques

It has already been mentioned that dry etching involves the interaction of gas phase molecules/atoms with a surface. More specifically, dry etching utilizes either plasmas that generate reactive species, or energetic ion beams to etch surfaces. Dry etching is particularly important to GaAs processing because, unlike silicon, there are no wet etching methods that result in negligible undercutting. Dry etching techniques can be characterized by either chemical or physical etching mechanisms. The chemical mechanisms tend to be more selective, i.e. more anisotropic, and tend to depend strongly on the specific material being etched. Several dry etch techniques will be briefly discussed below. For a more comprehensive description of these and other techniques, the reader is referred to the texts by Williams [26] or Sugawara [27].

Ion milling is a dry etch technique that uses a physical etching mechanism. In ion milling, ions of an inert gas are generated and then accelerated to impinge on a surface. The etching mechanism is simply the bombardment of these energetic ions on the surface, resulting in erosion. The energy of the ions can be controlled by varying the accelerating voltage, and it may be possible to change the selectivity by varying the angle of incidence.

Plasma etching is a term used to describe any dry etching process that utilizes reactive species generated from a gas plasma. For semiconductor processing, a low-pressure plasma, also called a glow discharge, is used. The glow discharge is characterized by pressures in the range 0.1–5 Torr and electron energies of 1–10 eV. The simplest type of plasma reactor consists of two parallel plates in a vacuum chamber filled with a gas at low pressure. A radio frequency (RF) voltage is applied between the two plates, generating plasma that emits a characteristic glow. Reactive radicals are produced by the plasma, resulting in a collection of gas phase species that are the products of collisions between photons, electrons, ions and atoms or molecules. These chemically reactive species can then collide with a nearby surface and react to form a volatile surface species, thereby etching the surface.

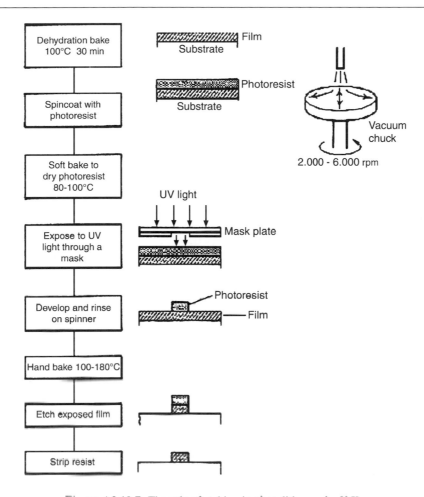

Figure A3.10.7. The role of etching in photolithography [25].

Reactive ion etching (RIE) is distinguished from plasma etching by the fact that the surface reactions are enhanced by the kinetic energy of the incoming reactive species. This type of chemical mechanism is referred to as a kinetically assisted chemical reaction, and very often results in highly anisotropic etching. RIE is typically performed at low pressures (0.01–0.1 Torr) and is used industrially to etch holes in GaAs.

Dry etching is a commonly used technique for creating highly anisotropic, patterned surfaces. The interaction of gas phase etchants with surfaces is of fundamental interest to understanding such phenomena as undercutting and the dependence of etch rate on surface structure. Many surface science studies aim to understand these interactions at an atomic level, and the next section will explore what is known about the etching of silicon surfaces.

A3.10.4.3 Atomic view of etching

On the atomic level, etching is composed of several steps: diffusion of the etch molecules to the surface, adsorption to the surface, subsequent reaction with the surface and, finally, removal of the reaction products. The third step, that of reaction between the etchant and the surface, is of considerable interest to the understanding of surface reactions on an atomic scale. In recent years, STM has given considerable insight into the

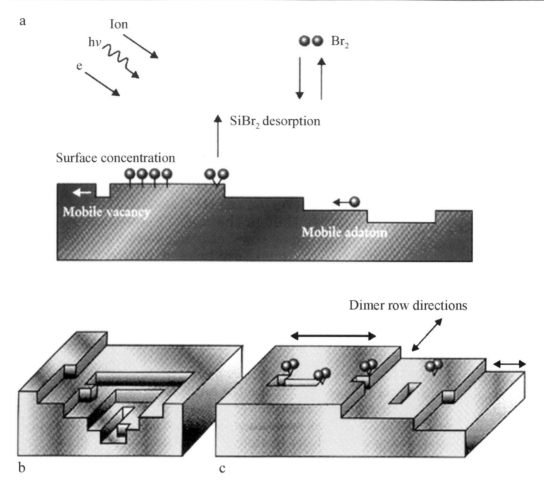

Figure A3.10.8. Depiction of etching on a Si(100) surface. (a) A surface exposed to Br$_2$ as well as electrons, ions and photons. Following etching, the surface either becomes highly anisotropic with deep etch pits (b), or more regular (c), depending on the relative desorption energies for different surface sites [28].

nature of etching reactions at surfaces. The following discussion will focus on the etching of silicon surfaces [28].

Figure A3.10.8 schematically depicts a Si(100) surface (a) being etched to yield a rough surface (b) and a more regular surface (c). The surfaces shown here are seen to consist of steps, terraces and kinks, and clearly have a three-dimensional character, rather than the two-dimensional character of an ideally flat, smooth surface. The general etching mechanism is based on the use of halogen molecules, the principal etchants used in dry etching. Upon adsorption on silicon at room temperature, Br$_2$ dissociates to form bromine atoms, which react with surface silicon atoms. Then, if an external source of energy is provided, for example by heating Si(100) to 900 K, SiBr$_2$ forms and desorbs, revealing the silicon atom(s) beneath and completing the etching process. Depending upon the relative desorption energies from various surface sites, the surface could be etched quite differently, as seen in figures A3.10.8(b) and (c).

Semiconductors such as silicon often undergo rearrangements, or reconstructions, at surface boundaries to lower their surface free energy. One way of lowering the surface free energy is the reduction of dangling

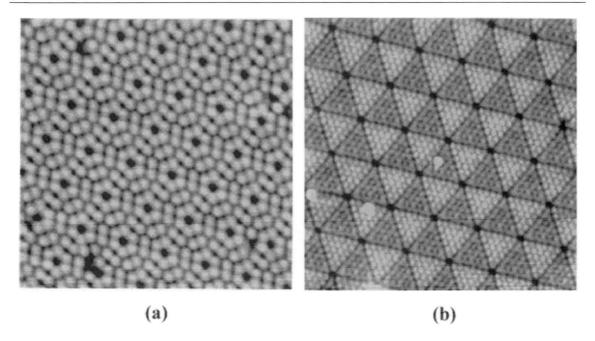

(a)   **(b)**

Figure A3.10.9. STM images of Si(111) surfaces before (a) and after (b) etching by bromine at 675 K. In (a) the (7×7) reconstructed surface is seen. In (b), the rest layer consisting of triangular arrays of Si atoms has been exposed by etching [28]. Both images show a 17×17 nm^2 area.

bonds, which are non-bonding orbitals that extend (dangle) into the vacuum. Si(111) undergoes a complex (7×7) reconstruction that was ultimately solved using STM. Figure A3.10.9(a) shows an STM image of the reconstructed Si(111) surface [29]. This reconstruction reduces the number of dangling bonds from 49 to 19 per unit cell.

The (7×7) reconstruction also affects the second atomic layer, called the rest layer. The rest layer is composed of silicon atoms arranged in triangular arrays that are separated from one another by rows of silicon dimers. Figure A3.10.9(b) shows the exposed rest layer following bromine etching at 675 K [29]. It is noteworthy that the rest layer does not reconstruct to form a new (7×7) surface. The stability of the rest layer following etching of (7×7)-Si(111) is due to the unique role of the halogen. The silicon adlayer is removed by insertion of bromine atoms into Si–Si dimer bonds. Once this silicon adlayer is gone, the halogen stabilizes the silicon rest layer by reacting with the dangling bonds, effectively inhibiting surface reconstruction to a (7×7) phase. Unfortunately, the exposure of the rest layer makes etching more difficult because to form SiBr$_2$, bromine atoms must insert into stronger Si–Si bonds.

Si(100) reconstructs as well, yielding a (1×2) surface phase that is formed when adjacent silicon atoms bond through their respective dangling bonds to form a more stable silicon dimer. This reconstructed bonding results in a buckling of the surface atoms. Furthermore, because Si–Si dimer bonds are weaker than bulk silicon bonds, the reconstruction actually facilitates etching. For a comprehensive discussion on STM studies of reconstructed silicon surfaces, see [30].

Si(100) is also etched by Br$_2$, although in a more dramatic fashion. Figure A3.10.10 shows a STM image of a Si(100) surface after etching at 800 K [28]. In this figure, the dark areas are etch pits one atomic layer deep. The bright rows running perpendicular to these pits are silicon dimer chains, which are composed of silicon atoms that were released from terraces and step edges during etching. The mechanism by which Si(100) is etched has been deduced from STM studies. After Br$_2$ dissociatively adsorbs to the surface, a bromine

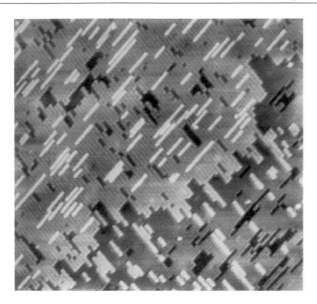

Figure A3.10.10. STM image (55×55 nm^2) of a Si(100) surface exposed to molecular bromine at 800 K. The dark areas are etch pits on the terraces, while the bright rows that run perpendicular to the terraces are Si dimer chains. The dimer chains consist of Si atoms released from terraces and step edges during etching [28].

atom bonds to each silicon atom in the dimer pairs. SiBr$_2$ is the known desorption product and so the logical next step is the formation of a surface SiBr$_2$ species. This step can occur by the breaking of the Si–Si dimer bond and the transfer of a bromine atom from one of the dimer atoms to the other. Then, if enough energy is available to overcome the desorption barrier, SiBr$_2$ will desorb, leaving behind a highly uncoordinated silicon atom that will migrate to a terrace and eventually re-dimerize. On the other hand, if there is not enough energy to desorb SiBr$_2$, then the Br atom would transfer back to the original silicon atom, and a silicon dimer bond would again be formed. In this scenario, SiBr$_2$ desorption is essential to the etching process.

Another view of the Si(100) etching mechanism has been proposed recently [28]. Calculations have revealed that the most important step may actually be the escape of the bystander silicon atom, rather than SiBr$_2$ desorption. In this way, the SiBr$_2$ becomes trapped in a state that otherwise has a very short lifetime, permitting many more desorption attempts. Preliminary results suggest that indeed this vacancy-assisted desorption is the key step to etching Si(100) with Br$_2$.

The implementation of tools such as the STM will undoubtedly continue to provide unprecedented views of etching reactions and will deepen our understanding of the phenomena that govern these processes.

A3.10.5 Catalytic reactions

A3.10.5.1 Introduction

A catalyst is a material that accelerates a reaction rate towards thermodynamic equilibrium conversion without itself being consumed in the reaction. Reactions occur on catalysts at particular sites, called 'active sites', which may have different electronic and geometric structures than neighbouring sites. Catalytic reactions are at the heart of many chemical industries, and account for a large fraction of worldwide chemical production. Research into fundamental aspects of catalytic reactions has a strong economic motivating factor: a better understanding of the catalytic process may lead to the development of a more efficient catalyst. While the implementation of a new catalyst based on surface science studies has not yet been realized, the investigation of catalysis using surface science methods has certainly shaped the current understanding of catalytic reactions. Several recommended texts on catalysis can be found in [31–33].

Fundamental studies in catalysis often incorporate surface science techniques to study catalytic reactions at the atomic level. The goal of such experiments is to characterize a catalytic surface before, during and after a chemical reaction; this is no small task. The characterization of these surfaces is accomplished using a number of modern analytical techniques. For example, surface compositions can be determined using x-ray photoelectron spectroscopy (XPS) or Auger electron spectroscopy (AES). Surface structures can be probed using low-energy electron diffraction (LEED) or STM. In addition, a number of techniques are available for detecting and identifying adsorbed species on surfaces, such as infrared reflection absorption spectroscopy, high-resolution electron energy-loss spectroscopy (HREELS) and sum frequency generation (SFG).

As with the other surface reactions discussed above, the steps in a catalytic reaction (neglecting diffusion) are as follows: the adsorption of reactant molecules or atoms to form bound surface species, the reaction of these surface species with gas phase species or other surface species and subsequent product desorption. The global reaction rate is governed by the slowest of these elementary steps, called the rate-determining or rate-limiting step. In many cases, it has been found that either the adsorption or desorption steps are rate determining. It is not surprising, then, that the surface structure of the catalyst, which is a variable that can influence adsorption and desorption rates, can sometimes affect the overall conversion and selectivity.

Industrial catalysts usually consist of one or more metals supported on a metal oxide. The supported metal can be viewed as discrete single crystals on the support surface. Changes in the catalyst structure can be achieved by varying the amount, or 'loading', of the metal. An increased loading should result in a particle size increase, and so the relative population of a particular crystal face with respect to other crystal faces may change. If a reaction rate on a per active site basis changes as the metal loading changes, then the reaction is deemed to be structure sensitive. The surface science approach to studying structure-sensitive reactions has been to examine the chemistry that occurs over different crystal orientations. In general, these studies have shown that close-packed, atomically smooth metal surfaces such as (111) and (100) fcc and (110) bcc surfaces are less reactive than more open, rough surfaces such as fcc(110) and bcc(111). The remaining task is then to relate the structure sensitivity results from single-crystal studies to the activity results over real-world catalysts.

Surface science studies of catalytic reactions certainly have shed light on the atomic-level view of catalysis. Despite this success, however, two past criticisms of the surface science approach to catalysis are that the pressure regimes (usually 10^{-10} Torr) and the materials (usually low-surface-area single crystals) are far removed from the high pressures and high-surface-area supported catalysts used industrially. These criticisms have been termed the 'pressure gap' and the 'materials gap'. To combat this criticism, much research in the last 30 years has focused on bridging these gaps, and many advances have been made that now suggest these criticisms are no longer warranted.

A3.10.5.2 *Experimental*

(a) Bridging the pressure gap

The implementation of high-pressure reaction cells in conjunction with UHV surface science techniques allowed the first true *in situ* postmortem studies of a heterogeneous catalytic reaction. These cells permit exposure of a sample to ambient pressures without any significant contamination of the UHV environment. The first such cell was internal to the main vacuum chamber and consisted of a metal bellows attached to a reactor cup [34]. The cup could be translated using a hydraulic piston to envelop the sample, sealing it from the surrounding UHV by means of a copper gasket. Once isolated from the vacuum, the activity of the enclosed sample for a given reaction could be measured at elevated pressures. Following the reaction, the high-pressure cell was evacuated and then retracted, exposing the sample again to the UHV environment, at which point any number of surface science techniques could be used to study the 'spent' catalyst surface.

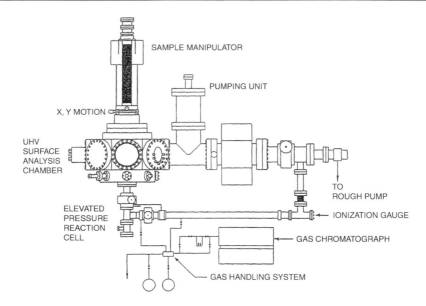

Figure A3.10.11. Side view of a combined high-pressure cell and UHV surface analysis system [37].

Shortly thereafter, another high-pressure cell design appeared [35]. This design consisted of a sample mounted on a retractable bellows, permitting the translation of the sample to various positions. The sample could be retracted to a high-pressure cell attached to the primary chamber and isolated by a valve, thereby maintaining UHV in the primary chamber when the cell was pressurized for catalytic studies. The reactor could be evacuated following high-pressure exposures before transferring the sample back to the main chamber for analysis.

A modification to this design appeared several years later (figure A3.10.11) [36, 37]. In this arrangement, the sample rod can be moved easily between the UHV chamber and the high-pressure cell without any significant increase in chamber pressure. Isolation of the reaction cell from UHV is achieved by a differentially pumped sliding seal mechanism (figure A3.10.12) whereby the sample rod is pushed through the seals until it is located in the high-pressure cell. Three spring-loaded, differentially pumped Teflon seals are used to isolate the reaction chamber from the main chamber by forming a seal around the sample rod. Differential pumping is accomplished by evacuating the space between the first and second seals (on the low-pressure side) by a turbomolecular pump and the space between the second and third seals (on the high-pressure side) by a mechanical (roughing) pump. Pressures up to several atmospheres can be maintained in the high-pressure cell while not significantly raising the pressure in the attached main chamber.

The common thread to these designs is that a sample can be exposed to reaction conditions and then studied using surface science methods without exposure to the ambient. The drawback to both of these designs is that the samples are still being analysed under UHV conditions *before* and *after* the reaction under study. The need for *in situ* techniques is clear.

Two notable *in situ* techniques are at the forefront of the surface science of catalysis: STM and SFG. STM is used to investigate surface structures while SFG is used to investigate surface reaction intermediates. The significance of both techniques is that they can operate over a pressure range of 13 orders of magnitude, from 10^{-10} to 10^3 Torr, i.e. they are truly *in situ* techniques. STM has allowed the visualization of surface structures under ambient conditions and has shed light on adsorbate-induced morphological changes that occur at surfaces, for both single-crystal metals and metal clusters supported on oxide single crystals. Studies of surface reactions with SFG have given insight into reaction mechanisms previously investigated under non-ideal

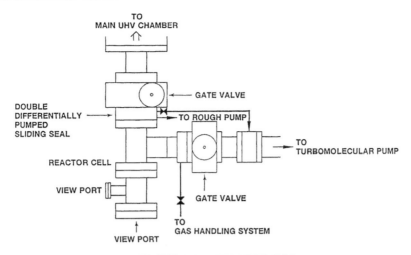

Figure A3.10.12. Side view of the high-pressure cell showing the connections to the UHV chamber, the turbomolecular pump and the gas handling system. The differentially pumped sliding seal is located between the high-pressure cell and the UHV chamber [37].

pressure or temperature constraints. Both SFG and STM hold promise as techniques that will contribute greatly to the understanding of catalytic reactions under *in situ* conditions.

(b) Bridging the materials gap

Single crystals are traditionally used in UHV studies because they provide an opportunity to well characterize a surface. However, as discussed above, single crystals are quite different from industrial catalysts. Typically, such catalysts consist of supported particles that can have multiple crystal orientations exposed at the surface. Therefore, an obstacle in attempting surface science studies of catalysis is the preparation of a surface in such a way that it mimics a real-world catalyst.

One criterion necessary for using charged-particle spectroscopies such as AES and EELS is that the material being investigated should be conductive. This requisite prevents problems such as charging when using electron spectroscopies and ensures homogeneous heating during thermal desorption studies. A problem then with investigating oxide surfaces for use as metal supports is that many are insulators or semiconductors. For example, alumina and silica are often used as oxide supports for industrial catalysts, yet both are insulators at room temperature, severely hindering surface science studies of these materials. However, thin-films of these and other oxides can be deposited onto metal substrates, thus providing a conductive substrate (*via* tunnelling) for use with electron spectroscopies and other surface science techniques.

Thin oxide films may be prepared by substrate oxidation or by vapour deposition onto a suitable substrate. An example of the former method is the preparation of silicon oxide thin-films by oxidation of a silicon wafer. In general, however, the thickness and stoichiometry of a film prepared by this method are difficult to control. On the other hand, vapour deposition, which consists of evaporating the parent metal in an oxidizing environment, allows precise control of the film thickness. The extent of oxidation can be controlled by varying the O_2 pressure (lower O_2 pressures can lead to lower oxides) and the film thickness can be controlled by monitoring the deposition rate. A number of these thin metal oxide films have been prepared by vapour deposition, including SiO_2, Al_2O_3, MgO, TiO_2 and NiO [38].

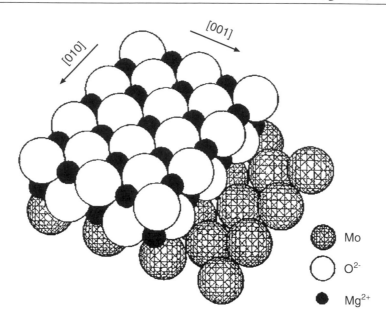

Figure A3.10.13. Ball model illustration of an epitaxial MgO overlayer on Mo(100) [38].

MgO films have been grown on a Mo(100) substrate by depositing Mg onto a clean Mo(100) sample in O_2 ambient at 300 K [39, 40]. LEED results indicated that MgO grows epitaxially at an optimum O_2 pressure of 10^{-7} Torr, with the (100) face of MgO parallel to the Mo(100) surface. Figure A3.10.13 shows a ball model illustration of the MgO(100) overlayer on Mo(100). The chemical states of Mg and O were also probed as a function of the O_2 pressure during deposition by AES and XPS. It was found that as the O_2 pressure was increased, the metallic Mg^0 ($L_{2,3}VV$) Auger transition at 44.0 eV decreased while a new transition at 32.0 eV increased. The transition at 32.0 eV was assigned to a Mg^{2+} ($L_{2,3}VV$) transition due to the formation of MgO. When the O_2 pressure reached 10^{-7} Torr, the Mg^{2+} feature dominated the AES spectrum while the Mg^0 feature completely diminished. XPS studies confirmed the LEED and AES results, verifying that MgO was formed at the optimal O_2 pressure. Furthermore, the Mg 2p and O 1s XPS peaks from the MgO film had the same binding energy (BE) and peak shape as the Mg 2p and O 1s peaks from an MgO single crystal. Both AES and XPS indicated that the stoichiometry of the film was MgO. Further annealing in O_2 did not increase the oxygen content of the film, which supports the fact that no evidence of Mg suboxides was found. This MgO film was successfully used to study the nature of surface defects in Li-doped MgO as they relate to the catalytic oxidative coupling of methane.

The deposition of titanium oxide thin-films on Mo(110) represents a case where the stoichiometry of the film is sensitive to the deposition conditions [41]. It was found that both TiO_2 and Ti_2O_3 thin-films could be made, depending on the Ti deposition rate and the O_2 background pressure. Lower deposition rates and higher O_2 pressures favoured the formation of TiO_2. The two compositionally different films could be distinguished in several ways. Different LEED patterns were observed for the different films: TiO_2 exhibited a (1×1) rectangular periodicity, while Ti_2O_3 exhibited a (1×1) hexagonal pattern. XPS Ti 2p data clearly differentiated the two films as well, showing narrow peaks with a Ti $2p_{3/2}$ BE of 459.1 eV for TiO_2 and broad peaks with a Ti $2p_{3/2}$ BE of 458.1 eV for Ti_2O_3. From LEED and HREELS results, it was deduced that the surfaces grown on Mo(110) were $TiO_2(100)$ and $Ti_2O_3(0001)$. Therefore, it is clear that vapour deposition allows control over thickness and extent of oxidation and is certainly a viable method for producing thin oxide films for use as model catalyst supports.

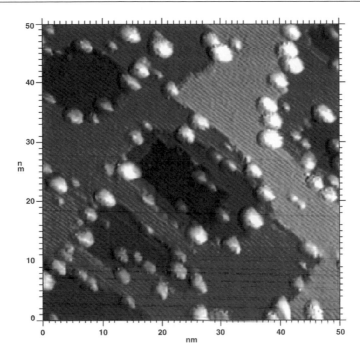

Figure A3.10.14. STM image of 0.25 ML Au vapour-deposited onto TiO$_2$(110). Atomic resolution of the substrate is visible as parallel rows. The Au clusters are seen to nucleate preferentially at step edges.

Metal vapour deposition is a method than can be used to conveniently prepare metal clusters for investigation under UHV conditions. The deposition is accomplished using a doser constructed by wrapping a high-purity wire of the metal to be deposited around a tungsten or tantalum filament that can be resistively heated. After sufficient outgassing, which is the process of heating the doser to remove surface and bulk impurities, then a surface such as an oxide can be exposed to the metal emanating from the doser to yield a model oxide-supported metal catalyst.

Model catalysts such as Au/TiO$_2$(110) have been prepared by metal vapour deposition [42]. Figure A3.10.14 shows a STM image of 0.25 ML (1 ML = 1.387×10^{15} atoms cm^{-2}) Au/TiO$_2$(110). These catalysts were tested for CO oxidation to compare to conventional Au catalysts. It is well known that for conventional Au catalysts there is an optimal Au cluster size (~3 nm) that yields a maximum CO oxidation rate. This result was duplicated by measuring the CO oxidation rate over model Au/TiO$_2$(110), where the cluster sizes were varied by manipulating the deposition amounts. There is a definite maximum in the CO oxidation activity at a cluster size of approximately 3.5 nm. Furthermore, investigation of the cluster electronic properties using scanning tunnelling spectroscopy (STS) revealed a correlation between the cluster electronic structure and the maximum in CO oxidation activity. Pd/SiO$_2$/Mo(100) model catalysts were also prepared and were found to have remarkably similar kinetics for CO oxidation when compared to Pd single crystals and conventional silica-supported Pd catalysts [43]. These results confirm that metal vapour deposition on a suitable substrate is a viable method for producing model surfaces for UHV studies.

Another method by which model-supported catalysts can be made is electron beam lithography [44]. This method entails spin-coating a polymer solution onto a substrate and then using a collimated electron beam to damage the polymer surface according to a given pattern. Next, the damaged polymer is removed, exposing the substrate according to the electron beam pattern, and the sample is coated with a thin metal film. Finally, the polymer is removed from the substrate, taking with it the metal film except where the metal

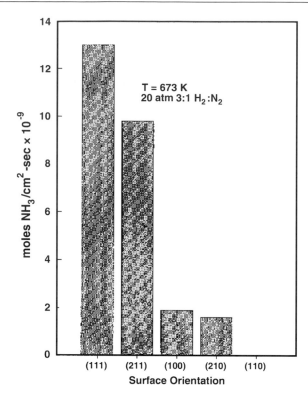

Figure A3.10.15. NH₃ synthesis activity of different Fe single-crystal orientations [32]. Reaction conditions were 20 atm and 600–700 K.

was bound to the substrate, leaving behind metal particles of variable size. This technique has been used to prepare Pt particles with 50 nm diameters and 15 nm heights on an oxidized silicon support [44]. It was found that ethylene hydrogenation reaction rates on the model catalysts agreed well with turnover rates on Pt single crystals and conventional Pt-supported catalysts.

A3.10.5.3   Atomic-level views of catalysis

(a) NH₃ synthesis: N₂+3H₂ ↔ 2NH₃

Ammonia has been produced commercially from its component elements since 1909, when Fritz Haber first demonstrated the viability of this process. Bosch, Mittasch and co-workers discovered an excellent promoted Fe catalyst in 1909 that was composed of iron with aluminium oxide, calcium oxide and potassium oxide as promoters. Surprisingly, modern ammonia synthesis catalysts are nearly identical to that first promoted iron catalyst. The reaction is somewhat exothermic and is favoured at high pressures and low temperatures, although, to keep reaction rates high, moderate temperatures are generally used. Typical industrial reaction conditions for ammonia synthesis are 650–750 K and 150–300 atm. Given the technological importance of the ammonia synthesis reaction, it is not surprising that surface science techniques have been used to thoroughly study this reaction on a molecular level [45, 46].

As mentioned above, a structure-sensitive reaction is one with a reaction rate that depends on the catalyst structure. The synthesis of ammonia from its elemental components over iron surfaces is an example of a structure-sensitive reaction. Figure A3.10.15 demonstrates this structure sensitivity by showing that the rate

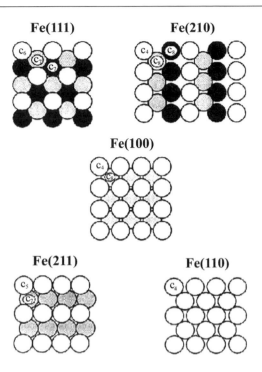

Figure A3.10.16. Illustrations of the surfaces in figure A3.10.15 for which ammonia synthesis activity was tested. The coordination of the surface atoms is noted in the figure [32].

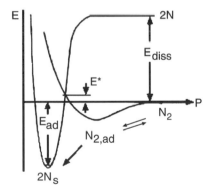

Figure A3.10.17. Potential energy diagram for the dissociative adsorption of N_2 [46].

of NH_3 formation at 20 atm and 600–700 K has a clear dependence on the surface structure [47]. The (111) and (211) Fe faces are much more active than the (100), (210) and (110) faces. Figure A3.10.16 depicts the different Fe surfaces for which ammonia synthesis was studied in figure A3.10.15. The coordination of the different surface atoms is denoted in each drawing. Surface roughness is often associated with higher catalytic activity, however in this case the (111) and (210) surfaces, both of which can be seen to be atomically rough, have distinctly different catalytic activities. Closer inspection of these surfaces reveals that the (111) and (211) faces have a C_7 site in common, i.e. a surface Fe atom with seven nearest neighbours. The high catalytic activity of the (111) and (211) Fe faces has been proposed to be due to the presence of these C_7 sites.

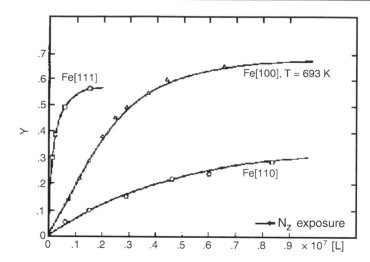

Figure A3.10.18. Surface concentration of nitrogen on different Fe single crystals following N_2 exposure at elevated temperatures in UHV [48].

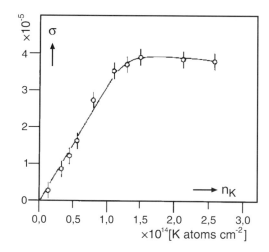

Figure A3.10.19. Variation of the initial sticking coefficient of N_2 with increasing potassium surface concentration on Fe(100) at 430 K [50].

It is widely accepted that the rate-determining step in NH_3 synthesis is the dissociative adsorption of N_2, depicted in a Lennard-Jones potential energy diagram in figure A3.10.17 [46]. This result is clearly illustrated by examining the sticking coefficient (the adsorption rate divided by the collision rate) of N_2 on different Fe crystal faces (figure A3.10.18) [48]. The concentration of surface nitrogen on the Fe single crystals at elevated temperatures in UHV was monitored with AES as a function of N_2 exposure. The sticking coefficient is proportional to the slope of the curves in figure A3.10.18. The initial sticking coefficients increase in the order $(110) < (100) < (111)$, which is the same trend observed for the ammonia synthesis catalytic activity at high-pressure (20 atm). This result indicates that the pressure gap for ammonia synthesis can be overcome: the kinetics results obtained in UHV conditions can be readily extended to the kinetics results obtained under high-pressure reaction conditions.

Further work on modified Fe single crystals explored the role of promoters such as aluminium oxide and potassium [49–51]. It was found that the simple addition of aluminium oxide to Fe single crystal surfaces decreased the ammonia synthesis rate proportionally to the amount of Fe surface covered, indicating no favourable interaction between Fe and aluminium oxide under those conditions. However, by exposing an aluminium-oxide-modified Fe surface to water vapour, the surface was oxidized, inducing a favourable interaction between Fe and the Al_xO_y. This interaction resulted in a 400-fold increase in ammonia synthesis activity for Al_xO_y/Fe(110) as compared to Fe(110) and an activity for Al_xO_y/Fe(110) comparable to that of Fe(111). Interestingly, aluminium-oxide-modified Fe(111) showed no change in activity. The increase in activity for Al_xO_y/Fe(110) to that of Fe(111) suggests a possible reconstruction of the catalyst surface, in particular that Fe(111) and Fe(211) surfaces may be formed. These surfaces have C_7 sites and so the formation of crystals with these orientations could certainly lead to an enhancement in catalytic activity. Thus, the promotion of Fe ammonia synthesis catalysts by Al_xO_y appears to be primarily a geometric effect.

The addition of potassium to Fe single crystals also enhances the activity for ammonia synthesis. Figure A3.10.19 shows the effect of surface potassium concentration on the N_2 sticking coefficient. There is nearly a 300-fold increase in the sticking coefficient as the potassium concentration reaches $\sim 1.5 \times 10^{14}$ K atoms cm^{-2}. Not only does the sticking coefficient increase, but with the addition of potassium as a promoter, N_2 molecules are bound more tightly to the surface, with the adsorption energy increasing from 30 to 45 kJ mol^{-1}. A consequence of the lowering of the N_2 potential well is that the activation energy for dissociation (E^* in figure A3.10.17) also decreases. Thus, the promotion of Fe ammonia synthesis catalysts by potassium appears to be primarily an electronic effect.

(b) Alkane hydrogenolysis

Alkane hydrogenolysis, or cracking, involves the dissociation of a larger alkane molecule to a smaller alkane molecule. For example, ethane hydrogenolysis in the presence of H_2 yields methane:

$$C_2H_6 + H_2 \rightarrow 2CH_4.$$

Cracking (or hydrocracking, as it is referred to when carried out in the presence of H_2) reactions are an integral part of petroleum refining. Hydrocracking is used to lower the average molecular weight (MW) of a higher MW hydrocarbon mixture so that it can then be blended and sold as gasoline. The interest in the fundamentals of catalytic cracking reactions is strong and it has been thoroughly researched.

Ethane hydrogenolysis has been shown to be structure sensitive over nickel catalysts [43], as seen in figure A3.10.20, where methane formation rates are plotted for both nickel single crystals and a conventional, supported nickel catalyst. There is an obvious difference in the rates over Ni(111) and Ni(100), and it is evident that the rate also changes as a function of particle size for the supported Ni catalysts. In addition, differences in activation energy were observed: for Ni(111) the activation energy is 192 kJ mol^{-1}, while for Ni(100) the activation energy is 100 kJ mol^{-1}. It is noteworthy that there is overlap between the hydrogenolysis rates over supported Ni catalysts with the Ni single crystals. The data suggest that small Ni particles are composed primarily of Ni(100) facets while large Ni particles are composed primarily of Ni(111) facets. In fact, this has been observed for fcc materials where surfaces with a (111) orientation are more commonly observed after thermally induced sintering. The structure sensitivity of this reaction over Ni surfaces has been clearly demonstrated.

The initial step in alkane hydrogenolysis is the dissociative adsorption, or 'reactive sticking' of the alkane. One might suspect that this first step may be the key to the structure sensitivity of this reaction over Ni surfaces. Indeed, the reactive sticking of alkanes has been shown to depend markedly on surface structure [52]. Figure A3.10.21 shows the buildup of surface carbon due to methane decomposition ($P_{\text{methane}} = 1.00$ Torr) over three single-crystal Ni surfaces at 450 K. The rate of methane decomposition is obviously dependent upon the surface structure with the decomposition rate increasing in the order (111) < (100) < (110). It can

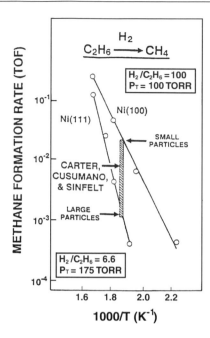

Figure A3.10.20. Arrhenius plot of ethane hydrogenolysis activity for Ni(100) and Ni(111) at 100 Torr and $H_2/C_2H_6 = 100$. Also included is the hydrogenolysis activity on supported Ni catalysts at 175 Torr and $H_2/C_2H_6 = 6.6$ [43].

be seen that, initially, the rates of methane decomposition are similar for Ni(100) and (110), while Ni(111) has a much lower reaction rate. With increasing reaction time, i.e. increasing carbon coverage, the rate over Ni(110) continues to increase linearly while both Ni(111) and (100) exhibit a nonlinear dependence. This linear dependence over Ni(111) may be due to either the formation of carbon islands or a reduced carbon coverage dependence as compared to Ni(111) and (100).

Hydrogenolysis reactions over Ir single crystals and supported catalysts have also been shown to be structure sensitive [53–55]. In particular, it was found that the reactivity tracked the concentration of low-coordination surface sites. Figure A3.10.22 shows ethane selectivity (selectivity is reported here because both ethane and methane are products of butane cracking) for *n*-butane hydrogenolysis over Ir(111) and the reconstructed surface Ir(110)-(1 × 2), as well as two supported Ir catalysts. There are clear selectivity differences between the two Ir surfaces, with Ir(110)-(1 × 2) having approximately three times the ethane selectivity of Ir(111). There is also a similarity seen between the ethane selectivity on small Ir particles and Ir(110)-(1 × 2), and between the ethane selectivity on large Ir particles and Ir(111).

The mechanisms by which *n*-butane hydrogenolysis occurs over Ir(110)-(1 × 2) and Ir(111) are different. The high ethane selectivity of Ir(110)-(1 × 2) has been attributed to the 'missing row' reconstruction that the (110) surface undergoes (figure A3.10.22). This reconstruction results in the exposure of a highly uncoordinated C_7 site that is sterically unhindered. These C_7 sites are capable of forming a metallocyclopentane (a five-membered ring consisting of four carbons and an Ir atom) which, based on kinetics and surface carbon coverages, has been suggested as the intermediate for this reaction [56, 57]. It has been proposed that the crucial step in this reaction mechanism over the reconstructed (110) surface is the reversible cleavage of the central C–C bond. On the other hand, the hydrogenolysis of *n*-butane over Ir(111) is thought to proceed by a different mechanism, where dissociative chemisorption of *n*-butane and hydrogen are the first steps. Then, the adsorbed hydrocarbon undergoes the irreversible cleavage of the terminal C–C bond. It is

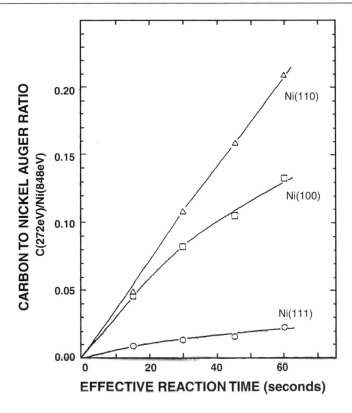

Figure A3.10.21. Methane decomposition kinetics on low-index Ni single crystals at 450 K and 1.00 Torr methane [43].

evident that surface structure plays an important role in hydrogenolysis reactions over both nickel and iridium surfaces.

(c) CO oxidation: $2CO+O_2 \rightarrow 2CO_2$

The oxidation of CO to CO_2, which is essential to controlling automobile emissions, has been extensively studied because of the relative simplicity of this reaction. CO oxidation was the first reaction to be studied using the surface science approach and is perhaps the most well understood heterogeneous catalytic reaction [58]. The simplicity of CO oxidation by O_2 endears itself to surface science studies. Both reactants are diatomic molecules whose adsorption on single-crystal surfaces has been widely studied, and presumably few steps are necessary to convert CO to CO_2. Surface science studies of CO and O_2 adsorption on metal surfaces have provided tremendous insight into the mechanism of the $CO-O_2$ reaction. The mechanism over platinum surfaces has been unequivocally established and the reaction has shown structure insensitivity over platinum [59], palladium [55, 59, 60] and rhodium surfaces [61, 62].

Although dissociative adsorption is sometimes observed, CO adsorption on platinum group metals typically occurs molecularly and this will be the focus of the following discussion. Figure A3.10.23 illustrates schematically the donor–acceptor model (first proposed by Blyholder [63]) for molecular CO chemisorption on a metal such as platinum. The bonding of CO to a metal surface is widely accepted to be similar to bond formation in a metal carbonyl. Experimental evidence indicates that the 5σ highest occupied molecular orbital (HOMO), which is regarded as a lone pair on the carbon atom, bonds to the surface by donating charge to unoccupied density of states (DOS) at the surface. Furthermore, this surface bond can be strengthened

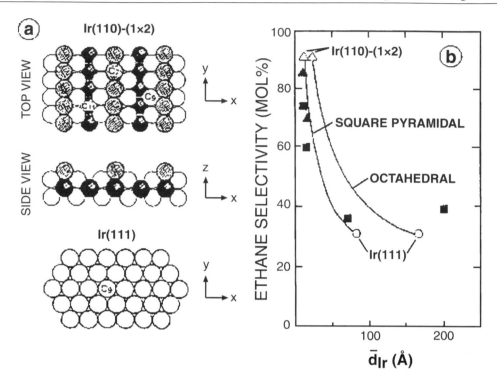

Figure A3.10.22. Relationship between selectivity and surface structure for *n*-butane hydrogenolysis on iridium. (a) Illustrations of the Ir(110)-(1 × 2) and Ir(111) surfaces. The *z*-axis is perpendicular to the plane of the surface. (b) Selectivity for C_2H_6 production (mol% total products) for *n*-butane hydrogenolysis on both Ni single crystals and supported catalysts at 475 K. The effective particle size for the single crystal surfaces is based on the specified geometric shapes [43]. △ Ir/Al$_2$O$_3$; □ Ir/SiO$_2$.

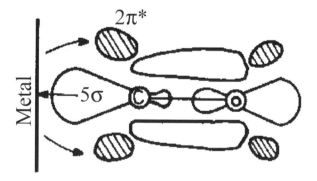

Figure A3.10.23. Schematic diagram of molecular CO chemisorption on a metal surface. The model is based on a donor–acceptor scheme where the CO 5σ HOMO donates charge to surface unoccupied states and the surface back-donates charge to the CO 2π LUMO [58].

by back-donation, which is the transfer of charge from the surface to the $2\pi^*$ lowest unoccupied molecular orbital (LUMO). An effect of this backbonding is that the C–O bond weakens, as seen by a lower C–O stretch frequency for adsorbed CO (typically <2100 cm^{-1}) than for gas phase CO (2143 cm^{-1}).

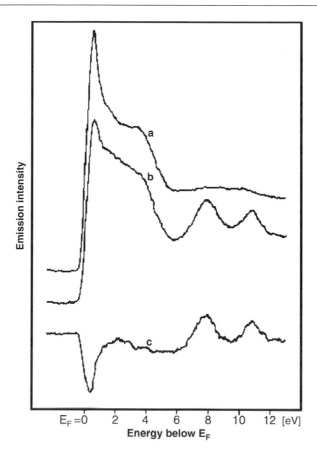

Figure A3.10.24. UPS data for CO adsorption on Pd(110). (a) Clean surface. (b) CO-dosed surface. (c) Difference spectrum (b-a). This spectrum is representative of molecular CO adsorption on platinum metals [58].

Ultraviolet photoelectron spectroscopy (UPS) results have provided detailed information about CO adsorption on many surfaces. Figure A3.10.24 shows UPS results for CO adsorption on Pd(110) [58] that are representative of molecular CO adsorption on platinum surfaces. The difference result in (c) between the clean surface and the CO-covered surface shows a strong negative feature just below the Fermi level (E_F), and two positive features at ~ 8 and 11 eV below E_F. The negative feature is due to suppression of emission from the metal d states as a result of an anti-resonance phenomenon. The positive features can be attributed to the 4σ molecular orbital of CO and the overlap of the 5σ and 1π molecular orbitals. The observation of features due to CO molecular orbitals clearly indicates that CO molecularly adsorbs. The overlap of the 5σ and 1π levels is caused by a stabilization of the 5σ molecular orbital as a consequence of forming the surface–CO chemisorption bond.

The adsorption of O_2 on platinum surfaces is not as straightforward as CO adsorption because molecular and dissociative adsorption can occur, as well as oxide formation [58]. However, molecular adsorption has been observed only at very low temperatures, where CO oxidation rates are negligible, hence this form of adsorbed oxygen will not be discussed here. UPS data indicate dissociative adsorption of O_2 on platinum surfaces at temperatures >100 K, and isotopic exchange measurements support this finding as well. The oxygen atoms resulting from O_2 dissociation can be either chemisorbed oxygen or oxygen in the form of an oxide. The two types of oxygen are distinguished by noting that oxide oxygen is located beneath the

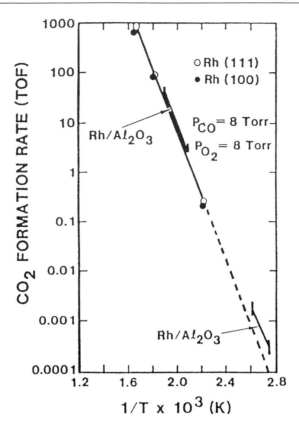

Figure A3.10.25. Arrhenius plots of CO oxidation by O_2 over Rh single crystals and supported Rh/Al_2O_3 at PCO = PO$_2$ = 0.01 atm [43]. The dashed line in the figure is the predicted behaviour based on the rate constants for CO and O_2 adsorption and desorption on Rh under UHV conditions.

surface ('subsurface') while chemisorbed oxygen is located on the surface. Experimentally, the two types of oxygen are discernible by AES, XPS and UPS. In general, it has been found that as long as pressure and temperature are kept fairly low, the most likely surface oxygen species will be chemisorbed. Therefore, when formulating a mechanism for reaction under these general conditions, only chemisorbed oxygen needs to be considered.

The mechanism for CO oxidation over platinum group metals has been established from a wealth of data, the analysis of which is beyond the scope of this chapter. It is quite evident that surface science provided the foundation for this mechanism by directly showing that CO adsorbs molecularly and O_2 adsorbs dissociatively. The mechanism is represented below (* denotes an empty surface site):

$$CO + {}^* \leftrightarrow CO_{ad}$$

$$O_2 + 2^* \rightarrow 2O_{ad}$$

$$O_{ad} + CO_{ad} \rightarrow CO_2 + 2^*.$$

The first step consists of the molecular adsorption of CO. The second step is the dissociation of O_2 to yield two adsorbed oxygen atoms. The third step is the reaction of an adsorbed CO molecule with an adsorbed oxygen atom to form a CO_2 molecule that, at room temperature and higher, desorbs upon formation. To simplify matters, this desorption step is not included. This sequence of steps depicts a Langmuir–Hinshelwood mechanism, whereby reaction occurs between two adsorbed species (as opposed to an Eley–Rideal mechanism, whereby reaction occurs between one adsorbed species and one gas phase species). The role of surface science studies in formulating the CO oxidation mechanism was prominent.

CO oxidation by O_2 is a structure-insensitive reaction over rhodium catalysts [61, 62]. Figure A3.10.25 illustrates this structure insensitivity by demonstrating that the activation energies over supported Rh catalysts and a Rh(111) single crystal (given by the slope of the line) were nearly identical. Furthermore, the reaction rates over both supported Rh/Al_2O_3 and single crystal Rh (111) surfaces were also remarkably similar. Thus, the reaction kinetics were quite comparable over both the supported metal particles and the single crystal surfaces, and no particle size effect (structure sensitivity) was observed.

The study of catalytic reactions using surface science techniques has been fruitful over the last 30 years. Great strides have been made towards understanding the fundamentals of catalytic reactions, particularly by bridging the material and pressure gaps. The implementation of *in situ* techniques and innovative model catalyst preparation will undoubtedly shape the future of catalysis.

References

[1] National Transportation Statistics 1998 (US Department of Transportation) table 4-3
[2] Roth A 1990 *Vacuum Technology* 3rd edn (New York: Elsevier)
[3] O'Hanlon J F 1989 *A User's Guide to Vacuum Technology* 2nd edn (New York: Wiley)
[4] Atkins P W 1990 *Physical Chemistry* 4th edn (New York: Freeman)
[5] West J M 1986 *Basic Corrosion and Oxidation* 2nd edn (Chichester: Ellis Horwood) p 220
[6] Marcus P 1998 Surface science approach of corrosion phenomena *Electrochim. Acta* **43** 109
[7] Itaya K 1998 *In situ* scanning tunneling microscopy in electrolyte solutions *Prog. Surf. Sci.* **58** 121
[8] Maurice V, Kitakatsu N, Siegers M and Marcus P 1997 Low coverage sulfur induced reconstruction of Ni(111) *Surf. Sci.* **373** 307
[9] Marcus P, Teissier A and Oudar J 1984 The influence of sulfur on the dissolution and the passivation of a nickel-iron alloy. 1. Electrochemical and radiotracer measurements *Corrosion Sci.* **24** 259
[10] Moffat T P, Fan F R F and Bard A 1991 Electrochemical and scanning tunneling microscopic study of dealloying of Cu_3Au *J. Electrochem. Soc.* **138** 3224
[11] Chen S J, Sanz F, Ogletree D F, Hallmark V M, Devine T M and Salmeron M 1993 Selective dissolution of copper from Au-rich Au-Cu alloys: an electrochemical STS study *Surf. Sci.* **292** 289
[12] Dahmen U 1982 Orientation relationships in precipitation systems *Acta Metall.* **30** 63
[13] Hahn E, Kampshoff E, Wälchli N and Kern K 1995 Strain driven fcc-bct phase transition of pseudomorphic Cu films on Pd(100) *Phys. Rev. Lett.* **74** 1803
[14] Zangwill A 1988 *Physics at Surfaces* (Cambridge: Cambridge University Press) pp 427–8
[15] Foord J S, Davies G J and Tsang W S 1997 *Chemical Beam Epitaxy and Related Techniques* (New York: Wiley)
[16] Panish M B and Temkin H 1993 *Gas Source Molecular Beam Epitaxy* (New York: Springer)
[17] Stringfellow G B 1989 *Organometallic Vapor-Phase Epitaxy* (San Diego, CA: Academic)
[18] Zangwill A 1988 *Physics at Surfaces* (Cambridge: Cambridge University Press) pp 428–32
[19] Brune H 1998 Microscopic view of epitaxial metal growth: nucleation and aggregation *Surf. Sci. Rep.* **31** 121
[20] Lewis B and Anderson J C 1978 *Nucleation and Growth of Thin Films* (New York: Academic)
[21] Venables J A, Spiller G D T and Hanbucken M 1984 Nucleation and growth of thin-films *Rep. Prog. Phys.* **47** 399
[22] Stoyanov S and Kashchiev D 1981 Thin film nucleation and growth theories: a confrontation with experiment *Current Topics in Materials Science* vol 7, ed E Kaldis (Amsterdam: North-Holland) p 69
[23] Brune H, Röder H, Boragno C and Kern K 1994 Microscopic view of nucleation on surfaces *Phys. Rev. Lett.* **73** 1955
[24] Sze S M 1985 *Semiconductor Devices* (New York: Wiley) p 456
[25] Cooke M J 1990 *Semiconductor Devices* (New York: Prentice-Hall) p 181
[26] Williams R 1990 *Modern GaAs Processing Methods* (Norwood: Artech House)
[27] Sugawara M 1998 *Plasma Etching: Fundamentals and Applications* (New York: Oxford University Press)
[28] Boland J J and Weaver J H 1998 A surface view of etching *Phys. Today* **51** 34
[29] Boland J J and Villarrubia J S 1990 Formation of Si(111)-(1 × 1)Cl *Phys. Rev. B* **41** 9865
[30] Wiesendanger R 1994 *Scanning Probe Microscopy and Spectroscopy* (Cambridge: Cambridge University Press)
[31] Thomas J M and Thomas W J 1996 *Principles and Practice of Heterogeneous Catalysis* (Weinheim: VCH)
[32] Somorjai G A 1993 *Introduction to Surface Chemistry and Catalysis* (New York: Wiley)
[33] Masel R I 1996 *Principles of Adsorption and Reaction on Solid Surfaces* (New York: Wiley)

[34] Blakely D W, Kozak E I, Sexton B A and Somorjai G A 1976 New instrumentation and techniques to monitor chemical surface reactions over a wide pressure range (10^{-8} to 10^5 Torr) in the same apparatus *J. Vac. Sci. Technol.* **13** 1091

[35] Goodman D W, Kelley R D, Madey T E and Yates J T Jr 1980 Kinetics of the hydrogenation of CO over a single crystal nickel catalyst *J. Catal.* **63** 226

[36] Campbell R A and Goodman D W 1992 A new design for a multitechnique ultrahigh vacuum surface analysis chamber with high-pressure capabilities *Rev. Sci. Instrum.* **63** 172

[37] Szanyi J and Goodman D W 1993 Combined elevated pressure reactor and ultrahigh vacuum surface analysis system *Rev. Sci. Instrum.* **64** 2350

[38] Goodman D W 1996 Chemical and spectroscopic studies of metal oxide surfaces *J. Vac. Sci. Technol.* A **14** 1526

[39] Wu M-C, Estrada C A, Corneille J S, He J-W and Goodman D W 1991 Synthesis and characterization of ultrathin MgO films on Mo(100) *Chem. Phys. Lett.* **472** 182

[40] He J-W, Corneille J S, Estrada C A, Wu M-C and Goodman D W 1992 CO interaction with ultrathin MgO films on a Mo(100) surface studied by IRAS, TPD, and XPS *J. Vac. Sci. Technol.* A **10** 2248

[41] Guo Q and Goodman D W Vanadium oxide thin-films grown on rutile $TiO_2(110)$-(1×1) and (1×2) surfaces *Surf. Sci.* **437** 38

[42] Valden M, Lai X and Goodman D W 1998 Onset of catalytic activity of gold clusters on titania with the appearance of nonmetallic properties *Science* **281** 1647

[43] Goodman D W 1995 Model studies in catalysis using surface science probes *Chem. Rev.* **95** 523

[44] Jacobs P W and Somorjai G A 1997 Conversion of heterogeneous catalysis from art to science: the surface science of heterogeneous catalysis *J. Mol. Catal.* A **115** 389

[45] Ertl G 1983 *Catalysis: Science and Technology* vol 4, ed J R Anderson and M Boudart (Heidelberg: Springer)

[46] Ertl G 1990 Elementary steps in heterogeneous catalysis *Agnew. Chem., Int. Ed. Engl.* **29** 1219

[47] Strongin D R, Carrazza J, Bare S R and Somorjai G A 1987 The importance of C_7 sites and surface roughness in the ammonia synthesis reaction over iron *J. Catal.* **103** 213

[48] Ertl G 1991 *Catalytic Ammonia Synthesis: Fundamentals and Practice, Fundamentals and Applied Catalysis* ed J R Jennings (New York: Plenum)

[49] Bare S R, Strongin D R and Somorjai G A 1986 Ammonia synthesis over iron single crystal catalysts—the effects of alumina and potassium *J. Phys. Chem.* **90** 4726

[50] Ertl G, Lee S B and Weiss M 1982 Adsorption of nitrogen on potassium promoted Fe(111) and (100) surfaces *Surf. Sci.* **114** 527

[51] Paàl Z, Ertl G and Lee S B 1981 Interactions of potassium, nitrogen, and oxygen with polycrystalline iron surfaces *Appl. Surf. Sci.* **8** 231

[52] Beebe T P, Goodman D W, Kay B D and Yates J T Jr 1987 Kinetics of the activated dissociation adsorption of methane on low index planes of nickel single crystal surfaces *J. Chem. Phys.* **87** 2305

[53] Wu M-C, Estrada C A, Corneille J S and Goodman D W 1996 Model surface studies of metal oxides: adsorption of water and methanol on ultrathin MgO films on Mo(100) *J. Chem. Phys.* **96** 3892

[54] Xu X and Goodman D W 1992 New approach to the preparation of ultrathin silicon dioxide films at low temperature *Appl. Phys. Lett.* **61** 774

[55] Szanyi J, Kuhn W K and Goodman D W 1994 CO oxidation on palladium: 2. A combined kinetic-infrared reflection absorption spectroscopic study of Pd(100) *J. Phys. Chem.* **98** 2978

[56] Engstrom J R, Goodman D W and Weinberg W H 1986 Hydrogenolysis of n-butane over the (111) and (110)-(1×2) surfaces of iridium: a direct correlation between catalytic selectivity and surface structure *J. Am. Chem. Soc.* **108** 4653

[57] Engstrom J R, Goodman D W and Weinberg W H 1988 Hydrogenolysis of ethane, propane, n-butane and neopentane over the (111) and (110)-(1×2) surfaces of iridium *J. Am. Chem. Soc.* **110** 8305

[58] Engel T and Ertl G 1978 Elementary steps in the catalytic oxidation of carbon monoxide on platinum metals *Adv. Catal.* **28** 1

[59] Berlowitz P J and Goodman D W 1988 Kinetics of CO oxidation on single crystal Pd, Pt, and Ir *J. Phys. Chem.* **92** 5213

[60] Szanyi J and Goodman D W 1994 CO oxidation on palladium: 1. A combined kinetic-infrared reflection absorption spectroscopic study of Pd(111) *J. Phys. Chem.* **98** 2972

[61] Oh S H, Fisher G B, Carpenter J E and Goodman D W 1986 Comparative kinetic studies of CO-O_2 and CO-NO reactions over single crystal and supported rhodium catalysts *J. Catal.* **100** 360

[62] Berlowitz P J, Goodman D W, Peden C H F and Blair D S 1988 Kinetics of CO oxidation by O_2 or NO on Rh(111) and Rh(100) single crystals *J. Phys. Chem.* **92** 1563

[63] Blyholder G 1964 Molecular orbital view of chemisorbed carbon monoxide *J. Phys. Chem.* **68** 2772

A3.11
Quantum mechanics of interacting systems: scattering theory

George C Schatz

A3.11.1 Introduction

Quantum scattering theory is concerned with transitions between states which have a continuous energy spectrum, i.e., which are unbound. The most common application of scattering theory in chemical physics is to collisions involving atoms, molecules and/or electrons. Such collisions can produce many possible results, ranging from elastic scattering to reaction and fragmentation. Scattering theory can also be used to describe collisions of atoms, molecules and/or electrons with solid surfaces and it also has application to many kinds of dynamical process in solids. These latter include collisions of conduction electrons in a metal with impurities or with particle surfaces, or collisions of collective wave motions such as phonons with impurities, or adsorbates. Scattering theory is also involved in describing the interaction of light with matter, including applications to elastic and inelastic light scattering, photoabsorption and emission. Additionally, there are many processes where continuum states of particles are coupled to continuum states of electromagnetic radiation, including photodissociation of molecules and photoemission from surfaces.

While the basic formalism of quantum scattering theory can be found in a variety of general physics textbooks [1–7] and textbooks that are concerned with scattering theory in a broad sense [8–11], many problems in chemical physics require special adaptation of the theory. For example, in collisions of particles with surfaces, angular momentum conservation is not important, but linear momentum conservation can be crucial. Also, in many collision problems involving atoms and molecules, the de Broglie wavelength is short compared to the distances over which the particles interact strongly, making classical or semiclassical theory useful. One especially important feature associated with scattering theory applications in chemical physics is that the forces between the interacting particles can usually be determined with reasonable accuracy (in principle to arbitrary accuracy), so explicit forms for the Hamiltonian governing particle motions are available. Often these forces are quite complicated, so it is not possible to develop analytical solutions to the scattering theory problem. However, numerical solutions are possible, so a significant activity among researchers in this field is the development of numerical methods for solving scattering problems. There are a number of textbooks which consider scattering theory applications of more direct relevance to problems in chemical physics [12–18], as well as numerous monographs that have a narrower focus within the field [19–29].

Much of what one needs to know about scattering theory can be understood by considering a particle moving in one dimension governed by a potential that allows it to move freely except for a range of coordinates where there is a feature such as a barrier or well that perturbs the free particle motion. Our discussion will therefore begin with this simple problem (section A3.11.2). Subsequently (section A3.11.3) more complete versions of scattering theory will be developed that apply to collisions involving particles having internal degrees of freedom. There are both time dependent and time independent versions of scattering theory, and both of these theories will be considered. In section A3.11.4, the numerical methods that are used to calculate scattering theory properties are considered, including time dependent and independent approaches. Also presented (section A3.11.5) are scattering theory methods for determining information that has been summed and averaged over many degrees of freedom (such as in a Boltzmann distribution). Finally, in section A3.11.6, scattering theory methods based on classical and semiclassical mechanics are described.

There are a variety of topics that will not be considered, but it is appropriate to provide references for further reading. The development in this paper assumes that the Born–Oppenheimer approximation applies in collisions between atoms and molecules and thus the nuclear motion is governed by a single potential energy surface. However there are many important problems where this approximation breaks down and multiple coupled potential energy surfaces are involved, with nonadiabatic transitions taking place during the scattering process. The theory of such processes is described in many places, such as [14, 15, 23, 25].

Other topics that have been omitted include the description of scattering processes using Feynman path integrals [18, 19] and the description of scattering processes with more than two coupled continua (i.e., where three or more independent particles are produced, as in electron impact ionization [30] or collision induced dissociation) [31]. Our treatment of resonance effects in scattering processes (i.e., the formation of metastable intermediate states) is very brief as this topic is commonly found in textbooks and one monograph is available [26]. Finally, it should be mentioned that the theory of light scattering is not considered; interested readers should consult textbooks such as that by Newton [8].

A3.11.2 Quantum scattering theory for a one-dimensional potential function

A3.11.2.1 Hamiltonian; boundary conditions

The problem of interest in this section is defined by the simple one-dimensional Hamiltonian

$$\hat{H} = \frac{\hat{P}^2}{2m} + V(x) \qquad\qquad (A3.11.1)$$

where $V(x)$ is the potential energy function, examples of which are pictured in figure A3.11.1. The potentials shown are of two general types: those which are constant in the limit of $x \to \pm\infty$ (figures A3.11.1(a) and A3.11.1(b)), and those which are constant in the limit of $x \to \infty$ and are infinite in the limit of $x \to -\infty$ (figure A3.11.1(c)) (of course this potential could be flipped around if one wants). In the former case, one can have particles moving at constant velocity in both asymptotic limits ($x \to \pm\infty$), so there are two physically distinct processes that can be described, namely, scattering in which the particle is initially moving to the right in the limit $x \to -\infty$, and scattering in which the particle is initially moving to the left in the limit $x \to \infty$. In the latter case (figure A3.11.1(c)), the only physically interesting situation involves the particle initially moving to the left in the limit $x \to \infty$. The former case is appropriate for describing a chemical reaction where there is either a barrier (figure A3.11.1(a)) or a well (figure A3.11.1(b)). It is also relevant to the scattering of an electron from the surface of a metal, where either transmission or reflection can occur. In figure A3.11.1(c), only reflection can occur, such as happens in elastic collisions of atoms, or low energy collisions of molecules with surfaces.

The physical question to be answered for figures A3.11.1(a) and (b) is: what is the probability P that a particle incident with an energy E from the left at $x \to -\infty$ will end up moving to the right at $x \to +\infty$? In the case of figure A3.11.1(c), only reflection can occur. However the change in phase of the wavefunction that occurs in this reflection is often of interest. In the following treatment the detailed theory associated with figures A3.11.1(a) and (b) will be considered. Eventually we will see that figure A3.11.1(c) is a subset of this theory.

The classical expression for the transmission probability associated with figures A3.11.1(a) or (b) is straightforward, namely

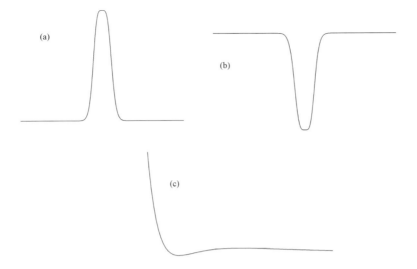

Figure A3.11.1. Potential associated with the scattering of a particle in one dimension. The three cases shown are (a) barrier potential, (b) well potential and (c) scattering off a hard wall that contains an intermediate well.

(1) $P(E) = 0$ if $V(x) \geq E$ for any x

(2) $P(E) = 1$ if $V(x) < E$ for all x.

The quantum solution to this problem is much more difficult for a number of reasons. First, it is important to know how to define what we mean by a particle moving in a given direction when $V(x)$ is constant. Secondly, one must determine the probability that the particle is moving in any specified direction at any desired location and, third, we need to be able to solve the Schrödinger equation for the potential $V(x)$.

A3.11.2.2   Wavepackets in one dimension

To understand how to describe a particle moving in a constant potential, consider the case of a free particle for which $V(x) = 0$. In this case the time-dependent Schrödinger equation is

$$i\hbar \frac{\partial \psi}{\partial t} = -\frac{\hbar^2}{2m} \frac{\partial^2 \psi}{\partial x^2} \tag{A3.11.2}$$

and if one invokes the usual procedure for separating the time and spatial parts of this equation, it can readily be shown that one possible solution is

$$\psi_k(x, t) = e^{-iEt/\hbar} \, e^{ikx} \tag{A3.11.3}$$

where

$$E = \frac{\hbar^2 k^2}{2m} \tag{A3.11.4}$$

is the particle's energy and $\hbar k$ is its linear momentum. Note that both energy and momentum of the particle are exactly specified in this solution. As might be expected from the uncertainty principle, the location of the particle is therefore completely undetermined. As a result, this solution to the Schrödinger equation, even though correct, is not useful for describing the scattering processes of interest.

In order to localize the particle, it is necessary to superimpose wavefunctions ψ_k with different momenta k. A very general way to do this is to construct a wavepacket, defined through the integral

$$\psi_{\text{wp}}(x,t) = \int_{-\infty}^{\infty} dk\, C(k)\psi_k(x,t)$$
$$= \int_{-\infty}^{\infty} dk\, C(k)\, e^{ikx}\, e^{-i\hbar k^2 t/2m} \tag{A3.11.5}$$

where $C(k)$ is a function which tells us how much of each momentum $\hbar k$ is contained in the wavepacket. If the particle is to move with roughly a constant velocity, $C(k)$ must be peaked at some k which is taken to be k_0. One function which accomplishes this is the Gaussian

$$C(k) = \sqrt{\frac{a}{2\pi^{3/2}}} \exp[-a^2(k-k_0)^2/2] \tag{A3.11.6}$$

where a measures the width of the packet. Substituting this into equation (A3.11.5), the result is:

$$\psi_{\text{wp}}(x,t) = \pi^{-1/4}[a(1+i\hbar t/ma^2)]^{-1/2} \exp\left[-\frac{(x-\hbar k_0 t/m)^2}{2a^2(1+i\hbar t/ma^2)} + ik_0 x - \frac{i\hbar t}{2ma^2}\right]. \tag{A3.11.7}$$

The absolute square of this wavefunction is $|\psi_{\text{wp}}(x,t)|^2$

$$|\psi_{\text{wp}}|^2 = \pi^{-1/2}a^{-1}(1+\hbar^2 t^2/m^2 a^4)^{-1/2} \exp\left[-\frac{(x-\hbar k_0 t/m)^2}{a^2(1+\hbar^2 t^2/m^2 a^4)}\right]. \tag{A3.11.8}$$

This is a Gaussian function which peaks at $x = \hbar k_0 t/m$, moving to the right with a momentum $\hbar k_0$. The width of this peak is

$$\Delta = a[(\ln 2)(1+\hbar^2 t^2/m^2 a^4)]^{1/2} \tag{A3.11.9}$$

which starts out at $\Delta = a(\ln 2)^{1/2}$ at $t=0$ and increases linearly with time for large t. This increase in width means that wavepacket spreads as it moves. This is an inevitable consequence of the fact that the wavepacket was constructed with a distribution of momentum components, and is a natural consequence of the uncertainty principle. Note that the wavefunction in equation (A3.11.7) still satisfies the Schrödinger equation (equation (A3.11.2)).

One can show that the expectation value of the Hamiltonian operator for the wavepacket in equation (A3.11.7) is:

$$\langle \hat{H} \rangle = \frac{\hbar^2 k_0^2}{2m} + \frac{\hbar^2}{4ma^2}. \tag{A3.11.10}$$

The first term is what one would expect to obtain classically for a particle of momentum $\hbar k_0$, and it is much bigger than the second term provided $k_0 a \gg 1$. Since the de Broglie wavelength λ is $2\pi/k_0$, this condition is equivalent to the statement that the size of the wavepacket be much larger than the de Broglie wavelength.

It is also notable that the spreading of the wavepacket can be neglected for times t such that $t \ll ma^2/\hbar$. In this time interval the centre of the wavepacket will have moved a distance $(k_0 a)a$. Under the conditions noted above for which $k_0 a \gg 1$, this distance will be many times larger than the width of the packet.

A3.11.2.3 Wavepackets for the complete scattering problem

The generalization of the treatment of the previous section to the determination of a wavepacket for the Hamiltonian in equation (A3.11.1) is accomplished by writing the solution as follows:

$$\psi_{\text{wp}}(x,t) = \int_{-\infty}^{\infty} dk\, C(k)\psi_k(x)\, e^{-iE_k t/\hbar} \tag{A3.11.11}$$

where ψ_k is the solution of the time-independent Schrödinger equation

$$\hat{H}\psi_k = E_k\psi_k \tag{A3.11.12}$$

for an energy E_k. By substituting equation (A3.11.11) into the time-dependent Schrödinger equation one can readily show that ψ_{wp} is a solution.

However, it is important to make sure that ψ_{wp} satisfies the desired boundary conditions initially and finally. Part of this is familiar already, since we have already demonstrated in equations (A3.11.3), (A3.11.5) and (A3.11.7) that use of $\psi_k = \mathrm{e}^{\mathrm{i}kx}$ and a Gaussian $C(k)$ gives a Gaussian wavepacket which moves with momentum $\hbar k_0$. This is the behaviour that is of interest initially ($t \to -\infty$) in the limit of $x \to -\infty$.

At the end of the collision ($t \to +\infty$) one expects to see part of the wavepacket moving to the right for $x \to \infty$ (the transmitted part) and part of it moving to the left for $x \to -\infty$ (the reflected part). Both this and the $t \to -\infty$ boundary condition can be satisfied by requiring that

$$\psi_k(x) \underset{x \to -\infty}{=} \mathrm{e}^{\mathrm{i}kx} + R\,\mathrm{e}^{-\mathrm{i}kx} \tag{A3.11.13a}$$

$$\underset{x \to +\infty}{=} T\,\mathrm{e}^{-\mathrm{i}\bar{k}x} \tag{A3.11.13b}$$

where R and T are as yet undetermined coefficients that will be discussed later and $\bar{k} = (2m(E - V_0)/\hbar^2)^{1/2}$ where V_0 is the value of the potential in the limit $x \to \infty$. Note that V_0 specifies the energy difference between the potential in the right and left asymptotic limits, and it has been assumed that $E > V_0$, as otherwise there could not be travelling waves in the $x \to \infty$ limit.

To prove that equation (A3.11.13b) gives a wavepacket which satisfies the desired boundary conditions, we note that substitution of equation (3.11.13a) into equation (A3.11.11) gives us two wavepackets which roughly speaking are given by

$$\psi_{\mathrm{wp}} \underset{x \to -\infty}{\approx} \mathrm{e}^{-(x-\hbar k_0 t/m)^2/2a^2} + R\,\mathrm{e}^{-(x+\hbar k_0 t/m)^2/2a^2}. \tag{A3.11.14a}$$

In the $t \to -\infty$ limit, only the first term, representing a packet moving to the right, has a peak in the $x \to -\infty$ region (the left asymptotic region). The second term peaks in the right asymptotic region but this is irrelevant as equation (A3.11.14a) does not apply there. Thus, in the left asymptotic region the second term is negligible and all we have is a packet moving to the right. For $t \to +\infty$, equation (A3.11.14a) still applies in the left asymptotic region, but now it is the second term which peaks and this packet moves to the left.

Now substitute equation (A3.11.13b) into equation (A3.11.11). Ignoring various unimportant terms, we obtain

$$\psi_{\mathrm{wp}} \underset{x \to +\infty}{\approx} T\,\mathrm{e}^{-(x-\hbar \bar{k}_0 t/m)^2/2a^2}. \tag{A3.11.14b}$$

This formula represents a packet moving to the right centred at $x = \hbar \bar{k}_0 t/m$. For $t \to -\infty$, this is negligible in the right asymptotic region, so the wavefunction is zero there, while for $t \to +\infty$ this packet is large for $x \to +\infty$ just as we wanted.

A3.11.2.4 Fluxes and probabilities

Now let us use the wavepackets just discussed to extract the physically measurable information about our problem, namely, the probabilities of reflection and transmission. As long as the wavepackets do not spread much during the collision, these probabilities are given by the general definition:

$$\text{probability} = \frac{|\text{total flux outgoing for process of interest}|}{|\text{total flux incident}|} \tag{A3.11.15}$$

where the flux is the number of particles per unit time that cross a given point (that cross a given surface in three dimensions), and the total flux is the spatial integral of the instantaneous flux. Classically the flux is just ρv where ρ is the density of particles (particles per unit length in one dimension) and v is the velocity of the particles. In quantum mechanics, the flux I is defined as

$$I = \text{Re}[\psi^* \hat{v} \psi] \tag{A3.11.16}$$

where $\hat{v}$ is the velocity operator ($\hat{v} = (-i\hbar/m)\partial/\partial x$ in one dimension) and Re implies that only the real part of $\psi^* \hat{v} \psi$ is to be used.

To see how equation (A3.11.16) works, substitute equation (A3.11.7) into (A3.11.16). Under the condition that wavepacket spreading is small (i.e., $\hbar t/ma^2 \ll 1$) we obtain

$$I = \left(\frac{\hbar k_0}{m}\right) \pi^{-1/2} a^{-1} \exp[-(x - \hbar k_0 t/m)^2/a^2] \tag{A3.11.17}$$

which is just $v_0|\psi_{\text{wp}}|^2$ where v_0 is the initial most probable velocity ($v_0 = \hbar k_o/m$). In view of equation (A3.11.14a), this is just the incident flux. The integral of this quantity over all space (the total flux) is $I_{\text{tot}}^{\text{inc}} = v_0$.

For the reflected wave associated with equation (A3.11.13a), the total outgoing flux is $I_{\text{tot}}^{\text{out}} = |R|^2 v_0$ so the reflection probability P_R is

$$P_R = |R|^2. \tag{A3.11.18a}$$

A similar calculation of the transmission probability gives

$$P_T = \frac{\overline{v}_0}{v_0}|T|^2 \tag{A3.11.18b}$$

where

$$\overline{v}_0 \equiv \frac{\hbar \overline{k}}{m}. \tag{A3.11.19}$$

A3.11.2.5   Time-independent approach to scattering

Note from equations (A3.11.18a, b) that all of the physically interesting information about the scattering process involves the coefficients R and T which are properties of the time *independent* wavefunction ψ_k obtained from equations (A3.11.12) with the boundary conditions in equations (A3.11.13). As a result, we can use scattering theory completely in a time independent picture. This picture can be thought of as related to the time dependent picture by the superposition of many Gaussian incident wavepackets to form a plane wave. The important point to remember in using time independent solutions is that the asymptotic solution given by equations (A3.11.13) involves waves moving to the left and right that should be treated *separately* in calculating fluxes since these solutions do not contribute at the same time to the evolution of $\psi_{\text{wp}}(x, t)$ in the $t \to \pm\infty$ limits. As a result, fluxes are evaluated by substituting either the left or right moving wavepacket parts of equations (A3.11.13) into (A3.11.16).

A3.11.2.6   Scattering matrix

It is useful to rewrite the asymptotic part of the wavefunction as

$$\psi_k(x) \underset{x \to -\infty}{=} e^{ikx} + S_{11} e^{-ikx} \tag{A3.11.20a}$$

$$\underset{x \to +\infty}{=} S_{12}(k/\overline{k})^{1/2} e^{-i\overline{k}x} \tag{A3.11.20b}$$

where the coefficients S_{11} and S_{12} are two elements of a 2×2 matrix known as the scattering (S) matrix. The other two elements are associated with a different scattering solution in which the incident wave at $t \to -\infty$ moves to the left in the $x \to +\infty$ region. The boundary conditions on this solution are

$$
\begin{aligned}
\psi_{\bar{k}} &\underset{x \to +\infty}{=} e^{-i\bar{k}x} + S_{22} e^{+i\bar{k}x} \\
&\underset{x \to -\infty}{=} S_{21}(\bar{k}/k)^{1/2} e^{-i\bar{k}x}.
\end{aligned}
\tag{A3.11.21}
$$

The S matrix has a number of important properties, one of which is that it is *unitary*. Mathematically this means that $\mathbf{S}^+\mathbf{S} = 1$ where $\mathbf{S}^+$ is the Hermitian conjugate (transpose of complex conjugate) of $\mathbf{S}$. This property comes from the equation of continuity, which says that for any solution ψ to the time dependent Schrödinger equation,

$$
\frac{\partial |\psi|^2}{\partial t} + \frac{\partial I}{\partial x} = 0
\tag{A3.11.22}
$$

where I is the flux from equation (A3.11.16). Equation (A3.11.22) can be proved by substitution of equation (A3.11.16) and the time dependent Schrödinger equation into (A3.11.22).

If $\psi = \psi_k(x) e^{-iEt/\hbar}$, $|\psi|^2$ is time independent, so equation (A3.11.22) reduces to $\partial I/\partial x = 0$, which implies I is a constant (i.e., flux is conserved), independent of x. If so then the evaluation of I at $x \to +\infty$ and at $x \to -\infty$ should give the same result. By directly substituting equations (A3.11.13) into (A3.11.16) one finds

$$
\begin{aligned}
I &\underset{x \to -\infty}{=} \frac{\hbar k}{m}(1 - |S_{11}|^2) \\
&\underset{x \to +\infty}{=} \frac{\hbar k}{m}|S_{12}|^2
\end{aligned}
\tag{A3.11.23}
$$

and since these two have to be equal, we find that

$$
|S_{11}|^2 + |S_{12}|^2 = 1
\tag{A3.11.24}
$$

which indicates that the sum of the reflected and transmitted probabilities has to be unity. This is one of the equations that is implied by unitarity of the S matrix. The other equations can be obtained by using the solution $\psi_{\bar{k}}$ (equation (A3.11.21)) and by using a generalized flux that is defined by

$$
I_{k\bar{k}} = \mathrm{Re}(\psi_k^* \hat{v} \psi_{\bar{k}}).
\tag{A3.11.25}
$$

Another useful property of the S matrix is that it is *symmetric*. This property follows from conservation of the fluxlike expression

$$
\tilde{I}_{k\bar{k}} = \mathrm{Re}(\psi_k v \psi_{\bar{k}})
\tag{A3.11.26}
$$

which differs from equation (A3.11.25) in the absence of a complex conjugate in the wavefunction ψ_k. The symmetry property of $\mathbf{S}$ implies that S_{12} in equation (A3.11.20b) equals S_{21} in equation (A3.11.21). Defining the probability matrix $\mathbf{P}$ by the relation

$$
P_{ij} = |S_{ij}|^2
\tag{A3.11.27}
$$

we see that symmetry of $\mathbf{S}$ implies equal probabilities for the $i \to j$ and $j \to i$ transitions. This is a statement of the principle of *microscopic reversibility* and it arises from the time reversal symmetry associated with the Schrödinger equation.

The probability matrix plays an important role in many processes in chemical physics. For chemical reactions, the probability of reaction is often limited by tunnelling through a barrier, or by the formation of metastable states (resonances) in an intermediate well. Equivalently, the conductivity of a molecular wire is related to the probability of transmission of conduction electrons through the junction region between the wire and the electrodes to which the wire is attached.

A3.11.2.7   Green's functions for scattering

Now let us write down the Schrödinger equation (equation (A3.11.12)) using equation (A3.11.1) for H and assuming that V_0 in figure A3.11.1 is zero. The result can be written

$$\left(\frac{\mathrm{d}^2}{\mathrm{d}x^2} + k^2\right)\psi_k(x) = \frac{2m}{\hbar^2}V(x)\psi_k(x). \tag{A3.11.28}$$

One way to solve this is to invert the operator on the left hand side, thereby converting this differential equation into an integral equation. The general result is

$$\psi_k(x) = \varphi_k(x) + \frac{2m}{\hbar^2}\int_{-\infty}^{\infty} G_0(x, x')V(x')\psi_k(x')\,\mathrm{d}x' \tag{A3.11.29}$$

where G_0 is called the Green function associated with the operator $\mathrm{d}^2/\mathrm{d}x^2 + k^2$ and φ_k is a solution of the homogeneous equation that is associated with equation (A3.11.28), namely

$$\left(\frac{\mathrm{d}^2}{\mathrm{d}x^2} + k^2\right)\varphi_k(x) = 0. \tag{A3.11.30}$$

To determine $G_0(x, x')$, it is customary to reexpress equation (A3.11.28) in a Fourier representation. Let $F_k(k')$ be the Fourier transform of $\psi_k(x)$. Taking the Fourier transform of equation (A3.11.28), we find

$$\frac{1}{\sqrt{2\pi}}\int_{-\infty}^{\infty} \mathrm{e}^{\mathrm{i}k'x}(k^2 - k'^2)F_k(k')\,\mathrm{d}k' = \frac{1}{\sqrt{2\pi}}\int_{-\infty}^{\infty} \mathrm{e}^{\mathrm{i}k'x}B(k')\,\mathrm{d}k' \tag{A3.11.31}$$

where

$$B(k') = \frac{1}{\sqrt{2\pi}}\int_{-\infty}^{\infty} \mathrm{e}^{-\mathrm{i}k'x}\frac{2m}{\hbar^2}V(x)\psi_k(x)\,\mathrm{d}x. \tag{A3.11.32}$$

Equation (A3.11.31) implies

$$F_k(k') = \frac{B(k')}{k^2 - k'^2} \tag{A3.11.33}$$

and upon inverting the Fourier transform we find

$$\psi_k(x) = \frac{1}{\sqrt{2\pi}}\int_{-\infty}^{\infty} \mathrm{e}^{\mathrm{i}k'x}\frac{B(k')}{k^2 - k'^2}\,\mathrm{d}k' = \frac{2m}{\hbar^2}\int_{-\infty}^{\infty} G_0(x, x')V(x')\psi_k(x')\,\mathrm{d}x' \tag{A3.11.34}$$

where the Green function is given by

$$G_0(x, x') = \frac{1}{2\pi}\int_{-\infty}^{\infty} \mathrm{e}^{\mathrm{i}k'(x-x')}(k^2 - k'^2)^{-1}\,\mathrm{d}k'. \tag{A3.11.35}$$

The evaluation of the integral in equation (A3.11.35) needs to be done carefully as there is a pole at $k' = \pm k$. A standard trick to do it involves replacing k by $k \pm \mathrm{i}\varepsilon$ where ε is a small positive constant that will be set to zero in the end. This reduces equation (A3.11.35) to

$$G_0(x, x') = \frac{1}{4\pi k}\lim_{\varepsilon \to 0}\int_{-\infty}^{\infty} \exp[\mathrm{i}k'(x - x')]\left(\frac{1}{k - k' \pm \mathrm{i}\varepsilon} + \frac{1}{k + k' \pm \mathrm{i}\varepsilon}\right)\mathrm{d}k'. \tag{A3.11.36}$$

This integral can be done by contour integration using the contours in figure A3.11.2. For the $+\mathrm{i}\varepsilon$ choice, the contour in figure A3.11.2(a) is appropriate for $x < x'$ as the circular part has a negative imaginary k' which

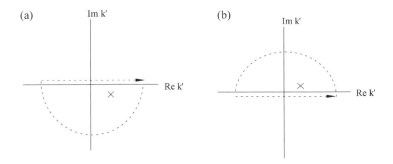

Figure A3.11.2. Integration contours used to evaluate equation (A3.11.36) (a) for $x < x'$, (b) for $x > x'$.

makes $e^{ik'(x-x')}$ vanish for $|k'| \to \infty$. Likewise for $x > x'$, we want to use the contour in figure A3.11.2(b) as this makes the imaginary part of k' positive along the circular part. In either case, the integral along the real axis equals the full contour integral, and the latter is determined by the residue theorem to be $2\pi i$ times the residue at the pole which is encircled by the contour.

The pole is at $k' = -k - i\varepsilon$ for the contour in figure A3.11.2(a) and at $k' = k + i\varepsilon$ for figure A3.11.2(b). This gives us

$$G_0^+(x, x') = \begin{cases} \left(\frac{-i}{2k}\right) e^{-ik(x-x')} & \text{for } x < x' \\ \left(\frac{-i}{2k}\right) e^{ik(x-x')} & \text{for } x > x' \end{cases} \tag{A3.11.37}$$

which we will call the 'plus' wave free particle Green function G_0^+. A different Green function ('minus' wave) is obtained by using $-i\varepsilon$ in the above formulas. It is

$$G_0^-(x, x') = \begin{cases} \left(\frac{i}{2k}\right) e^{ik(x-x')} & \text{for } x < x' \\ \left(\frac{i}{2k}\right) e^{-ik(x-x')} & \text{for } x > x'. \end{cases} \tag{A3.11.38}$$

Upon substitution of G_0 into equation (A3.11.29) we generate the following integral equation for the solution ψ_k^+ that is associated with G_0^+:

$$\psi_k^+(x) = \varphi_k(x) - \int_{-\infty}^{x} \left(\frac{i}{2k}\right) e^{ik(x-x')} \frac{2m}{\hbar^2} V(x')\psi_k^+(x')\, dx' - \int_{x}^{\infty} \left(\frac{i}{2k}\right) e^{-ik(x-x')} \frac{2m}{\hbar^2} V(x')\psi_k^+(x')\, dx'. \tag{A3.11.39}$$

For $x \to \pm\infty$, it is possible to make ψ_k^+ look like equation (A3.11.20) by setting $\varphi_k(x) = e^{ikx}$. This shows that the plus Green function is associated with scattering solutions in which outgoing waves move to the right in the $x \to \infty$ limit. For $x \to -\infty$, equation (A3.11.39) becomes

$$\psi_k^+ \underset{x \to -\infty}{=} e^{ikx} - e^{-ikx} \int_{-\infty}^{\infty} \left(\frac{i}{2k}\right) e^{ikx'} \frac{2m}{\hbar^2} V(x')\psi_k^+(x')\, dx'. \tag{A3.11.40}$$

By comparison with equation (A3.11.20a), we see that

$$S_{11} = -\frac{i}{2k} \int_{-\infty}^{\infty} e^{ikx'} \frac{2m}{\hbar^2} V(x')\psi_k^+(x')\, dx' \tag{A3.11.41}$$

which is an integral that can be used to calculate S_{11} provided that ψ_k^+ is known. One can similarly show that

$$S_{12} = 1 - \frac{i}{2k} \int_{-\infty}^{\infty} e^{ikx'} \frac{2m}{\hbar^2} V(x')\psi_k^+(x')\, dx'. \tag{A3.11.42}$$

The other S matrix components S_{21} and S_{22} can be obtained from the G_0^- Green function.

A3.11.2.8 Born approximation

If $V(x)$ is 'small', ψ_k^+ will not be perturbed much from what it would be if $V(x) = 0$. If so, then we can approximate $\psi_k^+ = e^{ikx}$ and obtain

$$S_{11} = -\frac{i}{2k} \int_{-\infty}^{\infty} e^{2ikx'} \frac{2m}{\hbar^2} V(x')\, dx' \tag{A3.11.43a}$$

$$S_{12} = 1 - \frac{i}{2k} \int_{-\infty}^{\infty} \frac{2m}{\hbar^2} V(x')\, dx'. \tag{A3.11.43b}$$

This is the one dimensional version of what is usually called the *Born approximation* in scattering theory. The transition probability obtained from equation (A3.11.43a) is

$$P_{11} = \frac{m^2}{\hbar^2 p^2} \left| \int_{-\infty}^{\infty} e^{2ikx'} V(x')\, dx' \right|^2 \tag{A3.11.44}$$

where $p = \hbar k$ is the momentum. Note that this approximation simplifies the evaluation of transition probabilities to performing an integral.

A number of improvements to the Born approximation are possible, including *higher order* Born approximations (obtained by inserting lower order approximations to ψ_k^+ into equation (A3.11.40), then the result into (A3.11.41) and (A3.11.42)), and the *distorted wave* Born approximation (obtained by replacing the free particle approximation for ψ_k^+ by the solution to a Schrödinger equation that includes part of the interaction potential). For chemical physics applications, the distorted wave Born approximation is the most often used approach, as the approximation of ψ_k^+ by a plane wave is rarely of sufficient accuracy to be even qualitatively useful. However, even the distorted wave Born approximation is poorly convergent for many applications, so other exact and approximate methods need to be considered.

A3.11.2.9 Variational methods

A completely different approach to scattering involves writing down an expression that can be used to obtain **S** directly from the wavefunction, and which is stationary with respect to small errors in the wavefunction. In this case one can obtain the scattering matrix element by variational theory. A recent review of this topic has been given by Miller [32]. There are many different expressions that give **S** as a functional of the wavefunction and, therefore, there are many different variational theories. This section describes the Kohn variational theory, which has proven particularly useful in many applications in chemical reaction dynamics. To keep the derivation as simple as possible, we restrict our consideration to potentials of the type plotted in figure A3.11.1(c), where the wavefunction vanishes in the limit of $x \to -\infty$, and where the S matrix is a scalar property so we can drop the matrix notation.

The Kohn variational approximation states that for a trial wavefunction $\tilde{\Psi}$ which has the asymptotic form

$$\tilde{\Psi} \underset{x \to \infty}{\sim} -v^{-1/2} (e^{-ikx} - e^{-ikx}\tilde{S}) \tag{A3.11.45}$$

that the quantity

$$S = \tilde{S} + \frac{i}{\hbar} \langle \tilde{\Psi} | \hat{H} - E | \tilde{\Psi} \rangle \tag{A3.11.46}$$

is stationary with respect to variations in $\tilde{\Psi}$, and $S = S_{ex}$ where S_{ex} is the exact scattering matrix when $\tilde{\Psi} = \Psi_{exact}$. Note that $\tilde{\Psi}$ is not complex conjugated in calculating $\langle \tilde{\Psi} |$.

To prove this we expand $\tilde{\Psi}$ about the exact wavefunction Ψ_{ex}, that is, we let

$$\tilde{\Psi} = \Psi_{ex} + \delta\Psi. \tag{A3.11.47}$$

Ψ_{ex} here is assumed to have the asymptotic form

$$\Psi_{\text{ex}} \underset{x\to\infty}{\sim} -v^{-1/2}(e^{-ikx} - e^{-ikx}S_{\text{ex}}). \tag{A3.11.48}$$

This means that

$$S\Psi \underset{x\to\infty}{\sim} v^{-1/2}e^{-ikx}\delta S \tag{A3.11.49}$$

where $\delta S = S - S_{\text{ex}}$. Then we see that

$$S = S_{\text{ex}} + \delta S + \frac{i}{\hbar}\langle\Psi_{\text{ex}} + \delta\Psi|\hat{H} - E|\Psi_{\text{ex}} + \delta\Psi\rangle = S_{\text{ex}} + \delta S + \frac{i}{\hbar}\langle\Psi_{\text{ex}}|\hat{H} - E|\delta\Psi\rangle + O(\delta\Psi^2) \tag{A3.11.50}$$

since $\hat{H} - E|\Psi_{\text{ex}}\rangle = 0$.

Now use integration by parts twice to show that

$$\left\langle\Psi_{\text{ex}}\left|\frac{d^2}{dx^2}\right|\delta\Psi\right\rangle = +(\Psi_{\text{ex}}\delta\Psi' - \Psi'_{\text{ex}}\delta\Psi)\Big|_{-\infty}^{\infty} + \left\langle\delta\Psi\left|\frac{d^2}{dx^2}\right|\Psi_{\text{ex}}\right\rangle \tag{A3.11.51}$$

which means that

$$\langle\Psi_{\text{ex}}|\hat{H} - E|\delta\Psi\rangle = \frac{-\hbar^2}{2\mu}(\Psi_{\text{ex}}\delta\Psi' - \Psi'_{\text{ex}}\delta\Psi)\Big|_{-\infty}^{\infty} + \langle\delta\Psi|\hat{H} - E|\Psi_{\text{ex}}\rangle. \tag{A3.11.52}$$

The last term vanishes, and so does the first at $x = -\infty$. The nonzero part is then

$$\frac{-\hbar^2}{2\mu}(-v^{-1/2}(e^{-lkx} - e^{ikx}S_{\text{ex}})v^{-1/2}(ik)e^{-ikr}\delta S + v^{-1/2}(-ik)(e^{-ikx} + e^{ikx}S_{\text{ex}})v^{-1/2}e^{ikx}\delta S)$$

$$= \frac{-\hbar^2}{2\mu}\left(2\frac{-ik}{v}\delta S\right) = i\hbar\,\delta S. \tag{A3.11.53}$$

So overall

$$S = S_{\text{ex}} + \delta S + \frac{i}{\hbar}(i\hbar\,\delta S) + O(\delta\Psi^2) = S_{\text{ex}} + O(\delta\Psi^2) \tag{A3.11.54}$$

which means that the deviations from the exact result are of second order. This means that S is stationary with respect to variations in the trial function. Later (section A3.11.4) we will show how the variational approach can be used in practical applications where the scattering wavefunction is expanded in terms of basis functions.

A3.11.3 Multichannel quantum scattering theory; scattering in three dimensions

In this section we consider the generalization of quantum scattering theory to problems with many degrees of freedom, and to problems where the translational motion takes place in three dimensions rather than one. The simplest multidimensional generalization is to consider two degrees of freedom, and we will spend much of our development considering this, as it contains the essence of the complexity that can arise in what is called 'multichannel' scattering theory. Moreover, models containing two degrees of freedom are of use throughout the field of chemical physics. For example, this model can be used to describe the collision of an atom with a diatomic molecule with the three atoms constrained to be collinear so that only vibrational motion in the diatomic molecule needs to be considered in addition to translational motion of the atom relative to the molecule. This model is commonly used in studies of *vibrational energy transfer* [29] where the collision causes changes in the vibrational state of the molecule. In addition, this model can be used to describe *reactive* collisions wherein an atom is transferred to form a new diatomic molecule [23, 24]. We will discuss both of these processes in the following two sections (A3.11.3.1 and A3.11.3.2).

The treatment of translational motion in three dimensions involves representation of particle motions in terms of plane waves $e^{ik \cdot r}$ where the wavevector k specifies the direction of motion in addition to the magnitude of the velocity. For problems involving the motion of isolated particles, i.e., gas phase collisions, all problems can be represented in terms of eigenfunctions of the total angular momentum, which is a conserved quantity. The relationship between these eigenfunctions and the plane wave description of particle motions leads to the concept of a *partial wave expansion*, something that is used throughout the field of chemical physics. This is described in the third part of this section (A3.11.3.3).

Problems in chemical physics which involve the collision of a particle with a surface do not have rotational symmetry that leads to partial wave expansions. Instead they have two dimensional translational symmetry for motions parallel to the surface. This leads to expansion of solutions in terms of diffraction eigenfunctions. This theory is described in the literature [33].

A3.11.3.1 Multichannel scattering—coupled channel equations

Consider the collision of an atom (denoted A) with a diatomic molecule (denoted BC), with motion of the atoms constrained to occur along a line. In this case there are two important degrees of freedom, the distance R between the atom and the centre of mass of the diatomic, and the diatomic internuclear distance r. The Hamiltonian in terms of these coordinates is given by:

$$\hat{H} = \frac{\hat{P}_R^2}{2\mu_{A,BC}} + \frac{\hat{P}_r^2}{2\mu_{BC}} + V(R, r) \tag{A3.11.55}$$

$$= -\frac{\hbar^2}{2\mu_{A,BC}} \frac{\partial^2}{\partial R^2} - \frac{\hbar^2}{2\mu_{BC}} \frac{\partial^2}{\partial r^2} + V(R, r) \tag{A3.11.56}$$

where $\mu_{A,BC}$ is the reduced mass associated with motion in the R coordinate, and μ_{BC} is the corresponding diatom reduced mass. Note that this Hamiltonian can be derived by starting with the Hamiltonian of the independent atoms and separating out the motion of the centre of mass. The second form (A3.11.56) arises by replacing the momentum operators by their usual quantum mechanical expressions.

We concentrate in this section on solving the time-independent Schrödinger equation, which, as we learned from section A3.11.2.5, is all we need to do to generate the physically meaningful scattering information. If BC does not dissociate then it is reasonable to use the BC eigenfunctions as a basis for expanding the scattering wavefunction. Assume that as $R \to \infty$, $V(R, r) \to V_{BC}(r)$. Then the BC eigenfunctions are solutions to

$$\left(\frac{-\hbar^2}{2\mu_{BC}} \frac{d^2}{dr^2} + V_{BC}(r) \right) \varphi_v(r) = \varepsilon_v \varphi_v(r) \tag{A3.11.57}$$

where ε_v is the vibrational eigenvalue. The expansion of Ψ in terms of the BC eigenfunctions is thus given by

$$\Psi(R, r) = \sum_v \varphi_v(r) g_v(R) \tag{A3.11.58}$$

where the g_v are unknown functions to be determined. This equation is called a *coupled channel* expansion.

Substituting this into the Schrödinger equation, we find

$$
\frac{-\hbar^2}{2\mu_{A,BC}} \sum_v \varphi_v(r) \frac{d^2 g_v}{dR^2} + \sum_v g_v(R) \left[\frac{-\hbar^2}{2\mu_{BC}} \frac{d^2}{dr^2} + V_{BC}(r) \right] \varphi_v(r) + \sum_v (V(R,r) - V_{BC}) g_v(R) \varphi_v(r)
$$

$$
= E \sum_v g_v(R) \varphi_v(r). \tag{A3.11.59}
$$

Now rearrange, multiply by φ_v, and integrate to obtain

$$
\frac{-\hbar^2}{2\mu_{A,BC}} \frac{d^2 g_v}{dR^2} = (E - \epsilon_v) g_v - \sum_{v'} \langle \phi_v | V - V_{BC} | \phi_{v'} \rangle g_{v'} \tag{A3.11.60}
$$

or

$$
\frac{d^2 g_v(R)}{dR^2} = \sum_{v'} U_{vv'}(R) g_{v'}(R) \tag{A3.11.61}
$$

where

$$
U_{vv'} = \frac{2\mu_{A,BC}}{\hbar^2}(E - \epsilon_v)\, \delta_{vv'} + \frac{2\mu_{BC}}{\hbar^2} \langle \phi_v | V - V_{BC} | \phi_{v'} \rangle. \tag{A3.11.62}
$$

In matrix–vector form these *coupled-channel* equations are

$$
\frac{d^2 g}{dR^2} = \mathbf{U} g \tag{A3.11.63}
$$

where g is the vector formed using the g_v as elements and $\mathbf{U}$ is a matrix whose elements are $U_{vv'}$. Note that the internal states may be either *open* or *closed*, depending on whether the energy E is above or below the internal energy ϵ_v. Only the open states (often termed *open channels*) have measureable scattering properties, but the closed channels can be populated as intermediates during the collision, sometimes with important physical consequences. In the following discussion we confine our discussion to the open channels.

The boundary conditions on the open channel solutions are:

$$
\mathbf{g}(R) \to 0 \text{ as } R \to 0 \tag{A3.11.64}
$$

provided that the potential is repulsive at short range, and

$$
\mathbf{g}(R) \to \mathbf{v}^{-1/2}(e^{-ikR} - e^{ikR} S) \text{ as } R \to \infty. \tag{A3.11.65}
$$

Here we have collected the N independent g that correspond to different incoming states for N open channels into a matrix $\mathbf{g}$ (where the *sans serif* bold notation is again used to denote a square matrix). Also we have the matrices

$$
(\mathbf{v})_{vv'} = v_v\, \delta_{vv'} \tag{A3.11.66}
$$

$$
(\mathbf{k})_{vv'} = k_v\, \delta_{vv'} \tag{A3.11.67}
$$

where

$$
k_v = \frac{\sqrt{2\mu_{A,BC}}}{\hbar^2}(E - \varepsilon_v) \tag{A3.11.68}
$$

$$
v_v = \frac{\hbar k_v}{\mu_{A,BC}} \tag{A3.11.69}
$$

$$
\mathbf{S} = S_{vv}. \tag{A3.11.70}
$$

S is the *scattering matrix*, analogous to that defined earlier. As before, the probabilities for transitions between states v and v' are

$$P_{vv'} = |S_{vv'}|^2. \tag{A3.11.71}$$

Often in numerical calculations we determine solutions $\mathbf{g}(R)$ that solve the Schrödinger equations but do not satisfy the asymptotic boundary condition in (A3.11.65). To solve for **S**, we rewrite equation (A3.11.65) and its derivative with respect to R in the more general form:

$$\mathbf{g} = (\mathbf{I} - \mathbf{OS})\mathbf{A} \tag{A3.11.72}$$
$$\mathbf{g}' = (\mathbf{I}' - \mathbf{O}'\mathbf{S})\mathbf{A} \tag{A3.11.73}$$

where the incoming and outgoing asymptotic solutions are:

$$\mathbf{I} = \kappa^{-1/2}\, e^{-i\kappa R} \tag{A3.11.74}$$
$$\mathbf{O} = \kappa^{-1/2}\, e^{i\kappa R}. \tag{A3.11.75}$$

A is a coefficient matrix that is designed to transform between solutions that obey arbitrary boundary conditions and those which obey the desired boundary conditions. **A** and **S** can be regarded as unknowns in equations (A3.11.72) and (A3.11.73). This leads to the following expression for **S**:

$$\mathbf{S} = \mathbf{W}^{-1}(\mathbf{I}'\mathbf{g} - \mathbf{I}\mathbf{g}')(\mathbf{O}'\mathbf{g} - \mathbf{O}\mathbf{g}')^{-1}\mathbf{W} \tag{A3.11.76}$$

where

$$\mathbf{W} = \mathbf{O}'\mathbf{I} - \mathbf{O}\mathbf{I}'. \tag{A3.11.77}$$

The present derivation can easily be generalized to systems with an arbitrary number of internal degrees of freedom, and it leads to coupled channel equations identical with equation (A3.11.63), where the coupling terms (A3.11.62) are expressed as matrix elements of the interaction potential using states which depend on these internal degrees of freedom. These internal states could, in principle, have a continuous spectrum but, in practice, if there are multiple continuous degrees of freedom then it is most useful to reformulate the problem to take this into account. One particularly important case of this sort arises in the treatment of reactive collisions, where the atom B is transferred from C to A, leading to the formation of a new arrangement of the atoms with its own scattering boundary conditions. We turn our attention to this situation in the next section.

A3.11.3.2   *Reactive collisions*

Let us continue with the atom–diatom collinear collision model, this time allowing for the possibility of the reaction $A + BC \rightarrow AB + C$. We first introduce *mass-scaled* coordinates, as these are especially convenient to describe rearrangements, using

$$R' = \left(\frac{\mu_{A,BC}}{m}\right)^{1/2} R \tag{A3.11.78}$$

$$r' = \left(\frac{\mu_{BC}}{m}\right)^{1/2} r. \tag{A3.11.79}$$

The choice of m in these formulas is arbitrary, but it is customary to take either $m = 1$ or

$$m = \sqrt{\frac{m_A m_B m_C}{m_A + m_B + m_C}} = \sqrt{\mu_{A,BC}\mu_{BC}}. \tag{A3.11.80}$$

Either choice is invariant to permutation of the atom masses.

In terms of these coordinates, the Hamiltonian of equation (A3.11.56) becomes

$$H = \frac{\hat{P}_R^2}{2\mu_{A,BC}} + \frac{\hat{P}_r^2}{2\mu_{BC}} + V = \frac{\hat{P}_R'^2 + \hat{P}_r'^2}{2m} + V. \tag{A3.11.81}$$

One nice thing about H in mass-scaled coordinates is that it is identical to the Hamiltonian of a mass point moving in two dimensions. This is convenient for visualizing trajectory motions or wavepackets, so the mass-scaled coordinates are commonly used for plotting data from scattering calculations.

Another reason why mass-scaled coordinates are useful is that they simplify the transformation to the Jacobi coordinates that are associated with the products $AB + C$. If we define S as the distance from C to the centre of mass of AB, and s as the AB distance, mass scaling is accomplished via

$$S' = \sqrt{\frac{\mu_{C,AB}}{m}} S \tag{A3.11.82}$$

$$s' = \sqrt{\frac{\mu_{AB}}{m}} s. \tag{A3.11.83}$$

The Hamiltonian in terms of product coordinates is

$$\hat{H} = \frac{\hat{P}_{S'}^2 + \hat{P}_{s'}^2}{2m} + V \tag{A3.11.84}$$

and the transformation between reagent and product coordinates is given by:

$$\begin{aligned} S' &= r' \sin \beta + R' \cos \beta \\ s' &= -r' \cos \beta + R' \sin \beta \end{aligned} \tag{A3.11.85}$$

where the angle β is defined by:

$$\tan \beta = \sqrt{\frac{m_B(m_A + m_B + m_C)}{m_A m_C}}. \tag{A3.11.86}$$

Equation (A3.11.85) implies that the $R', r' \rightarrow S', s'$ transformation is orthogonal, a point which is responsible for the similarities between the Hamiltonian expressed in terms of reagent and product mass-scaled coordinates ((A3.11.81) and (A3.11.84)). In fact, the reagent to product transformation can be thought of as a rotation by an angle β followed by a flip in the sign of s'. The angle β is sometimes called the 'skew' angle, and it can vary between 0 and 90°, as determined by equation (A3.11.86). If $m_A = m_B = m_C$ (i.e., all three masses are identical, as in the reaction $H + H_2$), then $\beta = 60°$, while for $m_B \gg m_A, m_C$, $\beta \rightarrow 90°$ and $m_B \ll m_A m_C$ gives $\beta \rightarrow 0$.

Although the Schrödinger equation associated with the $A + BC$ reactive collision has the same form as for the nonreactive scattering problem that we considered previously, it *cannot* be solved by the coupled-channel expansion used then, as the reagent vibrational basis functions cannot directly describe the product region (for an expansion in a finite number of terms). So instead we need to use alternative schemes of which there are many.

One possibility is to use *hyperspherical coordinates*, as these enable the use of basis functions which describe reagent and product internal states in the same expansion. Hyperspherical coordinates have been extensively discussed in the literature [34–36] and in the present application they reduce to polar coordinates (ρ, η) defined as follows:

$$\rho = \sqrt{R'^2 + r'^2} = \sqrt{S'^2 + s'^2} \qquad 0 \le \rho \le \infty \tag{A3.11.87}$$

$$\eta = \tan^{-1} \frac{r'}{R'} \qquad 0 \le \eta \le \beta. \tag{A3.11.88}$$

Hyperspherical coordinates have the properties that η motion is always bound since $\eta = 0$ and $\eta = \beta$ correspond to cases where two of the three atoms are on top of one another, yielding a very repulsive potential. Also, $\rho \to 0$ is a repulsive part of the potential, while large ρ takes us to the reagent and product valleys.

To develop coupled-channel methods to solve the Schrödinger equation, we first transform the Hamiltonian (A3.11.81) to hyperspherical coordinates, yielding:

$$\hat{H} = \frac{-\hbar^2}{2m}\left(\frac{\partial^2}{\partial\rho^2} + \frac{1}{\rho}\frac{\partial}{\partial\rho} + \frac{1}{\rho^2}\frac{\partial^2}{\partial\eta^2}\right) + V. \tag{A3.11.89}$$

Now define a new wavefunction $\chi = \rho^{+1/2}\psi$. Then

$$\hat{H}\psi = \frac{-\hbar^2}{2m}\left(\frac{\partial^2}{\partial\rho^2} + \frac{1}{\rho}\frac{\partial}{\partial\rho} + \frac{1}{\rho^2}\frac{\partial^2}{\partial\eta^2}\right)\rho^{-1/2}\chi + V\rho^{-1/2}\chi. \tag{A3.11.90}$$

After cancelling out a factor $\rho^{-1/2}$ and regrouping, we obtain a new version of the Schrödinger equation in which the first derivative term has been eliminated.

$$\left\{\frac{-\hbar^2}{2m}\left[\frac{\partial^2}{\partial\rho^2} + \frac{1}{4\rho^2} + \frac{1}{\rho^2}\frac{\partial^2}{\partial\eta^2}\right] + V\right\}\chi = E\chi. \tag{A3.11.91}$$

Now select out the η-dependent part to define vibrational functions at some specific ρ which we call $\bar{\rho}$.

$$-\frac{\hbar^2}{2m\bar{\rho}^2}\frac{d^2}{d\eta^2}\varphi_n + V(\bar{\rho},\eta)\varphi_n = \varepsilon_n\varphi_n \tag{A3.11.92}$$

with the boundary condition that $\varphi_n \to 0$ as $\eta \to 0$ and $\eta \to \beta$. For large $\bar{\rho}$ the φ_n will become eigenfunctions of the reagent and product diatomics.

To set up coupled-channel equations, use the expansion

$$\psi = \rho^{-1/2}\sum_n \varphi_n(\eta)g_n(\rho). \tag{A3.11.93}$$

This leads to

$$\frac{-\hbar^2}{2m}\frac{d^2}{d\rho^2}\sum_n \varphi_n g_n + \frac{-\hbar^2}{2m\rho^2}\sum_n g_n\frac{d^2\varphi_n}{d\eta^2} + \left(V - \frac{\hbar^2}{8m\rho^2}\right)\sum_n \varphi_n g_n = E\sum_n \varphi_n g_n. \tag{A3.11.94}$$

Now substitute the $\rho = \bar{\rho}$ solution for $d^2\varphi_n/d\eta^2$ from above, multiply by φ_n and integrate over η. This gives

$$\frac{d^2\mathbf{g}}{d\rho^2} = \mathbf{U}\mathbf{g} \tag{A3.11.95}$$

where

$$U_{nn'} = \frac{2m}{\hbar^2}\left\langle\varphi_n\left|V(\rho,\eta) - \frac{\bar{\rho}^2}{\rho^2}V(\bar{\rho},\eta)\right|\varphi_{n'}\right\rangle + \frac{2m}{\hbar^2}\delta_{nn'}\left(\varepsilon_n\frac{\bar{\rho}^2}{\rho^2} - \frac{\hbar^2}{8m\rho^2} - E\right). \tag{A3.11.96}$$

This equation may be solved by the same methods as used with the nonreactive coupled-channel equations (discussed later in section A3.11.4.2). However, because $V(\rho,\eta)$ changes rapidly with ρ, it is desirable to periodically change the expansion basis set φ_n. To do this we divide the range of ρ to be integrated into 'sectors' and within each sector choose a $\bar{\rho}$ (usually the midpoint) to define local eigenfunctions. The coupled-channel equations just given then apply within each sector, but at sector boundaries we change basis sets. Let $\bar{\rho}_1$ and $\bar{\rho}_2$ be the $\bar{\rho}$ associated with adjacent sectors. Then, at the sector boundary ρ_b we require

$$\Psi_1(\eta, \rho_b) = \Psi_2(\eta, \rho_b) \tag{A3.11.97}$$

or

$$\sum_n \varphi_n(\eta, \bar{\rho}_1) g_{nn_i}^{(1)}(\rho_b) = \sum_n \varphi_n(\eta, \bar{\rho}_2) g_{nn_i}^{(2)}(\rho_b). \tag{A3.11.98}$$

Multiply by $\varphi_n(\eta, \bar{\rho}_2)$ and integrate to obtain

$$\mathbf{g}^{(2)} = \mathbf{S}^{21} \mathbf{g}^{(1)} \tag{A3.11.99}$$

where

$$S_{nn'}^{21} = \langle \varphi_n(\eta, \bar{\rho}_2) | \varphi_n(\eta, \bar{\rho}_1) \rangle. \tag{A3.11.100}$$

The corresponding derivative transformation is:

$$\frac{d\mathbf{g}^{(2)}}{d\rho} = \mathbf{S}^{21} \frac{d\mathbf{g}^{(1)}}{d\rho}. \tag{A3.11.101}$$

This scheme makes it possible to propagate $\mathbf{g}$ from small ρ where $\mathbf{g}$ should vanish to large ρ where an asymptotic analysis can be performed.

To perform the asymptotic analysis we need to first write down the proper asymptotic solution. Clearly we want some solutions with incoming waves in the reagents, then outgoing waves in both reagents and products and other solutions with the reagent and product labels interchanged. One way to do this is to define a matrix of incoming waves $\mathbf{I}$ and outgoing waves $\mathbf{O}$ such that

$$I_{\alpha v' \alpha' v'} = \delta_{\alpha\alpha'} \delta_{vv'} \begin{cases} e^{-ik_v R'} & \alpha = 1 \\ e^{-ik_v S'} & \alpha = 2 \end{cases} \tag{A3.11.102a}$$

$$O_{\alpha v \alpha' v'} = \delta_{\alpha\alpha'} \delta_{vv'} \begin{cases} e^{ik_v R'} & \alpha = 1 \\ e^{ik_v S'} & \alpha = 2 \end{cases} \tag{A3.11.102b}$$

where α is an arrangement channel label such that $\alpha = 1$ and 2 correspond to the 'reagents' and 'products'. Also let φ_{av} be the reagent or product vibrational function. Then the asymptotic solution is

$$\Psi_{R'S' \to \infty} \sim \sum_{\alpha v} \varphi_{\alpha v} (v_\alpha)^{-1/2} \left(I_{\alpha v \alpha' v'} - \sum_{\alpha'' v''} S_{\alpha v \alpha'' v''} O_{\alpha'' v'' \alpha v} \right). \tag{A3.11.103}$$

We have expressed Ψ in terms of Jacobi coordinates as this is the coordinate system in which the vibrations and translations are separable. The separation does not occur in hyperspherical coordinates except at $\rho = \infty$, so it is necessary to interrelate coordinate systems to complete the calculations. There are several approaches for doing this. One way is to project the hyperspherical solution onto Jacobi's before performing the asymptotic analysis, i.e.

$$\rho^{-1/2} \sum_v \varphi_v(\eta) g_{vv'}(\rho) = \sum_v \varphi_v(r) G_{vv'}(R). \tag{A3.11.104}$$

The $\mathbf{G}$ matrix is then obtained by performing the quadrature

$$G_{vv'} = \int dr \varphi_v(r) \sum_{v''} \rho^{-1/2} \varphi_{v''}(\eta) g_{v''v'}(\rho) \tag{A3.11.105}$$

where $\rho(R, r), \eta(R, r)$ are to be substituted as needed into the right hand side.

A3.11.3.3   *Scattering in three dimensions*

All the theory developed up to this point has been limited in the sense that translational motion (the continuum degree of freedom) has been restricted to one dimension. In this section we discuss the generalization of this to three dimensions for collision processes where space is isotropic (i.e., collisions in homogeneous phases, such as in a vacuum, but not collisions with surfaces). We begin by considering collisions involving a single particle in three dimensions; the multichannel case is considered subsequently.

The biggest change associated with going from one to three dimensional translational motion refers to asymptotic boundary conditions. In three dimensions, the initial scattering wavefunction for a single particle is represented by a plane wave $e^{i\mathbf{k}\cdot\mathbf{r}}$ moving in a direction which we denote with the wavevector $\mathbf{k}$. Scattering then produces outgoing spherical waves as $t \to \infty$ weighted by an amplitude $f_k(\theta)$ which specifies the scattered intensity as a function of the angle θ between $\mathbf{k}$ and the observation direction. Mathematically the time independent boundary condition analogous to equations (A3.11.13a, b) is:

$$\psi_k(r) \underset{r\to\infty}{=} e^{i\mathbf{k}\cdot\mathbf{r}} + f_k(\theta)\frac{e^{ikr}}{r}. \tag{A3.11.106}$$

Note that for potentials that depend only on the scalar distance r between the colliding particles, the amplitude $f_k(\theta)$ does not depend on the azimuthal angle associated with the direction of observation.

The measurable quantity in a three dimensional scattering experiment is the differential cross section $d\sigma_k(\theta)/d\Omega$. This is defined as

$$\frac{d\sigma_k(\theta)}{d\Omega} = \frac{|\text{outgoing radial flux}|}{|\text{total incident flux}|} \tag{A3.11.107}$$

where outgoing flux refers to the radial velocity operator $\hat{v}_r = -i\hbar\partial/\partial r$. Substitution of equation (A3.11.106) into (A3.11.107) using (A3.11.16) yields

$$d\sigma_k(\theta)/d\Omega = |f_k(\theta)|^2. \tag{A3.11.108}$$

It is convenient to expand $f_k(\theta)$ in a basis of Legendre polynomials $P_\ell(\cos\theta)$ (as these define the natural angular eigenfunctions associated with motion in three dimensions). Here we write:

$$f_k(\theta) = \sum_\ell a_\ell^k P_\ell(\cos\theta). \tag{A3.11.109}$$

We call this a *partial wave expansion*. To determine the coefficients a_ℓ^k, one matches asymptotic solutions to the radial Schrödinger equation with the corresponding partial wave expansion of equation (A3.11.106). It is customary to write the asymptotic radial Schrödinger equation solution as

$$\psi_{\ell m}(r, \theta, \varphi) \underset{r\to\infty}{=} \frac{1}{r}Y_{\ell m}(\theta, \varphi)(e^{-i(kr-\ell\pi/2)} - S_\ell\, e^{i(kr-\ell\pi/2)}) \tag{A3.11.110}$$

where S_ℓ is the scattering matrix for the ℓth partial wave and m is the projection quantum number associated with ℓ. Unitarity of the scattering matrix implies that S_ℓ can be written as $\exp(2i\,\delta_\ell)$ where δ_ℓ is a real quantity known as the *phase shift*.

The asymptotic partial wave expansion of equation (A3.11.106) can be developed using the identity

$$e^{i\mathbf{k}\cdot\mathbf{r}} = e^{ikr\cos\theta} = \sum_{\ell=0}^{\infty} i^\ell (2\ell + 1) j_\ell(kr) P_\ell(\cos\theta) \tag{A3.11.111}$$

where $j_\ell(kr)$ is a spherical Bessel function. At large r, the spherical Bessel function reduces to

$$j_\ell(kr) \underset{r\to\infty}{=} \frac{\sin(kr - \ell\pi/2)}{kr}. \tag{A3.11.112}$$

If equation (A3.11.112) is then used to evaluate (A3.11.111) after substitution of the latter into (A3.11.106) and if equation (A3.11.109) is also substituted into (A3.11.106) and the result for each ℓ and m is equated to (A3.11.110), one finds that only $m = 0$ contributes, and that

$$a_\ell^k = \frac{(2\ell + 1)}{2ik}(S_\ell - 1). \qquad (A3.11.113)$$

From equations (A3.11.108), (A3.11.109) and (A3.11.113) one then finds

$$\frac{d\sigma_k(\theta)}{d\Omega} = \frac{1}{4k^2}\left|\sum_\ell (2\ell + 1)P_\ell(\cos\theta)(S_\ell - 1)\right|^2 \qquad (A3.11.114a)$$

$$= \frac{1}{k^2}\left|\sum_\ell (2\ell + 1)P_\ell(\cos\theta)\,e^{i\delta_\ell}\sin\delta_\ell\right|^2. \qquad (A3.11.114b)$$

This differential cross section may be integrated over scattering angles to define an integral cross section σ as follows:

$$\sigma = 2\pi\int_0^\pi \frac{d\sigma_k(\theta)}{d\Omega}\sin\theta\,d\theta = \frac{\pi}{k^2}\sum_\ell (2\ell + 1)|S_\ell - 1|^2 \qquad (A3.11.115a)$$

$$= \frac{4\pi}{k^2}\sum_\ell (2\ell + 1)\sin^2\delta_\ell. \qquad (A3.11.115b)$$

Equations (A3.11.114b) and (A3.11.115b) are in a form that is convenient to use for potential scattering problems. One needs only to determine the phase shift δ_ℓ for each ℓ, then substitute into these equations to determine the cross sections. Note that in the limit of large ℓ, δ_ℓ must vanish so that the infinite sum over partial waves ℓ will converge. For most potentials of interest to chemical physics, the calculation of δ_ℓ must be done numerically.

Equation (A3.11.115a) is also useful as a form that enables easy generalization of the potential scattering theory that we have just derived to multistate problems. In particular, if we imagine that we are interested in the collision of two molecules A and B starting out in states n_A and n_B and ending up in states n'_A and n'_B, then the asymptotic wavefunction analogous to equation (A3.11.106) is

$$\psi_{n_A n_B \to n'_A n'_B} = \exp(ik_{n_A n_B} \cdot r)|n_A n_B\rangle + r^{-1}\sum_{n'_A n'_B} f_{n_A n_B \to n'_A n'_B}(\theta)\exp(ik_{n'_A n'_B}r)|n'_A n'_B\rangle \qquad (A3.11.116)$$

where the scattering amplitude f is now labelled by the initial and final state indices. Integral cross sections are then obtained using the following generalization of equation (A3.11.115a):

$$\sigma_{n_A n_B \to n'_A n'_B} = \frac{\pi}{k_{n_A n_B}^2}\sum_J (2J + 1)|S_{n_A n_B \to n'_A n'_B}^J - \delta_{n_A n_B,\,n'_A n'_B}|^2 \qquad (A3.11.117)$$

where $\mathbf{S}$ is the multichannel scattering matrix, δ is the Kronecker delta function and J is the total angular momentum (i.e., the vector sum of the orbital angular momentum ℓ plus the angular momenta of the molecules A and B). Here the sum is over J rather than ℓ, because ℓ is not a conserved quantity due to coupling with angular momenta in the molecules A and B.

A3.11.4 Computational methods and strategies for scattering problems

In this section we present several numerical techniques that are commonly used to solve the Schrödinger equation for scattering processes. Because the potential energy functions used in many chemical physics problems are complicated (but known to reasonable precision), new numerical methods have played an important role in extending the domain of application of scattering theory. Indeed, although much of the formal development of the previous sections was known 30 years ago, the numerical methods (and computers) needed to put this formalism to work have only been developed since then.

This section is divided into two sections: the first concerned with time-dependent methods for describing the evolution of wavepackets and the second concerned with time-independent methods for solving the time independent Schrödinger equation. The methods described are designed to be representative of what is in use, but not exhaustive. More detailed discussions of time-dependent and time-independent methods are given in the literature [37, 38].

A3.11.4.1   Time-dependent wavepacket methods

(a) Overall strategy

The methods described here are all designed to determine the time evolution of wavepackets that have been previously defined. This is only one of several steps for using wavepackets to solve scattering problems. The overall procedure involves the following steps:

(a)   First, choose an initial wavepacket $\psi(x, t)$ that describes the range of energies and initial conditions that we want to simulate and which is numerically as well behaved as possible. Typically, this is chosen to be a Gaussian function of the translational coordinate x, with mean velocity and width chosen to describe the range of interest. In making this choice, one needs to consider how the spatial part of the Schrödinger equation is to be handled, i.e., whether the dependence of the wavepacket on spatial coordinates is to be represented on a grid, or in terms of basis functions.

(b)   Second, one propagates this wavepacket in time using one of the methods described below, for a sufficient length of time to describe the scattering process of interest.

(c)   Third, one calculates the scattering information of interest, such as the outgoing flux.

Typically, the ratio of this to the incident flux determines the transition probability. This information will be averaged over the energy range of the initial wavepacket, unless one wants to project out specific energies from the solution. This projection procedure is accomplished using the following expression for the energy resolved (time-independent) wavefunction in terms in terms of its time-dependent counterpart:

$$\psi^+(E) = \frac{1}{a(E)} \int_{-\infty}^{\infty} e^{iEt} \psi(t) \, dt \tag{A3.11.118}$$

where

$$a(E) = \frac{1}{v^{1/2}} \langle e^{-ikR} | \psi(0) \rangle. \tag{A3.11.119}$$

(b) Second order differencing

A very simple procedure for time evolving the wavepacket is the second order differencing method. Here we illustrate how this method is used in conjunction with a fast Fourier transform method for evaluating the spatial coordinate derivatives in the Hamiltonian.

If we write the time-dependent Schrödinger equation as $\partial \psi / \partial t = -(i/\hbar)\hat{H}\psi$, then, after replacing the time derivative by a central difference, we obtain

$$\frac{\partial x}{\partial t} = \frac{\psi(t + \Delta t) - \psi(t + \Delta t)}{2\Delta t}. \tag{A3.11.120}$$

After rearranging this becomes

$$\psi(t+\Delta t) = \psi(t-\Delta t) - \frac{2i\Delta t}{\hbar}\hat{H}\psi(t).$$
(A3.11.121)

To invoke this algorithm, we need to evaluate $\hat{H}\psi = (\hat{T}+\hat{V})\psi$. If ψ is represented on a uniform grid in coordinate x, an effective scheme is to use fast Fourier transforms (FFTs) to evaluate $\hat{T}\psi$. Thus, in one dimension we have, with n points on the grid,

$$\tilde{\psi}(k_j,t) = \sum_{m=1}^{n} e^{ik_j x_m}\psi(x_m,t)$$
(A3.11.122)

where the corresponding momentum grid is

$$k_j = \frac{2\pi j}{n\Delta x}.$$
(A3.11.123)

Differentiation of (A3.11.122) then gives

$$\hat{T}\tilde{\psi}(k_j,t) = \frac{p_j^2}{2m}\tilde{\psi}(k_j,t)$$
(A3.11.124a)

where $p_j = \hbar k_j/m$. This expression can be inverted to give

$$\hat{T}\psi(x_\ell) = \frac{1}{n}\sum_{j=-n/2}^{n/2-1} e^{-ik_j x_\ell}\frac{p_j^2}{2m}\tilde{\psi}(k_j,t).$$
(A3.11.124b)

This expression, in combination with (A3.11.122), determines the action of the kinetic energy operator on the wavefunction at each grid point. The action of $\hat{V}$ is just $V(x_\ell)\psi(x_\ell)$ at each grid point.

(c) Split-operator or Feit–Fleck method

A more powerful method for evaluating the time derivative of the wavefunction is the split-operator method [39]. Here we start by formally solving $i\hbar\partial\psi/\partial t = \hat{H}\psi$ with the solution $\psi(t) = e^{-i\hat{H}t/\hbar}\psi(0)$. Note that H is assumed to be time-independent. Now imagine evaluating the propagator $e^{-i\hat{H}t/\hbar}$ over a short time interval.

$$e^{-i\hat{H}\Delta t/\hbar} = e^{-i(\hat{T}+\hat{V})\Delta t/\hbar} \approx e^{-i\hat{V}\Delta t/2\hbar}e^{-i\hat{T}\Delta t/\hbar}e^{-i\hat{V}\Delta t/2\hbar}.$$
(A3.11.125)

Evidently, this formula is not exact if $\hat{T}$ and $\hat{V}$ do not commute. However for short times it is a good approximation, as can be verified by comparing terms in Taylor series expansions of the middle and right-hand expressions in (A3.11.125). This approximation is intrinsically unitary, which means that scattering information obtained from this calculation automatically conserves flux.

The complete propagator is then constructed by piecing together N time steps, leading to

$$e^{-i\hat{H}\Delta t/\hbar} = e^{-i\hat{V}\Delta t/2\hbar}e^{-i\hat{T}\Delta t/\hbar}e^{-\hat{V}\Delta t/\hbar}\cdots e^{-i\hat{V}\Delta t/\hbar}e^{-i\hat{T}\Delta t/\hbar}e^{-i\hat{V}\Delta t/2\hbar}.$$
(A3.11.126)

To evaluate each term we can again do it on a grid, using FFTs as described above to evaluate $e^{-i\hat{T}\Delta t/\hbar}$.

(d) Chebyshev method

Another approach [40] is to expand $e^{-i\hat{H}\Delta t/\hbar}$ in terms of Chebyshev polynomials and to evaluate each term in the polynomial at the end of the time interval. Here a Chebyshev expansion is chosen as it gives the most uniform convergence in representing the exponential over the chosen time interval. The time interval Δt is typically chosen to be several hundred of the time steps that would be used in the second order differencing or split-operator methods. Although the Chebyshev method is not intrinsically unitary, it is capable of much higher accuracy than the second order differencing or split-operator methods [41].

In order to apply this method it is necessary to scale $\hat{H}$ to lie in a certain finite interval which is usually chosen to be $(-1, 1)$. Thus, if V_{max} and V_{min} are estimates of the maximum and minimum potentials and T_{max} is the maximum kinetic energy, we use

$$\hat{H}_{norm} = [\hat{H} - (R + V_{min})]/R = \frac{\hat{H} - V_{min}}{R} - 1 \qquad (A3.11.127)$$

where

$$R = (T_{max} + V_{max} - V_{min})/2. \qquad (A3.11.128)$$

This choice restricts the range of values of $\hat{H}_{norm}$ to the interval $(0,1)$. Then the propagator becomes

$$e^{-i\hat{H}\Delta t/\hbar} = e^{-i\hat{H}_{norm}R\Delta t/\hbar}\, e^{-i(R+V_{min})\Delta t/\hbar}. \qquad (A3.11.129)$$

Now we replace the first exponential in the right-hand side of (A3.11.129) by a Chebyshev expansion as follows:

$$e^{-i\hat{H}\Delta t/\hbar} = e^{-i(R+V_{min})\Delta t/\hbar} \sum_{k=0}^{N} C_k J_k(R\Delta t/\hbar) i^k T_k(-\hat{H}_{norm}) \qquad (A3.11.130)$$

where T_k is a Chebyshev polynomial, and J_k is a Bessel function. The coefficients C_k are fixed at $C_0 = 1$, $C_k = 2, k > 0$.

To apply this method, the J_k are calculated once and stored while the T_k are generated using the recursion formula:

$$T_{k+1}(x) = 2x T_k(x) - T_{k-1}(x) \qquad (A3.11.131)$$

with $T_0 = 1$, $T_1 = x$. Actually the T_k are never explicitly stored, as all we really want is T_k operating onto a wavefunction. However, the recursion formula is still used to generate this, so the primary computational step involves $\hat{H}_{norm}$ operating onto wavefunctions. This can be done using FFTs as discussed previously.

(e) Short iterative Lanczos method

Another approach involves starting with an initial wavefunction ψ_0, represented on a grid, then generating $\hat{H}\psi_0$, and consider that this, after orthogonalization to ψ_0, defines a new state vector. Successive applications $\hat{H}$ can now be used to define an orthogonal set of vectors which defines as a *Krylov space* via the iteration: $(n = 0, \ldots, N)$

$$\beta_{n+1}|\psi_{n+1}\rangle = (\hat{H} - \alpha_n)|\psi_n\rangle - \beta_n|\psi_{n-1}\rangle \qquad (A3.11.132)$$

where

$$\alpha_n = \langle\psi_n|\hat{H}|\psi_n\rangle \qquad \beta_{n+1} = \langle\psi_n|\hat{H}|\psi_{n+1}\rangle. \qquad (A3.11.133)$$

The Hamiltonian in this vector space is

$$\mathbf{H} = \begin{pmatrix} \alpha_0 & \beta_1 & 0 & 0 & 0 \\ \beta_1 & \alpha_1 & \beta_2 & 0 & 0 \\ 0 & \beta_2 & \alpha_2 & \beta_3 & 0 \\ 0 & 0 & \beta_2 & \alpha_3 & \beta_4 \\ . & . & . & . & . \end{pmatrix}. \qquad (A3.11.134)$$

Here **H** forms an $N \times N$ matrix, where N is the dimensionality of the space and is generally much smaller than the number of grid points.

Now diagonalize **H**, calling the eigenvalues λ_k and eigenvectors $T_{\ell k}$. Numerically, this is a very efficient process due to the tridiagonal form of (A3.11.134). The resulting eigenvalues and eigenvectors are then used to propagate for a short time Δt via

$$[\psi(t+\Delta t)]_j = [\psi_0^+(t)\psi_0(t)]_j^{1/2} \sum_{\ell=0}^{N} \sum_{k=0}^{\ell} \frac{(\psi(t)_\ell)_j}{[\psi(t)_\ell^+ \psi(t)_\ell]_j^{1/2}} T_{\ell k}\, e^{-i\lambda_k \Delta t/\hbar} T_{0k} \qquad (A3.11.135)$$

where $(\psi_\ell)_j$ are coefficients that transfer between the jth grid point and the ℓth order Krylov space in (A3.11.132).

A3.11.4.2   Time-independent methods

Here we discuss several methods that are commonly used to propagate coupled-channel equations, and we also present a linear algebra method for applying variational theory. The coupled-channel equations are coupled ordinary differential equations, so they can in principle be solved using any one of a number of standard methods for doing this (Runge–Kutta, predictor–corrector etc). However these methods are very inefficient for this application and a number of alternatives have been developed which take advantage of specific features of the problems being solved.

(a) Gordon-type methods

In many kinds of atomic and molecular collision problem the wavefunction has many oscillations because the energy is high (i.e., $g(R) \approx e^{ikR}$ and k is large). In this case it useful to expand $g(R)$ in terms of oscillatory solutions to some reference problem that is similar to the desired one and then regard the expansion coefficients as the quantity being integrated, thereby removing most or all of the oscillations from the time dependence of the coefficients.

For example, suppose that we divide coordinate space into steps ΔR, then evaluate **U** (in equation (A3.11.61)) at the middle of each step and regard this as the reference for propagation within this step. Further, let us diagonalize **U**, calling the eigenvalues u_k and eigenvectors **T**. Then, as long as the variation in eigenvalues and eigenvectors can be neglected in each step, the Schrödinger equation solution within each step is easily expressed in terms of sin and cos $w_k \Delta R$, where $w_k = \sqrt{-u_k}$ (or exponential solutions if $u_k > 0$). In particular, if $g(R_0)$ is the solution at the beginning of each step, then **Tg** transforms into the diagonalized representation and **Tg**′ is the corresponding derivative. The complete solution at the end of each step would then be

$$\mathbf{g}(R_1) = \mathbf{T}^{-1}(\mathbf{w}^{-1}\sin(\mathbf{w}\Delta R)\mathbf{Tg}'(R_0) + \cos(\mathbf{w}\Delta R)\mathbf{Tg}(R_0)) \qquad (A3.11.136)$$

$$\mathbf{g}'(R_1) = \mathbf{T}^{-1}(\cos(\mathbf{w}\Delta R)\mathbf{Tg}'(R_0) - \mathbf{w}\sin(\mathbf{w}\Delta R)\mathbf{Tg}(R_0)). \qquad (A3.11.137)$$

In principle, one can do better by allowing for R-dependence to **U** and **T**. If we allow them to vary linearly with R, then we have Gordon's method [42]. However, the higher order evaluation in this case leads to a much more cumbersome theory that is often less efficient even though larger steps can be used.

One problem with using this method (or any method that propagates ψ) is that in regions where $u_k > 0$, the so-called 'closed' channels, the solutions increase exponentially. If such solutions exist for some channels while others are still open, the closed-channel solutions can become numerically dominant (i.e., so much bigger that they overwhelm the open-channel solutions to within machine precision and, after a while, all channels propagate as if they are closed). To circumvent this, it is necessary to 'stabilize' the solutions periodically. Typically this is done by multiplying $\mathbf{g}(R)$ and $\mathbf{g}'(R)$ by some matrix **h** that 'orthogonalizes' the solutions as best one can. For example, this can be done using $\mathbf{h} = \mathbf{g}^{-1}(R_s)$ where R_s is the value of R at the end of the 'current' step. Thus, after stabilization, the new $\mathbf{g}$ and $\mathbf{g}'$ are:

$$\mathbf{g}_{\text{new}}(R_s) = \mathbf{g}_{\text{old}}(R_s)\mathbf{g}_{\text{old}}^{-1}(R_4) = \mathbf{I} \tag{A3.11.138}$$

$$\mathbf{g}'_{\text{new}}(R_s) = \mathbf{g}'_{\text{old}}(R_s)\mathbf{g}_{\text{old}}^{-1}(R_4). \tag{A3.11.139}$$

One consequence of performing the stabilization procedure is that the initial conditions that correspond to the current $\mathbf{g}(R)$ are changed each time stabilization is performed. However this does not matter as long the initial $\mathbf{g}(R)$ value corresponds to the limit $R \to 0$ as then all one needs is for $\mathbf{g}(R)$ to be small (i.e., the actual value is not important).

(b) Log derivative propagation

One way to avoid the stabilization problem just mentioned is to propagate the log derivative matrix $\mathbf{Y}(R)$ [43]. This is defined by

$$\mathbf{Y}(R) = \mathbf{g}'(R)\mathbf{g}^{-1}(R) \tag{A3.11.140}$$

and it remains well behaved numerically even when $\mathbf{g}(R)$ grows exponentially. The differential equation obeyed by $\mathbf{Y}$ is

$$\mathbf{Y}'(R) = \mathbf{g}''(R)\mathbf{g}^{-1}(R) - \mathbf{g}'(R)\mathbf{g}^{-1}(R)\mathbf{g}'(R)\mathbf{g}^{-1}(R) = \mathbf{U}\mathbf{g}(R)\mathbf{g}^{-1}(R) - \mathbf{Y}^2(R) = \mathbf{U} - \mathbf{Y}^2(R). \tag{A3.11.141}$$

It turns out that one cannot propagate $\mathbf{Y}$ using standard numerical methods because $|Y|$ blows up whenever $|g|$ is zero. To circumvent this one must propagate $\mathbf{Y}$ by 'invariant imbedding'. The basic idea here is to construct a propagator $\mathbf{Y}$ which satisfies

$$\begin{pmatrix} \mathbf{g}'(R') \\ \mathbf{g}'(R'') \end{pmatrix} = \begin{pmatrix} \mathbf{Y}_1(R', R'') & \mathbf{Y}_2(R', R'') \\ \mathbf{Y}_3(R', R'') & \mathbf{Y}_4(R', R'') \end{pmatrix} \begin{pmatrix} -\mathbf{g}(R') \\ \mathbf{g}(R'') \end{pmatrix} \tag{A3.11.142}$$

where R' and R'' might form the beginning and end of a propagation step. Assuming for the moment that we know what the $\mathbf{Y}_i$ are, then the evolution of $\mathbf{Y}$ is as follows

$$\mathbf{Y}(R'') = \mathbf{g}'(R'')\mathbf{g}^{-1}(R'') = -\mathbf{Y}_3\mathbf{g}(R')\mathbf{g}^{-1}(R'') + \mathbf{Y}_4\mathbf{g}(R'')\mathbf{g}^{-1}(R'') \tag{A3.11.143}$$

$$\mathbf{Y}(R') = \mathbf{g}'(R')\mathbf{g}^{-1}(R') = -\mathbf{Y}_1\mathbf{g}(R')\mathbf{g}^{-1}(R') + \mathbf{Y}_2\mathbf{g}(R'')\mathbf{g}^{-1}(R'). \tag{A3.11.144}$$

So the overall result is

$$\mathbf{Y}(R'') = \mathbf{Y}_4 - \mathbf{Y}_3[\mathbf{Y}(R') + \mathbf{Y}_1]^{-1}\mathbf{Y}_2. \tag{A3.11.145}$$

To solve for the $\mathbf{Y}$, we begin by solving a reference problem wherein the coupling matrix is assumed diagonal with constant couplings within each step. (These could be accomplished by diagonalizing $\mathbf{U}$, but it would be better to avoid this work and use the diagonal $\mathbf{U}$ matrix elements.) Then, in terms of the reference $\mathbf{U}$ (which we call $\mathbf{U}_d$), we have

$$\mathbf{g}(R_1) = \mathbf{U}_d^{-1}\sin(\mathbf{U}_d\Delta R)\mathbf{g}'(R_0) + \cos(\mathbf{U}_d\Delta R)\mathbf{g}(R_0) \tag{A3.11.146}$$

$$\mathbf{g}'(R_1) = \cos\mathbf{U}_d\Delta R\mathbf{g}'(R_0) - \mathbf{U}_d\sin\mathbf{U}_d\Delta R\mathbf{g}(R_0) \tag{A3.11.147}$$

Now rearrange these to:

$$\mathbf{g}'(R_0) = \mathbf{U}_d(\sin^{-1}\mathbf{U}_d\Delta R)^{-1}\mathbf{g}(R_1) - \mathbf{U}_d\cot\mathbf{U}_d\Delta R\mathbf{g}(R_0) \tag{A3.11.148}$$

$$\mathbf{g}'(R_1) = \mathbf{U}_d\cot\mathbf{U}_d\Delta R\mathbf{g}(R_1) - \mathbf{U}_d(\sin\mathbf{U}_d\Delta R)^{-1}\mathbf{g}(R_0) \tag{A3.11.149}$$

which can be written:

$$\begin{pmatrix} \mathbf{g}'(R_0) \\ \mathbf{g}'(R_1) \end{pmatrix} = \begin{pmatrix} \mathbf{U}_d \cot \mathbf{U}_d \Delta R & \mathbf{U}_d (\sin \mathbf{U}_d \Delta R)^{-1} \\ \mathbf{U}_d (\sin \mathbf{U}_d \Delta R)^{-1} & \mathbf{U}_d \cot \mathbf{U}_d \Delta R \end{pmatrix} \begin{pmatrix} -\mathbf{g}(R_0) \\ \mathbf{g}(R_1) \end{pmatrix}. \tag{A3.11.150}$$

Note that $|\mathbf{U}_d \Delta R| < 0$ is required for meaningful results and thus ΔR cannot be too large. By comparing equations (A3.11.142) and (A3.11.150), we find:

$$\mathbf{Y}_1 = \mathbf{Y}_4 = \mathbf{U}_d \cot \mathbf{U}_d \Delta R \tag{A3.11.151}$$

$$\mathbf{Y}_2 = \mathbf{Y}_3 = \mathbf{U}_d (\sin \mathbf{U}_d \Delta R)^{-1}. \tag{A3.11.152}$$

The standard log-derivative propagator now corrects for the difference between $\mathbf{U}$ and $\mathbf{U}_d$ using a Simpson-rule integration. The specific formulas are

$$\mathbf{y}_1 \rightarrow \mathbf{y}_1 + \mathbf{Q}(R_0) \tag{A3.11.153}$$

$$\mathbf{y}_2 \rightarrow \mathbf{y}_2 \tag{A3.11.154}$$

$$\mathbf{y}_3 \rightarrow \mathbf{y}_3 \tag{A3.11.155}$$

$$\mathbf{y}_4 \rightarrow \mathbf{y}_4 + \mathbf{Q}(R_1). \tag{A3.11.156}$$

Then for a step divided into two halfsteps, at $R = a, c, b$, write $c = 1/2(a+b)$, $\Delta R = (b-a)/2$, $\Delta \mathbf{u} = \mathbf{U} - \mathbf{U}_d$, $a = R_0$, $b = R_1$. This leads to the following expression for $\mathbf{Q}$:

$$\mathbf{Q}(a) = \frac{\Delta R}{e} \Delta \mathbf{u}(a) \tag{A3.11.157}$$

$$\mathbf{Q}(c) = \frac{1}{2} \left(\mathbf{I} - \frac{\Delta R^2}{6} \Delta \mathbf{u}(c) \right)^{-1} \frac{4 \Delta R}{3} \Delta \mathbf{u}(c) \tag{A3.11.158}$$

$$\mathbf{Q}(b) = \frac{\Delta R}{3} \Delta \mathbf{u}(b). \tag{A3.11.159}$$

Propagation then proceeds from $R \rightarrow 0$ to large R, then the scattering matrix is easily connected to $\mathbf{Y}$ at large R.

(c) Variational calculations

Now let us return to the Kohn variational theory that was introduced in section A3.11.2.8. Here we demonstrate how equation (A3.11.46) may be evaluated using basis set expansions and linear algebra. This discussion will be restricted to scattering in one dimension, but generalization to multidimensional problems is very similar.

To construct $\tilde{\Psi}$, we use the basis expansion

$$\tilde{\Psi} = -u_0 + \sum_{t=1}^{N} u_t(r) C_t \tag{A3.11.160}$$

where $u_0(r)$ is a special basis function which asymptotically looks like $u_0(r) \sim v^{-1/2} e^{-ikr}$, and $u_1 = u_0^*$ is the outgoing wavepart, multiplied by a coefficient C_1 which is $\tilde{S}$. Typically the complete form of u_0 is chosen to be

$$u_0 = v^{-1/2} f(r) e^{-ikr} \tag{A3.11.161}$$

where $f(r)$ is a function which is unity at larger r and vanishes at small r. The functions $u_2, u_3, \ldots$ are taken to be square integrable and the coefficients $C_1, \ldots, C_N$ are to be variationally modified. Now substitute $\tilde{\Psi}$

into the expression for S. This gives

$$S = \tilde{S} + \frac{i}{\hbar}\left\langle -u_0 + \sum_t u_t C_t \middle| \hat{H} - E \middle| -u_0 + \sum_t u_t C_t \right\rangle = \tilde{S} + \frac{i}{\hbar}\langle u_0|\hat{H} - E|u_0\rangle - \frac{i}{\hbar}\sum_t \langle u_0|\hat{H} - E|u_t\rangle C_t$$

$$- \frac{i}{\hbar}\sum_t \langle u_t|\hat{H} - E|u_0\rangle C_t + \frac{i}{\hbar}\sum_t\sum_{t'} \langle u_t|\hat{H} - E|u_{t'}\rangle C_t C_{t'}. \qquad (A3.11.162)$$

Let us define a matrix **M** via

$$M_{0,0} = \langle u_0|\hat{H} - E|u_0\rangle \qquad (A3.11.163)$$
$$(M_0)_t = \langle u_t|\hat{H} - E|u_0\rangle \qquad (A3.11.164)$$
$$(M)_{tt'} = \langle u_t|\hat{H} - E|u_{t'}\rangle. \qquad (A3.11.165)$$

Also, employ integration by parts to convert (plus $\tilde{S} = C_1$), yielding

$$\frac{-i}{\hbar}\sum_t \langle u_0|\hat{H} - E|u_t\rangle C_t = \frac{-i}{\hbar}\sum_t \langle u_t|\hat{H} - E|u_0\rangle C_t - \frac{-i}{\hbar}(i\hbar\tilde{S}). \qquad (A3.11.166)$$

This replaces (A3.11.162) with

$$S = \frac{i}{\hbar}\left(M_{0,0} - 2\sum_t C_t M_{t,0} + \sum_{t,t'} C_t C_{t'} M_{t,t'}\right). \qquad (A3.11.167)$$

Now apply the variational criterion as follows:

$$\frac{\partial}{\partial C_t}S = 0 = \frac{i}{\hbar}\left(-2M_{t,0} + 2\sum_{t'} C_{t'} M_{t,t'}\right). \qquad (A3.11.168)$$

This leads to:

$$\mathbf{M}C = M_0 \qquad (A3.11.169)$$

and thus:

$$C = \mathbf{M}^{-1}M_0 \qquad (A3.11.170)$$

and the **S** matrix is given by:

$$S = \frac{i}{\hbar}(M_{0,0} - 2M_0^+\mathbf{M}^{-1}M_0 + M_0^+\mathbf{M}^{-1}\mathbf{M}\mathbf{M}^{-1}M_0) = \frac{i}{\hbar}(M_{0,0} - M_0^+\mathbf{M}^{-1}M_0). \qquad (A3.11.171)$$

This converts the calculation of S to the evaluation of matrix elements together with linear algebra operations. Generalizations of this theory to multichannel calculations exist and lead to a result of more or less the same form.

A3.11.5 Cumulative reaction probabilities

A special feature of quantum scattering theory as it applies to chemical reactions is that in many applications it is only the cumulative reaction probability (CRP) that is of interest in determining physically measurable properties such as the reactive rate constant. This probability P_{cum} (also denoted $N(E)$) is obtained from the S matrix through the formula:

$$P_{cum} = \sum_i \sum_f |S_{if}|^2. \qquad (A3.11.172)$$

Note that the sums are restricted to the portion of the full S matrix that describes reaction (or the specific reactive process that is of interest). It is clear from this definition that the CRP is a highly averaged property where there is no information about individual quantum states, so it is of interest to develop methods that determine this probability directly from the Schrödinger equation rather than indirectly from the scattering matrix. In this section we first show how the CRP is related to the physically measurable rate constant, and then we discuss some rigorous and approximate methods for directly determining the CRP. Much of this discussion is adapted from Miller and coworkers [44, 45].

A3.11.5.1 Rate constants

Consider first a gas phase bimolecular reaction (A + B $\rightarrow$ C + D). If we consider that the reagents are approaching each other with a relative velocity v, then the total flux of A moving toward B is just vC_A where C_A is the concentration of A (number of A per unit volume (or per unit length in one dimension)). If σ is the integral cross section for reaction between A and B for a given velocity v (σ is the reaction probability in one dimension), then for every B, the number of reactive collisions per unit time is $\sigma v C_A$. The total number of reactive collisions per unit time per unit volume (or per unit length in one dimension) is then $\sigma v C_A C_B$ where C_B is the concentration of B. Equating this to the rate constant k times $C_A C_B$ leads us to the conclusion that

$$k = \sigma v. \tag{A3.11.173}$$

This rate constant refers to reactants which all move with a velocity v whereas the usual situation is such that we have a Boltzmann distribution of velocities. If so then the rate constant is just the average of (A3.11.173) over a Boltzmann distribution P_B:

$$k(T) = \int_0^\infty P_B(v)v\sigma(v)\,dv. \tag{A3.11.174}$$

This expression is still oversimplified, as it ignores the fact that the molecules A and B have internal states and that the cross section σ depends on these states; σ depends also on the internal states of the products C and D. Letting the indices i and f denote the internal states of the reagents and products respectively, we find that σ in equation (A3.11.174) must be replaced by $\sum_f \sigma_{if}$ and the Boltzmann average must now include the internal states. Thus, equation (A3.11.174) becomes:

$$k(T) = \sum_i p_B(i) \int_0^\infty P_B(v_i)v_i \sum_f \sigma_{if}(v_i)\,dv_i \tag{A3.11.175}$$

where $p_B(i)$ is the internal state Boltzmann distribution.

Now let us write down explicit expressions for $p_B(i)$, $P_B(v_i)$ and σ_{if}. Denoting the internal energy for a given state i as ε_i and the relative translational energy as $E_i = 1/2\mu v_i^2$, we have (in three dimensions)

$$p_B(i) = e^{-\varepsilon_i/kT}/Q_{int} \tag{A3.11.176}$$

and

$$P_B = 4\pi(\mu/2\pi kT)^{3/2}v_i^2 \exp(-\mu v_i^2/2kT) \tag{A3.11.177}$$

where Q_{int} is the internal state partition function.

The cross section σ_{if} is related to the partial wave reactive scattering matrix S_{if}^J through the partial wave sum (i.e., equation (A3.11.117) evaluated for $n_A n_B \neq n_{A'} n_{B'}$).

$$\sigma_{if} = \frac{\pi}{k_i^2} \sum_J (2J+1)|S_{if}^J|^2 \tag{A3.11.178}$$

where $k_i = \mu v_i / \hbar$. Now substitute equations (A3.11.176)–(A3.11.178) into (A3.11.175). Replacing the integral over v_i by one over E_i leads us to the expression

$$k(T) = \frac{(2\pi\hbar)^2}{Q_{\text{int}} (2\pi\mu kT)^{3/2}} \sum_i e^{-\varepsilon_i/kT} \int_0^\infty e^{-E_i/kT} \sum_J (2J+1) \sum_f |S_{if}^J|^2 \, dE_i. \tag{A3.11.179}$$

If we now change the integration variable from E_i to the total energy $E = E_i + \varepsilon_i$, we can rewrite equation (A3.11.179) as

$$k(T) = \frac{kT}{h} \frac{1}{Q_{\text{int}} Q_{\text{trans}}} \int_0^\infty e^{-E/kT} P_{\text{cum}}(E) \, dE/kT \tag{A3.11.180}$$

where Q_{trans} is the translational partition function per unit volume:

$$Q_{\text{trans}} = \left(\frac{2\pi\mu kT}{h^2} \right)^{3/2} \tag{A3.11.181}$$

and P_{cum} is the cumulative reaction probability that we wrote down in equation (A3.11.172), but generalized to include a sum over the conserved total angular momentum J weighted by the usual $2J + 1$ degeneracy:

$$P_{\text{cum}}(E) = \sum_J (2J+1) \sum_i \sum_f |S_{if}^J|^2. \tag{A3.11.182}$$

Note that in deriving equation (A3.11.180), we have altered the lower integration limit in equation (A3.11.182) from zero to $-\varepsilon_i$ by defining S_{if}^J to be zero for $E_i < 0$.

In one physical dimension, equation (A3.11.180) still holds, but Q_{trans} is given by its one dimensional counterpart and (A3.11.172) is used for the CRP.

A3.11.5.2   Transition state theory

The form of equation (A3.11.182) is immediately suggestive of statistical approximations. If we assume that the total reaction probability $\sum_f |S_{if}|^2$ is zero for $E < E^\ddagger$ and unity for $E \geq E^\ddagger$ where $E^\ddagger$ is the energy of a critical bottleneck (commonly known as the transition state) then

$$P_{\text{cum}}^\ddagger = \sum_J (2J+1) \sum_i h(E - E_i^\ddagger) \tag{A3.11.183}$$

where h is a Heaviside (step) function which is unity for positive arguments and zero for negative arguments, and we have added the subscript i to $E_i^\ddagger$ since the bottleneck energies will in general be dependent on internal state.

Equation (A3.11.183) is simply a formula for the number of states energetically accessible at the transition state and equation (A3.11.180) leads to the thermal average of this number. If we imagine that the states of the system form a continuum, then $P_{\text{cum}}^\ddagger(E)$ can be expressed in terms of a density of states ρ as in

$$P_{\text{cum}}^\ddagger(E) = \int_0^E \rho^\ddagger(\varepsilon) \, d\varepsilon. \tag{A3.11.184}$$

Substituting this into the integral in equation (A3.11.180) and inverting the order of integration, one obtains

$$\int_0^\infty e^{-E/kT} \left(\int_0^E \rho^\ddagger(\varepsilon) \, d\varepsilon \right) dE/kT = \int_0^\infty \rho^\ddagger(\varepsilon) \left(\int_\varepsilon^\infty e^{-E/kT} \, dE/kT \right) d\varepsilon. \tag{A3.11.185}$$

The inner integral on the right-hand side is just $e^{-\varepsilon/kT}$, so equation (A3.11.185) reduces to the transition state partition function (leaving out relative translation):

$$Q^{\ddagger} = \int_0^{\infty} e^{-\varepsilon/kT} \rho^{\ddagger}(\varepsilon)\, d\varepsilon. \qquad (A3.11.186)$$

Using this in equation (A3.11.180) gives the following

$$k(T) = \frac{kT}{h}\frac{Q^{\ddagger}}{Q_{\text{int}} Q_{\text{trans}}}. \qquad (A3.11.187)$$

This is commonly known as the *transition state theory* approximation to the rate constant. Note that all one needs to do to evaluate (A3.11.187) is to determine the partition function of the reagents and transition state, which is a problem in statistical mechanics rather than dynamics. This makes transition state theory a very useful approach for many applications. However, what is left out are two potentially important effects, tunnelling and barrier recrossing, both of which lead to CRPs that differ from the sum of step functions assumed in (A3.11.183).

A3.11.5.3 Exact quantum expressions for the cumulative reaction probability

An important development in the quantum theory of scattering in the last 20 years has been the development of exact expressions which directly determine either $P_{\text{cum}}(E)$ or the thermal rate constant $k(T)$ from the Hamiltonian H. Formally, at least, these expressions avoid the determination of scattering wavefunctions and any information related to the internal states of the reagents or products. The fundamental derivations in this area have been presented by Miller [44] and by Schwartz *et al* [45].

The basic expression of $P_{\text{cum}}(E)$ is

$$P_{\text{cum}}(E) = \tfrac{1}{2}(2\pi\hbar)^2 \text{Tr}[\hat{F}\,\delta(E - \hat{H})\hat{F}\,\delta(E - \hat{H})] \qquad (A3.11.188)$$

where $\hat{F}$ is the symmetrized flux operator:

$$\hat{F} = \frac{1}{2}\left\{\frac{\hat{p}}{m}\delta(s) + \delta(s)\frac{\hat{p}}{m}\right\}. \qquad (A3.11.189)$$

Note that equation (A3.11.188) includes a quantum mechanical trace, which implies a sum over states. The states used for this evaluation are arbitrary as long as they form a complete set and many choices have been considered in recent work. Much of this work has been based on wavepackets [46] or grid point basis functions [47].

An exact expression for the thermal rate constant is given by:

$$k = Q^{-1}\int_0^{\infty} dt\, C_{\text{f}}(t) \qquad (A3.11.190)$$

where $C_{\text{f}}(t)$ is a flux–flux correlation function

$$C_{\text{f}}(t) = \text{Tr}(\hat{F}\,e^{i\hat{H}t_{\text{c}}^*/\hbar}\,\hat{F}\,e^{-i\hat{H}t_{\text{c}}/\hbar}). \qquad (A3.11.191)$$

Here t_{c} is a complex time which is given by $t_{\text{c}} = t - i\hbar/2kT$. Methods for evaluating this equation have included path integrals [45], wavepackets [48, 49] and direct evaluation of the trace in square integrable basis sets [50].

A3.11.6   Classical and semiclassical scattering theory

Although the primary focus of this article is on quantum scattering theory, it is important to note that classical and semiclassical approximations play an important role in the application of scattering theory to problems in chemical physics. The primary reason for this is that the de Broglie wavelength associated with motions of atoms and molecules is typically short compared to the distances over which these atoms and molecules move during a scattering process. There are exceptions to this of course, in the limits of low temperature and energy, and for light atoms such as hydrogen atoms, but for a very broad sampling of problems the dynamics is close to the classical limit.

A3.11.6.1   Classical scattering theory for a single particle

Consider collisions between two molecules A and B. For the moment, ignore the structure of the molecules, so that each is represented as a particle. After separating out the centre of mass motion, the classical Hamiltonian that describes this problem is

$$H = \tfrac{1}{2}\mu\,\dot{r}^2 + V(r) \tag{A3.11.192}$$

where the reduced mass is $\mu = m_A m_B/(m_A + m_B)$ and the potential V only depends on the distance r between the particles. Because of the spherical symmetry of the potential, motion of the system is confined to a plane. It is convenient to use polar coordinates, r, θ, φ and to choose the plane of motion such that $\varphi = 0$. In this case the orbital angular momentum is:

$$|L| = |r \times p_r| = \mu r^2\,\dot{\theta}. \tag{A3.11.193}$$

Since angular momentum is conserved, equation (A3.11.192) may be rearranged to give the following implicit equation for the time dependence of r:

$$\int_{r_1}^{r_2} \frac{\mathrm{d}r}{\sqrt{(E - L^2/2\mu r^2 - V)2/\mu}} = t_2 - t_1. \tag{A3.11.194}$$

The time dependence of θ can then be obtained by integrating (A3.11.193).

The physical situation of interest in a scattering problem is pictured in figure A3.11.3. We assume that the initial particle velocity v is coincident with the z axis and that the particle starts at $z = -\infty$, with $x = b = $ impact parameter, and $y = 0$. In this case, $L = \mu v b$. Subsequently, the particle moves in the x, z plane in a trajectory that might be as pictured in figure A3.11.4 (here shown for a hard sphere potential). There is a point of closest approach, i.e., $r = r_2$ (inner turning point for r motions) where

$$E = \frac{L^2}{2\mu r_<^2} + V(r_<). \tag{A3.11.195}$$

If we define $t_1 = 0$ at $r = r_<$, then the explicit trajectory motion is determined by

$$t = \int_{r_<}^{r(t)} \frac{\mathrm{d}r}{\sqrt{(E - L^2/2\mu r^2 - V(r))2/\mu}} \tag{A3.11.196}$$

$$\theta(t) = \pi + L \int_{-\infty}^{t} \frac{\mathrm{d}r}{\mu r(t')^2}. \tag{A3.11.197}$$

The final scattering angle θ is defined using $\theta = \theta(t = \infty)$. There will be a correspondence between b and θ that will tend to look like what is shown in figure A3.11.5 for a repulsive potential (here given for the special case of a hard sphere potential).

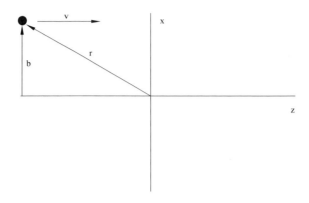

Figure A3.11.3. Coordinates for scattering of a particle from a central potential.

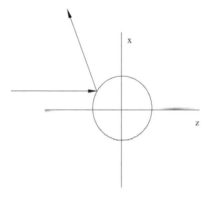

Figure A3.11.4. Trajectory associated with a particle scattering off a hard sphere potential.

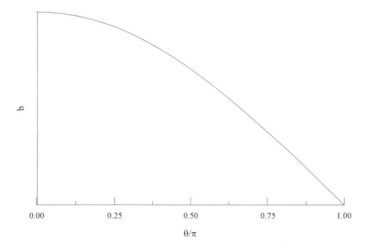

Figure A3.11.5. Typical dependence of b on θ (shown for a hard sphere potential).

In an ensemble of collisions, the impact parameters are distributed randomly on a disc with a probability distribution $P(b)$ that is defined by $P(b)\,db = 2\pi b\,db$. The *cross section* $d\sigma$ is then defined by

$$d\sigma = 2\pi b\,db. \tag{A3.11.198}$$

Now $d\sigma = (d\sigma/d\Omega)\,d\Omega$ or $(d\sigma/2\pi d\cos\theta)2\pi d\cos\theta = [d\sigma/2\pi d(\cos\theta)]2\pi \sin\theta\,d\theta = I(\omega)2\pi \sin\theta\,d\theta$ where $I(\omega)$ is the differential cross section. Therefore

$$I(\omega) = \frac{b\,db}{\sin\theta\,d\theta} = \frac{b}{\sin\theta}\left|\frac{db}{d\theta}\right| \tag{A3.11.199}$$

where the absolute value takes care of the case when $db/d\theta < 0$. The integral cross section is

$$\sigma = 2\pi \int_0^{\pi} I \sin\theta\,d\theta = 2\pi \int_0^{b_{\max}} b\,db \tag{A3.11.200}$$

where $b_{\max}$ is the value of the impact parameter that is associated with a scattering angle of 0 (i.e., scattering in the forward direction). Note that for a potential with infinite range (one that does not go to zero until $r = \infty$), the cross section predicted by (A3.11.200) is infinite. This is not generally correct, except for a Coulomb $(1/r)$ potential. This is a classical artifact; the corresponding quantum mechanical result is finite.

A simple example of a finite range potential is the *hard sphere*, for which $V(r) = 0$ for $r > a$, $V(r) = \infty$ for $r < a$. By geometry one can show that $2\varphi + \theta = \pi$ and $\sin\varphi = b/a$. Therefore

$$b = a \sin\left(\frac{\pi}{2} - \frac{\theta}{2}\right) = a\cos\theta/2 \tag{A3.11.201}$$

$$\frac{db}{d\theta} = \frac{-a}{2}\sin\theta/2 \tag{A3.11.202}$$

$$I(\omega) = \frac{a\cos\theta/2(a/2)\sin\theta/2}{\sin\theta} = \frac{a^2}{4} \tag{A3.11.203}$$

and

$$\sigma = 2\pi\frac{a^2}{4}\int \sin\theta\,d\theta = \pi a^2. \tag{A3.11.204}$$

This shows that the differential cross section is independent of angle for this case, and the integral cross section is, just as expected, the area of the circle associated with the radius of the sphere. More generally it is important to note that there can be many trajectories which give the same θ for different b'. In this case the DCS is just the sum over trajectories.

$$I(\theta) = \sum_n \frac{b_n}{\sin\theta}\left(\frac{db}{d\theta}\right)_n. \tag{A3.11.205}$$

An explicit result for the differential cross section (DCS) can be obtained by substituting $L = pb = \mu vb$ into the following expression:

$$\frac{\dot{\theta}}{\dot{r}} = \frac{d\theta}{dr} = \frac{L/\mu r^2}{\sqrt{(2/\mu)(E - V - L^2/2\mu r^2)}} = \frac{b/r^2}{\sqrt{(1 - V/E - (2\mu E)b^2/2\mu E r^2)}} = \frac{b/r^2}{\sqrt{1 - V/E - b^2/r^2}}. \tag{A3.11.206}$$

To integrate this expression, we note that θ starts at π when $r = \infty$, then it decreases while r decreases to its turning point, then r retraces back to ∞ while θ continues to evolve back to π. The *total change* in θ is then *twice* the integral

$$\theta = \pi - 2b\int_{r_<}^{\infty}\frac{dr}{r^2}\left(1 - \frac{V}{E} - \frac{b^2}{r^2}\right)^{-1/2}. \tag{A3.11.207}$$

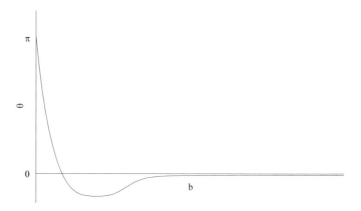

Figure A3.11.6. Dependence of scattering angle θ on impact parameter for a 6–12 potential.

Note that θ obtained this way can be negative. Because of cylindrical symmetry, only $|\theta|$ (or θ mod π) means anything.

For a typical interatomic potential such as a 6–12 potential, $\theta(b)$ looks like figure A3.11.6 rather than A3.11.5. This shows that for some θ there are three b (one for positive θ and two for negative θ) that contribute to the DCS. The θ where the number of contributing trajectories changes value are sometimes called *rainbow angles*. At these angles, the classical differential cross sections have singularities.

A3.11.6.2   Classical scattering theory for many interacting particles

To generalize what we have just done to reactive and inelastic scattering, one needs to calculate numerically integrated trajectories for motions in many degrees of freedom. This is most convenient to develop in space-fixed Cartesian coordinates. In this case, the classical equations of motion (Hamilton's equations) are given by:

$$\dot{x}_i = \frac{\partial H}{\partial p_i} = \frac{p_i}{m_i} \tag{A3.11.208}$$

$$\dot{p}_i = \frac{\partial H}{\partial x_i} = -\frac{\partial V}{\partial x_i} \tag{A3.11.209}$$

where m_i is the mass associated with the ith degree of freedom and the second equality applies to Cartesian coordinates. Methods for solving these equations of motion have been described in review articles, as have procedures for defining initial conditions [27]. Note that for most multidimensional problems it is necessary to average over initial conditions that represent the internal motions of the species undergoing collision. These averages are often determined by Monte Carlo integration (i.e., randomly sampling the coordinates that need to be averaged over). The initial conditions may be chosen from canonical or microcanonical ensembles, or they may be chosen to mimic an initially prepared quantum state. In the latter case, the trajectory calculation is called a 'quasiclassical' trajectory calculation.

A3.11.6.3   Semiclassical theory

The obvious defect of classical trajectories is that they do not describe quantum effects. The best known of these effects is tunnelling through barriers, but there are others, such as effects due to quantization of the reagents and products and there are a variety of interference effects as well. To circumvent this deficiency, one can sometimes use semiclassical approximations such as WKB theory. WKB theory is specifically for motion of a particle in one dimension, but the generalizations of this theory to motion in three dimensions are known and will be mentioned at the end of this section. More complete descriptions of WKB theory can be found in many standard texts [1–5, 18].

(a) WKB theory

In WKB theory, one generates a wavefunction that is valid in the $\hbar \to 0$ limit using a linear combination of exponentials of the form

$$\psi(x) = A(x)\, e^{iS(x)/\hbar} \tag{A3.11.210}$$

where $A(x)$ and $S(x)$ are real (or sometimes purely imaginary) functions that are derived from the Hamiltonian. This expression is, of course, very familiar from scattering theory applications described above (A3.11.2), where $A(x)$ is a constant, and $S(x)$ is kx. More generally, by substituting (A3.11.210) into the time independent Schrödinger equation in one dimension, and expanding $A(x)$ and $S(x)$ in powers of $\hbar$, one can show that the leading terms representing $S(x)$ have the form:

$$A(x) = [E - V(x)]^{-1/4}$$

$$S(x) = \pm \int \sqrt{\frac{2m}{\hbar^2}(E - V(x))}\, dx. \tag{A3.11.211}$$

Note that the integrand in $S(x)$ is just the classical momentum $p(x)$, so $S(x)$ is the classical *action* function. In addition, $A(x)$ is proportional to $p^{-1/2}$, which means that $|\psi|^2$ is proportional to the inverse of the classical velocity of the particle. This is just the usual classical expression for the probability density. Note that $A(x)$ and $S(x)$ are real as long as motion is classically allowed, meaning that $E > V(x)$. If $E < V(x)$, then $S(x)$ becomes imaginary and $\psi(x)$ involves real rather than complex exponentials. At the point of transition between allowed and forbidden regions, i.e., at the so-called *turning points* of the classical motion, $A(x)$ becomes infinite and the solutions above are not valid. However, it is possible to 'connect' the solutions on either side of the turning point using 'connection formulas' that are determined from exact solutions to the Schrödinger equation near the turning point. The reader should consult the standard textbooks [1–5, 18] for a detailed discussion of this.

   In applications to scattering theory, one takes linear combinations of functions of the form (A3.11.210) to satisfy the desired boundary conditions and one uses the connection formulas to determine wavefunctions that are valid for all values of x. By examining the asymptotic forms of the wavefunction, scattering information can be determined. For example, in applications to scattering from a central potential, one can solve the radial Schrödinger equation using WKB theory to determine the phase shift for elastic scattering. The explicit result depends on how many turning points there are in the radial motion.

(b) Scattering theory for many degrees of freedom

For multidimensional problems, the generalization of WKB theory to the description of scattering problems is often called Miller–Marcus or classical S-matrix theory [51]. The reader is referred to review articles for a more complete description of this theory [52].

   Another theory which is used to describe scattering problems and which blends together classical and quantum mechanics is the semiclassical wavepacket approach [53]. The basic procedure comes from the fact that wavepackets which are initially Gaussian remain Gaussian as a function of time for potentials that are constant, linear or quadratic functions of the coordinates. In addition, the centres of such wavepackets evolve in time in accord with classical mechanics. We have already seen one example of this with the free particle wavepacket of equation (A3.11.7). Consider the general quadratic Hamiltonian (still in one dimension but the generalization to many dimensions is straightforward)

$$\hat{H} = -\frac{\hbar^2}{2m}\frac{\partial^2}{\partial x^2} + V_0 + V_x(x - x_1) + \frac{1}{2}V_{xx}(x - x_t)^2. \tag{A3.11.212}$$

The Gaussian wavepacket is written as

$$\psi(x, t) = \exp[(i/\hbar)\alpha_t(x - x_t)^2 + (i/\hbar)p_t(x - x_t) + (i/\hbar)\gamma_t]. \tag{A3.11.213}$$

Here x_t and p_t are real time dependent quantities that specify the average position and momentum of the wavepacket ($p_t = \langle p \rangle$, $x_t = \langle x \rangle$) and α_t and γ_t are complex functions which determine the width, phase and normalization of the wavepacket.

Inserting equation (A3.11.213) into $\hat{H}\psi = i\hbar\partial\psi/\partial t$, and using equation (A3.11.212), leads to the following relation:

$$-\dot{\alpha}_t(x - x_t)^2 + (2\alpha_t\dot{x}_t - \dot{p}_t)(x - x_t) - \dot{\gamma}_t + p_t\dot{x}_t]\psi = \{[(2/m)\alpha_t^2 + \frac{1}{2}V_{xx}](x - x_t)^2$$
$$+ (2\alpha_t p_t/m + V_x)(x - x_t) + V_0 - i\hbar\alpha_t/m + p_t^2/2m\}\psi. \tag{A3.11.214}$$

Comparing coefficients of like powers of $(x - x_t)$ then gives us three equations involving the four unknowns:

$$\dot{\alpha}_t = -(2/m)\alpha_t^2 - V_{xx}/2 \tag{A3.11.215a}$$
$$2\alpha_t\dot{x}_t - \dot{p}_t = 2\alpha_t p_t/m + V_x \tag{A3.11.215b}$$
$$\dot{\gamma}_t = i\hbar\alpha_t/m + p_t\dot{x}_t - V_0 - p_t^2/2m. \tag{A3.11.215c}$$

To develop an additional equation, we simply make the *ansatz* that the first term on the left-hand side of equation (3.11.215b) equals the first term on the right-hand side and similarly with the second term. This immediately gives us Hamilton's equations

$$\dot{x}_t = p_t/m \tag{A3.11.216a}$$
$$\dot{p}_t = -V_x \tag{A3.11.216b}$$

from which it follows that x_t and p_t are related through the classical Hamiltonian function

$$H = p_t^2/2m + V_0 = E. \tag{A3.11.217}$$

Equations (A3.11.216) can then be cast in the general form

$$\dot{x}_t = \partial H/\partial p_t \tag{A3.11.218a}$$
$$-\dot{p}_t = \partial H/\partial x_t \tag{A3.11.218b}$$

and the remaining two equations in equation (A3.11.215) become

$$\dot{\alpha}_t = -(2/m)\alpha_t^2 - \frac{1}{2}V_{xx} \tag{A3.11.219a}$$
$$\dot{\gamma}_t = i\hbar\alpha_t/m + p_t\dot{x}_t - E. \tag{A3.11.219b}$$

It is not difficult to show that, for a constant potential, equations (A3.11.218) and (A3.11.219) can be solved to give the free particle wavepacket in equation (A3.11.7). More generally, one can solve equations (A3.11.218) and (A3.11.219) numerically for *any* potential, even potentials that are not quadratic, but the solution obtained will be exact only for potentials that are constant, linear or quadratic. The deviation between the exact and Gaussian wavepacket solutions for other potentials depends on how close they are to being *locally quadratic*, which means how well the potential can be approximated by a quadratic potential over the width of the wavepacket. Note that although this theory has many classical features, the $\hbar \to 0$ limit has not been used. This circumvents problems with singularities in the wavefunction near classical turning points that cause trouble in WKB theory.

References

[1]   Messiah A 1965 *Quantum Mechanics* (Amsterdam: North-Holland)
[2]   Schiff L I 1968 *Quantum Mechanics* (New York: McGraw-Hill)
[3]   Merzbacher E 1970 *Quantum Mechanics* (New York: Wiley)
[4]   Davydov A S 1976 *Quantum Mechanics* (Oxford: Pergamon)
[5]   Sakurai J J 1985 *Modern Quantum Mechanics* (Menlo Park: Benjamin-Cummings)
[6]   Adhi Kari S K and Kowolski K L 1991 *Dynamical Collision Theory and Its Applications* (New York: Academic)
[7]   Cohen-Tannouji C, Diu B and Laloë F 1977 *Quantum Mechanics* (New York: Wiley)
[8]   Newton R G 1982 *Scattering Theory of Waves and Particles* 2nd edn (New York: McGraw-Hill)
[9]   Rodberg L S and Thaler R M 1967 *Introduction to the Quantum Theory of Scattering* (New York: Academic)
[10] Roman P 1965 *Advanced Quantum Theory* (Reading, MA: Addison-Wesley)
[11] Simons J and Nichols J 1997 *Quantum Mechanics in Chemistry* (New York: Oxford University Press)
[12] Levine R D 1969 *Quantum Mechanics of Molecular Rate Processes* (London: Oxford University Press)
[13] Child M S 1974 *Molecular Collision Theory* (New York: Academic)
[14] Nikitin E E 1974 *Theory of Elementary Atomic and Molecular Processes in Gases* (Oxford: Clarendon)
[15] Massey H S W 1979 *Atomic and Molecular Collisions* (London: Taylor and Francis)
[16] Levine R D and Bernstein R B 1987 *Molecular Reaction Dynamics and Chemical Reactivity* (New York: Oxford University Press)
[17] Murrell J N and Bosanac S D 1989 *Introduction to the Theory of Atomic and Molecular Collisions* (New York: Wiley)
[18] Schatz G C and Ratner M A 1993 *Quantum Mechanics in Chemistry* (Englewood Cliffs, NJ: Prentice-Hall)
[19] Miller W H (ed) 1976 *Dynamics of Molecular Collisions* (New York: Plenum)
[20] Bernstein R B (ed) 1979 *Atom–Molecule Collision Theory. A Guide for the Experimentalist* (New York: Plenum)
[21] Truhlar D G (ed) 1981 *Potential Energy Surfaces and Dynamics Calculations* (New York: Plenum)
[22] Bowman J M (ed) 1983 *Molecular Collision Dynamics* (Berlin: Springer)
[23] Baer M (ed) 1985 *The Theory of Chemical Reaction Dynamics* (Boca Raton, FL: Chemical Rubber Company)
[24] Clary D C 1986 *The Theory of Chemical Reaction Dynamics* (Boston: Reidel)
[25] Baer M and Ng C-Y (eds) 1992 *State-Selected and State-to-State Ion–Molecule Reaction Dynamics Part 2. Theory (Adv. Chem. Phys. 72)* (New York: Wiley)
[26] Truhlar D G (ed) 1984 *Resonances in Electron–Molecule Scattering, van der Waals Complexes, and Reactive Chemical Dynamics (ACS Symp. Ser. 263)* (Washington, DC: American Chemical Society)
[27] Thompson D L 1998 *Modern Methods for Multidimensional Dynamics Computations in Chemistry* (Singapore: World Scientific)
[28] Hase W L (ed) 1998 *Comparisons of Classical and Quantum Dynamics (Adv. in Classical Trajectory Methods III)* (Greenwich, CT: JAI Press)
[29] Mullin A S and Schatz G C (eds) 1997 *Highly Excited Molecules: Relaxation, Reaction and Structure (ACS Symp. Ser. 678)* (Washington, DC: American Chemical Society)
[30] Rost J M 1998 Semiclassical s-matrix theory for atomic fragmentation *Phys. Rep.* **297** 272–344
[31] Kaye J A and Kuppermann A 1988 Mass effect in quantum-mechanical collision-induced dissociation in collinear reactive atom diatomic molecule collisions *Chem. Phys.* **125** 279–91
[32] Miller W H 1994 S-matrix version of the Kohn variational principle for quantum scattering theory of chemical reactions *Adv. Mol. Vibrations and Collision Dynamics* vol 2A, ed J M Bowman (Greenwich, CT: JAI Press) pp 1–32
[33] Mayne H R 1991 Classical trajectory calculations on gas-phase reactive collisions *Int. Rev. Phys. Chem.* **10** 107–21
[34] Delves L M 1959 Tertiary and general-order collisions *Nucl. Phys.* **9** 391–9
      Delves L M 1960 Tertiary and general-order collisions (II) *Nucl. Phys.* **20** 275–308
[35] Smith F T 1962 A symmetric representation for three-body problems. I. Motion in a plane *J. Math. Phys.* **3** 735–48
      Smith F T and Whitten R C 1968 Symmetric representation for three body problems. II. Motion in space *J. Math. Phys.* **9** 1103–13
[36] Kuppermann A 1997 Reactive scattering with row-orthonormal hyperspherical coordinates. 2. Transformation properties and Hamiltonian for tetraatomic systems *J. Phys. Chem.* A **101** 6368–83
[37] Kosloff R 1994 Propagation methods for quantum molecular-dynamics *Annu. Rev. Phys. Chem.* **45** 145–78
[38] Balint-Kurti G G, Dixon R N and Marston C C 1992 Grid methods for solving the Schrödinger equation and time-dependent quantum dynamics of molecular photofragmentation and reactive scattering processes *Int. Rev. Phys. Chem.* **11** 317–44
[39] Feit M D and Fleck J A Jr 1983 Solution of the Schrödinger equation by a spectral method. II. Vibrational energy levels of triatomic molecules *J. Chem. Phys.* **78** 301–8
[40] Tal-Ezer H and Kosloff R 1984 An accurate and efficient scheme for propagating the time dependent Schrödinger equation *J. Chem. Phys.* **81** 3967–71
[41] Leforestier C *et al* 1991 Time-dependent quantum mechanical methods for molecular dynamics *J. Comput. Phys.* **94** 59–80
[42] Gordon R G 1969 Constructing wave functions for bound states and scattering *J. Chem. Phys.* **51** 14–25
[43] Manolopoulos D E 1986 An improved log derivative method for inelastic scattering *J. Chem. Phys.* **85** 6425–9
[44] Miller W H 1974 Quantum mechanical transition state theory and a new semiclassical model for reaction rate constants *J. Chem. Phys.* **61** 1823–34
[45] Miller W H, Schwartz S D and Tromp J W 1983 Quantum mechanical rate constants for bimolecular reactions *J. Chem. Phys.* **79** 4889–98

[46] Zhang D H and Light J C 1996 Cumulative reaction probability via transition state wave packets *J. Chem. Phys.* **104** 6184–91

[47] Manthe U, Seideman T and Miller W H 1993 Full-dimensional quantum-mechanical calculation of the rate-constant for the $H_2 + OH \rightarrow H_2O + H$ reaction *J. Chem. Phys.* **99** 10 078–81

[48] Wahnstrom G and Metiu H 1988 Numerical study of the correlation function expressions for the thermal rate coefficients in quantum systems *J. Phys. Chem.* **92** 3240–52

[49] Thachuk M and Schatz G C 1992 Time dependent methods for calculating thermal rate coefficients using flux correlation functions *J. Chem. Phys.* **97** 7297–313

[50] Day P N and Truhlar D G 1991 Benchmark calculations of thermal reaction rates. II. Direct calculation of the flux autocorrelation function for a canonical ensemble *J. Chem. Phys.* **94** 2045–56

[51] Miller W H 1970 Semiclassical theory of atom–diatom collisions: path integrals and the classical S matrix *J. Chem. Phys.* **53** 1949–59
Marcus R A 1970 Extension of the WKB method to wave functions and transition probability amplitudes (S-matrix) for inelastic or reactive collisions *Chem. Phys. Lett.* **7** 525–32

[52] Miller W H 1971 Semiclassical nature of atomic and molecular collisions *Accounts Chem. Res.* **4** 161–7
Miller W H 1974 Classical-limit quantum mechanics and the theory of molecular collisions *Adv. Chem. Phys.* **25** 69–177
Miller W H 1975 Classical S-matrix in molecular collisions *Adv. Chem. Phys.* **30** 77–136

[53] Heller E 1975 Time dependent approach to semiclassical dynamics *J. Chem. Phys.* **62** 1544–55

Further Reading

Basic quantum mechanics textbooks that include one or more chapters on scattering theory as applied to physics:

Messiah A 1965 *Quantum Mechanics* (Amsterdam: North-Holland)
Schiff L I 1968 *Quantum Mechanics* (New York: McGraw-Hill)
Merzbacher E 1970 *Quantum Mechanics* (New York: Wiley)
Davydov A S 1976 *Quantum Mechanics* (Oxford: Pergamon)
Cohen-Tannouji C, Diu B and Laloë F 1997 *Quantum Mechanics* (New York: Wiley)
Sakurai J J 1985 *Modern Quantum Mechanics* (Menlo Park: Benjamin-Cummings)

Books that are entirely concerned with scattering theory (only the Levine text is concerned with applications in chemical physics):

Roman P 1965 *Advanced Quantum Theory* (Reading, MA: Addison-Wesley)
Rodberg L S and Thaler R M 1967 *Introduction to the Quantum Theory of Scattering* (New York: Academic)
Levine R D 1969 *Quantum Mechanics of Molecular Rate Processes* (London: Oxford)
Newton R G 1982 *Scattering Theory of Waves and Particles* 2nd edn (New York: McGraw-Hill)
Adhi Kari S K and Kowolski K L 1991 *Dynamical Collision Theory and Its Applications* (New York: Academic)
Murrell J N and Bosanac S D 1989 *Introduction to the Theory of Atomic and Molecular Collisions* (New York: Wiley)

Books with a more chemical bent that include chapters on scattering theory and related issues.

Child M S 1974 *Molecular Collision Theory* (New York: Academic)
Nikitin E E 1974 *Theory of Elementary Atomic and Molecular Processes in Gases* (Oxford: Clarendon)
Massey H S W 1979 *Atomic and Molecular Collisions* (London: Taylor and Francis)
Levine R D and Bernstein R B 1987 *Molecular Reaction Dynamics and Chemical Reactivity* (New York: Oxford University Press)
Schatz G C and Ratner M A 1993 *Quantum Mechanics in Chemistry* (Englewood Cliffs, NJ: Prentice-Hall)
Simons J and Nichols J 1997 *Quantum Mechanics in Chemistry* (New York: Oxford)

A3.12
Statistical mechanical description of chemical kinetics: RRKM

William L Hase

A3.12.1 Introduction

As reactants transform to products in a chemical reaction, reactant bonds are broken and reformed for the products. Different theoretical models are used to describe this process ranging from time-dependent classical or quantum dynamics [1, 2], in which the motions of individual atoms are propagated, to models based on the postulates of statistical mechanics [3]. The validity of the latter models depends on whether statistical mechanical treatments represent the actual nature of the atomic motions during the chemical reaction. Such a statistical mechanical description has been widely used in unimolecular kinetics [4] and appears to be an accurate model for many reactions. It is particularly instructive to discuss statistical models for unimolecular reactions, since the model may be formulated at the elementary microcanonical level and then averaged to obtain the canonical model.

Unimolecular reactions are important in chemistry, physics, biochemistry, materials science, and many other areas of science and are denoted by

$$A^* \rightarrow \text{products} \tag{A3.12.1}$$

where the asterisk denotes that the unimolecular reactant A contains sufficient internal vibrational/rotational energy to decompose. (Electronic excitation may also promote decomposition of A, but this topic is outside the purview of this presentation.) The energy is denoted by E and must be greater than the unimolecular decomposition threshold energy E_0. There are three general types of potential energy profiles for unimolecular reactions (see figure A3.12.1). One type is for an isomerization reaction, such as cyclopropane isomerization

$$CH_2\text{---}CH_2 \diagdown \diagup CH_2 \longrightarrow CH_3\text{---}CH\text{=}CH_2$$

for which there is a substantial potential energy barrier separating the two isomers. The other two examples are for unimolecular dissociation. In one case, as for formaldehyde dissociation

$$H_2CO \rightarrow H_2 + CO$$

there is a potential energy barrier for the reverse association reaction. In the other, as for aluminium cluster dissociation

$$Al_n \rightarrow Al_{n-1} + Al$$

there is no barrier for association.

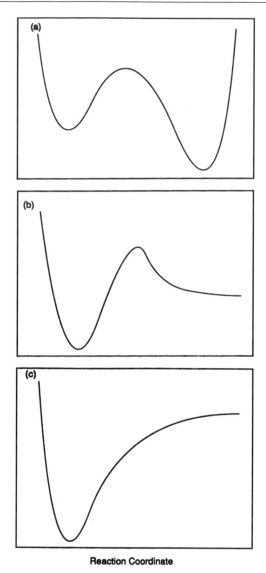

Figure A3.12.1. Schematic potential energy profiles for three types of unimolecular reactions. (a) Isomerization. (b) Dissociation where there is an energy barrier for reaction in both the forward and reverse directions. (c) Dissociation where the potential energy rises monotonically as for rotational ground-state species, so that there is no barrier to the reverse association reaction. (Adapted from [5].)

A number of different experimental methods may be used to energize the unimolecular reactant A. For example, energization can take place by the potential energy release in chemical reaction, i.e.

$$^1CH_2 + CH_3SiH_3 \rightarrow CH_3SiH_2CH_3^*$$

or by absorption of a single photon,

$$CH_3NC + h\nu \rightarrow CH_3NC^*.$$

Extensive discussions of procedures for energizing molecules are given elsewhere [5].

Quantum mechanically, the time dependence of the initially prepared state of A* is given by its wavefunction $\Psi(t)$, which may be determined from the equation of motion

$$i\hbar \frac{d\Psi(t)}{dt} = \hat{H}\Psi(t).$$ (A3.12.2)

At the unimolecular threshold of moderate to large size molecules (e.g. C_2H_6 to peptides), there are many vibrational/rotational states within the experimental energy resolution dE and the initial state of A* may decay by undergoing transitions to other states and/or decomposing to products. The former is called intramolecular vibrational energy redistribution (IVR) [6]. The probability amplitude versus time of remaining in the initially prepared state is given by

$$C(t) = \langle \Psi(0)|\Psi(t)\rangle$$ (A3.12.3)

and is comprised of contributions from both IVR and unimolecular decomposition. The time dependence of the unimolecular decomposition may be constructed by evaluating $|\Psi(t)|^2$ inside the potential energy barrier, within the reactant region of the potential energy surface.

In the statistical description of unimolecular kinetics, known as Rice–Ramsperger–Kassel–Marcus (RRKM) theory [4, 7, 8], it is assumed that complete IVR occurs on a timescale much shorter than that for the unimolecular reaction [9]. Furthermore, to identify states of the system as those for the reactant, a dividing surface [10], called a transition state, is placed at the potential energy barrier region of the potential energy surface. The assumption implicit in RRKM theory is described in the next section.

A3.12.2   Fundamental assumption of RRKM theory: microcanonical ensemble

RRKM theory assumes a microcanonical ensemble of A* vibrational/rotational states within the energy interval $E \rightarrow E + dE$, so that each of these states is populated statistically with an equal probability [4]. This assumption of a microcanonical distribution means that the unimolecular rate constant for A* only depends on energy, and not on the manner in which A* is energized. If $N(0)$ is the number of A* molecules excited at $t = 0$ in accord with a microcanonical ensemble, the microcanonical rate constant $k(E)$ is then defined by

$$\left.\frac{-dN(t)}{dt}\right|_{t=0} = k(E)N(t)|_{t=0}.$$ (A3.12.4)

The rapid IVR assumption of RRKM theory means that a microcanonical ensemble is maintained as the A* molecules decompose so that, at any time t, $k(E)$ is given by

$$\frac{-dN(t)}{dt} = k(E)N(t).$$ (A3.12.5)

As a result of the fixed time-independent rate constant $k(E)$, $N(t)$ decays exponentially, i.e.

$$N(t) = N(0)\exp[-k(E)t].$$ (A3.12.6)

A RRKM unimolecular system obeys the ergodic principle of statistical mechanics [11].

The quantity $-dN(t)/[N(t)\,dt]$ is called the lifetime distribution $P(t)$ [12] and according to RRKM theory is given by

$$P(t) = k(E)\exp[-k(E)t].$$ (A3.12.7)

Figure A3.12.2(a) illustrates the lifetime distribution of RRKM theory and shows random transitions among all states at some energy high enough for eventual reaction (toward the right). In reality, transitions between quantum states (though coupled) are not equally probable: some are more likely than others. Therefore, transitions between states must be sufficiently rapid and disorderly for the RRKM assumption to be mimicked,

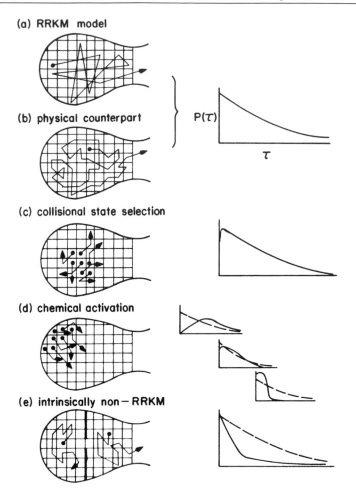

Figure A3.12.2. Relation of state occupation (schematically shown at constant energy) to lifetime distribution for the RRKM theory and for various actual situations. Dashed curves in lifetime distributions for (d) and (e) indicate RRKM behaviour. (a) RRKM model. (b) Physical counterpart of RRKM model. (c) Collisional state selection. (d) Chemical activation. (e) Intrinsically non-RRKM. (Adapted from [9].)

as qualitatively depicted in figure A3.12.2(b). The situation depicted in these figures, where a microcanonical ensemble exists at $t = 0$ and rapid IVR maintains its existence during the decomposition, is called *intrinsic* RRKM behaviour [9].

   The lifetime distribution will depend in part on the manner in which the energy needed for reaction is supplied. In many experiments, such as photoactivation and chemical activation, the molecular vibrational/rotational states are excited non-randomly. Regardless of the pattern of the initial energizing, the RRKM model of rapid IVR requires the distribution of states to become microcanonical in a negligibly short time. Three different possible lifetime distributions are represented by figure A3.12.2(d). As shown in the middle, the lifetime distribution may be similar to that of RRKM theory. In other cases, the probability of a short lifetime with respect to reaction may be enhanced or reduced, depending on the location of the initial excitation within the molecule. These are examples of *apparent* non-RRKM behaviour [9] arising from the initial non-random excitation. If there are strong internal couplings, $P(t)$ will become that of RRKM theory, equation (A3.12.5), after rapid IVR. A classic example of apparent non-RRKM behaviour is described below in section A3.12.8.1.

A situation that arises from the intramolecular dynamics of A* and completely distinct from apparent non-RRKM behaviour is *intrinsic* non-RRKM behaviour [9]. By this, it is meant that A* has a non-random $P(t)$ even if the internal vibrational states of A* are prepared randomly. This situation arises when transitions between individual molecular vibrational/rotational states are slower than transitions leading to products. As a result, the vibrational states do not have equal dissociation probabilities. In terms of classical phase space dynamics, slow transitions between the states occur when the reactant phase space is *metrically decomposable* [13, 14] on the timescale of the unimolecular reaction and there is at least one *bottleneck* [9] in the molecular phase space other than the one defining the transition state. An intrinsic non-RRKM molecule decays non-exponentially with a time-dependent unimolecular rate constant or exponentially with a rate constant different from that of RRKM theory.

The above describes the fundamental assumption of RRKM theory regarding the intramolecular dynamics of A*. The RRKM expression for $k(E)$ is now derived.

A3.12.3   The RRKM unimolecular rate constant

A3.12.3.1   Derivation of the RRKM $k(E)$

As discussed above, to identify states of the system as those for the reactant A*, a dividing surface is placed at the potential energy barrier region of the potential energy surface. This is a classical mechanical construct and classical statistical mechanics is used to derive the RRKM $k(E)$ [4].

In the vicinity of the dividing surface, it is assumed that the Hamiltonian for the system may be separated into the two parts

$$H = H_1 + H' \qquad (A3.12.8)$$

where H_1 defines the energy for the conjugate coordinate and momentum pair q_1, p_1 and H' gives the energy for the remaining conjugate coordinates and momenta. This special coordinate q_1 is called the *reaction coordinate*. Reactive systems which have a total energy $H = E$ and a value for q_1 which lies between $q_1^\ddagger$ are $q_1^\ddagger + dq_1^\ddagger$ called microcanonical transition states. The reaction coordinate potential at the transition state is E_0. The RRKM $k(E)$ is determined from the rate at which these transition states form products.

The hypersurface formed from variations in the system's coordinates and momenta at $H(p, q) = E$ is the microcanonical system's phase space, which, for a Hamiltonian with $3n$ coordinates, has a dimension of $6n - 1$. The assumption that the system's states are populated statistically means that the population density over the whole surface of the phase space is uniform. Thus, the ratio of molecules at the dividing surface to the total molecules $[dN(q_1^\ddagger, p_1^\ddagger)/N]$ may be expressed as a ratio of the phase space at the dividing surface to the total phase space. Thus, at any instant in time, the ratio of molecules whose reaction coordinate and conjugate momentum have values that range from $q_1^\ddagger$ to $q_1^\ddagger + dq_1^\ddagger$ and from $p_1^\ddagger$ to $p_1^\ddagger + dp_1^\ddagger$ to the total number of molecules is given by

$$\frac{dN(q_1^\ddagger, p_1^\ddagger)}{N} = \frac{dq_1^\ddagger\, dp_1^\ddagger \int \ldots \int_{H=E-E_1^\ddagger-E_0} dq_2^\ddagger \ldots dq_{3n}^\ddagger\, dp_2^\ddagger \ldots dp_{3n}^\ddagger}{\int \ldots \int_{H=E} dq_1 \ldots dq_{3n}\, dp_1 \ldots dp_{3n}} \qquad (A3.12.9)$$

where $E_1^\ddagger$ is the translational energy in the reaction coordinate. One can think of this expression as a reactant–transition state equilibrium constant for a microcanonical system. The term $dq_1^\ddagger\, dp_1^\ddagger$ divided by Planck's constant is the number of translational states in the reaction coordinate and the surface integral in the numerator divided by h^{3n-1} is the density of states for the $3n - 1$ degrees of freedom orthogonal to the reaction coordinate. Similarly, the surface integral in the denominator is the reactant density of states multiplied by h^{3n}.

To determine $k(E)$ from equation (A3.12.9) it is assumed that transition states with positive $p_1^{\ddagger}$ form products. Noting that $p_1^{\ddagger} = \mu_1 \, dq_1^{\ddagger}/dt$, where μ_1 is the reduced mass of the separating fragments, all transition states that lie within $q_1^{\ddagger}$ and $q_1^{\ddagger} + dq_1^{\ddagger}$ with positive $p_1^{\ddagger}$ will cross the transition state toward products in the time interval $dt = \mu_1 \, dq_1^{\ddagger}/p_1^{\ddagger}$. Inserting this expression into equation (A3.12.9), one finds that the reactant-to-product rate (i.e. flux) through the transition state for momentum $p_1^{\ddagger}$ is

$$\frac{dN(q_1^{\ddagger}, p_1^{\ddagger})}{dt} = \frac{N \frac{p_1^{\ddagger} \, dp_1^{\ddagger}}{\mu_1} \int \dots \int_{H=E-E_1^{\ddagger}-E_0} dq_2^{\ddagger} \dots dq_{3n}^{\ddagger} \, dp_2^{\ddagger} \dots dp_{3n}^{\ddagger}}{\int \dots \int_{H=E} dq_1 \dots dq_{3n} \, dp_1 \dots dp_{3n}}. \tag{A3.12.10}$$

Since the energy in the reaction coordinate is $E_1^{\ddagger} = p_1^{\ddagger 2}/2\mu_1$, its derivative is $dE_1^{\ddagger} = p_1^{\ddagger} \, dp_1^{\ddagger}/\mu_1$ so that equation (A3.12.10) can be converted into

$$\frac{dN(q_1^{\ddagger}, p_1^{\ddagger})}{dt} = \frac{N \, dE_1^{\ddagger} \int \dots \int_{H=E-E_1^{\ddagger}-E_0} dq_2^{\ddagger} \dots dq_{3n}^{\ddagger} \, dp_2^{\ddagger} \dots dp_{3n}^{\ddagger}}{\int \dots \int_{H=E} dq_1 \dots dq_{3n} \, dp_1 \dots dp_{3n}}. \tag{A3.12.11}$$

This equation represents the reaction rate at total energy E with a fixed energy in the reaction coordinate $E_1^{\ddagger}$ and may be written as

$$dN(E, E_1^{\ddagger})/dt = k(E, E_1^{\ddagger}) N \, dE_1^{\ddagger} \tag{A3.12.12}$$

where $k(E, E_1^{\ddagger})$ is a unimolecular rate constant. As discussed above, the integrals in equation (A3.12.11) are densities of states ρ, so $k(E, E_1^{\ddagger})$ becomes

$$k(E, E_1^{\ddagger}) = \frac{\rho^{\ddagger}(E - E_0 - E_1^{\ddagger})}{h\rho(E)}. \tag{A3.12.13}$$

To find the total reaction flux, equation (A3.12.12) must be integrated between the limits $E_1^{\ddagger}$ equal to 0 and $E - E_0$, so that

$$\frac{dN}{dt} = \frac{\int_0^{E-E_0} dN(E, E_1^{\ddagger})}{dt} = N \int_0^{E-E_0} k(E, E_1^{\ddagger}) \, dE_1^{\ddagger} = k(E)N \tag{A3.12.14}$$

where, using equation (A3.12.13), $k(E)$ is given by

$$k(E) = \frac{\int_0^{E-E_0} \rho^{\ddagger}(E - E_0 - E_1^{\ddagger}) \, dE_1^{\ddagger}}{h\rho(E)} = \frac{N^{\ddagger}(E - E_0)}{h\rho(E)}. \tag{A3.12.15}$$

The term $N^{\ddagger}(E - E_0)$ is the sum of states at the transition state for energies from 0 to $E - E_0$. Equation (A3.12.15) is the RRKM expression for the unimolecular rate constant.

Only in the high-energy limit does classical statistical mechanics give accurate sums and densities of state [15]. Thus, in general, quantum statistical mechanics must be used to calculate a RRKM $k(E)$ which may be compared with experiment [16]. A comparison of classical and quantum harmonic (see below) RRKM rate constants for $C_2H_5 \rightarrow H + C_2H_4$ is given in figure A3.12.3 [17]. The energies used for the classical calculation are with respect to the reactant's and transition state's potential minima. For the quantum calculation the energies are with respect to the zero-point levels. If energies with respect to the zero-point levels were used in the classical calculation, the classical $k(E)$ would be appreciably smaller than the quantum value [16].

RRKM theory allows some modes to be uncoupled and not exchange energy with the remaining modes [16]. In quantum RRKM theory, these uncoupled modes are not *active*, but are *adiabatic* and stay in fixed quantum states n during the reaction. For this situation, equation (A3.12.15) becomes

$$k(E, n) = \frac{N^{\ddagger}[E - E_0(n), n]}{h\rho(E, n)}. \tag{A3.12.16}$$

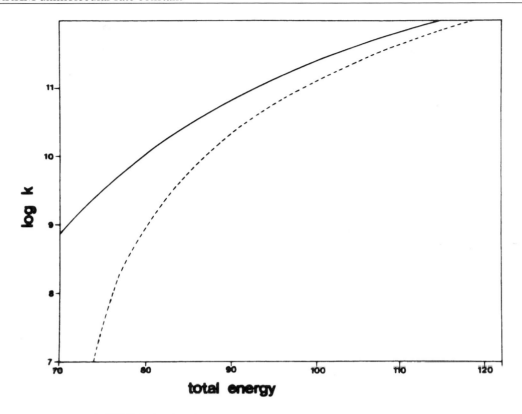

Figure A3.12.3. Harmonic RRKM unimolecular rate constants for $C_2H_5 \rightarrow H + C_2H_4$ dissociation: classical state counting (solid curve), quantal state counting (dashed curve). Rate constant is in units of s^{-1} and energy in kcal mol^{-1}. (Adapted from [17].)

In addition to affecting the number of active degrees of freedom, the fixed n also affects the unimolecular threshold $E_0(n)$. Since the total angular momentum j is a constant of motion and quantized according to

$$j = \sqrt{J(J+1)}\hbar \qquad (A3.12.17)$$

the quantum number J is fixed during the unimolecular reaction. This may be denoted by explicitly including J in equation (A3.12.16), i.e.

$$k(E, J, n) = \frac{N^{\ddagger}(E, J, n)}{\rho(E, J, n)}. \qquad (A3.12.18)$$

The treatment of angular momentum is discussed in detail below in section A3.12.4.3.

A3.12.3.2   k(E) as an average flux

The RRKM rate constant is often expressed as an average classical flux through the transition state [18–20]. To show that this is the case, first recall that the density of states $\rho(E)$ for the reactant may be expressed as

$$\rho(E) = \frac{d}{dE} \int \ldots \int dq_1 \ldots dq_{3n} \, dp_1 \ldots dp_{3n} \theta(E - H)/h^{3n} \qquad (A3.12.19)$$

where θ is the Heaviside function, i.e. $\theta(x) = 1$ for $x > 0$ and $\theta(x) = 0$ for $x < 0$. Since the delta and Heaviside functions are related by $\delta(x) = \mathrm{d}\theta(x)/\mathrm{d}x$, equation (A3.12.19) becomes

$$\rho(E) = \int \ldots \int \mathrm{d}q_1 \ldots \mathrm{d}q_{3n}\, \mathrm{d}p_1 \ldots \mathrm{d}p_{3n} \delta(E - H)/h^{3n}. \qquad (A3.12.20)$$

From equations (A3.12.11)–(A3.12.15) and the discussion above, the RRKM rate constant may be written as

$$k(E) = \frac{\int_0^{E-E_0} \left[\int \ldots \int \mathrm{d}q_2^{\ddagger} \ldots \mathrm{d}q_{3n}^{\ddagger}\, \mathrm{d}p_2^{\ddagger} \ldots \mathrm{d}p_{3n}^{\ddagger} \delta(E' - H) \right] \mathrm{d}H}{\int \ldots \int \mathrm{d}q_1 \ldots \mathrm{d}q_{3n}\, \mathrm{d}p_1 \ldots \mathrm{d}p_{3n} \delta(E - H)}. \qquad (A3.12.21)$$

The inner multiple integral is the transition state's density of states at energy E', and also the numerator in equation (A3.12.13), which gives the transition states sum of states $N^{\ddagger}(E - E_0)$ when integrated from $E' = 0$ to $E' = E - E_0$. Using Hamilton's equation $\mathrm{d}H/\mathrm{d}p_1 = \dot{q}_1$, $\mathrm{d}H$ in the above equation may be replaced by $\dot{q}_1 \mathrm{d}p_1$. Also, from the definition of the delta function

$$\int \delta(q_1 - q_1^{\ddagger})\, \mathrm{d}q_1 = 1. \qquad (A3.12.22)$$

This expression may be inserted into the numerator of the above equation, without altering the equation.

Making the above two changes and noting that $\delta(q_1 - q_1^{\ddagger})$ specifies the transition state, so that the $\ddagger$ superscript to the transition state's coordinates and momenta may be dropped, equation (A3.12.21) becomes

$$k(E) = \frac{\int \ldots \int \dot{q}_1\, \mathrm{d}q_1 \ldots \mathrm{d}q_{3n}\, \mathrm{d}p_1 \ldots \mathrm{d}p_{3n} \delta(q_1 - q_1^{\ddagger}) \delta(E - H)}{\int \ldots \int \mathrm{d}q_1 \ldots \mathrm{d}q_{3n}\, \mathrm{d}p_1 \ldots \mathrm{d}p_{3n} \delta(E - H)}. \qquad (A3.12.23)$$

The rate constant is an average of $\dot{q}_1 \delta(q_1 - q_1^{\ddagger})$, with positive $\dot{q}_1$, for a microcanonical ensemble $H = E$ and may be expressed as

$$k(E) = \langle \dot{q}_1 \delta(q_1 - q_1^{\ddagger}) \rangle. \qquad (A3.12.24)$$

The RRKM rate constant written this way is seen to be an average flux through the transition state.

A3.12.3.3 *Variational RRKM theory*

In deriving the RRKM rate constant in section A3.12.3.1, it is assumed that the rate at which reactant molecules cross the transition state, in the direction of products, is the same rate at which the reactants form products. Thus, if any of the trajectories which cross the transition state in the product direction return to the reactant phase space, i.e. recross the transition state, the actual unimolecular rate constant will be smaller than that predicted by RRKM theory. This one-way crossing of the transition state, with no recrossing, is a fundamental assumption of transition state theory [21]. Because it is incorporated in RRKM theory, this theory is also known as microcanonical transition state theory.

As a result of possible recrossings of the transition state, the classical RRKM $k(E)$ is an upper bound to the correct classical microcanonical rate constant. The transition state should serve as a bottleneck between reactants and products, and in *variational RRKM theory* [22] the position of the transition state along q_1 is varied to minimize $k(E)$. This minimum $k(E)$ is expected to be the closest to the truth. The quantity actually minimized is $N^{\ddagger}(E - E_0)$ in equation (A3.12.15), so the operational equation in variational RRKM theory is

$$\frac{\mathrm{d}N^{\ddagger}[E - E_0(q_1)]}{\mathrm{d}q_1} = 0 \qquad (A3.12.25)$$

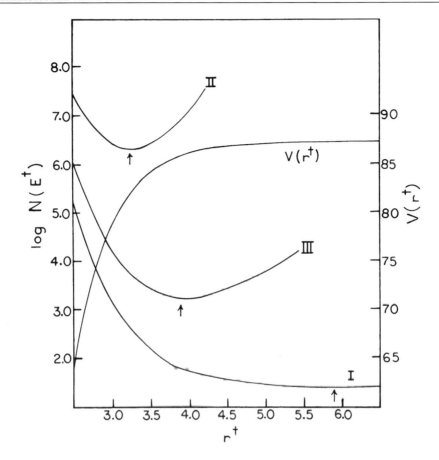

Figure A3.12.4. Plots of $N^{\ddagger}[E - E_0(q_1)]$ versus q_1 for three models of the $C_2H_6 \rightarrow 2CH_3$ potential energy function. r^{+} represents q_1 and the term on the abscissa represents $N^{\ddagger}$. (Adapted from [23].)

where $E_0(q_1)$ is the potential energy as a function of q_1. The minimum in $N^{\ddagger}[E - E_0(q_1)]$ is identified by $q_1 = q_1^{\ddagger}$ and this value for q_1, with the smallest sum of states, is expected to be the best bottleneck for the reaction.

For reactions with well defined potential energy barriers, as in figures A3.12.1(a) and (b), the variational criterion places the transition state at or very near this barrier. The variational criterion is particularly important for a reaction where there is no barrier for the reverse association reaction: see figure A3.12.1(c). There are two properties which gave rise to the minimum in $N^{\ddagger}[E - E_0(q_1)]$ for such a reaction. As q_1 is decreased the potential energy $E_0(q_1)$ decreases and the energy available to the transition state $E - E_0(q_1)$ increases. This has the effect of increasing the sum of states. However, as q_1 is decreased, the intermolecular anisotropic forces between the dissociating fragments increase, which has the effect of decreasing the available phase space and, thus, the sum of states. The combination of these two effects gives rise to a minimum in $N^{\ddagger}[E - E_0(q_1)]$. Plots of the sum of states versus q_1 are shown in figure A3.12.4 for three model potentials of the $C_2H_6 \rightarrow 2CH_3$ dissociation reaction [23].

Variational RRKM theory is particularly important for unimolecular dissociation reactions, in which vibrational modes of the reactant molecule become translations and rotations in the products [22]. For $CH_4 \rightarrow CH_3+H$ dissociation there are three vibrational modes of this type, i.e. the C- - -H stretch which is the reaction coordinate and the two degenerate H- - -CH$_3$ bends, which first transform from high-frequency to low-frequency vibrations and then hindered rotors as the H---C bond ruptures. These latter two degrees of freedom are called transitional modes [24, 25]. $C_2H_6 \rightarrow 2CH_3$ dissociation has five transitional modes, i.e. two pairs of degenerate CH_3 rocking/rotational motions and the CH_3 torsion.

To calculate $N^\ddagger(E - E_0)$, the non-torsional transitional modes have been treated as vibrations as well as rotations [26]. The former approach is invalid when the transitional mode's barrier for rotation is low, while the latter is inappropriate when the transitional mode is a vibration. Harmonic frequencies for the transitional modes may be obtained from a semi-empirical model [23] or by performing an appropriate normal mode analysis as a function of the reaction path for the reaction's potential energy surface [26]. Semiclassical quantization may be used to determine anharmonic energy levels for the transitional modes [27].

The intermolecular Hamiltonian of the product fragments is used to calculate the sum of states of the transitional modes, when they are treated as rotations. The resulting model [28] is nearly identical to phase space theory [29], if the distance between the product fragments' centres-of-mass is assumed to be the reaction coordinate [30]. A more complete model is obtained by using a generalized reaction coordinate, which may contain contributions from different motions, such as bond stretching and bending, as well as the above relative centre-of-mass motion [31].

Variational RRKM calculations, as described above, show that a unimolecular dissociation reaction may have two variational transition states [32–36], i.e. one that is a tight vibrator type and another that is a loose rotator type. Whether a particular reaction has both of these variational transition states, at a particular energy, depends on the properties of the reaction's potential energy surface [33–35]. For many dissociation reactions there is only one variational transition state, which smoothly changes from a loose rotator type to a tight vibrator type as the energy is increased [26].

A3.12.4 Approximate models for the RRKM rate constant

A3.12.4.1 *Classical harmonic oscillators: RRK theory*

The classical mechanical RRKM $k(E)$ takes a very simple form, if the internal degrees of freedom for the reactant and transition state are assumed to be harmonic oscillators. The classical sum of states for s harmonic oscillators is [16]

$$N(E) = \frac{E^s}{s! \prod_{i=1}^{s} h\nu_i}. \tag{A3.12.26}$$

The density $\rho(E) = dN(E)/dE$ is then

$$\rho(E) = \frac{E^{s-1}}{(s-1)! \prod_{i=1}^{s} h\nu_i}. \tag{A3.12.27}$$

The reactant density of states in equation (A3.12.15) is given by the above expression for $\rho(E)$. The transition state's sum of states is

$$N^\ddagger(E - E_0) = \frac{(E - E_0)^{s-1}}{(s-1)! \prod_{i=1}^{s-1} h\nu_i^\ddagger}. \tag{A3.12.28}$$

Inserting equations (A3.12.27) and (A3.12.28) into equation (A3.12.15) gives

$$k(E) = \frac{\prod_{i=1}^{s} \nu_i}{\prod_{i=1}^{s-1} \nu_i^\ddagger} \left(\frac{E - E_0}{E}\right)^{s-1}. \tag{A3.12.29}$$

If the ratio of the products of vibrational frequencies is replaced by ν, equation (A3.12.29) becomes

$$k(E) = \nu \left(\frac{E - E_0}{E}\right)^{s-1}. \tag{A3.12.30}$$

which is the Rice–Ramsperger–Kassel (RRK) unimolecular rate constant [16, 37]. Thus, the $k(E)$ of RRK theory is the classical harmonic limit of RRKM theory.

A3.12.4.2   Quantum harmonic oscillators

Only in the high-energy limit does classical statistical mechanics give accurate values for the sum and density of states terms in equation (A3.12.15) [3, 14]. Thus, to determine an accurate RRKM $k(E)$ for the general case, quantum statistical mechanics must be used. Since it is difficult to make anharmonic corrections, both the molecule and transition state are often assumed to be a collection of harmonic oscillators for calculating the sum $N^{\ddagger}(E - E_0)$ and density $\rho(E)$. This is somewhat incongruous since a molecule consisting of harmonic oscillators would exhibit intrinsic non-RRKM dynamics.

With the assumption of harmonic oscillators, the molecule's quantum energy levels are

$$E(n) = \sum_{i=1}^{s} n_i h \nu_i. \tag{A3.12.31}$$

The same expression holds for the transition state, except that the sum is over $s - 1$ oscillators and the frequencies are the $\nu_i^{\ddagger}$. The Beyer–Swinehart algorithm [38] makes a very efficient direct count of the number of quantum states between an energy of zero and E. The molecule's density of states is then found by finite difference, i.e.

$$\rho(E) = \frac{N(E + \Delta E/2) - N(E - \Delta E/2)}{\Delta E} \tag{A3.12.32}$$

where $N(E + \Delta E/2)$ is the sum of states at energy $E + \Delta E/2$. The transition state's harmonic $N^{\ddagger}(E - E_0)$ is counted directly by the Beyer–Swinehart algorithm. This harmonic model is used so extensively to calculate a value for the RRKM $k(E)$ that it is easy to forget that RRKM theory is not a harmonic theory.

A3.12.4.3   Overall rotation

Regardless of the nature of the intramolecular dynamics of the reactant A*, there are two constants of the motion in a unimolecular reaction, i.e. the energy E and the total angular momentum j. The latter ensures the rotational quantum number J is fixed during the unimolecular reaction and the quantum RRKM rate constant is specified as $k(E, J)$.

(a) Separable vibration/rotation

For a RRKM calculation without any approximations, the complete vibrational/rotational Hamiltonian for the unimolecular system is used to calculate the reactant density and transition state's sum of states. No approximations are made regarding the coupling between vibration and rotation. However, for many molecules the exact nature of the coupling between vibration and rotation is uncertain, particularly at high energies, and a model in which rotation and vibration are assumed separable is widely used to calculate the quantum RRKM $k(E, J)$ [4, 16]. To illustrate this model, first consider a linear polyatomic molecule which decomposes via a linear transition state. The rotational energy for the reactant is assumed to be that for a rigid rotor, i.e.

$$E_r = J(J + 1)\hbar^2/2I. \tag{A3.12.33}$$

The same expression applies to the transition state's rotational energy $E_r^{\ddagger}(J)$ except that the moment of inertia I is replaced by $I^{\ddagger}$. Since the quantum number J is fixed, the active energies for the reactant and transition state are $[E - E_r(J)]$ and $[E - E_0 - E_r^{\ddagger}(J)]$, respectively. The RRKM rate constant is denoted by

$$k(E, J) = \frac{N[E - E_0 - E_r^{\ddagger}(J)]}{h\rho[E - E_r(J)]} \tag{A3.12.34}$$

where $N^{\ddagger}$ and ρ are the sum and density of states for the vibrational degrees of freedom. Each J level is $(2J+1)^2$ degenerate, which cancels for the sum and density.

(b) The K quantum number: adiabatic or active

The degree of freedom in equation (A3.12.18), which has received considerable interest regarding its activity or adiabaticity, is the one associated with the K rotational quantum number for a symmetric or near-symmetric top molecule [39, 40]. The quantum number K represents the projection of J onto the molecular symmetry axis. Coriolis coupling can mix the $2J+1$ K levels for a particular J and destroy K as a good quantum number. For this situation K is considered an active degree of freedom. On the other hand, if the Coriolis coupling is weak, the K quantum number may retain its integrity and it may be possible to measure the unimolecular rate constant as a function of K as well as of E and J. For this case, K is an adiabatic degree of freedom.

It is straightforward to introduce active and adiabatic treatments of K into the widely used RRKM model which represents vibration and rotation as separable and the rotations as rigid rotors [41, 42]. For a symmetric top, the rotational energy is given by

$$E_r(J, K) = \frac{J(J+1)\hbar^2}{2I_a} + \left(\frac{1}{I_c} - \frac{1}{I_a}\right)K^2\hbar^2. \tag{A3.12.35}$$

If K is adiabatic, a molecule containing total vibrational–rotational energy E and, in a particular J, K level, has a vibrational density of states $\rho[E - E_r(J, K)]$. Similarly, the transition state's sum of states for the same E, J, and K is $N^{\ddagger}[E - E_0 - E_r^{\ddagger}(J, K)]$. The RRKM rate constant for the K adiabatic model is

$$k(E, J, K) = \frac{N^{\ddagger}[E - E_0 - E_r^{\ddagger}(J, K)]}{h\rho[E - E_r(J, K)]}. \tag{A3.12.36}$$

Mixing the $2J+1$ K levels, for the K active model, results in the following sums and densities of states:

$$N^{\ddagger}(E, J) = \sum_{K=-J}^{J} N^{\ddagger}[E - E_0 - E_r^{\ddagger}(J, K)] \tag{A3.12.37}$$

$$\rho(E, J) = \sum_{K=-J}^{J} \rho[E - E_r(J, K)]. \tag{A3.12.38}$$

The RRKM rate constant for the K active model is

$$k(E, J) = \frac{\sum_{K=-J}^{J} N^{\ddagger}[E - E_0 - E_r^{\ddagger}(J, K)]}{h\sum_{K=-J}^{J} \rho[E - E_r(J, K)]}. \tag{A3.12.39}$$

In these models the treatment of K is the same for the molecule and transition state. It is worthwhile noting that *mixed mode RRKM models* are possible in which K is treated differently in the molecule and transition state [39].

A3.12.5   Anharmonic effects

In the above section a harmonic model is described for calculating RRKM rate constants with harmonic sums and densities of states. This rate constant, denoted by $k_h(E, J)$, is related to the actual anharmonic RRKM rate constant by

$$k(E, J) = f_{anh}(E, J)k_h(E, J) = f_{anh}(E, J)\frac{N_h^{\ddagger}(E, J)}{h\rho_h(E, J)} \tag{A3.12.40}$$

where $N_h^\ddagger(E, J)$ and $\rho_h(E, J)$ are the harmonic approximations to the sum and density of states. The anharmonic correction, $f_{anh}(E, J)$, is obviously

$$f_{anh}(E, J) = \frac{N_{anh}^\ddagger(E, J)/N_h^\ddagger(E)}{\rho_{anh}(E, J)/\rho_h(E, J)} = \frac{f_{anh, N^\ddagger}(E, J)}{f_{anh, \rho}(E, J)} \quad (A3.12.41)$$

the ratio of anharmonic corrections for the sum and density. For energies near the unimolecular threshold, where the transition state energy $E - E_0$ is small, anharmonicity in the transition state may be negligible, so that $f_{anh}(E)$ may be well approximated by $1/f_{anh, \rho(E)}$ [43]. However, for higher energies, anharmonicity is expected to become important also for the transition state.

There is limited information available concerning anharmonic corrections to RRKM rate constants. Only a few experimental studies have investigated the effect of anharmonicity on the reactant's density of states (see below). To do so requires spectroscopic information up to very high energies. It is even more difficult to measure anharmonicities for transition state energy levels [44, 45]. If the potential energy surface is known for a unimolecular reactant, anharmonic energy levels for both the reactant and transition state may be determined in principle from large-scale quantum mechanical variational calculations. Such calculations are more feasible for the transition state with energy $E - E_0$ than for the reactant with the much larger energy E. Such calculations have been limited to relatively small molecules [4].

The bulk of the information about anharmonicity has come from classical mechanical calculations. As described above, the anharmonic RRKM rate constant for an analytic potential energy function may be determined from either equation (A3.12.4) [13] or equation (A3.12.24) [46] by sampling a microcanonical ensemble. This rate constant and the one calculated from the harmonic frequencies for the analytic potential give the anharmonic correction $f_{anh}(E, J)$ in equation (A3.12.41). The transition state's anharmonic classical sum of states is found from the phase space integral

$$N_{anh}^\ddagger(E, J) = \int \ldots \int dq_2 \ldots dq_{3n} \, dp_2 \ldots dp_{3n} \theta(E - H)/h^{3n-1} \quad (A3.12.42)$$

which may be combined with the harmonic sum $N_h^\ddagger(E, J)$ to give $f_{anh, N^\ddagger}(E, J)$. The classical anharmonic correction to the reactant's density of states, $f_{anh, \rho}(E, J)$, may be obtained in a similar manner.

A3.12.5.1 Molecules with a single minimum

Extensive applications of RRKM theory have been made to unimolecular reactions, for which there is a single potential energy minimum for the reactant molecule [4, 47]. For such reactions, the anharmonic correction $f_{anh}(E, J)$ is usually assumed to be unity and the harmonic model is used to calculate the RRKM $k(E, J)$. Though this is a widely used approach, uncertainties still remain concerning its accuracy. Anharmonic densities of states for formaldehyde [48] and acetylene [49], obtained from high-resolution spectroscopic experiments at energies near their unimolecular thresholds, are 11 and 6 times larger, respectively, than their harmonic densities. From calculations with analytic potential energy functions at energies near the unimolecular thresholds, the HCN quantum anharmonic density of states is 8 times larger than the harmonic value [50] and the classical anharmonic density of states for the model alkyl radical HCC is 3–5 times larger than the harmonic value [51]. There is a sense that the anharmonic correction may become less important for larger molecules, since the average energy per mode becomes smaller [4]. However, as shown below, this assumption is clearly not valid for large fluxional molecules with multiple minima.

Analytic expressions have been proposed for making anharmonic corrections for molecules with a single potential minimum [52–56]. Haarhoff [52] derived an expression for this correction factor by describing the molecules' degrees of freedom as a collection of Morse oscillators. One of the limitations of this model is that it is difficult to assign Morse parameters to non-stretching degrees of freedom. Following the spirit of Haarhoff's work, Troe [54] formulated the correction factor

$$f_{\text{anh},\rho}(E) = \prod_{i=1}^{m} \left(1 + \frac{E/D_i}{2s - 3}\right) \tag{A3.12.43}$$

for a molecule with s degrees of freedom, m of which are Morse stretches. The remaining $s - m$ degrees of freedom are assumed to be harmonic oscillators. The D_i are the individual Morse dissociation energies. To account for bend–stretch coupling, i.e. the attenuation of bending forces as bonds are stretched [51], Troe amended equation (A3.12.43) to give [55]

$$f_{\text{anh},\rho}(E) = \sum_{i=1}^{m} \left(1 + \frac{E/D_i}{2s - 3}\right)^2. \tag{A3.12.44}$$

The above expressions are empirical approaches, with m and D_i as parameters, for including an anharmonic correction in the RRKM rate constant. The utility of these equations is that they provide an analytic form for the anharmonic correction. Clearly, other analytic forms are possible and may be more appropriate. For example, classical sums of states for H–C–C, H–C=C, and H–C≡C hydrocarbon fragments with Morse stretching and bend–stretch coupling anharmonicity [51] are fit accurately by the exponential

$$f_{\text{anh},N}(E) = \exp(bE). \tag{A3.12.45}$$

The classical anharmonic density of states is then [56]

$$f_{\text{anh},\rho}(E) = \exp(bE)[1 + bE/s]. \tag{A3.12.46}$$

Modifying equation (A3.12.45) to represent the transition state's sum of states, the anharmonic correction to the RRKM rate constant becomes

$$f_{\text{anh}}(E) = \frac{\exp[b^{\ddagger}(E - E_0)]}{\exp(bE)[1 + bE/s]}. \tag{A3.12.47}$$

This expression, and variations of it, have been used to fit classical anharmonic microcanonical $k(E, J)$ for unimolecular decomposition [56].

A3.12.5.2   Fluxional molecules with multiple minima

Anharmonic corrections are expected to be very important for highly fluxional molecules such as clusters and macromolecules [30]. Figure A3.12.5 illustrates a possible potential energy curve for a fluxional molecule. There are multiple minima (i.e. conformations) separated by barriers much lower than that for dissociation. Thus, a moderately excited fluxional molecule may undergo rapid transitions between its many conformations, and all will contribute to the molecule's unimolecular rate constant. Many different conformations are expected for the products, but near the dissociation threshold E_0, only one set of product conformations is accessible. As the energy is increased, thresholds for other product conformations are reached. For energies near E_0, there is very little excess energy in the transition state and the harmonic approximation for the lowest energy product conformation should be very good for the transition state's sum of states. Thus, for $E \approx E_0$ the anharmonic correction in equation (A3.12.40) is expected to primarily result from anharmonicity in the reactant density of states.

The classical anharmonic RRKM rate constant for a fluxional molecule may be calculated from classical trajectories by following the initial decay of a microcanonical ensemble of states for the unimolecular reactant,

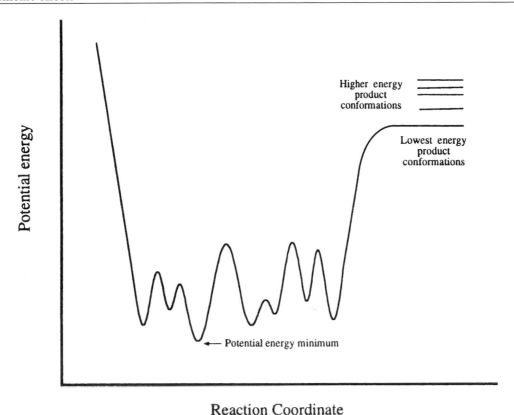

Figure A3.12.5. A model reaction coordinate potential energy curve for a fluxional molecule. (Adapted from [30].)

as given by equation (A3.12.4). Such a calculation has been performed for dissociation of the Al_6 and Al_{13} clusters using a model analytic potential energy function written as a sum of Lennard–Jones and Axelrod–Teller potentials [30]. Structures of some of the Al_6 minima, for the potential function, are shown in figure A3.12.6. The deepest potential minimum has C_{2h} symmetry and a classical $Al_6 \rightarrow Al_5 + Al$ dissociation energy E_0 of 43.8 kcal mol^{-1}. For energies 30–80 kcal mol^{-1} in excess of this E_0, the value of f_{anh} determined from the trajectories varies from 200 to 130. The harmonic RRKM rate constants are based on the deepest potential energy minima for the reactant and transition state, and calculated for a reaction path degeneracy of 6. As discussed above, even larger corrections are expected at lower energies, particularly for $E \approx E_0$, where anharmonicity in the transition state does not contribute $f_{anh}(E)$. However, because of the size of Al_6 and its long unimolecular lifetime, it becomes impractical to simulate the classical dissociation of Al_6 for energies in excess of E_0 much smaller than 30 kcal mol^{-1}. For the bigger cluster Al_{13}, the anharmonic correction varies from 5500 to 1200 for excess energies in the range of 85–185 kcal mol^{-1} [30]. These calculations illustrate the critical importance of including anharmonic corrections when calculating accurate RRKM rate constants for fluxional molecules.

In the above discussion it was assumed that the barriers are low for transitions between the different conformations of the fluxional molecule, as depicted in figure A3.12.5, and therefore the transitions occur on a timescale much shorter than the RRKM lifetime. This is the rapid IVR assumption of RRKM theory discussed in section A3.12.2. Accordingly, an initial microcanonical ensemble over all the conformations decays exponentially. However, for some fluxional molecules, transitions between the different conformations

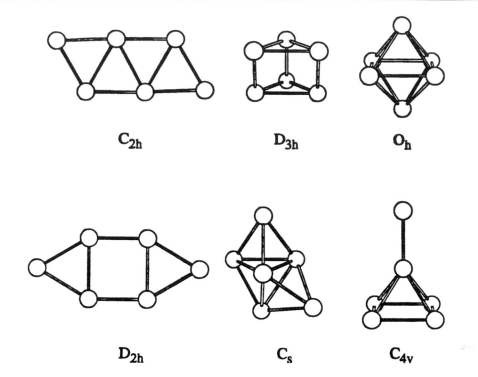

Figure A3.12.6. Structures for some of the potential energy minima for Al_6. The unimolecular thresholds for the C_{2h}, D_{3h}, C_5, O_h, C_{4v}, and D_h minima are 43.8, 40.0, 39.6, 38.8, 31.4 and 20.9 kcal mol^{-1}, respectively. (Adapted from [40].)

may be slower than the RRKM rate, giving rise to bottlenecks in the unimolecular dissociation [4, 57]. The ensuing lifetime distribution, equation (A3.12.7), will be non-exponential, as is the case for *intrinsic* non-RRKM dynamics, for an initial microcanonical ensemble of molecular states.

A3.12.6 Classical dynamics of intramolecular motion and unimolecular decomposition

A3.12.6.1 Normal-mode Hamiltonian: Slater theory

The classical mechanical model of unimolecular decomposition, developed by Slater [18], is based on the normal-mode harmonic oscillator Hamiltonian. Although this Hamiltonian is rigorously exact only for small displacements from the molecular equilibrium geometry, Slater extended it to the situation where molecules are highly vibrationally energized, undergo large amplitude motions and decompose. Since there are no couplings in the normal-mode Hamiltonian, the energies in the individual normal modes do not vary with time. This is the essential difference from the RRKM theory which treats a molecule as a collection of coupled modes which freely exchange energy.

The normal-mode harmonic oscillator classical Hamiltonian is

$$H = \sum_{i=1}^{s} \frac{(P_i^2 + \lambda_1 Q_i^2)}{2} \tag{A3.12.48}$$

where $\lambda_1 = 4\pi^2 v_i^2$. Solving the classical equations of motion for this Hamiltonian gives rise to quasiperiodic motion [58] in which each normal-mode coordinate Q_i varies with time according to

$$Q_i = Q_i^0 \cos(2\pi v_i t + \delta_i) \tag{A3.12.49}$$

where Q_i^0 is the amplitude and δ_i the phase of the motion. Thus, if an energy $E_i = (P_i^2 + \lambda_i Q_i^2)/2$ and phase δ_i are chosen for each normal mode, the complete intramolecular motion of the energized molecule may be determined for this particular initial condition.

Reaction is assumed to have occurred if a particular internal coordinate q, such as a bond length, attains a critical extension $q^\ddagger$. In the normal-mode approximation, the displacement d of internal coordinates and normal-mode coordinates Q are related through the linear transformation

$$d = LQ. \tag{A3.12.50}$$

The transformation matrix L is obtained from a normal-mode analysis performed in internal coordinates [59, 60]. Thus, as the evolution of the normal-mode coordinates versus time is evaluated from equation (A3.12.49), displacements in the internal coordinates and a value for q are found from equation (A3.12.50). The variation in q with time results from a superposition of the normal modes. At a particular time, the normal-mode coordinates may phase together so that q exceeds the critical extension $q^\ddagger$, at which point decomposition is assumed to occur.

The preceding discussion gives the essential details of the Slater theory. Energy does not flow freely within the molecule and attaining the critical reaction coordinate extension is not a statistically random process as in RRKM theory, but depends on the energies and phases of the specific normal modes excited. If a microcanonical ensemble is chosen at $t = 0$, Slater theory gives an initial decay rate which agrees with the RRKM value. However, Slater theory gives rise to intrinsic non-RRKM behaviour [12, 13]. The trajectories are quasiperiodic and each trajectory is restricted to a particular type of motion and region of phase space. Thus, as specific trajectories react, other trajectories cannot fill up unoccupied regions of phase space. As a result, a microcanonical ensemble is not maintained during the unimolecular decomposition. In addition, some of the trajectories may be unreactive and trapped in the reactant region of phase space.

Overall, the Slater theory is unsuccessful in interpreting experiments. Many unimolecular rate constants and reaction paths are consistent with energy flowing randomly within the molecule [4, 36]. If one considers the nature of classical Hamiltonians for actual molecules, it is not surprising that the Slater theory performs so poorly. For example, in Slater theory, the intramolecular and unimolecular dynamics of the molecule conform to the symmetry of the molecular vibrations. Thus, if normal modes of a particular symmetry type are excited (e.g. in-plane vibrations) a decomposition path of another symmetry type (e.g. out-of-plane dissociation) cannot occur. This path requires excitation of out-of-plane vibrations. Normal modes of different symmetry types for actual molecules are coupled by *Coriolis* vibrational–rotational interactions [61]. Similarly, nonlinear resonance interactions couple normal modes of vibration, allowing transfer of energy [62, 63]. Not including these effects is a severe shortcoming of Slater theory. Clearly, understanding the classical intramolecular motion of vibrationally excited molecules requires one to go beyond the normal-mode model.

A3.12.6.2 Coupled anharmonic Hamiltonians

The first classical trajectory study of unimolecular decomposition and intramolecular motion for realistic anharmonic molecular Hamiltonians was performed by Bunker [12, 13]. Both intrinsic RRKM and non-RRKM dynamics was observed in these studies. Since this pioneering work, there have been numerous additional studies [9, 17, 30, 64–67] from which two distinct types of intramolecular motion, chaotic and quasiperiodic [14], have been identified. Both are depicted in figure A3.12.7. Chaotic vibrational motion is not regular as predicted by the normal-mode model and, instead, there is energy transfer between the modes. If all the modes of the molecule participate in the chaotic motion and energy flow is sufficiently rapid, an initial microcanonical ensemble is maintained as the molecule dissociates and RRKM behaviour is observed [9]. For non-random excitation initial apparent non-RRKM behaviour is observed, but at longer times a microcanonical ensemble of states is formed and the probability of decomposition becomes that of RRKM theory.

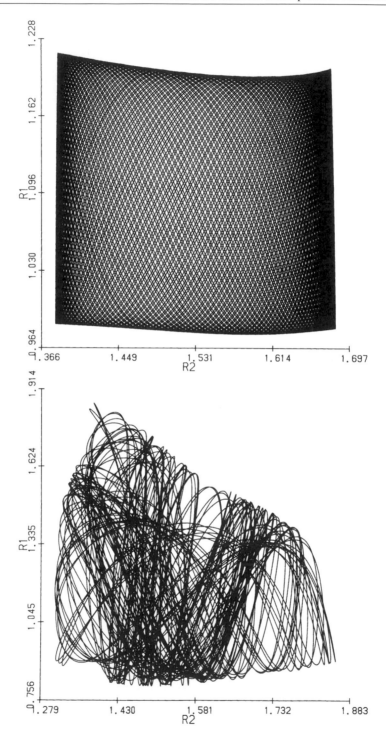

Figure A3.12.7. Two trajectories for a model HCC Hamiltonian. Top trajectory is for $n_{HC} = 0$ and $n_{CC} = 2$, and is quasiperiodic. Bottom trajectory is for $n_{HC} = 5$ and $n_{CC} = 0$, and is chaotic. R_1 is the HC bond length and R_2 the CC bond length. Distance is in Ångstroms (Å). (Adapted from [121] and [4].)

Quasiperiodic motion is regular as assumed by the Slater theory. The molecule's vibrational motion may be represented by a superposition of individual modes, each containing a fixed amount of energy. For some cases these modes resemble normal modes, but they may be identifiably different. Thus, although actual molecular Hamiltonians contain potential and kinetic energy coupling terms, they may still exhibit regular vibrational motion with no energy flow between modes as predicted by the Slater theory. The existence of regular motion for coupled systems is explained by the Kolmogorov–Arnold–Moser (KAM) theorem [4, 68].

Extensive work has been done to understand quasiperiodic and chaotic vibrational motion of molecular Hamiltonians [14, 58, 63, 68, 69]. At low levels of excitation, quasiperiodic normal mode motion is observed. However, as the energy is increased, nonlinear resonances [14, 62, 63, 68] result in the flow of energy between the normal modes, giving rise to chaotic trajectories. In general, the fraction of the trajectories, which are chaotic at a fixed energy, increases as the energy is increased. With increase in energy the nature of the quasiperiodic trajectories may undergo a transition from normal mode to another type of motion, e.g. local mode [70, 71]. In many cases the motion becomes totally chaotic before the unimolecular threshold energy is reached, so that the intramolecular dynamics is ergodic. Though this implies intrinsic RRKM behaviour for energies above the threshold, the ergodicity must occur on a timescale much shorter than the RRKM lifetime for a system to be intrinsically RRKM.

For some systems quasiperiodic (or nearly quasiperiodic) motion exists above the unimolecular threshold, and intrinsic non-RRKM lifetime distributions result. This type of behaviour has been found for Hamiltonians with low unimolecular thresholds, widely separated frequencies and/or disparate masses [12, 62, 65]. Thus, classical trajectory simulations performed for realistic Hamiltonians predict that, for some molecules, the unimolecular rate constant may be strongly sensitive to the modes excited in the molecule, in agreement with the Slater theory. This property is called *mode specificity* and is discussed in the next section.

It is of interest to consider the classical/quantal correspondence for the above different types of classical motion. If the motion within the classical phase space is ergodic so that the decomposing molecules can be described by a microcanonical ensemble, classical RRKM theory will be valid. However, the classical and quantal RRKM rate constants may be in considerable disagreement. This results from an incorrect treatment of zero-point energy in the classical calculations [17, 72] and is the reason quantum statistical mechanics is needed to calculate an accurate RRKM rate constant: see the discussion following equation (A3.12.15). With the energy referenced at the bottom of the well, the total internal energy of the dissociating molecule is $E = E^* + E_z^*$ where E^* is the internal energy of the molecule and E_z^* is its zero-point energy. The classical dissociation energy is D_e, and the energy available to the dissociating molecule at the classical barrier is $E - D_e$. Because the quantal threshold is $D_e + E_z^{\ddagger}$ where $E_z^{\ddagger}$ is the zero-point energy at the barrier, the classical threshold is lower than the quantal one by $E_z^{\ddagger}$. For large molecules with a large $E_z^{\ddagger}$ and/or for low levels of excitation the classical RRKM rate constant is significantly larger than the quantal one. Only in the high-energy limit are they equal; see figure A3.12.3.

Quasiperiodic trajectories, with an energy greater than the unimolecular threshold, are trapped in the reactant region of phase space and will not dissociate. These trajectories correspond to quantum mechanical compound-state resonances $|n\rangle$ (discussed in the next section), which have complex eigenvalues. Applying semiclassical mechanics to the trajectories [73–76] gives energies E_n, wavefunctions ψ_n, and unimolecular rate constants k_n for these resonances.

A classical microcanonical ensemble for an energized molecule may consist of quasiperiodic, chaotic, and 'vague tori' trajectories [77]. The lifetimes of trajectories for the latter may yield correct quantum k_n for resonance states [4].

A3.12.7 State-specific unimolecular decomposition

The quantum dynamics of unimolecular decomposition may be studied by solving the time-dependent Schrödinger equation, i.e. equation (A3.12.2). For some cases the dissociation probability of the molecule is sufficiently small that one can introduce the concept of quasi-stationary states. Such states are commonly referred to as resonances, since the energy of the unimolecular product(s) in the continuum is in resonance with (i.e. matches) the energy of a vibrational/rotational level of the unimolecular reactant. For unimolecular reactions there are two types of resonance states. A shape resonance occurs when a molecule is temporarily trapped by a fairly high and wide potential energy barrier and decomposes by tunnelling. The second type of resonance, called a Feshbach or compound-state resonance, arises when energy is initially distributed between molecular vibrational/rotational degrees of freedom which are not strongly coupled to the decomposition reaction coordinate motion, so that there is a time lag for unimolecular dissociation.

In a time-dependent picture, resonances can be viewed as localized wavepackets composed of a superposition of continuum wavefunctions, which qualitatively resemble bound states for a period of time. The unimolecular reactant in a resonance state moves within the potential energy well for a considerable period of time, leaving it only when a fairly long time interval τ has elapsed; τ may be called the lifetime of the almost stationary resonance state.

Solving the time-dependent Schrödinger equation for resonance states [78] one obtains a set of complex eigenvalues, which may be written in the form

$$E_n^0 = E_n - i\Gamma_n/2 \tag{A3.12.51}$$

where E_n and Γ_n are positive constants. The constant E_n, the real component to the eigenvalue, gives the position of the resonance in the spectrum. It is easy to see the physical significance of complex energy values. The time factor in the wavefunction of a quasi-stationary state is of the form

$$\exp[-(i/\hbar)E_n^0 t] = \exp[-(i/\hbar)E_n t]\exp[-(\Gamma_n/2\hbar)t]. \tag{A3.12.52}$$

Hence, all probabilities given by the squared modulus of the wavefunction decrease as $\exp[-(\Gamma_n/\hbar)t]$ with time, that is

$$|\psi_n(t)|^2 = |\psi_n(0)|^2 \exp[-(\Gamma_n/\hbar)t]. \tag{A3.12.53}$$

In particular, the probability of finding the unimolecular reactant within its potential energy well decreases according to this law. Thus Γ_n determines the lifetime of the state and the state specific unimolecular rate constant is

$$k_n = \Gamma_n/\hbar = 1/\tau_n \tag{A3.12.54}$$

where τ_n is the state's lifetime.

The energy spectrum of the resonance states will be quasi-discrete; it consists of a series of broadened levels with Lorentzian lineshapes whose full-width at half-maximum Γ is related to the lifetime by $\Gamma = \hbar/\tau$. The resonances are said to be isolated if the widths of their levels are small compared with the distances (spacings) between them, that is

$$\Gamma_n \ll 1/\rho(E) \tag{A3.12.55}$$

where $\rho(E)$ is the density of states for the energized molecule. A possible absorption spectrum for a molecule with isolated resonances is shown in figure A3.12.8. Below the unimolecular threshold E_0, the absorption lines for the molecular eigenstates are very narrow and are only broadened by interaction of the excited molecule with the radiation field. However, above E_0 the excited states leak toward product space, which gives rise to widths for the resonances in the spectrum. Each resonance has its own characteristic width (i.e. lifetime). As the linewidths broaden and/or the number of resonance states in an energy interval increases, the spectrum may no longer be quasi-discrete since the resonance lines may overlap, that is

$$\Gamma_n \gg 1/\rho(E). \tag{A3.12.56}$$

It is of interest to determine when the linewidth $\Gamma(E)$ associated with the RRKM rate constant $k(E)$ equals the average distance $\rho(E)^{-1}$ between the reactant energy levels. From equation (A3.12.54) $\Gamma(E) = \hbar k(E)$

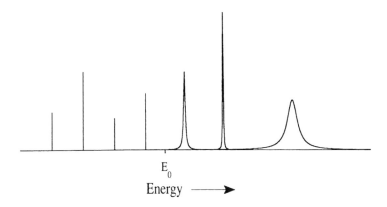

Figure A3.12.8. Possible absorption spectrum for a molecule which dissociates via isolated compound-state resonances. E_0 is the unimolecular threshold. (Adapted from [4].)

and from the RRKM rate constant expression equation (A3.12.15) $\rho(E)^{-1} = \hbar 2\pi K(E)/N^{\ddagger}(E - E_0)$. Equating these two terms gives $N^{\ddagger}(E - E_0) = 2\pi$, which means that the linewidths, associated with RRKM decomposition, begin to overlap when the transition state's sum of states exceeds six.

The theory of isolated resonances is well understood and is discussed below. Mies and Krauss [79, 80] and Rice [81] were pioneers in treating unimolecular rate theory in terms of the decomposition of isolated Feshbach resonances.

A3.12.7.1 Isolated reactant resonance states mode specificity

The observation of decomposition from isolated compound-state resonances does not necessarily imply mode-specific unimolecular decomposition. Nor is mode specificity established by the presence of fluctuations in state-specific rate constants for resonances within a narrow energy interval. What is required for mode-specific unimolecular decomposition is a distinguishable and, thus, assignable pattern (or patterns) in the positions of resonance states in the spectrum. Identifying such patterns in a spectrum allows one to determine which modes in the molecule are excited when forming the resonance state. It is, thus, possible to interpret particularly large or small state-specific rate constants in terms of mode-specific excitations. Therefore, mode specificity means there are exceptionally large or small state-specific rate constants depending on which modes are excited.

The ability to assign a group of resonance states, as required for mode-specific decomposition, implies that the complete Hamiltonian for these states is well approximated by a zero-order Hamiltonian with eigenfunctions $\phi_i(m)$ [58]. The ϕ_i are product functions of a zero-order orthogonal basis for the reactant molecule and the quantity m represents the quantum numbers defining ϕ_i. The wavefunctions ψ_n for the compound state resonances are given by

$$\psi_n = \sum_i c_{in}\phi_i(m). \tag{A3.12.57}$$

Resonance states in the spectra, which are assignable in terms of zero-order basis $\phi_i(m)$, will have a predominant expansion coefficient c_{in}. Hose and Taylor [58] have argued that for an assignable level $c_{in}^2 > 0.5$ for one of the expansion coefficients. The quasiperiodic and 'vague tori' trajectories for energies above the unimolecular threshold, discussed in the previous section, are the classical analogue of these quantum mode specific resonance states.

Mode specificity has been widely observed in the unimolecular decomposition of van der Waals molecules [82], e.g.

$$HF{-}HF \rightarrow 2HF.$$

A covalent bond (or particular normal mode) in the van der Waals molecule (e.g. the I_2 bond in I_2–He) can be selectively excited, and what is usually observed experimentally is that the unimolecular dissociation rate constant is orders of magnitude smaller than the RRKM prediction. This is thought to result from weak coupling between the excited high-frequency intramolecular mode and the low-frequency van der Waals intermolecular modes [83]. This coupling may be highly mode specific. Exciting the two different HF stretch modes in the $(HF)_2$ dimer with one quantum results in lifetimes which differ by a factor of 24 [84]. Other van der Waals molecules studied include $(NO)_2$ [85], NO–HF [86], and $(C_2H_4)_2$ [87].

There are fewer experimental examples of mode specificity for the unimolecular decomposition of covalently bound molecules. One example is the decomposition of the formyl radical HCO, namely

$$HCO \rightarrow H + CO.$$

Well defined progressions are seen in the stimulated emission pumping spectrum (SEP) [88, 89] so that quantum numbers may be assigned to the HCO resonance states, and lifetimes for these states may be associated with the degree of excitation in the HC stretch, CO stretch and HCO bend vibrational modes, denoted by quantum numbers v_1, v_2, and v_3, respectively. States with large v_1 and large excitations in the HC stretch have particularly short lifetimes, while states with large v_2 and large excitations in the CO stretch have particularly long lifetimes. Short lifetimes for states with a large v_1 might be expected, since the reaction coordinate for dissociation is primarily HC stretching motion. The mode specific effects are illustrated by the nearly isoenergetic (v_1, v_2, v_3) resonance states $(0, 4, 5)$, $(1, 5, 1)$ and $(0, 7, 0)$ whose respective energies (i.e. position in spectrum) are 12373, 12487 and 12544 cm^{-1} and whose respective linewidths Γ are 42, 55 and 0.72 cm^{-1}.

Time-dependent quantum mechanical calculations have also been performed to study the HCO resonance states [90, 91]. The resonance energies, linewidths and quantum number assignments determined from these calculations are in excellent agreement with the experimental results.

Mode specificity has also been observed for HOCl→Cl+OH dissociation [92–94]. For this system, many of the states are highly mixed and unassignable (see below). However, resonance states with most of the energy in the OH bond, e.g. $v_{OH} = 6$, are assignable and have unimolecular rate constants orders of magnitude smaller than the RRKM prediction [92–94]. The lifetimes of these resonances have a very strong dependence on the J and K quantum numbers of HOCl.

(a) Statistical state specificity

In contrast to resonance states which may be assigned quantum numbers and which may exhibit mode-specific decomposition, there are states which are intrinsically unassignable. Because of extensive couplings, a zero-order Hamiltonian and its basis set cannot be found to represent the wavefunctions ψ_n for these states. The spectrum for these states is irregular without patterns, and fluctuations in the k_n are related to the manner in which the ψ_n are randomly distributed in coordinate space. Thus, the states are intrinsically unassignable and have no good quantum numbers apart from the total energy and angular momentum. Energies for these resonance states do not fit into a pattern, and states with particularly large or small rate constants are simply random occurrences in the spectrum. For the most statistical (i.e. non-separable) situation, the expansion coefficients in equation (A3.12.56) are random variables, subject only to the normalization and orthogonality conditions

$$\sum_n c_{in}^2 = 1 \qquad \text{and} \qquad \sum_i c_{in}c_{im} = 0. \qquad (A3.12.58)$$

If all the resonance states which form a microcanonical ensemble have random ψ_n, and are thus intrinsically unassignable, a situation arises which is called *statistical state-specific behaviour* [95]. Since the wavefunction coefficients of the ψ_n are Gaussian random variables when projected onto ϕ_i basis functions for any zero-order representation [96], the distribution of the state-specific rate constants k_n will be as statistical as possible. If these k_n within the energy interval $E \rightarrow E + \mathrm{d}E$ form a continuous distribution, Levine [97] has argued that the probability of a particular k is given by the Porter–Thomas [98] distribution

$$P(k) = \frac{\nu}{2\bar{k}} \left(\frac{\nu k}{2\bar{k}} \right)^{((\nu-2)/2)} \frac{\exp(-\nu k/2\bar{k})}{\Gamma(1/2\nu)} \qquad (A3.12.59)$$

where $\bar{k}$ is the average state-specific unimolecular rate constant within the energy interval $E \rightarrow E + \mathrm{d}E$,

$$\bar{k} = \int_0^\infty k P(k) \, \mathrm{d}(k) \qquad (A3.12.60)$$

and ν is the 'effective number of decay channels'. Equation (A3.12.59) is derived in statistics as the probability distribution

$$X_\nu^2 = x_1^2 + x_2^2 + \cdots + x_\nu^2 \qquad (A3.12.61)$$

where the ν x_i are each independent Gaussian distributions [96]. Increasing ν reduces the variance of the distribution $P(k)$.

The connection between the Porter–Thomas $P(k)$ distribution and RRKM theory is made through the parameters $\bar{k}$ and ν. Waite and Miller [99] have studied the relationship between the average of the statistical state-specific rate constants k and the RRKM rate constant $k(E)$ by considering a separable (uncoupled) two-dimensional Hamilton, $H = H_x + H_y$, whose decomposition path is tunnelling through a potential energy barrier along the x-coordinate. They found that the average of the state-specific rate constants for a microcanonical ensemble $\bar{k}$ is the same as the RRKM rate constant $k(E)$. Though insightful, this is not a general result since the tunnelling barrier defines the dividing surface, with no recrossings, which is needed to derive RRKM from classical (not quantum) mechanical principles (see section A3.12.3). For state-specific decomposition which does not occur by tunnelling, a dividing surface cannot be constructed for a quantum calculation. However, the above analysis is highly suggestive that $\bar{k}$ may be a good approximation to the RRKM $k(E)$.

The parameter ν in equation (A3.12.59) has also been related to RRKM theory. Polik et al [80] have shown that for decomposition by quantum mechanical tunnelling

$$\nu = \left[\sum_n \kappa(E - E_n^\ddagger) \right]^2 \bigg/ \sum_n \kappa(E - E_n^\ddagger)^2 \qquad (A3.12.62)$$

where $\kappa(E - E_n^\ddagger)$ is a one-dimensional tunnelling probability through a potential barrier and $E_n^\ddagger$ is the vibrational energy in the $3N - 7$ modes orthogonal to the tunnelling coordinate. If the energy is sufficiently low that all the tunnelling probabilities are much less than 1 and one makes a parabolic approximation to the tunnelling probabilities [96, 100], equation (A3.12.62) becomes

$$\nu = \prod_{k=1}^{3N-7} \coth(\pi \omega_k^\ddagger / \omega_b) \qquad (A3.12.63)$$

where the $\omega_k^\ddagger$ are the $3N - 7$ frequencies for the modes orthogonal to the tunnelling coordinate and ω_b is the barrier frequency. The interesting aspect of equation (A3.12.63) is that it shows ν to be energy independent in the tunnelling region. On the other hand, for energies significantly above the barrier so that $\kappa(E - E_n^\ddagger) = 1$, it is easy to show [96, 100] that

$$\nu = N^\ddagger(E) \qquad (A3.12.64)$$

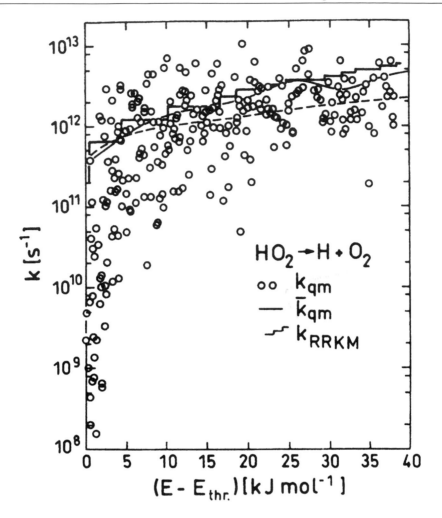

Figure A3.12.9. Comparison of the unimolecular dissociation rates for $HO_2 \rightarrow H+O_2$ as obtained from the quantum mechanical resonances (k_{qm}, open circles) and from variational transition state RRKM (k_{RRKM}, step function). E_{thr} is the threshold energy for dissociation. Also shown is the quantum mechanical average (solid line) as well as the experimental prediction for $J = 0$ derived from a simplified SACM analysis of high pressure unimolecular rate constants. (Adapted from [101].)

where $N^{\ddagger}(E)$ is the sum of states for the transition state. In this energy region, ν rapidly increases with increase in energy and the $P(k)$ distribution becomes more narrowly peaked.

Statistical state-specific behaviour has been observed in experimental SEP studies of $H_2CO \rightarrow H_2+CO$ dissociation [44, 48] and quantum mechanical scattering calculations of $HO_2 \rightarrow H+O_2$ dissociation [101, 102]. The state-specific rate constants for the latter calculation are plotted in figure A3.12.9. For both of these dissociations the RRKM rate constant and the average of the state-specific quantum rate constants for a small energy interval ΔE are in good agreement. Similarly, the fluctuations in the resonance rate constants are well represented by the Porter–Thomas distribution. That HO_2 dissociation is statistical state-specific, while HCO dissociation is mode specific, is thought to arise from the deeper potential energy well and associated greater couplings and density of states for HO_2.

A microcanonical ensemble of isolated resonances decays according to

$$N(t, E) = \sum_n \exp(-k_n t).$$ (A3.12.65)

If the state-specific rate constants are assumed continuous, equation (A3.12.65) can be written as [103]

$$N(t, E) = N_0 \int_0^\infty \exp(-kt) P(k) \, d(k)$$ (A3.12.66)

where N_0 is the total number of molecules in the microcanonical ensemble. For the Porter–Thomas $P(k)$ distribution, $N(t, E)$ becomes [103, 104]

$$N(t, E)/N_0 = (1 + 2\bar{k}t/\nu)^{-\nu/2}.$$ (A3.12.67)

The expression for $N(t, E)$ in equation (A3.12.67) has been used to study [103, 104] how the Porter–Thomas $P(k)$ affects the collision-averaged monoenergetic unimolecular rate constant $k(\omega, E)$ [105] and the Lindemann–Hinshelwood unimolecular rate constant $k_{uni}(\omega, T)$ [47]. The Porter–Thomas $P(k)$ makes $k(\omega, E)$ pressure dependent [103]. It equals $\bar{k}$ in the high-pressure $\omega \to \infty$ limit and $[(\nu-2)/\nu]\bar{k}$ in the $\omega \to 0$ low-pressure limit. $P(k)$ only affects $k_{uni}(\omega, T)$ in the intermediate pressure regime [40, 104], and has no affect on the high- and low-pressure limits. This type of analysis has been applied to $HO_2 \to H+O_2$ resonance states [106], which decay in accord with the Porter–Thomas $P(k)$. Deviations between the $k_{uni}(\omega, T)$ predicted by the Porter–Thomas and exponential $P(k)$ are more pronounced for the model in which the rotational quantum number K is treated as adiabatic than the one with K active.

A3.12.7.2 Individual transition state levels

The prediction of RRKM theory is that at low energies, where $N^\ddagger(E)$ is small, there are incremental increases in $N^\ddagger(E)$ and resulting in steps in $k(E)$. The minimum rate constant is at the threshold where $N^\ddagger(E) = 1$, i.e. $k(E_0) = 1/\rho(E_0)$. Steps are then expected in $k(E)$ as $N^\ddagger(E)$ increases by unit amounts. This type of behaviour has been observed in experiments for $NO_2 \to NO+O$ [107, 108], $CH_2CO \to CH_2+CO$ [44] and $CH_3CHO \to CH_3+HCO$ [109] dissociation. These experiments do not directly test the rapid IVR assumption of RRKM theory, since steps are expected in $N^\ddagger(E)$ even if all the states of the reactant do not participate in $\rho(E)$. However, if the measured threshold rate constant $k(E_0)$ equals the inverse of the accurate anharmonic density of states for the reactant (difficult to determine), RRKM theory is verified.

If properly interpreted [110], the above experiments provide information about the energy levels of the transition state, i.e. figure A3.12.10. For $NO_2 \to NO+O$ dissociation, there is no barrier for the reverse association reaction, and it has been suggested that the steps in its $k(E)$ may arise from quantization of the transition state's O- - -NO bending mode [107, 108]. Ketene (CH_2CO) dissociates on both singlet and triplet surfaces. The triplet surface has a saddlepoint, at which the transition state is located, and the steps in $k(E)$ for this surface are thought to result from excitation in the transition state's CH_2 wag and C- - -CO bending vibrations [44]. The singlet ketene surface does not have a barrier for the reverse $^1CH_2+CO$ association and the small steps in $k(E)$ for dissociation on this surface are attributed to CO free rotor energy levels for a loose transition state [44]. The steps for acetaldehyde dissociation [109] have been associated with the torsional and C- - -CO bending motions at the transition state.

Detailed analyses of the above experiments suggest that the apparent steps in $k(E)$ may not arise from quantized transition state energy levels [110, 111]. Transition state models used to interpret the ketene and acetaldehyde dissociation experiments are not consistent with the results of high-level *ab initio* calculations [110, 111]. The steps observed for NO_2 dissociation may originate from the opening of electronically excited dissociation channels [107, 108]. It is also of interest that RRKM-like steps in $k(E)$ are not found from

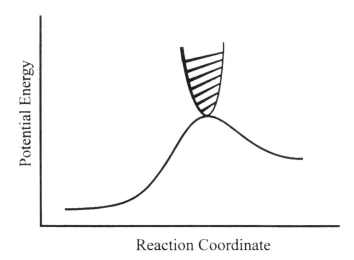

Figure A3.12.10. Schematic diagram of the one-dimensional reaction coordinate and the energy levels perpendicular to it in the region of the transition state. As the molecule's energy is increased, the number of states perpendicular to the reaction coordinate increases, thereby increasing the rate of reaction. (Adapted from [4].)

detailed quantum dynamical calculations of unimolecular dissociation [91, 101, 102, 112]. More studies are needed of unimolecular reactions near threshold to determine whether there are actual quantized transition states and steps in $k(E)$ and, if not, what is the origin of the apparent steps in the above measurements of $k(E)$.

A3.12.8   Examples of non-RRKM decomposition

A3.12.8.1   *Apparent non-RRKM*

Apparent non-RRKM behaviour occurs when the molecule is excited non-randomly and there is an initial non-RRKM decomposition before IVR forms a microcanonical ensemble (see section A3.12.2). Reaction pathways, which have non-competitive RRKM rates, may be promoted in this way. Classical trajectory simulations were used in early studies of apparent non-RRKM dynamics [113, 114].

To detect the initial apparent non-RRKM decay, one has to monitor the reaction at short times. This can be performed by studying the unimolecular decomposition at high pressures, where collisional stabilization competes with the rate of IVR. The first successful detection of apparent non-RRKM behaviour was accomplished by Rabinovitch and co-workers [115], who used chemical activation to prepare vibrationally excited hexafluorobicyclopropyl-d_2:

$$CH_2 + CF_2\!\!\overset{\displaystyle\diagup}{\underset{\displaystyle CH_2}{\diagdown}}\!\!CF\!-\!CF = CF_2 \longrightarrow CF_2\!\!\overset{\displaystyle\diagup}{\underset{\displaystyle CH_2}{\diagdown}}\!\!CF\!-\!CF\!\!\overset{\displaystyle\diagup}{\underset{\displaystyle CD_2}{\diagdown}}\!\!CF_2^*$$

The molecule decomposes by elimination of CF_2, which should occur with equal probabilities from each ring when energy is randomized. However, at pressures in excess of 100 Torr there is a measurable increase in the fraction of decomposition in the ring that was initially excited. From an analysis of the relative product yield *versus* pressure, it was deduced that energy flows between the two cyclopropyl rings with a rate of only

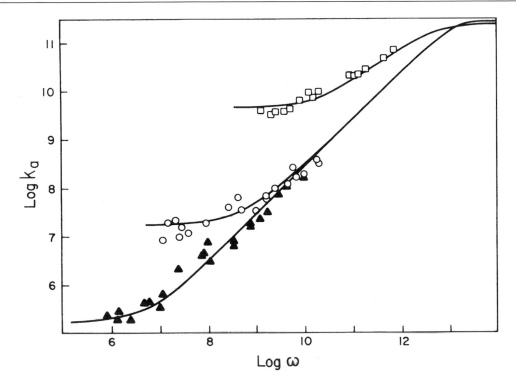

Figure A3.12.11. Chemical activation unimolecular rate constants *versus* ω for fluoroalkyl cyclopropanes. The $\square$, $\circ$, and $\blacktriangle$ points are for $R = CF_3$, C_3F_7, and C_5F_{11}, respectively. (Adapted from [116].)

3×10^9 s^{-1}. In a related set of experiments Rabinovitch *et al* [116] studied the series of chemically activated fluoroalkyl cyclopropanes:

$$R, \quad R = CF_3, C_3F_7, C_5F_{11}.$$

The chemically activated molecules are formed by reaction of $^1CH_2$ with the appropriate fluorinated alkene. In all these cases apparent non-RRKM behaviour was observed. As displayed in figure A3.12.11, the measured unimolecular rate constants are strongly dependent on pressure. The large rate constant at high pressure reflects an initial excitation of only a fraction of the total number of vibrational modes, i.e. initially the molecule behaves smaller than its total size. However, as the pressure is decreased, there is time for IVR to compete with dissociation and energy is distributed between a larger fraction of the vibrational modes and the rate constant decreases. At low pressures each rate constant approaches the RRKM value.

Apparent non-RRKM dynamics has also been observed in time-resolved femtosecond (fs) experiments in a collision-free environment [117]. An experimental study of acetone illustrates this work. Acetone is dissociated to the CH_3 and CH_3CO (acetyl) radicals by a fs laser pulse. The latter which dissociates by the channel

$$CH_3CO \rightarrow CO + CH_3$$

is followed in real time by fs mass spectrometry to measure its unimolecular rate constant. It is found to be 2×10^{12} s^{-1} and ~ 10 times smaller than the RRKM value, which indicates the experimental excitation process does not put energy in the C–C reaction coordinate and the rate constant value and short timescale reflects restricted IVR and non-RRKM kinetics.

A3.12.8.2   Intrinsic non-RRKM

As discussed in section A3.12.2, intrinsic non-RRKM behaviour occurs when there is at least one bottleneck for transitions between the reactant molecule's vibrational states, so that IVR is slow and a microcanonical ensemble over the reactant's phase space is not maintained during the unimolecular reaction. The above discussion of mode-specific decomposition illustrates that there are unimolecular reactions which are intrinsically non-RRKM. Many van der Waals molecules behave in this manner [4, 82]. For example, in an initial microcanonical ensemble for the $(C_2H_4)_2$ van der Waals molecule both the C_2H_4- - -C_2H_4 intermolecular modes and C_2H_4 intramolecular modes are excited with equal probabilities. However, this microcanonical ensemble is not maintained as the dimer dissociates. States with energy in the intermolecular modes react more rapidly than do those with the C_2H_4 intramolecular modes excited [85]. Furthermore, IVR is not rapid between the C_2H_4 intramolecular modes and different excitation patterns of these modes result in different dissociation rates. As a result of these different timescales for dissociation, the relative populations of the vibrational modes of the C_2H_4 dimer change with time.

Similar behaviour is observed in both experiments and calculations for HCO→H+CO dissociation [88–91] and in calculations for the X^-- - -CH_3Y ion–dipole complexes, which participate in S_N2 nucleophilic substitution reactions [118]. HCO states with HC excitation dissociate more rapidly than do those with CO excitation and, thus, the relative population of HC to CO excitation decreases with time. The unimolecular dynamics of the X^-- - -CH_3Y complex is similar to that for van der Waals complexes. There is weak coupling between the X^-- - -CH_3Y intermolecular modes and the CH_3Y intramolecular modes, and the two sets of modes react on different timescales.

Definitive examples of intrinsic non-RRKM dynamics for molecules excited near their unimolecular thresholds are rather limited. Calculations have shown that intrinsic non-RRKM dynamics becomes more pronounced at very high energies, where the RRKM lifetime becomes very short and dissociation begins to compete with IVR [119]. There is a need for establishing quantitative theories (i.e. not calculations) for identifying which molecules and energies lead to intrinsic non-RRKM dynamics. For example, at thermal energies the unimolecular dynamics of the Cl^-- - -CH_3Cl complex is predicted to be intrinsically non-RRKM [118], while experiments have shown that simply replacing one of the H-atoms of CH_3Cl with a CN group leads to intrinsic RRKM dynamics for the Cl^-- - -$ClCH_2CN$ complex [120]. This difference is thought to arise from a deeper potential energy well and less of a separation between vibrational frequencies for the Cl^-- - -$ClCH_2CN$ complex. For the Cl^-- - -CH_3Cl complex the three intermolecular vibrational frequencies are 64(2) and 95 cm^{-1}, while the lowest intramolecular frequency is the C–Cl stretch of 622 cm^{-1} [118]. Thus, very high-order resonances are required for energy transfer from the intermolecular to intramolecular modes. In contrast, for the Cl^-- - -$ClCH_2CN$ complex there is less of a hierarchy of frequencies, with 44, 66 and 118 cm^{-1} for the intermolecular modes and 207, 367, 499, 717, . . . for the intramolecular ones [120]. Here there are low-order resonances which couple the intermolecular and intramolecular modes. It would be very useful if one could incorporate such molecular properties into a theoretical model to predict intrinsic RRKM and non-RRKM behaviour.

Acknowledgments

The author wishes to thank the many graduate students, post-doctorals and collaborators who have worked with him on this topic. Of particular importance are the many discussions the author has had with Tom Baer, Reinhard Schinke and Jürgen Troe concerning unimolecular dynamics. Special thanks are given to Fleming Crim, Martin Quack and Kihyung Song for their valuable comments in preparing this chapter.

References

[1] Levine R D and Bernstein R B 1987 *Molecular Reaction Dynamics and Chemical Reactivity* (New York: Oxford University Press)

[2] Schinke R 1993 *Photodissociation Dynamics* (New York: Cambridge University Press)

[3] McQuarrie D A 1973 *Statistical Thermodynamics* (New York: Harper and Row)

[4] Baer T and Hase W L 1996 *Unimolecular Reaction Dynamics. Theory and Experiments* (New York: Oxford University Press)

[5] Steinfeld J I, Francisco J S and Hase W L 1999 *Chemical Kinetics and Dynamics* 2nd edn (Upper Saddle River, NJ: Prentice-Hall)

[6] Uzer T 1991 Theories of intramolecular vibrational energy transfer *Phys. Rep.* **199** 73–146

[7] Marcus R A 1952 Unimolecular dissociations and free radical recombination reactions *J. Chem. Phys.* **20** 359–64

[8] Rosenstock H M, Wallenstein M B, Wahrhaftig A L and Erying H 1952 Absolute rate theory for isolated systems and the mass spectra of polyatomic molecules *Proc. Natl Acad. Sci. USA* **38** 667–78

[9] Bunker D L and Hase W L 1973 On non-RRKM unimolecular kinetics: molecules in general and CH_3NC in particular *J. Chem. Phys.* **59** 4621–32

[10] Miller W H 1976 Importance of nonseparability in quantum mechanical transition-state theory *Acc. Chem. Res.* **9** 306–12

[11] Chandler D 1987 *Introduction to Modern Statistical Mechanics* (New York: Oxford University Press)

[12] Bunker D L 1964 Monte Carlo calculations. IV. Further studies of unimolecular dissociation *J. Chem. Phys.* **40** 1946–57

[13] Bunker D L 1962 Monte Carlo calculation of triatomic dissociation rates. I. N_2O and O_3 *J. Chem. Phys.* **37** 393–403

[14] Gutzwiller M C 1990 *Chaos in Classical and Quantum Mechanics* (New York: Springer)

[15] Slater J C 1951 *Quantum Theory of Matter* (New York: McGraw-Hill)

[16] Robinson P J and Holbrook K A 1972 *Unimolecular Reactions* (New York: Wiley)

[17] Hase W L and Buckowski D G 1982 Dynamics of ethyl radical decomposition. II. Applicability of classical mechanics to large-molecule unimolecular reaction dynamics *J. Comp. Chem.* **3** 335–43

[18] Slater N B 1959 *Theory of Unimolecular Reactions* (Ithaca, NY: Cornell University Press)

[19] Miller W H 1976 Unified statistical model for 'complex' and 'direct' reaction mechanisms *J. Chem. Phys.* **65** 2216–23

[20] Doll J D 1980 A unified theory of dissociation *J. Chem. Phys.* **73** 2760–2

[21] Truhlar D G and Garrett B C 1980 Variational transition-state theory *Acc. Chem. Res.* **13** 440–8

[22] Hase W L 1983 Variational unimolecular rate theory *Acc. Chem. Res.* **16** 258–64

[23] Hase W L 1972 Theoretical critical configuration for ethane decomposition and methyl radical recombination *J. Chem. Phys.* **57** 730–3

[24] Lin Y N and Rabinovitch B S 1970 A simple quasi-accommodation model of vibrational energy transfer *J. Phys. Chem.* **74** 3151–9

[25] Wardlaw D M and Marcus R A 1984 RRKM reaction rate theory for transition states of any looseness *Chem. Phys. Lett.* **110** 230–4

[26] Hase W L and Wardlaw D M 1989 *Bimolecular Collisions* ed M N R Ashfold and J E Baggott (London: Royal Society of Chemistry) p 171

[27] Song K, Peslherbe G H, Hase W L, Dobbyn A J, Stumpf M and Schinke R 1995 Comparison of quantum and semiclassical variational transition state models for $HO_2 \rightarrow H+O_2$ microcanonical rate constants *J. Chem. Phys.* **103** 8891–900

[28] Wardlaw D M and Marcus R A 1987 On the statistical theory of unimolecular processes *Adv. Chem. Phys.* **70** 231–63

[29] Chesnavich W J and Bowers M T 1979 *Gas Phase Ion Chemistry* vol 1, ed M T Bowers (New York: Academic) p 119

[30] Peslherbe G H and Hase W L 1996 Statistical anharmonic unimolecular rate constants for the dissociation of fluxional molecules. Application to aluminum clusters *J. Chem. Phys.* **105** 7432–47

[31] Klippenstein S J 1992 Variational optimizations in the Rice–Ramsperger–Kassel–Marcus theory calculations for unimolecular dissociations with no reverse barrier *J. Chem. Phys.* **96** 367–71

[32] Chesnavich W J, Bass L, Su T and Bowers M T 1981 Multiple transition states in unimolecular reactions: a transition state switching model. Application to $C_4H_8^+$ *J. Chem. Phys.* **74** 2228–46

[33] Hu X and Hase W L 1989 Properties of canonical variational transition state theory for association reactions without potential energy barriers *J. Phys. Chem.* **93** 6029–38

[34] Song K and Chesnavich W J 1989 Multiple transition states in chemical reactions: variational transition state theory studies of the HO_2 and HeH_2^+ systems *J. Chem. Phys.* **91** 4664–78

[35] Song K and Chesnavich W J 1990 Multiple transition state in chemical reactions. II. The effect of angular momentum in variational studies of HO_2 and HeH_2^+ systems *J. Chem. Phys.* **93** 5751–9

[36] Klippenstein S J, East A L L and Allen W D A 1994 First principles theoretical determination of the rate constant for the dissociation of singlet ketene *J. Chem. Phys.* **101** 9198–201

[37] Kassel L S 1928 Studies in homogeneous gas reactions. II. Introduction of quantum theory *J. Phys. Chem.* **32** 1065–79

[38] Beyer T and Swinehart D R 1973 Number of multiply-restricted partitions *ACM Commun.* **16** 379

[39] Zhu L, Chen W, Hase W L and Kaiser E W 1993 Comparison of models for treating angular momentum in RRKM calculations with vibrator transition states. Pressure and temperature dependence of $Cl+C_2H_2$ association *J. Phys. Chem.* **97** 311–22

[40] Hase W L 1998 Some recent advances and remaining questions regarding unimolecular rate theory *Acc. Chem. Res.* **31** 659–65

[41] Quack M and Troe J 1974 Specific rate constants of unimolecular processes. II. Adiabatic channel model *Ber. Bunsenges. Phys. Chem.* **78** 240–52

[42] Miller W H 1979 Tunneling corrections to unimolecular rate constants, with applications to formaldehyde *J. Am. Chem. Soc.* **101** 6810–14

[43] Bunker D L and Pattengill M 1968 Monte Carlo calculations. VI. A re-evaluation of the RRKM theory of unimolecular reaction rates *J. Chem. Phys.* **48** 772–6

[44] Green W H, Moore C B and Polik W F 1992 Transition states and rate constants for unimolecular reactions *Ann. Rev. Phys. Chem.* **43** 591–626

[45] Ionov S I, Brucker G A, Jaques C, Chen Y and Wittig C 1993 Probing the $NO_2 \rightarrow NO+O$ transition state via time resolved unimolecular decomposition *J. Chem. Phys.* **99** 3420–35

[46] Viswanathan R, Raff L M and Thompson D L 1984 Monte Carlo random walk calculations of unimolecular dissociation of methane *J. Chem. Phys.* **81** 3118–21

[47] Gilbert R G and Smith S C 1990 *Theory of Unimolecular and Recombination Reactions* (London: Blackwell)

[48] Polik W F, Guyer D R and Moore C B 1990 Stark level-crossing spectroscopy of S_0 formaldehyde eigenstates at the dissociation threshold *J. Chem. Phys.* **92** 3453–70

[49] Abramson E, Field R W, Imre D, Innes K K and Kinsey J L 1985 Fluorescence and stimulated emission $S_1 \rightarrow S_0$ spectra of acetylene: regular and ergodic regions *J. Chem. Phys.* **85** 453–65

[50] Wagner A F, Kiefer J H and Kumaran S S 1992 The importance of hindered rotation and other anharmonic effects in the thermal dissociation of small unsaturated molecules: application to HCN *Twenty-Fourth Symposium on Combustion* (Combustion Institute) 613–19

[51] Bhuiyan L B and Hase W L 1983 Sum and density of states for anharmonic polyatomic molecules. Effect of bend-stretch coupling *J. Chem. Phys.* **78** 5052–8

[52] Haarhoff P C 1963 The density of vibrational energy levels of polyatomic molecules *Mol. Phys.* **7** 101–17

[53] Forst W 1971 Methods for calculating energy-level densities *Chem. Rev.* **71** 339–56

[54] Troe J 1983 Specific rate constants $k(E, J)$ for unimolecular bond fissions *J. Chem. Phys.* **79** 6017–29

[55] Troe J 1995 Simplified models for anharmonic numbers and densities of vibrational states. I. Application to NO_2 and H_3^+ *Chem. Phys.* **190** 381–92

[56] Song K and Hase W L 1999 Fitting classical microcanonical unimolecular rate constants to a modified RRK expression: anharmonic and variational effects *J. Chem. Phys.* **110** 6198–207

[57] Duffy L M, Keister J W and Baer T 1995 Isomerization and dissociation in competition. The pentene ion story *J. Phys. Chem.* **99** 17 862–71

[58] Hose G and Taylor H A 1982 A quantum analog to the classical quasiperiodic motion *J. Chem. Phys.* **76** 5356–64

[59] Wilson E B Jr, Decius J C and Cross P C 1955 *Molecular Vibrations* (New York: McGraw-Hill)

[60] Califano S 1976 *Vibrational States* (New York: Wiley)

[61] Herzberg G 1945 *Molecular Spectra and Molecular Structure. II. Infrared and Raman Spectra of Polyatomic Molecules* (New York: Van Nostrand-Reinhold)

[62] Oxtoby D W and Rice S A 1976 Nonlinear resonance and stochasticity in intramolecular energy exchange *J. Chem. Phys.* **65** 1676–83

[63] W L Hase (ed) 1992 *Advances in Classical Trajectory Methods. 1. Intramolecular and Nonlinear Dynamics* (London: JAI)

[64] Sloane C S and Hase W L 1977 On the dynamics of state selected unimolecular reactions: chloroacetylene dissociation and predissociation *J. Chem. Phys.* **66** 1523–33

[65] Wolf R J and Hase W L 1980 Quasiperiodic trajectories for a multidimensional anharmonic classical Hamiltonian excited above the unimolecular threshold *J. Chem. Phys.* **73** 3779–90

[66] Viswanathan R, Thompson D L and Raff L M 1984 Theoretical investigations of elementary processes in the chemical vapor deposition of silicon from silane. Unimolecular decomposition of SiH_4 *J. Chem. Phys.* **80** 4230–40

[67] Sorescu D, Thompson D L and Raff L M 1994 Statistical effects in the skeletal inversion of bicyclo(2.1.0)pentane *J. Chem. Phys.* **101** 3729–41

[68] Lichtenberg A J and Lieberman M A 1992 *Regular and Chaotic Dynamics* 2nd edn (New York: Springer)

[69] Brickmann J, Pfeiffer R and Schmidt P C 1984 The transition between regular and chaotic dynamics and its influence on the vibrational energy transfer in molecules after local preparation *Ber. Bunsenges. Phys. Chem.* **88** 382–97

[70] Jaffé C and Brumer P 1980 Local and normal modes: a classical perspective *J. Chem. Phys.* **73** 5646–58

[71] Sibert E L III, Reinhardt W P and Hynes J T 1982 Classical dynamics of energy transfer between bonds in ABA triatomics *J. Chem. Phys.* **77** 3583–94

[72] Marcus R A 1977 Energy distributions in unimolecular reactions *Ber. Bunsenges. Phys. Chem.* **81** 190–7

[73] Miller W H 1974 Classical-limit quantum mechanics and the theory of molecular collisions *Adv. Chem. Phys.* **25** 69–177

[74] Child M S 1991 *Semiclassical Mechanics with Molecular Applications* (New York: Oxford University Press)

[75] Marcus R A 1973 Semiclassical theory for collisions involving complexes (compound state resonances) and for bound state systems *Faraday Discuss. Chem. Soc.* **55** 34–44

[76] Heller E J 1983 The correspondence principle and intramolecular dynamics *Faraday Discuss. Chem. Soc.* **75** 141–53

[77] Shirts R B and Reinhardt W P 1982 Approximate constants of motion for classically chaotic vibrational dynamics: vague tori, semiclassical quantization, and classical intramolecular energy flow *J. Chem. Phys.* **77** 5204–17

[78] Landau L D and Lifshitz E M 1965 *Quantum Mechanics* (London: Addison-Wesley)

[79] Mies F H and Krauss M 1966 Time-dependent behavior of activated molecules. High-pressure unimolecular rate constant and mass spectra *J. Chem. Phys.* **45** 4455–68

[80] Mies F H 1969 Resonant scattering theory of association reactions and unimolecular decomposition. II. Comparison of the collision theory and the absolute rate theory *J. Chem. Phys.* **51** 798–807

[81] Rice O K 1971 On the relation between unimolecular reaction and predissociation *J. Chem. Phys.* **55** 439–46

[82] Miller R E 1988 The vibrational spectroscopy and dynamics of weakly bound neutral complexes *Science* **240** 447–53

[83] Ewing G E 1980 Vibrational predissociation in hydrogen-bonded complexes *J. Chem. Phys.* **72** 2096–107

[84] Huang Z S, Jucks K W and Miller R E 1986 The vibrational predissociation lifetime of the HF dimer upon exciting the 'free-H' stretching vibration *J. Chem. Phys.* **85** 3338–41

[85] Casassa M P, Woodward A M, Stephenson J C and Kind D S 1986 Picosecond measurements of the dissociation rates of the nitric oxide dimer $v_1 = 1$ and $v_4 = 1$ levels *J. Chem. Phys.* **85** 6235–7

[86] Lovejoy C M and Nesbitt D J 1989 The infrared spectra of NO–HF isomers *J. Chem. Phys.* **90** 4671–80

[87] Fischer G, Miller R E, Vohralik P F and Watts R O 1985 Molecular beam infrared spectra of dimers formed from acetylene, methyl acetylene and ethene as a function of source pressure and concentration *J. Chem. Phys.* **83** 1471–7

[88] Tobiason J D, Dunlap J R and Rohlfing E A 1995 The unimolecular dissociation of HCO: a spectroscopic study of resonance energies and widths *J. Chem. Phys.* **103** 1448–69

[89] Stock C, Li X, Keller H-M, Schinke R and Temps F 1997 Unimolecular dissociation dynamics of highly vibrationally excited DCO ($\tilde{X}\,^2A'$). I. Investigation of dissociative resonance states by stimulated emission pumping spectroscopy *J. Chem. Phys.* **106** 5333–58

[90] Wang D and Bowman J M 1995 Complex L^2 calculations of bound states and resonances of HCO and DCO *Chem. Phys. Lett.* **235** 277–85

[91] Stumpf M, Dobbyn A J, Mordaunt D H, Keller H-M, Fluethmann H and Schinke R 1995 Unimolecular dissociations of HCO, HNO, and HO_2: from regular to irregular dynamics *Faraday Discuss.* **102** 193–213

[92] Barnes R J, Dutton G and Sinha A 1997 Unimolecular dissociation of HOCl near threshold: quantum state and time-resolved studies *J. Phys. Chem.* A **101** 8374–7

[93] Callegari A, Rebstein J, Muenter J S, Jost R and Rizzo T R 1999 The spectroscopy and intramolecular vibrational energy redistribution dynamics of HOCl in the $v(OH) = 6$ region, probed by infrared-visible double resonance overtone excitation *J. Chem. Phys.* **111** 123–33

[94] Skokov S and Bowman J M 1999 Variation of the resonance width of HOCl ($6v_{OH}$) with total angular momentum: comparison between *ab initio* theory and experiment *J. Chem. Phys.* **110** 9789–92

[95] Hase W L, Cho S-W, Lu D-H and Swamy K N 1989 The role of state specificity in unimolecular rate theory *Chem. Phys.* **139** 1–13

[96] Polik W F, Guyer D R, Miller W H and Moore C B 1990 Eigenstate-resolved unimolecular reaction dynamics: ergodic character of S_0 formaldehyde at the dissociation threshold *J. Chem. Phys.* **92** 3471–84

[97] Levine R D 1987 Fluctuations in spectral intensities and transition rates *Adv. Chem. Phys.* **70** 53–95

[98] Porter C E and Thomas R G 1956 Fluctuations of nuclear reaction widths *Phys. Rev.* **104** 483–91

[99] Waite B A and Miller W H 1980 Model studies of mode specificity in unimolecular reaction dynamics *J. Chem. Phys.* **73** 3713–21

[100] Miller W H, Hernandez R, Moore C B and Polik W F A 1990 Transition state theory-based statistical distribution of unimolecular decay rates with application to unimolecular decomposition of formaldehyde *J. Chem. Phys.* **93** 5657–66

[101] Stumpf M, Dobbyn A J, Keller H-M, Hase W L and Schinke R 1995 Quantum mechanical study of the unimolecular dissociation of HO_2: a rigorous test of RRKM theory *J. Chem. Phys.* **102** 5867–70

[102] Dobbyn A J, Stumpf M, Keller H-M and Schinke R 1996 Theoretical study of the unimolecular dissociation $HO_2 \rightarrow H+O_2$. II. Calculation of resonant states, dissociation rates, and O_2 product state distributions *J. Chem. Phys.* **104** 8357–81

[103] Lu D-H and Hase W L 1989 Monoenergetic unimolecular rate constants and their dependence on pressure and fluctuations in state-specific unimolecular rate constants *J. Phys. Chem.* **93** 1681–3

[104] Miller W H 1988 Effect of fluctuations in state-specific unimolecular rate constants on the pressure dependence of the average unimolecular reaction rate *J. Phys. Chem.* **92** 4261–3

[105] Rabinovitch B S and Setser D W 1964 Unimolecular decomposition and some isotope effects of simple alkanes and alkyl radicals *Adv. Photochem.* **3** 1–82

[106] Song K and Hase W L 1998 Role of state specificity in the temperature- and pressure-dependent unimolecular rate constants for $HO_2 \rightarrow H+O_2$ dissociation *J. Phys. Chem.* A **102** 1292–6

[107] Ionov S I, Brucker G A, Jaques C, Chen Y and Wittig C 1993 Probing the $NO_2 \rightarrow NO+O$ transition state via time resolved unimolecular decomposition *J. Chem. Phys.* **99** 3420–35
[108] Miyawaki J, Yamanouchi K and Tsuchiya S 1993 State-specific unimolecular reaction of NO_2 just above the dissociation threshold *J. Chem. Phys.* **99** 254–64
[109] Leu G-H, Huang C-L, Lee S-H, Lee Y-C and Chen I-C 1998 Vibrational levels of the transition state and rate of dissociation of triplet acetaldehyde *J. Chem. Phys.* **109** 9340–50
[110] King R A, Allen W D and Schaefer H F III 2000 On apparent quantized transition-state thresholds in the photofragmentation of acetaldehyde *J. Chem. Phys.* **112** 5585–92
[111] Gezelter J D and Miller W H 1996 Dynamics of the photodissociation of triplet ketene *J. Chem. Phys.* **104** 3546–54
[112] Schinke R, Beck C, Grebenshchikov S Y and Keller H-M 1998 Unimolecular dissociation: a state-specific quantum mechanical perspective *Ber. Bunsenges. Phys. Chem.* **102** 593–611
[113] Hase W L 1976 *Modern Theoretical Chemistry. 2. Dynamics of Molecular Collisions* part B, ed W H Miller (New York: Plenum) p 121
[114] Hase W L 1981 *Potential Energy Surfaces and Dynamics Calculations* ed D G Truhlar (New York: Plenum) p 1
[115] Rynbrandt J D and Rabinovitch B S 1971 Direct demonstration of nonrandomization of internal energy in reacting molecules. Rate of intramolecular energy relaxation *J. Chem. Phys.* **54** 2275–6
[116] Meagher J F, Chao K J, Barker J R and Rabinovitch B S 1974 Intramolecular vibrational energy relaxation. Decomposition of a series of chemically activated fluoroalkyl cyclopropanes *J. Phys. Chem.* **78** 2535–43
[117] Kim S K, Guo J, Baskin J S and Zewail A H 1996 Femtosecond chemically activated reactions: concept of nonstatistical activation at high thermal energies *J. Phys. Chem.* **100** 9202–5
[118] Hase W L 1994 Simulations of gas-phase chemical reactions: applications to S_N2 nucleophilic substitution *Science* **266** 998–1002
[119] Shalashilin D V and Thompson D L 1996 Intrinsic non-RRK behavior: classical trajectory, statistical theory, and diffusional theory studies of a unimolecular reaction *J. Chem. Phys.* **105** 1833–45
[120] Wladkowski B D, Lim K F, Allen W D and Brauman J I 1992 The S_N2 exchange reaction $ClCH_2CN+Cl^- \rightarrow Cl^-+ClCH_2CN$: experiment and theory *J. Am. Chem. Soc.* **114** 9136–53
[121] Hase W L 1982 *J. Phys. Chem.* **86** 2873–9

Further Reading

Forst W 1973 *Theory of Unimolecular Reactions* (New York: Academic)
Hase W L 1976 *Modern Theoretical Chemistry, Dynamics of Molecular Collisions* part B, ed W H Miller (New York: Plenum) p 121
Gilbert R G and Smith S C 1990 *Theory of Unimolecular and Recombination Reactions* (London: Blackwell Scientific)
Uzer T 1991 *Phys. Rep.* **199** 73–146
Baer T and Hase W L 1996 *Unimolecular Reaction Dynamics. Theory and Experiments* (New York: Oxford University Press)
Thompson D L 1999 *Int. Rev. Phys. Chem.* **17** 547–69
Quack M and Troe J 1981 *Theoretical Chemistry: Advances and Perspectives* vol 6B, ed E Henderson (New York: Academic) p 199
Holbrook K A, Pilling M J and Robertson S H 1996 *Unimolecular Reactions* 2nd edn (Chichester: Wiley)

A3.13
Energy redistribution in reacting systems

Roberto Marquardt and Martin Quack

A3.13.1 Introduction

Energy redistribution is the key primary process in chemical reaction systems, as well as in reaction systems quite generally (for instance, nuclear reactions). This is because many reactions can be separated into two steps:

(a) activation of the reacting species R, generating an energized species R^*:

$$R \rightharpoonup R^* \qquad (A3.13.1)$$

(b) reaction of the energized species to give products.

$$R^* \rightharpoonup P. \qquad (A3.13.2)$$

The first step (A3.13.1) is a general process of energy redistribution, whereas the second step (A3.13.2) is the genuine reaction step, occurring with a specific rate constant at energy E. This abstract reaction scheme can take a variety of forms in practice, because both steps may follow a variety of quite different mechanisms. For instance, the reaction step could be a barrier crossing of a particle, a tunnelling process or a nonadiabatic crossing between different potential hypersurfaces to name just a few important examples in chemical reactions.

The first step, which is the topic of the present chapter, can again follow a variety of different mechanisms. For instance, the energy transfer could happen within a molecule, say from one initially excited chemical bond to another, or it could involve radiative transfer. Finally, the energy transfer could involve a collisional transfer of energy between different atoms or molecules. All these processes have been recognized to be important for a very long time. The basic idea of collisional energization as a necessary primary step in chemical reactions can be found in the early work of van't Hoff [1] and Arrhenius [2, 3], leading to the famous Arrhenius equation for thermal chemical reactions (see also chapter A3.4)

$$k(T) = A(T) \exp\left(-\frac{E_A(T)}{RT}\right). \qquad (A3.13.3)$$

This equation results from the assumption that the actual reaction step in thermal reaction systems can happen only in molecules (or collision pairs) with an energy exceeding some threshold energy E_0 which is close, in general, to the Arrhenius activation energy defined by equation (A3.13.3). Radiative energization is at the basis of classical photochemistry (see e.g. [4–7] and chapter B2.5) and historically has had an interesting sideline in the radiation theory of unimolecular reactions [8], which was later superseded by the collisional Lindemann mechanism [9]. Recently, radiative energy redistribution has received new impetus through coherent and incoherent multiphoton excitation [10].

In this chapter we shall first outline the basic concepts of the various mechanisms for energy redistribution, followed by a very brief overview of collisional intermolecular energy transfer in chemical reaction systems. The main part of this chapter deals with true intramolecular energy transfer in polyatomic molecules, which is a topic of particular current importance. Stress is placed on basic ideas and concepts. It is not the aim of this chapter to review in detail the vast literature on this topic; we refer to some of the key reviews and books [11–32] and the literature cited therein. These cover a variety of aspects of the topic and further, more detailed references will be given throughout this review. We should mention here the energy transfer processes, which are of fundamental importance but are beyond the scope of this review, such as electronic energy transfer by mechanisms of the Förster type [33, 34] and related processes.

A3.13.2   Basic concepts for inter- and intramolecular energy transfer

The processes summarized by equation (A3.13.1) can follow quite different mechanisms and it is useful to classify them and introduce the appropriate nomenclature as well as the basic equations.

A3.13.2.1   *Processes involving interaction with the environment (bimolecular and related)*

(a) The first mechanism concerns bimolecular, collisional energy transfer between two molecules or atoms and molecules. We may describe such a mechanism by

$$M + R \rightarrow M + R^* \tag{A3.13.4}$$

or more precisely by defining quantum energy levels for both colliding species, e.g.

$$\{M(E_{Mi}) + R(E_{Ri})\}_I \rightarrow \{M(E_{Mf}) + R(E_{Rf})\}_F. \tag{A3.13.5}$$

This is clearly a process of intermolecular energy transfer, as energy is transferred between two molecular species. Generally one may, following chapter A.3.4.5, combine the quantum labels of M and R into one level index (I for initial and F for final) and define a cross section σ_{FI} for this energy transfer. The specific rate constant $k_{FI}(E_{t,I})$ for the energy transfer with the collision energy $E_{t,I}$ is given by:

$$k_{FI}(E_{t,I}) = \sigma_{FI}(E_{t,I})\sqrt{\frac{2E_{t,I}}{\mu}} \tag{A3.13.6}$$

with the reduced mass:

$$\mu = \left(\frac{1}{m_M} + \frac{1}{m_R}\right). \tag{A3.13.7}$$

We note that, by energy conservation, the following equation must hold:

$$E_{Mi} + E_{Ri} + E_{t,I} = E_{Mf} + E_{Rf} + E_{t,F}. \tag{A3.13.8}$$

Some of the internal (rovibronic) energy of the atomic and molecular collision partners is transformed into extra translational energy $\Delta E_t = E_{t,F} - E_{t,I}$ (or consumed, if ΔE_t is negative). If one averages over a thermal distribution of translational collision energies, one obtains the thermal rate constant for collisional energy transfer:

$$k_{FI}(T) = \left(\frac{8k_B T}{\pi\mu}\right)^{1/2} \int_0^\infty x \exp(-x)\sigma_{FI}(k_B T x)\,dx. \tag{A3.13.9}$$

We note here that the quantum levels denoted by the capital indices I and F may contain numerous energy eigenstates, i.e. are highly degenerate, and refer to chapter A3.4 for a more detailed discussion of these equations. The integration variable in equation (A3.13.9) is $x = E_{t,I}/k_B T$.

(b) The second mechanism, which is sometimes distinguished from the first although it is similar in kind, is obtained when we assume that the colliding species M does not change its internal quantum state. This special case is frequently realized if M is an inert gas atom in its electronic ground state, as the energies needed to generate excited states of M would then greatly exceed the energies available in ordinary reaction systems at modest temperatures. This type of mechanism is frequently called *collision induced intramolecular energy transfer*, as internal energy changes occur only *within the molecule* R. One must note that in general there is transfer of energy between *intermolecular translation* and intramolecular rotation and vibration in such a process, and thus the nomenclature 'intramolecular' is somewhat unfortunate. It is, however, widely used, which is the reason for mentioning it here. In the following, we shall not make use of this nomenclature and shall summarize mechanisms (a) and (b) as one class of bimolecular, intermolecular process. We may also note that, for mechanism (b) one can define a cross section σ_{fi} and rate constant k_{fi} between individual, non-degenerate quantum states i and f and obtain special equations analogous to equations (A3.13.5)–(A3.13.3), which we shall not repeat in detail. Indeed, one may then have cross sections and rates between different individual quantum states i and f of the same energy and thus no transfer of energy to translation. In this very special case, the redistribution of energy would indeed be entirely 'intramolecular' within R.

(c) The third mechanism would be transfer of energy between molecules and the radiation field. These processes involve absorption, emission or Raman scattering of radiation and are summarized, in the simplest case with one or two photons, in equations (A3.13.10)-(A3.13.12):

$$R_i + h\nu \rightarrow R_f \qquad \text{(absorption)} \qquad (A3.13.10)$$

$$R_{i'} \rightarrow R_{f'} + h\nu \qquad \text{(emission)} \qquad (A3.13.11)$$

$$R_{i''} + h\nu_i \rightarrow R_{f''} + h\nu_f \qquad \text{(Raman scattering)}. \qquad (A3.13.12)$$

In the case of polarized, but otherwise incoherent statistical radiation, one finds a rate constant for radiative energy transfer between initial molecular quantum states i and final states f:

$$k_{fi} = \frac{8\pi^3}{h^2} \frac{I_\nu^z}{(4\pi\epsilon_0)c} |M_{fi}^z|^2 \qquad (A3.13.13)$$

where $I_\nu^z = dI^z/d\nu$ is the intensity per frequency bandwidth of radiation and M_{fi}^z is the electric dipole transition moment in the direction of polarization. For unpolarized random spatial radiation of density $\rho(\nu)$ per volume and frequency, I_ν^z/c must be replaced by $\rho(\nu)/3$, because of random orientation, and the rate of induced transitions (absorption or emission) becomes:

$$k_{fi}^{\text{induced}} = B_{fi}\rho(\nu) \qquad (A3.13.14)$$

$$= \frac{8\pi^3}{3h^2(4\pi\epsilon_0)}\rho(\nu)|M_{fi}|^2.$$

B_{fi} is the Einstein coefficient for induced emission or absorption, which is approximately related to the absolute value of the dipole transition moment $|M_{fi}|$, to the integrated cross section G_{fi} for the transition and to the Einstein coefficient A_{fi} for spontaneous emission [10]:

$$B_{fi} = \frac{c}{h}G_{fi} = \frac{c^3}{8\pi h\nu_{fi}^3}A_{fi} \qquad (A3.13.15)$$

with

$$G_{fi} = \int_{\text{line}} \sigma_{fi}(\nu)\nu^{-1}\,d\nu \qquad (A3.13.16)$$

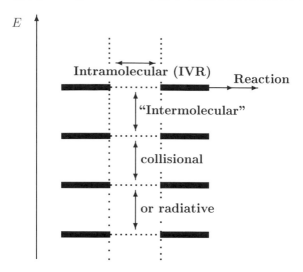

Figure A3.13.1. Schematic energy level diagram and relationship between 'intermolecular' (collisional or radiative) and intramolecular energy transfer between states of isolated molecules. The fat horizontal bars indicate thin energy shells of nearly degenerate states.

and $\sigma_{\text{fi}}(\nu)$ the frequency dependent absorption cross section. In equation (A3.13.15), $\nu_{\text{fi}} = |E_{\text{f}} - E_{\text{i}}|/h$. Equation (A3.13.17) is a simple, useful formula relating the integrated cross section and the electric dipole transition moment as dimensionless quantities, in the electric dipole approximation [10, 100]:

$$\frac{G_{\text{fi}}}{\text{pm}^2} \approx 41.624 \left| \frac{M_{\text{fi}}}{\text{Debye}} \right|^2. \tag{A3.13.17}$$

From these equations one also finds the rate coefficient matrix for thermal radiative transitions including absorption, induced and spontaneous emission in a thermal radiation field following Planck's law [35]:

$$k_{\text{fi}} = A_{\text{fi}} \frac{\text{sign}(E_{\text{f}} - E_{\text{i}})}{\exp((E_{\text{f}} - E_{\text{i}})/k_{\text{B}}T)) - 1}. \tag{A3.13.18}$$

Finally, if one has a condition with incoherent radiation of a small band width $\Delta\nu$ exciting a broad absorption band with $\sigma(\nu \pm \Delta\nu) \approx \sigma(\nu)$, one finds:

$$k_{\text{fi}}^{\text{induced}} = \frac{\sigma(\nu)}{h\nu} I \tag{A3.13.19}$$

where I is the radiation intensity. For a detailed discussion refer to [10]. The problem of coherent radiative excitation is considered in sections A3.13.4 and A3.13.5 in relation to intramolecular vibrational energy redistribution.

(d) The fourth mechanism is purely intramolecular energy redistribution. It is addressed in the next section.

A3.13.2.2 Strictly monomolecular processes in isolated molecules

Purely intramolecular energy transfer occurs when energy migrates within an isolated molecule from one part to another or from one type of motion to the other. Processes of this type include the vast field of molecular electronic radiationless transitions which emerged in the late 1960s [36], but more generally any type of intramolecular motion such as intramolecular vibrational energy redistribution (IVR) or intramolecular vibrational–rotational energy redistribution (IVRR) and related processes [37–39]. These processes will be discussed in section A3.13.5 in some detail in terms of their full quantum dynamics. However, in certain situations a statistical description with rate equations for such processes can be appropriate [38].

Figure A3.13.1 illustrates our general understanding of intramolecular energy redistribution in isolated molecules and shows how these processes are related to 'intermolecular' processes, which may follow any of the mechanisms discussed in the previous section. The horizontal bars represent levels of nearly degenerate states of an isolated molecule.

Having introduced the basic concepts and equations for various energy redistribution processes, we will now discuss some of them in more detail.

A3.13.3 Collisional energy redistribution processes

A3.13.3.1 The master equation for collisional relaxation reaction processes

The fundamental kinetic master equations for collisional energy redistribution follow the rules of the kinetic equations for all elementary reactions. Indeed an energy transfer process by inelastic collision, equation (A3.13.5), can be considered as a somewhat special 'reaction'. The kinetic differential equations for these processes have been discussed in the general context of chapter A3.4 on gas kinetics. We discuss here some special aspects related to collisional energy transfer in reactive systems. The general master equation for relaxation and reaction is of the type [11–13, 15, 25, 40, 41]:

$$\frac{dc_j(t)}{dt} = F(\{c_k(t)\}) \tag{A3.13.20}$$

$$c_j(t = 0) = c_{j0}. \tag{A3.13.21}$$

The index j can label quantum states of the same or different chemical species. Equation (A3.13.20) corresponds to a generally stiff initial value problem [42, 43]. In matrix notation one may write:

$$\frac{d\mathbf{c}(t)}{dt} = F[\mathbf{c}(t)] \tag{A3.13.22}$$

$$\mathbf{c}(t = 0) = \mathbf{c}_0. \tag{A3.13.23}$$

There is no general, simple solution to this set of coupled differential equations, and thus one will usually have to resort to numerical techniques [42, 43] (see also chapter A3.4).

A3.13.3.2 The master equation for collisional and radiative energy redistribution under conditions of generalized first-order kinetics

There is one special class of reaction systems in which a simplification occurs. If collisional energy redistribution of some reactant occurs by collisions with an excess of 'heat bath' atoms or molecules that are considered kinetically structureless, and if furthermore the reaction is either unimolecular or occurs again with a reaction partner M having an excess concentration, then one will have generalized first-order kinetics for populations p_j of the energy levels of the reactant, i.e. with

$$\frac{dp_j}{dt} = \sum_{k \neq j} (K'_{jk} p_k - K'_{kj} p_j) - k_j p_j \tag{A3.13.24}$$

$$p_j = c_j \left(\sum_k c_k \right)^{-1}. \tag{A3.13.25}$$

In equation (A3.13.24), k_j is the specific rate constant for reaction from level j, and K'_{jk} are energy transfer rate coefficients. With appropriate definition of a rate coefficient matrix $\mathbf{K}$ one has, in matrix notation,

$$\frac{\mathrm{d}\boldsymbol{p}}{\mathrm{d}t} = \mathbf{K}\boldsymbol{p} \tag{A3.13.26}$$

where for $j \neq i$

$$K_{ji}(T) = \left(\frac{8k_\mathrm{B}T}{\pi\mu}\right)^{1/2} [M] \int_0^\infty x \exp(-x)\sigma_{ji}(k_\mathrm{B}T\,x)\,\mathrm{d}x. \tag{A3.13.27}$$

(see equation (A3.13.9)) and

$$-K_{jj}(T) = k_j + \sum_{k \neq j} K_{kj}(T). \tag{A3.13.28}$$

The master equation (A3.13.26) applies also, under certain conditions, to radiative excitation with rate coefficients for radiative energy transfer being given by equations (A3.13.13)–(A3.13.19), depending on the case, or else by more general equations [10]. Finally, the radiative and collisional rate coefficients may be considered together to be important at the same time in a given reaction system, if time scales for these processes are of the appropriate order of magnitude.

The solution of equation (A3.13.26) is given by:

$$p(t) = \exp(\mathbf{K}t)p(0). \tag{A3.13.29}$$

This solution can be obtained explicitly either by matrix diagonalization or by other techniques (see chapter A3.4 and [42, 43]). In many cases the discrete quantum level labels in equation (A3.13.24) can be replaced by a continuous energy variable and the populations by a population density $p(E)$, with replacement of the sum by appropriate integrals [11]. This approach can be made the starting point of useful analytical solutions for certain simple model systems [11, 19, 44–46].

While the time dependent populations $p_j(t)$ may generally show a complicated behaviour, certain simple limiting cases can be distinguished and characterized by appropriate parameters:

(a) The long time steady state limit (formally $t \to \infty$) is described by the largest eigenvalue λ_1 of $\mathbf{K}$. Since all λ_j are negative, λ_1 has the smallest absolute value [35, 47]. In this limit one finds [47] (with the reactant fraction $F_\mathrm{R} = \sum_j p_j$):

$$-\frac{\mathrm{d}\ln(F_\mathrm{R}(t))}{\mathrm{d}t} = -\frac{\mathrm{d}\ln(\sum_j p_j(t))}{\mathrm{d}t} = k_\mathrm{uni} = -\lambda_1. \tag{A3.13.30}$$

Thus, this eigenvalue λ_1 determines the unimolecular steady-state reaction rate constant.

(b) The second largest eigenvalue λ_2 determines ideally the relaxation time towards this steady state, thus:

$$\tau_\mathrm{relax}^{-1} = -\lambda_2. \tag{A3.13.31}$$

More generally, further eigenvalues must be taken into account in the relaxation process.

(c) It is sometimes useful to define an incubation time τ_inc by the limiting equation for steady state:

$$-\ln(F_\mathrm{R}^{st}(t)) = -\lambda_1(t - \tau_\mathrm{inc}). \tag{A3.13.32}$$

Figure A3.13.2 illustrates the origin of these quantities. Refer to [47] for a detailed mathematical discussion as well as the treatment of radiative laser excitation, in which incubation phenomena are important. Also refer to [11] for some classical examples in thermal systems.

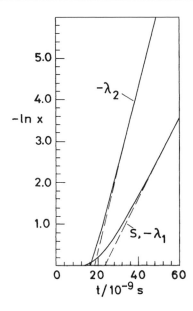

Figure A3.13.2. Illustration of the analysis of the master equation in terms of its eigenvalues λ_1 and λ_2 for the example of IR-multiphoton excitation. The dashed lines give the long time straight line limiting behaviour. The full line to the right-hand side is for $x = F_R(t)$ with a straight line of slope $-\lambda_1 = k_{uni}$. The intercept of the corresponding dashed line (F_R^{St}) indicates τ_{inc} (see equation (A3.13.32)). The left-hand line is for $x = |F_R - F_R^{St}|$ with limiting slope $-\lambda_2 = \tau_{relax}^{-1}$ (see text and [47]).

As a rule, in thermal unimolecular reaction systems at modest temperatures, λ_1 is well separated from the other eigenvalues, and thus the time scales for incubation and 'relaxation' are well separated from the steady-state reaction time scale $\tau_{reaction} = k_{uni}^{-1}$. On the other hand, at high temperatures, k_{uni}, τ_{relax}^{-1} and τ_{inc}^{-1} may merge. This is illustrated in figure A3.13.3 for the classic example of thermal unimolecular dissociation [48–51]:

$$N_2O + Ar \rightarrow N_2 + O + Ar. \qquad (A3.13.33)$$

Note that in the 'low pressure limit' of unimolecular reactions (chapter A3.4), the unimolecular rate constant k_{uni} is entirely dominated by energy transfer processes, even though the relaxation and incubation rates (τ_{relax}^{-1} and τ_{inc}^{-1}) may be much faster than k_{uni}.

The master equation treatment of energy transfer in even fairly complex reaction systems is now well established and fairly standard [52]. However, the rate coefficients K_{ij} for the individual energy transfer processes must be established and we shall discuss some aspects of this matter in the following section.

A3.13.3.3 *Mechanisms of collisional energy transfer*

Collisional energy transfer in molecules is a field in itself and is of relevance for kinetic theory (chapter A3.1), gas phase kinetics (chapter A3.4), RRKM theory (chapter A3.12), the theory of unimolecular reactions in general, as well as the kinetics of laser systems [53]. Chapters C3.3–C3.5 treat these subjects in detail. We summarize those aspects that are of importance for mechanistic considerations in chemically reactive systems.

We start from a model in which collision cross sections or rate constants for energy transfer are compared with a reference quantity such as average Lennard-Jones collision cross sections or the usually cited

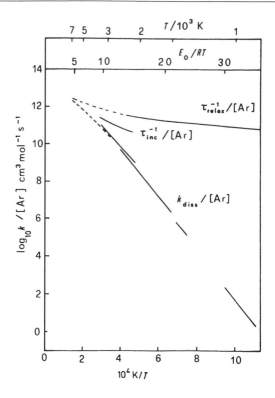

Figure A3.13.3. Dissociation ($k_{uni} = k_{diss}$), incubation (τ_{inc}^{-1}) and relaxation (τ_{relax}^{-1}) rate constants for the reaction $N_2O \rightarrow N_2 + O$ at low pressure in argon (from [11], see discussion in the text for details and references to the experiments).

Lennard-Jones collision frequencies [54]

$$Z_{LJ} = \pi \sigma_{AB}^2 \left(\frac{8 k_B T}{\pi \mu_{AB}} \right)^{1/2} \Omega_{AB}^{(2,2)^*} \tag{A3.13.34}$$

where σ_{AB} is the Lennard-Jones parameter and $\Omega_{AB}^{(2,2)^*}$ is the reduced collision integral [54], calculated from the binding energy ϵ and the reduced mass μ_{AB} for the collision in the Lennard-Jones potential

$$V(r) = 4\epsilon \left[\left(\frac{\sigma_{AB}}{r} \right)^{12} - \left(\frac{\sigma_{AB}}{r} \right)^6 \right]. \tag{A3.13.35}$$

Given such a reference, we can classify various mechanisms of energy transfer either by the probability that a certain energy transfer process will occur in a 'Lennard-Jones reference collision', or by the average energy transferred by one 'Lennard-Jones collision'.

With this convention, we can now classify energy transfer processes either as resonant, if $|\Delta E_t|$ defined in equation (A3.13.8) is small, or non-resonant, if it is large. Quite generally the rate of resonant processes can approach or even exceed the Lennard-Jones collision frequency (the latter is possible if other long-range potentials are actually applicable, such as by permanent dipole–dipole interaction).

Resonant processes of some importance include resonant electronic to electronic energy transfer (E–E), such as the pumping process of the iodine atom laser

$$\text{(E–E)} \quad O_2(^1\Delta) + I(^2P_{3/2}) \rightarrow O_2(^3\Sigma_g^-) + I(^2P_{1/2}). \tag{A3.13.36}$$

Another near resonant process is important in the hydrogen fluoride laser, equation (A3.13.37), where vibrational to vibrational energy transfer is of interest:

$$\text{(V–V)} \quad \text{HF}(v') + \text{HF}(v'') \rightarrow \text{HF}(v' + \Delta v) + \text{HF}(v'' - \Delta v), \tag{A3.13.37}$$

where Δv is the number of vibrational quanta exchanged. If HF were a harmonic oscillator, ΔE_t would be zero (perfect resonance). In practice, because of anharmonicity, the most important process is exothermic, leading to increasing excitation v' of some of the HF molecules with successive collisions [55, 56], because the exothermicity drives this process to high v' as long as plenty of HF(v'') with low v'' are available.

Resonant rotational to rotational (R–R) energy transfer may have rates exceeding the Lennard-Jones collision frequency because of long-range dipole–dipole interactions in some cases. Quasiresonant vibration to rotation transfer (V–R) has recently been discussed in the framework of a simple model [57].

'Non-resonant' processes include vibration–translation (V–T) processes with transfer probabilities decreasing to very small values for diatomic molecules with very high vibrational frequencies, of the order of 10^{-4} and less for the probability of transferring a quantum in a collision. Also, vibration to rotation (V–R) processes frequently have low probabilities, of the order of 10^{-2}, if ΔE_t is relatively large. Rotation to translation (R–T) processes are generally fast, with probabilities near 1. Also, the R–V–T processes in collisions of large polyatomic molecules have high probabilities, with average energies transferred in one Lennard-Jones collision being of the order of a few kJ mol^{-1} [11, 25], or less in collisions with rare gas atoms. As a general rule one may assume collision cross sections to be small, if ΔE_t is large [11, 58, 59].

In the experimental and theoretical study of energy transfer processes which involve some of the above mechanisms, one should distinguish processes in atoms and small molecules and in large polyatomic molecules. For small molecules a full theoretical quantum treatment is possible and even computer program packages are available [60–63], with full state to state characterization. A good example are rotational energy transfer theory and experiments on He + CO [64]:

$$\text{He} + \text{CO}(J') \rightarrow \text{He} + \text{CO}(j''). \tag{A3.13.38}$$

On the experimental side, small molecule energy transfer experiments may use molecular beam techniques [65–67] (see also chapter C3.3 for laser studies).

In the case of large molecules, instead of the detailed quantum state characterization implied in the cross sections σ_{fi} and rate coefficients K_{fi} of the master equation (A3.13.24), one derives more coarse grained information on 'levels' covering a small energy bandwidth around E and E' (with an optional notation $K_{FI}(E', E)$) or finally energy transfer probabilities $P(E', E)$ for a transition from energy E to energy E' in a highly excited large polyatomic molecule where the density of states $\rho(E')$ is very large, for example in a collision with a heat bath inert gas atom [11]. Such processes can currently be modelled by classical trajectories [68–70].

Experimental access to the probabilities $P(E', E)$ for energy transfer in large molecules usually involves techniques providing just the first moment of this distribution, i.e. the average energy $\langle \Delta E \rangle$ transferred in a collision. Such methods include UV absorption, infrared fluorescence and related spectroscopic techniques [11, 28, 71–74]. More advanced techniques, such as kinetically controlled selective ionization (KCSI [74]) have also provided information on higher moments of $P(E', E)$, such as $\langle (\Delta E)^2 \rangle$.

The standard mechanisms of collisional energy transfer for both small and large molecules have been treated extensively and a variety of scaling laws have been proposed to simplify the complicated body of data [58, 59, 75]. To conclude, one of the most efficient special mechanisms for energy transfer is the quasi-reactive process involving chemically bound intermediates, as in the example of the reaction:

$$\text{O}_2(v', j') + \text{O} \rightarrow \text{O}_3^* \rightarrow \text{O}_2(v'', j'') + \text{O}. \tag{A3.13.39}$$

Such processes transfer very large amounts of energy in one collision and have been treated efficiently by the statistical adiabatic channel model [11, 19, 30, 76–79]. They are quite similar mechanistically to chemical activation systems. One might say that in such a mechanism one may distinguish three phases:

(a)   Formation of a bound collision complex AB:

$$A(v', j') + B \rightarrow AB^*. \tag{A3.13.40}$$

(b)   IVRR in this complex:

$$AB^* \rightarrow AB^{**}. \tag{A3.13.41}$$

(c)   Finally, dissociation of the internally, statistically equilibrated complex:

$$AB^{**} \rightarrow A(v'', j'') + B. \tag{A3.13.42}$$

That is, rapid IVR in the long lived intermediate is an essential step. We shall treat this important process in the next section, but mention here in passing the observation of so-called 'supercollisions' transferring large average amounts of energy $\langle \Delta E \rangle$ in one collision [80], even if intermediate complex formation may not be important.

A3.13.4   Intramolecular energy transfer studies in polyatomic molecules

In this section we review our understanding of IVR as a special case of intramolecular energy transfer. The studies are based on calculations of the time evolution of vibrational wave packets corresponding to middle size and large amplitude vibrational motion in polyatomic molecules. An early example for the investigation of wave packet motion as a key to understanding IVR and its implication on reaction kinetics using experimental data is given in [81]. Since then, many other contributions have helped to increase our knowledge using realistic potential energy surfaces, mainly for two- and three-dimensional systems, and we give a brief summary of these results below.

A3.13.4.1   IVR and classical mechanics

Before undergoing a substantial and, in many cases, practically irreversible, change of geometrical structure within a chemical reaction, a molecule may often perform a series of vibrations in the multidimensional space around its equilibrium structure. This applies in general to reactions that take place entirely in the bound electronic ground state and in many cases to reactions that start in the electronic ground state near the equilibrium structure, but evolve into highly excited states above the reaction threshold energy. In the latter case, within the general scheme of equation (A3.13.1) a reaction is thought to be induced by a sufficiently energetic pulse of electromagnetic radiation or by collisions with adequate high-energy collision partners. In the first case, a reaction is thought to be the last step after a chain of excitation steps has transferred enough energy into the molecule to react either thermally, by collisions, or coherently, for instance by irradiation with infrared laser pulses. These pulses can be tuned to adequately excite vibrations along the reaction coordinate, the amplitudes of which become gradually larger until the molecule undergoes a sufficiently large structural change leading to the chemical reaction.

Vibrational motion is thus an important primary step in a general reaction mechanism and detailed investigation of this motion is of utmost relevance for our understanding of the dynamics of chemical reactions. In classical mechanics, vibrational motion is described by the time evolution $\tilde{q}(t)$ and $\tilde{p}(t)$ of general internal position and momentum coordinates. These time dependent functions are solutions of the classical equations of motion, e.g. Newton's equations for given initial conditions $\tilde{q}(t_0) = q_0$ and $\tilde{p}(t_0) = p_0$. The definition of initial conditions is generally limited in precision to within experimental uncertainties Δq_0 and Δp_0, more fundamentally related by the Heisenberg principle $\Delta q_0 \Delta p_0 \gtrsim = h/4\pi$. Therefore, we need to consider an initial distribution $F_0(q - q_0, p - p_0)$, with widths Δq_0 and Δp_0 and the time evolution $F_t(q - \tilde{q}(t), p - \tilde{p}(t))$, which may be quite different from the initial distribution F_0, depending on the integrability of the dynamical system. Ideally, for classical, integrable systems, vibrational motion may be understood as the motion of narrow, well localized distributions $F(q - \tilde{q}(t), p - \tilde{p}(t))$ (ideally δ-functions in a strict mathematical sense), centred around the solutions of the classical equations of motion. In this picture we wish to consider initial conditions that correspond to localized vibrational motion along specific manifolds, for instance a vibration that is induced by elongation of a single chemical bond (local mode vibrations) as a result of the interaction with some external force, but it is also conceivable that a large displacement from equilibrium might be induced along a single normal coordinate. Independent of the detailed mechanism for the generation of localized vibrations, harmonic transfer of excitation may occur when such a vibration starts to extend into other manifolds of the multidimensional space, resulting in trajectories that draw Lissajous figures in phase space, and also in configuration space [82] (see also [83]). Furthermore, if there is anharmonic interaction, IVR may occur. In [84, 85] this type of IVR was called classical intramolecular vibrational redistribution (CIVR).

A3.13.4.2 IVR and quantum mechanics

In time-dependent quantum mechanics, vibrational motion may be described as the motion of the wave packet $|\psi(q, t)|^2$ in configuration space, e.g. as defined by the possible values of the position coordinates q. This motion is given by the time evolution of the wave function $\psi(q, t)$, defined as the projection $\langle q|\psi(t)\rangle$ of the time-dependent quantum state $|\psi(t)\rangle$ on configuration space. Since the quantum state is a complete description of the system, the wave packet defining the probability density can be viewed as the quantum mechanical counterpart of the classical distribution $F(q - \tilde{q}(t), p - \tilde{p}(t))$. The time dependence is obtained by solution of the time-dependent Schrödinger equation

$$i\frac{h}{2\pi}\frac{d|\psi(t)\rangle}{dt} = \hat{H}|\psi(t)\rangle \tag{A3.13.43}$$

where h is the Planck constant and $\hat{H}$ is the Hamiltonian of the system under consideration. Solutions depend on initial conditions $|\psi(t_0)\rangle$ and may be formulated using the time evolution operator $\hat{U}(t, t_0)$:

$$|\psi(t)\rangle = \hat{U}(t, t_0)|\psi(t_0)\rangle. \tag{A3.13.44}$$

Alternatively, in the case of incoherent (e.g. statistical) initial conditions, the density matrix operator $\hat{P}(t) = |\psi(t)\rangle\langle\psi(t)|$ at time t can be obtained as the solution of the Liouville–von Neumann equation:

$$\hat{P}(t) = \hat{U}(t, t_0)\hat{P}(t_0)\hat{U}^\dagger(t, t_0) \tag{A3.13.45}$$

where $\hat{U}^\dagger(t, t_0)$ is the adjoint of the time evolution operator (in strictly conservative systems, the time evolution operator is unitary and $\hat{U}^\dagger(t, t_0) = \hat{U}^{-1}(t, t_0) = \hat{U}(t_0, t)$).

The calculation of the time evolution operator in multidimensional systems is a formidable task and some results will be discussed in this section. An alternative approach is the calculation of semi-classical dynamics as demonstrated, among others, by Heller [86–88], Marcus [89, 90], Taylor [91, 92], Metiu [93, 94] and coworkers (see also [83] as well as the review by Miller [95] for more general aspects of semiclassical dynamics). This method basically consists of replacing the δ-function distribution in the true classical calculation by a Gaussian distribution in coordinate space. It allows for a simulation of the vibrational quantum dynamics to the extent that interference effects in the evolving wave packet can be neglected. While the application of semi-classical methods might still be of some interest for the simulation of quantum dynamics in large polyatomic molecules in the near future, as a natural extension of classical molecular dynamics calculations [68, 96], full quantum mechanical calculations of the wave packet evolution in smaller polyatomic molecules are possible with the currently available computational resources. Following earlier spectroscopic work and three-dimensional quantum dynamics results [81, 97–100], Wyatt and coworkers have recently demonstrated applications of full quantum calculations to the study of IVR in fluoroform, with nine degrees of freedom [101, 102] and in benzene [103], considering all 30 degrees of freedom [104]. Such calculations show clearly the possibilities in the computational treatment of quantum dynamics and IVR. However, remaining computational limitations restrict the study to the lower energy regime of molecular vibrations, when all degrees of freedom of systems with more than three dimensions are treated. Large amplitude motion, which shows the inherently quantum mechanical nature of wave packet motion and is highly sensitive to IVR, cannot yet be discussed for such molecules, but new results are expected in the near future, as indicated in recent work on ammonia [105, 106], formaldehyde and hydrogen peroxide [106–108], and hydrogen fluoride dimer [109–111] including all six internal degrees of freedom.

A key feature in quantum mechanics is the dispersion of the wave packet, i.e. the loss of its Gaussian shape. This feature corresponds to a delocalization of probability density and is largely a consequence of anharmonicities of the potential energy surface, both the 'diagonal' anharmonicity, along the manifold in which the motion started, and 'off diagonal', induced by anharmonic coupling terms between different manifolds in the Hamiltonian. Spreading of the wave packet into different manifolds is thus a further important feature of IVR. In [84, 85] this type of IVR was called delocalization quantum intramolecular vibrational redistribution (DIVR). DIVR plays a central role for the understanding of statistical theories for unimolecular reactions in polyatomic molecules [84, 97], as will be discussed below.

A3.13.4.3   IVR within the general scheme of energy redistribution in reactive systems

As in classical mechanics, the outcome of time-dependent quantum dynamics and, in particular, the occurrence of IVR in polyatomic molecules, depends both on the Hamiltonian and the initial conditions, i.e. the initial quantum mechanical state $|\psi(t_0)\rangle$. We focus here on the time-dependent aspects of IVR, and in this case such initial conditions always correspond to the preparation, at a time t_0, of superposition states of molecular (spectroscopic) eigenstates involving at least two distinct vibrational energy levels. Strictly, IVR occurs if these levels involve at least two distinct vibrational manifolds in terms of which the total (vibrational) Hamiltonian is not separable [84]. In a time-independent view, this requirement states that the wave functions belonging to the two spectroscopic states are spread in a non-separable way over the configuration space spanned by at least two different vibrational modes. The conceptual framework for the investigation of IVR may be sketched within the following scheme, which also mirrors the way we might investigate IVR in the time-dependent approach, both theoretically and experimentally:

$$|\psi(t_{-1})\rangle \xrightarrow{\hat{U}_{\text{prep}}(t_0, t_{-1})} |\psi(t_0)\rangle \xrightarrow{\hat{U}_{\text{free}}(t, t_0)} |\psi(t)\rangle. \qquad (A3.13.46)$$

In a first time interval $[t_{-1}, t_0]$ of the scheme (A3.13.46), a superposition state is prepared. This step corresponds to the step in equation (A3.13.1). One might think of a time evolution $|\psi(t_{-1})\rangle \rightarrow |\psi(t_0)\rangle = \hat{U}_{\text{prep}}(t_0, t_{-1})|\psi(t_{-1})\rangle$, where $|\psi(t_{-1})\rangle$ may be a molecular eigenstate and $\hat{U}_{\text{prep}}$ is the time evolution operator obtained from the interaction with an external system, to be specified below. The probability distribution $|\psi(q, t_0)|^2$ is expected to be approximatively localized in configuration space, such that $|\psi(q^*, t_0)|^2 > 0$ for position coordinates $q^* \in \mathcal{M}^*$ belonging to some specific manifold $\mathcal{M}^*$ and $|\psi(q, t_0)|^2 \approx 0$ for coordinates $q \in \mathcal{M}$ belonging to the complementary manifold $\mathcal{M} = \overline{\mathcal{M}^*}$. In a second time interval $[t_0, t_1]$, the superposition state $|\psi(t_0)\rangle$ has a free evolution into states $|\psi(t)\rangle = \hat{U}_{\text{free}}(t, t_0)|\psi(t_0)\rangle$. This step corresponds to the intermediate step equation (A3.13.47), occurring between the steps described before by equations (A3.13.1) and (A3.13.2) (see also equation (A3.13.41)):

$$R^* \rightarrow R^{*'}. \qquad (A3.13.47)$$

IVR is present if $|\psi(q, t)|^2 > 0$ is observed for $t > t_0$ also for $q \in \mathcal{M}$. IVR may of course also occur during the excitation process, if its time scale is comparable to that of the excitation.

In the present section, we concentrate on coherent preparation by irradiation with a properly chosen laser pulse during a given time interval. The quantum state at time t_{-1} may be chosen to be the vibrational ground state $|\phi_0^{(g)}\rangle$ in the electronic ground state. In principle, other possibilities may also be conceived for the preparation step, as discussed in sections A3.13.1–A3.13.3. In order to determine superposition coefficients within a realistic experimental set-up using irradiation, the following questions need to be answered: (1) What are the eigenstates? (2) What are the electric dipole transition matrix elements? (3) What is the orientation of the molecule with respect to the laboratory fixed (linearly or circularly) polarized electric field vector of the radiation? The first question requires knowledge of the potential energy surface, or the Hamiltonian $\hat{H}_0(p, q)$ of the isolated molecule, the second that of the vector valued surface $\vec{\mu}(q)$ of the electric dipole moment. This surface yields the operator, which couples spectroscopic states by the impact of an external irradiation field and thus directly affects the superposition procedure. The third question is indeed of great importance for comparison with experiments aiming at the measurement of internal wave packet motion in polyatomic molecules and has recently received much attention in the treatment of molecular alignment and orientation [112, 113], including non-polar molecules [114, 115]. To the best of our knowledge, up to now explicit calculations of multidimensional wave packet evolution in polyatomic molecules have been performed upon neglect of rotational degrees of freedom, i.e. only internal coordinates have been considered, although calculations on coherent excitation in ozone level structures with rotation exist [116, 117], which could be interpreted in terms of wave packet evolution. A more detailed discussion of this point will be given below for a specific example.

A3.13.4.4 Concepts of computational methods

There are numerous methods for solving the time dependent Schrödinger equation (A3.13.43), and some of them were reviewed by Kosloff [118] (see also [119, 120]). Whenever projections of the evolving wave function on the spectroscopic states are useful for the detailed analysis of the quantum dynamics (and this is certainly the case for the detailed analysis of IVR), it is convenient to express the Hamiltonian based on spectroscopic states $|\phi_n\rangle$:

$$\hat{H}_0 = \sum_n \frac{h}{2\pi} \omega_n |\phi_n\rangle \langle \phi_n| \tag{A3.13.48}$$

where ω_n are the eigenfrequencies. For an isolated molecule $\hat{H} = \hat{H}_0$ in equation (A3.13.43) and the time evolution operator is of the form

$$\hat{U}(t, t_0) = \sum_n \exp(-i\omega_n(t - t_0))|\phi_n\rangle \langle \phi_n|. \tag{A3.13.49}$$

The time-dependent wave function is then given by the expression:

$$\psi(q, t) = \sum_n c_n^0 \exp(-i\omega_n t)\phi_n(q). \tag{A3.13.50}$$

Here, $\phi_n(q) = \langle q|\phi_n\rangle$ are the wave functions of the spectroscopic states and the coefficients c_n^0 are determined from the initial conditions

$$\psi(q, t_0) = \sum_n c_n^0 \phi_n(q), \qquad c_n^0 = \langle \phi_n|\psi(t_0)\rangle. \tag{A3.13.51}$$

Equation (A3.13.49) describes the spectroscopic access to quantum dynamics. Clearly, when the spectral structure becomes too congested, i.e. when there are many close lying frequencies ω_n, calculation of all spectroscopic states becomes difficult. However, often it is not necessary to calculate all states when certain model assumptions can be made. One assumption concerns the separation of time scales. When there is evidence for a clear separation of time scales for IVR, only part of the spectroscopic states need to be considered for fast evolution. Typically, these states have large frequency separations, and considering only such states means neglecting the fine-grained spectral structure as a first approximation. An example for separation of time scales is given by the dynamics of the alkyl CH chromophore in CHXYZ compounds, which will be discussed below. This group span a three-dimensional linear space of stretching and bending vibrations. These vibrations are generally quite strongly coupled, which is manifested by the occurrence of a Fermi resonance in the spectral structure throughout the entire vibrational energy space. As we will see, the corresponding time evolution and IVR between these modes takes place in less than 1 ps, while other modes become involved in the dynamics on much longer time scales (10 ps to ns, typically). The assumption for time scale separation and IVR on the subpicosecond time scale for the alkyl CH chromophore was founded on the basis of spectroscopic data nearly 20 years ago [98, 121]. The first results on the nature of IVR in the CH chromophore system and its role in IR photochemistry were also reported by that time [122, 123], including results for the acetylenic CH chromophore [124] and results obtained from first calculations of the wave packet motion [81]. The validity of this assumption has recently been confirmed in the case of CHF_3 both experimentally, from the highly resolved spectral structure of highly excited vibrational overtones [125, 126], and theoretically, including all nine degrees of freedom for modestly excited vibrational overtones up to 6000 cm^{-1} [102].

A3.13.4.5   IVR during and after coherent excitation: general aspects

Modern photochemistry (IR, UV or VIS) is induced by coherent or incoherent radiative excitation processes [4–7]. The first step within a photochemical process is of course a preparation step within our conceptual framework, in which time-dependent states are generated that possibly show IVR. In an ideal scenario, energy from a laser would be deposited in a spatially localized, large amplitude vibrational motion of the reacting molecular system, which would then possibly lead to the cleavage of selected chemical bonds. This is basically the central idea behind the concepts for a 'mode selective chemistry', introduced in the late 1970s [127], and has continuously received much attention [10, 117, 122, 128–135]. In a recent review [136], IVR was interpreted as a 'molecular enemy' of possible schemes for mode selective chemistry. This interpretation is somewhat limited, since IVR represents more complex features of molecular dynamics [37, 84, 134], and even the opposite situation is possible. IVR can indeed be selective with respect to certain structural features [85, 97] that may help mode selective reactive processes after tailored laser excitation [137].

To be more specific, we assume that for a possible preparation step the Hamiltonian might be given during the preparation time interval $[t_{-1}, t_0]$ by the expression:

$$\hat{H} = \hat{H}_0 + \hat{H}_i(t) \tag{A3.13.52}$$

where $\hat{H}_0$ is the Hamiltonian of the isolated molecule and $\hat{H}_i$ is the interaction Hamiltonian between the molecule and an external system. In this section, we limit the discussion to the case where the external system is the electromagnetic radiation field. For the interaction with a classical electromagnetic field with electric field vector $\vec{E}(t)$, the interaction Hamiltonian is given by the expression:

$$\hat{H}_i(t) = -\hat{\vec{\mu}}\vec{E}(t). \tag{A3.13.53}$$

where $\hat{\vec{\mu}}$ is the operator of the electric dipole moment. When we treat the interaction with a classical field in this way, we implicitly assume that the field will remain unaffected by the changes in the molecular system under consideration. More specifically, its energy content is assumed to be constant. The energy of the radiation field is thus not explicitly considered in the expression for the total Hamiltonian and all operators acting on states of the field are replaced by their time-dependent expectation values. These assumptions are widely accepted, whenever the number of photons in each field mode is sufficiently large. For a coherent, monochromatic, polarized field with intensity $I = \sqrt{\epsilon_0/\mu_0}|\vec{E}|^2 \approx 1 \text{ MW cm}^{-2}$ *in vacuo*, which is a typical value used in laser chemical experiments in the gas phase at low pressures, the number N_ν of mid infrared photons existing in a cavity of volume $V = 1 \text{ m}^3$ is [138, p 498]:

$$N_\nu = \frac{IV}{c_0 h\nu} \approx 10^{21}. \tag{A3.13.54}$$

Equation (A3.13.54) legitimates the use of this semi-classical approximation of the molecule–field interaction in the low-pressure regime.

Since $\hat{H}_i(t)$ is explicitly time dependent, the time evolution operator is more complicated than in equation A3.13.49. However, the time-dependent wave function can still be written in the form

$$\psi(q,t) = \sum_n c_n(t)\phi_n(q) \tag{A3.13.55}$$

with time-dependent coefficients that are obtained by solving the set of coupled differential equations

$$i\frac{dc_n(t)}{dt} = \sum_{n'}\{W_{nn'} + V_{nn'}(t)\}c_{n'}(t) \tag{A3.13.56}$$

where $W_{nn'} = \delta_{nn'}\omega_n$ ($\delta_{nn'}$ is the Kronecker symbol, ω_n were defined in equation (A3.13.48)) and

$$\begin{aligned}V_{nn'}(t) &= \frac{2\pi}{h}\langle\phi_n|\hat{H}_i(t)|\phi_{n'}\rangle \\ &= -\frac{2\pi}{h}\langle\phi_n|\hat{\vec{\mu}}|\phi_{n'}\rangle\vec{E}(t),\end{aligned} \tag{A3.13.57}$$

The matrix elements $\langle\phi_n|\hat{\vec{\mu}}|\phi_{n'}\rangle$ are multidimensional integrals $\int \phi_n^*(q)\vec{\mu}(q)\phi_{n'}(q)\,d\tau$ of the vector valued dipole moment surface. The time-independent part of the coupling matrix elements in equation (A3.13.57) can also be cast into the practical formula

$$V_{nn'}^0/(2\pi c_0 \text{ cm}^{-1}) = -0.46093\langle\phi_n|\hat{\mu}_\alpha/\text{Debye}|\phi_{n'}\rangle\sqrt{I_0/\text{MW cm}^{-2}}, \tag{A3.13.58}$$

where α is the direction of the electric field vector of the linearly polarized radiation field with maximal intensity I_0. The solution of equation (A3.13.56) may still be quite demanding, depending on the size of the system under consideration. However, it has become a practical routine procedure to use suitable approximations such as the QRA (quasiresonant approximation) or Floquet treatment [35, 122, 129] and programmes for the numerical solution are available [139, 140].

A3.13.4.6   *Electronic excitation in the Franck–Condon limit and IVR*

At this stage we may distinguish between excitation involving different electronic states and excitation occurring within the same electronic (ground) state. When the spectroscopic states are located in different electronic states, say the ground (g) and excited (e) states, one frequently assumes the Franck–Condon approximation to be applicable:

$$\langle\phi_n^{(g)}|\hat{\vec{\mu}}|\phi_{n'}^{(e)}\rangle \approx \vec{\mu}_{ge}\langle\phi_n^{(g)}|\phi_{n'}^{(e)}\rangle. \tag{A3.13.59}$$

Such electronic excitation processes can be made very fast with sufficiently intense laser fields. For example, if one considers monochromatic excitation with a wavenumber in the UV region (60 000 cm^{-1}) and a coupling strength $(\vec{\mu}_{ge}\vec{E})/hc \approx 4000\ \mathrm{cm}^{-1}$ (e.g. $\mu_{ge} \approx 1$ Debye in equation (A3.13.59), $I \approx 50\ \mathrm{TW\ cm^{-2}}$), excitation occurs within 1 fs [141]. During such a short excitation time interval the relative positions of the nuclei remain unchanged (Franck approximation). Within these approximations, if one starts the preparation step in the vibrational ground state $|\phi_0^{(g)}\rangle$, the resulting state $|\psi(t_0)\rangle$ at time t_0 has the same probability distribution as the vibrational ground state. However, it is now transferred into the excited electronic state where it is no longer stationary, since it is a superposition state of vibrational eigenstates in the excited electronic state:

$$|\psi(t_0)\rangle = \sum_n \langle \phi_n^{(e)}|\phi_0^{(g)}\rangle |\phi_n^{(e)}\rangle. \qquad (A3.13.60)$$

Often the potential energy surfaces for the ground and excited states are fairly different, i.e. with significantly different equilibrium positions. The state $|\psi(t_0)\rangle$ will then correspond to a wave packet, which has nearly a Gaussian shape with a centre position that is largely displaced from the minimal energy configuration on the excited surface and, since the Franck approximation can be applied, the expectation value of the nuclear linear momentum vanishes. In a complementary view, the superposition state of equation (A3.13.60) defines the manifold $\mathcal{M}^*$ in configuration space. It is often referred to as the 'bright' state, since its probability density defines a region in configuration space, the Franck–Condon region, which has been reached by the irradiation field through mediation by the electric dipole operator. After the preparation step, the wave packet most likely starts to move along the steepest descent path from the Franck–Condon region. One possibility is that it proceeds to occupy other manifolds, which were not directly excited. The occupation of the remaining, 'dark' manifolds (e.g. $\overline{\mathcal{M}^*}$) by the time-dependent wave packet is a characteristic feature of IVR.

Studies of wave packet motion in excited electronic states of molecules with three and four atoms were conducted by Schinke, Engel and collaborators, among others, mainly in the context of photodissociation dynamics from the excited state [142–144] (for an introduction to photodissociation dynamics, see [7], and also more recent work [145–149] with references cited therein). In these studies, the dissociation dynamics is often described by a time-dependent displacement of the Gaussian wave packet in the multidimensional configuration space. As time goes on, this wave packet will occupy different manifolds (from where the molecule possibly dissociates) and this is identified with IVR. The dynamics may be described within the Gaussian wave packet method [150], and the vibrational dynamics is then of the classical IVR type (CIVR [84]). The validity of this approach depends on the dissociation rate on the one hand, and the rate of delocalization of the wave packet on the other hand. The occurrence of DIVR often receives less attention in the discussions of photodissociation dynamics mentioned above. In [148], for instance, details of the wave packet motion by means of snapshots of the probability density are missing, but a delocalization of the wave packet probably takes place, as may be concluded from inspection of figure 5 therein.

A3.13.5 IVR in the electronic ground state: the example of the CH chromophore

A3.13.5.1 Redistribution during and after coherent excitation

A system that shows IVR with very fast spreading of the wave packet, i.e. DIVR in the subpicosecond time range, is that of the infrared alkyl CH chromophore, which will be used in the remaining part of this chapter to discuss IVR as a result of a mode specific excitation within the electronic ground state. The CH stretching and bending modes of the alkyl CH chromophore in CHXYZ compounds are coupled by a generally strong Fermi resonance [100, 151]. Figure A3.13.4 shows the shape of the potential energy surface for the symmetrical compound CHD_3 as contour line representations of selected one- and two-dimensional sections (see figure caption for a detailed description). The important feature is the curved shape of the $V(Q_s, Q_{b_1})$ potential section ($V(Q_s, Q_{b_2})$ being similarly curved), which indicates a rather strong anharmonic coupling. This feature is characteristic for compounds of the type CHXYZ [84, 100, 151–153]. Q_s, Q_{b_1} and Q_{b_2} are (mass weighted) normal coordinates of the CH stretching and bending motion, with symmetry A_1 and E,

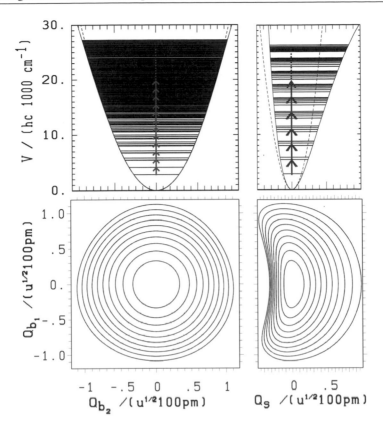

Figure A3.13.4. Potential energy cuts along the normal coordinate subspace pertaining to the CH chromophore in CHD_3. Q_{b_1} is the A' coordinate in C_s symmetry, essentially changing structure along the x-axis (see also figure A3.13.5), and Q_{b_2} is the A'' coordinate, essentially changing structure along the y-axis. Contour lines show equidistant energies at wave number differences of 3000 cm^{-1} up to $30\,000$ cm^{-1}. The upper curves are one-dimensional cuts along Q_{b_2} (left) and Q_s (right). The dashed curves in the two upper figures show harmonic potential curves (from [154]).

respectively, in the C_{3v} point group of symmetrical CHD_3. A change of Q_s is a concerted motion of all atoms along the z-axis, defined in figure A3.13.5. However, displacements along Q_s are small for the carbon and deuterium atoms, and large for the hydrogen atom. Thus, this coordinate essentially describes a stretching motion of the CH bond (along the z-axis). In the same way, Q_{b_1} and Q_{b_2} describe bending motions of the CH bond along the x- and y-axis, respectively (see figure A3.13.5). In the one-dimensional sections the positions of the corresponding spectroscopic states are drawn as horizontal lines. On the left-hand side, in the potential section $V(Q_{b_2})$, a total of 800 states up to an energy equivalent wave number of $25\,000$ cm^{-1} has been considered. These energy levels may be grouped into semi-isoenergetic shells defined by multiplets of states with a constant chromophore quantum number $N = v_s + \frac{1}{2}v_b = 0, \frac{1}{2}, 1, \frac{3}{2}, 2, \frac{5}{2}, \ldots$, where v_s and v_b are quantum numbers of effective basis states ('Fermi modes' [97, 152, 154]) that are strongly coupled by a 2:1 Fermi resonance. These multiplets give rise to spectroscopic polyads and can be well distinguished in the lower energy region, where the density of states is low.

In the potential section $V(Q_s)$, shown on the right hand side of Figure A3.13.4, the subset of A_1 energy states is drawn. This subset contains only multiplets with integer values of the chromophore quantum number $N = 0, 1, 2, \ldots$. This reduction allows for an easier visualization of the multiplet structure and also represents

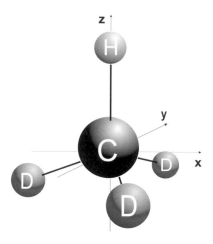

Figure A3.13.5. Coordinates and axes used to describe the wave packet dynamics of the CH chromophore in CHX_3 or CHXYZ compounds.

the subset of states that are strongly coupled by the parallel component of the electric dipole moment (see discussion in the following paragraph). The excitation dynamics of the CH chromophore along the stretching manifold can indeed be well described by restriction to this subset of states [97, 154].

Excitation specificity is a consequence of the shape of the electric dipole moment surface. For the alkyl CH chromophore in CHX_3 compounds, the parallel component of the dipole moment, i.e. the component parallel to the symmetry axis, is a strongly varying function of the CH stretching coordinate, whereas it changes little along the bending manifolds [155, 156]. Excitation along this component will thus induce preparation of superposition states lying along the stretching manifold, preferentially. These states thus constitute the 'bright' manifold in this example. The remaining states define the 'dark' manifolds and any substantial population of these states during or after such an excitation process can thus be directly linked to the existence of IVR. On the other hand, the perpendicular components of the dipole moment vector are strongly varying functions of the bending coordinates. For direct excitation along one of these components, states belonging to the bending manifolds become the 'bright' states and any appearance of a subsequent stretching motion can be interpreted as arising from IVR.

The following discussion shall illustrate our understanding of structural changes along 'dark' manifolds in terms of wave packet motion as a consequence of IVR. Figure A3.13.6 shows the evolution of the wave packet for the CH chromophore in CHF_3 during the excitation step along the parallel (stretching) coordinate [97]. The potential surface in the CH chromophore subspace is similar to that for CHD_3 (figure A3.13.4 above), with a slightly more curved form in the stretching–bending representation (figures are shown in [97, 151]). The laser is switched on at a given time t_{-1}, running thereafter as a continuous, monochromatic irradiation up to time t_0, when it is switched off. Thus, the electric field vector is given as

$$\vec{E}(t) = h(t - t_{-1})h(t_0 - t)\vec{E}_0 \cos(\omega_L t). \qquad (A3.13.61)$$

where $h(t)$ is the Heaviside unit step function, $\vec{E}_0$ is the amplitude of the electric field vector and $\omega_L = 2\pi c\tilde{\nu}_L$ its angular frequency. Excitation parameters are the irradiation intensity $I_0 = 30\,TW\,cm^{-2}$, which corresponds to a maximal electric field strength $E_0 \approx 3.4 \times 10^{10}\,V\,m^{-1}$, and wave number $\tilde{\nu}_L = 2832.42\,cm^{-1}$, which lies in the region of the fundamental for the CH stretching vibration (see arrows in the potential cut $V(Q_s)$ of figure (A3.13.4)). The figure shows snapshots of the time evolution of the wave packet between 50 and 70 fs after the beginning of the irradiation ($t_{-1} = 0$ here). On the left-hand side, contour maps of the

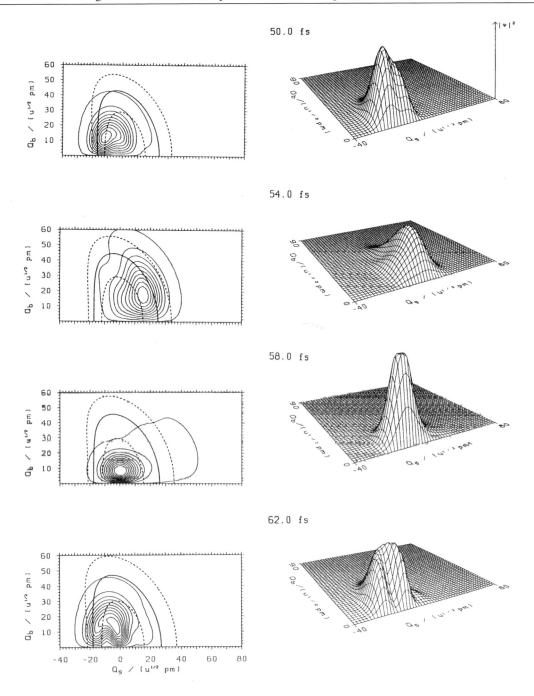

Figure A3.13.6. Time evolution of the probability density of the CH chromophore in CHF_3 after 50 fs of irradiation with an excitation wave number $\tilde{v}_L = 2832.42$ cm^{-1} at an intensity $I_0 = 30$ TW cm^{-2}. The contour lines of equiprobability density in configuration space have values 2×10^{-5} u^{-1} pm^{-2} for the lowest line shown and distances between the lines of 24, 15, 29 and 20×10^{-5} u^{-1} pm^{-2} in the order of the four images shown. The averaged energy of the wave packet corresponds to 6000 cm^{-1} (roughly 3100 cm^{-1} absorbed) with a quantum mechanical uncertainty of ± 3000 cm^{-1} (from [97]).

time-dependent, integrated probability density

$$|\psi(Q_s, Q_b, t)|^2 = \int_{\varphi_b} |\psi(Q_s, Q_b, \varphi_b, t)|^2 \, d\varphi_b \qquad (A3.13.62)$$

are shown, where Q_s is the coordinate for the stretching motion and $Q_b = \sqrt{Q_{b_1}^2 + Q_{b_2}^2}$, $\varphi_b = \arctan(Q_{b_2}/Q_{b_1})$ are polar representations of the bending coordinates Q_{b_1} and Q_{b_2}. Additionally, contour curves of the potential energy surface are drawn at the momentary energy of the wave packet. This energy is defined as:

$$E(t) = \sum_n E_n p_n(t) \qquad (A3.13.63)$$

where

$$p_n(t) = c_n^*(t) c_n(t) \qquad (A3.13.64)$$

are the time-dependent populations of the spectroscopic states during the preparation step (the complex coefficients $c_n(t)$ in equation (A3.13.64) are calculated according to equation (A3.13.55), the spectroscopic energies $E_n = \frac{h}{2\pi} \omega_n$ are defined in equation (A3.13.48); the dashed curves indicate the quantum mechanical uncertainty which arises from the superposition of molecular eigenstates). The same evolution is repeated on the right-hand side of the figure as a three-dimensional representation.

In the treatment adopted in [97], the motion of the CF_3 frame is implicitly considered in the dynamics of the normal modes. Indeed, the integrand $|\psi(Q_s, Q_b, \varphi_b, t)|^2$ in equation (A3.13.62) is to be interpreted as probability density for the change of the CHF_3 structure in the subspace of the CH chromophore, as defined by the normal coordinates Q_s, Q_{b_1} and Q_{b_2}, irrespective of the molecular structure and its change in the remaining space. This interpretation is also valid beyond the harmonic approximation, as long as the structural change in the CH chromophore space can be dynamically separated from that of the rest of the molecule. The assumption of dynamical separation is well confirmed, both from experiment and theory, at least during the first 1000 fs of motion of the CH chromophore.

When looking at the snapshots in figure A3.13.6, we see that the position of maximal probability oscillates back and forth along the stretching coordinate between the walls at $Q_s = -20$ and $+25 \sqrt{u}$ pm, with an approximate period of 12 fs, which corresponds to the classical oscillation period $\tau = 1/\nu$ of a pendulum with a frequency $\nu = c_0 \tilde{\nu} \approx 8.5 \times 10^{13}$ s^{-1} and wave number $\tilde{\nu} = 2850$ cm^{-1}. Indeed, the motion of the whole wave packet approximately follows this oscillation and, when it does so, the wave packet motion is semiclassical. In harmonic potential wells the motion of the wave packet is always semiclassical [157–159]. However, since the potential surface of the CH chromophore is anharmonic, some gathering and spreading out of the wave packet is observable on top of the semiclassical motion. It is interesting to note that, at this 'initial' stage of the excitation step, the motion of the wave packet is nearly semiclassical, though with modest amplitudes of the oscillations, despite the anharmonicity of the stretching potential.

The later time evolution is shown in figure A3.13.7, between 90 and 100 fs, and in figure A3.13.8, between 390 and 400 fs, after the beginning of the excitation (time step t_{-1}). Three observations are readily made: first, the amount of energy absorbed by the chromophore has increased, from 3000 cm^{-1} in figure A3.13.6, to 6000 cm^{-1} in figure A3.13.7 and 12 000 cm^{-1} in figure A3.13.8. Second, the initially semiclassical motion has been replaced by a more irregular motion of probability density, in which the original periodicity is hardly visible. Third, the wave packet starts to occupy nearly all of the energetically available region in configuration space, thus escaping from the initial, 'bright' manifolds into the 'dark' manifolds. From these observations, the following conclusions may be directly drawn: IVR of the CH chromophore in fluoroform is fast (in the subpicosecond time scale); IVR sets in already during the excitation process, i.e. when an external force field is driving the molecular system along a well prescribed path in configuration space (the 'bright' manifold); IVR is of the delocalization type (DIVR). Understanding these observations is central for the understanding of IVR and they are discussed as follows:

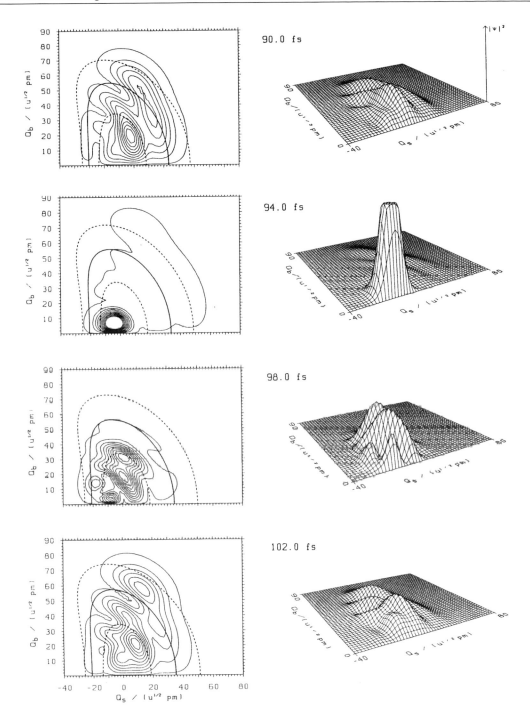

Figure A3.13.7. Continuation of the time evolution for the CH chromophore in CHF_3 after 90 fs of irradiation (see also figure A3.13.6). Distances between the contour lines are 10, 29, 16 and 9×10^{-5} u^{-1} pm^{-2} in the order of the four images shown. The averaged energy of the wave packet corresponds to 9200 cm^{-1} (roughly 6300 cm^{-1} absorbed) with a quantum mechanical uncertainty of ± 5700 cm^{-1} (from [97]).

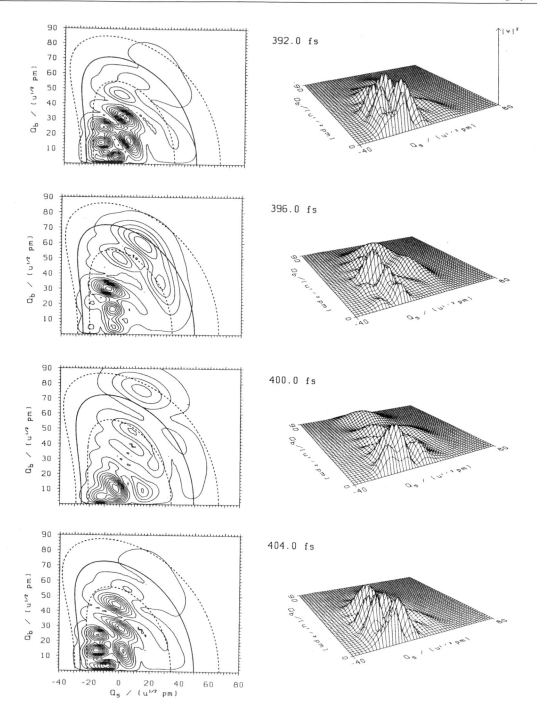

Figure A3.13.8. Continuation of the time evolution for the CH chromophore in CHF_3 after 392 fs of irradiation (see also figures A3.13.6 and A3.13.7). Distances between the contour lines are 14, 12, 13 and 14×10^{-5} u^{-1} pm^{-2} in the order of the four images shown. The averaged energy of the wave packet corresponds to 15 000 cm^{-1} (roughly 12 100 cm^{-1} absorbed) with a quantum mechanical uncertainty of ± 5800 cm^{-1} (from [97]).

(a) A more detailed analysis of quantum dynamics shows that the molecular system, represented by the group of vibrations pertaining to the CH chromophore in this example, absorbs continuously more energy as time goes on. Let the absorbed energy be $E_{abs} = N_{v,abs}(h/2\pi)\omega_L$, where $N_{v,abs}$ is the mean number of absorbed photons. Since the carrier frequency of the radiation field is kept constant at a value close to the fundamental of the stretching oscillation, $\omega_L \approx \omega_{N=1} - \omega_{N=0}$ (N being the chromophore quantum number here), this means that the increase in absorbed energy is a consequence of the stepwise multiphoton excitation process, in which each vibrational level serves as a new starting level for further absorption of light after it has itself been significantly populated. This process is schematically represented, within the example for CHD_3, by the sequence of upright arrows shown on the right-hand side of figure A3.13.4. $N_{v,abs}$ is thus a smoothly increasing function of time.

(b) The disappearance of the semiclassical type of motion and, thus, the delocalization of the wave packet, is understood to follow the onset of dephasing. With increasing energy, both the effective anharmonic couplings between the 'bright' stretching mode and the 'dark' bending modes, as well as the diagonal anharmonicity of the 'bright' mode increase. The larger the anharmonicity, the larger the deviation from a purely harmonic behaviour, in which the wave packet keeps on moving in a semiclassical way. In quantum mechanics, the increase in anharmonicity of an oscillator leads to an effective broadening $\Delta \nu_{eff} > 0$ in the distribution of frequencies of high-probability transitions—for transitions induced by the electric dipole operator usually those with a difference of ± 1 in the oscillator quantum number (for the harmonic oscillator $\Delta \nu_{eff} = 0$). On the other hand, these are the transitions which play a major role in the stepwise multiphoton excitation of molecular vibrations. A broadening of the frequency distribution invariably leads to a broadening of the distribution of relative phases of the time-dependent coefficients $c_n(t)$ in equation (A3.13.55). Although the sum in equation (A3.13.55) is entirely coherent, one might introduce an effective coherence time defined by:

$$\tau_{c,eff} = 1/\Delta \nu_{eff}. \tag{A3.13.65}$$

For the stretching oscillations of the CH chromophore in CHF_3 $\tau_{c,eff} \approx 100\,\text{fs}$. Clearly, typical coherence time ranges depend on both the molecular parameters and the effectively absorbed amount of energy during the excitation step, which in turn depends on the coupling strength of the molecule–radiation interaction. A more detailed study of the dispersion of the wave packet and its relationship with decoherence effects was carried out in [106]. In [97] an excitation process has been studied for the model of two anharmonically coupled, resonant harmonic oscillators (i.e. with at least one cubic coupling term) but under similar conditions as for the CH chromophore in fluoroform discussed here. When the cubic coupling parameter is chosen to be very small compared with the diagonal parameters of the Hamilton matrix, the motion of the wave packet is indeed semiclassical for very long times (up to 600 ps) and, moreover, the wave packet does probe the bending manifold without significantly changing its initial shape. This means that, under appropriate conditions, IVR can also be of the classical type within a quantum mechanical treatment of the dynamics. Such conditions require, for instance, that the band width $h\Delta \nu_{eff}$ be smaller than the resonance width (power broadening) of the excitation process.

(c) The third observation, that the wave packet occupies nearly all of the energetically accessible region in configuration space, has a direct impact on the understanding of IVR as a rapid promotor of microcanonical equilibrium conditions. Energy equipartition preceding a possible chemical reaction is the main assumption in quasiequilibrium statistical theories of chemical reaction dynamics ('RRKM' theory [161–163], 'transition state' theory [164, 165] but also within the 'statistical adiabatic channel model' [76, 77]; see also chapter A3.12 and further recent reviews on varied and extended forms of statistical theories in [25, 166–172]). In the case of CHF_3 one might conclude from inspection of the snapshots at the later stage of the excitation dynamics (see figure A3.13.8) that after 400 fs the wave packet delocalization is nearly complete. Moreover, this delocalization arises here from a fully coherent, isolated evolution of a system consisting of one molecule and a coherent radiation field (laser). Of course, within the common interpretation of the wave packet as a probability distribution in configuration space, this result means that, for an ensemble of identically prepared molecules, vibrational motion is essentially delocalized at this stage and vibrational energy is nearly equipartitioned.

However, the wave packet does not occupy all of the energetically accessible region. A more detailed analysis of populations [97, table IV] reveals that, during the excitation process, the absorbed energy is inhomogeneously distributed among the set of molecular eigenstates of a given energy shell (such a shell is represented by all nearly iso-energetic states belonging to one of the multiplets shown on the right-hand side of figure A3.13.4). Clearly, equipartition of energy is attained, if all states of an energy shell are equally populated. The microcanonical probability distribution in configuration space may then be represented by a typical member of the microcanonical ensemble, defined e.g. by the wave function

$$\psi_{\text{micro}} \approx \frac{1}{\sqrt{N_{\text{shell}}}} \sum_{n \in \text{shell}} \exp(-\mathrm{i}\varphi_n^{\text{random}})\phi_n \qquad (A3.13.66)$$

where N_{shell} denotes the number of nearly iso-energetic states ϕ_n of a shell and $\varphi_n^{\text{random}}$ is a random phase. Such a state is shown in figure A3.13.9. When comparing this state with the state generated by multiphoton excitation, the two different kinds of superposition that lead to these wave packets must, of course, be distinguished. In the stepwise multiphoton excitation, the time evolved wave packet arises from a superposition of many states in several multiplets (with roughly constant averaged energy after some excitation time and a large energy uncertainty). The microcanonical distribution is given by the superposition of states in a single multiplet (of the same averaged energy but much smaller energy uncertainty). In the case of the CH chromophore in CHF_3 studied in this example, the distribution of populations within a molecular energy shell is not homogeneous during the excitation process because the multiplets are not ideally centred at the multiphoton resonance levels and their energy range is effectively too large when compared to the resonance width of the excitation process (power broadening). If molecular energy shells fall entirely within the resonance width of the excitation, such as in the model systems of two harmonic oscillators studied in [97], population distribution within a shell becomes more homogeneous [97, table V]. However, as discussed in that work, equidistribution of populations does not imply that the wave packet is delocalized. Indeed, the contrary was shown to occur. If the probability distribution in configuration space is to delocalize, the relative phases between the superposition states must follow an irregular evolution, such as in a random phase ensemble, in addition to equidistribution of population. Thus, one statement would therefore be that IVR is not complete, although very fast, during the multiphoton excitation of CHF_3. Excitation and redistribution are indeed two concurring processes. In the limit of weak field excitation, in the spectroscopic regime, the result is a superposition of essentially two eigenstates (the ground and an excited state, for instance). Within the 'bright' state concept, strong IVR will be revealed by an instantaneous delocalization of probability density, both in the 'bright' and the 'dark' manifolds, as soon as the excited state is populated, because the excited state is, of course, a superposition state of states from both manifolds. On the other hand, strong field stepwise IR multiphoton excitation promotes, in a first step, the deposition of energy in a spatially localized, time-dependent molecular structure. Simultaneously, IVR starts to induce redistribution of this energy among other modes. The redistribution becomes apparent after some time has passed and is expected to be of the DIVR type, at least on longer time scales. DIVR may lead to a complete redistribution in configuration space, if the separation between nearly iso-energetic states is small compared to the power broadening of the excitation field. However, under such conditions, at least during an initial stage of the dynamics, CIVR will dominate.

In view of the foregoing discussion, one might ask what is a typical time evolution of the wave packet for the *isolated* molecule, what are typical time scales and, if initial conditions are such that an entire energy shell participates, does the wave packet resulting from the coherent dynamics look like a microcanonical distribution? Such studies were performed for the case of an initially pure stretching 'Fermi mode' (v_s, $v_b = 0$), with a high stretching quantum number, e.g. $v_s = 6$. It was assumed that such a state might be prepared by irradiation with some hypothetical laser pulse, without specifying details of the pulse. The energy of that

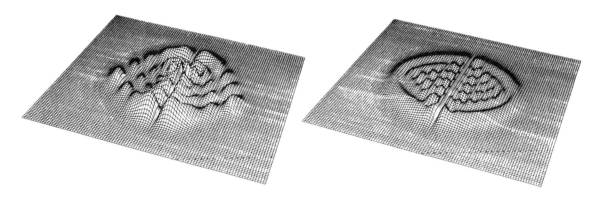

Figure A3.13.9. Probability density of a microcanonical distribution of the CH chromophore in CHF_3 within the multiplet with chromophore quantum number $N = 6$ ($N_{shell} = N + 1 = 7$). Representations in configuration space of stretching (Q_s) and bending (Q_b) coordinates (see text following equation (A3.13.62) and figure A3.13.10). Left-hand side: typical member of the microcanonical ensemble of the multiplet with $N = 6$ (random phases, equation (A3.13.66)). Right-hand side: microcanonical density $P_{micro} = \frac{1}{N_{shell}} \sum_{n \in shell} |\phi_n|^2$ for the multiplet with $N = 6$ ($N_{shell} = 7$). Adapted from [81].

state is located at the upper end of the energy range of the corresponding multiplet [81, 152, 154], which has a total of $N_{shell} = 7$ states. Such a state couples essentially to all remaining states of that multiplet. The corresponding evolution of the isolated system is shown as snapshots after the preparation step ($t_0 = 0$) in figure A3.13.10. The wave packet starts to spread out from the initially occupied stretching manifold (along the coordinate axis denoted by Q_s) into the bending manifold (Q_b) within the first 30–45 fs of evolution (left-hand side). Later on, it remains delocalized most of the time (as shown at the time steps 80, 220 and 380 fs, on the right-hand side) with exceptional partial recovery of the initial conditions at some isolated times (such as at 125 fs). The shape of the distribution at 220 fs is very similar to that of a typical member of the microcanonical ensemble in figure A3.13.9 above. However, in Figure A3.13.9, the relative phases between the seven superposition states were drawn from a random number generator, whereas in figure A3.13.10 they result from a fully coherent and deterministic propagation of a wave function.

IVR in the example of the CH chromophore in CHF_3 is thus at the origin of a redistribution process which is, despite its coherent nature, of a statistical character. In CHD_3, the dynamics after excitation of the stretching manifold reveals a less complete redistribution process in the same time interval [97]. The reason for this is a smaller effective coupling constant k'_{sbb} between the 'Fermi modes' of CHD_3 (by a factor of four) when compared to that of CHF_3. In [97] it was shown that redistribution in CHD_3 becomes significant in the picosecond time scale. However, on that time scale, the dynamical separation of time scales is probably no longer valid and couplings to modes pertaining to the space of CD_3 vibrations may become important and have additional influence on the redistribution process.

A3.13.5.2 IVR and time-dependent chirality

IVR in the CH chromophore system may also arise from excitation along the bending manifolds. Bending motions in polyatomic molecules are of great importance as primary steps for reactive processes involving isomerization and similar, large amplitude changes of internal molecular structure. At first sight, the one-dimensional section of the potential surface along the out-of-plane CH bending normal coordinate in CHD_3, shown in figure A3.13.4, is clearly less anharmonic than its one-dimensional stretching counterpart, also shown in that figure, even up to energies in the wave number region of $30\,000$ cm^{-1}. This suggests that coherent sequential multiphoton excitation of a CH bending motion, for instance along the x-axis in Figure A3.13.5,

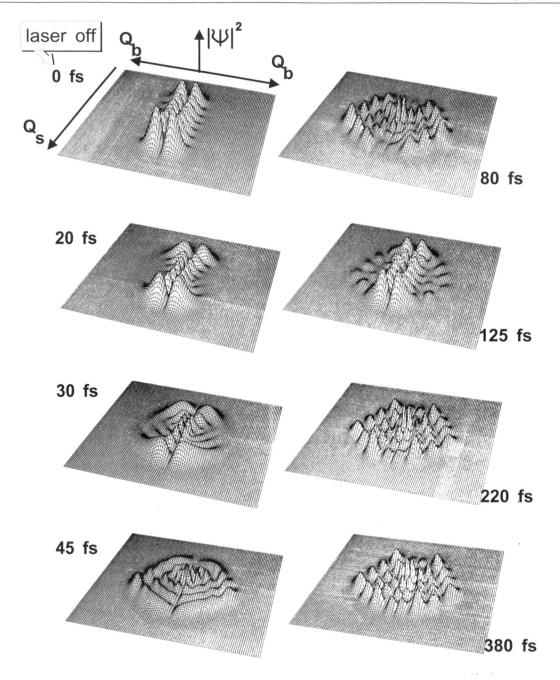

Figure A3.13.10. Time-dependent probability density of the isolated CH chromophore in CHF$_3$. Initially, the system is in a 'Fermi mode' with six quanta of stretching and zero of bending motion. The evolution occurs within the multiplet with chromophore quantum number $N = 6$ ($N_{shell} = N + 1 = 7$). Representations are given in the configuration space of stretching (Q_s) and bending (Q_b) coordinates (see text following equation (A3.13.62): Q_b is strictly a positive quantity, and there is always a node at $Q_b = 0$; the mirrored representation at $Q_b = 0$ is artificial and serves to improve visualization). Adapted from [81].

may induce a quasiclassical motion of the wave packet along that manifold [159, 160], which is significantly longer lived than the motion induced along the stretching manifold under similar conditions (see discussion above). Furthermore, the two-dimensional section in the CH bending subspace, spanned by the normal coordinates in the lower part of figure A3.13.4, is approximately isotropic. This corresponds to an almost perfect $C_{\infty v}$ symmetry with respect to the azimuthal angle φ (in the xy plane of figure A3.13.5), and is related to the approximate conservation of the bending vibration angular momentum ℓ_b [152, 173]. This implies that the direct anharmonic coupling between the degenerate bending manifolds is weak. However, IVR between these modes might be mediated by the couplings to the stretching mode. An interesting question is then to what extent such a coupling scheme might lead to a motion of the wave packet with quasiclassical exchange of vibrational energy between the two bending manifolds, following paths which could be described by classical vibrational mechanics, corresponding to CIVR. Understanding quasiclassical exchange mechanisms of large amplitude vibrational motion opens one desirable route of exerting control over molecular vibrational motion and reaction dynamics. In [154] these questions were investigated by considering the CH bending motion in CHD_3 and the asymmetric isotopomers CHD_2T and $CHDT_2$. The isotopic substitution was investigated with the special goal of a theoretical study of the coherent generation of dynamically chiral, bent molecular structures [174] and of the following time evolution. It was shown that IVR is at the origin of a coherent racemization dynamics which is superposed to a very fast, periodic exchange of left- and right-handed chiral structures ('stereomutation' reaction, period of roughly 20 fs, comparable to the period of the bending motion) and sets in after typically 300–500 fs. The main results are reviewed in the discussion of figures A3.13.11–A3.13.13.

The wave packet motion of the CH chromophore is represented by simultaneous snapshots of two-dimensional representations of the time-dependent probability density distribution

$$|\psi_{sb}(t, Q_s, Q_{b_i})|^2 = \int_{Q_{b_j}} dQ_{b_j} |\psi(t, Q_s, Q_{b_i}, Q_{b_j})|^2 \qquad (i \neq j) \qquad \text{(A3.13.67)}$$

and

$$|\psi_{bb}(t, Q_{b_1}, Q_{b_2})|^2 = \int_{Q_s} dQ_s |\psi(t, Q_s, Q_{b_1}, Q_{b_2})|^2. \qquad \text{(A3.13.68)}$$

Such a sequence of snapshots, calculated in intervals of 4 fs, is shown as a series of double contour line plots on the left-hand side of figure A3.13.11 (the outermost row shows the evolution of $|\psi_{bb}|^2$, equation (A3.13.68), the innermost row is $|\psi_{sb}|^2$, equation (A3.13.67), at the same time steps). This is the wave packet motion in CHD_3 for excitation with a linearly polarized field along the the x-axis at 1300 cm^{-1} and 10 TW cm^{-2} after 50 fs of excitation. At this point a more detailed discussion regarding the orientational dynamics of the molecule is necessary. Clearly, the polarization axis is defined in a laboratory fixed coordinate system, while the bending axes are fixed to the molecular frame. Thus, exciting internal degrees of freedom along specific axes in the internal coordinate system requires two assumptions: the molecule must be oriented or aligned with respect to the external polarization axis, and this state should be stationary, at least during the relevant time scale for the excitation process. It is possible to prepare oriented states [112, 114, 115] in the gas phase, and such a state can generally be represented as a superposition of a large number of rotational eigenstates. Two questions become important then: How fast does such a rotational superposition state evolve? How well does a purely vibrational wave packet calculation simulate a more realistic calculation which includes rotational degrees of freedom, i.e. with an initially oriented rotational wave packet? The second question was studied recently by full dimensional quantum dynamical calculations of the wave packet motion of a diatomic molecule during excitation in an intense infrared field [175], and it was verified that rotational degrees of freedom may be neglected whenever vibrational–rotational couplings are not important for intramolecular rotational–vibrational redistribution (IVRR) [84]. Regarding the first question, because of the large rotational constant of methane, the time scales on which an initially oriented state of the free molecule is maintained are likely to be comparatively short and it would also be desirable to carry out calculations that

CHD$_3$

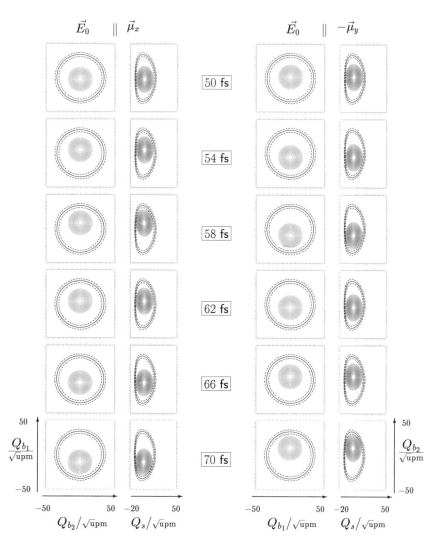

Figure A3.13.11. Illustration of the time evolution of reduced two-dimensional probability densities $|\psi_{bb}|^2$ and $|\psi_{sb}|^2$, for the excitation of CHD$_3$ between 50 and 70 fs (see [154] for further details). The full curve is a cut of the potential energy surface at the momentary absorbed energy corresponding to 3000 cm^{-1} during the entire time interval shown here ($\approx$6000 cm^{-1}, if zero point energy is included). The dashed curves show the energy uncertainty of the time-dependent wave packet, approximately 500 cm^{-1}. Left-hand side: excitation along the x-axis (see figure A3.13.5). The vertical axis in the two-dimensional contour line representations is the Q_{b_1}-axis, the horizontal axes are Q_{b_2} and Q_s, for $|\psi_{bb}|^2$ and $|\psi_{sb}|^2$, respectively. Right-hand side: excitation along the y-axis, but with the field vector pointing into the negative y-axis. In the two-dimensional contour line representations, the vertical axis is the Q_{b_2}-axis, the horizontal axes are Q_{b_1} and Q_s, for $|\psi_{bb}|^2$ and $|\psi_{sb}|^2$, respectively. The lowest contour line has the value 44×10^{-5} u^{-1} pm^{-2}, the distance between them is 7×10^{-5} u^{-1} pm^{-2}. Maximal values are nearly constant for all the images in this figure and correspond to 140×10^{-5} u^{-1} pm^{-2} for $|\psi_{bb}|^2$ and 180×10^{-5} u^{-1} pm^{-2} for $|\psi_{sb}|^2$.

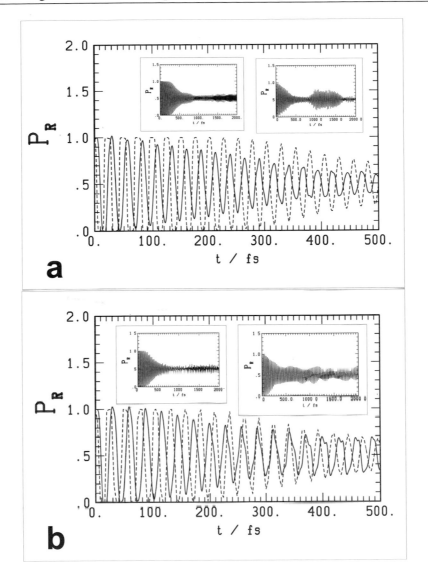

Figure A3.13.12. Evolution of the probability for a right-handed chiral structure $P_R(t)$ (full curve, see equation (A3.13.69)) of the CH chromophore in CHD_2T (a) and $CHDT_2$ (b) after preparation of chiral structures with multiphoton laser excitation, as discussed in the text (see also [154]). For comparison, the time evolution of P_R according to a one-dimensional model including only the Q_{b_2} bending mode (dashed curve) is also shown. The left-hand side insert shows the time evolution of P_R within the one-dimensional calculations for a longer time interval; the right-hand insert shows the P_R time evolution within the three-dimensional calculation for the same time interval (see text).

include rotational states explicitly. Such calculations were done, for instance, for ozone at modest excitations [116, 117], but they would be quite difficult for the methane isotopomers at the high excitations considered in the present example.

The multiphoton excitation scheme corresponding to excitation along the x-axis is shown by the upright arrows on the left-hand side of figure A3.13.4. In the convention adopted in [154], nuclear displacements

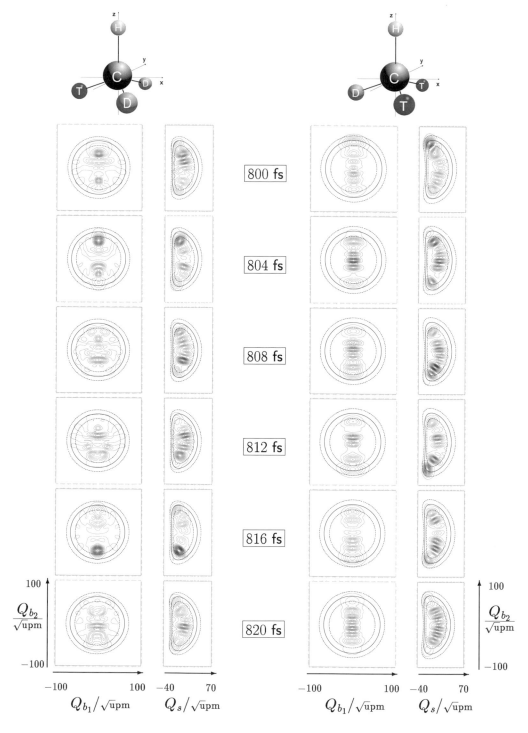

Figure A3.13.13. Illustration of the time evolution of reduced two-dimensional probability densities $|\psi_{\mathrm{bb}}|^2$ and $|\psi_{\mathrm{sb}}|^2$, for the isolated CHD_2T (left-hand side) and $CHDT_2$ (right-hand side) after 800 fs of free evolution. At time 0 fs the wave packets corresponded to a localized, chiral molecular structure (from [154]). See also text and figure A3.13.11.

along Q_{b_1} occur along the x-axis, displacements along Q_{b_2} are directed along the y-axis. One observes a semiclassical, nearly periodic motion of the wave packet along the excited manifold with a period of approximately 24 fs, corresponding to the frequency of the bending vibrations in the wave number region around 1500 cm^{-1}. At this stage of the excitation process, the motion of the wave packet is essentially one-dimensional, as seen from the trajectory followed by the maximum of the probability distribution and its practically unchanged shape during the oscillations back and forth between the turning points. The latter lie on the potential energy section defined by the momentary energy $E(t)$ of the wave packet, as described above, and describe the classically accessible region in configuration space. These potential energy sections are shown by the continuous curves in the figures, which are surrounded by dotted curves describing the energy uncertainty.

The sequence on the right-hand side of figure A3.13.11 shows wave packets during the excitation along the y-axis. Here, excitation was chosen to be antiparallel to the y-axis ($E_0 \| -\mu_y$). This choice induces a phase shift of π between the two wave packets shown in the figure, in addition to forcing oscillations along different directions. Excitation along the y-axis can be used to generate dynamically chiral structures. If the excitation laser field is switched off, e.g. at time step 70 fs after beginning the excitation, the displacement of the wave packet clearly corresponds to a bent molecular structure with angle $\vartheta \approx 10°$ (e.g. in the xy plane of figure A3.13.5). This structure will, of course, also change with time for the isolated molecule, and one expects this change to be oscillatory, like a pendulum, at least initially. Clearly, IVR will play some role, if not at this early stage, then at some later time. One question is, will it be CIVR or DIVR? When studying this question with the isotopically substituted compounds CHD$_2$T and CHDT$_2$, the y-axis being perpendicular to the C_s mirror plane, a bent CH chromophore corresponds to a chiral molecular structure with a well defined chirality quantum number, say R. As time evolves, the wave packet moves to the other side of the symmetry plane, $Q_{b_2} = 0$, implying a change of chirality. In this context, the enantiomeric excess can be defined by the probability

$$P_R(t) = \int_{-\infty}^{0} dQ_{b_2} |\psi_{b_2}(t, Q_{b_2})|^2 \tag{A3.13.69}$$

for right-handed ('R') chiral structures ($P_L(t) = 1 - P_R(t)$ is the probability for left-handed ('L') structures), where

$$|\psi_{b_2}(t, Q_{b_2})|^2 = \int_{Q_s} \int_{Q_{b_1}} dQ_s \, dQ_{b_1} |\psi(t, Q_s, Q_{b_1}, Q_{b_2})|^2. \tag{A3.13.70}$$

The time evolution of $P_R(t)$ is shown in figure A3.13.12 for the field free motion of wave packets for CHD$_2$T and CHDT$_2$ prepared by a preceding excitation along the y-axis.

In the main part of each figure, the evolution of P_R calculated within the stretching and bending manifold of states for the CH chromophore is shown (full curve). The dashed curve shows the evolution of P_R within a one-dimensional model, in which only the Q_{b_2} bending manifold is considered during the dynamics. Within this model there is obviously no IVR, and comparison of the full with the dashed curves helps to visualize the effect of IVR. The insert on the left-hand side shows a survey of the evolution of P_R for the one-dimensional model during a longer time interval of 2 ps, while the insert on the right-hand side shows the evolution of P_R for the calculation within the full three-dimensional stretching and bending manifold of states during the same time interval of 2 ps. The three-dimensional calculations yield a fast, initially nearly periodic, evolution, with an approximate period of 20 fs, which is superimposed by a slower decay of probability corresponding to an overall decay of enantiomeric excess $|D_{abs}(t)| = |1 - 2P_R(t)|$ on a time scale of 300–400 fs for both CHD$_2$T and CHDT$_2$. The decay is clearly more pronounced for CHD$_2$T (figure A3.13.12(a)). The first type of evolution corresponds to a stereomutation reaction, while the second can be interpreted as racemization. A further question is then related to the origin of this racemization.

Figure A3.13.14 shows the wave packet motion for CHD$_2$T and CHDT$_2$, roughly 800 fs after the initially localized, chiral structure has been generated. Comparison with the wave packet motion allows for

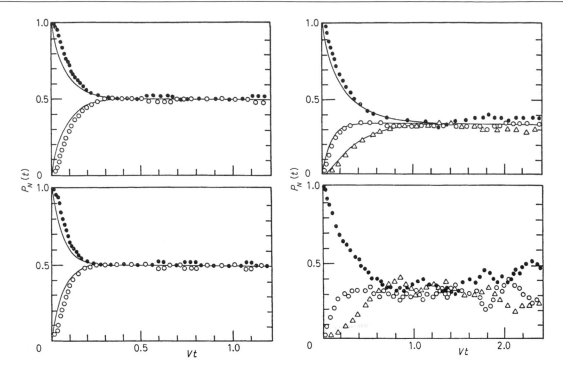

Figure A3.13.14. Illustration of the quantum evolution (points) and Pauli master equation evolution (lines) in quantum level structures with two levels (and 59 states each, left-hand side) and three levels (and 39 states each, right-hand side) corresponding to a model of the energy shell IVR (horizontal transition in figure A3.13.1). From [38]. The two-level structure (left) has two models: $|V_{ij}|^2 = $ const and random signs (upper part), random V_{ij} but $-V_m \leq V_{ij} \leq V_m$ (lower part). The right-hand side shows an evolution with initial diagonal density matrix (upper part) and a single trajectory (lower part).

the conclusion that racemization is induced by the presence of DIVR between all vibrational modes of the CH chromophore. However, while DIVR is quite complete for CHD_2T, after excitation along the y axis, it is only two-dimensional for $CHDT_2$. A localized exchange of vibrational energy in terms of CIVR has not been observed at any intermediate time step. Racemization is stronger for CHD_2T, for which DIVR occurs in the full three-dimensional subspace of the CH chromophore, under the present conditions. It is less pronounced for $CHDT_2$, which has a higher degree of localization of the wave packet motion. In comparison with the one-dimensional calculations in figure A3.13.12, it becomes evident that there is a decay of the overall enantiomeric excess for CHD_2T, as well as for $CHDT_2$, also in the absence of IVR. The decay takes place on a time scale of 500–1000 fs and is a consequence of the dephasing of the wave packet due to the diagonal anharmonicity of the bending motion. This decay may, of course, also be interpreted as racemization. However, it is much less complete than racemization in the three-dimensional case and clearly of secondary importance for the enantiomeric decay in the first 200 fs.

A3.13.6 Statistical mechanical master equation treatment of intramolecular energy redistribution in reactive molecules

The previous sections indicate that the full quantum dynamical treatment of IVR in an intermediate size molecule even under conditions of 'coherent' excitation shows phenomena reminiscent of relaxation and

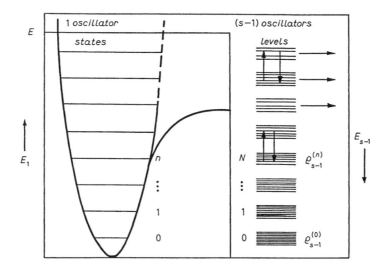

Figure A3.13.15. Master equation model for IVR in highly excited C_2H_6. The left-hand side shows the quantum levels of the reactive CC oscillator. The right-hand side shows the levels with a high density of states from the remaining 17 vibrational (and torsional) degrees of freedom (from [38]).

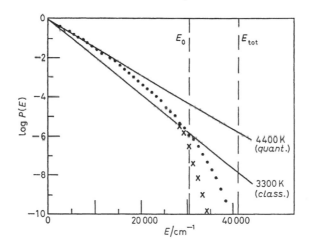

Figure A3.13.16. Illustration of the level populations (corresponding to the C–C oscillator states) from various treatments in the model of figure A3.13.15 for C_2H_6 at a total energy $E = (hc)\ 41\,000\ \mathrm{cm}^{-1}$ and a threshold energy $E_0 = (hc)\ 31\,000\ \mathrm{cm}^{-1}$. The points are microcanonical equilibrium distributions. The crosses result from the solution of the master equation for IVR at steady state and the lines are thermal populations at the temperatures indicated (from [38]: quant. is calculated with quantum densities of states, class. with classical mechanical densities.).

equilibration. This suggests that, in general, at very high excitations in large polyatomic molecules with densities of states easily exceeding the order of $10^{10}\ \mathrm{cm}^{-1}$ (or about 10^9 molecular states in an energy interval corresponding to 1 J mol^{-1}), a statistical master equation treatment may be possible [38, 122]. Such an approach has been justified by quantum simulations in model systems as well as analytical considerations [38], following early ideas in the derivation of the statistical mechanical Pauli equation [176]. Figure A3.13.14 shows the kinetic behaviour in such model systems. The 'coarse grained' populations of groups of quantum states ('levels' with less than 100 states, indexed by capital letters I and J) at the same total energy show very similar behaviour if calculated from the Schrödinger equation, e.g. equation (A3.13.43), or the Pauli equation (A3.13.71),

$$p(t) = \mathbf{Y}(t)p(0),\tag{A3.13.71}$$

with $\mathbf{Y}$ being given by:

$$\mathbf{Y}(t) = \exp(\mathbf{K}t),\tag{A3.13.72}$$

and the rate coefficient matrix elements in the limit of perturbation theory

$$K_{IJ} = 2\pi |V_{IJ}|^2/\delta_I.\tag{A3.13.73}$$

In equation (A3.13.73), δ_I is the average angular frequency distance between quantum states within level I and $|V_{IJ}|^2$ is the average square coupling matrix element (as angular frequency) between the quantum states in levels I and J (of total number of states N_I and N_J, respectively) and is given by:

$$|V_{IJ}|^2 = \frac{1}{N_I}\frac{1}{N_J}\sum_{i=1}^{N_I}\sum_{j=1}^{N_J}|V_{ij}|^2.\tag{A3.13.74}$$

Figure A3.13.14 seems to indicate that the Pauli equation is a strikingly good approximation for treating IVR under these conditions, involving even relatively few quantum states. This is, however, only true in this simple manner because we assume in the model that all the couplings are randomly distributed around their 'typical' values. This excludes any symmetry selection rules for couplings within the set of quantum states considered [177]. More generally, one has to consider only sets of quantum states with the same set of good (conserved) quantum numbers in such a master equation treatment. It is now well established that even in complex forming collisions leading to maximum energy transfer by IVR (section A3.13.3.3), conserved quantum numbers such as nuclear spin symmetry and parity lead to considerable restrictions [177, 178]. More generally, one has to identify approximate symmetries on short time scales, which lead to further restrictions on the density of strongly coupled states [179]. Thus, the validity of a statistical master equation treatment for IVR in large polyatomic molecules is not obvious *a priori* and has to be established individually for at least classes of molecular systems, if not on a case by case basis.

Figure A3.13.15 shows a scheme for such a Pauli equation treatment of energy transfer in highly excited ethane, e.g. equation (A3.13.75), formed at energies above both thresholds for dissociation in chemical activation:

$$H + C_2H_5 \rightarrow C_2H_6^* \rightarrow 2CH_3.\tag{A3.13.75}$$

The figure shows the migration of energy between excited levels of the ultimately reactive C–C oscillator, the total energy being constant at $E/hc = 41\,000$ cm^{-1} with a CC dissociation threshold of $31\,000$ cm^{-1}. The energy balance is thus given by:

$$E_{\text{tot}} = E_{\text{C–C}} + \sum_{i=1}^{17} E_i.\tag{A3.13.76}$$

The microcanonical equilibrium distributions are governed by the densities ρ_{s-1} in the $(s-1) = 17$ oscillators (figure A3.13.15):

$$p_{micro}(n) = \rho_{s-1}^{(n)}\left\{\sum_k \rho_{s-1}^{(k)}\right\}^{-1}\tag{A3.13.77}$$

where the 17 remaining degrees of freedom of ethane form essentially a 'heat bath'. The kinetic master equation treatment of this model leads to steady-state populations shown in figure A3.13.16. This illustrates that the steady-state populations under conditions where reaction equation (A3.13.75) competes with IVR differ from the microcanonical equilibrium populations at high energy, and both differ from thermal distributions shown as lines (quantum or classical). Whereas the deviation from a thermal distribution is well understood and handled by standard statistical theories such as RRKM (chapter A3.12) and the statistical adiabatic channel model [76], the deviation from the microcanonical distribution would lead to an intramolecular nonequilibrium effect on the rates of reaction which so far has not been well investigated experimentally [37–39].

A3.13.7 Summarizing overview on energy redistribution in reacting systems

It has been understood for more than a century that energy redistribution is a key process in chemical reactions, including in particular the oldest process of chemical technology used by mankind: fire or combustion, where both radiative and collisional processes are relevant. Thus one might think that this field has reached a stage of maturity and saturation. Nothing could be further from the truth. While collisional energy transfer is now often treated in reaction systems in some detail, as is to some extent routine in unimolecular reactions, there remain plenty of experimental and theoretical challenges. In the master equation treatments, which certainly should be valid here, one considers a statistical, macroscopic reaction system consisting of reactive molecules in a mixture, perhaps an inert gas heat bath.

The understanding of the second process considered in this chapter, intramolecular energy redistribution within a single molecular reaction system, is still in its infancy. It is closely related to the challenge of finding possible schemes to control the dynamics of atoms in molecules and the related change of molecular structure during the course of a chemical reaction [10, 117, 154, 175], typically in the femtosecond time scale, which has received increasing attention in the last few decades [180–182]. The border between fully quantum dynamical treatments, classical mechanical theories and, finally, statistical master equations for IVR type processes needs to be explored further experimentally and theoretically in the future. Unravelling details of the competition between energy redistribution and reaction in individual molecules remains an important task for the coming decades [37–40].

References

[1] van't Hoff J H 1884 *Études de Dynamique Chimique* (Amsterdam: F. Müller)
[2] Arrhenius S 1889 Über die Reaktionsgeschwindigkeit bei der Inversion von Rohrzucker durch Säuren *Z. Phys. Chem.* **4** 226–48
[3] Arrhenius S 1899 Zur Theorie der chemischen Reaktionsgeschwindigkeiten *Z. Phys. Chem.* **28** 317–35
[4] Calvert J G and Pitts J N 1966 *Photochemistry* (New York: Wiley)
[5] Simons J P 1977 The dynamics of photodissociation *Gas Kinetics and Energy Transfer* vol 2, ed P G Ashmore and R J Donovan (London: The Chemical Society) pp 58–95
[6] Turro N J 1978 *Modern Molecular Photochemistry* (Menlo Park, CA: Benjamin-Cummings)
[7] Schinke R 1993 *Photodissociation Dynamics* (Cambridge: Cambridge University Press)
[8] Perrin J 1922 On the radiation theory of chemical action *Trans. Faraday Soc.* **17** 546–72
[9] Lindemann F A 1922 Discussion contributions on the radiation theory of chemical action *Trans. Faraday Soc.* **17** 598–9
[10] Quack M 1998 Multiphoton excitation *Encyclopedia of Computational Chemistry* vol 3, ed R P von Schleyer *et al* (New York: Wiley) pp 1775–91
[11] Quack M and Troe J 1976 Unimolecular reactions and energy transfer of highly excited molecules *Gas Kinetics and Energy Transfer* vol 2, ch 5, ed P G Ashmore and R J Donovan (London: The Chemical Society) pp 175–238 (a review of the literature published up to early 1976)
[12] Nikitin E E 1974 *Theory of Elementary Atomic and Molecular Processes in Gases* (Oxford: Clarendon)
[13] Troe J 1975 Unimolecular reactions: experiment and theory *Physical Chemistry. An Advanced Treatise* vol VIB, ed H Eyring, D Henderson and W Jost (New York: Academic) pp 835–929
[14] Rice S A 1975 Some comments on the dynamics of primary photochemical processes *Excited States* ed E C Lim (New York: Academic) pp 111–320
[15] Light J C, Ross J and Shuler K E 1969 Rate coefficients, reaction cross sections and microscopic reversibility *Kinetic Processes in Gases and Plasmas* ed A R Hochstim (New York: Academic) pp 281–320
[16] Stockburger M 1973 *Organic Molecular Photophysics* vol 1, ed J Birks (New York: Wiley) p 57
[17] Montroll E W and Shuler K E 1958 The application of the theory of stochastic processes to chemical kinetics *Adv. Chem. Phys.* **1** 361–99

[18] Pritchard H O 1975 *Reaction Kinetics* vol 1, ed P G Ashmore (London: The Chemical Society)
[19] Quack M and Troe J 1981 Current aspects of unimolecular reactions *Int. Rev. Phys. Chem.* **1** 97–147
[20] Golden D M and Benson S W 1975 *Physical Chemistry. An Advanced Treatise* vol VII (New York: Academic) p 57
[21] Tardy D C and Rabinovitch B S 1977 Intermolecular vibrational energy transfer in thermal unimolecular systems *Chem. Rev.* **77** 369–408
[22] Troe J 1978 Atom and radical recombination reactions *Ann. Rev. Phys. Chem.* **29** 223–50
[23] Hippler H and Troe J 1989 *Advances in Gas Phase Photochemistry and Kinetics* ed M N R Ashfold and J E Baggott (London: Royal Society of Chemistry) pp 209–62
[24] Krajnovitch D J, Parmenter C S and Catlett D L Jr 1987 State-to-state vibrational transfer in atom-molecule collisions. Beams vs. bulbs *Chem. Rev.* **87** 237–88
[25] Gilbert R G and Smith S C 1990 *Theory of Unimolecular and Recombination Reactions* (London: Blackwell)
[26] Lambert J D 1977 *Vibrational and Rotational Relaxation in Gases* (Oxford: Oxford University Press)
[27] Weitz E and Flynn G W 1981 Vibrational energy flow in the ground electronic states of polyatomic molecules *Adv. Chem. Phys.* **47** 185–235
[28] Oref I and Tardy D C 1990 Energy transfer in highly excited large polyatomic molecules *Chem. Rev.* **90** 1407–45
[29] Weston R E and Flynn G W 1992 Relaxation of molecules with chemically significant amounts of vibrational energy: the dawn of the quantum state resolved era *Ann. Rev. Phys. Chem.* **43** 559–89
[30] Howard M J and Smith I W M 1983 The kinetics of radical-radical processes in the gas phase *Prog. Reaction Kin.* **12** 57–200
[31] Orr B J and Smith I W M 1987 Collision-induced vibrational energy transfer in small polyatomic molecules *J. Phys. Chem.* **91** 6106–19
[32] Flynn G W, Parmenter C S and Wodtke A M 1996 Vibrational energy transfer *J. Phys. Chem.* **100** 12 817–38
[33] Förster Th 1948 Zwischenmolekulare Energiewanderung und Fluoreszenz *Ann. Phys.* **2** 55–75
[34] Juzeliunas G and Andrews D L 2000 Quantum electrodynamics of resonance energy transfer *Adv. Chem. Phys.* **112** 357–410
[35] Quack M 1982 Reaction dynamics and statistical mechanics of the preparation of highly excited states by intense infrared radiation *Adv. Chem. Phys.* **50** 395–473
[36] Jortner J, Rice S A and Hochstrasser R M 1969 Radiationless transitions in photochemistry *Adv. Photochem.* **7** 149
[37] Quack M and Kutzelnigg W 1995 Molecular spectroscopy and molecular dynamics: theory and experiment *Ber. Bunsenges. Phys. Chem.* **99** 231–45
[38] Quack M 1981 Statistical mechanics and dynamics of molecular fragmentation *Nuovo Cimento* B **63** 358–77
[39] Quack M 1995 Molecular femtosecond quantum dynamics between less than yoctoseconds and more than days: experiment and theory *Femtosecond Chemistry* ed J Manz and L Woeste (Weinheim: Verlag Chemie) pp 781–818
[40] Oppenheim I, Shuler K E and Weiss G H 1977 *Stochastic Processes in Chemical Physics, The Master Equation* (Cambridge, MA: MIT Press)
[41] van Kampen N G 1981 *Stochastic Processes in Physics and Chemistry* (Amsterdam: North-Holland)
[42] Gear C W 1971 *Numerical Initial Value Problems in Ordinary Differential Equations* (Englewood Cliffs, NJ: Prentice-Hall)
[43] Shampine S 1994 *Numerical Solutions of Ordinary Differential Equations* (New York: Chapman and Hall)
[44] Troe J 1977 Theory of thermal unimolecular reactions at low pressures. I. Solutions of the master equation *J. Chem. Phys.* **66** 4745–57
[45] Troe J 1977 Theory of thermal unimolecular reactions at low pressures. II. Strong collision rate constants. Applications *J. Chem. Phys.* **66** 4758
[46] Troe J 1983 Theory of thermal unimolecular reactions in the fall-off range. I. Strong collision rate constants *Ber. Bunsenges. Phys. Chem.* **87** 161–9
[47] Quack M 1979 Master equations for photochemistry with intense infrared light *Ber. Bunsenges. Phys. Chem.* **83** 757–75
[48] Dove J E, Nip W and Teitelbaum H 1975 *Proc. XVth Int. Symp. on Combustion* (The Combustion Institute) p 903
[49] Olschewski H A, Troe J and Wagner H G 1966 Niederdruckbereich und Hochdruckbereich des unimolekularen N_2O-Zerfalls *Ber. Bunsenges. Phys. Chem.* **70** 450
[50] Martinengo A, Troe J and Wagner H G 1966 *Z. Phys. Chem. (Frankfurt)* **51** 104
[51] Johnston H S 1951 Interpretation of the data on the thermal decomposition of nitrous oxide *J. Chem. Phys.* **19** 663–7
[52] Venkatesh P K, Dean A M, Cohen M H and Carr R W 1999 Master equation analysis of intermolecular energy transfer in multiple-well, multiple-channel unimolecular reactions. II. Numerical methods and application to the mechanism of the $C_2H_5 + O_2$ reaction *J. Chem. Phys.* **111** 8313
[53] Smith K and Thomson R M 1978 *Computer Modelling of Gas Lasers* (New York: Plenum)
[54] Hirschfelder J D, Curtiss C F and Bird R B 1964 *Molecular Theory of Gases and Liquids* (New York: Wiley)
[55] Tabor M, Levine R D, Ben-Shaul A and Steinfeld J I 1979 Microscopic and macroscopic analysis of non-linear master equations: vibrational relaxation of diatomic molecules *Mol. Phys.* **37** 141–58
[56] Treanor C E, Rich J W and Rehm R G 1968 Vibrational relaxation of anharmonic oscillators with exchange-dominated collisions *J. Chem. Phys.* **48** 1798–807
[57] McCaffery A J 1999 Quasiresonant vibration–rotation transfer: a kinematic interpretation *J. Chem. Phys.* **111** 7697
[58] Levine R D and Bernstein R B 1987 *Molecular Reaction Dynamics and Chemical Reactivity* (New York: Oxford University Press)

[59] Steinfeld J I and Klemperer W 1965 Energy-transfer processes in monochromatically excited iodine molecules. I. Experimental results *J. Chem. Phys.* **42** 3475–97

[60] Gianturco F A 1979 *The Transfer of Molecular Energies by Collision* (Heidelberg: Springer)

[61] Bowman J M (ed) 1983 *Molecular Collision Dynamics* (Berlin: Springer)

[62] Hutson J M and Green S 1994 MOLSCAT computer code, version 14, distributed by Collaborative Computational Project No 6 of the Engineering and Physical Sciences Research Council (UK)

[63] Alexander M H and Manolopoulos D E 1987 A stable linear reference potential algorithm for solution of the quantum close-coupled equations in molecular scattering theory *J. Chem. Phys.* **86** 2044–50

[64] Bodo E, Gianturco F A and Paesani F 2000 Testing intermolecular potentials with scattering experiments: He–CO rotationally inelastic collisions *Z. Phys. Chem., NF* **214** 1013–34

[65] Fluendy M A D and Lawley K P 1973 *Applications of Molecular Beam Scattering* (London: Chapman and Hall)

[66] Faubel M and Toennies J P 1977 Scattering studies of rotational and vibrational excitation of molecules *Adv. Atom. Mol. Phys.* **13** 229

[67] Faubel M 1983 Vibrational and rotational excitation in molecular collisions *Adv. Atom. Mol. Phys.* **19** 345

[68] Bunker D L 1971 *Methods in Computational Physics* vol 10, ed B Alder (New York: Academic)

[69] Lenzer T, Luther K, Troe J, Gilbert R G and Lim K F 1995 Trajectory simulations of collisional energy transfer in highly excited benzene and hexafluorobenzene *J. Chem. Phys.* **103** 626–41

[70] Grigoleit U, Lenzer T and Luther K 2000 Temperature dependence of collisional energy transfer in highly excited aromatics studied by classical trajectory calculations *Z. Phys. Chem., NF* **214** 1065–85

[71] Barker J R 1984 Direct measurement of energy transfer in rotating large molecules in the electronic ground state *J. Chem. Phys.* **88** 11

[72] Miller L A and Barker J R 1996 Collisional deactivation of highly vibrationally excited pyrazine *J. Chem. Phys.* **105** 1383–91

[73] Hippler H, Troe J and Wendelken H J 1983 Collisional deactivation of vibrationally highly excited polyatomic molecules. II. Direct observations for excited toluene *J. Chem. Phys.* **78** 6709

[74] Hold U, Lenzer T, Luther K, Reihs K and Symonds A C 2000 Collisional energy transfer probabilities of highly excited molecules from kinetically controlled selective ionization (KCSI). I. The KCSI technique: experimental approach for the determination of $P(E', E)$ in the quasicontinuous energy range *J. Chem. Phys.* **112** 4076–89

[75] Steinfeld J I, Ruttenberg P, Millot G, Fanjoux G and Lavorel B 1991 Scaling laws for inelastic collision processes in diatomic molecules *J. Phys. Chem.* **95** 9638–47

[76] Quack M and Troe J 1974 Specific rate constants of unimolecular processes II. Adiabatic channel model *Ber. Bunsenges. Phys. Chem.* **78** 240–52

[77] Quack M and Troe J 1998 Statistical adiabatic channel model *Encyclopedia of Computational Chemistry* vol 4, ed P von Ragué Schleyer *et al* (New York: Wiley) pp 2708–26

[78] Quack M and Troe J 1975 Complex formation in reactive and inelastic scattering: statistical adiabatic channel model of unimolecular processes III *Ber. Bunsenges. Phys. Chem.* **79** 170–83

[79] Quack M 1979 Quantitative comparison between detailed (state selected) *relative* rate data and averaged (thermal) *absolute* rate data for complex forming reactions *J. Phys. Chem.* **83** 150–8

[80] Clary D C, Gilbert R G, Bernshtein V and Oref I 1995 Mechanisms for super collisions *Faraday Discuss. Chem. Soc.* **102** 423–33

[81] Marquardt R, Quack M, Stohner J and Sutcliffe E 1986 Quantum-mechanical wavepacket dynamics of the CH group in the symmetric top X_3CH compounds using effective Hamiltonians from high-resolution spectroscopy *J. Chem. Soc., Faraday Trans. 2* **82** 1173–87

[82] Herzberg G 1966 *Molecular Spectra and Molecular Structure III. Electronic Spectra and Electronic Structure of Polyatomic Molecules* (New York: Van Nostrand-Reinhold) (reprinted in 1991)

[83] Rice S A 1981 An overview of the dynamics of intramolecular transfer of vibrational energy *Adv. Chem. Phys.* **47** 117–200

[84] Beil A, Luckhaus D, Quack M and Stohner J 1997 Intramolecular vibrational redistribution and unimolecular reactions: concepts and new results on the femtosecond dynamics and statistics in CHBrClF *Ber. Bunsenges. Phys. Chem.* **101** 311–28

[85] Quack M 1993 Molecular quantum dynamics from high resolution spectroscopy and laser chemistry *J. Mol. Struct.* **292** 171–95

[86] Heller E J 1975 Time-dependent approach to semiclassical dynamics *J. Chem. Phys.* **62** 1544–55

[87] Heller E J 1981 The semiclassical way to molecular spectroscopy *Acc. Chem. Res.* **14** 368–78

[88] Heller E J 1983 The correspondence principle and intramolecular dynamics *Faraday Discuss. Chem. Soc.* **75** 141–53

[89] Noid D W, Koszykowski M L and Marcus R A 1981 Quasiperiodic and stochastic behaviour in molecules *Ann. Rev. Phys. Chem.* **32** 267–309

[90] Marcus R A 1983 On the theory of intramolecular energy transfer *Faraday Discuss. Chem. Soc.* **75** 103–15

[91] Hose G and Taylor H S 1982 A quantum analog to the classical quasiperiodic motion *J. Chem. Phys.* **76** 5356–64

[92] Wyatt R E, Hose G and Taylor H S 1983 Mode-selective multiphoton excitation in a model system *Phys. Rev. A* **28** 815–28

[93] Heather R and Metiu H 1985 Some remarks concerning the propagation of a Gaussian wave packet trapped in a Morse potential *Chem. Phys. Lett.* **118** 558–63

[94] Sawada S and Metiu H 1986 A multiple trajectory theory for curve crossing problems obtained by using a Gaussian wave packet representation of the nuclear motion *J. Chem. Phys.* **84** 227–38

[95] Miller W H 1974 Classical-limit quantum mechanics and the theory of molecular collisions *Adv. Chem. Phys.* **25** 69–177

[96] van Gunsteren W F and Berendsen H J C 1990 Computer simulation of molecular dynamics: methodology, applications, and perspectives in chemistry *Angew. Chem. Int. Ed. Engl.* **29** 992–1023

[97] Marquardt R and Quack M 1991 The wave packet motion and intramolecular vibrational redistribution in CHX_3 molecules under infrared multiphoton excitation *J. Chem. Phys.* **95** 4854–67

[98] Dübal H-R and Quack M 1980 Spectral bandshape and intensity of the C–H chromophore in the infrared spectra of CF_3H and C_4F_9H *Chem. Phys. Lett.* **72** 342–7
    Dübal H-R and Quack M 1981 High resolution spectroscopy of fluoroform *Chem. Phys. Lett.* **80** 439–44

[99] Dübal H-R and Quack M 1984 Tridiagonal Fermi resonance structure in the IR spectrum of the excited CH chromophore in CF_3H *J. Chem. Phys.* **81** 3779–91

[100] Quack M 1990 Spectra and dynamics of coupled vibrations in polyatomic molecules *Ann. Rev. Phys. Chem.* **41** 839–74

[101] Maynard A T, Wyatt R E and Iung C 1995 A quantum dynamical study of CH overtones in fluoroform. I. A nine-dimensional *ab initio* surface, vibrational spectra and dynamics *J. Chem. Phys.* **103** 8372–90

[102] Maynard A T, Wyatt R E and Iung C 1997 A quantum dynamical study of CH overtones in fluoroform. II. Eigenstates of the $v_{CH} = 1$ and $v_{CH} = 2$ regions *J. Chem. Phys.* **106** 9483–96

[103] Wyatt R E, Iung C and Leforestier C 1992 Quantum dynamics of overtone relaxation in benzene. II. Sixteen-mode model for relaxation from $CH(v = 3)$ *J. Chem. Phys.* **97** 3477–86

[104] Minehardt T A, Adcock J D and Wyatt R E 1999 Quantum dynamics of overtone relaxation in 30-mode benzene: a time-dependent local mode analysis for $CH(v = 2)$ *J. Chem. Phys.* **110** 3326–34

[105] Gatti F, Iung C, Leforestier C and Chapuisat X 1999 Fully coupled 6D calculations of the ammonia vibration–inversion–tunneling states with a split Hamiltonian pseudospectral approach *J. Chem. Phys.* **111** 7236–43

[106] Luckhaus D 2000 6D vibrational quantum dynamics: generalized coordinate discrete variable representation and (a)diabatic contraction *J. Chem. Phys.* **113** 1329–47

[107] Fehrensen B, Luckhaus D and Quack M 1999 Mode selective stereomutation tunnelling in hydrogen peroxide isotopomers *Chem. Phys. Lett.* **300** 312–20

[108] Fehrensen B, Luckhaus D and Quack M 1999 Inversion tunneling in aniline from high resolution infrared spectroscopy and an adiabatic reaction path Hamiltonian approach *Z. Phys. Chem., NF* **209** 1–19

[109] Zhang D H, Wu Q, Zhang J Z H, von Dirke M and Bačić Z 1995 Exact full dimensional bound state calculations for $(HF)_2$, $(DF)_2$ and HFDF *J. Chem. Phys.* **102** 2315–25

[110] Quack M and Suhm M A 1998 Spectroscopy and quantum dynamics of hydrogen fluoride clusters *Advances in Molecular Vibrations and Collision Dynamics, Vol. III Molecular Clusters* ed J Bowman and Z Bačić (JAI Press) pp 205–48

[111] Qiu Y and Bačić Z 1998 Vibration–rotation–tunneling dynamics of $(HF)_2$ and $(HCl)_2$ from full-dimensional quantum bound state calculations *Advances in Molecular Vibrations and Collision Dynamics, Vol. I–II Molecular Clusters* ed J Bowman and Z Bačić (JAI Press) pp 183–204

[112] Loesch H J and Remscheid A 1990 Brute force in molecular reaction dynamics: a novel technique for measuring steric effects *J. Chem. Phys.* **93** 4779–90

[113] Seideman T 1995 Rotational excitation and molecular alignment in intense laser fields *J. Chem. Phys.* **103** 7887–96

[114] Friedrich B and Herschbach D 1995 Alignment and trapping of molecules in intense laser fields *Phys. Rev. Lett.* **74** 4623–6

[115] Kim W and Felker P M 1996 Spectroscopy of pendular states in optical-field-aligned species *J. Chem. Phys.* **104** 1147–50

[116] Quack M and Sutcliffe E 1983 Quantum interference in the IR-multiphoton excitation of small asymmetric-top molecules: ozone *Chem. Phys. Lett.* **99** 167–72

[117] Quack M and Sutcliffe E 1984 The possibility of mode-selective IR-multiphoton excitation of ozone *Chem. Phys. Lett.* **105** 147–52

[118] Kosloff R 1994 Propagation methods for quantum molecular dynamics *Ann. Rev. Phys. Chem* **45** 145–78

[119] Dey B D, Askar A and Rabitz H 1998 Multidimensional wave packet dynamics within the fluid dynamical formulation of the Schrödinger equation *J. Chem. Phys.* **109** 8770–82

[120] Quack M 1992 Time dependent intramolecular quantum dynamics from high resolution spectroscopy and laser chemistry *Time Dependent Quantum Molecular Dynamics: Experiment and Theory. Proc. NATO ARW 019/92 (NATO ASI Ser. Vol 299)* ed J Broeckhove and L Lathouwers (New York: Plenum) pp 293–310

[121] Quack M 1981 *Faraday Discuss. Chem. Soc.* **71** 309–11, 325–6, 359–64 (Discussion contributions on flexible transition states and vibrationally adiabatic models; statistical models in laser chemistry and spectroscopy; normal, local, and global vibrational states)

[122] Quack M 1982 The role of intramolecular coupling and relaxation in IR-photochemistry *Intramolecular Dynamics, Proc. 15th Jerusalem Symp. on Quantum Chemistry and Biochemistry (Jerusalem, Israel, 29 March–1 April 1982)* ed J Jortner and B Pullman (Dordrecht: Reidel) pp 371–90

[123] von Puttkamer K, Dübal H-R and Quack M 1983 Time-dependent processes in polyatomic molecules during and after intense infrared irradiation *Faraday Discuss. Chem. Soc.* **75** 197–210

[124] von Puttkamer K, Dübal H R and Quack M 1983 Temperature-dependent infrared band structure and dynamics of the CH chromophore in $C_4F_9–C\equiv C–H$ *Chem. Phys. Lett.* **4–5** 358–62

[125] Segall J, Zare R N, Dübal H R, Lewerenz M and Quack M 1987 Tridiagonal Fermi resonance structure in the vibrational spectrum of the CH chromophore in CHF_3. II. Visible spectra *J. Chem. Phys.* **86** 634–46

[126] Boyarkin O V and Rizzo T R 1996 Secondary time scales of intramolecular vibrational energy redistribution in CF_3H studied by vibrational overtone spectroscopy *J. Chem. Phys.* **105** 6285–92

[127] Schulz P A, Sudbo A S, Krajnovitch D R, Kwok H S, Shen Y R and Lee Y T 1979 Multiphoton dissociation of polyatomic molecules *Ann. Rev. Phys. Chem.* **30** 395–409

[128] Zewail A H 1980 Laser selective chemistry—is it possible? *Phys. Today* Nov, 27–33

[129] Quack M 1978 Theory of unimolecular reactions induced by monochromatic infrared radiation *J. Chem. Phys.* **69** 1282–307

[130] Quack M, Stohner J and Sutcliffe E 1985 Time-dependent quantum dynamics of the picosecond vibrational IR-excitation of poly-atomic molecules *Time-Resolved Vibrational Spectroscopy, Proc. 2nd Int. Conf. Emil-Warburg Symp. (Bayreuth-Bischofsgrün, Germany, 3–7 June 1985)* ed A Laubereau and M Stockburger (Berlin: Springer) pp 284–8

[131] Mukamel S and Shan K 1985 On the selective elimination of intramolecular vibrational redistribution using strong resonant laser fields *Chem. Phys. Lett.* **5** 489–94

[132] Lupo D W and Quack M 1987 IR-laser photochemistry *Chem. Rev.* **87** 181–216

[133] von Puttkamer K and Quack M 1989 Vibrational spectra of $(HF)_2$, $(HF)_n$ and their D-isotopomers: mode selective rearrangements and nonstatistical unimolecular decay *Chem. Phys.* **139** 31–53

[134] Quack M 1991 Mode selective vibrational redistribution and unimolecular reactions during and after IR-laser excitation *Mode Selective Chemistry* ed J Jortner, R D Levine and B Pullman (Dordrecht: Kluwer) pp 47–65

[135] Crim F F 1993 Vibrationally mediated photodissociation: exploring excited state surfaces and controlling decomposition pathways *Ann. Rev. Phys. Chem.* **44** 397–428

[136] Nesbitt D J and Field R W 1996 Vibrational energy flow in highly excited molecules: role of intramolecular vibrational redistribution *J. Phys. Chem.* **100** 12 735–56

[137] He Y, Pochert J, Quack M, Ranz R and Seyfang G 1995 Discussion contributions on unimolecular reactions dynamics *J. Chem. Soc. Faraday Discuss.* **102** 358–62, 372–5

[138] Siegman A E 1986 *Lasers* (Oxford: Oxford University Press)

[139] Quack M and Sutcliffe E 1986 Program 515. URIMIR: unimolecular reactions induced by monochromatic infrared radiation *QCPE Bull.* **6** 98

[140] Marquardt R, Quack M and Stohner J, at press

[141] Marquardt R and Quack M 1996 Radiative excitation of the harmonic oscillator with applications to stereomutation in chiral molecules *Z. Phys. D* **36** 229–37

[142] Cotting R, Huber J R and Engel V 1993 Interference effects in the photodissociation of FNO *J. Chem. Phys.* **100** 1040–8

[143] Schinke R and Huber J R 1995 Molecular dynamics in excited electronic states—time-dependent wave packet studies *Femtosecond Chemistry: Proc. Berlin Conf. Femtosecond Chemistry (Berlin, March 1993)* (Weinheim: Verlag Chemie)

[144] Schinke R and Huber J R 1993 Photodissociation dynamics of polyatomic molecules. The relationship between potential energy surfaces and the breaking of molecular bonds *J. Phys. Chem.* **97** 3463

[145] Meier C and Engel V 1995 Pump–probe ionization spectroscopy of a diatomic molecule: sodium molecule as a prototype example *Femtosecond Chemistry: Proc. Berlin Conf. Femtosecond Chemistry (Berlin, March 1993)* (Weinheim: Verlag Chemie)

[146] Meyer S and Engel V 1997 Vibrational revivals and the control of photochemical reactions *J. Phys. Chem.* **101** 7749–53

[147] Flöthmann H, Beck C, Schinke R, Woywod C and Domcke W 1997 Photodissociation of ozone in the Chappuis band. II. Time-dependent wave packet calculations and interpretation of diffuse vibrational structures *J. Chem. Phys.* **107** 7296–313

[148] Loettgers A, Untch A, Keller H-M, Schinke R, Werner H-J, Bauer C and Rosmus P 1997 *Ab initio* study of the photodissociation of HCO in the first absorption band: three-dimensional wave packet calculations including the $\tilde{X}^2 A' - \tilde{A}^2 A''$ Renner–Teller coupling *J. Chem. Phys.* **106** 3186–204

[149] Keller H-M and Schinke R 1999 The unimolecular dissociation of HCO. IV. Variational calculation of Siegert states *J. Chem. Phys.* **110** 9887–97

[150] Braun M, Metiu H and Engel V 1998 Molecular femtosecond excitation described within the Gaussian wave packet approximation *J. Chem. Phys.* **108** 8983–8

[151] Dübal H-R, Ha T-K, Lewerenz M and Quack M 1989 Vibrational spectrum, dipole moment function, and potential energy surface of the CH chromophore in CHX_3 molecules *J. Chem. Phys.* **91** 6698–713

[152] Lewerenz M and Quack M 1988 Vibrational spectrum and potential energy surface of the CH chromophore in CHD_3 *J. Chem. Phys.* **88** 5408–32

[153] Marquardt R, Sanches Gonçalves N and Sala O 1995 Overtone spectrum of the CH chromophore in CHI_3 *J. Chem. Phys.* **103** 8391–403

[154] Marquardt R, Quack M and Thanopoulos I 2000 Dynamical chirality and the quantum dynamics of bending vibrations of the CH chromophore in methane isotopomers *J. Phys. Chem. A* **104** 6129–49

[155] Ha T-K, Lewerenz M, Marquardt R and Quack M 1990 Overtone intensities and dipole moment surfaces for the isolated CH chromophore in CHD_3 and CHF_3: experiment and *ab initio* theory *J. Chem. Phys.* **93** 7097–109

[156] Hollenstein H, Marquardt R, Quack M and Suhm M A 1994 Dipole moment function and equilibrium structure of methane in an analytical, anharmonic nine-dimensional potential surface related to experimental rotational constants and transition moments by quantum Monte Carlo calculations *J. Chem. Phys.* **101** 3588–602

[157] Schroedinger E 1926 Der stetige Übergang von der Mikro- zur Makromechanik *Naturwissenschaften* **14** 664–6

[158] Kerner E H 1958 Note on the forced and damped oscillator in quantum mechanics *Can. J. Phys.* **36** 371–7

[159] Marquardt R and Quack M 1989 Infrared-multiphoton excitation and wave packet motion of the harmonic and anharmonic oscillators: exact solutions and quasiresonant approximation *J. Chem. Phys.* **90** 6320–7

[160] Marquardt R and Quack M 1989 Molecular motion under the influence of a coherent infrared-laser field *Infrared Phys.* **29** 485–501

[161] Rice O K and Ramsperger H C 1927 Theories of unimolecular gas reactions at low pressures *J. Am. Chem. Soc.* **49** 1617–29

[162] Kassel L S 1928 Studies in homogeneous gas reactions I *J. Phys. Chem.* **32** 225–42

[163] Marcus R A and Rice O K 1951 The kinetics of the recombination of methyl radicals and iodine atoms *J. Phys. Colloid. Chem.* **55** 894–908

[164] Evans M G and Polanyi M 1935 Some applications of the transition state method to the calculation of reaction velocities, especially in solution *Trans. Faraday Soc.* **31** 875–94

[165] Eyring H 1935 The activated complex in chemical reactions *J. Chem. Phys.* **3** 107–15

[166] Hofacker L 1963 Quantentheorie chemischer Reaktionen *Z. Naturf.* A **18** 607–19

[167] Robinson P J and Holbrook K A 1972 *Unimolecular Reactions* (New York: Wiley)

[168] Quack M and Troe J 1981 Statistical methods in scattering *Theor. Chem. Adv. Perspect.* B **6** 199–276

[169] Truhlar D G, Garrett B C and Klippenstein S J 1996 Current status of transition state theory *J. Phys. Chem.* **100** 12 771–800

[170] Baer T and Hase W L 1996 *Unimolecular Reaction Dynamics. Theory and Experiment* (New York: Oxford University Press)

[171] Holbrook K A, Pilling M J and Robertson S H 1996 *Unimolecular Reactions* 2nd edn (New York: Wiley)

[172] Quack M 1990 The role of quantum intramolecular dynamics in unimolecular reactions *Phil. Trans. R. Soc. Lond.* A **332** 203–20

[173] Luckhaus D and Quack M 1993 The role of potential anisotropy in the dynamics of the CH chromophore in CHX$_3$ (C_{3v}) symmetric tops *Chem. Phys. Lett.* **205** 277–84

[174] Quack M 1989 Structure and dynamics of chiral molecules *Angew. Chem. Int. Ed. Engl.* **28** 571–86

[175] Hervé S, Le Quéré F and Marquardt R 2001 Rotational and vibrational wave packet motion during the IR multiphoton excitation of HF *J. Chem. Phys.* **114** 826–35

[176] Pauli W 1928 Über das H-Theorem vom Anwachsen der Entropie vom Standpunkt der neuen Quantenmechanik *Probleme der Modernen Physik (Festschrift zum 60. Geburtstage A. Sommerfelds)* ed P Debye (Leipzig: Hirzel) pp 30–45

[177] Quack M 1977 Detailed symmetry selection rules for reactive collisions *Mol. Phys.* **34** 477–504

[178] Cordonnier M, Uy D, Dickson R M, Kew K E, Zhang Y and Oka T 2000 Selection rules for nuclear spin modifications in ion-neutral reactions involving H$_3^+$ *J. Chem. Phys.* **113** 3181–93

[179] Quack M 1985 On the densities and numbers of rovibronic states of a given symmetry species: rigid and nonrigid molecules, transition states and scattering channels *J. Chem. Phys.* **82** 3277–83

[180] Manz J and Wöste L (ed) 1995 *Femtosecond Chemistry: Proc. Berlin Conf. Femtosecond Chemistry (Berlin, March, 1993)* (Weinheim: Verlag Chemie)

[181] Gaspard P and Burghardt I (ed) 1997 *XXth Solvay Conf. on Chemistry: Chemical Reactions and their Control on the Femtosecond Time Scale (Adv. Chem. Phys. 101)* (New York: Wiley)

[182] *Femtochemistry 97* 1998 (The American Chemical Society) (*J. Phys. Chem.* **102** (23))

Further Reading

Ashmore P G and Donovan R J (ed) 1980 *Specialists Periodical Report: Gas Kinetics and Energy Transfer* vol 1 (1975), vol 2 (1977), vol 3 (1978), vol 4 (1980) (London: The Royal Society of Chemistry)

Bunsen. Discussion on 'Molecular spectroscopy and molecular dynamics. Theory and experiment *Ber. Bunsengesellschaft Phys. Chem.* **99** 231–582

Bunsen. Discussion on 'Intramolecular processes' *Ber. Bunsenges. Phys. Chem.* **92** 209–450

Bunsen. Discussion on 'Unimolecular reactions' *Ber. Bunsenges. Phys. Chem.* **101** 304–635

Faraday Discuss. Chem. Soc. 1983 Intramolecular kinetics, No 75

Faraday Discuss. Chem. Soc. 1986 Dynamics of molecular photofragmentation, No 82

Faraday Discuss. Chem. Soc. 1995 Unimolecular dynamics, No 112

Gilbert R G and Smith S C 1990 *Theory of Unimolecular and Recombination Reactions* (Oxford: Blackwell)

Holbrook K, Pilling M J and Robertson S H 1996 *Unimolecular Reactions* (New York: Wiley)

Levine R D and Bernstein R B 1987 *Molecular Reaction Dynamics and Chemical Reactivity* (New York: Oxford University Press)

Quack M 1982 *Adv. Chem. Phys.* **50** 395–473

Quack M and Troe J 1981 Statistical methods in scattering *Theor. Chem.: Adv. Perspect.* B **6** 199–276

A3.14
Nonlinear reactions, feedback and self-organizing reactions

Stephen K Scott

A3.14.1 Introduction

A3.14.1.1 *Nonlinearity and feedback*

In the reaction kinetics context, the term 'nonlinearity' refers to the dependence of the (overall) reaction rate on the concentrations of the reacting species. Quite generally, the rate of a (simple or complex) reaction can be defined in terms of the rate of change of concentration of a reactant or product species. The variation of this rate with the extent of reaction then gives a 'rate–extent' plot. Examples are shown in figure A3.14.1. In the case of a first-order reaction, curve (i) in figure A3.14.1(a), the rate–extent plot gives a straight line: this is the only case of 'linear kinetics'. For all other concentration dependences, the rate–extent plot is 'nonlinear': curves (ii) and (iii) in figure A3.14.1(a) correspond to second-order and half-order kinetics respectively. For all the cases in figure A3.14.1(a), the reaction rate is maximal at zero extent of reaction, i.e. at the beginning of the reaction. This is characteristic of 'deceleratory' processes. A different class of reaction types, figure A3.14.1(b), show rate–extent plots for which the reaction rate is typically low for the initial composition, but increases with increasing extent of reaction during an 'acceleratory' phase. The maximum rate is then achieved for some non-zero extent, with a final deceleratory stage as the system approaches complete reaction (chemical equilibrium). Curves (i) and (ii) in figure A3.14.1(b) are idealized representations of rate–extent curves observed in isothermal processes exhibiting 'chemical feedback'. Feedback arises when a chemical species, typically an intermediate species produced from the initial reactants, influences the rate of (earlier) steps leading to its own formation. Positive feedback arises if the intermediate accelerates this process; negative feedback (or inhibition) arises if there is a retarding effect. Such feedback may arise chemically through 'chain-branching' or 'autocatalysis' in isothermal systems. Feedback may also arise through thermal effects: if the heat released through an exothermic process is not immediately lost from the system, the temperature of the reacting mixture will rise. A reaction showing a typical overall Arrhenius temperature dependence will thus show an increase in overall rate, potentially giving rise to further self-heating. Curve (iii) in figure A3.14.1(b) shows the rate–extent curve for an exothermic reaction under adiabatic conditions. Such feedback is the main driving force for the process known as combustion: endothermic reactions can similarly show self-cooling and inhibitory feedback. Specific examples of the origin of feedback in a range of chemical systems are presented below.

The branching cycle involving the radicals H, OH and O in the H_2+O_2 reaction involves the three elementary steps

$$H + O_2 \rightarrow OH + O \tag{A3.14.1}$$

$$O + H_2 \rightarrow OH + H \tag{A3.14.2}$$

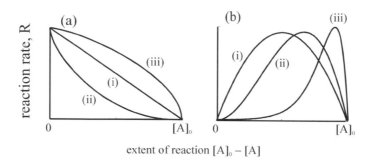

Figure A3.14.1. Rate-extent plots for (a) deceleratory and (b) acceleratory systems.

$$OH + H_2 \rightarrow H_2O + H. \tag{A3.14.3}$$

In steps (1) and (2) there is an increase from one to two 'chain carriers'. (For brevity, step (x) is used to refer to equation $(A3.14.x)$ throughout.) Under typical experimental conditions close to the first and second explosion limits (see section A3.14.2.3), steps (2) and (3) are fast relative to the rate determining step (1). Combining $(1) + (2) + 2 \times (3)$ gives the overall stoichiometry

$$H + 3H_2 + O_2 \rightarrow 3H + 2H_2O$$

so there is a net increase of 2 H atoms per cycle. The rate of this overall step is governed by the rate of step (1), so we obtain

$$d[H]/dt = +2k_1[O_2][H]$$

where the + sign indicates that the rate of production of H atoms increases proportionately with that concentration.

In the bromate–iron clock reaction, there is an autocatalytic cycle involving the species intermediate species $HBrO_2$. This cycle is comprised of the following non-elementary steps:

$$HBrO_2 + BrO_3^- + H^+ \rightarrow 2BrO_2 + H_2O \tag{A3.14.4}$$

$$BrO_2 + Fe^{2+} + H^+ \rightarrow HBrO_2 + Fe^{3+}. \tag{A3.14.5}$$

Step (5) is rapid due to the radical nature of BrO_2, so the overall stoichiometric process given by $(4) + 2 \times (5)$, has the form

$$HBrO_2 + BrO_3^- + 2Fe^{2+} + 3H^+ \rightarrow 2HBrO_2 + 2Fe^{3+} + H_2O$$

and an effective rate law

$$d[HBrO_2]/dt = +k_4[BrO_3{}^-][H^+][HBrO_2]$$

again showing increasing rate of production as the concentration of $HBrO_2$ increases.

In 'Landolt'-type reactions, iodate ion is reduced to iodide through a sequence of steps involving a reductant species such as bisulfite ion (HSO_3^-) or arsenous acid (H_3AsO_3). The reaction proceeds through two overall stoichiometric processes. The Dushman reaction involves the reaction of iodate and iodide ions

$$IO_3^- + 5I^- + 6H^+ \rightarrow 3I_2 + 3H_2O \tag{A3.14.6}$$

with an empirical rate law $R_\alpha = (k_1 + k_2[I^-])[IO_3^-][I^-][H^+]^2$.

The iodine produced in the Dushman process is rapidly reduced to iodide via the Roebuck reaction

$$I_2 + H_3AsO_3 + H_2O \rightarrow 2I^- + H_3AsO_4 + 2H^+. \tag{A3.14.7}$$

If the initial concentrations are such that $[H_3AsO_3]_0/[IO_3^-]_0 > 3$, the system has excess reductant. In this case, the overall stoichiometry is given by $(6) + 3 \times (7)$ to give

$$3H_3AsO_3 + IO_3^- + 5I^- \rightarrow 3H_3AsO_4 + 6I^-$$

i.e. there is a net production of one iodide ion, with an overall rate given by R_α. At constant pH and for conditions such that $k_2[I^-] \gg k_1$, this can be approximated by

$$d[I^-]/dt = k[IO_3^-][I^-]^2$$

where $k = k_2[H^+]^2$. This again has the autocatalytic form, but now with the growth proportional to the square of the autocatalyst (I^-) concentration. Generic representations of autocatalytic processes in the form

$$\text{quadratic autocatalysis} \quad A + B \rightarrow 2B \qquad \text{rate} = kab$$
$$\text{cubic autocatalysis} \quad A + 2B \rightarrow 3B \qquad \text{rate} = kab^2$$

where a and b are the concentrations of the reactant A and autocatalyst B respectively, are represented in figure A3.14.1(b) as curves (i) and (ii). The bromate–iron reaction corresponds to the quadratic type and the Landolt system to the cubic form.

A3.14.1.2 Self-organizing systems

'Self-organization' is a phrase referring to a range of behaviours exhibited by reacting chemical systems in which nonlinear kinetics and feedback mechanisms are operating. Examples of such behaviour include ignition and extinction, oscillations and chaos, spatial pattern formation and chemical wave propagation. There is a formal distinction between thermodynamically closed systems (no exchange of matter with the surroundings) and open systems [1]. In the former, the reaction will inevitably attain a unique state of chemical equilibrium in which the forward and reverse rates of every step in the overall mechanism become equal (detailed balance). This equilibrium state is temporally stable (the system cannot oscillate about equilibrium) and spatially uniform (under uniform boundary conditions). However, nonlinear responses such as oscillation in the concentrations of intermediate species can be exhibited as a transient phenomenon, provided the system is assembled with initial species concentrations sufficiently 'far from' the equilibrium composition (as is frequently the case). The 'transient' evolution may last for an arbitrary long (pehaps even a geological timescale), but strictly finite, period.

A3.14.2 Clock reactions, chemical waves and ignition

A3.14.2.1 Clock reactions

The simplest manifestation of nonlinear kinetics is the clock reaction—a reaction exhibiting an identifiable 'induction period', during which the overall reaction rate (the rate of removal of reactants or production of final products) may be practically indistinguishable from zero, followed by a comparatively sharp 'reaction event' during which reactants are converted more or less directly to the final products. A schematic evolution of the reactant, product and intermediate species concentrations and of the reaction rate is represented in figure A3.14.2. Two typical mechanisms may operate to produce clock behaviour.

The Landolt reaction (iodate + reductant) is prototypical of an autocatalytic clock reaction. During the induction period, the absence of the feedback species (here iodide ion, assumed to have virtually zero initial concentration and formed from the reactant iodate only via very slow 'initiation' steps) causes the reaction mixture to become 'kinetically frozen'. There is reaction, but the intermediate species evolve on concentration scales many orders of magnitude less than those of the reactant. The induction period depends on the initial

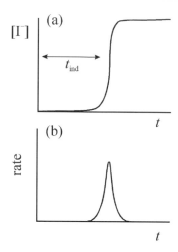

Figure A3.14.2. Characteristic features of a clock reaction, illustrated for the Landolt reaction, showing (a) variation of product concentration with induction period followed by sharp 'reaction event'; (b) variation of overall reaction rate with course of reaction.

concentrations of the major reactants in a manner predicted by integrating the overall rate cubic autocatalytic rate law, given in section A3.14.1.1.

The bromate–ferroin reaction has a quadratic autocatalytic sequence, but in this case the induction period is determined primarily by the time required for the concentration of the 'inhibitor' bromide ion to fall to a critical low value through the reactions

$$BrO_3^- + Br^- + 2H^+ \rightarrow HBrO_2 + HOBr \qquad (A3.14.8)$$

$$HBrO_2 + Br^- + H^+ \rightarrow 2HOBr. \qquad (A3.14.9)$$

Bromide ion acts as an inhibitor through step (9) which competes for $HBrO_2$ with the rate determining step for the autocatalytic process described previously, steps (4) and (5),

Steps (8) and (9) constitute a pseudo-first-order removal of Br^- with $HBrO_2$ maintained in a low steady-state concentration. Only once $[Br^-]<[Br^-]_{cr} = k_3[BrO_3^-]/k_2$ does step (3) become effective, initiating the autocatalytic growth and oxidation.

Clock-type induction periods occur in the spontaneous ignition of hydrocarbon–oxygen mixtures [2], in the setting of concrete and the curing of polymers [3]. A related phenomenon is the induction period exhibited during the self-heating of stored material leading to thermal runaway [4]. A wide variety of materials stored in bulk are capable of undergoing a slow, exothermic oxidation at ambient temperatures. The consequent self-heating (in the absence of efficient heat transfer) leads to an increase in the reaction rate and, therefore, in the subsequent rate of heat release. The Semenov and Frank–Kamenetskii theories of thermal runaway address the relative rates of heat release and heat loss under conditions where the latter is controlled by Newtonian cooling and by thermal conductivity respectively. In the Frank–Kamenetskii form, the heat balance equation shows that the following condition applies for a steady-state balance between heat transfer and heat release:

$$\kappa \nabla^2 T - (-\Delta H)c^n A e^{-E/RT} = 0$$

where κ is the thermal conductivity and ∇^2 is the Laplacian operator appropriate to the particular geometry. The boundary condition specifies that the temperature must have some fixed value equal to the surrounding temperature T_a at the edge of the reacting mass: the temperature will then exceed this value inside the reacting mass, varying from point to point and having a maximum at the centre. Steady-state solutions are only possible if the group of quantities $(-\Delta H)a_0^2 c^n E A\, e^{-E/RT_a}/\kappa R T_a^2$, where a_0 is the half-width of the pile, is less than some critical value. If this group exceeds the critical value, thermal runaway occurs. For marginally supercritical situations where thermal balance is almost achieved, the runaway is preceded by an induction period as the temperature evolves on the Fourier time scale, $t_F = c_p a_0^2/\kappa$, where c_p is the heat capacity. For large piles of low thermal conductivity, this may be of the order of months.

A3.14.2.2   Reaction–diffusion fronts

A 'front' is a thin layer of reaction that propagates through a mixture, converting the initial reactants to final products. It is essentially a clock reaction happening in space. If the mixture is one of fuel and oxidant, the resulting front is known as a flame. In each case, the unreacted mixture is held in a kinetically frozen state due to the virtual absence of the feedback species (autocatalyst or temperature). The reaction is initiated locally to some point; for example, by seeding the mixture with the autocatalyst or providing a 'spark'. This causes the reaction to occur locally, producing a high autocatalyst concentration/high temperature. Diffusion/conduction of the autocatalyst/heat then occurs into the surrounding mixture, initiating further reaction there. Front/flames propagate through this combination of diffusion and reaction, typically adopting a constant velocity which depends on the diffusion coefficient/thermal diffusivity and the rate coefficient for the reaction [5]. In each case, the speed c has the form $c \propto \sqrt{Dk}$. In gravitational fields, convective effects may arise due to density differences between the reactants ahead and the products behind the front. This difference may arise from temperature changes due to an exothermic/endothermic reaction or due to changes in molar volume between reactants and products—in some cases the two processes occur and may compete. In solid-phase combustion systems, such as those employed in self-propagating high-temperature synthesis (SHS) of materials, the steady flame may become unstable and a pulsing or oscillating flame develop—a feature also observed in propagating polymerization fronts [6].

A3.14.2.3   Ignition, extinction and bistability

In flow reactors there is a continuous exchange of matter due to the inflow and outflow. The species concentrations do not now attain the thermodynamic chemical equilibrium state—the system now has steady states which constitute a balance between the reaction rates and the flow rates. The steady-state concentrations (and temperature if the reaction is exo/endothermic) depend on the operating conditions through experimental parameters such as the flow rate. A plot of this dependence gives the steady-state locus, see figure A3.14.3. With feedback reactions, this locus may fold back on itself, the fold points corresponding to critical conditions for ignition or extinction—the plot is also then known as a 'bifurcation diagram'. Between these points, the system exhibits bistability, as either the upper or the lower branch can be accessed; so the system may have different net reaction rates for identical operating conditions. Starting with a long residence time (low flow rates), the system lies on the 'thermodynamic branch', with a steady-state composition close to the equilibrium state. As the residence time is decreased (flow rate is increased), so the steady-state extent of conversion decreases, but at the turning point in the locus a further decrease in residence time causes the system to drop onto the lower, flow branch. This jump is known as 'washout' for solution-phase reactions and 'extinction' in combustion. The system now remains on the flow branch, even if the residence time is increased: there is hysteresis, with the system jumping back to the thermodynamic branch at the 'ignition' turning point in the locus. Many reactions exhibiting clock behaviour in batch reactors show ignition and extinction in flow systems. The determination of such bifurcation diagrams is a classic problem in chemical reactor engineering and of great relevance to the safe and efficient operation of flow reactors in the modern chemical industry. More complex steady-state loci, with isolated branches or multiple fold points leading to three accessible competing states have been observed in systems ranging from autocatalytic solution-phase

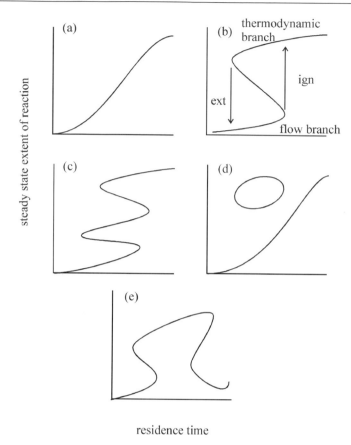

Figure A3.14.3. Example bifurcation diagrams, showing dependence of steady-state concentration in an open system on some experimental parameter such as residence time (inverse flow rate): (a) monotonic dependence; (b) bistability; (c) tristability; (d) isola and (e) mushroom.

reactions, smouldering combustion and in catalytic reactors [7]. Bistability has been predicted from certain models of atmospheric chemistry [8]. In the $H_2 + O_2$ and other branched-chain reactions, a balance equation for the radical species expresses the condition for a steady-state radical concentration. The condition for an 'ignition limit', i.e. for the marginal existence of a steady state, is that the branching and termination rates just balance. This can be expressed in terms of a 'net branching factor', $\phi = k_b - k_t$ where k_b and k_t are the pseudo-first-order rate constants for branching and termination respectively. For the hydrogen–oxygen system at low pressures, this has the form

$$2k_b[O_2] = k_{t1}[O_2][M] + k_{t2}$$

where k_{t1} corresponds to a three-body termination process (with [M] being the total gas concentration) and k_{t2} to a surface removal of H-atoms. This condition predicts a folded curve on the pressure–temperature plane—the first and second explosion limits, see figure A3.14.4.

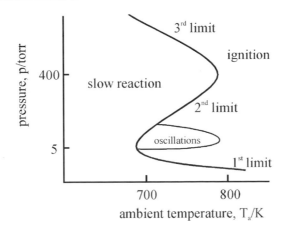

Figure A3.14.4. $P-T_a$ ignition limit diagram for $H_2 + O_2$ system showing first, second and third limits as appropriate to a closed reactor. The first and second limits have similar positions in a typical flow reactor, for which there is also a region of oscillatory ignition as indicated.

A3.14.3 Oscillations and chaos

Despite previous worries about restrictions imposed by thermodynamics, the ability of homogeneous isothermal chemical systems to support long-lived (although strictly transient) oscillations in the concentrations of *intermediate* species even in closed reactors is now clearly established [9]. The reaction system studied in greatest detail is the Belousov–Zhabotinsky (BZ) reaction [10–12], although the CIMA/CDIMA system involving chlorine dioxide, iodine and malonic acid is also of importance [13]. In flow reactors, oscillations amongst the concentrations of all species, including the reactants and products, are possible.

A3.14.3.1 The Belousov–Zhabotinsky reaction

The BZ reaction involves the oxidation of an organic molecule (citric acid, malonic acid (MA)) by an acidified bromate solution in the presence of a redox catalyst such as the ferroin/ferriin or Ce^{3+}/Ce^{4+} couples. For a relatively wide range of initial reactant concentrations in a well-stirred beaker, the reaction may exhibit a short induction period, followed by a series of oscillations in the concentration of several intermediate species and also in the colour of the solution. The response of a bromide-ion-selective electrode and of a Pt electrode (responding to the redox couple) for such a system is shown in figure A3.14.5. Under optimal conditions, several hundred excursions are observed. In the redox catalyst concentrations, the oscillations are of apparently identical amplitude and only minutely varying period: the bromide ion concentration increases slowly with each complete oscillation and it is the slow build-up of this inhibitor, coupled with the consumption of the initial reactants, that eventually causes the oscillations to cease (well before the system approaches its equilibrium concentration).

The basic features of the oscillatory mechanism of the BZ reaction are given by the Field–Koros–Noyes (FKN) model [14]. This involves three 'processes'—A, B and C. Process A involves steps (8) and (9) from section A3.14.2.1, leading to removal of 'inhibitor' bromide ion. Process B involves steps (3) and (4) from Section A3.14.1.1 and gives the autocatalytic oxidation of the catalyst. This growth is limited partly by the disproportionation reaction

$$2HBrO_2 \rightarrow BrO_3^- + HOBr + H^+. \tag{A3.14.10}$$

The 'clock' is reset through process C. Bromomalonic acid, BrMA, is a by-product of processes A and B

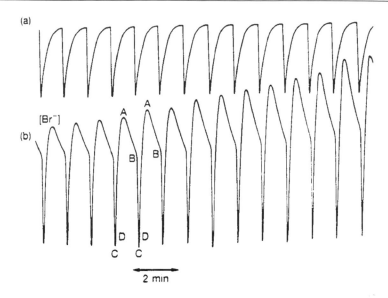

Figure A3.14.5. Experimental records from Pt and Br$^-$-ion-sensitive electrode for the BZ reaction in batch showing regular oscillatory response.

(possibly from HOBr) which reacts with the oxidized form of the redox catalyst. This can be represented as

$$2M_{ox} + MA + BrMA \rightarrow 2M_{red} + f\,Br^-. \qquad (A3.14.11)$$

Here, f is a stoichiometric factor and represents the number of bromide ions produced through this overall process for each two catalyst ions reduced. Because of the complex nature of this process, involving various radical species, f can lie in a range between 0 and ~3 depending on the $[BrO_3^-]/[MA]$ ratio and the $[H^+]$ concentration (note that these may change during the reaction).

The behaviour of the BZ system can be modelled semi-quantitatively by the 'oregonator' model [15]:

$$\frac{dx}{dt} = \frac{1}{\varepsilon}\left\{ x(1-x) - fz\frac{(x-q)}{(x+q)} \right\}$$

$$\frac{dz}{dt} = x - z$$

where x and z are (scaled) concentrations of $HBrO_2$ and M_{ox} respectively, and $\varepsilon = k_{11}[Org]/k_5[BrO_3^-][H^+]$ and $q = 2k_8k_{10}/k_9k_4$ are parameters depending on the rate coefficients and the initial concentrations [16], with [Org] being the total concentration of organic species (MA + BrMA). Oscillations are observed in this model for $0.5 < f < 1 + \sqrt{2}$. More advanced models and detailed schemes account for the difference between systems with ferrion and cerium ion catalysts and for the effect of oxygen on the reaction [17].

Under some conditions, it is observed that complex oscillatory sequences develop even in batch systems, typically towards the end of the oscillatory phase of the reaction. Transient 'chaos'—see section A3.14.3.3—appears to be established [18].

In flow reactors, both simple period-1 oscillations (every oscillation has the same amplitude and period as its predecessor) and more complex periodic states can be established and sustained indefinitely. The first report of chemical chaos stemmed from the BZ system [19] (approximately contemporaneously with observations in the biochemical peroxidase reaction [20]). These observations were made in systems with relatively high flow rate and show complexity increasing through a sequence of 'mixed-mode' wave forms comprising one large excursion followed by n small peaks, with n increasing as the flow rate is varied. Subsequent period-doubling and other routes to chaos have been found in this system at low flow rates [21]. A relatively simple two-variable extension of the oregonator model can adequately describe these complex oscillations and chaos.

A3.14.3.2   The CIMA/CDIMA system

The reaction involving chlorite and iodide ions in the presence of malonic acid, the CIMA reaction, is another that supports oscillatory behaviour in a batch system (the chlorite–iodide reaction being a classic 'clock' system: the CIMA system also shows reaction–diffusion wave behaviour similar to the BZ reaction, see section A3.14.4). The initial reactants, chlorite and iodide are rapidly consumed, producing ClO_2 and I_2 which subsequently play the role of 'reactants'. If the system is assembled from these species initially, we have the CDIMA reaction. The chemistry of this oscillator is driven by the following overall processes, with the empirical rate laws as given:

$$\begin{array}{llll} (1) & MA + I_2 \rightarrow IMA + I^- + H^+ & r_1 = k_1[MA] & (A3.14.12a) \\ (2) & ClO_2 + I^- \rightarrow ClO_2^- + \frac{1}{2}I_2 & r_2 = k_2[ClO_2][I^-] & (A3.14.12b) \\ (3) & ClO_2^- + 4I^- + 4H^+ \rightarrow 2I_2 + Cl^- + 2H_2O & r_3 = k_3 \dfrac{[ClO_2^-][I_2][I^-]}{\alpha + [I^-]^2}. & (A3.14.12c) \end{array}$$

The concentrations of the major reactants ClO_2 and I_2, along with H^+, are treated as constants, so this is a two-variable scheme involving the concentrations of ClO_2^- and I^-. Step ($12c$) constitutes the main feedback process, which here is an inhibitory channel, with the rate decreasing as the concentration of iodide ion increases (for large $[I^-]$ the rate is inversely proportional to the concentration). Again, exploiting dimensionless terms, the governing rate equations for u (a scaled $[I^-]$) and v (scaled $[ClO_2^-]$) can be written as [22, 23]:

$$\frac{du}{dt} = a - u - \frac{4uv}{1 + u^2} \qquad \frac{dv}{dt} = b\left(u - \frac{uv}{1 + u^2}\right)$$

where a and b are constants depending on the rate coefficients and the initial concentrations of the reactants.

Another important reaction supporting nonlinear behaviour is the so-called FIS system, which involves a modification of the iodate–sulfite (Landolt) system by addition of ferrocyanide ion. The Landolt system alone supports bistability in a CSTR: the addition of an extra feedback channel leads to an oscillatory system in a flow reactor. (This is a general and powerful technique, exploiting a feature known as the 'cross-shaped diagram', that has led to the design of the majority of known solution-phase oscillatory systems in flow reactors [25].) The FIS system is one member of the important class of pH oscillators in which H^+ acts as an autocatalyst in the oxidation of a weak acid to produce a strong acid. Elsewhere, oscillations are observed in important chemical systems such as heterogeneously catalysed reactions or electrochemical and electrodissolution reactions [26].

A3.14.3.3   Combustion systems

Oscillatory behaviour occurs widely in the oxidation of simple fuels such as H_2, CO and hydrocarbons. Even in closed reactors, the $CO + O_2$ reaction shows a 'lighthouse effect', with up to 100 periodic emissions of chemiluminescence accompanying the production of electronically excited CO_2. Although also described as 'oscillatory ignition', each 'explosion' is associated with less than 1% fuel consumption and can be effectively isothermal, even for this strongly exothermic reaction. Many hydrocarbons exhibit 'cool flame' oscillations

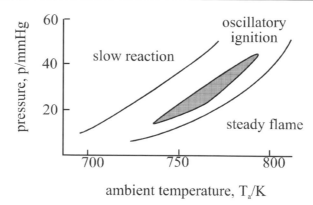

Figure A3.14.6. P–T_a ignition limit diagram for $CO + O_2$ system in a flow reactor showing location of ignition limits and regions of simple and complex (shaded area) oscillations.

in closed reactors, with typically between two and seven 'bursts' of light emission (from excited HCHO) and reaction, accompanied by self-heating of the reacting mixture. In continuous flow reactors, these modes can be sustained indefinitely. Additionally, true periodic ignitions occur for both the $CO + O_2$ and $H_2 + O_2$ systems [27, 28].

The p–T_a 'ignition' diagram for the $CO + O_2$ reaction under typical experimental conditions is shown in figure A3.14.6. Example oscillations observed at various locations on this diagram are displayed in figure A3.14.7. Within the region marked 'complex oscillations' the simple period-1 oscillation is replaced by waveforms that have different numbers of excursions in the repeating unit. The complexity develops through a 'period-doubling' sequence, with the waveform having 2^n oscillations per repeating unit, with n increasing with T_a. The range of experimental conditions over which the higher-order periodicities exist decreases in a geometric progression, with $n \to \infty$ leading to an oscillation with no repeating unit at some finite ambient temperature. Such *chaotic* responses exist over a finite range of experimental conditions and differ fundamentally from stochastic responses. Plotting the amplitude of one excursion against the amplitude of the next gives rise to a 'next-maximum map' (figure A3.14.8). This has a definite structure—a single-humped maximum—characteristic of a wide class of physical systems showing the period-doubling route to chaos.

The mechanistic origin of simple and complex oscillation in the $H_2 + O_2$ system is well established. The basic oscillatory clockwork involves the self-acceleration of the reaction rate through the chain-branching cycle, steps (1)–(3) in section A3.14.1.1 and the 'self-inhibitory' effect of H_2O production. Water is an inhibitor of the $H_2 + O_2$ system under these pressure and temperature conditions through its role in the main chain-termination step

$$H + O_2 + M \to HO_2 + M \tag{A3.14.13}$$

where M is a 'third-body' species which removes energy, stabilizing the newly-formed HO_2 bond. H_2 and O_2 have third body efficiencies of 1 and 0.3 respectively (these are measured relative to H_2), but H_2O is substantially more effective, with an efficiency of ~6.3 relative to H_2. Following an ignition, then, the rate of step (A3.14.13) is enhanced relative to the branching cycle, due to the now high concentration of H_2O, and further reaction is inhibited. The effect of the flow to the reactor, however, is to replace H_2O with H_2 and O_2, thus lowering the overall third-body effectiveness of the mixture in the reactor. Eventually, the rate of step (A3.14.13) relative to the branching rate becomes sufficiently small for another ignition to occur. Complex oscillations require the further feedback associated with the self-heating accompanying ignition. A 'minimal' complex oscillator mechanism for this system has been determined [28].

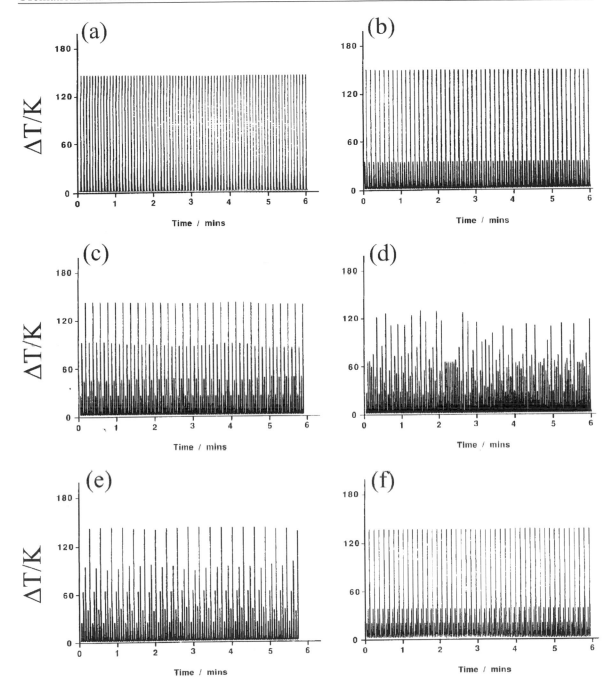

Figure A3.14.7. Example oscillatory time series for $CO + O_2$ reaction in a flow reactor corresponding to different $P-T_a$ locations in figure A3.14.6: (a) period-1; (b) period-2; (c) period-4; (d) aperiodic (chaotic) trace; (e) period-5; (f) period-3.

Surprisingly, the origin of the complex oscillations and chaos in the $CO + O_2$ system (where trace quantities of H-containing species have a major influence on the reaction and, consequently, many of the reactions of the $H_2 + O_2$ system predominate) are far from established to date.

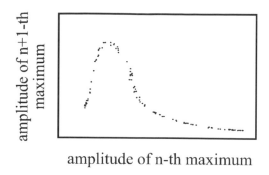

Figure A3.14.8. Next-maximum map obtained by plotting maximum temperature in one ignition against maximum in next ignition from trace (d) of figure A3.14.7.

The 'cool flames' associated with the low-temperature oxidation of hydrocarbon have great relevance, being the fundamental cause of knock in internal combustion engines. Their mechanistic origin arises through a thermokinetic feedback [2]. The crucial feature is the reaction through which O_2 reacts by addition to an alkyl radical $R^{\cdot}$

$$R^{\cdot} + O_2 = RO_2^{\cdot}. \tag{A3.14.14}$$

Under typical operating conditions, in the absence of self-heating from the reaction, the equilibrium for this step lies in favour of the product $RO_2^{\cdot}$. This species undergoes a series of intramolecular hydrogen-abstraction and further O_2-addition steps before fragmentation of the carbon chain. This final step produces three radical species, leading to a delayed, but overall branching of the radical chain ('degenerate branching'). This channel is overall an exothermic process, and the acceleration in rate associated with the branching leads to an increase in the gas temperature. This increase causes the equilibrium of step (A3.14.14) to shift to the left, in favour of the $R^{\cdot}$ radical. The subsequent reaction channel for this species involves H-atom abstraction by O_2, producing the conjugate alkene. This is a significantly less exothermic channel, and the absence of branching means that the overall reaction rate and the rate of heat release fall. The temperature of the reacting mixture, consequently, decreases, causing a shift of the equilibrium back to the right. Complex oscillations have been observed in hydrocarbon oxidation in a flow reactor, although chaotic responses have not yet been reported.

A3.14.3.4 *Controlling chaos*

The simple shape of the next-maximum map has been exploited in approaches to 'control' chaotic systems. The basic idea is that a system in a chaotic state is coexisting with an infinite number of unstable periodic states—indeed the chaotic 'strange attractor' is comprised of the period-1, period-2, period-4 and all other periodic solutions which have now become unstable. Control methods seek to select one of these unstable periodic solutions and to 'stabilize' it by applying appropriate but very small perturbations to the experimental operating conditions. Such control methods can also be adapted to allow an unstable state, such as the period-1 oscillation, to be 'tracked' through regions of operating conditions for which it would be naturally unstable. These techniques have been successfully employed for the BZ chaos as well as for chaos in lasers and various other physical and biological systems. For a full review and collection of papers see [29].

A3.14.4 **Targets and spiral waves**

The BZ and other batch oscillatory systems are capable of supporting an important class of reaction–diffusion structures. As mentioned earlier, clock reactions support one-off travelling wave fronts or flame, converting

Figure A3.14.9. Reaction–diffusion structures for an excitable BZ system showing (a) target and (b) spiral waves. (Courtesy of A F Taylor.)

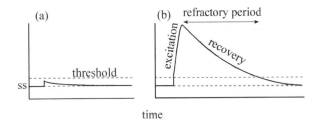

Figure A3.14.10. Schematic representation of important features of an excitable system (see the text for details).

reactants to products. In an oscillatory system, the 'resetting' process can be expected to produce a 'wave back' following the front, giving rise to a propagating wave pulse. Furthermore, as the system is then returned more or less to its initial state, further wave initiation may be possible. A series of wave pulses travelling one after the other forms a wave train. If the solution in spread as a thin film, for example in a Petri dish, and initiation is from a point source, the natural geometry will be for a series of concentric, circular wave pulses—*a target 'pattern'* [16, 30, 31]. An example of such reaction–diffusion structures in the BZ system is shown in figure A3.14.9(a). For such studies, the reactant solution is typically prepared with initial composition such that the system lies just outside the range for which it is spontaneously oscillatory (i.e. for f marginally in excess of $1 + \sqrt{2}$, see section A3.14.3.1) by increasing the initial malonic acid concentration relative to bromate. The system then sits in a stable steady state corresponding to the reduced form of the catalyst, and has the property of being *excitable*. An excitable system is characterized by (i) having a steady state; (ii) the steady state is stable to small perturbations and (iii) if the perturbation exceeds some critical or threshold value, the system responds by exhibiting an *excitation event*. For the BZ system, this excitation event is the oxidation of the redox catalyst, corresponding to process B with a local colour change in the vicinity of the perturbation (initiation) site. This response is typically large compared with the critical stimulus, so the system acts as a 'nonlinear amplifier' of the perturbing signal. Following the excitation, the system eventually returns to the initial steady state and recovers its excitability. There is, however, a finite period, the *refractory period*, between the excitation and the recovery during which the system is unresponsive to further stimuli. These basic characteristics are summarized in figure A3.14.10. Excitability is a feature not just of the BZ system, but is found widely throughout physical and, in particular, biological systems, with important examples in nerve signal transmission and co-ordinated muscle contraction [32].

The target structures shown in figure A3.14.9(a) reveal several levels of detail. Each 'pacemaker site' at the centre of a target typically corresponds to a position at which the system exhibits some heterogeneity, in

Figure A3.14.11. Spiral waves imaged by photoelectron electron microscopy for the oxidation of CO by O_2 on a Pt(110) single crystal under UHV conditions. (Reprinted with permission from [35], © The American Institute of Physics.)

some cases due to the presence of dust particles or defects of the dish surface. It is thought that these alter the local pH, so as to produce a composition in that vicinity such that the system is locally oscillatory, the spontaneous oscillations then serving as initiation events. Different sites have differing natural oscillatory frequencies, leading to the differing observed wavelengths of the various target structures. The speed of the waves also depends on the frequency of the pacemaker, through the so-called dispersion relation. The underlying cause of this for the BZ system is that the speed of a given front is dependent on the bromide ion (inhibitor) concentration into which it is propagating: the higher the pacemaker frequency, the less time the bromide ion concentration has to fall, so high-frequency (low-period) structures have lower propagation speeds.

If a wave pulse is broken, for example through mechanical disturbance, another characteristic feature of excitable media is that the two 'ends' then serve as sites around which the wave may develop into a pair of counter-rotating spirals, see figure A3.14.9(b). Once created, the spiral core is a persistent structure (in contrast to the target, in which case removal of the pacemaker heterogeneity prevents further initiation). The spiral structures have a wavelength determined by the composition of the bulk solution rather than the local properties at the core (although other features such as the *meandering* of the core may depend more crucially on local properties).

Targets and spirals have been observed in the CIMA/CDIMA system [13] and also in dilute flames (i.e. flames close to their lean flammability limits) in situations of enhanced heat loss [33]. In such systems, substantial fuel is left unburnt. Spiral waves have also been implicated in the onset of cardiac arrhythmia [32]: the normal contractive events occurring across the atria in the mammalian heart are, in some sense, equivalent to a wave pulse initiated from the sino-atrial node, which acts as a pacemaker. If this pulse becomes fragmented, perhaps by passing over a region of heart muscle tissue of lower excitability, then spiral structures (in 3D, these are *scroll waves*) or 're-entrant waves' may develop. These have the incorrect sequencing of

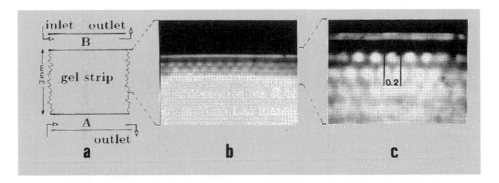

Figure A3.14.12. The first experimental observation of a Turing pattern in a gel strip reactor. Solutions containing separate components of the CIMA/CDIMA reaction are flowed along each edge of the strip and a spatial pattern along the horizontal axis develops for a range of experimental conditions. (Reprinted with permission from [38], © The American Physical Society.)

contractions to squeeze blood from the atria to the ventricles and impair the operation of the heart. Similar waves have been observed in neuronal tissue and there are suggested links to pathological behaviour such a epilepsy and migraine [34]. Spirals and targets have also been observed accompanying the oxidation of CO on appropriate single-crystal catalysts, such as Pt(110), and in other heterogeneously catalysed systems of technological relevance [35] (see figure A3.14.11). The light-sensitive nature of the $Ru(bipy)_2$-catalysed BZ system has been exploited in many attempts to 'control' or influence spiral structures (for example to remove spirals). The excitable properties of the BZ system have also been used to develop generic methods for devising routes through complex mazes or to construct chemical equivalents of logic gates [36].

A3.14.5 Turing patterns and other structures

A3.14.5.1 Turing patterns

Diffusive processes normally operate in chemical systems so as to disperse concentration gradients. In a paper in 1952, the mathematician Alan Turing produced a remarkable prediction [37] that if *selective* diffusion were coupled with chemical feedback, the opposite situation may arise, with a spontaneous development of sustained spatial distributions of species concentrations from initially uniform systems. Turing's paper was set in the context of the development of form (morphogenesis) in embryos, and has been adopted in some studies of animal coat markings. With the subsequent theoretical work at Brussels [1], it became clear that oscillatory chemical systems should provide a fertile ground for the search for experimental examples of these Turing patterns.

The basic requirements for a Turing pattern are:

(i)   the chemical reaction must exhibit feedback kinetics;

(ii) the diffusivity of the feedback species must be less than those of the other species;

(iii) for the patterns to be sustained, the system must be open to the inflow and outflow of reactants and products.

Requirement (i) is met particularly well by the BZ and CIMA/CDIMA reactions, although many chemical reactions with feedback are known. Requirements (ii) and (iii) were met almost simultaneously through the use of 'continuous flow unstirred reactors' (CFURs) in which the reaction is carried out in a dilute gel or membrane, with reactant free flows across the edges or faces of the gel. The incorporation of large indicator molecules such as starch into the gel is the key. This indicator is used with the CIMA/CDIMA system for

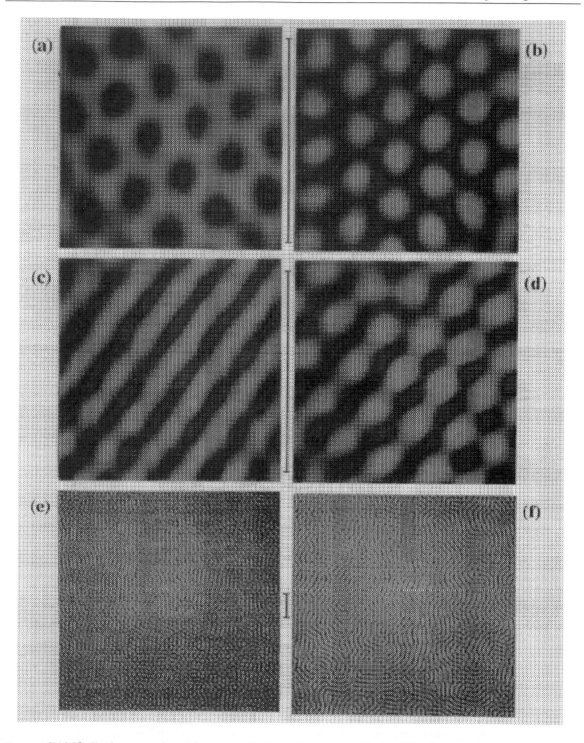

Figure A3.14.13. Further examples of the various Turing patterns observable in a 2D gel reactor. (a) and (b) spots, (c) and (d) stripes, (e) and (f): wider field of view showing long-range defects in basic structure. The scale bar alongside each figure represents 1 mm. (Reprinted with permission from [39], © The American Institute of Physics.)

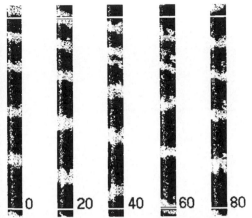

Figure A3.14.15. The differential flow-induced chemical instability (DIFICI) in the BZ reaction. (Reprinted with permission from [44], © The American Physical Society.)

Figure A3.14.14. A cellular flame in butane oxidation on a burner. (Courtesy of A C McIntosh.)

which I_3^- is formed where the I^- concentration is high, and this binds as a complex to the starch to produce the characteristic blue colour. The complexed ion is temporarily immobilized compared with the free ion, thus reducing the effective diffusion coefficient in a kind of 'reactive chromatography' [24]. In this way the first laboratory examples of Turing patterns were produced in Bordeaux [38] and in Texas [39]: examples are shown in figures A3.14.12 and A3.14.13. Turing patterns have not been unambiguously observed in the BZ system as no similar method of reducing the diffusivity of the autocatalytic species $HBrO_2$ has been devised.

A3.14.5.2   Cellular flames

Such 'diffusion-driven instabilities' have been observed earlier in combustion systems. As early as 1892, Smithells reported the observation of 'cellular flames' in fuel-rich mixtures [40]. An example is shown in figure A3.14.14. These were explained theoretically by Sivashinsky in terms of a 'thermodiffusive' mechanism [41]. The key feature here involves the role played by the Lewis number, Le, the ratio of the thermal to mass diffusivity. If $Le < 1$, which may arise with fuel-rich flame, for which H-atoms are the relevant species, of relatively low thermal conductivity (due to the high hydrocarbon content), a planar flame is unstable to spatial perturbations along the front. This mechanism has also been shown to operate for simple one-off chemical wave fronts, such as the iodate–arsenite system [42] and for various pH-driven fronts [43], if the diffusivity of I^- or H^+ are reduced via complexing strategies similar to that described above for the CIMA system.

A3.14.5.3   Other reaction–diffusion structures

The search for Turing patterns led to the introduction of several new types of chemical reactor for studying reaction–diffusion events in feedback systems. Coupled with huge advances in imaging and data analysis capabilities, it is now possible to make detailed quantitative measurements on complex spatiotemporal behaviour. A few of the reactor configurations of interest will be mentioned here.

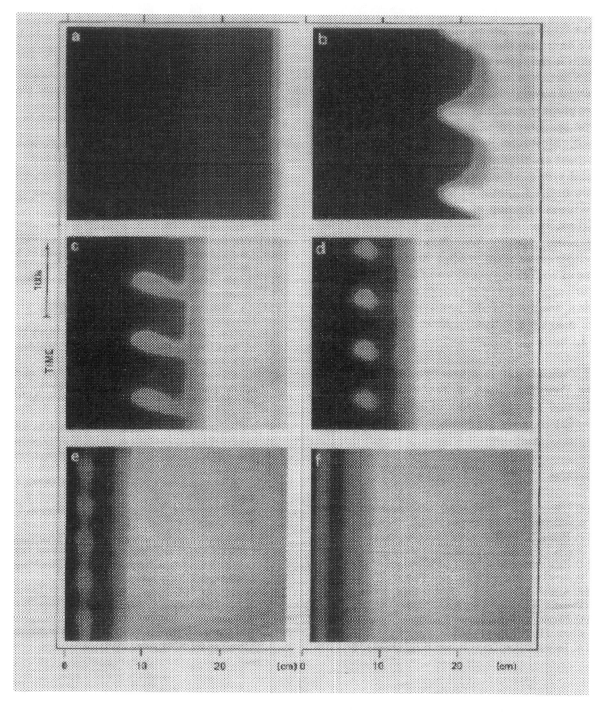

Figure A3.14.16. Spatiotemporal complexity in a Couette reactor: space–time plots showing the variation of position with time of fronts corresponding to high concentration gradients of I_3^- in the CIMA/CDIMA reaction. (Reprinted with permission from Ouyang *et al* [45], © Elsevier Science Publishers 1989.)

Laboratory experiment

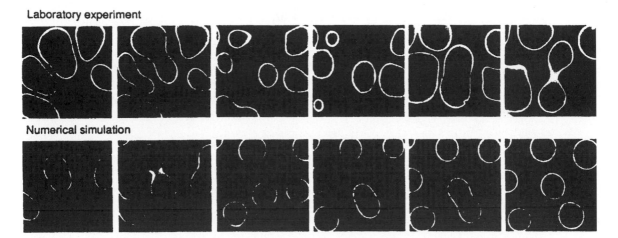

Numerical simulation

Figure A3.14.17. Self-replicating spots in the FIS reaction in a CFUR, comparing an experimental time sequence with numerical simulation based on a simple autocatalytic scheme. (Reprinted with permission from Lee *et al* [46], © Macmillan Magazines Ltd. 1994.)

The Turing instability is specific in requiring the feedback species to be selectively immobilized. An related instability, the differential flow-induced chemical instability or DIFICI requires only that one active species be immobilized relative to the others [44]. The experimental configuration is simple: a column of ion exchange beads is loaded with one chemical component, for example, ferrion for the BZ system. The remaining species are prepared in solution and flowed through this column. Above some critical flow rate, a travelling spatial structure with narrow bands of oxidized reagent (in the BZ system) separated by a characteristic wavelength and propagating with a characteristic velocity (not equal to the liquid flow velocity) is established. This effect has been realized experimentally—see figure A3.14.15.

If a fluid is placed between two concentric cylinders, and the inner cylinder rotated, a complex fluid dynamical motion known as Taylor–Couette flow is established. Mass transport is then by exchange between eddy vortices which can, under some conditions, be imagined as a substantially enhanced diffusivity (typically with 'effective diffusion coefficients several orders of magnitude above molecular diffusion coefficients') that can be altered by varying the rotation rate, and with all species having the same diffusivity. Studies of the BZ and CIMA/CDIMA systems in such a Couette reactor [45] have revealed bifurcation through a complex sequence of front patterns, see figure A3.14.16.

The FIS reaction (section A3.14.3.2) has been studied in a CFUR and revealed a series of structures known as 'serpentine patterns'; also, the birth, self-replication and death of 'spots', corresponding to regions of high concentration of particular species (see figure A3.14.17) have been observed [46].

A3.14.6 Theoretical methods

Much use has been, and continues to be, made of simplified model schemes representative of general classes of chemical or thermal feedback. The oregonator and Lengyel–Epstein models for the BZ and CDIMA systems have been given earlier. Pre-eminent among the more abstracted caricature models is the brusselator introduced by Prigogine and Lefever [47] which has the following form:

$$A \rightarrow X \qquad\qquad (A3.14.15a)$$

$$B + X \rightarrow Y + D \qquad\qquad (A3.14.15b)$$

$$Y + 2X \rightarrow 3X \tag{A3.14.15c}$$

$$X \rightarrow E. \tag{A3.14.15d}$$

Here, A and B are regarded as 'pool chemicals', with concentrations regarded as imposed constants. The concentrations of the intermediate species X and Y are the variables, with D and E being product species whose concentrations do not influence the reaction rates. The reaction rate equations for [X] and [Y] can be written in the following dimensionless form:

$$dx/dt = A - Bx + yx^2 - x \qquad dy/dt = Bx - yx^2.$$

Oscillations are found in this model if $B < B^* = 1 + A^2$.

A variation on this theme introduced by Gray and Scott, known as the 'autocatalator', is also widely exploited. This is often written in the form

$$P \rightarrow A \tag{A3.14.16a}$$

$$A \rightarrow B \tag{A3.14.16b}$$

$$A + 2B \rightarrow 3B \tag{A3.14.16c}$$

$$B \rightarrow C \tag{A3.14.16d}$$

so here A and B are equivalent to the Y and X in the brusselator and the main clockwork again involves the cubic autocatalytic step (15c) or (16c). The dimensionless equations here are

$$da/dt = \mu - \kappa a - ab^2 \qquad db/dt = \kappa a + ab^2 - b \tag{A3.14.17}$$

where μ is a scaled concentration of the reactant P and κ is a dimensionless rate coefficient for the 'uncatalysed' conversion of A to B in step (16b). In this form, the model has oscillatory behaviour over a range of experimental conditions:

$$\mu_1^* < \mu < \mu_2^* \qquad \text{with} \quad (\mu_{1,2}^*)^2 = \tfrac{1}{2}[1 - 2\kappa \pm (1 - 8\kappa)^{1/2}]. \tag{A3.14.18}$$

Outside this range, the system approaches the steady state obtained by setting $da/dt = db/dt = 0$:

$$(a_{ss}, b_{ss}) = (\mu/(\mu^2 + \kappa), \mu). \tag{A3.14.19}$$

The existence of an upper and a lower limit to the range of oscillatory behaviour is more typical of observed behaviour in chemical systems.

The autocatalytic driving step (16c) can also be taken on its own, or with the 'decay' step (16d) in models of open systems such as a CSTR, with an inflow of species A and, perhaps, of B. This system is then one of the simplest to show bistability and more complex steady-state loci of the type described in section A3.14.2.3. Also, generic features of wave front propagation can be studied on the basis of this scheme [7]. A comprehensive account can be found in the book by Gray and Scott (see Further Reading). Essentially, this model has been used in the context of modelling the broad features of oscillations in glycolysis and, with some modification, for animal coat patterning through Turing-like mechanisms.

The main theoretical methods have in common the determination of the *stability* of steady-state or other simple solutions to the appropriate form of the governing mass balance equations. Bifurcations from simple to more complex responses occur when such a solution loses its stability. Thus, the steady state given in equation (A3.14.19) does not cease to exist in the oscillatory region defined by (A3.14.18), but is now unstable, so that, if the system is perturbed, it departs from the steady state and moves to the (stable) oscillatory state which is also a solution of the reaction rate equations (A3.14.17).

In the generalized representation of the rate equations

$$da/dt = f(a,b) \qquad db/dt = g(a,b)$$

where f and g are functions of the species concentrations, a determining role is played by the eigenvalues of the Jacobian matrix J defined by

$$J = \begin{pmatrix} \partial f/\partial a & \partial f/\partial b \\ \partial g/\partial a & \partial g/\partial b \end{pmatrix}$$

evaluated with the steady-state concentrations. (This is readily generalized to n-variable systems, with J then being an $n \times n$ matrix.) Bifurcations corresponding to a turning or fold point in a steady-state locus (ignition or extinction point) occur if a real eigenvalue passes through zero. Equivalently, this arises if the determinant $\det J = 0$. This is knows as a 'saddle-node' bifurcation. The oscillatory instability, or Hopf bifurcation, occurs if the real part of an imaginary pair of eigenvalues passes through zero (provided all other eigenvalues are negative or have negative real parts). For a two-variable system, this occurs if the trace $\mathrm{Tr}\, J = 0$, and is the origin of the result in equation (A3.14.18).

The autocatalator model is in many ways closely related to the FONI system, which has a single first-order exothermic reaction step obeying an Arrhenius temperature dependence and for which the role of the autocatalyst is taken by the temperature of the system. An extension of this is the Sal'nikov model which supports 'thermokinetic' oscillations in combustion-like systems [48]. This has the form:

$$
\begin{aligned}
&P \rightarrow A && \text{rate} = k_0 p \\
&A \rightarrow B + \text{heat} && \text{rate} = k_1(T)a.
\end{aligned}
$$

The reactant P is again taken as a pool chemical, so the first step has a constant rate. The rate of the second step depends on the concentration of the intermediate A and on the temperature T and this step is taken as exothermic. (In the simplest case, k_0 is taken to be independent of T and the first step is thermoneutral.) Again, the steady state is found to be unstable over a range of parameter values, with oscillations being observed.

The approach to investigating spatial structure is similar—usually some simple solutions, such as a spatially uniform steady state exists and the condition for instability to spatial perturbations is determined in terms of eigenvalues of an extended Jacobian matrix. For conditions marginally beyond a bifurcation point (whether the instability is temporal or spatial), amplitude equations, such as the complex Ginsburg–Landau equation, are exploited [49]. For conditions far from bifurcation points, however, recourse to numerical integration is generally required. Frequently these will involve reaction–diffusion (and perhaps advection) equations, although representations of such systems in terms of cellular automata or gas-lattice models can be advantageous [50].

References

[1] Nicolis G and Prigogine I 1977 *Self-organization in Nonequilibrium Systems* (New York: Wiley)

[2] Griffiths J F 1986 The fundamentals of spontaneous ignition of gaseous hydrocarbons and related organic compounds *Adv. Chem. Phys.* **64** 203–303

[3] Epstein I R and Pojman J A 1999 Overview: nonlinear dynamics related to polymeric systems *Chaos* **9** 255–9

[4] Gray P and Lee P R 1967 Thermal explosion theory *Combust. Oxid. Rev.* **2** 1–183

[5] Scott S K and Showalter K 1992 Simple and complex propagating reaction-diffusion fronts *J. Phys. Chem.* **96** 8702–11

[6] Pojman J A, Curtis G and Ilyashenko V M 1996 Frontal polymerisation in solution *J. Am. Chem. Soc.* **118** 3783–4

[7] Gray P and Scott S K 1983 Autocatalytic reactions in the isothermal continuous, stirred-tank reactor: isolas and other forms of multistability *Chem. Eng. Sci.* **38** 29–43

[8] Johnson B R, Scott S K and Tinsley M R 1998 A reduced model for complex oscillatory responses in the mesosphere *J. Chem. Soc. Faraday Trans.* **94** 2709–16

[9] Epstein I R and Showalter K 1996 Nonlinear chemical dynamics: oscillation, patterns and chaos *J. Phys. Chem.* **100** 13 132–47

[10] Winfree A T 1984 The prehistory of the Belousov–Zhabotinsky oscillator *J. Chem. Educ.* **61** 661–3
[11] Zhabotinsky A M 1991 A history of chemical oscillations and waves *Chaos* **1** 379–86
[12] Tyson J J 1976 *The Belousov–Zhabotinsky Reaction (Lecture Notes in Biomathematics vol 10)* (Berlin: Springer)
[13] De Kepper P, Boissonade J and Epstein I R 1990 Chlorite–iodide reaction: a versatile system for the study of nonlinear dynamical behaviour *J. Phys. Chem.* **94** 6525–36
[14] Field R J, Koros E and Noyes R M 1972 Oscillations in chemical systems, part 2: thorough analysis of temporal oscillations in the bromate–cerium–malonic acid system *J. Am. Chem. Soc.* **94** 8649–64
[15] Field R J and Noyes R M 1974 Oscillations in chemical systems, part 4: limit cycle behavior in a model of a real chemical reaction *J. Chem. Phys.* **60** 1877–84
[16] Tyson J J 1979 Oscillations, bistability and echo waves in models of the Belousov–Zhabotinskii reaction *Ann. New York Acad. Sci.* **316** 279–95
[17] Zhabotinsky A M, Buchholtz F, Kiyatin A B and Epstein I R 1993 Oscillations and waves in metal-ion catalysed bromate oscillating reaction in highly oxidised states *J. Phys. Chem.* **97** 7578–84
[18] Wang J, Sorensen P G and Hynne F 1994 Transient period doublings, torus oscillations and chaos in a closed chemical system *J. Phys. Chem.* **98** 725–7
[19] Schmitz R A, Graziani K R and Hudson J L 1977 Experimental evidence of chaotic states in Belousov–Zhabotinskii reaction *J. Chem. Phys.* **67** 4071–5
[20] Degn H, Olsen L F and Perram J W 1979 Bistability, oscillations and chaos in an enzyme reaction *Ann. New York Acad. Sci.* **316** 623–37
[21] Swinney H L, Argoul F, Arneodo A, Richetti P and Roux J-C 1987 Chemical chaos: from hints to confirmation *Acc. Chem. Res.* **20** 436–42
[22] Gyorgyi L and Field R J 1992 A three-variable model of deterministic chaos in the Belousov–Zhabotinsky reaction *Nature* **355** 808–10
[23] Lengyel I, Rabai G and Epstein I R Experimental and modelling study of oscillations in the chlorine dioxide–iodine–malonic acid reaction *J. Am. Chem. Soc.* **112** 9104–10
[24] Lengyel I and Epstein I R 1992 A chemical approach to designing Turing patterns in reaction–diffusion systems *Proc. Natl Acad. Sci.* **89** 3977–9
[25] Epstein I R, Kustin K, De Kepper P and Orban M 1983 Oscillatory chemical reactions 1983 *Sci. Am.* **248** 96–108
[26] Imbihl R and Ertl G 1995 Oscillatory kinetics in heterogeneous catalysis *Chem. Rev.* **95** 697–733
[27] Johnson B R and Scott S K 1990 Period doubling and chaos during the oscillatory ignition of the CO + O_2 reaction *J. Chem. Soc. Faraday Trans.* **86** 3701–5
[28] Johnson B R and Scott S K 1997 Complex and non-periodic oscillations in the H_2 + O_2 reaction *J. Chem. Soc. Faraday Trans.* **93** 2997–3004
[29] Ditto W L and Showalter K (eds) 1997 Control and synchronization of chaos: focus issue *Chaos* **7** 509–687
[30] Zaikin A N and Zhabotinsky A M 1970 Concentration wave propagation in two-dimensional liquid-phase self-oscillating system *Nature* **225** 535–7
[31] Winfree A T 1972 Spiral waves of chemical activity *Science* **175** 634–6
[32] Winfree A T 1998 Evolving perspectives during 12 years of electrical turbulence *Chaos* **8** 1–19
[33] Pearlman H 1997 Target and spiral wave patterns in premixed gas combustion *J. Chem. Soc. Faraday Trans.* **93** 2487–90
[34] Larter R, Speelman B and Worth R M 1999 A coupled ordinary differential equation lattice model for the simulation of epileptic seizures *Chaos* **9** 795–804
[35] Nettesheim S, von Oertzen A, Rotermund H H and Ertl G 1993 Reaction diffusion patterns in the catalytic CO-oxidation on Pt(110): front propagation and spiral waves *J. Chem. Phys.* **98** 9977–85
[36] Toth A and Showalter K 1995 Logic gates in excitable media *J. Chem. Phys.* **103** 2058–66
[37] Turing A M 1952 The chemical basis of morphogenesis *Phil. Trans. R. Soc.* B **641** 37–72
[38] Castets V, Dulos E, Boissonade J and De Kepper P 1990 Experimental evidence of a sustained standing Turing-type nonequilibrium structure *Phys. Rev. Lett.* **64** 2953–6
[39] Ouyang Q and Swinney H L 1991 Transition to chemical turbulence *Chaos* **1** 411–20
[40] Smithells A and Ingle H 1892 The structure and chemistry of flames *J. Chem. Soc. Trans.* **61** 204–16
[41] Sivashinsky G I 1983 Instabilities, pattern formation and turbulence in flames *Ann. Rev. Fluid Mech.* **15** 179–99
[42] Horvath D and Showalter K 1995 Instabilities in propagating reaction–diffusion fronts of the iodate–arsenous acid reaction *J. Chem. Phys.* **102** 2471–8
[43] Toth A, Lagzi I and Horvath D 1996 Pattern formation in reaction–diffusion systems: cellular acidity fronts *J. Phys. Chem.* **100** 14 837–9
[44] Rovinsky A B and Menzinger M 1993 Self-organization induced by the differential flow of activator and inhibitor *Phys. Rev. Lett.* **70** 778–81
[45] Ouyang Q, Boissonade J, Roux J C and De Kepper P 1989 Sustained reaction–diffusion structures in an open reactor 1989 *Phys. Lett.* A **134** 282–6
[46] Lee K-J, McCormick W D, Pearson J E and Swinney H L 1994 Experimental observation of self-replicating spots in a reaction–diffusion system *Nature* **369** 215–8
[47] Prigogine I and Lefever R 1968 Symmetry breaking instabilities in dissipative systems *J. Chem. Phys.* **48** 1695–700

[48] Gray P, Kay S R and Scott S K Oscillations of simple exothermic reactions in closed systems *Proc. R. Soc. Lond.* A **416** 321–41
[49] Borckmans P, Dewel G, De Wit A and Walgraef D Turing bifurcations and pattern selection *Chemical Waves and Patterns* eds R Kapral and K Showalter (Dordrecht: Kluwer) ch 10, pp 323–63
[50] Kapral R and Wu X-G Internal noise, oscillations, chaos and chemical waves *Chemical Waves and Patterns* eds R Kapral and K Showalter (Dordrecht: Kluwer) ch 18, pp 609–34

Further Reading

Scott S K 1994 *Oscillations, Waves and Chaos in Chemical Kinetics* (Oxford: Oxford University Press)

A short, final-year undergraduate level introduction to the subject.

Epstein I R and Pojman J A 1998 *An Introduction to Nonlinear Chemical Dynamics: Oscillations, Waves, Patterns and Chaos* (Oxford: Oxford University Press)

Introductory text at undergraduate/postgraduate level.

Gray P and Scott S K 1994 *Chemical Oscillations and Instabilities* (Oxford: Oxford University Press)

Graduate-level introduction mainly to theoretical modelling of nonlinear reactions

Scott S K 1993 *Chemical Chaos* (Oxford: Oxford University Press)

Graduate-level text giving detailed summary of status of chaotic behaviour in chemical systems to 1990.

Field R J and Burger M (eds) 1984 *Oscillations and Travelling Waves in Chemical Systems* (New York: Wiley)

Multi-author survey of nonlinear kinetics field to 1984, still a valuable introduction to researchers in this area.

Kapral R and Showalter K (eds) 1995 *Chemical Waves and Patterns* (Dordrecht: Kluwer)

Multi-author volume surveying chemical wave and pattern formation, an up-to-date introduction for those entering the field.